# ...AND THIS TEXT'S STUDY TOOLS CAN HELP!

## Do I understand what I've just read?

Take time out to test your understanding before moving on in the chapter.

- ***Concept Checkpoints*** are designed to help you test yourself with a variety of useful strategies including filling in tables, reorganization of information, drawing, and short answer responses.

## Can I interpret and understand what I see?

Visualization is critical in understanding biology concepts and processes. It's also a powerful technique for remembering information.

- ***Visual Thinking questions*** accompany many of the illustrations in this text. These questions ask you to demonstrate your understanding of what is being depicted in the figure, using short answer, drawing, and other strategies.



tions readily take place in water because so many kinds of compounds are water soluble and therefore move among water molecules as separate molecules or ions.

How do solutes such as sugar or salt dissolve in solvents such as water? Water molecules gather closely around any particle that exhibits an electrical charge, such as ions, and polar molecules, such as sugars. For example, sodium chloride (table salt) is made up of the positively charged sodium ($Na^+$) and negatively charged chloride ($Cl^-$) ions. These ions are attracted to one another and cluster in a regular pattern, forming crystals. When you put salt in water, some ions break away from the crystals because the positive ends of some water molecules are attracted to the $Cl^-$ ions, while the negative ends of other water molecules are attracted to the $Na^+$ ions. These attractions are shown in the right-hand portion of **Figure 3.10**. These attractions are stronger than the attraction between the ions that keep the crystal together. Therefore, the ions are pulled from their positions in the crystal as is shown on the left. Water molecules then surround each ion, forming a *hydration shell*, which keeps the ions apart. The salt is said to be *dissolved*. Hydration shells form around all polar molecules and ions when in water.

**Water excludes nonpolar molecules**

Remember the old saying that oil and water don't mix? This statement is true because oil is a nonpolar molecule and cannot form hydrogen bonds with water; it cannot dissolve in water. Instead, the water molecules form hydrogen bonds with each other, causing the water to exclude the nonpolar oil molecules. It is almost as if nonpolar molecules move away from contact with the water. For this reason, nonpolar molecules are referred to as hydrophobic (HI-dro-FO-bik). The word *hydrophobic* comes from Greek words meaning "water" (*hydros*) and "fearing" (*phobos*). This tendency for nonpolar molecules to group together in a water solution is called *hydrophobic bonding* (**Figure 3.11**). Hydrophobic forces determine the three-dimensional shapes of many biological molecules, which are usually surrounded by water within organisms.

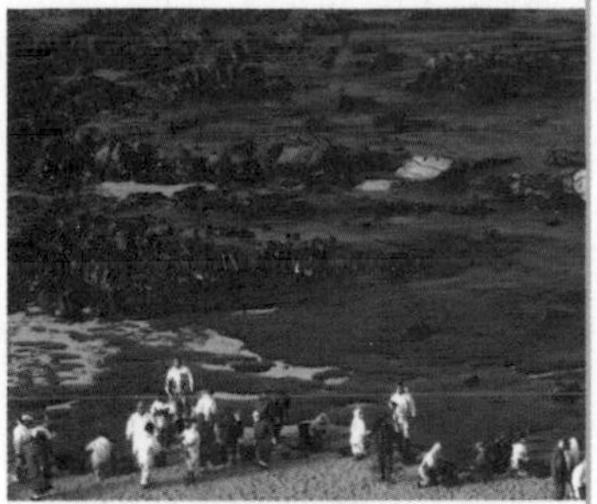

**Figure 3.11** Oil is hydrophobic. Crude petroleum from an oil spill floats on the surface of the ocean because it is hydrophobic and less dense (lighter) than water. The sinking of the oil tanker *Prestige* in November 2002 dumped thousands of tons of fuel oil into fish ing waters and onto beaches in Spain, France, and Portugal. This photo shows cleanup work on the shore of a creek in a northwestern village in Spain in December 2002.

CONCEPT CHECKPOINT

10. List and explain the chemical property of water that results in each of its following biologically important characteristics. Provide an example of why the characteristic is biologically important. (a) cohesion of water; (b) adhesion of water; (c) powerful solvent.

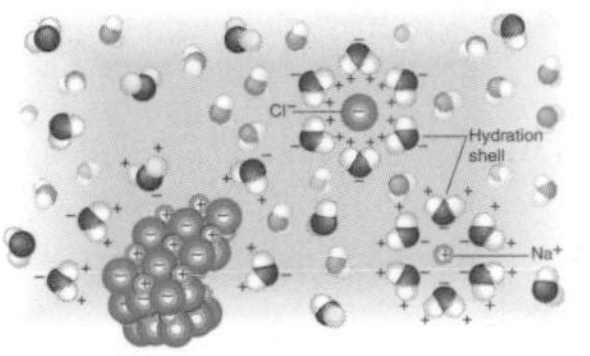


**Figure 3.10** How salt dissolves in water.

**Visual Thinking:** Would salt dissolve in a nonpolar liquid as it does in water? Why or why not?

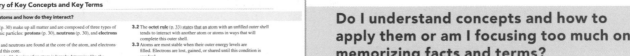

**3.8 Water, when ionized, can accept or donate protons in the reactions of life.**

Although water is commonly thought of as $H_2O$, its molecules often break spontaneously. When this happens, one hydrogen atom nucleus (a proton) dissociates, or separates, from the rest of the water molecule, leaving behind its electron. Because its positive charge is no longer balanced by an electron, it becomes a positively charged hydrogen ion, $H^+$. The remaining part of the water molecule now has an extra electron. It is therefore a negatively charged hydroxyl ion, $OH^-$. This process of spontaneous ion formation is called **ionization** (EYE-uh-nih-ZAY-shun):

$$H_2O \rightleftarrows OH^- + H^+$$

The ionization of water is critical to the chemical reactions that make up life processes. Why? Many reactions that take place in liv-

## Do I understand concepts and how to apply them or am I focusing too much on memorizing facts and terms?

The more you review material, the more you truly understand it. The ***Levels of Understanding*** review material at the end of each chapter uses a host of strategies to help ensure that you understand the chapter material from a variety of perspectives



CHAPTER REVIEW

**Summary of Key Concepts and Key Terms**

**What are atoms and how do they interact?**

**3.1 Atoms** (p. 30) make up all matter and are composed of three types of subatomic particles: **protons** (p. 30), **neutrons** (p. 30), and **electrons** (p. 30).
**3.1** Protons and neutrons are found at the core of the atom, and electrons surround this core.
**3.1** The chemical behavior of an atom is largely determined by the distribution of its electrons, particularly the number of electrons in its outermost **energy level** (**shell**, p. 31).
**3.1 Elements** (p. 30) are pure substances that are made up of a single kind of atom and cannot be separated into different substances by ordinary chemical methods.
**3.1** Atoms that have the same number of protons but different numbers of neutrons are called **isotopes** (EYE-suh-topes, p. 30).
**3.2 Molecules** (p. 31) are two or more atoms held together by sharing electrons.
**3.2** Molecules can be made up of atoms of the same element or atoms of different elements.
**3.2** Molecules made up of atoms of different elements are called **compounds** (p. 31). Compounds are also atoms of different elements that have lost or gained electrons (**ions**, p. 31) and that are held together by opposing electrical charges.
**3.2** The **octet rule** (p. 33) states that an atom with an unfilled outer shell tends to interact with another atom or atoms in ways that will complete this outer shell.
**3.3** Atoms are most stable when their outer energy levels are filled. Electrons are lost, gained, or shared until this condition is reached.
**3.3 Chemical bonds** (p. 33) are forces that hold atoms together.
**3.3** In **covalent** (p. 33) bonds, atoms share outer shell electrons. In **ionic** (eye-ON-ick, p. 33) bonds, atoms that have lost or gained electrons (ions) bond by the attraction of opposing electrical charges.
**3.4** An atom tends to lose (give) electrons if it has an outer shell needing many electrons to be complete. An atom tends to gain (take) electrons if it has a nearly completed outer shell.
**3.4** When an atom loses an electron, it is oxidized. When an atom gains an electron, it is reduced. Together, **oxidation** (OK-si-DAY-shun, p. 34) and **reduction** (p. 36) reactions are called **redox reactions** (p. 34).
**3.5** A covalent bond is the sharing of one or more pairs of electrons between atoms.
**3.5** Covalent bonds store energy as well as hold molecules together.

**What properties of water are important to life?**

**3.6** Water is a **polar molecule** (p. 36), having slightly positive and slightly negative ends, and attracts other polar molecules and charged particles.
**3.6** The cohesion of water molecules, which is their tendency to stick together by means of **hydrogen bonds** (p. 36), and adhesion, which is their tendency to be attracted to the polar molecules of other substances, are key to the movement of water in plants.
**3.7** Generally, a **solvent** (p. 37) is the substance in the solution that is present in the greater amount. It dissolves the **solute** (p. 37) to form the solution.
**3.7** Water dissolves polar substances and excludes nonpolar substances.
**3.7** Chemical interactions readily take place in water because so many kinds of molecules are water **soluble** (SOL-you-bul, p. 37) and therefore move among water molecules as separate molecules or ions.
**3.8** Water spontaneously ionizes (**ionization**, p. 38), forming hydrogen ions ($H^+$) and hydroxide ions ($OH^-$).
**3.8** An **acid** (p. 39) is any substance that dissociates to form $H^+$ ions when it is dissolved in water; a **base** (p. 39) is any substance that combines with $H^+$ ions and $OH^-$ ions do.
**3.8** The **pH scale** (p. 39) indicates the concentration of $H^+$ ions in a solution.

**Level 1 Learning Basic Facts and Terms**

**Multiple Choice**

1. An atom is a(n)
   a. subatomic particle.
   b. isotope with a single proton.
   c. submicroscopic particle that makes up matter.
   d. particle carrying a positive charge.
2. In covalent bonding
   a. ions give and take electrons.
   b. atoms share electron pairs.
   c. oxidation occurs.
   d. cohesion holds the molecule together.
3. Diatomic molecules
   a. are made up of two atoms.
   b. have double bonds.
   c. have two protons, two neutrons, and two electrons.
   d. have two electrons in each shell.



**Level 2 Learning Concepts**

1. a. Draw a diagram of an atom with an atomic number of 1. Label the nucleus and the subatomic particles, and show the electrical charge of each particle.
   b. Draw an diagram of the same atom.
2. List and explain three factors that influence how an atom interacts with other atoms. What is the significance of the octet rule?
3. Nitrogen gas is formed via a triple bond between two nitrogen atoms. What type of bonding is this? What does this bond have in common with the bonds forming water molecules? How is this bond different from the bonds forming water molecules?
4. Describe some of the chemical properties of the water molecule that make it biologically important. Relate these properties to water's molecular structure. How do these properties affect the interactions of other molecules dissolved or suspended in it?
5. The pH of the digestive juices within the human small intestine is between 7.5 and 8.5. Would you describe this environment as acidic or basic? Is the concentration of H+ ions in these digestive juices greater than or less than the concentration of H+ ions in a solution of neutral pH?
6. Water is discussed in this chapter as an essential component of life on Earth. Describe how the chemical properties and bonding characteristics of water relate to its biological importance.

**Level 3 Critical Thinking—Life Applications**

1. Based on what you learned in this chapter about the interactions of oil and water, suggest at least one reason why it is necessary to clean up an oil spill in the ocean.
2. Radon is a serious indoor air pollutant. Studies show that inhaling large amounts of radon increases the risk of lung cancer. Using Figure
3. Unpolluted rainwater has a pH of about 5. Acid precipitation has a pH of about 4. About how much more acidic is acid precipitation than unpolluted rainwater? Suggest one reason why you think acid precipitation might be harmful to living things.
4. In early 2004, the NASA probe "Spirit" landed successfully on Mars.

**You'll find more study suggestions and tools on this text's companion Web site at www.wiley.com/college/alters**

# Biology

## UNDERSTANDING LIFE

Sandra Alters, Ph.D.

Brian Alters, Ph.D.

*McGill University*

**WILEY**

John Wiley & Sons, Inc.

| ACQUISITIONS EDITOR | Rebecca Hope |
| MARKETING MANAGER | Clay Stone |
| DEVELOPMENT EDITOR | Len Neufeld |
| ASSOCIATE PRODUCTION MANAGER | Kelly Tavares |
| DESIGN DIRECTOR | Harry Nolan |
| COVER DESIGN | Sue Noli |
| PHOTO RESEARCHER | Jennifer MacMillan, Elyse Rieder |
| PHOTO RESEARCHER COORDINATOR | Joy Sikorski |
| PHOTO CREDIT | John Bracegirdle/Taxi/Getty Images |
| ILLUSTRATION EDITOR | Anna Melhorn |
| COVER PHOTO | John Bracegirdle/Taxi/Getty Images |

This book was set in 10/12 Times Roman by GGS Book Services, Atlantic Highlands and printed and bound by Von Hoffmann Press.

To order books or for customer service please, call 1(800)-CALL-WILEY (225-5945).

ISBN 0-471-43365-9

Printed in the United States of America

10 9 8 7 6 5 4 3 2 1

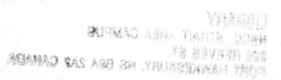

# PREFACE

As teachers, we can all probably pinpoint a time or instance that to us represents an optimal classroom experience: a well-planned lecture, useful visuals, a really effective example, perhaps a creative activity—students' faces indicated that whatever we had planned and executed was working. They got it. The classroom buzzed.

In our decades of teaching biology and training science teachers, we've strived to understand what creates this kind of connection and energy in the classroom. Based on what we learned from our experiences and by following research in teaching and learning theory, we began crafting approaches tailored specifically to the needs of a unique breed of student: the non-science major.

Non-majors present a host of unique challenges when they enter the biology classroom: Many are science shy. Some are academically underprepared; all have varying biological knowledge, a mix of learning styles and motivations for learning, and a host of preconceptions. And for many of these students, introductory biology is the only science course they'll take. Teaching this course is a special and serious responsibility—to teach students early in their college careers; we are charged with teaching students not just about science, but also about how to study and learn.

Our experience has reinforced for us that students' learning success depends on their generating links between relevant information they already know and new information. From such a constructivist perspective, learning is a social process in which students make sense of experience in terms of what they already know. It is a complex task, to tap into what students know in order to help them build on their knowledge, replace their misconceptions with scientifically useful conceptions, and construct meaning from their learning experiences.

We can probably all agree that a textbook alone isn't the answer. But an effective textbook can reinforce an instructor's efforts to create the optimal learning environment for making connections. A text can do this by encouraging efficient study skills and guiding the student to what's important. So, in developing this book we've examined our own classroom strategies for creating that optimal learning environment and have tried to incorporate them in the explanations and pedagogical features of this text. We hope that by focusing our efforts on the following key teaching challenges —those that hundreds of biology instructors repeatedly underscored in their responses to our surveys, reviews, and interviews—*Biology: Understanding Life* and its unique support package can help you create a buzz in your classroom.

## Challenge #1: Helping students relate biology to their own lives

First and foremost, students' successful learning is heavily dependent on their generating links between relevant information they already know and new information. What makes this more challenging, instructors tell us, is that they often feel they have to "convince" students that many aspects of biology are relevant to their lives. Features in this text that help address this challenge include:

▶ **Classroom-tested analogies** throughout the text help students visualize and understand concepts and processes by comparing them with readily recognizable, real-life examples.

▶ *Just Wondering* **boxes** answer questions from students across the country enrolled in introductory biology courses for non-majors. At the conclusion of each box, students are invited to submit their own questions to the authors by visiting the text's companion Web site.

▶ *In The News* **integrated features** place biology topics in the context of current events and students' lives in three ways:

- *In The News* chapter-opening story introduces each chapter by describing a news story related to chapter content and directing students to the related video on the text's companion Web site.

- *In The News* DVD (available free to adopters) presents the 42 real news stories on which the chapter-opening stories are based—this can be a great way to launch a lecture, and our Instructor's Resource Guide provides numerous suggestions for doing just that.

- *In The News* end-of-chapter feature asks students to reread the chapter-opening story and complete critical-thinking exercises based on it. Students are encouraged to view the video footage on the Web site, read a summary of the video, explore links relating to the story, and complete and submit answers to critical-thinking questions.

## Challenge #2: Addressing your students' varying backgrounds in science and study skills

Instructors routinely report that a key challenge in their teaching is accurately pitching their course to address their students' varied backgrounds—in familiarity with science as well as study skills, both specific to studying science and in general. Features in this text that help address this challenge include:

- **Question-and-answer organization** is introduced with this text's unique **Chapter Guide,** replacing the traditional chapter outline. The Chapter Guide presents the key questions and answers that organize the chapter. At the end of each chapter, key concepts and terms are reviewed in the same format.

- **Concept Checkpoint questions** are at the end of many chapter sections. These critical-thinking questions encourage students to check whether they remember and understand the concepts discussed, before continuing their reading.

- **Pronunciation guides** provide phonetic representations of pronunciation of key terms as appropriate, to help students who are not familiar with scientific vocabulary.

- **Visual Thinking questions** appear with many of the text's figures. The purpose of these short-answer questions is to help students understand the information in the figure and/or to connect the figure to the text.

- **Chapter Review** at the end of each chapter includes questions structured according to a three-level system, providing different types of questions designed to encourage students to think about different strategies for studying the chapter's material.

> **LEVEL 1** **Learning Basic Facts and Terms.** These questions focus on the knowledge level; they consist of basic recall questions.

> **LEVEL 2** **Learning Concepts.** These are short-answer questions that assess comprehension but that can be a bit more difficult as well, venturing into application.

> **LEVEL 3** **Critical Thinking and Life Applications.** These questions can run the gamut of the higher-order thinking questions in Bloom's taxonomy.

## Challenge #3: Bringing the classroom and subject to life

More and more, today's student requires heightened activity and visual stimulation to build knowledge successfully. Feedback from teachers indicates that a teacher can never have enough tools to address these needs, all in an effort to build energy and help students visualize concepts and consolidate information. Features in this text that help address this challenge include:

- **Instructor's Resource Guide.** This unique resource guide, with contributions from a panel of science educators, is designed to enhance classroom effectiveness. It provides hundreds of creative ideas for in-class and out-of-class activities, lecture launchers, interactive lectures, tips for using *In The News* videos, and guidelines for effectively integrating this text's support materials on a topic-by-topic basis.

- **Instructor's Resource CD.** PowerPoint presentations with embedded art, video clips, and animations, plus JPG files and PPT files of *all* line art, photos, and tables.

- In The News **DVD.** All *In The News* video news stories, each 1–3 minutes in duration and professionally produced by

ScienCentral, are provided in DVD format to ensure maximum efficiency and flexibility in using them in the classroom. Also available in VHS tape format on request.

- *Teaching Biology in Higher Education,* **by Brian and Sandra Alters.** This supplement is for all who teach in the life sciences at the postsecondary level. Covering a great variety of teaching scenarios, it concisely explains how to teach effectively whether lecturing to hundreds of students or conducting labs with a few students. Biologists will immediately recognize the practical applications.

## Challenge #4: Effectively testing and assessing student understanding

Through our surveys and reviews we have become acquainted with numerous examples of well-designed course goals. Many of these goals are aimed at building students' biological literacy, at ensuring they'll take with them an understanding of the principal concepts of biology, and at helping them use these concepts to make decisions in their own lives. Yet most of the instructors who formulated these goals admit that, as a result of their growing enrollments, lack of time, and students' inclination to rely heavily on memorization, their tests do not fully reflect or support the goals. Based on our understanding of how critical it is to test and assess knowledge in this course, we have included the following feature to help address this challenge:

- **Levels of Understanding Testing Program.** This is a completely reconceptualized testing program, prepared by science educators and reviewed by an assessment specialist. This collection of over 4,000 multiple-choice, short-answer, and essay items is divided into three banks of test questions: Learning Basic Facts and Terms, Learning Concepts, and Critical Thinking and Life Applications. The questions contained in these banks directly reinforce the Chapter Review materials and encourage students to view study and learning as a multitiered process.

## Challenge #5: Teaching scientific methods

The vast majority of students enroll in a non-majors biology course in order to fulfill a general education requirement, and for many of those students this will be the only college-level science course they will take. Accordingly, most instructors rank among their most important course goals to instill in students an understanding of how scientific knowledge is acquired and to encourage students to use the scientific way of thinking to view the world around them. But given many students' fear of science and their interest in applications of biology to the real world, instructors sometimes struggle with how to properly emphasize the importance of scientific methods. Features in this text that help address this challenge include:

- **A separate scientific methods chapter** helps students understand that scientific methods are used in many disciplines and are routinely applied to everyday life.

- **A "dissection" of an entire scientific research paper** in an appendix shows students how scientists communicate with one another and provides a model for developing similar reports.

▶ *How Science Works* **boxes** give students insight into the applications, processes, and methods of science as well as the tools and discoveries of scientists.

## Challenge #6: Communicating with students outside of class and administering your course—especially with large sections.

As budgets are cut and lecture sections balloon in size, instructors across the country struggle with the challenges of more students, less help in teaching and grading, and greater pressure to demonstrate achieved learning outcomes. The result, many instructors report, is their inability to communicate with and assess student understanding in the manner in which they would prefer. In hope of helping teachers increase their efficiency in this area of responsibility, we are pleased to offer with our text the following innovative tool:

▶ **eGrade Plus** provides an integrated suite of teaching and learning resources, including an online version of the text, in one easy-to-use Web site. Organized around the essential activities you perform in class, *eGrade Plus* helps you:

- **Prepare and present.** Create class presentations using a wealth of Wiley-provided resources, including an online version of the textbook, PowerPoint slides, *In The News* items, and more—making your preparation time more efficient. You may easily adapt, customize, and add to this content to meet the needs of your course.
- **Create assignments.** Automate the assigning and grading of homework or quizzes by using Wiley-provided question banks or by writing your own. Student results will be automatically graded and recorded in your gradebook. eGrade Plus can link homework problems to the relevant section of the online text, providing students with context-sensitive help.
- **Track student progress.** Keep track of your students' progress via an instructor's gradebook, which allows you to analyze individual and overall class results to determine their progress and level of understanding
- **Administer your course.** *eGrade Plus* can easily be integrated with another course management system, gradebooks, or other resources you are using in your class, providing you with the flexibility to build your course in your own way.

## ACKNOWLEDGMENTS

Textbook authors are only one part of the extensive team needed to produce a multifaceted, outstanding biology program such as this. We feel fortunate to have John Wiley & Sons as our publisher, for the company is a true leader in excellence in college publishing.

Each individual working on the team that produced this book is an outstanding professional. Keri Witman and and Kaye Pace first visited us in Montreal to discuss the possibility of this project and then welcomed us as new members of the Wiley family. David Brake and his staff at Content Connections helped us develop our initial plan for the book and provided our link to the hundreds of reviewers who gave us valuable input. Many thanks to David and the Content Connections staff.

Rebecca Hope was the driving force behind the project, providing insightful direction and a focus on excellence. We deeply appreciate her willingness to do whatever was needed to make this book and its supplements a cut above the rest in the field of introductory biology.

Johnna Barto gave initial editorial direction to the book, but after her retirement we were incredibly lucky to have Len Neufeld take over the editorial reins. Len's intelligence, dedication, considerable editorial skills, eye for detail, and easygoing nature made day-to-day work a pleasure and contributed immeasurably to the excellence of the project. Assisting editorially was Deborah Allen, a skilled biologist and editor, who helped develop the prose of the text while keeping the book to a manageable length. Len and Deborah also analyzed illustrations for their content accuracy and fit with the text, creating a situation in which the words and the art mesh with one another incredibly well and making this edition a superb learning tool. Barbara Heaney supervised the editorial team and kept all the wheels turning. A heartfelt thanks to such a skilled editorial team.

Harry Nolan is responsible for the design, which is not only visually stunning, but useful, uncluttered, and extremely helpful to students and professors alike. Kelly Tavares managed the incredibly complex production process and kept it running smoothly—as impossible as that might have seemed at times. Jennifer MacMillan and Elyse Rieder located the beautiful photographs used throughout the book, while Anna Melhorn coordinated the illustration program.

Merillat Staat found a top-notch group of biologists and science educators to develop the supplements that accompany the book. These resources are outstanding, and Merillat was key to coordinating these projects and maintaining their excellence.

Another key member of the Wiley team is Clay Stone. In his role as marketing manager, Clay helped us all to remain focused on the needs of our audience.

Many thanks go to authors James Quinn and Jane Sirdevan and to the journal *Biological Conservation* (Elsevier Science Ltd.) for allowing us to annotate and reprint the research article in Appendix A. We believe that this model will be useful to students and professors alike.

Sharon Shriver was instrumental in finding our ScienCentral link, which we think is a key aspect of our biology program in that it provides instructors with an important resource in developing interactive lectures and active learning environments. Sharon, thank you. John Peters contributed many of the thoughtful questions throughout the text; we appreciate his contribution immensely. Also, a personal thank you to Kam Yee for his consultation on fish identification.

Finally, a heartfelt thanks to the professors who have offered their comments, suggestions, and criticisms throughout this project. They remain a constant source of inspiration and guidance. We acknowledge the following professors who contributed their time and talent to review portions of the manuscript and those who participated in surveys conducted in the preparation of *Biology: Understanding Life*.

Marilyn Shopper, *Johnson County Community College*
Suzanne Simoneau, *Augusta State University*
Dianne Snyder
Eric Strauss, *Boston College*
AJ Strong, *Wichita State University*
Nathan Strong, *New Hampshire Technical Institute*
Delon Stultz, *Mesa Community College*
Aleta Sullivan, *Pearl River Community College*
Julie Sutherland, *College of DuPage*
Bradley J. Swanson, *Central Michigan University*
Samuel F. Tarsitano, *Texas State University*
Salvatore Tavormina, *Austin Community College*
Kathy Tehrani, *Cincinnati State Technical and Community College*
Tom Timmons, *Murray State University*
Sue Trammell, *John A. Logan College*
Eileen Underwood, *Bowling Green State University*
Pat Wadsworth, *University of Massachusetts*
Carol Wake, *South Dakota State University*
Hong Li Wang, *University of Arkansas–Little Rock*
Art Weiner, *University of Alaska–Anchorage*
Ruth Welti, *Kansas State University*
Jane Weston, *Genesee Community College*
Rachel Willard, *University of Denver*
Lawrence Williams, *University of Houston*
Mala wingerd, *San Diego State University*
Christopher J. Winslow, *Bowling Green State University*
Donald Wujek , *Oakland Community College*
Kenneth Wunch, *Sam Houston State University*

## ◢ ILLUSTRATION PROGRAM

*These reviewers viewed online selected, preliminary elements of the text's art program, submitting feedback addressing the illustrations' accuracy, quality, clarity of concept, necessity to text, and comparison to other text art programs available in the market.*

David C. Belt, *Johnson County Community College*
Brenda Blackwelder, *Central Piedmont Community College*
David Byres, *Florida Community College–Jacksonville*
Estella B. Chen, *Georgia State University*
Christine Curran, *University of Cincinnati*
Paul Decelles, *Johnson County Community College*
Anne Donnelly, *State University of New York–Cobleskill*
Carmen Eilertson, *Georgia State University*
Dianne M. Jedlicka, *Columbia College Chicago*
David Krauss, *Boston College*
Maureen Leupold, *Genesee Community College*
Tammy Liles, *Lexington Community College*
Joseph R. Mendelson III, *Utah State University*
Juliet Noor, *Louisiana State University*
Donald J. Reinhardt, *Georgia State University*
Dan Rogers, *Somerset Community College*
Carolyn Rost, *Cincinnati State Technical and Community College*
Kim Cleary Sadler, *Middle Tennessee State University*
A. Spencer Tomb, *Kansas State University*
Sue Trammell, *John A. Logan College*
Craig Tuerk, *Morehead State University*
Carol Wake, *South Dakota State University*
Kenneth Wunch, *Sam Houston State University*

## ◢ PRODUCT DESIGN AND VIABILITY

*These reviewers viewed online a detailed table of contents, a statement of objectives for the text and support package, and a preliminary designed*

sample chapter, submitting feedback regarding the text's vision and the chapter's execution.

Neil Baker, *The Ohio State University*
Donna Bivans, *Pitt Community College*
Mehdi Borhan, *Johnson County Community College*
Robert Boyd, *Auburn University*
David Byres, *Florida Community College–Jacksonville*
Jocelyn Cash, *Central Piedmont Community College*
Jan R. P. Coles, *Kansas State University*
Jerry L. Cook, *Sam Houston State University*
Deborah Dardis, *Southeastern Louisiana University*
Jean DeSaix, *University of North Carolina–Chapel Hill*
Anne Donnelly, *State University of New York–Cobleskill*
Ravi Gargesh, *Johnson County Community College*
Dana Haine, *Central Piedmont Community College*
William J. Higgins, *University of Maryland*
Brian Hoffman, *Park University*
Terry L. Hufford, *The George Washington University–Mount Vernon College*
Ari Jumpponen, *Kansas State University*
Peggy Lepley, *Cincinnati State Technical and Community College*
Maureen Leupold, *Genesee Community College*
Kimberly G. Lyle-Ippolito, *Anderson University*
Gordon L. Mendenhall, *University of Indianapolis*
Craig Milgrim, *Grossmont College*
Nancy H. Miller, *Diablo Valley College*
Robert Moldenhauer, *St. Clair County Community College*
Randy Moore, *University of Minnesota*
Sean O'Keefe, *Morehead State University*
Jack Pennington, *St. Louis Community College–Forest Park*
David K. Peyton, *Morehead State University*
George Pinchuk, *Mississippi State University*
Thoniot T. Prabhakaran, *Texas State University–San Marcos*
Donald J. Reinhardt, *Georgia State University*
Michael H. Renfroe, *James Madison University*
Kim Cleary Sadler, *Middle Tennessee State University*
Dave Sheldon, *St. Clair County Community College*
Marilyn Shopper, *Johnson County Community College*
Suzanne Simoneau, *Augusta State University*
Dianne Snyder, *Augusta State University*
Larry D. Spears, *John A. Logan Community College*
Kathy Tehrani, *Cincinnati State Technical and Community College*
T. G. Thomas, *Tarrant County College*
Teresa A. Thomas, *Southwestern College*
Carol Wake, *South Dakota State University*
Richard Webster, *University of Illinois*
Art Weiner, *University of Alaska–Anchorage*
Richard Whalen, *South Dakota State University*
Kenneth Wunch, *Sam Houston State University*

## ◢ CONTENT ASSESSMENT

*These reviewers focused on individual chapters or blocks of chapters, submitting feedback on the chapters' strengths and weaknesses, writing style, level of detail, organization, art program, and competitiveness with existing products in the market.*

Mark Ainsworth, *Seattle Central Community College*
Shylaja Akkaraju, *College of DuPage*
Corrie Andries, *Albuquerque Technical Vocational Institute*
Gregory Antipa, *San Francisco State University*
Kemuel Badger, *Ball State University*

Marilyn C. Baguinon, *Kutztown University*
Neil Baker, *The Ohio State University*
Marilyn Banta, *University of Northern Colorado*
Don Bard, *University of California–Santa Cruz*
David C. Belt, *Johnson County Community College*
Nick Bhattacharya, *Mesa Community College*
Donna Bivans, *Pitt Community College*
Brenda Blackwelder, *Central Piedmont Community College*
Lesley Blair, *Oregon State University*
Roman E. Boldyreff, *Cuyahoga Community College*
Robert Boyd, *Auburn University*
Marlies K. Boyd, *Modesto Junior College*
Agnello Braganza, *Chabot College*
Susan J. Brown, *Kansas State University*
Dana Brown-Haine, *Central Piedmont Community College*
David Bryan, *Cincinnati State Technical and Community College*
David Byres, *Florida Community College–Jacksonville*
Todd Carter, *Seward County Community College*
Jocelyn Cash, *Central Piedmont Community College*
Estella B. Chen, *Georgia State University*
Elisabeth A. Ciletti, *Pasadena City College*
Jan R. P. Coles, *Kansas State University*
Jerry L. Cook, *Sam Houston State University*
Mitchell B. Cruzan *Portland State University*
Christine Curran, *University of Cincinnati*
Michael Dabney, *Hawaii Pacific University*
Don C. Dailey, *Austin Peay State University*
Garry Davies, *University of Alaska–Anchorage*
Clementine A. deAngelis, *Tarrant County College*
Paul Decelles, *Johnson County Community College*
AnneDonnelly, *State University of New York–Cobles*kill
Carmen Eilertson, *Georgia State University*
Michelle Geary, *West Valley College*
Gisele Giorgi, *Skyline College*
Jack M. Goldberg, *University California–Davis*
Robert M. Goodman, *University of Wisconsin - Madison*
Bernard Hauser, *University of Florida*
Susan Herking, *Cincinnati State Technical and Community College*
Terry L. Hufford, *The George Washington University–Mount Vernon
    College*
Martha Jack, *Indiana University of Pennsylvania*
Dianne M. Jedlicka, *Columbia College Chicago*
Seema Jejurikar, *Bellevue Community College*
Mitrick A. Johns, *Northern Illinois University*
Timothy C. Johnston, *Murray State University*
Elaine Kent, *California State University–Sacramento*
Scott S. Kinnes, *Azusa Pacific University*
Thomas E. Kober, *Cincinnati State College*
Hal Kramer, *Fordham University*
David Krauss, *Boston College*
Kevin Krown, *San Diego State University*
Dale Lambert, *Tarrant County College*
Kaddee Lawrence, *Highline Community College*
Maureen Leupold, *Genesee Community College*
Anna Levin, *Diablo Valley College*
Tammy Liles, *Lexington Community College*
Matthew J. Linton, *University of Utah*
Paul Lonquich, *Los Angeles Valley College*
Melanie Loo, *California State University–Sacramento*
David Loring, *Johnson County Community College*
Ann S. Lumsden, *Florida State University*

William MacKay, *Edinboro University of Pennsylvania*
Godfred Masinde, *San Bernardino Valley College*
Amy Massengill, *Middle Tennessee State University*
Jacqueline S.McLaughlin, *The Pennsylvania State University–Berks-
    Lehigh Valley*
Malinda McMurry, *Morehead State University*
Joseph R. Mendelson III, *Utah State University*
Nancy H. Miller, *Diablo Valley College*
V. Christine Minor, *Clemson University*
Randy Moore, *University of Minnesota*
Jorge A. Moreno, *University of Colorado–Boulder*
Woody Moses, *Highline Community College*
Ann Murkowski, *North Seattle Community College*
Juliet Noor, *Louisiana State University*
Amanda N. Orenstein, *Saddleback College*
Robert Paterson, *North Carolina State University*
Jack Pennington, *St. Louis Community College–Forest Park*
Gary Pettibone, *Buffalo State College*
David K. Peyton, *Morehead State University*
Jay Phelan, *University of California–Los Angeles*
George Pinchuk, *Mississippi State University*
Kathryn Stanley, *Podwall Nassau Community College*
Robert Pope, *Miami-Dade College*
Jerry Purcell, *San Antonio College*
Walter Rast, *Texas State University–San Marcos*
Renee Redman, *University of North Carolina–Greensboro*
Mary Rees, *Catholic University*
Donald J. Reinhardt, *Georgia State University*
Erin Rempala-Kim, *Grossmont College*
Carol Rhodes, *College of San Mateo OF*
Dan Rogers, *Somerset Community College*
Carolyn Rost, *Cincinnati State Technical and Community College*
Peter Ruben, *Utah State University*
Michael L. Rutledge, *Middle Tennessee State University*
Kim Cleary Sadler, *Middle Tennessee State University*
Roger Raymond, *Saft University of Alaska–Anchorage*
Brian Saunders, *North Seattle Community College*
Louis Scala, *Kutztown University*
Erik Scully, *Towson University*
Sharon Shapiro, *Southwestern College*
Marilyn Shopper, *Johnson County Community College*
Suzanne Simoneau, *Augusta State University*
Howard Singer, *New Jersey City University*
Dianne Snyder, *Augusta State University*
Peter Svensson, *West Valley College*
Bradley J. Swanson, *Central Michigan University*
Samuel F. Tarsitano, *Texas State University*
Kathy Tehrani, *Cincinnati State Technical and Community College*
Tom Timmons, *Murray State University*
A. Spencer Tomb, *Kansas State University*
Sue Trammell, *John A. Logan College*
Carol Wake, *South Dakota State University*
Jerry Waldvogel, *Clemson University*
Hong Li. Wang, *University of Arkansas–Little Rock*
Randall Warwick, *Coastline Community College*
Lisa Weasel, *Portland State University*
Kathy Webb, *Bucks County Community College*
Art Weiner, *University of Alaska–Anchorage*
Janet Wolkenstein, *Hudson Valley Community College*
Kenneth Wunch, *Sam Houston State University*
Carol Wymer, *Morehead State University*
Gregory Zagursky, *Radford University*

## SPECIALIZED CONSULTANTS FOR
### *BIOLOGY: UNDERSTANDING LIFE*

*These reviewers focused on the text's parts, or blocks of chapters relating to an overarching area of biology, examining these chapters in line-by-line detail. Expert reviewers were asked to provide feedback to the authors on accuracy, currency, clarity of explanations, level of detail, applied and research examples included in the chapters, and writing style.*

### Margaret Beard, University of Wisconsin–Oshkosh

Margaret Beard earned her M.S. and Ph.D. from the University of Michigan, completing her postdoctoral experience at Albert Einstein College of Medicine. She has taught introductory biology, animal physiology, cellular biology, histology, human physiology, and neurobiology to undergraduates at Reed College, Columbia University, the University of Wisconsin–Oshkosh, and Holy Cross College, as well as medical students at Oregon Health Sciences University. She pursued cell biological/neurobiological research with an abiding interest in peroxisomes while at the University of Michigan, Albert Einstein College of Medicine, Reed College, and Columbia University and then at The Nathan Kline Institute (for neurobiological research), affiliated with New York University and the State of New York.

### Kyle Harms, Louisiana State University

Kyle Harms grew up in rural Iowa and attended Iowa State University (ISU) as an undergraduate student. He majored in biology and minored in Spanish at ISU. He pursued graduate studies in the Department of Ecology and Evolutionary Biology at Princeton University and received his Ph.D. from Princeton in 1997. His Ph.D. dissertation explored mechanisms that maintain plant species diversity in a tropical forest in Panama. After spending postdoctoral stints at the Smithsonian Tropical Research Institute in Panama, the University of California at Santa Barbara, and Cornell University, he joined the faculty of the Department of Biological Sciences at Louisiana State University. His current research interests concern the population, community, and evolutionary ecology of organisms inhabiting tropical forests and subtropical pine savannas.

### Jason Koontz, Augustana College

Jason Koontz earned his B.S. from Iowa State University, M.S. from Miami University (Ohio), and Ph.D. from Washington State University, all in botany. For four years he worked as an assistant research scientist in plant systematics with the Illinois Natural History Survey in Champaign, Ill., doing botanical surveys at Illinois Department of Transportation project sites. He also continued his research on the larkspur genus (*Delphinium*) and conservation genetics of rare plants. During that time, Dr. Koontz held an affiliate assistant professor position in the Department of Plant Biology at the University of Illinois at Urbana–Champaign where he co-taught Integrative Biology 100/101, a non-majors, general biology course. Currently, Dr. Koontz is an assistant professor of biology at Augustana College, Rock Island, Ill. where he teaches botany and cell biology.

### Michael Simmons, University of Minnesota–Twin Cities

Michael Simmons is a professor in the Department of Genetics, Cell Biology and Development at the University of Minneota–Twin Cities. He received his B.A. degree in biology from St. Vincent College, and his M.S. and Ph.D. degrees in genetics from the University of Wisconsin–Madison. Dr. Simmons has taught courses in general biology, genetics, population biology, population genetics, and molecular biology. His research activities are focused on the genetic significance of transposable genetic elements, especially those present in the genome of *Drosophila melanogaster*. He has served on advisory committees at the National Institutes of Health and is currently a member of the editorial board of *Genetics*, published by the Genetics Society of America. In 1986, Dr. Simmons received the Morse-Amoco teaching award from the University of Minnesota in recognition of his contributions to undergraduate education.

### William Zimmerman, Amherst College

William Zimmerman teaches a course on human sociobiology at Amherst College ("The Evolution of Human Nature") and has co-written a text with Timothy Goldsmith, *Biology, Evolution and Human Nature*, published by Wiley.

## ANCILLARY PROGRAM AUTHORS FOR
### *BIOLOGY: UNDERSTANDING LIFE*

*A textbook's support package is crucial to helping an instructor achieve the goal of creating the optimal course experience. In order to ensure that each component of this text's ancillary program reflects Biology:* Understanding Life*'s attention to teaching strategies and learning styles, we brought together a group of seasoned, highly talented educators to help us craft effective tools. Rather than supplying the "same old, same old" support products—many of which we toss aside and never use—each team member listed here brought to their task a dedication to creating innovative solutions to teaching and learning challenges. For complete descriptions of this text's ancillaries, including those already described in this preface, please consult the "supplements" link on this text's companion Web site at www.wiley.com/college/alters.*

### Jerry Cook, Sam Houston State University
**PowerPoint Presentations**

Jerry Cook has been an assistant professor of biology at Sam Houston State University for the past five years, teaching non-majors contemporary biology and several major's-level classes. Jerry believes that adding a visual component to lectures is an effective way to demonstrate the content of the topic and to increase student understanding.

### Lee Dorosz
**www.wiley.com/college/alters–Careers in Biology Site**

Lee Dorosz is now a retired prof. of biology, having taught at San Jose State University beginning in 1970 and having chaired the department for eight years. To prepare his students for a lifetime of education, he spends as much time exploring with them why they are learning as what they are learning. He focuses on linking the classroom closely to the realities of their young lives. He considers biology an absolutely wonderful vehicle with which to make such links.

### Anthony Gaudin, Ivy Tech State College
**Instructor's Resource Guide—Lecture Outlines**

Anthony Gaudin has been a professor of biology at Ivy Tech State College since 1998 and has been dean of academic affairs since 1999. He has an extensive background in education and has taught courses in biology, anatomy and physiology, vertebrate biology, and herpetology. He believes that properly structured classroom lectures enable instructors to communicate more effectively with students.

**Tom Kubiszyn, University of Houston**
**Assessment Consultant for**
**Levels of Understanding Testing Program**

Tom Kubiszyn taught assessment, statistics and applied psychology courses at the University of Texas at Austin from 1976 to 2003. He is currently professor and director of school psychology training at the University of Houston, and his introductory tests and measurements text is now in its seventh edition. The approach of *Biology: Understanding Life*'s text's testing program reflects Tom's conviction that technically sound, challenging, and fair tests and assessments can effectively enhance instruction and student motivation and satisfaction.

**Ann Lumsden, Florida State University**
**Teaching Assistant's Guide**

Ann Lumsden has taught biology at Florida State University for the past 21 years and is currently the coordinator for the Non-majors Biology Program and the Teaching and Learning Workshop for graduate students. Ann has worked extensively in training teaching assistants and feels that proper guidance and instruction in the teaching process can greatly benefit both teaching assistants and their students.

**Kimberly Lyle-Ippolito, Anderson University**
**Levels of Understanding Testing Program**

Kimberly Lyle-Ippolito has been teaching at the undergraduate level at Anderson University for four years, following a career in research and clinical medicine. In her four years at Anderson she has been awarded Professor of the Year twice. Kimberly feels that relevant test items serve to make biology more meaningful and enjoyable for students.

**Patricia Mancini, Bridgewater State College**
**Levels of Understanding Testing Program**

Patricia Mancini is an assistant professor of biological sciences at Bridgewater State College, where she teaches majors' and non-majors' courses in microbiology, parasitology, immunology, and general biology. Her teaching centers on critical thinking and understanding the conceptual basis of a subject before further independent exploration.

**Kathleen Marrs, Indiana University**
**Purdue University Indianapolis**
**Instructor's Resource Guide–***In the News*

Kathleen A. Marrs joined the Department of Biology at Indiana University Purdue University Indianapolis in 1998. She works to identify and target misconceptions that students bring to biology courses, create interactivity and active learning in large lecture courses, and enhance content knowledge in biology using a Web-based method known as "just-in-time teaching."

**Peter Ommundsen**
**Levels of Understanding Testing Program**

Peter Ommundsen has taught college-level biology for 32 years. He is an advocate of problem-based learning and favors test questions that incorporate realistic contexts and decisions that might confront citizens and employees.

**John Peters, College of Charleston**
**www.wiley.com/college/alters–Studying Biology Site**

John Peters is a senior instructor of biology and directs the supplemental instruction program for the natural sciences at the College of Charleston. He is currently researching the effect of problem-based learning on science literacy in non-science majors.

**Michael Rutledge, Middle Tennessee State University**
**Instructor's Resource Guide–Activities**

Michael Rutledge teaches introductory biology at Middle Tennessee State University. He seeks to promote meaningful learning in his class through the use of activities and assignments that make the course student centered and the content relevant to students.

**Kim Cleary Sadler, Middle Tennessee State University**
**Instructor's Resource Guide–Activities**

An advocate of cooperative learning in large lecture classes, Kim Cleary Sadler has taught non-majors introductory biology at Middle Tennessee State University for 15 years. Guided by current research findings on active learning, she creates mini learning communities within her large classes and uses multiple strategies to reinforce application and relevance of key biological concepts.

**Brian Shmaefsky, Kingwood College**
**www.wiley.com/college/alters–Self-Quizzes**
**Levels of Understanding Testing Program**

Brian Shmaefsky has been a professor of biology at Kingwood College for the past 12 years and coordinates the college's service learning program. He stresses concept application with his students and uses his biochemical industry experience to develop ways of incorporating "real world" science skills into his teaching.

**Marilyn Shopper, Johnson County Community College**
**Student Study Guide**

Marilyn Shopper has taught principles of biology and human anatomy and physiology at Johnson County Community College for twenty-four years. Marilyn believes the student must be active and involved in his or her learning. It is essential for the student to make connections among concepts and connect those concepts to real life.

**Sue Trammell, John A. Logan College**
**eGrade Plus – Assessment Exercises**
**Levels of Understanding Testing Program**

Sue Trammell has taught biology and botany at the community college level for over fifteen years and has been at John A. Logan College for the past three years. She believes biology students should be given opportunities to solve problems and do experiments in order to develop critical thinking skills about the natural world.

**Mark Walvoord, University of Oklahoma**
**Personal Response System Questions**

Mark Walvoord has taught introductory zoology at the University of Oklahoma for the past year and a half. He works to implement and experiment with the latest technological advances in the classroom, including on-line course management and personal response systems to encourage student participation, interest, and learning.

# BRIEF CONTENTS

# CONTENTS

---

## Chapter 17: SPECIATION AND EXTINCTION: HOW SPECIES ARISE AND DIE OUT    264

## CHAPTER 18 THE EVOLUTION OF LIFE ON EARTH    281

# PART 5: BIOLOGICAL DIVERSITY AND ITS EVOLUTION    307

## Chapter 19 VIRUSES AND BACTERIA    307

## CHAPTER 31 EXCRETION 532

## CHAPTER 32 NERVOUS SYSTEM COMMUNICATION 548

## Chapter 33 THE SENSES 572

## CHAPTER 34 PROTECTION, SUPPORT, AND MOVEMENT 591

## Chapter 35  HORMONES   610

## Chapter 36  SEX, REPRODUCTION, AND DEVELOPMENT   627

## PART 8: INTERACTIONS AMONG ORGANISMS AND WITH THE ENVIRONMENT   654

### Chapter 37  ANIMAL BEHAVIOR   654

**Chapter 42 THE BIOSPHERE: TODAY AND TOMORROW 739**

IN THE NEWS  Shark Test, 739

# Just Wondering & How Science Works Boxes

## just wondering . . .

## How Science Works

# BIOLOGY: UNDERSTANDING LIFE

## In The News | Galactic Real Estate

Our title, *Biology: Understanding Life*, poses a serious goal—comprehending and developing insight into how life works. This chapter takes the first steps toward achieving that goal by describing the characteristics of life. The video news story, "Galactic Real Estate," reveals how our Earth provides us with the requirements for life as we know it. You can view this video by visiting the *In The News* section of this text's companion Web site at www.wiley.com/college/alters.

What does it take for a planet to be habitable? Astrobiologist Guillermo Gonzalez of Iowa State University and his colleagues at the University of Washington have researched this question, and their results are in the news. Their research has produced a description of the Galactic Habitable Zone (GHZ), places that are most hospitable to life in the Milky Way galaxy. A galaxy is a large system of billions of stars, some with planets, plus various types of interstellar matter. Our solar system is situated within the Milky Way.

Prior to Gonzalez's work, astronomers usually focused on the Circumstellar Habitable Zone (CHZ) when thinking about life. The CHZ is the region around a star that has the right conditions for a planet to maintain surface water for at least a few billion years. Gonzalez and his team have studied a broader area of galactic real estate—the entire Milky Way—and a broader set of conditions. They have identified three necessary requirements for a planet to sustain life: an ocean and dry land, moderately high levels of oxygen and low levels of carbon dioxide, and long-term climate stability.

Conversely, the Gonzalez team notes that many conditions existing in certain regions of our galaxy negate the possibility of life there. First, the center of the galaxy often emits dangerous radiation that would kill any life nearby. At the far reaches of the galaxy there are not enough heavy elements to form stars and planets of a size able to support life. Not surprisingly, then, Gonzalez and his colleagues have determined that the GHZ is a ring halfway out on the disk of the Milky Way, but they note that not all places there can support life. Nevertheless, the Earth is in the GHZ and supports life quite well. What does the Earth have that makes it such a good place to live?

Gonzalez notes that the Earth has a moon, whose gravitational pull (along with that of the sun) helps stabilize the angle of the axis around which the Earth rotates. A stable rotational axis is important for maintaining a stable climate. The Earth is also the proper distance from the sun and has a nearly circular orbit, which maintains liquid water on the planet. The large planet Jupiter also helps shield the Earth from too frequent comet impacts. Yes, we on Earth have location, location, location!

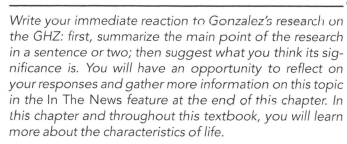

*Write your immediate reaction to Gonzalez's research on the GHZ: first, summarize the main point of the research in a sentence or two; then suggest what you think its significance is. You will have an opportunity to reflect on your responses and gather more information on this topic in the* In The News *feature at the end of this chapter. In this chapter and throughout this textbook, you will learn more about the characteristics of life.*

## CHAPTER GUIDE

### What is biology?

**1.1** Biology is the study of life.

**1.2** Biology reveals the themes and characteristics of life.

### What are the themes and characteristics of life?

**1.3** Organisms have a cellular composition and hierarchical organization.

**1.4** Organisms interact with one another and the environment.

**1.5** Organisms are more than the sum of their parts.

**1.6** Organisms reproduce, passing on biological information.

**1.7** Organisms use and transform energy.

**1.8** The structures of organisms fit their functions.

**1.9** Organisms exhibit diversity and unity.

**1.10** Organisms alive today descended from those that lived long ago.

### How do scientists organize the diversity of life?

**1.11** Scientists use classification systems to organize the living world.

**1.12** Scientists classify organisms based on common ancestry.

**1.13** An increasingly accepted classification scheme includes three domains.

# What is biology?

## 1.1 Biology is the study of life.

When you decided to take this course, you probably had a good notion of what biology was all about—organisms and how they work. You probably expected to learn about plants, animals, and microorganisms, and hopefully about yourself. However, you may not have realized the wide scope of the discipline we call biology. In fact, biology comprises many subdisciplines, which are narrower areas of study under the larger biology "umbrella."

For example, molecular biologists study life at the chemical level, probing the workings of the hereditary material and the molecular "chain of command" within cells. Cell biologists study life from a slightly different perspective. They study individual cells or groups of cells, often by growing them outside of organisms in cultures; they examine cell-to-cell interactions and the effects of the environment on cells. Many cancer researchers are also cell biologists because they study this disease on the cellular level. Some biologists work at the organism level, studying animals, plants, and other multicellular organisms. Some biologists study populations—individuals of the same species occurring together at one place and at one time—and are interested in interactions among them. These population biologists study topics such as the changes in population sizes and the causes of these fluctuations. Yet other biologists work with a global orientation and are therefore interested in questions that have worldwide impact. For example, some study what effects changes in global weather patterns and the burning of billions of acres of tropical rain forest have on the world of living things.

As you can see, biology comprises areas of study that focus on life at a variety of levels and from a diversity of perspectives. Biologists ask questions that probe the intricacies of life, calling on the knowledge and techniques of related fields. They answer questions that add to our knowledge base and investigate areas such as the transfer of hereditary material from organism to organism, the relationship between diet and disease, and the development of new food plants. This exploration advances knowledge in applied fields such as medicine, agriculture, and industry and results in the creation of products that enhance and lengthen lives. As you read *Biology: Understanding Life*, you will find out more about these and other topics that are a part of biology today. As the ancient Greeks put it, we will have a *bios logo*—a discourse on life.

## 1.2 Biology reveals the themes and characteristics of life.

Nonscientists—such as yourself—often view biology as an accumulation of facts, but it is much more than that. All sciences, including biology, are a way, or process, of understanding the world. Studying biology involves learning about problem solving, the way scientists go about their work, and worlds you may know little about—like the world of microbes. Studying biology involves learning how to protect the planet and how to take care of your body. Most of all, when you study biology you are learning about yourself—your evolutionary roots and your connectedness to all things living on this Earth.

As a result of observing living things, their environments, and the interactions between them, biologists have posed questions, put forth hypotheses based on their observations, and tested their predictions of the living world. Rising out of this abundance of tested predictions are the themes, or accepted explanations, of life that permeate the science of biology. Although the specific details of these themes may be updated or changed as biologists modify their hypotheses, the themes themselves transcend time, describing the characteristics of life and helping to answer the question, "What distinguishes the living from the nonliving?" However, as you read the following descriptions of these themes and related characteristics (also listed in **Table 1.1**), you may realize that nonliving things possess certain of these characteristics too. Taken together, however, these ideas embody life; and that is where we begin—with the idea that the whole is equal to more than the sum of its parts.

| TABLE 1.1 | The Characteristics of Life |
| --- | --- |

- Cellular composition and hierarchical organization (cells → tissues → organs → organ systems)
- Interaction with one another and the environment
- Emergent properties (the characteristics of an organism result from its organization, not simply from the sum of its parts)
- DNA as hereditary material, directing structure and function, which is passed on through reproduction
- Transformation of energy and maintenance of a steady state
- Form that fits function
- Diversity in type with unity in patterns among organisms
- Evolution of species

# What are the themes and characteristics of life?

## 1.3 Organisms have a cellular composition and hierarchical organization.

First, what constitutes the "parts" of living things? You have probably known the answer to this question since middle school—possibly before: All living things are composed of cells. **Figure 1.1** shows an example of the cellular structure of living things. Here, cells making up the outer layer of the human gallbladder are lined up in rows.

A **cell** is a microscopic mass of protoplasm. It is bounded by a membrane and contains a chemically active mixture of complex substances suspended in water, including hereditary material. Until the invention of the microscope around 1600, naturalists could not see or study this invisible level of organization of living things. It was not until the mid-1800s that advances in the technology of the microscope allowed botanist Matthias Schleiden and zoologist Theodor Schwann to determine, after examining many different types of organisms, that the cell is the smallest living unit of structure of organisms. Shortly thereafter, the German medical microscopist Rudolf Virchow argued that all cells could arise only from preexisting cells. The cells, he wrote, are "the last constant link in the great chain of mutually subordinated formations that form tissues, organs, systems, the individual. Below them is nothing but change."

Virchow's statement refers to the hierarchy, or levels, of organization seen in all living things. Living things, or **organisms**, are either *multicellular* (composed of many cells) or *unicellular* (composed of a single cell). Single-celled organisms that work and live together as a team are called *colonial organisms*, like the *Volvox* shown in **Figure 1.2**. Whether unicellular, multicellular, or colonial, every organism has the cell as its simplest level of structure and function.

Smaller units, such as atoms and molecules, make up cells but are not living units. The hereditary material, or DNA, present in almost every cell of an organism's body is a molecule composed of atoms. Although this huge molecule is critical to life and is passed on from parent to offspring, it is still not a living unit by itself.

Cells in multicellular organisms are organized to form the structures and perform the functions of the organism. The next more complex, or more inclusive, level of organization is the tissue. **Tissues** are groups of similar cells that work together to perform a function. Grouped together, various tissues form a structural and functional unit called an **organ**. An **organ system** is a group of organs that function together to carry out the principal activities of the organism—its most complex level of organization.

Interactions take place within an organism among its levels of organization. For example, when a chameleon sees an insect land on a nearby plant, a complex series of events occur within the chameleon. First, the nerve *cells* embedded in the back wall of its eyes (*organs*) conduct impulses to its brain (an *organ*). The vision center within the brain (*nervous tissue*) interprets these impulses, resulting in the chameleon seeing the insect. Impulses speed to other brain centers to coordinate a rapid response in the form of nerve impulses to the muscles (*tissues*) in its tongue (an *organ*). The muscles respond at incredible speed, allowing the tongue to extend to its prey, as shown in **Figure 1.3**. Had it been a better day for the insect, its nervous system would have reacted more quickly in perceiving and responding to the threat.

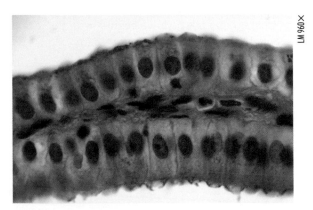

LM 960×

**Figure 1.1** Cellular structure of the outer layer of the human gallbladder. The gallbladder, where bile is stored and concentrated, is a sac attached to the underside of the liver. Bile helps digest fats.

**Visual Thinking:** Describe what the cells look like in this micrograph.

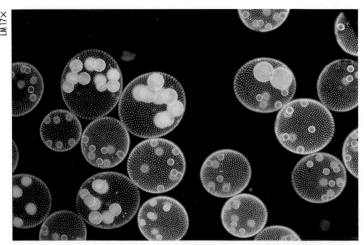

LM 17×

**Figure 1.2** The cells of *Volvox* work as a team. These algae live in ponds and streams. The outer spheres are each a colony of cells that work together. The dots on each sphere are individual cells, which are held together by strands of protoplasm. Within the spheres are newly developing colonies.

**Figure 1.3** The long tongue of the chameleon extends to snare its insect prey.

**Figure 1.4** The white growths on these roots are fungi, called mycorrhizae. The plant and the fungi live in a close association for the benefit of both.

## 1.4 Organisms interact with one another and the environment.

Interactions occur not only among the levels of organization within organisms but also between organisms and their external environments. Biologists usually classify the interactions between organisms and their environments into the following hierarchy: populations, communities, ecosystems, and the biosphere.

A **population** consists of the individuals of a given species that occur together at one place and at one time. Ernst Mayr, one of the leading evolutionary biologists of the past century, defines **species** (SPEE-shees or SPEE-sees) as "groups of interbreeding natural populations that are reproductively isolated from other such groups." Reproductive isolation means that members of one species do not breed with members of other species in the wild. Mayr's definition of species will be used throughout this book. Organisms that do not reproduce sexually such as bacteria and certain plants, animals, protists, and fungi are designated as species by means of their morphological characteristics (form and structure), and biochemical characteristics (chemical composition and processes). No definition of species is accepted by all biologists. There are many species concepts, as discussed in Chapter 17.

The foxes living in a small forest are an example of a population. The foxes interact with one another in a variety of ways. Sometimes they compete for the same limited resources and for mates. Conversely, individuals within animal populations may also work cooperatively for common purposes. Foxes often hunt together, for example, working with one another to overtake and trap prey.

Populations of different species that interact with one another make up a **community** of organisms. A forest community may be made up of populations of bacteria, fungi, earthworms, plant-eating and animal-eating insects, mice, deer, salamanders, frogs,

foxes, snakes, hawks, trees, and grasses. These populations within the forest community compete with one another for resources, as do organisms within populations. The foxes, snakes, and hawks, for example, compete with one another to capture the mice for food. Other types of interactions may also exist within the community, such as mutualistic relationships in which two different species live in a close association for the benefit of both. For example, fungi and plants are often interdependent. Some fungi live near the roots of many plants, such as trees in a forest. These fungi (seen as white in **Figure 1.4**) envelop the roots and send billions of minute cell extensions into the soil. These microscopic "fingers" of fungus absorb water and nutrients better than the plants' roots could alone, and they pass the substances to the plant. In turn, the fungus uses certain products that the plant makes by photosynthesis.

An **ecosystem** is a community and its physical environment, including all the interactions among organisms and between organisms and the environment. The forest ecosystem includes all the organisms previously discussed as well as nonliving components of the environment such as air and water, which contribute substances needed for the ecosystem to function.

The **biosphere** is the part of the Earth where biological activity exists. Most people refer to the biosphere as simply "the environment." Within this global environment, living things interact with each other and with nonliving resources in various ways. Organisms other than humans use only *renewable* (replaceable) *resources*. When they die, decomposers return the nutrients held within the dead organisms to the soil and air. Humans, however, use many *nonrenewable resources* and generate large amounts of waste. These levels of organization—from population to biosphere—build on the levels of organization of individual organisms from atoms to the organism. All these levels are depicted in **Figure 1.5**.

### CONCEPT CHECKPOINT

1. What is the hierarchy of structure and function in organisms, beginning with cells?
2. What is the hierarchy of interactions between organisms and the environment?

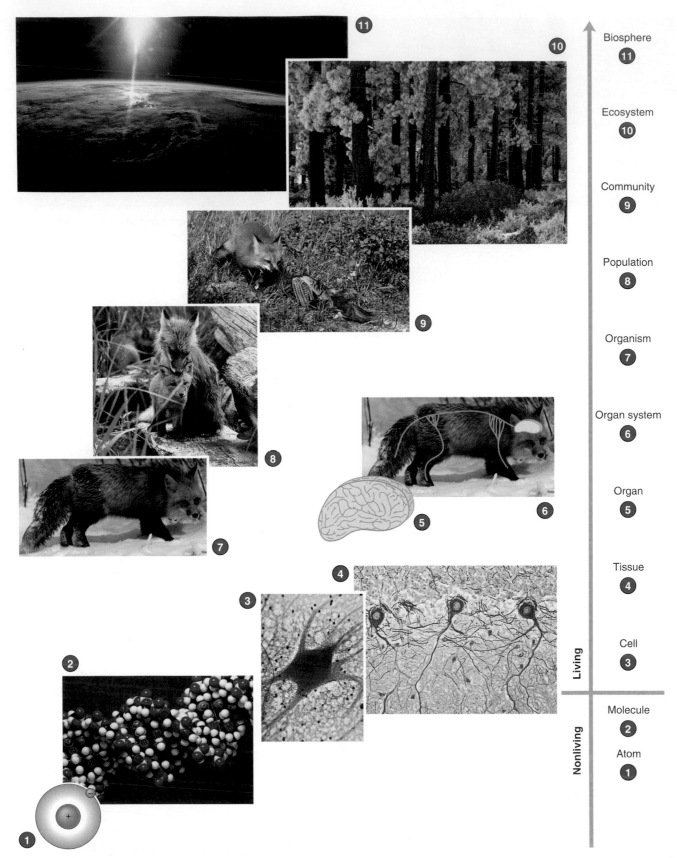

**Visual Summary** **Figure 1.5 The levels of biological organization.** The progression of organization from atoms to the biosphere. The "lowest," or least complex, level of biological structure and function is the cell. The "highest," or most complex, level of biological structure and function is the biosphere.

## 1.5 Organisms are more than the sum of their parts.

As you were reading about organisms and how they interact with one another, you may have gained insight into the earlier statement, "the whole is equal to more than the sum of its parts." It is an important concept in biology and is termed *emergent properties*. Cells are much more than the atoms and molecules that compose them. Tissues are much more than the cells that make up their structure, and so on. In other words, the workings of a cell, tissue, organ, system, or organism cannot be predicted simply by knowing which components make up its structure.

To illustrate further, you might decide to build a cabinet to house your CD player and CDs. If your neighbor came over and looked at the materials you bought to build the cabinet—wood, nails, hinges, and wood stain—he or she would not be able to tell what you were building and how it would function. Only when you organize the materials—build the cabinet—will your neighbor be able to see the properties of your CD cabinet. Pieces of wood, unable to function as doors by themselves, now take on the properties of doors when attached with hinges to the body of the cabinet. These new characteristics have emerged at this higher level of organization of the wood, nails, hinges, and stain. So, too, do unique properties of living things emerge at each level of their organization. These emergent properties cannot be predicted simply by knowledge of an organism's component parts.

## 1.6 Organisms reproduce, passing on biological information.

All living things possess biological information that directs their structure and function, and ultimately, therefore, their emergent properties. This biological information is the hereditary material, or genes.

Genes are made up of molecules of **DNA**, or **deoxyribonucleic acid** (de-ok-see-RYE-boh-new-KLAY-ick). In these molecules of DNA exist the "code of life," instructions that are translated into a working organism. These instructions take the form of molecular subunits of the DNA molecule called *nucleotides*. Using only four different nucleotides, DNA codes for all the structural and functional components of an organism. We will examine DNA, its structure, and its functioning, in detail in Chapters 9 and 10. A model of this complex molecule is shown in **Figure 1.6**.

The secret to DNA's ability to carry information regarding the variety of structural and functional components of all living things lies

**Figure 1.6** A model of DNA, the hereditary material.

in its code. When translated in a living organism, the code represented by a single gene results in the formation of a specific molecule designed to do a specific job.

DNA is passed from organism to organism during reproduction, as organisms give rise to others of their kind. Some organisms reproduce sexually and some asexually, whereas others reproduce using both methods. In sexual reproduction, two parents give rise to offspring; in asexual reproduction, a single parent gives rise to offspring. In either case, genes—the units of heredity—are passed from one generation to the next.

### CONCEPT CHECKPOINT

3. Develop an example of emergent properties to show that you understand this concept.
4. How does DNA pass on biological information from parent to offspring?

## 1.7 Organisms use and transform energy.

You need energy to get through the day, just as all organisms need energy to do the work of living. This work involves many processes, such as movement, cell repair, reproduction, and growth. It also involves maintaining a stable internal environment in spite of a differing external environment. Energy drives the chemical reactions that underlie all these activities.

What is the source of the energy that fuels life processes and helps organisms maintain an inner equilibrium? Ultimately, the energy for nearly all life comes from the sun. For example, certain organisms within our forest ecosystem such as the trees and grasses make their own food by capturing energy from the sun in a process called *photosynthesis* (fote-oh-SIN-thuh-sis). These organisms are called *producers*. During photosynthesis, producers convert the energy in sunlight into chemical energy by locking it within the bonds of the food molecules they synthesize. Organisms that cannot make their own food, such as the insects, deer, frogs, and hawks in the forest, are called *consumers*. They feed on the producers and on each other, passing along energy that was once captured from the sun (**Figure 1.7**). Both the producers and the consumers release the stored energy in food by breaking down its molecules, using much of the released energy to do work. Some of it is lost as heat and is therefore unusable. As a result, organisms need a continual input of energy to fuel the chains of chemical reactions that move, store, and free energy needed to perform the activities of life. *Decomposers*, such as many types of bacteria and fungi, break down the organic molecules of dead organisms, serving as the last link in the flow of energy through an ecosystem and contributing to the recycling of nutrients within the environment.

## 1.8 The structures of organisms fit their functions.

Would it make sense to try to turn a screw with a hammer or eat soup with a knife? Of course not—tools and kitchen utensils are structured in specific ways to do specific jobs. Just as the shape and structure of a

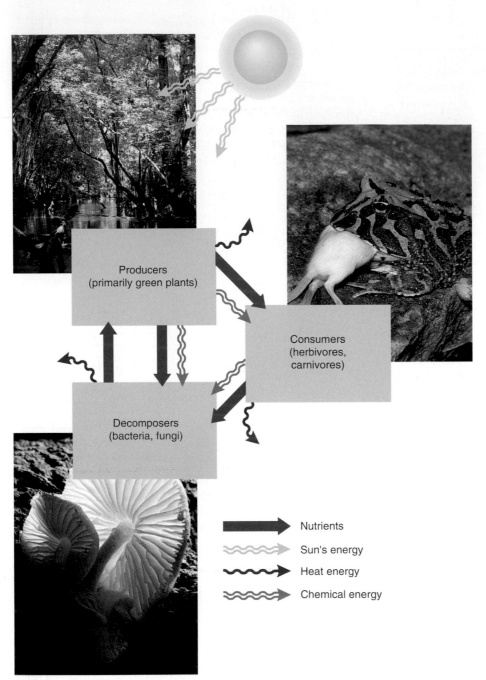

Producers
(primarily green plants)

Consumers
(herbivores,
carnivores)

Decomposers
(bacteria, fungi)

→ Nutrients

〰→ Sun's energy

〰→ Heat energy

〰→ Chemical energy

**Figure 1.7** Flow of energy through an ecosystem. Producers transform the sun's energy (yellow wavy lines) to chemical energy by means of photosynthesis. Nutrients (blue arrows) are then transferred from producer to consumer and from consumer to consumer. Decomposers such as fungi break down the organic molecules of dead organisms, making these nutrients available for reuse. Some energy is "lost" as heat (red wavy lines), thereby becoming unavailable to the ecosystem to do work; therefore, organisms need a constant input of energy to perform the activities of life.

 **Visual Thinking:** Describe what would happen if there were no longer an input of energy into the Earth's ecosystems.

For example, what types of feathers would make good insulators? As you ponder this question, think about your experiences with trying to keep warm when it's cold. You probably learned that you are warmer in cold weather when you wear many thin layers rather than a single thick layer. Layers of clothes trap air between the layers. Your body heat then warms the trapped air. So, getting back to the question of feathers—fluffy feathers would trap air and therefore make the best insulators. Down feathers are fluffy feathers (**Figure 1.8a**), and provide excellent insulation for birds. Birds have other types of feathers as well, structured for other functions, such as flight, camouflage, waterproofing, and display. Figure 1.8*b* shows the brightly colored display feathers on the head of a cockatoo, and Figure 1.8*c* shows contour feathers, which are flattened feathers on the wings and which aid in flight.

### CONCEPT CHECKPOINT

**5.** Examine Figure 1.7, which depicts the flow of energy through an ecosystem. What do you think would happen to an ecosystem if you removed one component involved in energy flow (i.e., decomposers, producers, sunlight, or consumers)?

**6.** Pick two organisms and a structure of each. Describe how form fits function.

## 1.9 Organisms exhibit diversity and unity.

The diversity, or variety, of living things is astounding. Biologists estimate that from 5 million to 30 million different species exist on Earth. As you can see from this range of numbers, scientists are unsure how many species exist; they have discovered, described, and catalogued probably fewer than half of them. Unfortunately, many species are becoming extinct, or dying out, as their habitats are destroyed, and before scientists can even study and classify them.

Only a tiny sampling of the array of the species humans have seen and categorized is shown in **Figure 1.9**. Although these organisms are very different from one another, each has characteristics common to all species. These shared characteristics, however, may or may not be visible to the naked eye. The other themes of life have described these characteristics, such as a cellular organization and the ability to transform energy to do the work of life. In the next sec-

screwdriver or a spoon fits its function, the structures of living things fit their functions. Biologists sum up this idea with the phrase "form fits function." By analyzing form, inferences can be made regarding function. Conversely, knowing function gives insight into form.

(a)                    (b)

(c)

**Figure 1.8** Different types of feathers: form fits function. (a) Down feathers act as insulators. (b) The brilliantly colored feathers on the head of this cockatoo are display feathers. (c) Quite different in form are the contour feathers used for flight.

(a)

LM 220×

(b)                    (c)

**Figure 1.9** The diversity of species. Although scientists have categorized many species, millions of species remain to be described. (a) Axolotl, an unusual amphibian. (b) Bird's nest fungi. (c) Pyrocystis, a bioluminescent single-celled organism.

tion we will explore how all organisms are related, which is the basis of why they are diverse in their types but unified in their patterns.

## 1.10 Organisms alive today are descended from those that lived long ago.

All organisms have common characteristics because they are related to one another. Just as you have a history and a family tree, so does the Earth's family of organisms. Yet all organisms have differences because as they changed over time, they diverged from one another. The Earth itself has changed from its beginnings some 4.6 billion years ago.

Scientists know little about what early Earth was like nearly 4 billion years ago, but they do agree that it was a harsh environment different from that of today. Under these conditions, scientists hypothesize that the elements and simple compounds of the primitive atmosphere reacted with one another, forming complex molecules. Biochemical change took place over time and resulted in the appearance of single-celled organisms approximately 3.5 billion years ago. The remains of these early cells (and of any organisms) preserved in rocks are called *fossils*.

Fossils provide scientists with a record of the history of life and document the changes in living things that have taken place (**Figure 1.10**). By about 2 billion years ago, the fossil record documents the existence of cells more complex than the first cells, and by 500 million years ago, an abundance of multicellular organisms—with members of groups similar to those that exist today—had appeared.

The fossil record is only one piece of evidence indicating that living things have changed over time. Using this and other types of

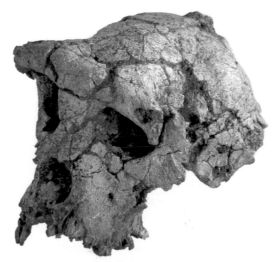

**Figure 1.10** Fossil links to human ancestors. This skull, discovered in Chad in central Africa by an international team of scientists, dates back six or seven million years. Thought to be the earliest known member of the human family, *Sahelanthropus tchadensis* was nicknamed "Toumai," which means "hope of life."

evidence, Charles Darwin, a nineteenth-century English naturalist, not only demonstrated to the scientific community the factuality of "descent with modification" (**evolution**) but also proposed a mechanism for the process. Since Darwin's time, the science of evolution has been consistently supported by an overwhelming amount of data and has therefore become the underpinning concept of biology. Evolution embodies the ideas that organisms alive today are descendants of organisms that lived long ago, and that organisms have changed and diverged from one another over billions of years. Scientists still ponder, examine, and develop hypotheses regarding details of the mechanisms of evolution, but they agree that evolution has been and is taking place.

How does evolution take place? A key idea is that some of the individuals within a population of organisms possess heritable characteristics that favor their survival. These individuals are more likely to live to reproductive age than are individuals that do not possess the favorable characteristics. These reproductively advantageous traits (called adaptive traits or adaptations) are passed on from surviving individuals to their offspring. Over time, the number of individuals carrying these traits will increase within the population, and the nature of the population as a whole will gradually change. This process of survival of the most reproductively fit organisms is called *natural selection*.

Scientists disagree regarding the pace of evolutionary change. Darwin suggested that new species develop slowly and gradually as an entire species changes over time. In the early 1970s, two American scientists, Niles Eldredge and Stephen Jay Gould, proposed that a species might remain relatively unchanged over long periods of time, punctuated with comparatively rapid periods of change. Chapters 15–18 discuss these ideas and the major concepts of evolution in more detail.

### CONCEPT CHECKPOINT

7. Pick two organisms that appear to be very different from one another and list three characteristics that they have in common.

8. For each organism that you chose, list two or three adaptations that make it well suited to survival in its environment.

# How do scientists organize the diversity of life?

## 1.11 Scientists use classification systems to organize the living world.

One reason scientists "organize" life is that there are millions of species; studying life is nearly impossible without bringing some order to the wide array and incredible number of living things. Not only does order help prevent chaos, but it helps scientists better understand organisms they study, providing a context of all life from which to view a narrow slice.

For example, you may have an interest in cars. You may spend time researching an old car you inherited—finding out about the original engine, paint, interior, and so forth. However, putting your inheritance into the context of car manufacturers (Ford, for example) and year (1965, for example) gives you additional information. Suddenly you realize that your car may be a classic. Through research you discover that it is one of the first Mustangs made— sporty cars that had their origins in the Ford Falcon, one of the first successful "small" cars. You know that the first Mustangs came "stock" with V-6 engines but that they could be ordered with a V-8. Which is yours? Does it have a hardtop or is it a convertible? If it is the famous Shelby GT 350, you know that restored it would be worth well over $100,000! Suddenly, by placing your car within the context of Ford vehicles and their manufacturing history, you have much more knowledge than without this context.

Notice, too, that having names for cars that everyone recognizes makes discussing cars easy. Just say Ford Mustang, Jeep Cherokee, or Honda Accord and everyone knows what you mean. So, too, scientists have developed names for organisms that other scientists recognize. Organisms have two scientific names, called **binomial nomenclature**, which literally means "two-name nam-

ing." We humans are *Homo sapiens*. Your housecat is *Felis domesticus*. American red squirrels are *Tamiasciurus hudsonicus*. Some of the bacteria in your intestines are *Escherichia coli*.

The Swedish botanist Carolus Linnaeus established the system of binomial nomenclature in the mid-1700s. The two names used in scientific naming, such as *Tamiasciurus hudsonicus*, are the genus name plus the species name. The "first name" or genus name is the same for all organisms in the same genus. A **genus** comprises closely related species, organisms with a common ancestor in the not-too-distant past.

For example, the American red squirrels mentioned previously are members of the genus *Tamiasciurus*. Other closely related species are also members of this genus, such as Douglas squirrels, *Tamiasciurus douglasii*. In another example, cabbage and rutabaga are closely related and are in the same genus, *Brassica*. However, their species names are different. Cabbage is *Brassica oleracea* and rutabaga is *Brassica rapus*. Notice that the genus name is capitalized and that the species name begins with a lowercase letter. Both are italicized.

Along with establishing the system of binomial nomenclature, Linnaeus developed a classification system on which contemporary classification is founded. **Classification** is the categorization of organisms into a coherent scheme. Linnaeus proposed that all organisms be grouped into two kingdoms, the plants and the animals, and established the beginnings of the modern hierarchical system of classification using the terms *kingdoms*, *classes*, *genera*, and *species*.

The classification scheme increasingly used today is shown in **Figure 1.11**—domain, kingdom, phylum, class, order, family, genus, and species. The level of **domain**, which has only recently come into use, is the most inclusive, and the level of species is the least inclusive. Thus, as shown in Figure 1.11, domains of organisms are

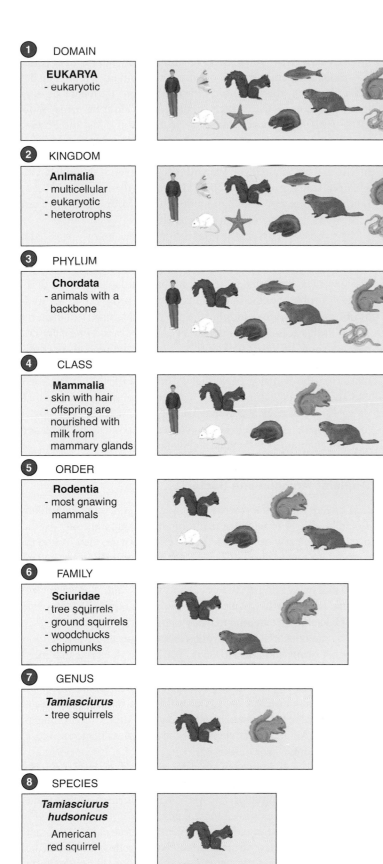

**1** DOMAIN

**EUKARYA**
- eukaryotic

**2** KINGDOM

**Animalia**
- multicellular
- eukaryotic
- heterotrophs

**3** PHYLUM

**Chordata**
- animals with a backbone

**4** CLASS

**Mammalia**
- skin with hair
- offspring are nourished with milk from mammary glands

**5** ORDER

**Rodentia**
- most gnawing mammals

**6** FAMILY

**Sciuridae**
- tree squirrels
- ground squirrels
- woodchucks
- chipmunks

**7** GENUS

*Tamiasciurus*
- tree squirrels

**8** SPECIES

*Tamiasciurus hudsonicus*

American red squirrel

**Figure 1.11 Classification scheme for a squirrel.** As we move from the most inclusive category, domain, to the least inclusive category, species, fewer organisms are included in each category, reflecting increasing relatedness.

subdivided into groupings that include fewer and fewer organisms. From domains through species, these groupings reflect increasing closeness in their evolutionary ancestry. Many types of evidence determine this closeness, including the similarities of organisms in their behavioral charcteristics (in the case of animals) and biochemical characteristics.

Each domain is subdivided into **kingdoms** and each kingdom is subdivided into **phyla** (singular, phylum [FYE-lum]). The animal kingdom, for example, has approximately 19 phyla, depending on the particular classification system. The top box in Figure 1.11 shows some members of the domain Eukarya **1** . The next box shows some representatives of the animal kingdom **2** . Note that the paramecium, the mushroom, and the tree are no longer shown—they are eukaryotes but not animals. One of the phyla within this kingdom is the phylum Chordata: animals with a backbone **3** . (See Chapter 22 and the glossary for a more complete and precise definition of chordates.) Notice that the sea star and the butterfly are not shown. These two organisms, though both animals, are not chordates. The chordates are more closely related to one another than to the sea star or butterfly.

The next subcategory is **class**. The class Mammalia **4** is one subgroup of the phylum Chordata. The snake and the fish are not included in this group because they are not mammals. Mammals have skin with hair and nourish their young with milk secreted by mammary glands. These are characteristics that fish and snakes do not have.

The next subcategory is **order**. One order of class Mammalia is the order Rodentia **5** , which includes most gnawing mammals. Humans are not gnawing animals and are not included in this subgroup. The last three subgroupings, in order of increasing relatedness, are **family 6** , genus **7** , and species **8** .

## 1.12 Scientists classify organisms based on common ancestry.

**Taxonomy** (tack-SAHN-uh-me) is the science of classifying organisms. It is not a static science because the criteria used for classification change over time as new technologies for studying life are added to the biologist's toolbox.

A brief history of taxonomy shows how scientists build on prior knowledge but adapt and refine their ideas as new ones emerge. Since Linnaeus's work predated Darwin, Linnaeus did not use evolutionary history as a criterion for classification. He classified organisms based on overall resemblances, which was called "natural" classification. His plant taxonomy, for example, was based solely on the number and arrangement of reproductive organs.

In 1859, Charles Darwin published *On the Origin of Species by Means of Natural Selection* in which he described the theory of evolution and the mechanism of natural selection (see Chapter 15). As biologists came to accept his work, they began to stress the significance of evolutionary relationships in classification. One of those scientists was zoologist Ernst Haeckel who, in 1866, proposed a third kingdom, the Protista. The Protista comprised all bacteria, protozoans (single-celled organisms more complex than bacteria), algae, and fungi.

As biologists began to focus on evolutionary history and evolutionary relationships in their work, the science of systematics developed. **Systematics** (sis-teh-MAH-ticks) is the study of the diversity of organisms, focusing not only on their comparative relationships but also on their evolutionary relationships. As part of their work, *systematists* (sis-TEH-mah-tists) classify organisms in ways that reflect their relatedness. Therefore, the term *systematics* is often used interchangeably with the term *taxonomy* since biologists today classify organisms with evolutionary relatedness in mind.

By the mid-1900s, evidence provided by genetics and physiology became increasingly important in determining evolutionary relationships. In 1969, American ecologist Robert H. Whittaker proposed a five-kingdom classification system, which was embraced for decades and is still widely used: bacteria, protists, plants, animals, and fungi. Whittaker and all modern biologists use many types of evidence to determine shared ancestry, such as organisms' biochemistry, cell structure, development, physiology, and behavior. They also study the fossil record.

In 1977, a discovery by American microbiologist Carl R. Woese (WOHS) resulted in a challenge to the five-kingdom system. By studying sequences of molecules in microorganisms' protein-making machinery (ribosomal RNA), Woese collected evidence that bacteria were of two fundamentally different types: the bacteria and the Archaebacteria (ARE-kee-back-TEAR-ee-ah). However, as his work progressed, Woese concluded that the Archaebacteria were not bacteria at all, and changed their name to Archaea (ARE-kee-uh). Many species of these simple microorganisms live in extreme environments such as the Dead Sea, hot springs, and acid lakes.

Woese's work supported the concept that there are three principal lines of evolution. In 1990, Woese proposed the domain as a classification category above kingdom, which reflects these three evolutionary lines. The Royal Swedish Academy of Sciences in 2003 awarded Woese the Crafoord Prize in Biosciences for this work. This prize acknowledges scientific accomplishments for which there is no category in the Nobel Prizes in science.

---

### CONCEPT CHECKPOINT

9. Are organisms belonging to a single phylum classified in the same kingdom? in the same family? Why or why not?

10. What are the primary differences between the three-domain system and the five-kingdom system? During class one of your classmates makes the following remark, "What difference does it make whether there are five kingdoms or three domains? Aren't we just arguing about silly names?" How would you respond to this student?

---

## 1.13 An increasingly accepted classification scheme includes three domains.

At this time, members of the biological community continue to debate which classification scheme most accurately depicts evolutionary relationships among all life. Woese's ideas are gaining acceptance, and we will use the three-domain system as the classifi-

cation framework for this book: Archaea, Bacteria, and Eukarya (you-CARE-yuh or you-CARE-ee-uh), as shown in **Figure 1.12**.

Scientific debate exists not only at the domain level but also at other levels of classification. Currently, there is debate on how many kingdoms exist in each domain. For example, some biologists, particularly those who study protists, think that the Protista should be divided into about 20 kingdoms!

In this book, we will discuss the Archaea and the Bacteria primarily at the domain level. We will recognize four kingdoms in the Eukarya domain: Protista (pro-TISS-tah), Plantae (PLAN-tee), Fungi (FUN-jeye), and Animalia (ah-nih-MAY-lee-ah). Figure 1.12 shows members of each of these groups.

As you can see from the brief history of taxonomy presented in the previous section, a sharp shift in taxonomy came with Darwin's work, which caused changes in classification schemes as technologies advanced that helped trace evolutionary relatedness. Similarly, Woese's proposal of the three-domain system arose out of his study of ribosomal RNA to determine the evolutionary history of bacteria.

Will this classification debate ever be resolved? Probably not, because some scientists will always favor certain methods over others to determine relatedness and will debate inferences that others make from the evidence. Nevertheless, one way of looking at life will rise above the others, becoming the most commonly accepted approach in the biological community. The three-domain system is on the rise today, although it contains many classification "wrinkles" that will continue to be debated and worked out in years to come.

### Domains Archaea and Bacteria

If you were asked to tell what you know about bacteria, you likely would be able to provide some information. You might suggest that bacteria are nearly everywhere in our environment: they live on the surfaces we touch, the clothes we wear, and even within our bodies! They are single cells that cause many common diseases. You might know that many can be killed with antibiotics, although some have become dangerously resistant. You might even mention that a variety of bacteria are essential to many important biological processes, such as the ones that take nitrogen from the air and convert it to a form that plants can use. However, you would likely be speechless when asked about archaeans, organisms in the domain Archaea.

Although they are simple cells like bacteria, archaeans do not cause common diseases such as strep throat or some sexually transmitted infections. They are usually found in extreme environments, such as the Antarctic, the depths of the ocean, volcanic areas, and swamp bottoms.

Bacteria and archaeans are primarily distinguished from one another by their ribosomal RNA. Other differences include chemical variations within their cell walls, within their cell membranes, and in certain of their physiological processes.

Archaeans and bacteria are alike, too: both lack membrane-enclosed intracellular structures called organelles. However, they do have some organelles that are not enclosed by membranes, and neither has a membrane that surrounds the hereditary material of the cell, forming a nucleus. Instead, their genetic material is located within a particular region within the cell called a nucleoid. Such unicellular organisms, having no membrane-bounded organelles and nuclei, are called **prokaryotes** (pro-KARE-ee-oats). This word means before (*pro-*) the nucleus (-*karyon*).

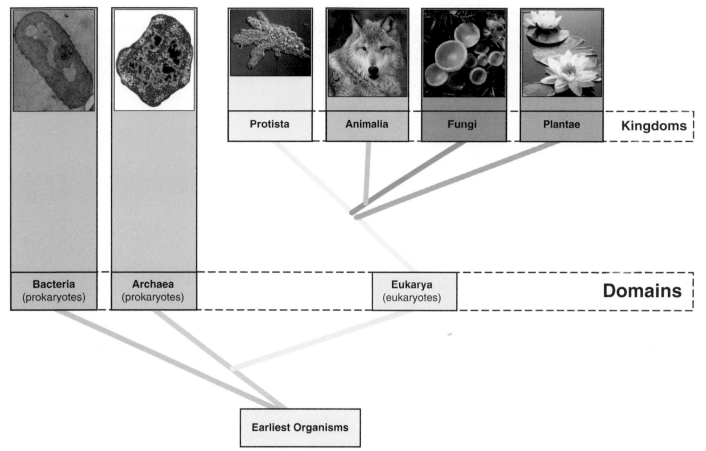

**Figure 1.12 Members of the three domains, showing the four kingdoms of domain Eukarya.** Left to right; *Escherichia coli*, a normal bacterial resident of the vertebrate large intestine; *Methanocuccus voltae*, a methane-producing archaebacterium; *Ameba proteus*, a protist commonly found in fresh water, the oceans, and the upper layers of the soil; *Canis lupus*, the timber wolf, which lives in packs in nothern climates; *Sowerbyella rhenana*, an orange cup fungus from the Willamete Valley in Oregon; and *Nymphaea alba*, the European white waterlily, which commonly grows in fresh water ponds and lakes. This "tree of life" shows that prokaryotes arose first, with eukaryotes evolving from prokaryotes. Protists were the first eukaryotes to develop, with plants, fungi, and animals arising from the protists.

## Domain Eukarya

The organisms comprising the domain Eukarya are either unicellular or multicellular. Each of their cells contains membrane-bounded organelles and a membrane-bounded nucleus. Such cells are called **eukaryotes** (you-KARE-ee-oats), a word that means true (*eu-*) nucleus (*-karyon*). Four kingdoms of eukaryotes comprise the domain Eukarya.

**Kingdom Protista** Protista, or protists, are single-celled eukaryotic organisms that are much more complex and much larger than bacteria and archaeans. You can easily view protists by collecting a small amount of pond water and looking at it under a microscope. You may be familiar with some common single-celled protists such as the bloblike ameba or the oval-shaped paramecium. Protists also include the algae, which have both multicellular and unicellular forms. Algae are classified as protists because they seem to be more closely related to this group than to the plants, animals, or fungi.

As mentioned previously, the classification of protists is in a state of flux because our knowledge of their evolutionary history at this time is incomplete. Protists represent an important step in early evolution. They evolved from prokaryotes, most likely from simple communities of prokaryotic cells. Eventually, they gave rise to the entire line of eukaryotes as shown in Figure 1.12.

**Kingdom Fungi** If you've ever seen mold growing on stale bread or on rotting fruit, or mushrooms growing in the forest, you are familiar with some members of the fungi kingdom. Fungi are decomposers and survive by breaking down substances and absorbing the breakdown products. Therefore, they are considered absorptive **heterotrophs**; that is, they obtain food by eating and digesting other organisms. Most fungi are multicellular; a few, such as the yeasts, are unicellular.

Most fungi are *saprophytes*; that is, they live on organisms that are no longer alive, such as fallen trees and leaves, and dead animals. Some fungi, however, are *parasites*—they live on living organisms. Certain human diseases and conditions such as athlete's foot and ringworm are caused by fungi, as are plant diseases such as potato blight.

**Kingdom Plantae** The plants are multicellular eukaryotic organisms that make their own food; such organisms are called **autotrophs**. (Autotrophs are producers.) Plants are photosynthetic autotrophs; that is, they make their own food by the process of photosynthesis, in which they use carbon dioxide from the air, light energy from the sun, and water. We discuss this process in depth in Chapter 7. Plants probably evolved from multicellular green algae. Examples of plants are mosses, ferns, pine trees, and flowering plants.

**Kingdom Animalia** Animals are multicellular eukaryotic organisms that cannot make their own food; such organisms are heterotrophs. (Heterotrophs are consumers or decomposers.)

The animal kingdom is an extraordinarily diverse group of organisms, which can be categorized in two groups: those without a backbone (the invertebrates) and those with a backbone (the chordates, mainly the vertebrates). Vertebrae are the series of bones that stacked one upon the other comprise the vertebral column, or backbone. Examples of invertebrates are worms, sea stars, spiders, crabs, and insects. Examples of vertebrates are dogs, cats, fish, birds, frogs, snakes, and, of course, humans.

### CONCEPT CHECKPOINT

**11.** What do all organisms of the domain Eukarya have in common?

**12.** Summarize how organisms in each of the kingdoms of the domain Eukarya acquire energy from their environments.

## CHAPTER REVIEW

## Summary of Key Concepts and Key Terms

### What is biology?

**1.1** Biology is the study of life.

**1.1** Biology comprises areas of study that focus on life at a variety of levels and from a diversity of perspectives.

**1.2** The themes of biology are science-based explanations regarding the structure, function, and behavior of living things as well as their interrelationships.

### What are the themes and characteristics of life?

**1.3** Living things (**organisms**, p. 4) are composed of **cells** (p. 4) and are hierarchically organized.

**1.3** **Tissues** (p. 4) are groups of similar cells that work together to perform a function; grouped together, various tissues form a structural and functional unit called an **organ** (p. 4); an **organ system** (p. 4) is a group of organs that function together to carry out the principal activities of the organism—its most complex level of organization.

**1.4** Living things interact with each other and their environments.

**1.4** A **species** (SPEE-shees or SPEE-sees; p. 5) is a group of interbreeding natural populations that are reproductively isolated from other such groups.

**1.4** A **population** (p. 5) consists of the individuals of a given species that occur together at one place and at one time.

**1.4** Populations of different species that interact with one another make up a **community** (p. 5) of organisms.

**1.4** An **ecosystem** (p. 5) is a community and its physical environment, including all the interactions among organisms and between organisms and the environment.

**1.4** The **biosphere** (p. 5) is the part of the Earth where biological activity exists.

**1.5** Living things have emergent properties.

**1.6** Living things reproduce and pass on biological information to their offspring.

**1.6** This biological information is the hereditary material, or genes; genes are made up of molecules of **DNA**, or deoxyribonucleic acid (de-ok-see-RYE-boh-new-KLAY-ick, p. 7).

**1.7** Living things transform energy and maintain a steady internal environment.

**1.8** Living things exhibit forms that fit their functions.

**1.9** Living things display both diversity and unity.

**1.10** Species change over time, or evolve (**evolution**; ev-oh-LOO-shun, p. 10).

### How do scientists organize the diversity of life?

**1.11** Organisms have two scientific names, which is referred to as **binomial nomenclature** (p. 10).

**1.11** Scientists use **classification** (p. 10) systems in which they categorize organisms into a coherent scheme to order life.

**1.11** The classification scheme increasingly used today is **domain** (p. 10), **kingdom** (p. 11), **phyla** (p. 11), **class** (p. 11), **order** (p. 11), **family** (p. 11), **genus** (p. 10), and species.

**1.12** Modern **taxonomy** (tack-SAHN-uh-me, p. 11), the science of classifying organisms, categorizes organisms based on their common ancestry (their evolutionary history).

**1.12** **Systematics** (sis-teh-MAH-ticks, p. 12) is the study of the diversity of organisms, focusing on their comparative relationships and their evolutionary relationships.

**1.12** In 1990, American microbiologist Carl Woese proposed the domain as a new category in the classification of life.

**1.13** This book uses the three-domain system of classification: Archaea, Bacteria, and Eukarya.

**1.13** Archaeans and bacteria are single-celled organisms that have no membrane-bounded organelles and nuclei; such organisms are called **prokaryotes** (pro-KARE- ee-oats, p. 12).

**1.13** **Eukaryotes** (you-KARE-ee-oats, p. 13) have cells with membrane-bounded organelles and a membrane-bounded nucleus.

**1.13** The four kingdoms of eukaryotes are protists, fungi, plants, and animals.

**1.13** **Heterotrophs** (p. 13) obtain food by eating and digesting other organisms; **autotrophs** (p. 13) make their own food.

## Level 1    Learning Basic Facts and Terms

**Multiple Choice**

1. The basic unit of structure of all living things is
   a. protein.  c. the atom.
   b. the cell.  d. DNA.
2. Organisms that interbreed freely in their natural settings and do not interbreed with other populations are called a
   a. genus.  c. species.
   b. phylum.  d. family.
3. An example of emergent properties in living things is
   a. an organism is more than a collection of cells.
   b. structures of organisms fit their functions.
   c. consumers feed on producers and other consumers.
   d. populations of organisms change over time.
4. An example of scientific binomial nomenclature is
   a. John Doe.  c. red squirrel.
   b. Honda Accord.  d. *Homo sapiens*.

5. The phrase "common ancestry" refers to
   a. organisms within one family.
   b. kingdoms of organisms subdivided into groupings.
   c. natural selection.
   d. evolutionary relationships.

**True–False**

6. _____ Organisms are diverse in their types but unified in their patterns because they are all related.
7. _____ Fossils provide a record of the history of life.
8. _____ The most recently proposed classification scheme is Whittaker's five-kingdom system.
9. _____ Systematists classify organisms in ways that reflect their evolutionary relatedness.
10. _____ The protists are a small group of prokaryotic organisms.

## Level 2    Learning Concepts

1. What are three topics that biologists might study?
2. List the eight unifying themes of biology (the characteristics of life) described in this textbook.
3. List and describe the four levels of internal organization in multicellular organisms.
4. Name and describe the different levels of interactions that occur between organisms and their environments.
5. How does human interaction with the biosphere differ from that of all other organisms?

6. From what ultimate source do you obtain the energy that keeps you alive? Explain how you obtain energy from that source.
7. What is DNA? Explain its role in living things.
8. What is evolution? How are fossils associated with evolution?
9. Name the three domains of living things. To which domain do you belong?
10. Place the following terms in their correct sequence, starting with the term that refers to the group containing the largest number of organisms: phylum, species, genus, domain, order, family, class, kingdom.

## Level 3    Critical Thinking and Life Applications

1. Two organisms are members of the same order. Are the following statements true or false? Explain your answers.
   a. The organisms must belong to the same class.
   b. The organisms must belong to the same genus.
   c. The organisms must have the same "first name" using binomial nomenclature.
2. If all living things are composed of one or more cells, it follows that a computer-driven machine cannot be alive. And yet such a machine can transform energy, do work, make copies of itself, and evolve over time to better suit the challenges of its environment. Can machines be considered living? From a biological perspective, what constitutes life?
3. Choose two organisms and state the common name and kingdom of each one. Describe three characteristics the organisms have in common. Describe three ways in which the organisms are different from one another.
4. Although you never thought you had a "green thumb," you recently developed an interest in gardening and took an adult-education course on gardening offered at your local library. You've just planted vegetable seeds and young plants. Assuming that your course will help

you manage your garden well, use your knowledge of the characteristics of life to predict five observations you will make over the growing season.
5. Structurally, most viruses are made up of hereditary material encased in protein. Some biologists suggest that viruses are a bridge between the living and the nonliving world. Other scientists suggest that viruses are simply nonliving things. Evaluate both points of view regarding viruses using your personal knowledge of having had a viral infection, such as the flu, your knowledge of the characteristics of life, and the information stated in this question.
6. Based on your knowledge of life thus far, which of the following would you consider alive? Defend your answer.
   • a carrot that has been in your refrigerator for a week.
   • the wood in a fallen tree branch
   • an apple seed
   • a human embryo frozen at an in vitro fertilization clinic
   • a sperm cell
7. Write down 10 issues or problems confronting your generation today.
8. How many of the issues you listed in question #7 are related directly or indirectly to the science of biology?

In The News | **Critical Thinking**

**GALACTIC REAL ESTATE**

Now that you understand more about the characteristics of life, reread this chapter's opening story about research into the galactic habitable zone (GHZ). To understand this research better, it may help you to follow these steps:

1. Review your immediate responses to Gonzalez's research on the GHZ that you wrote when you began reading this chapter.
2. Based on your current understanding, again summarize the main point of the research in a sentence or two.
3. What questions do you now have about this research that the chapter's opening story does not answer?
4. Collect new information about the research. Visit the *In The News* section of this text's companion Web site at www.wiley.com/college/alters and watch the "Galactic Real Estate" video. Then use the "summary" link to read the accompanying story and access related links. Use this information, the links provided, and other online and library resources to answer your questions and find updates about this research topic. State the sources of your information. Explain why you think the information is accurate. Also determine whether the information expresses a particular point of view or is biased in any way.
5. What in your view is the most significant aspect of this research? Give reasons for your opinion and for any changes in your ideas based on the additional information you have collected and the analysis you have done.

# HOW SCIENTISTS DO THEIR WORK

## Publish *and* Perish

You may have heard the phrase "publish *or* perish" in academic or scientific circles. It means that scientists and other academicians at research universities and colleges must publish their research and other scholarly work in order to achieve tenure and keep their jobs. Recently, scientists have been uttering a twist on that phrase: "publish *and* perish." This phrase refers to scientists' fear that someone in their ranks might publish discoveries or information in scientific journals that terrorists might use to help wreak their havoc. To avoid this real possibility, scientists have recently decided to police what is published in their journals.

After a workshop and discussion with national security officials, editors and publishers of the major scientific journals made a somber announcement at a meeting of the American Association for the Advancement of Science (AAAS). They decided not to publish scientific material that might be at odds with national security interests and are anxious to monitor scientific publications to avoid governmental intervention and censorship.

The video news story, "Publish and Perish," provides an example of the type of information that could be worrisome. Researcher Eckard Wimmer and his colleagues at the State University of New York at Stony Brook published a now-famous paper in the journal *Science*, in which they described their methods of synthesizing live poliovirus from scratch, using pieces of DNA they bought over the Internet. Could terrorists use these methods to make other viruses that could be tools of bioterror, such as the smallpox virus? Wimmer contends that viruses such as smallpox are too complex to be synthesized using their methods and that scientists do not yet possess the technological ability to synthesize such viruses as they did the simpler polio virus. You can view this video by visiting the *In The News* section of this text's companion Web site at www.wiley.com/college/alters.

Although no one wants terrorists to have access to information that could prove dangerous to public security, scientists still want to preserve open communication among the members of their research community. Achieving this end may involve a balancing act, but scientists feel confident that they will recognize potentially dangerous information when they see it, and so they will work with authors to modify their manuscripts for "safe" publication.

*Write your immediate reaction to the AAAS announcement noted above. First, summarize the main part of the announcement in a sentence or two; then suggest what you think its significance is. You will have an opportunity to reflect on your responses and gather more information on this topic in the* In The News *feature at the end of this chapter. In this chapter, you will learn more about how scientists do their work and share their methods, data, and conclusions with the rest of the scientific community.*

## What is the day-to-day work of science and how is it conducted?

**2.1** Scientists study the natural world.

**2.2** Scientists develop tentative explanations, or hypotheses, to guide inquiry.

**2.3** Scientists test hypotheses.

**2.4** Scientists use diverse methods to test hypotheses.

**2.5** Scientists develop theories from consistently supported and related hypotheses.

## How do scientists report their work?

**2.6** Scientists report their work in journals, in books, and at conferences.

# What is the day-to-day work of science and how is it conducted?

## 2.1 Scientists study the natural world.

On any given day in your local newspaper, you are likely to find a variety of news articles related to science. You might find articles on severe acute respiratory syndrome (SARS) a contagious infection that was first described in 2003. Other articles may include topics such as genetically altered foods, human cloning, water or air pollution, new fossil finds, and global warming. A recent article in the *New York Times*, for example, was titled "Doing Science at the Top of the World." Such a headline might make you wonder what the scientists were doing at the North Pole and how they were doing it. You might want to read their story to see how these scientists studied changes in the polar ice, ocean, and atmosphere (**Figure 2.1**).

Although scientists may differ in the focus and location of their work, the aim of all science is to better understand the *natural world*. In doing so, scientists share scientific methods that often include the following key features: observation, hypothesis development, and hypothesis testing. A **hypothesis** (hi-POTH-uh-sis) is a tentative explanation of a phenomenon that guides inquiry. Put another way, scientists typically explain the things they observe with testable statements. For example, scientists may observe that the summer ice mass at the North Pole has decreased from what it was the decade prior. They might develop a hypothesis that the reduced ice mass is a temporary shift in the range of normal variation. Then they test this hypothesis to determine whether data support or refute it.

Scientists test their hypotheses in a variety of ways, producing results that are open to testing and verification by others in their disciplines. Because a hallmark of science is the testable hypothesis, science does not explore supernatural explanations to understand how the natural world works. For example, scientists would not hypothesize that supernatural beings melted ice at the North Pole, nor would they explore this idea.

As just noted, observation is an important process of science, but direct observation alone does not constitute science. For example,

some scientists infer from evidence the causes of extinctions that occurred millions of years ago, but they cannot directly observe those extinctions or their causes currently happening. Theoretical physicists analyze experimental data produced by experimental physicists, but they typically do not conduct the experiments and observations.

> CONCEPT CHECKPOINT
>
> **1.** How would you describe what the work of a scientist entails?

## 2.2 Scientists develop tentative explanations, or hypotheses, to guide inquiry.

The specific pathways of thinking that scientists follow in developing hypotheses vary greatly. In general, however, a scientist considers the available information that might help answer a question and uses this information to develop possible explanations to answer the question. Scientists may gather information regarding their questions by searching the scientific literature and synthesizing the information they find. They may make observations of their own or find anecdotal information (individual occurrences) that relate to their question. They develop their hypotheses based on this information and their observations.

Hypothesis testing isn't just for scientists; it can be used to solve everyday problems and answer everyday questions. For example, you may observe that many of your friends grow healthy and robust houseplants, while you struggle to keep yours alive. You may be asking yourself, "Why can't I grow healthy houseplants?" Some explanations that might come to mind immediately are: The plants are getting too much or too little water, the plants are getting too much or too little sun, or the plants need fertilizer. Your tentative explanations are preliminary alternative hypotheses.

You still need to refine your hypotheses and gather information pertinent to your question. You read articles in gardening magazines that discuss how to grow healthy houseplants, and you determine that most of your actions in nurturing your houseplants are proper. You are giving your plants the right amount of water and sunlight. However, you realize that you are not doing one important thing: You are not fertilizing

**Figure 2.1 Doing science at the top of the world.** Recently, researchers camped at the North Pole Environmental Observatory to retrieve and install instruments that record Arctic conditions year-round. Scientists are collecting data that will help them determine whether the sharp warming around the Arctic perimeter over the last decade is an early result of global warming or simply a temporary shift within a normal range of variation.

them. As you think about nutrients and plants, you also remember that you potted your plants using soil that you used to grow kitchen herbs last spring. You remember that overused soil can become depleted of nutrients. So from your research and current knowledge, you decide that the best hypothesis as to why your houseplants do not thrive is that they do not receive sufficient nutrients.

---

### CONCEPT CHECKPOINT

**2.** What steps would you take to develop a hypothesis to explain why the fish in the lake behind your house are dying?

---

## 2.3 Scientists test hypotheses.

As you refine your hypothesis on plant growth, you may come up with additional questions before you test it. What should you use as a source of nutrients for your plants? How much should you give them? Will a low dose work as well as a high dose? How often should the nutrients be added to the plants' containers? Gathering additional information may help you refine your hypothesis for testing.

### Gathering preliminary information

You remember reading advertisements stating that Brand X fertilizer causes houseplants to grow 30% taller and produce healthier, greener plants. You look up articles about fertilizing houseplants in gardening magazines and find that they suggest a specific proportion of nutrients as best for certain houseplants. You note that Brand X fertilizer has these specific proportions. Some of your friends may use Brand X fertilizer and say that it works well for their plants. From these observations and testimonials, you might decide to use Brand X fertilizer to test your hypothesis that your plants are dying due to lack of nutrients.

### Refining hypotheses

To test your idea, you state your hypothesis in an if/then format. Your newly refined hypothesis, which is based in part on the company advertising, might be the following: *If* houseplants are given Brand X fertilizer, *then* they will grow at least 30% taller and be greener than plants with no fertilizer. What you are actually doing is generating an argument that states an expected outcome.

### Testing hypotheses

Testing a hypothesis involves producing data or evidence that either supports or contradicts the hypothesis. Scientists test hypotheses in many ways. This "testing" process involves the use of *deductive reasoning*. Deduction begins with a general statement, such as Brand X fertilizer will increase growth of houseplants, and proceeds to a specific statement, such as Brand X fertilizer will increase growth of *my* houseplants.

While scientists are formulating hypotheses, they are also thinking about ways to test their hypotheses and what might happen as a result. Then, as scientists test their hypotheses, they compare their mental arguments for or against their hypotheses with what their testing actually shows. The ways in which a hypothesis is tested and the

ways in which data and evidence are collected depend on the hypothesis itself as well as on the scientist doing the testing. For example, paleontologists often test their hypotheses about prehistoric life by gathering evidence from living descendants of prehistoric species and from fossils, the preserved remains or other traces of organisms. Some population biologists test their hypotheses regarding interactions among populations using computer (mathematical) models. Biomedical researchers may test hypotheses by comparing two or more groups of subjects with regard to an intervention that occurs naturally, such as comparing the percentage of unimmunized persons contracting the flu with the percentage of immunized persons contracting the disease during the same time period.

### Controlling variables

Many scientists test hypotheses by means of controlled experiments. In a **controlled experiment**, a researcher manipulates, or changes, one factor and observes how unmanipulated factors respond. All other factors are kept constant throughout the experiment, that is, they are controlled. In analyzing the simplified hypothesis in our previous houseplant scenario, notice that the factor you are manipulating—the fertilizer—is stated after the "if" part of the hypothesis. This factor is called the **independent variable**, or manipulated variable. The word *variable* refers to a factor that changes, or varies. To test your hypothesis, you will need to add different amounts of Brand X fertilizer to groups of houseplants (**Figure 2.2**). Also notice that the factor stated after the "then" part of the hypothesis depends on the manipulated (independent) variable. This factor is called the **dependent variable**, or responding variable, and it varies in response to changes in the independent variable. In your experiment, plant growth and color are dependent variables.

The controlled variables are the factors in the experimental setup that are not allowed to change. In this case, they include the type of soil, the amount of time the plants are exposed to the light, the temperature in which the plants are grown, the amount of water the plants

**Figure 2.2** The independent variable in this experiment is the amount of fertilizer. From left to right, the amounts are 15 ml, 30 ml, 45 ml, and 60 ml.

---

 **Visual Thinking:** The amounts of fertilizer shown in the graduated cylinders increase from left to right, yet the cylinder third from the left (45 ml) has a higher level of fertilizer than the cylinder to the far right (60 ml). How do you explain this discrepancy?

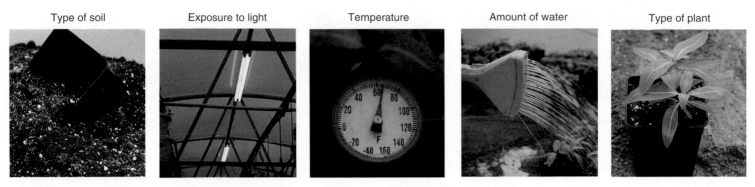

| Type of soil | Exposure to light | Temperature | Amount of water | Type of plant |

**Figure 2.3** The controlled variables in this experiment are type of soil, time of exposure to light, temperature, amount of water, and type of plant.

are given, and the type of plant tested as shown in **Figure 2.3**. Can you think of any other variables that may need to be controlled?

As you design a controlled experiment to test your hypothesis, you should first outline how you would manipulate the independent variable. In this case, you may decide to give one group of plants no fertilizer and another group of plants the dose of fertilizer suggested on the fertilizer container. You also may choose to give some plants slightly more fertilizer than the suggested dose and others slightly less. If the suggested dose is 30 milliliters (ml) of properly diluted liquid fertilizer, you may decide to give a second group of plants 15 ml, a third group 45 ml, and a fourth group 60 ml as shown in Figure 2.2. Each group of plants should be the same type of plant, such as all periwinkle plants of approximately the same size. You should pot them all in the same soil in the same type and size of pot, place them approximately the same distance from the same light source, and water them all with the same amount of water at the same time. These plants will receive the *treatment*—in this case, the fertilizer. They are the experimental plants. A fifth group of plants should be treated the same as the others except that it will not receive the treatment (fertilizer). This group serves as the control. The **control** is the standard against which the experimental effects on the treatment plants can be compared. Any changes in the control are the result of factors other than the treatment. **Figure 2.4** shows the setup of the controlled experiment.

## Collecting and organizing data

Over a preselected time, possibly based on the advertised length of time needed for the fertilizer to cause change, you should observe your plants and collect *data* regarding leaf color and plant height. You could also choose other indicators of plant growth such as

number of leaves or number of flowers per plant. The data regarding the height of your plants are called **quantitative data** because these data are based on numerical measurements, such as inches or centimeters. The data regarding leaf color in our example are called **qualitative data** because these data are descriptive and not based on numerical measurements.

At the beginning of your experiment, you should measure the height of each plant in each group and record these measurements in a data table. Find the *mean* (average) height of the plants in each group by summing the heights of the plants and dividing each sum by the number of plants. Record your data for each group. At the end of your predetermined growing period—six weeks, for example, you measure plant height again, determine the mean heights for each group, and record these data. Your summary data table may look like this:

**Mean Plant Height in Centimeters (cm)**

| | Amount of Fertilizer (ml) | | | | |
| --- | --- | --- | --- | --- | --- |
| | **0** | **15** | **30** | **45** | **60** |
| 0 weeks | 145 | 150 | 148 | 151 | 149 |
| 6 weeks | 166 | 180 | 181 | 179 | 184 |
| Increase in plant height | 21 | 30 | 33 | 28 | 35 |

Regarding the collection of qualitative data, you should determine whether the color of the leaves of each plant is light green, medium green, or dark green. Record this information in a data table as well. Find the *mode,* the value that occurs most often, for each group. Record the mode for each group in a summary data table. Do the same at the end of your predetermined growing period. Your table may look like this:

**Description of Green Color of Leaves (mode)**

| | Amount of Fertilizer (ml) | | | | |
| --- | --- | --- | --- | --- | --- |
| | **0** | **15** | **30** | **45** | **60** |
| 0 weeks | medium | medium | medium | medium | medium |
| 6 weeks | light | medium | dark | dark | dark |
| Change in leaf color | negative | none | positive | positive | positive |

| 15 ml | 30 ml | 45 ml | 60 ml | Control (no fertilizer) |

**Figure 2.4** Setup of the controlled experiment showing the different amounts of fertilizer used.

In both tables, the control plants are those receiving 0 ml of fertilizer (no treatment); the initial heights and colors of plants are those for 0 weeks.

### Drawing conclusions from data

Remember your hypothesis: If houseplants are given Brand X fertilizer, then they will grow at least 30% taller and be greener than plants with no fertilizer. Do your data support this? Looking at the qualitative data, we see that the plants with fertilizer were greener than the control with no fertilizer. In this experiment, 30 ml to 60 ml produced the greenest plants.

Looking at the quantitative data, can you determine whether the plants were at least 30% taller than the control? First, let's compare the plants treated with 15 ml of fertilizer with the control plants. The control plants grew 21 centimeters, and the test plants grew 30 centimeters—9 centimeters more than the control. Since 9 is 43% of 21, the plants receiving 15 ml of Brand X fertilizer grew 43% taller than the control! Similarly, the plants receiving 30 ml of Brand X fertilizer grew 57% taller, the plants receiving 45 ml grew 33% taller, and the plants receiving 60 ml grew 67% taller. In summary, these data do support your hypothesis. In fact, your test data suggest that Brand X fertilizer may help houseplants grow even greener and taller than the manufacturer suggests.

### Understanding limitations to experimentation

Why do you think the manufacturer of Brand X fertilizer is not making greater claims regarding this product? Your experiment certainly shows much greater than a 30% increase in growth. The answer lies in the limitations of your experiment. First, you experimented with one type of houseplant. Brand X fertilizer is probably sold for use on a variety of houseplants. Not all plants may grow as vigorously in response to the fertilizer as yours did. Second, you probably experimented on five small groups of plants. Scientists often perform experiments over and over again, in **repeated trials**, before drawing conclusions from their data. Using repeated trials increases the reliability of the results by reducing the effects of chance or error that may occur in a single trial. The Brand X fertilizer manufacturing company probably performs hundreds of experiments on a variety of houseplants before drawing conclusions from their data and generalizing to all houseplants.

What if the fertilized plants had not grown better than the control? Evidence that contradicts a hypothesis weakens the likelihood that it is accurate, and in many cases, especially in the physical sciences, it disconfirms the hypothesis. However, evidence that supports a hypothesis does *not* establish that further testing will also produce supporting evidence. Unforeseen factors may affect the outcome of a future experiment. In your experiment, for example, the data collected using your plants supported your hypothesis. But data collected using other plants in someone else's home may not support your hypothesis. In summary, then, scientists cannot actually *prove* hypotheses; they can only *support* hypotheses with evidence.

---

**CONCEPT CHECKPOINT**

3. Revisit Concept Checkpoint 2. Suppose that you notice that owners of lake property are putting large amounts of fertilizer on their lawns and gardens, and you suspect that this has something to do with fish in the lake dying. By gathering information about fresh-water fish kills, you learn that fertilizers can stimulate algal growth. When the fertilizer is consumed, much of the algae dies, and bacteria decompose the dead material. The bacterial population grows and oxygen consumption by the bacteria rises. You suspect that the fish kill in your lake is due to reduced levels of oxygen in the water. Develop a testable hypothesis about the cause of the fish kill based on this information.
4. Outline a design for a controlled experiment that would test this hypothesis.

---

## 2.4 Scientists use diverse methods to test hypotheses.

There is no "one" scientific method—no single way in which scientists go about their work. Although some scientists find it quite useful to first make direct observations, then develop hypotheses, and finally test their hypotheses by means of controlled experiments, not all scientists proceed in this manner. One reason for this diversity of process is that not all sciences are based on the accumulation of observational data, so the starting point is different. For example, a chemist cannot directly observe the chemical bonds he or she may be studying.

Another reason for the diversity of scientific process is that testing hypotheses by means of controlled experiments is often difficult, if not impossible, in some scientific disciplines. For example, paleontologists cannot conduct controlled experiments to see what happened in the past.

Another example of a field in which the testing of hypotheses by controlled experimentation is often difficult is in biomedical research. It is impossible to control all factors when dealing with humans. When feasible, subjects are chosen randomly for both a "control" group and a "test" group in an effort to obtain groups of people that are as similar to one another as possible. In addition, scientists cannot conduct on humans controlled experiments in which a treatment might be harmful. In such cases, biomedical researchers may use methods other than controlled experiments to test their hypotheses. Such investigators often conduct observational studies in which they do not manipulate an independent variable, which is often called an *intervention* in biomedical research. Instead, the researchers may report observations that have biomedical importance, such as the numbers of persons receiving flu vaccinations during September of a particular year. Or they may test a hypothesis by comparing two or more groups of subjects with regard to an intervention that occurs naturally. For example, researchers may compare the death rates from lung cancer of cigarette smokers and nonsmokers.

When research with human subjects involves a planned intervention (a controlled experiment), the investigator must be sure that it is ethical to conduct an experimental study to answer the research question and test the hypothesis. In other words, the researcher must be sure that participants are not subjected to undue risk and that they are informed as to the possible benefits and risks of the experiment; the researcher must also have written consent of the participants in the study. Using human subjects in research often raises many ethical questions that might not come up when other forms of life such as bacteria or plants are studied.

For these reasons and others, scientists who study the workings of the human body often use tissues, chemicals, and mathematics to model human systems rather than using humans. These model systems are useful when a problem must be studied under simplified, well-controlled conditions and when a controlled study in humans is unethical. However, such approaches are limited in their applications, because nonliving systems cannot mimic the actual workings of a living organism. Often, therefore, certain

# How Science Works

## PROCESSES AND METHODS

### Is there bias in science?

People usually think about the work of scientists as being objective and controlled, and they envision scientists as designing experiments that eliminate any personal biases they may have. Although scientists usually strive for these goals, achieving total objectivity and eliminating all bias may be impossible.

Stephen Jay Gould, the late world-renowned scientist and Harvard professor, provides an interesting example that probes the power of subjectivity and bias in his book *The Mismeasure of Man*. He relates the story of Paul Broca, one of the many nineteenth-century scientists who were engaged in craniometry, the measurement of the skull and facial bones, and thus measurement of the contents of the cranium. The nineteenth-century sketch in **Figure 2.A** shows a measuring device used in this practice.

During the nineteenth century, European leaders and intellectuals viewed some races, especially their own, as being superior to others and believed that those races at the bottom of their ranking had inferior characteristics of anatomy and/or physiology, such as inferior intelligence. Accepting this view of a racial hierarchy, Broca, a professor of clinical surgery in Paris, believed that brain size was related to intelligence, with the more intelligent races (in his view) having larger brains and the less intelligent races having smaller brains. When another scientist challenged his belief, Broca set out to defend himself. He hypothesized that white Europeans would have larger brains than other races, thus explaining their (perceived) higher intelligence. He meticulously gathered quantitative data on cranial ca-

pacity to test his hypothesis. He carefully designed controls for his procedures so that error would not affect his measurements. Broca concluded that the data supported his hypothesis.

A modern-day reevaluation of Broca's data suggests that while his individual measurements are reliable, Broca's conclusions are invalid. Bias influenced both the formulation of his hypothesis and the collection of his "supporting" data. As Gould points out, "Broca's cardinal bias lay in his assumption that human races could be ranked in a linear

**Figure 2.A** Nineteenth-century device used to measure the skull and facial bones, and thus the contents of the cranium. The device is bottom left, while a section of a skull is upper right.

scale of mental worth." In fact, brain size is only one of many factors that affect intelligence, some of which are hard to quantify. Brain size alone does not predict intelligence with any accuracy. Furthermore, Gould suggests that this same bias influenced the type of data Broca collected, as he *subconsciously* searched for those data that would support his hypothesis. If the data clearly did not support his hypothesis, he would develop other criteria that would

supersede the importance of the data that didn't "fit."

Broca's story is not just a historical curiosity. This example shows that although data may be reliable—collected meticulously and flawlessly—the interpretation of those data may not be objective if affected by factors such as prejudice and bias. Scientists today usually try to eliminate as much bias in their work as possible in a variety of ways. To remove the subjective element from the evaluation of the effects of a new drug, for example, medical researchers often choose a double-blind methodology. In such a study, half of the participants are randomly chosen to receive the new drug. The other half of the participants receive a placebo, a pill that resembles the drug but contains no medication. Neither the researchers nor the participants know which pill is the drug (thus, double-blind); the drug company stamps the pills with code letters and reveals the code only after the study ends.

Publishing the results of scientific studies—making them immediately available to scientists and laypeople of many different backgrounds and points of view—is another very important way to control for bias. If another researcher or team cannot reproduce the results of the published research, the validity of the results becomes questionable and may eventually be refuted. If the interpretation of data is heavily biased, as in Broca's case, those in the scientific community will often perceive that bias and reject the study's conclusions. Thus, though it is not perfect, the scientific process has built-in self-monitoring systems that eventually succeed in reducing the amount of bias in research. ●

animals are used in limited numbers as experimental models. The use of animals in biomedical research has helped scientists make advances in knowledge regarding a wide array of human diseases.

In addition to the fact that there is no "one" scientific method, many people refer to science as "organized common sense" and suggest that the methods scientists use have no special link with science. People across disciplines and in their everyday lives engage in these generalized processes. Think about the steps that you might take to solve a problem. First, you have to identify the problem, figure out how to solve the problem, and try that method; then you have to attempt another approach and see if you get the same results; and finally, you need to compare your results with those of others or with your previous knowledge. Are you thinking like a scientist in this case, or do scientists simply use organized, logical processes to solve their problems and answer their questions as you might?

### CONCEPT CHECKPOINT

5. Why do scientists use diverse methods to test hypotheses? Why don't they all use a single approach?
6. When you answer questions in your everyday life, are you "doing science"?

## 2.5 Scientists develop theories from consistently supported and related hypotheses.

You've probably heard someone say, "Oh, that's *just* a theory." However, using the word "theory" in a scientific context has a much different meaning than using it in an everyday context. A scientific theory is not just a hunch or guess as in: "I have a theory why my roommate doesn't get up for 8 o'clock classes!" That theory *is* just a theory, or hunch, or guess. Scientific theories are much, much more.

When scientists communicate in a scientifically exacting manner in research journals, they normally use **theory** to mean an explanation of phenomena that has been rigorously tested. A scientific theory is a powerful concept that helps scientists make predictions about the world and explain particular phenomena. It can incorporate facts, laws, inferences, and tested hypotheses. In science, theories can be considered factual if they have been repeatedly confirmed and never refuted. For example, in the mid-1800s, evolution was considered to be an unconfirmed theory, but due to the amount of confirming evidence since that time, and the lack of counterevidence, the theory is now considered as fact. Theories supported by

## *just wondering . . .*

### Questions students ask

### Why is it that there's no "Truth" in science?

To most people, *truth* means something that is always accurate. In science, however, hypotheses and theories may change as new data emerge. Scientists collect data and gather evidence that may support or contradict a theory. If the data contradict the theory, it may need to be modified or even discarded. Therefore, the very nature of the scientific process is one of an openness to change. In other words, science is tentative, whereas most people consider truth to be absolute.

Some hypotheses and theories have not changed significantly in hundreds of years, and they may not change in the future. However, we cannot know what evidence may come to light to affect a particular theory. For example, prior to the time of Copernicus (1473–1543), the widely accepted view was that the sun orbited the Earth (**Figure 2.B**). Copernicus published a book in which he pointed out the logical reasons why the correct view is that the Earth orbits the sun. Later, Galileo (1564–1642) defended this theory, which still stands today after continued testing as technology advanced over the centuries. In another example, the theory of evolution by natural selection proposed by Charles Darwin in 1859 remains valid today, though scientists continue to argue about the mode and tempo of evolution. You can read about these arguments in Chapter 17. ●

**Figure 2.B** Ptolemy's model of an Earth-centered universe. Ptolemy was an astronomer, mathematician, and geographer who lived in the second century. He provided a mathematical model of the universe that accounted for the movements of the sun, the moon, and the five planets known at the time. This illustration depicts Ptolemy's model by showing the Earth in the center surrounded by the "spheres" of each celestial body. This geocentric model was accepted for about 1500 years until Copernicus, also a mathematician and astronomer, proposed the sun-centered (heliocentric) universe in the 16th century.

such an overwhelming weight of evidence are accepted as scientifically valid statements; they are well-substantiated explanations.

Both laws and theories have equal potential for factuality. But scientific theories are different from **scientific laws** because scientific laws *describe* patterns of regularity with respect to natural phenomena; they do not *explain* them as do hypotheses and theories. For example, the law of inertia states that an object in motion tends to stay in motion and an object at rest tends to stay at rest. In general, theories explain what laws describe. However, a satisfactory explanation for a law may not exist. For example, when Isaac Newton was asked to give a theory for his law of gravity, he admitted he had none.

The work of scientists focuses on generating and testing hypotheses and theories. However, analyses of data from different scientific perspectives can result in opposing theories based on the same evidence. Continued analyses over time often result in one scientific theory prevailing over others. In addition, the possibility always remains that future evidence will cause a theory to be revised, since scientific knowledge grows and changes as new data are collected, analyzed, and then synthesized with previous information. This idea is discussed in more detail in the "Just Wondering" box.

### CONCEPT CHECKPOINT

**7.** Often people will say, "Well, you know that evolution is *just a theory.*" What might you say to explain the difference between a scientific theory and what people mean when they say this?

# How do scientists report their work?

## 2.6 Scientists report their work in journals, in books, and at conferences.

You might not think that scientists need to communicate with one another, and maybe you envision the lone scientist in his or her lab coat toiling away in the laboratory. However, scientists do not usually work in isolation. Communication among scientists is crucial in order to replicate and build on each other's work. Scientists communicate with one another just like nonscientists do: via telephone, surface mail, and the Internet. However, the two primary "formal" avenues of communication among scientists are scientific conferences and scientific journals.

Professional organizations exist for nearly every subdiscipline of science, and most of these organizations hold meetings or conferences in which members come together to discuss the latest research findings in their particular field. Usually, scientists present papers in a lecture format, outlining their work and their results or they may discuss their work with others one-on-one or in a small group in poster sessions (**Figure 2.5**).

Research papers that scientists write for submission to journals in their fields, whether electronic or hard-copy journals, follow a format that is common to scientific disciplines and usually includes these parts: title, abstract, introduction, methods, results, discussion/conclusion, and references. Similarly, most student science competitions require students to report their research in this format. Science courses in which there is a laboratory component and in which professors require laboratory reports often follow this format or an adaptation of it as well.

You can examine an annotated reprint of a scientific paper in Appendix A. The article describes research conducted by James Quinn and Jane Sirdevan of McMaster University in Hamilton, Ontario (Canada), near Niagara Falls. These researchers tested a variety of nesting materials called *substrates* to see on which types of materials Caspian terns, a species of waterbird (**Figure 2.6**), were most likely to nest. The researchers also investigated whether the survival of chicks, the birds' offspring, varied with substrate type. This information was used to prepare nesting sites for the terns on artificial islands built in an effort to relocate them. The land on which the terns were currently nesting was privately owned and likely to be developed.

The paper was published in 1998 in the journal *Biological Conservation*, a peer-reviewed journal. Virtually all scientific journals are peer reviewed, which means that articles appearing

**Figure 2.5** Scientists report on their work at scientific conferences.

**Figure 2.6** A Caspian tern.

in them have been sent to other scientists knowledgeable in the field for their comments on the work and their determination of whether the research should be published. *Biological Conservation* is a British publication and therefore uses British spellings, which differ occasionally from American spellings. The paper, written in a format that is standard for most scientific papers, is an example of scientific process and how scientists conduct and report their work. In Appendix A, the side notes next to the article explain its organization. Annotations and comments help explain its content.

### CONCEPT CHECKPOINT

8. When conducting scientific research, scientists often build on the work of previous studies. In what section of the scientific paper by James Quinn and Jane Sirdevan (see Appendix A) do the authors discuss other scientists' work that contributed to their study?

9. Provide an example of background work that contributed to Quinn and Sirdevan's study.

# CHAPTER REVIEW

## Summary of Key Concepts and Key Terms

### What is the day-to-day work of science and how is it conducted?

**2.1** Scientists study the natural world using a variety of scientific methods.

**2.1** The methods scientists use often, but not always, include direct observation, hypothesis formation, and hypothesis testing.

**2.1** A **hypothesis** (hi-POTH-uh-sis; p. 19) is a testable statement or an explanation that guides inquiry.

**2.2** Scientists usually formulate hypotheses by synthesizing information from various sources.

**2.3** Testing a hypothesis involves producing data or evidence that either supports or does not support the hypothesis.

**2.3** Many scientists test hypotheses by means of **controlled experiments** (p. 20), in which one factor is manipulated while others are observed that respond to the change.

**2.3** The factor that is manipulated in a controlled experiment is called the **independent variable** (p. 20); those that respond are called the **dependent variables** (p. 20).

**2.3** Controlled variables (**control**, p. 21) are factors kept constant in the experimental setup.

**2.3** **Quantitative data** (p. 21) are based on numerical measurements, such as inches or centimeters, while **qualitative data** (p. 21) are descriptive and not based on numerical measurements.

**2.3** Scientists often perform experiments over and over again, in **repeated trials** (p. 22), before drawing conclusions from their data.

**2.4** Scientists test hypotheses in a variety of ways.

**2.4** The processes scientists use are diverse for many reasons, including the following: Not all sciences are based on the accumulation of observational data, and testing hypotheses by means of controlled experiments is difficult in some fields of scientific inquiry.

**2.4** Many methods used by scientists are common to other disciplines.

**2.5** **Scientific theories** (p. 24) are explanations of phenomena that have been rigorously tested—powerful concepts that explain particular phenomena.

**2.5** Some scientific theories are upheld over time so consistently that they are considered factual.

**2.5** **Scientific laws** (p. 25) describe patterns of regularity with respect to natural phenomena.

### How do scientists report their work?

**2.6** The two primary avenues of communication among scientists are scientific conferences and scientific journals.

**2.6** Research papers describing scientific studies usually include these parts: title, abstract, introduction, methods, results, discussion/conclusion, and references.

## Level 1    Learning Basic Facts and Terms

**Matching**

1. _____ a tentative explanation that guides inquiry
2. _____ one method of testing a hypothesis
3. _____ factor that is manipulated in a controlled experiment
4. _____ based on numerical measurements
5. _____ describes patterns of regularity of natural phenomena

    a. independent variable
    b. dependent variable
    c. controlled experiment
    d. hypothesis
    e. quantitative data
    f. scientific law

**True–False**

6. _____ There is only one true scientific method.
7. _____ Gathering information on a question before developing a hypothesis is like cheating on an exam.
8. _____ Scientists often perform experiments over and over again, in repeated trials.
9. _____ If scientists cannot confirm their hypotheses, their methods are faulty.
10. _____ Scientific research papers follow a format that is common to scientific disciplines.

## Level 2    Learning Concepts

1. List three key features of scientific inquiry. Are these features always present?
2. Define the term *hypothesis*. The text states that scientists cannot prove hypotheses. Explain this statement.
3. What is a scientific theory? What is the difference between a scientific theory and a scientific law?
4. In the context of a controlled experiment, define the term *variable*. What is the difference between an independent variable and a dependent variable? Explain what is meant by the phrase "controlled experiment."
5. Describe the difference between quantitative data and qualitative data.
6. Do all scientists perform controlled experiments as part of their work? Why or why not?
7. Why do scientists conduct repeated trials within an experiment? What purpose do repeated trials serve?
8. Read Quinn and Sirdevan's paper in Appendix A. What is the purpose of the abstract?
9. Your friend is "hooked" on telephoning for psychic readings. Are psychic readings considered scientific? Why or why not?
10. Which of the following questions is within the realm of scientific investigation?
    - Is classical music more artistic than rock music?
    - Do girls develop motor skills earlier in development than boys?
    - Does listening to Mozart affect learning in children?
    - Is capital punishment just?
    - Do nonhuman animals have self-awareness?
    - How are whooping cranes able to migrate long distances?
11. Can the data from a scientific experiment support a scientific hypothesis? Can the data prove the hypothesis? Explain.

## Level 3    Critical Thinking and Life Applications

1. You've been hearing about how regular exercise (say, walking for one hour, four days a week) could help people lose weight. You decide to do a series of controlled experiments to test this idea. You persuade 50 students at your school to participate. State your hypothesis and identify the independent and dependent variables.
2. What is a control? How would you set up the control in the preceding experiment?
3. Enzymes help chemical reactions take place in living systems. These biological molecules are present in a variety of toiletry items and laundry detergents. Susan noticed that her contact lens cleaner listed "enzymes" in the ingredient list. She wondered whether the cleaner contained protease, which degrades protein. She decided to test this by determining whether the cleaner degraded milk protein. Using an if/then format, state a hypothesis for Susan's investigation. Describe a simple controlled experiment that would test the hypothesis.
4. Deductive reasoning is often used in both scientific and criminal investigations. In this context, how is science sometimes like a crime investigation? Give specific examples.
5. State two examples in which the study of biology could help you become a more enlightened citizen.

## In The News    Critical Thinking

### PUBLISH *AND* PERISH

Now that you understand more about how scientists conduct and communicate their work, reread this chapter's opening story about the scientific journal editors' and publishers' decision not to publish scientific information that might threaten national security. In thinking through the significance of this announcement, it may help you to follow these steps:

1. Review your immediate response to the announcement described in this chapter's opening story.
2. Based on your current understanding, again summarize the main point of the opening story in a sentence or two.
3. What questions do you now have about the announcement made at the AAAS meeting that the opening story does not answer?
4. Collect new information about this topic. Visit the *In The News* section of this text's companion Web site at www.wiley.com/college/alters and watch the "Publish and Perish" video. Then use the "summary" link to read the accompanying story and access related links. Use this information, the links provided, and other online and library resources to answer your questions and find updates about this topic. State the sources of your information. Explain why you think the information is accurate. Also determine whether the information expresses a particular point of view or is biased in any way.
5. In your view, what is the most significant aspect of this announcement? Give reasons for your opinion and for any changes in your ideas based on the additional information you have collected and the analysis you have done.

# THE CHEMISTRY OF LIFE

## In The News | *E. coli* vs. *E. coli*

On any given day, approximately 10 million Americans consume restaurant hamburgers. In fact, 77% of all American restaurants have this "all-American meal" on the menu. Nevertheless, several incidents of contamination by a deadly strain of *E. coli* bacteria in undercooked hamburgers have spurred debate about how to protect consumers from dangerous food microorganisms without creating new safety hazards. Irradiating food—treating food with radiation—is one answer to this problem, and learning more about chemistry will help you better understand this process. The video news story "*E. coli* vs. *E. coli*" tells of another possible solution. You can view this video by visiting the *In The News* section of this text's companion Web site at www.wiley.com/college/alters.

Incidents of *E. coli* poisoning that occurred within the last two decades initiated the food irradiation debate. Illness or death from *E. coli* infections had been rare until 1986, when two elderly women in Washington State succumbed to *E. coli* (strain 0157:H7) after consuming tainted ground beef at a fast-food restaurant. In 1993, another outbreak of 0157:H7 sickened 600 people and killed four children who had eaten contaminated fast-food hamburgers. In 1996, over 6000 children in Japan were sickened by *E. coli* found in beef in their school lunches. Reports of meat suppliers recalling ground beef because the meat may be tainted with *E. coli* 0157:H7 are still periodically in the news.

*E. coli* is only one of many organisms that contaminate the food supply. Salmonella bacteria are a common contaminant of poultry. Certain insects and their larvae contaminate flour. A wide range of organisms cause not only foodborne illness but also the spoilage of food. What can be done? Inspection by touch, sight, and smell is one way to help ensure the safety of the food we eat. Proper storage and handling is another. Killing bacteria and pathogens is yet a third.

So what is the debate about? Supporters point out that irradiation has been shown to be the only way to rid ground beef of *E. coli* 0157:H7 before cooking. Irradiation also kills other bacteria, as well as insects and fungi that can make people sick or cause food spoilage. In addition, food irradiation can be used to inhibit the sprouting of vegetables and delay the ripening of fruits. Food irradiation, proponents hold, would make the food supply safer, would improve the quality of food, and would extend the "shelf life" of food.

Opponents point out that irradiation induces chemical changes in the nutrients that compose food. Therefore, they argue, these changes can affect the color, odor, and texture of food, and may change the food's chemistry to create toxins that are harmful. They also charge that irradiation lowers the nutritional value of food; it affects some vitamins, for example.

*Write your immediate reaction to the issue raised: first, summarize the issue in a sentence or two; then, suggest how you think this issue could be or should be resolved. You will have an opportunity to reflect on your responses and gather more information on this issue in the* In The News *feature at the end of this chapter. In this chapter, you will learn more about the chemistry of life.*

## CHAPTER GUIDE

### What are atoms and how do they interact?

**3.1** Atoms are submicroscopic units of matter that interact by means of their electrons.

**3.2** Atoms interact to form molecules and compounds.

**3.3** Chemical bonds hold atoms together.

**3.4** Ionic bonds are attractions between ions of opposite charge.

**3.5** Covalent bonds are a sharing of electrons between atoms.

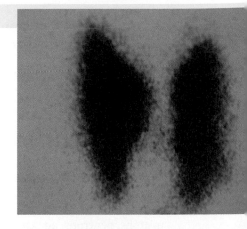

### What properties of water are important to life?

**3.6** Water molecules "stick to" other water molecules.

**3.7** Water is a powerful solvent.

**3.8** Water, when ionized, can accept or donate protons in the reactions of life.

# What are atoms and how do they interact?

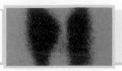

## 3.1 Atoms are submicroscopic units of matter that interact by means of their electrons.

**Atoms** are submicroscopic particles that make up all matter. Matter is the physical material that makes up everything in the universe; it is anything that takes up space and has a measurable amount of substance, or mass. Chemistry is the science of matter.

Although scientists know a great deal about atoms, a simplified explanation provides a good starting point in understanding their complex structures. Every atom is made up of particles tinier than the atom itself. These subatomic particles are of three types: **protons, neutrons**, and **electrons**. Protons and neutrons are found at the core, or nucleus, of the atom; electrons surround the nucleus.

Protons have mass and carry a positive (+) charge. Neutrons, though similar to protons in mass, are neutral and carry no charge. Electrons have very little mass and carry a negative (−) charge; for this reason, the *atomic mass* of an atom is defined as the combined mass of all its protons and neutrons without regard to its electrons.

**Elements** are pure substances that are made up of a single kind of atom and cannot be separated into different substances by ordinary chemical methods. The identity of an atom is determined by the number of its protons. For example, an atom containing two protons is helium, a gas you have probably seen used to blow up balloons. An atom possessing seven protons is nitrogen. **Table 3.1** lists the most common elements in the human body and in the Earth's crust. The number of protons in each type of element is also listed in this table and is called the *atomic number*.

The number of protons in an atom is the same as the number of electrons in that atom. Atoms are therefore electrically neutral; the positive charges of the protons are balanced by the negative charges of the electrons. The number of neutrons in an atom, however, may or may not equal the number of protons. Atoms that have the same number of protons but different numbers of neutrons are called **isotopes** (EYE-suh-topes).

The isotopes of an element differ in atomic mass but have similar chemical properties. In general, electrons determine the chemical properties of an element because atoms interact by means of their electrons, not their protons or neutrons. For example, the three naturally occurring isotopes of hydrogen are shown in **Figure 3.1**. Each isotope has a single proton in its nucleus but a different number of neutrons. Deuterium, tritium, and hydrogen therefore, differ in their atomic masses but have the same chemical properties as one another because they all have the same number of electrons.

| TABLE 3.1 | | | The Most Common Elements in the Human Body and in the Earth's Crust* | | |
|---|---|---|---|---|---|
| **Element** | **Symbol** | **Atomic Number** | **Percent of Human Body by Weight** | **Approximate Percent of Earth's Crust by Weight** | **Biological Importance or Function** |
| Oxygen | O | 8 | 65.0 | 46.6 | Necessary for cellular respiration, component of water |
| Carbon | C | 6 | 18.5 | 0.03 | Backbone of organic molecules |
| Hydrogen | H | 1 | 9.5 | 0.14 | Electron carrier, component of water and most organic molecules |
| Nitrogen | N | 7 | 3.3 | Trace | Component of all proteins and nucleic acids |
| Calcium | Ca | 20 | 1.5 | 3.6 | Component of bones and teeth, trigger for muscle contraction |
| Phosphorus | P | 15 | 1.0 | 0.07 | Backbone of nucleic acids, important in energy transfer |
| Potassium | K | 19 | 0.4 | 2.6 | Principal positive ion in cells, important in nerve function |
| Sulfur | S | 16 | 0.3 | 0.03 | Component of most proteins |
| Chlorine | Cl | 17 | 0.2 | 0.01 | Principal negative ion bathing cells |
| Sodium | Na | 11 | 0.2 | 2.8 | Principal positive ion bathing cells, important in nerve function |
| Magnesium | Mg | 12 | 0.1 | 2.1 | Critical component of many energy-transferring enzymes |
| Iron | Fe | 26 | Trace | 5.0 | Critical component of hemoglobin in the blood |

*Many other elements, such as copper, zinc, molybdenum, iodine, silicon, aluminum, and manganese, occur in trace amounts in the human body. Other common elements in the Earth's crust include silicon (27.7%) and aluminum (8.1%).

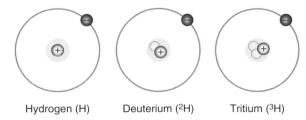

Hydrogen (H)  Deuterium ($^2$H)  Tritium ($^3$H)

**Figure 3.1** The three naturally occurring isotopes of hydrogen. The yellow balls represent neutrons.

The chemical behavior of atoms is due to not only the number of electrons but also their arrangement. Although scientists cannot precisely locate the position of any individual electron at a particular time, they can predict where an electron is most likely to be. This volume of space around a nucleus where an electron is most likely to be found is called the **shell**, or **energy level**, of that electron.

Atoms can have many electron shells. Atoms with more than two electrons have more than one electron shell. In **Figure 3.2a**, the atomic nucleus is shown as a small circle surrounded by spheres. These spheres represent electron shells. In Figure 3.2b, electrons are shown within the shells that are depicted here two dimensionally as circles. Notice that the atom of nitrogen pictured in Figure 3.2b has electrons occupying two shells. The innermost shell contains two electrons; the second and outermost shell has five electrons.

Because the energy of electrons increases as their distance from the attractive force of the nucleus increases, the various electron shells of atoms are also called *energy levels*. Electrons occupying increasingly distant shells from the nucleus have a stepwise increase in their levels of energy. The number of protons in the nucleus also influences the energy of electrons at the various energy levels, since the attractive force of the nucleus increases as its number of protons increases. Therefore, atomic nuclei with more protons have a greater attractive force than atomic nuclei with fewer protons.

---

### CONCEPT CHECKPOINT

1. Describe the structure of an atom.
2. Which type of subatomic particle is key in identifying an atom? Which type largely determines its chemical properties?
3. What is an isotope?

---

## 3.2 Atoms interact to form molecules and compounds.

The atoms of most elements interact with one another, a key condition to life. Without interaction among atoms, nothing could live, breathe, and grow. The hereditary information—DNA—could not exist. Metabolism could not take place. The universe would be simply a collection of elements.

Electrons, then, are crucial to life processes since atoms interact by means of their electrons. Some atoms share their electrons, while others attract one another via opposing electrical charges. In either case, the interacting atoms are called **molecules**. Molecules can be made up of atoms of the same element or atoms of different elements. In molecules made up of atoms of the same element, the

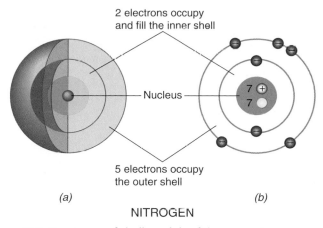

2 electrons occupy and fill the inner shell

Nucleus

5 electrons occupy the outer shell

(a)   (b)

NITROGEN

**Figure 3.2** Two types of shell models of the same atom.

atoms interact by sharing electrons. Molecules made up of atoms of different elements are called **compounds**. Compounds include not only atoms that interact by sharing electrons but also atoms that have lost or gained electrons and are held together by opposing electrical charges. Atoms that have lost or gained electrons are called **ions** (EYE-onz), or charged particles.

### Representing molecules and compounds

Chemists have devised methods to make it easier to talk about molecules and compounds. For example, the oxygen in the air we breathe consists of molecules made up of pairs of atoms of the same element—oxygen. These pairs are represented by the chemical formula $O_2$. A *chemical formula* is a type of "shorthand" used to describe the composition of a molecule. The atoms are represented by symbols, such as $O$ for oxygen. Chemical symbols are shown in Table 3.1. A subscript shows the number of these atoms present in the molecule. An example of a compound with which we are all familiar is water, $H_2O$. In this chemical formula, the symbol $H$ stands for the element hydrogen. The chemical formula $H_2O$ shows that water is made up of two atoms of hydrogen and one atom of oxygen.

Although chemical formulas represent the composition of molecules and compounds, they do not show their shapes. Chemists use models to represent the shapes of molecules. A space-filling model shows the space occupied by electrons. Space-filling models, such as the one depicting a molecule of water in **Figure 3.3**, are used throughout this chapter and the next.

### Factors that influence interaction among atoms

What influences this important process of interaction among atoms? Three factors affect whether an atom will interact with other atoms and also influence the type of interactions likely to take place. These factors are (1) the tendency of electrons to occur in pairs, (2) the tendency of atoms to balance positive and negative

**Figure 3.3** A space-filling model of one molecule of water.

# How Science Works

## The use of isotopes in medicine and food preservation

As mentioned earlier in this chapter, isotopes of an element differ from one another in their numbers of neutrons. Isotopes also differ in that some are *radioactive*. That is, some have unstable nuclei and emit charged particles and rays. Although radioactive isotopes have the potential to be deadly, they have important uses in many spheres of life, including medicine and food preservation.

Radioactive isotopes emit three types of charged particles and rays: (1) alpha ($\alpha$) particles, which are high-energy helium nuclei, (2) beta ($\beta$) particles, which are electron-like particles formed when neutrons are transformed into protons, and (3) gamma ($\gamma$) rays, which are a form of electromagnetic radiation similar to X rays.

Gamma rays are used in food irradiation. Food is irradiated within its packaging by exposing it to these rays or to high-energy electron beams produced by electron accelerators. The radiation kills living organisms within the food as the energy passes through it. Irradiating food does not make the food radioactive, just as a dentist's X ray does not make your teeth radioactive.

The food irradiation process was patented in the United States in 1921 but the Food and Drug Administration (FDA) did not approve it for use on the first food products (wheat, wheat flour, and white potatoes) until the early 1960s. Since then, food irradiation has been a contentious issue in the United States. Approximately 40 years after its approval, irradiation remains in limited use, although the FDA has approved its use on fresh produce, herbs, spices, pork, poultry, and red meat.

Radiation is also used in the diagnosis of disease. When used in diagnosis, radioactive isotopes are called *tracers* or *labels* because they can be used to follow the fate of certain substances in the body. The body is unable to distinguish between radioactive isotopes and stable ones of the same element. Therefore, physicians can use radioactive isotopes as substitutes for their stable counterparts.

For example, suppose your doctor wants to investigate the rate at which your

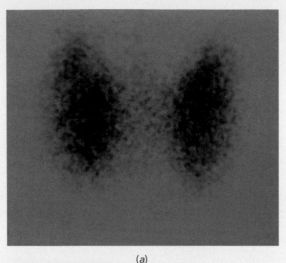

(a)

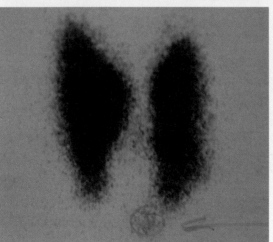

(b)

**Figure 3.A** Thyroid scans after consumption of radioactive sodium iodide. (*a*) A normal thyroid. (*b*) An enlarged thyroid.

thyroid gland is using iodine, a substance in your diet that your body uses to make thyroid hormone. To determine this rate,

you drink a solution of radioactive iodine, which your body will use in the same way it uses the unradioactive iodine you acquire in your food. Equipment that counts the emission of particles from the nuclei of the radioactive iodine is placed over your thyroid gland and measures how fast the tracer iodine is entering your thyroid. The doctor can then see *exactly* how fast your thyroid is using iodine by simply reading the information on the detector. A visual representation of radioactivity levels is shown in the photos in **Figure 3.A**.

Isotopes are also effective in treating cancer. Three methods are used: (1) aiming high-energy radiation at the cancer, (2) implanting seeds or beads of isotope within the cancerous region, or (3) ingesting or injecting the isotope. High-energy radiation is often used to treat deep tumors. Using intense, short irradiation helps reduce the damage to surrounding tissues. Implanting radioactive seeds is used to treat certain cancers such as prostate and cervical cancer. These cancers seem to respond well to a constant beam of radiation. The ingestion method is often used to treat thyroid conditions such as overactive thyroid. In this case, much larger doses of isotope are used than in diagnosis. The isotope accumulates in the gland and damages the hormone-producing cells, thereby lowering production of thyroid hormone.

Radiation also carries risks; it can be detrimental to health or even fatal. When alpha, beta, and gamma rays strike living cells, they can cause parts of their molecules to break apart and ionize. Radiation can also cause the formation of free radicals, which are uncharged but unstable and highly reactive atoms or compounds. Free radicals and ions can interact with other molecules in the body, causing their destruction. They may also produce chemical changes that may contribute to various disease conditions including cancer. Used properly, however, radiation can have many important applications in both food preservation and in medicine. ●

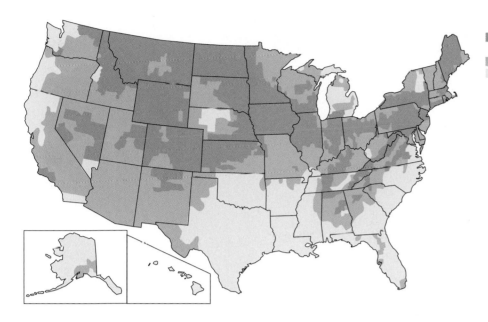

More than 4 picocuries
per liter

2–4 picocuries per liter

Less than 2 picocuries per liter

**Figure 3.4** Concentrations of radon gas throughout the United States.  Note: A curie is a unit of radioactivity. A picocurie is one trillionth of curie. On the map, the darker areas have higher concentrations of radioactivity, and the lighter areas have lower concentrations.

charges, and (3) the tendency of the outer shell, or energy level, of electrons to be full. This third factor is often called the **octet rule**.

The word *octet* means "eight objects" and refers to the fact that eight is a stable number for electrons in an outer shell. The outer electron shell of many atoms contains a maximum of eight electrons. (The first energy level is an exception to this rule because it contains a maximum of two electrons.) The octet rule states that an atom with an unfilled outer shell tends to interact with another atom or atoms in ways that will complete this outer shell. Although the octet rule does not apply to all atoms, it does apply to all biologically important ones—those involved in the structure, energy needs, and metabolism of living things. We will look more closely at these atoms and the molecules they make up later in this chapter and Chapter 4.

Atoms of elements that have equal numbers of protons and electrons and have full outer-electron energy levels are the only ones that exist as single atoms. Atoms with these characteristics are called *noble gases*, or inert gases. They do not react readily with other elements.

Most of the noble gases—helium, neon, argon, krypton, xenon, and radon—are rare. They are also relatively unreactive and unimportant in living systems. Helium and radon are probably the best known and the least rare noble gases. Radon, in fact, has gained more notice in recent years. Formed in rock or soil particles from the radioactive decay of radium, radon gas can seep through cracks in basement walls and remain trapped in homes that are not well ventilated. Prolonged exposure to radioactive radon in levels greater than those normally found in the atmosphere is thought to lead to lung cancer. **Figure 3.4** shows concentrations of radon gas throughout the United States.

CONCEPT CHECKPOINT

4. Most atoms interact with one another. They may form molecules. Define the term *molecule*.
5. What is the key difference between a molecule that is a compound and one that is not?
6. Name the three factors that influence whether and how an atom will interact with other atoms.

## 3.3  Chemical bonds hold atoms together.

How can atoms having incomplete outer shells satisfy the octet rule? There are three ways:

- Atoms can gain electrons from other atoms.
- Atoms can lose electrons to other atoms.
- Atoms can share electron pairs with other atoms.

Such interactions among atoms result in **chemical bonds**, forces that hold atoms together. The number and type of atoms and the way in which atoms are held together and arranged in space determine the properties of a molecule—whether the molecule is part of the wing of a fly, a hair on your head, or the hamburger on your plate. The making and breaking of chemical bonds are also crucial to life processes such as digesting your food or healing a cut.

As mentioned previously, some atoms bond by sharing outer shell electrons. This type of bond is called **covalent** (ko-VAY-lent). Other atoms that have lost or gained electrons (ions) bond by the attraction of opposing electrical charges. This type of bond is called **ionic** (eye-ON-ick). Other weaker kinds of bonds also occur.

The electrons in the outermost (highest) energy levels, or shells of atoms are key to bonding. Why? All electrons in an atom are attracted to the positive charge of the nucleus, but electrons far from the nucleus are not held as tightly as electrons closer to the nucleus. In addition, these "far out" electrons have more energy than "close in" electrons. Energy-rich, loosely held electrons interact with other atoms more easily than energy-poor, tightly held electrons.

## 3.4  Ionic bonds are attractions between ions of opposite charge.

So why do some atoms share electrons while others "give" and "take" electrons? An atom tends to lose (give) electrons if it has an outer shell needing many electrons to be complete. An atom tends to gain (take) electrons if it has a nearly completed outer shell. A "giver" tends to interact with a "taker," resulting in a completed outer shell for each.

For example, an atom with a nearly completed outer shell tends to take electrons from an atom that has only one or two electrons in its outer shell instead of the eight it needs for completion. Once these outer electrons are gone from the atom giving up electrons, the next shell in becomes its new, complete outer shell. As a result of this interaction, neither atom is electrically neutral; the number of protons no longer equals the number of electrons in either atom. The atom taking on electrons acquires a negative charge; the atom giving up electrons acquires a positive charge. They are not called atoms because they are no longer electrically neutral. They are charged and, as mentioned previously, such charged particles are called ions.

**Figure 3.5a** illustrates the formation of ions using table salt—sodium chloride—as an example. Notice in the upper left portion of the illustration that sodium (Na) is an element with 11 protons and 11 electrons. It is a soft, silver-white metal that occurs in nature as a part of ionic compounds. Of sodium's 11 electrons, 2 are in its innermost energy level (full with 2 electrons), 8 are at the next level, and 1 is at the outer energy level. Because of this distribution of electrons, sodium's outer energy level is not full, and therefore the octet rule is not satisfied.

Chlorine (Cl) is an element with 17 protons and 17 electrons, as shown in the upper right portion of the illustration. In its molecular form ($Cl_2$), it is a greenish-yellow gas that is poisonous and irritating to the nose and throat. In the ionic compound sodium chloride, however, it does not have these characteristics. Of chlorine's 17 electrons, 2 are at its innermost energy level, 8 at the next energy level, and 7 at the outer energy level. Chlorine, like sodium, has an outer energy level that is not full. Sodium and chlorine atoms can interact with one another in a way that results in both having full outer energy levels.

When placed together, the metal sodium and the gas chlorine react explosively. The single electron in the outer energy level of each sodium atom is lost to each chlorine atom as shown in the lower portion of Figure 3.5a. The result is the production of $Na^+$ and $Cl^-$ ions. These ions come together because their opposite charges attract one another. This type of attraction is called *electrostatic attraction* and results in ionic bonding of the sodium and chloride ions. As these ions are drawn to one another, they form geometrically perfect crystals of salt as shown in Figure 3.5b.

The transfer of electrons between atoms is an important chemical event—one type of chemical reaction. This type of chemical interaction has a special vocabulary. When an atom loses (gives up) an electron, it is *oxidized*. The process by which this occurs is called an **oxidation** (OK-si-DAY-shun), meaning "to combine with oxygen." The name reflects the fact that in biological systems, oxygen, which strongly attracts electrons, is the most frequent electron acceptor. Therefore, atoms that give up electrons to oxygen are "acted upon" by oxygen, or oxidized. Although not found in living systems, the formation of rust is an oxidation with which you are likely familiar. When

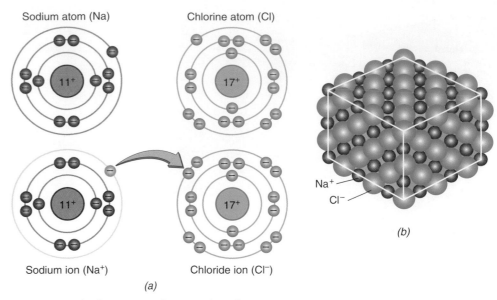

Sodium atom (Na)   Chlorine atom (Cl)

Sodium ion (Na⁺)   Chloride ion (Cl⁻)

Na⁺
Cl⁻

(b)

(a)

**Figure 3.5** The formation of an ionic bond.

**Visual Thinking:** Explain what the arrow in the bottom portion of (a) is showing.

iron combines with oxygen in the presence of moisture, it becomes oxidized. The product of this oxidation is commonly known as rust.

Conversely, when an atom gains an electron, it becomes *reduced*. The process is called a **reduction**. This name reflects the fact that the addition of an electron reduces the charge by one. For example, if a molecule had a charge of $+2$, the addition of an electron ($-1$) would reduce the molecule's charge to $+1$. A commonly held misconception is that a reduced molecule is somehow smaller (reduced in size). That is not the case.

Oxidation and reduction always take place together because every electron that is lost by an atom (oxidation) is gained by some other atom (reduction). Together they are therefore called **redox reactions**. In a redox reaction the charge of the oxidized atom is increased, and the charge of the reduced atom is lowered.

### CONCEPT CHECKPOINT

7. Which one of the following atoms would be most likely to ionize and form an ionic compound with another ion (a) radon, with an outer shell of eight electrons, (b) chlorine, with an outer shell of seven electrons, or (c) carbon, with an outer shell of four electrons? Why?
8. When oxidation takes place, such as when iron rusts, what is happening chemically?

## 3.5 Covalent bonds are a sharing of electrons between atoms.

As you can see in **Figure 3.6** (upper right), a hydrogen atom has a single electron and an unfilled outer electron shell. A filled outer shell at this energy level requires two electrons. When hydrogen atoms are close enough to one another, an interesting thing happens. They form pairs, with each of their single electrons moving around the two nuclei. These paired atoms of hydrogen are called *diatomic* (die-uh-TOM-ick) *molecules* and are represented by the chemical formula $H_2$ which is hydrogen gas. Diatomic molecules, the simplest molecules, contain only two atoms. Other examples of diatomic molecules are $O_2$ (oxygen gas), $N_2$ (nitrogen gas), and $Cl_2$ (chlorine gas).

As a result of this sharing of electrons, the diatomic hydrogen gas molecule is electrically balanced because it now contains two protons and two electrons. In addition, each hydrogen atom has two electrons in its outer shell, completing this shell. Thus, by sharing their electrons and forming a covalent bond, the two hydrogen atoms form a stable molecule.

Covalent bonds can be very strong, that is, difficult to break. *Double bonds*, those bonds in which two pairs of electrons are shared, are stronger than *single bonds* in which only one pair of electrons is shared. As you might expect, *triple bonds*, those bonds in which three pairs of electrons are shared, are the strongest of these three types of covalent bonds. In chemical formulas that show the structure of covalently bonded molecules, single bonds are represented by a single line between two bonded atoms, double bonds by two lines, and triple bonds by three lines. For example, the structural formula of hydrogen gas is H - H, oxygen gas is O = O, and nitrogen gas is N≡N.

An atom can also form covalent bonds with more than one other atom. Carbon (C), for example, contains six electrons: two in the inner shell and four in the outer shell. To satisfy the octet rule, it must gain four additional electrons by sharing its four outer-shell electrons with another atom or atoms, forming four covalent bonds. Because four covalent bonds may form in many ways, carbon atoms are versatile and are the basis for all four primary biological molecules: carbohydrates, lipids, proteins, and nucleic acids. We discuss these key molecules of life in Chapter 4.

The strength of a covalent bond refers to the amount of energy needed to make or break that bond. The energy that goes into making the bond is held within the bond and is released when the bond is broken. Therefore, covalent bonds are actually a storage place for

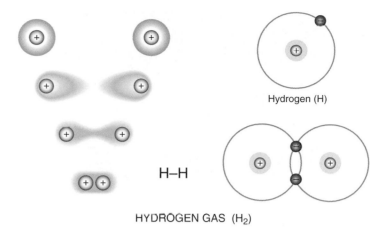

Hydrogen (H)

H–H

HYDROGEN GAS ($H_2$)

**Figure 3.6** Two types of shell models (left and right) and a structural formula (center) depicting the sharing of electrons in a diatomic molecule.

energy as well as a type of chemical "glue" that holds molecules together. Accordingly, covalent bonds hold another key to life—the ability of living things to store and use energy by making and breaking covalent bonds, using molecules as a type of energy currency.

---

 CONCEPT CHECKPOINT

**9.** Carbon (with an outer shell of four electrons), hydrogen (with an outer shell of one electron), and oxygen (with an outer shell of six electrons) typically form covalent bonds with one another. What is the maximum number of hydrogen atoms with which an atom of carbon can bond? What is the maximum number of oxygen atoms with which an atom of carbon can bond? Explain your answers.

---

# What properties of water are important to life?

## 3.6 Water molecules "stick to" other water molecules.

Water is the most abundant molecule in your body, making up about two-thirds of your body weight. Although it seems to be a simple covalently bonded molecule, water is vital to life. When life on Earth was beginning, this liquid provided a medium in which other molecules could move and interact. Life evolved as a result of these interactions. And life, as it evolved, maintained these ties to water (as illustrated in **Figure 3.7**). Three-fourths of the Earth's surface is covered by water. Where water is plentiful, such as in the tropical rain forests, the land abounds with life. Where water is scarce, such as in the desert, the land seems almost lifeless except after a rainstorm. No plant or animal can grow and reproduce without some amount of water. The chemistry of life, then, is water chemistry.

Water has a simple molecular structure: one oxygen atom bonded by single covalent bonds to two hydrogen atoms. The resulting molecule satisfies the octet rule, and its positive and negative charges are balanced. However, the electron pair shared in each covalent bond is more strongly attracted to the oxygen nucleus than to either of the hydrogen nuclei. Why?

The oxygen atom in each water molecule contains eight protons, and each hydrogen atom contains only one proton. Therefore, the negatively charged electrons are more strongly attracted to the oxygen nucleus with its eight positive charges than to the hydrogen

**Figure 3.7 Water is the cradle of life.** This mass of frog eggs is attached to a rock in the watery environment of a stream bottom.

nucleus with its single positive charge. Although the electrons surround both the oxygen and hydrogen nuclei, they are far more likely to be found near the oxygen nucleus at a given moment than near one of the hydrogen nuclei. Because of this situation, the oxygen end of the water molecule has a partial negative charge as shown in both the shell model and space-filling model in **Figure 3.8**. Molecules such as water that have opposite partial charges at different ends of the molecule are called **polar molecules**.

Any covalent bond between different atoms is polar to some degree because no two elements have identical numbers of protons in their nuclei. The term *electronegativity* refers to the extent to which an element attracts bonding electrons in a covalent bond.

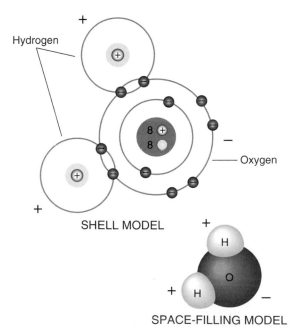

**Figure 3.8** A shell model and a space-filling model of one molecule of water.

**Visual Thinking:** Which model better helps you understand the structure of water? Why?

The greater the difference between the electronegativities of two elements bonded covalently, the more polar the bond.

The polarity of water contributes to its ability to attract other molecules and form special types of chemical bonds with them. These bonds between water molecules, called **hydrogen bonds**, have approximately 5 to 10% the strength of covalent bonds. Hydrogen bonds are shown as the dotted lines holding water molecules together in Figure 3.B in the *Just Wondering* box. As shown in the illustration of ice, all four possible hydrogen bonds have formed between each molecule of water and other water molecules, resulting in a rigid latticework of molecules held a considerable distance from one another. Therefore, ice is less dense than water; that is, there are fewer molecules per gram. Thus, ice floats on water, and bodies of water freeze from the top down. An ice covering acts as insulation for the water below, protecting bodies of water from freezing into a solid block, which allows aquatic life to survive beneath ice-covered surfaces.

Hydrogen bonds are weak, and therefore these bonds are constantly made and broken in water below its boiling point of 100°C (212°F) and above its freezing point of 0°C (32°F). Each bond lasts only 1/100,000,000,000 of a second! Nevertheless, the cumulative effect of very large numbers of hydrogen bonds is responsible for the many important physical properties of water. The ability of water molecules to form weak bonds among themselves and with other molecules is the reason for much of the organization and chemistry of living things.

All organisms have a considerable amount of water in their structures. In fact, the human body is roughly 60 to 70% water. Water modulates temperature in the bodies of living things. How does it do this? One of the properties of water is its ability to absorb or release much heat energy with little change in its temperature. Water is said to have a high *specific heat*, which is the amount of heat required to raise or lower the temperature of one gram of a substance by one degree Celsius. Hydrogen bonding in water is the key to this property. As water absorbs heat, hydrogen bonds are broken. Only after hydrogen bonds have been broken does heat absorption increase the motion of water molecules, raising water's temperature. Therefore, most organisms can exist in environments with wide temperature fluctuations and yet maintain steady internal temperatures, using other behavioral and physiological mechanisms as well.

Hydrogen bonding also results in a tendency of water molecules, or other like molecules, to "stick together." This property is called *cohesion*. The cohesion of water molecules plays an important role in upward water movement in plants. As molecules of water evaporate from the leaves of plants, they "pull" other molecules of water with them, which pull on the molecules next to them and so forth. This force is transmitted throughout the plant to the roots, from one water molecule to the next. In addition, water molecules are attracted to the molecules of the walls of the very narrow transport tubes within plants. This property—the weak electrical attraction of one type of molecule to another type—is known as *adhesion*. Water actually tends to climb the walls! The two forces of cohesion and adhesion link water molecules together within a plant in an unbroken stream. This resulting upward movement of water within narrow tubes against the force of gravity is called *capillary action*.

*Surface tension* refers to the cohesion of water molecules at its surface, making the water appear as though it is covered by a thin film or skin. It is the reason that water molecules can bear the weight of organisms such as the water strider shown in **Figure 3.9**.

## *just wondering . . .*

### If water (a liquid) is made up of two gases, hydrogen and oxygen, then why isn't water a gas?

You have made an interesting observation and questioned one of water's many fascinating properties: at room temperature, both hydrogen ($H_2$) and oxygen ($O_2$) are gases, but water is not. In addition, other substances whose molecules are similar in size to those of water, such as ammonia and methane, *are* gases at room temperature. Why not water?

The answer to your question and the underlying cause of many of water's properties such as its surface tension and its expansion upon freezing all have to do with the way in which the water molecule is constructed. Water is a polar molecule as described in Section 3.6. Because of this polarity, its positively charged portions are attracted to the negatively charged portions of other water molecules and vice versa. These attractive forces are called *hydrogen bonds*. Hydrogen bonding results in each water molecule being attracted to others, with each of those molecules attracted to others, and so forth, forming an extensive network among water molecules at room temperature. Water mole-

cules, therefore, do not move independently as in a gas but move together, forming a liquid.

At room temperature, water molecules are in constant motion, so their hydrogen bonds are constantly broken and remade. However, at other temperatures, the properties of water change because temperature affects hydrogen bonding. To illustrate: above 100°C (212°F) the hydrogen bonds weaken, and water molecules begin to break free from one another, forming steam, or water vapor. However, the bonds within the water molecules do not break, so steam is made up of water, not hydrogen and oxygen gases. Below 0°C (32°F) the water molecules are no longer moving vigorously enough to break and remake hydrogen bonds, so the water molecules become locked into a crystalline lattice with the hydrogen bonds holding each water molecule apart at bond's length as shown in **Figure 3.B**. Therefore, the water expands and becomes the solid we call ice. At the surface of ice, however, hydrogen bonds continue to be broken and remade, causing ice to be slippery, which allows us to skate, slide, and fall on this amazing compound. ●

---

*Are you wondering about a topic in biology and how it relates to your life? Submit your question by clicking the* Just Wondering *link in this text's companion Web site at www.wiley.com/college/alters.*

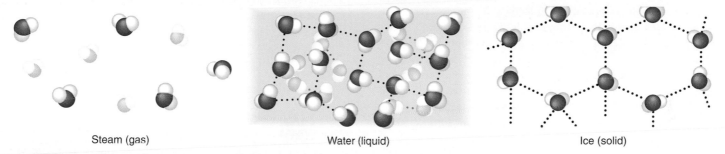

Steam (gas)          Water (liquid)          Ice (solid)

**Figure 3.B** Three forms of water.

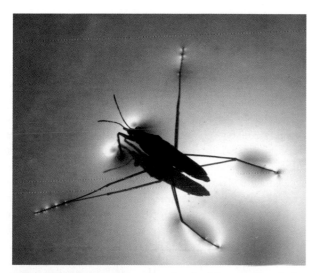

**Figure 3.9** Because water molecules "stick together," they can support the water strider.

### 3.7 Water is a powerful solvent.

Did you ever wonder what happens when the sugar in your coffee dissolves? The answer is that the sugar "goes into solution." But what does that mean?

A solution is a homogeneous mixture of two or more substances. *Homogeneous* means that the substances are uniform in composition. In other words, the solution looks the same throughout. You do not see sugar and coffee separate from one another, because the sugar molecules move among the water molecules as separate molecules when they are in solution.

Generally, a **solvent** is the substance in the solution that is present in the greater amount. It dissolves the **solute**—the substance that dissolves in the solvent to form the solution. In the case of a black coffee with sugar, the solvent is water and the solutes are sugar and coffee molecules.

Water is a powerful solvent. Compounds that dissolve in water are said to be **soluble** (SOL-you-bul) in water. Chemical interactions readily take place in water because so many kinds of compounds are water soluble and therefore move among water molecules as separate molecules or ions.

How do solutes such as sugar or salt dissolve in solvents such as water? Water molecules gather closely around any particle that exhibits an electrical charge, such as ions, and polar molecules, such as sugars. For example, sodium chloride (table salt) is made up of the positively charged sodium ($Na^+$) and negatively charged chloride ($Cl^-$) ions. These ions are attracted to one another and cluster in a regular pattern, forming crystals. When you put salt in water, some ions break away from the crystals because the positive ends of some water molecules are attracted to the $Cl^-$ ions, while the negative ends of other water molecules are attracted to the $Na^+$ ions. These attractions are shown in the right-hand portion of **Figure 3.10**. These attractions are stronger than the attraction between the ions that keep the crystal together. Therefore, the ions are pulled from their positions in the crystal as is shown on the left. Water molecules then surround each ion, forming a *hydration shell*, which keeps the ions apart. The salt is said to be *dissolved*. Hydration shells form around all polar molecules and ions when in water.

### Water excludes nonpolar molecules

Remember the old saying that oil and water don't mix? This statement is true because oil is a nonpolar molecule and cannot form hydrogen bonds with water; it cannot dissolve in water. Instead, the water molecules form hydrogen bonds with each other, causing the water to exclude the nonpolar oil molecules. It is almost as if nonpolar molecules move away from contact with the water. For this reason, nonpolar molecules are referred to as hydrophobic (HI-dro-FO-bik). The word *hydrophobic* comes from Greek words meaning "water" (*hydros*) and "fearing" (*phobos*). This tendency for nonpolar molecules to group together in a water solution is called *hydrophobic bonding* (**Figure 3.11**). Hydrophobic forces determine the three-dimensional shapes of many biological molecules, which are usually surrounded by water within organisms.

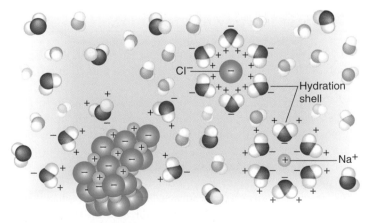

**Figure 3.10** How salt dissolves in water.

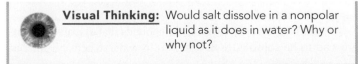

**Visual Thinking:** Would salt dissolve in a nonpolar liquid as it does in water? Why or why not?

**Figure 3.11** Oil is hydrophobic. Crude petroleum from an oil spill floats on the surface of the ocean because it is hydrophobic and less dense (lighter) than water. The sinking of the oil tanker *Prestige* in November 2002 dumped thousands of tons of fuel oil into fishing waters and onto beaches in Spain, France, and Portugal. This photo shows cleanup work on the shore of a creek in a northwestern village in Spain in December 2002.

### CONCEPT CHECKPOINT

10. List and explain the *chemical* property of water that results in each of its following biologically important characteristics. Provide an example of why the characteristic is biologically important. (a) cohesion of water; (b) adhesion of water; (c) powerful solvent.

## 3.8 Water, when ionized, can accept or donate protons in the reactions of life.

Although water is commonly thought of as $H_2O$, its molecules often break spontaneously. When this happens, one hydrogen atom nucleus (a proton) dissociates, or separates, from the rest of the water molecule, leaving behind its electron. Because its positive charge is no longer balanced by an electron, it becomes a positively charged hydrogen ion, $H^+$. The remaining part of the water molecule now has an extra electron. It is therefore a negatively charged hydroxyl ion, $OH^-$. This process of spontaneous ion formation is called **ionization** (EYE-uh-nih-ZAY-shun):

$$H_2O \rightleftharpoons OH^- + H^+$$

The ionization of water is critical to the chemical reactions that make up life processes. Why? Many reactions that take place in living systems involve the transfer of a proton from an acid to a base. An **acid** is any substance that dissociates to form $H^+$ ions when it is dissolved in water. A **base** is any substance that combines with or accepts $H^+$ ions, as $OH^-$ ions do. Water, when ionized, can act as either an acid or a base, accepting protons or donating protons. Therefore, water plays a vital role in the chemistry of life as either a reactant—a molecule that enters into a chemical reaction—or a product—a molecule that is produced in a chemical reaction.

In a volume of water, only very few water molecules are ionized at a single instant in time. Scientists calculate that the fraction of water molecules dissociated (ionized) at any given time in pure water is 0.0000001. This tiny number can be written another way by using exponential notation. This is done by counting the number of places to the right of the decimal point. Because there are seven places, this number is written as $10^{-7}$. The $-7$ is called an *exponent*; the minus sign means that the number is less than 1.

To indicate the concentration of $H^+$ ions in a solution, scientists have devised a scale based on the slight degree of spontaneous ionization of water. This scale is called the **pH scale**. The letters "pH" come from Latin words meaning "potential of hydrogen." This phrase means that acidity is due to a predominance of hydrogen ions in a water-containing solution.

The pH values of the scale generally range from 0 to 14. The pH of a solution is determined by taking the negative value of the exponent of its hydrogen ion concentration. For example, pure water has a hydrogen ion concentration of $10^{-7}$ and therefore a pH of 7. When water ionizes, hydroxyl ions ($OH^-$) are produced in a concentration equal to the concentration of hydrogen ions. In pure water the concentrations of both $H^+$ and $OH^-$ ions are equal to $10^{-7}$. Because these ions join spontaneously, water, at pH 7, is neutral.

An acid produces $H^+$ ions; the more hydrogen ions an acid produces, the stronger an acid it is. Although an acid produces a higher concentration of $H^+$ ions than pure water (0.00001 as opposed to 0.0000001, for example), its pH is lower. Using the above numbers to illustrate, we find that the first number (0.00001 or $10^{-5}$) represents the hydrogen ion concentration of an acid. The negative value of its exponent results in a pH of 5. The second number (0.0000001 or $10^{-7}$) represents the hydrogen ion concentration of water. The negative value of its exponent results in a pH of 7. Each single-unit decrease in pH, however, does not correspond to a one-fold increase in acidity. Depending on the acid and how completely it ionizes, a change of one unit may correspond to as much as a tenfold increase in acidity.

**Figure 3.12** shows many common substances and their pH values. The pH of champagne, for example, is about 4. This low pH is due to the dissolved carbonic acid that causes champagne to bubble. The hydrochloric acid (HCl) of your stomach ionizes completely to $H^+$ and $Cl^-$. It forms a strong acid with a pH of 0.1 but becomes diluted with other stomach fluids, resulting in a pH of 2–3. Rain, snow, fog, and even clouds can also be acids due to the mixing of moisture in the air with pollutants. Clouds hanging over the spruce-fir forests found on Clingmans Dome (**Figure 3.13**), the highest peak in Great Smoky Mountains National Park, have been measured to have a pH as low as 2.0. When this acidic moisture falls to the ground, it is called *acid precipitation*, and it can have a devastating effect on both living and nonliving things. We discuss acid precipitation in more detail in Chapter 42.

Any increase in the concentration of a base lowers the $H^+$ ion concentration because the base combines with $H^+$. Bases therefore have pH values higher than water's neutral value of 7. For example, the envi-

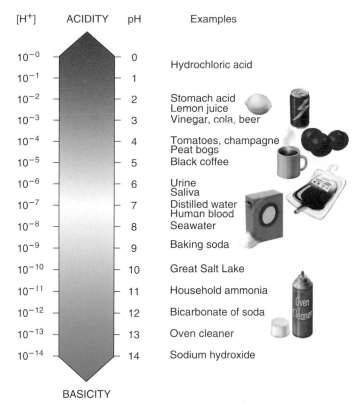

| [H⁺] | ACIDITY | pH | Examples |
|---|---|---|---|
| $10^{-0}$ | | 0 | Hydrochloric acid |
| $10^{-1}$ | | 1 | |
| $10^{-2}$ | | 2 | Stomach acid<br>Lemon juice |
| $10^{-3}$ | | 3 | Vinegar, cola, beer |
| $10^{-4}$ | | 4 | Tomatoes, champagne<br>Peat bogs |
| $10^{-5}$ | | 5 | Black coffee |
| $10^{-6}$ | | 6 | Urine<br>Saliva |
| $10^{-7}$ | | 7 | Distilled water<br>Human blood |
| $10^{-8}$ | | 8 | Seawater |
| $10^{-9}$ | | 9 | Baking soda |
| $10^{-10}$ | | 10 | Great Salt Lake |
| $10^{-11}$ | | 11 | Household ammonia |
| $10^{-12}$ | | 12 | Bicarbonate of soda |
| $10^{-13}$ | | 13 | Oven cleaner |
| $10^{-14}$ | | 14 | Sodium hydroxide |

BASICITY

**Figure 3.12** The pH scale.

ronment of your small intestine is kept at a basic pH of between 7.5 and 8.5. Strong bases such as sodium hydroxide (NaOH) have pH values of 12 or more. As with acids, a change of 1 in the pH value of a base may reflect up to a tenfold change in pH, depending on the base.

### CONCEPT CHECKPOINT

**11.** Why is the spontaneous ionization of water important to life?

**12.** Define the term *pH*.

**13.** What do low pH values indicate? What do high pH values indicate?

**Figure 3.13** Acid precipitation. Clouds hang over the spruce-fir forests at Clingmans Dome, Great Smoky Mountains National Park. The devastating effects of acid fog and other acid precipitation on these trees is hard to miss.

## CHAPTER REVIEW

## Summary of Key Concepts and Key Terms

### What are atoms and how do they interact?

**3.1 Atoms** (p. 30) make up all matter and are composed of three types of subatomic particles: **protons** (p. 30), **neutrons** (p. 30), and **electrons** (p. 30).

**3.1** Protons and neutrons are found at the core of the atom, and electrons surround this core.

**3.1** The chemical behavior of an atom is largely determined by the distribution of its electrons, particularly the number of electrons in its outermost **energy level** (**shell**, p. 31).

**3.1 Elements** (p. 30) are pure substances that are made up of a single kind of atom and cannot be separated into different substances by ordinary chemical methods.

**3.1** Atoms that have the same number of protons but different numbers of neutrons are called **isotopes** (EYE-suh-topes, p. 30).

**3.2 Molecules** (p. 31) are two or more atoms held together by sharing electrons.

**3.2** Molecules can be made up of atoms of the same element or atoms of different elements.

**3.2** Molecules made up of atoms of different elements are called **compounds** (p. 31). Compounds are also atoms of different elements that have lost or gained electrons (**ions**, p. 31) and that are held together by opposing electrical charges.

**3.2** The **octet rule** (p. 33) states that an atom with an unfilled outer shell tends to interact with another atom or atoms in ways that will complete this outer shell.

**3.3** Atoms are most stable when their outer energy levels are filled. Electrons are lost, gained, or shared until this condition is reached.

**3.3 Chemical bonds** (p. 33) are forces that hold atoms together.

**3.3** In **covalent** (p. 33) bonds, atoms share outer shell electrons. In **ionic** (eye-ON-ick, p. 33) bonds, atoms that have lost or gained electrons (ions) bond by the attraction of opposing electrical charges.

**3.4** An atom tends to lose (give) electrons if it has an outer shell needing many electrons to be complete. An atom tends to gain (take) electrons if it has a nearly completed outer shell.

**3.4** When an atom loses an electron, it is oxidized. When an atom gains an electron, it is reduced. Together, **oxidation** (OK-si-DAY-shun, p. 34) and **reduction** (p. 36) reactions are called **redox reactions** (p. 34).

**3.5** A covalent bond is the sharing of one or more pairs of electrons between atoms.

**3.5** Covalent bonds store energy as well as hold molecules together.

### What properties of water are important to life?

**3.6** Water is a **polar molecule** (p. 36), having slightly positive and slightly negative ends, and attracts other polar molecules and charged particles.

**3.6** The cohesion of water molecules, which is their tendency to stick together by means of **hydrogen bonds** (p. 36), and adhesion, which is their tendency to be attracted to the polar molecules of other substances, are key to the movement of water in plants.

**3.7** Generally, a **solvent** (p. 37) is the substance in the solution that is present in the greater amount. It dissolves the **solute** (p. 37) to form the solution.

**3.7** Water dissolves polar substances and excludes nonpolar substances.

**3.7** Chemical interactions readily take place in water because so many kinds of molecules are water **soluble** (SOL-you-bul, p. 37) and

therefore move among water molecules as separate molecules or ions.

**3.8** Water spontaneously ionizes (**ionization**, p. 38), forming hydrogen ions ($H^+$) and hydroxide ions ($OH^-$).

**3.8** An **acid** (p. 39) is any substance that dissociates to form $H^+$ ions when it is dissolved in water; a **base** (p. 39) is any substance that combines with $H^+$ ions, as $OH^-$ ions do.

**3.8** The **pH scale** (p. 39) indicates the concentration of $H^+$ ions in a solution.

**3.8** The ionization of water is critical to life processes because many reactions that take place in living systems involve the transfer of a proton from an acid to a base.

---

| **Level 1** | **Learning Basic Facts and Terms** |
|---|---|

**Multiple Choice**

1. An atom is a(n)
   a. subatomic particle.
   b. isotope with a single proton.
   c. submicroscopic particle that makes up matter.
   d. particle carrying a positive charge.
2. In covalent bonding
   a. ions give and take electrons.
   b. atoms share electron pairs.
   c. oxidation occurs.
   d. cohesion holds the molecule together.
3. Diatomic molecules
   a. are made up of two atoms.
   b. have double bonds.
   c. have two protons, two neutrons, and two electrons.
   d. have two electrons in each shell.

4. Oil and water don't mix because
   a. the surface tension of the oil is not great enough.
   b. oil molecules are too cohesive.
   c. hydrogen bonds will not allow the interaction.
   d. water does not dissolve nonpolar substances.
5. What term describes any substance that dissociates to form $H^+$ ions when it is dissolved in water?
   a. acid     b. base     c. isotope     d. molecule

**Matching**

6. ____ element    a. neutral and carries no charge
7. ____ noble gas    b. pure substance made up of a single kind of atom
8. ____ shell    c. volume of space around a nucleus where electrons are
9. ____ neutron    d. has full outer-electron energy levels
10. ____ reduction    e. process of gaining an electron

## Level 2    Learning Concepts

1. a. Draw a diagram of an atom with an atomic number of 1. Label the nucleus and the subatomic particles, and show the electrical charge of each particle.
   b. Draw an isotope of the same atom.
2. List and explain three factors that influence how an atom interacts with other atoms. What is the significance of the octet rule?
3. Nitrogen gas is formed via a triple bond between two nitrogen atoms. What type of bonding is this? What does this bond have in common with the bonds forming water molecules? How is this bond different from the bonds forming water molecules?

4. Describe some of the chemical properties of the water molecule that make it biologically important. Relate these properties to water's molecular structure. How do these properties affect the interactions of other molecules dissolved or suspended in it?
5. The pH of the digestive juices within the human small intestine is between 7.5 and 8.5. Would you describe this environment as acidic or basic? Is the concentration of H+ ions in these digestive juices greater than or less than the concentration of H+ ions in a solution of neutral pH?
6. Water is discussed in this chapter as an essential component of life on Earth. Describe how the chemical properties and bonding characteristics of water relate to its biological importance.

## Level 3    Critical Thinking—Life Applications

1. Based on what you learned in this chapter about the interactions of oil and water, suggest at least one reason why it is necessary to clean up an oil spill in the ocean.
2. Radon is a serious indoor air pollutant. Studies show that inhaling large amounts of radon increases the risk of lung cancer. Using Figure 3.4, determine whether you live in an area of high radon concentrations. Then find out what you can do to test your home for radon and what corrective actions you can take if high levels are found. Summarize that information here.

3. Unpolluted rainwater has a pH of about 5. Acid precipitation has a pH of about 4. About how much more acidic is acid precipitation than unpolluted rainwater? Suggest one reason why you think acid precipitation might be harmful to living things.
4. In early 2004, the NASA probe "Spirit" landed successfully on Mars. It was designed to look in rocks and sediments for evidence that liquid water existed on the surface of Mars in the past. The probe landed in what scientists believe is a dried lakebed. Why do you suppose NASA chose this site and designed the probe to accomplish such a goal?

## In The News    Critical Thinking

### E. COLI VS. E. COLI

Now that you understand more of the science behind food irradiation, reread this chapter's opening story. To develop an informed and thoughtful decision about your stand on this issue, it may help you to follow these steps:

1. Review your immediate responses to the issue that you wrote when you began reading this chapter.
2. Based on your current understanding, again summarize the issue, using either a statement or a question. An issue is a point on which people hold differing views.
3. Collect new information about the issue. Visit the *In The News* section of this text's companion Web site at www.wiley.com/college/alters and watch the "*E. coli* vs. *E. coli*" video. Then use the "summary" link to read the accompanying story and access related links. Use this information, the links provided, and other online and library resources to find updates about this issue. State the sources of that information. Explain why you think the information is accurate. Also determine if the information expresses a particular point of view or is biased in any way.
4. Determine which individuals, groups, or organizations have a stake in the issue. What does each stand to gain or lose depending on the outcome of the issue?
5. List possible outcomes (resolutions) of the issue. List the pros and cons of each outcome.
6. Which outcome do you think would be best? Why? Note whether your opinion differs from or is the same as what you wrote when you began reading this chapter. Give reasons for your opinion and for any changes in your ideas based on the additional information you have collected and the analysis you have done.

# BIOLOGICAL MOLECULES

## Mad Cow Clues

It wasn't until December 2003 that the United States experienced its first case of mad cow disease. "Mad cow" is the name commonly given to bovine spongiform encephalitis (BSE), a fatal disease of adult cattle in which the brain slowly degenerates. Why the concern? Scientists think that people can contract a disease similar to BSE by eating contaminated beef products. The exact cause of BSE is not known, but the general consensus in the scientific community is that BSE is caused by a prion (PREE-ahn)—a malformed protein that is infectious and pokes holes in the brain tissue of cows. If humans eat beef contaminated with BSE prions, they may develop a variant of a similar human disease called CJD (Creutzfeldt-Jacob disease). Amid all the concern over mad cow disease, one researcher is taking a different look at these misshapen biological molecules called prions.

Susan Lindquist, biologist and director of the Whitehead Institute for Biomedical Research has teamed up with physicists at the University of Chicago to determine whether prions can be useful. Lindquist notes that these misfolded proteins can set up a chain reaction in cells, in which the prions convert other proteins to the misfolded, prion form. Since proteins must be folded properly to do their jobs, they cease functioning when converted to misshapen prions. Nevertheless, during this chain reaction of

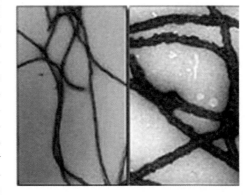

proteins to prions, the prions stick together, forming extremely tough fibers. Lindquist and her team thought these tough fibers might be used to do something useful, so they manipulated uninfective prions from yeast cells so that they would bond with gold and silver, forming extremely tiny, tough wires. In the photo, uncoated prion fibers are shown on the left and coated prion fibers on the right. The coated prion wires conduct electricity and are smaller than any that can be manufactured currently. Lindquist expects to use them to connect circuits. Might computers of the future conduct electricity with biological molecules? You can learn more about this research by viewing the video news story "Mad Cow Clues" at the *In The News* section of this text's companion Web site at www.wiley.com/college/alters.

*Write your immediate reaction to Lindquist's research on the prion wires: first, summarize the main point of the research in a sentence or two; then suggest what you think the significance of this research is. You will have an opportunity to reflect on your responses and gather more information on this topic in the* In The News *feature at the end of this chapter. In this chapter, you will learn more about the structure of biological molecules in general and proteins in particular.*

## CHAPTER GUIDE

### What are the general characteristics of biological molecules?

**4.1** Biological molecules are carbon based.

**4.2** Biological molecules interact by means of functional groups.

**4.3** Biological molecules are assembled or disassembled by adding or removing water.

### What are the major classes of biological molecules and their characteristics?

**4.4** Carbohydrates contain C, H, and O with an H-to-O ratio of 2:1.

**4.5** Lipids have an H-to-O ratio higher than 2:1.

**4.6** Proteins are chains of amino acids.

**4.7** Nucleic acids are composed of nucleotides.

# What are the general characteristics of biological molecules?

The differences between living and nonliving things relate to the characteristics of life that we discussed in Chapter 1, such as reproduction and the ability to transform energy. However, another characteristic distinguishes the living from the nonliving—molecular composition. Living or once-living things are composed of and produce biological molecules. What are biological molecules and how do they differ from the molecules that make up the nonliving world?

## 4.1 Biological molecules are carbon based.

With the exception of water, most molecules that are formed by living organisms and make up their structures contain the element *carbon*. This fascinating element lends itself to being the basis of living material because of its ability to form four stable bonds with other atoms. This ability allows carbon to interact with other molecules in myriad ways, forming a variety of large compounds.

Another atom with bonding properties similar to carbon is silicon. It is abundant in the Earth's crust but not in most living things. Nevertheless, some organisms use silicon in building outer skeletons or shells. An example of such organisms are the diatoms (**Figure 4.1**), one of the most abundant single-celled eukaryotes.

The carbon-containing molecules that make up living things are called **organic compounds**. Organic compounds are often very large and usually are held together by covalent bonds.

Inorganic compounds—substances not containing carbon—are often quite small and are usually held together by ionic bonds. A few carbon-containing, covalently bonded inorganic molecules exist (such as carbon dioxide and carbonic acid); they are considered inorganic molecules because they do not make up the structure of living or once-living things.

## 4.2 Biological molecules interact by means of functional groups.

It is often helpful to think of an organic molecule as a carbon-based core, or skeleton, with other special parts attached. Each of these special parts is really a group of atoms called a **functional group** and has distinct chemical properties.

A hydroxyl group (—OH), for example, is a functional group. Key functional groups are illustrated in **Figure 4.2**. These groups are important because many chemical reactions that occur within organisms involve the transfer of a functional group from one molecule to another. Other frequent chemical reactions involve the making and breaking of carbon-carbon bonds.

Some of the molecules of living things are simple organic molecules, often having only one functional group. Other far larger molecules, called *macromolecules*, contain thousands of atoms and may have many functional groups. The four major groups of biologically important macromolecules are carbohydrates, lipids, proteins, and nucleic acids (see **Table 4.1**).

| Functional Group | Structural Formula | Model |
|---|---|---|
| Hydroxyl | $-OH$ | |
| Carbonyl | $\begin{array}{c} -C- \\ \| \\ O \end{array}$ | |
| Carboxyl | $-C\begin{array}{c} \nwarrow O \\ \searrow OH \end{array}$ | |
| Amino | $-N\begin{array}{c} \nearrow H \\ \searrow H \end{array}$ | |
| Sulfhydryl | $-SH$ | |
| Phosphate | $\begin{array}{c} H \\ \| \\ -O-P-OH \\ \| \\ O \end{array}$ | |

**Figure 4.2** The chemical building blocks of life, functional groups. These are six of the most important functional groups involved in chemical reactions.

LM 150×

**Figure 4.1** Diatoms have silicon in the structure of their glasslike shells.

**Visual Thinking:** Which of the functional groups do not contain carbon?

| TABLE 4.1 | Biologically Important Macromolecules | | |
|---|---|---|---|
| **Macromolecule** | **Subunits** | **Function** | **Example** |
| **Carbohydrates** | | | |
| Starch, glycogen | Glucose | Stores energy | Potatoes |
| Cellulose | Glucose | Makes up cell walls in plants animals | Paper |
| Chitin | Modified glucose | Makes up the exterior skeleton in some animals | Crab shells |

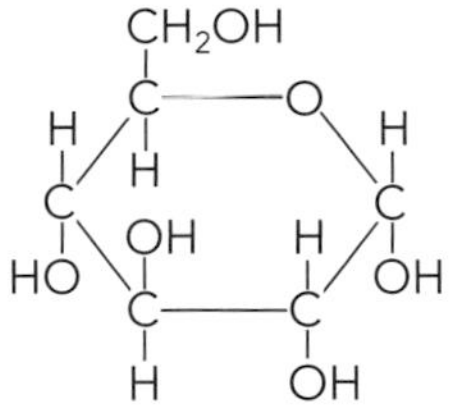

Glucose (ring structural formula)

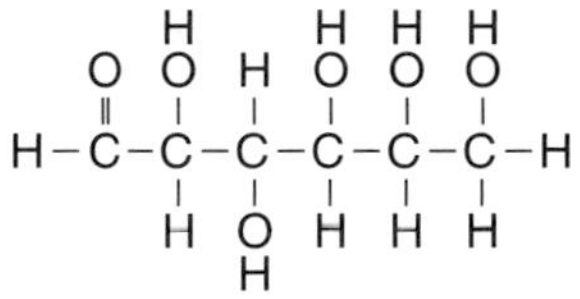

Glucose (linear structural formula)

| **Lipids** | | | |
|---|---|---|---|
| Fats (triglycerides) | Glycerol + three fatty acids | Store energy | Butter |
| Phospholipids | Glycerol + two fatty acids + phosphate | Make up cell membranes | All membranes |
| Steroids | Four carbon rings | Act as chemical messengers | Cholesterol, estrogen |

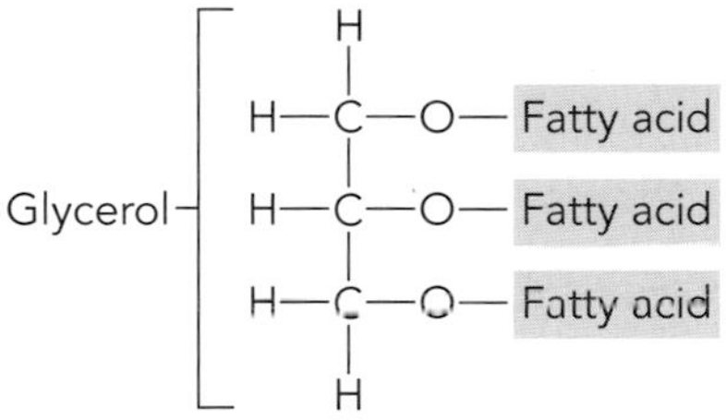

Generalized triglyceride

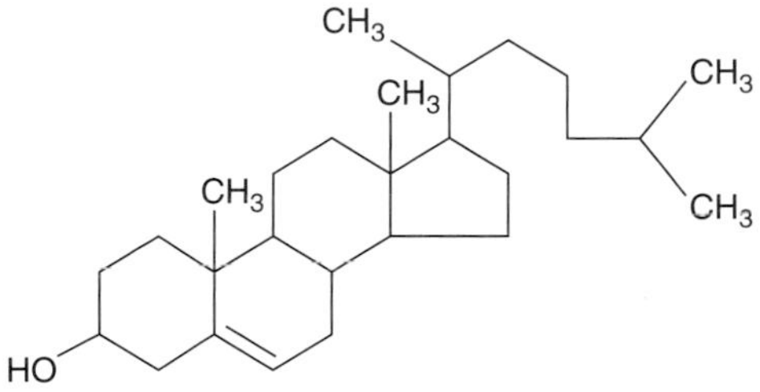

Cholesterol (a steroid)

| **Proteins** | | | |
|---|---|---|---|
| Enzymes | Amino acids | Help chemical reactions take place | Lysozyme in saliva |
| Peptides | Amino acids | Use as chemical messengers | Peptide neurotransmitters |
| Structural | Amino acids | Make up tissues that support body structures and provide movement | Muscle |

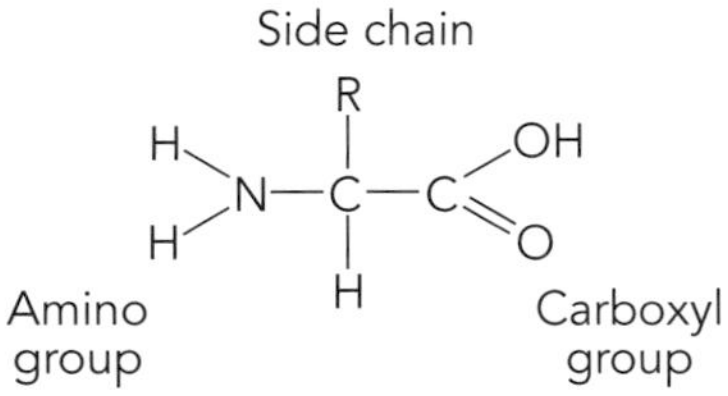

Generalized amino acid

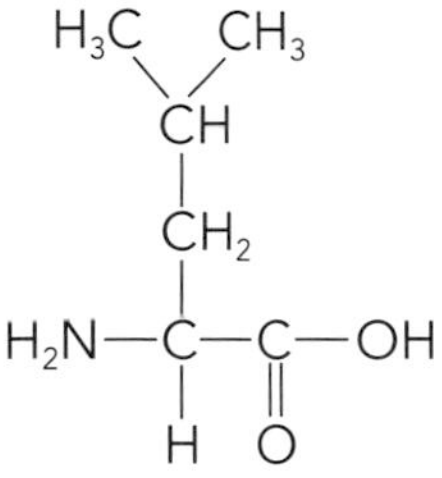

Leucine (amino acid)

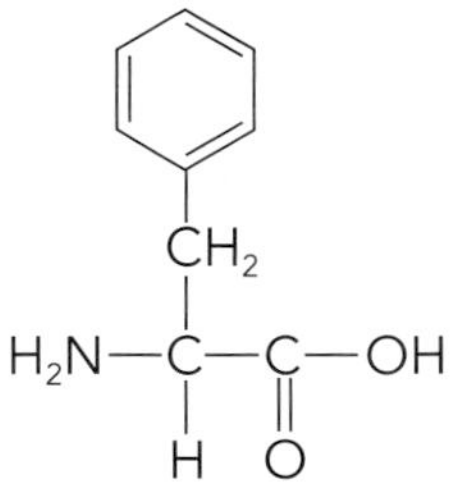

Phenylalanine (amino acid)

| **Nucleic Acids** | | | |
|---|---|---|---|
| DNA | Nucleotides | Encodes hereditary information | Chromosomes |
| RNA | Nucleotides | Helps decode hereditary information | Messenger RNA |

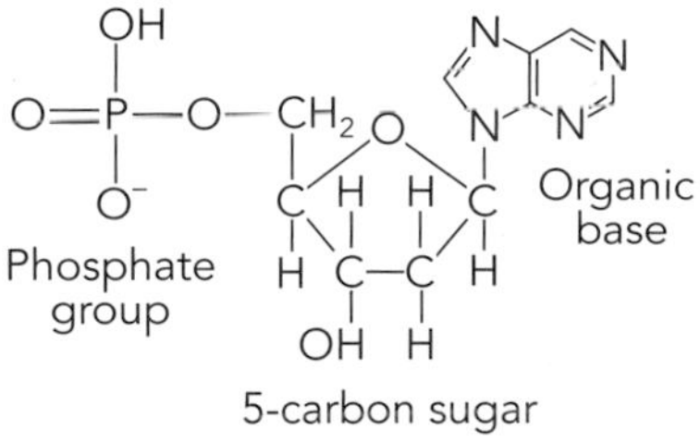

Generalized nucleotide

CONCEPT CHECKPOINT

1. Would an organic chemist study the chemical composition of rocks and soil? Why or why not?
2. Structurally, organic compounds consist of two primary parts, which also determine their properties. Name these two primary parts.

## 4.3 Biological molecules are assembled or disassembled by adding or removing water.

Two extremely important processes in organisms are the putting together and taking apart of biological molecules. If these processes could not be carried out, then life itself would cease. Assembling molecules is needed not only for growth and repair, but also for manufacturing molecules that are critical in helping essential chemical reactions take place. Similarly, disassembling molecules is not only necessary for digestion, to provide molecules that can be absorbed by organisms, but also to provide molecules that can enter cells and be used in reactions that take place there, where more disassembly and reassembly occur. Assembly and disassembly of biological molecules can be regarded as the work of life. As one molecule is broken down, another is built up in the give and take of metabolism, the sum of all the chemical reactions of a living thing.

### Assembly: dehydration synthesis

You might think that the four types of biologically important macromolecules would be assembled by four distinct processes, but this is not the case. Although each type is composed of different building blocks, the four types of biological molecules are all assembled by the same process. This assembly process is not unlike putting Lego pieces together to form larger structures from smaller ones. In biological molecules, the building blocks are called *monomers*. Monomers are small, similar building blocks that are linked together by covalent bonds to form large, chainlike molecules called *polymers*. Lipids are a bit different; they are composite molecules, which means that their building blocks are not all the same or similar; they are built from more than one type of molecule. Even so, the process of assembling both polymers and lipids is the same: dehydration synthesis.

During **dehydration synthesis**, one molecule of water is removed (dehydration) from each two monomers that are joined (synthesis). One monomer loses its hydroxyl group (—OH), and the other loses an atom of hydrogen (H). Water is formed. Having lost electrons they shared in covalent bonding, both monomers bond covalently with one another as shown in **Figure 4.3**. (Also see Figures 4.5, 4.8, and 4.11.) The process of dehydration synthesis uses energy, which is stored in the bond that is made, and takes place with the help of special molecules called *enzymes* (see Chapter 6).

### Disassembly: hydrolysis

Polymers and lipids are disassembled in an opposite process called **hydrolysis** (see Figure 4.11). During hydrolysis the covalent bonds are broken between monomers (and the building blocks of lipids) with the addition of water and in the presence of enzymes. The term *hydrolysis* literally means "to break apart" (*lysis*) "by means of water" (*hydro*). The hydroxyl group of a water molecule bonds to one monomer, and the hydrogen atom bonds to its neighbor. The energy held in the bond is released.

CONCEPT CHECKPOINT

3. List the four major classes of biologically important macromolecules.
4. Biological molecules are assembled by dehydration synthesis and disassembled by hydrolysis. How do these names describe the processes?

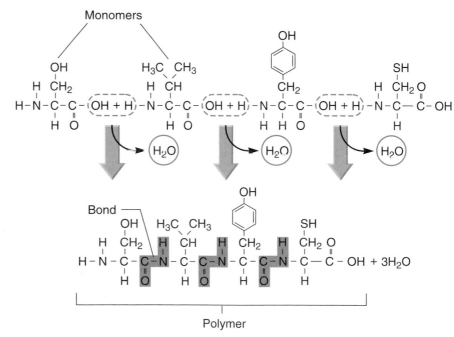

**Figure 4.3** Dehydration synthesis.

**Visual Thinking:** If you were explaining dehydration synthesis by using this illustration, to which parts would you point when explaining *dehydration* and *synthesis*?

# What are the major classes of biological molecules and their characteristics?

## 4.4 Carbohydrates contain C, H, and O with an H-to-O ratio of 2:1.

### Some carbohydrates supply energy

Have you ever heard of carbohydrate loading? This is the practice of eating a carbohydrate-based diet for a few days before a marathon or similar sports event. Such a carbohydrate based diet would include fruits, vegetables, and grain foods such as cereals, hearty breads, and pasta. During carbo-loading, athletes also cut down on pre-event exercise to allow their muscles time to store the carbohydrates as glycogen, the storage form of carbohydrates in animals. As you can see, a primary role of carbohydrates in the animal body is to provide energy.

**Carbohydrates** are molecules that contain carbon, hydrogen, and oxygen, with an H-to-O ratio of 2:1. Abundant energy is locked in their many carbon-hydrogen covalent bonds. Plants, algae, and some bacteria produce carbohydrates by the process of photosynthesis. Most organisms, such as humans, use carbohydrates as an important fuel, breaking these bonds and releasing energy to sustain life.

Among the least complex of the carbohydrates are the simple sugars or **monosaccharides** (MON-oh-SACK-uh-rides). This word comes from two Greek words meaning "single" (*monos*) and "sweet" (*saccharon*) and reflects the fact that monosaccharides are individual sugar molecules. Some of these sweet-tasting sugars have as few as three carbon atoms. The monosaccharides that play a central role in energy storage, however, have six carbon atoms. The primary energy-storage molecule used by living things is glucose ($C_6H_{12}O_6$), a six-carbon sugar. Notice in **Figure 4.4** that glucose exists as a straight chain or as a ring of atoms. This is the situation with other sugars as well.

Glucose is not the only sugar with the formula $C_6H_{12}O_6$. Other monosaccharides that have this same formula are fructose and galactose. Because these molecules have the same molecular formula as glucose but are put together slightly differently, they are called *isomers*, or alternative forms, of glucose. You can compare the structures of glucose and fructose in **Figure 4.5**, and your taste buds can tell the difference: fructose tastes much sweeter than glucose.

In living systems, two of these three sugars are often found covalently bonded to one another. Two monosaccharides linked to-

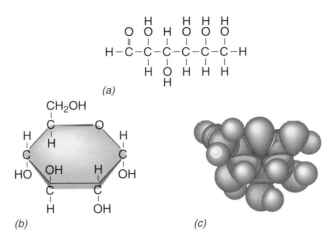

**Figure 4.4 Structure of a glucose molecule.** The structural formula of glucose (*a*) in its linear form and (*b*) as a ring structure. (*c*) Three-dimensional (space filling) model of glucose (hydrogen, blue; oxygen, red; carbon, black).

gether form a **disaccharide** (dye-SACK-uh-ride). Many organisms, such as plants, link monosaccharides together to form disaccharides, which are less readily broken down while being transported within the organism. Sucrose, or table sugar, is a disaccharide formed by linking a molecule of glucose to a molecule of fructose by dehydration synthesis as shown in Figure 4.5. It is the common transport form of sugar in plants.

Lactose, or milk sugar (glucose + galactose), is a disaccharide that many mammals produce to feed their young. Between 30 and 50 million Americans cannot digest lactose, a condition known as lactose intolerance. When lactose-intolerant individuals ingest lactose, they may experience bloating, cramping, nausea, and diarrhea.

Not only do organisms unlock and use the energy within carbohydrate molecules, but they also store this energy. To do so, however, organisms must convert soluble sugars such as glucose to an insoluble form to be stored. Sugars are made insoluble by joining them together into long polymers called **polysaccharides**, sometimes called

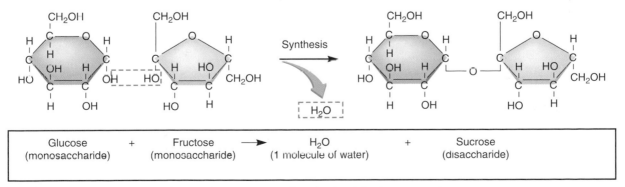

| Glucose (monosaccharide) | + | Fructose (monosaccharide) | → | H₂O (1 molecule of water) | + | Sucrose (disaccharide) |

**Figure 4.5** Disaccharides are formed from monosaccharides by dehydration synthesis.

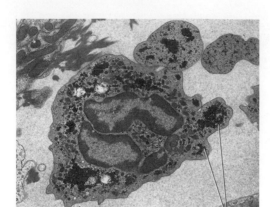

Glycogen
granules

**Figure 4.6** Glycogen, a polysaccharide often called animal starch. Glycogen storage granules can be seen in the cytoplasm of many types of cells, such as liver cells, muscle cells, and some white blood cells. The electron micrograph shows a neutrophil, a type of white blood cell that is abundant in the body; neutrophils ingest foreign material and store glycogen as granules in their cytoplasm.

*complex carbohydrates*. Plants store energy in polysaccharides called **starches**. The starch amylose, for example, is made up of hundreds of glucose molecules linked together in long, unbranched chains. Most plant starch is a branched version of amylose called *amylopectin* (AM-ih-low-PECK-tin). Animals store glucose in highly branched polysaccharides called **glycogen** (GLYE-ko-jen) (**Figure 4.6**).

### Some carbohydrates are used for structure

The chief component of plant cell walls is a polysaccharide called *cellulose*. A modified form of cellulose called *chitin* is found in in-sects, many fungi, and certain other organisms. Cellulose is chemically similar to amylose but is bonded in a way that most organisms cannot digest (**Figure 4.7**). For this reason, cellulose works well as a biological structural material and occurs widely in this role in plants. In your diet, this indigestible plant material is called *fiber*. Fiber provides some important health benefits such as helping to prevent constipation and to lower blood cholesterol. Lower blood cholesterol is associated with a lower risk of heart disease.

 CONCEPT CHECKPOINT

**5.** What are the two main uses of carbohydrates in living things?
**6.** What monomers are used to build carbohydrates? What is the general name for carbohydrate polymers? Name a specific monomer and specific polymer.

## 4.5 Lipids have an H-to-O ratio higher than 2:1.

Lipids include a wide variety of molecules, all of which can be dissolved in oil. Lipids cannot be dissolved in water because almost all the bonds in lipids are nonpolar carbon-carbon or carbon-hydrogen bonds. Three important categories of lipids are (1) oils, fats, and waxes; (2) phospholipids; and (3) steroids.

### Fats and oils supply a high concentration of energy

When organisms have glucose molecules to store for long periods, they usually convert them to **fats**. Fats are nonpolar, insoluble molecules, so they work well for storage. Many of us wish that we had less of this long-term storage material!

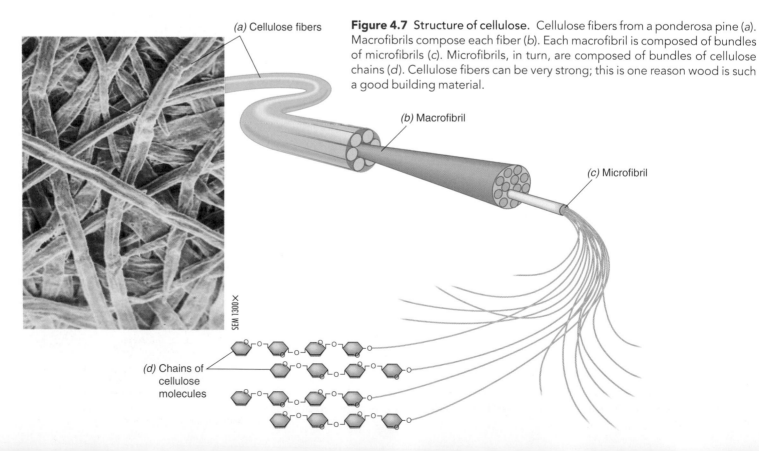

(a) Cellulose fibers

(b) Macrofibril

(c) Microfibril

SEM 1300×

(d) Chains of cellulose molecules

**Figure 4.7** Structure of cellulose. Cellulose fibers from a ponderosa pine (a). Macrofibrils compose each fiber (b). Each macrofibril is composed of bundles of microfibrils (c). Microfibrils, in turn, are composed of bundles of cellulose chains (d). Cellulose fibers can be very strong; this is one reason wood is such a good building material.

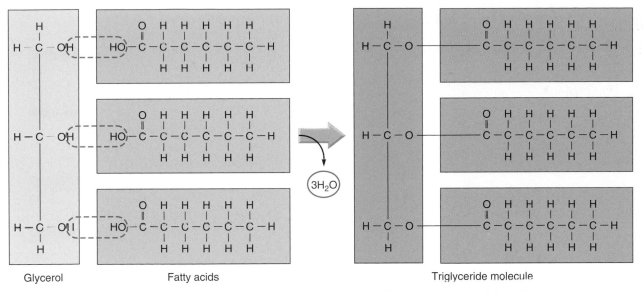

Glycerol          Fatty acids                                          Triglyceride molecule

**Figure 4.8** **Structure of a triglyceride.** Triglycerides are composite molecules made up of three fatty acid molecules bonded to a single glycerol molecule. This bonding takes place by dehydration synthesis.

Fats also provide more energy in our diets than do equivalent amounts of carbohydrates or proteins. Although we can gain weight from eating too many calories from any foods, fats provide more calories, gram for gram, than do proteins or carbohydrates. Why? The reason lies in the molecular composition of fats. Fats are large molecules made up of carbon, hydrogen, and oxygen, with a high hydrogen-to-oxygen ratio. These huge molecules contain more energy-storing carbon-hydrogen bonds than do either carbohydrates or proteins.

All lipids are composite molecules; that is, they are made up of more than one component. Oils and fats are built from two different kinds of subunits:

- *Glycerol*: Glycerol is a three-carbon molecule, with each carbon bearing a hydroxyl (—OH) group. The three carbons form the backbone of the fat molecule.
- *Fatty acids*: Fatty acids have long *hydrocarbon* chains (chains consisting only of carbon and hydrogen atoms) ending in a carboxyl (—COOH) group. Three fatty acids are attached to a single glycerol backbone (**Figure 4.8**). Because there are three fatty acids, the resulting fat molecule is called a **triglyceride**. Most dietary fat is in the form of triglycerides.

Triglycerides differ depending on the structure of their fatty acids. As **Figure 4.9** shows, a fatty acid with only single bonds between its carbon atoms can hold more hydrogen atoms than a fatty

**Figure 4.9 Saturated and polyunsaturated triglycerides.** (a) Palmitic acid, a fatty acid with only single bonds between its carbon atoms, has a maximum of hydrogen atoms and is a saturated fatty acid. The space-filling model shows a saturated triglyceride that consists of three saturated fatty acids chains attached to a glycerol backbone. (b) Linolenic acid, with three double bonds and thus fewer than the maximum number of hydrogen atoms bonded to the carbon chain, is a polyunsaturated fatty acid. The space-filling model shows a polyunsaturated triglyceride that consists of three unsaturated fatty acids chains attached to a glycerol backbone. The bars show where double bonds are located.

Palmitic acid (saturated)

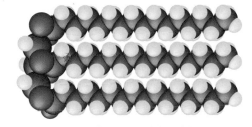

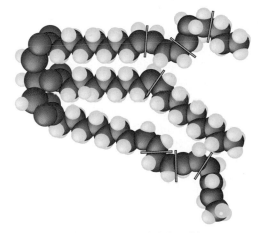

(a) Saturated triglyceride

Linolenic acid (polyunsaturated)

(b) Polyunsaturated triglyceride

| TABLE 4.2 | Fatty Acid Content of Common Fats and Oils | | |
|---|---|---|---|
| Type of fat/oil | % Saturated | % Monounsaturated | % Polyunsaturated |
| Canola oil | 7 | 59 | 29 |
| Corn oil | 13 | 24 | 59 |
| Olive oil | 13 | 73 | 8 |
| Peanut oil | 17 | 46 | 31 |
| Margarine (hard) | 20 | 45 | 33 |
| Vegetable shortening | 25 | 45 | 26 |
| Chicken fat | 30 | 45 | 21 |
| Pork fat | 39 | 45 | 12 |
| Beef fat | 50 | 42 | 4 |
| Butter | 65 | 30 | 4 |
| Coconut oil | 90 | 5 | 1 |

Note: Totals are not 100% because other lipids are present in the fat or oil.

acid with double bonds between its carbon atoms. A fatty acid that carries as many hydrogen atoms as possible, such as the fatty acid in Figure 4.9*a*, is *saturated*. Fats composed of fatty acids with double bonds are *unsaturated* because the double bonds replace some of the hydrogen atoms. If a fat has one double bond, it is *monounsaturated*. If it has more than one double bond, it is *polyunsaturated* (Figure 4.9*b*).

The difference between fats and oils has to do with the number of double bonds in their fatty acids—their degree of saturation. Generally, the more unsaturated the fatty acids of a fat are, the more soft or liquid the fat is at room temperature. Fats rich in monounsaturated triglycerides, such as olive oil, and polyunsaturated triglycerides, such as corn oil, have low melting points and are therefore liquid fats, or oils, at room temperature. In general, plant and fish oils are rich in polyunsaturates. Exceptions are the so-called tropical oils, which are highly saturated yet are still liquid at room temperature.

Animal fats, in contrast, are usually saturated and occur as hard fats. Human diets with large amounts of saturated fats may contribute to clogged arteries and raise the risk of developing diseases of the circulatory system. However, the fats found in foods contain mixtures of saturated and unsaturated fatty acids as shown in **Table 4.2**. You can usually tell the degree of saturation by determining the hardness of the fat. You've probably noticed that the fat on a steak at room temperature is harder than the fat of a chicken at room temperature. Chicken fat is less saturated than beef fat, so it is softer than beef fat. Chicken is recommended over beef for people avoiding saturated fats.

You have likely heard about trans fats in the news. Trans fats are processed fats, created by adding hydrogen to oil to harden it, as is done to produce margarine. Like saturated fats, trans fats tend to raise blood cholesterol levels. We discuss dietary fat, including trans fats, in more detail in Chapter 27. (See Table 27.1 for a listing of foods that commonly contain trans fats.)

Waxes differ from fats and oils by having a chemical backbone slightly different from glycerol and having only one carbon chain. All waxes are water resistant and serve as a protective coating on plant leaves and fruits, and on the outer covering of some insects. Waxes are also produced by honeybees to create their honeycombs.

## Phospholipids make up cell membranes, protect, and waterproof

Phospholipids are similar to fats and oils except that one of the fatty acid chains is replaced by a phosphate group attached to an alcohol molecule. Phospholipids play a key role in the structure of cell membranes and are discussed in more detail in Chapter 5 (see Figure 5.7).

## Steroids include cholesterol and hormones

Membranes often contain steroids, which are lipids whose structure is very different from fats and oils. Steroids are composed of four carbon rings, as shown in **Figure 4.10**. Humans naturally make a variety of steroids. For example, humans manufacture cholesterol, a component of cell membranes, and the male and female sex hormones, estrogen and testosterone, which are discussed in Chapter 36. In addition, many steroids have therapeutic value, such as inhaled steroids that calm inflamed airways as a treatment for asthma.

CONCEPT CHECKPOINT

7. What is the difference between a lipid and a fat?
8. Which is the most heart-healthy of the following foods: chicken, coconut oil, butter, and pork. Why?
9. Where are phospholipids and steroids found in the human body?

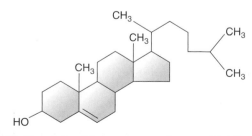

**Figure 4.10 Example of the structure of a steroid.** This particular molecule is cholesterol, a component of cell membranes, but a molecule also implicated in heart disease.

## 4.6 Proteins are chains of amino acids.

### Proteins have varied functions

**Proteins**, the third major group of macromolecules that make up the bodies of organisms, play diverse roles in living things. They transport other molecules, provide for muscle contraction, help protect the body from disease, play an important role in nerve transmission, and help control the growth and differentiation of cells. Perhaps the most important proteins are **enzymes**, proteins capable of speeding up specific chemical reactions. Other short proteins called **peptides** are used as chemical messengers within your brain and throughout your body. Collagen, a structural protein, is an important part of bones, cartilage, and tendons.

Despite these varied functions, all proteins have the same basic structure: a long chain of amino acids linked end to end. Amino acids are small molecules containing an amino group ($—NH_2$), a carboxyl group ($—COOH$), a hydrogen atom, a carbon atom, and a *side chain* that differs among amino acids. In a generalized formula for an amino acid, the side chain is shown as *R*. The identity and unique chemical properties of each amino acid are determined by the nature of the R group.

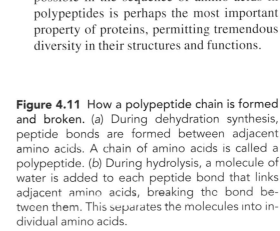

Only 20 different amino acids combine to make up the diverse array of proteins found in living things. Each protein differs according to the amount, type, and arrangement of amino acids that make up its structure. There are 20 common amino acids, each with the same chemical backbone but with different R groups.

Each amino acid has a free amino group ($—NH_2$) at one end and a free carboxyl group ($—COOH$) at the other end. During dehydration synthesis, each of these groups on separate amino acids loses a molecule of water between them, forming a covalent bond that links the two amino acids (**Figure 4.11**). This bond is called a *peptide bond*. A long chain of amino acids linked by peptide bonds is a *polypeptide*. Proteins are long, complex polypeptides. The great variability possible in the sequence of amino acids in polypeptides is perhaps the most important property of proteins, permitting tremendous diversity in their structures and functions.

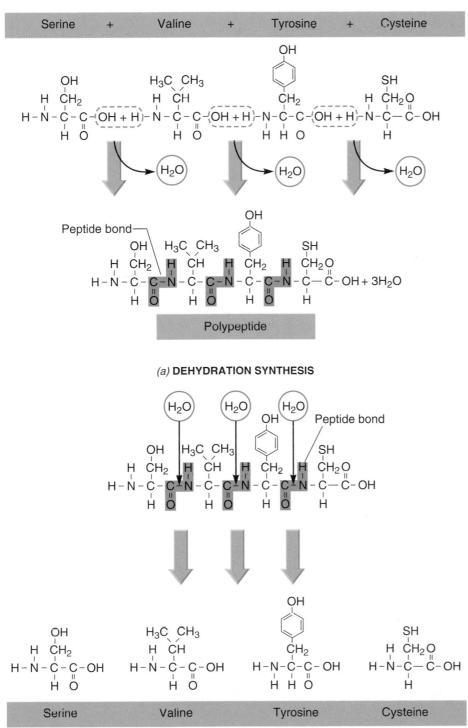

*(a)* **DEHYDRATION SYNTHESIS**

*(b)* **HYDROLYSIS**

**Figure 4.11 How a polypeptide chain is formed and broken.** (a) During dehydration synthesis, peptide bonds are formed between adjacent amino acids. A chain of amino acids is called a polypeptide. (b) During hydrolysis, a molecule of water is added to each peptide bond that links adjacent amino acids, breaking the bond between them. This separates the molecules into individual amino acids.

**Visual Thinking:** Draw a circle around the R group (side chain) in the amino acids serine, valine, tyrosine, and cysteine shown at the top of the illustration.

**I've been weight training at a gym and have discovered that many body builders take steroids. I'm thinking about taking them, too. Are they dangerous?**

You're probably talking about anabolic steroids, which may be called 'roids, juice, hype, or pump at the gym. These controversial drugs are really synthetic hormones, chemicals that affect the activity of specific organs or tissues. Anabolic steroids affect the body in ways similar to the male sex hormone testosterone and stimulate the buildup of muscles. The term *anabolic* simply means "building body tissue." However, along with building a championship body, anabolic steroids strikingly change the body's metabolism.

Female athletes on steroids often experience side effects such as shrinking breasts, a deepening voice, and an increase in body hair. Male athletes frequently find that their testicles shrink, their nipples and breasts increase in size, and they become impotent. Youngsters who take these drugs risk stunting their growth because anabolic steroids cause bones to stop growing prematurely. In addition, anabolic steroids may cause psychological effects such as "steroid rage," a state of mind in which users attack people and things around them.

If you are thinking about using anabolic steroids, first consider that they can be deadly. Second, realize that if they don't kill you, they will most likely have a negative effect on your health and can result in high blood pressure, heart disease, liver damage, cancer, stroke, blood clots, urinary and bowel problems, headaches, aching joints, muscle cramps, nausea, vomiting, sleep problems, severe acne, and baldness. Instead of putting your life and health in jeopardy, work out without steroids, practicing good nutrition, proper conditioning, and appropriate training. ●

*Are you wondering about a topic in biology and how it relates to your life? Submit your question by clicking the* Just Wondering *link in this text's companion Web site at www.wiley.com/college/alters.*

### Proteins have levels of structure

The sequence of amino acids that makes up a particular polypeptide chain is termed the *primary structure* of a protein (**Figure 4.12 ❶** ). This sequence determines the further levels of structure of the protein molecule resulting from bonds that form between these groups. Having the proper sequence of amino acids, then, is crucial to the functioning of a protein because protein structure is the basis of protein function. Proteins are unique among macromolecules in that they recognize and interact with a wide array of other molecules, carrying out the diverse activities we just discussed in this section. The specific shapes of proteins give them these capabilities. Therefore, if the protein does not assume its correct shape, it will not work properly or at all. Because different amino acid R groups have different chemical properties, the shape of a protein may be altered by a single amino acid change.

The R groups of the amino acids in a polypeptide chain interact with their neighbors, forming hydrogen bonds. In addition, portions of a protein chain with many nonpolar R groups tend to be shoved into the interior of the protein because of their hydrophobic properties. That is, polar water molecules "push away" these nonpolar molecules (see Chapter 3). Because of these interactions, polypeptide chains tend to fold spontaneously into sheets or wrap into coils. This folded or coiled shape is called its *secondary structure* ❷ . Proteins made up largely of sheets often form fibers such as keratin fibers in hair, fibrin in blood clots, and silk in spiders' webs. Proteins that have regions forming coils frequently fold into globular (globelike) shapes such as the globin subunits of hemoglobin in blood.

Hydrogen bonding and ionic bonds form within globular proteins, resulting in proteins with more complex shapes than the secondary structure. This level of structure is called *tertiary structure* ❸ . For proteins that consist of subunits (separate polypeptide chains), the way these subunits are assembled into a whole is called the *quaternary structure* ❹ .

The three-dimensional structure of a protein is crucial to its function. Although most bonds involved in determining the three-dimensional shape of a protein are weak, one type is strong: nonpolar covalent disulfide bridges. These bonds form between sulfur atoms found in the amino acid cysteine. Many proteins contain this amino acid, and the disulfide bridges formed in these proteins have a strong influence on the final three-dimensional shape of the protein. Nevertheless, protein structure can be easily degraded by heat or various chemicals. You see this type of irreversible change when you cook an egg. The white part of the egg becomes firm and opaque, much different from the runny substance of a raw egg. Similarly, if your body temperature rises too high, proteins involved in your metabolism can be degraded and no longer function, with possibly fatal consequences. Health care practitioners generally suggest that a fever be brought down when it reaches between 102°F and 104°F and that medical care should be sought immediately if it rises above 104°F.

---

### CONCEPT CHECKPOINT

10. An analogy can be made between protein structure and the use of letters, words, and sentences in a language. With respect to the chemical makeup of proteins, letters could be analogous to _____, words could be analogous to _____, and sentences could be analogous to _____. To what characteristic of proteins could the meaning of the words and sentences be analogous?

11. Why is the shape of a protein crucial to its proper functioning?

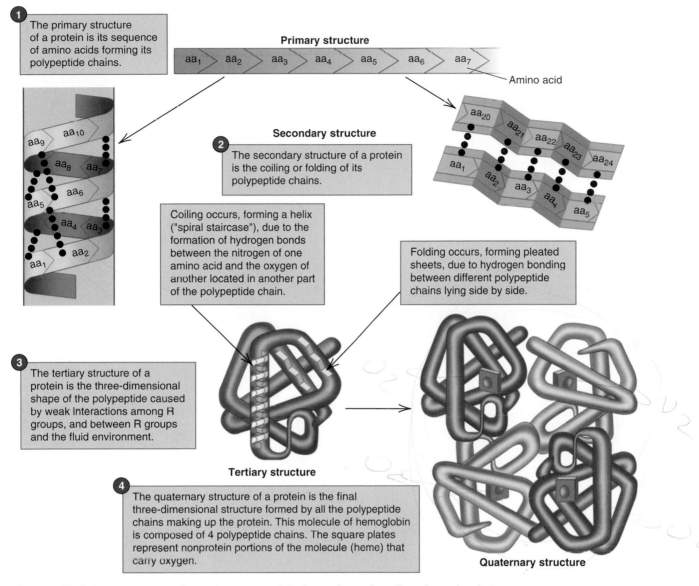

**1** The primary structure of a protein is its sequence of amino acids forming its polypeptide chains.

**Primary structure**

aa₁  aa₂  aa₃  aa₄  aa₅  aa₆  aa₇

Amino acid

**Secondary structure**

**2** The secondary structure of a protein is the coiling or folding of its polypeptide chains.

Coiling occurs, forming a helix ("spiral staircase"), due to the formation of hydrogen bonds between the nitrogen of one amino acid and the oxygen of another located in another part of the polypeptide chain.

Folding occurs, forming pleated sheets, due to hydrogen bonding between different polypeptide chains lying side by side.

**3** The tertiary structure of a protein is the three-dimensional shape of the polypeptide caused by weak interactions among R groups, and between R groups and the fluid environment.

**Tertiary structure**

**4** The quaternary structure of a protein is the final three-dimensional structure formed by all the polypeptide chains making up the protein. This molecule of hemoglobin is composed of 4 polypeptide chains. The square plates represent nonprotein portions of the molecule (heme) that carry oxygen.

**Quaternary structure**

**Figure 4.12** Primary structure determines a protein's shape due to bonding along the chain.

# 4.7 Nucleic acids are composed of nucleotides.

### Nucleic acids are molecules of heredity

Organisms store information about the structures of their proteins in macromolecules called **nucleic acids**. Nucleic acids hold the blueprint for living things, such as information about your hair color, skin color, and height. Along with other supplementary molecules, nucleic acids comprise the hereditary material, directing the formation and operation of living systems.

Nucleic acids are long polymers of repeating subunits called *nucleotides*. Each nucleotide is made up of three smaller building blocks:

- A five-carbon sugar.
- A phosphate group ($-PO_4^=$).
- An organic, nitrogen-containing molecule called a *nitrogenous base*.

To form the nucleic acid chain, the sugars and phosphate groups making up the nucleotides are linked; a nitrogenous base protrudes from each sugar.

### The nucleic acid DNA stores information for making proteins

Organisms have two forms of nucleic acid. One form, deoxyribonucleic acid (DNA), stores the information for making proteins in the sequence of its bases. This sequence is a biological code. The order in which the nucleotides are linked together forms the code that ultimately specifies the order of amino acids in a particular protein. The proteins coded by DNA not only make up much of the structure of organisms, but also perform diverse jobs such as transporting molecules, speeding up chemical reactions, and serving as chemical messengers.

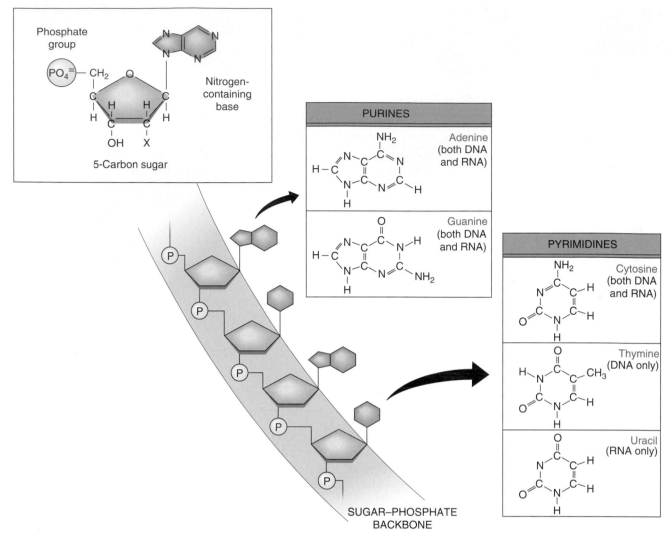

**Figure 4.13 The structure of a nucleotide and the formation of nucleic acids.** As shown in the inset, a nucleotide is composed of a five-carbon sugar, a phosphate group, and an organic nitrogenous (nitrogen-containing) base. In RNA, the five-carbon sugar is ribose, which contains a hydroxyl group (—OH) at the X. In DNA, the five-carbon sugar is deoxyribose, which contains a hydrogen atom (H) at the X. The sugar and phosphate groups make up the backbone of a nucleic acid chain, while the nitrogenous bases link the two sides of the chain. The five nitrogenous bases that occur in the nucleic acids of DNA and RNA are shown on the right.

## The nucleic acid RNA directs the production of proteins

The other form of nucleic acid is ribonucleic acid (RNA), which directs the manufacture of proteins using the code from DNA. Cells contain three types of RNA, and each plays a special role in the manufacture of proteins.

One type of RNA brings information from the DNA to structures called ribosomes to direct which protein is assembled. A second type of RNA transports amino acids, which are used to build the protein, to the ribosomes. In addition, these RNA molecules position each amino acid at the correct place on the elongating protein chain. The third type of RNA makes up part of the structure of the ribosomes.

**Figure 4.13** shows the structure of nucleotides and the formation of nucleic acids. Details of the structure of DNA and the ways it interacts with RNA are presented in Chapters 9 and 10.

### CONCEPT CHECKPOINT

**12.** An analogy can be made between the structure of nucleic acids, such as DNA, and the use of letters, words, and sentences in a language. With respect to the chemical makeup of DNA, letters could be analogous to _____, and words/sentences could be analogous to _____. To what characteristic of DNA could the meaning of the words and sentences be analogous?

**13.** What is the difference between the roles of DNA and RNA?

# CHAPTER REVIEW

## Summary of Key Concepts and Key Terms

### What are the general characteristics of biological molecules?

**4.1** Most molecules that are formed by living organisms and make up their structures contain the element carbon.

**4.1** The carbon-containing molecules that make up living things are called **organic compounds** (p. 44).

**4.2** Organic compounds have a carbon core, or skeleton, with various **functional groups** (p. 44) attached to the carbon atoms.

**4.2** There are four major groups of biologically important macromolecules: carbohydrates, lipids, proteins, and nucleic acids.

**4.3** Biological molecules are assembled by **dehydration synthesis** (p. 46), in which water is removed as building blocks are linked by covalent bonding.

**4.3** Biological molecules are disassembled by **hydrolysis** (p. 46), in which water is added and links between building blocks are broken.

### What are the major classes of biological molecules and their characteristics?

**4.4 Carbohydrates** (p. 47) are molecules that contain carbon, hydrogen, and oxygen, with an H-to-O ratio of 2:1.

**4.4** Most organisms use carbohydrates as an important fuel.

**4.4** The least complex carbohydrates are simple sugars, or **monosaccharides** (p. 47). Two monosaccharides linked together form a **disaccharide** (dye-SACK-uh-ride, p. 47).

**4.4** Plants store energy as **starches** (p. 48); animals store energy as **glycogen** (p. 48). Both are types of complex carbohydrates called **polysaccharides** (p. 47).

**4.4** Some carbohydrates are used for structure in plants but are indigestible in humans.

**4.5** Lipids have an H-to-O ratio higher than 2:1.

**4.5** Three important categories of lipids are oils, fats, and waxes; phopholipids; and steroids.

**4.5** Oils and **fats** (p. 48) are composed of a glycerol backbone and fatty acid chains; **triglycerides** (p. 49) are fats that have three fatty acid chains.

**4.5** The fats are important in the long-term storage of energy.

**4.5** Phospholipids are structured similarly to fats and oils; they are a major component of cell membranes.

**4.5** Steroids are composed of four carbon rings and are found in cell membranes and sex hormones.

**4.6 Proteins** (p. 51) are polymers of amino acids that play diverse roles in organisms.

**4.6** Perhaps the most important proteins are **enzymes** (p. 51), proteins capable of speeding up specific chemical reactions.

**4.6** Other short proteins called **peptides** (p. 51) are used as chemical messengers.

**4.6** Because the 20 amino acids that occur in proteins have side chains with differing chemical properties, the shape and therefore the functioning of a protein are critically affected by its particular sequence of amino acids.

**4.6** Proteins have up to four levels of structure.

**4.7 Nucleic acids** (p. 53) hold the blueprint for living things.

**4.7** Hereditary information is stored as a sequence of nucleotides in the nucleotide polymer DNA.

**4.7** A second form of nucleic acid, RNA, directs the production of proteins.

## Level 1 — Learning Basic Facts and Terms

### Multiple Choice

1. Most molecules that make up living and once-living things contain the element
   - a. silicon.
   - b. carbon.
   - c. water.
   - d. calcium.
2. Molecules containing thousands of atoms and many functional groups are
   - a. macromolecules.
   - b. supermolecules.
   - c. megamolecules.
   - d. hydroxymolecules.
3. A polysaccharide that is a primary component of plant cell walls is
   - a. starch.
   - b. glucose.
   - c. glycogen.
   - d. cellulose.
4. Lipids are made up of more than one type of component, so they are termed
   - a. hydrocarbon chains.
   - b. composite molecules.
   - c. fats.
   - d. polyunsaturated.
5. Proteins are chains of
   - a. monosaccharides.
   - b. triglycerides.
   - c. amino acids.
   - d. nucleic acids.

### True–False

6. ____ Peptide bonds are formed by dehydration synthesis.
7. ____ The three components of nucleic acids are a sugar, a phosphate group, and a nitrogenous base.
8. ____ RNA stores the information for making proteins, while DNA directs the manufacture of proteins using the RNA code.
9. ____ The building blocks of biological molecules are called polymers.
10. ____ Biological molecules interact by means of functional groups.

## Level 2    Learning Concepts

1. Distinguish between organic and inorganic molecules. Which would you primarily study if you wanted to learn more about the human body?
2. Distinguish among monosaccharides, disaccharides, and polysaccharides. To what group of biological macromolecules do they belong, and why are they important?
3. Why do plants and animals not store glucose as a simple sugar for future use? What forms of storage are most common in plants and animals?
4. Discuss the three classes of macromolecules taken in as food energy by humans. Which chemical process, dehydration synthesis or hydrolysis, do you think is essential in the digestion of food?
5. DNA is often called the code of life. Explain why.
6. How does the sequence of amino acids in a protein ultimately determine its levels of structure?

## Level 3    Critical Thinking and Life Applications

1. While doing your grocery shopping, you find this label on a product:

| Nutritional Information per Serving | |
| --- | --- |
| Calories | 270 |
| Calories from fat | 45 |
| Total fat | 5 grams |
| Saturated fat | 1.5 grams |
| Total carbohydrates | 48 grams |
| Sugar | 4 grams |
| Protein | 8 grams |

What macromolecules are present in this food? Would this food be part of a heart-healthy diet? Defend your answer with evidence.

2. If three molecules of a fatty acid each of which has the formula $C_{16}H_{22}O_2$ are joined to a molecule of glycerol ($C_3H_8O_3$), what chemical formula would the resulting molecule have? Explain your answer.
3. Would the triglyceride in question 2 be saturated or unsaturated? Explain your answer.
4. Name two functions of proteins in living systems. How might an error in the DNA of an organism affect protein function?
5. Sarah decided that she needed to lose weight, and planned to exercise daily and eat only 1500 calories per day. She was shocked to learn that one gram of fat contained more than twice as many calories as a gram of carbohydrate or protein. Based on what you know about the structures of these molecules, propose a reason to explain Sarah's finding.
6. Do you use nutrition labels? Find the nutrition label on a packaged product that you consume frequently. Which do you consider the most important piece of information on the label? Why?

## In The News    Critical Thinking

**MAD COW CLUES**

Now that you understand more about the structure of proteins, reread this chapter's opening story about research into prions. To understand this research better, it may help you to follow these steps:

1. Review your immediate responses to Lindquist's research on prions that you wrote when you began reading this chapter.
2. Based on your current understanding, again summarize the main point of the research in a sentence or two.
3. What questions do you now have about this research that this chapter's opening story does not answer?
4. Collect new information about the research. Visit the *In The News* section of this text's companion Web site at www.wiley.com/college/alters and

watch the "Mad Cow Clues" video. Then use the "summary" link to read the accompanying story and access related links. Use this information, the links provided, and other online and library resources to answer your questions and find updates about this research topic. State the sources of your information. Explain why you think the information is accurate. Also determine whether the information expresses a particular point of view or is biased in any way.

5. What in your view is the most significant aspect of this research? Give reasons for your opinion and for any changes in your ideas based on the additional information you have collected and the analysis you have done.

# CELL STRUCTURE AND FUNCTION

## In The News | Agre and Cells

Can you imagine the thrill of winning a Nobel Prize? Peter Agre, professor of biological chemistry at Johns Hopkins University School of Medicine in Baltimore, Maryland, and pictured here, knows this thrill and talks about his recent Nobel Prize-winning discovery in the video news story "Agre and Cells." You can see this video by visiting the *In The News* section of this text's companion Web site at www.wiley.com/college/alters.

Agre and fellow researcher Roderick MacKinnon of The Rockefeller University in New York City shared the 2003 Nobel Prize in Chemistry for discovering how aquaporins work. *Aquaporins* are water pores in cells. More specifically, they are specialized cell membrane proteins that control the flow of water into and out of cells without allowing other molecules to pass. Plants have 35 types of aquaporins, while mammals, including humans, have 10. They are all critical to proper cell functioning and health. For example, if aquaporins in human kidneys behave improperly, then water is not reabsorbed from kidney tubules and put back into the bloodstream. Therefore, a person urinates large quantities of dilute urine and is continually thirsty. This condition is a type of diabetes insipidus (which is different from the sugar-regulating disease diabetes mellitus). If persons with diabetes insipidus do not consume adequate fluids to compensate for their water losses, they become dehydrated and their blood levels of sodium rise, compromising health.

So how did researchers determine how aquaporins allow water to move into and out of cells while not allowing other molecules to tag along? Agre, MacKinnon, and their research teams used the Terascale Computing System at the Pittsburgh Supercomputing Center to simulate the process and see what happened. First, they simulated the structure of aquaporins, which had been previously determined. The aquaporins were placed within a simulated cell membrane, forming a channel, and water molecules were placed on both sides of the membrane. The supercomputer showed water molecules passing through the aquaporin channel single file, entering with their oxygen atoms facing "downstream," and leaving with their oxygen atoms facing "upstream." This inward orientation of the negatively charged oxygen atoms blocked them from dragging positively charged ions or protons through the channel with them. If water were to pull positive ions through the channels, then the electrical balance inside and outside of these cells would be upset, and the cells would cease to function.

*Write your immediate reaction to Agre's research on aquaporins: first, summarize the main point of the research in a sentence or two; then suggest what you think its significance is. You will have an opportunity to reflect on your responses and gather more information on this topic in the* In The News *feature at the end of this chapter. In this chapter, you will learn more about the structure and function of cells.*

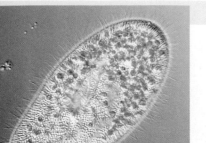

# CHAPTER GUIDE

## What are cells?

**5.1** Cells are the smallest living unit of structure and function.

**5.2** Most cells are microscopic.

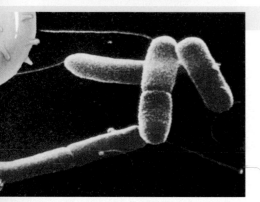

## What are the primary types of cells?

**5.3** Cells are of two types: prokaryotic and eukaryotic.

**5.4** Prokaryotic cells have a simpler structure than eukaryotic cells.

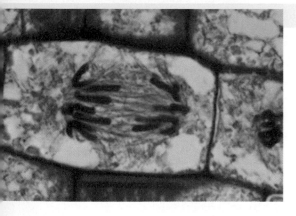

## What are the structures of eukaryotic cells and their functions?

**5.5** The plasma membrane encloses the cell.

**5.6** Cell walls provide rigid exteriors for plant cells, fungi, and many protists.

**5.7** The cytoplasm provides the viscous, fluid interior of the cell.

**5.8** The cytoskeleton helps support and shape the cell.

**5.9** Cilia and flagella provide motility.

**5.10** Membranous organelles form functional compartments within the cytoplasm.

**5.11** Bacterialike organelles release and store energy.

**5.12** Bacterial, animal, and plant cells differ in their structures.

## How do molecules and particles move into and out of cells?

**5.13** Passive transport is molecular movement that does not require energy.

**5.14** Active transport is molecular movement that requires energy.

**5.15** Cells move large molecules and particles into the cell by endocytosis.

**5.16** Cells move large molecules and particles out of the cell by exocytosis.

# What are cells?

When British microscopist Robert Hooke looked at a sliver of cork under his primitive homemade microscope in the mid-1600s, he thought it looked like row upon row of empty boxes. Hooke, of course, was looking at dead plant cells and saw only their outer layer—their cell walls. From his observations, however, he coined the word "cell," which comes from a Latin word meaning small compartment.

Scientists today know that cells are the smallest units of life that can exist independently. Cells can take in nutrients, break them down to release energy, and get rid of wastes. They can reproduce, react to stimuli, and maintain an internal environment different from their surroundings. In multicellular organisms such as humans, cells work together in groups to maintain life. In single-celled organisms such as the paramecium in **Figure 5.1a**, each cell survives independently. This chapter will help you become familiar with the structure of cells and how they work. By studying their structure and function, you will also see their tremendous diversity and complexity and will begin to understand the cellular level of organization of the human body.

## 5.1 Cells are the smallest living units of structure and function.

Nearly 200 years after Hooke's discovery, in 1839, botanist Matthias Schleiden and zoologist Theodor Schwann formulated the theory that all living things are made up of cells. In other words, they realized that cells make up the structure of such things as houseplants and people—but not of rocks or soil. It took another 50 years and the work of Rudolf Virchow for scientists to understand another basic concept about cells: living cells can only be produced by other living cells. As Virchow put it, "all cells from cells." Together, these concepts are called the **cell theory**.

The cell theory, which is today recognized as fact, is a profound statement regarding the nature of living things. It includes three basic principles:

- All living things are made up of one or more cells.
- The smallest *living* unit of structure and function of all organisms is the cell.
- All cells arise from preexisting cells.

When these statements were formulated in the mid-1800s, scientists finally discarded the idea of *spontaneous generation*: that living things could arise from the nonliving. The theory of spontaneous generation had suggested that frogs could be born of the mud in a pond and that rotting meat could spawn the larvae of flies. Various scientists attacked this idea from the mid-1600s onward; it was a debate that lasted for centuries. After the development of the cell theory, however, and groundbreaking experiments designed by French chemist Louis Pasteur, scientists recognized that life arose directly from the growth and division of single cells. Today, scientists know that life on Earth represents a continuous line of descent from the first cells that evolved on Earth.

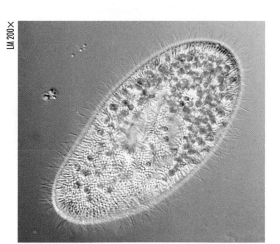

(a)

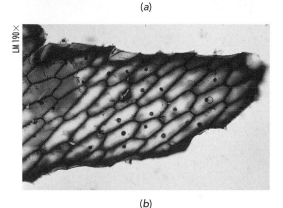

(b)

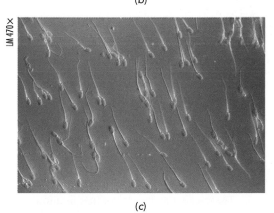

(c)

**Figure 5.1** Cells are diverse in shape, structure, and function. (a) *Paramecium*, a single-celled protozoan. (b) Onion cells. (c) Sperm cells.

## 5.2 Most cells are microscopic.

Most animal cells are extremely small, ranging in diameter from about 10 to 30 micrometers (μm). There are 1000 μm in 1 millimeter (mm)—the width of a paper clip's wire. So, as you might expect, most cells are invisible to the naked eye without the aid of a microscope. The How Science Works box will give you some insight into the tools scientists use to visualize cells. Your red blood cells, for example, are so small that it would take a row of about 2500 of

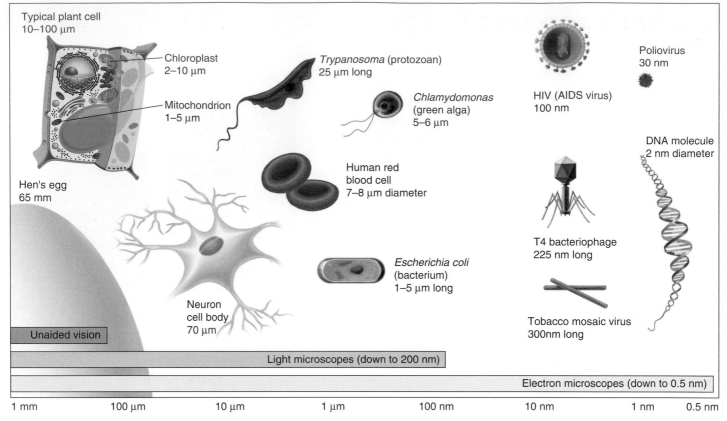

**Figure 5.2 Size comparisons.** The cells, viruses, and molecule shown in this illustration are not drawn to scale. Their size relationships are shown instead by the scale at the bottom of the figure. The largest are to the left and the smallest to the right. The hen's egg is a single cell. Some large plant cells can just be seen with the naked eye. A light microscope is needed to view most single cells, including bacteria. An electron microscope is needed to view viruses and the DNA molecule.

them to span the diameter of a dime. Only a few kinds of cells are large. If you had eggs for breakfast, you were eating single cells! Other kinds of cells are long and thin, such as nerve cells that run from your spinal cord to your toes or fingers. However, very few cells are as large as a hen's egg or as long as a nerve cell. **Figure 5.2** shows a comparison of cell sizes.

To understand why cells are so small, you must first realize that most cells are constantly working, doing such jobs as breaking down molecules for energy, producing substances that cells need, and getting rid of wastes. Each cell must move substances in and out across its boundary—the cell membrane—quickly enough to meet its needs. Therefore, the amount of membranous surface area

a cell has in relationship to the volume it encloses is crucial to its survival. In general, the smaller the cell, the larger its surface area–to–volume ratio (**Figure 5.3**).

### CONCEPT CHECKPOINT

1. When asked in a test "Why are cells so small?" a college freshman provided the following answer. Each underlined term might be an error. If it is, correct the error.

   "A small cell has a <u>smaller</u> surface area–to–volume ratio than a large cell, which <u>enhances</u> the uptake of molecules required for metabolism and the removal of accumulated wastes."

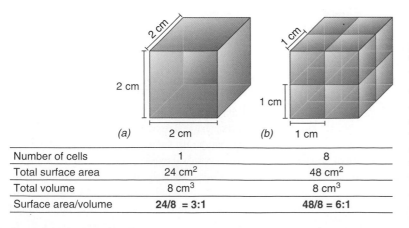

| Number of cells | 1 | 8 |
|---|---|---|
| Total surface area | 24 cm$^2$ | 48 cm$^2$ |
| Total volume | 8 cm$^3$ | 8 cm$^3$ |
| Surface area/volume | **24/8 = 3:1** | **48/8 = 6:1** |

**Figure 5.3 Cells maintain a large surface area-to-volume ratio.** The smaller the cell, the larger its ratio of surface area to volume. This illustration depicts a large cell (*a*) with the same volume as eight smaller cells (*b*). Volume is length × width × height = 2 × 2 × 2 = 8 cm$^3$. The surface area of the large cell is 6 times the area of one side, or 6 × 4 = 24 cm$^2$. The surface area of one of the smaller cells is also 6 times the area of one side, or 6 × 1 = 6 cm$^2$; thus, the combined surface area of the eight smaller cells is 8 × 6 = 48 cm$^2$. For the large cell, the ratio of surface area to volume is 24:8 = 3:1. For the eight smaller cells, the ratio of surface area to volume is 48:8 = 6:1. Thus, the surface area–to–volume ratio for the eight smaller cells is twice that for the one large cell!

# How Science Works

## PROCESSES AND METHODS

### *Exploring the microscopic world: the tools of scientists*

A world invisible to the naked eye had its roots of discovery in the late 1600s with the work of two pioneers in the field of microscopy: Anton van Leeuwenhoek and Marcello Malpighi. Microscopy is the use of a microscope—an optical instrument consisting of a lens or a combination of lenses for magnifying things that are too small to see clearly or at all. Leeuwenhoek's and Malpighi's inventions allowed them to observe such things as plant and animal tissues, blood cells, and sperm; little escaped their observant and technologically aided eyes.

Microscopes improved somewhat over the next two hundred years, but it was not until the beginning of the twentieth century that this technology began to advance in sophistication at a fast pace, resulting in the array of light microscopes and electron microscopes that scientists routinely use today. Photographs prepared from three types of microscopes— the compound light microscope, the transmission electron microscope, and the scanning electron microscope— are most often used in this book.

The compound light microscope (**Figure 5.A**) uses two lenses (therefore, compound) to magnify an object. A mirror focuses light from the room up to the eye from beneath the specimen, or a lamp is used for illumination (therefore, light). The compound light microscope visualizes eukaryotic cells (such as plant and animal cells) and prokaryotic cells (bacteria) and can magnify them up to 1500

times. Special stains are often used to visualize particular structures, as is shown in **Figure 5.B**. Brightfield microscopy, the technique used to visualize this stained specimen, passes light directly through the specimen. Unstained organisms can be seen quite well using a technique called phase-contrast microscopy. This technique allows the organism to be viewed while still alive (many types of stains kill cells) and uses direct and indirect lighting to intensify the variations in density within the cell as shown in **Figure 5.C**. A variety of other techniques can be used with the compound microscopes to view organisms, but these two are most commonly used.

The electron microscope does not use light to visualize specimens but instead uses a fine beam of electrons transmitted to a specimen in a vacuum. The transmitted electrons, after partial absorption by the object, are focused by magnets to form the image of the specimen. Electrons have a shorter wavelength than visible light; this difference gives the electron microscope a greater resolving power than the compound light microscope. Resolving power is the ability to distinguish two points as being separate from one another. Therefore, specimens can be magnified more highly with the electron microscope and still transmit a clear image. As with light microscopy, a variety of techniques can be used to prepare specimens. There are two types of electron microscopes: the transmission electron microscope (TEM) and the scanning electron microscope (SEM). **Figure 5.D** is a scanning electron microscope.

The transmission electron microscope is generally used to visualize slices of cells that have been specially prepared. The images are therefore of the interiors of cells, as is shown in **Figure 5.E**. In this photograph (called an electron micrograph), you can see a variety of organelles within this plant cell.

The scanning electron microscope is generally used to visualize surfaces of specimens that have been glued or taped to a metal slide and then covered with a microscopically thin layer of a metal, usually gold or platinum. The metal gives off secondary electrons when excited by a beam of electrons scanned across its surface. These secondary electrons are collected on a screen, which results in an image of the surface of the specimen. **Figure 5.F** is a scanning electron micrograph of the antenna of a moth. ●

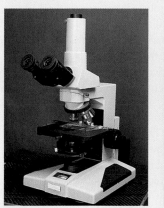

**Figure 5.A** Compound light microscope.

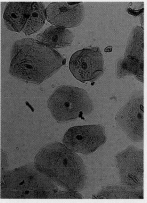
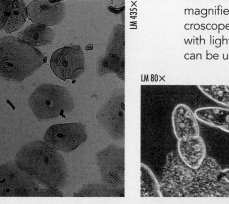

LM 435×

**Figure 5.B** Stained cheek cells as seen under the compound light microscope.

LM 80×

**Figure 5.C** Unstained paramecia as seen with phase-contrast microscopy.

**Figure 5.D** A scanning electron microscope.

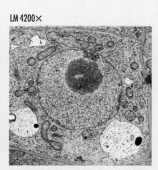

LM 4200×

**Figure 5.E** Color-enhanced transmission electron micrograph (TEM) of a plant cell.

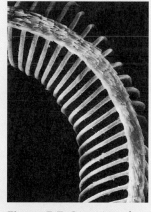

**Figure 5.F** Scanning electron micrograph of the antenna of a moth.

# What are the primary types of cells?

## 5.3 Cells are of two types: prokaryotic and eukaryotic.

All cells can be grouped into two broad categories: **prokaryotic cells** (PRO-kare-ee-OT-ick) and **eukaryotic cells** (YOO-kare-ee-OT-ick). Cells are placed into one of these categories based on their type of structure. All prokaryotes are single-celled organisms—they consist of a single prokaryotic cell. Eukaryotes can be single-celled or multicellular organisms—they consist of one or multiple eukaryotic cells.

All prokaryotic cells share features not found in eukaryotic cells, and all eukaryotic cells share features not found in prokaryotic cells. But because both types are living cells, they have some features in common. Almost all cells have the following four characteristics:

- A surrounding membrane.
- A thick fluid enclosed by this membrane, which, along with the other cell contents, is called *protoplasm*.
- Organelles, or little organs, located within the protoplasm that carry out certain cellular functions.
- A control center that contains the hereditary material, DNA.

## 5.4 Prokaryotic cells have a simpler structure than eukaryotic cells.

Prokaryotic cells were the first type of cell to exist as life arose on Earth more that 3 billion years ago (see Chapter 18). Eukaryotic cells evolved from these simpler cells. Two of the three domains of life—bacteria and archaeans—are prokaryotes. The third domain, the eukaryotes, comprises protists, fungi, plants, and animals. (Section 1.13 discusses the characteristics that distunguish archaeans from bacteria; Figure 1.12 shows the evolutionary relationships among the three domains.)

The two main characteristics of prokaryotic cells that make them less complex that eukaryotic cells are:

- Prokaryotic cells have no membrane-bounded nucleus containing the DNA; rather, there is a region of DNA concentration within the cell called a *nucleoid*.
- The organelles in prokaryotic cells are not bounded by membranes and thus do not compartmentalize the cell.

**Figure 5.4** (a bacterial cell), **Figure 5.5** ( an animal cell), and **Figure 5.6** ( a plant cell) will help you visualize prokaryotic and eukaryotic cell structures.

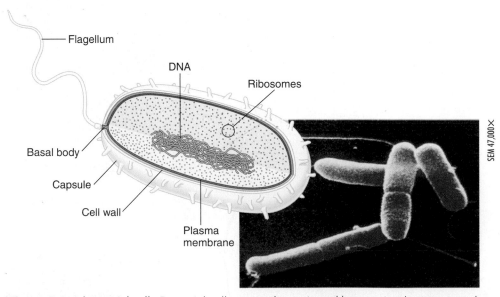

**Figure 5.4** A bacterial cell. Bacterial cells are prokaryotic and have a simpler structure than eukaryotic cells.

 **Visual Thinking:** Although bacteria (prokaryotes) have no membrane-bounded organelles, they do have organelles within their cytoplasm. What organelles are shown in this illustration? What is the function of these organelles?

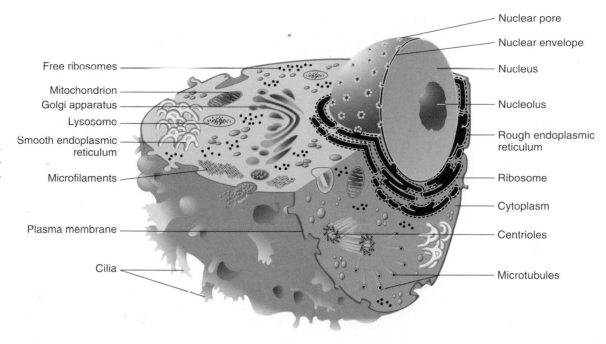

**Figure 5.5** An animal cell. A generalized representation of an animal cell showing the major organelles.

Labels: Nuclear pore, Nuclear envelope, Nucleus, Nucleolus, Rough endoplasmic reticulum, Ribosome, Cytoplasm, Centrioles, Microtubules, Free ribosomes, Mitochondrion, Golgi apparatus, Lysosome, Smooth endoplasmic reticulum, Microfilaments, Plasma membrane, Cilia

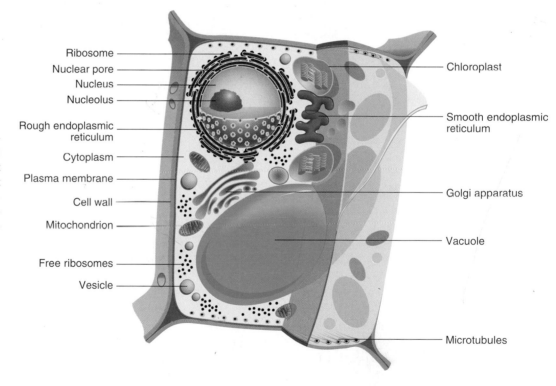

**Figure 5.6** A plant cell. A generalized representation of a plant cell showing the major organelles. Plant cells differ from animal cells in that they have a rigid cell wall, contain chloroplasts, and frequently have large, water-filled vacuoles occupying a major part of the cell volume.

Labels: Ribosome, Nuclear pore, Nucleus, Nucleolus, Rough endoplasmic reticulum, Cytoplasm, Plasma membrane, Cell wall, Mitochondrion, Free ribosomes, Vesicle, Chloroplast, Smooth endoplasmic reticulum, Golgi apparatus, Vacuole, Microtubules

# What are the structures of eukaryotic cells and their functions?

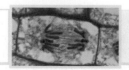

Although eukaryotic cells, such as plant and animal cells, are quite a diverse group, they share a basic architecture. They are all bounded by a membrane that encloses a semifluid material crisscrossed with a supporting framework of protein. All eukaryotic cells also possess many organelles (**Table 5.1**). These organelles are of two general kinds: (1) membranes or organelles derived from membranes and (2) bacterialike organelles.

Most biologists agree that the bacterialike organelles in eukaryotes were derived from ancient *symbiotic bacteria*. The word "symbiosis" means that two or more organisms live together in a close association.

| TABLE 5.1 | Eukaryotic Cell Structures and Their Functions | |
|---|---|---|
| **Structure** | **Description** | **Function** |
| **Exterior Structures** | | |
| Cell wall | Outer layer of cellulose or chitin, or absent | Protection, support |
| Plasma membrane | Lipid bilayer in which proteins are embedded | Regulation of what passes in and out of cell, cell-to-cell recognition |
| Flagella (cilia) | Cellular extensions with 9 + 2 arrangement of pairs of microtubules | Motility or moving fluids over surfaces |
| **Interior Structures and Organelles** | | |
| Endoplasmic reticulum (ER) | Network of internal membranes | Formation of compartments and vesicles; modification and transport of proteins; synthesis of carbohydrates and lipids |
| Ribosomes | Small, complex assemblies of protein and RNA, often bound to ER | Sites of protein synthesis |
| Nucleus | Spherical structure bounded by a double membrane, site of chromosomes | Control center of cell |
| Chromosomes | Long threads of DNA associated with protein | Sites of hereditary information |
| Nucleolus | Site within nucleus of rRNA synthesis | Synthesis and assembly of ribosomes |
| Golgi apparatus | Stacks of flattened vesicles | Packaging of proteins for export from cell |
| Lysosomes | Membranous sacs containing digestive enzymes found in animal cells | Digestion of various molecules |
| Cytoskeleton | Network of protein filaments, fibers, and tubules | Structural support, cell movement |
| Mitochondria | Bacterialike elements with inner membrane highly folded | "Power plant" of the cell |
| Chloroplasts | Bacterialike elements with inner membrane forming sacs containing chlorophyll, found in plant cells and algae | Site of photosynthesis |

Biologists hypothesize that in like fashion bacteria and the precursors to eukaryotic cells lived symbiotically hundreds of millions of years ago. This hypothesis, called the **endosymbiont theory**, explains how complex cells can evolve from two or more simpler cells living in a symbiotic relationship. Today, these bacterial endosymbionts are mitochondria, which occur in all but a very few eukaryotic organisms, and chloroplasts, which occur in algae and plants. Chapter 18 discusses the endosymbiont theory in more detail and presents some of the evidence that supports this theory.

## 5.5 The plasma membrane encloses the cell.

Every cell is bounded by a **plasma membrane**, so named because it encloses the *protoplasm*—the semifluid cell contents, including the cell organelles and nucleus (control center) of the cell. The electron micrograph in **Figure 5.7a** shows that the plasma membrane is a double-layered structure. The plasma membrane has two main components: *phospholipids* (shown as the blue balls with tails in Figure 5.7b) and *proteins* (shown as the purple globular structures).

Phospholipids have only two fatty acid chains as you may recall from Chapter 4; phosphoric acid takes the place of the third, as shown in Figure 5.7c. The kink in one of the fatty acid chains re-

sults from its being unsaturated, which means that it has one or more carbon–carbon double bonds. The straight fatty acid chain is saturated; it has no double bonds.

Lipids are nonpolar molecules and will not dissolve in water (i.e., form hydrogen bonds with water). Phospholipids, however, *are* polar and *do* interact with (form hydrogen bonds with) water via the phosphate group. Therefore, phospholipids have a portion of the molecule that is hydrophilic and a portion of the molecule that is hydrophobic, as shown in Figure 5.7c.

When a collection of phospholipid molecules is placed in water, their hydrophobic fatty acid chains are pushed away by the water molecules that surround them. The water molecules move toward the molecules that form hydrogen bonds with them—the phosphoric acid. As a result of these orientations of the two parts of phospholipids to water, the phospholipids form two layers of molecules, with their hydrophobic "tails" oriented inward (away from the water) and their hydrophilic "heads" oriented outward (toward the water, Figure 5.7b). This double-layered structure is called a *lipid bilayer* and is the foundation of the membranes of all living things.

Did you ever blow soap bubbles when you were a child? Can you remember how the soap film forming the bubble seemed to swirl and move as the bubble floated in the air? Lipid bilayers are fluid much like the film of the soap bubbles. The fluid nature of the bilayer is

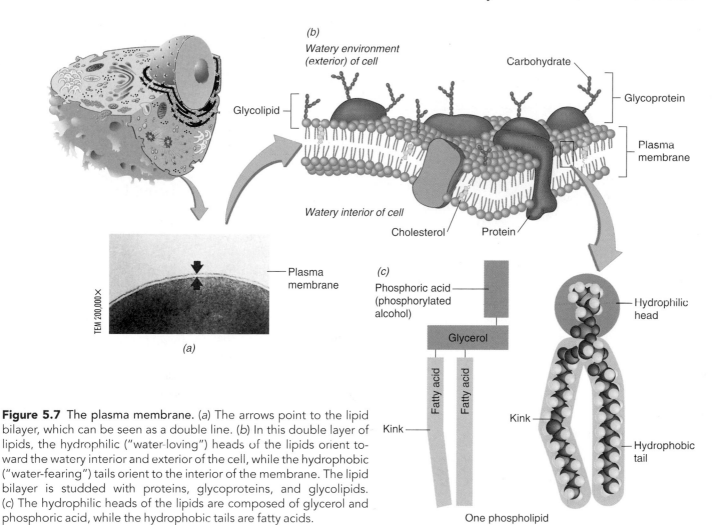

**Figure 5.7** The plasma membrane. (*a*) The arrows point to the lipid bilayer, which can be seen as a double line. (*b*) In this double layer of lipids, the hydrophilic ("water-loving") heads of the lipids orient toward the watery interior and exterior of the cell, while the hydrophobic ("water-fearing") tails orient to the interior of the membrane. The lipid bilayer is studded with proteins, glycoproteins, and glycolipids. (*c*) The hydrophilic heads of the lipids are composed of glycerol and phosphoric acid, while the hydrophobic tails are fatty acids.

caused by movement of the phospholipid molecules. Although water constantly forces the phospholipid molecules to form a bilayer, the hydrogen bonds between the water molecules and the hydrophilic heads of the phospholipids are constantly breaking and reforming. In addition, molecules of cholesterol are found unevenly distributed throughout the lipid bilayer. Sandwiched in between the phospholipid molecules, cholesterol molecules modulate the fluidity of the bilayer.

If cells were encased with pure lipid bilayers, however, they would be unable to survive because the interior of the lipid bilayer is completely nonpolar. It repels most water-soluble molecules that attempt to pass through it. As **Figure 5.8** shows, the lipid bilayer allows only a few small uncharged molecules, such as individual water ($H_2O$), carbon dioxide ($CO_2$), oxygen ($O_2$), and ammonia ($NH_3$) molecules, to pass through (Figure 5.8*a*). (Ammonia is a common cellular waste product formed from the breakdown of proteins.) Substances that are soluble in oil, such as vitamins A, D, and E and other hydrocarbons (molecules that contain only hydrogen and carbon), do move across the lipid bilayer (Figure 5.8*b*). However, the bilayer prevents any water-soluble substances such as sugars, amino acids, and proteins (Figure 5.8*c*) as well as charged particles such as calcium ions ($Ca^{++}$) and potassium ions ($K^+$) from entering or leaving the cell (Figure 5.8*d*). These molecules pass

through the cell membrane by means of proteins that are embedded in the lipid bilayer. Proposed in the early 1970s by Singer and Nicholson, this model of the cell membrane is widely accepted today. It is called the **fluid mosaic model**, a name that describes the fluid nature of a lipid bilayer studded with a mosaic of proteins.

A variety of proteins control the interactions of the cell with its environment. Notice in Figure 5.7*b* that membrane proteins often have sugars bonded to them as well as to some lipid molecules. These molecules are called glycoproteins and glycolipids.

Some membrane proteins are called *channels* or *transporters*. They act as doors that let specific molecules into and out of the cell, or actively move molecules in one direction or the other. Other membrane proteins, called *receptors*, recognize certain chemicals, such as hormones, that signal cells to respond in particular ways. Receptor proteins cause changes within the cell when they come in contact with such chemical messengers or signals. A third type of membrane proteins often the glycoproteins, identify a cell as being of a particular kind. These cell surface markers allow cells to distinguish one another, an ability that helps cells function correctly in a multicellular organism (such as your body). A fourth type of membrane proteins act as enzymes, catalyzing the production of substances within cells in response to signals from outside cells.

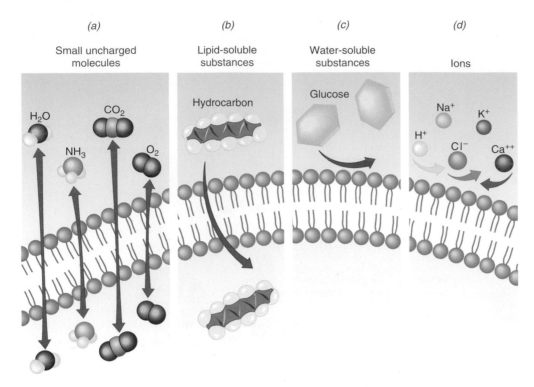

*(a)* Small uncharged molecules

*(b)* Lipid-soluble substances

*(c)* Water-soluble substances

*(d)* Ions

**Figure 5.8** Some substances can pass across the lipid bilayer; others cannot.

 **Visual Thinking:** Why does the lipid bilayer not allow ions to pass?

---

**CONCEPT CHECKPOINT**

2. Mitochondria and chloroplasts have their own DNA. What is a widely accepted scientific explanation for this fact?

3. Develop three analogies between the functions of molecules found in cell membranes and the functions of human body parts. For example, membrane receptor proteins could be likened to ears. Membrane receptors recognize certain chemicals that signal cells to respond in particular ways. Ears recognize sounds that often result in your responding in particular ways.

## 5.6 Cell walls provide rigid exteriors for plant cells, fungi, and many protists.

Plants, fungi, and many protists (especially algae) have a rigid, thick structure called a **cell wall** that surrounds the plasma membrane, as can be seen in the plant cell in **Figure 5.9**. Almost all plant cell walls are made up of cellulose, large molecules formed by the linking of glucose units, polysaccharides, and protein. The cellulose acts like a supporting mesh within the wall, with its fibers running in various directions. Fungi have cell walls that contain chitin, the same substance that is found in the shells of organisms such as grasshoppers and lobsters.

Cell walls perform many jobs for cells. In plants and fungi, they help impart stiffness to the tissues. They also provide some protection from a drying environment. In single-celled organisms, cell walls give shape to the organisms and help protect them.

## 5.7 The cytoplasm provides the viscous, fluid interior of the cell.

The **cytoplasm** (SYE-toe-PLAZ-um) of the cell is the viscous (thick) fluid containing all cell organelles *except* the nucleus. (The protoplasm includes the nucleus.) The word "cytoplasm" literally means living gel (*plasm*) of the cell (*cyto*). The major components of the cytoplasm are a gel-like fluid (the cytosol), storage substances, a network of interconnected filaments and fibers, and cell organelles.

The fluid part of the cytoplasm is made up of approximately 75% water and 25% proteins. The proteins, mostly enzymes and structural proteins, make the cytoplasm viscous—much like thickening gelatin. Nonprotein molecules involved in the various chemical reactions of the cell, such as ions and adenosine triphosphate, (ATP, an important energy-storing molecule), are also dissolved in this fluid.

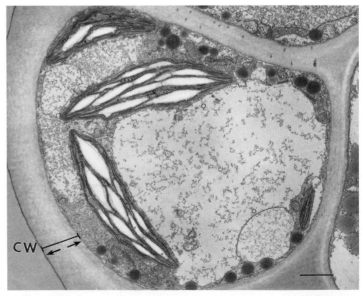

**Figure 5.9** The plant cell wall. This electron micrograph also shows the large vacuole of the plant cell and the chloroplasts next to it. The plasma membrane is visible as a thin line between the thick plant cell wall (CW and arrows) and the interior of the cell.

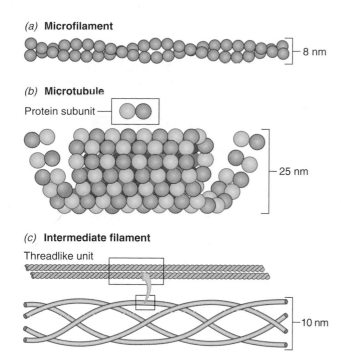

*(a)* **Microfilament**

⎬ 8 nm

*(b)* **Microtubule**

Protein subunit ⎯

⎬ 25 nm

*(c)* **Intermediate filament**

Threadlike unit

⎬ 10 nm

**Figure 5.10** Types of fibers found in the cytoskeleton.

## 5.8 The cytoskeleton helps support and shape the cell.

The cytoplasm also contains a *cytoskeleton*—a network of filaments and fibers that do many jobs. Although the term *skeleton* implies rigidity, the cytoskeleton is not a permanent, inflexible network; it undergoes rapid and continuous change.

The cytoskeleton is made up of three different types of fibers: *microfilaments, microtubules,* and *intermediate filaments.* The microfilaments are the thinnest—a twisted double chain of protein—and are not visible with an ordinary light microscope (**Figure 5.10a**). The microtubules, a chain of proteins wrapped in a spiral to form a tube, are the thickest members of the cytoskeleton (Figure 5.10b). As the name suggests, the intermediate filaments are an in-between size. These are threadlike protein molecules that wrap around one another to form ropes of protein (Figure 5.10c).

The microfilaments and intermediate filaments provide the protein network that lies beneath the plasma membrane to help support and shape the cell. This network of proteins extends into the cytoplasm and looks something like the web of proteins illustrated in **Figure 5.11**. Microtubules are also part of this skeleton of the cytoplasm. Together, these three types of protein fibers provide the cell with mechanical support and help anchor many of the organelles (mitochondria are an exception). They also help move substances from one part of the cell to another.

## 5.9 Cilia and flagella provide motility.

Single-celled eukaryotic cells are often motile—able to move within their environments. One way in which these cells move is by means of cell extensions that look somewhat like hairs. These structures, called **cilia** (SILL-ee-uh), are short and often cover a cell. In human cells, cilia are found only on sections of certain cells and are used to move substances across their surfaces. Figure 28.7 shows the cilia that line the trachea, or windpipe. They help sweep invading particles and organisms up the trachea and away from the lungs. Some cells have whiplike extensions called **flagella** (fluh-JELL-uh). Used strictly for movement, these structures are longer than cilia, but fewer are usually present on a cell. Sperm are the only human cells that have flagella. Normal human sperm have a single flagellum, as you can see in Figure 5.1c.

Bar = 0.25 μm

MT

IF

A

*(a)*

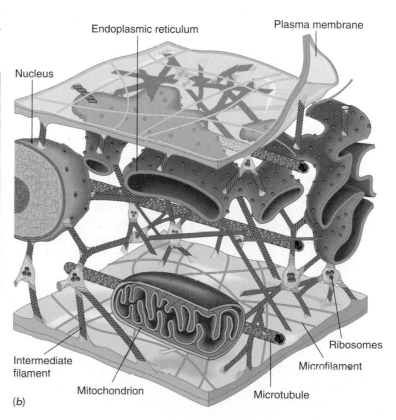

Endoplasmic reticulum

Plasma membrane

Nucleus

Intermediate filament

Mitochondrion

Ribosomes

Microfilament

Microtubule

*(b)*

**Figure 5.11** The cytoskeleton. (*a*) Micrograph of the fibers making up the cytoskeleton. A = actin filaments (microfilaments), IF = intermediate filaments, MT = microtubules. Notice that the intermediate filaments appear solid, while the microtubules have a hollow core. (*b*) A section of a eukaryotic cell and the network of proteins that helps support and shape the cell.

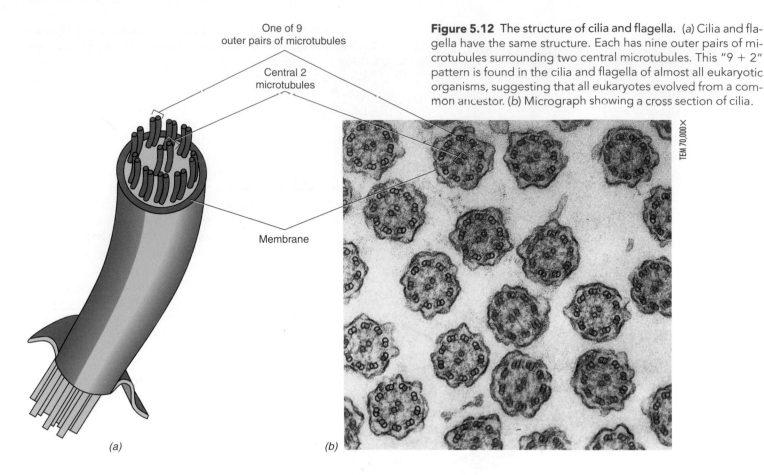

One of 9
outer pairs of microtubules

Central 2
microtubules

Membrane

(a)

(b)

TEM 70,000×

**Figure 5.12** **The structure of cilia and flagella.** (*a*) Cilia and flagella have the same structure. Each has nine outer pairs of microtubules surrounding two central microtubules. This "9 + 2" pattern is found in the cilia and flagella of almost all eukaryotic organisms, suggesting that all eukaryotes evolved from a common ancestor. (*b*) Micrograph showing a cross section of cilia.

Although cilia and flagella differ in length, they have the same structure: They are bundles of microtubules covered with the plasma membrane of the cell. As you can see in **Figure 5.12**, nine outer pairs of microtubules surround two central microtubules. As these microtubules dip into the cell beneath the level of the plasma membrane, they connect with another structure called the *basal body*. Also composed of microtubules, a basal body serves to anchor a cilium or flagellum to the cell as shown in Figure 5.4. Basal bodies are identical in structure to centrioles. **Centrioles** are pairs of cylindrical microtubular structures found in the cytoplasm of animal cells that play a role in cell division (see Chapter 12).

Cilia and flagella in eukaryotes beat in a whiplike fashion because of microtubules sliding past each other in a manner similar to the sliding of actin and myosin filaments in muscle contraction (see Chapter 34). The structure of the flagella and basal body in bacteria differs from eukaryotic flagellar and basal body structure. In addition, bacteria swim by rotating their flagella rather than by whipping them as eukaryotes do.

## 5.10 Membranous organelles form functional compartments within the cytoplasm.

Entangled within the cytoskeletal fibers, the cell organelles work constantly, each contributing in a special way to the life and well-being of the cell. Most organelles of eukaryotic cells are bounded by membranes and form organized compartments within the cytoplasm. Such compartments are an important characteristic of eu-

karyotic cells not present in prokaryotic cells. By means of compartments, enzymes and their substrates are kept in particular places in cells and at optimum concentrations. Metabolic pathways can therefore run more efficiently, with their components grouped together. Toxic waste products, when they arise, can also be separated from the rest of the cell, a safe distance from enzymes and other substances that might be damaged by toxins. The compartmentalized eukaryotic cell is an efficiently running living machine.

### The nucleus contains the hereditary material and controls cell activities

The **nucleus**, or control center of the cell, is made up of an outer, double membrane that encloses the chromosomes and one or more nucleoli. The word "eukaryote" means true nucleus and refers to the fact that the nucleus is a closed compartment bounded by a membrane. Other than large central vacuoles in plants, this compartment is the largest in eukaryotic cells.

The outer, double membrane of the nucleus is called the *nuclear envelope*. The inner of the two membranes actually forms the boundary of the nucleus, and the outer membrane is continuous with the endoplasmic reticulum. At various spots on its surface, the double membrane fuses to form openings called *nuclear pores*. **Figure 5.13a** shows the surface of the nuclear envelope of a cell that was frozen and then cracked during a special type of preparation for the electron microscope. The pores look like pockmarks on the surface of the membrane. These ringlike holes are lined with proteins and serve as passageways for molecules entering and leaving the nucleus (Figure 5.13*b*).

DNA is bound to proteins in the nucleus, forming a complex called *chromatin*. In a cell that is not dividing, the chromatin is strung out, looking like strands of microscopic pearls. But as a cell begins to divide, the DNA coils more tightly around the proteins, condensing to form shortened, thickened structures called **chromosomes** (**Figure 5.14**).

The **nucleolus** (noo-KLEE-oh-lus, plural *nucleoli*) is a darkly staining region within the nucleus (**Figure 5.15**). Most cells have two or more nucleoli. Making up the bulk of the nucleolar material is a special area of DNA that directs the synthesis of **ribosomal ribonucleic acid**, or **rRNA**. As rRNA is made at this DNA, it forms clumps of molecules that are structural components of **ribosomes** (RYE-bah-somes), which are spherical organelles found in the cytoplasm at which proteins are manufactured.

## The endoplasmic reticulum helps move substances within the cell

The **endoplasmic reticulum** (EN-doe-PLAZ-mik ri-TIK-yuh-lum), or ER, is an extensive system of interconnected membranes that forms flattened channels and tubelike canals within the cytoplasm, almost like a cellular subway system. The channels

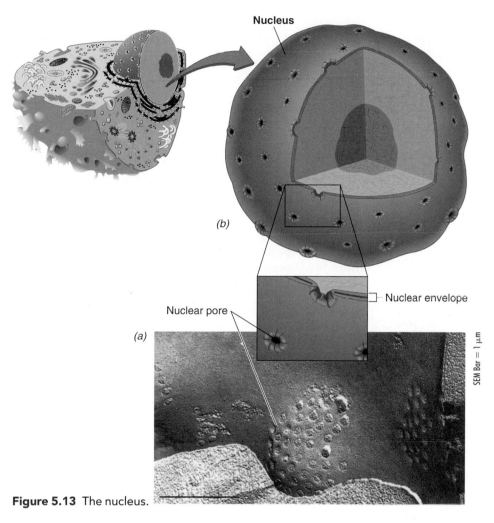

Nucleus

(b)

(a)

Nuclear pore

Nuclear envelope

SEM Bar = 1 μm

**Figure 5.13** The nucleus.

Within the nucleus is the hereditary material **deoxyribonucleic acid** (de-ok-see-RYE-boh-new-KLAY-ick), or **DNA**. DNA determines whether your hair is blond or brown or whether a plant flowers in pink or white. To accomplish such amazing feats, DNA performs one job: It directs the synthesis of ribonucleic acid, or RNA, which in turn directs the synthesis of proteins. The structure, interactions, and roles of both DNA and RNA are discussed further in Chapters 9 and 10.

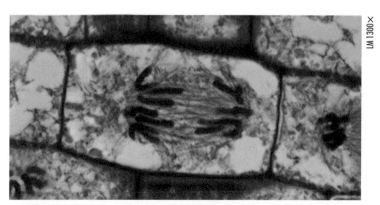

LM 1300×

**Figure 5.14** Chromosomes. As a cell begins to divide, DNA coils and supercoils, condensing to form shortened, thickened structures called chromosomes. The chromosomes in this cell have separated from one another in a way that will provide a complete set of hereditary material to each new daughter cell.

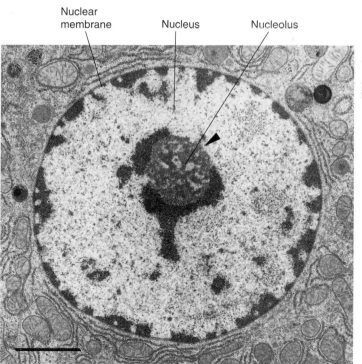

Nuclear membrane        Nucleus        Nucleolus

TEM Bar = 2 μm

**Figure 5.15** The nucleolus. Electron micrograph of a nucleolus within the nucleus of the cell. The nucleolus consists of DNA, ribosomal RNA (rRNA), and ribosomal proteins.

are used to move substances from one part of the cell to another. The name *endoplasmic reticulum* is descriptive of its location and appearance. The word "endoplasmic" means within the cytoplasm, and the word "reticulum" comes from a Latin word meaning a little net. That is exactly what the ER looks like—a net within the cytoplasm.

There are two types of endoplasmic reticula in cells: **rough ER** and **smooth ER**. *Rough* refers to the appearance of this type of ER. With ribosomes dotting their surfaces, rough ER membranes look like long sheets of sandpaper (**Figure 5.16**).

Ribosomes are the places where proteins are manufactured. Some ribosomes are attached to the rough ER, but others are in the cytoplasm bound to cytoskeletal fibers. The cytoplasmic ribosomes help produce proteins for use in the cell, such as those making up the structure of the cytoskeletal fibers or certain organelles. The proteins made on the ribosomes of the rough ER are most often destined to leave the cell. Cells specialized for secreting proteins, such as the pancreatic cells that manufacture the hormone insulin, contain large amounts of rough ER. After the proteins are manufactured on the ribosomes on the surface of the ER, they enter the inner space of the ER. Within this channel, the

proteins may be changed by enzymes bound to the inner surface of the ER membrane. Carbohydrate molecules are often added to them.

Smooth ER has no ribosomes attached to its surfaces. Therefore, it does not have the grainy appearance of rough ER and does not manufacture proteins. Instead, smooth ER has enzymes bound to its inner surfaces that help build carbohydrates and lipids. Cells specialized for the synthesis of these molecules, such as animal cells that produce male or female sex hormones, have abundant smooth ER.

When the proteins reach the end of their journey in the ER, they are encased in tiny membrane-bounded sacs called *vesicles*. These vesicles eventually fuse with the membranes of another organelle called the **Golgi apparatus** (GOL-gee).

### The Golgi apparatus modifies molecules and readies them for transport

First described by the physician Camillo Golgi in the last half of the nineteenth century, the Golgi apparatus (or Golgi for short) looks

---

## *just wondering . . .*

### *Questions students ask*

### How do scientists know what is in the nucleus of a cell?

The story of scientists' quests to understand the structure of the nucleus began in the early 1700s. Microscopes at that time, though primitive by today's standards, were sophisticated enough to allow scientists to see the nuclei of plant cells and some animal cells. The German microscope in **Figure 5.G** is an example of microscope technology of the late 1700s. However, it wasn't until 1833 that a nucleus was thought to be in every cell. Prior to that time, scientists held the view that some cells had no nuclei because they could not see them. Today we realize that scientists were simply observing some cells during cell division, when the nucleus loses its membrane and the nuclear material looks much different from that of a nondividing cell.

As time went on, microscope technology improved with the invention of the high-power oil immersion lens in 1870. The microtome, which could slice individual cells into thin sections, was invented in 1866. In addition, new methods of fixing and staining cells were developed. Some dyes were produced that were highly specific for staining certain cell parts, such as the nucleus. All of these factors allowed scientists to view the nuclei of cells with much greater magnification, detail, and clarity.

As the quality of optical equipment and cell preparation techniques improved, the precision with which cells "at rest" and undergoing nuclear division could be studied increased. The nucleus no longer looked like a granule-filled compartment that split indiscriminately during cell division. Scientists could now see a nucleus filled with threads and bands of material, which they named *chromatin* after the Greek word for color, since this material readily took on color when stained. The word *chromosome*, meaning "colored body," was proposed in 1888.

**Figure 5.G** Eighteenth-century German microscope.

During the 1800s, many scientists were also studying the principles of heredity and the nature of fertilization as well as the events of cell division. So, investigations into the function of the nuclear material paralleled the study of its structure. By 1866, Gregor Mendel had published his theories of inheritance; his work is described in Chapter 13. In the late 1870s, scientists described the process of nuclear division and the process of nuclear reduction-division that occurs in sex cells. By 1884, scientists realized that fertilization in both animals and plants consists of the fusion of maternal and paternal sex cells. Taking all of this information into account, an American graduate student named Walter Sutton hypothesized that the hereditary factors were on the chromosomes. At the turn of the twentieth century, scientists began to look toward the chemical nature of this hereditary material. Then, for more than half a century, chemists joined biologists in their quest to understand the structure of nuclear material.

The rest of this fascinating story of discovery is continued later in this book. Chapter 9 chronicles investigations that experimentally determined that the hereditary material is located in the nucleus. It goes on to discuss the discovery of the molecular makeup of the hereditary material, culminating with the Nobel Prize–winning work of two young scientists, James Watson and Francis Crick. As you can see, your question took scientists from many fields over one hundred years to answer. Today, using electron microscopes and highly sophisticated laboratory techniques, molecular biologists and others are still asking and answering questions about the nucleus and the nature of the hereditary material. ●

---

*Are you wondering about a topic in biology and how it relates to your life? Submit your question by clicking the* Just Wondering *link in this text's companion Web site at* www.wiley.com/college/alters.

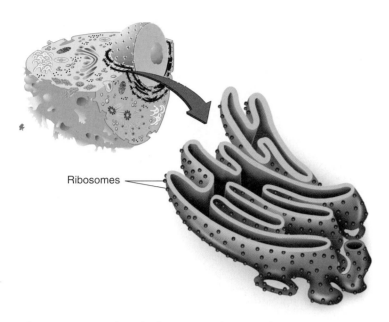

**Figure 5.16** Rough endoplasmic reticulum (rough ER).

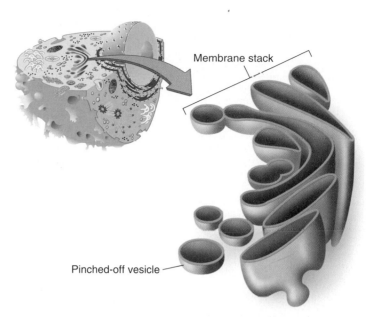

**Figure 5.17** A Golgi apparatus. A Golgi apparatus is a structure that collects, modifies, packages, and distributes molecular products. When these molecular products are ready for transport, they are pinched off into vesicles that move either to the cell membrane to exit the cell or to locations within the cell.

like a microscopic stack of pancakes in the cytoplasm (**Figure 5.17**). Animal cells each contain 10 to 20 sets of these flattened membranes. Plant cells may contain several hundred because the Golgi apparatus is involved in the synthesis and maintenance of plant cell walls (a structure animal cells do not have).

The Golgi can be thought of as the packaging and shipping station of the cell. Molecules come to a Golgi apparatus in vesicles pinched off from the ER (**Figure 5.18 ①**). The membranes of the vesicles fuse with the membranes of a Golgi apparatus **②**. Once inside the space formed by the Golgi membranes **③**, the molecules may be modified by the formation of new chemical bonds or by the addition of carbohydrates. For example, mucin, which is a protein with attached carbohydrates that forms a major part of the mucous secretions of the body, is put together in its final form in the Golgi.

When molecular products are ready for transport, they are sorted and pinched off in separate vesicles **④**. Each vesicle travels to its destination and fuses with another membrane. The vesicles containing those molecules that are to be secreted from the cell fuse with the plasma membrane **⑤**. In this way, the contents of the vesicle are liberated from the cell. Other vesicles fuse with the membranes of organelles such as lysosomes, delivering the new molecules to their interiors.

## Lysosomes digest large molecules

The new molecules delivered to **lysosomes** (LYE-so-somes) are digestive enzymes—molecules that help break large molecules into smaller molecules. Lysosomes are, in fact, membrane-bounded bags of many different digestive enzymes found in animal cells. Several hundred of these organelles may be present in one cell alone.

Lysosomes and their digestive enzymes are extremely important to the health of a cell. They help cells function by aiding in cell renewal, constantly breaking down old cell parts as they are replaced with new cell parts. During development, lysosomes help remodel

tissues, such as the reabsorption of the tadpole tail as the tadpole develops into a frog. In some cells, lysosomes also break down substances brought into the cell from the environment. For example, one job of certain white blood cells is to get rid of bacteria invading the body. The cytoplasm of these cells flows around their prey, engulfing them in a membrane-bounded sac called a *vacuole*. A lysosome then fuses with this vacuole, and digestion of the invader begins.

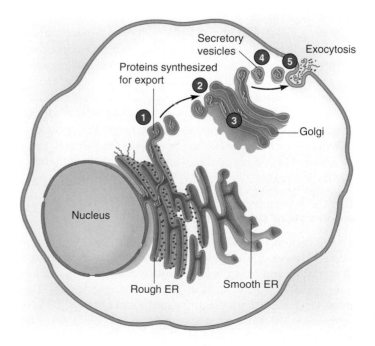

**Figure 5.18** A molecule is packaged for transport in the Golgi apparatus.

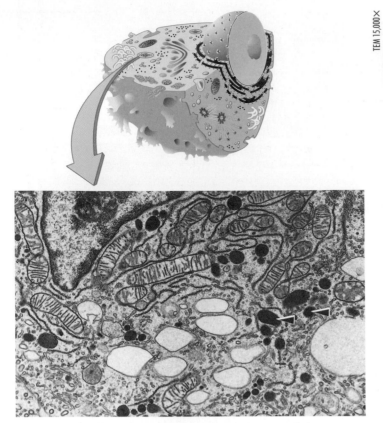

Figure 5.19 Lysosomes. Electron micrograph of lysosomes, pointed out by black arrowheads. The blackened areas within the lysosomes are digestive enzymes. These enzymes break down old cell parts and other materials brought into the cell.

What happens to the digestion products of lysosomes (**Figure 5.19**)? Substances such as parts of bacteria are packaged in a vesicle, transported to the plasma membrane, and exit the cell. Other molecules, such as the breakdown products of old cell parts, may simply be released into the cytoplasm. These cellular building blocks can be recycled—used once again to build new cell parts.

There are over 40 known inherited diseases called lysosomal storage disorders, in which an individual lacks a particular enzyme to break down substances within lysosomes, so they are stored there. As these waste products accumulate in the cells of affected individuals, they damage cells, especially in the brain, skeleton, skin, and heart. The symptoms are progressive and serious. Effective enzyme replacement therapies have been developed for only a few of these diseases.

### Vacuoles store substances

The word **vacuole** (VACK-yoo-ol) comes from a Latin word meaning empty. However, these membrane-bounded storage sacs only look empty, as you might note in **Figure 5.20**. Within vacuoles can be found such substances as water, food, and wastes. Their number, kind, and size vary in different kinds of cells.

Vacuoles are most often found in plant cells. These giant water-filled sacs play a major role in helping plant tissues stay rigid (see Section 5.13). They also may contain toxic chemicals that are released when the plant is harmed by plant-eating organisms. Plant

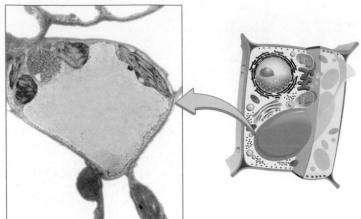

Figure 5.20 The plant cell vacuole. This colorized electron micrograph shows that vacuoles in plants (pale green) often occupy a majority of the cell space. Note that the nucleus (orange) and chloroplasts (bright green and red) are pushed against the plasma membrane.

vacuoles isolate toxic byproducts of chemical reactions that take place in the cell and digest large molecules, acting like the lysosomes of animal cells. In mature plant cells, the vacuole is often so large that it takes up most of the interior of the cell.

## 5.11 Bacteria-like organelles release and store energy.

### Mitochondria break down fuel molecules, releasing energy

The **mitochondria** (MITE-oh-KON-dree-uh) that occur in most eukaryotic cells are thought to have originated as symbiotic bacteria, as mentioned in Section 5.4. According to the endosymbiont theory, the bacteria that became mitochondria were engulfed by eukaryotic cells early in their evolutionary history. Before they acquired these bacteria, the host cells were unable to carry out chemical reactions necessary for living in an atmosphere that had increasing amounts of oxygen. The engulfed bacteria were able to carry out these reactions *using oxygen* and are considered to be the precursors to mitochondria.

Mitochondria are oval, sausage-shaped, or threadlike organelles about the size of bacteria that *have their own DNA*, suggesting that they were once free-living bacteria. They are bounded by a double membrane. The outer of the two membranes is smooth and defines the shape of the organelle. The inner membrane, however, has many folds called *cristae* that dip into the gel-like interior of the mitochondrion, termed the *matrix*. These cristae resemble the folded membranes that occur in various groups of bacteria. Notice in **Figure 5.21** that this arrangement of membranes forms two mitochondrial compartments.

The job of the mitochondria is to break down fuel molecules, releasing energy for cell work. The two most important fuels of cells are glucose and fatty acids. Some organisms, such as plants, make their fuel (glucose) using the raw materials of carbon dioxide, water, and sunlight. Other organisms, such as animals, eat food and digest it to produce glucose and fatty acids, which can then be

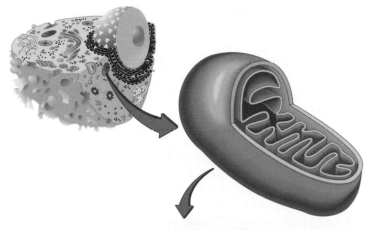

## Chloroplasts use energy from the sun to make sugars

Symbiotic events similar to those postulated for the origin of mitochondria also seem to have been involved in the origin of **chloroplasts**. They are thought to be derived from symbiotic photosynthetic bacteria. These energy-capturing organelles are found in the cells of plants and algae. In these organelles, the energy in sunlight is used to power the reactions that make the cellular fuel—glucose—by using molecules of carbon dioxide from the air. Together, the complex series of chemical reactions that perform these tasks is known as *photosynthesis* (see Chapter 7).

The glucose can be either broken down in the mitochondria to release its energy to ATP for immediate use or short-term storage, or stored as complex carbohydrates for later use. Chloroplasts have a structure similar to the mitochondria. Like mitochondria, chloroplasts have their own DNA and are bounded by a double membrane. The inner membrane encloses a space filled with a thick fluid called *stroma* and an extensive array of saclike structures called *thylakoids* (THIGH-leh-koidz). Stacks of thylakoids are called *grana* (**Figure 5.22**). Chlorophyll, an organic molecule that can absorb light energy from the sun and that allows photosynthesis to take place, is found within the thylakoids. As with the mitochondria, chloroplasts provide an orderly, closed compartment within the cell in which a series of reactions can occur.

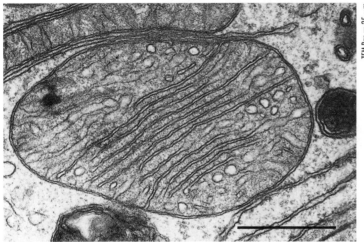

TEM Bar = 25 μm

**Figure 5.21** The mitochondrion.

transported to the cells. In both plants and animals, these fuels are broken down by cells, releasing energy by means of a series of oxygen-requiring reactions called *cellular respiration*. The energy released during these reactions is stored for later use in special molecules called ATP. (See Chapters 6 and 8 for a detailed description of ATP and cellular respiration.)

Cellular respiration begins in the cytoplasm, but most of the energy from the breakdown of glucose and fatty acids is generated in the mitochondria. The enzymes that are used in this breakdown are bound to the membranes of the mitochondrion. Within this closed compartment, the complex series of reactions needed to break apart glucose and to capture the liberated energy and store it are accomplished in an orderly and efficient way separated from the rest of the cell.

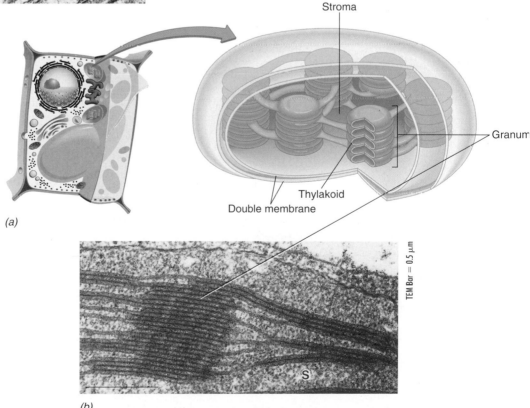

TEM Bar = 0.5 μm

**Figure 5.22** The chloroplast. (a) Chloroplasts have a double outer membrane and a system of interior membranes that form sacs called thylakoids. A stack of thylakoids is called a granum (pl. grana). (b) Electron micrograph of a chloroplast showing a granum, thylakoids, and stroma.

4. Analogies are helpful for remembering the basic functions of many of the organelles in a cell. For example, think of the cell as a large factory complex that makes a variety of widgets. Name the parts of the cell that are analogous to the following factory departments: (a) engineering department—houses blueprints and instructions for building widgets; (b) manufacturing department—makes widgets; (c) waste division—stores chemical wastes from the manufacturing process; (d) recycling center—deconstructs and recycles broken, damaged, or old widgets; (e) sorting division—sorts and labels widgets; (f) central receiving—boxes up the widgets and puts shipping tags on the boxes; and (g) power plant—provides electricity for the factory.

5. To which type of organic molecules in cells are the widgets in the above example analogous?

## 5.12 Bacterial, animal, and plant cells differ in their structures.

**Table 5.2** lists the differences among bacterial, animal (including human), and plant cells. As the table shows, prokaryotes (bacterial cells) and eukaryotes (plant and animal cells) differ greatly in structure and complexity. The table will help you clarify and classify the differences among these three cell types. The most distinctive structural difference between these two cell types is the extensive subdivision, or compartmentalization, of the interior of eukaryotic cells by membranes. Not shown in the table are differences in physiology in prokaryotes and eukaryotes. Eukaryotic cells, which are larger than prokaryotic cells, generally have lower metabolic and growth rates, and divide much less often. Prokaryotic cells are described in more detail in Chapter 19.

| TABLE 5.2 | A Comparison of Bacterial, Animal, and Plant Cells | | |
|---|---|---|---|
| | **Bacterium** | **Animal** | **Plant** |
| **Exterior Structures** | | | |
| Cell wall | Present (protein polysaccharide) | *Absent* | Present (cellulose) |
| Plasma membrane | Present | Present | Present |
| Flagella (cilia) | Sometimes present | Sometimes present | Sperm of a few species possess flagella |
| **Interior Structures and Organelles** | | | |
| Endoplasmic reticulum | *Absent* | Usually present | Usually present |
| Microtubules | *Absent* | Present | Present |
| Centrioles | *Absent* | Present | *Absent* |
| Golgi apparatus | *Absent* | Present | Present |
| Nucleus | *Absent* | Present | Present |
| Mitochondria | *Absent* | Present | Present |
| Chloroplasts | *Absent* | *Absent* | Present |
| Chromosomes | A single circle of naked DNA | Multiple units, DNA associated with protein | Multiple units, DNA associated with protein |
| Ribosomes | Present | Present | Present |
| Lysosomes | *Absent* | Present | Present |
| Vacuoles | *Absent* | *Absent* or small | Usually a large single vacuole in mature cell |

# How do molecules and particles move into and out of cells?

## 5.13 Passive transport is molecular movement that does not require energy.

Cells usually live in an environment where they are bathed in water. This fact may not seem to be true when you consider that bacteria live on your body and that protists live in the soil. However, all cells must move substances across their membranes to survive, and water is the liquid in which their molecules are dissolved. Water also provides a fluid environment within which molecules can move. Therefore, water must surround cells, even in microscopic amounts that may not be readily apparent.

How do molecules get where they are going? To answer this question, you must understand that molecules cannot move *purposefully* from one spot to another. Molecules move randomly. All molecules and small particles have a constant, inherent jiggling motion called *Brownian movement*. Because molecules are always jiggling, they tend to bump into things, such as other molecules. A bump may push a molecule in a particular direction until it bumps into another molecule and gets pushed again.

This random motion in all directions often results in a net movement of molecules in a particular direction. The term *net movement* means that, although individual members of a group of molecules are traveling in different directions, the resulting movement of the group is in one direction. This idea is similar to the net movement of your money into and out of your bank account. Money may move into your bank account from a variety of directions, such as student loans, gifts from your parents, and money from a part-time job. It may also move out of your bank account in a variety of directions, such as payments for your car, rent on your apartment, and bills for clothes. After adding movement in and subtracting movement out, you see the net movement. More of your money moves in one direction than the other.

The net movement of molecules in living systems occurs through differences in concentration, pressure, or electrical charge. These differences are referred to as a **gradient**. The net movement often requires no input of energy. In this type of movement, molecules spread out spontaneously, thus traveling from regions of high concentration to regions of low concentration or from regions of high pressure to regions of low pressure. In addition, ions (charged particles) move toward unlike charges or away from like charges.

As molecules or ions move from regions of high concentration to regions of low concentration, they are referred to as going *down* a concentration, pressure, or electrical gradient. Some molecules move into and out of cells as they travel down gradients. Molecular movement down a gradient but across a cell membrane is called **passive transport** because no work is being done. The three types of passive transport are diffusion, osmosis, and facilitated diffusion.

## Diffusion is the net movement of molecules from regions of higher concentration to regions of lower concentration

**Diffusion** eventually results in a uniform distribution of the molecules, as you've experienced. Did you ever wake up to the smell of freshly brewed coffee? Did you ever spill a bottle of a liquid that had a strong odor, such as ammonia or perfume? In any similar circumstance, the molecules that reached the smell receptors in your nose moved from an area of high concentration (the spill, for example) to an area of low concentration (your nose). They moved down the concentration gradient, eventually becoming evenly spread. This process is illustrated in **Figure 5.23** with a lump of sugar.

## Osmosis is the diffusion of water across a differentially permeable membrane

**Osmosis** (os-MOE-sis) is a special form of diffusion in which water molecules flow from a solution with a lower solute concentration to a solution with a higher solute concentration, across a differentially permeable membrane. A *differentially permeable membrane* allows only certain types of molecules to pass through it, or permeate it, freely. Differentially permeable membranes are also called *semipermeable* or *selectively permeable membranes*. Plasma membranes are differentially permeable membranes. Most types of molecules that occur in cells cannot pass freely across plasma membranes.

The cytoplasm of a cell consists of many different types of molecules and ions dissolved in water. The mixture of these molecules and water is called a *solution*. Water, the most common of the molecules in the mixture, is called the *solvent*. The other kinds of molecules dissolved in the water are called *solutes*.

The cytoplasm of living cells contains approximately 1% dissolved solutes. If a cell is immersed in pure water, interesting things

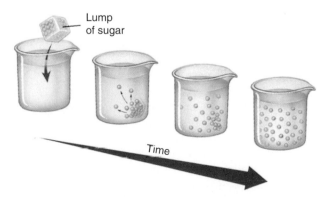

**Figure 5.23 Process of diffusion.** Sugar molecules move from an area of higher concentration (the cube) to an area of lower concentration (the water in the beaker); eventually, there will be an even distribution of sugar molecules throughout the water.

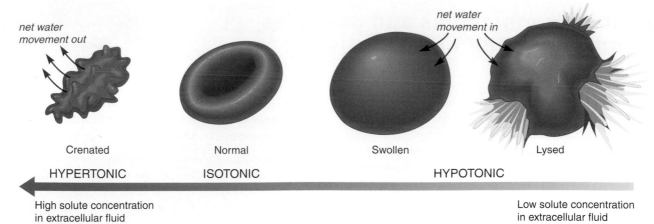

*net water movement out*

*net water movement in*

| Crenated | Normal | Swollen | Lysed |

**HYPERTONIC** **ISOTONIC** **HYPOTONIC**

High solute concentration in extracellular fluid

Low solute concentration in extracellular fluid

**Figure 5.24** Effects of hypertonic, isotonic, and hypotonic solutions on red blood cells. The normal shape of a red blood cell is that shown in the isotonic solution.

begin to happen. Because the water has a lower concentration of solutes than the cell, water molecules begin to move into the cell. The pure water is said to be **hypotonic** (Greek *hypo*, under) with respect to the cytoplasm of the cell because its concentration of solutes is less than the concentration of solutes in the cell.

As you can see in **Figure 5.24**, a cell in a hypotonic solution begins to blow up like a balloon. Similarly, if a cell is placed in a **hypertonic** (Greek *hyper* meaning over) solution, one with a solute concentration higher than the cytoplasm of the cell, water will move out of the cell, and the cell will shrivel. If a cell is placed in a solution with the same concentration of solutes as its cytoplasm, an **isotonic** (Greek *isos* meaning equal) solution, water will move into and out of the cell, but no net movement will take place.

Intuitively, you might think that as "new" water molecules diffuse into a cell placed in a hypotonic solution, the pressure of the cytoplasm pushing against the cell membrane will build. This is indeed what happens. The cell will eventually reach an equilibrium—a point at which the osmotic force driving water inward is counterbalanced exactly by the pressure outward of the cytoplasm. However, the pressure within the cell may become so great that the cell bursts like a balloon. The cell is said to *lyse*.

Single-celled and multicellular organisms have various mechanisms that work to keep their cells from swelling and bursting or shriveling like prunes. Single-celled organisms that live in fresh water, for example, battle a constant influx of water. Some of these organisms have one or more organelles that collect water from the cell's interior and transport it to the cell's surface. Plant cells avoid bursting in a hypotonic environment because of their strong cell walls. Instead, pressure builds up within the cell, called *turgor*. Turgor gives rigidity to plant cells. When in hypertonic surroundings, plant cells lose turgor, the plasma

membrane and protoplasm pull away from the cell wall in a process called *plasmolysis* (plaz-MAHL-eh-sis), and the plant wilts.

### Facilitated diffusion uses transport proteins for movement across a differentially permeable membrane

One of the most important properties of any cell is its ability to move substances necessary for survival into its interior and get rid of unnecessary or harmful substances. Your cells, for example, move glucose into the cytoplasm from the bloodstream to be used for fuel. Without the ability to move glucose into your cells, you would die.

Cells can perform these feats because they have differentially permeable membranes. Some of the channels in the cell membrane help certain molecules and ions enter or leave the cell and speed their movement by providing them with a passageway. These passageways are most likely transport proteins that extend from one side of the membrane to the other. After the transport protein binds with a solute molecule (**Figure 5.25 ❶**), its shape changes **❷**, and the molecule moves across the membrane **❸**. This type of transport process,

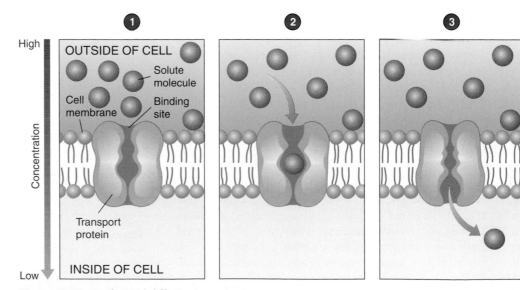

High

**OUTSIDE OF CELL**

Solute molecule

Cell membrane

Binding site

Concentration

Transport protein

Low

**INSIDE OF CELL**

**Figure 5.25** Facilitated diffusion.

in which molecules move down the concentration gradient by means of transport proteins but without an input of energy, is called **facilitated diffusion**. Facilitated diffusion helps rid the cell of certain molecules present in high concentrations and moves molecules into the cell that are present on the outside in high concentrations. Aquaporins (see this chapter's opening story) are transport proteins that facilitate the movement of water molecules across cell membranes.

Facilitated diffusion has two essential characteristics: (1) it is *specific*, with only certain molecules being able to traverse a given channel, and (2) it is *passive*, with the direction of net movement being determined by the relative concentrations of the transported molecule inside and outside the membrane.

---

### CONCEPT CHECKPOINT

6. A very overworked and tired first-year medical student is asked to do a red blood cell count on a patient. She extracts some blood and places a tiny drop on a slide that has several larger drops of distilled water on it. When she looks at the red blood cells under the microscope, she notices that they are swollen with water and some have burst. Explain to her why this has happened.

7. What would happen to cells from the leaf of a plant if they were placed in the distilled water on the slide?

---

## 5.14 Active transport is molecular movement that requires energy.

Cells often move substances into or out of the cell *against* the gradients of concentration, pressure, and electrical charge. A cell uses energy to move molecules against a gradient much like you might use energy to move something against gravity. For example, if you are driving downhill, you can put your car into neutral and coast (although that is not the safest way to drive downhill). The car will continue to move without a push from the engine. As soon as you come to a hill, however, you must put the car in gear and press on the accelerator or the car will soon come to a stop. Cells, too, need to use energy to move molecules uphill, or against a gradient, and cells also need to expend energy to move large molecules or particles into the cell that cannot move across the cell membrane.

A cell takes up or eliminates many molecules and ions against a concentration gradient. These molecules and ions enter

and leave cells by way of a variety of selectively permeable transport channels. In all these cases, a cell must expend energy to transport these molecules against the concentration gradient and maintain the concentration difference. This type of transport is called **active transport**. Active transport is one of the most important functions of any cell. Without it, the cells of your body would be unable to maintain the proper concentrations of substances they need for survival.

An example of an important active transport mechanism in animal cells is a plasma membrane protein called the *sodium-potassium pump*. (Such active transport proteins are often called pumps.) More than one-third of all the energy expended by an animal cell that is not dividing is used to actively transport sodium ions ($Na^+$) out of the cell and potassium ions ($K^+$) into the cell. A steep gradient of these ions is necessary for various cell functions, but most importantly for nerve cells to carry impulses and muscle cells to contract. The type of channel by which *both* ions are transported across the cell membrane in opposite directions is called a coupled channel, which has binding sites for both molecules on one membrane transport protein.

The sodium-potassium pump uses energy to power its shape changes that move $Na^+$ and $K^+$ ions across the cell membrane (**Figure 5.26 ❶**). The energy is supplied when a molecule of adenosine triphosphate (ATP) is broken down into adenosine diphosphate (ADP) and a phosphate group (shown as P). ATP is often used to power chemical reactions in cells; it is discussed in more detail in Chapter 6. Sodium ions are moved out of the cell to maintain a low internal concentration relative to the concentration outside the cell ❷. Conversely, potassium ions are moved into the cell to maintain a high internal concentration relative to the concentration outside the cell ❸. Three sodium ions are moved out for every two potassium ions that are moved in.

## 5.15 Cells move large molecules and particles into the cell by endocytosis.

Certain types of cells transport particles, small organisms, or large molecules such as proteins into their cells. In humans, for example, white blood cells police the body fluids and ingest substances as

**Figure 5.26** The sodium-potassium pump uses energy to move sodium ions ($Na^+$) out of an animal cell and potassium ions ($K^+$) into the cell.

**Visual Thinking:** How does the illustration show that energy is being used to move sodium and potassium ions?

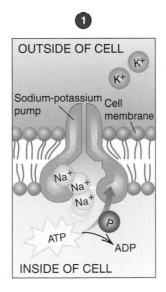
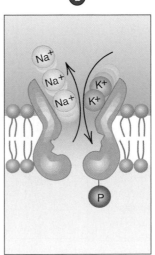
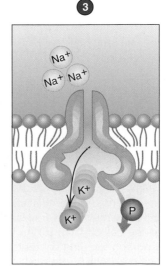

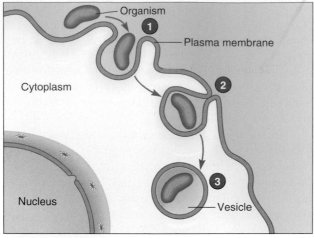

*(a)* PHAGOCYTOSIS

*(b)* PINOCYTOSIS

**Figure 5.27** Phagocytosis and pinocytosis.

large as invading bacteria. In nature, some single-celled organisms often eat other single-celled organisms whole. How can cells move such large substances into their interiors?

Cells such as these ingest particles or molecules that are too large to move across the membrane by a process called **endocytosis**, which literally means "into the cell." During endocytosis, a cell surrounds the particle or large molecules with fingerlike extensions of the membrane-covered cytoplasm (**Figure 5.27 ❶** and **❷**). The edges of the membrane eventually meet on the other side of the now enveloped substance. Because of the fluid nature of the lipid bilayer, the membranes fuse together, forming a vesicle around the particle **❸**.

If the material that is brought into the cell contains an organism or a fragment of organic matter as in Figure 5.27a, the endocytosis is called **phagocytosis** (FAG-oh-sye-TOE-sis) (Greek *phagein* meaning to eat, and *cytos* meaning cell). If the material brought into the cell is liquid—contains dissolved molecules as in Figure 5.27b—the endocytosis is referred to as **pinocytosis** (Greek *pinein* meaning to drink). Pinocytosis is common among the cells of multicellular animals. Human egg cells, for example, are nursed by surrounding cells that secrete nutrients that the maturing egg cell takes up by pinocytosis.

There are two types of endocytosis: *bulk-phase endocytosis* and *receptor-mediated endocytosis*. Bulk-phase endocytosis is nonspecific. That is, any particles or large molecules that happen to be present in the enveloped fluid gain entry into the cell. Receptor-mediated endocytosis is specific. Only molecules that bind to receptors on the outside surface of the plasma membrane are brought into the cell.

## 5.16 Cells move large molecules and particles out of the cell by exocytosis.

The reverse of endocytosis is **exocytosis** (**Figure 5.28**). During exocytosis, a cell discharges material by packaging it in a vesicle **❶** and moving the vesicle to the cell surface **❷**. The

**Figure 5.28** Exocytosis. (*a*) Proteins and other molecules are secreted from cells in small pockets called vesicles, whose membranes fuse with the cell membrane. When this fusion occurs, the contents of the vesicles are released to the cell surface. (*b*) An electron micrograph showing contents of a vesicle being expelled from a cell.

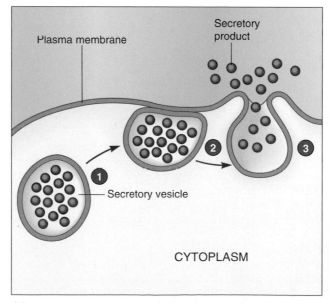

*(a)*

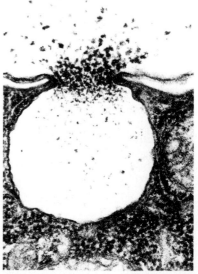

*(b)*

membrane of the vesicle fuses with the cell membrane, and the contents are expelled ❸. In plants, exocytosis is the main way that cells move the materials from the Golgi apparatus and out of the cytoplasm to construct the cell wall. In animals, many cells are specialized for secretion using the mechanism of exocytosis, including cells that produce and secrete digestive enzymes or hormones.

### CONCEPT CHECKPOINT

8. The following are types of molecular transportation within cells: active transport, passive transport, facilitated diffusion, pinocytosis, and exocytosis. For each of these modes of transport, list the following: (a) the direction of molecular transport (high concentration → low, or low concentraton → high), (b) the general characteristics of the molecules transported, and (c) an example of the type of molecules transported.

## CHAPTER REVIEW

## Summary of Key Concepts and Key Terms

### What are cells?

**5.1** A cell is a membrane-bounded unit containing hereditary material and cytoplasm; cells can release energy from fuel and use that energy to grow and reproduce.

**5.1** The **cell theory** (p. 59) states that all living things are made up of cells and that all cells arise from the growth and division of other cells.

**5.2** Most cells divide before they grow too large, maintaining a large surface area-to-volume ratio and the ability to move substances across their cell membranes quickly enough to meet their needs.

### What are the primary types of cells?

**5.3** All cells can be classified as either **prokaryotic** (PRO-kare-ee-OT-ick, p. 62) or **eukaryotic** (YOO-kare-ee-OT-ick, p. 62).

**5.4** Prokaryotic cells are the bacteria and archaeans; they have a relatively simple structure.

**5.4** Eukaryotic cells are more complex than prokaryotic cells and contain many organelles.

**5.4** Eukaryotic cells make up the bodies of plants, animals, protists, and fungi.

**5.4** The **endosymbiont theory** (p. 64) describes how eukaryotes arose through symbiotic relationships among prokaryotic cells.

### What are the structures of eukaryotic cells and their functions?

**5.5** All cells are bounded by a **plasma membrane** (p. 64), which is a phospholipid bilayer with embedded proteins.

**5.5** Phospholipids have a portion that is hydrophilic and a portion that is hydrophobic.

**5.5** Lipid bilayers allow only certain molecules to pass through them.

**5.5** The model of the cell membrane widely accepted today is called the **fluid mosaic model** (p. 65), a name that describes the fluid nature of a lipid bilayer studded with a mosaic of proteins.

**5.5** Membrane proteins facilitate the movement of many types of molecules into and out of the cell.

**5.6** Plants, fungi, and many protists have rigid **cell walls** (p. 66) surrounding the cell membrane; cell walls perform many jobs, including protecting the membrane from drying and imparting shape and stiffness.

**5.7** The **cytoplasm** (SYE-toe-PLAZ-um, p. 66) of the cell is the viscous fluid containing all cell organelles except the nucleus.

**5.8** The cytoskeleton is a network of filaments and fibers in the cytoplasm, which undergoes continuous change and performs many jobs for the cell.

**5.9** In eukaryotes, **cilia** (SILL-ee-uh, p. 67) and **flagella** (fluh-JELL-uh, p. 67) are composed of microtubules and help propel cells; in some eukaryotic cells, cilia are used to move substances across surfaces.

**5.9** **Centrioles** (p. 68) are pairs of cylindrical microtubular structures found in the cytoplasm of animal cells that play a role in cell division.

**5.10** Eukaryotic cells contain membrane-bounded organelles: **nucleus** (p. 68), **endoplasmic reticulum** (EN-doe-PLAZ-mik ri-TIK-yuh-lum, p. 69), **Golgi apparatus** (GOL-gee, p. 70), **lysosomes** (LYE-so-somes, p. 71), and **vacuoles** (VACK-yoo-olz, p. 72).

**5.10** Within the nucleus is the hereditary material **deoxyribonucleic acid** (de-ok-see-RYE-boh-new-KLAY-ick, p. 69), or **DNA**.

**5.10** As a cell begins to divide, the DNA condenses to form shortened, thickened structures called **chromosomes** (p. 69).

**5.10** The **nucleolus** (noo-KLEE-oh-lus, p. 69) is a darkly staining region within the nucleus; it has a special area of DNA that directs the synthesis of **ribosomal ribonucleic acid**, or **rRNA** (RYE-buh-SO-mull RYE-boh-new-KLAY-ick, p. 69).

**5.10** **Ribosomes** (RYE-bah-somes, p. 69), which are small, spherical organelles located in the cytoplasm or on the endoplasmic reticulum, are the sites of protein synthesis.

**5.10** There are two types of endoplasmic reticula in cells: **rough ER** (p. 70) and **smooth ER** (p. 70).

**5.11** Eukaryotic cells also contain bacterialike organelles: **mitochondria** (MITE-oh-KON-dree-uh, p. 72) and **chloroplasts** (p. 73).

**5.12** Bacterial, animal, and plant cells differ in a variety of significant ways

## How do molecules and particles move into and out of cells?

**5.13** Molecules move randomly with a constant jiggling motion.

**5.13** Molecules and ions undergo net movement in response to differences, or **gradients** (p. 75), in concentration, pressure, and electrical charge, resulting in an equal distribution of these molecules and ions.

**5.13** Movement down a gradient requires no added energy to take place.

**5.13** Molecular movement down a gradient but across a cell membrane is called **passive transport** (p. 75).

**5.13** The three types of passive transport are **diffusion** (p. 75), **osmosis** (os-MOE-sis, p. 75), and **facilitated diffusion** (p. 77).

**5.13** A solution is **hypotonic** (p. 76) with respect to another solution when its solute concentration is less than the solute concentration of the other, **hypertonic** (p. 76) when its solute concentration is higher

than the solute concentration of the other, and **isotonic** (p. 76) when their solute concentrations are the same.

**5.14** When cells move substances against a gradient, energy is required; this movement is called **active transport** (p. 77).

**5.15** The transport into the cell of particles or molecules that are too large to move across the membrane takes place by a process called **endocytosis** (p. 78), in which the cell engulfs large molecules or particles.

**5.15** If the material that is brought into the cell contains an organism or a fragment of organic matter, the endocytosis is **phagocytosis** (FAG-oh-sye-TOE-sis, p. 78).

**5.15** If the material brought into the cell is liquid, the endocytosis is **pinocytosis** (p. 78).

**5.16** The reverse of endocytosis is **exocytosis** (p. 78).

## Level 1  Learning Basic Facts and Terms

**Multiple Choice**

1. Which organelle/structure is common to both plant and animal cells?
   a. chloroplasts
   c. mitochondria
   b. cell wall
   d. a large central vacuole
2. What is the mechanism by which a small single-celled organism could ingest a smaller single-celled organism?
   a. pinocytosis
   c. osmosis
   b. exocytosis
   d. phagocytosis
3. The membrane of this organelle has pores that allow the passage of large molecules, such as those that make up ribosomes.
   a. smooth ER
   b. Golgi apparatus
   c. nucleus
   d. rough ER
   e. mitochondrion

4. Which of the following organelles is paired correctly with its function?
   a. mitochondrion . . . . . .photosynthesis
   b. Golgi apparatus . . . . . .receiving, tagging, and shipping proteins
   c. lysosomes . . . . . .storage
   d. vacuoles . . . . . .digestion

**True–False**

5. Microfilaments of the cell cytoskeleton are involved primarily in protein synthesis.
6. Membrane-bound organelles are advantageous because they allow toxic waste products of metabolism to be compartmentalized.
7. Cells of organs, such as the pancreas, that synthesize polypeptide hormones probably have extensive rough endoplasmic reticula.
8. Nucleic acids are likely to be found only in the nucleus.
9. Endosymbiosis explains the origin of all eukaryotic intracellular organelles.

## Level 2  Learning Concepts

1. What characteristics do diffusion and facilitated diffusion have in common? In what ways do they differ?
2. A human liver cell has produced a protein to be used outside of the cell. Using what you know about the endoplasmic reticulum and Golgi apparatus, trace the path of the protein from production to removal from the cell.
3. Draw a diagram of a cell engaging in endocytosis. Give an example of this process in the human body.
4. Many types of cells in the human body, such as cells of the pancreas and small intestine, secrete products such as hormones or enzymes. What method of transportation might such cells use to secrete large amounts of the proteins they produce? Give a rationale for your answer.

5. What makes biological membranes differentially permeable? Why is it necessary for the membranes of living cells to be differentially permeable?
6. Identify the type(s) of molecular movement involved in each of the following. Which are active and which are passive processes? For (b), identify the solute, solvent, and solution.
   a. Glucose molecules leave your bloodstream (where they are usually in higher concentrations than in your cells) and enter your cells by attaching to specialized carrier proteins.
   b. You dissolve a spoonful of instant coffee crystals in a mug of hot water.
   c. Your cells "import" sodium ions by transporting them against the concentration gradient.

## Level 3  Critical Thinking and Life Applications

1. Some of the skeletal muscle cells that allow you to move your legs can be as long as 30 to 40 centimeters. These cells have many nuclei. Explain why.
2. You're examining three cells under the electron microscope. You know that one is a bacterium, one is a plant cell, and the third is an animal cell. From their structures, how can you tell which is which?

3. Some protozoans (single-celled organisms) contain contractile vacuoles, which remove excess water from the cell. What does the presence of contractile vacuoles suggest about the environment in which these protozoans are found?
4. Many antibiotics kill bacteria because they hamper the function of prokaryotic ribosomes but have no effect on eukaryotic ribosomes. Which essential cellular process do antibiotics disrupt in bacteria?

## In The News | Critical Thinking

### AGRE AND CELLS

Now that you understand more about the structures and functions of cells, reread this chapter's opening story about aquaporin research. To better understand this research, it may help you to follow these steps:

1. Review your immediate responses to Agre's research on aquaporins that you wrote when you began reading this chapter.
2. Based on your current understanding, again summarize the main point of the research in a sentence or two.
3. What questions do you now have about this research that the opening story does not answer?
4. Collect new information about the research. Visit the *In The News* section of this text's companion Web site at www.wiley.com/college/alters and

watch the "Agre and Cells" video. Then use the "summary" link to read the accompanying story and access related links. Use this information, the links provided, and other online and library resources to answer your questions and find updates about this research topic. State the sources of your information. Explain why you think the information is accurate. Also determine whether the information expresses a particular point of view or is biased in any way.

5. What in your view is the most significant aspect of this research? Give reasons for your opinion and for any changes in your ideas based on the additional information you have collected and the analysis you have done.

# THE FLOW OF ENERGY WITHIN ORGANISMS

## In The News | Mighty Mini Motor

Machines the size of molecules? Seems impossible, doesn't it? However, molecular biologist Peixuan Guo and his research team at Purdue University's School of Veterinary Medicine are developing a molecular-sized "mighty mini motor" that someday might be able to travel within the body performing such tasks as carrying "healthy" genes to repair damaged ones.

About 1/3000 the width of a human hair, Guo's biomachine is made from RNA, a nucleic acid whose role is usually one of transcribing and translating the DNA hereditary message during the manufacture of proteins. RNA molecules can bind ATP, one of the most important energy-bearing molecules in living things. ATP performs the essential job of transferring energy during chemical reactions. Guo and his team discovered that on binding ATP, RNA moves. The team used this new, exciting knowledge to develop their motor.

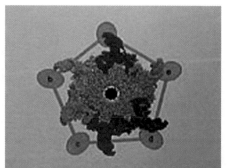

RNA is a flexible molecule, and it lends itself to being manipulated into the six-sided engine they devised, which is shown in the photo. Guo's team is currently exploring ways in which to use this biological engine, fueled by the same molecule that fuels your cells. You can view a video about this research by visiting the *In The News* section of this text's companion Web site at www.wiley.com/college/alters.

*Write your immediate reaction to Guo's research on the mini motor: first, summarize the main point of the research in a sentence or two; then suggest what you think the significance of this research is. You will have an opportunity to reflect on your responses and gather more information on this topic in the* In The News *feature at the end of this chapter. In this chapter, you will learn more about ATP and how it fuels the chemical reactions of life.*

# CHAPTER GUIDE

## How is energy released and stored within cells?

**6.1** Life involves a constant flow of energy.

**6.2** Energy is released and stored during chemical reactions.

## How are chemical reactions regulated in living systems?

**6.3** Enzymes lower the activation energy needed to get reactions started.

**6.4** Enzymes bind substrates at active sites, which brings the substrates together.

**6.5** Environmental conditions affect enzyme activity.

**6.6** Cofactors and coenzymes help enzymes regulate chemical reactions.

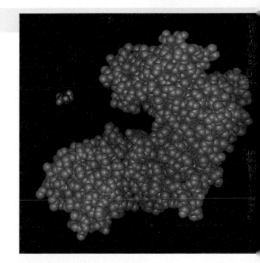

## How is energy captured and supplied in chemical reactions in living systems?

**6.7** Adenosine triphosphate (ATP) is the primary energy carrier of living systems.

**6.8** ATP stores, transfers, and releases energy.

# How is energy released and stored within cells?

## 6.1 Life involves a constant flow of energy.

Life can be viewed as a constant flow of energy that is channeled by organisms to do the work of living. Energy is the ability to do work, which encompasses many things, such as the pull of a cable on a rollercoaster car or the swift dash of a horse. Energy exists in many forms, including mechanical force, heat, sound, electricity, light, radioactivity, and magnetism.

Energy also exists in two states: potential and kinetic. Energy not actively doing work but having the capacity to do so is called **potential energy**. A rollercoaster car perched atop its first hill, for example, possesses this type of stored energy, as does a flying squirrel crouched in a tree. As the car rolls down the hill or the squirrel jumps (**Figure 6.1**), each is actively engaged in doing work. This form of energy is called **kinetic energy**, or the energy of motion. Much of the work performed by living organisms involves changing potential energy to kinetic energy, such as changing the potential energy in the food you eat to the kinetic energy of muscular activity.

All the changes in energy that take place in the universe, from nuclear explosions to the buzzing of bees, are governed by the two **laws of thermodynamics**. The **first law of thermodynamics** is also called the law of conservation of energy. It states that energy can neither be created nor destroyed. It can, however, change from one form to another and from one state to another. For example, chemical energy in the food you eat is converted to mechanical energy when you move your muscles. As chemical energy changes to mechanical energy, some is converted to heat energy as well. This heat helps keep your body warm.

The **second law of thermodynamics** states that disorder in a closed physical system tends to increase. Stated simply, disorder is more likely than order. You can relate this to keeping your personal environment in order. Your bedroom continually becomes messy unless you make an effort to keep it neat. In general, matter or energy in order tends to have higher energy associated with it, while matter or energy in disorder tends to have lower energy associated with it. Although the total amount of energy in the universe does not change

(first law of thermodynamics), the amount of useful energy available to do work decreases as progressively more useful energy is transformed to less useful energy (second law of thermodynamics).

**Entropy** (ENN-truh-pee) is a measure of the disorder (or randomness) of a physical system. The greater the disorder, the greater the entropy. The loss of available energy during a process is a result of a tendency for randomness, or disorder. Sometimes the second law of thermodynamics is simply stated as "entropy increases." So, although energy cannot be destroyed, the universe is constantly moving toward increasing entropy.

It takes energy to overcome increasing entropy. Plants need radiant energy from the sun, or they will die. Living systems are not closed but open systems; their entropy can decrease. In other words, organisms must have an input of energy so that their cell systems will continue to function, maintaining molecular order. Molecular disorder results in death. For example, you need to eat (take in energy) or you will die (your chemical pathways will become disordered and you will no longer function).

Although energy cannot come into or go out of the universe, the surface of the Earth is constantly receiving "new" energy primarily from the sun. Photosynthetic organisms such as green plants capture some of it. In photosynthesis, the energy from sunlight is changed to chemical energy, combining small molecules into more complex molecules by means of reactions that store energy. This stored energy can be shifted to other molecules by forming different chemical bonds. Thus, energy continuously flows into and through the biological world, with new energy from the sun constantly flowing, "replacing" energy that is no longer available to do work.

## 6.2 Energy is released and stored during chemical reactions.

**Figure 6.2** is a graph of the varying energy states of the molecules in a chemical reaction. Doesn't this graph look like the beginning hill of a rollercoaster? The **reactants** are the molecules entering into the chemical reaction—the rollercoaster cars. As the cars are perched at the foot of the first hill, energy is needed to push them up and over the top. In chemical reactions, that "push" is called *activation energy*. The activation energy is represented by the height of the curve. All chemical reactions require activation energy to get started.

Activation energy is needed for various reasons. In some chemical reactions, it helps break or destabilize old bonds, which must occur before new bonds are formed. In other reactions, it excites electrons, which then achieve a higher energy level prior to "pairing up" in covalent bonds. In certain chemical reactions, activation energy helps molecules overcome the mutual repulsion of their many electrons, allowing them to get close enough to react. In any chemical reaction, the activation energy performs one or more of these jobs and allows the chemical reaction to "go." The reactants undergo a chemical change, resulting in new bonding arrangements between the molecules. The changed reactants are called **products**.

**Figure 6.1 A flying squirrel and energy transformation.** A flying squirrel sitting on a tree branch has the capability of leaping from it. In other words, the squirrel possesses potential energy. As it leaps from the tree, potential energy is transformed into kinetic energy—the energy of motion.

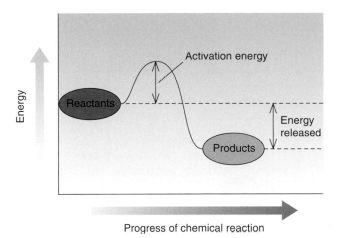

**Figure 6.2 An exergonic reaction.** In an exergonic reaction, when reactants are broken down, energy is released. The reaction products contain less energy than the reactants.

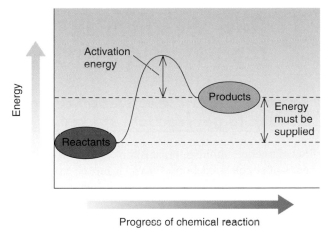

**Figure 6.3 An endergonic reaction.** In an endergonic reaction, when reactants are joined into products, energy must be supplied. The products contain more energy than the reactants.

## Reactions that break apart molecules release energy

Sometimes reactants are chemically broken down to yield products. When a reactant is broken down, energy is released from the chemical bonds that were holding it together. This type of reaction is called exergonic, meaning "energy out." In an **exergonic reaction**, the products contain less energy than the reactant. The excess energy is released to nearby molecules (see Figure 6.2).

Exergonic reactions proceed spontaneously once they get going because they are thermodynamically favored. That is, they "follow" the second law of thermodynamics. Therefore, they occur with no additional input of energy from another chemical reaction. Spontaneous reactions are often referred to as "downhill" reactions because, like a ball rolling down a hill, spontaneous processes occur without help other than the input of activation energy. With the ball analogy, the activation energy can be likened to the push the ball received to get it rolling down the hill.

Your body is a literal hotbed of exergonic reactions. For example, a series of exergonic reactions breaks down the glucose in your cells to supply your body with energy. Furthermore, when you take in less energy (calories) than you use, your body breaks down stored nutrients, such as fat, to supply your body with energy. The heat released from these reactions helps keep your body warm. Reactions such as these, which release energy by breaking down complex molecules into simpler molecules, are called **catabolic reactions** (CAT-uh-BOL-ick).

## Reactions that join molecules store energy

In some chemical reactions, reactants are chemically joined to yield a product. When reactants are joined, energy is used to build the chemical bonds holding the product together. Such processes are thermodynamically unfavorable. This type of reaction is called endergonic, meaning "energy in." In an **endergonic reaction**, the product contains more energy in its chemical bonds than the reactants (**Figure 6.3**).

Endergonic reactions *do not occur spontaneously*. Such nonspontaneous reactions are often referred to as "uphill" reactions because, like a ball rolling up a hill, nonspontaneous processes can only occur with help—an input of energy in addition to activation energy. Baking a cake is an example of an endergonic reaction. In order for batter to turn into cake, it must have an input of heat energy by baking in an oven. No matter how long it sits out on the counter at room temperature, it will not rise and become firm. It needs the added heat from the oven for its transformation to cake.

Endergonic reactions play important roles in your body, such as putting molecules together to build muscles, bones, and fat. Reactions like this, which use energy to build complex molecules from simpler molecules, are called **anabolic reactions** (AN-uh-BOL-ick).

In living things, exergonic and endergonic reactions are *coupled reactions*, meaning that they occur in conjunction with one another. In this way, the energy released when molecules are split in exergonic reactions is used to power the combining of molecules in endergonic reactions (**Figure 6.4**).

EXERGONIC REACTION      ENDERGONIC REACTION

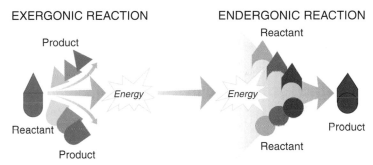

**Figure 6.4 Coupled reactions.** The energy released in an exergonic reaction is used to drive an endergonic reaction.

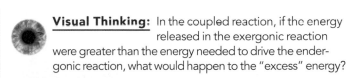

**Visual Thinking:** In the coupled reaction, if the energy released in the exergonic reaction were greater than the energy needed to drive the endergonic reaction, what would happen to the "excess" energy?

### CONCEPT CHECKPOINT

1. A dead tree in a forest is struck by lightning and bursts into flames. As the tree burns, the organic molecules that make up the tree are converted into inorganic molecules such as carbon dioxide, carbon monoxide, and water. (a) Is this process exergonic or endergonic? (b) Does it increase or decrease the entropy of the universe? (c) What provided the activation energy to get this reaction started? (d) Do the chemical reactants (the molecules of the tree) have higher or lower energy than the chemical products ($CO$, $CO_2$, $H_2O$, ash . . .)? What happened to this energy?

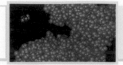

# How are chemical reactions regulated in living systems?

## 6.3 Enzymes lower the activation energy needed to get reactions started.

The flow of energy within an organism like yourself consists of a long series of coupled reactions. Energy is moved from one molecule to the next by means of exergonic and endergonic reactions. Some of this energy is stored in the bonds of molecules that make up the structure of your body. Some is freed to do cellular work. Some is transformed to heat energy.

These chains of reactions that move, store, and release energy are called **metabolic pathways**. Metabolic pathways accomplish jobs such as obtaining energy from the food you eat and repairing tissues that are worn out or damaged. By means of such pathways, your body and the bodies of all living things work to maintain order and avoid increasing entropy. Put simply, all the metabolic pathways in your body work to help you survive. However, for these chemical reactions to occur, they must be pushed over the hill of activation energy. In addition, they must be controlled so that they occur at the locations and times they are needed.

As you read earlier in this chapter, chemical reactions require activation energy to get started. Where does this energy come from? Heat can provide activation energy. If molecules are heated, they move very quickly and bump into one another forcefully. In such a situation, the kinetic energy of the bump can provide enough energy for activation. However, this situation rarely occurs in living organisms because heating would destroy important substances in organisms and would be a very nonspecific method of starting reactions. Instead, the chemical reactions of living things take place in the moderate temperatures of living cells. So how do reactions in living systems get pushed over the activation energy "hill?"

Living things have **enzymes** (ENN-zymes), that lower, or lessen, the activation energy needed, allowing chemical reactions to take place. Although RNA can act as an enzyme, most enzymes are proteins, and they act as catalysts. A *catalyst* is a substance that increases the rate of a chemical reaction but is not chemically changed by the reaction. Catalysts and enzymes lower the barrier of the activation energy (**Figure 6.5**). That is, they reduce the amount of energy needed for a reaction to occur by bringing together substrates (reactants) or by placing stress on the bonds of a single substrate, thereby making its bonds more reactive and easy to break. **Substrates** are the substances on which enzymes act. Enzymes are biological catalysts and control all the chemical reactions making up the metabolic pathways in living things. Life is therefore a process regulated by enzymes.

## 6.4 Enzymes bind substrates at active sites, which brings the substrates together.

If you were to look at the three-dimensional structure of an enzyme, you might notice several indentations on its surface. These surface depressions are called **active sites**, and they allow for the binding of other molecules, such as substrates, drug inhibitors, and vitamins. Three-dimensional shape is important for active site function, and it results from the tertiary or quaternary structure of the protein.

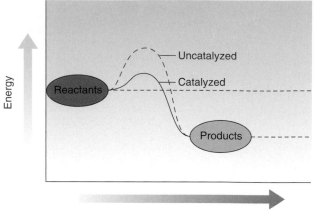

**Figure 6.5 The function of enzymes.** Enzymes catalyze particular reactions by lowering the amount of activation energy required to initiate the reaction.

 **Visual Thinking:** How does the graph show the difference in activation energy needed in a catalyzed reaction versus an uncatalyzed reaction?

The active sites are the locations on the enzyme where a reaction is catalyzed—where *catalysis* takes place. For catalysis to occur, a substrate must fit into the active site of an enzyme so that many of its atoms nudge up against atoms of the enzyme. The binding of a substrate causes the enzyme to adjust its shape slightly, allowing a better fit called an *induced fit* (**Figure 6.6**). Induced fit can be thought of as how a baseball (the substrate) fits into a catcher's mitt (the enzyme). The shapes are complementary at the site in the mitt where the ball is caught. The center of the mitt has roughly the shape of a ball but takes on that shape more exactly when the catcher catches a missed pitch. Just as only baseballs fit into catcher's mitts—not footballs or soccer balls, for example—substrates only fit into, or bind with, the active sites of certain enzymes.

Enzyme/substrate interactions are specific. Therefore, enzymes typically catalyze only one or a few different chemical reactions. Be-

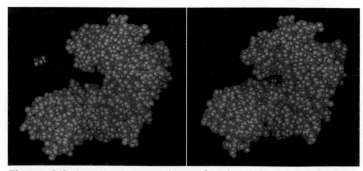

**Figure 6.6 A computer simulation for the induced fit of an enzyme** (hexokinase, shown in blue) and its substrate (glucose, shown in red). Note the differences in the shape of the enzyme before (left photo) and after (right photo) the substrate has bound.

cause of this specificity, each cell in your body contains 1000 to 4000 different types of enzymes.

Enzymes catalyze both endergonic and exergonic reactions. **Figure 6.7** shows how an enzyme catalyzes an exergonic reaction. In this example, a molecule of sucrose ① binds to the active site of an enzyme ②. After binding takes place ③, certain atoms within the active site of the enzyme chemically interact with the sucrose. This interaction causes a slight change in the shape of both the enzyme and the sucrose molecule. This change in shape places stress on the bonds joining the glucose and fructose subunits, reducing the amount of activation energy needed to be absorbed by the molecules for them to react. The bond then breaks ④, and the products are released from the enzyme ⑤. This reaction is a hydrolysis; note the addition of water.

Enzymes also catalyze reactions that bind two substrates together. In this situation, an enzyme acts like a dating service. It gets the substrates together so that their meeting does not occur simply by chance. Interestingly, Chinese biologists and chemists characterize enzymes in a similar way, calling them *tsoo mei*, or "marriage brokers."

**Figure 6.8** shows a reaction in which two reactants bond with one another. The energy needed for this reaction to take place is not shown. Without an enzyme, this reaction takes place only when the reactants collide by chance. Figure 6.8 shows how an enzyme facilitates this reaction. Both substrates bind at the active site on the enzyme. As binding occurs, the enzyme changes shape for a better fit ①. The binding of the substrates to the enzyme distorts and weakens their chemical bonds. In addition, their interaction with the enzyme orients the substrates ② so that their reactive sites are in contact with one another and react ③. The products from the reaction

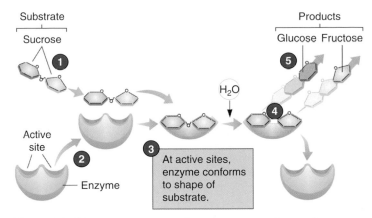

**Figure 6.7** How an enzyme catalyzes an exergonic reaction.

> **Visual Thinking:** How is the induced fit between the enzyme and the substrate depicted in this figure?

are released, and the enzyme returns to its normal shape ④. Moreover, enzymes do not get used up after one interaction takes place. They are only intermediaries in chemical reactions and are then released, available to catalyze yet another reaction ⑤.

### CONCEPT CHECKPOINT

2. List as many characteristics of enzymes as you can remember from your reading.

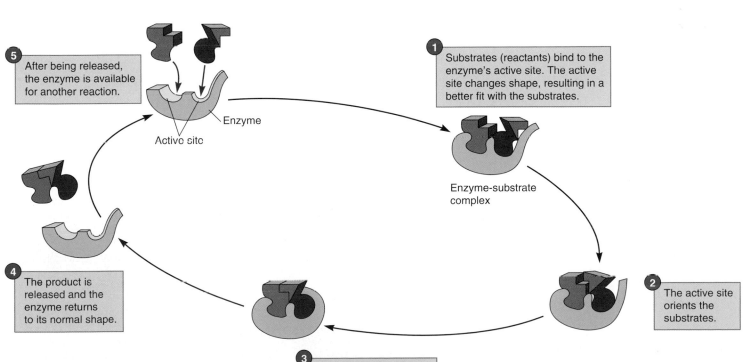

**Figure 6.8** Enzymes catalyze reactions between two substrates. Without an enzyme, reactants collide only by chance. With an enzyme, a reaction is more likely to occur.

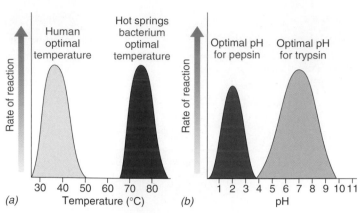

(a)

(b)

**Figure 6.9** The enzymes of living things work best in conditions specific to the organism.

 **Visual Thinking:** At about which temperature will reactions involving enzymes of the human body take place most quickly?

## 6.5 Environmental conditions affect enzyme activity.

The activity of an enzyme is affected by anything that changes its three-dimensional shape. If it loses its shape, or is denatured, an enzyme cannot bind with a substrate. Temperature is one factor that denatures protein. You can see the protein in an egg white become denatured when you cook it. The heat permanently changes the shape of the egg proteins, making the egg white firmer and more opaque. Two other environmental conditions that affect protein shape and therefore enzyme activity are pH and the binding of specific chemicals to the protein.

Most human enzymes function best between 36°C and 38°C, close to body temperature. Similarly, the enzymes of other living things work best at temperature ranges specific to the organism. Bacteria that live in hot springs, for example, have enzymes that function at temperatures between 74°C and 76°C (approximately 167°F) (**Figure 6.9a**). At temperatures colder or warmer than the enzyme's optimum range, the bonds between the amino acids that determine the enzyme's shape become weak or rigid, changing the shape of the active site and causing chemical reactions to stop.

Enzymes also work best within a particular range of pH values. As you may recall from Chapter 3, pH is a measure of the hydrogen ion concentration of a solution. Most enzymes work best within the range of pH 6 to 8, which is a medium to medium-low hydrogen ion concentration. Trypsin, for example, is a protein-digesting enzyme that functions well in the alkaline environment

of the human small intestine because the optimal pH of trypsin is 6 to 8. Some enzymes, however, function in environments with a high hydrogen ion concentration (low pH). Pepsin, another enzyme that digests proteins, functions well in the acidic environment of the human stomach because the optimal pH of pepsin is about 2 (Figure 6.9b). Pepsin is inactivated when it reaches the more basic environment of the small intestine, where trypsin takes over.

The activity of an enzyme is sensitive not only to pH and temperature but also to the presence of specific chemicals that bind to the enzyme and cause changes in its shape. By means of these specific chemicals, a cell is able to turn enzymes on and off. When the binding of a chemical changes the shape of the protein and the active site so that it can catalyze a chemical reaction, the chemical is called an *activator*. When the binding causes a change in the active site that shuts off enzyme activity, the chemical is called an *inhibitor*. Enzymes usually have special activator and inhibitor binding sites, and these binding sites are different from their active sites. The antibiotic penicillin is a common example of an enzyme inhibitor. Penicillin kills certain disease-causing bacteria by inhibiting the main enzyme involved in the synthesis of their cell walls.

Enzyme activity within an organism is often regulated by an end product acting as an inhibitor of its own synthesis. For example, suppose that an end product is synthesized via a series of chemical reactions (a metabolic pathway) involving multiple substrates, enzymes, and products (**Figure 6.10**). As the end product is synthesized and its concentration builds up in the cell, it binds to the inhibitor binding site of the first enzyme in the pathway (Enzyme 1). The first substrate (Substrate A) is no longer able to bind to the active site of Enzyme 1, and the pathway is shut down. This process is a type of **negative feedback**: In effect, the end product feeds

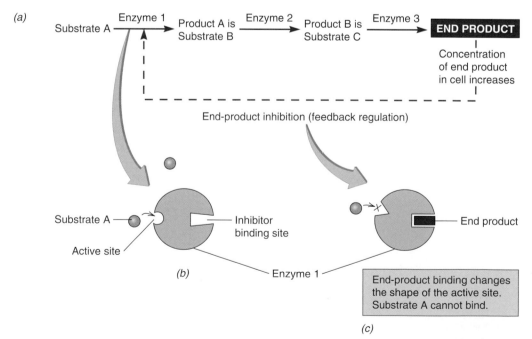

**Figure 6.10 End-product inhibition (feedback regulation).** An end product controls the rate of its own synthesis by acting on Enzyme 1, the catalyst for the first reaction in a pathway. (a) In this example, an end product is synthesized via a metabolic pathway that consists of three substrates and three enzymes. As the concentration of the end product in the cell increases, end-product inhibition increases. (b) Enzyme 1 has an active site where Substrate A binds and also has an inhibitor binding site. (c) As the end product accumulates, it binds to the inhibitor binding site of Enzyme 1, changing the shape of the active site, which prevents Substrate A from binding.

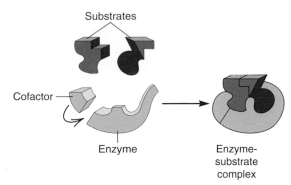

Substrates

Cofactor

Enzyme

Enzyme-
substrate
complex

**Figure 6.11** Cofactors help enzymes catalyze chemical reactions. Cofactors are nonprotein parts of enzymes. This illustration shows how coenzymes are often a part of the active site of the enzyme.

information back to the first enzyme in the pathway, shutting down the pathway when there is excess end product in the cell. When the concentration of end product in the cell decreases sufficiently, inhibition ends and synthesis resumes. Many enzyme-catalyzed processes within cells work in this way, temporarily increasing or decreasing the activity of enzymes in metabolic pathways and thus regulating the amount of end product that is produced.

## 6.6 Cofactors and coenzymes help enzymes regulate chemical reactions.

Enzymes often have additional parts in their structures that are made up of molecules other than proteins, as shown in **Figure**

**6.11**. These additional chemical parts are called **cofactors**. Cofactors help enzymes catalyze chemical reactions. For example, many enzymes have metal ions such as zinc, iron, or copper locked into their active sites. These ions help draw electrons from substrate molecules. One of your digestive enzymes, carboxypeptidase, breaks down proteins in foods by using a zinc ion to draw electrons away from the bonds being broken in the food. Many trace elements necessary for your health, such as manganese, help enzymes in this way.

When the cofactor is an organic (carbon-containing) molecule, it is called a **coenzyme**. Many of the vitamins that your body requires, such as members of the B-vitamin complex, are used to synthesize coenzymes that maintain health. These coenzymes perform many jobs in the body, playing key roles in the reactions of cellular respiration, amino acid synthesis, and protein metabolism. Table 27.2 lists the vitamins, their major functions, and rich food sources.

### CONCEPT CHECKPOINT

3. List three environmental factors that affect enzyme activity.
4. A thermostat in a home regulates heat by end-product inhibition, with sufficient heat being the end product. Develop an analogy between end-product inhibition in enzyme-catalyzed processes within cells and the thermostat in a house.
5. What roles do many of the trace elements and vitamins in your diet play in enzyme activity?

## *just wondering . . .*

*Questions students ask*

### What are "natural" vitamins? Are they better than other kinds of vitamins?

Your question is no surprise . . . whether shopping for vitamins on the Web, in health food stores, or at your local pharmacy, consumers are bombarded with claims that products are "all natural." Just what does that mean? Are other vitamins "unnatural?"

The word "natural," when applied to dietary supplements such as vitamins, usually means that the product came from a biological source. For example, some e-tailers of vitamins state that their products are derived from the livers of shark, halibut, and catfish. Are these vitamins better than those synthesized in the laboratory?

The answer is a simple "no." Vitamins from natural sources are the same as laboratory-made vitamins. In addition, the U.S. Food and Drug Administration (FDA) has long held that there is no nutritional difference between natural and synthetic vitamins. It has even barred companies that sell vitamins from making claims on their labels that natural vitamins are superior to synthetic vitamins.

Another question, and possibly a more important one, is whether you should take vitamin supplements. Nutritionists will tell you that if you follow the Food Guide Pyramid, which is illustrated and described in Chapter 27, then you should obtain enough vitamins from your diet if you are a healthy person with no underlying disease. People with certain digestive tract and metabolic diseases have higher needs for some or all vitamins, and there is increasing evidence that elderly persons also have higher needs. If you are a total vegetarian (a vegan), you may need more vitamins than your diet provides. In addition, pregnant women have higher nutritional needs and should seek advice from a health care provider. If you decide that your diet does not meet your vitamin needs, choose a multiple vitamin supplement that stays within the Dietary Reference Intake (DRI) for each vitamin because taking high doses can be harmful. DRIs can be found at the Web site of the Food and Nutrition Information Center of the U.S. Department of Agriculture. ●

*Are you wondering about a topic in biology and how it relates to your life? Submit your question by clicking the* Just Wondering *link in this text's companion Web site at* www.wiley.com/college/alters.

# How is energy captured and supplied in chemical reactions in living systems?

## 6.7 Adenosine triphosphate (ATP) is the primary energy carrier of living systems.

Many metabolic pathways in your body break down complex substances into simpler substances, releasing energy in the process. What happens to the energy that is released in these exergonic reactions? What if your body could only use this energy when it was released? You might run out of energy a few hours after lunch, leaving you without enough energy to eat another meal. Obviously, living things could not survive under such circumstances. They have evolved ways to capture the energy released from food after it is eaten and absorbed so that this energy can be used to do cell work in the future.

One way your body stores energy from the food you eat is by converting it to fat. Fat, however, cannot serve your cells' immediate energy needs. It is used for long-term storage. As a quick source of energy, you store excess carbohydrates as glycogen in your liver and skeletal muscles. Although glycogen can be more readily converted into energy than can body fat, it is still unable to meet immediate energy demands. The primary molecule used by cells to capture energy and supply it at a moment's notice is called **adenosine triphosphate** (uh-DEN-oh-SEEN try-FOS-fate), or **ATP**. Each ATP molecule is made up of three subunits:

- A five-carbon sugar called *ribose*.
- A double-ringed molecule called *adenine*.
- Three phosphate groups ($PO_4$) linked in a chain called a *triphosphate group*.

Together, the ribose sugar and the adenine rings are called *adenosine*. The "working end" of the molecule, however, is the triphosphate group, especially the terminal (end) phosphate (**Figure 6.12**).

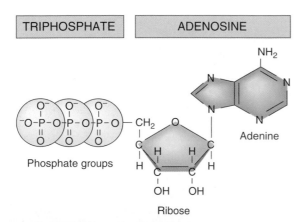

**Figure 6.12** Adenosine triphosphate (ATP) is the primary energy currency of the cell.

## 6.8 ATP stores, transfers, and releases energy.

The covalent bonds that link the phosphate groups in ATP are fairly unstable and require very little activation energy to be broken. Because of their instability and the negative charges that repel one another in the triphosphate group, ATP has a high *group transfer potential*—its terminal phosphate group easily transfers from one molecule to another during coupled reactions.

ATP transfers chemical energy as it transfers its terminal phosphate group. The entire ATP molecule contains energy "stored" in its molecular configuration. When ATP is coupled to endergonic reactions, the bond that links its terminal phosphate group to the rest of the ATP molecule is broken by hydrolysis and all of that energy is released. Some of the energy is transferred, along with the phosphate group, to another molecule in a process called **phosphorylation** (FOS-for-ih-LAY-shun). This process is extremely important in living systems because most cellular work depends on ATP energizing other molecules by phosphorylation.

ATP is such a universally used molecule to store and transfer energy in living systems that it is referred to as the energy "currency" of the cell. The analogy goes like this: ATP saves energy (like we save money) when it captures the energy that is released from exergonic reactions. Similarly, when cells need energy to drive endergonic reactions, the cell can "spend" ATP to provide this energy. In so doing, ATP connects the metabolic steps in the metabolic machinery of living systems.

**Figure 6.13** shows an example of an endergonic reaction coupled with the exergonic reaction of the hydrolysis of ATP.

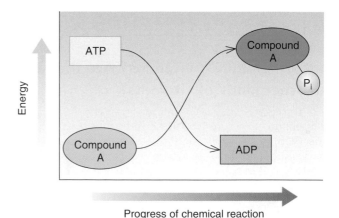

**Figure 6.13 A coupled reaction.** The energy that is released from conversion of ATP to ADP + $P_i$ (an exergonic reaction) is used to drive the conversion of Compound A to Compound A + $P_i$ (an endergonic reaction). The graph shows that ATP loses energy when it is converted to ADP, and Compound A gains energy when it is phosphorylated. The energy from one molecule is transferred to the other.

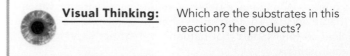

**Visual Thinking:** Which are the substrates in this reaction? the products?

**Figure 6.14 ATP plays a role in firefly light.** Fireflies, also called lightning bugs, produce light during a chemical reaction involving luciferin, a small organic molecule; luciferase, an enzyme; and ATP. When these molecules react in the presence of oxygen, luciferin glows. The chemical energy of ATP powers the reaction. Males actually "flash" females to attract a mate!

During this reaction, the terminal phosphate group of ATP is hydrolyzed and then bonded to Compound A. Energy is transferred to Compound A with the phosphate group. That is, Compound A *gains* energy when it acquires a phosphate group from ATP. Compound A might be a muscle fiber that contracts when phosphorylated, or an intermediate product formed as a plant stores the energy from sunlight in molecules of sugar during photosynthesis. ATP loses energy when its phosphate group is cleaved. Some of this energy is transformed into heat energy, which is unusable for work, but most of it is transferred to Compound A. The molecule that remains is called **adenosine diphosphate**, or **ADP**.

ATP is used to fuel a variety of cell processes. It is used, for example, when deep-water fishes, glow worms, and fireflies produce light (**Figure 6.14**). It is used to build larger, more complex molecules such as proteins from smaller, simpler molecules such as amino acids. It also provides energy for you to move, fueling the reactions that take place in your muscles causing muscle fibers to contract. Cells also use ATP to help move substances against a gradient—it is the fuel of active transport. Finally, the energy in ATP is changed into electrical energy, primarily in nerves.

ATP is constantly being split into ADP and phosphate to drive the endergonic, energy-requiring processes of the cell by means of coupled reactions. In addition, ATP is continually being made from ADP, phosphate, and energy during coupled exergonic reactions. The energy for these exergonic reactions comes from the breakdown of energy molecules such as fats and carbohydrates. (These processes are described in Chapter 8.) This recycling happens quickly and is depicted in **Figure 6.15**. If you could mark every ATP molecule in your body at one instant in time and then watch them, they would be gone in a flash. Most cells maintain a particular molecule of ATP for only a few seconds before using it.

Where does the energy come from that is captured in molecules of ATP? The ultimate source is energy from the sun that is captured by plants during the process of photosynthesis. Plants and the animals that eat them are food sources for other animals. Your body breaks down the food you eat into molecules such as glucose that are usable for fuel by your cells. Your cells break down these fuel molecules, releasing energy that is then captured in molecules of ATP.

---

**CONCEPT CHECKPOINT**

6. How do you supply your body with the energy required to make ATP?
7. For most life on earth, where does the energy needed to make ATP *ultimately* come from?
8. List at least four types of biological work that living organisms must perform in order to stay alive. Which of these requires energy from ATP?

---

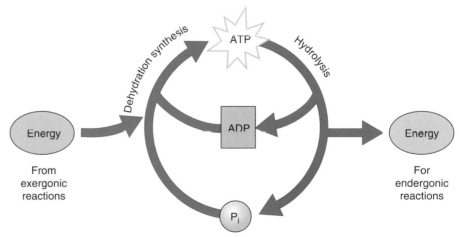

**Figure 6.15 The ATP–ADP cycle.** When ATP is used to drive the energy-requiring activities of living things, the high-energy bond that links the last phosphate group to the ATP molecule is broken, releasing energy. Cells always contain a pool of ATP, ADP, and phosphate ($P_i$), so ATP is continually being made from ADP, phosphate, and the energy released from exergonic reactions.

## CHAPTER REVIEW

## Summary of Key Concepts and Key Terms

### How is energy released and stored within cells?

**6.1** Energy exists as either **potential energy** (stored energy; p. 84) and **kinetic energy** (energy actively doing work; p. 84).

**6.1** The **laws of thermodynamics** (p. 84) describe patterns of regularity with respect to changes in energy that take place in the universe.

**6.1** Energy can be changed from one form to another during chemical reactions.

**6.1** The **first law of thermodynamics** (p. 84) states that energy cannot be created or destroyed; it can only be changed from one form or state to another.

**6.1** The **second law of thermodynamics** (p. 84) states that disorder in a closed physical system tends to increase; closed systems go from a state of higher useful energy to a state of lower useful energy.

**6.1** **Entropy** (ENN-truh-pee; p. 84) is a measure of the disorder (or randomness) of a physical system.

**6.2** Chemical reactions involve changing one substance into another.

**6.2** The substances entering a chemical reaction are **reactants** (p. 84). The changed reactants are **products** (p. 84).

**6.2** Energy is needed to initiate a chemical reaction. This energy is the activation energy.

**6.2** In **exergonic reactions** (EK-sur-GON-ick; p. 85), reactants are broken down to yield products. The products have less energy than the reactants; the excess energy is released.

**6.2** Reactions that release energy by breaking down complex molecules into simpler molecules are called **catabolic reactions** (CAT-uh-BOL-ick; p. 85).

**6.2** In **endergonic reactions** (ENN-der-GON-ick; p. 85), reactants are chemically joined to yield a product. The product has more energy than the reactants.

**6.2** Reactions that use energy to build complex molecules from simpler molecules are called **anabolic reactions** (AN-uh-BOL-ick; p. 85).

**6.2** In living systems, endergonic reactions are coupled with exergonic reactions to provide the energy needed to drive the endergonic process.

### How are chemical reactions regulated in living systems?

**6.3** Chains of chemical reactions that move, store, carry, and release energy in living systems are called **metabolic pathways** (p. 86).

**6.3** Special proteins called **enzymes** (p. 86) regulate the reactions in metabolic pathways.

**6.3** **Substrates** (p. 86) are the substances on which enzymes act.

**6.3** Living things control the time and place of chemical reactions by controlling the activity of the appropriate enzymes.

**6.3** Enzymes lower the activation energy and speed up reactions.

**6.4** Enzymes work by bringing substrates together so that they react more easily and by placing stress on certain bonds, which lowers the amount of activation energy that must be absorbed by the substrates to react.

**6.4** The **active sites** (p. 86) are the locations on the enzyme where a reaction is catalyzed.

**6.5** The activity of an enzyme is affected by anything that changes its three-dimensional shape, such as pH, temperature, and the binding of specific chemicals.

**6.5** The activity of enzymes is regulated by changes in enzyme shape; these changes result when activator and inhibitor molecules bind to specific enzymes.

**6.5** End-product inhibition is a type of **negative feedback** (p. 88): The end product feeds information back to the first enzyme in the pathway, shutting down the pathway when there is excess end product.

**6.6** Special nonprotein molecules called **cofactors** (p. 89), and **coenzymes** (p. 89) help enzymes catalyze chemical reactions.

### How is energy captured and supplied in chemical reactions in living systems?

**6.7** **Adenosine triphospate** (uh-DEN-o-seen try-FOS-fate; p. 90), or **ATP**, is made up of ribose, adenine, and three phosphate groups.

**6.8** ATP transfers chemical energy as it transfers its terminal phosphate group, a process called **phosphorylation** (FOS-for-ih-LAY-shun; p.

90). The molecule that remains is called **adenosine diphosphate** (uh-DEN-o-seen dye-FOS-fate; p. 91), or **ADP**.

**6.8** ATP is the energy currency of the cell in that it can capture energy from an exergonic reaction, carry this energy, and then "spend" it when needed.

## Level 1 — Learning Basic Facts and Terms

**Multiple Choice**

1. What role does ATP play in our cells?
   a. ATP provides energy to drive endergonic cellular processes.
   b. ATP provides energy to pump molecules into cells against their concentration gradients.
   c. ATP provides energy for muscle contraction.
   d. All of the above.

2. Why are enzymes important in metabolism?
   a. They provide energy to make ATP.
   b. They transport molecules across cell membranes.
   c. They shuttle phosphate groups from substrates to products.
   d. They speed up biochemical reactions.

3. Where does the energy come from to synthesize ATP from ADP in living cells?
   a. from the breakdown of ATP to ADP
   b. from chemical energy stored in food molecules like carbohydrates
   c. from digestion
   d. from chemical energy stored in carbon dioxide molecules

4. Frequently, an enzyme inhibitor is the last product produced in a biochemical pathway. This inhibitor binds to the enzyme catalyzing the first step in a series of chemical reactions and prevents it from binding with its substrate. This process is an example of:
   a. enzyme regulation.
   b. end-product inhibition.
   c. negative feedback.
   d. all of the above.

5. Which of the following is important in determining enzyme function?
   a. chemical structure of the enzyme
   b. three-dimensional structure of the active site
   c. temperature of the enzyme's environment
   d. all of the above

6. Enzymes
   a. can speed up reactions by eliminating the need for activation energy.
   b. are usually nucleic acids.
   c. often require coenzymes to function.
   d. are rarely very specific in the reactions they catalyze.

**True–False**

7. Enzymes are biological catalysts that lower the activation energy of a reaction by bringing the reactants into proper orientation for the reaction to take place.

8. Enzymes are biological catalysts that speed up reactions by making an endergonic reaction an exergonic reaction.

9. The first law of thermodynamics could be interpreted to mean that the total amount of energy in the universe is constant.

10. According to the laws of thermodynamics, it is *not* possible to create a system that is 100% efficient in transforming energy from one type to another.

## Level 2 — Learning Concepts

1. If one were to compare the action of enzymes in metabolic pathways to the assembly-line process of building an automobile,
   a. the automobile parts would be analogous to the _____.
   b. the finished auto would be analogous to the _____.
   c. assembly-line workers would be analogous to _____.
   d. the assembly-line tools used by the workers would be analogous to _____.

2. The hypothetical biochemical reaction A + B → C + D requires 3.2 kcal/mole of energy to proceed. (Think of a kcal/mole simply as a unit of energy.) If this reaction were coupled with the reaction ATP → ADP + P (which releases 7.2 kcal/mole of energy), what would happen to the "leftover" energy?

3. Explain why the cells of your body require so many different enzymes (1000 to 4000 per cell).

4. ATP can be thought of as the "everyday pocket change of energy currency," while food molecules, such as glucose, are the $500 bills. Explain this analogy.

5. What do enzyme activators and enzyme inhibitors have in common? Which plays a key role in the process of negative feedback?

6. One hears the word "metabolism" used in many ways, such as metabolic rate, metabolizing a drug, and high metabolism. In the most complete sense of the word, what is metabolism?

7. The graph below represents the changes in energy of a chemical reaction in the absence of an enzyme. What does "A" represent? Overall is this an exergonic or endergonic reaction? What would this graph look like in the presence of an enzyme?

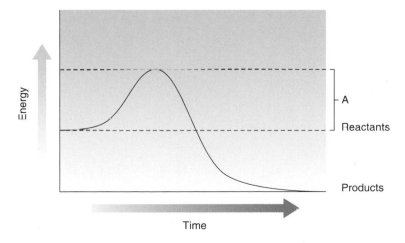

**Critical Thinking and Life Applications**

1. A student in your class comments that "Living organisms seem to contradict the second law of thermodynamics. After all . . . living organisms create highly complex ordered cells and tissues from less ordered organic molecules." How might you respond to this student?

2. The result of many genetic disorders is a defect in an enzyme. In what way would a defective enzyme affect an organism if the defective enzyme was involved in metabolic reactions that produce ATP?

3. The graph at right represents the effect of temperature on the activity of a human digestive enzyme. Explain this graph in terms of what you know about how enzymes function with respect to temperature. What temperature in degrees Celsius do you think corresponds to the peak of the graph? How might one measure enzyme activity?

4. Energy is continually entering the Earth's atmosphere from the sun. Explain the role played by photosynthetic organisms in capturing this energy and contributing to the biological flow of energy on Earth.

5. Industrial pollution can change the pH of a pond or river to make the water more acidic. How can this affect the metabolic pathways of the plants that live in the water?

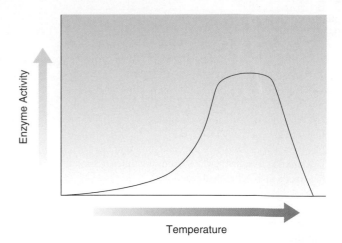

Enzyme Activity

Temperature

---

**In The News** | **Critical Thinking**

**MIGHTY MINI MOTOR**

Now that you understand more about how ATP powers chemical reactions, reread this chapter's opening story about research into the development of biological mini motors powered by ATP. To understand this research better, it may help you to follow these steps:

1. Review your immediate responses to Guo's research on the mini motor that you wrote when you began reading this chapter.

2. Based on your current understanding, again summarize the main point of the research in a sentence or two.

3. What questions do you now have about this research that the opening story leaves unanswered?

4. Collect new information about the research. Visit the *In The News* section of this text's companion Web site at www.wiley.com/college/alters and

watch the "Might Mini Motor" video. Then use the "summary" link to read the accompanying story and access related links. Use this information, the links provided, and other online and library resources to answer your questions and find updates about this research topic. State the sources of your information. Explain why you think the information is accurate. Also determine if the information expresses a particular point of view or is biased in any way.

5. What in your view is the most significant aspect of this research? Give reasons for your opinion and for any changes in your ideas based on the additional information you have collected and the analysis you have done.

# STORING ENERGY FROM THE SUN: PHOTOSYNTHESIS

## In The News | Plant Impostors

Are bacteria masquerading as plants? Not really . . . but Edward DeLong (shown in the photo) and his research team at the Monterey Bay Aquarium Research Institute in California have found something close to such a situation. The team discovered ocean-dwelling bacteria that have a rhodopsin-based mechanism for making food using the energy of sunlight, much like plants make food using chlorophyll. Rhodopsin is a red pigment found in the retina, which lines the back of our eyes. In our eyes, this pigment undergoes a chemical alteration when activated by light and sends signals to the brain, which are interpreted as images. However, these ocean-dwelling bacteria use rhodopsin in a different way: to capture light during photosynthesis. Photosynthesis is the process of transforming light energy into stored chemical energy.

Like the rhodopsin in our eyes, the rhodopsin in these bacteria can absorb various wavelengths of light. This ability is useful to the bacteria because not all wavelengths of light penetrate very deeply into the ocean. For example, nearly all red light is converted to heat in the first few meters of ocean water. Violet, blue, and green wavelengths of visible light reach the deepest—to about 100 meters below the surface. However, only 1% of the violet, blue, and green light that strikes the surface of the water makes it to this depth.

DeLong and his research team suggest interesting applications of this previously unrecognized photosynthetic pathway, such as developing microbial fuel cells that run on light. You can view the "Plant Impostors" video to learn more about these ideas and this research by visiting the *In The News* section of this text's companion Web site at www.wiley.com/college/alters.

*Write your immediate reaction to DeLong's discovery of a rhodopsin-based photosynthetic pathway in ocean-dwelling microbes: first, summarize the main point of the research in a sentence or two; then suggest what you think the significance of this research is. You will have an opportunity to reflect on your responses and gather more information on this topic in the* In The News *feature at the end of this chapter. In this chapter, you will learn more about the process of photosynthesis.*

## CHAPTER GUIDE

### What is the source of energy for life?

**7.1** Sunlight is electromagnetic energy.

**7.2** Pigments in plants, algae, and cyanobacteria absorb light energy.

**7.3** Autotrophs and heterotrophs release the energy in food via cellular respiration.

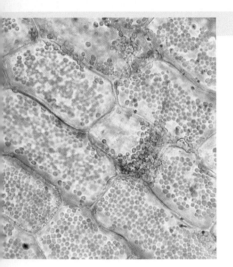

### How do autotrophs store energy from the sun?

**7.4** Photosynthesis stores energy from the sun in carbohydrates.

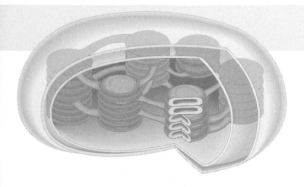

### What are the reactions that enable autotrophs to store energy from the sun?

**7.5** Light-dependent reactions produce ATP and NADPH.

**7.6** Light-independent reactions produce carbohydrates.

# What is the source of energy for life?

Imagine what might happen if the light of the sun were blocked from the Earth forever. You might predict that the temperature of the Earth's atmosphere would drop, and you would be correct. Obviously, you might also think that the Earth would become shrouded in darkness. That is true, too, but more devastating consequences would occur. Life on the surface of the Earth—as we know it today—would totally and completely end.

Most living organisms of Earth depend on the sun and have done so for billions of years. Life depends on a process called photosynthesis, which is powered by the sun. By means of photosynthesis, energy from the sun is captured and stored in carbohydrate molecules, which is a primary source of fuel for living things. A byproduct of photosynthesis is oxygen, another life requirement for all oxygen-breathing organisms, such as yourself. **Figure 7.1** is a photo of a waterweed, *Elodea canadensis*, releasing oxygen bubbles from its leaves as it photosynthesizes underwater.

## 7.1 Sunlight is electromagnetic energy.

How do photosynthetic organisms harvest light energy from the sun? How do they use this energy to create chemical bonds in food molecules? First, it is important to understand what light energy is.

As you may recall from Chapter 6, energy is able to create change—to do work. Sunlight is a form of energy known as *elec-*

*tromagnetic energy*, also called *radiation*. Electromagnetic energy (sunlight) travels as waves although it also behaves like individual particles—discrete packets of energy, which are called *photons*.

The array of light waves coming from the sun vary in length. Their lengths are measured from the crest of one wave to the crest of the next and are expressed in meters or in billionths of meters (nanometers, or nm). The shortest wavelengths are gamma rays; the longest are radio waves.

The full range of electromagnetic radiation in the universe is called the *electromagnetic spectrum* (**Figure 7.2**). Visible light is only a small part of this spectrum, but it, too, is made up of many different wavelengths. Remember the last rainbow you saw? Its array of colors was caused by the separation of the various wavelengths of visible light as they passed through tiny droplets of water in the air.

## 7.2 Pigments in plants, algae, and cyanobacteria absorb light energy.

We still have not answered the question, how do photosynthetic organisms harvest light energy from the sun? The simple answer is pigments.

Plants contain pigments in the chloroplasts of their cells. **Pigments** are molecules that absorb some visible wavelengths of light and transmit or reflect others. (Transmitting means allowing

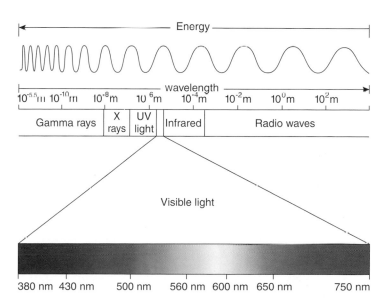

**Figure 7.1** Oxygen production. The waterweed *Elodea canadensis* releases oxygen bubbles from its leaves as it photosynthesizes underwater.

**Figure 7.2** The electromagnetic spectrum. Visible light represents only a small part of the electromagnetic spectrum, that between 380 and 750 nanometers (nm).

light to pass through.) Pigments are responsible for all the colors you see, including the greens, reds, yellows, and other colors of plants.

How do pigments absorb light? How do they absorb only certain wavelengths of light? It works like this: a pigment molecule captures light energy when a photon boosts one of the electrons in the pigment molecule to a higher energy state. This electron is then said to be excited. Boosting an electron to a higher energy level requires just the right amount of energy—no more and no less. For example, when you climb a ladder, you must raise your foot just so far to climb a rung, not one centimeter more or less. Similarly, specific atoms can absorb only certain photons of light—those that correspond to available energy levels. Therefore, a given atom or molecule has a characteristic range, or *absorption spectrum*, of photons it is capable of absorbing depending on the electron energy levels that are available in it.

For example, artists' red paint looks red because it is manufactured using a pigment that absorbs variable amounts of different wavelengths, including red, but reflects light enriched in red wavelengths. These red wavelengths bounce off the pigment and are reflected back to your eyes. Pigments located in special cells within your eyes then absorb the photons of this red light. The resulting electron excitations ultimately generate nerve impulses that are sent to the brain. The brain interprets these impulses as red. Similarly, white objects absorb no light and reflect all the wavelengths of visible light to your eye. A black object absorbs all wavelengths, reflecting none. You probably have noted that wearing white clothing is cooler in the summer sun than wearing black clothing.

Organisms have evolved a variety of pigments. Among the many groups of plant pigments, two groups that are important to

the process of photosynthesis are the **carotenoids** (kuh-ROT-uh-noids) and the **chlorophylls**. Carotenoids absorb photons of green, blue, and violet wavelengths and reflect red, yellow, and orange. Chlorophylls, as shown in **Figure 7.3**, absorb photons of violet-blue and red wavelengths and reflect green and yellow.

Chlorophyll is used as the primary light gatherer in all plants and algae and in certain photosynthetic bacteria—the cyanobacteria. The carotenoids are important, however, because they absorb wavelengths of light that chlorophyll cannot, and then pass the energy to chlorophyll, thereby increasing the spectrum of light that can be absorbed. Chlorophyll masks the presence of the carotenoids, which become visible only when chlorophyll breaks down. This breakdown occurs in autumn (in the areas of the world that have seasons) as the days shorten and the amount of sunlight decreases. As a result, the carotenoids become visible, providing the magnificent yellows and oranges of fall foliage.

---

### CONCEPT CHECKPOINT

1. Why are chlorophyll and the carotenoids called pigments?
2. How are photosynthetic pigments similar to the antenna on your car radio?

---

## 7.3 Autotrophs and heterotrophs release the energy in food via cellular respiration.

So which members of the living world are able to carry out photosynthesis? They are the plants, algae, certain protists, and certain prokaryotes. Organisms that produce their own food by *photosynthesis*, along with a few others that use chemical energy in a similar way, are called **autotrophs** (AW-toe-trofs). The word "autotroph" literally means "self-feeder." All organisms live on the food produced by autotrophs, including the autotrophs themselves.

Chemosynthetic autotrophs use energy derived from inorganic chemicals to make their own food. These prokaryotes live in places where the inorganic chemicals they use are abundant, such as near deep-sea thermal vents, acidic thermal springs, and acid mine drainage streams.

Photosynthetic autotrophs, which are the subject of this chapter, make their own food by harvesting light energy from the sun and changing it into stored chemical energy within food. More specifically, the energy from the sun powers the synthesis of glucose molecules, using the raw materials of carbon dioxide and water (**Figure 7.4 ❶**). Oxygen is a byproduct of this process ❷. Photosynthetic organisms are critical to life on Earth not only because they manufacture food for themselves as well as other organisms but also because they generate large amounts of oxygen.

The glucose produced during photosynthesis ❸ is a sugar that plants and other photosynthetic organisms use as a source of energy (from which they manufacture ATP; see Chapter 6) and as building blocks to construct larger molecules. From glucose, plants and other autotrophs manufacture cellulose (found in the

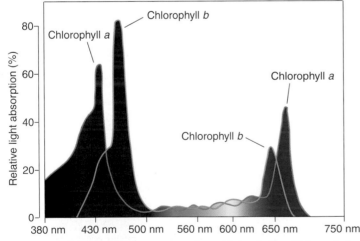

**Figure 7.3 Absorption spectra for chlorophylls.** Chlorophylls, such as chlorophyll *a* and chlorophyll *b*, absorb predominantly violet-blue and red light in two narrow bands of the spectrum.

**Visual Thinking:** A rose has green leaves during seasons of active growth and photosynthesis. Why? Use the graph to help you answer the question.

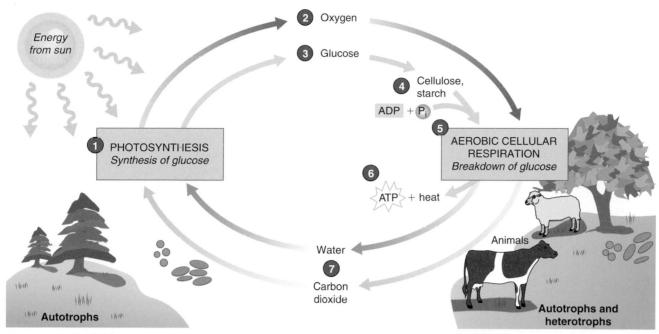

**Figure 7.4 The major metabolic pathways of autotrophs and heterotrophs.** Photosynthetic autotrophs make their own food by using carbon dioxide from the atmosphere, water, and energy from the sun, and changing it into stored chemical energy within food, shown here as glucose. Oxygen is a byproduct of those reactions. Au-totrophs and heterotrophs use the food made via photosynthesis as an energy source. They unlock the stored energy through the chemical reactions of cellular respiration, generating ATP, which can be used to do cell work. Byproducts of cellular respiration are water and carbon dioxide.

**Visual Thinking:** Which biochemical process do you think evolved first on primitive Earth: photosynthesis or cellular respiration. Why?

cell walls of many autotrophs) and starch **4** . Starch is a storage form of sugar, which autotrophs can convert to glucose when they need energy. Autotrophs also use inorganic nutrients such as phosphates and nitrates to build large organic molecules from glucose. This is why plants need fertilizer, which contains phosphates and nitrates.

Organisms, such as humans, that cannot produce their own food are called **heterotrophs** (HET-ur-oh-trofs)—literally, "other-feeders." Heterotrophs consume other organisms and, indirectly, the food produced by autotrophs. At least 95% of species—all animals, all fungi, and most protists and bacteria—are heterotrophs. They live by feeding on the chemical energy that is fixed (incorporated into carbohydrates) by photosynthesis. Like photosynthetic autotrophs, heterotrophs use glucose to make ATP.

The energy that organisms use to make ATP, then, is actually energy from the sun, captured during photosynthesis. By eating plants, and by eating animals that eat plants or other photosynthetic autotrophs such as algae, other animals can harvest the energy from the sun that was originally captured and stored by autotrophs.

The primary process by which organisms unlock the energy in food, which originally comes from the sun, is called *cellular respiration* **5** . During this complex series of chemical reactions, ATP is made **6** by breaking down nutrient molecules such as glucose to release the energy stored in its bonds, capturing it in ATP. Some en-

ergy is released as heat. We can see the results of this energy transfer in a variety of situations, one of which is shown in **Figure 7.5**.

Aerobic cellular respiration is oxygen dependent. *All* organisms that use oxygen—including plants, which produce the oxygen that they (and we) need—break down nutrient molecules by

**Figure 7.5 Metabolic heat.** A skunk cabbage generates enough heat internally, through cellular respiration, to melt surrounding snow when it emerges in the spring.

means of *aerobic respiration*. The byproducts of aerobic respiration—carbon dioxide and water **7**—are the raw materials of photosynthesis. As Figure 7.4 shows, the two processes of photosynthesis and cellular respiration are intimately connected, with each fueling the other. These major metabolic pathways are the energy crossroads of cells. Cellular respiration is discussed in more detail in Chapter 8.

**CONCEPT CHECKPOINT**

**3.** Fill in the blanks.

Plants are _____, using energy from the _____ to manufacture _____, with _____ as a byproduct. Animals are _____; they consume plants (or other organisms that consume plants). By the process of _____, animals unlock the energy stored in their food, capturing it in molecules of _____, with _____ and _____ (the raw materials of photosynthesis) as byproducts.

# How do autotrophs store energy from the sun?

## 7.4 Photosynthesis stores energy from the sun in carbohydrates.

We still have one question left unanswered: How do plants use the energy from the sun to create chemical bonds in food molecules? This question is echoed above: How do autotrophs store energy from the sun? The answer is **photosynthesis**, the process whereby light energy is transformed into stored chemical energy. It is a complicated process, but this section provides a simplified overview. Sections 7.5 and 7.6 describe the process in more detail.

The process of photosynthesis is described chemically in the following equation:

$$6CO_2 + 6H_2O \xrightarrow[\text{energy}]{\text{Light}} C_2H_{12}O_6 + 6O_2$$

Carbon  Water          Glucose  Oxygen
dioxide

This chemical equation represents the beginning and end points in a series of chemical reactions and merely summarizes complex events, much as you might summarize the details of a book in a single paragraph. It shows that the net effect of photosynthesis is the production of one molecule of glucose and six molecules of oxygen from six molecules of carbon dioxide and six molecules of water in the presence of sunlight. The main product of photosynthesis is actually glyceraldehyde-3-phosphate (G3P), which can be synthesized into glucose, sucrose, complex carbohydrates, and other organic compounds. Glucose is used to simplify our discussion and to show the relationship between photosynthesis and respiration.

A more accurate summary of the chemical reactions of photosynthesis is represented by the following equation:

$$6CO_2 + 12H_2O \xrightarrow[\text{energy}]{\text{Light}} C_6H_{12}O_6 + 6H_2O + 6O_2$$

Here, water appears on both sides of the equation—as a substrate and as a product. Writing the equation in this way emphasizes that the water molecules on the left side of the equation and those on the right side are *not* the same water molecules. The six molecules of oxygen (12 atoms) produced by photosynthesis are derived from those in the water molecules on the left side of the equation. Plants, algae, and the

cyanobacteria produce oxygen in this way. In addition, the water molecules on the right side of the equation derive their oxygen atoms from the oxygen in carbon dioxide. Glucose, which is an organic molecule, is constructed from the carbon and oxygen in carbon dioxide and the hydrogen in water, which are both inorganic molecules. During the process of photosynthesis, inorganic molecules are broken apart and their atoms are reshuffled and put back together again to form organic molecules—molecules of living things. Photosynthesis, then, is a process during which the living and nonliving worlds meet.

### Light-dependent reactions take place in sunlight

The complex events of photosynthesis summarized by the previous equation involve two sets of chemical reactions as shown in **Figure 7.6**. During the first set of reactions **1**, adenosine triphosphate (ATP) and the reduced form of nicotinamide adenine dinucleotide phosphate (NADPH, an electron carrier, see next paragraph) are formed using energy captured from sunlight. These molecules store the energy harvested from the sun and are used in the synthesis of sugars in the second set of reactions. The reactions that produce ATP and NADPH take place only in the presence of sunlight and are therefore called the **light-dependent reactions**.

### Electron carriers transfer electrons

NADPH is an *electron carrier* and therefore is an important molecule in both the light-dependent and light-independent reactions of photosynthesis. Electron carriers transfer electrons as hydrogen atoms from one molecule to another during oxidation-reduction, or redox reactions (see Chapter 3). In biological systems, electrons often do not travel alone from one atom to another but instead take along a proton ($H^+$). Together the electron and the proton are a hydrogen atom. $NADP^+$ (the oxidized form of NADPH) is only one of many electron carriers that are key players in the processes of photosynthesis and cellular respiration. The $^+$ sign indicates that this molecule is oxidized; that is, it has given up electrons. It can accept two electrons and one proton ($H^+$) to become reduced to NADPH. Notice that NADPH bears no positive charge because it has been reduced; that is, it has accepted an electron. The extra $H^+$ and electron are shown by the addition of one H to NADP. The electrons that $NADP^+$ accepts contain a great deal of energy; thus, they produce molecules that carry energy in the form of energized electrons.

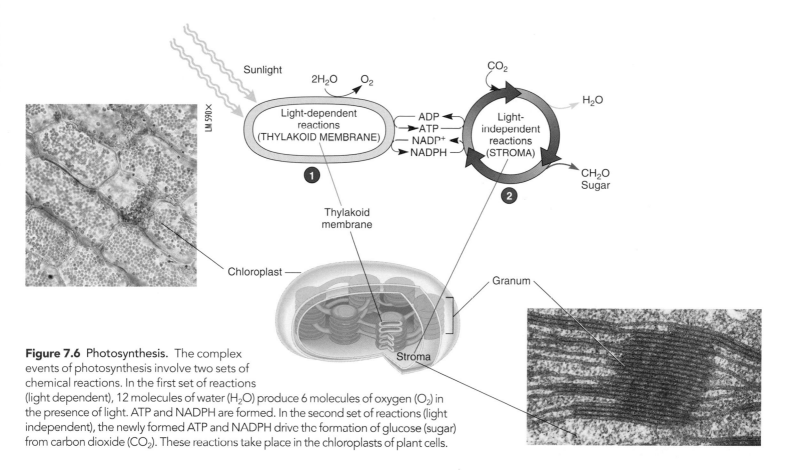

**Figure 7.6 Photosynthesis.** The complex events of photosynthesis involve two sets of chemical reactions. In the first set of reactions (light dependent), 12 molecules of water ($H_2O$) produce 6 molecules of oxygen ($O_2$) in the presence of light. ATP and NADPH are formed. In the second set of reactions (light independent), the newly formed ATP and NADPH drive the formation of glucose (sugar) from carbon dioxide ($CO_2$). These reactions take place in the chloroplasts of plant cells.

## Light-independent reactions do not need sunlight

A second series of reactions uses the ATP and NADPH formed during the light-dependent reactions to provide energy for the formation of glucose from carbon dioxide. These reactions are called the **light-independent reactions** (Figure 7.6 ❷ ) because there is no direct involvement of light in the reactions.

## Photosynthetic reactions take place on and in special structures

In photosynthetic prokaryotes, all photosynthetic reactions take place on infoldings of the cell membrane. In photosynthetic eukaryotes (plants, algae, and single-celled protists called euglenoids), all photosynthetic reactions take place within cell organelles called **chloroplasts**. The light-dependent reactions take place on membranes within the chloroplasts, and the light-independent reactions take place in the surrounding fluid.

Chloroplasts are shown in Figure 7.6. The photo to the left is a light micrograph of live plant cells; the small dots within the cells are chloroplasts. One chloroplast is illustrated to the right of the photo, and the inner chloroplast structures are shown to the right of the illustration in an electron micrograph.

Chloroplasts were probably derived evolutionarily from photosynthetic bacteria that were like the cyanobacterium shown in **Figure 7.7**. The internal membranes of these organelles are organized into a system of flattened sacs called **thylakoids** (THIGH-leh-koidz). Stacks of thylakoids are referred to as

**grana** (GRA-nuh). Each thylakoid is a closed compartment. Chorophyll, accessory pigments, and enzymes are embedded in the thylakoid membranes. The light-dependent reactions take place here. Surrounding the thylakoids is a fluid called the **stroma** (STROH-muh), which contains the enzymes of the light-independent reactions.

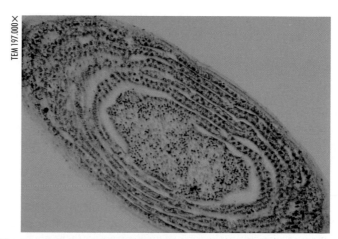

**Figure 7.7 Cyanobacteria.** Also called blue-green bacteria, cyanobacteria are a diverse and widely distributed group of photosynthetic bacteria. The cytoplasm of this cyanobacterium is filled with photosynthetic membranes, which are called thylakoids, even though they are not discrete organelles as in plants.

In summary, the main points of photosynthesis are as follows:

- The light-dependent reactions of photosynthesis perform the important job of changing the energy in sunlight into usable chemical energy.
- That energy is held in molecules of ATP and NADPH, which contribute the energy in the light-independent reactions to construct sugars from carbon dioxide.
- During the synthesis of sugars, chemical energy is transferred from ATP and NADPH to the sugars.
- The energy-rich sugars, which are the products of photosynthesis, are used to fuel the activities of life for both the autotrophs that produce the sugars and the heterotrophs that consume the autotrophs.

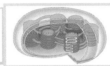

## CONCEPT CHECKPOINT

4. *LEO the lion says GER (grrrrrr)* is a good mnemonic (memory) device often used for remembering what happens to electrons during oxidation-reduction reactions. What do you think each of the letters in LEO and GER stands for?
5. What molecules produced in the light-dependent reactions are used in the light-independent reactions to make sugar?
6. Besides the molecules produced in the light-dependent reactions, what inorganic molecule obtained from the plant's environment is also needed in the light-independent reactions?

# What are the reactions that enable autotrophs to store energy from the sun?

## 7.5 Light-dependent reactions produce ATP and NADPH.

Now that you have the big picture of photosynthesis, we will discuss the process in more detail. First, we will discuss the light-dependent reactions, which involve the production of ATP and NADPH. The first event leading to the production of these molecules is the capturing of a photon of light by a pigment.

### Pigment molecules funnel energy to chlorophyll

In all but the most primitive prokaryotes, individual networks of chlorophylls, carotenoids, and other pigment molecules capture light. These networks, located in the thylakoid membranes of the chloroplasts, are called *photocenters*. The pigment molecules within the photocenters are arranged in such a way that they act like sensitive antennae. These antennae capture and funnel photon energy to special molecules of chlorophyll called **chlorophyll *a***. The variety of pigments in the photocenters allow a plant to capture energy from a range of wavelengths of light and to use it to power photosynthesis.

When light of the proper wavelength (photon) strikes any pigment molecule of the photocenter as shown in **Figure 7.8**, one of its electrons is boosted to a higher energy level. This energized pigment molecule passes the energy to a neighboring molecule within the photocenter. The energy is passed from one pigment molecule to the next, until it reaches the molecule of chlorophyll *a* that will participate in photosynthesis. This molecule is called either *P700* or *P680*. The *P* stands for *pigment*. The number refers to the wavelength at which this molecule of chlorophyll *a* absorbs light. Both P700 and P680 absorb wavelengths in the far-red portion of the light spectrum.

A simple analogy to the form of energy transfer in the photocenter can be shown with pool balls. If you place three pool balls in a straight line touching each other and hit the first one with the cue ball, shooting in the direction of the line, the last ball in the line will be propelled by the hit, but the other two will stay in place. The energy is

transferred through the first two balls to the most distant one. In a similar way, the pigment molecules of the photocenter channel energy in the form of excited electrons to a molecule of chlorophyll *a*.

### Plants and algae use a two-stage photosystem to carry out the light-dependent reactions

Plants as well as algae have two different types of photosystems that play a role in photosynthesis. These photosystems are named in order of their discovery: photosystem I and photosystem II. In **photosystem I**, energy is transferred to a molecule of P700. In **photosystem II**, energy is transferred to a molecule of P680.

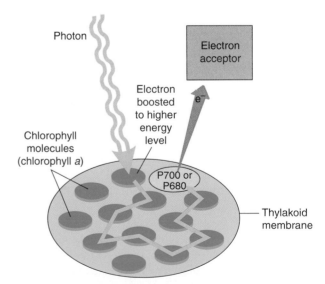

**Figure 7.8 How photocenters work.** When a photon strikes any pigment molecule of the photocenter, one of its electrons is boosted to a higher energy level. The energy is passed from one pigment molecule to the next, until it reaches the molecule of chlorophyll *a* that will participate in photosynthesis. This molecule is called either *P700* or *P680*.

## In noncyclic electron flow, both photosystems work together to harvest energy

Both photosystems absorb light at the same time (**Figure 7.9** ❶ and ⑫). However, it is easiest to first describe what happens in photosystem II to explain the events of the light-dependent reactions.

When four photons of light are absorbed by pigment molecules in photosystem II ❶, the energy is funneled to P680 ❷. This energy causes four electrons to be ejected from the P680 molecule ❸. The electrons are captured sequentially by an electron acceptor located within the thylakoid membrane ❹. However, an electron "hole" is now left in the P680 molecule. The P680 pulls four electrons from two molecules of water ❺, which is normally a hard task to accomplish. The P680 strongly attracts these electrons, however,

causing each molecule of water to be split into two hydrogen ions and an oxygen atom. Oxygen atoms from two split molecules of water quickly join to form a molecule of oxygen ❻—a byproduct of photosynthesis and the gas that helps keep us alive.

Meanwhile, the electron acceptor passes the four energized electrons from the P680 to the next in a series of electron acceptors and carrier proteins called an **electron transport chain** ❼. Here, electrons are passed from one carrier protein to another. As each of these carriers accepts electrons in turn, the electrons fall to lower energy levels, releasing energy in the process. The energy released from the electrons being passed along the chain pumps protons ($H^+$) from the stroma through the thylakoid membrane and into the interior of the thylakoid compartment. This proton pumping ❽ and the ATP formation described next are shown in the box at the bottom of

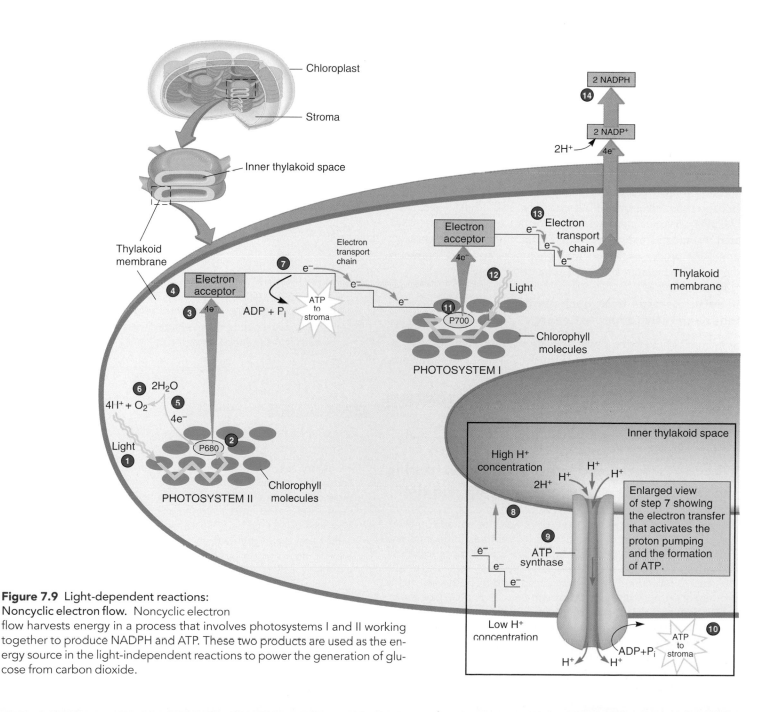

**Figure 7.9** Light-dependent reactions: Noncyclic electron flow. Noncyclic electron flow harvests energy in a process that involves photosystems I and II working together to produce NADPH and ATP. These two products are used as the energy source in the light-independent reactions to power the generation of glucose from carbon dioxide.

the illustration. A high concentration of protons builds up within this thylakoid space, and the protons begin to diffuse back across the membrane through a transport protein, ATP synthase **9**. As the protons travel through these channels, enzymes couple the energy from the movement of the protons to the reaction of bonding inorganic phosphate ($P_i$) to ADP, forming ATP **10**. This process in which a cell uses the potential energy in a concentration gradient to make ATP is called **chemiosmosis** (KEM-ee-oz-MOH-sis). In 1978, British biochemist Peter Mitchell was awarded the Nobel Prize for his description of this process.

When the electrons finish their journey through the electron transport chain (see **7**), they are then accepted by the chlorophyll *a* molecule of photosystem I—P700 **11**. This chlorophyll molecule, as mentioned previously, has its electrons energized by photons of light at the same time as P680 is energized **12**. So it, too, develops electron holes as energized electrons are ejected from their shells. The electrons of P680, now devoid of their boost of energy, fill these electron holes. Meanwhile, the energized electrons of P700 have been passed down the electron transport chain of photosystem I **13**. ATP molecules are not generated in this transport chain. Instead, the last acceptors are two molecules of $NADP^+$, each of which picks up a proton in addition to two electrons to become reduced to NADPH **14**. Because the electrons it accepts are still highly energized, NADPH serves as a storage depot for this energy and is considered a source of reducing power for the light-independent reactions. The processes that occur in photosystems I and II make up the light-dependent reactions of eukaryotic photocenters.

### In cyclic electron flow, each photosystem works independently to harvest energy

Sometimes photosystem I works on its own in a modified way, independent of photosystem II (**Figure 7.10**). In this situation,

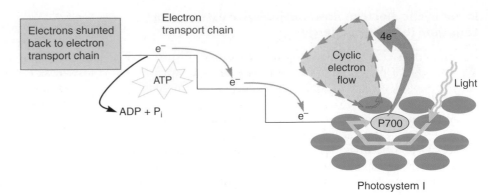

**Figure 7.10 Light-dependent reactions: Cyclic electron flow.** Cyclic electron flow harvests energy in a process that involves photosystem I working independently of photosystem II. This method of generating energy produces ATP, but it does not directly produce NADPH, as does noncyclic electron flow.

$NADP^+$ does not accept the excited electrons ejected from the P700 molecule. Instead, they are shunted to the electron transport chain that connects photosystems I and II. Here, the electrons are passed along the chain, triggering the events of ATP production. The electrons, no longer in an excited state, are then accepted back by P700 to complete a cycle. Cyclic electron flow provides eukaryotic organisms with ATP when metabolic demands are such that there is no need for additional NADPH.

---

### CONCEPT CHECKPOINT

7. Hydroelectric dams generate electricity by allowing lakeside water to flow over the dam, which provides the energy needed to turn huge electricity-generating turbines. How is the chemiosmotic process of making ATP during the light-dependent reactions analogous to a hydroelectric dam?

8. What is the final acceptor of high-energy electrons in noncyclic electron flow? What will this molecule shuttle to the light-dependent reactions?

---

## *just wondering . . .*

### *Questions students ask*

### Why do weeds and crabgrass grow during the hot summer while lawns dry up and die?

The answer to your question lies in the fact that not all plants undergo photosynthesis in the same way. Kentucky bluegrass and similar grasses use $C_3$ photosynthesis, the type of photosynthesis described in this chapter. It is termed $C_3$ because molecules of $CO_2$ are used to form the three-carbon compound 3PG, the first stable product of photosynthesis produced in the Calvin cycle (see Figure 7.11). Weeds and crabgrass, which thrive during a hot, dry summer, use a $C_4$ photosynthetic pathway.

$C_4$ photosynthesis is a cumbersome way to generate three-carbon molecules, but it is actually an advantage to plants growing in hot, dry, sunny regions. Why? $C_4$ plants use water more efficiently and use less energy than $C_3$ plants in high temperatures because their leaf anatomy is unique and they have the ability to fix $CO_2$ by mechanisms and under conditions that $C_3$ plants cannot. In essence, $C_4$ plants are able to grow and thrive while $C_3$ plants wither and die in dry heat and intense sunlight. But pesky weeds are not unique in having this adaptation. Corn and sugarcane, two extremely important crops, are also $C_4$ plants. ●

---

*Are you wondering about a topic in biology and how it relates to your life? Submit your question by clicking the Just Wondering link in this text's companion Web site at www.wiley.com/college/alters.*

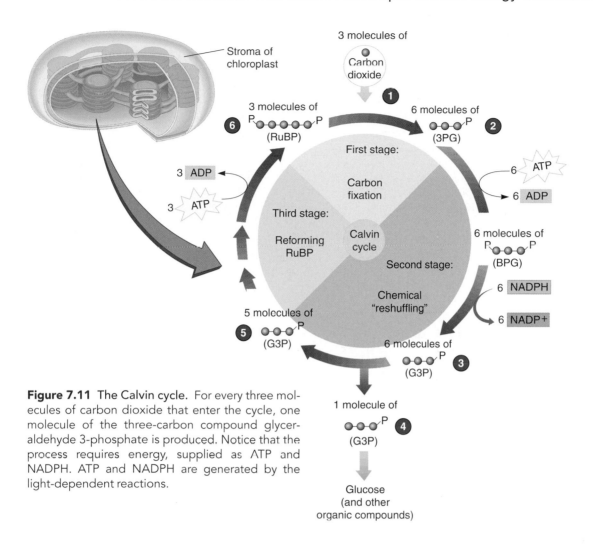

**Figure 7.11** The Calvin cycle. For every three molecules of carbon dioxide that enter the cycle, one molecule of the three-carbon compound glyceraldehyde 3-phosphate is produced. Notice that the process requires energy, supplied as ATP and NADPH. ATP and NADPH are generated by the light-dependent reactions.

## 7.6 Light-independent reactions produce carbohydrates.

The light-dependent reactions of photosynthesis described in the preceding section use light energy to produce (1) metabolic energy in the form of ATP and (2) reducing power in the form of NADPH. But this is only half the story. Photosynthetic organisms use the ATP and NADPH produced by the light-dependent reactions to build organic molecules from atmospheric carbon dioxide, a process called **carbon fixation** or the **Calvin cycle**, after American chemist Melvin Calvin who discovered its steps. These organic molecules are food for the plant and also provide the plant with raw materials for growth. This phase of photosynthesis does not need light energy to drive its reactions; therefore, it is referred to as the *light-independent reactions*. In plants, algae, and the euglenoids, these reactions are carried out by a series of enzymes located in the stroma of the chloroplast.

The Calvin cycle has three stages that produce one three-carbon sugar, *glyceraldehyde 3-phosphate (G3P)*, from three molecules of carbon dioxide ($CO_2$). Two molecules of G3P are used to manufacture one glucose molecule (and other carbohydrates) as shown at the bottom of the cycle. **Figure 7.11** depicts the Calvin cycle in simplified form; the grey balls stand

for carbon atoms and will help you follow the flow of these reactions.

In the first stage of the Calvin cycle, carbon dioxide joins a five-carbon molecule called *ribulose 1, 5 bisphosphate (RuBP)* from the "end" of the cycle ❶ in the process of carbon fixation. Thus, the inorganic molecule carbon dioxide is attached to a biological molecule, making $CO_2$ available for use to synthesize glucose and complex carbohydrates. For three turns of the cycle, three $CO_2$ molecules join with three RuBP molecules and form six three-carbon molecules (3PG) ❷. In the second stage of the cycle, these molecules are chemically reshuffled to produce glyceraldehyde 3-phosphate molecules ❸ Only one of these six G3P molecules is used in the manufacture of carbohydrates ❹. The other five molecules of G3P ❺ are used to resynthesize three molecules of RuBP, the third stage of the cycle, ❻. These molecules of RuBP are needed to keep the cycle going. In this way, RuBP is used as a handle to fix carbon dioxide, and the three-carbon molecules produced are used to build sugars for energy and storage and to reconstruct or recycle molecules of RuBP.

Referring back to the equation of photosynthesis on page 100, we see that it takes six molecules of carbon dioxide to produce one molecule of glucose. For one molecule of G3P to be produced, the cycle must take place three times, and it must take place six times to produce one molecule of glucose.

CHAPTER REVIEW

## Summary of Key Concepts and Key Terms

### What is the source of energy for life?

**7.1** The sun is the ultimate source of energy for virtually all life on Earth.

**7.1** Photosynthesis is a process whereby energy from the sun is captured by living organisms and used to produce molecules of food.

**7.1** Conversely, the byproduct of photosynthesis—oxygen—and its product of carbohydrates are the substrates of aerobic respiration.

**7.2** Two important types of pigments that play a role in photosynthesis are the **chlorophylls** (p. 99) and the **carotenoids** (kuh-ROT-uh-noids; p. 99).

**7.2** Chlorophyll and carotenoid pigments are organized into photocenters and serve to capture light energy and change it into chemical energy useful in photosynthesis.

**7.2** The energy from the sun is electromagnetic radiation and behaves as both waves and discrete packets of energy called photons.

**7.2** Special molecules called **pigments** (p. 97) absorb some wavelengths of light and reflect others.

**7.3** Organisms capable of making their own food by photosynthesis—plants, algae, and certain protists and certain prokaryotes—are called photosynthetic **autotrophs** (AW-toe-trofs; p. 97).

**7.3** Organisms not capable of making their own food, called **heterotrophs** (HET-ur-oh-trofs; p. 97), depend on the food produced by autotrophs.

**7.3** The primary process by which organisms unlock the energy in food is cellular respiration.

**7.3** The byproducts of aerobic cellular respiration—carbon dioxide and water—are the substrates of photosynthesis.

### How do autotrophs store energy from the sun?

**7.4** During **photosynthesis** (p. 100), atmospheric carbon dioxide is used to produce molecules of sugar, with oxygen as a byproduct.

**7.4** Photosynthesis takes place in the **chloroplasts** (p. 101) of photosynthetic eukaryotic cells and on infoldings of the cell membrane in photosynthetic prokaryotes.

**7.4** The internal membranes of chloroplasts are flattened sacs called **thylakoids** (THIGH-luh-koidz; p. 101). Stacks of thylakoids are

**grana** (GRA-nuh, p. 101). Surrounding the thylakoids is a fluid called the **stroma** (STROH-muh, p. 101).

**7.4** Photosynthesis involves two sets of reactions: the synthesis of ATP and NADPH, which are the **light-dependent reactions** (p. 100), and the use of ATP and NADPH, which are the **light-independent reactions** (p. 101) to produce glucose.

### What are the reactions that enable autotrophs to store energy from the sun?

**7.5** Photocenters, which are arrays of pigment molecules located within the thylakoid membranes of chloroplasts, capture and direct photon energy toward single molecules of **chlorophyll a** (p.102) that participate in photosynthesis.

**7.5** Plants and algae use a two-stage photosystem to carry out the light-dependent reactions of photosynthesis.

**7.5** During noncyclic electron flow, energized electrons ejected from **photosystem II** (p. 102) are passed down an **electron transport chain** (p. 103), triggering events of ATP production.

**7.5** The electron hole in photosystem II is filled by electrons from the breakdown of water, a process that releases oxygen as a byproduct.

**7.5** Energized electrons ejected from **photosystem I** (p. 102), along with hydrogen ions, reduce NADP+ to NADPH.

**7.5** The process in which a cell uses the potential energy in a concentration gradient to make ATP is called **chemiosmosis** (KEM-ee-oz-MOH-sis, p. 104).

**7.5** The result of the light-dependent reactions is the synthesis of ATP and NADPH molecules as well as the byproduct, oxygen.

**7.6** Photosynthetic organisms use the ATP and NADPH produced by the light-dependent reactions to build organic molecules from atmospheric carbon dioxide, a process called **carbon fixation** or the **Calvin cycle** (p. 105).

**7.6** Three main events take place during the light-independent reactions of photosynthesis: the fixation of carbon, the production of glucose and other carbohydrates, and the reformation of an organic molecule that attaches to carbon dioxide to begin the cycle again.

## Level 1 | Learning Basic Facts and Terms

**Multiple Choice**

1. How is water used in photosynthesis?
   a. It helps ATP synthase make ATP.
   b. It provides electrons to fill electron "holes" in the P680 molecule.
   c. Water is not used at all in photosynthesis.
   d. It carries electrons to the light-independent reactions (Calvin cycle).

2. Which of the following groups contains organisms that are photosynthetic autotrophs?
   a. animals
   b. plants
   c. protists
   d. prokaryotes
   e. b, c, and d are correct

3. Which of the following molecules produced in the light-dependent reactions are used in the Calvin cycle (light-independent reactions)?
   a. ATP and NADPH
   b. ATP and $CO_2$
   c. ADP and $CO_2$
   d. NADPH and ADP

4. What is the role of NADPH in the Calvin cycle?
   a. Provides the high-energy electrons and $H^+$ ions necessary to produce sugars from $CO_2$
   b. Provides the energy necessary to turn ADP into ATP

c. Provides the carbon and oxygen necessary to make sugar
d. Provides the iron necessary for photosynthesis to occur

5. Carbohydrates produced during photosynthesis may be used for which of the following?
   a. transformed into starch
   b. transformed into a structural polysaccharide such as cellulose
   c. used to fuel cellular respiration
   d. combined with other plant nutrients and converted into proteins or DNA
   e. all of the above

**Matching**
Match events (6–11) with the stage(s) of photosynthesis (a–c) where each occurs.
   a. Cyclic electron flow
   b. Noncyclic electron flow
   c. Calvin cycle

6. Chemiosmosis occurs.
7. Carbon fixation occurs.
8. Oxidation-reduction reactions occur.
9. Water is oxidized.
10. NADP is reduced.
11. NADPH is oxidized.

## Level 2    Learning Concepts

1. Which of the following would be considered an accurate description of the differences between animal cells and plant cells?
   a. Plants produce sugars but don't use them in metabolism; animals consume and use sugars in metabolism but cannot produce them.
   b. Plant cells undergo only photosynthesis; animal cells only undergo cellular respiration.
   c. Plant cells produce only $O_2$ from metabolism; animal cells produce only $CO_2$ from metabolism.
   d. Plant cells have both mitochondria and chloroplasts; animal cells only have mitochondria.

2. Consider the following groups of organisms: aquatic algae, terrestrial plants, fungi, cyanobacteria. Which does not belong with the others? Explain your answer.

3. Of the various metabolic activities in a plant (Calvin cycle, cyclic electron flow, noncyclic electron flow, cellular respiration), which could not occur if the plant was placed in darkness for several days? Explain your reasoning.

4. In the overall summary reactions of photosynthesis, which of the starting reactants is being reduced? Which is being oxidized? What products do each of these reactants become when they are oxidized or reduced? Where does each of these oxidation-reduction reactions occur in photosynthesis?

5. Place the following in order of the progression of energy transformations that occur during photosynthesis.
   • light energy
   • electron flow
   • chemical energy stored in sugars
   • chemical energy stored in ATP and NADPH
   • proton (H+ ion) gradient

## Level 3    Critical Thinking and Life Applications

1. Imagine you place a potted plant in your classroom window after weighing it carefully. Pot and all, it weighs just 3 pounds. Watering it properly, you let the plant grow for a year and then reweigh it. Pot and all, it now weighs 4-1/2 pounds. Where precisely did the 1-1/2 pounds of extra plant mass come from?

2. The roots of plants are not exposed to sunlight—they are under the ground. How do they manufacture ATP?

3. Scientists argue that global warming is being caused by the release of large amounts of $CO_2$ into the atmosphere through the burning of fossil fuels, such as coal and oil. Carbon dioxide in the atmosphere traps heat close to the Earth, much like glass traps heat inside a greenhouse. Explain why cutting down trees of the world's rain forests is intensifying the problem of large amounts of $CO_2$ in the atmosphere.

4. With reference to question 3, which reactions of photosynthesis would be directly responsible for reducing atmospheric $CO_2$?

5. If you were to wrap one of your houseplants in transparent green cellophane and set it in the sunlight, how do you think this would affect photosynthesis? What if you were to do the same with blue cellophane? (*Hint*: Green cellophane filters out all colors of light except green, which it transmits to the plant.)

6. Scientists think that about 60 million years ago a large meteor struck Earth blanketing it with a thick haze of atmospheric dust lasting for years. Explain how this event might have affected life on Earth at this time. Relate your answer to metabolism.

## In The News | **Critical Thinking**

**PLANT IMPOSTORS**

Now that you understand more about the process of photosynthesis, reread this chapter's opening story about DeLong's discovery of a rhodopsin-based photosynthetic pathway in ocean-dwelling microbes. To understand this research better, it may help you to follow these steps:

1. Review your immediate reaction to DeLong's research that you wrote when you began reading this chapter.
2. Based on your current understanding, again summarize the main point of the research in a sentence or two.
3. What questions do you have about this research that this chapter's opening story does not answer?

4. Collect new information about the research. Visit the *In The News* section of this text's companion Web site at www.wiley.com/college/alters and watch the "Plant Impostors" video. Then use the "summary" link to read the accompanying story and access related links. Use this information, the links provided, and other online and library resources to answer your questions and find updates about this research topic. State the sources of your information. Explain why you think the information is accurate. Also determine whether the information expresses a particular point of view or is biased in any way.
5. What in your view is the most significant aspect of this research? Give reasons for your opinion and for any changes in your ideas based on the additional information you have collected and the analysis you have done.

# RELEASING ENERGY FROM FOOD: CELLULAR RESPIRATION

## In The News | Alarm Cells

Terrorists can strike in many ways, and one of those ways is with the use of biological agents—bacteria such as anthrax and toxins such as sarin gas. Bioweapons such as these can harm and kill humans. Generally, if the presence of biological agents is suspected, scientists test for specific agents to determine if risk exists. But testing for individual agents takes time, and if none of the tests proves positive, how do scientists know if a harmful unidentified agent is present? What if new, genetically engineered substances are used for which scientists have no tests?

David Cliffel and his team of scientists at Vanderbilt University have developed a new device to help defend against biological agents. The device is called a *microphysiometer*, and it tests human cells exposed to a biological agent to see if cellular respiration or other metabolic processes are affected, rather than testing for the agent itself. Cellular respiration is the process whereby fuel molecules (primarily the sugar glucose) are broken down, releasing energy. The microphysiometer measures changes in extracellular glucose, lactic acid, oxygen, and acidification rate. By measuring extracellular glucose and oxygen, for example, the team can see if the cells are using the glucose and the oxygen they need to burn this fuel during cellular respiration. Lactic acid and carbonic acid are also produced in cells as

byproducts of various chemical reactions, so production of these compounds and the consequent pH level of the fluid surrounding the cells tells whether various aspects of metabolism—the chemical reactions of the cell—are proceeding normally.

Using living cells as a detection mechanism for the effects of various chemical and biological agents has applications in areas other than bioterrorism. For example, the microphysiometer can also be used to detect environmental toxins in groundwater and to test for toxicity of compounds being screened as drugs for human use. You can learn more about this research and view the video "Alarm Cells" by visiting the *In The News* section of this text's companion Web site at www.wiley.com/college/alters.

*Write your immediate reaction to Cliffel's research on testing cells for the effects of bioagents: first, summarize the main point of the research in a sentence or two; then suggest what you think the significance of this research is. You will have an opportunity to reflect on your responses and gather more information on this topic in the* In The News *feature at the end of this chapter. In this chapter, you will learn more about cellular respiration and its role in the general metabolism of cells.*

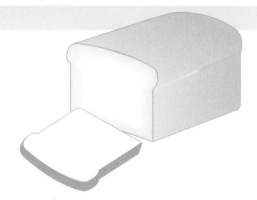

## How do organisms transform energy to maintain life?

**8.1** Organisms release energy from carbohydrates, capturing it to do cell work.

## How do organisms release energy from food?

**8.2** During aerobic cellular respiration, glucose is broken down, yielding energy.

**8.3** Food is converted into substrates that can enter the pathways of aerobic respiration.

## What are the chemical pathways that release energy from glucose?

**8.4** Some energy released from glucose is captured in ATP molecules during glycolysis.

**8.5** The remaining energy in glucose is captured during the Krebs cycle.

**8.6** The electron transport chain produces ATP from reduced electron carriers.

**8.7** Fermentation takes place without oxygen.

# How do organisms transform energy to maintain life?

## 8.1 Organisms release energy from carbohydrates, capturing it to do cell work.

Plants and other photosynthetic organisms play a major role in converting energy from the sun to chemical energy that can be used to maintain life. Using carbon dioxide from the atmosphere and the sun's energy, photosynthetic organisms build carbohydrates, which hold chemical energy locked in their bonds. However, that is only part of the story. To maintain life, organisms must extract the energy locked in carbohydrates and capture it in molecules of ATP that can be used to perform cell work. As the sugars that make up carbohydrates are broken down to release energy, the carbon atoms are incorporated into carbon dioxide, and this gas is returned to the environment. This cycling of carbon and transformation of energy is shown in Figure 7.4.

In organisms such as humans and plants, the sugars (most specifically glucose) that make up carbohydrates are broken down by a series of chemical reactions called **cellular respiration**. The term *cellular* means that this process takes place in the cells of living things. The term *respiration* describes the process: the breakdown of fuel molecules with a resulting release of energy in which an inorganic compound, such as oxygen, serves as the terminal electron acceptor. The terminal electron acceptor is the last compound to accept electrons in the series of chemical reactions that comprise cellular respiration. If the terminal electron acceptor is oxygen, as it is in organisms like humans and plants, the process is called aerobic cellular respiration. Aerobic (air-OH-bick) means "with oxygen."

Since aerobic cellular respiration uses oxygen and produces carbon dioxide, you might think that it has something to do with breathing. In fact, the two processes are intimately connected. The

oxygen in the air we and other organisms breathe in is used as the terminal electron acceptor during aerobic cellular respiration. The carbon dioxide we breathe out is an end product derived from the breakdown of sugar. We use the energy that is liberated from sugar and exhale the carbon atoms as $CO_2$.

Organisms that do not use oxygen as the terminal electron acceptor, such as some bacteria and yeasts, make ATP by either anaerobic electron transport or fermentation. During anaerobic electron transport (an-air-OH-bick; "without oxygen"), organisms such as certain prokaryotes use the pathways of aerobic respiration but use compounds such as nitrate or sulfate, rather than oxygen, as the terminal electron acceptor. For example, bacteria that live in your intestines and others that live in the soil use nitrate ($NO_3^-$) instead of oxygen as a terminal electron acceptor.

Fermentation, on the other hand, consists of **glycolysis**—an initial series of reactions of cellular respiration—and one or two additional reactions that take place anaerobically. As a result, the carbon and hydrogen atoms of the glucose molecule are not fully oxidized as in cellular respiration. Instead, the atoms in the glucose molecule are simply rearranged into a molecule containing less chemical energy than glucose. Put simply, fermentation does not extract as much energy from glucose molecules as does aerobic and anaerobic respiration. As a result, fermentative organisms use a great deal of glucose to derive the energy they need for life. However, they are able to thrive in environments devoid of oxygen.

Products of fermentation are a part of our daily lives. Baker's yeast, for example, can generate ATP by fermentation. During this process, the yeast produces carbon dioxide gas, the end product that causes bread dough to rise. Certain yeasts produce alcohol during fermentation and are used in the production of beers, wines, and other alcoholic beverages. In addition, muscle cells, when they are deprived of sufficient oxygen for aerobic respiration, undergo a fermentative process on a short-term basis, which produces lactic acid. The traditional process of soy sauce production involves a double fermentation by both molds and bacteria, which produces small amounts of alcohol and lactic acid (**Figure 8.1**). Fermentation is discussed in more detail in Section 8.7.

Organisms use either cellular respiration, anaerobic electron transport, or fermentative pathways to generate ATP, suggesting a strong evolutionary relationship among all living things.

**Figure 8.1** Fermentation of soy and wheat in large steel tanks during soy sauce production.

CONCEPT CHECKPOINT

1. For each of the following organisms, list their source of energy, the type of metabolic pathway they use, whether they are aerobic or anaerobic, and whether they use or produce $CO_2$ during metabolism: plants, animals, and baker's yeast.

# How do organisms release energy from food?

## 8.2 During aerobic cellular respiration, glucose is broken down, yielding energy.

So how does cellular respiration extract the energy locked in glucose and transform it into energy available for cell work? Cellular respiration is a complicated process, but this section provides a simplified overview. Sections 8.4 to 8.6 describe the process in more detail.

Cellular respiration is described chemically in the following formula:

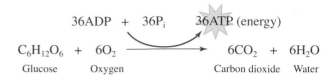

$$36ADP + 36P_i \longrightarrow 36ATP \text{ (energy)}$$
$$C_6H_{12}O_6 + 6O_2 \longrightarrow 6CO_2 + 6H_2O$$
Glucose　　Oxygen　　　　　　Carbon dioxide　Water

Notice the ATP that is generated. These molecules can enter into the chemical reactions of living things to power endergonic reactions. Put simply, ATP can do cell work.

Although the formula is stated as if glucose were broken down to carbon dioxide and water, generating ATP in a single step, this is not the case. This simplified formula shows that as aerobic respiration occurs, the *net effect* is the breakdown of one molecule of glucose in the presence of six molecules of oxygen to yield six molecules of carbon dioxide and six molecules of water. During this process, enough energy is liberated from the glucose as it is being cleaved to power many endergonic reactions: 36 molecules of ADP are bonded to atoms of inorganic phosphate groups—an energy yield of 36 ATPs. Therefore, the breakdown of glucose during aerobic respiration releases energy. In other words, the exergonic reactions of cellular respiration help generate molecules of ATP. The production of ATP takes place via endergonic reactions (see Chapter 6).

The complex series of reactions of cellular respiration can be divided into three parts: glycolysis (glye-KOL-uh-sis), the Krebs cycle, and the electron transport chain. The word "glycolysis" comes from Greek words meaning "to break apart" (*lysis*) a "sugar" (*glyco*). Glycolysis takes place in the cytosol, the gel-like fluid of the cytoplasm. It is the first stage of extracting energy from glucose (**Figure 8.2 ①**). No oxygen is needed for glycolysis to take place. It is an exergonic metabolic pathway in which ATP is generated, but the total yield of ATP molecules is small—only two ATPs for each original glucose molecule. Energy-rich NADH is also produced ②.

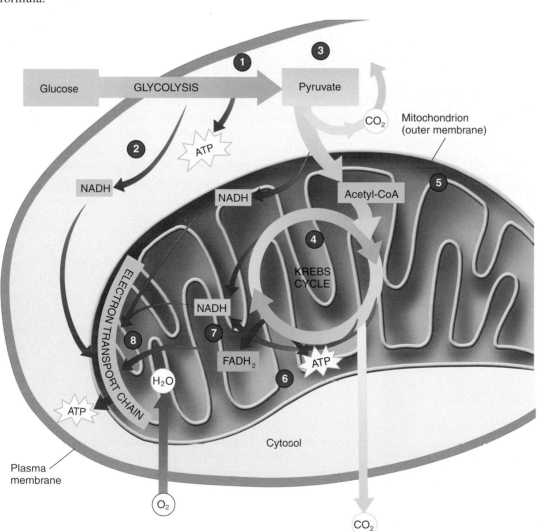

**Figure 8.2** An overview of cellular respiration. 36 ATPs are produced for every one glucose molecule.

 **Visual Thinking:** Glucose is shown entering the cell at the upper left. What is its role in cellular respiration? Oxygen is shown entering the cell at the lower left. What is its role? Carbon dioxide is shown leaving the cell at the lower right. Where do the carbon molecules of this byproduct come from?

When glycolysis is completed, the six-carbon glucose has been cleaved in half, yielding two three-carbon molecules called pyruvate, or pyruvic acid ③. The two pyruvate molecules still contain most of the energy that was present in the one original glucose molecule.

The **Krebs cycle** ④ takes place in the mitochondria and is the second stage of extracting energy from glucose. Named after the German biochemist Hans Krebs, this cycle of reactions begins with a two-carbon molecule called acetyl-CoA ⑤ produced by reactions prior to the cycle that remove one carbon dioxide molecule ($CO_2$) from each pyruvate molecule formed by glycolysis. Each of these two-carbon molecules combines with a four-carbon molecule to form a six-carbon molecule called citric acid (which is not shown in Figure 8.2). For this reason, the Krebs cycle is also often called the citric acid cycle. For every two pyruvate molecules entering the Krebs cycle as the result of the glycolytic breakdown of one glucose molecule, two more ATP molecules are made ⑥ and a large number of electrons are removed from the substrates in the cycle. These electrons pass to a series of electron carriers also located in the mitochondria ⑦, which form part of an **electron transport chain** ⑧ similar to the electron transport chains found in the thylakoid membranes in the chloroplasts of plants. ATPs are produced by means of chemical reactions and processes that take place along the chain.

### CONCEPT CHECKPOINT

**2.** List the molecules that enter into the reactions of glycolysis. List the molecules that are products of the reactions of glycolysis. What are the fates of these products?

**3.** List the molecules that enter into the reactions of the Krebs cycle. List the molecules that are products of the reactions of the Krebs cycle. What are the fates of these products?

## 8.3 Food is converted into substrates that can enter the pathways of aerobic respiration.

Food is a complex mixture of carbohydrates, lipids, proteins, and other molecules. If cellular respiration breaks down glucose derived from the digestion of carbohydrates, then what happens to all the noncarbohydrate foods that we and other organisms eat? Food is converted into substrates that can enter the pathways of aerobic respiration.

## *just wondering . . .*

### *Questions students ask*

### I really want to lose weight. I've read about foods that can "melt away" fat. Is that true?

Many diet programs and schemes promise quick weight loss by eating particular foods, taking "fat-burning" pills, or avoiding certain nutrients, such as carbohydrates. In general, these pills and programs will not turn your body into a fat-burning machine, although dieters across America would love to believe those claims.

In order to "melt away" fat, a substance or food would have to change your metabolism in profound ways. Although certain substances and nutrients in foods do increase metabolism, they won't help you lose weight. Caffeine, for example, is a naturally occurring stimulant drug. It can raise your heart rate and boost your energy level, but you won't melt away fat by drinking coffee or cola. Iodine, a mineral element found in iodized salt, also affects metabolism; it is needed for thyroid hormone production. Iodine deficiency has a significant negative effect on energy metabolism, but this deficiency rarely occurs in the United States. In addition, many vitamins and a few minerals (magnesium and iron, in particular) participate in chemical reactions that are involved in energy metabolism. However, taking more vitamins and minerals will not make you burn fat faster or slim you.

The American Dietetic Association refers to the notion that any food or food component can "burn up" the calories in food or melt away body fat as a myth and wishful thinking. Foods generally don't affect metabolism. The human body is adaptable and can use various mixes of biological fuels. Section 8.3 discusses how proteins, fats, and carbohydrates are converted into substrates that can enter the pathways of aerobic respiration at various points. If a person loses weight by eating a large proportion of one food, such as grapefruit, the loss is not because the grapefruit somehow magically changed the dieter's metabolism, rerouting biochemical pathways. It is likely due to substituting low-calorie grapefruit for other higher-calorie foods, helping the dieter maintain a total energy intake lower than the amount needed to support his or her weight.

You may be familiar with low-carbohydrate high-protein diets, such as the Atkins Diet and the Carbohydrate Addict's Diet, the most popular type of diet in the United States today. Such diets result in the need to deaminate (remove amino groups from) large amounts of amino acids for energy, which can cause dehydration and problems for people with unhealthy livers and kidneys. In the short term, low-carbohydrate diets cause a greater loss of body water than body fat. Moreover, low-carbohydrate diets are nutritionally inadequate according to a scientific review of popular diets conducted by researchers M. Freedman, J. King, and E. Kennedy from the U.S. Department of Agriculture. There is also concern that high-protein diets may contain high amounts of saturated fats, which may negatively affect health. Nevertheless, proponents of low-carbohydrate diets contend that diet composition is the key for weight loss. Scientific studies refute this claim and show that individuals on low-carbohydrate diets simply consume fewer calories than they did before the diet.

Regardless of whether a weight loss drug is available over-the-counter or by prescription, much or all of the weight one loses while taking the drug is usually regained when it's discontinued. Overweight and obesity is epidemic in the United States today. Looking for the quick fix or miracle "fat-burning" cure is not the answer to sustainable weight loss. What does work is a twofold approach: eating a balanced diet containing fewer calories than you burn and exercising regularly. These lifestyle practices will enhance your health as well as take inches off your thighs and hips. ●

*Are you wondering about a topic in biology and how it relates to your life? Submit your question by clicking the* Just Wondering *link in this text's companion Web site at* www.wiley.com/college/alters.

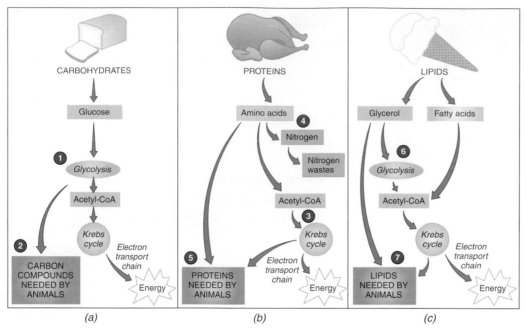

**Figure 8.3 Metabolism of food.** The complex molecules in food must be broken down into simpler molecules before cells can use them as fuel. Carbohydrates are broken down into simpler sugars; proteins are broken down into amino acids; and lipids are broken down into fatty acids and glycerol. The hamburger in the photo supplies all three types of complex molecules.

The first thing that happens in food's journey toward ATP production is that digestive system enzymes break down complex molecules into simple ones. Carbohydrates, such as those found in vegetables, pasta, and bread, for example, are split into simple sugars such as glucose or into sugars that are changed to glucose (**Figure 8.3a**). Proteins, which are found in foods such as meats

and nuts, are broken down to amino acids (Figure 8.3*b*). Lipids, which can be found in oil-based salad dressings and ice cream, are split into fatty acids and glycerol (Figure 8.3*c*). These steps taken by the digestive system yield no usable energy, but they change a diverse array of complex molecules into a small number of simpler molecules that your cells can use either as fuel or as building blocks to manufacture substances for growth, repair, and maintenance.

Once carbohydrates are changed to glucose, they are completely metabolized by glycolysis, the Krebs cycle, and the electron transport chain (Figure 8.3 ❶). Certain intermediary products formed in glycolysis may be used to build carbon compounds the body needs ❷. After the digestion of proteins to amino acids, their carbon portions are chemically modified and are then metabolized by the Krebs cycle ❸ and the electron transport chain, skipping over glycolysis. The nitrogen in proteins is eliminated in the urine ❹. Often, amino acids are used to build proteins rather than being used for energy ❺. Some of the breakdown products of the fats in your diet are converted to substances that can also be metabolized by glycolysis, the Krebs cycle and the electron transport chain ❻ or used to build other lipids the body needs ❼.

# What are the chemical pathways that release energy from glucose?

## 8.4 Some energy released from glucose is captured in ATP molecules during glycolysis.

As we have seen, aerobic respiration can be thought of as three integrated sets of chemical reactions: glycolysis, the Krebs cycle, and the electron transport chain. In this section, we'll describe the process of glycolysis in some detail.

Scientists think that glycolysis was one of the first metabolic processes to evolve. One reason is that this process uses no molecular oxygen and therefore occurs readily in an environment devoid of oxygen—a characteristic of the atmosphere of the primitive Earth. In addition, all of the reactions of glycolysis occur free in the cytosol of the cytoplasm; none is associated with any organelle or membrane structure. Early cells most certainly had no specialized structures such as organelles; thus, metabolic processes that evolved independently of membrane systems are probably evolutionarily older than those associated with membrane systems.

### Glycolysis generates a small amount of ATP

Living things possess glycolytic enzymes. However, most present-day organisms are able to extract considerably more energy from glucose molecules than glycolysis does. For example, of the 36 ATP molecules you obtain from each glucose molecule that you metabolize, only 2 are obtained by glycolysis. Why, then, is glycolysis still maintained even though its energy yield is comparatively meager? This simple question has an important answer: Evolution is a step-by-step process. Cells that were capable of glycolysis survived the early competition of life. Later changes in catabolic metabolism built on this process. Glycolysis was not discarded during the course of evolution but rather was used as the starting point for the further extraction of chemical energy.

The catabolic pathway of glycolysis is shown in **Figure 8.4**. The gray balls represent carbon molecules. Its reactions can be divided into four stages:

- *Stage A—Glucose mobilization*. Glycolysis begins as glucose is changed into a compound that can be split into two molecules.

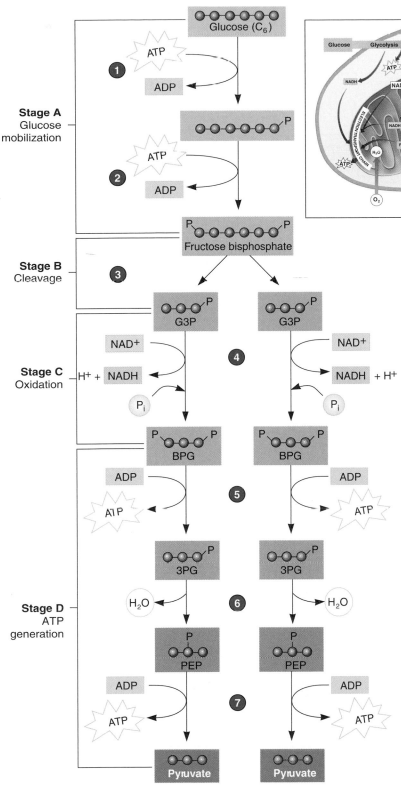

**Stage A**
Glucose
mobilization

**Stage B**
Cleavage

**Stage C**
Oxidation

**Stage D**
ATP
generation

- *Stage C—Oxidation.* During reaction **4**, two H atoms are removed from each three-carbon molecule. $NAD^+$ accepts one hydrogen ion and two electrons, forming two molecules of NADH for each original glucose molecule. A hydrogen ion ($H^+$) is left over. As this chemical change takes place, an inorganic phosphate molecule is bonded to each three-carbon molecule, resulting in the three-carbon molecule BPG. $NAD^+$ is a major electron acceptor (carrier) in cellular respiration.

- *Stage D—ATP generation.* During reaction **5**, a phosphate group on each three-carbon molecule (BPG) is removed and bonded to an ADP molecule, producing two ATP molecules for each original glucose molecule. Then, during reaction **6**, a molecule of water is removed from each 3PG molecule, producing two PEP molecules. In reaction **7**, the phosphate group is removed from each PEP molecule and bonded to an ADP molecule, producing two ATP molecules for each original glucose molecule. The end product of glycolysis is two three-carbon molecules called pyruvate. (G3P = Glyceraldehyde 3-phosphate, BPG = 1,3-Bisphosphoglycerate, 3PG = 3-Phosphoglycerate, PEP = Phosphoenolpyruvate).

In glycolysis, because each glucose molecule is split into two three-carbon molecules, the overall net reaction sequence yields two ATP molecules, as well as two molecules of pyruvate:

$$
\begin{array}{ccc}
-2\,ATP & +\ \ 2\,(2\,ATP) & =\quad +2\,ATP \\
\text{Stage A} & \text{Stage D} & \text{net gain}
\end{array}
$$

Although this is not a great amount of energy—only 2% of the energy in the glucose molecule—glycolysis does generate ATP. During the first anaerobic, or airless, stages of life on Earth, this reaction sequence was the only way for living things to extract energy from food molecules.

The three most significant changes that take place during glycolysis are as follows:

- Glucose is converted to pyruvate.
- ADP + $P_i$ is converted to ATP.
- $NAD^+$ is converted to NADH.

These three products can be formed continually as long as the substrates used to produce them are available. The glucose is supplied by the food you eat. ADP and $P_i$ continually become available as ATP is broken down to do cellular work. But cells contain only a small amount of $NAD^+$, and the supply is quickly depleted unless NADH passes along its hydrogen ion and two electrons to another electron carrier. In this way, NADH is oxidized to form $NAD^+$ once again. Cells recycle NADH in one of two ways:

- *By means of the electron transport chain as part of cellular respiration.* NADH passes its hydrogen ion and two electrons to a molecule that shuttles them into the mitochondria, ultimately

**Figure 8.4 Glycolysis.** Glycolysis produces a net yield of two ATP molecules for each original glucose molecule, as well as two molecules of NADH and two molecules of pyruvate.

During each of the reactions **1** and **2**, ATP transfers chemical energy to glucose as it transfers its terminal phosphate group. Before energy can be released from glucose, your body must spend two ATP molecules to "prime" the glucose "pump."

- *Stage B—Cleavage.* During reaction **3**, the six-carbon product of glucose mobilization is split into two three-carbon molecules.

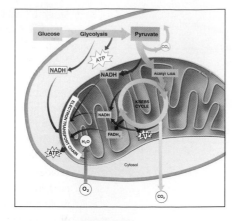

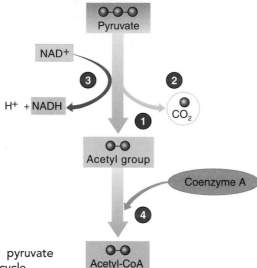

**Figure 8.5** How pyruvate enters the Krebs cycle.

passing them to an electron carrier in the electron transport chain. $NAD^+$ is thus reformed.

- *By means of fermentation.* Certain bacteria and yeasts produce an organic molecule from pyruvate that will accept the hydrogen ion and two electrons from NADH and thus re-form $NAD^+$. We will consider fermentation in more detail later in this chapter.

### Pyruvate is the end product of glycolysis

In all aerobic organisms—like you—the oxidation of glucose continues with the further oxidation of the product of glycolysis, which is pyruvate. In eukaryotic organisms, aerobic metabolism takes place in the mitochondria.

Before entering the reactions of the Krebs cycle, each pyruvate molecule is cleaved into a two-carbon molecule called an *acetyl group* (**Figure 8.5 1** ). The leftover carbon atom from each pyruvate is split off as carbon dioxide gas ($CO_2$) **2** . In addition, one $H^+$ and two electrons reduce $NAD^+$ to NADH **3** . In the course of these reactions, the two-carbon acetyl fragment removed from pyruvate is added to a carrier molecule called coenzyme A, or CoA, forming a compound called acetyl-CoA **4** . These reactions take place in the matrix of the mitochondrion as shown in the inset figure.

The reactions that link glycolysis and the Krebs cycle produce a molecule of NADH inside the mitochondrion, the energy of which is later used to produce ATP molecules. Of greater importance, however, is the acetyl-CoA, the energy-rich first substrate in the Krebs cycle. Acetyl-CoA is produced not only during the break-

down of glucose but also during the metabolic breakdown of proteins, fats, and other lipids. This molecule is the point at which many of the catabolic processes involved in the breakdown of proteins, lipids, and carbohydrates converge (see Figure 8.3).

## 8.5 The remaining energy in glucose is captured during the Krebs cycle.

The second series of reactions comprising aerobic cellular respiration is the Krebs cycle. This cyclic chain of eight enzyme-mediated reactions oxidizes acetyl-CoA. These reactions are shown in **Figure 8.6**. The cycle has two stages:

- *Stage A—Preparation reactions.* These two reactions set the scene. In the first reaction, acetyl-CoA joins a four-carbon molecule from the end of the cycle to form the six-carbon molecule citric acid **1** . In the next reaction **2** , chemical groups are rearranged.
- *Stage B—Energy extraction.* Four of the remaining six reactions are oxidations in which hydrogen ions and electrons are removed from the intermediate compounds in the cycle to form three NADH molecules for each acetyl-CoA, six for each original glucose molecule. Reaction **3** is an oxidation that produces NADH and is followed by the release of carbon dioxide. During reaction **4** oxidation occurs again, producing NADH and causing the release of a second carbon dioxide molecule. During the cycle **5** , a molecule called GTP is produced, which is quickly used to produce an ATP; two ATPs are generated for each original molecule of glucose. In addition, one molecule of $FADH_2$ (an electron carrier similar to NADH) is formed **6** , two for each original glucose.

Together, the eight reactions make up a cycle that begins and ends with the same four-carbon molecule **7** , **8** . At every turn of the cycle, acetyl-CoA enters and is oxidized to $CO_2$ and $H_2O$, and the hydrogen ions and electrons are donated to electron carriers.

In the process of aerobic respiration, the glucose molecule has been consumed entirely. Its six carbons were first split into three-carbon units during glycolysis. One of the carbons of each three-carbon unit was then lost as $CO_2$ in the conversion of pyruvate to acetyl-CoA, and the other two were lost during the oxidations of the Krebs cycle. Part of the glucose molecule's energy and its electrons, which are preserved in four ATP molecules and the reduced state of 12 electron carriers, are all that is left.

---

**CONCEPT CHECKPOINT**

4. Fill in the table below showing the net gain of each of these molecules produced per glucose molecule.

| | ATP | NADH | FADH$_2$ |
|---|---|---|---|
| Glycolysis | | | |
| Krebs cycle | | | |
| Total net gain | | | |

5. After the complete oxidation of glucose via glycolysis and the Krebs cycle, *most* of the energy once stored in a glucose molecule is now stored in the molecules shown in the above table. What has happened to the remaining energy?

**Figure 8.6** The Krebs cycle. The Krebs cycle produces three molecules of NADH, one molecule of FADH$_2$, and one molecule of ATP for each acetyl-CoA molecule that enters the cycle.

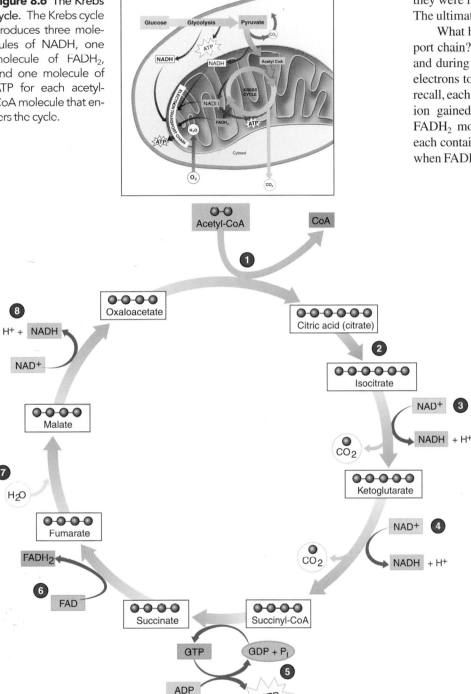

they were in the electron transport chains of photosynthesis. The ultimate result is the production of ATP.

What happens during the reactions of the electron transport chain? The NADH molecules formed during glycolysis and during the subsequent oxidation of pyruvate carry their electrons to the inner mitochondrial membrane. As you may recall, each NADH contains a pair of electrons and a hydrogen ion gained when NADH was formed from NAD$^+$. The FADH$_2$ molecules are already attached to this membrane, each containing two hydrogen ions and two electrons gained when FADH$_2$ was formed from FAD (see Figure 8.7).

As each of the carriers in the electron transport chain accepts electrons in turn, the electrons fall to lower energy levels, releasing energy in the process. The energy released from the electrons being passed along the chain pumps the protons (hydrogen ions, H$^+$) from the inner mitochondrial compartment (matrix) through the inner mitochondrial membrane proteins and into the outer mitochondrial compartment. As a high concentration of protons builds up in the outer compartment, the protons begin to flow back across the membrane through a channel formed by the membrane protein ATP synthase. As the protons travel inward through these channels, proteins associated with the channels couple the energy from the movement of the protons to the reaction of bonding P$_i$ to ADP, forming ATP (chemiosmosis).

Oxygen, an excellent electron acceptor, forms the last link in the chain in aerobic respiration. A molecule other than oxygen, such as nitrate or sulfate, forms the last link in the chain in anaerobic electron transport. In aerobic respiration—the process on which we are focusing—each atom of oxygen accepts two hydrogen ions and two electrons to form one molecule of water. In addition, three molecules of ATP are formed during various chemical processes that take place. Water is therefore an end product of cellular respiration, with one molecule of water produced for each NADH or FADH$_2$ oxidized.

Because three of the electron carriers in the chain act as "proton pumps," most NADH molecules produced during cellular respiration ultimately result in the production of three ATP molecules. Each FADH$_2$ activates two of the pumps and leads to the production of two ATP molecules. However, eukaryotes carry out glycolysis in the cytoplasm and the Krebs cycle within the mitochondria. This separation of the two processes within the cell requires transporting the electrons of the NADH created during glycolysis across the mitochondrial membrane. Only the electrons of NADH are shuttled across the membrane, not the entire molecule. The electron transfers that are made as these electrons are shuttled into the mitochondria result in the formation of only two ATPs, not three as is usually the case. Thus, each glycolytic NADH produces

## 8.6 The electron transport chain produces ATP from reduced electron carriers.

How can the energy preserved in 12 electron carriers, NADH and FADH$_2$, be transformed into energy that is able to do cell work? In other words, how can this energy be converted to ATP? The process of generating ATP from reduced electron carriers takes place within the inner mitochondrial membranes. Here, embedded within these membranes, are many series of electron carrier proteins. Each series is known as an electron transport chain (**Figure 8.7**). Electrons are passed from one carrier protein to another in the same way only two ATP molecules instead of three in the final total. **Figure 8.8** describes the total energy yield from the aerobic metabolism of one molecule of glucose. **Figure 8.9** shows the relationships between photosynthesis and cellular respiration, adding detail to the generalized overview of these two processes shown in Figure 7.4.

Interestingly, poisons such as certain pesticides, cyanide, and carbon monoxide work by interfering with chemical reactions that occur in the chain. For example, cyanide blocks the reaction in which the final electron acceptor passes its electrons to oxygen. Therefore, the system gets "backed up." Since the final electron acceptor cannot pass its electrons to oxygen, it cannot receive electrons. The entire system stops working all the way back to NADH, and no ATP is produced. For this same reason, most cells of complex organisms cannot live long without oxygen. If there is no terminal electron acceptor in the chain (oxygen), then the chain will shut down. The amount of energy produced without the reactions of the electron transport chain is inadequate to sustain life processes. In summary, without the reactions of the electron transport chain, most of the ATP production of cellular respiration stops . . . and life itself ends.

**Figure 8.7 The electron transport chain.** The NADH and FADH$_2$ formed during the Krebs cycle and glycolysis pass their electrons to electron receptors located in the inner membrane of the mitochondrion.

**Visual Thinking:** The mitochondrion is nicknamed "the powerhouse of the cell." What is happening in this illustration that justifies that nickname?

### CONCEPT CHECKPOINT

6. If you could measure the pH in a mitochondrion actively undergoing chemiosmosis, where would you expect more acidic conditions—in the outer compartment or the inner compartment? Why?
7. Once NADH and FADH$_2$ release their electrons to the electron transport chain, what do you think happens to these molecules?

**Visual Summary** **Figure 8.8 The energy extracted from the oxidation of glucose.** A net total of 2 ATPs are produced during glycolysis. Two NADH molecules are also produced, which each yield only 2 (rather than 3) molecules of ATP after their electrons are shuttled from the cytoplasm to the inner mitochondrial membrane. Two NADH, yielding 6 ATPs, are produced as pyruvate is oxidized to acetyl-CoA. In the Krebs cycle, 2 ATPs are produced directly; 6 NADH are produced, yielding 18 ATPs; and 2 FADH$_2$ are produced, yielding 4 ATPs.

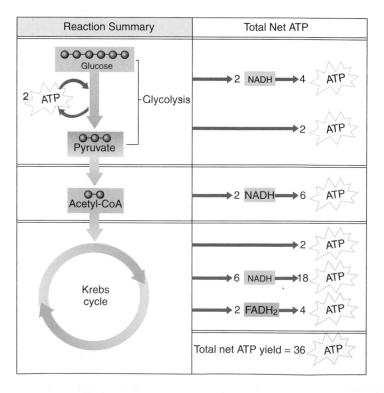

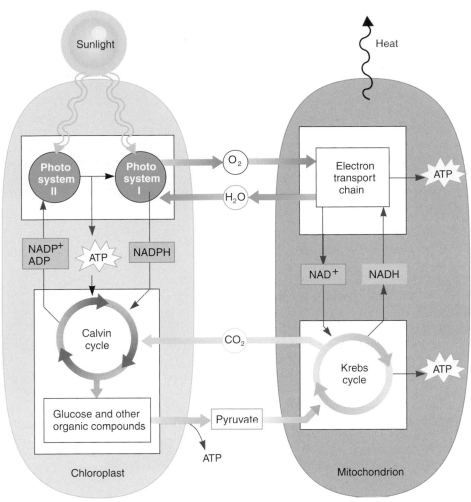

**Visual Summary**    **Figure 8.9** The metabolic machine. The diagram shows how the processes of photosynthesis and aerobic respiration are connected. The products of photosynthesis (glucose and other organic compounds) are converted to pyruvate and used by the nonphotosynthesizing parts of the plant to obtain energy. Leaf cells (the primary photosynthesizing parts of the plant) use the products of photosynthesis to obtain energy in the dark (when the light-dependent reactions are not taking place).

 **Visual Thinking:** Which types of organisms have both kinds of cells? Which types of organisms have only one of these two kinds of cells? Does the "metabolic machine" work if these cells are found in different organisms? How?

## 8.7 Fermentation takes place without oxygen.

As previously mentioned, aerobic cellular respiration cannot take place without oxygen because oxygen is necessary for the electron transport chain to function. Not only will the reactions of the electron transport chain come to a halt, but the Krebs cycle reactions will also stop because NADH will not be recycled to $NAD^+$, the needed electron acceptors in the cycle.

Although humans and other air-breathing organisms have no alternatives except death if deprived of oxygen for more than a few minutes, some organisms have more flexible metabolisms. Some organisms, such as certain bacteria, have enzymes that carry out both aerobic respiration and an alternative pathway called fermentation. Some organisms, such as other bacteria and yeasts, have the enzymes for fer-

mentation only. **Fermentation** comprises glycolysis and the production of an organic molecule from pyruvate that will accept the hydrogen ion and two electrons from NADH and thus re-form $NAD^+$. The end products of fermentation depend on the organic molecule that is produced from pyruvate.

What do you think of when you hear the word "fermentation?" You might think of alcoholic beverages, such as beer and wine, or fermented foods, such as olives, vinegar, and hard cider. Bacteria and yeasts carry out many types of fermentation. Often, the resulting reduced compound is an acid. In some organisms, such as yeasts, an organic molecule called acetaldehyde is formed from pyruvate and then reduced, producing an alcohol (**Figure 8.10a**). This particular type of fermentation is of great interest to people because it is the source of the ethyl alcohol in wine and beer. However, ethyl alcohol is an undesirable end product for yeast because it becomes toxic to the yeast when it reaches high levels. That is why wine contains only about 12% alcohol—12% is approximately the amount it takes to kill the yeast fermenting the sugars.

Have your muscles ever ached when you ran for miles or exercised vigorously? That muscle pain was likely due to a similar fermentative process muscles undergo when they produce ATP without sufficient oxygen (Figure 8.10b). During strenuous exercise, muscle cells break down large amounts of glucose to produce ATP. However, because muscle cells are packed with mitochondria, the pace of glucose breakdown far outstrips the blood's ability to deliver oxygen for aerobic respiration. Therefore, the muscle cells switch from aerobic respiration to fermentation. Muscle cells do not change pyruvate to another organic molecule to be reduced as bacteria and yeasts do, but instead directly reduce pyruvate to lactic acid. This acid builds up in the muscle and tends to produce a sensation of muscle fatigue. Gradually, however, the blood carries the lactic acid to the liver and certain other cells, where it is broken down aerobically. This need for oxygen to break down the lactic acid produces what is termed an *oxygen debt* and is the reason a person continues to pant after completing strenuous exercise.

### CONCEPT CHECKPOINT

8. Account for all of the energy that was once stored in a glucose molecule after it is completely metabolized through aerobic cellular respiration (glycolysis, Krebs cycle, and the electron transport chain).

9. Account for all of the energy that was once stored in a glucose molecule after it is fermented in a muscle cell.

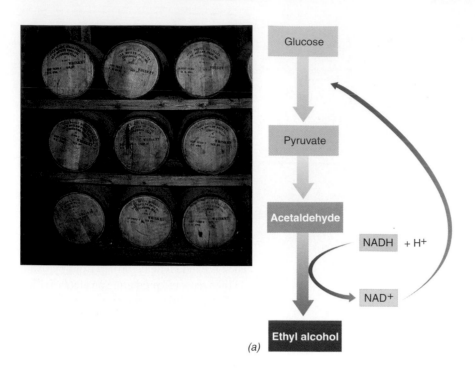

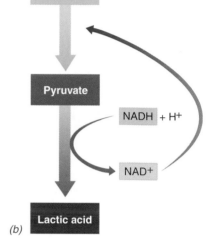

(a)

(b)

**Figure 8.10 Anaerobic metabolism.** (*a*) Yeasts form acetaldehyde from pyruvate. This molecule accepts the hydrogen ion and two electrons from NADH, re-forming NAD to be used in glycolysis. The result is the ethyl alcohol of wine, beer, and other alcoholic beverages. (*b*) When animal cells do not have sufficient oxygen, pyruvate accepts the hydrogen ion and two electrons from NADH to re-form NAD. This converts pyruvate into lactic acid, which builds up in the muscles until oxygen becomes available. This is what happens when you exercise strenuously. The ache in the arm muscles of the wheelchair racers, like the ache in the leg muscles of a runner, is the result of a buildup of lactic acid.

# CHAPTER REVIEW

## Summary of Key Concepts and Key Terms

### How do organisms transform energy to maintain life?

**8.1** Organisms release energy from glucose during **cellular respiration** (p. 111).

**8.1** If the terminal electron acceptor in respiration is oxygen, the process is called aerobic cellular respiration.

**8.1** Organisms that do not use oxygen as the terminal electron acceptor make ATP by either anaerobic cellular respiration or fermentation.

**8.1** Fermentation consists of **glycolysis** (p. 111)—an initial series of reactions of cellular respiration—and one or two additional reactions that take place anaerobically.

### How do organisms release energy from food?

**8.2** During aerobic respiration, glucose is broken down in the presence of oxygen, capturing energy in molecules of ATP; the byproducts of this process are carbon dioxide and water.

**8.2** Cellular respiration is composed of three series of chemical reactions: glycolysis, the **Krebs cycle** (p. 113), and the **electron transport chain** (p. 113).

**8.3** In the human body, most carbohydrates are changed to glucose and completely metabolized by aerobic respiration; parts of protein and lipid molecules are also metabolized by components of this metabolic pathway.

**What are the chemical pathways that release energy from glucose?**

**8.4** During glycolysis, glucose is split in half to pyruvate, and ATP and NADH are formed.
**8.4** Before entering the Krebs cycle, pyruvate is converted to acetyl-CoA.
**8.5** The Krebs cycle oxidizes acetyl-CoA, producing ATP and NADH.
**8.6** The NADH and $FADH_2$ molecules formed during the breakdown of glucose are themselves oxidized via the electron transport chain; these oxidations power the chemiosmotic production of ATP.
**8.6** Aerobic respiration yields 36 ATP molecules from glycolysis, the Krebs cycle, and the electron transport chain.

**8.7** Certain anaerobic organisms release energy from glucose during **fermentation** (p. 119).
**8.7** Organisms that metabolize nutrient molecules anaerobically by the process of fermentation use pyruvate or a molecule derived from pyruvate to accept the electrons produced during the glycolytic breakdown of glucose.
**8.7** Fermentation produces end products that are frequently acids or alcohols.

## Level 1 — Learning Basic Facts and Terms

**Multiple Choice**

1. What is ATP used for in cells?
   a. ATP provides the energy to drive endergonic (uphill) cellular processes.
   b. ATP provides the energy to pump molecules into cells against their concentration gradients.
   c. ATP provides the energy for muscle cells to contract.
   d. All of the above.
2. Which of the following is a product of glycolysis?
   a. NADPH
   b. $CO_2$
   c. Glucose
   d. NADH
3. Each glucose molecule oxidized completely by aerobic cellular respiration yields approximately
   a. 36 ATPs.
   b. 2 ATPs.
   c. 200 ATPs.
   d. 686 ATPs.

4. During the conversion of pyruvate (a 3-carbon molecule) to acetyl CoA (a 2 carbon molecule), _____ is given off as a byproduct of cellular respiration.
5. During cellular respiration, the energy necessary to recycle ADP back into ATP in cells ultimately comes from _____?

**Matching**

6. Occurs in the cytosol of the cell
7. Oxygen used directly
8. The pathway where most of the oxidation of the carbon intermediates derived from glucose occurs during cellular respiration
9. The *main* energy-producing pathway during aerobic cellular respiration (i.e., where most of the ATP is made)
10. Requires some ATP to get it started

a. Glycolysis
b. Fermentation
c. Krebs cycle
d. Electron transport chain

## Level 2 — Learning Concepts

1. What specific role does the oxygen we breathe play in metabolism?
2. You stop at a fast-food restaurant to eat a burger and fries. Explain how your body converts these foods into usable energy.
3. In your muscles, why does ATP production change with oxygen concentration? What might be one way the body would naturally increase oxygen supply to muscles?
4. We have discussed how the chemistry of life is integrally involved with water. What is the link between water and the electron transport chain?
5. In many respects, photosynthesis and cellular respiration are opposites of one another. How is the Krebs cycle the "opposite" of the Calvin cycle?
6. How is the electron transport chain the "opposite" of the noncyclic electron flow of the light independent reactions?

## Level 3 — Critical Thinking and Life Applications

1. Blood boosting*, a very dangerous practice that is often abused by athletes, involves removing a couple of pints of blood from the body several weeks before a competition and then re-infusing it back into the circulatory system the day before a big event. The expected metabolic benefits of this would be to
   a. increase ATP production.
   b. keep muscle cells out of the more inefficient anaerobic pathways longer during strenuous exercise.
   c. prevent buildup of lactic acid during strenuous activity.
   d. All of the above.
2. Which of the following would you predict would occur in the mitochondria of a person with a disorder that prevents $H^+$ ions from passing through ATP synthase as quickly as in an unaffected person?
   a. The outer compartment would be more acidic than normal.
   b. The amount of ATP produced in the cell would be higher than normal.
   c. The amount of oxygen produced by the cell would diminish.
3. One day after eating way too much over the holidays you decide to go on a diet and exercise program to lose weight. After two months of eating smaller portions, eating low fat food, and exercising daily you manage to lose 20 pounds—most of it fat. In what chemical form did most of that fat leave your body?
4. Is the following statement true or false? The process of glucose breakdown during cellular respiration has both endergonic and exergonic elements. Support your answer with evidence.
5. Chapter 5 briefly described the endosymbiont theory, which suggests that mitochondria originated as bacteria that were engulfed by pre-eukaryotic cells. What have you learned in this chapter that supports that theory?

*Note: This practice is *very* dangerous; it can lead to strokes and cardiac arrest!

## In The News | Critical Thinking

### ALARM CELLS

Now that you understand more about cellular respiration, reread this chapter's opening story about research into the development of a device to test cells for the effects of bioagents. To understand this research better, it may help you to follow these steps:

1. Review your immediate reaction to Cliffel's research on the development of a microphysiometer to test human cells exposed to biological agents that you wrote when you began reading this chapter.

2. Based on your current understanding, again summarize the main point of the research in a sentence or two.

3. What questions do you now have about this research that the opening story does not answer?

4. Collect new information about the research. Visit the *In the News* section of this test's companion Web site at www.wiley.com/college/alters and watch the "Alarm Cells" video. Then use the "summary" link to read the accompanying story and access related links. Use this information, the links provided, and other online and library resources to answer your questions and find updates about this research topic. State the sources of your information. Explain why you think the information is accurate. Also determine if the information expresses a particular point of view or is biased in any way.

5. What in your view is the most significant aspect of this research? Give reasons for your opinion and for any changes in your ideas based on the additional information you have collected and the analysis you have done.

# DNA STRUCTURE AND REPLICATION

## In The News    Lifespan Gene

Genes are our units of heredity. They hold the instructions for making proteins and other molecules crucial to our structure and function. This chapter describes genes and the hereditary material that comprises them—DNA. These molecules are the amazing "code of life." But do they also hold the key to longevity? Is there a lifespan gene?

Scientists are now studying a gene called Silent Information Regulator 2, or SIR2. Leonard Guarente, a professor of biology at Massachusetts Institute of Technology (MIT), studies SIR2 genes and their link to longevity in a variety of organisms. One of the organisms he studies is the free-living soil nematode, or roundworm, shown in the photo. To the naked eye, these worms are about the size of the head of a pin, but their well-developed organ systems provide a good model for organ systems such as our own.

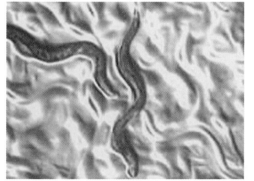

Can manipulating the SIR2 gene prolong life? Currently, the only known way to lengthen the lifespan is a calorie-restricted diet. However, when these roundworms were stripped of their SIR2 genes, calorie restriction did not extend their lives. Guarente thinks that SIR2 regulates metabolism, reproduction, and aging when caloric intake is low to help the body survive.

Might a gene that regulates aging in worms also be found in humans? Guarente and his colleague Cynthia Kenyons at the University of California San Francisco think so. They have also found the SIR2 gene in yeast cells (a type of fungus), and it appears to control lifespan in yeast as it does in roundworms. Since roundworms and yeast are only distantly related, the researchers think it is highly probable that this gene is present in most living things. Might it be manipulated to control the lifespan? You can learn more about this research and view a video on the lifespan gene by visiting the *In The News* section of this text's companion Web site at www.wiley.com/college/alters.

*Write your immediate reaction to Guarente's research on the lifespan gene: first, summarize the main point of the research in a sentence or two; then suggest what you think the significance of this research is. You will have an opportunity to reflect on your responses and gather more information on this topic in the In The News feature at the end of this chapter. In this chapter, you will learn more about genes and the hereditary material that comprises them—DNA.*

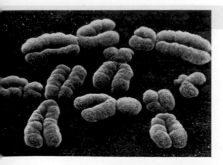

## Where is hereditary material located?

**9.1** Chromosomes contain hereditary material.

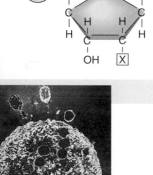

## How did scientists discover the characteristics of hereditary material?

**9.2** A century of research provided information about hereditary material.

## What is hereditary material?

**9.3** Hereditary material is DNA, a double-stranded helix.

**9.4** DNA is a self-replicating molecule.

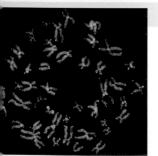

## What are genes?

**9.5** Genes are sequences of nucleotides that direct primarily the synthesis of polypeptides.

**9.6** DNA contains nucleotide sequences that are not expressed.

**9.7** DNA contains repeated sequences.

## Can DNA change?

**9.8** Environmental factors can mutate, or alter, the structure of DNA.

**9.9** Transposable genetic elements are a source of genetic change.

# Where is hereditary material located?

## 9.1 Chromosomes contain hereditary material.

Many first-time parents no doubt wish that their newborn child would come with a set of instructions. Indeed, each child does, but this information has little do with parenting. The instructions come from the parents themselves, hidden in the *chromosomes* each contributed to the child upon conception. Chromosomes, the brightly stained bodies seen in **Figure 9.1**, contain all the information necessary for the development and maintenance of a human being. This information is a highly specialized biochemical "language," somewhat like a complex computer code that can be understood by certain other molecules in the body. For every living organism, it is the hereditary information contained within chromosomes that determines which traits will be exhibited. For children, chromosomes determine which combined traits they will inherit from their parents—such as whether the hair is blond or brown and whether the child is male or female. Most importantly, chromosomes hold the key to the mystery of what makes any one of us unique.

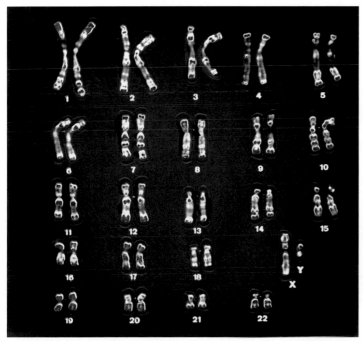

**Figure 9.1** A human karyotype. The chromosomes are stained to visualize the patterns of bands within them. The sex chromosomes are labeled on the karyotype as *X* and *Y*.

Figure 9.1 shows a *karyotype* (CARE-ee-oh-type), the set of chromosomes found in an individual cell in their condensed form during cell division. A karyotype is often represented by an arrangement of chromosomes according to their lengths. Figure 9.1 shows the complete set of 46 chromosomes that are characteristic of human cells. Each species of eukaryotic organism has its own distinctive number of chromosomes. These chromosomes are linear, as in the photo. Nearly all bacteria have a single, circular chromosome.

Scientists develop karyotypes by specially treating and staining a cell just as it is about to divide. The cell is then photographed, and the chromosomes are either cut out of the photograph and physically arranged on a piece of paper or, more commonly today, are manipulated on a computer. Each chromosome has a mate, so the mates are matched and then the matched pairs are placed in order of decreasing size. One purpose of developing a karyotype is to screen for abnormalities in the chromosomes, a topic that is discussed in Chapter 14.

Since the discovery of chromosomes, scientists have learned a great deal about their structure and function. Scientists know that the chromosomes of eukaryotes are made up of two principal components: **deoxyribonucleic** (de-ok-see-RYE-boh-noo-KLAY-ick) **acid** (DNA) and proteins. The DNA contains the hereditary information, commonly called the *code of life*. The proteins, called histone proteins, are bound to the DNA at regularly spaced intervals and help wrap the DNA molecule into a more compact fiber, as shown in **Figure 9.2**. This complex of DNA and histone proteins is called **chromatin** (KRO-muh-tin).

DNA exists as one very long, double-stranded molecule that extends unbroken through the chromosome's entire length. DNA strands are coiled within cells, much as you might coil up your garden hose. The coiling of DNA within cells, however, is much more complicated. Figure 9.2 shows how this coiling is accomplished.

First, the double strand of DNA ❶ winds around separate clusters of eight histone proteins ❷. Each DNA–histone complex is called a *nucleosome*. The nucleosomes look similar to pearls on a string (see the micrograph of rat liver DNA next to the drawing). This string of pearls then wraps up into larger coils ❸, which wrap again into even larger coils called *supercoils* ❹. These supercoils are then looped and packaged with other proteins to form the condensed chromosome ❺.

This process of coiling and condensing of chromatin takes place just before and during cell division, which is described in Chapter 12. This process of DNA compaction is so effective that if all the DNA of your body were strung out and placed end to end, it would extend from the sun to the planet Jupiter and back again about 118 times!

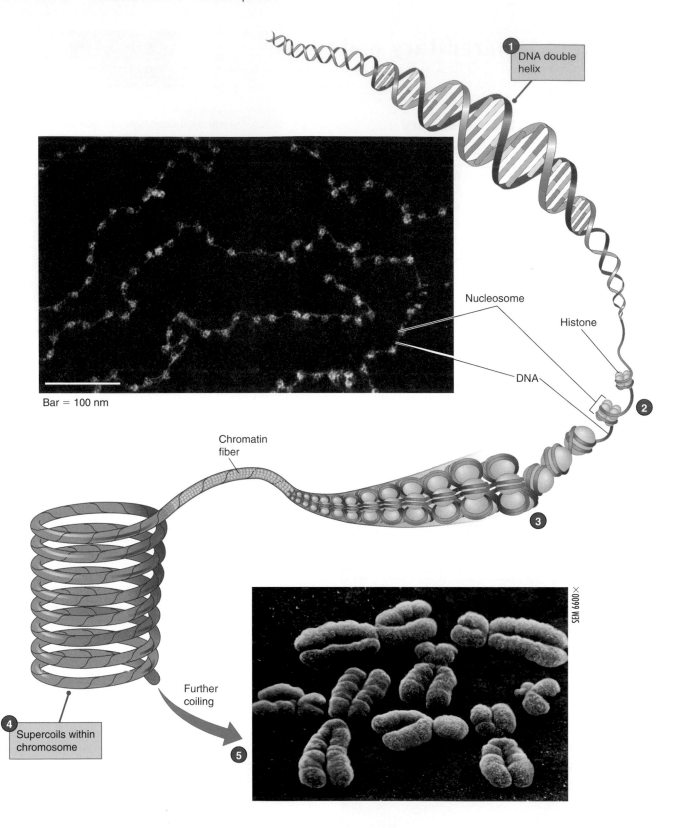

Bar = 100 nm

**Visual Summary** Figure 9.2 How DNA is coiled within chromosomes.

 **Visual Thinking:** The chromatin material condenses to form chromosomes (as shown in the bottom photo) only at a certain time in the "life" of a cell. When does it condense to form chromosomes? What is the benefit to the cell of chromatin being condensed into chromosomes at that time rather than being "strung out"?

# How did scientists discover the characteristics of hereditary material?

## 9.2 A century of research provided information about hereditary material.

Although scientists have been observing and studying chromosomes since the late 1800s, they did not know that DNA was the hereditary material until 1944, with the work of Oswald Avery, which we will describe shortly. Today, scientists understand—in considerable detail—the structure of DNA and the way that information in the DNA of a developing organism directs the formation of organs and tissues. However, it took about a century of research by a succession of scientific investigators to gain this knowledge. **Table 9.1** presents a timeline of significant events in DNA research. Now let's look at the story behind the discovery of DNA as the hereditary material and the determination of its structure and function.

### Miescher isolated "nuclein" from the nuclei of cells

By the early 1800s, scientists had located and named the nucleus of the cell but had no notion of the hereditary material contained within. In 1869, a Swiss chemist named Friedrich Miescher iso-

lated the compound contained within the nuclei of a variety of cells. He noted that his isolate was acidic, phosphorus rich, and made of large molecules. He did not call it DNA; this designation would not come until the 1920s. Miescher called this material *nuclein* because it was specifically associated with the cell nucleus.

### Morgan determined that hereditary material is on chromosomes

Gregor Mendel, an Austrian monk trained in botany and mathematics, conducted pioneering work in genetics in the late 1800s. He suggested that traits are inherited as discrete factors. A bit later, at the beginning of the twentieth century, Walter Sutton suggested that these factors were on the chromosomes, but he had no direct evidence to support this hypothesis. In 1910, however, Thomas Hunt Morgan's experiments provided the first clear evidence upholding Sutton's theory. Morgan was able to show that a specific sex chromosome, the X chromosome, is associated with eye color in fruit flies. We will look more closely at the work of Mendel, Sutton, and Morgan in Chapter 13.

| TABLE 9.1 | Some Important Events Leading to the Discovery of the Genetic Code | |
|---|---|---|
| **Date** | **Researchers** | **Discoveries** |
| 1869 | Friedrich Miescher | Isolated DNA (called it nuclein). |
| Late 1800s | Gregor Mendel | Discovered that traits are inherited as discrete factors. |
| Early 1900s | Walter Sutton | Hypothesized that hereditary information is located on the chromosomes. |
| 1910 | Thomas Hunt Morgan | Provided the first clear evidence supporting Sutton's hypothesis that the hereditary information is located on the chromosomes. |
| 1920s | Phoebus Levene | Discovered that there are two types of nucleic acid within cells, and that both types have nearly identical structures. |
| Early 1940s | Oswald Avery | Provided preliminary evidence that DNA is the hereditary material. |
| Late 1940s | Erwin Chargaff | Showed that the proportion of the bases varies in the nucleic acids of different types of organisms. Also noted that the proportions of any one particular base is always roughly equal to the proportion of another particular base. |
| Early 1950s | George Beadle and Edward Tatum | Developed the one-gene–one-enzyme theory, which states that the production of a given enzyme is under the control of a particular unit of heredity, or gene. |
| 1952 | Alfred Hershey and Martha Chase | Confirmed that DNA (not protein) is the hereditary material. |
| 1953 | Rosalind Franklin and Maurice Wilkins | Analyzed DNA by X-ray diffraction and concluded that it is shaped like a helix (winding staircase). |
| 1953 | James Watson and Francis Crick | Developed double-helix model for the structure DNA. |

# How Science Works

**PROCESSES AND METHODS**

## Is DNA or protein the genetic material?

Students studying biology today sometimes take the information presented in textbooks for granted. For instance, it seems a given that all the genetic information about you—your appearance, how your cells function, your gender—is contained in your DNA. As little as 65 years ago, many scientists thought that the genetic information was contained in and passed along to offspring by proteins located in chromosomes. In 1952, Alfred Hershey and Martha Chase performed a series of important experiments that helped determine that DNA and not protein was the genetic material. Building on an increasing body of experimental evidence that pointed to DNA, Hershey and Chase used viruses to carry out a clear either/or test. They showed that an infection by a virus involved an injection of the virus DNA, not protein, into a cell.

Hershey and Chase reasoned that if DNA were indeed the genetic carrier, then bacteria infected with a virus would contain DNA from the virus inside their cells. They reached this conclusion because they knew that when viruses infect cells, they somehow direct the cell to manufacture more virus particles. On the other hand, if protein were the material of which genes are made, then virus-infected bacteria would contain viral proteins from the original infecting viruses within their cells.

The challenge for Hershey and Chase was to find a way to look inside an infected cell and see what parts of the infecting virus could be found. Hershey and Chase chose to work with a bacteriophage (or simply phage), a type of virus that infects bacteria. Hershey and Chase knew from the previous experiments of other researchers that the phage was not engulfed and taken into the host bacterium during infection, but that part of it remained outside the cell, where it could be seen with an electron microscope.

A key part of Hershey and Chase's experiments was the way they distinguished between DNA and protein—they used radioactive labels. Proteins contain sulfur, but DNA does not, so radioactive sulfur served as a label for protein. DNA contains phosphorus, but protein does not, so radioactive phosphorus served as a label for DNA. Hershey and Chase performed two experiments (**Figure 9.A**). In experiment 1, they labeled the protein coat of the phage with $^{35}$S, a radioactive isotope of sulfur. (See Chapter 3 for a discussion of isotopes.) Then they mixed the labeled phage with *Escherichia coli* bacteria, which are found in abundance in your digestive tract, and allowed the viruses to infect the bacteria.

After infecting the bacteria with phage, Hershey and Chase separated whatever remained outside the bacteria from the infected bacterial cells by agitating the cells with a quick spin in a blender. They then centrifuged the mixture (spun it rapidly). Centrifugation pulled the heavier bacteria to the bottom of a test tube, leaving the lighter material in the solution on top, which is called the supernatant.

In the supernatant was viral (phage) material that was not injected into the bacterium upon infection. Hershey and Chase found all the $^{35}$S-labeled protein in the supernatant, not in the bacteria. Thus, the phage protein did not enter the bacteria during infection.

In experiment 2, Hershey and Chase labeled the phage DNA with an isotope of phosphorus, $^{32}$P, and performed the same agitating and centrifugation as they did in experiment 1. The $^{32}$P was found in the bottom of the test tube, with the bacteria. Clearly, $^{32}$P-labeled DNA had entered the bacteria.

These experiments went a long way toward convincing the scientific community that DNA, not protein, was the genetic material. It confirmed Avery's conclusion that DNA is the transforming, heritable substance. Within the year, Watson and Crick's discovery of DNA's structure showed how easily DNA could fulfill its suggested genetic role. ●

**Figure 9.A** The experiments of Hershey and Chase showing that DNA, not protein, is the genetic material.

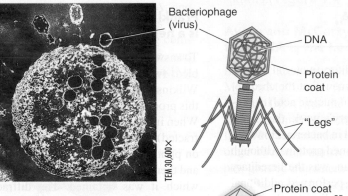

Bacteriophage (virus)

DNA

Protein coat

"Legs"

TEM 30,600×

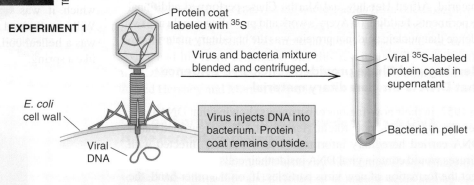

**EXPERIMENT 1**

Protein coat labeled with $^{35}$S

*E. coli* cell wall

Viral DNA

Virus and bacteria mixture blended and centrifuged.

Virus injects DNA into bacterium. Protein coat remains outside.

Viral $^{35}$S-labeled protein coats in supernatant

Bacteria in pellet

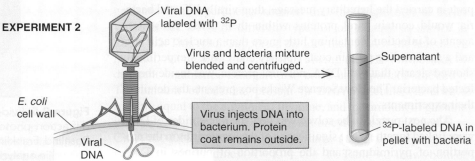

**EXPERIMENT 2**

Viral DNA labeled with $^{32}$P

*E. coli* cell wall

Viral DNA

Virus and bacteria mixture blended and centrifuged.

Virus injects DNA into bacterium. Protein coat remains outside.

Supernatant

$^{32}$P-labeled DNA in pellet with bacteria

## Watson and Crick determined the structure of DNA

Learning informally of Franklin and Wilkins's results before they were published in 1953, James Watson and Francis Crick, two young scientists at Cambridge University in England, quickly worked out a likely structure of the DNA molecule. They built models of the nucleotides, assembled them into molecular structures, and then tested each to see whether its structure fit with what they knew from Chargaff's, Franklin's, and Wilkins's work. They finally hit on the idea that the molecule might be a double helix (two springs twisted together) in which the uprights of the ladder were alternating sugar molecules and phosphate molecules of the nucleotides, bonded together to form long chains. Pointing inward toward one another from these two long chains were the nucleotide bases: purines and pyrimidines (**Figure 9.6**). Pairing a purine (which is large) with a pyrimidine (which is small) resulted in "lad-

der rungs" of uniform length, which gives the molecule stability. Always pairing the purine adenine (A) with the pyrimidine thymine (T) and the purine guanine (G) with the pyrimidine cytosine (C) yielded a molecule in which the amount of adenine was always equal to the amount of thymine, and the amount of guanine was always equal to the amount of cytosine, a molecule consistent with Chargaff's observations. The bases were held together by hydrogen bonds—two between adenine and thymine, and three between guanine and cytosine. These pairings—A with T, G with C—are referred to as the complementarity of the bases. For their discoveries concerning the molecular structure of DNA and its significance for information transfer in living material, Watson, Crick, and Wilkins shared the 1962 Nobel Prize in Physiology or Medicine. Franklin died before nominations were made, so she was unable to share the award.

### CONCEPT CHECKPOINT

3. The structure of DNA might be thought of as analogous to that of a twisted rope ladder with rigid wooden rungs. In this analogy, which part of the DNA molecule do the rigid rungs of the ladder represent? What does the flexible rope portion of the ladder represent?

## 9.4 DNA is a self-replicating molecule.

Each cell in our bodies, with the exception of the sex cells and red blood cells, carries a full complement of hereditary material. This fact holds true for other living things as well. As organisms grow and new cells are formed, exact copies of the DNA must be generated and "parceled out" to them. How might this occur? A key part of the answer involves the complementarity of the bases.

The sequence of bases on one side of the DNA ladder is a code that determines the sequence of bases on the other side of the ladder. If the sequence on one side were ATTGCAT, for example, the complementary sequence on the other side would have to be TAACGTA.

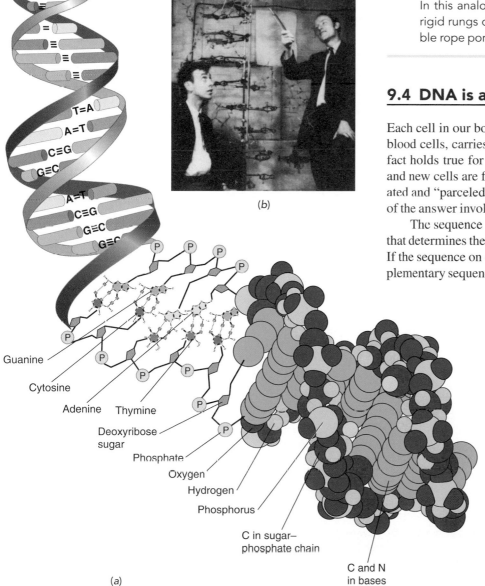

Guanine
Cytosine
Adenine   Thymine
Deoxyribose sugar
Phosphate
Oxygen
Hydrogen
Phosphorus
C in sugar–phosphate chain
C and N in bases

(a)

(b)

**Figure 9.6 Structure of DNA.** (*a*) In the upper portion of this illustration, the purple bands represent the sugar–phosphate "uprights" of the DNA "ladder." The bases are denoted by the colored cylinders and—a little further down in the model—the letters A (adenine), C (cytosine), T (thymine), and G (guanine). A G C base pair is bonded by three hydrogen bonds, and an A-T base pair is bonded by two hydrogen bonds. The middle part of the illustration shows more detail in the structures of the nucleotides and in how they link together. In the bottom portion of the model, each atom is represented by a sphere. (*b*) Watson and Crick with their first model of the DNA double helix.

Before a cell divides, an enzyme breaks the bonds between the complementary bases in short sections of the double-stranded DNA molecules, and the complementary strands separate from one another. This process is commonly referred to as *unwinding* or *unzipping*. Replication begins at various places in the DNA strand and proceeds in both directions from each replication point, creating replication bubbles as shown in **Figure 9.7a**.

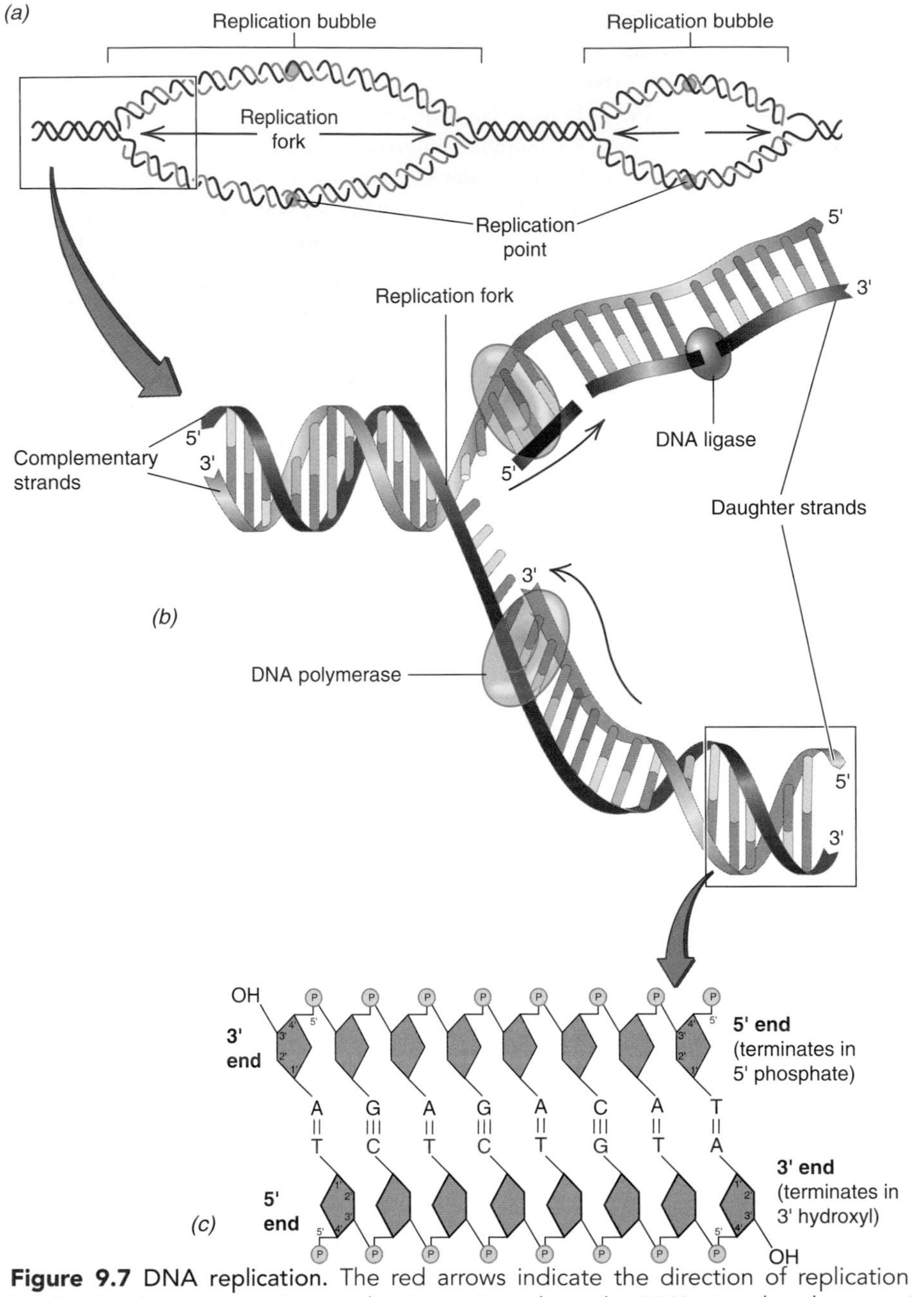

Figure 9.7b shows one end of a replication bubble, which is called a *replication fork*. At these forks, each separated strand serves as a template for the synthesis of a new, complementary strand. In a process directed by the enzyme DNA polymerase, bases in free nucleotide units, which are present along with the DNA in the nucleus, link to complementary bases on each of the DNA strands.

Each strand of DNA has a 3′ (3-prime) end and a 5′ (5-prime) end. Find these designations on the DNA molecule shown in Figure 9.7b. These names refer to positions of particular carbon atoms in the sugar molecules of the ladder uprights. More specifically, note that in Figure 9.7c, the 5′ end of each DNA strand terminates with a phosphate group, which is attached to the carbon occupying the 5′ position. The 3′ end of each DNA strand terminates with a hydroxyl group (–OH), which is attached to the carbon occupying the 3′ position.

DNA polymerase can add nucleotides only to a 3′–OH group. Put simply, it works at the 3′ end of the molecule, adding nucleotides in the 5′ to 3′ direction on the growing strand. Therefore, replication proceeds *toward* the replication fork on one strand—the upper strand in Figure 9.7b. On the other strand, replication proceeds *away from* the replication fork—the lower strand in Figure 9.7b. An enzyme called *DNA ligase* assists bonding during replication.

Thus, the process of DNA replication begins with one double-stranded DNA molecule and ends with two double-stranded DNA molecules. Each of the two new molecules contains one strand from the original molecule plus a new, complementary strand assembled from free nucleotides. In this process, termed *semiconservative replication*, each of the new double strands of DNA retains half the original DNA molecule (*semi-* means half). The two new double-stranded molecules are identical to each other and to the original parent molecule. In 1959, American biochemist and physician Arthur Kornberg received the Nobel Prize in Physiology or Medicine for discovering the process by which DNA is replicated in bacterial cells.

**Figure 9.7** DNA replication. The red arrows indicate the direction of replication. (a) Replication begins at various replication points along the DNA strand and proceeds in both directions simultaneously, forming replication bubbles. (b) The double-stranded DNA molecule separates at its bases, forming a split, or replication fork. Each separated strand acts as a template. The enzyme DNA polymerase adds free nucleotides to the growing, complementary daughter strands, and the enzyme DNA ligase catalyzes bonding. Replication proceeds from the 5′ end to the 3′ end on each growing strand. (c) Each of the complementary DNA strands has a 3′ end and a 5′ end. The 5′ end of one strand is opposite the 3′ end of the other strand.

**Visual Thinking:** In part (b) of this illustration, describe how replication is shown to be occurring in opposite directions on the unzipped DNA molecule. Why does the process occur in this fashion?

# What are genes?

## 9.5 Genes are sequences of nucleotides that direct primarily the synthesis of polypeptides.

Before Watson and Crick's discovery of the structure of DNA, scientists were working to determine how the hereditary material directs the growth and development of an organism. They asked the question, "What is the hereditary message and how is that message expressed?" Experiments conducted in the late 1940s and early 1950s by two geneticists, George Beadle and Edward Tatum, answered this question. These researchers studied the biochemical characteristics of the expression of what we now know as genes. They worked with the common bread mold *Neurospora* (ner-ROS-per-ah).

At the same time Beadle and Tatum were doing their work, various biochemists were also studying the manufacture and breakdown of organic molecules within cells. They determined that cells build and degrade molecules by sequences of steps in which each step is catalyzed by a specific enzyme. These enzymatically controlled sequences of steps are called biochemical pathways.

Beadle and Tatum studied mold cultures that they had exposed to X rays. The X rays induced *mutations*, or changes, in the hereditary material of the organism, resulting in a variety of mutants, each unable to manufacture certain amino acids. Amino acids, as noted in Chapter 4, are the building blocks of proteins. Using known information about the biochemical pathways of *Neurospora*, they hypothesized that specific enzymes must be involved in the manufacture of these amino acids and that some or all of the enzymes in each biochemical pathway must not be doing their jobs.

To test their hypothesis, Beadle and Tatum chose mutants unable to synthesize the amino acid arginine. They supplied each mutant with various compounds intermediate in the arginine pathway and observed whether the mutant was then able to synthesize argi-

nine. Using this method, Beadle and Tatum were able to infer the presence of defective enzymes in this biochemical pathway.

From these studies, Beadle and Tatum proposed the **one-gene–one-enzyme theory**, which states that the production of a given enzyme is under the control of a specific gene, a unit of heredity (**Figure 9.8**). If the gene mutates, the enzyme will not be synthesized properly or will not be made at all. Therefore, the reaction it catalyzes will not take place, and the product of the reaction will not be produced.

For example, suppose the product of a biochemical pathway is arginine (pictured as an oval with two tails), as shown in Figure 9.8. In order to produce that product, acetylornithine **1** is converted to ornithine **2**. You can think of this change as enzyme A changing the rectangle to a square. Next, enzyme B adds a tail to the square, forming citrulline **3**. Enzyme C changes the square to the oval arginosuccinine **4**, and enzyme D adds another tail to the oval representing arginine **5**. Notice how blocking one step would also block all the other steps "down the line." For their discovery that genes act by regulating definite chemical events, Beadle and Tatum were awarded the Nobel Prize in Physiology or Medicine in 1958.

Scientists later refined the one-gene–one-enzyme hypothesis to say that the production of a given polypeptide (a *portion* of an enzyme or other protein) is under the control of a single gene. This was commonly known as "one gene–one polypeptide." Today, scientists know that in most cases one gene codes for multiple polypeptides, depending on factors involved in gene expression. (See Chapter 10.) Some genes code for proteins directly, because some proteins are composed of only one or a few polypeptides, and some genes code for certain types of RNA.

In summary, then, a unit of heredity is a **gene**—a sequence of nucleotides that codes for (scientists say encodes) the amino acid sequences of polypeptides or for RNA.

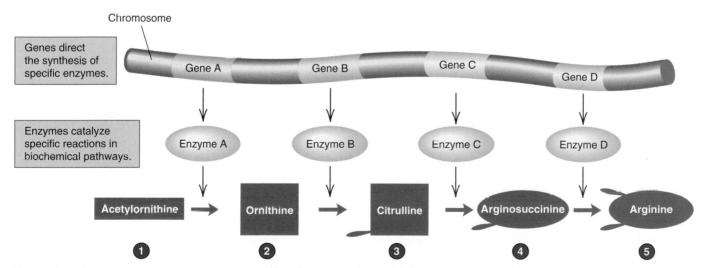

**Figure 9.8** The one-gene-one-enzyme theory. This illustration shows the biochemical pathway for the synthesis of the amino acid arginine. Each gene directs the synthesis of one enzyme that catalyzes a specific step in this pathway.

## 9.6 DNA contains nucleotide sequences that are not expressed.

You might think that your entire complement of DNA is used to produce polypeptides or RNA, but that is not the case. The **genome** of an organism, all the genetic information in one complete set of hereditary material, contains nucleotide sequences that are not expressed, such as promoters, introns, repetitive DNA, and telomeres.

*Promoters* can be thought of as "start here" messages to decode genes. The nucleotide sequences that comprise genes can be likened to letters in a word. When you read words, you are aided by spaces between the words that signal the beginning of a new word. So, too, the sequences of nucleotides that comprise genes must be set off in some way. There must be a cue that tells where each gene begins. This is the job of promoters. They are key in translating a gene to a polypeptide and will be discussed in more detail in Chapter 10.

*Introns* are extra sequences of nucleotides that intervene between the polypeptide-specifying portions of the gene. They are called introns because they intrude into the gene but are not expressed. The remaining segments of the gene—the nucleotide sequences that encode the amino acid sequence of the polypeptide—are called *exons*, those portions of a gene that will be expressed. Introns interrupt most, but not all, eukaryotic genes. They vary in length from about 50 nucleotide pairs to thousands of nucleotide pairs. For example, the gene that codes for the connective tissue collagen in chickens is comprised of 37,000 nucleotide pairs, but only 4600 nucleotides code for the protein. Introns play several roles in gene regulation and gene expression, which are topics of Chapter 10.

## 9.7 DNA contains repeated sequences.

The DNA of eukaryotic organisms also contains repeated sequences, called *repetitive DNA*. From 20 to 50 percent of eukaryotic genomes contains repeated sequences. The repeats may be short—just a few nucleotides—or may be long—up to a few hundred nucleotides. Scientists hypothesize that repeats may play structural or organizational roles in chromosomes, may be involved in certain processes during gamete production, may protect other DNA sequences, and may serve no current purpose.

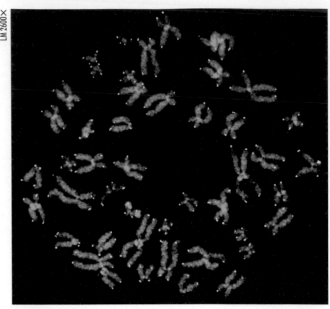

**Figure 9.9** Telomeres (yellow) at the ends of human chromosomes.

One type of repetitive DNA is located at the ends of eukaryotic chromosomes. These repetitive DNA sequences that "cap" the ends of chromosomes are called *telomeres* and are shown in **Figure 9.9**. Every time a cell replicates, the telomeres shorten. Cells cannot function when telomeres become too short; they allow cells to distinguish between chromosome ends and the ends of broken DNA. When the chromosome ends become too short for the cell to function, it dies. However, in some cells the enzyme telomerase helps rebuild the repetitive sequences after cell division. The repetitive sequences, then, preserve the life of a cell, conferring a type of cell immortality. Telomerase is only found in certain cells, such as eggs, sperm, and cancer cells. In their quest to find new cancer therapies, scientists are investigating ways of inhibiting the telomerase function in cancer cells as a means of causing cancer cell death.

---

**CONCEPT CHECKPOINT**

4. Sequences of DNA nucleotides have a variety of functions in cells. List and describe those functions.

---

## Can DNA change?

## 9.8 Environmental factors can mutate, or alter, the structure of DNA.

The answer to the question "Can DNA change?" is both yes and no. Generally speaking, an individual's DNA does not change enough over the lifespan to show any noticeable differences when it is analyzed in the laboratory, such as when DNA is analyzed in criminal cases. However, environmental factors, which are discussed in this section, *can* change DNA in significant ways.

A change in the genetic message of a chromosome caused by a change in the DNA molecule is called a **point mutation** or **gene mutation**. Through mutation, a new gene is produced. Changes can also take place in sections of chromosomes, in which chromosomes break and reattach in new ways. Humans may inherit chromosomes having structural changes (chromosomal rearrangements) such as these. We will discuss the occurrence and inheritance of chromosomal rearrangements in more detail in Section 14.15.

# just wondering . . .

## I've heard that DNA analysis can help solve crimes. How is DNA analyzed?

The use of DNA analysis in criminal investigations is called DNA profiling. The term for the analytic process, called *DNA fingerprinting*, was coined in 1985 by British geneticist Alec Jeffreys. He was the first to use the technique in paternity, immigration, and murder cases.

DNA fingerprints were first allowed as evidence in U.S. courts in 1990. Today, DNA evidence is used in fewer than 1% of prosecutions, but these cases usually involve murder or rape, in which the DNA of the perpetrator has been left behind in either blood or semen.

One way scientists analyze DNA is by amplifying certain fragments with the polymerase chain reaction (PCR), separating the fragments using electrophoresis, and then visualizing the bands formed. These techniques are described in Chapter 11. The resulting banding patterns, such as those shown in **Figure 9.B**, create the DNA fingerprint. This fingerprint is unique to every individual, just as the fingerprints on your hands are unique. DNA fingerprinting can identify an individual with virtual certainty—even if the DNA sample and the individual have aged from when the crime was committed.

DNA fingerprinting is an important tool to help solve crimes, and a huge backlog exists in processing samples. In late 2003, about 350,000 DNA samples from crime scenes and victims awaited analysis nationwide. About 300,000 samples from convicted criminals needed processing as well. One method for tackling this backlog is a new computer chip that can process eight DNA samples in 20 minutes, compared to the current method that processes one sample in several hours. At this writing, the House of Representatives had passed the Advancing Justice Through DNA Technology Act, which provided $105 million over four years from Congress to speed DNA analysis, eliminate state and federal DNA analysis backlogs, and train nurses to treat sexual assault cases. ●

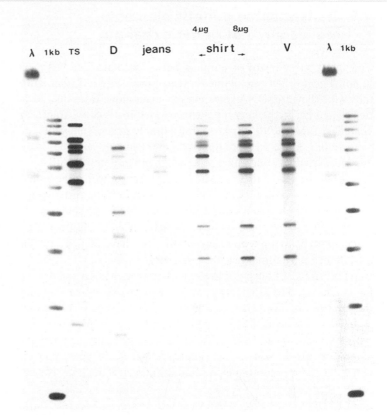

**Figure 9.B** DNA fingerprinting of the blood from a defendant, a victim, and the victim's clothing. A stabbing victim's DNA profile is shown in column V; the defendant's is shown in column D. The banding patterns of the blood taken from the defendant's shirt and jeans match those of the victim, not the defendant.

*Are you wondering about a topic in biology and how it relates to your life? Submit your question by clicking the* Just Wondering *link in this text's companion Web site at* www.wiley.com/college/alters.

There is evidence that chromosomes break naturally, sometimes through the activity of transposable genetic elements, which are DNA sequences that can move from one place in the DNA to another. Occasionally, transposable elements induce breaks in chromosomes as they move. Transposable genetic elements are discussed in the next section.

Ionizing radiation, ultraviolet radiation, and chemicals are also sources of change in DNA. **Ionizing radiation**, such as X rays and nuclear radiation, is a form of energy known as *electromagnetic energy*. Sunlight is also a type of electromagnetic energy. Ionizing radiation, however, has a higher level of energy than ordinary light does. When ionizing radiation reaches a cell, it can create charged molecular fragments, or *free radicals*, which are highly reactive. They may interact with DNA, producing chromosomal breaks or changes (mutations) in the nucleotide structure.

Ultraviolet (UV) radiation, the component of sunlight and of the lights used in tanning beds that leads to suntan and sunburn, is much lower in energy than X rays but still higher in energy than ordinary light. Certain molecules within the structure of chromosomes absorb UV radiation, developing chemical bonds among them that are not normally present. These unusual bonds produce a kink in the molecular structure of the DNA of chromosomes. Normally, a chromosome is able to be repaired by enzymes removing the affected molecules and synthesizing new, undamaged molecules. Sometimes, however, mistakes can take place in some part of the repair process and a change may occur in a gene.

Sometimes chemicals damage DNA directly. LSD (the hallucinogenic compound lysergic acid diethylamide), marijuana (leaves of the *Cannabis sativa* plant), cyclamates (compounds widely used as nonnutritive sugar substitutes), and certain pesticides (compounds used to kill insects on crops and other plants) are a few examples of chemicals known to damage DNA. In general, chemicals alter the molecular structure of nucleotides, resulting in a mispairing of bases. As with UV radiation, mistakes can take place as the cell works to repair the DNA, and changes may then occur in a gene.

## 9.9 Transposable genetic elements are a source of genetic change.

You've probably heard of jumping beans, but have you heard of jumping *genes*? This nickname for **transposable genetic elements** tells what they are: DNA sequences that can move from one position to another in the genome. First reported in 1948 by American geneticist Barbara McClintock, transposable genetic elements, or *transposons*, make up more than 40% of the human genome. They are widespread in other eukaryotes and in prokaryotes as well.

When transposons move from one location to another, they may break chromosomes or mutate genes. Therefore, these jumping genes have profound significance, since they contribute considerably to the rate of mutation in an organism. Their widespread distribution in the living world suggests that they have played a role in evolution, contributing to genetic diversity.

Insertions of transposable elements are the cause of many visible mutations. McClintock discovered transposons when studying the mottling of the kernels in maize (corn) as shown in **Figure 9.10**. Few scientists appreciated the significance of McClintock's work initially, since scientists viewed genes as occupying fixed positions on chromosomes. Her evidence that some genes moved about the genome contradicted what was known at the time. Never-

**Figure 9.10** Transposable genetic elements are responsible for the mottling in kernels of maize.

theless, in the 1960s and 1970s, jumping genes were discovered in bacteria and in fruit flies, and other scientists began basing their work on McClintock's findings. She was awarded the Nobel Prize for her work on transposable genetic elements in 1983, about 35 years after her first publication on the topic.

### CONCEPT CHECKPOINT

5. Describe three cellular events that can mutate DNA.

## CHAPTER REVIEW

### Summary of Key Concepts and Key Terms

**Where is hereditary material located?**

**9.1** Chromosomes contain hereditary material.
**9.1** A karyotype is the set of chromosomes found in an individual cell in their condensed form during cell division.
**9.1** Proteins and **deoxyribonucleic** (de-ok-see-RYE-boh-noo-KLAY-ick) **acid** (DNA, p. 125) make up the **chromatin** (p. 125) of the cell.

**9.1** Chromatin forms chromosomes by coiling and condensing prior to cell division.

**How did scientists discover the characteristics of hereditary material?**

**9.2** Miescher isolated "nuclein" from the nuclei of cells.
**9.2** Morgan determined that hereditary material is on chromosomes.
**9.2** Levene discovered two types of **nucleic acid** (p. 128): deoxyribonucleic acid (DNA) and **ribonucleic acid** (**RNA**, p. 128).
**9.2** Both DNA and RNA are made up of **nucleotides** (p. 133), which each contain a sugar, a phosphate group, and a base.
**9.2** There are five different bases in nucleotides: adenine (A), guanine (G), cytosine (C), thymine (T), and uracil (U); A, G, and C are found in both DNA and RNA; T is found only in DNA, and U is found only in RNA.

**9.2** A and G are **purines** (PYOOR-eens, p. 128); C, T, and U are **pyrimidines** (pih-RIM-uh-deens, p. 128).
**9.2** Avery and McCarty provided preliminary evidence that DNA is the hereditary material.
**9.2** Chargaff provided additional evidence that DNA is the hereditary material
**9.2** Chargaff also noted these rough proportions in the amounts of the bases in DNA molecules: A = T, G = C, and A + G = C + T.

**What is hereditary material?**

**9.3** Hershey and Chase provided conclusive evidence that DNA, not protein, is the hereditary material.
**9.3** Using X-ray diffraction analysis, Franklin and Wilkins provided evidence that DNA is a helical coil.

**9.3** Watson and Crick determined the structure of DNA.
**9.3** DNA is a double-stranded, helical molecular "ladder," with rungs of uniform length and with alternating sugar–phosphate units forming the ladder uprights.

**9.3** Each "rung" pairs a purine with a pyrimidine—A always pairs with T, and C always pairs with G; this is known as the complementarity of the bases.

**9.4** During DNA replication, DNA polymerase bonds free nucleotides to the exposed bases of unzipped DNA, producing two exact copies of the original DNA strand.

## What are genes?

**9.5** The units of heredity are **genes** (p. 133).

**9.5** Beadle and Tatum proposed the **one-gene–one-enzyme theory** (p. 133), which states that the production of a given enzyme is under the control of a specific gene.

**9.5** Today, scientists know that in most cases one gene codes for multiple polypeptides depending on factors involved in gene expression.

**9.5** Although most genes code for a string of amino acids (polypeptides), there are also genes devoted to the production of certain types of RNA.

**9.6** The **genome** (p. 134) of an organism contains nucleotide sequences that are not expressed, such as promoters and introns.

**9.7** DNA contains repeated sequences of nucleotides, including telomeres.

## Can DNA change?

**9.8** A change in the genetic message of a chromosome caused by a change in the DNA molecule is called a **point mutation** or **gene mutation** (p. 134).

**9.8** Chromosomes may break, and the broken pieces may reattach in new ways.

**9.8** Chromosome breaks may occur naturally by means of transposable genetic elements.

**9.8** Environmental factors such as high-energy radiation (**ionizing radiation**, p. 135), low-energy radiation (UV radiation), and certain chemicals can damage DNA.

**9.9** **Transposable genetic elements** (p. 136), or transposons, are DNA sequences that can move from one position to another in the genome.

**9.9** Insertions of transposable elements are the cause of many visible changes in organisms.

---

## Level 1 — Learning Basic Facts and Terms

**Matching**

Match the scientist with his/her discovery:

1. Alfred Hershey and Martha Chase
2. Gregor Mendel
3. George Beadle and Edward Tatum
4. James Watson and Francis Crick
5. Barbara McClintock
6. Thomas Hunt Morgan
7. Erwin Chargaff
8. Rosalind Franklin and Maurice Wilkins

a. Transposable genetic elements
b. Amount of A=T and C=G within a cell
c. DNA rather than protein is the genetic material
d. Determined the structure of DNA
e. One gene–one enzyme hypothesis
f. Genes are located on chromosomes
g. Traits are inherited as discrete factors
h. X-ray diffraction suggested DNA is a helical molecule

9. Which of the following is characteristic of DNA replication?
   a. It occurs in the same manner on both strands of the double helix.
   b. It always produces three identical copies of DNA.
   c. It is prone to making frequent errors.
   d. It involves complementary base pairing.

10. How is DNA different from RNA?
   a. RNA is protein; DNA is nucleic acid.
   b. RNA contains ribose sugar; DNA contains deoxyribose sugar.
   c. Uracil takes the place of guanine in RNA.
   d. RNA lacks the phosphate group found in DNA.

---

## Level 2 — Learning Concepts

1. Explain how one of your unique characteristics, such as your hair color, is a result of specific chemical instructions. Where do these instructions originate?
2. What are the two types of nucleic acid found in your cells? Describe the structures of each.
3. Explain the term *double helix* and describe its structure.
4. What characteristics of DNA did Chargaff's experiments reveal? Why was this significant?
5. At the end of Watson and Crick's landmark paper in *Science*, in which they proposed the structure of the DNA molecule, they made the following statement: "It has not escaped our notice that the pairing . . . immediately suggests a possible copying mechanism for the genetic material." What were they referring to, and what do you suppose they meant by this?
6. Beadle and Tatum's "one-gene–one enzyme" hypothesis is only partially accurate. Explain why this is so.
7. Many of the pioneering discoveries in molecular genetics were made by women. Summarize one of their discoveries that you feel was insightful or interesting.

## Critical Thinking and Life Applications

1. As cells replicate their DNA before they divide, the ends of the strands of DNA are not completely replicated. Over time, this nonreplication leads to shortening of the DNA. Why might this pose a problem to cells? What characteristic of the DNA provides some protection against this problem?

2. You may know that Dolly, the first cloned mammal (a sheep), died. Dolly was produced by inserting the nucleus of a fully formed adult udder cell into a denucleated egg cell. This fused cell began to divide into an embryo, was implanted in a surrogate mother, and ultimately a lamb (Dolly) was born. Although Dolly died relatively young, she had many physical ailments that were characteristic of much older sheep. Speculate on why this was the case.

3. Viruses replicate by inserting their DNA into host cells. This DNA contains the instructions for building more viruses inside host cells. Often, this viral DNA is spliced into the DNA of the host cell, where it is copied and passed to daughter cells along with the rest of the host cell's DNA. Scientists have hypothesized that some of this viral DNA is still in our cells, having been passed down through the generations, but is no longer expressed. What characteristic of DNA is consistent with this hypothesis?

4. The nineteenth- and twentieth-century race to find the molecule of inheritance and determine its structure is a classic example of the process of science. Explain what this statement means.

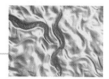

## In The News | Critical Thinking

### LIFESPAN GENE

Now that you understand more about genes and DNA, reread this chapter's opening story about research into a lifespan gene. To understand this research better, it may help you to follow these steps:

1. Review your immediate reaction to Guarente's research on the SIR2 gene that you wrote when you began reading this chapter.

2. Based on your current understanding, again summarize the main point of the research in a sentence or two.

3. What questions do you now have about this research that are not answered by the opening story?

4. Collect new information about the research. Visit the *In The News* section of this text's companion Web site at www.wiley.com/college/alters and

watch the "Lifespan Gene" video. Then use the "summary" link to read the accompanying story and access related links. Use this information, the links provided, and other online and library resources to answer your questions and find updates about this research topic. State the sources of your information. Explain why you think the information is accurate. Also determine if the information expresses a particular point of view or is biased in any way.

5. What in your view is the most significant aspect of this research? Give reasons for your opinion and for any changes in your ideas based on the additional information you have collected and the analysis you have done.

# GENE EXPRESSION AND REGULATION

## In The News | Mouse of a Different Color

Scientists have found direct links between specific genes and the development of certain diseases and conditions, such as Type I diabetes, Alzheimer's disease, and breast cancer. However, having a gene that is linked to a disease is not the same as having the disease. A gene must be turned on and expressed. That is, the polypeptides for which genes code must by synthesized and exert their effects on the organism.

Scientists use mice as one model system to study gene expression and its control in mammals. Until recently, scientists have been able to control *where* in mice particular genes are expressed but not *when* the genes are expressed. Heidi Scrable and her colleagues at the University of Virginia School of Medicine are *In the News*, having developed a way to turn gene expression on and off in mice.

Scrable and her team have been able to take a sequence of DNA found in *Escherichia coli* bacteria, a common resident of the human large intestine, and transfer it to mice, incorporating it into their DNA. This sequence of DNA is a genetic switch that controls the metabolism of lactose, commonly known as milk sugar. The gene switch turns on the gene to which it is attached in the presence of lactose and turns off the gene in its absence. Scrable attached the lactose switch to a gene in mice that produces pigment. When lactose is not present in the drinking water of the mice, the pig

ment gene is not turned on, and the mice are albino (without pigmentation). When lactose is present in the drinking water of the mice, the switch turns on the pigment gene, and these same mice develop brown pigment in their eyes, skin, and fur.

Scrable used this visible system for reversible control of gene expression so that she and her team were able to tell easily whether they were successful in transferring the gene control system into the mouse. She hopes that other scientists will use the technique for controlling genes linked to human disease, allowing them more experimental control over the timing of gene expression. You can learn more about this research and view the video "Mouse of a Different Color" by visiting the *In The News* section of this text's companion Web site at www.wiley.com/college/alters.

---

*Write your immediate reaction to Scrable's research on a system for reversible control of gene expression: first, summarize the main point of the research in a sentence or two; then suggest what you think the significance of this research is. You will have an opportunity to reflect on your responses and gather more information on this topic in the* In The News *feature at the end of this chapter. In this chapter, you will learn more about gene expression and its regulation.*

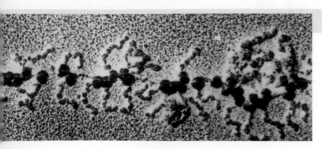

## How is the DNA code expressed?

**10.1** The DNA code is expressed primarily as the production of polypeptides.

**10.2** The flow of genetic information is from DNA to RNA to polypeptide.

**10.3** The DNA code is transcribed into the more mobile code of RNA.

**10.4** The RNA code is translated into polypeptides.

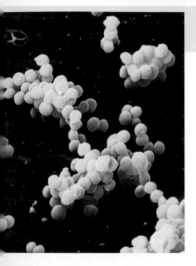

## How is the production of polypeptides, or gene expression, regulated?

**10.5** In prokaryotes, gene expression is regulated primarily by controlling transcription.

**10.6** In eukaryotes, gene expression is regulated by controlling transcription, mRNA processing, and translation.

## What are master control genes, and how is gene expression activated?

**10.7** Master control genes trigger the development of organs and other structures.

**10.8** Gene expression is activated by environmental factors and signaling molecules.

# How is the DNA code expressed?

## 10.1 The DNA code is expressed primarily as the production of polypeptides.

Knowing about the structure of DNA does not tell you how it works. However, knowing that the structure of DNA is directly related to its function is a good start toward that understanding.

DNA functions as the universal code of life; it is the genetic material of all organisms and most viruses. As we described in Chapter 9, DNA is a double-stranded helix comprising a sequence of nucleotide pairs. It is replicated as organisms grow and develop, assuring that each new body cell contains an exact copy of the hereditary code. Occasional errors in the DNA code, which are due to mutations or errors in DNA replication that are not repaired, result in genetic variation. In fact, heredity and variation—its sources, nature, and implications—is what **genetics** is all about.

How does DNA fulfill its role as the code of life? It's actually quite simple—DNA primarily directs the synthesis of **polypeptides**, which are chains of amino acids linked by peptide bonds. **Proteins** are large molecules made up of one to several polypeptides, so in some instances, polypeptides *are* proteins. Some DNA also directs the synthesis of types of RNA that are *not* translated into polypeptides, often called noncoding RNA (ncRNA). Each unit of DNA that codes for a polypeptide or strand of ncRNA is called a **gene**. In other words, genes carry codes as sequences of DNA bases. The code of life, then, is a code primarily for making polypeptides but also for making ncRNA. These polypeptides are used to assemble proteins responsible for regulating body functions, such as building tissues or digesting food. **Gene expression** is the process by which genes produce polypeptides and RNA, and by which these products exert their effects on an organism. If the regulation of gene expression malfunctions, then certain polypeptides needed to carry on life activities are not made properly, are made in deficit or excess, or are not made at all. The result may be minor and go unnoticed, or it may be devastating.

## 10.2 The flow of genetic information is from DNA to RNA to polypeptide.

DNA does not make polypeptides directly. It uses another nucleic acid, called ribonucleic acid, or RNA, to carry the code from the nucleus to the cytoplasm of the cell. There, polypeptides are manufactured on the **ribosomes** (RYE-buh-somes) (see Chapter 5).

As we described in Chapter 9, RNA is similar to DNA, made up of nucleotides containing a sugar, a nitrogenous base, and a phosphate group (see Figure 9.3). In DNA the sugar is deoxyribose; in RNA it is ribose, a sugar with one more oxygen atom than deoxyribose. Both DNA and RNA have four nitrogenous bases in their structures. Both have the bases guanine, adenine, and cytosine. However, DNA contains only thymine, while RNA contains only uracil (see Figure 9.4).

Cells contain three types of RNA, and each plays a special role in the production of polypeptides. In general, polypeptide synthesis occurs like this: **Messenger RNA (mRNA)** is made in the nucleus using DNA as a template. It then travels to the ribosomes to direct the synthesis of a polypeptide. Put simply, mRNA takes the message of

DNA to the cytoplasm to be translated into a polypeptide. A second type of RNA, called **transfer RNA (tRNA)**, is found in the cytoplasm. During polypeptide synthesis, tRNA molecules transport amino acids, which are used to build the polypeptide, to the ribosomes. In addition, tRNA molecules position each amino acid at the correct place in the growing polypeptide chain. So transfer RNA gets its name because it transfers amino acids from the cytoplasm to the ribosomes. The third type of RNA is the primary constituent of ribosomes and is therefore called **ribosomal RNA (rRNA)**. These complex polypeptide-making factories contain more than 50 different proteins in their structure along with RNA. **Figure 10.1** illustrates

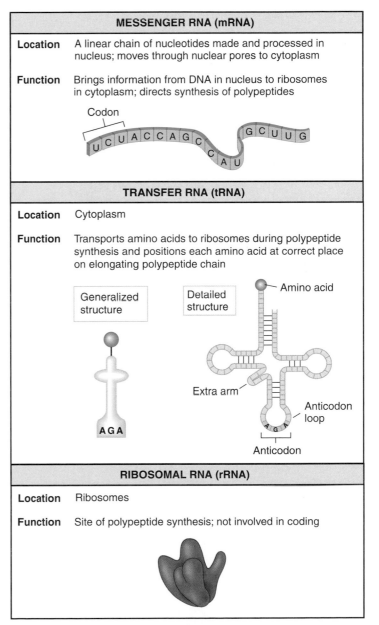

**Figure 10.1 The three types of RNA.** The diagrammatic representations of the shapes of these molecules are used to depict each type of RNA throughout this chapter.

the three types of RNA, gives their locations, describes their functions, and shows diagrammatic representations of their shapes.

**1.** There are three types of RNA involved in gene expression. Name the type that carries out each of the following roles: (1) A molecular shuttle that carries the components of a polypeptide to the site of protein synthesis. (2) A molecular messenger that relays the genetic information to the site of protein synthesis. (3) The molecular "workbench" of protein synthesis.

## 10.3 The DNA code is transcribed into the more mobile code of RNA.

More specifically, how does the process of polypeptide synthesis (gene expression) work? The first step in this process is the copying of the gene into a strand of mRNA, a process called **transcription**.

Transcription begins when the enzyme RNA polymerase (pol-IH-meh-race) binds to a sequence of nucleotides that makes up a promoter site on a single DNA strand (**Figure 10.2 ❶** ). A *promoter site* is a sequence of nucleotides located at the beginning of the gene that is being expressed. You can think of such a sequence of nucleotides as a signal that says, "Start here."

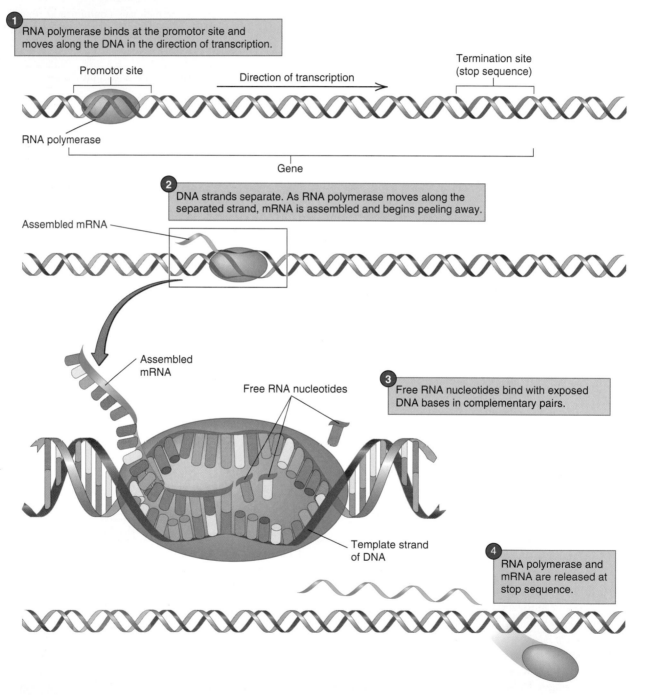

**❶** RNA polymerase binds at the promotor site and moves along the DNA in the direction of transcription.

Promotor site

Direction of transcription

Termination site (stop sequence)

RNA polymerase

Gene

**❷** DNA strands separate. As RNA polymerase moves along the separated strand, mRNA is assembled and begins peeling away.

Assembled mRNA

Assembled mRNA

Free RNA nucleotides

**❸** Free RNA nucleotides bind with exposed DNA bases in complementary pairs.

Template strand of DNA

**❹** RNA polymerase and mRNA are released at stop sequence.

**Figure 10.2 Transcription.** One of the strands of DNA functions as a template on which free RNA nucleotides are assembled into a strand of mRNA.

Next, the enzyme helicase causes the DNA base-pair hydrogen bonds to break, freeing the DNA strand that is to be transcribed ❷. The RNA polymerase begins moving along the separated DNA strand and binding free RNA nucleotides with the now-exposed DNA bases in a sequence complementary to the sequence of the DNA bases. Each RNA nucleotide is incorporated successively into a growing mRNA chain ❸. The complementary pairs of RNA and DNA bases are: RNA adenine pairs with DNA thymine, RNA uracil pairs with DNA adenine, RNA guanine pairs with DNA cytosine, and RNA cytosine pairs with DNA guanine. The assembled sections of mRNA then peel away from the template strand of DNA ❹.

The RNA polymerase continues to move along the DNA strand and bind RNA nucleotides until it arrives at a stop sequence located at the end of the gene. At that point, it disengages from the DNA, and the newly assembled mRNA strand floats free ❹.

A processing step then removes segments of RNA from the original transcript that are not used in polypeptide synthesis (see Figure 10.7). These extra sequences of nucleotides that intrude between the polypeptide-specifying portions of the gene are called *introns* because they *intervene* into the gene but are not expressed. The remaining segments of the gene—the nucleotide sequences that encode the amino acid sequence of the polypeptide—are called *exons*, those portions of a gene that will be *expressed*. Cutting out introns and splicing together exons results in the final mRNA strand. After cutting and splicing occur, the mRNA strands leave the nucleus through the nuclear pores and travel to the ribosomes in the cytoplasm of the cell.

Cutting out introns and splicing together exons do not always occur in the same way along an mRNA transcript. Such **alternative splicing** of mRNA transcripts is the reason that one gene can code for multiple polypeptides. If a gene has multiple introns, they may be removed separately or as a unit. If they are removed as a unit, and if one or more exons lies between the introns, then the exons are removed as well. Thus, various possibilities exist for a processed mRNA transcript, depending on whether exons were cut out of the transcript along with introns.

For example, **Figure 10.3** illustrates alternative splicing in the *Sex-lethal (Sxl)* gene, which regulates sex determination in the fruit fly, *Drosophila*. The exons and introns in the gene are shown at the top. Introns are the horizontal lines between the numbered exons. During the development of some fly embryos, exon 3 is cut out of the transcript along with the introns. Therefore, the processed mRNA transcript has the sequence of exons shown in the middle. This transcript results in the production of a functional polypeptide. When this polypeptide is present, a set of genes is expressed that causes these fly embryos to develop into females. Conversely, during the development of other fly embryos, exon 3 is not cut out of the transcript. Therefore, the processed mRNA transcript has the sequence of exons shown at the bottom. This transcript contains a "stop" message that results in the production of a nonfunctional polypeptide. Without the functional polypeptide, a different set of genes is expressed that causes these fly embryos to develop into males.

## 10.4 The RNA code is translated into polypeptides.

How is mRNA used in the production of polypeptides? You can think of mRNA as computerized instructions or a template for an assembly line. After the instructions are copied from the master plans (the genes) during transcription, they are modified during processing, and then they are used to build products. The products are polypeptides, and the process of making polypeptides is called **translation**.

What exactly is the DNA code that must be translated in living things to produce polypeptides? In 1961, as a result of experiments led by Francis Crick, one of the researchers who worked out the structure of DNA, scientists learned that the DNA code is made up of sequences of three nucleotide bases. Each sequence is called a triplet—for example, the sequence of DNA bases AGA is a triplet, as is CTC. As described in Section 10.3, the DNA sequence of bases is transcribed into the complementary sequence of mRNA bases. Thus, each mRNA triplet is the complement of the corresponding original DNA triplet—for example, the mRNA triplets UCU and GAG are the complements of the DNA triplets AGA and CTC. Each mRNA

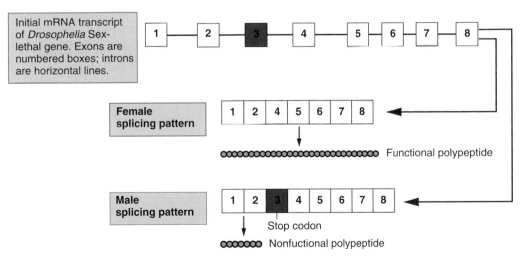

**Figure 10.3** Alternative splicing in the *Sex-lethal (Sxl)* gene in the fruit fly, *Drosophila*. The Sex-lethal gene regulates sex determination in the fruit fly. In the female splicing pattern, exon 3 is deleted, resulting in an mRNA transcript that produces a functional polypeptide. In the male splicing pattern, exon 3 is retained, resulting in a nonfunctional polypeptide. Fly embryos with the functional polypeptide develop into females; embryos with the nonfunctional polypeptide develop into males.

 **Visual Thinking:** If a mistake occurred and the introns between exons 2 and 3, 3 and 4, 4 and 5, and 5 and 6 were deleted as a unit, what splicing pattern would result? Do you think the polypeptide produced from this gene would be functional?

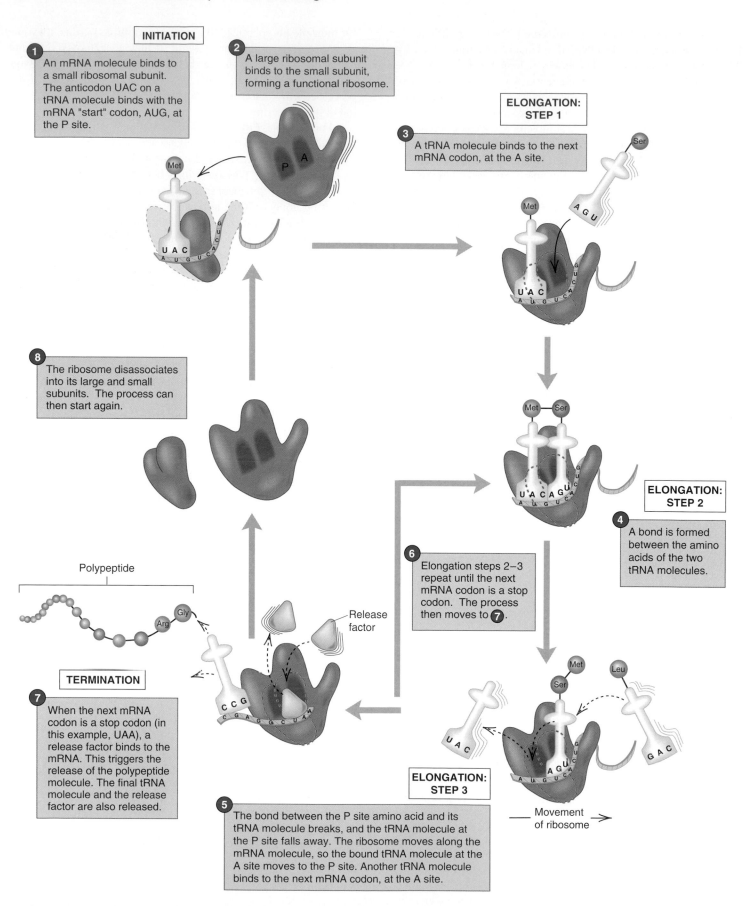

**INITIATION**

**1** An mRNA molecule binds to a small ribosomal subunit. The anticodon UAC on a tRNA molecule binds with the mRNA "start" codon, AUG, at the P site.

**2** A large ribosomal subunit binds to the small subunit, forming a functional ribosome.

**ELONGATION: STEP 1**

**3** A tRNA molecule binds to the next mRNA codon, at the A site.

**8** The ribosome disassociates into its large and small subunits. The process can then start again.

**ELONGATION: STEP 2**

**4** A bond is formed between the amino acids of the two tRNA molecules.

**6** Elongation steps 2–3 repeat until the next mRNA codon is a stop codon. The process then moves to **7**.

Polypeptide

Release factor

**TERMINATION**

**7** When the next mRNA codon is a stop codon (in this example, UAA), a release factor binds to the mRNA. This triggers the release of the polypeptide molecule. The final tRNA molecule and the release factor are also released.

**ELONGATION: STEP 3**

**5** The bond between the P site amino acid and its tRNA molecule breaks, and the tRNA molecule at the P site falls away. The ribosome moves along the mRNA molecule, so the bound tRNA molecule at the A site moves to the P site. Another tRNA molecule binds to the next mRNA codon, at the A site.

Movement of ribosome

**Figure 10.4** Translation and the termination of protein synthesis. A polypeptide is formed after elongation has been repeated many times. The small ribosomal subunit is shown as a transparent structure in all but the initiation step so that the process of translation can be viewed more easily.

triplet is called a **codon** (KOH-don). In most cases, a codon stands for (i.e., codes for, or *encodes*), a specific amino acid. Researchers broke this code and learned which codons stand for which amino acids, the building blocks of polypeptides (**Table 10.1** on page 146).

Since each codon is a sequence of three mRNA bases, and there are four bases (A, C, G, U), the total number of possible codons is 64 ($4 \times 4 \times 4 = 4^3 = 64$), as shown in Table 10.1. As you can see in the table, three codons are *stop* codons: they act as "stop" signals for the process of translation. The other 61 of the codons stand for amino acids; one of these codons (AUG) does double duty—it can encode the amino acid methionine, and it can also function as a "start" signal for translation.

Table 10.1 also shows that most of the amino acids are encoded by more than one codon (the exceptions are methionine and tryptophan). This is what you would expect, given that there are 64 codons and that only 20 amino acids are used in polypeptide production.

The molecules that bond the mRNA codons are molecules of tRNA. Each tRNA molecule has an **anticodon** loop—a portion of the molecule with a sequence of three base pairs complementary to a specific mRNA codon (see Figure 10.1). On the end opposite to the anticodon loop, each tRNA molecule carries the amino acid corresponding to its anticodon sequence. For example, a tRNA molecule having the anticodon AGA, which is complemenatary to the mRNA sequence UCU, carries the amino acid serine, which is encoded by UCU (see Table 10.1). A special family of enzymes links the amino acids to the tRNA molecules.

During the process of translation, the genetic code—the sequence of codons in mRNA—is deciphered and polypeptides are synthesized. This process takes place on the ribosomes, which consist of two RNA-protein particles: a small subunit and a large subunit. These subunits come together in polypeptide synthesis.

To initiate the process, the start codon AUG of the mRNA binds to a small ribosomal subunit. A tRNA molecule with the anticodon

UAC and carrying the amino acid methionine detects the start codon and binds to it (**Figure 10.4** ❶ ). Then, a large ribosomal subunit binds to the small subunit, resulting in a functional ribosome ❷ . This functional ribosome has an A (aminoacyl) site and a P (peptidyl) site. The tRNA already present occupies the P site of the ribosome.

Next, the elongation of the polypeptide begins. *Elongation* consists of three steps that take place over and over until the entire mRNA is "read." In Step 1, a tRNA with an anticodon complementary to the next mRNA codon "in line" binds to the mRNA at the A site ❸ . This tRNA carries the next amino acid that will be added to the chain. In the figure, the tRNA has the anticodon AGU and carries the amino acid serine; it binds to the complementary codon UCA on the mRNA. In Step 2, a peptide bond is formed between the two adjacent amino acids ❹ . In Step 3, the tRNA at the P site breaks away from its amino acid and the mRNA, leaving the ribosome. The ribosome moves so that the tRNA originally located at the A site shifts   with its attached growing amino acid chain—to the P site ❺ . Elongation steps 2–3 then repeat many times, as, one by one, tRNA molecules bind to codons at the A site on the mRNA, and amino acids line up in an order determined by the sequence of mRNA codons that passed through the ribosome. This continues until a stop codon is encountered on the mRNA ❻ .

When a stop codon is encountered (see Table 10.1), no tRNA exists to bind to it. Instead, it is recognized by a release factor, a protein that brings about the release of the newly made polypeptide from the ribosome. The final tRNA molecule is also released, as is the release factor ❼ . The small and large ribosomal sugbunits disassociate, and the process of translation can then begin again ❽ .

**Figure 10.5** shows that more than one ribosome at a time reads and translates the mRNA message to synthesize a polypeptide. Here, a group of many ribosomes, called a *polyribosome*, is reading along an mRNA molecule taken from a cell of a fly. Although the process is shown in a fly, it occurs in a similar fashion in all organisms. The polypeptides being made can be seen dangling behind each ribosome.

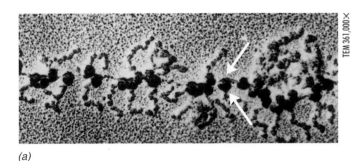

TEM 361,000×

*(a)*

*(b)*

**Figure 10.5 Protein synthesis in a midge.** (a) An electron micrograph of ribosomes in the midge *Chironomus tentans*. (A midge is a tiny, non-biting, mosquito-like fly.) These ribosomes are reading along an mRNA molecule from left to right, assembling polypeptides that dangle behind them like tails. Visible are the two subunits (*arrows*) of each ribosome translating the mRNA. (b) An illustration depicting the same process described above. The mRNA molecule is shown in blue, the ribosome subunits are shown in two shades of brown, and the polypeptides are shown in shades of green.

### CONCEPT CHECKPOINT

2. The following is a section of a template strand of DNA for the gene that codes for the CFTR protein, which regulates the water content of mucous secretions by respiratory cells.

   GATTACAGATTACACCTAGCTATC

   Transcribe and translate this sequence of DNA into a polypeptide. Show the mRNA sequence, and the amino acid sequence of the polypeptide.

3. Suppose the TACAGA sequence in the above gene was an intron. How would the mRNA sequence change? What effect would this have on the amino acid sequence of the polypeptide?

4. Using the information in Table 10.1 and your understanding of the relationship between codons and anticodons, can you determine which amino acid a tRNA molecule with the anticodon GGG carries?

| TABLE 10.1 | | | The Genetic Code | | | | | | | | | | |

**Second Nucleotide in Codon**

| | | U | | | C | | | A | | | G | | |
|---|---|---|---|---|---|---|---|---|---|---|---|---|---|
| **U** | UUU | Phe | F | *Phenylalanine* | UCU | Ser | S | *Serine* | UAU | Tyr | Y | *Tyrosine* | UGU | Cys | C | *Cysteine* | **U** |
| | UUC | Phe | F | *Phenylalanine* | UCC | Ser | S | *Serine* | UAU | Tyr | Y | *Tyrosine* | UGC | Cys | C | *Cysteine* | **C** |
| | UUA | Leu | L | *Leucine* | UCA | Ser | S | *Serine* | UAA | | Stop codon | | UGA | | Stop codon | | **A** |
| | UUG | Leu | L | *Leucine* | UCG | Ser | S | *Serine* | UAG | | Stop codon | | UGG | Trp | W | *Tryptophan* | **G** |
| **C** | CUU | Leu | L | *Leucine* | CCU | Pro | P | *Proline* | CAU | His | H | *Histidine* | CGU | Arg | R | *Arginine* | **U** |
| | CUC | Leu | L | *Leucine* | CCC | Pro | P | *Proline* | CAC | His | H | *Histidine* | CGC | Arg | R | *Arginine* | **C** |
| | CUA | Leu | L | *Leucine* | CCA | Pro | P | *Proline* | CAA | Gln | Q | *Glutamine* | CGA | Arg | R | *Arginine* | **A** |
| | CUG | Leu | L | *Leucine* | CCG | Pro | P | *Proline* | CAG | Gln | Q | *Glutamine* | CGG | Arg | R | *Arginine* | **G** |
| **A** | AUU | Ile | I | *Isoleucine* | ACU | Thr | T | *Threonine* | AAU | Asn | N | *Asparagine* | AGU | Ser | S | *Serine* | **U** |
| | AUC | Ile | I | *Isoleucine* | ACC | Thr | T | *Threonine* | AAC | Asn | N | *Asparagine* | AGC | Ser | S | *Serine* | **C** |
| | AUA | Ile | I | *Isoleucine* | ACA | Thr | T | *Threonine* | AAA | Lys | K | *Lysine* | AGA | Arg | R | *Arginine* | **A** |
| | AUG | Met | M | *Methionine* Start codon | ACG | Thr | T | *Threonine* | AAG | Lys | K | *Lysine* | AGG | Arg | R | *Arginine* | **G** |
| **G** | GUU | Val | V | *Valine* | GCU | Ala | A | *Alanine* | GAU | Asp | D | *Aspartic acid* | GGU | Gly | G | *Glycine* | **U** |
| | GUC | Val | V | *Valine* | GCC | Ala | A | *Alanine* | GAC | Asp | D | *Aspartic acid* | GGC | Gly | G | *Glycine* | **C** |
| | GUA | Val | V | *Valine* | GCA | Ala | A | *Alanine* | GAA | Glu | E | *Glutamic acid* | GGA | Gly | G | *Glycine* | **A** |
| | GUG | Val | V | *Valine* | GCG | Ala | A | *Alanine* | GAG | Glu | E | *Glutamic acid* | GGG | Gly | G | *Glycine* | **G** |

*First nucleotide in codon (5' end)* / *Third Nucleotide in Codon (3' end)*

In this table, the first nucleotide in the codon identifies the row of the amino acid. The second nucleotide in the codon identifies the column of the amino acid. This process identifies a box of four amino acids. The third nucleotide in the codon identifies the specific amino acid in the box of four. Each triplet (e.g., UUU) is a codon. The table also shows the amino acids encoded by the codons and the three-letter and single-letter abbreviations for each amino acid. In many cases, as shown in the table, codons that encode the same amino acid differ in the third nucleotide only (e.g., CCU, CCC, CCA, and CCG all encode proline). UAA, UAG, and UGA are stop codons. AUG encodes methionine and is also a start codon.

# How is the production of polypeptides, or gene expression, regulated?

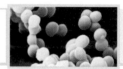

## 10.5 In prokaryotes, gene expression is regulated primarily by controlling transcription.

Transcription and translation occur somewhat differently in prokaryotes and eukaryotes, because of the differences between prokaryotic and eukaryotic cells. In prokaryotes, such as bacteria (**Figure 10.6**), the DNA is not contained within a nucleus. Therefore, transcription and translation occur simultaneously in the cytoplasm. Also, introns occur only rarely in prokaryotic DNA, and then only among the Archea. Bacteria and archeans are two of the three domains of life (see Figure 1.12).

As a result of these differences, prokaryotes regulate gene expression mainly by controlling transcription. Control of gene expression can be positive and result in turning a gene on, or it can be negative and result in turning a gene off. If a gene is turned on, a gene product will be synthesized; if a gene is turned off, a gene product will not be synthesized.

Prokaryotes need certain gene products all the time, such as tRNA, rRNA, and enzymes that catalyze reactions that must take place almost continually to maintain life. The genes that encode these products are called **constitutive** (con-STI-tu-tive) **genes**, and they are nicknamed "housekeeping genes." In general, constitutive genes are expressed continually in prokaryotes.

Other types of genes encode gene products that are not produced continually but only when needed. This ability to turn genes on and off in response to diverse and suddenly changing environmental conditions allows prokaryotes to save energy by not producing substances that might be useless to them in a particular environment at a particular time. For example, the *Escherichia coli* bacteria found in your large intestine live on whatever undigested material comes their way. Depending on what you have been eating, their type of food may change.

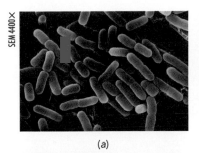

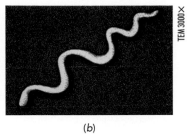

(a)

(b)

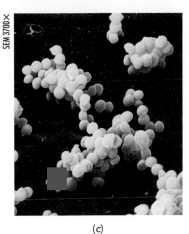

(c)

**Figure 10.6** Various types of bacteria. (a) *Escherichia coli*, a rod-shaped bacterium that is a normal inhabitant of the intestines. (b) *Treponema pallidum*, a spiral-shaped bacterium that causes the sexually transmitted infection syphilis. (c) *Staphylococcus aureus*, a spherical bacterium that usually lives harmlessly on the skin, inside the nose, or in the throat.

To get more specific with our example, *E. coli* can use several carbohydrates for food, such as glucose, sucrose, galactose, and lactose. If the bacterium has glucose, it will use it. If it is in an environment containing only lactose, it will use that. However, *E. coli* must synthesize two enzymes to metabolize lactose, and it has the ability to do just that. When the enzymes are needed, it turns on the genes that encode those enzymes. When the enzymes are not needed, it turns those genes off. Such genes are called **structural genes**—those that code for enzymes or structural proteins.

Turning genes on in response to the presence of a substrate is called *induction*; turning genes off is called *repression*. Prokaryotes regulate both induction and repression by positive and negative control mechanisms, which involve regulator genes.

**Regulator genes** code for products that control the expression of other genes. In a positive control mechanism, the product of the regulator gene turns on the expression of one or more structural genes. In negative control mechanisms, the product of the regulator gene shuts off the expression of structural genes. Positive control systems require the binding of regulator proteins to turn genes on. These proteins are called *activators*. Negative control systems require the binding of regulator proteins to turn genes off. These proteins are called *repressors*. Activators and repressors regulate gene expression by binding to regulator protein sites next to the promoter sites of structural genes.

Bacteria appear to be quite efficient in their control of gene expression. Genes with related functions are frequently grouped in the bacterial genome in units called **operons**. Gene expression in an operon is regulated as a unit. For example, genes that produce enzymes involved in the metabolism of lactose occur together and are regulated together. This particular operon is called the *lac* operon, short for lactose operon.

Each operon contains a promoter site, an operator site (which is the regulator protein binding site), and a set of structural genes.

Although some operons are regulated by positive control mechanisms, most operons are regulated by negative control, so the regulator protein is a repressor. When the repressor binds to the operator, RNA polymerase cannot transcribe the structural genes in the operon. When the operator is free of a repressor, RNA polymerase can bind and transcription takes place.

---

### CONCEPT CHECKPOINT

5. The preceding section states (italics added): "*Prokaryotes* regulate both *induction* and *repression* by *positive and negative control mechanisms*, which involve *regulator genes*. Explain what this statement means, being sure to define all the italicized words and phrases.

---

## 10.6 In eukaryotes, gene expression is regulated by controlling transcription, mRNA processing, and translation.

Gene expression in eukaryotes is much more complicated than in prokaryotes because eukaryotic cells and organisms are much more complex. For example, although prokaryotes have a single, circular strand of DNA containing anywhere from a few hundred to a few thousand genes, eukaryotes have numerous chromosomes containing many thousands of genes. Humans, for example, have about 30,000 genes in their genomes.

Along with having more genetic material, eukaryotic cells are compartmentalized by membranous organelles, unlike prokaryotic cells. Therefore, the events of gene expression are separated: transcription of RNA and its processing take place in the nucleus, while translation occurs in the cytoplasm at the ribosomes. This physical separation of the events of gene expression makes it possible for regulation to occur in different places within cells (see Figure 10.7).

In addition to having more complex cells than prokaryotes, many eukaryotes are multicellular. Multicellular organisms are composed of a variety of tissues and undergo developmental processes unlike any in unicellular organisms. In multicellular organisms, not every gene product is produced by or needed in every tissue. In humans, for example, certain cells of the stomach, intestines, and pancreas produce digestive enzymes for use in the digestive tract. Red blood cells—not other types of cells—synthesize hemoglobin to carry oxygen to all parts of the body. As shown by these examples, the expression of some genes depends on the tissue in which they are found; regulation of gene expression is tissue-specific. In addition, regulation of gene expression is time-specific in multicellular eukaryotes. During fetal development, specific proteins play crucial roles at certain times, directing the series of biochemical reactions that result in differentiation of tissues and growth. As this growth and development take place, genes are transcribed in a specific order, each gene for a specified time.

### Regulation of gene expression at transcription

Transcriptional control mechanisms are more complex in eukaryotes than in prokaryotes. In both, transcriptional regulation occurs by means of proteins interacting with DNA. In eukaryotes, such proteins are called **transcription factors**. Unlike prokaryotic

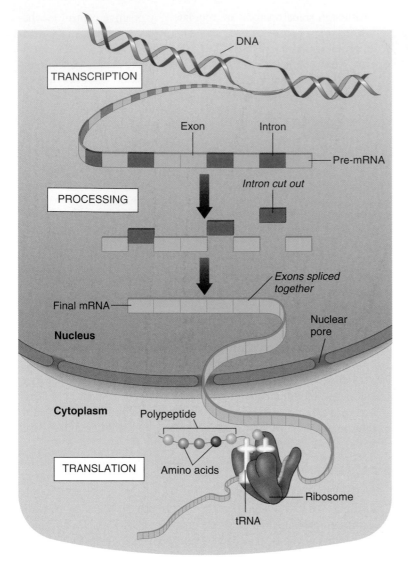

**Figure 10.7 Eukaryotic gene expression.** Expression can be regulated during transcription, processing, and translation.

 **Visual Thinking:** In what part of the cell do transcription and mRNA processing take place in eukaryotes? in prokaryotes? In what part of the cell does translation take place in eukaryotes? in prokaryotes? In what part of the cell are tRNA and amino acids located in eukaryotes? in prokaryotes?

transcriptional control, eukaryotic control is generally positive because genes are usually inactive. In general, eukaryotic genes must be turned on.

Eukaryotic chromosome structure also plays a role in the regulation of transcription. In order to be transcribed, DNA must have its protein binding sites accessible to transcription factors and RNA polymerase. Figure 9.2 shows that DNA does not exist as naked strands in the nucleus. The wrapping of DNA around histone proteins to form nucleosomes has an inhibitory effect on transcription. In other words, the packaging of DNA gets in the way of its doing

its job. How do eukaryotic cells overcome such physical transcriptional barriers? The answer is by means of a molecular crowbar. Among the proteins that collaborate in transcription, some pry open regions of the DNA blocked by nucleosomes, allowing access to the DNA. Histone-like proteins exist in prokaryotic cells as well, but relatively little compaction of DNA takes place in prokaryotic cells as compared to eukaryotic cells.

### Regulation of gene expression at mRNA processing

As mentioned previously, eukaryotic cells process mRNA within the nucleus by cutting out introns—noncoding portions of DNA—and splicing together exons—portions that encode a polypeptide. By controlling the removal of introns, eukaryotic cells control the polypeptides that are made. For example, if two introns border an exon or exons, and the introns are removed as a unit, the exons in between will also be cut out. The polypeptide formed from the mRNA will be different from the one encoded by the mRNA retaining those exons. Such alternative splicing of transcripts enables a single gene to code for various polypeptides.

### Regulation of gene expression at translation

During translation, as described in Section 10.4, mRNA transcripts are used to manufacture polypeptides at the ribosomes. Looking back at Figure 10.5 will help remind you that more than one ribosome at a time reads and translates a strand of mRNA. Polyribosomes, or groups of ribosomes, move along the mRNA—a polypeptide assembly line of sorts. This assembly line continues manufacturing molecules of polypeptide until the mRNA is degraded, or broken down.

Control of gene expression at translation relies on the time it takes until mRNA is degraded. If an mRNA is broken down swiftly, then it cannot support many rounds of polypeptide synthesis. On the other hand, if an mRNA is long-lived, then it can support the production of many more copies of the polypeptide.

Translational control is a mechanism that often operates during fetal development, when only certain amounts of products are needed. Messenger RNA can be degraded rapidly, preventing undesired polypeptide synthesis.

**Figure 10.7** summarizes the sequence of transcription, mRNA processing, and translation described in Sections 10.3 and 10.4.

### CONCEPT CHECKPOINT

6. Briefly describe the mechanisms by which eukaryotes regulate gene expression at each of the following potential points of regulation: transcription, mRNA processing, translation. Which of these mechanisms can also be found in prokaryotes? You may find it helpful to organize this information in a table.

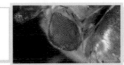

# What are master control genes, and how is gene expression activated?

## 10.7 Master control genes trigger the development of organs and other structures.

The fly in **Figure 10.8** gives new meaning to the phrase "having eyes in the back of your head," for it has eyes on its antennae and eyes on its wings. This mutant *Drosophila* (fruit fly) is only one of many types produced in the laboratory of Dr. Walter Gehring and his colleagues at the University of Basel in Switzerland. These mutants hold a major key to the understanding of how eyes develop and how the genes that control the development of organs are regulated.

The Gehring group discovered that by turning on a gene called *eyeless* in parts of a fruit fly where it is normally turned off, they could produce flies with eyes in unusual places—their legs, antennae, and wings. Some flies grew as many as 14 eyes. The results of their research indicate that the *eyeless* gene is a master control gene—a gene that, by itself, triggers the development of an organ or other structure. To do this, master control genes produce transcription factors, which are protein switches that bind to genes or groups of genes and turn them on and off. They coordinate the action of a number of genes, which, in turn, determine the development of a large region or body segment. Master control genes are also called **homeotic genes** and were discovered in fruit flies in 1983.

Homeotic genes encode transcription factors, the protein switches that bind to genes or groups of genes and turn them on and off. In this case, *eyeless* encodes a transcription factor that switches on a developmental pathway involving several thousand genes.

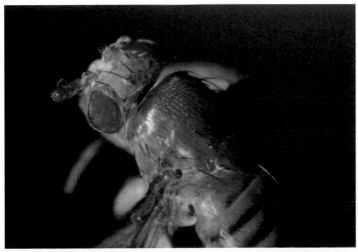

**Figure 10.8** Induction of eyes on the antennae and wings of *Drosophila* by targeted expression of the *eyeless gene* in these structures.

Prior to the *eyeless* discovery, scientists had found other genes that trigger the development of different cell types in the eye such as cells and pigments that detect light. Since the *eyeless* discovery, they have found a second master control gene. This gene, called *twin of eyeless*, is required for the initiation of *eyeless* expression in the embryo and acts through *eyeless* to activate the eye developmental pathway.

Researchers have discovered that homeotic genes have been conserved (are the same) in organisms that are very different from one another, such as yeast, fruit flies, mice, and humans. The presence of the same homeotic genes in such different animals suggests that these master control genes have a crucial function in many—and possibly all—animals.

The degree to which gene sequence and organization are conserved among organisms also gives scientists information about the time that has passed since they diverged during their evolution. For example, scientists have found genes that are functionally equivalent to the *Drosophila eyeless* gene in mice. This discovery suggests that these homeotic genes date back to the common ancestor of fruit flies and mammals.

## 10.8 Gene expression is activated by environmental factors and signaling molecules.

By now you probably have a vision of prokaryotic life and can picture the unicellular bacterium, such as the *E. coli* described earlier, living in an ever-changing environment like the human large intestine. In the environments of prokaryotes, nutrient availability is probably one of the most important factors affecting gene expression, and environmental factors, such as the nutrient lactose, are inducers of transcription. Thus, changes in environmental conditions are closely followed by changes in gene expression, so that only needed gene products are made.

The cells of multicellular eukaryotic organisms live in much less variable environments. The expression of genes is infrequently a response to environmental factors. Two examples of known environmental inducers of transcription in eukaryotes are high heat and light. To illustrate, when organisms are subjected to high temperatures, transcription is initiated to manufacture a group of proteins that help stabilize the internal cellular environment. These proteins are called heat-shock proteins and are found in both prokaryotes and eukaryotes. Similarly, light induces the production of ribulose 1,5 bisphosphate carboxylase (RBC) in plants when they are exposed to light. RBC plays an important role in photosynthesis in green plants.

Most frequently, eukaryotic gene expression is induced by hormones, growth factors, and related compounds. These substances are called **signaling molecules** because they convey signals within and between cells of a multicellular organism, stimu-

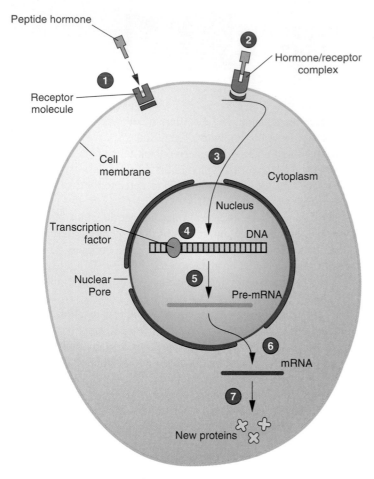

**Figure 10.9** How peptide hormones work.

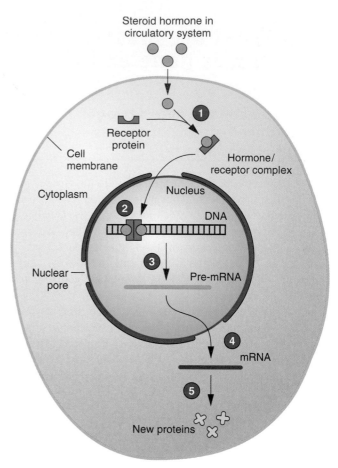

**Figure 10.10** How steroid hormones work.

lating coordinated changes in gene expression. Two major classes of hormones work within animals: peptide hormones and steroid hormones.

## Peptide hormones

**Peptide hormones** are made of amino acids, but the amino acid chain length varies greatly from hormone to hormone. Polypeptide hormones have chain lengths of several dozen or more amino acids, such as the hormone insulin. Even larger are protein hormones that may have over 200 amino acids with carbohydrates attached at several positions.

Peptide hormones are unable to pass through the cell membrane. Instead, as shown in **Figure 10.9**, the peptide hormone binds to a receptor molecule embedded in the cell membrane of a target cell **1**, forming a hormone/receptor complex **2**. The binding of a hormone to a receptor activates a cytoplasmic protein that triggers a cascade of intracellular changes **3**. These changes transmit the signal to the nucleus, inducing a transcription factor to bind to DNA **4**. The bound transcription factor stimulates transcription **5**. The transcript is processed and transported to the cytoplasm **6**, where the mRNA is translated into proteins **7**. An example of a peptide hormone is prolactin, which stimulates cells of the mammary glands to produce milk.

## Steroid hormones

**Steroid hormones** are all made from cholesterol, a lipid synthesized by the liver and present in many foods, such as eggs, dairy products, and beef. A characteristic of steroid hormones is their set of carbon rings.

Steroid hormones, being lipid soluble, pass freely through the lipid bilayer of a target cell membrane and bind to receptor proteins located within the cytoplasm of target cells (**Figure 10.10 1**). The hormone-receptor complex moves into the nucleus of the cell where it binds to a hormone response element in the DNA **2**, triggering transcription **3**. The transcript is processed and transported to the cytoplasm **4**. The mRNA is translated into proteins **5**. Examples of steroid hormones are estrogens and testosterone. The cells these hormones target produce proteins involved in such processes as the development and maintenance of female and male sexual characteristics.

---

### CONCEPT CHECKPOINT

**7.** State the evolutionary implication of the discovery of functionally equivalent homeotic genes for eye development in fruit flies and mice.

**8.** Summarize the main difference in the mechanisms by which peptide hormones and steroid hormones elicit a response in a cell.

# just wondering . . .

## What happens if the body makes a "mistake" in regulating gene expression?

Gene expression involves the processes of transcription and translation, with the end products being RNA and polypeptides. When the regulation of this process goes awry, it is possible that too much, too little, or no end product is made. It is also possible that a product is made at the wrong time. All of these situations can have deadly consequences or can produce structural or functional problems. One example of a malfunction of gene expression is Rett syndrome (RTT).

RTT is a serious disorder of the nervous system that affects approximately 1 in 10,000 girls but rarely affects boys. Girls with RTT develop normally until 6 to 18 months of age with no indication that anything is wrong. Suddenly, development regresses, and the child loses any ability to speak. She also loses hand skills, with repetitive hand movements taking the place of formerly coordinated movements. The child may also develop irregular breathing, motor control problems, and slowed head growth. Although the RTT patient may live to adulthood, she never regains the ability to speak or use her hands.

The cause of the majority of cases of RTT is a defect in the MECP2 gene on the X chromosome. The X chromosome is a sex chromosome and is instrumental in determining gender; females have two X chromosomes, and males have a single X and single Y sex chromosome.

The MECP2 gene codes for MeCP2 protein, which is thought to be a repressor of transcription. RTT mutations in the MECP2 gene cause no functional MeCP2 protein to be produced. It appears that RTT is caused by the expression of genes in some tissues when those genes should be turned off. As of this writing, scientists had not yet isolated those genes. Researchers expect that identifying the genes that are erroneously expressed may be a first step toward a treatment for this disease. Researchers must also determine the function of the abnormal MeCP2 protein. ●

*Are you wondering about a topic in biology and how it relates to your life? Submit your question by clicking the* Just Wondering *link in this text's companion Web site at www.wiley.com/college/alters.*

## CHAPTER REVIEW

## Summary of Key Concepts and Key Terms

### How is the DNA code expressed?

**10.1 Genetics** (p. 141) is the study of heredity and of the sources, nature, and implications of heritable variation.

**10.1 Genes** (p. 141) are units of DNA that encode ncRNA and, primarily, **polypeptides** (p. 141).

**10.1 Proteins** (p. 141) are large molecules comprised of one to several polypeptides; therefore, in some instances, polypeptides are proteins.

**10.1 Gene expression** (p. 141) is the process by which genes produce polypeptides and RNA, and by which these products exert their effects on an organism.

### How are polypeptides made using the code of DNA?

**10.2** DNA uses RNA to carry the genetic code to the cytoplasm of the cell, where polypeptides are manufactured on **ribosomes** (RYE-buh-somes, p. 141).

**10.2** Cells contain three types of RNA, which all play a role in polypeptide synthesis: **Messenger RNA (mRNA**, p. 141), **transfer RNA (tRNA**, p. 141), and **ribosomal RNA (rRNA**, p. 141).

**10.2** DNA acts as a template to manufacture mRNA in the nucleus; the mRNA then travels to the ribosomes in the cytoplasm; in the cytoplasm, tRNA transports amino acids into the correct positions to form polypeptides; this whole process takes place on the ribosomes, which contain rRNA.

**10.3** In **transcription (p. 142)**, the DNA code for a gene is copied onto a strand of mRNA.

**10.3** RNA polymerase binds to the DNA at a promoter site; the enzyme helicase causes the DNA strands to separate; RNA polymerase moves along a separated strand, binding free RNA nucleotides to the exposed bases on the strand; the RNA nucleotides bind together to form a growing chain of mRNA, which peels away from the DNA strand, until the mRNA is complete.

**10.3** mRNA is then processed in the nucleus; introns are removed and exons are spliced.

**10.3** Cutting out introns and splicing together exons do not always occur in the same way along an mRNA transcript. Such **alternative splicing** (p. 143) of mRNA transcripts is the reason that one gene can code for multiple polypeptides.

**10.4** The DNA code copied onto mRNA is used to make polypeptides, in a process called **translation** (p. 143).

**10.4** The DNA code is made up of sequences of three nucleotides; each sequence is called a triplet and functions as a **codon** (p. 144); a codon stands for (encodes) a specific amino acid.

**10.4** There are 64 possible codons: 61 of the codons encode amino acids, and one of these also functions as a "start" signal; the other three codons function as "stop" signals.

**10.4** Since there are only 20 amino acids used in polypeptide synthesis, and there are 61 codons for amino acids, most amino acids are encoded by more than one codon.

**10.4** Each tRNA molecule has an **anticodon** (p. 144) loop that binds with the complementary codon on mRNA.

**10.4** The process of translation consists of these steps: (1) the mRNA binds to a small ribosomal subunit, and then a special tRNA molecule binds to the mRNA "start" codon; (2) a large ribosomal subunit binds to the small ribosomal subunit, forming a functional ribosomal complex; (3) tRNA molecules carrying amino acids corresponding to their anticodons bind to the complementary mRNA codons as the ribosomal complex moves along the mRNA strand; (4) the amino acids bind together to form a polypeptide that is released from the ribosomal complex when a "stop" codon is reached on the mRNA.

**10.4** More than one ribosome at a time reads and translates the mRNA message to synthesize a polypeptide.

## How is the production of polypeptides, or gene expression, regulated?

**10.5** Prokaryotes regulate gene expression mainly by controlling transcription.

**10.5** **Constitutive** (con-STI-tu-tive) **genes** (p. 146) code for products needed almost all the time.

**10.5** **Structural genes** (p. 147) code for enzymes or structural proteins.

**10.5** Control of gene expression can be positive and result in turning a gene on, or it can be negative and result in turning a gene off. If a gene is turned on, a gene product will be synthesized; if a gene is turned off, a gene product will not be synthesized.

**10.5** **Regulator genes** (p. 147) code for products that control the expression of other genes.

**10.5** Genes with related functions are frequently grouped in the bacterial genome in units called **operons** (p. 147).

**10.6** Control of gene expression is much more complicated in eukaryotes than in prokaryotes.

**10.6** In eukaryotes, gene expression is regulated by controlling transcription, mRNA processing, and translation.

**10.6** In eukaryotes, protein **transcription factors** (p. 147), which interact with DNA, control transcriptional regulation.

## How is gene expression activated?

**10.7** **Homeotic genes** (master control genes, p. 149) encode transcription factors, the protein switches that bind to genes or groups of genes and turn them on and off.

**10.7** Researchers have discovered that homeotic genes have been conserved (are the same) in organisms that are very different from one another, such as yeast, fruit flies, mice, and humans.

**10.8** In prokaryotes, environmental factors are inducers of transcription.

**10.8** Eukaryotic gene expression is induced by hormones, growth factors, and related compounds called **signaling molecules** (p. 149) because they convey signals within and between cells of a multicellular organism, stimulating coordinated changes in gene expression.

**10.8** Two major classes of hormones work within animals: **peptide hormones** (p. 150) and **steroid hormones** (p. 150).

---

## Level 1 | Learning Basic Facts and Terms

**Multiple Choice**

1. Genes that are turned on or off in response to specific environmental changes are referred to as
   a. constitutive genes.
   c. regulator genes.
   b. structural genes.
   d. operons.

2. Gene regulation in eukaryotes includes all of the following levels of control except:
   a. controls that affect the rate of gene transcription.
   b. controls that govern cutting and splicing of mRNA transcripts.
   c. controls that affect the precision of mRNA transcription.
   d. controls that regulate which mRNA transcripts will be translated into polypeptides.

3. RNA differs from DNA in which of the following ways?
   a. RNA is double stranded, while DNA is single stranded.
   b. Uracil takes the place of thymine in RNA.
   c. DNA has four different nitrogenous bases, while RNA has five.
   d. RNA is only found in the cytoplasm; DNA is only found in the nucleus.

4. The cytoplasm of eukaryotic cells contains large numbers of tRNA molecules which attach themselves to free
   a. amino acids.
   d. anticodons.
   b. polypeptides.
   e. regulator genes.
   c. nucleotides.

5. Which of the following is not directly involved in the process of translation?
   a. ribosomes
   d. amino acids
   b. tRNA
   e. RNA polymerase
   c. mRNA

**Matching**

6. operon
7. homeotic genes
8. structural genes
9. steroid hormones
10. transcription factors
11. regulator genes
12. polyribosome

a. Genes that are grouped and regulated together
b. Encode products that control the expression of other genes
c. Molecules that bind to DNA and regulate transcription
d. Signaling molecules made up of cholesterol
e. Coordinate the actions of a large number of genes
f. Encode enzymes and structural proteins
g. Multiple ribosomes attached to and transcribing a single mRNA transcript

## Level 2 | Learning Concepts

1. Distinguish among rRNA, tRNA, and mRNA. What does each abbreviation stand for, and what are the respective functions of each?

2. Most cells in your body have the same number of chromosomes (notable exceptions are red blood cells, sex cells, and liver cells), and the exact same set of genes, and yet there are a multitude of different types of cells in the body. How is this possible given that most of them possess the same genetic information?

3. If stretched out, the DNA in the nucleus of a single human cell would be well over a meter in length, thousands of times larger than the diameter of the cell's nucleus. Briefly describe how a human cell packages all this DNA in a tiny nucleus. What is the significance of this DNA packaging to the regulation of gene expression?

4. Hypothesize what might occur in the process of transcribing and translating the following hypothetical DNA template strand if each of the mutations listed below occurred.

GATTACAGATTACACCTAGCTATC

Suggest a possible sequence of subsequent events for each. Which mutation would change the structure of the polypeptide product the most? You will need to consult Table 10.1 to determine your answer.

a. T is substituted for the first C.
b. An additional C is added after the first C.

## Level 3 | Critical Thinking and Life Applications

1. New molecular technologies in biology have made it possible to insert the gene for human growth hormone (HGH) into the DNA of bacteria. The bacteria that take up the HGH gene are able to transcribe and translate this gene into small quantities of HGH, which can be harvested and given to people with HGH-related growth disorders. What fundamental similarities must bacteria and humans share in order to make this technology possible?

2. People who have a form of diabetes called type II diabetes (or noninsulin-dependent diabetes) develop a lack of cellular receptivity to insulin. Based on your knowledge of peptide hormone action, what do you think this means?

3. Which of the following may be a predicted effect of a mutation in the regulator gene resulting in a nonfunctional repressor protein in the lac operon? Explain your answer.
a. continuous transcription of the genes that metabolize lactose
b. accumulation of large quantities of lactose in the cell
c. continuous blocking of expression of the genes that metabolize lactose

## In The News | Critical Thinking

### MOUSE OF A DIFFERENT COLOR

Now that you understand more about gene expression and its regulation, reread this chapter's opening story about research into the development of a system for reversible control of gene expression. To understand this research better, it may help you to follow these steps:

1. Review your immediate reaction to Scrable's research on the development of a visible, reversible system for the control of gene expression that you wrote when you began reading this chapter.

2. Based on your current understanding, again summarize the main point of the research in a sentence or two.

3. What questions do you now have about this research that the opening story does not answer?

4. Collect new information about the research. Visit the *In The News* section of this text's companion Web site at www.wiley.com/college/alters and

watch the "Mouse of a Different Color" video. Then use the "summary" link to read the accompanying story and access related links. Use this information, the links provided, and other online and library resources to answer your questions and find updates about this research topic. State the sources of your information. Explain why you think the information is accurate. Also determine whether the information expresses a particular point of view or is biased in any way.

5. What in your view is the most significant aspect of this research? Give reasons for your opinion and for any changes in your ideas based on the additional information you have collected and the analysis you have done.

# BIOTECHNOLOGY AND GENETIC ENGINEERING

## Human Cloning—The Science

Cloning has been in the news for many years, most noticeably since Dolly the sheep was revealed to the public in 1997. Born in 1996, Dolly was the first mammal to be cloned from an adult cell. To develop the sheep clone, scientists removed the nucleus from an unfertilized egg and replaced it with a nucleus from a mammary gland cell. The egg cell then developed into an embryo, which was implanted into the uterus of a sheep to complete its development. Scientists call this cloning process *nuclear transfer*. The clone is the genetic duplicate of the "parent," with the exception of the mitochondrial DNA found in the egg cell. Since the cloning of Dolly, a variety of organisms have been cloned, such as calves, mice, monkeys, and pigs. Now, human cloning is in the news.

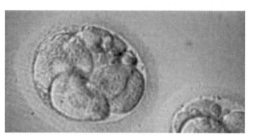

There are two types of human cloning: therapeutic cloning and reproductive cloning. In therapeutic cloning, a cloned cell is allowed to develop into a blastocyst, which is a hollow ball of about 200 cells. One portion of the blastocyst contains cells that have the ability to differentiate (change) into any cell type, such as a nerve cell or muscle cell. These cells are called stem cells, and they can be used to treat certain diseases. The stem cells are removed and the blastocysts destroyed. (See *In The News*, Chapter 12, for more information on stem cells.)

In reproductive cloning, the blastocyst would be implanted in a surrogate mother and allowed to develop until birth. Rudolf Jaenisch, of the Whitehead Institute for Biomedical Research and the Department of Biology at the Massachusetts Institute of Technology (MIT), is a leader in nuclear transfer research and opposes human reproductive cloning. Jaenisch and his team have conducted studies on reproductive cloning in mice. Results of their studies suggest that cloned mammals are abnormal. You can view the video news story, "Human Cloning—The Science," which describes Jaenisch's work in more detail. Visit the *In The News* section of this text's companion Web site at www.wiley.com/college/alters.

Should human reproductive cloning be allowed or banned in the United States? What are the arguments against human reproductive cloning? Health risks to the cloned child are a major focus. Beyond the problems Jaenisch and his team suggest, which focus on an unacceptable extent of abnormal gene expression in clones, there are concerns that a cell from an adult might carry accumulated genetic mutations that could predispose the cloned individual to cancers and other diseases. Another concern is that if cells are used from deceased persons, then the person who "donates" the cells cannot give consent. Yet others are concerned about the mental and emotional life of a cloned individual, who may feel that his or her individuality has been compromised.

What are arguments in favor of human reproductive cloning? Supporters suggest that this technology could aid infertile and same-sex couples in having children biologically related to one parent. In addition, supporters suggest that parents could have children without the risk of known genetic diseases. Many think that human reproductive cloning expands reproductive freedom.

*Many more arguments than those mentioned here can be made for and against human reproductive cloning. Write your immediate reaction to the issue raised. First, summarize the issue in a sentence or two. How do you think this issue could be or should be resolved? You will have an opportunity to reflect on your responses and gather more information on this issue in the* In The News *feature at the end of this chapter. In this chapter, you will learn more about molecular biotechnology, including cloning.*

## CHAPTER GUIDE

### What is biotechnology?

**11.1** Classical biotechnology involves manipulating organisms.

**11.2** Molecular biotechnology involves manipulating genes.

### How are genes manipulated?

**11.3** Genes are transferred naturally among certain bacteria.

**11.4** Scientists can engineer gene transfer.

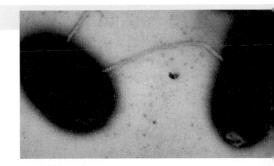

### What are some applications of biotechnology and genetic engineering?

**11.5** Proteins can be produced for therapeutic use.

**11.6** Genetically engineered vaccines can be produced.

**11.7** DNA fingerprinting is done by copying, decoding, and "reading" DNA.

**11.8** Results of research on genome structure and function have a wide variety of applications.

**11.9** Gene therapy can be used to treat disease.

**11.10** Biotechnology can be used to modify food.

### What are some controversies regarding genetic engineering?

**11.11** The use of transgenic plants and animals has potential benefits and dangers.

# What is biotechnology?

*Biotechnology* is a word used frequently in scientific circles and is often discussed on the evening news and in the newspaper. What does it mean? The word "biotechnology" has been used longer than you might expect. It was used as far back as 1917, when it referred to the large-scale production of substances synthesized by bacteria grown in vats. Because the roots of the biological manipulation of organisms are traceable back 10,000 years or so, many refer to these "old" technologies as classical biotechnology. Molecular biotechnology is the "new" biotechnology. **Biotechnology** is the use of scientific and engineering principles to manipulate organisms or their genes, producing one or more of the following:

- Organisms with specific biochemical, morphological, and/or growth characteristics.
- Organisms that produce useful products.
- Information about an organism or tissue that would otherwise not be known.

## 11.1 Classical biotechnology involves manipulating organisms.

As far back as 10,000 years ago, humans selected plants and animals with specific characteristics to propagate from the wild. In a simple way, this process is considered biotechnological: Humans were using selection to produce organisms with specific characteristics. For more than 8000 years, bacteria and yeasts have been used to produce products such as beer, vinegar, yogurt, and cheese, although the processes involved were not understood at the time. Loaves of yeast breads have even been found in Egyptian pyramids built 6000 years ago. And more than 3000 years ago, the Chinese and the Central American Indians used products produced by molds and fungi to treat infections. Today we know that certain molds and bacteria are natural sources of bacteria-fighting antibiotics.

Since ancient times, plant and animal breeders have selectively bred organisms to develop hybrids having certain desired characteristics. *Hybrids* are the offspring produced by crossing two genetically dissimilar varieties of a species. For example, the cauliflower, broccoli, and cabbage shown in **Figure 11.1** have all been bred from one species of wild mustard. By selecting certain characteristics of each plant they wished to cultivate, breeders were able to develop hybrid plants that are quite different from one another. (In other contexts, the word *hybrid* may also refer to the cross between two different species of organisms. A mule, for example, is the hybrid offspring of a horse and a donkey.)

At the turn of the nineteenth century, scientists began developing hybrids using scientific principles brought to light by Gregor Mendel: selecting and recombining hereditary factors now known as genes. Beginning in the late 1920s, scientists induced genetic mutations with X rays to obtain plants with new characteristics. After World War II, the list of mutagens they used increased. These techniques of selection, hybridization, and mutation are the basis of **classical biotechnology**.

Using carefully planned and controlled breeding programs, scientists found that they could produce hybrids that were stronger or improved in certain ways from their parents, a feature termed *hybrid vigor*. During the 1930s, a worldwide effort was initiated to increase food production in developing countries. By developing hybrid varieties of food crops using the techniques of selection and recombination, agricultural researchers began a *green revolution* in the 1940s in Mexico and the 1960s in India and Pakistan, dramatically increasing the yields of food crops in these countries. For example, hybrid corn developed in the United States helped double crop yields, and new dwarf varieties of wheat introduced to farmers in Mexico resulted in a 300% increase in wheat yields there.

At about the same time that Mendel was conducting genetic research with pea plants in the mid-1800s, strides were also being made in classical biotechnology in the field of microbiology. For example, processes of fermentation that had been used without being fully understood for thousands of years were explained during the 1800s. In 1837–1838, researchers concluded that yeasts are alive. Almost 30 years later, Louis Pasteur confirmed that certain bacteria and yeasts formed molecules such as acetic acid, lactic acid, butyric acid, alcohol, and carbon dioxide. He determined why and how wine often turned to vinegar, and he developed a process to kill the microorganisms causing this fermentative change. This process, termed *pasteurization*, is still used to help preserve wine and other foods such as milk.

Knowledge of fermentative processes was also put to work during World War I. Because of blockades during the war, the Germans were unable to obtain the vegetable oils from which they extracted the glycerol necessary for manufacturing explosives. To get around this obstacle, the Germans devised a biological way to produce glycerol, using the fermentative abilities of yeast. Similarly, the British developed biological methods to produce acetone and butanol, using the bacterium *Clostridium acetobutylicum*. Acetone was used in manufacturing ammunition, and butanol was used in producing artificial rubber.

The use of classical biotechnology to fight disease progressed rapidly. In the late 1800s, Robert Koch developed the germ theory of disease, based on the idea that microbes cause infection. He discovered both the bacterium that causes anthrax, a disease of cattle that sometimes affects humans, and the bacterium that causes tuberculosis. During this time, Koch also de-

**Figure 11.1** Broccoli, cauliflower, and cabbage. These three species of vegetables have all been bred from one species of wild mustard.

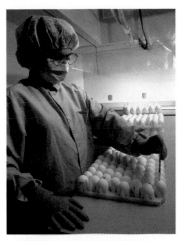

**Figure 11.2 Growing viruses in eggs.** Many viruses can be cultured in eggs, but today this technique is used only to culture the flu virus. A hole is drilled in the shell at 5 to 14 days after fertilization, and the virus is injected into the site where it will replicate.

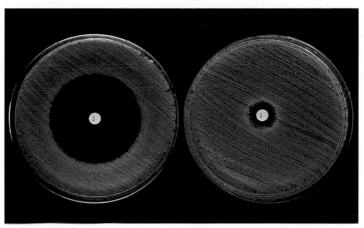

**Figure 11.4** Penicillin resistance (*right*) and sensitivity (*left*) in cultures of bacteria *Staphylococcus aureus*. Bacteria were spread on the nutrient surface of each of these two Petri plates. A disk containing the antibiotic penicillin was then placed in the center of each plate. The bacteria in the plate on the right carry a gene that confers resistance to penicillin. Therefore, these bacteria grew quite close to the disk. The bacteria in the plate on the left do not carry this resistance gene; the antibiotic inhibited the bacteria from growing in a large area surrounding the disk.

veloped *pure culture* methods. Using these methods, scientists could isolate and work with a specific organism, free from contamination by other organisms.

Microorganisms were also used to produce various disease-fighting products such as vaccines. Although Edward Jenner did not understand the role of microorganisms in disease, he is credited with developing the first successful vaccine against smallpox in 1796. Jenner injected volunteers with the relatively harmless *Vaccinia* virus, which conferred resistance against the deadly smallpox virus. Almost 100 years later, having a greater understanding than Jenner had of the disease process, Louis Pasteur and others developed the first cholera, diphtheria, and tetanus vaccines. By the early 1900s, mammalian cell culture techniques had been developed. By the mid-1900s, scientists were able to replicate viruses within these cultures (**Figure 11.2**), harvest and attenuate (weaken) them, and use them in vaccine preparations. Thus, microbes were themselves being used to fight human disease.

An important step in the development of disease-fighting products occurred quite by accident in 1927, when Alexander Fleming (**Figure 11.3**) discovered antibiotics. He noticed that a mold, which had contaminated his bacterial cultures, inhibited the growth of surrounding bacteria (**Figure 11.4**). That mold was *Penicillium notatum*, which naturally produces the antibacterial product we now call penicillin. By 1940, the first purified preparations of penicillin were available. This antibiotic played a major role in preventing the death

of soldiers from infection during World War II. Pharmaceutical companies began an intensive search for new antibiotics, screening bacteria and molds for their antibiotic potential.

## 11.2 Molecular biotechnology involves manipulating genes.

During the 1950s, a revolution took place that set the stage for the birth of molecular biotechnology: the discovery of the molecular structure of the hereditary material deoxyribonucleic acid (DNA). This revolution involved more than 50 years of work by a succession of scientists building on the work of Gregor Mendel (see Table 9.1). These scientific discoveries are described in Chapter 9.

Locating and characterizing the hereditary material were the first steps in what might be termed a *gene revolution*. Genes and genetic diversity then could be studied at the molecular level. Scientists began to manipulate genes in bacteria, intervening in and directing their natural methods of genetic recombination. **Genetic recombination** is the exchange of DNA sequences between molecules. In the rest of this chapter we look at what scientists have discovered.

**Figure 11.3** Alexander Fleming.

---

**CONCEPT CHECKPOINT**

1. Describe some of the basic differences between classical biotechnology and molecular biotechnology. Give two examples of each.
2. Describe some of the basic scientific discoveries that have made molecular biotechnology possible.

# How are genes manipulated?

## 11.3 Genes are transferred naturally among certain bacteria.

Bacteria reproduce asexually, when one cell splits into two. However, certain bacteria transfer genetic material from one cell to another. Genetic material is exchanged between bacteria by three mechanisms: transformation, transduction, and conjugation (**Figure 11.5**). In **transformation**, pieces of DNA from a donor bacterial cell that has lysed (ruptured) are released into the surrounding medium ➊. Under certain conditions, some bacterial cells can become recipient cells and take up DNA from their surroundings and incorporate it into their chromosome ➋, thereby becoming transformed, or changed. Not all bacteria can be transformed; the ability to take up DNA fragments is an inherited characteristic. Genetic engineers, however, have found a way to induce certain bacteria to take up DNA by cultivating them in the presence of specific chemicals.

During **transduction**, DNA from a donor bacterium is transferred to a recipient bacterium by a virus. For this transfer to occur, the virus must first inject its DNA ➌ and combine its genetic material with that of the donor bacterium it has infected ➍. The donor bacterium then manufactures new virus particles containing viral and donor bacterial DNA ➎. These particles burst from the

donor cell and infect a recipient cell ➏. The DNA from the virus and from the donor bacterium is then incorporated into the DNA of the recipient bacterium ➐. Only certain lysogenic viruses (viruses that infect a cell but do not immediately replicate) or damaged viruses are capable of combining their genetic material with bacterial DNA. Additionally, not all bacteria are capable of being infected by a virus carrying bacterial genes.

During **conjugation** (con-juh-GAY-shun), a donor and a recipient bacterium make contact, and the DNA from the donor is transferred to the recipient. This transfer takes place in bacteria having extra chromosomal pieces of DNA called **plasmids** (PLAZ-mids). Plasmids replicate independently of the main chromosome and make up only about 0.04% of the DNA in bacteria.

Some plasmids have several special genes that promote the transfer of the plasmid to other cells. This set of genes is referred to as a fertility factor, and such plasmids are called *F plasmids*. Fertility genes code for proteins that form a tubelike mating bridge called an *F pilus* (or *sex pilus*) on the surface of the bacterial cell (**Figure 11.6**). When the F pilus of an F⁺ cell (a bacterial cell with an F plasmid) makes contact with the surface of an F⁻ cell (a cell without an F plasmid), (see Figure 11.5 ➑ ), the F plasmid replicates ➒ and is then transferred to the F⁻ cell ➓.

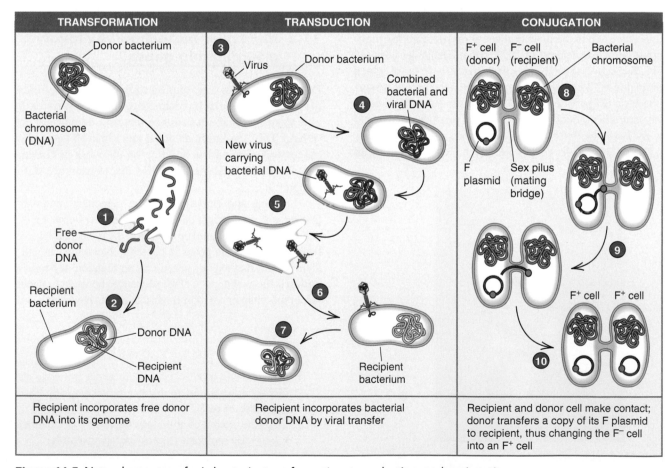

| TRANSFORMATION | TRANSDUCTION | CONJUGATION |
|---|---|---|
| Recipient incorporates free donor DNA into its genome | Recipient incorporates bacterial donor DNA by viral transfer | Recipient and donor cell make contact; donor transfers a copy of its F plasmid to recipient, thus changing the F⁻ cell into an F⁺ cell |

**Figure 11.5** Natural gene transfer in bacteria: transformation, transduction, and conjugation.

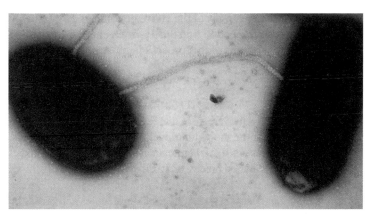

**Figure 11.6** Genetic recombination between *E. coli* bacteria. The bacteria are purple, and the F pilus joining them is green.

 **Visual Thinking:** The F pilus looks like a bridge connecting the two bacteria. What is the function of this bridge?

## 11.4 Scientists can engineer gene transfer.

Although scientists were able to manipulate the natural mechanisms of bacterial (prokaryotic) gene transfer in the 1950s and 1960s, it was not until the mid-1970s that they were able to combine DNA from different species of organisms and manipulate genes in eukaryotes. The techniques of molecular biotechnology that involve the manipulation of genes themselves, and not just entire organisms, are called **genetic engineering**, or **recombinant DNA technology**. *Recombinant DNA* contains DNA (hereditary material) originating from two or more sources.

In the 1980s, molecular biotechnology became a major area of growth in business, and the business of manufacturing began to look different from the stereotyped notion of an assembly line in a noisy factory. The laboratory technician in **Figure 11.7** is putting bacteria to work to create products for human use. The bacteria are growing in the small, table-top fermentors on the laboratory benches. These fermentors are only an intermediate step. Eventually, the bacteria used in this process will be grown in huge vats, like those shown in **Figure 11.8**.

To turn bacteria into living "factories," scientists isolate individual genes and transfer them from one kind of organism—a human, for example—to another, such as a bacterium. The bacteria are then propa-

gated in vast quantities, producing the substance they are now genetically programmed to manufacture.

This new technology for isolating individual genes and transferring them from one kind of organism to another has revolutionized scientists' ability not only to create new products for therapeutic use, but also to improve the characteristics of plants and animals. Gene manipulation offers enormous potential for use in agriculture and medicine, and it has already produced many important applications. Genetic engineering is also an important tool that helps scientists learn about gene structure, function, and regulation. Thus, these powerful techniques are used not only to produce new products but also to gain a better understanding of the molecular basis of life. The first commercial application of recombinant DNA technology was the production of proteins for therapeutic use, as we will see in the next section.

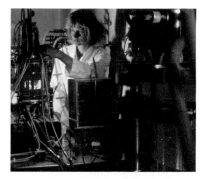

**Figure 11.7** A laboratory technician growing bacteria in table-top fermentors. Here, the technician is taking a sample from a fermentor during recombinant DNA experiments.

**Figure 11.8** Large fermentors are used for the commercial production of genetically engineered bacterial products. Large tanks such as these hold hundreds—and some hold thousands—of gallons of bacterial culture.

### CONCEPT CHECKPOINT

3. For each of the methods by which bacteria exchange genetic information—transformation, transduction, and conjugation—describe the form in which the DNA is acquired and give a brief description of the process.

# What are some applications of biotechnology and genetic engineering?

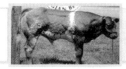

## 11.5 Proteins can be produced for therapeutic use.

The genetic engineer uses a variety of complex technological methods and laboratory equipment to study and manipulate genes. As scientists learn more about the process of genetic engineering, they

develop new methods and equipment to advance their studies. The field of DNA technology is constantly expanding and changing. Let's look at some of the basic methods employed today in the context of current applications.

One of the first applications of recombinant DNA technology was the insertion of human genes into bacteria and the manipula-

muscle of the person being vaccinated. The DNA becomes incorporated into the muscle cells of the recipient and causes those cells to produce the protein for which the DNA codes. The person's immune system then develops antibodies against this protein (antigen), which confers resistance against the pathogen.

DNA vaccines have many advantages over other types of vaccine preparations:

- No risk of infection by the pathogen.
- No risk of illness from contamination of the vaccine.
- A long-lasting immune system response.
- The ability to administer vaccines for a variety of diseases in a single shot.

DNA vaccines will be less costly than traditional vaccines and will not need to be refrigerated; thus, they can be more easily used in developing countries where many of the diseases they will protect against are prevalent. Researchers are currently working on DNA vaccines for a variety of diseases including malaria, rabies, epidemic (viral) diarrhea, herpes, tuberculosis, and the human papillomaviruses (HPVs) that cause cervical cancer. However, scientists are still uncertain how well they will work long term and what side effects might be encountered.

## 11.7 DNA fingerprinting is done by copying, decoding, and "reading" DNA.

The televising of the O. J. Simpson trial in 1994 and the detailed reporting of Monica Lewinsky's "stained" dress in 1998 have made DNA evidence and DNA fingerprinting familiar ideas. In the early days of DNA fingerprinting, it took many weeks, a team of people, and a sample of blood at least the size of a dime to perform the procedure. Today, DNA fingerprinting is automated and can be performed on minute samples, such as those obtained from saliva on a licked envelope or a spot of blood on a fence gate. The sample can be from 30-year-old evidence files or even biological specimens preserved for hundreds of years. Moreover, a national database exists of the DNA of adults convicted of crimes in the United States, to which police and forensics experts can compare known samples to identify criminals or corpses. Currently, the U.S. government is seeking to widen that database to include DNA samples from adults who have been arrested but not convicted and from juvenile offenders.

What is a DNA fingerprint? British geneticist Alec Jeffreys coined the term in 1985 to allude to the traditional use of fingerprints as a means of human identification. However, DNA fingerprinting has nothing to do with the unique patterns of whorls and ridges on fingertips. Instead, it uses unique patterns of genotypic information for human identification.

Five DNA fingerprints are shown in **Figure 11.13**. Each is a distinctive column of dark and light bands. These bands represent DNA fragments that were amplified using a technique called the **polymerase chain reaction (PCR)**. The fragments are drawn through a gel using an electrical field, which separates the fragments into bands of differing lengths. This process is called **electrophoresis** and is discussed later in this section. The bands are visualized with dyes.

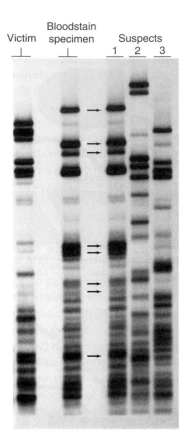

**Figure 11.13** DNA fingerprints. These were prepared from DNA isolated from a bloodstain at the site of a crime, from blood obtained from three individuals suspected of committing the crime, and from blood obtained from the victim of the crime. The arrows denote DNA fragments from suspect 1 that are not present in suspect 2 or 3 and that match those from the bloodstain at the crime scene.

**Visual Thinking:**
Could the bloodstain specimen be from the victim and not from suspect 1? How can you tell?

### Making copies of DNA fragments: the polymerase chain reaction

Developed in 1983 by American scientist Kary B. Mullis, the polymerase chain reaction (PCR) is a method used to make unlimited numbers of copies of segments of DNA. This technique is extremely important in genetic engineering today—not just in DNA fingerprinting, but in a variety of other applications as well. The DNA used for PCR can be in a mixture with other molecules and can come from anywhere, such as a cloned cell, a bloodstain, or even an organism that has been long dead, such as the preserved remains of extinct organisms, to determine their relatedness with other extinct or present-day organisms. Not only has PCR revolutionized genetic research, it has also enabled the analysis of DNA from extremely small samples at crime scenes that were previously impossible to analyze. Moreover, PCR has applications in health care. For example, PCR has allowed a more accurate HIV screening of blood samples than tests that screen for antibodies, since PCR can detect the HIV viral genome before the body has mounted an immune response.

The polymerase chain reaction is illustrated in **Figure 11.14**. "Polymerase" refers to DNA polymerase, which bonds nucleotides to DNA template strands, synthesizing DNA. "Chain reaction" means that the DNA necessary to the reaction is itself produced by the reaction, thus causing more of the same reaction.

The first step in the reaction is heating DNA fragments to separate the double DNA strands into single-stranded DNA **1**. After the DNA cools, *primers* a few nucleotides in length are hybridized with a portion of the DNA at the places where the "copying" is to begin **2**. The primers attach to the DNA, providing initiation sites

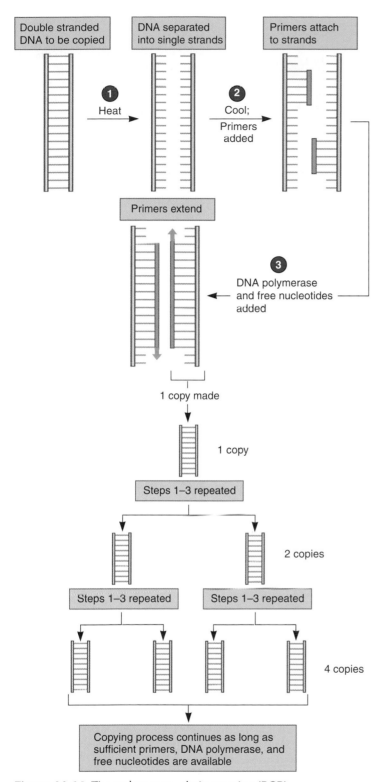

**Figure 11.14** The polymerase chain reaction (PCR).

stranded DNA fragment is separated into two single strands; each strand is hybridized with a nucleotide primer; and DNA polymerase adds free nucleotides to complete the strands, forming two new, double-stranded fragments from one original. Thus, one strand produces two, which produce four, which produce eight—the numbers quickly escalate. In a few hours, one original DNA fragment can be used to make a billion copies!

Researchers who want to amplify a particular gene choose two primers, one with a sequence that is specific to the beginning of the gene and the other with a sequence specific to the end of the gene. Scientists conducting DNA fingerprinting generally use primers with sequences that are randomly generated; thus, the locations where the primers bind are random.

PCR is performed by machines such as the one shown in **Figure 11.15**. In addition, researchers at Lawrence Livermore National Laboratory have developed a hand-held PCR instrument. This technology has many applications in medical diagnostics and in the military. For example, soldiers in the field can use this portable device to detect low levels of microorganisms contaminating water or supplies and to detect the spores of organisms used in biological warfare.

## Separating DNA fragments via electrophoresis

After DNA fragments have been amplified using random primers, they must be separated via electrophoresis, which is done as follows. A gel is heated until it becomes a liquid and is then poured

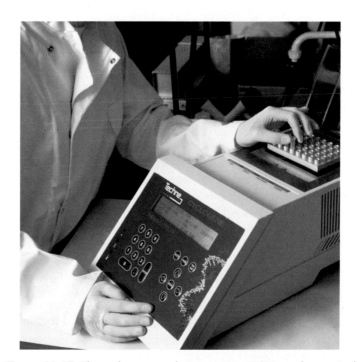

**Figure 11.15** The polymerase chain reaction (PCR) can be carried out by machine. The PCR machine is a thermal cycler—it goes through cycles of heating and cooling to carry out this chain of reactions. The laboratory technician simply loads the DNA to be copied, primer, free nucleotides, and DNA polymerase in tubes, which she is placing in the top of the machine. A special type of DNA polymerase that is resistant to destruction by high temperature is used. The reactions will take place in a cyclic fashion as long as sufficient reactants are available.

for the elongation of new DNA fragments. The enzyme DNA polymerase and free nucleotides (adenine, cytosine, guanine, and thymine) are added to the solution ③. DNA polymerase extends each primer, adding nucleotides in sequence from the site of the primer to the end of the fragment. Each new DNA fragment is double stranded once again and is an exact replica of the original fragment. This process is repeated over and over again: A double-

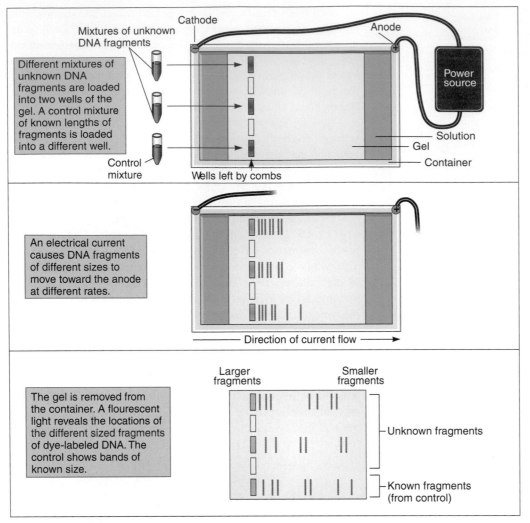

Cathode

Mixtures of unknown DNA fragments

Anode

**Different mixtures of unknown DNA fragments are loaded into two wells of the gel. A control mixture of known lengths of fragments is loaded into a different well.**

Power source

Solution

Gel

Control mixture

Wells left by combs

Container

**An electrical current causes DNA fragments of different sizes to move toward the anode at different rates.**

Direction of current flow

Larger fragments

Smaller fragments

**The gel is removed from the container. A flourescent light reveals the locations of the different sized fragments of dye-labeled DNA. The control shows bands of known size.**

Unknown fragments

Known fragments (from control)

**Figure 11.16** Electrophoresis of DNA fragments (overhead view). Two mixed samples of unknown DNA restriction fragments and one known sample, the control, are electrophoresed. The two unknowns are compared to the control.

into a shallow glass or plastic container. "Combs" are placed at one end of the gel, and as the gel cools and hardens, notches called *wells* form around the combs. Once the gel has cooled, the combs are removed and a solution (through which an electric current will run) is poured over the gel. A negative electrode (a cathode) is placed at the end of the gel containing the wells, and a positive electrode (an anode) is placed at the opposite end. The electrodes are connected to a source of electric power. This setup is illustrated in **Figure 11.16**.

The DNA fragments to be electrophoresed are placed into the wells, and the power source is turned on, creating an electric current in the gel between the negatively charged cathode and the positively charged anode. DNA is negatively charged; therefore, the fragments are repelled by the cathode and attracted to the anode. The fragments begin moving through the gel toward the anode. The fragments move at different rates, however, depending on their size, because the gel acts like a strainer, with porelike spaces that the fragments must pass through. Smaller fragments move through these spaces more easily and quickly than larger fragments, which have to "squeeze" through.

Because of these differences in rate of movement, the fragments separate into bands—groups of similar fragments that have moved similar distances. The largest fragments remain closest to the cathode, while the smallest fragments move farthest toward the anode, as shown in Figure 11.16. These bands can then be visualized with the use of dyes.

Looking back at Figure 11.13, you can see how the sequence of the dark bands of the victim and all three suspects differs. These bands are DNA fragments unique to each individual. Each is compared to the bloodstain found at the scene of a crime. The blood specimen clearly did not come from the victim, suspect 2, or suspect 3. It came from suspect 1.

Important components of DNA fingerprints are *variable number tandem repeats*, or *VNTRs*. What are VNTRs? As mentioned in Chapter 9, much of the DNA of eukaryotic organisms contain repeated sequences of nucleotides, called repetitive DNA. These repeated DNA sequences occur in tandem, that is, one right after the other. Repeats of varied lengths occur at several chromosomal locations and differ from person to person. In DNA fingerprinting using VNTRs, some DNA sequences containing VNTRs are first found. Then primers are designed to amplify the VNTR regions along with some flanking DNA sequences using PCR. The amplified fragments are then separated using gel electrophoresis so that differences in length, due to differences in the numbers of the tandem repeats, become apparent.

DNA fingerprints are also used to identify the biological father of a child in paternity cases. **Figure 11.17** shows the DNA fingerprints of a mother, her child, and two men, each claiming to be the child's father. Many of the bands in the DNA fingerprint of possible father no. 2 match those in the child's fingerprint, as denoted by the arrows, while many of the mother's bands match as well. Possible father no. 2 is the father of the child.

CONCEPT CHECKPOINT

5. What "tool" of biotechnology has made it possible to obtain enough genetic material for a DNA fingerprint when very little DNA is available?

6. How is a DNA vaccine different from a genetically engineered vaccine?

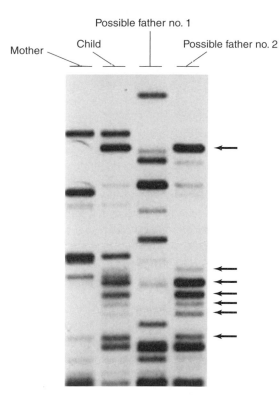

**Figure 11.17** DNA fingerprints of a mother, her child, and two men. Each of the men claimed to be the child's father. Arrows mark bands in the DNA of possible father no. 2 that are the same as bands in the child's DNA. The DNA of possible father no. 1 lacks bands similar to the child's. Possible father no. 2 is the biological father.

## 11.8 Results of research on genome structure and function have a wide variety of applications.

Analyzing the human genome is a huge task. In 1990, however, scientists undertook this task and the **Human Genome Project (HGP)** was launched. The HGP was a worldwide research effort that determined the DNA sequence of the human genome—all 3 billion base pairs and 30,000 to 40,000 genes! This sequencing was completed in April 2003. Contributors to the HGP included the National Institutes of Health (NIH), the U.S. Department of Energy (DOE), numerous universities throughout the United States, and international partners in the United Kingdom, France, Germany, Japan, and China.

The highest priority of the Human Genome Project was to provide a complete sequence of human DNA to the research community as a publicly available resource, and it accomplished this goal. This resource is significant because it will give scientists information they need to study the genes that cause various diseases as well as study the relationships between the structure of genes and the proteins they produce.

The HGP also included efforts to characterize and sequence the entire genomes of several other organisms, many of which are used extensively in biological research. Therefore, the HGP research holds enormous significance not only in medicine, but also in industry, agriculture, and livestock breeding. This new genetic knowledge raises many ethical, legal, and social questions. Therefore, a major component of ongoing research is devoted to analysis of such questions and the development of policy options for the public to consider.

The HGP heralded a new subdiscipline of genetics called genomics. **Genomics** involves mapping, sequencing, and analyzing the functions of entire genomes. This science can be divided into two areas: *structural genomics*, which is the study of the three-dimensional organization of proteins, and *functional genomics*, which is the study of genes, their resulting proteins, and the roles played by those proteins in metabolism.

To study the expression of entire genomes or large sets of genes, scientists use a technology called **DNA microarrays** or **DNA chips**. DNA microarrays allow scientists to screen thousands of genes to determine which genes or gene products are present in a sample.

To produce a DNA microarray, scientists first generate DNA fragments from the cells they are studying by cloning the DNA using techniques such as those described in Section 11.5. Each DNA fragment represents an individual gene. The fragments are placed on a solid surface in a specific pattern, or array. High-density arrays are produced by robotlike machinery that can place thousands of DNA fragment samples at known locations on the surface of a glass or silicon wafer, much like a computer chip. The result is the DNA chip.

Next, complementary DNAs (cDNAs) are made from the mRNAs present in the cells being studied (see Section 11.5). The cDNAs are labeled with fluorescent dyes. The labeled cDNAs are incubated on the chip so that base pairing (hybridization) can take place. The surface of the array is then scanned for fluorescence to determine which hybridizations have taken place. If a cDNA pairs with a DNA fragment, that is evidence of expression of that gene in that fragment in the cell.

**Figure 11.18** shows a DNA microarray. The cDNAs in this microarray are from yeast cells grown in either the presence or absence of glucose. The green spots represent mRNAs transcribed

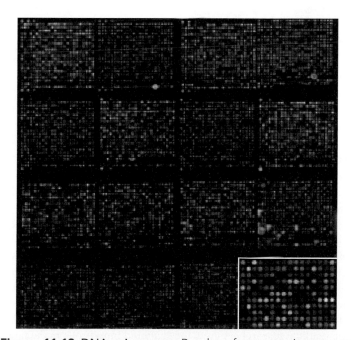

**Figure 11.18** DNA microarray. Results of an experiment performed using a mixture of cDNAs representing mRNAs transcribed from yeast cells growing in the presence of glucose (green-labeled cDNAs) or without glucose (red-labeled DNAs). Yellow-labeled DNAs correspond to genes that are expressed under both growth conditions. The inset at the lower right shows a close-up of a small portion of the microarray.

from yeast cells growing in the presence of glucose. The red spots represent mRNAs transcribed from yeast growing without glucose. The yellow spots represent genes that are expressed under both growth conditions. The inset at the lower right shows a close-up of a small portion of the microarray. This example shows how scientists can tell which genes are expressed under certain conditions and which ones are not.

Use of the DNA microarray technique has wide-ranging and important applications in medicine that are just beginning to be realized. For example, scientists are studying the genes of cells as they are transformed from "normal" cells into cancerous cells. Using DNA microarrays, researchers can tell which genes are turned off or turned on as transformation occurs. Microarrays are also being used to develop correlations between responses to medicinal drugs and the genetic profiles of patients.

---

### CONCEPT CHECKPOINT

**7.** State the purpose of generating a cDNA microarray. Why would the microarray from a heart muscle cell be different from that of a pancreatic cell?

**8.** It is one thing to read a book, but it is another to comprehend its meaning. How do you think this concept applies to the Human Genome Project? What further research must be done before we comprehend the "book of life"?

---

## 11.9 Gene therapy can be used to treat disease.

Recombinant DNA technology entered a new phase of application on September 14, 1990, when four-year-old Ashanthi De-Silva became the first person to undergo gene therapy. **Gene therapy** is the treatment of a genetic disorder by the insertion of "normal" copies of a gene into the cells of a patient carrying "defective" copies of the gene. This young girl suffered from the rare genetic disorder severe combined immunodeficiency (SCID). Key immune system cells called *T cells* (see Chapter 30) were not working because they lacked the enzyme adenosine deaminase, or ADA. Without these cells, Ashanthi had no defense against infection.

To treat Ashanthi, Drs. Michael Blaese and W. French Anderson, along with their colleagues at the National Institutes of Health (NIH) in Bethesda, Maryland, removed some of her white blood cells, which carry the defective gene, and cultured them. To the cell culture they then added bioengineered viruses into which they had incorporated working copies of the ADA gene. When the viruses infected the blood cells, they inserted the ADA genes into the cells. The medical researchers grew these altered blood cells in the laboratory until they numbered in the billions and then injected them into Ashanthi's blood. Thinking that these cells could live for only a few months, Drs. Blaese and Anderson repeated the gene-insertion process quite often over a two-year period. A few months into this process in 1991, another young ADA patient, Cynthia Cutshall, was included in this pioneering research.

The gene treatment ended after two years, but the girls' T cells continued to express the ADA gene; their cellular and humoral immunity had been restored. The results indicated that gene transfer into long-lasting "parent" T cells was successful and that new T cells carried a working copy of the gene. Although medical researchers suggest that they still must perfect certain components of this treatment, they conclude that gene therapy can be a safe and effective addition to other treatments for SCID. To date, a small number of ADA-deficient patients have been treated by gene therapy, and the therapy has resulted in improved immune function.

Genetic diseases most likely to be cured with gene therapy in the future are those, like SCID, which are caused by a single defective gene. In addition, genetic disorders that result in the deficiency of a cellular product are more likely to be curable than diseases that result in an overabundance of a product or a harmful substance. Scientists have learned how to turn genes on to produce a product but have not yet determined how to turn them off so that they no longer produce an unwanted product.

A more recent approach to gene therapy is *targeted gene repair*. The process of repairing a gene involves locating the gene and then changing its nucleotide sequence, which changes the function of the gene. To change genes within cells, targeted gene repair uses DNA molecules synthesized in the laboratory to activate and direct the cells' own DNA repair systems. The advantages of gene repair over gene replacement are that the gene remains under the control of its own promoter, and the permanent change is inheritable by daughter cells. Therefore, an effective repair may only require one intervention. Various techniques of targeted gene repair are currently being researched, with particular focus on methods to deliver the correcting molecule to the targeted nucleus.

## 11.10 Biotechnology can be used to modify food.

Biotechnology is being used in a variety of ways in the food industry. It focuses on increasing the yields of various foods and creating products with superior qualities. The most widespread use of biotechnology in the food industry is in agriculture. Through use of gene splicing, scientists are working to improve crops and forest trees by making them more resistant to disease, frost, and herbicides (chemicals that kill weeds). Scientists expect that plant improvement through the use of biotechnology will be an important part of increasing the food supply for the burgeoning world population, which is predicted to reach 10.7 billion by the year 2030.

### Transgenic plants

In early 1994, the FDA approved the first genetically engineered food: a tomato containing a gene that allows it to ripen longer on the vine and yet reach the supermarket without softening. Later that year, many genetically altered foods were approved, such as a squash that resists viruses, cotton and soybean plants that resist certain herbicides, and a potato that produces a pesticide that kills Colorado potato beetles. Such genetically altered plants, which carry genes from other organisms, are called **transgenic** plants (tranz-GEE-nik).

Since 1980, a plasmid of the bacterium *Agrobacterium tumefaciens*, which causes a tumorlike disease called *crown gall* in plants, has been the main vehicle used to introduce foreign genes into broadleaf plants such as tomatoes, tobacco, and soybeans. However, in order to use the plasmid to genetically engineer plants, scientists first had to remove *A. tumefaciens'* disease-causing genes—a technique known as disarming—while leaving its natural ability to transfer DNA intact.

After *A. tumefaciens* has been disarmed, it can be used as a vector to shuttle desired genes into plant cells. A part of its plasmid integrates into the plant DNA, carrying whichever genes the genetic engineer has inserted into the plasmid genome. **Figure 11.19** (right) diagrams the steps of this process. New plants can be micropropagated, or grown from these transformed cells in tissue culture. The cells develop into plantlets, which can be grown in conventional ways.

Many plant species are not natural hosts for the *Agrobacterium* organisms; thus, scientists have been looking for other ways to insert genes into plants. One commonly used method shoots microscopic metal pellets coated with DNA into plant cells, as shown in Figure 11.19 (left). The DNA on the pellets enters the nuclei of the plant cells, which are then cultured and propagated.

## Transgenic and cloned animals

In 1997, when cloning produced a sheep named Dolly, many people thought that the cloning of organisms was a new scientific breakthrough. But cloning is not new: plants have been cloned for

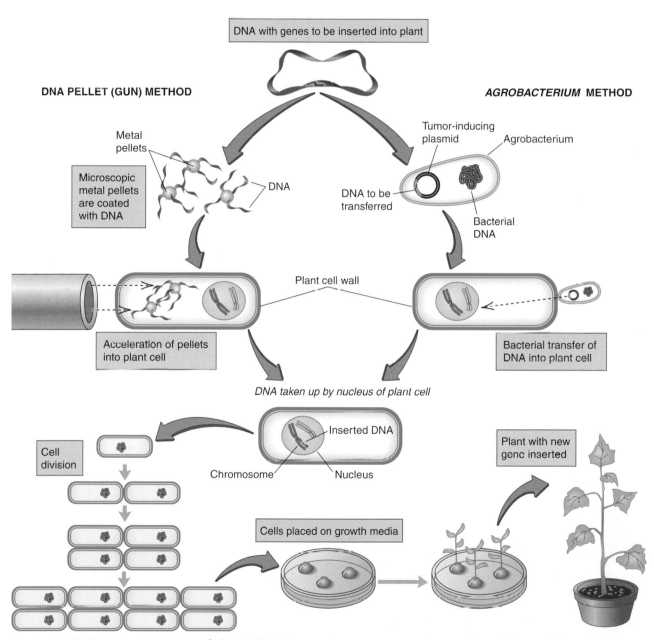

**Figure 11.19** Genetic engineering of plants. This diagram shows two ways to insert a gene of interest into a plant.

**Visual Thinking:** With the *Agrobacterium* method of inserting genes into the plant, what is the mechanism by which the bacterium transfers the DNA into the plant cell?

hundreds of years; you may have even done it yourself. Whenever a cutting from a plant is rooted, the new plant is a clone—that is, it is an individual plant genetically identical to another plant.

Animals have been cloned since the 1970s. Scientists have taken embryos at early stages of development and split them in two. The split embryos were clones of one another. What was different about the cloning of Dolly was that the genetic material was taken from an *adult* cell of Dolly's 6-year-old "parent." The nucleus from a single mammary gland cell was inserted into an egg cell cleared of its hereditary material, in a process called *nuclear transfer*. The resulting embryo was implanted in the oviduct of a surrogate mother. That embryo grew to be Dolly.

The technique of nuclear transfer provides another way for scientists to produce transgenic animals (those containing genes from other animals) that may be useful in a variety of ways. For example, transgenic animals might be engineered to produce substances in their milk for treating human diseases or to produce organs for transplantation. Nexia Biotechnologies in Quebec, Canada, for example, is currently cloning goats that have the genetic information to produce synthetic spider silk in their milk. This material, called BioSteel, will be used to produce artificial tendons and ligaments; it has a variety of other medical applications as well.

The beef industry may also use bioengineering in years to come to produce better and more inexpensive cuts of meat. In 1998, scientists discovered the gene that controls "double muscling" in cattle. Double muscling refers to a trait in cattle that provides more meat per animal on the same food intake as ordinary animals. Double muscling was first reported in 1807 in a strain of cattle called Belgian Blues. Notice how heavily muscled the Belgian Blue bull is in **Figure 11.20**. Unlike other types of cattle, whose meat gets tough when its muscles hyperdevelop, the Belgian Blue's muscles stay tender and yet are relatively low in fat.

When researchers isolated the double muscling gene in 1998, they discovered that it was a mutant of a gene that makes a protein called myostatin. Myostatin normally limits muscle growth, but the mutation blocks its effects and muscles grow larger. Scientists are not pursuing the development of herds of Belgian Blues with the double muscling mutation at this time because the offspring of the cows are so large that they must be delivered by cesarean section. Researchers are searching for a less extreme myostatin mutation and are also trying to identify another gene with a less drastic influence on muscle mass so that calves can be delivered without surgery.

**Figure 11.20** A heavily muscled, low-fat Belgian Blue bull.

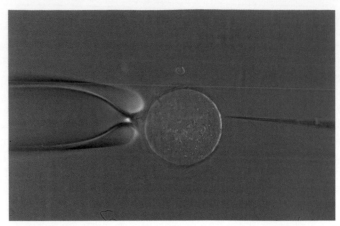

**Figure 11.21** Introducing foreign genes into the nucleus of a mouse egg. The blunt tip of the glass micropipet supporting the egg with a slight suction can be seen on the left. The egg is in the center, and the microneedle delivering the DNA is on the right.

Since the 1980s, transgenic animals have been produced by the microinjection of DNA fragments into single-celled zygotes. **Figure 11.21** shows a mouse egg undergoing this process. However, producing transgenic animals by microinjection of DNA is inefficient and cost-prohibitive. Therefore, aspects of the nuclear transfer technique used to clone Dolly may provide a more viable approach.

One technique being used at this time is as follows: fibroblasts (precursors of connective tissue cells) are taken from an animal fetus. The fibroblasts are placed in cell culture and allowed to grow and multiply **Figure 11.22** ❶. Foreign DNA is then mixed with the fibroblasts along with certain chemicals, and some of the cells take up the foreign DNA ❷. The fibroblasts are then grown under special conditions that allow scientists to select those cells that have successfully incorporated the foreign DNA ❸. The nuclei of these transgenic fibroblasts are transferred to cow oocytes (eggs) that have had their nuclei removed. To produce each embryo, a transgenic fibroblast is injected into the zona pellucida (outer coating) of a bovine oocyte ❹. These two cells are placed on a special microslide, and an electrical current is passed across them. The current fuses the cells and stimulates them to divide, forming many identical embryos ❺. The embryos are implanted in surrogate cows, and after the gestation period, transgenic cows are born ❻. These cows will produce (usually in their milk) the product for which the foreign DNA codes.

### The use of genetically engineered hormones in milk production

In the dairy industry, the focus has been on the production of genetically engineered hormones that, when injected into cows, would increase their milk production. In 1993, the FDA approved genetically engineered bovine somatotropin (bST). Recombinant bST is made by a process similar to that used in the production of synthetic human insulin and is identical to the hormone naturally produced by cows. Before giving its approval for commercial use, the FDA conducted more than 120 studies on bST and concluded that milk and meat from bST-treated cows are safe to consume. In addition, pasteurization destroys approximately 90% of the bST in milk. A variety of regulatory agencies, such as the National Institutes of Health in the United States as well as those in Canada and abroad, have confirmed these conclusions.

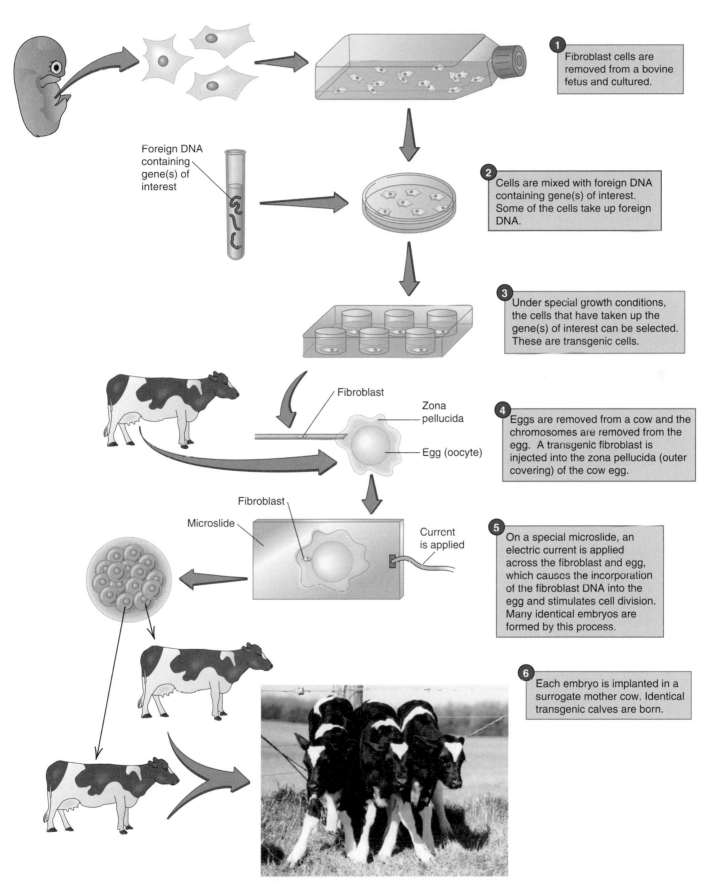

1 Fibroblast cells are removed from a bovine fetus and cultured.

Foreign DNA containing gene(s) of interest

2 Cells are mixed with foreign DNA containing gene(s) of interest. Some of the cells take up foreign DNA.

3 Under special growth conditions, the cells that have taken up the gene(s) of interest can be selected. These are transgenic cells.

Fibroblast

Zona pellucida

Egg (oocyte)

4 Eggs are removed from a cow and the chromosomes are removed from the egg. A transgenic fibroblast is injected into the zona pellucida (outer covering) of the cow egg.

Fibroblast

Microslide

Current is applied

5 On a special microslide, an electric current is applied across the fibroblast and egg, which causes the incorporation of the fibroblast DNA into the egg and stimulates cell division. Many identical embryos are formed by this process.

6 Each embryo is implanted in a surrogate mother cow. Identical transgenic calves are born.

**Figure 11.22** A nuclear transfer procedure to clone transgenic cows.

**Can human cloning really be done? Do you get an exact copy of yourself—a second you—or is it more like having a twin? Would the cloned person be somehow "weaker" or defective? I know Dolly the sheep was sickly and died young.**

It is likely that human cloning can really be done. If human cloning proceeds in the United States, scientists plan to use the nuclear transfer method described in Figure 11.22 and in this chapter's opening story, which is the same procedure that was used to create Dolly the sheep. With animal cloning, this procedure has only a 1 to 2% success rate, and there is no reason to think that the success rate with humans would be higher. Dolly was the sole survivor of 277 cloning attempts. At this writing, no scientifically verified human clone has been born.

Would a clone of you be a "second" you? No. A clone would be similar to a twin born at another time or a child born with the same genetic complement as your own, rather than with half your genes. However, your clone would not be your twin or child but would have a new type of biological relationship with you—the new individual would be your clone.

Are all clones sickly? There is controversy about this question in the scientific community. Some suggest that it is impossible to make generalizations about the health of clones based on the animals that have been cloned. Others suggest that clones do have survival and health problems that are probably brought about by difficulties in "developmental reprogramming." That is, nuclei that are transferred from cells into enucleated eggs must stop previous patterns of gene activity and start up new ones that drive embryo growth. Scientists are currently researching developmental reprogramming. In addition, you can learn more about the research of MIT researcher Rudolf Jaenisch and his team on human reproductive cloning by reading in the opening story of this chapter. ●

*Are you wondering about a topic in biology and how it relates to your life? Submit your question by clicking the Just Wondering link in this text's companion Web site at www.wiley.com/college/alters.*

# What are some controversies regarding genetic engineering?

## 11.11 The use of transgenic plants and animals has potential benefits and dangers.

There has been considerable discussion in the scientific community and among nonscientists as well about the possibility of inadvertently creating undesirable or potentially dangerous organisms in the course of a recombinant DNA experiment. For example, what if someone fragmented the DNA of a cancer cell and then incorporated these DNA fragments at random into viruses that are propagated within bacterial cells? Might there not be a danger that one of the resulting bacteria or viruses could be capable of infecting humans and causing a disease—even cancer?

Even though most recombinant DNA experiments are not dangerous, such concerns are taken seriously. It was the scientists themselves who first realized the possible dangers associated with genetic engineering technology. These scientists, along with government agencies, such as the National Institutes of Health, have drawn up formal guidelines that govern all such experimentation in the United States. Many other countries have done the same. Many scientific organizations have ethics committees to deal with such issues, especially as they relate to gene therapy in humans.

Scientists and governmental regulatory agencies not only monitor experiments but also study products to make sure they do not cause increased health risks to consumers or pose environmental hazards. Genetically engineered products on the market today, especially pharmaceutical and agricultural products, have been shown to be safe and effective in scientific studies, but they have aroused much public concern and controversy. Two major areas of controversy are genetically modified foods and the release of transgenic animals into the environment.

Genetically modified (GM) foods are those whose genes have been altered by genetic engineering techniques. The list of GM foods is lengthy and includes giant pumpkins, strains of wheat with twice the usual yield, and crops grown from "Roundup Ready" seeds. Roundup is a chemical that kills plants. Its manufacturer has genetically modified soybeans and other crop plants to resist the herbicide, so that farmers can spray the crops to kill weeds while not killing the crop.

Genetically modified foods are common in the United States but are opposed by many Europeans. In July 2003 the European

Parliament approved legislation requiring that food and feed made with genetically altered ingredients be labeled as such. The law would require that GM foods be traced from the moment of their inception, such as when genetically engineered wheat is used to make bread.

What are the controversies surrounding GM foods? Arguments in favor of GM foods include disease resistance, improved characteristics, and higher yields. One argument against GM foods is that the long-term safety of eating them has not been established. Although recent studies show GM foods to be safe, they might contain proteins that could trigger allergic reactions in a small percentage of people. It has also been shown that pollen from genetically modified crops has transferred to similar species in surrounding fields, resulting in unknown genetic recombinations.

The spread of bioengineered genes to native populations is a potential threat with transgenic animals. Transgenic salmon have been produced in response to the near extinction of Atlantic salmon. These GM fish have been given a growth hormone coding sequence from Chinook salmon and a regulatory sequence from the ocean pout, a bottom-dwelling fish. The result is a salmon growth hormone gene (GH gene) that is activated in the winter and shut down in the summer by the regulatory sequence. This gene is complementary to the native GH gene, which is inactive in the winter. As a result, GM salmon typically mature about one year more quickly than native salmon (**Figure 11.23**).

To prevent the spread of bioengineered genes from these salmon, which are all female, scientists have subjected the salmon to a pressure shock treatment that makes them sterile. However, there can be no guarantee that this treatment is 100% effective. If nonsterile GM salmon were to be released to the environment, they could transfer their GM gene to native populations. It is possible that the GM fish could outcompete any native Atlantic salmon left, but it is not clear that they would do so. The GM fish have disadvantages compared to native fish in that they use more oxygen, swim more slowly, and avoid predators less well. Whether these disadvantages would be outweighed by earlier maturation is still unknown.

Genetic engineering may sound rather ominous to some, and it does raise legitimate concerns, but it is important to remember that

**Figure 11.23** Alantic salmon (*Salmo salar*), both about two years of age. The top fish is a GM variety and is harvestable. The smaller "standard" fish will need another year of growth before it is commercially harvestable.

scientists have been manipulating genes for centuries using classical biotechnological techniques—techniques with which most people are comfortable. Molecular biotechnology is simply more specific, quick, and versatile.

---

### CONCEPT CHECKPOINT

9. In order for human gene therapy to permanently cure a genetic disorder such as diabetes or hemophilia, into what general kind of cells must the corrected gene be inserted? Explain.
10. Describe some potential benefits and risks of developing transgenic animals or plants. Do you think the benefits outweigh the risks?

---

## CHAPTER REVIEW

## Summary of Key Concepts and Key Terms

### What is biotechnology?

**11.1 Biotechnology** (p. 156) is the manipulation of organisms or their genes to yield specific characteristics, useful products, and information that would otherwise not be known.

**11.1 Classical biotechnology** (p. 156) uses the techniques of selection, hybridization, and mutation.

**11.2** Molecular biotechnology, which involves the manipulation of genes themselves, began with the discovery of the structure of the genetic material DNA in the 1950s.

**11.2 Genetic recombination** (p. 157) is the exchange of DNA sequences between molecules.

### How are genes manipulated?

**11.3** Certain bacteria can naturally transfer genetic material from one cell to another by **transformation** (p. 157), **transduction** (p. 157), and **conjugation** (p. 157).

**11.3.** Conjugation takes place in bacteria having extra chromosomal pieces of DNA called **plasmids** (PLAZ-mids, p. 157).

**11.4** Scientists were able to manipulate these gene transfer mechanisms in the 1950s and 1960s and, by the 1970s, discovered how to put eukaryotic genes into prokaryotes to amplify and study those genes.

**11.4** Techniques of gene manipulation are called **genetic engineering** (p. 159) or **recombinant DNA technology** (p. 159).

## What are some applications of biotechnology and genetic engineering?

**11.5** One of the first applications of recombinant DNA technology was the insertion of human genes into bacteria, causing these organisms to produce human proteins for therapeutic use.

**11.5** To produce human proteins by means of recombinant DNA technology, the genes that code for the desired protein must first be identified within the human genome.

**11.5** One method of finding a gene is **shotgun cloning** (p. 160), which researchers use when they know very little about the gene they are trying to find; this process results in a complete **genomic library** (p. 160): a collection of copies, or **clones** (p. 160), of DNA fragments that represent the entire genome of an organism.

**11.5** Shotgun cloning involves cutting the DNA of the entire genome into pieces with **restriction enzymes** (p. 160).

**11.5** The base sequence at which a particular restriction enzyme makes a cut is called a **restriction site** (p. 160).

**11.5** Carefully using restriction enzymes, scientists can produce a range of sizes of fragmented DNA, called **restriction fragments** (p. 160).

**11.5** The restriction fragments, each of which may contain one or more genes, are inserted into bacteria or yeast cells using either plasmids or viruses.

**11.5** Plasmids and viruses used to insert restriction fragments of foreign DNA into cells are called **cloning vectors** (p. 161).

**11.5** Two other methods of finding a gene, used under different circumstances, are **complementary DNA (cDNA) cloning** (p. 161) and **gene synthesis cloning** (p. 162).

**11.5** Cloned genes must be screened in order to select the gene(s) of interest.

**11.5** Most screening techniques involve DNA hydridization using molecular **probes** (p. 162), molecules that bind to specific genes, or nucleotide sequences.

**11.5** Products that have been developed by genetic engineering to treat a variety of human disorders include human insulin, used to treat diabetes; human growth hormone, used to treat children who do not produce enough of this hormone for normal growth; and the human interferons, proteins that interfere with the ability of a virus to invade a cell.

**11.6** Scientists are beginning to develop genetically engineered vaccines.

**11.6** The development of DNA vaccines is a promising avenue of research to develop a variety of effective, low-cost vaccines that will be safe.

**11.7** DNA fingerprints can be developed from small samples for medical and forensic applications.

**11.7** The **polymerase chain reaction** (**PCR**, p. 164) is a process used to produce multiple copies of DNA fragments for DNA fingerprinting and for a wide variety of other applications in molecular biotechnology.

**11.7** DNA fingerprints are banding patterns formed by DNA fragments drawn through a filtering gel material using an electrical field, a process called **electrophoresis** (p. 164).

**11.7** Important components of DNA fingerprints are restriction fragments of variable number tandem repeats, or VNTRs; VNTRs are repeated DNA sequences that occur one behind the other in the genome.

**11.8** **The Human Genome Project** (**HGP**, p. 167) was a worldwide effort to sequence the DNA of the human genome; the sequencing was completed in 2003.

**11.8** The HGP heralded a new subdiscipline of genetics called **genomics** (p. 167), which involves mapping, sequencing, and analyzing the functions of entire genomes.

**11.8** The HGP will provide information that scientists can use to help clarify the relationships between the structure of genes and the proteins they produce.

**11.8** To study the expression of entire genomes or large sets of genes, scientists use a technology called **DNA microarrays**, or **DNA chips** (p. 167).

**11.9** Genetic engineering has also been used to treat genetic disorders by means of a technique called **gene therapy** (p. 168), which involves the insertion of "normal" genes into the cells of a patient to replace "defective" genes.

**11.9** The process of targeted gene repair involves changing the nucleotide sequence of a particular gene, which changes the function of the gene.

**11.10** Genetic engineering is being used in a variety of ways in the food industry, where it focuses on increasing the yields of various foods and creating products with superior qualities.

**11.10** Organisms that carry genes from other organisms are called **transgenic** (tranz-GEE-nik, p. 168).

**11.10** In agriculture, scientists are improving crops and forest trees by making them more resistant to disease, frost, and herbicides.

**11.10** Regarding animals, research is being conducted to develop varieties of quickly growing salmon as well as transgenic animals that may be able to produce products useful to humans.

**11.10** In the dairy industry, genetically engineered hormones have been produced that increase the milk production of cows.

## What are some controversies regarding genetic engineering?

**11.11** Scientists and governmental agencies regulate genetic engineering experimentation and the products that result from the research.

**11.11** Two major areas of controversy in biotechnology are genetically modified foods and the release of transgenic animals into the environment.

## Level 1 | Learning Basic Facts and Terms

**Multiple Choice**

1. Which of the following could be considered tools used by biotechnologists?
   a. restriction enzymes
   b. PCR
   c. plasmids
   d. cloning vectors
   e. chemicals that stimulate transformation in bacteria
   f. all of the above

2. Recombinant DNA technology would most accurately be described as
   a. transferring genes from one species to another.
   b. cloning of embryos.
   c. using restriction enzymes to cleave DNA.
   d. separating DNA via gel electrophoresis.
   e. all of the above.

3. Which of the following is an example of classical biotechnology?
   a. creating transgenic organisms
   b. cloning embryos
   c. PCR
   d. DNA fingerprinting
   e. selective breeding of cattle

4. Restriction enzymes
   a. are made synthetically by biotechnologists.
   b. function to cleave foreign DNA as a protection against viruses in humans.
   c. bind consistently to a specific DNA recognition sequence.
   d. cut DNA in a very unpredictable manner.
   e. are used in PCR to amplify DNA.
5. Transformation is a process whereby
   a. bacteria utilize restriction enzymes to cleave DNA.
   b. bacteria take up DNA from their environment.
   c. bacteria duplicate their DNA.
   d. bacteria produce plasmids.

**Matching**

Pair the biotechnology procedure with its use in genetic engineering.

6. gel electrophoresis
7. PCR
8. gene synthesis cloning
9. DNA hybridization
10. cDNA cloning

a. a method of amplifying DNA
b. a method of separating fragments of DNA by size for a DNA fingerprint
c. allows for molecular probes to locate genes
d. cloning a synthetically produced DNA sequence
e. cloning genes made from mRNA transcripts

---

## Level 2 — Learning Concepts

1. Define the term *biotechnology*. Distinguish between biotechnology and genetic engineering.
2. What are the differences between classical biotechnology and molecular biotechnology? Are both being practiced today? Support your answer with evidence.
3. Describe one way in which scientists insert human genes into bacteria using the natural ability of certain bacteria to transfer genetic material.
4. What is the role of restriction enzymes in genetic engineering?

5. What is a molecular probe? What is its function in screening clones?
6. Why is the polymerase chain reaction (PCR) an important technique in genetic engineering today?
7. How does the process of gene therapy treat disease? Is this technique in widespread use today? Support your answer with evidence.
8. Why is more screening required in shotgun cloning than in cDNA cloning? What cloning procedure requires the least screening?
9. Why would the DNA fingerprints of two brothers not be identical? Revisit this question after you've read Chapter 12.

---

## Level 3 — Critical Thinking and Life Applications

1. If scientists can engineer bacteria to produce human proteins, why can't they genetically engineer bacteria to synthesize gold or other precious metals?
2. In contrast to genetic disorders caused by a nonfunctional protein product (as is the case in SCID), some genetic disorders occur because the gene mutation *produces* something that is toxic to a cell. How could gene therapy be used to treat such a disorder?
3. Imagine that you've created a genomic library of the entire human chromosome that contains the insulin gene. You then make hybrid plasmids with these DNA fragments, and you induce bacteria to take up these plasmids through transformation. Your goal now is to find the bacteria that took up the insulin gene among a variety of other bacterial cells. Characterize what occurred in this population of bacteria exposed to the genomic library that requires you to screen them for the insulin gene.
4. Here is part of the DNA sequence for the ADA gene.

ATCCATAGTCAAATTTTGGATCCGGATCCGGGCGCAAAATGGGGC

If you wanted to create a genomic library that contained this gene, which of the following restriction enzymes would you *not* want to use? The arrows show where each enzyme cuts base pairs. Explain your answer.

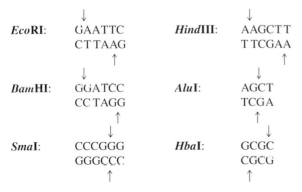

| | | | |
|---|---|---|---|
| ***Eco*RI**: | GAATTC<br>CTTAAG | ***Hind*III**: | AAGCTT<br>TTCGAA |
| ***Bam*HI**: | GGATCC<br>CCTAGG | ***Alu*I**: | AGCT<br>TCGA |
| ***Sma*I**: | CCCGGG<br>GGGCCC | ***Hba*I**: | GCGC<br>CGCG |

5. What advantages do you think exist for a person using genetically engineered human insulin versus insulin extracted and purified from an animal source? Discuss a recipient's possible reaction to human (bacterial) insulin versus animal-derived insulin.
6. One night in the town of Micksup, Mississippi, three girls were born in the same hospital at about the same time. Coincidentally, the three sets of parents were neighbors. The nurse in the maternity ward that night misplaced the infants' ID bracelets, and when he finally found them, he forgot which bracelet went with which infant. He notified the parents and suggested they do DNA fingerprinting to determine which child belonged to which parents. The results of the DNA fingerprinting are below. Match the children (Ruth, Sally, and Lane) with their correct parents.

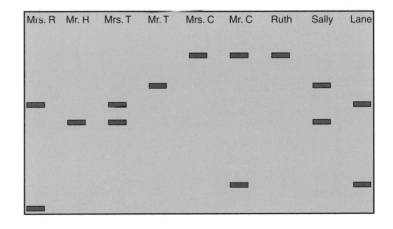

## In The News | **Critical Thinking**

**HUMAN CLONING—THE SCIENCE**

Now that you understand more of the science behind the process, reread this chapter's opening story about the issue of human cloning. To make an informed and thoughtful decision about your stand on this issue, it may help you to follow these steps:

1. Review your immediate reaction to the issue that you wrote when you began reading this chapter.

2. Based on your current understanding, identify the issue using either a statement or a question. An issue is a point on which people hold differing views.

3. Collect new information about the issue. Visit the *In The News* section of this text's companion Web site at www.wiley.com/college/alters and watch the "Human Cloning—The Science" video. Then use the "summary" link to read the accompanying story and access related links. Use this information, the links provided, and other online and library resources to find updates about this issue. State the sources of that information. Explain why you think the information is accurate. Also determine whether the information expresses a particular point of view or is biased in any way.

4. Determine which individuals, groups, or organizations have a stake in the issue. What does each stand to gain or lose depending on how the issue is resolved?

5. List possible outcomes (resolutions) of the issue. List the pros and cons of each outcome.

6. Which outcome do you think would be best? Why? Note whether your opinion differs from or is the same as what you wrote when you began reading this chapter. Give reasons for your opinion and for any changes in your ideas based on the additional information you have collected and the analysis you have done.

# CELL REPRODUCTION

## In The News | Stem Cell Shakes

Can cell reproduction help provide a new "magic bullet" in medicine? Can making unfertilized egg cells divide as if they were fertilized be a way of generating stem cells without the use of embryos? What are stem cells, and how can they be used to treat disease?

Stem cells are clusters of cells found within an embryo when it is merely a hollow ball called a blastocyst. Scientists are learning how to stimulate stem cells to develop into a variety of cell types that may be used to treat diseases and conditions such as Parkinson's disease, Alzheimer's disease, Type I diabetes, certain types of arthritis, some forms of blindness, and damage from strokes, spinal cord injuries, and burns. Ping Wu, Richard Coggeshall, and their research teams at the University of Texas Medical Branch at Galveston, Texas, have successfully induced human stem cells to develop into neurons (shown in photo) when implanted in the brains and spinal cords of rats. You can view the video news story, "Stem Cell Shakes," which describes Wu and Coggeshall's work in more detail, by visiting the *In The News* section of this text's companion Web site at www.wiley.com/college/alters.

Although the use of stem cells to treat human disease holds much promise, there is heated controversy surrounding the research on and the use of tissue from human embryos. Most agree that tissue obtained from a miscarriage is acceptable to use for these purposes. However, if a problem within the developing embryo caused the miscarriage, the tissue may be unusable. The controversy focuses on tissue derived from elective abortions.

Opponents of harvesting embryonic stem cells worry that the use of such tissue to treat human disease will en-courage abortion. For example, a woman who wants to help treat her father's Parkinson's disease might conceive and then abort an embryo to use its tissue. Opponents also contend that economic pressures may be imposed on abortion clinics to obtain and sell embryonic tissue. And some worry that doctors, families, or economic need could pressure women to become "embryo factories."

Proponents of stem cell research point to the need to heal the sick and argue that stem cells represent a highly promising means of treating a variety of diseases and conditions. They think that aborted embryonic tissues should be used for such purposes rather than being buried or discarded, and they suggest that ethical guidelines can be established to prevent any abuse of the tissue. They also point out that human embryos created in the course of fertility treatment and donated by couples for research can be used as a source of stem cells.

Researcher Kent Vrana and his team at Wake Forest University Baptist Medical Center have been researching a way to solve the embryo "problem." They have successfully stimulated eggs to divide to the blastocyst stage without being fertilized. Is this a solution?

*Write your immediate reaction to this information about the issue of stem cell research: first, summarize the issue in a sentence or two; then, suggest how you think this issue could be or should be resolved. You will have an opportunity to reflect on your responses and gather more information on this issue in the* In The News *feature at the end of this chapter. In this chapter, you will learn more about cell reproduction.*

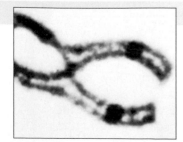
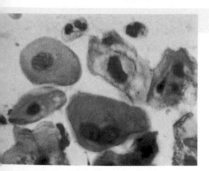

# How do eukaryotic cells reproduce?

## 12.1 Eukaryotic cells reproduce by the processes of mitosis and meiosis.

When you think about your uniqueness, you might think about your interests, your behavior, and how you look. You probably never think about it in terms of molecules. Yet your uniqueness is due in large part to the highly ordered and orchestrated process of DNA replication, transcription, and translation, which results in the expression of your heredity. A tremendous amount of information resides in the DNA packaged neatly in chromosomes in the nucleus of each cell of your body, and your cells are able to separate this genetic material in an organized way during cell division.

Your body uses two methods to divide nuclear material: **mitosis** (my-TOE-sis) and **meiosis** (my-OH-sis). Mitosis is a process of nuclear division that, when coupled with division of the cytoplasm, produces two cells, each of which is identical to the original parent cell. Mitosis is preceded by replication of the cell's DNA, so each daughter cell will have a full complement of genetic material. In animals, mitosis is used for growth and repair of *somatic* (body) cells. Meiosis is a process of nuclear division that, when coupled with division of the cytoplasm, produces four cells from an original parent cell. However, these cells are not identical to the parent cell because each has only half of the parent cell's complement of genetic material. In animals, meiosis produces **gametes**, or sex cells, and is used for reproduction.

To illustrate, look at the human life cycle diagrammed in **Figure 12.1**, which is representative of all animal life cycles. A **life cycle** is the progression of stages an organism passes through from its conception until it conceives another similar organism. The animal life cycle begins with the fusion of gametes from a male and a female of the same species. The female gamete is the egg, and the male gamete is the sperm. The diploid (2N) zygote in the diagram is the single cell formed by the fusion of these gametes; the zygote will develop into the new individual. **Diploid** (2N) means that the cell or individual has the full complement of hereditary material characteristic of that species. In the case of humans, the diploid (full) complement of hereditary material is 46 chromosomes (23 pairs). You can see the full complement of human chromosomes in the karyotype shown in Figure 9.1.

After a person (or other animal) grows to sexual maturity, the sex organs begin to produce gametes by meiosis (see Figure 12.1). During meiosis, four sex cells develop from a single parent cell, but each sex cell is **haploid** (N); that is, it contains only one chromosome from each pair in the parent cell, half of the full complement of hereditary material. In the case of humans, the haploid complement of chromosomes is 23—that is, 23 single chromosomes. Because of this reduction in chromosome number, one sex cell from each of two parent organisms can join together in a process called **fertilization** to form the first cell (the zygote) of a new individual with a full complement of hereditary material. This new cell is diploid. That is, it contains two of each chromosome—double the haploid amount of genetic information. This type of reproduction, which involves the fusion of gametes to produce the first cell of a new individual, is called **sexual reproduction**.

The life cycles of both plants and animals have two phases: a haploid phase and a diploid phase. In animals, however, the diploid phase of the life cycle predominates. The haploid phase consists only of single-celled gametes. In most animals, gametes live only hours to about six days after meiosis is complete, unless fertilization occurs. During fertilization, two haploid gametes join to form the first cell in the diploid phase of the life cycle. Plant life cycles differ from animal life cycles because most plants have a distinct multicellular haploid phase as well as a multicellular diploid phase. One phase usually dominates over the other, but both phases are always present. (See Chapter 23 for details regarding the variations in plant life cycles.)

Compare the generalized plant life cycle in **Figure 12.2** with the animal (human) life cycle in Figure 12.1. In plant life cycles, the multicellular haploid phase is the *gametophyte* (geh-MEE-toe-fite) (gamete-plant), and the diploid phase is the *sporophyte* (SPOR-ih-fite) (spore-plant). The existence of a multicellular haploid phase in plants means that there are some striking differences in the life cycles of animals and plants. In almost all animals, gametes are produced in diploid individuals by the process of meiosis. In plants, gametes are produced in haploid individuals (gametophytes) by the process of mitosis. The product of fertilization in plants is the first cell (the zygote) of a diploid sporophyte (spore-producing plant),

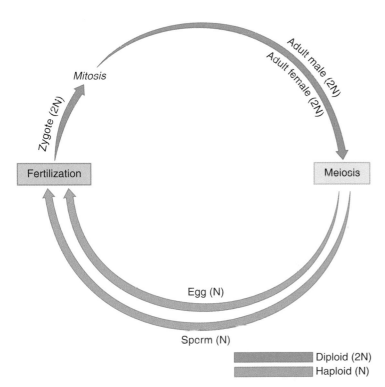

**Figure 12.1** The human (animal) life cycle.

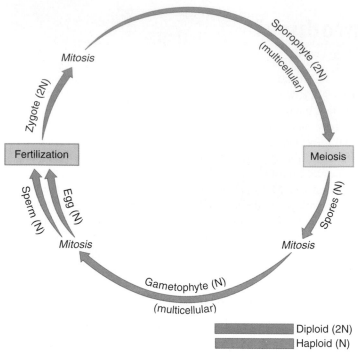

Diploid (2N)
Haploid (N)

**Figure 12.2** The generalized plant life cycle.

 **Visual Thinking:** Mitosis is shown three times in the generalized plant life cycle, yet only once in the animal life cycle (Figure 12.1). Explain this difference.

not a diploid gamete-producing organism as in animals. Spores are haploid, single-celled bodies capable of growing via mitosis, into the haploid gametophyte (see Figure 12.2). Because sporophytes are diploid, they produce spores by meiosis, reducing the number of chromosomes by half during this process. **Table 12.1** summarizes these differences in the life cycles of animals and plants. Throughout their life cycles, both plant and animals use mitosis for growth and repair of tissues.

Single-celled eukaryotes (such as many of the protists) generally reproduce by mitosis, in a process called **asexual reproduction**. During reproduction, the parent organism divides by mitosis, producing two identical organisms. Mitosis plays no role in the growth of single-celled organisms but does in the growth of populations of single-celled organisms. Some simple animals such as the hydra, protists such as *Paramecium*, and many plants such as the maternity plant (*Kalanchoe digremontiana*) and the strawberry plant (**Figure 12.3**) are capable of reproducing asexually by mitosis. Many of the plants that can reproduce asexually also reproduce sexually.

CONCEPT CHECKPOINT

**1.** The term *ploidy* refers to the number of sets of chromosomes in a cell. A haploid cell has one set; a diploid cell has two. Does the ploidy of a cell change when it divides by mitosis? by meiosis? Why or why not?

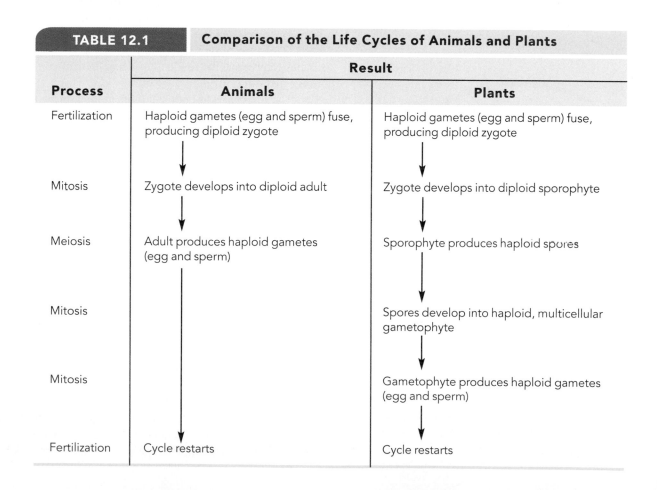

| TABLE 12.1 | Comparison of the Life Cycles of Animals and Plants | |
|---|---|---|
| | **Result** | |
| **Process** | **Animals** | **Plants** |
| Fertilization | Haploid gametes (egg and sperm) fuse, producing diploid zygote ↓ | Haploid gametes (egg and sperm) fuse, producing diploid zygote ↓ |
| Mitosis | Zygote develops into diploid adult ↓ | Zygote develops into diploid sporophyte ↓ |
| Meiosis | Adult produces haploid gametes (egg and sperm) ↓ | Sporophyte produces haploid spores ↓ |
| Mitosis | | Spores develop into haploid, multicellular gametophyte ↓ |
| Mitosis | | Gametophyte produces haploid gametes (egg and sperm) ↓ |
| Fertilization | Cycle restarts | Cycle restarts |

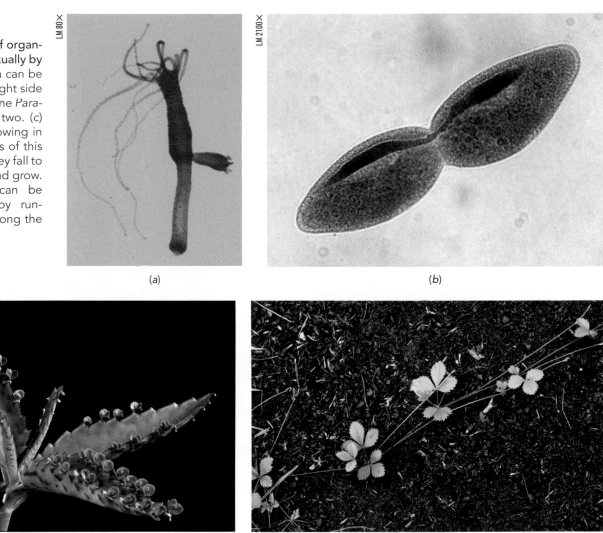

**Figure 12.3** Examples of organisms that reproduce asexually by mitosis. (a) A new hydra can be seen budding from the right side of the parent hydra. (b) One *Paramecium* is splitting into two. (c) Many new plants are growing in the notches of the leaves of this maternity plant. When they fall to the soil, they take root and grow. (d) Strawberry plants can be propagated asexually by runners. New plants arise along the runners.

LM 80×

LM 2100×

(a)

(b)

(c)

(d)

# What do eukaryotic cells do when they are not reproducing?

## 12.2 When not reproducing, eukaryotic cells are growing and carrying out other life functions.

**Figure 12.4** illustrates the **cell cycle**, the "life cycle" of the cell. The duration of the cell cycle varies among organisms. Some organisms, such as some prokaryotes, have cell cycles as short as 10 minutes, while other organisms have cycles that are much longer—days, months, or even years. As you can see in the illustration, only a small portion of a cell's life is involved with mitosis. The activities of **interphase** occupy most of the cell cycle. Interphase is the part of the cell cycle during which a cell grows, replicates cell organelles, replicates DNA, assembles the "machinery" of mitosis, and condenses its DNA.

During the $G_1$ stage of interphase—the time gap ($G_1$ for gap) between the last cell division and the start of DNA replication—the cell is growing. For many organisms, this growth period occupies the major portion of the cell's lifespan. The cell doubles its size and carries out its normal life functions.

During the S (for "synthesis"—DNA replication) stage of interphase, a complete replica of the cell's DNA is made in preparation for cell division. By the end of the S stage, the cell contains two complete, identical copies of hereditary information.

During the $G_2$ stage of interphase—the time gap between the end of DNA replication and the beginning of cell division—the supercoils of DNA condense into tightly compacted bodies that become visible as chromosomes during mitosis. Each chromosome contains two copies of hereditary information (the original and its replica synthesized during the S stage), which can be seen as two strands of chromosomal material held together at a single point. The strands are called **sister chromatids** (CROW-mah-tidz).

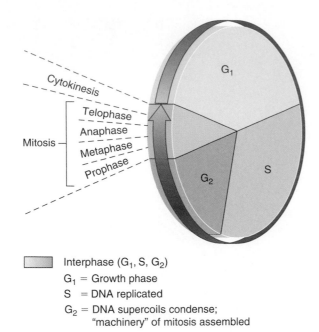

Interphase (G₁, S, G₂)

G₁ = Growth phase
S = DNA replicated
G₂ = DNA supercoils condense;
     "machinery" of mitosis assembled

**Figure 12.4** The cell cycle.

 **Visual Thinking:** The cell cycle is depicted here in a manner similar to a pie chart. The sizes of the pieces of the pie show the approximate proportion of time of each phase with respect to the entire cycle. If this figure depicts a six-hour cell cycle, approximately how much time would be devoted to DNA replication? to mitosis and cytokinesis?

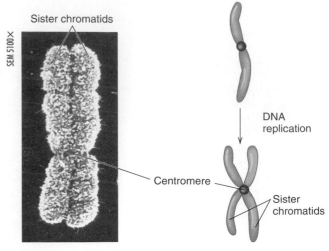

**Figure 12.5** Sister chromatids. An original and a replicated chromosome are compacted and held together at the centromere.

The sister chromatids are connected by a centromere, as shown in **Figure 12.5**.

Mitosis is a continuous sequence of events that occurs just after interphase and that results in the separation of the sister chromatids. In other words, the two sets of hereditary material (the original and the replica) are separated so each new cell can receive one complete set of hereditary material. To more easily understand this continuous process, scientists divide it into four phases: prophase, metaphase, anaphase, and telophase. We look more closely at these stages of mitosis in the next section. Cytokinesis, division of the cytoplasm, which occurs after mitosis, is discussed in Section 12.4.

# What happens during the process of mitosis?

## 12.3 During mitosis, duplicated chromosomes move to opposite poles of the cell.

### Prophase

The cell is in the first stage of mitosis, **prophase (Figure 12.6 ❶ )**, when the chromosomes (the sister chromatids) have condensed to the point where they become visible under a light microscope. As prophase continues, the chromosomes continue to shorten and thicken (see micrograph in Figure 12.5). The nucleolus disappears because the cell is unable to make ribosomal RNA (rRNA) when the chromosome bearing the rRNA genes is condensed. It is rRNA that makes up most of the substance of the nucleolus (see Chapter 5).

While the chromosomes are condensing, another series of equally important events is also occurring: Special microtubules (thin, tubelike protein structures) called *spindle fibers* are being assembled. In animal cells, these spindle fibers extend from a pair of related microtubular structures called *centrioles* (see Figure 12.6). The microtubules of the spindle appear to form from granules surrounding the centrioles called the *microtubule-organizing center (MTOC)*,

or *centrosome*, which functions throughout the cell cycle to organize the cell's microtubules. Plant cells do not contain centrioles, but spindle fibers are formed inside the plant cell's cytoskeleton.

In early prophase, the centrosomes of animal cells begin to move away from one another. By the end of prophase, each member of the pair has moved to an opposite end, or pole, of the cell. As the spindle fibers form, the nuclear membrane breaks down. The spindle fibers form a bridge between the centrosomes, spanning the distance from one pole to the other. Some spindle fibers extending from each pole attach to their side of the centromere of each sister chromatid. By this time, each chromatid has developed a *kinetochore* (kih-NET-ih-core) at its centromere. Several pole-to-centromere microtubules attach to each kinetochore, so that one sister chromatid is attached to one pole and the other sister chromatid to the other pole. The proper attachment of spindle fibers to kinetochores is critical to the separation and movement of sister chromatids during later stages of mitosis.

When the centrosomes reach the poles of the cell in animal mitosis, they radiate an array of microtubules called an *aster*. The function of the aster is not well understood, but evidence suggests that it probably acts as a support during the movement of the sister chromatids.

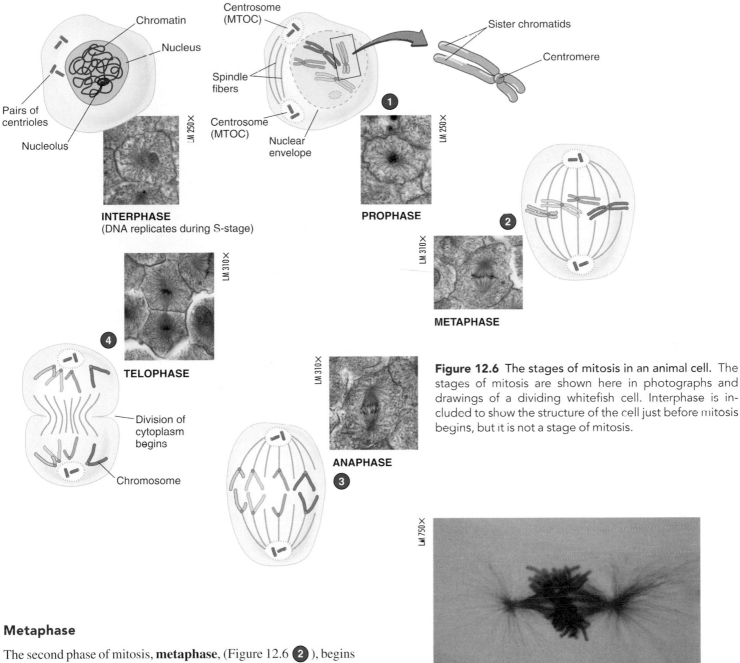

**Figure 12.6 The stages of mitosis in an animal cell.** The stages of mitosis are shown here in photographs and drawings of a dividing whitefish cell. Interphase is included to show the structure of the cell just before mitosis begins, but it is not a stage of mitosis.

## Metaphase

The second phase of mitosis, **metaphase**, (Figure 12.6 ② ), begins when the pairs of sister chromatids align in one plane at the center of the cell. When viewed two dimensionally with a light microscope, the pairs may seem to be in a line, as in the micrograph of metaphase in Figure 12.6. However, three dimensionally, they are not in a line, but form a circle perpendicular to the direction of the spindle fibers, as indicated in **Figure 12.7**. Positioned by the microtubules attached to the kinetochores at their centromeres, all the chromosomes are equidistantly arranged between the two poles at the "equator" of the cell. The region of this circular arrangement, called the *metaphase plate*, is not a physical structure but indicates approximately where the future plane of cell division will be.

## Anaphase

At the beginning of **anaphase** (Figure 12.6 ③ ), the sister chromatids separate from each other at the centromere. Before this split, the chromatids are pulled in two directions at once by opposing mi-

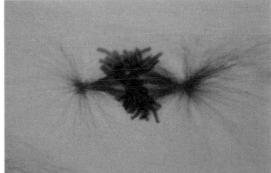

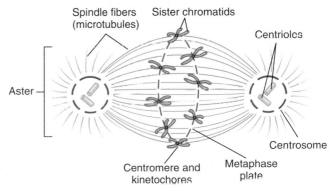

**Figure 12.7 The metaphase plate.** In metaphase, the chromosomes form a circular array around the spindle midpoint.

crotubules—somewhat like a cellular tug-of-war. After separation, the chromatids (now called chromosomes) move rapidly toward opposite poles of the cell, each pulled at its kinetochore by attached, shortening microtubules.

The separation of the sister chromatids in anaphase equally divides the duplicate sets of hereditary material, so each of the two new cells that are forming will receive a complete set of chromosomes. This partitioning of the genetic material is the essence of the process of mitosis.

### Telophase

The events of the last phase of mitosis, **telophase** (Figure 12.6 **4** ), ready the cell for *cytokinesis*, or division of the cytoplasm. The spindle fibers are chemically disassembled and disappear. A nuclear envelope reforms around each set of chromosomes. These chromosomes begin to uncoil, and the nucleolus reappears as rRNA is once again made. The events of telophase are very much like the events of prophase, but in reverse order.

## 12.4 After mitosis, the cytoplasm of the cell divides.

At the end of telophase, mitosis is complete. The eukaryotic cell has divided its duplicated hereditary material, which is now contained in two nuclei positioned at opposite ends of the cell. While this process has been going on, the cytoplasmic organelles, such as the mitochondria (which replicate through interphase), have been about equally distributed to each side of the cytoplasm.

At this point, the process of cell division is still not complete. The division of the cytoplasm—that portion of the cell outside the nucleus—began during telophase but is completed at a stage of the cell cycle called **cytokinesis**. This separation of the one cell into two cells usually occurs shortly after mitosis is complete. Cytokinesis generally involves the division of the cell into approximately equal halves. Sometimes, however, the division is unequal and the daughter cells are not the same size.

### Cytokinesis in animal cells

In human cells and in the cells of other eukaryotes that lack cell walls, cytokinesis occurs by a pinching of the cell in two. A belt of microfilaments that encircles the cell at the metaphase plate accomplishes this pinching. These microfilaments contract, forming a *cleavage furrow* around the circumference of the cell (**Figure 12.8**). As contraction proceeds, the furrow deepens until the opposing edges of the cell membrane make contact with one another. Then the membranes fuse, separating the one cell into two cells.

### Cytokinesis in plant cells

In plants and some algae, a rigid wall surrounds the cell and its membrane (see Figure 5.9). Cytokinesis involves the laying down of a new cell wall between the two daughter cells rather than a pinching in of the cytoplasm. Plants manufacture new sections of

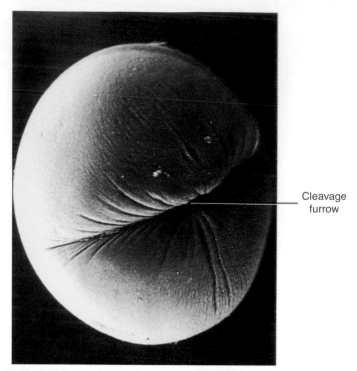

**Figure 12.8 Cytokinesis in an animal cell.** A cleavage furrow is beginning to encircle this dividing sea urchin egg.

membrane and wall from tiny vesicles most likely derived from the Golgi complex. These sections accumulate at the metaphase plate and fuse, beginning the formation of a partition called the *cell plate*. A new cell wall forms between the two membranes of the cell plate. **Figure 12.9** shows cytokinesis in a plant cell and the formation of a cell plate.

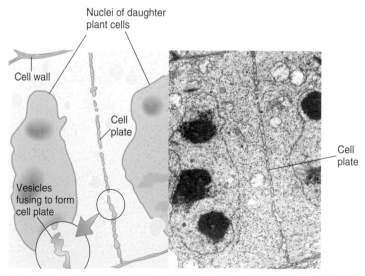

**Figure 12.9 Cytokinesis in plant cells.** In this photograph and companion drawing, a cell plate is forming between daughter nuclei. The cell plate is composed of pieces of cell membrane (vesicles) that fuse together.

CONCEPT CHECKPOINT

**2.** Complete the following table from memory. Reread Sections 12.2–12.4 to check your work and fill in any missing information.

| | **Description of Cellular Events** |
|---|---|
| **Interphase** | G1 |
| | DNA replication |
| **Prophase** | |
| **Metaphase** | |
| **Anaphase** | |
| **Telophase** | Spindle fibers are disassembled; nuclear membrane reforms; nucleolus reforms. |

# What happens if the cell cycle becomes deregulated?

## 12.5 Cancer arises as a result of deregulation of the cell cycle and loss of control over mitosis.

Mutations in the genes that regulate cell growth and cell division can result in cancer. That is, cancer is a deregulation of the cell cycle and consequent loss of control over mitosis.

Although cancer is often thought of as a single disease, more than 200 types exist. **Figure 12.10** shows the cancer death rates for the most prevalent cancers in the United States for both males and females over the last seven decades. Although cancer death rates decreased in the 1990s and cancer incidence rates stabilized, experts predict that the total number of cancer cases will double by 2050 as an increasing proportion of the U.S. population reaches older ages.

What is cancer? Cancer, no matter what type, always has four characteristics: (1) uncontrolled cell growth, (2) loss of cell differentiation (specialization), (3) invasion of normal tissues, and (4) **metastasis** (muh-TAS-tuh-sis), or spread, to multiple sites.

Cancer eventually causes death because the cancer cells continually increase in number while spreading to vital areas of the body, occupying the space in which normal cells would otherwise reside and carry out normal body functions. Cancer cells put pressure on normal tissues and replace them, causing pain and bleeding because these cells invade nerves and blood vessels. In sum, cancer disrupts normal cellular activities that are essential for life.

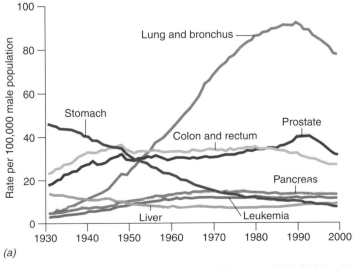

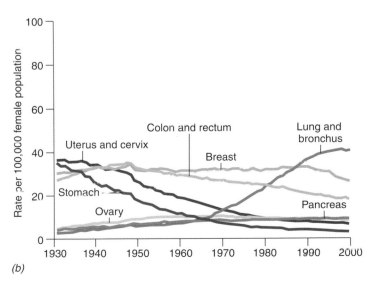

**Figure 12.10** Cancer death rates, United States, 1930–2000. (a) Males. (b) Females.

## *just wondering . . .*

### What can I do to decrease my risk of getting cancer?

Researchers suggest that 90% of all cancers are environmentally induced; that is, they are not inherited but result from external factors. **Table 12.2** lists external factors that are initiators and/or promoters of cancer. Many of these factors are also implicated in other diseases or conditions that endanger health, so avoiding these factors may increase your chances for a longer, healthier life. Although your heredity is not a factor you can control, knowledge about your hereditary background can help you control other factors important to your health.

One way to decrease your risk of cancer is to determine which cancer risks apply to you and to change your behavior accordingly.

Another way is to follow the recommendations of the American Cancer Society (ACS) regarding diagnostic tests (your local chapter of the ACS will send these recommendations to you on request). In addition, follow the general dietary recommendations listed in **Table 12.3** and look for the danger signs of cancer listed in **Table 12.4.** If you notice any of these signs, see your doctor immediately. Otherwise, the ACS recommends a cancer-related checkup by a physician every three years for persons aged 20 to 39 years and annually for those 40 years of age and older. Persons at risk for particular cancers may need to see their physician more often. ●

*Are you wondering about a topic in biology and how it relates to your life? Submit your question by clicking the* Just Wondering *link in this text's companions Web site at* www.wiley.com/college/alters.

| TABLE 12.2 | Major Factors That Increase the Risk of Cancer | |
|---|---|---|
| **Factor** | **Examples of Implicated Cancers** | **Comments** |
| Heredity | Retinoblastoma (childhood eye cancer) Osteosarcoma (childhood bone cancer) | Most cancers are not caused by heredity alone. Persons having family histories of certain cancers should follow physicians' recommendations. |
| Tumor viruses | Liver cancer Adult T cell leukemia/lymphoma Cervical cancer | Five viruses are initiators of certain cancers. (See Table 19.1) |
| Tobacco use | Lung cancer Cancers of the oral cavity, esophagus, and larynx Cancers of the kidney and bladder | Cigarette smoking is responsible for approximately one-third of all cancers. Nonsmokers have an increased risk of smoking-related cancers if they regularly breathe in sidestream smoke. |
| Alcohol consumption | Cancers of the oral cavity, esophagus, and larynx Breast cancer | The combined use of alcohol and tobacco leads to a greatly increased risk of these cancers. The mechanism of action in breast cancer is not yet known. |
| Industrial hazards | Lung cancer | Certain fibers, such as asbestos, chemicals such as benzene and arsenic, and wood and coal dust are prominent industrial hazards. |
| Ultraviolet radiation from the sun | Skin cancers | Those at greatest risk are fair-skinned persons who burn easily. However, everyone is at risk and should wear sunscreens and protective clothing when in the sun for extended periods of time. All types of UV radiation in tanning beds (UVA, UVB, & UVC) are harmful and may lead to skin cancer. |
| Ionizing radiation | Related to location and type of exposure | Eliminate unnecessary medical X rays to lower cancer risk. Infants and children are particularly susceptible to the damaging effects of ionizing radiation. Check your home to detect high levels of radon gas. |
| Hormones (estrogen and possibly testosterone) | Breast, cervical, ovarian, and prostate cancers | Estrogen-only and estrogen-progesterone hormone replacement therapies both increase the risk of breast cancer. Oral contraceptives increase the risk of breast cancer and cervical cancer, while reducing the risk of ovarian cancer. The role of testosterone in prostate cancer is unclear. |
| Diet | Breast and prostate cancers (weak association with high-fat diets), stomach and esophageal cancers (nitrites). | Nitrites found in salt-cured, salt-pickled, and smoked foods increase the risk of cancer. |

| **TABLE 12.3** | **General Dietary Recommendations to Reduce the Risk of Cancer** |
| --- | --- |

1. Avoid obesity by balancing caloric intake with exercise.
2. Choose foods that are low in fat and that help maintain a healthful weight.
3. Limit consumption of meats, especially high-fat and processed meats.
4. Eat five or more servings of fruits and vegetables each day.
5. Eat other foods from plant sources, such as breads, cereals, grain products, soy products, rice, pasta, or beans, several times each day.
6. If alcohol is consumed, limit intake to one mixed drink or glass of wine per day for women and two for men.

Adapted from *Cancer Facts & Figures 2004*, the American Cancer Society.

| **TABLE 12.4** | **The Seven Warning Signs of Cancer** |
| --- | --- |

**C**hange in bowel or bladder functions
**A** sore that does not heal
**U**nusual bleeding or discharge
**T**hickening or lump in any tissue
**I**ndigestion (chronic) or difficulty in swallowing
**O**bvious change in a wart or mole
**N**agging hoarseness or cough

The most significant characteristic of all cancer cells is their uncontrolled growth, which reflects a loss of regulation of the cell cycle. Normally, external and internal chemical signals control the length of the cell cycle and the duration of each stage. The current view of cell-cycle control is that transitions from one stage in the cycle to the next are regulated by checkpoints, which are mechanisms that stop the cycle until a critical process is completed or until damaged DNA is repaired. In cancer cells, checkpoints in the cell cycle typically have become deregulated due to any of a variety of genetic defects.

Cancer develops through a process that can be divided into three stages—initiation, promotion, and progression—as shown in **Figure 12.11**.

### Initiation

The genetic defects that cause cancer arise as a result of damage to the DNA in normal body cells. Typically, this damage occurs through a series of gene mutations. The process of change that a normal cell undergoes as it becomes a cancer cell is called **transformation**. The first stage in the process is the development of mutations in proto-oncogenes and/or tumor-suppressor genes. **Proto-oncogenes** are normal and necessary genes that play important roles in the regulation of biochemical activities within cells, including those activities related to cell division. Proto-oncogenes become **oncogenes** (ON-ko-jeens, cancer-causing genes) only if some part of their DNA undergoes mutation. Oncogenes can be thought of as "on" switches in the development of cancer, signaling cells to speed up their growth and decrease their levels of differentiation. The word "oncogenes" is derived from the Greek word *onkos*, meaning "mass" or "tumor."

The full development of a cancerous state usually requires additional mutations. Typically, these mutations affect **tumor-suppressor genes** which are normally involved in the restraint of cell growth. These genes can be regarded as "off" switches in the development of cancer, signaling cells to slow their growth and increase their levels of differentiation. The expression of tumor-suppressor genes is required for the normal functioning of cells; they appear to induce normal cells to differentiate into mature cell types with reduced or no growth potential. These genes are inactivated when they undergo mutation, thereby allowing cancerous growth.

Mutations in proto-oncogenes and in tumor-suppressor genes may be inherited, or they can be caused by viruses, chemicals, or radiation. All four of these mutagenic agents are referred to as *initiators*, and the process of proto-oncogene or tumor-suppressor gene mutation is called *initiation*. Unfortunately, initiation may occur after only brief exposure to an initiator. Initiation does not directly result in cancer; rather, as shown in Figure 12.11 (top), it results in a mutated cell (precancerous cell) that may or may not look abnormal and that gives rise to other initiated cells when it divides.

### Promotion

For cancer to occur, initiated cells must undergo *promotion*, the second stage in the development of cancer. The same agent that caused initiation may cause promotion in the same cell, or another agent may cause it. Promotion is a process by which initiated cells are stimulated to grow and divide when they normally would not. It is a gradual process and happens over a long time, as opposed to the short time required for initiation. Research suggests that if the promoter—the agent causing promotion—is withdrawn in the early stages in some situations, then cancer development can be reversed. For example, if a smoker stops smoking, that person's risk of lung cancer may return eventually to that of a nonsmoker, depending on how long and how much smoking has occurred. A somewhat different effect is observed with respect to sun exposure. If you stop lying in the sun to get a tan, you will not erase the effects of prior years of sun exposure, but you will avoid adding to your risk of developing skin cancer.

Evidence suggests that promoters may cause genes to mutate. Those agents that can both initiate and promote cancer are called complete **carcinogens** (kar-SIN-uh-jens), or cancer-causing substances. Most substances linked with the development of cancer are complete carcinogens. A few, however, act only as an initiator or as a promoter. Heredity, for example, acts only as an initiator. Conversely, asbestos acts only as a promoter in cells initiated by other agents, such as cigarette smoke and air pollution.

As promotion proceeds, the oncogenes begin to be transcribed, producing polypeptides. These polypeptides cause cells to grow and divide when normal cells would not and cause a variety of other changes, such as modifications in cell shape and structure, cell-to-cell interactions, membrane properties, cytoskeletal structure, and protein secretion. The mechanisms by which a particular polypeptide interacts with a cell to produce one or more of these effects vary. Oncogene-transcribed polypeptides may affect cell growth, for example, by mimicking growth factors and binding to receptor sites on the surfaces of certain cells, activating specific enzymes within them. The activated enzymes cause these cells to grow and divide when normally they would not.

When damage to the DNA of a promoted cell is not drastic, most of the normal components of the cell are produced and it still responds normally to growth-inhibiting factors. Sometimes,

however, cells in the early stages of promotion grow and divide abnormally, forming a benign (noncancerous) tumor (Figure 12.11, middle). **Benign tumors** (buh-NINE) are growths or masses made up of partially transformed cells; these growths are confined to one location and are encapsulated, shielding them from surrounding tissues. Such tumors are not life threatening in themselves, but some exhibit growth patterns that are characteristic of the development of cancer cells. These cells, which are said to exhibit *dysplasia*, may acquire additional mutations to change from harmless to cancerous. **Figure 12.12** shows normal cells and dysplastic cells (color differences are due to the use of different stains). The normal cells are all approximately the same size and have nuclei that look similar (Figure 12.12*a*). The dysplastic cells are irregular in size and in the appearance of their nuclei (Figure 12.12*b*).

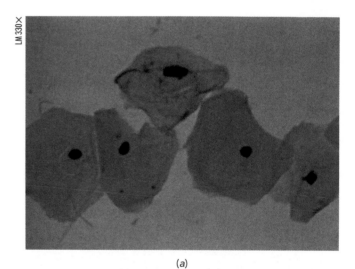

(a)

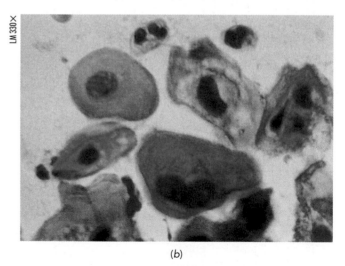

(b)

**Figure 12.12** (*a*) Normal cells. (*b*) Dysplastic cells.

**Visual Thinking:** Describe the differences you see between normal and dysplastic cells in these micrographs.

**Visual Summary**  **Figure 12.11** Stages in the development of cancer.

**INITIATION (Stage I)**
Mutagenic agents—inherited factors, viruses, chemicals, or radiation (initiators)

Normal cell

Mutation

Gene

Precancerous cell

**PROMOTION (Stage II)**
Mutagenic agents—viruses, chemicals, or radiation (promoters)

Oncogenes activated and expressed

Benign cells (partially transformed)

(Continued promotion)

**PROGRESSION (Stage III)**

Malignant cells (cancer)

## Progression

If promoters continue to affect partially transformed cells, the cells reach a point where they irreversibly become cancer cells. This point marks the beginning of the third stage in the development of cancer, called *progression* (Figure 12.11, bottom). During progression, the transformed cells usually become less differentiated than benign cells and increase their rate of growth and division without regard to the body's needs. In addition, these cancer cells have the ability to (1) invade and kill other tissues and (2) metastasize, or move to other areas of the body. Tumors made up of cells with these properties are **malignant** (muh-LIG-nunt).

The spread of cancer during progression is a multistage process, like the development of the cancer cells themselves. **Figure 12.13** shows the stages in the development of a carcinoma, a cancer of the epithelial tissues that tends to invade surrounding tissue. Skin, breast, lung, and colon cancers are examples of carcinomas. In ❶ a partially transformed tumor cell is shown in the epidermis of the skin. This cell multiplies ❷, forming a mass of dysplastic cells. As the DNA of these cells is further damaged, they irreversibly become cancer cells. Their growth rate increases, and they form a cancerous tumor that is localized ❸, referred to as *in situ*, meaning "in place." These are small, localized tumors that have not invaded the surrounding normal tissue. During the next stage of cancer progression ❹, cancer cells invade the surrounding tissue by secreting chemicals that break down the intercellular matrix—the substances that hold cells together. Other secretions cause the cells in the invaded tissue to break apart. Cancer cells then invade the underlying dermis, entering blood and lymph vessels and traveling throughout the body.

### CONCEPT CHECKPOINT

3. For the three stages of cancer development—initiation, promotion, and progression—give some examples of factors that may cause each stage to occur.

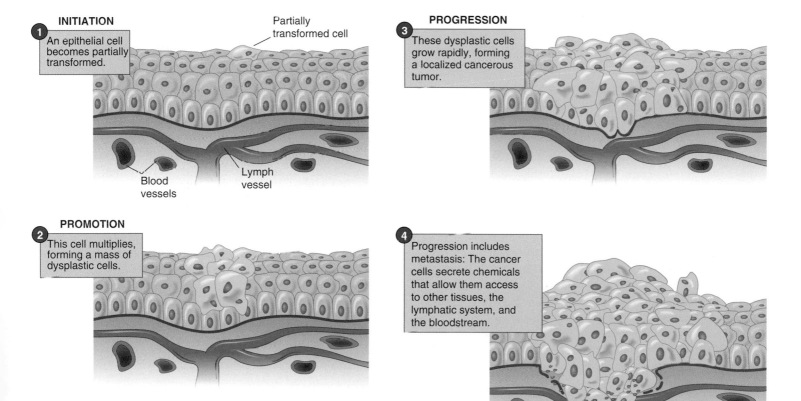

**Figure 12.13** How cancer cells multiply and spread.

# How Science Works

## Genetically programmed cell death and cancer treatment

Cell death is genetically programmed. Some cells of the human body, such as skin cells, glandular cells, and red blood cells, automatically self-destruct after a limited lifespan. They body then replaces these cells with new cells formed by cell division of the survivors. However, not all cells are programmed to die after a short life. In fact, some cells are programmed to last the lifetime of the organism, such as the muscle cells making up the heart and skeletal muscles, and the nerve cells of the central nervous system. Nevertheless, there can be a glitch in the system—some cells that should die do not, such as cancer cells, some of which develop in part from a failure of programmed cell death. These cells that evade death also continue to divide uncontrollably.

As we have seen, the most distinctive feature of cancer cells is their unregulated growth.

Cell biologists call one type of programmed cell death apoptosis (AHP-op-TOE-sis). In this type of cell suicide programming, the cell membrane remains intact as the cell dies, so that it does not release its contents and trigger a local inflammatory reaction, such as you might experience when your cells die due to a cut or abrasion. During apoptosis, the dying cell splits into small membrane-bound bodies that are engulfed and digested by white blood cells.

Genetically programmed cell death and apoptosis is currently the subject of considerable research activity. Apoptosis already plays a pivotal role in the mechanism of action of most anticancer drugs. Many chemotherapy agents target a tumor suppressor gene called p53. In many cancers, this gene is mutated and does not do its job of suppressing tumor growth. The drugs re-

store tumor suppressor function, and the cancer shrinks.

In a recent breakthrough in cell death research, scientists tested various compounds on normal cells and on a variety of cancers, including cancers of the breast, lung, and skin. As they had hoped, one of the compounds triggered apoptosis in the cancer cells, with little effect on normal cells. The other compound was not as effective when used on whole cells, but showed potency when tested in other ways. The researchers speculate that the less effective drug had difficulty penetrating cancer cell membranes. The researchers believe that the work is promising but emphasize that it could take years to develop useful drugs based on these compounds. In the meantime, scientists around the world continue their work on cell death programming, working to develop drugs that cause cancer cells to self-destruct. ●

# What happens during the process of meiosis?

## 12.6 During meiosis, the number of chromosomes is halved.

Most animals, plants, algae, and fungi, as well as certain protists, reproduce sexually. In sexual reproduction, gametes of opposite sexes or mating types (sometimes just termed + cells and − cells) unite in the process of fertilization, producing the first cell (zygote) of a new individual.

Humans have 46 chromosomes in each of their body cells—the diploid number. If the two cells that joined in fertilization each contained 46 chromosomes, however, the first cell of the future offspring would have 92 chromosomes. An individual born after 10 generations would have more than 47,000 chromosomes! Such a continued doubling of the number of chromosomes in each new individual would obviously be unworkable. Even early investigators realized that there must be some way in which the number of chromosomes is reduced by half during gamete formation. They reasoned that if sex cells had half the number of chromosomes of body cells, then the fusion of gametes during fertilization would produce a zygote with the full complement of chromosomes. Investigators soon observed this special type of cell division and named it *meiosis*, from a Greek word meaning "less."

During meiosis, the diploid number of chromosomes is reduced by half, forming haploid cells. For this reason, meiosis is often called *reduction division*. First, a diploid parent cell replicates its DNA and then divides, splitting its chromosomes between two

daughter cells. Then, each of these cells divides, resulting in an end product of four haploid daughter cells.

Meiosis occurs in diploid parent cells that each contain pairs of chromosomes. These pairs are referred to as **homologous chromosomes** (hoe-MOL-uh-gus), or **homologues** (HOME-uh-logs). "Homologous" is derived from a Greek word meaning "agreeing." The two chromosomes in a pair contain the same linear sequence of genes. The corresponding genes on the two chromosomes are called *alleles*, alternative forms of the same gene.

In humans, the diploid cells that undergo meiosis each contain 23 pairs of homologues (see the karyotype in Figure 9.1). When the haploid gametes (egg and sperm, each containing 23 chromosomes) that result from meiosis fuse during fertilization to form a human zygote, the zygote receives one allele for each gene on the 23 chromosomes in the mother's egg and one allele for each gene on the 23 chromosomes in the father's sperm. Thus, the zygote receives a full complement of 46 chromosomes. All organisms produced by sexual reproduction receive half their chromosomes from one parent and half from the other.

The process of meiosis, which follows DNA replication, consists of two sets of stages, called *meiosis I* and *meiosis II*. In **meiosis I**, homologous pairs of chromosomes line up in two rows at the equator of the diploid parent cell, which then divides, splitting the pairs and forming two daughter cells. In these daughter cells, each chromosome still consists of two sister chromatids. In **meiosis II**, the chromosomes line up at the equator of each of the daughter cells, which then divide, splitting the sister chromatids and forming four haploid daughter cells.

## Meiosis I—splitting homologous pairs of chromosomes

In the first stage of meiosis I (**Figure 12.14**), called prophase I, the individual chromosomes condense as their DNA coils more tightly; at this point, they become visible under a light microscope. Homologous chromosomes line up side by side in a process called **synapsis** (sih-NAP-sis). Because each chromosome is made up of two sister chromatids, the paired homologous chromosomes together have four chromatids. These synapsed homologues are therefore called *bivalents* (meaning "two strong"), or *tetrads* (meaning "groups of four").

After synapsis, a process called **crossing over** begins. During crossing over, homologous *nonsister* chromatids cross over one another. At the place where they cross over, pieces break away from each chromatid and attach to the other chromatid. The chromatids exchange segments (**Figure 12.15**). Once crossing over is complete, the nuclear envelope dissolves and the homologues begin to move apart. The four chromatids cannot separate from one another, however, because (1) the sister chromatids are held together at their centromeres and (2) the paired homologues are held together at the points where crossing over occurred. As the chromosomes move apart somewhat, the points of crossing over can be seen under a light microscope as X shaped structures called **chiasmata** (keye-AZ-muh-tuh; singular *chiasma* [keye-AZ-muh]; see micrograph in Figure 12.15).

When sister chromatids cross over and exchange segments, the event is not genetically significant because the sisters and the exchanged segments are identical. Crossing over of nonsister chromatids, however, is a significant event in meiosis because it produces new combinations of alleles. Notice in Figure 12.15 that before the exchange of segments, there are only two types of chromatids: the identical purple sister chromatids and the identical green sister chromatids. The green chromatids and the purple

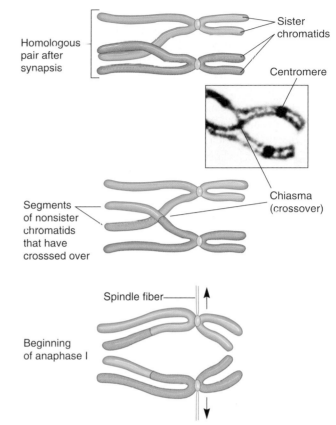

**Figure 12.15 Crossing over.** Replicated homologous chromosomes line up side by side (synapsis) during prophase I. Crossing over occurs when nonsister chromatids exchange segments at locations where they cross. Once crossing over is complete, chromosomes move apart slightly, forming X-shaped structures called chiasmata (photograph). Crossing over produces new combinations of alleles. In anaphase I, the homologous pairs are pulled completely apart.

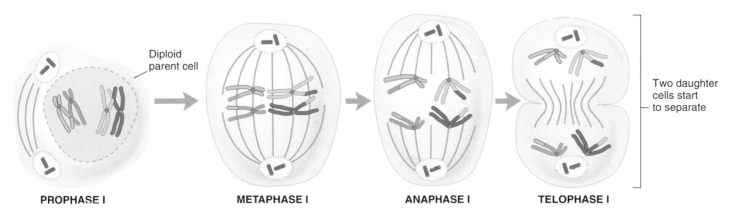

**Figure 12.14 Meiosis I in an animal cell.** Meiosis ensures the reassortment of genetic material. It is preceded by interphase, during which the chromosomes are replicated. An interphaselike period of variable length often occurs between meiosis I and meiosis II, allowing the completion of cytokinesis. During this time, there is no replication of chromosomes. Meiosis II is shown in Figure 12.16.

chromatids are homologous. After crossing over, there are four types of chromatids: one purple chromatid and its sister with a green segment, and one green chromatid and its sister with a purple segment. Thus, the process of crossing over provides one way in which offspring can have a genetic makeup different from that of either parent.

In metaphase I, the next phase of meiosis, the microtubles form a spindle as in mitosis, but with a crucial difference: The chromosomes line up with their homologues double file in meiosis, not single file as in mitosis. For each pair of homologues the orientation on the metaphase plate is random; which homologue is oriented toward which pole is a matter of chance.

After spindle attachment to the kinetochore of each homologous pair is complete, the homologues begin to move toward opposite poles of the cell. As this movement occurs, the homologues pull apart at their cross over points. During anaphase I, one member of each chromosome pair is pulled closer to each pole. Each chromosome still has two sister chromatids, but these chromatids are no longer identical, due to crossing over. During the last stage of meiosis I (telophase I), the two groups of chromosomes are gathered together at their respective poles, forming two chromosome clusters. A nuclear membrane forms around each group of chromosomes, and the cell begins to cleave.

An interphase-like period of variable length called *interkinesis* often occurs between meiosis I and meiosis II. *During this time, there is no replication of DNA.* Cytokinesis occurs, and the cells separate completely into two daughter cells.

## Meiosis II—splitting sister chromatids

At the end of meiosis I and cytokinesis, each daughter cell has only one member of each homologous pair. Each chromosome, however, still consists of two (now nonidentical) sister chromatids. Meiosis II separates these sister chromatids (**Figure 12.16**).

The two cells formed during meiosis I divide during meiosis II. In prophase II, the chromatin condenses to chromosomes, the nuclear envelopes disappear, and the spindle fibers form. Then the chromosomes line up (metaphase II), and their sister chromatids separate, and move toward opposite poles of each cell (anaphase II). In telophase II, the chromatids (now called chromosomes, since they are no longer attached to each other) are gathered together at each pole, the nucleoli reappear, and nuclear envelopes form around each set of chromosomes. Then each of the two cells begins to divide.

Meiosis II results in the production of four daughter cells, each with a haploid number of chromosomes. The cells that contain these haploid nuclei function as gametes for sexual reproduction in most animals and as spores in plants. In the females of many species of animals, including humans, only one of the four gametes is functional (see Chapter 36). **Figure 12.17** and **Table 12.5** compare mitotic and meiotic cell division.

## The importance of meiotic recombination

The reassortment of genetic material that occurs during meiosis generates variability in the hereditary material of the offspring.

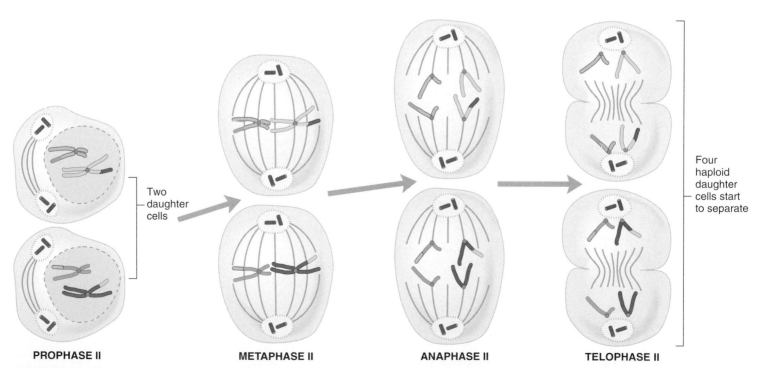

**PROPHASE II**      **METAPHASE II**      **ANAPHASE II**      **TELOPHASE II**

Two daughter cells

Four haploid daughter cells start to separate

**Figure 12.16** Meiosis II in an animal cell. Meiosis II is preceded by cytokinesis (daughter cells separate completely) and, sometimes, by an interphaselike period called *interkinesis*.

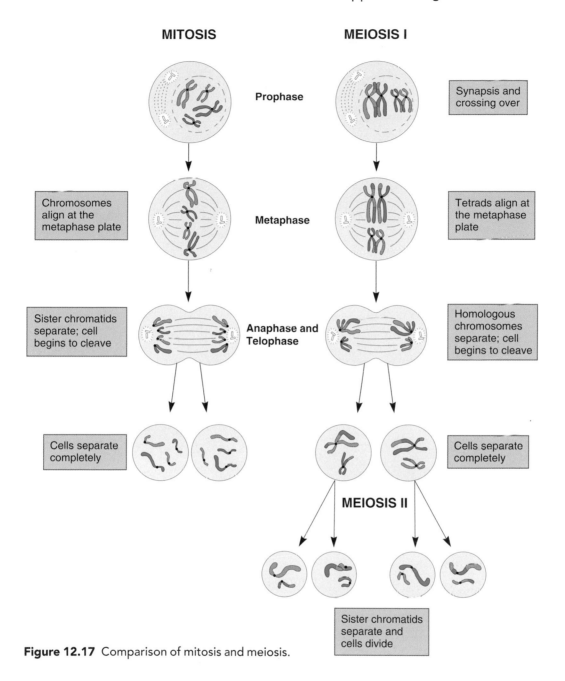

**Figure 12.17** Comparison of mitosis and meiosis.

For example, the copy of chromosome 14 that a particular human gamete receives has no influence on which copy of chromosome 5 it will receive. Each of the 23 pairs of chromosomes goes through meiosis independently of all the others, so there are $2^{23}$ (more than 8 million) different possibilities for the kinds of gametes that can be produced, with no two of them alike. In addition, crossing over adds even more variability. The subsequent union of two gametes thus creates a unique individual, a new combination of 23 pairs of chromosomes that probably has never occurred before and probably will never occur again. When you study evolution in Part 4 of this textbook, you will see that variability within species is essential to the process of evolution by natural selection. Meiosis is one source of variability.

**CONCEPT CHECKPOINT**

4. What are the similarities and differences between the following stages of mitosis and meiosis? Figures 12.6, 12.14, 12.16, 12.17, and Table 12.5 will aid you with this process. You may also want to draw a diagram of each stage to visually highlight the similarities and differences.
   - Prophase of mitosis and prophase I of meiosis
   - Anaphase of mitosis and anaphase I of meiosis
   - Telophase of mitosis and telophase I of meiosis
   - Prophase of mitosis and prophase II of meiosis
   - Anaphase of mitosis and anaphase II of meiosis
   - Telophase of mitosis and telophase II of meiosis

| TABLE 12.5 | Comparison of Mitosis and Meiosis | | |
|---|---|---|---|

| | Mitosis | Meiosis | |
|---|---|---|---|
| | | Meiosis I | Meiosis II |
| | • Occurs in diploid or haploid cells.<br>• Preceded by DNA replication during interphase, resulting in chromosomes that each consist of two sister chromatids. | • Occurs in diploid cells.<br>• Preceded by DNA replication during interphase, resulting in chromosomes that each consist of two sister chromatids. | • Occurs in cells from meiosis I.<br>• Not preceded by DNA replication. |
| | **1. PROPHASE** | **1. PROPHASE I** | **5. PROPHASE II** |
| | • Chromosomes condense.<br>• Nuclear membrane disintegrates.<br>• Spindle fibers form. | • Chromosomes condense.<br>• Homologous pairs of chromosomes align side by side (synapsis).<br>• Crossing over occurs.<br>• Nuclear membrane disintegrates.<br>• Spindle fibers form. | • Chromosomes condense.<br>• Nuclear membrane disintegrates.<br>• Spindle fibers form. |
| | **2. METAPHASE** | **2. METAPHASE I** | **6. METAPHASE II** |
| | • Spindle fibers attach to centromeres.<br>• Chromosomes align along metaphase plate. | • Spindle fibers attach to centromeres.<br>• Homologous pairs align along metaphase plate. | • Spindle fibers attach to centromeres.<br>• Chromosomes align along metaphase plate. |
| | **3. ANAPHASE** | **3. ANAPHASE I** | **7. ANAPHASE II** |
| | • Sister chromatids separate and move toward opposite poles. | • Homologous pairs separate and move toward opposite poles (each chromosome in each pair still consists of two sister chromatids). | • Sister chromatids separate and move toward opposite poles. |
| | **4. TELOPHASE** | **4. TELOPHASE I** | **8. TELOPHASE II** |
| | • Separated sister chromatids gather near each pole (separated chromatids are called chromosomes).<br>• Nuclear membrane forms around each group of chromosomes.<br>• Cytoplasmic division (cytokinesis) begins. | • Separated homologues gather near each pole.<br>• Nuclear membrane forms around each group of homologues.<br>• Cytoplasmic division (cytokinesis) begins. | • Separated sister chromatids gather near each pole (separated chromatids are called chromosomes).<br>• Nuclear membrane forms around each group of chromosomes.<br>• Cytoplasmic division (cytokinesis) begins. |
| | **followed by . . .** | **followed by . . .** | **followed by . . .** |
| | • Completion of cytokinesis.<br>• Result is two daughter cells, each with chromosomes identical to those in the original cell. | • Completion of cytokinesis.<br>• Result is two daughter cells, each with chromosomes different from those in the original cell because of crossing over.<br>• Meiosis II then occurs in these two daughter cells. | • Completion of cytokinesis.<br>• Result is four haploid daughter cells, each with its own unique set of chromosomes, different from those in the original cell and different from each other. |

# CHAPTER REVIEW

## Summary of Key Concepts and Key Terms

### How do eukaryotic cells reproduce?

**12.1** Hereditary material is passed on to new cells by **mitosis** (my-TOE-sis, p. 179) or **meiosis** (my-OH-sis, p. 179).

**12.1** Mitosis, cell division that occurs as a part of growth in animals and plants, and as a part of reproduction in plants, produces two identical cells from an original parent cell.

**12.1** Meiosis, cell division that occurs as a part of reproduction in both plants and animals, produces four cells from one parent cell, each with half the full complement of genetic material.

**12.1** In animals, meiosis produces **gametes** (p. 179), or sex cells. In plants, meiosis produces spores.

**12.1** A **life cycle** (p. 179) is the progression of stages an organism passes through from its conception until it conceives another similar organism.

**12.1** **Diploid** (2N) (p. 179) means that the cell or individual has the full complement of hereditary material characteristic of that species.

**12.1** **Haploid** (N) (p. 179) means that the cell or individual has only one chromosome from each pair in the parent cell; that is, it contains half the full complement of hereditary material characteristic of that species.

**12.1** **Sexual reproduction** (p. 179) involves the fusion of gametes (**fertilization**, p. 179) to produce a new individual.

**12.1** **Asexual reproduction** (p. 180) does not involve the union of gametes to produce a new individual.

## What do eukaryotic cells do when they are not reproducing?

**12.2 Interphase** (p. 181) is the major portion of the **cell cycle** (p. 181). During the $G_1$ stage of interphase, the cell grows and carries out normal life functions.

**12.2** During the S stage of interphase, the cell replicates its DNA.

**12.2** During the $G_2$ phase of interphase, the supercoils of DNA condense into tightly compacted bodies, which can be seen as two strands of chromosomal material called **sister chromatids** (p. 181), held together at a single point.

**12.2** Mitosis is a continuous sequence of events that occurs just after interphase and that results in the separation of the sister chromatids.

## What happens during the process of mitosis?

**12.3** Mitosis is made up of **prophase** (p. 182), **metaphase** (p. 183), **anaphase** (p. 183), and **telophase** (p. 184).

**12.4 Cytokinesis** (p. 184) is the physical division of the cytoplasm of a eukaryotic cell into two daughter cells.

## What happens if the cell cycle becomes deregulated?

**12.5** Cancer arises as a result of deregulation of the cell cycle and loss of control over mitosis.

**12.5** The most significant characteristic of all cancer cells is their uncontrolled growth.

**12.5 Metastasis** (muh-TAS-tuh-sis, p. 185) is the spread of cancer cells to multiple sites.

**12.5** Cancer cells arise as a result of damage to a cell's DNA, typically through a series of gene mutations.

**12.5** The process of change that a normal cell undergoes as it becomes a cancer cell is called **transformation** (p. 186).

**12.5 Proto-oncogenes** (p 186) become **oncogenes** (ON-ko-jeens, cancer-causing genes, p. 186) only if some part of their DNA undergoes mutation.

**12.5 Tumor-suppressor genes** (p. 186) can be thought of as "off" switches in the development of cancer, signaling cells to slow their growth and increase their levels of differentiation.

**12.5** Those agents that can both initiate and promote cancer are called complete **carcinogens** (kar-SIN-uh-jens, p. 186), or cancer-causing substances.

**12.5 Benign tumors** (buh-NINE, p. 187) are growths or masses that are made up of partially transformed cells, are confined to one location, and are encapsulated, shielding them from surrounding tissues.

**12.5** Tumors made up of cells that have the ability to invade and kill other tissues and metastasize, or move to other areas of the body, are **malignant** (muh-LIG-nunt, p. 189).

## What happens during the process of meiosis?

**12.6** Meiosis is a process of nuclear division in which the number of chromosomes in cells is halved, forming gametes in most animals and spores in plants.

**12.6 Homologous chromosomes** (hoe-MOL-uh-gus), or **homologues** (HOME-uh-logs, p. 190) are the members of a chromosome pair.

**12.6** Meiosis consists of two sets of mitosis-like divisions called **meiosis I** (p. 190) and **meiosis II** (p. 190).

**12.6** In meiosis I, homologous pairs of chromosomes are split between two daughter cells. In meiosis II, sister chromatids are split between daughter cells.

**12.6** In prophase I, homologous chromosomes line up side by side in a process called **synapsis** (sih-NAP-sis, p. 191). After synapsis, a process called **crossing over** (p. 191) begins.

**12.6** During crossing over, homologous nonsister chromatids exchange segments, creating new combinations of genes.

**12.6** The points of crossing over can be seen under a light microscope as X-shaped structures called **chiasmata** (keye-AZ-muh-tuh; singular *chiasma* [keye-AZ-muh], p. 191).

**12.6** The reassortment of DNA segments that occurs during meiosis generates variability in the hereditary material of offspring.

**12.6** Variability within species is essential to the process of evolution by natural selection; meiosis is one source of variability.

---

| Level 1 | Learning Basic Facts and Terms |
|---------|-------------------------------|

For questions 1–6 choose from the following:

a. meiosis I only
b. meiosis II only
c. mitosis, meiosis I and II
d. meiosis I and mitosis only
e. meiosis II and mitosis only

1. Chromatin condenses into visible chromosomes; the nuclear membrane is disassembled, and chromosomes are composed of two chromatids.
2. Chromosomes are composed of two chromatids, and each chromosome is lined up individually with its centromere on the metaphase plate.
3. Separation of homologous chromosomes occurs.
4. Separation of sister chromatids occurs.
5. Synapsis and crossing over occur.
6. Preceded by replication of DNA.

**True–False**

7. _____ Haploid sex cells have half the number of chromosomes of the original parent cell.
8. _____ Homologous pairs are normally present in haploid cells.
9. _____ In animals, a fertilized sex cell is diploid.
10. _____ The series of cellular events that occur during prophase of mitosis, meiosis I, and meiosis II are identical.
11. _____ For human somatic cells (nonsex cells), 2N = 46. This means that each of our cells has 23 homologous pairs of chromosomes.
12. _____ Tumor-suppressor genes are inactivated when they undergo mutation, thereby allowing cancerous growth.
13. _____ In cancer, the cell cycle and mitosis become deregulated.

## Level 2    Learning Concepts

1. What events occur during meiosis that contribute to genetic recombination? Explain the importance of the genetic recombination that occurs during meiosis.
2. Place these steps in the correct sequence. What process is being described?
   a. Chromosomes line up in center of cell.
   b. Nuclear envelope forms and chromosomes uncoil.
   c. Chromosomes condense and spindle fibers form.
   d. Sister chromatids separate and move to opposite poles of cell.
3. Place these steps in the correct sequence. What process is being described?
   a. Homologous chromosomes move toward opposite poles of cell; chromatids do not separate.
   b. Chromosomes gather together at two poles of cell, and nuclear membranes reform.
   c. Homologous chromosomes pair and exchange segments.
   d. Homologous chromosomes align on a central plane.
   e. Haploid cells separate completely.
4. Compare the cells that result from mitosis with those that result from meiosis. How are they different?
5. Which of the following is most similar in terms of DNA nucleotide sequence? Explain your answer.
   • two nonhomologous chromosomes
   • two nonsister chromatids of homologous chromosomes
   • sister chromatids
6. An organism has a diploid chromosome number of 2N = 4. The chromosomes are called A, B, C, and D, and the homologous pairs are AC and BD. What possible chromosome combinations could exist in the sex cells of this organism?

## Level 3    Critical Thinking and Life Applications

1. The amount of DNA per nucleus was measured on several hundred cells from the tip of the root of a plant. The amount of DNA ranged from 10 to 20 picograms (pg). (A picogram is a trillionth of a gram.) Describe the cell-cycle stage(s) (including cell division) that each of these cells is in:
   • A cell with 20 pg of DNA
   • A cell with 10 pg of DNA
   • A cell with 15 pg of DNA
2. The amount of DNA per nucleus was measured on several hundred cells in the testes of a Rhesus monkey. The amount of DNA ranged from 20 to 80 pg. Describe the cell-cycle stage(s) that each of these cells is in:
   • A cell with 40 pg of DNA        • A cell with 20 pg of DNA
   • A cell with 49.3 pg of DNA      • A cell with 80 pg of DNA
3. The cells in the ovaries of an organism contain four total chromosomes (labeled AA, aa, BB, bb) prior to meiosis. Each letter corresponds to a chromatid, and paired letters correspond to a chromosome with two chromatids. The homologous pairs are AA–aa and BB–bb. What possible chromosome combinations could exist in the sex cells she produces at the end of meiosis?
4. If cells in interphase are subjected to colchicine, a drug that interferes with the functioning of the spindle apparatus, at which stage will mitosis be arrested (stopped)?
5. While looking at an unknown cell under a high-powered microscope you observe that it is in metaphase, but you don't know whether it is metaphase of mitosis or metaphase I or II of meiosis. Could you decide which it is? Explain.
6. The meiotic division of a human sperm-producing cell produces two normal haploid sperm cells, one cell that has an extra chromosome, and another cell that is lacking the same chromosome. What may have gone wrong in the meiotic process to produce these four cells? What would result if each of the above cells were to fertilize an egg cell that had the normal complement of chromosomes?
7. Scientists have discovered a number of drugs that prevent angiogenesis (the growth and infiltration of blood vessels into body tissues). How might these drugs be useful in fighting cancer?

## In The News    Critical Thinking

**STEM CELL SHAKES**

Now that you understand more about cell reproduction and the cell cycle, reread this chapter's opening story about stem cells. To develop an informed and thoughtful stand on this issue, it may help you to follow these steps:

1. Review your immediate reaction to the opening story that you wrote when you began reading this chapter.
2. Based on your current understanding, identify the issue concerning stem cell research, using either a statement or a question. An issue is a point on which people hold differing views.
3. Collect new information about the issue. Visit the *In The News* section of this text's companion Web site at www.wiley.com/college/alters and watch the "Stem Cell Shakes" video. Then use the "summary" link to read the accompanying story and access related links. Use this information, the

links provided, and other online and library resources to find updates about this issue. State the sources of your information. Explain why you think the information is accurate. Also determine whether the information expresses a particular point of view or is biased in any way.

4. Determine which individuals, groups, or organizations have a stake in the issue. What does each stand to gain or lose depending on the outcome?
5. List possible outcomes (resolutions) of the issue. List the pros and cons of each outcome.
6. Which outcome do you think would be best? Why? Note whether your opinion differs from or is the same as what you wrote when you began reading this chapter. Give reasons for your views and for any changes in them based on the additional information you have collected and the analysis you have done.

# PATTERNS OF INHERITANCE

## In The News | Ancestry and Obesity

Gregor Mendel, a key figure in this chapter, studied patterns of inheritance in the mid-1800s, and today scientists are still studying the topic. Present-day scientists, however, are able to study patterns of inheritance that are a bit more complicated than those of flower color and seed shape that Mendel probed. One current investigation into patterns of inheritance involves the relationship between ancestry and the risk of obesity.

Mark Shriver, a professor of anthropology and genetics at The Pennsylvania State University, and geneticists at the University of Alabama at Birmingham are investigating whether genes important in certain diseases are inherited with genes related to ancestry. Shriver and the research team are working to answer questions such as why black women have a twofold higher risk of obesity than white women and why white women have a higher risk of developing osteoporosis than black women. Osteoporosis is a condition in which bones lose density, becoming weak and breaking easily.

To investigate the genetic aspects of the relationship between ancestry and obesity and other conditions such as osteoporosis, Shriver and the research team are looking at hundreds of DNA sequences called genetic markers to determine the genetic ancestry of individuals—that is, the percentage of genes they have that are Native American, West African, or European. They then use this information to study the association between genetic ancestry and disease, particularly obesity. Thus far, the researchers have found significant associations between West African ancestry, body mass index (BMI), and chromosomes 1, 11, and 12. The BMI is a relationship between weight and height that is associated with body fat and health risk (see Chapter 27).

Research on genetic ancestry appears to help researchers find genes that affect individuals' risks of diseases. The researchers expect that future investigations will help develop methods to study certain genetic markers within a person's genome and determine their predispositions to various diseases and conditions. Armed with this information, health care providers can then develop intervention strategies that will work with a person's genetic makeup to help avoid conditions such as obesity. You can learn more about this research by viewing the video news story "Ancestry and Obesity" at the *In The News* section of this text's companion Web site at www.wiley.com/college/alters.

*Write your immediate reaction to Shriver's research on genetic ancestry and obesity: first, summarize the main point of the research in a sentence or two; then suggest what you think its significance is. You will have an opportunity to reflect on your responses and gather more information on this topic during the* In The News *feature at the end of this chapter. In this chapter, you will learn more about patterns of inheritance.*

# CHAPTER GUIDE

## When did Mendel establish a theory of inheritance?

**13.1** Views on inheritance varied widely over centuries.

**13.2** Mendel established a theory of inheritance in the 1860s.

## How did Mendel investigate inheritance?

**13.3** Mendel studied the inheritance of individual traits.

**13.4** Mendel suggested that traits are inherited as discrete factors.

**13.5** Mendel showed that outward appearance is due to inherited factors.

**13.6** Mendel tested his conclusions.

**13.7** Mendel studied the inheritance of pairs of factors.

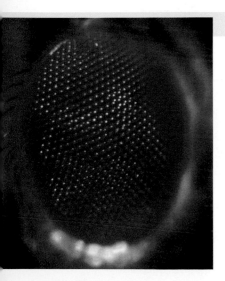

## How did Mendel's work influence further research on inheritance?

**13.8** Sutton suggested that Mendel's factors (genes) are on chromosomes.

**13.9** Morgan provided evidence that genes are on chromosomes.

# When did Mendel establish a theory of inheritance?

## 13.1 Views on inheritance varied widely over centuries.

Although genetic inheritance seems obvious today, this fact was not always apparent to scientists, naturalists, and philosophers. Hippocrates (460–377 B.C.) believed that a child inherited traits, or characteristics, from "particles" given off by all parts of the bodies of the father and mother. These particles, he suggested, traveled to the sex organs. During intercourse, the father's particles merged with the mother's particles to form the child. Many held this idea of inheritance until the mid-1800s.

Another commonly held idea about inheritance was that the male and female traits blended in the offspring. Thus, according to the theory of blending inheritance, a parent with red hair and a parent with brown hair would be expected to produce children with reddish-brown hair, and a tall parent and a short parent would produce children of intermediate height. However, taken to its logical conclusion, this theory suggests that all individuals within a species would eventually look like one another as their traits continually blended together.

Other ideas regarding inheritance were formulated after the invention of a simple, hand-held microscope. Anton van Leeuwenhoek (1632–1723) observed sperm for the first time with the microscope, drew pictures of them, and developed hypotheses regarding inheritance based on his observations. Before the nineteenth century it was widely held that each sperm contained a tiny but whole human called a *homunculus*, as shown in **Figure 13.1**. The miniature human was supposedly implanted in the uterus during fertilization, where it grew into a baby. Because the tiny human was fully formed, its body was thought to contain sperm that also each encased another preformed individual, and so forth . . . ad infinitum. Another theory held that it was the eggs of the mother that contained minute humans.

Not until the mid-1800s, with the work of Schleiden, Schwann, and Virchow (see Chapter 5), did scientists realize that new life arose from old life in the form of new cells arising from old cells. By means of the growth and division of these cells, new organisms developed from individual cells of parent organisms. Scientists of the late 1800s studied the nuclei of cells to uncover the mysteries of cell growth and division. They observed the complex process of mitosis (see Chapter 12) and wondered why cells went through this intricate process. Would it not be more efficient for a cell simply to pinch in two along its middle? By 1883, scientists knew that mitosis ensured an equal distribution of the nuclear material to two daughter cells. Not only did each daughter cell receive the same amount of nuclear material; each received a complete amount of the nuclear material.

At this same time, scientists also observed that an even more complex series of nuclear events, now known as *meiosis* (see Chapter 12), preceded the formation of eggs and sperm. By 1885, several scientists had independently concluded that the nuclear material was the physical bond that linked generations of organisms. However, scientists still did not understand how nuclear material regulated the development of fertilized eggs or how it was related to heredity and variation. Around 1900, scientists began to answer this question by piecing together re-

**Figure 13.1** A homunculus. Before the 1800s, it was widely held that each sperm contained a miniature human. During fertilization the tiny human was implanted in the uterus where it grew into a baby.

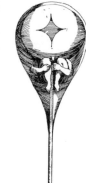

search of the day with research that had long been ignored: the work of Gregor Mendel, an Austrian monk who had trained in botany and mathematics at the University of Vienna.

## 13.2 Mendel established a theory of inheritance in the 1860s.

In the mid-1850s, approximately 30 years before scientists discovered the link between heredity and the complex processes of mitosis and meiosis, Mendel began his work with the garden pea. Mendel chose the pea plant because it was an annual plant that was small and easy to grow. In addition, pea plants are well suited to studies of inheritance because each pea flower contains both female parts and male parts. The female parts are the stigma, style, and ovules, where fertilization occurs and the fertilized egg develops. The male parts are the filaments that support anthers on which pollen is found. Both the female and male parts are enclosed and protected by the petals as shown in **Figure 13.2**. The petals remain closed until after pollination and

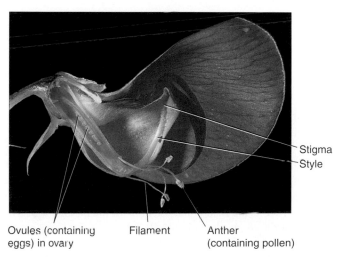

Ovules (containing eggs) in ovary    Filament    Anther (containing pollen)

Stigma
Style

**Figure 13.2** A cutaway of a flower of a pea plant.

**Visual Thinking:** Where are sperm located in this flower?

fertilization take place, unlike many other species of plants. The gametes, or sex cells, produced within each flower—sperm within pollen grains on the anthers, and eggs within the ovules—fuse to produce the embryos that are inside the seeds of new plants. Reproduction in angiosperms such as pea plants is illustrated in Figure 23.8.

Fertilization of this sort, called *self-fertilization*, takes place naturally within individual pea flowers if they are not disturbed. As a result, the offspring of the self-fertilization of garden peas are derived from one pea plant, not two. After generations of self-fertilization, some plants produce offspring consistently identical to the parent with respect to certain defined characteristics; these plants are said to be **true-breeding**.

Mendel selected a number of different true-breeding *varieties* of pea plants. Differing varieties, or strains, of an organism belong to the same species, but each has one or more distinctive character-istics that are passed from parent to offspring. To study the inheritance patterns of these characteristics, Mendel took true-breeding plants and artificially cross-pollinated them, which led to cross-fertilization. *Cross-fertilization* occurs when the sperm of one plant fertilizes the eggs of another plant. To do this, Mendel removed the anthers from a flower before they shed pollen and then dusted the stigma of the flower with pollen from another plant, cross-pollinating them. In this way, Mendel was able to perform experimental crosses between two different true-breeding varieties of pea plants that exhibited differences regarding particular traits. The offspring, or progeny, of the cross between two different varieties of plants of the same species are called *hybrids*. (In other contexts, the word "hybrid" may also refer to the cross between two different species of organisms. A mule, for example, is the hybrid offspring of a horse and a donkey.)

# How did Mendel investigate inheritance?

## 13.3 Mendel studied the inheritance of individual traits.

Mendel first designed a set of experiments that involved crossing varieties of pea plants differing from one another in a single characteristic. Mendel chose seven different characteristics, or **traits**, to study over the course of his work, which are listed and shown in **Figure 13.3**. Although Mendel was not the first to perform such experiments, he was the first to count and classify the plants that resulted from the crosses and compare the proportions using mathematical models. Mendel planned to use his data to deduce the laws by which traits are passed from generation to generation.

Mendel crossed true-breeding plants, each having contrasting forms of the single traits listed in Figure 13.3, by artificially polli-nating one with the other. These plants are the **parental (P) generation**, and their hybrid offspring are the **first filial (F₁) generation**. (The word *filial* is from Latin words meaning "son" and "daughter.") The F₁ offspring, or progeny, are called *monohybrids* because they are the product of two plants that differ from one another in a single trait.

When Mendel crossed two contrasting varieties, such as tall plants with dwarf plants (P generation) as shown in **Figure 13.4** ❶, the hybrid offspring (F₁) that he obtained were not intermedi-ate in size, as the theory of blending inheritance predicted. Instead, each hybrid offspring resembled *only one* of the parents ❷. Thus, in a cross of dwarf plants with tall plants, *all* the F₁ offspring were tall. In a different cross of purple-flowered plants with white-flow-ered plants, *all* the F₁ offspring had purple flowers.

Mendel referred to the form of a trait that was expressed in the F₁ plants as **dominant** and to the alternative form that was not expressed in the F₁ plants as **recessive**. He chose the word "recessive" because this form of the trait receded (disappeared) in the hybrids. For each of the seven contrasting pairs of traits that Mendel examined, one mem-

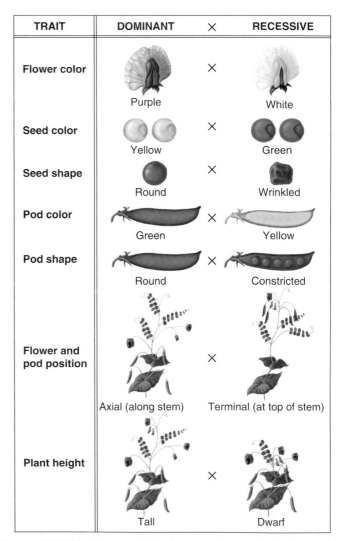

| TRAIT | DOMINANT | × | RECESSIVE |
|---|---|---|---|
| Flower color | Purple | × | White |
| Seed color | Yellow | × | Green |
| Seed shape | Round | × | Wrinkled |
| Pod color | Green | × | Yellow |
| Pod shape | Round | × | Constricted |
| Flower and pod position | Axial (along stem) | × | Terminal (at top of stem) |
| Plant height | Tall | × | Dwarf |

**Figure 13.3** The seven pairs of contrasting traits in the garden pea studied by Mendel.

**P generation** (Dominant)        (Recessive)

**1** Tall        ×        Dwarf

**Figure 13.4 The F₃ generation allowed Mendel to determine which of the F₂ plants were true-breeding.** Mendel allowed the F₂ generation to self-pollinate. He then reasoned from the F₃ offspring that the F₂ generation exhibited the ratio of nontrue-breeding dominant (left), to two nontrue-breeding dominant (middle), to one true-breeding recessive (right).

**Visual Thinking:** The F₁ generation of plants were self-pollinated. What type of pollination was used with the P generation? What is the difference between pollination and fertilization?

**F₁ generation**

**2** All tall (dominant trait)

*Self-fertilization*

**F₂ generation**

**3**

Tall        Tall        Tall        Dwarf

*Self-fertilization  Self-fertilization  Self-fertilization  Self-fertilization*

**4**

**F₃ generation**

**6**        **5**

**7**

When Mendel counted the traits he observed in the F₂ generation, he discovered that the ratio of plants exhibiting the dominant form of the trait to plants exhibiting the recessive form was approximately 3 to 1. That is, for each trait, about three-fourths of the plants exhibited the dominant form and one-fourth exhibited the recessive form. Notice that in his experiments, Mendel used hundreds and sometimes thousands of plants. Using large numbers of plants is part of a good experimental design because it increases the reliability of the results by reducing the effects of chance or error that may occur with small numbers of plants. You can see the same principle at work if you toss a coin and count heads and tails: with only a small number of tosses, heads or tails might dominate, but over many tosses the ratio will get closer and closer to 1:1.

Mendel went on to examine what happened when each F₂ plant was allowed to self-fertilize (Figure 13.4 **4**). He found that the recessive one-fourth was always true-breeding. For example, self-fertilizing dwarf F₂ plants reliably produced only dwarf offspring **5**. By contrast, only one-third of the tall F₂ individuals (one-fourth of the total offspring) proved true-breeding **6**, whereas two-thirds of the tall plants were not true-breeding. This last class of plants produced dominant and recessive F₃ individuals in a ratio of 3:1 **7**, just as the F₁ generation did. Thus, the ratio of individuals in the F₂ population was 1:2:1: (**1** true-breeding dominant : **2** not-true-breeding dominant : **1** true-breeding recessive).

ber of each pair was dominant and the other was recessive. Figure 13.3 shows which traits Mendel found to be dominant and which traits he found to be recessive.

Then Mendel went a step further. He allowed each F₁ plant to mature and self-fertilize. He collected and planted seeds from each plant, which produced the **second filial (F₂) generation**. He found that most of the F₂ plants exhibited the dominant form of the trait and looked like the F₁ plants, but some exhibited the recessive form **3**. None of the plants had blended characteristics. The clear-cut differences enabled Mendel to count the number of each of the two contrasting varieties of F₂ progeny and compare the results. His counts for experiments with all seven traits are shown in **Figure 13.5**.

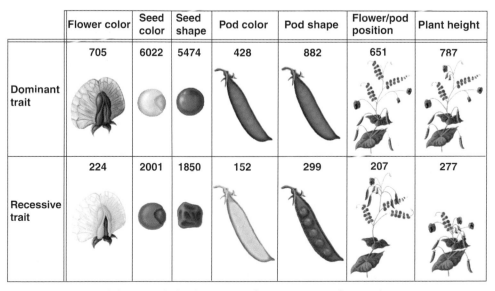

| | Flower color | Seed color | Seed shape | Pod color | Pod shape | Flower/pod position | Plant height |
|---|---|---|---|---|---|---|---|
| **Dominant trait** | 705 | 6022 | 5474 | 428 | 882 | 651 | 787 |
| **Recessive trait** | 224 | 2001 | 1850 | 152 | 299 | 207 | 277 |

**Figure 13.5** Mendel's monohybrid crosses of seven traits of pea plants: F₂ generation after self-fertilization of F₁ generation. The numbers reflect an approximate 3:1 ratio of dominant-to-recessive forms in the F₂ generation for each trait.

CONCEPT CHECKPOINT

1. Describe the difference between a true-breeding individual and a hybrid.
2. What evidence led Mendel to the conclusion that one trait seems to be dominant over another?

## 13.4 Mendel suggested that traits are inherited as discrete factors.

What did Mendel do with experimental data he collected? He drew conclusions regarding the nature of heredity—conclusions that have withstood further experimentation over time. In fact, Mendel's work is historically looked upon as the birth of genetics, and four of his conclusions are considered to be the first established principles of genetics. In modern terms, these four principles can be summarized as follows:

- Traits are inherited as genes.
- Alleles are alternative forms of genes.
- Gametes receive only one allele of each pair.
- Alleles may differ or may be identical.

### Traits are inherited as genes

As mentioned previously, Mendel observed that the plants exhibiting the traits he studied did not produce progeny of intermediate appearance when crossed. These observations did not uphold the theory of blending inheritance but suggested instead that traits are inherited as discrete factors that are either present or absent in a particular generation. These factors, hypothesized Mendel, act later in the offspring to produce the trait. Today, scientists call these factors **genes,** the units of hereditary.

### Alleles are alternative forms of genes

For each pair of traits that Mendel examined, one form was not expressed in the $F_1$ hybrids, although it reappeared in some $F_2$ individuals. He inferred from these observations that each individual, with respect to each trait, contains two factors. The members of each pair of factors may contain information for (be a code for) the same form of a trait or for the two different forms of the trait. Today, scientists call each member of a factor pair an **allele** (uh-LEEL). That is, an allele is a particular form of a gene. Each human, for example, receives one allele for each gene from the mother's egg and one allele for each gene from the father's sperm. As we saw in Section 13.3, Mendel used the term *dominant* to refer to the form of a trait that was expressed in the $F_1$ generation and the term *recessive* for the form that was not expressed in that generation.

### Gametes receive only one allele of each pair

Since the two factors (alleles) that coded for a trait in his pea hybrids remained "uncontaminated" (not blending with one another) and since experiments on various traits showed similar results, Mendel concluded that his hybrids must form egg cells and pollen cells (gametes) in which the factors for each trait separate from one another "in equal shares" during gamete formation. This concept is referred to as **Mendel's law of segregation**. Today, scientists state this law as follows: Each gamete receives only one of an organism's pair of alleles, and which one it receives is determined by chance. This random segregating process takes place during the process of meiosis (see Chapter 12).

### Alleles may differ or may be identical

Mendel made three crucial observations about the offspring from self-fertilization of his monohybrid crosses of pea plants (see Figure 13.4):

- In the $F_2$ generation, there was a 3:1 ratio of plants exhibiting the dominant form of a trait versus plants exhibiting the recessive form.
- Plants exhibiting the dominant form of a trait in the $F_2$ generation would either breed true in the $F_3$ generation or would produce offspring exhibiting the dominant versus recessive traits in the same 3:1 ratio.
- Plants exhibiting the recessive form of a trait would always breed true.

These observations suggested to Mendel that true-breeding plants receive *only* the dominant factor (allele) or *only* the recessive factor from each parent, whereas nontrue-breeding plants are hybrids, receiving the dominant factor from one parent and the recessive factor from the other parent. Today, scientists call an individual having two identical alleles for a trait **homozygous** (hoe-muh-ZYE-gus) for that trait. The prefix *homo* means "the same"; the suffix *zygous* refers to the zygote, or fertilized egg. An individual having two different alleles for a trait is said to be **heterozygous** for that trait (*hetero* means "different").

Mendel obtained clear results because he was studying alleles that exhibited complete dominance. However, many alleles exhibit incomplete dominance (see Chapter 14 for discussion). In that case, heterozygotes will have intermediate phenotypes. In fact, the "blending" of genes observed by some early investigators might have been visible expressions of incomplete dominance. Since Mendel's time, many examples of incomplete dominance have been found for various traits in both plants and animals, such as pink flowers in snapdragons resulting from crosses of red-flowered snapdragons with white-flowered snapdragons.

 CONCEPT CHECKPOINT

3. Using modern genetic terminology, describe each of the following principles of inheritance discovered by Mendel: traits are inherited as genes; alleles are alternative forms of genes; gametes receive only one allele of each pair; alleles may differ or may be identical.

## 13.5 Mendel showed that outward appearance is due to inherited factors.

Looking back on Mendel's experiments, you can use the conclusions he drew to further analyze and understand his experiments and his data. Mendel used letters to represent alleles. The modern convention is to name the gene for the trait associated with the recessive allele. This pattern is used most of the time, but there are exceptions. In addition, an uppercase letter is commonly used to denote the dominant allele, and the lowercase version of the same letter to denote the recessive allele. In the parental generation, the cross of true-breeding (therefore homozygous) tall pea plants (the dominant trait) and true-breeding dwarf pea plants (the recessive trait) can be represented as DD × dd, as shown in **Figure 13.6** ❶ (the × denotes a cross between two plants). Because the tall parent can produce only D gametes and the dwarf parent can produce only d gametes, the union of an egg and a sperm from these parents can produce only heterozygous Dd offspring in the $F_1$ generation ❷. Because the D allele is dominant, all of the $F_1$ individuals are tall. The d allele, though present, is not visibly expressed.

To distinguish between the presence of an allele and its expression, scientists use the term **genotype** (JEEN-uh-type) to refer to an organism's allelic makeup and the term **phenotype** (FEE-nuh-

type) to refer to the expression of those alleles. The phenotype—the organism's outward appearance—is the end result of the functioning of the enzymes and other proteins coded by an organism's genotype. Figure 13.6 shows the difference between these two terms at the top of the illustration ❸.

The genotype of the $F_1$ generation of pea plants from the cross of true-breeding tall plants (DD) and true-breeding dwarf plants (dd) is Dd, but the phenotype of the hybrids is tall because D is the dominant allele. When these $F_1$ plants are allowed to self-fertilize, the D and d alleles segregate (separate from one another) randomly during gamete formation. Their subsequent union at fertilization to form $F_2$ individuals is also random. Figure 13.6 ❹ shows the possible combinations of the gametes formed by the $F_1$ plants through self-fertilization. Their random combination produces plants in an approximate ratio of 1 DD : 2 Dd : 1 dd ❺ (Dd and dD are equivalent and are both represented by Dd). The ratio of phenotypes is 3 tall : 1 dwarf ❻.

Using a simple diagram called a **Punnett** (PUN-et) **square** is another way to visualize the possible combinations of genes in a cross. Named after its originator, English geneticist Reginald Punnett, the Punnett square is used to show in an orderly way how possible female gametes combine with possible male gametes. The male gametes are shown at the top of the square, and the female gametes are shown along the left side. As in **Figure 13.7**, the square is divided into smaller squares that form columns and rows, providing a "cell" for each possible combination of gametes.

Whether using a Punnett square or visualizing the gametes and their combinations as in Figure 13.6, you can see the expected ratios of the three kinds of $F_2$ plants: one-fourth are true-breeding dd dwarf plants, two-fourths are heterozygous Dd tall plants, and one-fourth are true-breeding DD tall plants. The 3:1 phenotypic ratio is really a 1:2:1 genotypic ratio.

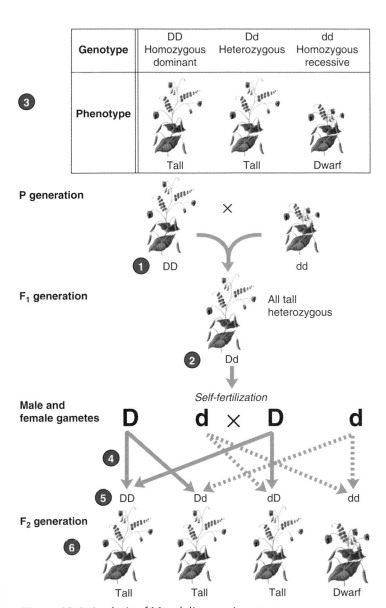

**Figure 13.6** Analysis of Mendel's experiments.

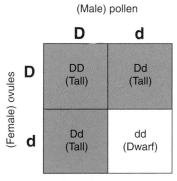

**Figure 13.7** Punnett square analysis of Mendel's $F_1$ generation. The genotypes in each square are determined by the segregated alleles for the trait from the female at the left of the square and from the male at the top of the square. Thus, the top left cell represents the combination of the D at the top and the D at the left. The top right cell represents the combination of the d at the top and the D at the left. The genotypes in the lower squares are represented in the same way. Each cell contains one possible $F_2$ combination from this cross of $F_1$ plants. The phenotypic ratio is 3:1 (3 tall phenotypes to 1 dwarf phenotype), but the genotypic ratio is 1:2:1. Look at Figure 13.6 and see that the genotypic ratio matches the expected $F_2$ ratio of true-breeding dominants to nontrue-breeding dominants to true-breeding recessives.

## 13.6 Mendel tested his conclusions.

How could Mendel be certain his conclusions were accurate? To test his conclusions further and to distinguish between homozygous dominant and heterozygous phenotypes, Mendel devised two procedures. The first test, already mentioned, is to self-fertilize a plant having a dominant phenotype. If the plant is homozygous dominant, it breeds true: tall plants produce only tall plants, purple-flowered plants produce only progeny with purple flowers, and so forth. But if the plant is heterozygous, it produces dominant and recessive offspring in a 3:1 ratio when self-fertilized.

The other procedure Mendel used is called a *testcross* and is illustrated in **Figure 13.8**. In this procedure, a phenotypically dominant test plant whose genotype is unknown is crossed with a known homozygous recessive plant. Mendel predicted that if the test plant is homozygous for the dominant trait, all the progeny will be hybrids and will therefore look like the test plant. This is shown in the cross at the right in Figure 13.8. Alternatively, if the test plant is heterozygous, then *half* the progeny will be heterozygous and look like the test plant, and half will be homozygous recessive and therefore will exhibit the recessive characteristic. This is shown in the cross at the left in Figure 13.8. Figure 13.8 uses the seed characteristics that Mendel actually employed in his first test-cross experiments.

What were Mendel's actual results? When he performed the testcross using hybrid $F_1$ plants having smooth seeds (Ww) and crossed them with plants having wrinkled seeds (ww), he obtained 208 seeds: 106 smooth seeds, and 102 wrinkled seeds—a 1:1 ratio, just as he predicted. These data confirmed the primary conclusion Mendel drew from earlier work: Alternative factors (alleles) segregate from one another in the formation of gametes, coming together in the progeny in a random manner.

CONCEPT CHECKPOINT

4. Below are several possible monohybrid crosses. Determine the predicted genotypic and phenotypic ratios for each cross. Feel free to use a Punnett square to help you determine these ratios. After you have completed the table, cover your answers and try to determine the ratios without using a Punnett square.

| | Phenotypic ratio (dominant : recessive) | Genotypic ratio (CC:Cc:cc) |
|---|---|---|
| Cc x Cc | 3:1 | 1:2:1 |
| CC x Cc | | |
| CC x cc | | |
| Cc x cc | | |

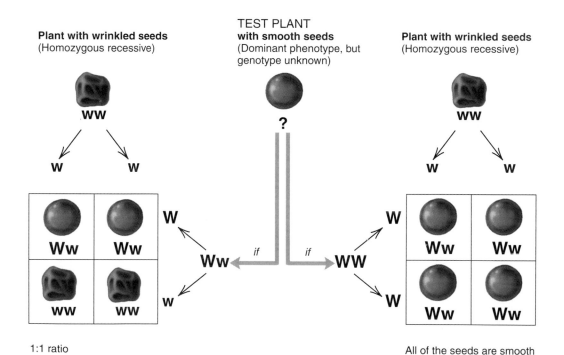

**Plant with wrinkled seeds** (Homozygous recessive)

**TEST PLANT with smooth seeds** (Dominant phenotype, but genotype unknown)

**Plant with wrinkled seeds** (Homozygous recessive)

1:1 ratio
Half of the seeds are smooth
Half of the seeds are wrinkled

All of the seeds are smooth

**Figure 13.8 A testcross.** This cross tests for the genotype of a plant having a dominant phenotype but an unknown genotype. Is the genotype homozygous dominant or heterozygous? To find out, Mendel crossed the "test" plant with a known homozygous recessive plant. If the test plant is homozygous dominant for the trait, as in the cross on the right, then the cross yields plants all having the dominant phenotype. If the test plant is heterozygous for the trait, as in the cross on the left, the cross yields plants with the dominant phenotype and plants with the recessive phenotype, in a 1:1 ratio (half and half). ("w" stands for the recessive allele for wrinkled seeds; "W" stands for the dominant allele for smooth seeds.)

## 13.7 Mendel studied the inheritance of pairs of factors.

Expanding on his law of segregation, Mendel asked a new question: Do the factor pairs (alleles) that determine particular traits segregate independently of factor pairs that determine other traits? In other words, does the segregation of one factor pair influence the segregation of another?

To answer his question, Mendel first developed a series of true-breeding lines of peas that differed from one another in two of the seven pairs of characteristics with which he had worked in his monohybrid (single trait) studies. He then crossed pairs of plants that exhibited contrasting forms of the two characteristics and that bred true. For example, he crossed plants having smooth, yellow seeds with plants having wrinkled, green seeds as shown in **Figure 13.9 ①**. From his monohybrid studies, Mendel knew that the traits "smooth seeds" (W) and "yellow seeds" (G) are dominant to "wrinkled seeds" (w)

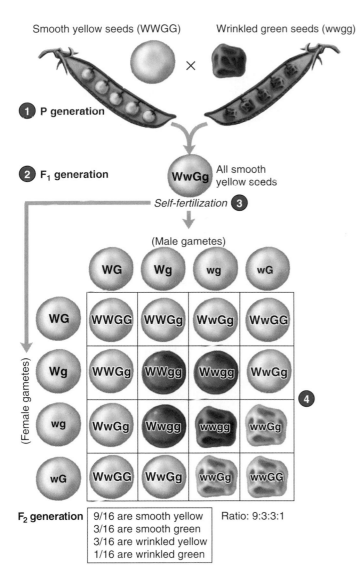

**Figure 13.9 A Punnett square showing the results of Mendel's self-cross of dihybrid smooth yellow-seeded plants.** The plants in the F₁ generation are the progeny of a cross between a homozygous dominant parent and a homozygous recessive parent. As shown in the Punnett square in the F₂ generation, the approximate ratio of the four possible combinations of phenotypes resulting from the self-fertilization of the F₁ generation is predicted to be 9:3:3:1, the ratio that Mendel found.

and "green seeds" (g). Therefore, the genotypes of the true-breeding parental (P) plants were WWGG and wwgg.

Mendel's F₁ progeny are *dihybrids*—the product of two plants that differ from one another in two traits. As in his monohybrid crosses, all the F₁ progeny had smooth, yellow seeds—the dominant phenotypes ❷. Mendel then allowed the F₁ dihybrids to self-fertilize ❸. The seeds he obtained from these self-crosses ❹ were 315 smooth, yellow seeds; 101 wrinkled, yellow seeds; 108 smooth, green seeds; and 32 wrinkled, green seeds—an approximate ratio of 9:3:3:1. (Mendel's numbers are not shown in the Punnett square.) Other dihybrid crosses also produced offspring having the same approximate ratio.

Mendel reasoned that if the factors (alleles) for seed color and seed shape segregated into gametes independently of one another and

were therefore inherited independently of one another, then the outcome for each trait would exhibit the 3:1 ratio of a monohybrid cross. For smooth versus yellow seeds, we see that there were 423 smooth seeds (315 + 108) and 133 wrinkled seeds (101 + 32), a 3:1 ratio. What about "seed color"—what is the approximate ratio of yellow seeds to green seeds? The answer is 416 yellow seeds and 140 green seeds, also a ratio of 3:1. Mendel's results were strong evidence that the contrasting factors (alleles) for seed shape and seed color assort independently of one another during gamete formation. Today, this concept is referred to as **Mendel's law of independent assortment** and can be stated as follows: The transmission of alleles for one trait into the gametes does not affect the transmission of alleles for other traits.

The Punnett square in Figure 13.9 shows visually how the two 3:1 ratios combine to produce a 9:3:3:1 ratio. Assuming that the alleles segregate randomly and assort independently as the gametes are formed, we see that the four female gametes (eggs) and the four male gametes (sperm) of the two parents can combine in 16 ways. Smooth, yellow seeds (WWGG, WWGg, WwGg, WwGG, and WwGg) result from 9 of the 16 combinations. Smooth, green seeds (WWgg and Wwgg) result from 3 of the 16 combinations. Wrinkled, yellow seeds (wwGG and wwGg) also result from 3 of the 16 combinations. Wrinkled, green seeds (wwgg) result from only 1 of the 16 combinations. Thus, Mendel's actual results (an approximate 9:3:3:1 ratio) confirm the analysis shown in the Punnett square for F₂ progeny.

Punnett squares can also be used to show how these ratios are governed by the laws of *probability*, or chance. **Figure 13.10** (left) shows the Punnett square results of the self-cross of hybrid plants having smooth seeds, one of the traits shown in the dihybrid cross in Fig-

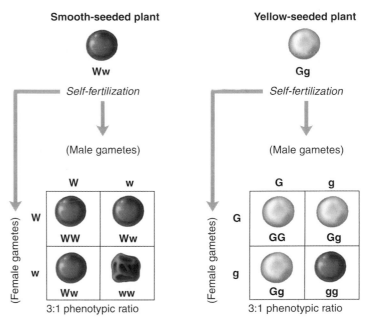

**Figure 13.10 Mendel's law of independent assortment.** These Punnett squares show the results of the self-cross of hybrid plants having smooth seeds (left) and yellow seeds (right), the factors shown in the dihybrid cross in Figure 13.9. Mendel noted that the probability that a trait will appear in the dihybrid cross is the probability that one of the traits will appear multiplied by the probability that the other trait will appear. This reasoning led him to the understanding that the contrasting traits in garden peas with which he worked assort independently from one another during gamete formation.

| TABLE 13.1 | Probabilities in a Dihybrid Cross | |
|---|---|---|
| | **Probability of smooth seeds = 3/4** | **Probability of wrinkled seeds = 1/4** |
| **Probability of yellow seeds = 3/4** | Probability of smooth, yellow seeds = 3/4 × 3/4 = **9/16** | Probability of wrinkled, yellow seeds = 1/4 × 3/4 = **3/16** |
| **Probability of green seeds = 1/4** | Probability of smooth, green seeds = 3/4 × 1/4 = **3/16** | Probability of wrinkled, green seeds = 1/4 × 1/4 = **1/16** |

ure 13.9. The genotype of these plants is Ww. The female gametes and the male gametes can combine in four ways. The probability that this cross will produce a plant having wrinkled seeds (ww) is one in four (1/4), because there are four possible combinations and only one of them is ww. Similarly, there is a three-in-four (3/4) probability that the cross will produce plants having smooth seeds (WW, Ww, and Ww), because three of the four combinations yield smooth seeds. Thus, the laws of probability predict that there will be a 3:1 phenotypic ratio of smooth seeds to wrinkled seeds (3/4:1/4 is the same as 3:1).

Figure 13.10 (right) shows the Punnett square results of the self-cross of hybrid plants having yellow seeds, the other trait shown in the dihybrid cross in Figure 13.9. The genotype of these plants is Gg. By the same reasoning we just applied to seed shape, we can see that the probability of producing a plant with green seeds is 1/4 and the probability of producing a plant with yellow

seeds is 3/4. Thus, probability predicts a 3:1 phenotypic ratio of yellow seeds to green seeds.

Now look at these traits (seed shape and seed color) as they occur together in a dihybrid cross. The probability that one event (e.g., a plant with wrinkled seeds) will occur at the same time as another, independent event (e.g., a plant with green seeds) is simply the product of their individual probabilities. Thus, the probability that a plant with wrinkled, green seeds will appear in the $F_2$ generation is equal to the probability of observing a plant with wrinkled seeds (1/4) times the probability of observing a plant with green seeds (1/4), or 1/16. **Table 13.1** shows this probability along with the probabilities that the other possible $F_2$ offspring will appear (plants with smooth, yellow seeds; plants with wrinkled, yellow seeds; and plants with smooth, green seeds). As you can see from the table, it makes no difference whether you calculate the expected ratio of offspring in a dihybrid cross by using a Punnett square or by using mathematics—in both cases, the prediction is the same, an expected ratio of 9:3:3:1.

**CONCEPT CHECKPOINT**

5. Below are eight possible testcrosses on pea plants using the seed color (yellow vs. green) and the seed shape (round vs. wrinkled) traits. Yellow (G) is dominant to green (g), and round (W) is dominant to wrinkled (w). For each cross, determine the predicted phenotypic ratios (yellow-round: green-round: yellow-wrinkled: green-wrinkled) and the probability of producing each phenotype: GGWW x ggww; GgWW x ggww; GgWw x ggww; GGWw x ggww; ggWw x ggww; ggWW x ggww; GGww x ggww; Ggww x ggww.

6. What pattern do you observe in the testcrosses in #5?

# How did Mendel's work influence further research on inheritance?

## 13.8 Sutton suggested that Mendel's factors (genes) are on chromosomes.

During the years Mendel was experimenting with pea plants to determine the principles of heredity, other scientists were studying the structure of cells. Although Mendel published his theories of inheritance in 1866, few other scientists read his work or understood its significance. In 1888, scientists gave the name *chromosomes* to the discrete, threadlike bodies that form as the nuclear material condenses during these processes. However, they still had no idea of the link between Mendel's factors of inheritance and the newly named chromosomes.

In 1900, three biologists, Carl Correns, Hugo de Vries, and Eric von Tschermak, independently worked out Mendel's principles of heredity. However, they knew nothing of Mendel's work until they searched the literature before publishing their results. This rediscovery of Mendel's work—his hypotheses supported by

the independent work of others—helped scientists begin to make connections between Mendel's ideas and chromosomes.

By the late 1800s, scientists had observed the process of fertilization and knew that sexual reproduction required the union of an egg and a sperm. However, they did not know how each contributed to the development of a new individual. An American graduate student, Walter Sutton, suggested that if Mendel's hypotheses were correct, then each gamete must make equal hereditary contributions. Sutton also suggested that because sperm contain little cytoplasm, the hereditary material must reside within the nuclei of the gametes. He noted that chromosomes are in pairs and segregate during meiosis, as did Mendel's factors. In fact, the behavior of chromosomes during the meiotic process paralleled the behavior of the hereditary factors. Using this line of reasoning, Sutton suggested that Mendel's factors were located on chromosomes. Years later, a single type of small fly—the fruit fly—provided the most conclusive evidence to uphold Sutton's chromosomal theory of inheritance.

## 13.9 Morgan provided evidence that genes are on chromosomes.

In 1910, Thomas Hunt Morgan, studying the fruit fly *Drosophila melanogaster*, detected a male fly that differed strikingly from normal flies of the same species. This fly had white eyes, as shown in **Figure 13.11a**. Normal fruit fly eyes are red (Figure 13.11*b*). Morgan quickly designed experiments to determine whether this new trait was inherited in a Mendelian fashion.

Morgan first crossed the white-eyed male fly to a normal female to see whether red or white eyes were dominant (**Figure 13.12 ❶**). All $F_1$ progeny had red eyes ❷, and Morgan therefore concluded that red eye color was dominant over white. Following the experimental procedure that Mendel had established long ago, Morgan then crossed flies from the $F_1$ generation with each other. Eye color did indeed segregate among the $F_2$ progeny as predicted by Mendel's theory. Of 4252 $F_2$ progeny that Morgan examined, 782 had white eyes—an imperfect 3:1 ratio but one that nevertheless provided clear evidence of segregation. Something was strange about Morgan's result, however—something totally unpredicted by Mendel's theory: All of the white-eyed $F_2$ flies were males! ❸

How could this strange result be explained? The solution to this puzzle involves sex. In *Drosophila* (as in most animals), the sex of the fly is determined by specific chromosomes called X and Y chromosomes. Both female and male flies have four pairs of chromosomes. In females, one of those pairs consists of two X chromosomes. In males, however the corresponding pair is an X chromosome and a Y chromosome. To explain his result, Morgan hypothesized that the white-eyed factor (allele) is located on the X chromosome. Because the white-eye allele is recessive to the red-eye allele, Morgan's result was a natural consequence of the Mendelian segregation of alleles. To see why, look again at Morgan's experiment in Figure 13.12 and note the chromosomes. In the P generation ❶, white-eyed male flies are crossed with normal red-eyed females. The white-eyed allele is shown on the male's X chromosome, while the Y chromosome does not contain an allele for eye color. Because the male has only one allele for eye color, that trait is expressed in the phenotype, regardless of whether it is dominant or recessive. The red-eyed allele is shown on each of the female's X chromosomes.

The $F_1$ generation flies all exhibit red eyes ❷. The male flies have inherited the chromosome carrying the red-eyed allele from their mothers and the Y chromosome from their fathers. Therefore, all the males will be red-eyed. The females have inherited the X chromosome carrying the red-eyed allele from their mothers and the X chromosome carrying the white-eyed allele from their fathers. Since the red-eyed allele is dominant to white, all females in the $F_1$ generation will be red-eyed as well.

In the $F_2$ generation ❸, males can inherit either the red-eyed or white-eyed allele on an X chromosome from their mothers and the Y chromosome from their fathers. Therefore, half the males will have red eyes and half will have white eyes. The females can inherit either the red-eyed or white-eyed allele from their mothers but only the red-eyed allele from their fathers. Therefore, females can be either homozygous red-eyed or heterozygous; in either case they will be red-eyed.

Morgan's experiment is important in the history of genetics because it presented the first clear evidence upholding Sutton's theory

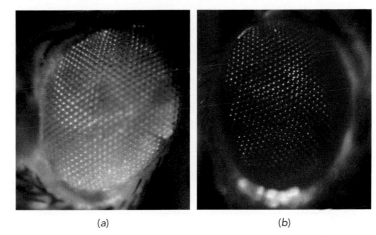

**Figure 13.11** Eyes of *Drosophila melanogaster*. (a) White eye. (b) Red eye.

that the factors (alleles) determining Mendelian traits are located on chromosomes. When Mendel observed the segregation of alternative factors in pea plants, he was observing the outward reflection of the meiotic segregation of homologous chromosomes. Morgan's work also revealed that certain factors were linked to the X chromosome, which is now called *X-linked inheritance*.

Throughout human history, scientists' attempts to understand the nature of heredity had been highly speculative, until

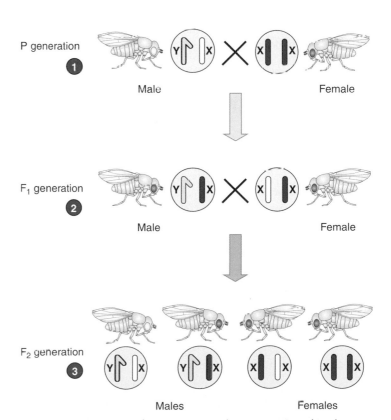

**Figure 13.12** Morgan's experiment demonstrating the chromosomal basis of sex linkage in *Drosophila melanogaster*.

the work of Mendel, Sutton, and Morgan. These early geneticists took the first real steps toward solving the puzzles of inheritance and laid the foundation for one of the great scientific advancements of the twentieth century: an understanding of the nature of genetic material and how it is transmitted from generation to generation. Their basic outline of heredity led to a long chain of questions and continues to be modified and refined as scientists provide additional experimental data and evidence.

## just wondering . . .

*Questions students ask*

### Why do scientists study tiny organisms like fruit flies? Why don't they choose something larger?

*Drosophila melanogaster* (**Figure 13.A**), known as the fruit fly to most people, is an excellent organism for geneticists to study for many reasons. First, fruit flies are easy to breed, and their life cycles are short. A single female can lay several hundred eggs, which develop into adults within 12 days. Therefore, a geneticist can study multiple generations of flies within a few months. Second, fruit flies are easy to maintain and take up little room in the laboratory. Populations of flies can be kept in small containers with easily prepared media. Third, fruit flies exhibit variations in certain inherited traits such as eye color and wing formation that are easy to see under a dissecting microscope. The geneticist can easily anesthetize the population, study the flies, and place them back in the bottle before they revive. Fourth, fruit flies have only four pairs of chromosomes. Experiments can often be simpler using organisms with few chromosomes rather than organisms with a large number of chromosomes.

Underlying the use of particular types of organisms for genetic studies is the assumption that genetic principles are universal. Thus,

**Figure 13.A** The fruit fly, *Drosophila melanogaster*.

Mendel and Morgan were able to articulate laws of genetics that apply to all living things, even though they studied only fruit flies and peas. ●

*Are you wondering about a topic in biology and how it relates to your life? Submit your question by clicking the* Just Wondering *link in this text's companions Web site at* www.wiley.com/college/alters.

## CHAPTER REVIEW

### Summary of Key Concepts and Key Terms

#### When did Mendel establish a theory of inheritance?

**13.1** Prior to the nineteenth century, speculations about inheritance varied widely.

**13.1** Hippocrates believed that parents gave off particles from their bodies that traveled to their sex organs during reproduction and were merged in the offspring.

**13.1** A common idea was that male and female traits blended in the offspring.

**13.1** Some scientists and philosophers held that either sperm or eggs carried tiny preformed beings within them.

**13.1** Not until the mid-1800s did scientists realize that new life arose from old life in the form of new cells arising from old cells.

**13.1** In the mid-1850s, Gregor Mendel, an Austrian monk trained in botany and mathematics, began studying patterns of inheritance using garden pea plants.

**13.2** Mendel cross-fertilized **true-breeding** (p. 200) varieties of pea plants that exhibited alternative forms of seven different traits.

**13.2** Mendel established a theory of inheritance in the 1860s.

#### How did Mendel investigate inheritance?

**13.3** Mendel studied the inheritance of individual **traits** (p. 200).

**13.3** The purpose of Mendel's experiments was to observe the offspring from the crossing of each pair of plants and to look for patterns in the transmission of single traits.

**13.3** Mendel crossed true-breeding plants, each having contrasting forms of a single trait, by artificially pollinating one with the other; these plants are the **parental (P) generation** (p. 200), and their hybrid offspring are the **first filial (F$_1$) generation** (p. 200).

**13.3** Mendel found that one of the alternative traits was expressed in the $F_1$ plants, but the other was not; he referred to the form of a trait that was expressed in the $F_1$ plants as **dominant** (p. 200) and to the unexpressed form as **recessive** (p. 200).

**13.3** Mendel allowed each $F_1$ plant to mature and self-fertilize. He collected and planted seeds from each plant, which produced the **second filial ($F_2$) generation** (p. 201). From counting progeny types in the $F_2$ generation, Mendel found that the dominant forms of traits appeared in 75% of $F_2$ plants, while the recessive forms appeared in 25% of $F_2$ plants—a 3:1 ratio.

**13.3** When Mendel allowed the $F_2$ plants to self-fertilize, he found that the ratio of true-breeding $F_2$ plants to hybrid $F_2$ plants was 1 true-breeding dominant to 2 hybrid dominant to 1 true-breeding recessive.

**13.3** Mendel deduced from the 3:1 ratio that traits are specified by discrete factors that do not blend in the offspring.

**13.4** Today, scientists refer to Mendel's factors as **genes** (p. 202) and to alternative forms of his factors as **alleles** (uh-LEELZ; p. 202).

**13.4** Because Mendel observed that traits did not blend in the offspring, Mendel hypothesized that factors (alleles) separate from one another during gamete formation; we call this concept **Mendel's law of segregation** (p. 202).

**13. 4** An individual having two identical alleles for a trait is **homozygous** (hoe-muh-ZYE-gus; p. 202) for that trait, while an individual having two different alleles for a trait is said to be **heterozygous** (p. 202) for that trait.

**13.5** Mendel showed that outward appearance is due to inherited factors (alleles).

**13.5** To distinguish between the presence of an allele and its outward expression, scientists use the term **phenotype** (FEE-nuh-type; p. 203) for the outward appearance of an individual and **genotype** (JEEN-uh-type; p. 203) for the genetic makeup of an individual.

**13.5** A **Punnett square** (p. 203) is one way to visualize the genotypes of progeny in simple Mendelian crosses and to illustrate their expected ratios.

**13.6** To determine the genotype of a phenotypically dominant test plant, it is crossed with a known homozygous recessive plant, which is a procedure known as a testcross.

**13.6** If the test plant used in a testcross is homozygous dominant, all the progeny will have the dominant phenotype; if the test plant is heterozygous, half the progeny will have the dominant phenotype and half the progeny will have the recessive phenotype.

**13.7** Working with pea plants that differed from one another in two characteristics, Mendel discovered that alleles of different genes assort independently during gamete formation, a concept referred to as **Mendel's law of independent assortment** (p. 205).

### How did Mendel's work influence further research on inheritance?

**13.8** Near the turn of the twentieth century, after three biologists independently upheld Mendel's ideas, scientists began to make links between chromosomes and inheritance.

**13.9** Thomas Hunt Morgan provided the first clear evidence that genes reside on chromosomes.

**13.9** Morgan demonstrated that the segregation of the white-eye trait in *Drosophila melanogaster* was associated with the segregation of the X chromosome, now called X-linked inheritance.

---

| Level 1 | **Learning Basic Facts and Terms** |

**Multiple Choice**

1. What contributed to Mendel's success in determining basic principles of genetics?
   a. He used plants that could self-pollinate.
   b. He used a simple experimental design and quantified his experimental results.
   c. He understood how chromosomes moved during meiosis.
   d. He was a monk.

2. The cross Aa × aa is an example of
   a. a dihybrid cross.
   b. a linkage cross.
   c. a cross between two true-breeding lines.
   d. a testcross.
   e. It has no special name given to it.

3. Which part of the theory of inheritance did Mendel establish?
   a. independent assortment of alleles
   b. genes are located on chromosomes
   c. DNA is the genetic material
   d. blending inheritance

4. An allele that has no noticeable effect on the appearance of an organism unless it is found in the homozygous condition is
   a. X-linked.
   b. dominant.
   c. codominant.
   d. recessive.
   e. heterozygous.

5. What is a testcross?
   a. A mating between an individual of unknown genotype and an individual that is homozygous recessive for the trait of interest.
   b. A mating between an individual of unknown genotype with an individual that is homozygous dominant for the trait of interest.
   c. A mating between an individual of unknown genotype with an individual that is heterozygous for the trait of interest.
   d. A mating between two individuals that are heterozygous for the trait of interest.
   e. A mating between an individual that is homozygous recessive with an individual that is homozygous dominant for the trait of interest.

6. True-breeding individuals
   a. are always homozygous recessive.
   b. are always homozygous dominant.
   c. are heterozygous.
   d. when self-fertilized, always produce offspring with the same phenotype as the parent.

**True–False**

7. Sexual reproduction provides less variation among individuals in a population than does asexual reproduction.

8. The law of segregation means that alleles separate during sex-cell formation.

9. When one is homozygous for a given gene, it means that two identical alleles were inherited.

10. Mendel's parental crosses were examples of cross-fertilization, while the F1 crosses were examples of self-fertilization.

## Level 2 | Learning Concepts

1. Assume that "S" represents the dominant trait of having long leaves, and "s" represents the recessive short-leafed trait in a plant. In the parental generation, you cross a homozygous long-leafed plant with a homozygous short-leafed plant. Draw a Punnett square illustrating this cross, and give the genotypes and phenotypes of the $F_1$ generation.

2. Using a Punnett square, show the genotypes and phenotypes of the $F_2$ generation if the $F_1$ plants in question 1 are self-fertilized.

3. In question 2, what is the probability that the $F_2$ plants will have short leaves?

4. What is Mendel's law of independent assortment?

5. In a dihybrid cross between organisms that are heterozygous for both traits, what is the probability that their offspring will exhibit the phenotype for both recessive traits? both dominant traits?

6. How did the work of Walter Sutton and Thomas Hunt Morgan change the way scientists viewed the role of egg and sperm in heredity and reproduction?

7. All seven of the garden pea traits studied by Mendel obeyed the principle of independent assortment. Based on your knowledge of inheritance thus far, what does this mean about the chromosomal location of these genes with respect to one another?

## Level 3 | Critical Thinking—Life Applications and Genetics Problems

1. An individual has genotype AaBb. What are the chances of producing a sex cell with a "B" allele? an "a" allele? an "a" and a "B" allele?

2. Both Gregor Mendel and Thomas Hunt Morgan chose organisms to study that were well suited to their particular investigations. Provide evidence to uphold this claim.

3. Can you think of a distinctive human trait that is not inherited? Explain.

4. Draw cells that show how two genes located on the same homologous pair would assort during meiosis. What gametes would result with respect to these two genes? What meiotic event would have to occur for this individual to produce all four different types of gametes?

5. If purple-flowered pea plants were not dominant to white-flowered pea plants, and if the inheritance of these traits did "blend," what would you expect the phenotype of $F_1$ plants to be if produced from a cross of true-breeding purple-flowered plants and true-breeding white-flowered plants? If the $F_1$ plants were allowed to self-fertilize, what would you expect would be the phenotype(s) of the $F_2$ generation of plants? If more than one phenotype would occur, give the ratio of the phenotypes.

GENETICS PROBLEMS

6. Among Hereford cattle, there is a dominant allele called *polled*; the individuals that have this allele lack horns. After college, you become a cattle baron and stock your spread entirely with polled cattle. You have many cows and few bulls. You personally make sure that each cow has no horns. Among the calves that year, however, some grow horns. Angrily, you dispose of them and make certain that no horned adult has gotten into your pasture. The next year, however, more horned calves are born. What is the source of your problem? What should you do to rectify it?

7. Many animals and plants bear recessive alleles for albinism, a condition in which homozygous individuals completely lack any pigments. An albino plant lacks chlorophyll and is white. An albino person lacks melanin. If two normally pigmented persons heterozygous for the same albinism allele have children, what proportion of their children would be expected to be albino?

8. In *Drosophila*, the allele for dumpy wings (d) is recessive to the normal long-wing allele (D). The allele for black body (b) is recessive to the normal gray body allele (B). In a cross of DdBb x ddbb, 200 offspring are produced. What phenotypes would result, and how many of each would you predict?

9. The results of the cross in question 8 were 55 Long-Gray, 48 Long-Black, 45 Dumpy-Gray, and 52 Dumpy-Black. Since these results are not identical to the expected, what would you conclude?

10. In some families, children are born who exhibit recessive traits and who therefore must be homozygous for the recessive allele specifying the trait even though neither parent exhibits the trait. What can account for this occurrence?

11. Two individuals that are heterozygous for five independently assorting genes are crossed. *Without using a Punnett square*, determine the probability of producing a child that is homozygous recessive at each gene. (*Hint*: Since the genes assort independently, consider them separately. A law of probability states that if you wish to determine the probability of several independent events occurring together, you multiply their individual probabilities.)

12. In question 11, what would be the probability of producing a child that expresses the dominant trait for all five genes?

## In The News | Critical Thinking

### ANCESTRY AND OBESITY

Now that you understand more about patterns of inheritance, reread this chapter's opening story about research into ancestry and obesity. To assist you in understanding this research better, it may help you to follow these steps:

1. Review your immediate reaction to Shriver's research on ancestry and obesity that you wrote when you began reading this chapter.

2. Based on your current understanding, again summarize the main point of the research in a sentence or two.

3. What questions do you now have about this research that this chapter's opening story does not answer?

4. Collect new information about the research. Visit the *In The News* section of this text's companion Web site at www.wiley.com/college/alters, and

watch the "Ancestry and Obesity" video. Then use the "summary" link to read the accompanying story and access related links. Use this information, the links provided, and other online and library resources to answer your questions and find updates about this research topic. State the sources of your information. Explain why you think the information is accurate. Also determine whether the information expresses a particular point of view or is biased in any way.

5. What in your view are the most significant aspects of this research? Give reasons for your opinion and for any changes in your ideas based on the additional information you have collected and the analysis you have done.

# HUMAN GENETICS

## In The News | Stop That Gene!

It is rare that genes alone can devastate your life. Nevertheless, this is the case with Huntington's disease, a degenerative brain disorder caused by a mutant, dominant gene. You can view the video news story, "Stop That Gene," about how Huntington's has affected the life of a young man, by visiting the *In The News* section of this text's companion Web site at www.wiley.com/college/alters.

Early symptoms of Huntington's disease include forgetfulness, clumsiness, involuntary twitching, and lack of coordination. As the disease progresses, walking, speaking, and swallowing become more difficult. There is no cure for Huntington's, and death often occurs from complications such as choking, infection, or heart failure. However, promising research results of today may change the deadly outcome of having inherited this gene—and possibly other disease-producing genes—in the future.

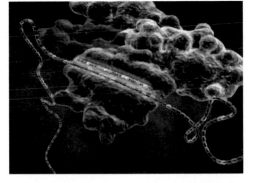

Medical researchers at the University of Iowa School of Medicine are experimenting with techniques to shut down the expression of the Huntington's gene. They describe their search in the "Stop That Gene" video news story. First some background: As with most genes, the Huntington's gene is used as a template to produce messenger RNA (mRNA), which then carries the genetic code from the nucleus to the cytoplasm, where proteins are synthesized from the code. In the case of the Huntington's gene, the proteins that are produced ravage the brain, so shutting down the ability of the RNA to code for the synthesis of these "bad" proteins would effectively cure the disease.

Dr. Henry Paulson and his team are using a recently discovered cell mechanism called RNA interference (RNAi) to block the expression of the mRNA transcribed on the Huntington's gene. Cells naturally regulate the expression of genes and the replication of viruses via RNAi. Scientists are trying to harness the process to fight disease. To do this job, they use pieces of RNA called small interfering RNAs (siRNA).

The illustration on this page, which is also shown in the video, is a computer-generated representation of a protein complex carrying the siRNA to the target RNA. The siRNA binds to the target only, allowing the protein complex to cut the target RNA. After the RNA is cut, cellular nucleases (enzymes that degrade nucleic acids) break apart the target RNA molecule, rendering it useless.

So far, however, scientists have not developed a method to introduce siRNA into living brains, where it would be needed to do its work to treat Huntington's disease. Nevertheless, excitement is high among scientists that this technique might ultimately lead to a treatment that will save lives and stop the devastation of Huntington's disease and possibly other inherited conditions.

*Write your immediate reaction to Paulson's research on treating Huntington's disease: first, summarize the main point of the research in a sentence or two; then suggest what you think its significance is. You will have an opportunity to reflect on your responses and gather more information on this topic in the* In The News *feature at the end of this chapter. In this chapter, you will learn more about the inheritance of human traits and genetic disorders.*

# CHAPTER GUIDE

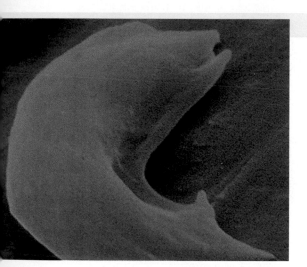

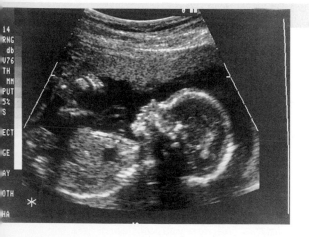

# How are human traits inherited?

Did you ever wonder how you got your curly hair and fair skin? Or why you are color blind and your siblings are not? Or why your brother or sister looks so much like you . . . or not like you at all? In the sections that follow, we will examine the basics of how human traits are inherited to help you understand some of the mechanisms that resulted in many of your distinguishing traits. Meanwhile, try to roll your tongue into a tube. You have heredity to credit for your ability to do it (if you can), not expertise or practice!

## 14.1 Some human traits are caused by dominant or recessive alleles.

Some human traits are inherited in a Mendelian fashion. That is, they are inherited showing the same patterns of dominance and recessiveness Gregor Mendel saw when he studied pea plants. Usually, every individual who carries the dominant allele manifests the trait. Only individuals who carry two recessive alleles exhibit the recessive trait. **Table 14.1** lists some of these traits, or characteristics, that are caused by dominant or recessive alleles.

### CONCEPT CHECKPOINT

1. Unattached earlobes are a dominant trait. If *A* denotes the allele for unattached earlobes, and *a* denotes the allele for attached earlobes, what is (are) the possible genotype(s) of a person who has unattached earlobes?
2. Could both parents of a person with unattached earlobes have attached earlobes? Why or why not?

## 14.2 Some human traits are caused by incomplete dominance of alleles.

Alleles do not always exhibit clear-cut dominance or recessiveness. The pea plants that Mendel worked with exhibited complete dominance, and for this reason his work challenged the concept of blending inheritance (see Chapter 13). However, since Mendel's time, researchers have discovered many cases of **incomplete dominance** in which alternative alleles are not dominant over or recessive to other alleles governing a particular trait. Instead, heterozygotes are phenotypic intermediates. Inheritance patterns follow Mendelian patterns, but the results differ when alleles exhibit incomplete dominance rather than complete dominance and recessiveness.

Humans have traits that exhibit incomplete dominance, such as the extent of curliness of the hair. A curly-haired individual and a straight-haired individual will have children with wavy hair. Persons with wavy hair will have children who are either straight haired, wavy haired, or curly haired.

**Figure 14.1** shows a cross between two wavy-haired people. *H* denotes the functional hair allele. The genotype *HH* is necessary for the curly trait to be expressed. With only one functional *H* allele, the person with the *HH'* genotype has wavy hair. The *H'H'* person without the functional *H* allele has straight hair.

## 14.3 Some human traits are caused by codominant alleles.

A slightly different situation that Mendel did not encounter but that builds on his work occurs with alleles that are codominant. With

---

| TABLE 14.1 | Common Dominant and Recessive Human Traits | |
| --- | --- | --- |
| **Trait** | **Dominant** | **Recessive** |
| Front hairline | Widow's peak (V-shaped front hairline)<br>*WW* or *Ww* | Straight front hairline<br>*ww* |
| Tongue rolling | Can roll tongue into tubelike shape<br>*RR* or *Rr* | Cannot roll tongue into tubelike shape<br>*rr* |
| Thumb crossing | Natural placement of left thumb over right when hands are folded (fingers interlocked)<br>*CC* or *Cc* | Natural placement of right thumb over left when hands are folded (fingers interlocked)<br>*cc* |
| Pinky finger shape | Bent pinky—last segment of pinky finger bends toward the ring finger<br>*BB* or *Bb* | Straight pinky<br>*bb* |
| Earlobe attachment | Earlobes are unattached—they hang free<br>*AA* or *Aa* | Earlobes are attached to the side of the head<br>*aa* |
| Thumb shape | When straightening out the thumb as in hitchhiking, the last segment (tip) of the thumb bends back to a 30 degree angle or less<br>*HH* or *Hh* | Hitchhiker's thumb—when straightening out the thumb as in hitchhiking, the last segment (tip) of the thumb bends back to nearly a 90 degree angle<br>*hh* |

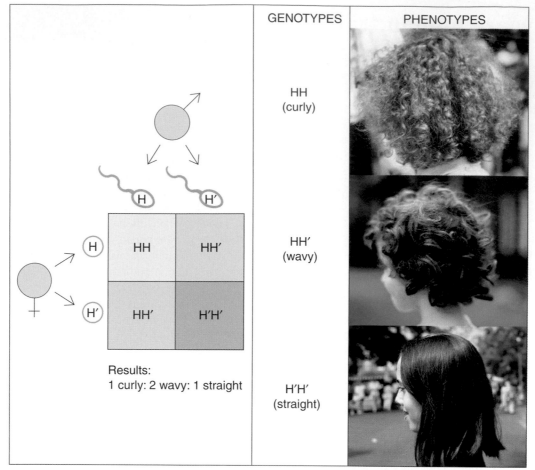

| GENOTYPES | PHENOTYPES |
|---|---|
| HH (curly) | |
| HH' (wavy) | |
| H'H' (straight) | |

Results:
1 curly: 2 wavy: 1 straight

**Figure 14.1** Incomplete dominance. Sometimes, alternative alleles are not dominant or recessive. For example, neither straight nor curly hair is dominant in Caucasians. The predicted ratio of the offspring of two wavy-haired people will be one curly-haired individual, one straight-haired individual, and two wavy-haired individuals. Wavy hair is the phenotypic intermediate. Note that uppercase letters are used for both alleles rather than upper- case and lowercase letters to illustrate the point that neither allele is dominant over or recessive to the other.

**codominant alleles**, neither allele is dominant; in a heterozygote, each allele contributes equally.

In humans the alleles that determine the A, B, and AB blood types are a good example of codominance. Each *A* and *B* allele codes for the production of different enzymes. These enzymes add certain sugar molecules to lipids on the surface of red blood cells. The enzyme produced by allele *A* results in the addition of one type of sugar; allele *B* adds a different sugar. The sugars act as recognition markers for the immune system and are called cell surface antigens. Persons having only *A* alleles will have only A antigens on their red blood cells and have blood type A, whereas persons having only *B* alleles will have only B antigens on their red blood cells and have blood type B. Both A and B antigens will be expressed in heterozygous individuals, which will give them the AB blood type.

Persons with type A blood respond to B antigens as foreign. That is, their plasma contains antibodies that respond to B antigens. Antibodies are proteins that bind to antigens. In the case of blood cells, if antibodies bind to cell surface antigens, they cause the blood cells to clump, which is a serious condition. Put simply, having a foreign antigen (a foreign blood type) in the body stimulates a response to destroy the "invader." Therefore, persons with type A blood cannot be given type B blood and vice versa. However, persons with type AB blood can be given either A blood, B blood, or AB blood; they produce no antibodies to these antigens because they are "normal" or "self" to that person.

In addition to being coded by codominant alleles, blood types are coded by more than the two alleles *A* and *B*. Humans can also have blood type O. The next section discusses this topic.

## 14.4 Many human genes have multiple alleles.

Thus far we have only discussed genes that exist in two allelic forms. However, a gene may be represented by more than two alleles. Some genes consist of a system of alleles, or **multiple alleles**.

The ABO blood types are determined by three alleles: $I^A$, $I^B$, and $i$. In this system of multiple alleles, the alleles $I^A$ and $I^B$ exhibit codominance (as described in Section 14.3) and the $i$ allele is recessive. Four phenotypes (A, B, AB, and O blood types) are determined by these alleles. Different combinations of the three possible alleles occur in different individuals, with each individual having two alleles. Since the *A* and *B* alleles are not dominant over one

| TABLE 14.2 | **Human ABO Blood Group Characteristics** | | | |
|---|---|---|---|---|
| Genotype | Blood Type | Frequency of Blood Type in U.S. | Antigens Present on Red Blood Cells | Antibodies Present in Plasma |
| $I^A I^A$ or $I^A i$ | A | 40% | A | B |
| $I^B I^B$ or $I^B i$ | B | 10% | B | A |
| $I^A I^B$ | AB | 4% | A and B | neither A nor B |
| $ii$ | O | 46% | neither A nor B | A and B |

 **Visual Thinking:** Why can O blood be given to anyone in emergencies, without threat of an A B antigen/antibody reaction, yet a person with O blood can only receive O blood?

another, they are both shown by the letter *I*, which denotes the gene responsible for producing the A and B antigens.

A person having an $I^A I^A$ genotype produces only the A cell surface antigen, whereas a person having an $I^A I^B$ genotype produces both A and B antigens. The *A* and *B* alleles are codominant, and both are expressed. A person having an $I^A i$ phenotype will produce only the A antigen (and so is said to have type A blood). The *i* allele produces neither A nor B antigens, so type O blood (*ii*) is known as the universal donor. However, this term is misleading because other blood groups found in human blood can cause reactions between donated blood and its recipients. Therefore, type O blood is given to persons with other blood types only in life-or-death situations. **Table 14.2** summarizes information about the ABO blood types.

## 14.5 Some human traits are caused by multiple genes.

Many traits in humans (and in other organisms) are controlled by more than one gene. The determination of a trait by alleles of two or more genes is called **polygenic inheritance**. This type of inheritance can be identified by its seemingly continuous nature rather than by the presence of two alternative forms of a trait. For example, with the inheritance of dominant and recessive traits, a person has either one form of a trait or its alternative: hitchhiker's thumb or straight thumb, attached earlobes or unattached earlobes, widow's peak or straight front hairline. With Mendel's pea plants, flowers are either purple or white, and seeds are either yellow or green. In polygenic inheritance, however, the additive effect of two or more genes determines the phenotype. No either/or situation exists, and a range of expressions of the trait is found in the population. Examples of such traits are height, eye color, and skin color.

Each gene in polygenic inheritance contributes to a single phenotype. For example, scientists currently think that three genes (six alleles) contribute to skin color. We could call those genes A, B, and C and denote their alleles as *A* and *a*, *B* and *b*, and *C* and *c*, where *A*, *B*, and *C* contribute to dark skin color and *a*, *b*, and *c* contribute to light skin color.

In skin color, alleles code for the production of the skin pigment melanin, and the effect of each allele is additive. Therefore, a

person with the genotype *AABBCC* has six alleles that code for a high production of melanin, so the skin would be very dark. A person whose genotype was *aabbcc* has six alleles that code for a low production of melanin, so the skin would be very light. **Figure 14.2** shows the range of skin colors possible in the offspring of medium-skinned parents.

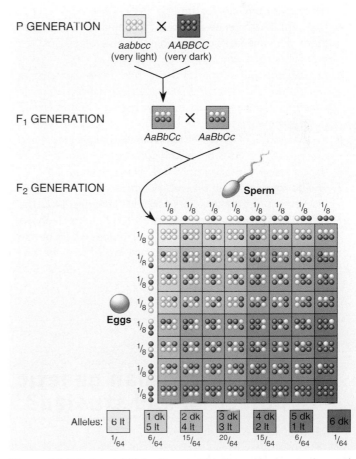

**Figure 14.2** Polygenic inheritance. This Punnett square shows the possible skin colorations of offspring of medium-skinned parents. The colored squares beneath the Punnett square show the frequency of genotypes having the same phenotype out of the 64 genotypic possibilities.

## 14.6 The environment influences genetic expression.

Sometimes the phenotype of an individual is not completely due to his or her genotype. Often, environmental conditions can influence the phenotypic expression of a gene. One example is the effect of ultraviolet radiation on skin color. If your genotype was *Aabbcc* as described in Section 14.5, how might the phenotypic expression of that genotype vary if you lived in Miami, Florida, vs. Bangor, Maine?

Environmental conditions also interact with the genes to affect human health. For example, one form of anemia is caused by an interaction of certain environmental factors and the inheritance of a mutant form of an enzyme called G6PD. Persons who inherit the gene that codes for the mutant enzyme have a condition known as G6PD deficiency. The G6PD enzyme is important to the structural integrity of red blood cell membranes. Certain chemicals trigger the destruction of red blood cells in persons with G6PD deficiency. When people with this condition eat fava beans (hearty beans cultivated in the Mediterranean) or take specific antibiotics or other drugs, chemicals in these substances trigger reactions that break down their red blood cell membranes and they become anemic. That is, they have fewer red blood cells than normal and so may look pale, have shortness of breath, and feel tired much of the time.

Sometimes the interplay between genes and the environment is not as clear-cut as in G6PD deficiency. Genetic factors may predispose a person to a disease. That is, a person may contract a particular disease or develop a particular condition more easily than other people in the general population exposed to the same environmental factors. Asthma, diabetes mellitus, migraine headaches, cardiovascular disease, and cancer are common conditions in which environmental factors modify the expression of certain genes.

### CONCEPT CHECKPOINT

3. In traits that show incomplete dominance, can persons heterozygous for a trait be distinguished from persons homozygous dominant for the trait? Why or why not?

4. Explain the difference between codominance and incomplete dominance. Give examples of phenotypes of heterozygotes of each.

5. Explain the difference between inheritance patterns in which traits are caused by multiple alleles and patterns in which traits are caused by multiple genes. Phenotypically, how might you be able to distinguish between the two?

6. Keisha and Tamika are sisters with the same father and mother, yet Keisha is quite light skinned and Tamika is very dark skinned. What could account for this difference?

7. The bar graph at right graphically represents the frequency distribution of the skin colors as shown in Figure 14.2. The bellshaped curve superimposed on the bar graph shows what the likely distribution of the phenotypes would be in a real population. Propose an explanation of why the bell curve is more likely than the bar graph to accurately represent the frequency distribution of skin color in a real population.

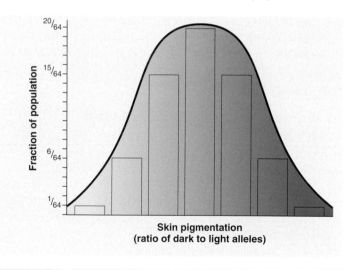

Skin pigmentation
(ratio of dark to light alleles)

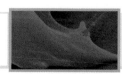

## How are human genetic disorders inherited and studied?

## 14.7 A few human genetic disorders are caused by dominant alleles.

As mentioned in Section 14.1, some inherited human traits are caused by dominant alleles. Individuals who carry the dominant allele manifest the trait, and the children of such affected heterozygous individuals each have a 50 percent chance of inheriting it. The human dominant genetic disorders that exist often kill persons with the disorders before they reach reproductive age. Therefore, inherited forms of these traits are not prevalent in the population.

## *just wondering . . .*

**Can good health habits overcome "bad genes"? Can eating fiber, drinking red wine, avoiding meat, and running 10 miles a day, for example, compensate for a family history of cancer and heart disease?**

Your question focuses on the extent to which environment (which includes lifestyle factors) can affect the expression of genes. We must first remember that the environment does not affect the expression of some genes. With genetic diseases such as cystic fibrosis, Duchenne muscular dystrophy, Down syndrome, Huntington's disease, and many others, there is a direct relationship between inheriting the gene and the expression of the disease. Living a healthy lifestyle will not directly affect whether a person gets the disease—only genes will. However, treating the disease may help affected individuals live longer and more comfortable lives.

In other diseases or disorders, environmental factors modify the expression of certain genes. Diseases or disorders of this type include asthma, diabetes mellitus, migraine headaches, cardiovascular disease, and cancer. Asthma, for example, is a disease of the airways. The mechanism of inheritance of asthma is not well understood; recent data suggest that inheritance may involve an autosomal codominant gene. Environmental factors such as air pollution, respiratory infections, and tobacco smoke often trigger asthma attacks in persons with the disease. Can persons with asthma compensate for their inheritance of the disease by avoiding factors that trigger attacks? All that asthma patients can do is lessen the severity and frequency of attacks, but they cannot make the disease "go away" with an asthma-friendly lifestyle.

Your "bad genes" vs. healthy lifestyle question is often asked with respect to cardiovascular disease and cancer. These are two diseases in which environment plays a significant role in their development. American Cancer Society statistics show that about three-fourths of all cancers are caused by environmental factors alone, such as cigarette smoking and unhealthy diets, which can cause gene mutations that lead to disease. However, scientists have discovered inherited abnormalities of genes for several cancers that tend to run in families, and they refer to them as "inherited cancers." These inherited abnormalities include certain types of cancers of the colon, breast, skin, kidney, and brain and account for only a small proportion of cancers in the United States. Can living a healthy lifestyle change the expression of these genes? No, but having a healthy lifestyle can significantly reduce the chance of developing cancer in general.

Recently, British researchers discovered that family history was a significant risk factor for two common types of stroke. A risk factor is a characteristic that increases an individual's chance of developing a health problem—stroke in this case. Half of all stroke risk is due to high blood pressure, smoking, diabetes, and high cholesterol. Having a healthy lifestyle can significantly reduce your chance of stroke, but it cannot change your family history.

Of all the risk factors for cardiovascular disease (including stroke), family history is only one factor. Elevated blood cholesterol, cigarette smoking, high blood pressure, physical inactivity, obesity, and diabetes mellitus are all risk factors for all types of cardiovascular disease. Although individuals cannot change the genes they inherit, they can reduce their other risk factors. Heart-healthy behaviors such as not smoking, maintaining an appropriate height/weight ratio, exercising regularly, eating a diet rich in fruits and vegetables, and getting regular medical checkups dramatically reduce the risk of cardiovascular disease. ●

*Are you wondering about a topic in biology and how it relates to your life? Submit your question by clicking the* Just Wondering *link in this text's companion Web site at* www.wiley.com/college/alters.

## Huntington's disease

*Huntington's disease* is a dominant disorder that killed folk singer and songwriter Woody Guthrie (**Figure 14.3**), who inherited the disease from his mother. Guthrie, who died in 1967 at the age of 55, often wrote songs about unemployment and social injustice. His most famous work is "This Land Is Your Land." His son is composer and singer Arlo Guthrie (**Figure 14.4**), who wrote the 1967 hit song *Alice's Restaurant*. Arlo did not inherit the disease from his father and was entertaining audiences around the world at the time of this writing at age 56.

Huntington's disease has persisted in the human population because it is caused by a mutant dominant allele that usually does not show its effects until individuals are older than 30 years of age—after they may have passed the allele to their children. This disorder causes progressive deterioration of brain cells and eventually death. About 30,000 Americans currently have Huntington's disease. Woody Guthrie suffered with the disease, which affected his behavior and mental abilities, for at least the last 13 years of his life.

**Figure 14.3** Woody Guthrie. This promotional photo was taken in 1944.

**Figure 14.4** Arlo Guthrie. During a 2003 benefit for the Woody Guthrie Foundation and Archive at the Ryman Auditorium in Nashville, Arlo sings one of his father's songs.

**Figure 14.5** Teen with progeria. John Tacket, of Bay City, Michigan, has progeria, a disease of premature aging. This photo was taken when John, then 15, spoke during a 2003 news conference announcing the discovery of the gene that causes this rare and fatal genetic condition. Dr. Francis Collins, director of the National Human Genome Research Institute, looks on.

## Progeria

Another dominant genetic disorder is *progeria*, which means "prematurely old." Progeria differs from the inheritance of Huntington's disease in that it appears to be the result of a spontaneous mutation that occurs during meiosis; neither parent carries the mutation. This rare mutation results in dwarfism, baldness, a small face and jaw relative to head size, and aged-looking skin in children, as shown in **Figure 14.5**. In addition, children affected with progeria suffer from joint stiffness, hip dislocation, and cardiovascular disease. The symptoms of progeria are not related to aging; persons with this disorder only look old and die at a young age.

## Other dominant genetic disorders

Other dominant genetic disorders include:

- *Marfan's syndrome*, which results in skeletal, eye, and cardiovascular defects. A hallmark of Marfan's syndrome is unusually long limbs. Many contend that America's sixteenth president, Abraham Lincoln, was likely affected with the disease; yet others note that his hands were highly muscular, not slender as in Marfan's.
- *Polydactyly* (PAHL-ee-DAK-tih-lee), which results in extra fingers or toes.
- *Achondroplasia* (ay-KON-dro-PLAY-zee-ah), which results in a form of dwarfism.
- *Hypercholesterolemia* (HI-per-ko-LESS-ter-ol-EE-mee-ah), which results in high blood cholesterol levels and a higher than average risk of developing coronary artery disease.

## 14.8 Most human genetic disorders are caused by recessive alleles.

Most inherited genetic disorders are recessive. The mutant alleles causing these diseases are able to persist in the population among carriers, persons who inherit an allele from only one parent and do not have the disease. Persons receiving a mutant allele from each of their parents are affected with the disease and, in the case of some diseases, may die before reaching adulthood.

### Cystic fibrosis

Often, recessive genetic disorders are seen primarily within specific populations or races of people unless many members of the population marry and have children outside of their population or race. Among white Americans of European descent, for example, the most common fatal genetic disorder is *cystic fibrosis* (*CF*). About 1 in 20 in this population has one mutant allele and shows no symptoms. Approximately 1 in 1800 has two mutant alleles and therefore has the disease.

The cause of cystic fibrosis is a defect in the transport of chloride ions across the membranes of glands that secrete mucus and sweat, which affects water movement across these membranes as well. Affected individuals secrete thick mucus that clogs the airways of their lungs and the passages of the pancreas and liver. Lung disease accounts for most of the illness in CF since multiple disorders of the lungs arise when mucus blocks the airways.

With the development of new therapies, the life expectancy for persons with CF has improved dramatically in recent decades, increasing from about 18 years in 1980 to nearly 33 years in 2000. One advance in therapy is a vest that uses a technology called high-frequency chest wall oscillation (HFCWO). The vest is inflatable and is connected by tubes to a generator (**Figure 14.6**). During therapy, the vest inflates and deflates rapidly, applying gentle pressure to the chest wall. This vacillating pressure works to loosen and

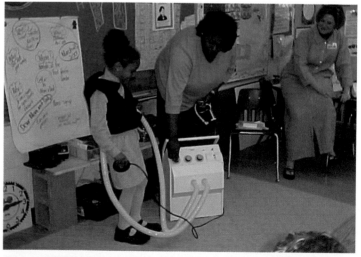

**Figure 14.6** High-frequency chest wall oscillation vest used for cystic fibrosis therapy. In this photo, a second grader with cystic fibrosis shows her classmates how the vest works. Attached to an air pulse generator by connecting hoses, the vest provides focused pressure and oscillations over the chest, helping patients clear their airways of mucus on their own.

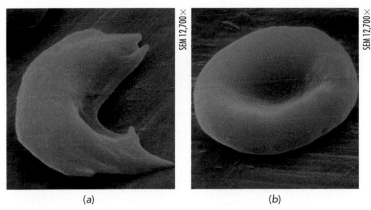

SEM 12,700×   SEM 12,700×

(a)                    (b)

**Figure 14.7** (*a*) Sickle-shaped red blood cell from person afflicted with sickle cell disease. (*b*) Normal human red blood cell.

thin mucus and move it toward the larger airways where it can be cleared by coughing or suctioning.

## Sickle cell disease

*Sickle cell disease* is a recessive disorder most common among African blacks and their descendants. This disorder is caused by a mutant HBB gene, which provides instructions for making a part of the hemoglobin molecule in red blood cells. In the United States, about 9 percent of blacks are heterozygous for this mutant allele; about 0.2 percent are homozygous and therefore have sickle cell disease. In some groups of people in Africa, up to 45 percent of the individuals are heterozygous for this allele. Heterozygous carriers of the sickle cell (mutant HBB) gene do not have the disease but can pass on the gene to their children.

Individuals affected with sickle cell disease are unable to transport oxygen to their tissues properly because the oxygen-carrying hemoglobin molecules are defective. Red blood cells that contain large proportions of such defective molecules become sickle shaped (**Figure 14.7a**) and stiff; normal red blood cells are disk shaped (Figure 14.7*b*) and much more flexible. As a result of their stiffness and irregular shape, the sickle-shaped red blood cells are unable to move easily through capillaries. Therefore, they tend to accumulate in blood vessels, reducing the blood supply to the organs they serve and causing pain, tissue destruction, and an early death.

The gene for sickle cell disease is most prevalent in the regions of Africa where malaria is prevalent. Malaria is a disease caused by microorganisms that live in a person's red blood cells. These microbes are injected into a person's bloodstream by the bite of a female *Anopheles* mosquito. A long-lasting disease, malaria affects the physical and mental development of its victims, causing damage to many body organs. Scientists have discovered that the defective hemoglobin molecules of the person with sickle cell disease produce conditions that are unfavorable to the growth of the malaria organism, but these persons eventually die of their disease. However, persons heterozygous for the sickle cell gene, though unaffected by sickle cell disease, are more resistant to malaria than persons with no sickle cell gene. Heterozygotes, then, have a survival advantage with respect to malaria and do not have sickle cell disease, although their offspring may be affected by the disease if they reproduce with persons also carrying the defective gene.

## Tay-Sachs disease

*Tay-Sachs disease* is an incurable, fatal recessive disorder caused by a mutation in a gene called hex A that codes for the production of the enzyme hexosaminidase A. This enzyme is necessary for breaking down certain fatty substances in brain and nerve cells. In persons with the mutation, these substances build up and and gradually destroy brain and nerve cells, until the entire central nervous system stops working.

Although rare in most human populations, Tay-Sachs has a high incidence among Jews of Eastern and Central Europe and among American Jews, 90 percent of whom are descendants of Eastern and Central Europeans. Geneticists estimate that, within these Jewish populations, 1 in 28 individuals carries the allele for this disease and approximately 1 in 3600 infants is born with the disorder.

The most common form of the disease affects infants. Affected children appear unaffected at birth but begin to show signs of mental deterioration at about 8 months of age. As the brain begins to deteriorate, affected children become blind; they usually die by the age of 5 years.

---

**CONCEPT CHECKPOINT**

8. Explain why parents who are heterozygous for a dominant disorder could be expected to pass on the trait to 50 percent of their children.
9. Draw a Punnett square to show the inheritance of sickle cell disease in the offspring of parents who are both carriers of the sickle cell allele. Use *s* to show the sickle cell allele and *S* to show the normal allele. What is the probability of their having a child with the disease?

---

## 14.9 Some human genetic disorders are caused by sex-linked alleles.

Humans have 23 pairs of chromosomes: a pair of **sex chromosomes** and 22 pairs of **autosomes** (**AW**-tuh-somes). The sex chromosomes determine a person's sex; a female has two X chromosomes and a male has an X and a Y chromosome, just as Morgan discovered in fruit flies (see Chapter 13). Autosomes, or non-sex chromosomes, carry the majority of an individual's genetic information but do not determine sex. They are the same in both sexes. Thus far in this chapter, the genes coding for the traits and disorders discussed have all been autosomal.

Almost all **sex-linked alleles** and disorders are really *X-linked alleles and disorders*, because the alleles for these disorders are found on the X chromosome. Y-linked disorders are rare. The reason is that the human Y chromosome is much smaller than the X chromosome and carries few genes. Therefore, recessive X-linked disorders appear more frequently in males than in females because males need to inherit only one mutant allele, not two, for the trait to be expressed. Some recessive X-linked disorders are red-green color blindness, hemophilia, and Duchenne muscular dystrophy.

Persons with *red-green color blindness* cannot distinguish red from green. This problem is due to inheriting a mutant allele located on the X chromosome that causes a defect in either the red or

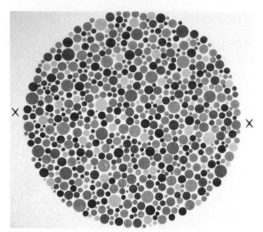

**Figure 14.8** A test for red-green color blindness. If you have normal color vision, you will see a winding path of green dots between the points marked X. If you have red-green color blindness, you will not be able to distinguish the green dots from red and therefore will not see the green path.

green color receptors in the cone cells of the retina of the eye. The pattern in **Figure 14.8** tests for an inability to distinguish red from green. Do you have red-green color blindness?

Individuals with *X-linked hemophilia* have inherited a mutant allele that results in their being unable to produce a factor needed in blood clotting. Therefore, the wounds of persons with this clotting disorder continue to bleed unless they receive treatment to stop it. Such a condition is life threatening.

*Duchenne muscular dystophy (DMD)* is a fatal disease in which the muscles gradually weaken and degenerate. Boys are affected with the disease when they receive the DMD allele from their mothers, who can carry the mutant allele yet remain unaffected. Girls are unaffected because they would need to receive a DMD allele from both their mother and their father to be affected by the disease. However, a boy with the DMD allele does not live long enough to father children.

As the muscles in a child with DMD weaken, the spine begins to curve and the abnormal body posture interferes with the functioning of internal organs, especially the lungs. DMD patients usually die in their teens or twenties of respiratory infections or respiratory failure when the diaphragm, a sheetlike muscle essential to breathing, becomes affected.

## 14.10 Pedigrees trace dominant, recessive, and sex-linked traits over generations.

**Pedigrees** are diagrams of genetic relationships among family members over several generations. By studying which relatives exhibit a trait, it is often possible to say if the gene producing the trait is sex linked or autosomal. Pedigree analysis also helps geneticists determine whether a trait is a dominant or a recessive characteristic. In many cases, it is also possible to infer which individuals are homozygous or heterozygous for the allele carrying the trait.

### Conventions used in pedigrees

**Figure 14.9** is a pedigree that shows the history of red-green color blindness in a family. As with all pedigrees, the circles represent females, and the squares represent males. Solid-colored symbols stand for individuals who exhibit the trait being studied. In this case, solid color individuals are color blind. Partially colored symbols stand for individuals carrying a mutant allele but not having the disorder—that is, they carry one allele for color blindness but are not color blind. Horizontal lines connecting a circle and a square denote mating. Vertical lines coming from two parents indicate their children, arranged along a horizontal line in birth order.

**Figure 14.9** A pedigree showing red-green color blindness. From this pedigree, it can be determined that the carriers of the color-blindness allele are the female in generation I, the three females in generation II, and the first and third females in generation III. The status of the other females is unknown.

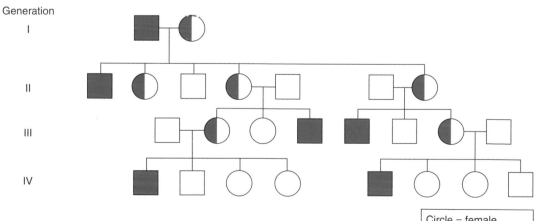

## How to analyze a pedigree

In studying a pedigree to determine whether a trait is sex linked or autosomal, first notice whether the trait is expressed more frequently in males than in females. If so, the trait may be a sex-linked trait. If the trait is expressed equally in both males and females, it is most likely an autosomal trait (whether dominant or recessive). Is red-green color blindness a sex-linked (X-linked) trait or an autosomal trait?

Second, to determine whether a trait is dominant or recessive, notice whether each person expressing the trait has a parent who expressed the trait. In our example, if color blindness is dominant to normal color vision, then each color-blind person will have a color-blind parent. However, if the trait is recessive, persons expressing the trait can have parents who are heterozygous (carry the mutant allele), but who do not express the trait. If each parent passes on the recessive gene to a child, the child is then homozygous recessive and the trait will be expressed. In sex-linked traits, however, males receiving the recessive trait from their mothers on their only X chromosome will express the trait. Is color blindness a dominant or a recessive characteristic?

Next, determine which individuals are carriers of the color-blindness gene. A color-blind male always contributes an X chromosome with the defective allele to his daughters; it is the only X he can contribute. His sons, however, get a Y chromosome from him and an X chromosome from their mother. A color-blind female can contribute only a chromosome with the defective allele to her children because both her X chromosomes carry the mutant allele. Women who are carriers of the trait, however, may contribute either a normal or a defective X to her children. The Punnett square in **Figure 14.10** illustrates the contributions of alleles by a normal mother and a color-blind father (Figure 14.10a) and a color-blind mother and normal father (Figure 14.10b).

## Pedigree analysis practice

Queen Victoria's family pedigree for hemophilia is shown in **Figure 14.11**. Using the strategy outlined in the preceding paragraphs, analyze the pedigree to determine whether hemophilia is sex linked or autosomal and whether it is a dominant or recessive trait. The answer is in the figure caption.

As shown in the royal pedigree, in the six generations since Queen Victoria, 10 of her male descendants have had hemophilia. The British royal family escaped the disorder because Queen Victoria's son King Edward VII did not inherit the defective allele. Three of Victoria's nine children did receive the defective allele, however, and carried it by marriage into many of the royal families of Europe. It is still being transmitted to future generations among these family lines, except in Russia, where the five children of Alexandra, Victoria's granddaughter, were killed soon after the Russian Revolution.

### CONCEPT CHECKPOINT

11. Analyze the following pedigree and determine whether the gene producing the trait is sex linked or autosomal and whether it is dominant or recessive. Give a rationale for your answer.

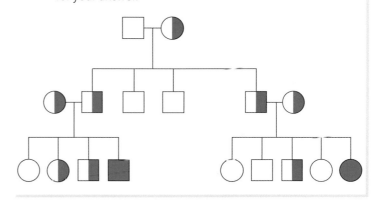

## 14.11 Karyotypes show errors in chromosome structure or number in individuals.

The 23 pairs of human chromosomes (including the two sex chromosomes) are shown in **Figure 14.12**. Arranged in this manner according to size, shape, and other characteristics, the chromosome pairs make up a **karyotype** (KAIR-ee-oh-type). Notice the small Y chromosome in pair 23 that is characteristic of males.

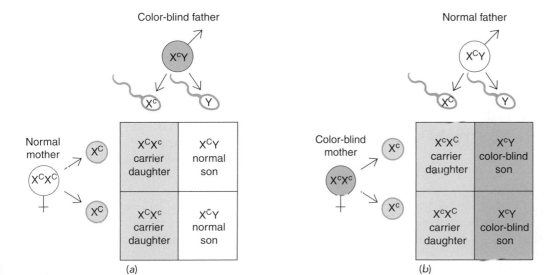

(a)

(b)

**Figure 14.10 Punnett squares depicting the outcomes for the sex-linked disorder color blindness.** (a) A Punnett square for a normal mother and color-blind father. (b) A Punnett square for a color-blind mother and normal father.

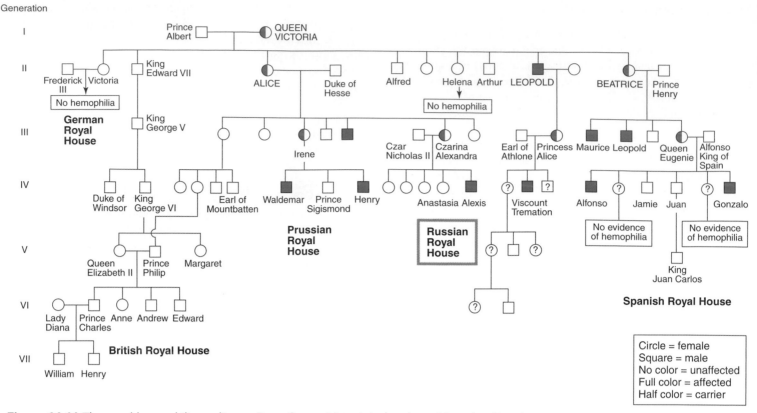

Generation

**Figure 14.11 The royal hemophilia pedigree.** From Queen Victoria's daughter, Alice, the disorder was introduced into the Prussian and Russian royal houses, and from her daughter Beatrice, it was introduced into the Spanish royal house. Victoria's son Leopold, himself a victim, also transmitted the disorder in a third line of descent.

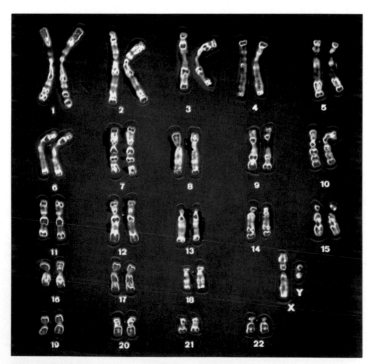

**Figure 14.12** A normal karyotype of a human male.

To make a karyotype, scientists use cells from blood, bone marrow, amniotic fluid, or placental tissue. The last two sources of cells are used to analyze the chromosomes of fetuses. The harvested cells are then grown in cell culture, and as they are dividing, their cell division is stopped with chemicals. In this way, scientists can produce cells having the condensed chromosomes of mitosis and can stain them. The dye stains regions of the chromosomes rich in the base pair adenine–thymine, which results in the banding patterns you see in Figure 14.12. Even the thinnest bands contain over a million base pairs—hundreds of genes! After being stained, the chromosomes are photographed and maniplulated digitally to pair homologues (the pairs you see in Figure 14.12) and order the pairs from 1 to 23. Prior to manipulating chromosomes with computers, scientists would simply cut the chromosomes out of the photograph, and physically pair and order them.

Analyzing karyotypes helps scientists identify chromosome abnormalities as the cause of disorders or diseases. They compare chromosomes for length, placement of centromeres, and locations and size of their bands. They also look for characteristics such as the loss of all or part of a chromosome, or the addition of extra chromosomes or chromosome fragments. Mutations in single genes (see Chapter 9) cannot be seen. Therefore, karyotypes are useful in studying diseases or disorders caused only by gross abnormalities

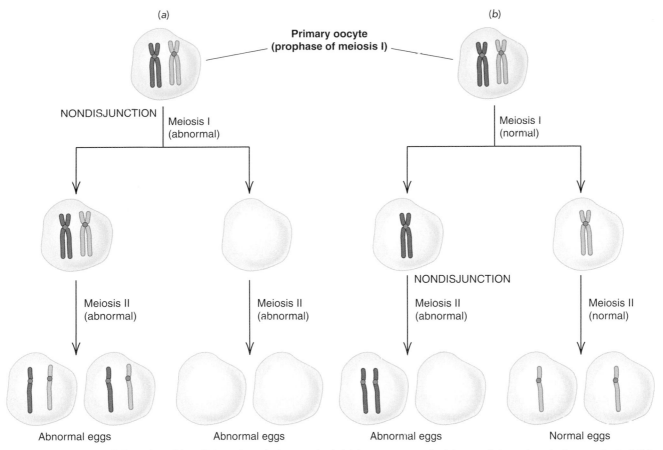

(a)                                                                     (b)

**Primary oocyte
(prophase of meiosis I)**

NONDISJUNCTION    Meiosis I
(abnormal)                                    Meiosis I
(normal)

Meiosis II
(abnormal)          Meiosis II
(abnormal)                    NONDISJUNCTION

Meiosis II
(abnormal)          Meiosis II
(normal)

Abnormal eggs        Abnormal eggs              Abnormal eggs        Normal eggs

**Figure 14.13** Nondisjunction. Nondisjunction during meiosis I (*a*) as compared with nondisjunction during meiosis II (*b*).

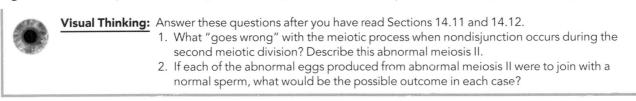

**Visual Thinking:** Answer these questions after you have read Sections 14.11 and 14.12.
1. What "goes wrong" with the meiotic process when nondisjunction occurs during the second meiotic division? Describe this abnormal meiosis II.
2. If each of the abnormal eggs produced from abnormal meiosis II were to join with a normal sperm, what would be the possible outcome in each case?

in chromosomes. They are not useful in studying inherited disorders caused by mutations in the chemical structure of the DNA, such as the diseases and disorders discussed thus far in this chapter. Such mutations are invisible in a karyotype.

## 14.12 Errors in chromosome number may result from errors in meiosis.

Humans can inherit an abnormal number of autosomes or sex chromosomes, which can be seen in a karyotype. How does such a situation arise? In humans, it comes about almost exclusively as a result of errors during meiosis, a type of cell division that results in gametes in animals (see Chapter 12).

Early in meiosis, homologous chromosomes line up side by side in a process called *synapsis*. At this point in meiosis, each chromosome is made up of two sister chromatids, so each pair of homologous chromosomes has four chromatids. Gametes can gain or lose chromosomes at this time if two homologous chromosomes

fail to separate (disjoin) in a process called **nondisjunction** (non-dis-JUNK-shun).

**Figure 14.13** shows what happens if nondisjunction occurs at either the first or the second meiotic division. Figure 14.13*a* shows nondisjunction occurring during meiosis I at the top of the figure. Compare with normal meiosis I in Figure 14.13*b*. If a homologous pair of chromosomes does not separate during meiosis I, both chromosomes will appear in one daughter cell, and the other daughter cell will receive one chromosome less than normal—no chromosomes for our example. When these cells undergo meiosis II, none of the resultant cells is normal. Two will contain an extra chromosome, and two will be missing a chromosome.

If nondisjunction occurs during meiosis II when two sister chromatids fail to separate, the cells resulting from this division will be abnormal, as seen in Figure 14.13*b*. One will have an extra chromosome, and one will receive one chromosome less than normal—no chromosomes for our example. The cell in which sister chromatids separate normally will yield normal daughter cells (right side).

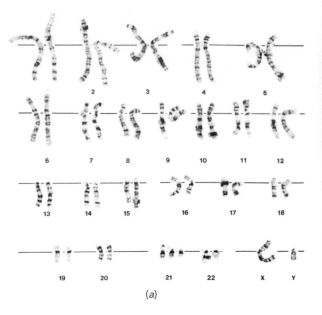

**Figure 14.14** Down syndrome. (*a*) Karyotype of a male with Down syndrome. (*b*) Down syndrome child with his mother.

(a)

(b)

## 14.13 Humans may inherit abnormal numbers of autosomal chromosomes.

Look at the karyotype in **Figure 14.14a** and compare it with that in Figure 14.12. What differences do you see? Carefully examine chromosome 21. One karyotype contains an extra copy of this chromosome, a situation called *trisomy 21*. A **trisomy** (TRI-so-me)

is a condition in which an individual has inherited an extra chromosome, resulting in three copies of a particular chromosome instead of the usual two copies. Therefore, the individual would have 47 instead of 46 chromosomes. (See **Table 14.3**.)

J. Langdon Down first described the developmental problem produced by trisomy 21 in 1866. For this reason, it is called *Down syndrome*. The chromosomal basis of Down syndrome was

| TABLE 14.3 | Summary of Chromosomal Abnormalities in Structure or Number | |
|---|---|---|
| **Condition** | **Characteristics** | **Difference** |
| Trisomy 13 | Multiple defects, including severe mental retardation and deafness; death before 6 months of age for 90% of those who survive birth | Three of chromosome 13 |
| Trisomy 18 | Facial deformities; heart defects; death before 1 year of age for 90% of those who survive birth | Three of chromosome 18 |
| Down syndrome (trisomy 21) | Developmental delay of skeletal system; mental retardation | Three of chromosome 21 |
| Trisomy 22 mosaic | Similar features to Down syndrome; more severe skeletal deformities | Three of chromosome 22 in some cells of the body |
| Cri du chat | Moon face; severe mental retardation | Deletion on chromosome 5 |
| Triple X female | Underdeveloped female characteristics | XXX; XXXX; XXXXX |
| Turner syndrome | Sterile female; webbed neck; broad chest | XO |
| Klinefelter syndrome | Sterile male; male and female characteristics | XXY |
| XYY syndrome | Fertile male; usually slightly taller then average; usually severe acne | XYY |

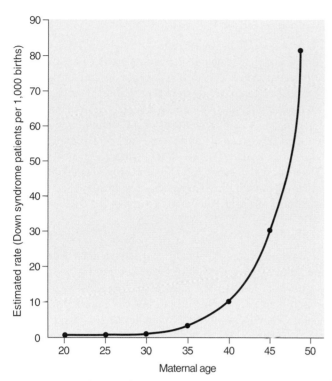

**Figure 14.15 Incidence of Down syndrome vs. maternal age.** This graph shows that few Down syndrome babies are born to mothers under age 30. As maternal age rises to 40 years, 10 out of every 1000 births are of Down syndrome children (1 percent). As a woman approaches menopause, her chances of bearing a Down syndrome child are over 8 percent (80 out of 1000 births).

At the time of this writing, understanding about some aspects of oocyte development was in flux. Recent research results suggest that our current understanding is incorrect and that primary oocytes die relatively quickly—over three to five years—but are replenished with new primary oocytes. However, the production rate of new primary oocytes lags behind the death rate of older primary oocytes, so the pool of primary oocytes decreases gradually as a woman ages until she reaches approximately 50 years. Theories about aging eggs and nondisjunction may soon be revised.

The incidence of nondisjunction of chromosomes other than those involved in Down syndrome also rises with maternal age. However, babies with other serious autosomal chromosome abnormalities are rare. Fertilized eggs with the improper number of chromosomes are almost always inviable; that is, they are unable to survive. These eggs do not begin normal development or implantation and are cast out of the body—a process called *spontaneous abortion*. This increase in the incidence of nondisjunction and its negative impact on the viability of zygotes is one reason older women often have a harder time conceiving than do younger women.

### CONCEPT CHECKPOINT

12. Explain why a pregnant woman over the age of 35 years might choose to have cells from the amniotic fluid or placental tissue used to produce a karyotype of her fetus. What information would the karyotype provide?

worked out nearly a century after Down identified the syndrome. In individuals with Down syndrome, maturation of the skeletal system is delayed, so they generally have a short stocky build, short hands, flattened facial features, and poor muscle tone. In addition, those with Down syndrome are mentally retarded.

Nondisjunction of chromosome 21 can occur in either parent. However, it is more likely to occur in females, and it increases with maternal age. In mothers younger than 20 years of age, the occurrence of Down syndrome children is only about 1 per 1700 births. In mothers 20 to 30 years old, the incidence is only slightly greater—about 1 per 1400. However, in mothers 30 to 35 years old, the incidence almost doubles to 1 per 750. In fact, the occurrence of nondisjunction of chromosome 21 increases sharply in the female population at about age 35 as shown in **Figure 14.15**. In mothers older than 45 years of age, the incidence of Down syndrome babies is approximately 1 in 32 births.

Why does this rise in the incidence of nondisjunction occur as women age? The short answer is that as a woman ages, so do her eggs. The longer explanation is that a woman is born with all the eggs she will ever have. The eggs begin meiosis during fetal development, but the process is arrested in prophase I of meiosis I. At this stage of meiosis, homologous chromosomes are lined up side by side (see Chapter 12). Over time, paired chromosomes may become unpaired, setting the stage for nondisjunction. The more time passes, the greater the chance that unpairing will occur.

## 14.14 Humans may inherit abnormal numbers of sex chromosomes.

Nondisjunction can occur with sex chromosomes as well as with autosomes. However, persons inheriting an extra X or Y chromosome or one X too few do not have the severe developmental problems that persons with too many or too few autosomes do. Nevertheless, persons who inherit abnormal numbers of sex chromosomes often exhibit some physical features that differ from XX or XY individuals.

### Triple X females

When X chromosomes fail to separate in meiosis, some gametes are produced that possess both of the X chromosomes; other gametes have no sex chromosome and are designated O. If an XX gamete joins an X gamete during fertilization, the result is an XXX (triple X) zygote. Even though triple X females usually have underdeveloped breasts and genital organs, they can often bear children. In addition, a small number of XXX people have lower-than-average intelligence. This condition occurs in about 1 out of every 700 births.

Although rare, a few individuals have been discovered to have tetra X (XXXX) and penta X (XXXXX) genotypes. Individuals having these genotypes are similar phenotypically to triple X individuals but usually suffer mental retardation.

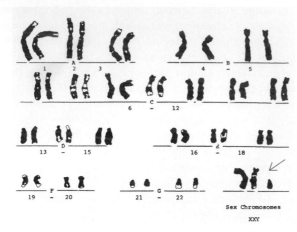

**Figure 14.16** A karyotype of a person with Klinefelter syndrome. Note the XXY set of sex chromosomes shown at the arrow.

## Klinefelter syndrome

If the XX gamete joins a Y gamete, the XXY zygote (**Figure 14.16**) develops into a sterile male. In addition to having male genitalia and characteristics, the individual has some female characteristics, such as breasts and a high-pitched voice. In some cases, XXY individuals have lower-than-average intelligence. This condition, called *Klinefelter syndrome*, occurs in about 1 out of every 500 male births.

## Turner syndrome

If an O gamete (no X) from the mother fuses with a Y gamete, the resulting OY zygote is nonviable and fails to develop further. If, on the other hand, an O gamete from either the mother or the father fuses with an X gamete to form an XO zygote, the result is a sterile female individual of short stature, webbed skin in the neck region, low-set ears, a broad chest, and sex organs that do not mature during puberty. The mental abilities of an XO individual may be slightly below normal. This condition, called Turner syndrome, occurs roughly once in every 2500 female births. Some of the ways in which nondisjunction in gametes can result in abnormal numbers of sex chromosomes are shown in **Figure 14.17**.

## XYY males

The Y chromosome occasionally fails to separate from its sister chromatid in meiosis II. Failure of the Y chromosome to separate leads to the formation of YY gametes. Viable XYY zygotes that develop into males are unusually tall but have normal fertility. The frequency of XYY among newborn males is about 1 per 1000.

The XYY syndrome has some interesting history associated with it. In the 1960s, the frequency of XYY males in penal and mental health institutions was reported to be approximately 2 percent (that is, 20 per 1000)—20 times higher than in the general population. This observation led to the suggestion that XYY males are inherently antisocial and violent. Further studies revealed, however, that XYY males have a higher probability of coming into conflict with the law than XY ("normal") males, but their crimes are usually nonviolent. Most XYY males lead normal lives and cannot be distinguished from other males.

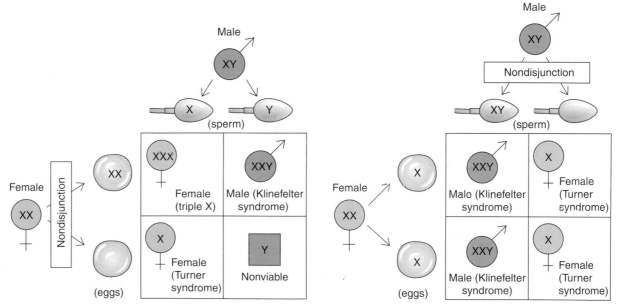

**Figure 14.17** How nondisjunction can result in abnormalities in the number of sex chromosomes. Both Klinefelter syndrome and Turner syndrome result from nondisjunction of either the male or female gamete. These Punnett squares also show how nondisjunction in female gametes can result in a triple X female.

 **Visual Thinking:** The text describes XYY males, but they are not shown in this illustration. Explain in detail how a YY gamete can result from an original XY cell during meiosis.

**13.** Explain the events that could lead to a person inheriting an improper number of sex chromosomes.

## 14.15 Humans may inherit chromosomes having structural changes.

Although a person may have inherited the proper number of chromosomes, he or she may develop a structural defect in a chromosome if one breaks and the cell does not repair the break or repairs the break incorrectly. Some chromosomal information may be added, lost, or moved from one location on the chromosome to another. These broken or misrepaired chromosomes can then be passed on in the gametes of the parents to produce disorders in the offspring.

How do chromosomes break? There is evidence that chromosomes break naturally, sometimes through the activity of transposable genetic elements, which are DNA sequences that can move from one place in the DNA to another (see Chapter 9). Occasionally, transposable elements induce breaks in a chromosome as they move. In addition, scientists can induce breaks in the laboratory by irradiating cells with X rays.

Chromosomal rearrangement changes the way that the genetic message is organized, interpreted, and expressed. Geneticists have identified many types of chromosome rearrangements. Some types are duplications, translocations, inversions, and deletions.

### Duplications

Newly added chromosomal information can be caused by a duplication of a section of a chromosome. Seen in a karyotype, a chromosome with a duplication appears longer than its homologue. In **Figure 14.18a** you can see that one red section is shown as duplicated, which results in a longer chromosome than normal.

### Translocations

A translocation could also produce an abnormally long chromosome. Translocations involve breaks in two or more nonhomologous chromosomes, followed by reattachments in new combinations. Depending on the lengths of the translocated pieces, some chromosomes could appear shorter than usual, and some longer, as shown in Figure 14.18b.

### Inversions

If a chromosomal inversion occurs, the broken piece of chromosome reattaches to the same chromosome but in a reversed direction (Figure 14.18c). Put another way, a chromosome piece is detached, flipped around 180°, and then reattached.

### Deletions

Chromosomes may become shortened because of deletions. Deletions may be due to a break near a chromosome tip, with the small piece becoming lost (Figure 14.18d). The most common situation, however, is the loss of internal pieces.

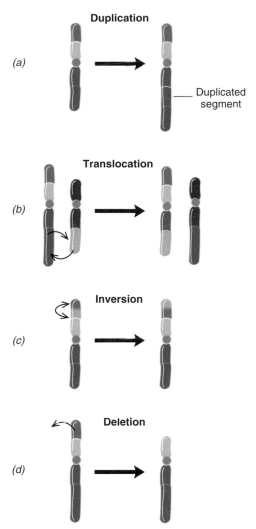

**Figure 14.18** Types of chromosomal rearrangement. (*a*) Duplication. (*b*) Translocation. (*c*) Inversion. (*d*) Deletion.

**Visual Thinking:** Which of these changes in chromosome structure would be visible in a karyotype? Give a rationale for your answer.

A disorder associated with a chromosomal deletion in humans is the cri du chat (kree-du-shah) syndrome. First described in 1963 by a French geneticist, cri du chat means "cat cry" and describes the catlike cry made by some cri du chat babies. Other symptoms of this disorder include varying degrees of mental impairment, a round "moon face," and wide-set eyes. Table 14.3 lists inheritance patterns of differences in chromosome number and structure discussed in Sections 14.13 to 14.15, along with additional trisomies.

**14.** Describe the difference between a duplication and a deletion. How would each of these changes in chromosome structure alter the physical appearance of the affected chromosome?

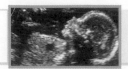

# How can individuals learn more about their genetic makeup?

"What are the chances that I'll develop Huntington's disease like my grandfather?" "I'm a carrier for sickle cell anemia, and my husband is not sure whether he is. Can our children inherit the disease?" "I'm having a baby and I'm 43 years old. Should I have tests to see if the baby is OK?" These are typical types of questions people ask who seek genetic testing and counseling. What is a genetic test and why might you want one?

## 14.16 Genetic information is available through genetic testing and counseling.

Genetic testing is the examination of a person's DNA for genes implicated in diseases or disorders. Genetic testing may also be performed to determine the identity of a deceased person or the perpetrator of a crime, who has left behind "genetic evidence," such as semen, saliva, or skin cells.

In **genetic testing**, the DNA is often taken from white blood cells in a blood sample. However, it can be taken from other DNA-containing body fluids, such as the amniotic fluid that surrounds a fetus in a pregnant woman, or from other tissues, such as bone marrow or cheek cells. Scientists may use karyotypes to detect abnormalities such as extra chromosomes or changes in chromosome structure, as described in Section 14.15. Scientists can also test for mutations on the molecular level, by examining the DNA molecule itself, or look for the presence or absence of gene products, such as certain proteins. In forensic or criminal cases, identity can be determined by comparing DNA samples to those of possible suspects or relatives of the deceased.

Genetic testing may be done in "high-risk" situations, such as in families with inherited diseases. Predictive tests are now available for many genetic diseases and for cancers in which heredity is a major factor, such as certain forms of breast and colon cancer.

**Genetic counselors** are specially trained health professionals who provide support for individuals when they consider genetic testing, when they receive test results, and as they deal with the information in the months that follow. Couples contemplating pregnancy may choose genetic testing to learn if they carry recessive alleles for inherited disorders such as cystic fibrosis, sickle cell anemia, or Tay-Sachs disease, which they could pass on to their children. Usually, geneticists first prepare family pedigrees for such couples to determine their chances of carrying deleterious alleles.

Newborns are also routinely screened for certain genetic diseases within days of birth. For example, many states require testing for phenylketonuria, or PKU, a genetic disorder in which the newborn does not metabolize protein properly. Untreated, infants accumulate toxic substances in their brains that impair brain development. However, infants fed special diets have little or no impairment.

## 14.17 During pregnancy, various procedures can assess fetal genetic makeup.

When a couple is expecting a child and both parents have a significant probability of being carriers of a serious recessive allele, the pregnancy is said to be a high-risk pregnancy. Another class of high-risk pregnancies occurs with mothers who are older than 35 years of age.

When a pregnancy is diagnosed as being high risk, many women elect to undergo *amniocentesis*, a procedure that permits the prenatal diagnosis of many genetic disorders (**Figure 14.19a**).

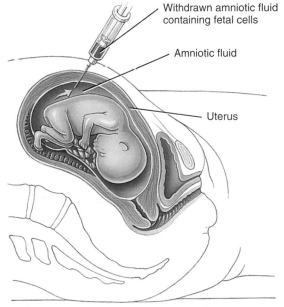

(a)

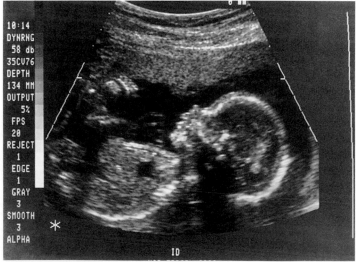

(b)

**Figure 14.19** Amniocentesis and ultrasound. (a) In amniocentesis, a needle is inserted into the amniotic cavity, and a sample of amniotic fluid containing some free cells derived from the fetus is drawn into a syringe. The fetal cells are then grown in tissue culture so that DNA testing can be performed. Ultrasound is used to guide the positioning of the needle. (b) Ultrasound image of a 20-week-old fetus.

In the fourth month of pregnancy, a hypodermic needle is used to obtain a small sample of amniotic fluid from the mother. The amniotic fluid, which bathes the fetus, contains free-floating cells derived from the fetus. Once removed, these cells can be cultured in the laboratory and used to test for many of the most common genetic disorders.

During amniocentesis, the position of the needle in relationship to the fetus is observed by means of a technique called *ultrasound*. Pulses of high-frequency sound waves are sent into the body and are reflected back in various patterns depending on the tissues or fluids the waves hit. These patterns of sound-wave reflections are then mapped to produce a picture of inner tissues (Figure 14.19*b*). This allows the person performing the amniocentesis to avoid damaging the fetus. Ultrasound also allows the fetus to be examined for the presence of major abnormalities.

Today, using ultrasound to guide them, physicians are able to sample fetal skin and certain other tissues. Cells from the outermost fetal membrane, the chorion, can also be sampled by a technique known as *chorionic villus sampling* (*CVS*). To perform CVS, a physician inserts a thin tube or a needle into the vagina and up into the uterus, vacuuming up fingerlike projections of the chorion called villi. This procedure can be performed in the second to third month of pregnancy.

The options for treating fetal abnormalities, though increasing, are still limited. Open fetal surgery was first performed at the University of California at San Francisco in 1981. Various surgical procedures are conducted today to treat fetal urinary obstructions, certain respiratory problems, and certain growths. The use of laser scalpels and new fetal imaging technologies such as fetoscopic cameras are allowing surgeons to treat such disorders less invasively. In some cases, surgery can be performed without opening the uterus. Gene therapy, the treatment of a genetic disorder by the insertion of "normal" copies of a gene into the cells of a patient carrying "defective" copies of the gene (see Chapter 11), may also be used.

At this time, however, most of the problems that can be diagnosed by pedigree analysis, amniocentesis, chorionic villus sampling, and ultrasound techniques cannot be treated. Sometimes the only options available to a couple are to continue the pregnancy and deal with the problems after birth or to have a therapeutic abortion. Continued progress in developing gene therapies and laser technologies may, in the future, have a dramatic effect on a physician's ability to treat genetic disorders before birth.

## CONCEPT CHECKPOINT

**15.** When a genetic test is performed, for what does it test?

**16.** How is amniocentesis performed? How does the use of ultrasound help in the procedure? Is amniocentesis a genetic test?

# CHAPTER REVIEW

## Summary of Key Concepts and Key Terms

### How are human traits inherited?

**14.1** Some human traits are caused by dominant or recessive alleles and are inherited in a Mendelian fashion.

**14.2** In traits exhibiting **incomplete dominance** (p. 213), alternative forms of a trait are neither dominant nor recessive; heterozygotes are phenotypic intermediates.

**14.3** In traits exhibiting codominance (**codominant alleles**, p. 214), alternative forms of an allele are both dominant; heterozygotes exhibit both phenotypes.

**14.4** Some genes consist of a system of alleles, or **multiple alleles** (p. 214), which are usually codominant to one another.

**14.4** The ABO system of human blood types is an example of multiple alleles, with two of the alleles (*A* and *B*) codominant to one another and one allele (*O*) recessive to the other two.

**14.5** Many traits are controlled by more than one gene, which is called **polygenic inheritance** (p. 215).

**14.5** Polygenic inheritance can be identified by its continuous nature and includes traits such as skin color, height, and eye color.

**14.6** Environmental conditions can influence the phenotypic expression of a gene.

### How are human genetic disorders inherited and studied?

**14.7** A few human genetic disorders are caused by dominant alleles.

**14.7** Dominant alleles that lead to severe genetic disorders are less common; in some of the more frequent ones, the expression of the alleles does not occur until after the individuals have reached their reproductive years.

**14.8** Many of the most common genetic disorders are associated with recessive alleles, the functioning of which may lead to the production of defective versions of enzymes that normally perform critical functions.

**14.8** Because recessive traits are expressed only in homozygotes, the alleles are not eliminated from the human population.

**14.9** Some human genetic disorders are caused by **sex-linked (X-linked) alleles** (p. 219).

**14.9** Human cells contain 46 chromosomes: 44 **autosomes** (AW-tuh-somes, p. 219) and 2 **sex chromosomes** (p. 219); the autosomes form 22 pairs of homologous (matched) chromosomes.

**14.10** Patterns of inheritance observed in family histories, or **pedigrees** (PED-uh-greez, p. 220), can be used to determine the mode of inheritance of a particular trait.

**14.10** Pedigree analysis can often determine whether a trait is associated with a dominant or a recessive allele and whether the gene determining the trait is sex linked.

**14.11** A **karyotype** (KAIR-ee-oh-type, p. 221) is a picture of chromosomes paired with their homologues and arranged according to size.

**14.11** Researchers often use karyotypes to study the inheritance of genetic disorders caused by the loss of all or part of a chromosome or by the addition of extra chromosomes or chromosome fragments.

**14.12** Errors in chromosome number usually result from **nondisjunction** (non-dis-JUNK-shun, p. 223) in meiosis.

**14.12** Nondisjunction occurs at either the first or second meiotic division when homologous chromosomes fail to separate.

**14.13** In humans, the inheritance of one autosome too few usually results in a nonviable zygote.

**14.13** The inheritance of one autosome too many, or a **trisomy** (TRI-so-me, p. 224), often results in a nonviable zygote (with the exception of chromosomes 13, 18, 21, and 22).

**14.13** Individuals with an extra copy of chromosome 21 have Down syndrome, a disorder much more frequent among children born to mothers older than 35 years.

**14.14** Persons who inherit abnormal numbers of sex chromosomes often have abnormal features and may exhibit intellectual deficits.

**14.14** Examples of genetic disorders caused by abnormal numbers of sex chromosomes are triple X females, XXY males (Klinefelter syndrome), XO females (Turner syndrome), and XYY males.

**14.15** Humans may inherit chromosomes having structural changes, such as duplications, translocation, inversions, and deletions.

### How can individuals learn more about their genetic makeup?

**14.16** **Genetic testing** (p. 228) is the examination of a person's DNA to check for a disease or disorder, or to determine identity.

**14.16** **Genetic counselors** (p. 228) provide support for individuals considering or undergoing genetic testing.

**14.17** Cells for genetic testing of fetuses can be collected by amniocentesis and chorionic villus sampling.

**14.17** Although most genetic disorders cannot yet be cured, research scientists are making progress in developing gene therapies and other technologies that may, in the future, have an important effect on a physician's ability to treat genetic disorders before birth.

## Level 1 — Learning Basic Facts and Terms

**Multiple Choice**

1. With incomplete dominance
   a. traits show either dominance or recessiveness.
   b. heterozygotes are phenotypic intermediates.
   c. phenotypic traits have a continuous nature.
   d. genetic counseling is useful.

2. In codominance
   a. a gene is represented by more than two alleles.
   b. a "double dose" of the gene is needed for a trait to be expressed.
   c. each allele contributes equally in the heterozygote.
   d. the phenotype is not completely due to the genotype.

3. Polygenic inheritance of a trait means that
   a. the trait is controlled by more than one gene.
   b. many genes are controlled by more than one trait.
   c. many alleles code for a single trait.
   d. genes for a trait are inherited from more than one parent.

**True–False**

4. _____ Most human genetic disorders arise from the inheritance of dominant alleles.

5. _____ In recessive disorders, mutant alleles are able to persist in the population among carriers.

6. _____ Tay-Sachs disease has a high incidence among Native American populations.

**Matching**

7. _____ autosome — a. complete complement of chromosomes, paired and ordered

8. _____ pedigree — b. procedure for prenatal diagnosis of a genetic disorder

9. _____ karyotype — c. chromosome that does not determine gender

10. _____ trisomy — d. diagram of genetic relationships among family members

   e. inheritance of an extra copy of a chromosome

## Level 2 — Learning Concepts

1. What happens when human offspring inherit abnormal numbers of sex chromosomes? Summarize the four examples discussed in the chapter, and give the genotype of each.

2. What would you conclude about the inheritance pattern of each human trait in each of the following situations?
   a. The trait is expressed more frequently in males than in females.
   b. Offspring who exhibit this trait have at least one parent who exhibits the same trait.
   c. Offspring can exhibit this trait even though the parents do not.
   d. The trait is expressed equally in both males and females.

3. Why do most genetic disorders in humans result from recessive genes? Name several examples.

4. Distinguish between incomplete dominance and codominance. Describe the phenotype of a heterozygote in each case.

5. If a couple wants to have a child but suspects that they may be at risk for a genetic disorder, what can they do? If a pregnancy turns out to be high risk, what options are available?

## Level 3    Critical Thinking—Life Applications and Human Genetics Problems

1. Familial hypercholesterolemia is an inherited genetic condition that leads to the development of cardiovascular disease at a relatively young age. Affected men typically have heart attacks in their 40s to 50s, and affected women in their 50s and 60s. Your father has this condition and his mother had it as well, but his father did not. Your mother does not have the condition, and your older brother appears unaffected. What is the inheritance pattern of this condition? What are the chances you inherited this disorder?

2. Suppose that you and your partner were interested in having children. What information would you be able to give a genetic counselor to help him or her identify or exclude you or your offspring from genetically based diseases or birth defects?

3. You were thrilled to have a new baby sister, but excitement turned to anxiety when she began to have digestive problems with persistent vomiting. Doctors quickly learned that she had an autosomal recessive disorder called *galactosemia*. Your sister lacked the enzyme that breaks down the milk sugar galatose, so her pediatrician put her on a special diet free of lactose and galactose. In a short time, your sister was fine.
   a. Neither of your parents is affected with galactosemia. If your sister's genotype is *gl/gl*, what are your parents' genotypes?
   b. You are not affected with galactosemia. What is your genotype or possible genotypes?
   c. Will your sister automatically pass on this disease to her children? Why or why not?

4. The extra chromosome 21 that is found in persons with Down syndrome is the cause of multiple developmental defects. What might this tell you about the interaction of genes on a particular chromosome?

5. A woman whose blood type is AB marries a man with the same blood type. Draw a Punnett square to illustrate the possible genotypes of their children. What blood type will each genotype have?

6. George has the same type of hemophilia as did Queen Victoria and some of her descendants. He marries his mother's sister's daughter Patricia. His maternal grandfather also had hemophilia. George and Patricia have five children: Two daughters are normal, and two sons and one daughter develop hemophilia. Draw the pedigree.

7. A couple with a newborn baby are troubled that the child does not appear to resemble either of them. Suspecting that a mix-up occurred at the hospital, they check the blood type of the infant. It is type O. Because the father is type A and the mother is type B, they conclude that a mistake must have been made. Are they correct?

8. How many chromosomes would you expect to find in the karyotype of a person with Turner syndrome?

9. A woman is married for the second time. Her first husband was blood type A, and her child by that marriage was type O. Her new husband is type B, and their child is type AB. What is the woman's ABO genotype and blood type?

10. Total color blindness is a rare hereditary disorder among humans in which no color is seen, only shades of gray. It occurs in individuals homozygous for a recessive allele and is not sex linked. A non-color-blind man whose father is totally color blind intends to marry a non-color-blind woman whose mother was totally color blind. What are the chances that they will produce offspring who are totally color blind?

11. This pedigree is of a rare trait in which children have extra fingers and toes.

Generation

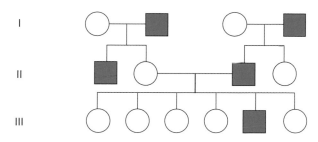

Which one of the following patterns of inheritance is consistent with this pedigree?
a. autosomal recessive
b. autosomal dominant
c. X-linked recessive
d. X-linked dominant
e. Y-linked

---

| In The News | **Critical Thinking** |

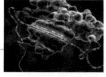

### STOP THAT GENE!

Now that you understand more about human genetics, reread this chapter's opening story about research into treatments for devastating genetic diseases such as Huntington's. To understand this research better, it may help you to follow these steps:

1. Review your immediate reaction to Paulson and his team's research that you wrote when you began reading this chapter.
2. Based on your current understanding, again summarize the main point of the research in a sentence or two.
3. What questions do you now have about this research that this chapter's opening story does not answer?
4. Collect new information about the research. Visit the *In The News* section of this text's companion Web site at www.wiley.com/college/alters and

watch the "Stop That Gene!" video. Then use the "summary" link to read the accompanying story and access related links. Use this information, the links provided, and other online and library resources to answer your questions and find updates about this research topic. State the sources of your information. Explain why you think the information is accurate. Also determine whether the information expresses a particular point of view or is biased in any way.

5. What in your view is the most significant aspect of this research? Give reasons for your opinion and for any changes in your ideas based on the additional information you have collected and the analysis you have done.

# THE HISTORY AND DEVELOPMENT OF EVOLUTIONARY THEORY

## In The News | Fossil Fraud

Science may not be the first thing you think of when you hear the words fraud, deceit, deception, and trickery. Although science is not immune from deception, the scientific community has safeguards in place to help avert the publication and acceptance of fraudulent work and claims. The peer review process is an important part of these safeguards and allows scientists the opportunity to read and critique colleagues' work prior to its publication. After publication, other researchers often replicate the work and then build on it to progress in scientific understanding. Sometimes, however, the system breaks down and leaves room for fraud. In legal terms, fraud is a deception deliberately practiced to secure unfair or unlawful gain. Recently, a fossil fraud was perpetrated on the scientific community, but scientists were not duped for long.

Fossils are preserved remains, impressions, or other traces of organisms that lived in the past. Fossils provide a record of the past and are part of the evidence of evolution. They can also be scarce and expensive commodities. A few years ago, a fossil described as the "missing link" between dinosaurs and birds appeared at one of the world's largest gem and mineral shows in Tucson, Arizona. Part dinosaur and part bird, the "missing link" called archaeoraptor (see photo) sold for $80,000! Subsequently, the fossil was shown to be a fake. How was this forgery discovered? How can scientists be certain that fossils are true representations of life from the past and not fakes built to look authentic?

Scientists at the University of Texas in Austin used a CAT (computed axial tomography) scanner, similar to that used in medical procedures, to visualize archaeoraptor from the inside out. Medical CAT scanners use X rays to create three-dimensional images of the inner body. The scientific scanner analyzes materials denser and smaller than the human body; it provided a three-dimensional view of the interior of the fossil and enabled scientists to see what materials made up the fossil, without destroying it. A scientific CAT scanner also shows whether a fossil's materials are natural and whether pieces have been glued together. It can not only find fossil frauds but can also document legitimate fossil finds. The CAT scan of archaeoraptor showed that the fossil had been pieced together from several specimens, primarily the front end of a bird and the tail of a nonflying dinosaur. You can learn more about detecting fossil frauds by visiting the *In The News* section of this text's companion Web site at www.wiley.com/college/alters.

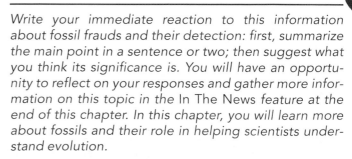

*Write your immediate reaction to this information about fossil frauds and their detection: first, summarize the main point in a sentence or two; then suggest what you think its significance is. You will have an opportunity to reflect on your responses and gather more information on this topic in the* In The News *feature at the end of this chapter. In this chapter, you will learn more about fossils and their role in helping scientists understand evolution.*

## CHAPTER GUIDE

### How was the theory of evolution by natural selection developed?

**15.1** Darwin developed his theory based on his observations and other evidence.

**15.2** Various factors influenced Darwin's thinking.

### What is natural selection?

**15.3** Natural selection is a mechanism of evolution.

**15.4** Darwin's finches provide classic examples of natural selection.

### Was Darwin the first to suggest evolution?

**15.5** Lamarck proposed evolutionary ideas before Darwin.

**15.6** Darwin published his theory in 1859.

### What other lines of evidence support an evolutionary view?

**15.7** Fossils provide a record of the past.

**15.8** Fossil dating shows that time existed for organisms to develop from ancestral forms.

**15.9** Comparative anatomy shows relatedness among organisms.

**15.10** Comparative embryology shows relatedness among organisms.

**15.11** Patterns of change in the molecular record provide strong, direct evidence for evolution.

### How do scientists depict evolutionary relationships?

**15.12** Scientists construct phylogenetic trees that show patterns of relationships among organisms.

# How was the theory of evolution by natural selection developed?

## 15.1 Darwin developed his theory based on his observations and other evidence.

The story of Charles Darwin and the development of his ideas about evolution (briefly introduced in Section 1.10) begins in the early 1800s, when he was a medical student at Edinburgh. While there, his interest in medicine waned, but his interest in natural history grew through interactions with his mentor, Robert Edmund Grant, an expert on sponges. After leaving Edinburgh, Darwin attended Cambridge University. In 1831, his professors recommended him to be the on-board naturalist for a five-year voyage around the world (**Figure 15.1**) on H.M.S. *Beagle*, more because of his enthusiasm for natural history than for his mediocre academic achievement. From 1831 to 1836, Darwin had the chance to study plants and animals on continents, islands, and seas in many different parts of the world. He was able to experience firsthand the remarkable diversity of living things on the Galapagos Islands, off the west coast of South America. Such an opportunity played an important role in the development of his thinking about the nature of life on Earth.

When the *Beagle* set sail, Darwin was fully convinced that species were unchanging. At this time in his career, Darwin's concept of species was remarkably close to this modern definition: A **species** (SPEE-sheez or SPEE-seez) is a group of interbreeding natural populations that are reproductively isolated from other such groups. In other words, species are defined by their reproductive isolation from one another. Organisms that do not reproduce sexually are designated as species by means of their morphological and biochemical characteristics. Scientists today know that species change over time, a concept that Darwin rejected at that point in his thinking.

Darwin wrote that it was not until two or three years after his return that he began to consider seriously the possibility that species could change. Then he began to formulate a theory integrating his observations of the trip with his understanding of geology, population biology, and the fossil record. In 1838, Darwin began to write his explanation of the diversity of life on Earth and the ways in which living things are related to one another.

During his five years on the *Beagle*, Darwin observed many phenomena that were of central importance to developing his theory of evolution. While in southern South America, for example, Darwin observed fossils of extinct armadillos that were similar to armadillos still living in that area (**Figure 15.2**). He found it interesting that such similar yet distinct living and fossil organisms were found in this same small geographical area. This observation suggested to Darwin that the fossilized armadillos were related to the present-day armadillos—that they were distant relatives that had died out.

Darwin also observed that geographical areas with similar climates, such as the desert areas of Australia, South Africa, and Chile, are each populated by *different* species of plants and animals. These differences suggested to Darwin that factors other than or in addition to climate must play a role in plant and animal diversity. Otherwise, all lands having the same climate would have the same species of animals and plants. However, he noted that these organisms are often similar to one another, shaped by environmental similarities.

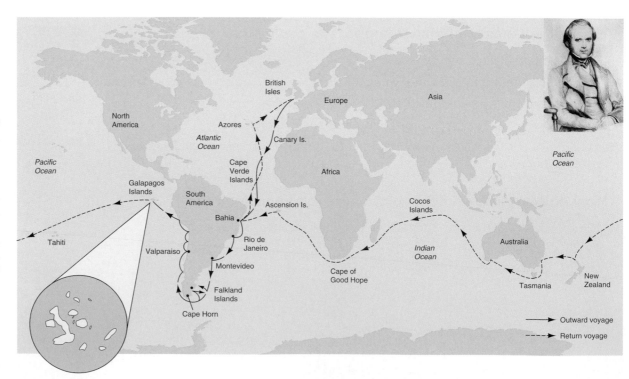

**Figure 15.1** Darwin and the voyage of H.M.S. Beagle. The voyage begins and ends in the British Isles. Before reaching the Galapagos Islands, Darwin made observations in South America that were critical in the development of some of his ideas.

The fact that the geographically isolated and relatively young Galapagos Islands (formed by undersea volcanoes) were home to a profusion of living organisms resembling plants and animals that lived on the nearby coast of South America also struck Darwin. Notice, for example, the similarity of the two birds in **Figure 15.3**. The bird on the left is a medium ground finch and is found on the Galapagos Islands. The blue-black grassquit, shown on the right, is found in grasslands along the Pacific coast from Mexico to Chile. These observations suggested to Darwin that ancestors of the Galapagos organisms had long ago flown, swam, or hitchhiked (were transported by other organisms) to the islands from the mainland. In summary, Darwin came to the following realizations:

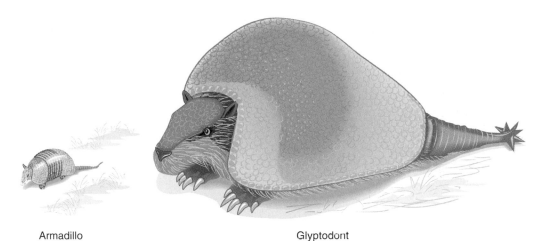

Armadillo · Glyptodont

**Figure 15.2 Distant relatives.** A reconstruction of a glyptodont, an extinct 2-ton South American armadillo (right), compared with a modern armadillo (left), which averages about 10 pounds. Darwin observed glyptodont fossils and conjectured that present-day armadillos living in the area were related to these giant armadillos that lived long ago in the same locale.

- Organisms of the past and present are related to one another.
- Factors other than or in addition to climate play a role in the development of plant and animal diversity.
- Members of the same species often change slightly in appearance after becoming geographically isolated from one another.
- Organisms living on oceanic islands often resemble organisms living on a nearby mainland.

## 15.2 Various factors influenced Darwin's thinking.

As Darwin studied the data he collected during his voyage, he reflected on their significance in the context of what was known about

(a) (b)

**Figure 15.3 Similar birds from different places.** The medium ground finch (left) lives on the Galapagos Islands, while the similar-looking blue-black grassquit (right) lives along the Pacific coast, from Mexico to South America. Darwin thought that the Galapagos birds might be related to ancestors that originally lived on the mainland.

geology, the breeding of domesticated animals, and population biology. The next sections discuss the state of knowledge in those areas at that time.

### The Earth has changed over time

In the late eighteenth and early nineteenth centuries, geologists (scientists who study the rock layers of the Earth) saw evidence that the Earth had gradually changed over time, acted on by natural forces such as wind, rain, heat, and cold. This concept of gradual change in the Earth influenced Darwin greatly. His understanding of the age of the Earth was derived from one of the most significant works on this topic at this time, Charles Lyell's *Principles of Geology*, published from 1830 to 1833. In this multivolume work, Lyell described how the physical structure of the Earth that we see today largely resulted from gradual change over millions of years. This view of gradual change, called *uniformitarianism*, challenged the prevailing view of catastrophic change. This view, called *catastrophism*, held that catastrophic changes such as volcano eruptions, earthquakes, and great floods alone accounted for the changes in the Earth's topography, resulting in the Earth being dated as only hundreds of thousands or a few million years old at most. Lyell and other geologists argued that gradual changes as well as catastrophic changes were responsible for the landscape of the planet. It was because of Lyell's book that Darwin realized that the Earth could be hundreds of millions of years old, which would give enough time for evolution to occur in large-scale ways.

Geologists also noticed that the fossils found within the Earth's rock layers were similar to but different in many ways from living organisms—an observation Darwin himself had made on his voyage. Not only had the Earth changed, scientists thought, but evidence existed that the organisms living on Earth had also changed.

(a)                              (b)

**Figure 15.4** Artificial selection: another clue for Darwin. Breeders can select for traits they wish to perpetuate, resulting in different varieties of a given species. These decorative chickens were produced after many generations of selecting chickens that exhibited showy head plumage.

## Artificial selection can create new varieties of organisms

As he pondered these ideas that the Earth and its organisms may have changed over time, Darwin reflected on the results of a process that he called *artificial selection*. In artificial selection, a breeder selects for desired characteristics, such as those of the chickens shown in **Figure 15.4**. These very different varieties of chickens come from the same ancestral stock, but through artificial breeding over successive generations, they have changed dramatically.

Artificial selection is based on the *natural variation* seen in all organisms. Although individuals within any population are similar, they possess characteristics that vary from individual to individual. Since prehistoric times, farmers and animal breeders have taken advantage of the natural variation within a population to select for characteristics they find valuable or useful, such as higher milk yield in cows. By choosing organisms that naturally exhibit a particular trait and then breeding that organism with another of the same species exhibiting the same trait, breeders are able (over successive generations) to produce animals or plants that all have the desired inherited trait. This trait will be retained and enhanced in successive generations when these organisms are bred with one another.

For example, different breeds of dogs have been selectively bred for different traits for thousands of years. Although your collie may look quite different from your neighbor's terrier, both animals belong to the same species, but each has been selected to retain traits characteristic of its breed (**Figure 15.5**). Even the turkey you eat on Thanksgiving has been selectively bred for a large cavity for stuffing.

## The growth of populations has limits

Pondering his observations, Darwin began to study Thomas Malthus's *Essay on the Principles of Population*. Malthus, an economist who lived from 1766 to 1834, focused much of his work on human populations. He pointed out that populations of plants and animals (including humans) tend to increase exponentially, but food

(a)                              (b)

**Figure 15.5** Artificial selection leads to different breeds of the same species. All dogs belong to the species *Canis familiaris*, but through artificial selection, breeders have been able to choose which traits each breed should retain. The collie in (a) has long, straight hair and a pointed snout. The terrier in (b) has short, curly hair and a less pointed snout.

supplies increase only arithmetically, as farmers gradually use more and better agricultural methods to grow crops. In an exponential progression, a population (for example) increases as its number is multiplied by a constant factor. For instance, in the exponential progression 2, 4, 8, 16, and so forth, each number is two times the preceding one. These numbers can also be expressed as $2^1$, $2^2$, $2^3$, and $2^4$. The number, or power, to which 2 is raised is termed the *exponent*, from which the term *exponential growth* is derived. An arithmetic progression, in contrast, to an exponential progression, is one in which the numbers increase by a constant difference, as in the progression 2, 4, 6, 8, and so forth. In this progression, each number is 2 greater than the preceding one. **Figure 15.6** shows the difference in how the numbers in these two types of mathematical progressions increase.

Thus, Malthus realized that factors existed to limit the exponential growth of populations. If populations were to grow unchecked, organisms would cover the entire surface of the Earth within a surprisingly short time. However, the world is not covered by ants, spiders, and poi-

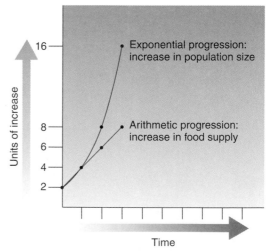

**Figure 15.6** Two types of mathematical progressions. Malthus noted that population size increases exponentially, while food supply increases only arithmetically. After three time intervals, population size has increased twice as much as food supply.

son ivy. Instead, populations of organisms vary in number within a certain limited range. Space and food are limiting factors of population growth; death limits infinite population growth. Malthus noted that in human populations, death was caused by famine, disease, and war.

After reading Malthus's work on populations, Darwin concluded that under circumstances of conflict between a population's limited natural resources and its continued growth, favorable variations would tend to be preserved and unfavorable ones would not.

## What is natural selection?

### 15.3 Natural selection is a mechanism of evolution.

Darwin made associations between the processes of artificial breeding, factors limiting survival, and reproduction within natural populations. He realized that environmental factors affecting natural populations could influence which organisms lived and which ones died, naturally selecting traits in ways similar to those in which breeders artificially select traits.

More specifically, Darwin suggested that those individuals that possess physical, behavioral, or other attributes well suited to their environments are more likely to survive than those that possess physical, behavioral, or other attributes less suited to their environments. The survivors have the opportunity to pass on their favorable characteristics to their offspring. The key ideas here are that (1) variation exists among individual traits within a population, (2) these traits are passed on to offspring within the population, (3) every generation produces more offspring than can survive, and (4) individuals who survive long enough to reproduce, or who reproduce the most, are those with the most favorable traits. (Traits are also inherited that are neither favorable nor unfavorable to survival.) Naturally occurring heritable traits that are favorable to survival and reproduction within populations are called **adaptations**.

To illustrate, let us use an example of a species of berry-eating birds living in an area with two food sources: berries and seeds. If bushes bore few berries during a particular season, then berry-eating birds able to crush seeds and eat them as an alternative food source would survive longer than birds that did not have seed-crushing beaks. Many non–seed-eating birds would not live to reproductive age. The seed-eating birds, on the other hand, would survive the bad berry season and likely would reproduce. Many of their progeny would have seed-crushing beaks. If another season of few berries followed, birds with seed-crushing beak adaptations would continue to have a survival advantage over other birds that were unable to live on this or another alternative food source.

Predators can also influence survival and reproduction. For example, if a new predator that fed on sand flies immigrated to a light-colored sandy beach area, then the predator would be able to see dark sand flies better than light sand flies because the dark flies would contrast with the light sand. Because the predator would eat more dark sand flies than light ones, more light than dark sand flies would survive and reproduce. Over successive generations, the population would change to consist of more light than dark flies. In this situation, light coloration has an adaptive advantage over dark.

Notice that in evolutionary contexts, the term *adaptation* is used differently than in its everyday sense or in other scientific contexts. In

an evolutionary context, adaptations are naturally occurring heritable traits, present in some members of a population of organisms, that confer a reproductive advantage to the portion of the population possessing them.

As adaptive, or reproductively advantageous, traits are passed on from surviving individuals to their offspring, the individuals carrying these traits increase in number within the population, and the *nature of the population as a whole changes*. Darwin called this process, in which organisms having adaptive traits reproduce in greater numbers than those without such traits, **natural selection**.

Change in populations of organisms therefore occurs over time because of natural selection. The environment imposes conditions that determine the results of the selection and thus the direction of change. This driving force of change—natural selection—is often referred to as *survival of the fittest*. The term *fittest* does not have the everyday meaning of the healthiest, strongest, or most intelligent. Fitness in the context of natural selection refers to **reproductive fitness**—the comparative ability of an organism to survive to reproductive age in a particular environment and to produce viable offspring, thereby passing on its genes. This idea is often called *Darwinian fitness*.

Natural selection provides a simple and direct explanation of biological diversity—why animals are different in different places. Environments differ; thus, organisms are favored by natural selection differently in different places. The nature of a population changes as more individuals are born that possess the selected traits. **Evolution** by means of natural selection is this process of change over time by which existing populations of organisms develop from ancestral forms through modification of their characteristics. It is the descent, with modification, of different lineages from common ancestors.

### CONCEPT CHECKPOINT

1. The following are observations that Darwin made or knowledge he acquired during and after his voyage on the HMS *Beagle*. How does each relate to evolution?
   a. Fossils found in a place resemble modern organisms living in that place.
   b. Geographically distinct places have similar climates and are home to similar but distinct organisms.
   c. Organisms living on oceanic islands often resemble organisms found living on a nearby mainland.
   d. Earth is much older than hundreds of thousands or a few million years.
   e. Artificial selection can be used to breed organisms with desired traits.
   f. Population growth outpaces resource availability.

## 15.4 Darwin's finches provide classic examples of natural selection.

The process of evolution by natural selection can be observed if it takes place relatively quickly—over a period of years to several thousand years, for example—resulting in the existence of multiple species that are closely related and that developed from a single ancestral species. What can be seen are clusters of these closely related species found living near one another. Such clusters of species are often found on a group of islands, in a series of lakes, or in other environments that are close to but separated from one another. Organisms living in such sharply discontinuous habitats are said to be geographically isolated from one another. A *habitat* is a place where an organism lives or grows.

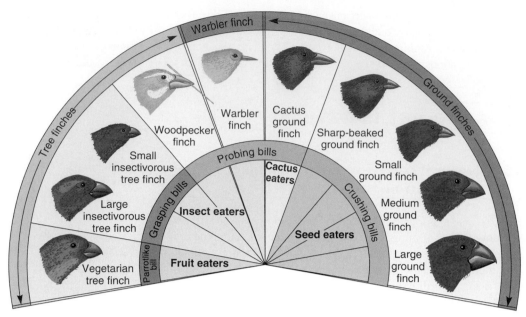

**Figure 15.7 Darwin's finches.** These are 10 species of Darwin's finches from Indefatigable Island, one of the Galapagos Islands, showing differences in bills and feeding habits. The bills of several of these species resemble those of different, distinct families of birds on the mainland. All of these birds are thought to be derived from a single common ancestor.

The Galapagos Islands are a particularly striking example of sharply discontinuous habitats, providing a natural laboratory in which to view the results of natural selection. The islands are all relatively young in geological terms (several million years) and have never been connected with the adjacent mainland of South America or with any other area. Made up of 13 major islands and some very tiny islands, the Galapagos are separated from one another by distances of up to 100 miles and are 600 miles from the South American mainland (see Figure 15.1). As a group, they exhibit diverse habitats. For example, the lowlands of the Galapagos are covered with thorn scrub. At higher elevations, attained only on the larger islands, there are moist, dense forests.

Formed by undersea volcanoes, the Galapagos Islands were uninhabited when they appeared above the surface of the water. The ancestors of all the organisms found on the Galapagos today reached these islands by crossing the sea, some on their own, some in or on the bodies of other organisms, and some blown by the wind. Only eight species of land birds reached the islands. One of these species was the finch, which fascinated Darwin. Hypothetically, the ancestor of Darwin's finches reached these islands earlier than any of the other birds. If so, all the types of habitats where birds occur on the mainland were unoccupied on the Galapagos—and the ancestral finches were able to take advantage of them all!

As the finches moved into these vacant habitats, the birds best suited to each particular habitat were selected for by nature. In other words, those birds possessing naturally occurring varia-

tions in their characteristics that were beneficial to survival in each habitat lived to reproduce. Their offspring also possessed these heritable traits. Over time, the population of finches occupying each habitat changed—the ancestral finches had split into a number of diverse populations. This phenomenon, by which a population of a species changes as it occupies different habitats within a region, is referred to as **adaptive radiation**. Some of these populations become so changed from the others that interbreeding is no longer possible: On the Galapagos, new species of finches were formed.

The evolution of Darwin's finches on the Galapagos Islands provides one of the classic examples of natural selection at work. The descendants of the original finches that reached the Galapagos Islands now occupy many different kinds of habitats on the islands (**Figure 15.7**) and are found nowhere else in the world. Among the 13 species of Darwin's finches that inhabit the Galapagos (10 are shown in Figure 15.7), there are three main groups: ground finches, tree finches, and warbler finches. The ground finches feed on seeds of different sizes. The size of their bills is related to the size of the seeds on which the birds feed. The tree finches, as their name suggests, eat insects, buds, or fruit found in the trees. Again, the size and shape of their bills are related to their food. The most unusual member of this group is the woodpecker finch. This bird carries a twig or a cactus spine, which it uses to probe for insects in deep crevices. It is an extraordinary example of a bird that uses a tool. Last, the warbler finches search continually with their slender beaks over leaves and branches for insects.

# How Science Works

## DISCOVERIES

### Natural selection is being observed today in Darwin's finches

If natural selection can be observed, you might wonder if anyone is observing and chronicling the process today. Since 1973, Princeton University researchers Rosemary and Peter Grant (**Figure 15.A**) have been doing just that—conducting a long-term study in natural selection on the Galapagos Islands. The Grants are studying a few species of Darwin's finches (see Figure 15.7) and have amassed a wealth of data regarding the evolution of beak size in these birds.

The Grants' laboratory is the island of Daphne Major (**Figure 15.B**). They chose this island because populations of birds on the island are small, there is little disturbance of the environment by humans, and environmental conditions vary greatly from year to year. Annual rainfall is erratic, resulting in major differences in the growth and characteristics of plants from year to year. For example, there was no rain in 1988, while 1987 experienced 236 inches (600 cm) of rain! During times of little rain, large, tough seeds predominate. Con-

**Figure 15.A** Rosemary and Peter Grant on-site at Daphne Major.

**Figure 15.B** Daphne Major, a partly collapsed remnant crater of a volcano and a natural laboratory in the Galapagos Islands.

versely, when rain is abundant, small, soft seeds predominate.

As they observed, tagged, and measured the birds, and documented the environment

from year to year, the Grants noted how environmental changes, especially changes in the types of seeds available, favored certain individuals within populations of birds. For example, during a severe drought in 1977, the plants withered, and seeds of all types were scarce. The bird population quickly ate all of the small, soft seeds available. With only the large, tough seeds left, birds that had beaks unable to crush these seeds died quickly. Those who survived were larger birds with larger, stronger beaks capable of crushing this usually less preferred type of seed. As the Grants observed and measured offspring the next season, they noted that the predominant progeny were large birds with large, strong beaks. Conversely, during a period of heavy rains, from 1984 to 1985, the predominant seed type was small and soft. The birds with smaller beaks that could best feed on these seeds were the ones to survive in greater numbers and produce more progeny. Over the years, the predominance of particular characteristics in the bird population shifted based on a natural selection process in which those best adapted to the environment lived to reproductive age in greater numbers and produced more offspring than those less well adapted. ●

# Was Darwin the first to suggest evolution?

## 15.5 Lamarck proposed evolutionary ideas before Darwin.

French biologist Jean-Baptiste Lamarck (1744–1829), who coined the word "biologist," is arguably the best known of the early evolutionists and was the first to present a coherent view of evolution. Lamarck's approach to evolution was embodied in his theory of the inheritance of acquired characteristics. The first part of this theory was based on the observation that changes in the environment cause changes in the needs of organisms living in that environment. Changed needs result in changed behaviors. For example, cheetahs (large African cats) eat animals such as impalas and gazelles. If the cheetahs' prey developed the ability to outrun cheetahs, then cheetahs would need to run faster to overtake their prey. These ideas are accurate. However, Lamarck inaccurately concluded that this *need* would cause cheetahs to run faster than they previously did.

The theory goes on to state that changed behaviors lead to greater or lesser use of a given structure or organ, and this may, in-

deed, be true. But Lamarck held that greater use causes the structure to increase in size; conversely, disuse causes it to shrink or even disappear. To a certain extent, these concepts are true with muscles. Nevertheless, muscles do not disappear with disuse, nor to organs such as your brain grow with use and shrink or disappear with disuse. Lamarck went on to state, erroneously, that these acquired characteristics are heritable traits and would be passed on over generations. For example, cheetahs that ran faster would develop larger, stronger muscles. Their progeny would inherit these larger muscles, and continued muscle use would increase muscle mass even more. This increase would be passed on, in turn, to their offspring.

In summary, Lamarck's theory of acquired characteristics makes three main points. First, he suggested that environmental pressures caused organisms to change their behaviors, which may be true. Second, he stated that changed behaviors led to the use or disuse of structures, which caused them to enlarge or shrink and possibly disappear. Parts of this point may be accurate in some cases. Third, he stated that such acquired characteristics would be passed on from generation to generation—they were heritable traits. As a result, all organisms changed gradually, driven by their needs in a changing environment. This third point is inaccurate.

We know today that *acquired traits are not inherited*. If that were true, the more that track stars trained for their events, the better their children would be at track! Heritable traits are those encoded in genes, whether the trait be a large, strong beak that crushes hard seeds, a colorful display of feathers that attracts a mate, or an outer capsule that helps a bacterium invade the body and cause disease. The driving force of evolution *is* environmental change, but evolution *is not* driven by needs caused by environmental change, as Lamarck suggested. Instead, it is driven by the selection of organisms already possessing heritable adaptive traits—those traits that make organisms better suited to the environment than others with less adaptive traits. Adaptive traits confer reproductive advantage; that is, adaptive traits help organisms live to reproductive age and generate progeny. Those progeny reflect the genetic composition of their parents. Thus the genetic composition of the population—geneticists call it the *gene pool*—will begin to shift from genes that code for less adaptive heritable traits to genes that code for more adaptive heritable traits.

Although Lamarck's theory of the inheritance of acquired characteristics is known to be false today, his ideas were reasonable hypotheses at the time because no one understood the mechanisms of heredity. Lamarck held yet other notions no longer held today. For example, he viewed evolution as a process of increasing complexity and perfection. Today scientists recognize that evolution is a process driven by natural selection, a process that favors alleles that improve the relative reproductive success of their carriers. Thus, evolution does not move toward increased complexity and perfection, but toward increased relative reproductive success. During this process, adaptive traits are perpetuated within populations, and nonadaptive traits are generally eliminated. So too, as environments change, traits that were once adaptive may become nonadaptive and vice versa.

## 15.6 Darwin published his theory in 1859.

Darwin drafted the overall argument for evolution by natural selection in 1842 and continued to refine it for many years. The stimulus that finally brought it into print was an essay that he received in 1858 from a young English naturalist, Alfred Russel Wallace (1823–1913; **Figure 15.8**). Wallace sent the essay to Darwin from Malaysia; it concisely set forth the theory of evolution by means of natural selection. Like Darwin, Wallace had been influenced greatly in his development of this theory by reading Malthus's 1798 essay. After Darwin received Wallace's essay, a joint scientific presentation of their ideas at a seminar in London was arranged in 1858. Darwin then proceeded to complete his own book, on which he had been working for some time, and submitted it for publication in what he considered an abbreviated version.

Darwin's book, *On the Origin of Species by Means of Natural Selection*, appeared in November 1859 and caused an immediate sensation. Some called the book "glorious" and were in complete agreement with Darwin's theories. Others criticized the book, admiring some parts and asserting that other parts were false. Still others attacked Darwin on religious grounds. Many of the clergy, however, openly agreed with Darwin, saying that the theory of evolution did not deny the existence of God.

**Figure 15.8** Alfred Russel Wallace. Darwin and Wallace jointly presented the theory of evolution by natural selection at a meeting in London in 1858.

One of the many debates that raged within the scientific community at the time Darwin published his work regarded the mechanism of evolutionary change—natural selection. As mentioned earlier, no one at this time had any concept of genes or of how heredity works, and so it was impossible for Darwin to explain completely how evolution occurs. Gregor Mendel (see Chapter 13) had begun his groundbreaking work in the study of inheritance but did not publish it until 1866. In fact, Mendel's work was not well understood, and the science of genetics was not established until the beginning of the twentieth century—40 years after the first publication of *Origin*.

Within 20 years or so after the publication of *Origin*, however, the concept that species have changed over time was well accepted. More than a century has elapsed since Charles Darwin's death in 1882. During this period, the evidence supporting his theory has grown progressively stronger. Evolution is a theory that has been upheld countless times as it is tested and retested. Scientists consider evolution, like the Earth traveling around the Sun, to be a fact. Scientists are continuously learning more, however, about the complexities of the mechanisms of evolution and the history of the evolution of life on Earth, topics that will be discussed in more detail in Chapters 16–18.

## What other lines of evidence support an evolutionary view?

## 15.7 Fossils provide a record of the past.

**Fossils** are preserved remains, impressions, or other traces of organisms that lived in the past. Fossils may be nearly complete impressions of organisms or merely burrows, tracks, molecules, or other traces of their existence.

What types of fossils are most abundant? Fossils of the hard parts of organisms are the type most often found (**Figure 15.9**), rather than fossils formed from soft body parts, which usually decay quickly and leave no trace. Sometimes, however, soft-bodied animals are preserved in exceptionally fine-grained mud, in conditions in which the supply of oxygen was poor while the mud was

**Figure 15.9** A fossil skeleton. This paleontologist is in his lab with the hind foot of a Colombian mammoth on his workbench. This specimen is an excellent example of a fossil formed from the hard parts of an organism.

being deposited, thus slowing the decomposition of the organism. Eventually, the soft parts of the organism decay completely, leaving behind a mold, or impression, of its body. Molds may become filled with minerals, such as lime or silica found in underground water, forming casts that resemble the original organism or body part

(**Figure 15.10**). In general, however, fossils of soft-bodied organisms like worms are rare. Even though soft-bodied animals evolved before their hard-bodied counterparts, there is comparatively little evidence of their history in the **fossil record**—the history of life as preserved in fossils. Only a minute fraction of the organisms living at any one time are preserved as fossils.

Fossils create an ordered record of life because they appear in an ordered arrangement within the Earth's crust. Fossils are found in rock in the order of their existence because most fossils are preserved in sedimentary rocks, which are laid down layer upon layer over time. In general, the oldest layers of rock are at the bottom and the youngest are at the top. How does this occur? Sedimentary rocks are made up of particles of other rocks, cast off as they weather and disintegrate. Running water, such as a river or stream, picks up these pieces of rock and carries them to lakes or oceans where they are deposited as sediment, better known as mud, sand, or gravel. Over time, the sediment hardens into rock. While some sediment is hardening, other sediment is still being deposited, thus creating layers, or strata, of rock formed one on top of the other. As a result of this process, most sedimentary rock has a stratified (layered) appearance, such as that seen in the photo of the Grand Canyon in **Figure 15.11**.

Using their understanding of the formation of sedimentary rock, scientists can determine the age of fossils and use this information to establish the broad patterns of the progression of life on Earth. This process is called *relative dating* because it yields information about the ages of rocks and fossils in comparison to one another, but it does not reveal their exact ages. In other words, it reveals an order of events without dates.

Is relative dating accurate? Some of the strongest evidence of the accuracy of this system comes from the mineral and petroleum industries. Geologists in these industries use fossils to determine where to mine and drill. If, for example, a petroleum company finds oil in a particular rock stratum in one part of the world, it looks for oil in that same stratum in another location. This method is much more successful in finding oil than is random drilling.

To determine relative dating of rock strata, scientists who study fossils (paleontologists) compare the sequences in which fossils appear in the different layers of sedimentary rock. Fossils found in the same strata (layer) are assumed to be of the same age. Since

**Figure 15.10** A fossil "mold." Shown here is a fossil crinoid, an echinoderm having a small, cup-shaped body with branched arms attached to the body by a stalk. This fossil illustrates the preservation of soft-bodied fossils in fine-grained sedimentary rocks.

**Figure 15.11** The Grand Canyon. Note the stratified appearance of the sedimentary rock making up the walls of the canyon. The Grand Canyon exhibits a wide variety of clearly exposed layers of rock.

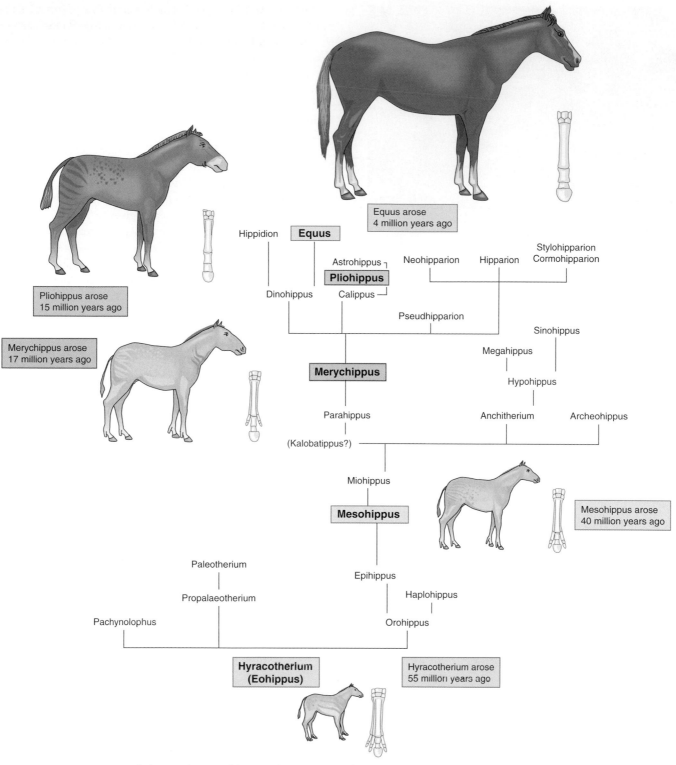

**Figure 15.12 Reconstruction of the evolution of horses based on fossils.** The phylogenetic tree of horses is an excellent example of how the evolution of a particular vertebrate group can be traced when fossils are abundant. The tree shows the relative dating of the fossil forms: The most recent forms are at the top of the tree (*Equus* is the modern horse); the most ancient forms are at the bottom. Fossils of forms at the same horizontal level were found in the same rock strata. The absolute ages of the pictured forms were established by dating the age of the rock layers or by dating the fossils themselves. The relative sizes of the pictured forms are accurate. Note that the ancestral tree of the horse is not linear, like a ladder, but instead branches like a bush.

 **Visual Thinking:** Which organism is the most recent common ancestor to Dinohippus, Calippus, and Pseudhipparion?

the youngest layers of rock are at the top and the oldest are at the bottom, the fossils in upper layers mostly represent more recent species than the fossils in lower layers. From this information and other data, paleontologists can construct ancestral trees, called *phylogenetic trees* or *evolutionary trees*, which show the history of an organism over time (**Figure 15.12**). These trees are discussed in more detail in Section 15.12.

By piecing together various types of information when examining rocks and fossils, paleontologists can develop an accurate understanding of where past organisms lived and gain some appreciation of the environments in which they lived, in addition to learning when they lived. For example, limestone that contains corals would have been deposited in a marine environment. Oak leaves found in sandstone suggest that a location was once a land mass. As you can see from these examples, fossils and the rock in which they are embedded provide information about the history of an area, which gives scientists clues to the location of continents, ponds, lakes, and oceans and how their location has changed over time. Thus, by preserving past life in their layers, fossil-containing sedimentary rocks provide some of the strongest evidence of evolution.

 CONCEPT CHECKPOINT

2. Some argue that evolution is "just a theory," implying that it is just a hunch or guess. What might you say to explain the difference between a scientific theory and a guess?

## 15.8 Fossil dating shows that time existed for organisms to develop from ancestral forms.

Direct methods of dating fossils first became available in the late 1940s. These methods use naturally occurring isotopes of certain elements found in rocks or fossils to determine their ages. This dating process is called *absolute dating* or *radiometric dating*. Although radiometric dating does have sources of error, it has yielded a great deal of consistent data. In addition, the relative ages of the rocks correlate well with the absolute ages determined by the use of isotopes. Scientists widely accept radiometric dating as a reliable way to determine the exact ages of rock strata.

Many different isotopes are used in radiometric dating. Isotopes are atoms of an element that have the same number of protons but different numbers of neutrons in their nuclei (see Chapter 3). They therefore differ from one another in their atomic numbers. *Radioactive isotopes* are unstable; their nuclei decay, or break apart, at a steady rate, producing other isotopes and emitting energy. After decay, some radioactive isotopes may give rise to other elements.

Some methods of radiometric dating give scientists information about the age of rocks; others measure the length of time since the death of an organism. One method of dating organisms that died less than 50,000 years ago is called *carbon dating*, which estimates the relative amount of the different isotopes of carbon present in a fossil (or other organic material). The process of carbon dating is described in the *Just Wondering* box.

With the use of radiometric dating methods, knowledge of the ages of various rocks has become more precise. The oldest rocks on Earth that have been dated (using isotopes with half-lifes much longer than carbon) include rocks from South Africa, southwestern Greenland, and Minnesota, which are approximately 3.9 billion years old. Meteorites have been dated at about 4.6 billion years old. Rocks brought back to Earth from the moon have been dated from 3.3 to 4.6 billion years old. These pieces of evidence suggest that the Earth and the moon, which were most likely formed from the same processes at the same time, are about 4.6 billion years old.

What significance does the age of the Earth hold for the science of evolution? The accumulation of adaptations and the development of new species usually take thousands, if not millions, of years. Until Darwin's time, most people, including most scientists, thought that the Earth was approximately 6000 to 10,000 years old. This would not allow enough time for the process of evolution to take place. Not until the present techniques of radiometric dating were developed could scientists begin to solve the time problem of evolution and accurately measure the age of the Earth and its fossils.

## *just wondering . . .*

*Questions students ask*

### How does carbon dating work?

The carbon that is incorporated into the bodies of living organisms consists of the same fixed proportion of $^{14}C$ and $^{12}C$ that occurs in the atmosphere. After an organism dies, however, and is no longer incorporating carbon, the $^{14}C$ in it gradually decays to nitrogen because it is a radioactive isotope. It takes 5,730 years for half of the $^{14}C$ present in a sample to be converted by this process; this length of time is called the *half-life* of the $^{14}C$ isotope. By measuring the amount of $^{14}C$ in a fossil, scientists can estimate the proportion of the $^{14}C$ to all other carbon that is still present and compare that with the ratio of these isotopes as they occur in the atmosphere. In this way, scientists can then estimate the length of time over which the $^{14}C$ has been decaying, which is the same as the length of time since the organism died.

For fossils older than 50,000 years, the amount of $^{14}C$ remaining is so small that it is not possible to measure it precisely enough to provide accurate estimates of age. Rocks in which older fossils are embedded, or rocks in the same relative position as older fossils, may be dated by using other radioactive isotopes, such as the isotope thorium-230, which has a half-life of 75,000 years, or the isotope lead-206, which has a half-life of 4.5 billion years. ●

*Are you wondering about a topic in biology and how it relates to your life? Submit your question by clicking the Just Wondering link in this text's companion Web site at www.wiley.com/college/alters.*

## 15.9 Comparative anatomy shows relatedness among organisms.

Comparative studies of animal anatomy provide strong evidence for evolution. During Darwin's time, scientists began looking for evolutionary relationships in the anatomical structures of organisms. If derived from the same ancestor, organisms should possess similar structures, with modifications reflecting adaptations to their environments. Such relationships have been shown most clearly in vertebrate animals.

Within the subphylum Vertebrata, the classes of organisms, such as birds, mammals, and amphibians, have the same basic anatomical plan of groups of bones (as well as nerves, muscles, and other organs and systems), but these bones are put to different uses among the classes. For example, the forelimbs seen in **Figure 15.13** are all constructed from the same basic array of bones, modified in one way in the wing of a bat, in another way in the leg of a horse, and in yet another way in the arm of a human. The bones are said to be **homologous** in the different vertebrates—that is, having the same embryological origin and the same underlying anatomy but now differing in function. These vertebrates all use the same bones in the same relative positions derived from the same embryological tissues to do different jobs.

In some cases, homologous structures exist among organisms, but they are no longer useful. These structures have diminished in size over time. **Figure 15.14** shows the tiny leg bones of the python, which no longer serve a purpose. Such structures that are present in an organism in a diminished size but are no longer useful are called *vestigial* structures.

Organisms that share a way of life often have similar structures with similar functions, but in contrast to homologous structures, these structures do *not* have the same embryological origin and do

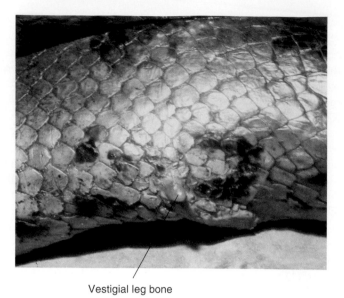

Vestigial leg bone

**Figure 15.14** A vestigial structure in a python. Pythons possess tiny leg bones that serve no purpose in locomotion.

*not* share the same underlying anatomy. Such body parts are called **analogous** structures.

The wings of birds and insects, which are similar in shape and function, are analogous structures. Plants also show analogous structures. For example, three different families of flowering plants—cacti, euphorbia, and milkweeds—have all developed thick, barrellike, fleshy stems that store water, which is an adaptive characteristic in a desert environment. These plants look so much alike that the casual North American observer might think they were all cacti. However, they evolved independently of one another in different parts of the world (southwest North America, Africa, and the Mediterranean, respectively). **Figure 15.15** shows the striking similarity between cacti and euphorbia.

The presence of analogous structures shows how in similar habitats, natural selection can lead to similar but not identical anatomical structures. Change over time among different species of organisms having different ancestors that results in similar structures and adaptations is called **convergent evolution**.

## 15.10 Comparative embryology shows relatedness among organisms.

Embryologists—scientists who study the development of organisms from conception to birth—noticed as early as the nineteenth century (around the time of Darwin) that various groups of organisms, though different as adults, possessed early developmental stages that were quite similar. For example, the embryological development of all vertebrate animals is similar in that all vertebrate embryos have gill arches, seen as pouches below the head. Only fish, however, actually develop gills; the pouches develop into different structures in other animals. Similarly, the embryological development of the backbone is similar in vertebrates, but some organisms develop a tail, while others such as humans do not. The similarities among embryos are shown in **Figure 15.16**. Tailbones of humans—the fused coccyx bones at the end of the spine—are vestigial structures.

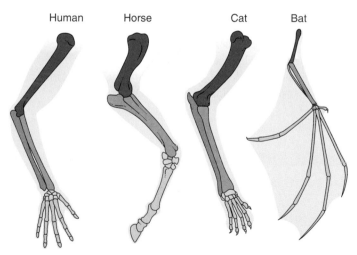

Human    Horse    Cat    Bat

**Figure 15.13** Homology among vertebrate limbs. Homologies among the forelimbs of four mammals show the ways that the proportions of the bones have changed in relation to the particular way of life of the organism, but that all the forelimbs have the same basic bone structure.

**Visual Thinking:** As shown here, the wing of a bat is homologous to the arm and hand of a human. Is the wing of a bat homologous to the wing of an insect? Why or why not?

(a)　　　　　　　　　　　　　　(b)

**Figure 15.15** An example of convergent evolution. (a) North American cactus. (b) *Euphorbiaceae*, found in Africa.

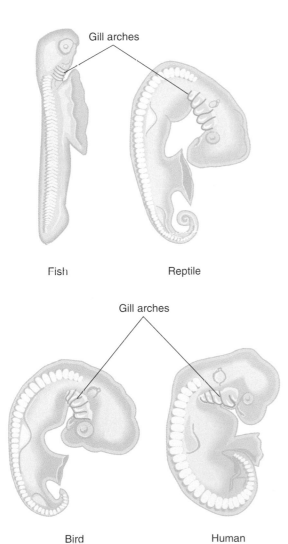

**Figure 15.16** The embryos of various groups of vertebrate animals. These embryos show the primitive features that all vertebrate animals share early in development, such as gill arches and a tail.

These similar developmental forms tell scientists that similar genes are at work during the early developmental stages of related organisms. The genes active during development have been passed on to distantly related organisms from a common ancestor. Over time, new genetic instructions are added to the old. Old and new instructions are expressed at different times during development, and sometimes old instructions are deleted. Evolution acts on all stages of an organism, so there is no one universal stage through which all vertebrates pass. Nevertheless, similarities certainly exist. The similar embryos in Figure 15.16 develop into adult organisms that are quite different from one another.

## 15.11 Patterns of change in the molecular record provide strong, direct evidence for evolution.

Today, using the tools of molecular biology gives scientists additional evidence for evolution and new insights into the evolutionary relationships among organisms. One way molecular biologists study the evolution of organisms is by looking at their hereditary material (DNA). Some evolutionary change involves the formation of new alleles from the old by mutation; favorable new alleles persist because of natural selection. Similarly, unfavorable alleles are "weeded out" from the population in that they confer a reproductive disadvantage. Significant evolutionary change also takes place by natural selection acting on variation created by genetic (sexual) recombination alone—that is, without the introduction of new alleles by mutation. Genetic recombination occurs during crossing over in meiosis (see Chapter 12).

During the process of natural selection, therefore, changes in species involve an accumulation of genetic change in the DNA. Organisms that are more distantly related will have accumulated a greater number of changes in their DNA than organisms that more recently evolved from a common ancestor. Put simply, species with more similar DNA are more closely related; species with less similar DNA are less closely related.

Changes in DNA also result from the accumulation of neutral mutations—those that confer neither a reproductive advantage nor a reproductive disadvantage. Neutral mutations accumulate at a constant rate. Evidence indicates that most of the divergence in the DNA of organisms is due to the accumulation of neutral mutations, not to the accumulation of new, adaptively superior mutant alleles. Thus, the overall difference between the DNA of two organisms is a measure of the time since they diverged (had a common ancestor) but may *not* reflect the degree to which they have diverged morpho-

logically. A good example of this is gorillas, chimpanzees, and humans. Chimpanzees and gorillas are very similar in their morphology, skeletal structures, dentition, and to a lesser extent behaviors, but they are very different in all of these ways from humans. Nonetheless, the DNA of humans and chimpanzees are more similar (have more gene sequences in common) than either is to the DNA of gorillas. This is why it is thought that chimpanzees and humans diverged from a common ancestor 6 to 8 million years ago, whereas gorillas diverged some 12 to 15 million years ago.

# How do scientists depict evolutionary relationships?

## 15.12 Scientists construct phylogenetic trees that show patterns of relationships among organisms.

As you can see from this chapter, various lines of inquiry help scientists determine the evolutionary relationships among organisms. In the mid-1800s, Ernst Haeckel, an artist and a scientist, pictorially represented the taxonomic groups of organisms as a tree that depicted these groups' evolutionary relationships (**Figure 15.17**). Scientists still use this technique today, but today's evolutionary trees are structured more like graphs, not modeled on actual trees.

The evolutionary history of an organism is called its **phylogeny**, which comes from Greek words meaning the origin (*-geny*) of a race or tribe (*phylo-*). Therefore, a depiction of the phylogeny of organisms—their pattern of evolutionary relationships—is called a **phylogenetic tree** (as we saw in Figure 15.12). Many phylogenetic trees place information about the pattern of relationships among organisms on the horizontal axis and information about the degree of genetic divergence or time on the vertical axis. In this orientation, the origin of the tree is at the base.

Phylogenetic trees have three main parts: *tips, branches,* and *nodes*. The tips, branches, and nodes can represent populations of any **taxon** (pl. *taxa*), or named group of organisms, such as a phylum, family, or order. **Figure 15.18** shows these three main tree parts. It represents two ways to show the same information. In fact, this information can be shown in a variety of ways. However, once you understand how to read phylogenetic trees using these models, you will be able to read other trees.

The branches are similar to branches on a tree; they represent the evolution of populations over time. The tips of the branches represent either extinct taxa or taxa living today. In Figure 15.18, all the branch tips reach the top of the time line, meaning that they are present-day taxa. If tips end prior to the top of the time line, which is prior to the present day, then they represent taxa that died out (became

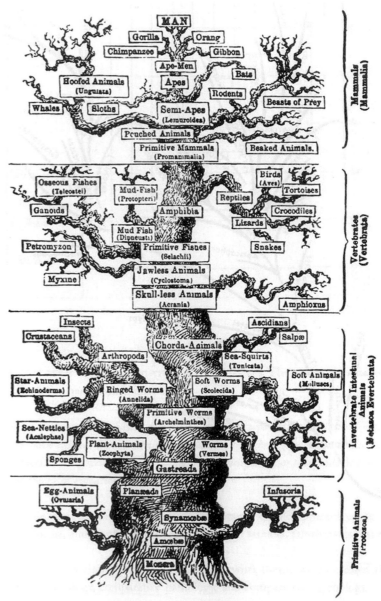

**Figure 15.17** An evolutionary tree of Ernst Haeckel's time.  Ernst Haeckel is thought to have been the first to describe evolutionary relationships among living organisms as analogous to a tree. Haeckel also postulated a common origin for all life, as shown by the common trunk at the base of the tree.

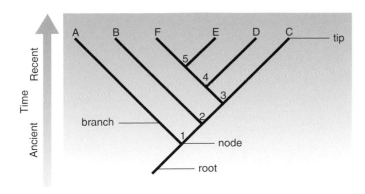

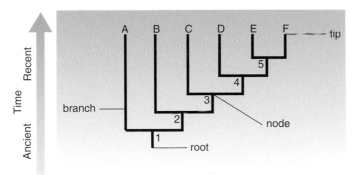

**Figure 15.18** Two phylogenetic trees showing the same information in different ways. Phylogenetic trees show evolutionary relationships among organisms. These vertical trees are read from bottom to top, but they could be oriented horizontally and read from left to right.

extinct). The nodes are the places where one branch diverges from another. They represent points where one population splits into two or more descendant populations. Notice that each of these trees has a root that identifies the origin, that is, the population ancestral to all the taxa on the tree.

To read the trees in Figure 15.18, start at the bottom and read up. The population at node 1 is ancestral to taxa A–E. At node 1, this population split into two branches, which represent two populations. One group evolved into taxon A, and the other evolved into the population at node 2, which is ancestral to taxa B, C, D, E, and F. The node 2 population then split into two populations. One group evolved into taxon B, and the other evolved into the population at node 3, which is ancestral to taxa C, D, E, and F. Continue in like manner to finish reading the tree.

Scientists use a wide variety of data to construct phylogenetic trees. Some scientists may use only a single set of data and a particular type of data. For example, the phylogenetic tree in **Figure 15.19** is constructed using comparisons between the sequences of genes in present-day organisms to make inferences about the evolutionary history of these organisms. Organisms are positioned on this tree on the basis of the differences in the nucleotide base sequence of the gene that codes for an enzyme called *cytochrome C oxidase*. Underlying the development of the tree is the theory that these differences exist because mutations in the bases accumulate independently over time in each evolving species after the split from a common ancestor. The higher the number of differences among the bases, the greater the number of mutations and therefore the longer ago in time these organisms had a common ancestor (the more distantly they are related). The

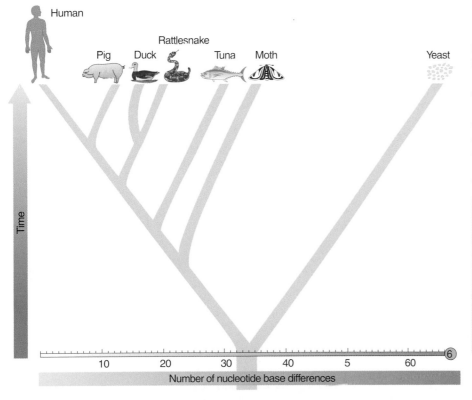

**Figure 15.19** A phylogenetic tree constructed using information from molecular biology. All of the organisms shown in this illustration have an enzyme called cytochrome C oxidase, which helps transfer electrons in the electron transfer system (see Chapter 8). The genes that code for this enzyme in these organisms are not the same, owing to changes over time as organisms diverged from one another. This phylogenetic tree shows the nucleotide base differences in these genes between humans and other organisms.

**Visual Thinking:** Which two organisms shown in the tree appear to be the most distantly related based on their differences in cytochrome C oxidase? the most closely related? Explain your answer. To which organism shown on the tree are humans most closely related? Explain your answer.

fewer the number of nucleotide base differences, the more recently these organisms had a common ancestor (the more closely they are related).

Phylogenetic trees constructed from analyses of molecular differences are similar to those built by use of anatomical studies. For example, both approaches show whales, dolphins, and porpoises clustering together, as do the primates and the hoofed animals.

 **CONCEPT CHECKPOINT**

**3.** Describe and provide an example of how each of the following lines of evidence support evolution: molecular evidence, embryological evidence, homologies, analogies, fossil evidence, age of the Earth.

# CHAPTER REVIEW

## Summary of Key Concepts and Key Terms

### How was the theory of evolution by natural selection developed?

**15.1** While studying **species** (p. 234) on oceanic islands, Darwin accumulated a wealth of evidence that organisms have changed over time.

**15.2** Other factors that influenced Darwin's thinking were geological evidence that the Earth had changed over time, the changes in organisms that breeders of domesticated animals were able to attain by using artificial breeding methods, and Malthus's ideas on population dynamics.

### What is natural selection?

**15.3** Darwin proposed that evolution occurs as a result of **natural selection** (p. 237).

**15.3** The mechanism of natural selection works in the following manner: Some individuals have naturally occurring heritable traits (**adaptations**, p. 237) that make them better suited to a particular environment, allowing more of them to survive to reproductive age and produce more offspring than individuals lacking these traits.

**15.3** **Reproductive fitness** (p. 237) is a measure of the tendency of some organisms to leave more offspring than competing members of the same population.

**15.3** Genetic traits possessed by fitter individuals will appear in greater proportions among members of succeeding generations.

**15.3** Traits allowing greater reproduction will increase in frequency over time.

**15.3** The environment imposes conditions that determine the direction of selection and thus the direction of change.

**15.3** **Evolution** (p. 237) is the descent, with modification, of different lineages from common ancestors.

**15.4** Isolated populations subjected to unique combinations of selective pressures (the conditions imposed by nature) diverge biologically from one another and may ultimately become so different that they are distinct species; this process is called **adaptive radiation** (p. 238).

### Was Darwin the first to suggest evolution?

**15.5** Jean-Baptiste Lamarck, who predated Darwin, was the first to present a coherent view of evolution; his theory of the inheritance of acquired characteristics has been shown to be false; acquired traits are not inherited.

**15.6** Alfred Russel Wallace sent an essay to Charles Darwin that set forth a theory of evolution by means of natural selection similar to Darwin's own theory; the two men presented the theory jointly in 1858.

**15.6** Darwin published his theory in 1859 in a book titled *On the Origin of Species by Means of Natural Selection*.

**15.6** A wealth of evidence since Darwin's time has supported his proposals that evolution occurs and that natural selection is the method that produces adaptation.

### What other lines of evidence support an evolutionary view?

**15.7** The **fossil record** (p. 241), which lets us see how change is correlated with age, provides direct evidence that upholds the science of evolution.

**15.8** By measuring the amounts of certain radioactive isotopes found within rocks and **fossils** (p. 240), scientists can determine the age of fossils and use this information to establish patterns of life's progression.

**15.9** Comparative studies of animal anatomy show that many organisms have **homologous** (p. 244) structures: groups of bones, nerves, muscles, and organs with the same underlying anatomical plan and embryological origin but with different forms and functions, such as the wing of a bat and the arm of a human.

**15.9** Homologous structures provide evidence of evolutionary relatedness.

**15.9** **Analogous** (p. 244) structures are often present within organisms that share a way of life and are similar in form and function, but these structures do not have the same embryological origin and do not share the same underlying anatomy.

**15.9** Analogous structures provide evidence for natural selection and **convergent evolution** (p. 244).

**15.10** Comparative embryological studies show that many organisms have early developmental stages that are similar, which provides evidence of evolutionary relatedness.

**15.11** The pattern of change seen in the molecular record is also evidence for evolution and provides strong, direct evidence for change over time.

**15.11** Molecular biologists study the evolution of organisms by comparing their hereditary material and quantifying the differences.

**15.11** The molecular record exhibits accumulated changes among organisms; the amount of change correlates with age as determined in the fossil record.

## How do scientists depict evolutionary relationships?

**15.12** The evolutionary history of an organism is called its **phylogeny** (p. 246).

**15.12** The depiction of the phylogeny of organisms is called a **phylogenetic tree** (p. 246).

**15.12** Phylogenetic trees depict patterns of relationships among major named groups of organisms called taxa (**taxon**, p. 246).

**15.12** Scientists who develop phylogenetic trees use a wide variety of data to construct trees.

## Level 1    Learning Basic Facts and Terms

**Multiple Choice**

1. What is meant by the term *Darwinian fitness*?
   a. Physical strength
   b. The ability to survive and reproduce
   c. A strong, healthy appearance
   d. High aggressiveness
   e. All of the above

2. Which of the following is true of adaptations that contribute to the evolution of a population?
   a. They increase an organism's evolutionary fitness.
   b. They are heritable.
   c. They occur naturally.
   d. They are encoded in an individual's genes.
   e. All of the above.

3. Which of the following is an acceptable definition of evolution?
   a. A change in the physical characteristics of an individual
   b. A change in the genetic makeup of an individual due to a mutation
   c. A change in environmental conditions
   d. A change over time by which existing populations develop from ancestral forms through modification of their characteristics
   e. All of the above.

4. Evolution is a biological theme that ties all others together. This is because the process of evolution
   a. explains how organisms become adapted to their environment.
   b. explains the diversity of organisms.
   c. explains why organisms have common characteristics.

   d. explains why distantly related organisms sometimes resemble one another.
   e. All of the above

5. Which of the following factors can affect the evolution of a population?
   a. Climate
   b. Mutations
   c. Food availability
   d. Introduction of a new predator
   e. All of the above

**Fill in the Blank**

6. Two species live in a very similar environment but on different continents. Although these two species are not at all closely related, they may appear quite similar as a result of (convergent/divergent) _____ evolution.

7. Ichthyosaur was an aquatic dinosaur. Fossils show that it had a dorsal fin and a tail just like fish do, even though its closest relatives were terrestrial reptiles that had neither dorsal fins nor aquatic tails. The dorsal fins and tails of ichthyosaurs (compared to the fins and tail of fishes) are (analogous/homologous) _____.

8. The emergence of many new species, all from a common ancestor, is an example of (adaptive radiation/reproductive fitness) _____.

9. Lamarck proposed the inheritance of (acquired/inherited) _____ characteristics.

## Level 2    Learning Concepts

1. What is wrong with the statement, "species are perfectly adapted to their environment"?

2. A dachshund and a Siberian husky are both dogs, but they look very different from each other. By what process did the two breeds come to look so different? On what observable genetic principle is this process based?

3. Evolution by natural selection holds the status of a scientific theory because diverse lines of evidence support it. Briefly summarize those lines of evidence.

4. What evolutionary concepts described below are exemplified by the Tasmanian Wolf, a marsupial from Australia that is similar in body form to the North American wolf?
   a. The body forms of these two mammals evolved independently of one another.

   b. The forearms of these two species have identical skeletal arrangements, as did an ancient common ancestor.

5. How would Mendel's insights have helped Darwin develop and argue his theory of evolution by natural selection?

6. Explain how scientists can use the $^{14}C$ method to date a fossil.

7. How old is the Earth according to radioactive dating methods? Why is the age of the Earth significant with respect to evolution?

8. What are vestigial structures? Give an example.

9. Explain how studies of comparative anatomy and comparative embryology support evolutionary science.

## Critical Thinking and Life Applications

1. If a population were composed of only a few organisms, what effects might natural selection have on its continued survival? Would a larger population have a better chance of survival? Why or why not?
2. Humans shape their environment in ways that other organisms cannot. Are humans subject to the same pressures of natural selection as other organisms? Why or why not?
3. Organisms are often referred to as "progressing up the ladder of evolution." Can evolution be accurately compared to a ladder? Can some organisms accurately be referred to as being "higher" on the ladder than others? Why or why not?
4. One problem that we are facing worldwide today is the appearance of antibiotic-resistant strains of bacteria. Diseases that were once easily treated with antibiotics are becoming difficult or impossible to treat. Using your understanding of evolutionary processes from this chapter, explain how bacteria can develop resistance to antibiotics.
5. Consider the following ads from matchmaking.com. Which of the following love seekers has the greatest "fitness" among this population in a strictly evolutionary sense? Support your argument.
   - Single white male (SWM) age 30–35 who is interested in lifting weights, working out, and a childless life (and has done something about it), seeks a gorgeous SWF who knows what a man really needs.
   - Feisty DWF age 65 seeking an elderly gentlemen with lots of money to keep me happy and comfortable in my retirement.
   - Tall, handsome, and rich DBM, two children, seeking an attractive, kind, amorous SBF who is willing to sign a prenuptial agreement and have more babies.
   - Attractive SBF, never been married, cannot have kids, nonsmoker, age 25–30, seeking a tall, good-looking SWM (my knight in shining armor).
6. Some moles have eyes that are so reduced in size that they function only minimally in light perception. Develop a plausible argument for why structures like these might become minimized to the point of being vestigial (diminished to the point of having little or no real function).
7. Oxford University scientist Richard Dawkins' book on the theories behind evolution is entitled *The Blind Watchmaker*. What do you suppose is the significance of this rather ironic title?

## In The News | Critical Thinking

**FOSSIL FRAUD**

Now that you understand more about fossils, reread this chapter's opening story about detecting fossil frauds. To better understand this information, it may help you to follow these steps:

1. Review your immediate reaction to the detection of fossil frauds that you wrote when you began reading this chapter.
2. Based on your current understanding, again summarize the main point of the research in a sentence or two.
3. What questions do you now have about this information that this chapter's opening story does not answer?
4. Collect new information about the research. Visit the *In The News* section of this text's companion Web site at www.wiley.com/college/alters and watch the "Fossil Frauds" video. Then use the "summary" link to read the accompanying story and access related links. Use this information, the links provided, and other online and library resources to answer your questions and find updates about this topic. State the sources of your information. Explain why you think the information is accurate. Also determine whether the information expresses a particular point of view or is biased in any way.
5. What in your view are the most significant aspects of this information? Give reasons for your opinion and for any changes in your ideas based on the additional information you have collected and the analysis you have done.

# BEYOND DARWIN: A GENETIC BASIS OF EVOLUTION

## In The News | Snake Origins

Scientists know that snakes evolved from ancient lizards. But lizards have legs and snakes do not. How did snakes lose their legs? Scientists are using genetics to find answers to this evolutionary question.

Two hypotheses have been advanced to explain why snakes have no legs. One hypothesis is that snakes evolved on land and "lost" their legs because having small legs or no legs was adaptive for burrowing creatures; having large legs was not. Snakes with large legs could not burrow as well or as quickly as those with smaller or no legs to escape heat or predators. Thus, natural selection "weeded out" snakes with legs; snakes with legs tended to die before they reproduced. Conversely, snakes with small or no legs survived, reproduced, and passed on the alleles for this trait to their offspring.

The other hypothesis is that snakes evolved in the water and lost their legs there. If this were the case, snakes would be closely related to extinct marine lizards called mosasaurs. The closest living relatives of mosasaurs are the monitor lizards, such as the Komodo dragon.

For years, scientists favored the land hypothesis. However, in 1997, researchers described several marine snake fossils with tiny hind limbs. These fossils suggested that snakes might indeed have had an aquatic origin. And that is where the genetic sleuthing begins.

Blair Hedges, a professor of biology at The Pennsylvania State University, and Nicolas Vidal, a postdoctoral student, collected DNA from all living families of lizards and most living families of snakes. They analyzed the DNA and developed an evolutionary tree from these analyses. The tree clearly shows that snakes are not closely related to monitor lizards and therefore did not arise, or lose their legs, in the sea. Hedges and Vidal's genetic data support the land hypothesis for the origin of snakes and where they lost their legs. You can learn more about this research and view the video "Snake Origins" by visiting the *In The News* section of this text's companion Web site at www.wiley.com/college/alters.

*Write your immediate reaction to Hedges and Vidal's research on the origin of snakes: first, summarize the main point in a sentence or two; then suggest what you think its significance is. You will have an opportunity to reflect on your responses and gather more information on this topic in the In The News feature at the end of this chapter. In this chapter, you will learn more about the genetic basis of evolution.*

## Is evolution still accepted in the form Darwin proposed?

**16.1** Today's view of evolution includes a genetic basis and additional mechanisms of change.

## How does evolution occur?

**16.2** Genetic variation is the raw material of evolution.

**16.3** Natural selection acts on phenotypes, affecting genotypes of populations.

**16.4** Natural selection affects populations in three major ways.

**16.5** Genetic variation arises by random mutation and recombination.

## What factors contribute to changes in gene frequencies within populations?

**16.6** Random genetic drift, gene flow, and nonrandom mating change gene frequencies within populations.

**16.7** Populations are rarely in genetic equilibrium.

# Is evolution still accepted in the form Darwin proposed?

## 16.1 Today's view of evolution includes a genetic basis and additional mechanisms of change.

Within 20 years or so after the 1859 publication of *On the Origin of Species by Means of Natural Selection*, the scientific community had accepted the theory of evolution proposed in that work by Charles Darwin (**Figure 16.1**): that organisms are related and have descended, with modification, from common ancestors. Today's scientific literature does not challenge the idea that life has evolved and considers as fact the concept that species have changed over time.

Nevertheless, scientists' ideas about the mechanism of evolution have changed over the years, especially in the context of genetics. As mentioned in Chapter 15, no one during Darwin's time had an understanding of genes or of how heredity works, so it was impossible for Darwin to explain completely how evolution occurs. Although Gregor Mendel published his work on the study of inheritance in 1866, his work was not well understood and the science of genetics was not established until the beginning of the twentieth century, well after Darwin's death in 1882.

Darwin's theory of evolution by natural selection is also called *Darwinism*, or **Darwinian evolution**. Darwin's theory comprises more than one concept; the main tenets of Darwinism are summarized in **Table 16.1**. In the early part of the twentieth century, a genetics-based view of natural selection was developed, which was called *neo-Darwinism*, or the **neo-Darwinian theory of evolution**. "Neo-" is a prefix from the Greek meaning "new" or "modern." In neo-Darwinism, concepts from the emerging science of population genetics were used to extend Darwin's ideas. **Population genetics** is the science of the frequency and distribution of genes in populations of organisms. A

**Figure 16.1** Charles Darwin in 1855.

*population* consists of the individuals of a given species that occur together at one place and at one time.

In viewing evolution through the lens of population genetics, scientists now saw natural selection as a process that alters gene frequencies in a population—a concept Darwin was unable to consider. Darwin did propose that heritable variation within populations was the basis for natural selection, but he had no understanding of how that variation is inherited. Once genes and gene mutations were discovered in the early twentieth century, evolutionists began to recognize the importance of genetic variation within a population to the process of natural selection.

A few decades after the development of neo-Darwinism, scientists determined that mechanisms of evolution existed in addition to natural selection. Within the scientific community, this combination of mechanisms is called the **modern synthesis**, or the **evolutionary synthesis**. Concepts that compose the evolutionary synthesis will be described throughout this chapter.

| TABLE 16.1 | Darwin's Theory |
|---|---|
| **Tenets** | **Explanations** |
| Species change over time. | Species are *not* constant and unchanging. The concept that species change over time is the basic theory of evolution. |
| All organisms have descended from common ancestors. | The history of life takes the form of a branching tree. |
| Evolution is gradual. | The evolutionary process is continuous, with no abrupt changes. |
| Species multiply. | "Old" species give rise to "new" species; speciation occurs. |
| Natural selection takes place. | Organisms having adaptive traits survive in greater numbers than those without such traits. |

Adapted from Mayr, E. (2001). *What Evolution Is*. New York: Basic Books (p. 86).

# How does evolution occur?

## 16.2 Genetic variation is the raw material of evolution.

**Genetic variation** refers to heritable differences among individuals in a population. You can see the effects of genetic variation in the population of flamingos shown in **Figure 16.2**. Some have

longer legs than others, and others differ slightly from their neighbors in their coloration patterns. What other variations can you see?

Think of any other population you might encounter and identify variations among its individual members. For example, what differences do you see among a population of humans riding the subway at rush hour . . . or among a population of redwoods in a California

**Figure 16.2 Variation within a species.** How many differences among these flamingos can you determine from this photo?

forest . . . or among a population of cats at a local cat show? Your lists of variations would likely be extremely long. These observable variations among organisms are part of their *phenotypes*, as you may recall from Chapter 13. Moreover, phenotypes are not just outward appearances but also include observable biochemical characteristics, such as the ability of a bacterium to metabolize certain amino acids, or behaviors, such as the courtship display of the male sage grouse.

What is the source of phenotypic variation? Mendel suggested that the characteristics of organisms are inherited as discrete factors, which are either present or absent in a particular generation. A eukaryotic organism can have only two different alleles of a particular gene (one on each member of a chromosome pair), but many different alleles of that gene can exist within a population. Scientists know that variation within a species, as well as variation among species, is due to the expression of differences in their genes (alleles). The genetic composition of an individual is called its *genotype*.

What role does phenotypic variation play in evolution? It plays a huge and integral role—variation is the necessary raw material of evolution. It is the basis on which gene frequencies are altered in a population by natural selection. If all traits (behaviors, physical features, and other characteristics) were the same in every member of a species, then all members of the population would have the same level of fitness. That is, each member's traits would confer no more and no less of a reproductive advantage than the traits of any other member of the population (see Chapter 15). If those traits were **adaptive traits**—if they allowed individuals to survive to reproductive age and bear offspring—then the population would be more likely to survive. If they were not adaptive, then the population would not survive. Variation within a population provides adaptive opportunity for future environmental changes.

**Figure 16.3** shows two species of pine trees that have adaptive traits relating to fires. The scraggly-looking jack pines have cones that are held closed by resin for as much as 20 years. They open only in the intense heat of a fire, dispersing their seeds, which fall to the ground or are blown elsewhere by the wind. Jack pines in which this mechanism works best will be the ones most likely to reproduce and have their genes remain in the gene pool after a fire. Ponderosa pines have fire-resistant bark, but not all Ponderosa pines survive fires: larger, healthier ones are more likely to survive and reproduce. Thus, not all members of each population have the same level of fitness even though all members of each population have adaptive traits relating to fires. Why? Because variation in alleles that encode fire-resistant traits and traits of general health and size exists among members of each population.

## 16.3 Natural selection acts on phenotypes, affecting genotypes of populations.

In the context of genetics, how does natural selection work? Natural selection acts on the phenotype, which reflects the underlying genotype. It works like this: In most populations, more offspring are produced than can be supported by available resources. The individuals that survive long enough to reproduce, or that reproduce the most, are those with alleles and combinations of alleles that confer an adaptive advantage. Thus, natural selection acts on individuals, but populations evolve.

(a)

(b)

**Figure 16.3 Adaptive traits and variation.** These two species of pine trees have adaptive traits relating to fires. Variations of the traits within populations of each are the basis on which gene frequencies are altered in a population by natural selection. (a) The jack pine has pine cones sealed shut until opened by the intense heat of a fire.

(b) The stately Ponderosa pine has fire-resistant bark. The largest and healthiest trees will have bark most able to survive heat and flames. Jack and Ponderosa pines in which these mechanisms work best for seed dispersal will contribute their genes to the gene pools of the next generation of each species.

## just wondering . . .

**I've noticed that some male animals have conspicuous features, such as brightly colored feathers, that attract females. But don't predators notice them as well? If so, aren't the males that females find most attractive the ones most likely to be killed? How can such traits be adaptive?**

It is true in that the males of some species have exaggerated traits, sometimes called ornaments, to attract females of that species. These traits are involved in one type of *sexual selection*, in which organisms are selected for mating based on their morphology (form and structure) or behavior. Sexual selection can involve individuals of one sex—usually females—choosing mates with certain preferred characteristics that advertise their quality. It can also involve individuals of one sex—generally males—"winning" mates by being victorious in fights with one another. In both cases, the male is advertising his quality, which is ultimately his genetic quality—sperm that carry genes conferring greater fitness than the genes of other males. However, in both cases, there are costs to the organism to produce the structures needed, such as bright plumage to woo a female or large horns to fight effectively.

What are these costs? One cost is energy. An organism must direct a tremendous amount of energy to developing showy, bright plumage or sizable horns. Another cost is the increased likelihood of death from predators. For example, not only might a brightly plumed male be more noticeable to predators, but his plumage also might hinder a quick escape. And this is just your point . . . how does the adaptive advantage of showy or bright features that elicit female choice, such as the extensive plumage on the peacock shown in **Figure 16.A**, balance their disadvantage, or cost? And why do females, such as the peahen shown in the foreground of the same photo, choose males with these showy features?

A few hypotheses have been proposed that suggest answers to these questions. One hypothesis is referred to as the *handicap theory*. It suggests that the cost of developing and having ornaments varies among males. If a male is weak or sickly, his body uses

**Figure 16.A Sexual selection.** Adult peacocks (males) have a train, which they spread out like a fan when they are courting females. Peahens (females; foreground) prefer males with elaborate trains.

its energy for survival. Instead of developing bright colors, extensive plumage, and a long tail, for example, the animal might develop dull colors, minimal plumage, and a short tail. Nevertheless, for a weak, sickly male, escape from predators might be difficult even without a long tail or plumage getting in the way. In contrast, a strong, healthy male will develop brighter colors, extensive plumage, and a long tail. The healthy male has energy to devote to ornaments and still escape easily from predators. Having these characteristics, then, is a clear, reliable signal to females of his species that he is a high-quality male, and not having these characteristics is a clear, reliable signal of a low-quality male. For the high-quality male, the cost becomes an advantage. That is, the adaptive advantage of developing showy or bright features to attract females, mate, and produce offspring outweighs the disadvantage.

Another hypothesis is called *runaway sexual selection*. It suggests that characteristics involved in sexual selection might have originated as characteristics conferring an adaptive advantage to males in their early evolution and having nothing to do with sexual selection. For example, males with long tails might somehow have been better able to survive predators and thus live to reproduce. If a mutation occurred in one or some females that caused them to prefer males with long tails, these females would mate with males exhibiting long tails and would produce sons with long tails and daughters who preferred a long tail in males. As the mutation for female preference appeared in greater numbers in the gene pool, males with long tails would gain an advantage in mating. If tail length in males evolved beyond the optimal length for survival against predators, then mating preference alone would begin to drive the evolution of the trait. The male trait of a long tail and the female trait of a preference for a long tail would reinforce each other in a runaway process.

Other hypotheses exist, but these two explanations provide ideas of scientifically plausible answers to your questions. Scientists are still gathering evidence to support or refute the various hypotheses that seek to explain various facets of sexual selection and the role of female choice in the mating systems of organisms. ●

*Are you wondering about a topic in biology and how it relates to your life? Submit your question by clicking the* Just Wondering *link in this text's companion Web site at www.wiley.com/college/alters.*

---

For example, remember the research Peter and Rosemary Grant are conducting on Daphne Major in the Galapagos? (See Chapter 15.) They found that in periods of little rain, larger birds with larger, stronger beaks than others in their species were the ones to survive. Why? The medium ground finch (*Geospiza fortis*)—one of the species studied by the Grants—is an excellent ex-

ample to illustrate what happens. This finch species has a blunt beak, well suited for eating small soft seeds of perennials. Larger birds in this species, however, have larger, stronger beaks that can break open the larger, harder seeds produced by a weed called the caltrop. In times of drought, many perennials die, yet the caltrop survives. Birds able to feed on caltrop seeds survive to reproductive age, while birds unable to feed on these large, hard seeds are more likely to die before they reproduce. As the Grants observed and measured offspring the next season, they noted an increasing proportion of large birds with large, strong beaks.

This phenotype of a large, strong beak reflects a particular genotype made up of alleles encoding a large, strong beak. This genotype is associated with higher reproductive fitness. As selection acts on this adaptive trait (the phenotype) and on the alleles encoding this trait (the genotype), the alleles increase in frequency in the population.

This example shows how the relative frequencies of alleles can change in a gene pool. A **gene pool** is all the alleles of all the genes of all the individuals in a population. The frequency of an allele is the proportion (fraction) of a given allele to all variants of that allele in a population. For example, if two alleles—red and white—existed for flower color, and if each allele occurred 20 times in a population of 20 petunias, then the frequency of each allele in the population would be 20 out of 40 total flower color alleles, or 1 out of 2, or 0.5.

What would happen if a population lacked variation in its gene pool? Genetic variation is key to the survival of a species because the alleles that confer reproductive fitness at one time are not necessarily the alleles that confer fitness at another time. If little genetic variation exists in a population and environmental conditions change, then the population may die out if its gene pool contains no adaptive alleles. Variation in the gene pool provides the potential for the survival of the population as the environment changes.

Today, population geneticists define evolution as change in the frequencies of alleles in the gene pool over time. Small-scale change, such as an increase in alleles for large beaks and a decrease in alleles for small beaks in the gene pool over a few generations, is called **microevolution**. It is evolution at or below the species level. Conversely, **macroevolution** is large-scale evolution. It is evolution above the species level, such as at the level of genera and families.

### CONCEPT CHECKPOINT

1. Do you think that it is possible to determine the frequency of a recessive allele for a given gene in a population simply by counting the individuals that express the recessive phenotype and those that do not? Explain.
2. An allele that is currently adaptive in Atlantic salmon might someday become detrimental. What could cause this situation to occur?

## 16.4 Natural selection affects populations in three major ways.

Let's go back to the Grants' research on Daphne Major. The microevolution of larger birds with larger, stronger beaks than others in their species is an example of **directional selection**. In directional selection, an extreme phenotype is fittest. With Darwin's finches on Daphne Major, you may remember, selection favored the "extreme" phenotype of large, strong beaks, during years with little rain. That is, directional selection occurs when the environment of a population changes and only the extreme phenotype in one direction happens to be best adapted for the new conditions. In this case, the new conditions were large, tough seeds as a food source, since not enough rain had fallen for most plants that produced small, soft seeds to survive.

Prior to selection of this trait, most of Darwin's finches on Daphne Major had the smaller, less strong type of beak that was suited to eating the predominant food—small, soft seeds. This initial situation is depicted in graph ❶ in **Figure 16.4**. The birds with larger, stronger beaks survived the food shift and were parents to the next generation. As the Grants observed and measured offspring the next season, they noted an increased proportion of large birds with large, strong beaks. What had been a larger, stronger beak in the previous year's population had now become almost the average beak ❷. Within this new population, some individuals—those to the right of the average—had even larger, stronger beaks. If the drought continued, selection pressure for larger, stronger beaks would continue, and selection would continue to move in that direction, as shown in the next generation ❸. Thus, a change in food size and toughness would select for a shift in mean beak size and strength in the population.

What if environmental conditions changed on Daphne Major so that large, strong beaks were still adaptive, but so were small, thin beaks? For example, if the only two food sources were large, tough seeds and soft seeds found in the crevices of rocks, then birds with small, thin beaks might survive as well, since they could forage for food in narrow crevices. If that were the case, then individuals having beaks of each type in the initial situation ❹ would survive and would be parents to the next generation of birds ❺. With continued selection pressure for these two beak types, **disruptive selection** would continue, diversifying the population into two extremes ❻. Disruptive selection (diversifying selection) occurs when two or more "extreme" phenotypes are fitter than the intermediates between them. New species may form as a single species is split in this way.

A third type of selection, called **stabilizing selection**, can affect a population if the environment does not change. For example, if the rain remained steady over the years on Daphne Major and small, soft seeds were abundant, average-sized beaks would be favored in the initial situation ❼. "Extreme" phenotypes would become less frequent in the population because they would be less fit than those with the optimal, average-sized beak. Birds with average-sized beaks would survive in greater numbers to reproductive age and would be the predominant parents of the next generation ❽. Over time, the average phenotype would not change, and variability in the population would be reduced ❾.

## 16.5 Genetic variation arises by random mutation and recombination.

### Mutation

Why do genes differ within a species? One relatively small source of genetic variation is mutation. **Mutations** are permanent changes in the genetic material, and they are the *only* source of new alleles.

How do mutations happen? They may arise spontaneously or may be caused by ionizing radiation (such as X rays), ultraviolet radiation (UV rays), or chemicals (such as LSD or pesticides). Such factors can alter molecules within the DNA, producing new alleles of genes. Another source of mutation is transposons, genes that move from place to place on chromosomes (see Chapter 9). When transposons move, they may break chromosomes or mutate genes. In fact, these jumping genes contribute significantly to the rate of mutation in organisms.

It is a misconception to think that mutations arise out of need. For example, an animal's genes will not mutate to confer faster run-

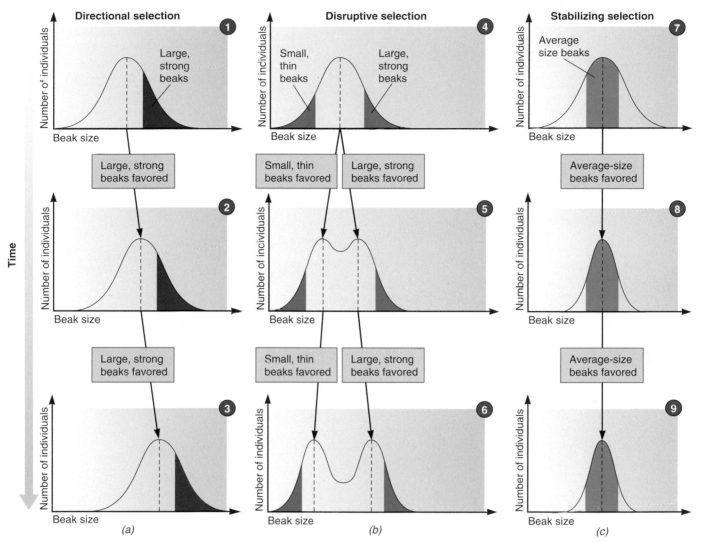

**Figure 16.4** The three basic types of selection and their effects. Beak size is indicated by shading, from small (lighter shading) to large (darker shading); the average beak size in the population is indicated by the dotted lines; and the curves show how many individual birds have each beak size. (a) In directional selection, the average phenotype moves toward an extreme (in this case, toward a larger beak size). (b) In disruptive selection, the population separates into two increasingly different groups, with each extreme phenotype becoming an average (in this case, there are more small-beak birds, more large-beak birds, and fewer birds with average-size beaks). (c) In stabilizing selection, the average phenotype does not change over time, but the range of variation in the population diminishes (in this case, there are fewer small-beak birds and fewer large-beak birds).

 **Visual Thinking:** Disruptive selection is also called diversifying selection. Look at the graphs of this process and explain why both names make sense.

ning ability just because it needs to run faster. Some animals are genetically fast runners and others are not. Although humans can run faster through proper training, they cannot change their genes and cannot pass on this acquired characteristic to their children. In other words, mutations are not goal directed—they happen randomly.

Mutations may be harmful, beneficial, or neutral. Harmful mutations are nonadaptive—reproductively disadvantageous. Helpful mutations are adaptive—reproductively advantageous. Neutral mutations are neither. Harmful mutations are weeded out of a population via selection, because organisms with these mutations tend to die prior to reproduction or are unable to reproduce. Beneficial

mutations persist. Mutations that are neutral—neither adaptive nor nonadaptive—are not directly affected by selection and may persist in the gene pool, providing variation within the population and adaptive "opportunity."

Although mutation is a relatively small source of genetic variation, it is *the* source of new alleles. Without mutation, ultimately there would be no variation among organisms and therefore no evolution. Even mutation in only one nucleotide may have important consequences for an organism.

An example of a single nucleotide mutation in the human population is the sickle cell mutation. Sickle cell disease is a recessive

**Figure 16.5 Areas of the world in which malaria is prevalent.** Malaria is a serious, often fatal disease. Persons with a single nucleotide mutation on the hemoglobin gene are resistant to one particularly deadly form of the malaria parasite. Homozygotes for this mutated allele have sickle cell disease as well as resistance. Heterozygotes do not have the disease, yet retain their resistance.

 **Visual Thinking:** Find where you live on the map. Is malaria prevalent there? Is the sickle cell allele likely to be prevalent in the country where you live? Why or why not?

Areas at risk from malaria

disorder found in human populations in equatorial Africa, the Middle East, and the Far East. A person homozygous for this allele has sickle cell disease and is unable to transport oxygen to his or her tissues properly. Why? In people with this condition, molecules of the protein hemoglobin, which carry oxygen in the blood, are defective due to one amino acid substitution. These defective molecules cause red blood cells to become sickle shaped, as shown in Figure 14.7a. The sickled, stiff cells do not pass easily through capillaries and clump there, reducing the blood supply to the tissues they serve. The result is pain, tissue destruction, and early death.

However, persons who are heterozygous for this allele and who live in regions of the world in which malaria is prevalent (**Figure 16.5**) actually benefit from it. Malaria is one of the 10 most prevalent and deadly diseases in the world. Both persons with sickle cell disease and carriers (heterozygotes) are resistant to one particularly deadly species of malaria parasite. This beneficial effect of being a carrier for sickle cell disease is the reason that the sickle cell allele has persisted in certain human populations.

### Recombination

Recombination is a major source of genetic variation. Three processes result in new combinations of existing genes: crossing over; segregation and independent assortment; and fertilization. Crossing over is the process whereby sister chromatids swap parts during meiosis. Segregation means that homologous chromosomes, which carry alternative forms of genes (alleles), separate from one another during meiosis. Independent assortment means that the distribution of alleles for one trait does not affect the distribution of alleles for other traits in the formation of gametes during meiosis unless the alleles are on the same chromosome. The processes of crossing over, and segregation and independent assortment, shuffle alleles into new combinations in the gametes that form potential new individuals at fertilization. Fertilization creates new allele combinations as egg and sperm unite (**Figure 16.6**). These processes are described in Chapter 12.

### CONCEPT CHECKPOINT

3. Which of the following are always characteristics of mutations? Which might be characteristics of certain mutations only?

   random, heritable, beneficial, neutral, purposeful, predictable, rare, nonadaptive, adaptive, lethal

4. Explain why the frequency of sickle cell disease is greater in Africa than it is in North America.

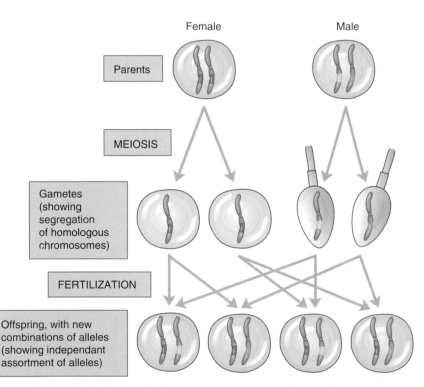

**Figure 16.6 Segregation and independent assortment of genes.** These processes are a major source of genetic variation. In this example, the female parent has only red alleles for a particular gene, while the male parent has one yellow allele and one green allele for that gene. After the homologous chromosomes segregate (separate) during meiosis, the eggs all have the red allele, while half the sperm have the yellow allele and half the green allele. When fertilization takes place, these alleles are independently assorted, and two types of zygotes may be formed: those with the red and yellow alleles and those with the red and green alleles. Neither combination existed in the parents. (Crossing over, which also results in new combinations of alleles, is illustrated in Figure 12.15.)

# What factors contribute to changes in gene frequencies within populations?

## 16.6 Random genetic drift, gene flow, and nonrandom mating change gene frequencies within populations.

In addition to mutation and natural selection, other factors contribute to change in a population's gene pool over time: random genetic drift, gene flow, and nonrandom mating.

### Random genetic drift

**Random genetic drift** is a change in the frequencies of alleles in the gene pool of a finite population due to chance events. With a finite (limited) population size, **sampling error** ensures that evolu-

tion will occur. What does this mean? Sampling refers to taking a random selection of alleles from the population, and error refers to the possibility that the frequencies of the alleles in the sample will not be exactly the same as the frequencies of the alleles in the total population. In other words, sampling error is the deviation of a sample from the larger population from which it was drawn.

A simple experiment illustrates how sampling error causes random genetic drift. **Figure 16.7** shows the results when we performed this experiment:

1. Put 100 small red squares and 100 small green squares into a covered box. This represents a population of 100 individual organisms, each with two alleles for color, a red allele and a green allele. A red square

| (a) POPULATION OF 100 INDIVIDUALS | In original population | First random pick | In new population based on first random pick | Second random pick |
|---|---|---|---|---|
| Total number of alleles | 200 | 100 | 200 | 100 |
| Number of red alleles | 100 | 52 | 104 | 53 |
| Number of green alleles | 100 | 48 | 96 | 47 |
| Frequency of red allele | 0.5 | 0.52 | 0.52 | 0.53 |
| Frequency of green allele | 0.5 | 0.48 | 0.48 | 0.47 |
| Deviation of the frequency of each allele from the frequency in the original population | — | 0.02 (2%) | — | 0.03 (3%) |

| (b) POPULATION OF 50 INDIVIDUALS | In original population | First random pick | In new population based on first random pick | Second random pick |
|---|---|---|---|---|
| Total number of alleles | 100 | 50 | 100 | 50 |
| Number of red alleles | 50 | 28 | 56 | 29 |
| Number of green alleles | 50 | 22 | 44 | 21 |
| Frequency of red allele | 0.5 | 0.56 | 0.56 | 0.58 |
| Frequency of green allele | 0.5 | 0.44 | 0.44 | 0.42 |
| Deviation of the frequency of each allele from the frequency in the original population | — | 0.06 (6%) | — | 0.08 (8%) |

| (c) POPULATION OF 10 INDIVIDUALS | In original population | First random pick | In new population based on first random pick | Second random pick |
|---|---|---|---|---|
| Total number of alleles | 20 | 10 | 20 | 10 |
| Number of red alleles | 10 | 4 | 8 | 3 |
| Number of green alleles | 10 | 6 | 12 | 7 |
| Frequency of red allele | 0.5 | 0.4 | 0.4 | 0.3 |
| Frequency of green allele | 0.5 | 0.6 | 0.6 | 0.7 |
| Deviation of the frequency of each allele from the frequency in the original population | — | 0.1 (10%) | — | 0.2 (20%) |

**Figure 16.7** Random genetic drift through sampling error in three population sizes

represents a gamete with an allele for red color; a green square represents a gamete with an allele for green color. The frequency of each color allele in this population is 100/200, or 0.5 (50%).

2. Shake the box and then uncover it. Without looking, pick 100 squares out of the box—this represents the random selection of gametes involved in reproduction in this population, leading to the next generation. Each of the 100 individuals contributes one gamete to fertilization.

3. Count the number of red squares and the number of green squares that you picked from the box. We got 52 reds and 48 greens (Figure 16.7*a*, first random pick). Thus, the frequency of color alleles in the first random pick is 0.52 (52%) reds and 0.48 (48%) greens. The frequency of each allele deviates by 0.02 (2%) from the frequency in the original population.

4. Now represent the new, next-generation population based on this allele frequency: place 104 red squares (52 × 2) and 96 green squares (48 × 2) in the box (Figure 16.7*a*, new population based on first random pick).

5. Repeat the procedure: pick 100 squares, and count the reds and the greens. We got 53 reds and 47 greens (Figure 16.7*a*, second random pick)—that is, frequencies of 0.53 (53%) reds and 0.47 (47%) greens. The deviation of allele frequencies is now slightly larger (0.03, or 3%). With continued repetitions, the deviation is likely to increase further.

Figures 16.7*b* and *c* show the results of this same experiment starting with smaller population sizes—50 red and 50 green alleles in Figure 16.7*b* (population of 50 individuals) and 10 red and 10 green alleles in Figure 16.7*c* (population of 10 individuals). As you can see from the figures, the smaller the original population, the larger the deviation:

- When we began with 50 of each allele (Figure 16.7*b*), the allele frequencies after the first random pick were 0.56 (56%) reds and 0.44 (44%) greens, or a deviation of 0.06 (6%). After the second random pick, the allele frequencies were 0.58 (58%) reds and 0.42 (42%) greens, or a deviation of 0.08 (8%).

- When we began with 10 of each allele (Figure 16.7*c*), the allele frequencies after the first random pick were 0.40 (40%) reds and 0.60 (60%) greens, or a deviation of 0.10 (10%). After the second random pick, the allele frequencies were 0.30 (30%) reds and 0.70 (70%) greens, or a deviation of 0.20 (20%).

This inverse relationship between the size of the population and the size of the sampling error means that genetic drift is more effective as a force of evolution when the population is small.

How does sampling error work in actual populations of organisms? If a population is large and if large numbers of organisms in the population consistently reproduce, there is a strong likelihood that the gene pool of the progeny will be similar to the gene pool of the parents. (In our experiment, all the simulated populations were much smaller than most natural populations.) Conversely, if a population is small or if only small numbers of organisms reproduce over many generations, there is a strong likelihood that the gene pool of the progeny will deviate from the gene pool of the parents. That is, the frequencies of various alleles may vary widely from their frequencies in the gene pool of the previous generation.

Genetic drift can occur over the course of one or a few generations, rather than over many generations, if a small population separates from a much larger population and begins a new population elsewhere. This extreme type of genetic drift is called the **founder effect**. The gene pool of the founder population comprises the alleles of only a small portion of the original population, which means that a large sampling error is likely. That is, the frequencies of the alleles in the small population are likely to be quite different from the frequencies of the alleles in the larger population. In this case, the gene pool of the founder population will contain only a small portion of the total genetic variation of the larger population, and over subsequent generations, allelic frequencies in the gene pool based on the founder population are likely to become increasingly different from the allelic frequencies in the original population.

The Amish people of eastern Pennsylvania provide an excellent example of the founder effect in human populations. This population, which currently totals about 20,000 was established by approximately 200 German immigrants in the 1700s. Because this population has been almost completely closed to immigration and to intermarriage with the surrounding American population, it carries an unusual concentration of gene mutations that cause a number of inherited disorders. These disorders, such as abnormalities of the teeth and nails, forms of dwarfism, and polydactyly (extra fingers and toes; **Figure 16.8**), are rare in large populations. They show up more frequently in the Amish because their children, being the result of intermarriage within a small population with little genetic diversity, are more likely to inherit two copies of recessive genes that cause these conditions.

In the extreme case, deleterious genes can become fixed within small populations. A gene is *fixed* within a population when it reaches a frequency of 1.00. In other words, 100% of the alleles for that trait are one particular allele—no variation of the trait exists in that population. How can this occur? The short answer is that random genetic drift is not evolution by natural selection. That is, random genetic drift does not involve selection pressure for alleles that are adaptive. Rather, random fluctuations result in the replacement of old alleles by new ones, and the new ones are not necessarily beneficial. Also, in a small population, natural selection may be insufficient to overcome the effects of random drift toward fixation.

Genetic drift can also occur because of a **population bottleneck**, an evolutionary event in which a significant percentage of a population—usually 50% or more—dies before reaching reproductive age or is otherwise prevented from reproducing. A popula-

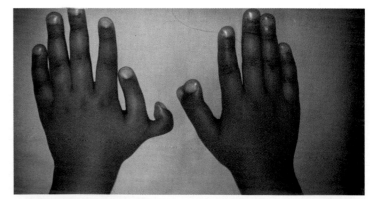

**Figure 16.8** Polydactyly—extra fingers or toes—is caused by an inherited gene mutation. This condition arises more frequently in the Amish than in the general population because their small, closed population has little genetic diversity. In such a population, children are more likely to inherit two copies of recessive genes that cause such conditions than in large populations with greater genetic diversity.

tion's loss of genetic variation from a population bottleneck can cause the genetic distance between the current and the ancestral populations to increase very quickly.

Northern elephant seals, which inhabit coastal waters from the Gulf of Alaska to Baja California, have experienced a population bottleneck. Between 1818 and 1860 these huge sea mammals shown in **Figure 16.9** were hunted for their blubber, which was processed into oil and used as a machine lubricant, lamp fuel, waterproofing, and an ingredient in paint and soap. By the late 1800s, seal hunters had turned to other species for blubber because they believed elephant seals to be extinct. However, a population estimated at 20 to 100 northern elephant seals had survived and that population has given rise to a current population of about 175,000. Since members of the current population are all descended from the 20 to 100 survivors, they are all genetically similar. Biologists are concerned that because of this limited genetic variability, northern elephant seals may be vulnerable to future selection pressures, such as new diseases to which no members of the population may be immune.

Accidental events can also cause genetic drift and population bottlenecks. For example, if a small population of white-flowered (genotype ww) and purple-flowered (genotypes WW and Ww) pea plants was growing at the foot of a mountain and most of the white-flowered plants were killed by a mudslide (but not many of the purples), then the recessive allele for white flower color would disproportionately be taken from the gene pool. In a large population, this chance event would not have much effect, but in a small population it could lead to the disappearance of white flowers.

### Gene flow

The gene pools of populations tend to diverge from one another by genetic drift. Conversely, gene flow—if unopposed by other factors—tends to keep the gene frequencies of neighboring populations of a species the same. **Gene flow** is the exchange of genes among neighboring populations due to immigration and emigration. Immigration is the movement of organisms into a population, and emigration is the movement of organisms out of a population.

**Figure 16.9** Male and female northern elephant seals (*Mirounga angustirostris*) along the shoreline. The northern elephant seals living today are all descended from survivors of a population bottleneck due to hunting. Therefore, the current population has little variation in its gene pool and may be susceptible to diseases and conditions that may arise in the future.

The gene content of every local population is strongly affected by gene flow. It contributes to the stability of widespread species.

Gene flow is generally a stabilizing process, but there is concern that gene flow could bring genetic modifications from transgenic plants to neighboring populations of wild relatives. For example, one fear is that plants engineered to resist herbicides could pass their genes on to weeds, producing populations of plant pests that herbicides wouldn't be able to kill. However, in order for gene flow to occur between neighboring populations, the plants must be close enough to exchange pollen via insects or the wind, they must flower at the same time, fertilization must take place, and the resultant seeds must survive and germinate into fertile plants. Several technologies are being discussed to decrease the risk of gene flow between transgenic and wild plants. One idea is to render the pollen of transgenic plants sterile, a so-called terminator technology.

### Nonrandom mating

**Nonrandom mating** is inbreeding (mating between closely related organisms or, in plants, self-pollination) or the selection of mates for specific characteristics, both of which reduce genetic variability within a gene pool. Nonrandom mating may also occur if a population is clumped, a situation in which some organisms in the population are close to one another but far from others. Most mating would likely take place within clumps, with few instances of mating between individuals of different clumps.

---

### CONCEPT CHECKPOINT

5. Imagine that you had a small-necked bottle full of 300 marbles with 1/3 red, 1/3 white, and 1/3 blue. The marbles can fit through the neck of the bottle only one at a time. If you turned the bottle over and shook once, only a few marbles would come out before the marbles would get stuck in the bottle's neck. When you shook the bottle, 6 marbles emerged: 1 red, 2 white, and 3 blue. How are the marbles that emerged from the bottle analogous to a founder population in a gene pool? How do they illustrate the founder effect?

---

## 16.7 Populations are rarely in genetic equilibrium.

Evolution is a change in the frequencies of genotypes and alleles in the gene pool over time, so a nonevolving population would be one in which genotype frequencies and allele frequencies remain constant over generations. Such a population would be in genetic equilibrium. Can such a population exist?

In 1908 and 1909, respectively, English mathematician Godfrey Hardy and German physician Wilhelm Weinberg independently developed a model that is considered to be one of the foundations of the genetic theory of evolution. Called the **Hardy-Weinberg equilibrium**, it states that genotype frequencies and allele frequencies will remain constant across generations (i.e., evolution will *not* take place) when the following five conditions are met:

- Mating is random, meaning that individuals of different sexes can pair without limitations.
- The population is very large.
- Genes are not added from outside the population—that is, there is no gene flow.

- Genes do not change from one allelic state to another—that is, no mutation occurs.
- All individuals have equal probabilities of survival and of reproduction—that is, there is no selection of certain alleles.

The idealized population meeting all these conditions is called an *equilibrium population* because it will remain in genetic equilibrium as long as the conditions are met. But clearly, it is rare for the gene pool of any real population to meet all these conditions, especially those of no mutations and random reproduction. Thus, evolutionary change, based principally on natural selection and to a lesser extent on genetic drift, mutation pressure, and gene flow, is a part of virtually all populations.

 **CONCEPT CHECKPOINT**

6. Match the conditions of the Hardy-Weinberg equilibrium (using the numbers in the list above) with the opposing conditions of evolution: genetic drift; gene flow between populations; mutations; natural selection; nonrandom mating.

## CHAPTER REVIEW

## Summary of Key Concepts and Key Terms

### Is evolution still accepted in the form Darwin proposed?

**16.1** The field of evolution did not progress much beyond Darwin's theory until the twentieth century because of the lack of a suitable explanation of how hereditary traits are transmitted.

**16.1** Darwin's theory of evolution by natural selection is also called *Darwinism*, or **Darwinian evolution** (p. 253).

**16.1** In the early part of the twentieth century, a genetics-based view of natural selection was developed, which was called *neo-Darwinism*, or the **neo-Darwinian theory of evolution** (p. 253).

**16.1** The **modern synthesis**, or the **evolutionary synthesis** (p. 253), which has a foundation in **population genetics** (p. 253), includes mechanisms of evolution in addition to natural selection.

### How does evolution occur?

**16.2** **Genetic variation** (p. 253) refers to heritable differences among individuals in a population.

**16.2** Variation due to the expression of differences in alleles of genes is the necessary raw material of evolution.

**16.2** Individuals that survive long enough to reproduce, or that reproduce the most, are those with alleles and combinations of alleles that confer adaptive advantage.

**16.2** Alleles associated with **adaptive traits** (p. 254), and with consequent reproductive fitness, increase in frequency from one generation to the next.

**16.3** Genetic variation is key to survival of a species because alleles conferring reproductive fitness at one time are not necessarily alleles that confer fitness at another time.

**16.3** **Microevolution** (p. 255) is small-scale change in the frequencies of alleles in the **gene pool** (p. 255) over time; it is evolution at or below the species level.

**16.3** **Macroevolution** (p. 255) is large-scale evolution; it is evolution above the species level.

**16.4** Natural selection affects populations in three major ways.

**16.4** In **directional selection** (p. 255), one extreme phenotype is fittest when the environment of a population changes.

**16.4** **Disruptive selection** (p. 255), also called *diversifying selection*, occurs when two or more extreme phenotypes are fitter than intermediate phenotypes when the environment of a population changes.

**16.4** When the environment of a population does not change, the average phenotype does not change and variability in the population becomes reduced via **stabilizing selection** (p. 255).

**16.5** A small source of genetic variation in populations is **mutation** (p. 256), which creates new alleles; mutations may arise spontaneously or through the action of radiation, chemicals, or transposons.

**16.5** Mutations may be harmful (nonadaptive—reproductively disadvantageous), beneficial (adaptive—reproductively advantageous), or neutral (neither).

**16.5** Recombination is the major source of genetic variation, resulting in new combinations of existing genes through the processes of crossing over, and segregation and independent assortment, which occur during gamete formation, and fertilization, which results in new combinations of alleles.

### What factors contribute to changes in gene frequencies within populations?

**16.6** Factors in addition to mutation and natural selection that contribute to change in genotype frequencies over time are random genetic drift, gene flow, and nonrandom mating.

**16.6** **Random genetic drift** (p. 258) is a change in the frequencies of alleles in the gene pool of a finite population due to chance events, such as sampling error.

**16.6** **Sampling error** (p. 258) is the possible deviation of a sample from the larger population from which it was drawn.

**16.6** Large populations are likely to experience less genetic drift than small populations.

**16.6** Genetic drift can occur in one or a few generations if a small (founder) population separates from a much larger population and begins a new population elsewhere founder effect (p. 259).

**16.6** A **population bottleneck** (p. 260) is an evolutionary event in which a significant percentage of a population or species is prevented from reproducing.

**16.6** **Gene flow** (p. 260) is the exchange of genes among neighboring populations due to immigration and emigration; it is the major reason for the stability of widespread species.

**16.6** **Nonrandom mating** (p. 260) is inbreeding or the selection of mates for specific characteristics, which may sometimes reduce genetic variability within a gene pool.

**16.7** Populations that are not evolving are called equilibrium populations.

**16.7** Natural populations are rarely in genetic equilibrium.

**16.7** The **Hardy-Weinberg equilibrium** (p. 261) states that evolution will not take place if a population has random mating, is very large, experiences no gene flow, has no mutations occurring, and is not affected by natural selection.

## Level 1    Learning Basic Facts and Terms

**Multiple Choice**

1. Changes in allelic frequencies in populations over many generations might occur for which of the following reasons?
   a. New mutations
   b. Migration
   c. Natural selection
   d. All of the above
2. Small changes in allelic frequencies caused by genetic drift in populations from one generation to the next would be considered a form of
   a. macroevolution.
   b. microevolution.
   c. Darwinian evolution.
   d. Hardy-Weinberg equilibrium.
3. New alleles can arise through
   a. spontaneous mutations.
   b. chemically-induced mutations.
   c. radiation-induced mutations.
   d. transposon-induced mutations.
   e. All of the above.
4. Natural selection acts on individuals, but _____ evolve.
   a. populations
   b. families of organisms
   c. genes
   d. alleles

**True–False**

Which of the following is an acceptable definition of evolution? Indicate true or false for each.

5. _____ A change in the gene pool of a population.
6. _____ A change in the genetic makeup of an individual.
7. _____ A change in environmental conditions.
8. _____ A change in the physical characteristics of an individual.
9. _____ A change in the allelic or genotypic frequencies of a population.

## Level 2    Learning Concepts

1. Is it possible for the entire gene pool of a natural population to be in genetic equilibrium? Explain.
2. What would have to be true of an allele and the population it is in for its frequency to change *exclusively* by genetic drift?
3. Describe the founder effect. Give an example of somewhere in the world where this effect may have occurred. (Do not use the examples of the Amish and of northern elephant seals, discussed in the text.)
4. Describe what is meant by a population bottleneck.

## Level 3    Critical Thinking and Life Applications

1. For an incompletely dominant genetic trait, there are two alleles in the population. Is it possible for the allelic frequencies to be determined simply by counting all the phenotypes with respect to this trait observed in the population?
2. DDT is an insecticide that kills mosquitoes, a host that transmits malaria to humans. In fact, DDT is so effective at killing mosquitoes that the number of malaria deaths in places where it was sprayed dropped to near zero. However, DDT causes significant damage to the ecosystem. Since the ban on DDT, the number of deaths from malaria has risen dramatically. What do you think would happen to the frequency of the sickle cell allele in Africa if DDT use again became widespread? Why?
3. Due to loss of habitat, the California condor population was reduced to just a few individuals held in captivity in zoos. Breeding programs have increased their numbers to the point where many are being reintroduced into the wild, although biologists are concerned that this reintroduced population will have limited ability to survive future environmental change. What is the reason for this concern?
4. Human development has resulted in fragmented patches of natural habitat harboring small, isolated populations of native species. How is this situation analogous to the founder effect? Why would ecologists be concerned that these isolated populations are more susceptible to extinction?
5. For each of the following, indicate what type of selection is occurring.
   a. An African butterfly exists in two strikingly different color patterns, each of which closely resembles other species that are distasteful to insect-eating birds.
   b. The mean tail feather length of male grackles (birds found east of the Rocky Mountains) is increasing because females prefer to mate with males with longer tails.
   c. Disease-causing bacteria found in many hospitals are increasingly becoming resistant to antibiotics.
   d. Most mountain lions produce a litter of two to three kittens.

## In The News    Critical Thinking

**SNAKE ORIGINS**

Now that you understand more about the genetic basis of evolution, reread this chapter's opening story about research into snake origins. To better understand this research, it may help you to follow these steps:

1. Review your immediate reaction to Hedges and Vidal's research on the origin of snakes that you wrote when you began reading this chapter.
2. Based on your current understanding, again summarize the main point of the research in a sentence or two.
3. What questions do you now have about this research that this chapter's opening story does not answer?
4. Collect new information about the research. Visit the *In The News* section of this text's companion Web site at www.wiley.com/college/alters and

watch the "Snake Origins" video. Then use the "summary" link to read the accompanying story and access related links. Use this information, the links provided, and other online and library resources to answer your questions and find updates about this topic. State the sources of your information. Explain why you think the information is accurate. Also determine if the information expresses a particular point of view or is biased in any way.

5. In your view, what is the most significant aspect of this research? Give reasons for your opinions and for any changes in your ideas based on the additional information you have collected and the analysis you have done.

# SPECIATION AND EXTINCTION: HOW SPECIES ARISE AND DIE OUT

## In The News | Eagle Expansion

In 1782, the bald eagle was adopted as a symbol of the United States for its independence and strength. At that time, an estimated 50,000 to 200,000 of these birds inhabited what are now the contiguous 48 states. By the 1970s, however, the bald eagle population had dropped so low—to less than 1000 birds—that it was deemed an endangered species in 43 of the "lower 48" states. **Endangered species** are those that face a very high risk of extinction (dying out) in the wild in the near future. Since the 1970s, the bald eagle has made a comeback, with an estimated 20,000 birds, including 5000 breeding pairs, now inhabiting the lower 48. In 1995, the status of the bald eagle was upgraded to threatened. **Threatened species** are those with populations so low that they are likely to become endangered in the future. However, in May 2004 a top governmental official said the American bald eagle will be removed from the threatened species list. The birds will instead be protected under the federal Bald Eagle Protection Act of 1940.

Why did the bald eagle population decline to endangered status? And how did it make a comeback? The bald eagle became an endangered species as a result of human activity. Logging and land development reduced its habitats, the areas in which it lived. Then, in the 1950s and 1960s, the widespread use of pesticides, most notably DDT, resulted in reproductive impairment of the birds. They ingested these chemicals by consuming fish from contaminated waters. As a result, their eggs had thin shells and often broke during incubation. In 1972 DDT was banned. Although reproduction rebounded in the bald eagle population, humans continued to encroach on their habitat and numbers remained low.

In 1976, the New York State Department of Environmental Conservation launched the first North American bald eagle hacking project. As used here, "hacking" is not a computer term but is environmental conservation jargon referring to a technique in which humans assist the development of young birds and then release them to the wild to populate the surrounding area. You can learn more about the project and how it helped increase bald eagle population numbers by viewing the video and reading the story "Eagle Expansion" in the *In The News* section of this text's companion Web site at www.wiley.com/college/alters.

Write your immediate reaction to this information about the bald eagle: first, summarize the main point in a sentence or two; then suggest what you think its significance is. You will have an opportunity to reflect on your responses and gather more information on this topic in the In The News feature at the end of this chapter. In this chapter, you will learn more about speciation and extinction processes.

# CHAPTER GUIDE

## What are species and how do new species arise?

**17.1** Scientists recognize various definitions of species.

**17.2** Species arise by genetic isolation and divergence.

## How is reproductive isolation maintained during the speciation process?

**17.3** Premating and postmating reproductive isolating mechanisms maintain genetic isolation.

## How do scientists characterize the mode and tempo of speciation and evolutionary change?

**17.4** Speciation and evolutionary change can take place slowly and gradually, or suddenly and rapidly.

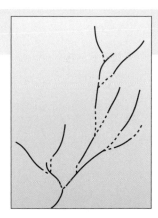

## Why do organisms become extinct?

**17.5** The adaptive capability of a species is one factor in the extinction process.

**17.6** Environmental change is another factor in the extinction process.

# What are species and how do new species arise?

## 17.1 Scientists recognize various definitions of species.

A *concept* is a notion, idea, or mental picture. In the case of species, it is a mental picture of a relationship among organisms. Defining the concept of species is a difficult task. In fact, there is no universal definition accepted by all biologists; there are many species concepts. On some level, though, we all have an idea of what a species is. We can distinguish a cat from a fish from a rose bush from a mushroom. When describing different types of organisms, we use what biologists might call the **morphological species concept**. It emphasizes pattern and recognizes species as groups of organisms that share certain morphological (anatomical) and biochemical traits. Defining a species, however, is not this simple.

Sometimes organisms look like one another due to convergent evolution, but they are actually different species. For example, the Australian echidna, or spiny anteater, looks similar to the North American porcupine (**Figure 17.1**). They are not only different *species*, but they are also different *orders* of mammals. The echidna is a monotreme (a mammal that lays eggs), while the porcupine is a rodent. They look alike because they evolved in very similar environments, though in different parts of the world. Conversely, organisms sometimes look very different from one another even though they are the same species. The fish shown in **Figure 17.2** look much different from one another, but are members of the same species. They simply look different at different stages of growth. Dogs are another example of organisms that can look very different from one another (think of a miniature poodle and a Great Dane) yet be members of the same species, *Canis familiaris*. Thus, organisms are generally recognized as species by means of sophisticated morphological and biochemical analyses, not just simple comparisons of external features.

The **biological species concept** provides another approach to the definition of a species. It recognizes **species** (SPEE-sheez or SPEE-seez) as groups of interbreeding natural populations that are reproductively isolated from other such groups. In other words, members of one species do not breed with members of other species in the wild. Of course, populations of the same species may be isolated from one another and may not be actively interbreeding, though the potential for their interbreeding exists.

As you can see, the biological species concept relates to populations and is applicable only to sexually reproducing organisms. When discussing organisms that have unknown reproductive characteristics, reproduce asexually, or are extinct, we must go back to the morphological species concept or to another species concept perhaps based on evolutionary or ecological criteria.

The biological species concept is the most widely accepted species concept and is particularly useful when studying how new species arise and are maintained in sexually reproducing populations. In fact, the definition of speciation—the origination of new species—is based on this view of a species. **Speciation** is the evolution of reproductive isolation between two or more groups of organisms.

## 17.2 Species arise by genetic isolation and divergence.

Under what conditions do new species arise? Let's revisit the definition of speciation: the evolution of reproductive isolation between two or more groups of organisms. Reproductive isolation really means genetic isolation in which there is no gene flow, or exchange of genes among neighboring populations. You may recall from Chapter 16 that gene flow—if unopposed by other factors—tends to keep the gene frequencies of neighboring populations of a species the same. It is the major reason for the stability of widespread species. If new species are to arise, then, there must be little gene flow among populations.

(a)                                              (b)

**Figure 17.1 Organisms may look alike but may be members of different species.** The North American porcupine (*Erethizon dorsatum*) (a) and the Australian echidna (*Tachyglossus aculeatus*) (b) are members not only of different species but also of different orders; they look alike due to convergent evolution.

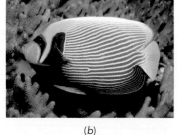

(a)                                              (b)

**Figure 17.2 Organisms may look different but may be members of the same species.** Fish grow throughout their lives and often change color during various phases of growth. These fish look very different, but they are members of the same species (*Pomacanthus imperator*). An adult is shown in (a) and a juvenile in (b).

Notice that speciation also involves evolution, in that a change occurs in the frequencies of alleles in the gene pool over time (see Chapter 16). This means that populations will become separate species only if they diverge genetically from the original population of which they were a part. The genetic divergence (difference) must be so great that it provides a reproductive barrier. New species arise only when interbreeding does not occur between populations whose individuals have access to one another or, if mating occurs and offspring are produced, the offspring are infertile.

Two basic mechanisms by which species may achieve genetic isolation and genetic divergence are allopatric (al-ah-PAH-trick) speciation and sympatric (sim-PAH-trick) speciation.

## Allopatric speciation

The word "allopatric" comes from Greek words meaning "other fatherland" and means living in different geographical areas. **Allopatric speciation** is characterized by the separation of members of a population by a physical barrier that blocks gene flow, which sets the stage for genetic and ecological divergence as the populations respond to different selection pressures.

What physical barriers can isolate some members of a population from others? Water may act as a barrier for terrestrial organisms. Some, however, can swim, fly, drift, or be transported across water from one island to another, from one continent to another, or between a continent and an island. You may recall from Chapter 16 that Darwin hypothesized this type of geographic isolation prior to the adaptive radiation of finches on the Galapagos Islands. He conjectured that many of the Galapagos plants and animals found their way to the islands from the nearby coast of South America. The 13 species of finches that now occupy the Galapagos Islands all arose from one South American species. By adaptive radiation the finch populations changed as they dispersed within the different habitats on the various islands.

Geological changes such as emergence of a mountain range or a land bridge can divide terrestrial or aquatic populations. In the case of aquatic populations, lowered water levels may result in lakes that were previously continuous bodies of water. *Continental drift* is the movement of the continents over time. Hundreds of millions of years ago, the continents did not exist as they do today. Instead, they were all joined in one giant continent called Pangea. This supercontinent began to break up about 225 million years ago, separating both terrestrial and aquatic populations. This slow movement continues even today.

Researchers have determined that continental drift was an isolating mechanism triggering speciation in a family of flightless birds called ratites (RAT-tights). Members of this family are two species of rhea (REE-ah) from South America, the ostrich from Africa, at least four species of kiwi from New Zealand, the emu (E-mew) from Australia, and three species of cassowary (KA-SAH-wheh-ree) from Australia and Papua New Guinea (**Figure 17.3**). Molecular studies of proteins show that ratites are more closely related to one another than to any other bird. Since they are flightless, these birds are unable to fly from continent to continent. Scientists compared the bones of a 25-million-year-old fossil ratite to those of the modern emu and cassowary to determine the specific times of divergence of the various species in the ratite family. Their conclusion was that continental drift explained most,

but not all, of the present distribution of the ratites. That is, the timing of the divergence of each species from the ancestral bird coincided with the splitting of the continent on which they presently reside from the supercontinent. The two ratites whose divergence data did not fit with the timing of continental drift are the kiwis and the ostrich. Scientists are working on various hypotheses to explain these data.

A geographic barrier increases the opportunity for speciation, but separated populations must also diverge genetically and ecologically if they are to evolve into separate species. At least four factors

(a)

(b)

(c)

(d)

(e)

**Figure 17.3** Continental drift as an isolating mechanism. Separation of the continents played a key role in the allopatric speciation of this family of ratites, which are flightless birds with small wings and flat breastbones. (*a*) rhea, (*b*) ostrich, (*c*) kiwi, (*d*) emu, and (*e*) cassowary.

make geographically separated populations diverge over time: (1) Geographically separated populations will likely have different allelic frequencies from the start. It is highly unlikely that the geographic barrier will split the population into two groups with identical gene pools. (2) If a small population becomes isolated from a larger population, genetic drift alone may cause genetic divergence. You may recall that random genetic drift is a change in the frequencies of alleles in the gene pool of a finite population due to chance events (see Chapter 16). Genetic drift can occur in one or a few generations if a small population (founder population) leaves a much larger population and establishes a new population elsewhere. (3) Separated populations experience different mutations. With no gene flow between the populations, new mutant genes stay within the population in which they arise. (4) Different selection pressures will likely exist in the separate geographical areas that the split populations inhabit. Therefore, genetic differences will likely accumulate.

Eventually, as genetic differences increase between the two populations, interbreeding may become impossible, even if the two populations are brought together again. **Figure 17.4a** shows a diagrammatic representation of allopatric speciation. Allopatric speciation is the most common source of new species, especially in animals.

## Sympatric speciation

The word "sympatric" comes from Greek words that mean to live in the same geographical area. **Sympatric speciation** begins as members of a population occupy different ecological zones within the same geographical area. For example, individuals may occupy special habitats (areas in which they live) or use different food sources. As with allopatric speciation, the splitting of a population may act as a trigger for genetic and ecological divergence. Sympatric speciation has been particularly important in the evolution of plants.

The most common type of sympatric speciation occurs with polyploid plants. The term *ploidy* refers to the number of sets of chromosomes in a cell. A typical cell in a species contains two full sets of chromosomes and is called diploid (2N). Cells with only one set of chromosomes—as usually occurs in gametes—are called haploid (N). Cells with an extra set of chromosomes are called polyploid; having an extra set is called **polyploidy**.

How do cells become polyploid? In one scenario, a cell in the meristematic tissue (undifferentiated growth tissue) may replicate its chromosomes and undergo mitosis, but cytokinesis does not occur. That is, the chromosomes replicate and separate,

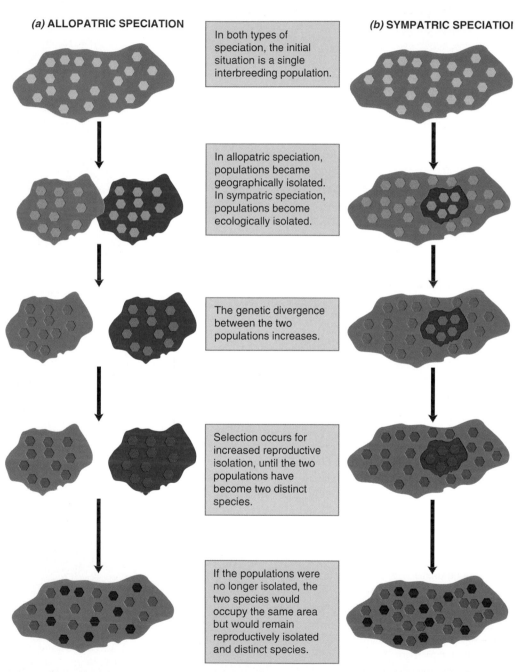

*(a)* **ALLOPATRIC SPECIATION**

In both types of speciation, the initial situation is a single interbreeding population.

In allopatric speciation, populations became geographically isolated. In sympatric speciation, populations become ecologically isolated.

The genetic divergence between the two populations increases.

Selection occurs for increased reproductive isolation, until the two populations have become two distinct species.

If the populations were no longer isolated, the two species would occupy the same area but would remain reproductively isolated and distinct species.

*(b)* **SYMPATRIC SPECIATION**

**Figure 17.4 Allopatric and sympatric speciation.** (a) Populations that are geographically separated and occupy different ranges are said to be allopatric; speciation triggered by geographical separation is termed *allopatric speciation*. (b) Populations that are not geographically separated and do not occupy different ranges are said to be sympatric. Speciation in organisms occupying different ecological zones while still inhabiting the same geographical area is termed *sympatric speciation*.

but the cell does not divide. Therefore, the resulting cell has twice the usual number of chromosomes. If such a cell gives rise to flowers, then diploid (not haploid) gametes will be produced and self-pollination will produce tetraploid offspring. The tetraploid offspring will be reproductively isolated from their diploid ancestors.

In a similar situation, cross-pollination may occur between different but closely related species of plants. Usually such hybrid plants are sterile because their chromosomes cannot pair during meiosis. However, if a cell in the meristematic tissue of this hybrid also becomes polyploid as just described, then the plant can reproduce because it has two full sets of chromosomes. Pairing can occur and fertility is restored. These plants *are* new species. Polyploid plants may arise in other similar ways, and they occur at low frequencies in many natural populations.

How are polyploid plants genetically isolated from the population of which they are a part? Although tetraploid plants can reproduce successfully by self-fertilization, they do not reproduce successfully by mating with diploid plants. The offspring of such a mating is triploid, caused by diploid gametes from the tetraploid plant combining with the haploid gametes from the diploid plant. Most triploid plants have trouble during growth and development. Even if triploid plants survive, they cannot reproduce. Having an odd number of chromosomes makes chromosome pairing impossible and meiosis cannot take place. Therefore, tetraploid plants growing within a population of diploid plants form a genetically isolated population. Thus, speciation can occur in just one generation. Figure 17.4*b* illustrates sympatric speciation.

In the early 1900s, Danish scientist Öjvind Winge provided initial hypotheses on the role of polyploidy in plant speciation, and Dutch botanist Hugo de Vries discovered sympatric speciation by polyploidy in evening primroses as he studied mutations in plants. Today, scientists estimate that 25 to 50% of all plants, particularly angiosperms (flowering plants) are polyploid species. Polyploidy is common in crop species; some polyploid plants are bread wheat, oats, peanuts, sugar cane, Irish potatoes, tobacco, cotton, and apples. Polyploid plants not only have larger cells than the nonpolyploid plant, but the plants themselves are often larger and have higher yields. Thus, polyploid plants often respond favorably to diverse environmental conditions, which has led to the deliberate creation of polyploid varieties of useful plants. Although polyploidy does not generate genetic diversity simply because the genome is doubled, it is associated with gene rearrangement and altered gene expression patterns.

Polyploidy occurs rarely in animals because they are generally unable to survive with chromosomal aberrations. Although it is found in some insects, amphibians, and reptiles, polyploidy is better tolerated in plants. The high frequency of polyploid lineages in plants indicates that this method of speciation is highly successful.

It is not clear whether there are examples of sympatric speciation that do not involve chromosomal changes such as polyploidy. Some scientists suggest that small changes in habitat without a true geographic barrier can be a sympatric trigger for genetic and ecological divergence. An example often used is the treehopper, *Enchenopa binotata*, shown in **Figure 17.5**.

(a)          (b)

**Figure 17.5** The treehopper, *Enchenopa binotata*, and sympatric speciation. The two-spotted treehopper in (*a*) is in its preferred habitat, the bittersweet plant. The treehopper in (*b*), though the same species, prefers hickory trees. These insects tend to mate on the plants on which they feed, so these sympatric subsets of the population may become genetically isolated.

There are over 3000 species of treehoppers, which are insects that feed on plants. They are most abundant in forest or savanna habitats, particularly in the tropics. Two subsets of the population of the treehopper *E. binotata* have adapted to different host plants while living in the same geographical area, splitting the population. One group lives on the bittersweet plant, while the other lives on hickory trees. Since treehoppers tend to mate on the plants on which they feed, there will likely be little gene flow between these two groups of *E. binotata*. Over time, these population subsets may accumulate genetic differences. Although not a geographic barrier, adaptation to different host plants may act as a barrier to interbreeding. Eventually, this split population may evolve into different species. The differing choices of host plants in *E. binotata* reflects small-scale change in the frequencies of alleles governing this trait. Such small-scale change is called *microevolution*, as was mentioned in Chapter 16.

## CONCEPT CHECKPOINT

1. Apple maggot flies lay their eggs on the apples of domestic apple trees or on the small applelike fruit of hawthorn trees. Both trees grow in northern temperate climates. Two hundred years ago, the ancestors of apple maggot flies laid their eggs only on hawthorn fruit. Today, females generally lay their eggs on the type of fruit they grew up in, and males tend to look for mates on the type of fruit they grew up in. So "hawthorn" flies generally end up mating with other "hawthorn" flies and "apple" flies generally end up mating with other "apple" flies. Explain what has likely happened to gene flow within populations of apple maggot flies over the past 200 years. Is allopatiric or sympatric speciation possible? Explain.

## How is reproductive isolation maintained during the speciation process?

### 17.3 Premating and postmating reproductive isolating mechanisms maintain genetic isolation.

With few exceptions (such as polyploidy in plants), speciation is a slow process and is usually not observable in one's lifetime. How is reproductive (genetic) isolation maintained during the slow process of speciation? All reproductive isolating mechanisms restrict gene flow. If gene flow is not restricted between split populations and if significant gene flow resumes, then split populations no longer diverge genetically. They become a single population sharing a single gene pool. Speciation will not result.

### Premating reproductive isolating mechanisms

**Premating reproductive isolating mechanisms** are physical barriers, physiological functions, behaviors, or anatomical structures that prevent organisms from mating. There are selective pressures for such isolating mechanisms across species, since mating between members of different species usually results in offspring that are sterile or that have a low level of reproductive fitness. Premating reproductive isolating mechanisms include:

- *Maintenance of geographic isolation*. If a physical barrier remains over time between members of a split population, then they cannot reach each other to mate and their gene pools will remain separate (**Figure 17.6**).
- *Maintenance of habitat isolation*. If members of a split population continue to occupy different habitats within the same geographical locale over time, their gene pools will remain separate (**Figure 17.7**).
- *Temporal (time-related) isolation*. If members of a split population live in the same geographical locale but breed at different times, mating will be prevented and their gene pools will remain

(a)

(b)

**Figure 17.7** Habitat isolation. Occupying different habitats while living within the same range maintains reproductive isolation between two species of wild lilacs. On the California coast *Ceanothus thyrsiflorus* (a) grows on wooded slopes and canyons having moist, good soil, while *Ceanothus dentatus* (b) grows on sandy hills and flats having dry, poor soil.

(a)

(b)

**Figure 17.6** Geographic isolation. Geographic barriers maintain reproductive isolation among Darwin's finches on the Galapagos Islands. The Cocos finch (a) and the cactus ground finch (b) are two of the 14 species of finches that reside on different islands in the Galapagos chain.

(a)　　　　　　　　　　(b)

**Figure 17.8** Temporal isolation. Breeding at different times while living within the same range maintains reproductive isolation between two species of frogs. (a) *Rana pipiens*, the Northern leopard frog, inhabits a range from southern Labrador (Canada) south to Pennsylvania and Kentucky, and west into the Pacific states. (b) The range of *Rana catesbeiana*, the North American bullfrog, overlaps that of the leopard frog. The bullfrog inhabits a range from Nova Scotia (Canada) to central Florida and west into the Pacific states. The period of most active mating is April for the leopard frog and early July for the bullfrog.

separate. If breeding times overlap, different breeding sites are generally used (**Figure 17.8**).

- *Behavioral isolation.* If members of a split population live in the same geographical locale, but have developed different courtship behaviors, mating will be prevented and their gene pools will remain separate (**Figure 17.9**).
- *Mechanical isolation.* If members of a split population develop differences in their male and female structures, such as in their genitalia or flowers, so that copulation or pollen transfer cannot take place, then mating will be prevented and their gene pools will remain separate (**Figure 17.10**).

**Figure 17.9 Behavioral isolation.** Differences in courtship displays maintain reproductive isolation among various species of fiddler crabs. Male fiddler crabs have one enormous claw, which is held in front of the body much like a fiddle (hence the name). This claw often contrasts in color with the rest of the body and is used by the male at certain times of the year to attract females in a complicated courtship display. Each species of the fiddler crab genus *Uca* has distinctive courtship displays.

Reproductive isolation is crucial to speciation and to the maintenance of separate species in closely related individuals. Often, several premating reproductive isolating mechanisms, not just one, exist between species and help maintain genetic isolation. An excellent example of a pair of species having multiple premating reproductive isolating mechanisms is *Gilia capitata chamissoni* and *Gilia millefoliata*, two small annuals with purple flowers. The first lives on sand dunes and the second in meadows. The first flowers late in the year and the second flowers early in the year. Finally, the first has large flowers and is pollinated by bees, while the second has small flowers and relies on self-pollination.

(a)

(b)

**Figure 17.10 Mechanical isolation.** A common type of mechanical isolation in plants occurs when flowers are adapted for different groups of pollinators as seen in these two species of flowering plants from the North American columbine genus *Aquilegia*. (a) *Aquilegia canadensis* is pollinated by hummingbirds with beaks slightly longer than the 1/2-inch to 3/4-inch flower spur. The spur is the slender, hollow projection from the columbine, and it is a reservoir for nectar, the sugary fluid that pollinators are really after. (b) *Aquilegia flavescens* has a longer spur and is pollinated by moths with a long proboscis. The hummingbirds cannot reach the nectar of *Aquilegia flavescens*. Not all flowering plants have spurs, but botanists have determined that floral nectar spurs have enabled columbines and other flowering plants with spurs to specialize on different pollinators.

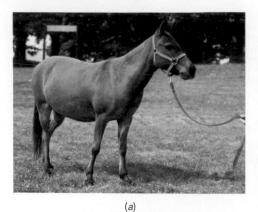
(a)

(b)

(c)

**Figure 17.11 The mule is the classic example of hybrid sterility.** The mule (a) is the sterile offspring of a donkey (b) and a horse (c).

### Postmating reproductive isolating mechanisms

When members of two species mate, their offspring are **hybrids**. Although hybrids can occur in nature, mating between different species is not common owing to premating reproductive isolating mechanisms. Nevertheless, if such mating does occur, postmating reproductive isolating mechanisms exist that help maintain genetic isolation between species. **Postmating reproductive isolating mechanisms** are structures, functions, or developmental abnormalities that, once mating has occurred, prevent organisms of two species or diverging populations from producing vigorous, fertile offspring. Postmating reproductive isolating mechanisms include:

- *Hybrid inviability*. After mating, fertilization may not occur if the female reproductive tract is an unsuitable environment for the sperm or if the organisms have a genetic incompatibility. Even if fertilization occurs, development may stop if the organisms have differing developmental systems. Hybrid inviability is common in plants; hybrid seeds may not germinate or may die soon after germination. An example of hybrid inviability in animals is the combination of sheep and goats. If hybrid embryos are artificially created, they die early in development.

- *Hybrid sterility*. If the hybrid organism develops into an adult, it will likely be unable to produce functional gametes. The classic example of hybrid sterility is the mule (**Figure 17.11**.) A mule is the offspring of a horse and a donkey; it cannot reproduce.

- *Hybrid breakdown*. Sometimes first-generation hybrids appear vigorous and are fertile. If these individuals mate with each other or with individuals of either parent species, the offspring have decreased fitness, are not well adapted, and are soon eliminated. For example, hybrid breakdown occurs between certain species of cotton plants. Hybrids appear vigorous, but their offspring die as seeds, die early in development, or develop into weak plants unable to reproduce.

# How do scientists characterize the mode and tempo of speciation and evolutionary change?

## 17.4 Speciation and evolutionary change can take place slowly and gradually, or suddenly and rapidly.

Well, which is it, you might ask . . . slow and gradual, or sudden and rapid? The answer is simply too complicated to choose one mechanism or the other. Both theories have been useful in interpreting the fossil record.

Darwin proposed the theory of **phyletic gradualism**, which asserts that new species develop slowly and gradually as an entire species changes over time. The phrase "phyletic gradualism" was not Darwin's terminology. It is a modern phrase used to refer to these ideas of Darwin. Phyletic gradualism involves species developing gradually, leaving numerous transitional forms; the entire geographic population generally speciates. In 1972, a new perspective was introduced regarding the manner and pace of evolutionary change. This theory, developed by American paleontologists Niles Eldredge and Stephen Jay Gould, is called **punctuated equilibrium**.

In contrast to phyletic gradualism, punctuated equilibrium involves geologically rapid speciation with long periods of stasis, therefore leaving relatively few transitional fossil forms; speciation occurs in only a small subpopulation of the entire geographic population. The punctuationalist idea of rapid change means that the period of time during which speciation occurs is short with respect to the period of equilibrium or *stasis*, during which small variations do not lead to speciation. The term *rapid* does not imply that speciation takes any particular (short) length of time. That time may vary

| TABLE 17.1 | Phyletic Gradualism vs. Punctuated Equilibrium |
|---|---|
| **Phyletic Gradualism**<br>**Principal Proponent: Darwin** | **Punctuated Equilibrium**<br>**Principal Proponents: Eldredge and Gould** |
| New species develop gradually and slowly with little evidence of stasis (no significant change). | New species develop rapidly and then experience long periods of stasis. |
| The fossil record should contain numerous transitional forms within the lineage of any one type of organism. | The fossil record should contain few transitional forms with the maintenance of given forms for long periods of time. |
| New species arise via the transformation of an ancestral population. | New species arise as lineages are split. |
| The entire ancestral form usually transforms into the new species. | A small subpopulation of the ancestral form gives rise to the new species. |
| Speciation usually involves the entire geographical range of the species (sympatry). | The subpopulation is in an isolated area at the periphery of the range (allopatry). |

from species to species and has been estimated to fall between 50,000 and 500,000 years—incredibly short in geological terms. In part, Eldredge and Gould developed punctuated equilibrium to help explain the appearance of portions of the fossil record that included long periods of unchanged fossils followed rapidly by new types of fossils.

**Table 17.1** presents the main ideas embodied in both phyletic gradualism and punctuated equilibrium. The phylogenetic trees shown in **Figure 17.12** illustrate these ideas. In Figure 17.12*a*, the dots depict apparent gaps in the fossil record from a gradualist point of view: gradual change of a species but no speciation event. *Apparent gaps* are areas in which few, if any, intermediate forms in the fossil record have been found. Speciation is depicted in the gradualist model by one line splitting from another. In Figure 17.12*b*, the dots show the same apparent gaps from a punctuationalist point of view: a speciation event occurring within a short period of time.

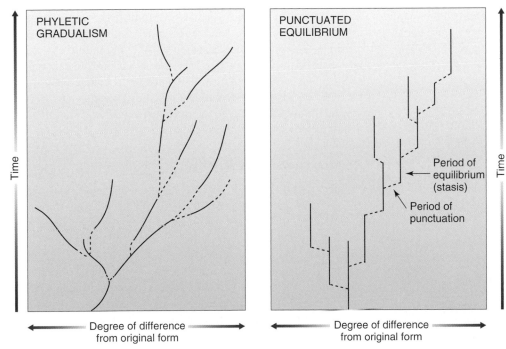

**Figure 17.12 Phyletic gradualism and punctuated equilibrium.** (*a*) Phyletic gradualism: time versus morphology. (*b*) Punctuated equilibrium: time versus morphology. In both representations, time progresses up the Y-axis. Morphology, however, does *not* progress to the right along the X-axis. This axis simply denotes morphological change, whether to the left or the right.

 **Visual Thinking:** In the phyletic gradualism diagram, the solid lines are slanted; in the punctuated equilibrium diagram, they are straight up and down. What does each type of line denote?

CONCEPT CHECKPOINT

2. Which type of reproductive isolating mechanism has probably occurred in each of the following examples? (a) Because of a single mutation in a developmental gene, two closely related species of wasps have a slightly different anatomy of their sex organs making interbreeding impossible. (b) Two species of snakes are morphologically very similar but breed during different seasons. (c) Two similar species of treehoppers live and feed on two separate species of trees. (d) Two similar species of orchids flower at different times during the spring. (e) Due to a geographic barrier, a small peripheral portion of a population remains isolated and its gene pool diverges from that of the original population.

3. Which of the examples above fits with the punctuated equilibrium view of speciation?

# Why do organisms become extinct?

## 17.5 The adaptive capability of a species is one factor in the extinction process.

Species **extinction**—death of all members of a species—is part of the natural course of evolution. It has been estimated that billions of species of plants and animals have become extinct since life arose on Earth.

What causes extinctions? Environmental change is a major culprit, and this factor is discussed in the next section. Environmental change alone, however, generally does not cause extinctions unless the change is catastrophic. Rather, it is the interplay between environmental change and the adaptive capability of a species that can lead to extinction. Three major factors render a species vulnerable to extinction when the environment changes: localized distribution, overspecialization, and reduced genetic variability.

### Localized distribution

**Localized distribution** means that a species lives within a small geographic range. The *range* is the area over which the species is distributed. Therefore, if change occurs in that small area in which the species lives, then it is in danger of being eliminated. For example, the Maya mouse (*Peromyscus mayensis*) occurs naturally only on one isolated mountaintop in western Guatamala. If a dramatic environmental change were to occur there, the Maya mouse could easily become extinct. The Maya mouse is considered an endangered species.

### Overspecialization

**Overspecialization** means that a species is adapted to a narrow range of environmental conditions. One example is the Karner Blue butterfly (**Figure 17.13a**), which is most widespread in Wisconsin and feeds only on leaves of the wild lupine plant (Figure 17.13b) when it is in its caterpillar stage. If the lupines become unavailable in the butterflies' range, then the species will die out. The Karner Blue was listed as an endangered species in 1992.

Another example of overspecialization comes with an interesting story from the tropical island of Mauritius, which is located in the Indian Ocean east of Africa and Madagascar (**Figure 17.14a**). A few tambalacoque trees grow on this island. Hundreds of years ago, tambalacoque trees were abundant there and provided timber for island residents. Flightless dodo birds (Figure 17.14b) were also abundant then and ate the fruit of the tambalacoque trees (Figure 17.14c). In 1505, the island became a stopover for ships engaged in the spice trade. The dodo bird served as a source of fresh meat for sailors, and its population declined. Years later when the island was used as a Dutch penal colony, rats, pigs, and monkeys were introduced to the island. These predators ate dodo eggs that were laid in ground nests. Prior to this time, the dodo bird had no such predators. By 1681, the dodo bird had become extinct, and by 1973 only 13 tambalacoque trees remained. For some reason, the seeds produced by these 300-year-old trees would not sprout. When tackling this problem, researchers discovered that when the dodo bird ate the fruit of the tambalacoque trees, its digestive system "processed" the pitlike seeds, readying them for sprouting upon elimination. The tambalacoque tree had overspecialized—its seeds were able to sprout only after a journey through a dodo and possibly a few other now-extinct species, such as a large-billed parrot and a giant tortoise. Now the dodos and the few other tambalacque seed-processing species are gone. Researchers thought that turkeys might be able to do the same job, and encouraged the birds to eat tambalacoque pits, even though they would not eat the whole fruit. Since then, a few pits have sprouted into healthy, young trees—some say the first new tambalacoque trees on Mauritius Island in more than three centuries.

### Reduced genetic variability

**Reduced genetic variability** means that a species has a narrow range of genotypes. Genetic diversity within the gene pool of a species, rather than genetic uniformity, is adaptive. That is, it helps species survive in the face of environmental change or inbreeding, which may occur among close relatives.

Environments can change in a variety of ways. Not only can the local weather change, but also long-term climates can change, introducing changes in conditions that affect organisms. New diseases, predators, and parasites can emerge. Populations of competitive species can increase in size. Food supplies can diminish or change. Droughts or floods can occur. In such variable environments, differences among individuals of a species increase the probability that some individuals of the species will survive any changes that occur and will live to reproduce. For example, if all the plants in a species have a limited tolerance for drought, then a single drought can destroy an entire population. If, however, some plants within the population have a greater drought tolerance, they might live to reproduce and the species would survive.

(a)           (b)

**Figure 17.13** The Karner Blue butterfly has overspecialized and has become an endangered species. (a) Karner Blue butterfly. (b) The wild lupine. The Karner Blue caterpillar will feed only on the leaves of the wild lupine.

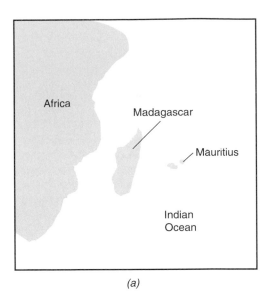

*(a)*               *(b)*

**Figure 17.14 Extinct and endangered species on the island of Mauritius.** (a) Mauritius is a tropical island in the Indian Ocean east of Africa and Madagascar. (b) The dodo bird is an extinct species from the island. Its extinction played a role in the decline of island tambalacoque trees because the trees overspecialized in using the dodo bird as a "processor" of its seeds as a condition for their sprouting. (c) The tambalacoque tree.

*(c)*

In small populations with little genetic diversity, offspring are more likely to inherit two copies of recessive genes that cause deleterious conditions. Therefore, populations with little genetic variation produce offspring who are more likely to have inherited traits that weaken individuals or make them less likely to survive to reproduce. Low genetic variability is one threat to the cheetah's survival.

Cheetahs are spotted, wild cats (**Figure 17.15**) and are the world's fastest-running land animals. Scientists have determined that the small populations of cheetahs in sub-Saharan Africa and in northern Iran are highly inbred. The most widely accepted hypothesis on how their genetic variability became reduced is that cheetahs went through a genetic bottleneck about 10,000 years ago due to a natural disaster that wiped out many mammalian species. The number of cheetahs to survive that cataclysmic event is estimated at 100 to 300, and the species has expanded only through inbreeding of this tiny population. Estimates place the current populations at 10,000 to 20,000. Nevertheless, the lack of genetic variability in addition to other threats cheetahs face, such as habitat destruction, illegal hunting, and predation by lions make the cheetah's future uncertain. The cheetah is an endangered species along with many other species of wild cats.

## 17.6 Environmental change is another factor in the extinction process.

Environmental change is a major factor in the extinction process. The primary types of environmental change that contribute to the extinction process are habitat loss; environmental contamination; competition, predation, and parasitism from introduced or nonnative species; overkilling; and commercial use and abuse.

### Habitat loss

**Habitat loss** is the shrinking and fragmentation of the ranges in which species live. For example, approximately half the forests that once covered the Earth are now gone. Why is this happening? Many processes contribute to habitat loss, and they are all the result of human activity. Some of these are the development of land for residential use, oil and gas exploration and extraction, the conversion of land to agricultural uses, and the use of natural resources, such as forests, for a variety of purposes.

When species lose their habitats, they lose not only the places in which they live and reproduce, but they also lose their sources of food and water. Fragmenting habitats involves breaking one large

**Figure 17.15 Cheetahs are the world's fastest running land animals.** These large cats may reach speeds in excess of 65 miles per hour in short hunting bursts. This species is endangered due to low genetic variability, loss of habitat, illegal hunting, and predation by lions.

## *just wondering . . .*

### I've heard that we are on the brink of a sixth great extinction. Is that true?

Your question is an interesting one, and there is no consensus on the answer. Most scientists knowledgeable in this area would answer yes. However, many would say no.

When speciation and extinction are in balance—that is, when the number of new species that arise equals the number of old species that die out—biodiversity is maintained. *Biodiversity*, or biological diversity, refers to the variety and abundance of organisms living in a particular area or region. Biodiversity is important for many reasons, but one primary reason is that organisms interact with and are dependent on one another; each plays a role in the web of life. When some organisms become extinct, these interactions and relationships are changed in ways that may have negative effects on the species that remain. At the extreme, some extinctions may trigger further extinctions. Extinctions may also have positive effects; they may open up niches previously occupied and result in the diversification and radiation of other organisms.

Biodiversity is also important to the survival of humans. Nature provides raw materials for every aspect of human survival, including food, water, shelter, and medicines, and it forms the basis for our global economy.

Since the beginning of life on Earth, there have been at least 10 *mass extinctions*. A mass extinction is a catastrophic event that causes large numbers of species to become extinct. The five greatest mass extinctions occurred: (1) Near the end of the Ordovician period about 443 million years ago. (2) Near the end of the Devonian period about 360 million years ago. (3) At the end of the Permian period about 250 million years ago. (4) At the end of the Triassic period about 210 million years ago. (5) At the end of the Cretaceous period about 65 million years ago.

Scientists study mass extinctions by analyzing fossils and databases and by computer modeling. Many hypotheses have been developed to explain mass extinction, but scientists think that in general they are caused by drastic environmental changes that follow catastrophic events such as meteorite impacts or massive volcanic eruptions. A *How Science Works* box in Chapter 18 discusses the development of hypotheses to explain mass extinctions, using the mass extinction at the end of the Cretaceous as an example.

Why do some scientists think that we are on the brink of a sixth catastrophic extinction event? Research scientists from a variety of disciplines are gathering increasing evidence suggesting that the balance of life on Earth is being irreversibly altered. Some scientists predict that if things do not change, 30 to 70% of all species now living will disappear within the next 50 years or so.

Extinction is a normal part of evolution. Species arise and species die out. In birds, for example, the normal "background" extinction rate is one species every 100 years or so. Since 1800, however, the Earth has lost 106 species of birds when it should have lost only about two. This is approximately 50 times the background extinction rate.

Why is this occurring? The short answer is human activity. Humans cause large environmental disruptions that result in the extinction of species. Some of these disruptions are the clear-cutting of forests and the deposition of contaminated sediments in lakes and along shorelines. Disruptions such as these lead to the fragmentation, degradation, and loss of habitats.

Other scientists do not agree that we are on the brink of a sixth catastrophic extinction event. They argue that it is extremely difficult to determine extinction rates with reasonable accuracy, especially when we are not sure how many species there are at the start. They suggest that many estimates of species loss are simply speculations with little supporting evidence. For example, geologist A. J. van Loon argues that no scientific "proof" exists for a current mass extinction (in his 2003 article "The Dubious Role of Man in a Questionable Mass Extinction," *Earth Science Reviews*, 62:177–186) **(Figure 17.A**, left).

Countering these arguments, other scientists note that satellite photography provides accurate data on changing land-use patterns and deforestation rates. Moreover, biologists in the field are documenting the effects of the fragmented forests and habitat loss on species diversity. An example of this type of research can be found in the July 24, 2003 issue of the journal *Nature* ( B. W. Brook et al., "Catastrophic Extinctions Follow Deforestation in Singapore," *Nature*, 424:420–423) (Figure 17.A, right). ●

*Are you wondering about a topic in biology and how it relates to your life? Submit your question by clicking the* Just Wondering *link in this text's companion Web site at* www.wiley.com/college/alters.

**Figure 17.A**
Journal articles arguing each side of the "sixth great extinction" issue.

area of habitat into several smaller patches separated by nonhomogeneous areas. This often reduces living spaces to parcels too small to allow large, genetically diverse populations to survive, and it creates living spaces unsuitable for species that need sizeable areas to roam. Risk of predation is generally higher in fragmented habitats as well. In the United States, for example, populations of predators such as raccoons, foxes, coyotes, house cats, and crows tend to increase in fragmented habitats where humans live because these organisms are adapted to feeding on such substances as garbage, pet food, and agricultural crops. In fragmented habitats, individuals of many species face a high probability of death by predation due to high predator populations. Moreover, small parcels of habitat are easily penetrated by the predators.

Scientists think that one example of death by predation due to habitat fragmentation can be seen in the decline of populations of neotropical migrant songbirds in the United States over the past decade. These birds breed in North America in the summer and migrate to Central and South America for the winter. Examples of species experiencing declines are the ovenbird, the wood thrush, and the cerulean warbler (**Figure 17.16**).

One technique used to try to counter the effects of habitat fragmentation is the establishment of wildlife corridors. These corridors are strips of homogeneous habitat that connect the fragments so that animals can move from one fragment to another (**Figure 17.17**). They may take the form of rows of trees as has been used in agricultural regions of Australia, woodland corridors running parallel to or over highways to link forests in Florida, or protected areas linking wildlife reserves in Africa. Corridors are used also for maintaining populations of insect-pollinated plants. Studies show, however, that some corridor designs can actually promote extinctions while others prevent extinctions. More research is needed to determine the best way to prevent extinctions due to habitat fragmentation.

## Environmental contamination

**Environmental contamination** involves the presence of toxic (poisonous) substances in a habitat. Such substances vary widely from raw sewage to chemicals to organisms that cause toxic red tides (see Chapter 20). Toxic substances not only kill organisms directly, but also affect their developmental processes, reducing fitness. In addition, some organisms accumulate certain chemicals in their bodies. As organisms higher on the food chain eat organisms lower on the food chain, toxins accumulate in predators, a concept called biological magnification (see Chapter 42).

## Nonnative species

**Nonnative species** are organisms that are alien to a particular *ecosystem*. (An ecosystem includes all the living and nonliving factors within a certain area.) Because these alien species evolved separately from the ecosystem they invade, the native plants and animals there are vulnerable to the effects of competition, predation, or parasitism that the newcomers may bring. In the last section, we discussed the extinction of the dodo bird in the context of overspecialization of tambalacoque trees. The dodo bird also illustrates the role of competition or predation

(a)

(b)

(c)

**Figure 17.16** Declines in neotropical songbirds appear to be due to habitat fragmentation and resultant predation. Examples of species in decline in the United States in the past decade are (a) the ovenbird, (b) the wood thrush, and (c) the cerulean warbler.

from introduced or nonnative species. The dodo population flourished on Mauritius Island until rats, pigs, and monkeys were introduced to their island habitat. These nonnative species were predators and ate dodo eggs until the native species was eliminated.

We must be very careful when introducing nonnative organisms into an environment because we cannot always predict their effect. The prickly pear cactus was brought into Australia with the intention of using it to create fence-like hedges to keep livestock

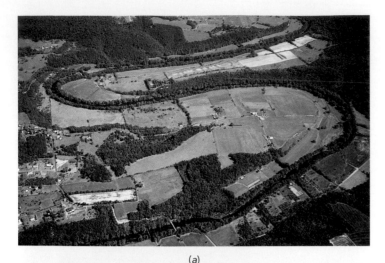

(a)

(b)

**Figure 17.17.** Wildlife corridors connect habitat fragments. Animal movement between fragments is possible through the corridors. (a) Corridor along the Shenandoah River in Virginia. (b) Timber cuts that have preserved corridors in Washington state.

 **Visual Thinking:** Wildlife corridors give plants, as well as animals, the opportunity to survive and thrive in ways they cannot within habitat fragments, yet plants are immobile. How might plants benefit from wildlife corridors?

within certain areas. Its population quickly soared, and it nearly wiped out other species of plants through fierce competition. Australia eventually rid itself of the cactus by introducing a nonnative predator, the cactus-eating South American moth, *Cactoblastis*. The moth lays its eggs in the flesh of the cactus, and the young eat their way to adulthood, destroying the plant. Since *Cactoblastis* saved plant species from extinction in Australia, it was introduced to the island of Nevis in the West Indies to control species of pest cacti. It spread to surrounding islands and eventually reached the Florida Keys. There it nearly rendered extinct the native semaphore cactus. Today, for this cactus to survive, it must be grown in a cage that *Cactoblastis* cannot invade.

**Figure 17.18.** Confiscated endangered species products.

## Overkilling and commercial abuse

There are many types of overkilling; all types involve harvesting at a rate that depletes numbers dramatically and may threaten the survival of species. Overharvesting of trees for wood products and timber affects populations of trees and affects habitats as well. Overhunting, overfishing, and commercial abuse, including illegal trade in endangered species (**Figure 17.18**), threaten species with extinction because populations can be depleted to such low levels that they lack genetic diversity. For example, several species of fishes including Atlantic cod, haddock, and bluefin tuna have been overfished, and their populations are considered depleted. Salmon have been overfished in the past, but their populations are recovering. King crab has been overharvested as well, depleting this species' populations. In many places, heavy fishing pressure and environmental problems have forced governments to limit or halt fishing until fish populations can recover.

Not only does overfishing and overharvesting of shellfish deplete populations, it changes marine ecosystems and has a domino effect on other organisms. The overharvesting of oysters in the Chesapeake Bay, for example, is one factor in the alteration of the water chemistry there because fewer of these mollusks exist to filter the water. According to the Chesapeake Bay Foundation, overharvesting of the oysters, combined with loss of oyster reef habitat, pollution, and disease, has resulted in the dramatic decline of these organisms. This change has made Chesapeake Bay inhospitable to the once abundant populations of manatees, giant sturgeon, whales and alligators. Chesapeake Bay is now undergoing ecological restoration.

### CONCEPT CHECKPOINT

4. For each of the factors involved in the extinction process described in this section, provide one or two examples of your own that you've experienced, learned in school, or read about in the news. Which of these factors could be the result of human activity?

# CHAPTER REVIEW

## Summary of Key Concepts and Key Terms

### In The News: Eagle Expansion

**Endangered species** (p. 264) are species that face a very high risk of extinction (dying out) in the wild in the near future.

**Threatened species** (p. 264) are species whose populations are so low that they are likely to become endangered in the future.

### What are species and how do new species arise?

**17.1** Many species concepts have been formulated; there is no universal definition accepted by all biologists.

**17.1** The **morphological species concept** (p. 266) recognizes species as groups of organisms that share certain morphological (anatomical) and biochemical traits.

**17.1** The **biological species concept** (p. 266) recognizes **species** (SPEE-sheez or SPEE-seez, p. 266) as groups of interbreeding natural populations that are reproductively isolated from other such groups.

**17.1** **Speciation** (p. 266) is the evolution of reproductive isolation between two or more groups of organisms.

**17.1** Other species concepts apply evolutionary, ecological, or other criteria.

**17.2** **Allopatric speciation** (al-ah-PAH-trick, p. 267) occurs when members of a population are separated by a physical barrier, triggering genetic and ecological divergence.

**17.2** **Sympatric speciation** (sim-PAH-trick, p. 268) begins as members of a population occupy different ecological zones within the same geographical area; individuals may occupy special habitats or use different food sources; the splitting of a population may act as a trigger for genetic and ecological divergence.

**17.2** Cells with an extra set of chromosomes are called polyploid, and having an extra set is called **polyploidy** (p. 268).

### How is reproductive isolation maintained during the speciation process?

**17.3** Reproductive isolating mechanisms are restrictions to gene flow.

**17.3** **Premating reproductive isolating mechanisms** (p. 270) are physical barriers, physiological functions, behaviors, or anatomical structures, that prevent organisms from mating.

**17.3** Often, several premating reproductive isolating mechanisms, not just one, exist between species and help maintain genetic isolation.

**17.3** Although **hybrids** (p. 272) can occur in nature, mating between different species is not common due to premating reproductive isolating mechanisms.

**17.3** **Postmating reproductive isolating mechanisms** (p. 272) are structures, functions, or developmental abnormalities that, once mating has occurred, prevent organisms of two species or diverging populations from producing vigorous, fertile offspring.

### How do scientists characterize the mode and tempo of speciation and evolutionary change?

**17.4** Speciation and evolutionary change can take place slowly and gradually, or suddenly and rapidly.

**17.4** The theory of **phyletic gradualism** (p. 272) asserts that new species develop slowly and gradually as an entire population changes over time.

**17.4** **Punctuated equilibrium** (p. 272) involves geologically rapid speciation with long periods of stasis, therefore leaving relatively few transitional fossil forms; speciation occurs in only a small subpopulation of the entire geographic population.

### Why do organisms become extinct?

**17.5** **Extinction** (p. 274) is the death of all members of a species.

**17.5** Extinctions are caused by the interplay between environmental change and the adaptive capability of a species.

**17.5** Three major factors render a species vulnerable to extinction when the environment changes: **localized distribution** (p. 274), **overspecialization** (p. 274), and **reduced genetic variability** (p. 274).

**17.6** The primary types of environmental change that contribute to the extinction process are **habitat loss** (p. 275); **environmental contamination** (p. 277); predation and parasitism from introduced or **nonnative species** (p. 277); overkilling; and commercial use and abuse.

| Level 1 | Learning Basic Facts and Terms |
|---|---|

### Multiple Choice

1. The prinicpal component of the biological species concept is
   a. reproductive isolation.
   c. genetic drift.
   b. anatomical similarities.
   d. evolutionary fitness.

2. A reproductive isolating mechanism
   a. might prevent mating.
   b. might act by preventing hybrids from producing fertile offspring.
   c. might not prevent mating but might prevent fertilization.
   d. might prevent potential mates from ever encountering each other.
   e. All of the above are true.

3. Sympatric speciation is
   a. the appearance of a new species in the same geographic area as the parent population.
   b. the process by which most animal species have evolved.
   c. initiated by the appearance of a geographical barrier.
   d. the emergence of many species from a single ancestor.

4. Speciation may arise from the evolution of different courtship displays and dances. This is an example of
   a. mechanical isolation.
   d. behavioral isolation.
   b. reduced hybrid viability.
   e. sexual selection.
   c. hybrid sterility.

5. The biological species concept is inadequate for grouping
   a. sexually reproducing organisms.
   b. extinct organisms.
   c. populations that produce hybrid offspring that are always nonviable.
   d. populations that produce hybrid offspring that are always infertile.
   e. All of the above are true.

**True–False**

6. Allopatric speciation generally produces a new species in a single generation.
7. Allopatric speciation produces new species due to differences in reproductive behaviors.
8. Failure of chromosomes to segregate properly during cell division may lead to sympatric speciation.
9. Geographic isolation is required for allopatric speciation.

## Level 2 | Learning Concepts

1. Which species concept is used in describing sympatric and allopatric speciation?
2. Although transitional fossils (fossils of organisms intermediate between ancestral forms and descendants) have been found, they are relatively rare. How would a supporter of punctuated equilibrium interpret this? How would a supporter of gradualism interpret it?

3. Which mode of speciation did Darwin observe on the Galapagos Islands? Explain.
4. Why are mass extinctions often followed by large-scale adaptive radiation?
5. Once a population is geographically isolated, a number of factors can act on it to cause genetic divergence. Describe two of these factors.

## Level 3 | Critical Thinking and Life Applications

1. Two populations of fish were isolated in separate lakes about 10,000 years ago because a drier climate gradually lowered water levels (i.e., what started as one population in one lake became subdivided). There are no distinct morphological differences between these two populations of fish, but when brought together in the same location, they will not interbreed because their mating seasons are different. What evolutionary concepts are represented in this scenario?
2. There is current debate centered on the pace of evolution. From which "school" of evolutionary pace is each of the following statements?
   a. The "sudden" appearance of new species in the fossil record is not really as sudden as the fossil record seems to imply. Rather, it occurred as a gradual accumulation of small changes over millions of years, many of which are not recorded in the fossil record.
   b. A new species forms most of its unique features as it comes into existence and then changes little for the duration of its existence.
   c. The periods of modification of species, measured in tens of thousands to millions of years, must have been very short in comparison with the periods during which they remained unchanged.

3. Approximately one-third of all species of fish live in fresh water, while the other two-thirds are saltwater (marine) species. Fresh water lakes and rivers, however, cover only about 1% of the Earth's surface compared to saltwater, which covers three-fourths of the Earth's surface. What might account for this?
4. The Hawaiian Islands are relatively young volcanoes that arose from the Pacific Ocean devoid of any living organisms. Now these islands support a number of species found nowhere else on Earth. What biological process could account for this phenomenon?
5. The Hawaiian Islands are also the site of numerous recent extinctions and loss of biodiversity. Present some reasons why organisms on these islands might be particularly susceptible to extinction.
6. The Florida Keys is an island archipelago that contains virtually no species that are native only to the Keys. Why might this be so?
7. In the next decade NASA is proposing a "Mars Sample Return Mission." This is a mission to send an unmanned probe to bring soil samples back to Earth to be analyzed for possible microbial life. Many scientists are concerned that if there is life on Mars, it could be extremely dangerous to life on this planet if it were ever accidentally introduced into the Earth's ecosystem. Do you think their concerns are valid? Support your answer.

## In The News | Critical Thinking

**EAGLE EXPANSION**

Now that you understand more about species extinction, reread this chapter's opening story about the decline and resurgence of the U.S. bald eagle population. To better understand this information, it may help you to follow these steps:

1. Review your immediate reaction to this information about the bald eagle that you wrote when you began reading this chapter.
2. Based on your current understanding, again summarize the main point of the research in a sentence or two.
3. What questions do you now have about this information that this chapter's opening story does not answer?

4. Collect new information about the research. Visit the *In The News* section of this text's companion Web site at www.wiley.com/college/alters and watch the "Eagle Expansion" video. Then use the "summary" link to read the accompanying story and access related links. Use this information, the links provided, and other online and library resources to answer your questions and find updates about this topic. State the sources of your information. Explain why you think the information is accurate. Also determine whether the information expresses a particular point of view or is biased in any way.
5. What in your view are the most significant aspects of this information? Give reasons for your opinion and for any changes in your ideas based on the additional information you have collected and the analysis you have done.

# THE EVOLUTION OF LIFE ON EARTH

## In The News | Walking Whales

Life on Earth evolved in the sea and then moved to land, but you may not realize that some land species moved back to the sea. One such evolutionary process occurred with whales, and a fossil find of "walking whales" is in the news.

Recently, J. G. M. "Hans" Thewissen of the North Eastern Ohio Universities Medical College and his colleagues discovered a full-skeleton fossil of a whale ancestor more ancient than the amphibious full-skeleton fossils previously discovered. This latest fossil find shows that ancient ancestors to the whales were terrestrial organisms. The photo shows an artist's rendition of the transitional amphibianlike whale relative.

This new fossil find not only shows the terrestrial heritage of whales, it also confirms DNA evidence showing that whales are very closely related to hippos. Previously, paleontologists thought that whales were most closely related to an extinct group of hyenalike organisms, but anatomical analyses of the fossil skeleton added evidence that the hippo is the whale's closest living relative.

The fossil record of whale evolution is quite extensive and more complete with this discovery, showing a series of intermediates between the land ancestor of whales and the fully aquatic whales. You can read more about Thewissen's research on the evolution of whales by viewing the video and reading the story "Walking Whales" in the *In The News* section of this text's companion Web site at www.wiley.com/college/alters.

*Write your immediate reaction to this research on whale ancestry: first, summarize the main point in a sentence or two; then suggest what you think its significance is. You will have an opportunity to reflect on your responses and gather more information on this topic in the In The News feature at the end of this chapter. In this chapter, you will learn more about the evolution of life on Earth.*

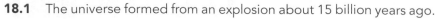

## CHAPTER GUIDE

### How did life on Earth originate?

**18.1** The universe formed from an explosion about 15 billion years ago.

**18.2** Various hypotheses explain the origin of organic molecules about 3.8 bya.

### When did cells arise?

**18.3** The first cells appeared in the Archean era about 3.5 bya.

**18.4** Eukaryotic cells appeared in the Proterozoic era about 2.1 bya.

### When did multicellular organisms arise?

**18.5** An "explosion" of multicellular life occurred in the Cambrian period (544–490 mya).

**18.6** Chordates likely evolved from early echinodermlike organisms in the Cambrian period.

**18.7** Fishes with bony skeletons appeared in the Devonian period (417–360 mya).

**18.8** Land plants became abundant in the Carboniferous period (360–285 mya).

**18.9** Reptiles dominated the Triassic and Jurassic periods (250–145 mya).

**18.10** Flowering plants appeared in the Cretaceous period (145–65 mya).

**18.11** Mammals dominated the Cenozoic era beginning 65 mya.

### What is the history of human evolution?

**18.12** Primates appeared in the early Cenozoic era about 65 mya.

**18.13** Anthropoids appeared about 50 mya.

**18.14** The oldest evidence of hominins is 6 to 7 million years old.

**18.15** The oldest human fossils are about 2 million years old.

**18.16** A now-extinct species of humans arose 1.8 mya.

**18.17** The earliest fossils of *Homo sapiens* are about 500,000 years old.

# How did life on Earth originate?

## 18.1 The universe formed from an explosion about 15 billion years ago.

Vast areas of gas and dust particles swirl in the Orion nebula M42, which is shown in **Figure 18.1**. Orion is a constellation, or group of stars, that can be seen in the night sky of the northern hemisphere. Surrounding some of its stars are cloudlike concentrations of gases and "stardust" called nebulae. Nebulae are found in many sections of the Milky Way galaxy and are part of a cosmic life cycle—the birth, life, and death of stars.

Nebulae are "born" from the dust and gases hurled from unstable stars as they explode. Over time, this material condenses, giving birth to second- and later-generation stars. Then, in turn, these stars become unstable and explode, spewing dust and gases into space. And so the nebula life cycle continues.

The American astronomer Edwin Hubble (1889–1953), for whom the Hubble telescope is named, observed that the galaxies in the universe are moving away from one another. In other words, the universe is expanding. From the rate of expansion of the universe, astronomers infer that it began approximately 15 billion years ago as an explosion from a single point of super-condensed matter, in an event called the **Big Bang**. The universe—as it is known today—did not exist before that time.

The gases and dust from the Big Bang produced an early generation of stars. Then, over billions of years, these stars exploded forming nebulae, and their "space debris" gave rise to other stars and planets. Our solar system probably formed in this way some 4.6 billion years ago. For almost a billion years or so, the molten Earth cooled, eventually forming a hardened, outer crust. Approximately 3.8 billion years ago, the oldest rocks on Earth formed. After that time, some 3.5 billion years ago, life began. (When referring to these distant time periods, we sometimes abbreviate "billion years ago" as simply "bya.")

**Figure 18.1** Orion nebula M42.

## 18.2 Various hypotheses explain the origin of organic molecules about 3.8 bya.

The environment of primitive Earth was an incubator for life. In trying to understand what the Earth was like 3.8 billion years ago, scientists often use information from events that are observable today as starting points to build hypotheses. For example, scientists have determined that the dust and gases ejected from unstable stars contain mostly hydrogen, as well as helium and varying amounts of other elements such as nitrogen, sodium, sulfur, and carbon. Some of these elements combine to form compounds such as hydrogen sulfide, methane, water, and ammonia. As the Earth coalesced, or came together into one mass, it may have contained some of these gases and vapors produced by the planet as it cooled. As the water vapor in this mix condensed, it could have caused millions of years of torrential rains extensive enough to form oceans. Other scientists suggest that water and gas deep within the Earth were vented to the surface by volcanoes.

Scientists agree that the environment of primitive Earth was harsh and violent, bathed in the energy of radiation from the sun, electrical storms, high temperature impact events, wave action, and constant volcanic eruptions. Many biochemists and geochemists hypothesize that under these conditions the elements and simple compounds of the primitive atmosphere reacted with one another to spontaneously form more complex molecules, capturing the surrounding energy within their bonds. To test this hypothesis, Stanley Miller and Harold Urey, working at the University of Chicago in 1953, created these atmospheric conditions in a glass vessel. They introduced four gases into the vessel—methane ($CH_4$), ammonia ($NH_3$), water vapor ($H_2O$), and hydrogen gas ($H_2$). Electrodes within the chamber shot sparks of electricity (like lightning) through the gas mixture while condensers cooled it.

Miller and Urey wondered whether complex organic molecules, of the sort that make up the structure of all living things, would form in their simulated atmosphere, dissolve in the water vapor, and fall into their "ocean." Within a week, they had their answer. A total of 15% of the carbon that was originally present as methane gas had been converted into other, more complex compounds of carbon.

Miller and Urey, joined later by many other investigators around the world, tried similar experiments with different combinations of simple compounds that might have made up the primitive atmosphere or that were likely present on the Earth's surface. In addition, a variety of energy sources were tried, such as ultraviolet light, heat, and radioactivity. More recent experiments have used updated data regarding the probable composition of the early atmosphere. The outcome was always the same: more complex organic molecules formed from simpler ones. These newly formed molecules included many that are essential components of living things: the building blocks of proteins (amino acids), nucleic acids (ribose sugars and purine and pyrimidine bases) lipids and membranes (fatty acids). From this array of experiments, scientists developed the "primordial soup theory" for the origin of life, namely, that complex organic

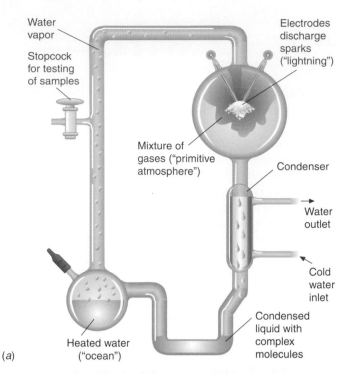

**Figure 18.3 Hydrothermal vents.** Located on the floor of deep oceans, hydrothermal vents are cracks in the Earth's crust that release hot gases and sulfur compounds.

(b)

**Figure 18.2 The Miller-Urey experiment.** (a) Miller and Urey's apparatus consisted of a closed tube connecting two chambers. The upper chamber contained a mixture of gases thought to resemble the Earth's atmosphere. Any complex molecules formed in the atmosphere chamber would be dissolved and carried in droplets to the lower "ocean" chamber, from which samples were withdrawn for analysis. (b) Stanley Miller with the apparatus in 1991.

molecules first formed from simpler ones in the ancient atmosphere. These molecules then accumulated in primitive seas and smaller lakes and lagoons, and spontaneously formed aggregates that could reproduce themselves, starting with nucleic acids, the only group of biological molecules that can make copies of themselves. The Miller-Urey experimental apparatus is shown in **Figure 18.2**.

Life may have also arisen in hydrothermal vents, openings in the deep ocean floor from which hot gases and sulfur compounds escape (**Figure 18.3**). In 1988, Günter Wächtershäuser of the Technical University of Munich proposed that this environment could have supplied the energy and nutrients needed for life to come about. The minerals deposited there appear to catalyze reactions that convert nitrogen into ammonia at the high temperatures and pressure of the vents. In addition, the hydrothermal vents in ex-

istence today provide nutrients for extensive communities of living organisms, such as tubeworms, clams, and bacteria.

Likely conditions of the primitive Earth can be created in the laboratory, and simple organic compounds are formed. But scientists cannot re-create the millions of years that passed and so far have been unable to synthesize molecules or aggregates of them that are capable of reproducing themselves under these conditions. Nonetheless, many biochemists continue to investigate many lines of reasoning about how the precursors to life, and life itself, originated.

The first step toward the evolution of life must have been the synthesis of even more complex organic molecules than the molecules previously described. Scientists know that polymers, such as polysaccharides and proteins, do not spontaneously develop from a mixture of their simpler building blocks, or monomers (in this case, sugars and amino acids). Most polymer synthesis reactions depend on the removal of water and the input of energy to chemically link one monomer to another. This process, called dehydration synthesis (see Chapter 3), could have taken place if condensing agents, which combine with water and release energy, were present with the monomers. If condensing agents were not present, heat and evaporation could also promote the dehydration synthesis of polymers.

In the 1950s, American biochemist Sidney Fox and his coworkers created laboratory conditions under which dry mixtures of amino acids could be heated to form polymers, which he called "proteinoids." In 1998, Günter Wächtershäuser and colleague Claudia Huber reported that they linked amino acids together into short, proteinlike chains (peptide chains) under hydrothermal vent conditions. The formation of early proteins is significant because proteins are not only structural molecules;

many are enzymes that act as catalysts to increase the rates of organic reactions. Metal ions and clays (composed chiefly of minerals) could also have served as early catalysts. When combined into sequences, enzymatic reactions can be thought of as the beginning of metabolic systems—pathways of chemical reactions that sustain life. To be considered life, metabolic systems must be organized within a cellular structure, be able to carry information about themselves, and have the ability to pass on this information by the process of replication.

Self-replicating RNA molecules are generally thought to have started the process of biological evolution because, like all polynucleotides and unlike other biological polymers such as proteins and carbohydrates, they can act as a template or guide for the synthesis of a second polynucleotide that has a complementary sequence of bases (see Chapters 9 and 10). Enclosed by a membrane, replicating RNA (or DNA) and enzymes could be organized to carry on life functions. Chemicals could be selectively allowed into or kept out of the cell's interior. Wastes could be eliminated from the cell, and hereditary material could be passed on from cell to cell.

How much time passed before biochemical evolution became biological evolution—the evolution of life? How did this evolution take place? There is yet no scientific consensus on these points.

### CONCEPT CHECKPOINT

1. Compare the basic research question that Miller and Urey were asking with the question that Fox and, later, Wächtershäuser and Huber were asking.

## *just wondering . . .*

*Questions students ask*

### How does science reconcile the differences between religious beliefs and scientific explanations such as evolution?

It is not within the scientific domain to reconcile science with other disciplines such as art, music, or religion. What you may really be asking is how individual scientists personally reconcile such differences.

Evolution fits into the belief systems of the majority of the world's religions; how it fits varies from religion to religion.

The minority of scientists who embrace one of the small number of religious beliefs in opposition to evolution usually hold their religion separate from their scientific work and do not try to reconcile the two. Science and religion are different ways of knowing. ●

*Are you wondering about a topic in biology and how it relates to your life? Submit your question by clicking the* Just Wondering *link in this text's companions Web site at* www.wiley.com/college/alters.

## When did cells arise?

### 18.3 The first cells appeared in the Archean era about 3.5 bya.

By using relative dating techniques (see Chapter 15), geologists and paleontologists have been able to study the layers of the Earth's crust to discover the time scale of Earth's history. They can tell the age of rocks relative to one another from the unique sets of fossils embedded within them. The Earth's crust can then be seen as a geological time line—a calendar of sorts—in which five main divisions of rock strata are recognized as associated with major **eras** of geological time. These time periods, from the formation of Earth until the present day, are shown in **Figure 18.4** on p. 286.

Eras are subdivided into shorter time units called **periods**. In addition, the periods of the Cenozoic era are subdivided into **epochs** (EP-uks *or* EE-poks). Early geologists defined and named these time units as they discovered them. Note that each unit does not stand for the same length of time. Distinctive events mark the beginnings and ends of time units, and these occurred over varying amounts of time. Figure 18.4 also shows a one-year geological "calendar" to help you relate the geological time line to a single year, making it easier to understand how long each geological time unit is in relation to the others.

Look at the bottom of the geological time line in Figure 18.4 and find the Archean era. It spans January through early June in our "calendar" of life. It is likely that the first cells evolved during this time, about 3.5 billion years ago (3500 million years ago) and fed on organic materials in the environment. The evolution of a pigment system that could capture the energy from sunlight and store it in chemical bonds most likely evolved as early as 2.5 to 3.3 billion years ago in ancient cyanobacteria (SIGH-an-oh-back-TEAR-ee-ah). These single-celled oxygen-producing bacteria became key figures in the evolution of life as it is known today. Their photosynthesis gradually added oxygen to the atmosphere and the oceans around 2 billion years ago, as evidenced by Archean and Proterozoic sediments. Scientists have also discovered fossils of the cyanobacteria.

**Visual Summary**    **Figure 18.4** Geological time.  A summary of the major biological and geological events in Earth's history. The time line isn't drawn to scale: if 0.1 inch represented 1 million years, the time line would be almost 40 feet high! The one-year geological calendar should help you get a sense of the relative length of the geological time divisions. (MYA = millions of years ago.)

| Era | Period | Epoch | MYA | Major Biological and Geological Events | One-Year Geological Calendar |
|---|---|---|---|---|---|
| Cenozoic | Quaternary | Pleistocene | | End of last ice age. (10,000 years ago) | December 31 (11:59 PM) |
| | | | | Modern humans appear (Cro-Magnons). (40,000 years ago) | December 31 (11:55 PM) |
| | | | | Neanderthals appear. (230,000 years ago) | December 31 (11:34 PM) |
| | | | | *Homo sapiens* (early human beings). (500,000 years ago) | December 31 (11:02 PM) |
| | | | 1.8 | | |
| | Tertiary | Pliocene | | *Homo erectus* (sophisticated tools; fire). (1.8 mya) | December 31 (late evening) |
| | | | | Human forms (genus *Homo*) appear. (2 mya) | December 31 (early evening) |
| | | | | *Australopithecus anamensis* (first human ancestor to be fully bipedal). (4 mya) | December 31 (midafternoon) |
| | | | 53 | | |
| | | Miocene | 26 | Hominins (humanlike forms) appear. (6–7 mya) | December 31 (mid-day) |
| | | Oligocene | | *Aegytopithecus* (ancestor of early hominids). (27 mya) | December 30 (early morning) |
| | | | | *Parapithecus* (ancestor of Old World monkeys and hominids). (30 mya) | December 29 (early afternoon) |
| | | | 38 | | |
| | | Eocene | | Anthropoid primates appear. (50 mya) | December 27 |
| | | | 54 | | |
| | | Paleocene | | Primates appear. | |
| | | | | Mammals become abundant. | |
| | | | 65 | | |
| Mesozoic | Cretaceous | | | **Major extinction (dinosaurs, many other animals, many plants, and most marine genera disappear). (65 mya)** | December 26 |
| | | | | Flowering plants appear. (140 mya) | December 20 |
| | | | 145 | | |
| | Jurassic | | | *Archeopteryx* (oldest fossil bird). (145 mya) | December 19 |
| | | | | Large dinosaurs dominate Earth. | |
| | | | | Birds appear. | |
| | | | | Mammals appear. (200 mya) | December 15 |
| | | | 210 | | |
| | Triassic | | | **Major extinction (most marine species and some terrestrial animal species disappear). (210 mya)** | December 14 |
| | | | | Pangea (supercontinent) forms. (240 mya) | December 12 |
| | | | | Dinosaurs appear. | |
| | | | 250 | | |
| Paleozoic | Permian | | | **Major extinction (most marine species disappear). (250 mya)** | December 11 |
| | | | | Conifers appear. | |
| | | | 285 | | |
| | Carboniferous | | | Reptiles appear. | |
| | | | | Coal deposits form. | |
| | | | | Horsetails, ferns, and seed-bearing plants become abundant. | |
| | | | 360 | | |
| | Devonian | | | **Major extinction (most marine invertebrates and many fishes disappear). (360 mya)** | December 2 |
| | | | | Amphibians appear. | |
| | | | | Bony fishes appear. | |
| | | | | Insects appear. | |
| | | | | Jawed fishes appear. (410 mya) | November 28 |
| | | | 417 | | |
| | Silurian | | | Land plants appear. | |
| | | | 443 | | |
| | Ordovician | | | **Major extinction (almost all corals and fish and many other species disappear). (443 mya)** | November 26 |
| | | | 490 | | |
| | Cambrian | | | Vertebrates appear. (500 mya) | November 21 |
| | | | | Chordates appear. (520 mya) | November 20 |
| | | | | "Explosion" of living forms, including origin of main invertebrate phyla. | |
| | | | 544 | | |
| Proterozoic | | | | Multicellular animals appear. (600 mya) | November 13 |
| | | | | Fungi appear. | |
| | | | | Multicellular organisms appear. (1200 mya) | September 27 |
| | | | | Eukaryotic cells appear. (2100 mya) | July 18 |
| | | | | Oxygenation of atmosphere and oceans. | |
| | | | 2600 | | |
| Archean | | | | Stromalites form. | |
| | | | | First living cells appear. (3500 mya) | March 29 |
| | | | | First rocks form. (3800 mya) | March 4 |
| | | | | Origin of Earth. (4600 mya) | January 1 |
| | | | 4600 | | |

(a)

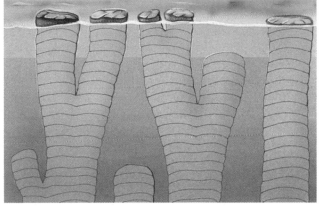

(b)

**Figure 18.5** Stromatolites. (*a*) These stromatolites, located in Shark Bay, Western Australia, formed about 4000 years ago. The largest structures are about 1.5 meters (approximately 4.5 feet) across. (*b*) This diagram shows that much of the stromatolite formation is underwater.

Although fossils of single-celled organisms are usually difficult to find, the cyanobacteria sometimes grew in "piles," creating fossilized columns of organisms and the sediments that collected around them. These ancient columns of cyanobacteria are called stromatolites (stroh-MAT-eh-lites) and date back 3.5 billion years. In **Figure 18.5a**, the tops of stromatolites are visible above the surface of the water in which they form. Like icebergs, however, much of the stromatolite is hidden under water, as shown in Figure 18.5*b*. Actively developing stromatolite formations can be observed even today in the warm, shallow waters of places such as the Gulf of California, western Australia, and San Salvador (Watling Island), Bahamas. The shapes and sizes of the bacteria within present-day stromatolites look very much like the bacteria found within fossil stromatolites (**Figure 18.6**).

## 18.4 Eukaryotic cells appeared in the Proterozoic era about 2.1 bya.

Once again, refer to Figure 18.4 and find the Proterozoic (PRAHT-er-eh-ZOE-ik) era, which spans from June through mid-November in our geological "calendar." The beginning of Proterozoic time, which extends from 2.6 billion to 544 million

years ago, is marked by the formation of a stable oxygen-containing atmosphere. The term *Proterozoic* is derived from two Greek words meaning "prior" (*proteros*) "life" (*zoe*) and refers to the fact that, until a few decades ago, no evidence of life during this time had been found. Once scientists began looking for microfossils, however, they found them in both this and the Archean era.

Until recently, it was thought that eukaryotes arose late in the Proterozoic era, but a discovery published by an Australian team of geoscientists in 1999 described evidence of eukaryotic life at least 2.1 billion years ago, much earlier during the Proterozoic. The evidence is the presence of biological lipids in shale deposits of this age from northwest Australia. Such "molecular fossils" are biochemicals left behind by cells unable to remain intact during the heat and pressure to which they were exposed over billions of years. These biochemical traces show that eukaryotes are 1 billion years older than previously thought. Although they appeared early in the fossil record, eukaryotes did not begin to diversify until 1.2 to 1.0 billion years ago, and it is thought that they did so only after the evolution of sexual reproduction, or after sufficient levels of oxygen developed in the oceans.

How did the eukaryotes arise? Lynn Margulis, in her 1981 book titled *Symbiosis in Cell Evolution*, proposed a controversial answer to that question, but one that has since been upheld by a great deal of evidence. This theory is the **serial endosymbiotic theory (SET)**. *Symbiosis* means "living together," and *endo-* means "within." According to the SET, eukaryotes arose through a succession of symbiotic relationships among various prokaryotic cells in which some prokaryotic cells became attached to or were engulfed by other, pre-eukaryotic (host prokaryotic) cells. These endosymbionts were precursors to present-day organelles of eukaryotes, namely mitochondria and chloroplasts, and quite possibly cilia and flagella as well.

One widely held hypothesis about how these symbiotic relationships first evolved is that predatory prokaryotes engulfed other prokaryotes as food, but some of the prokaryotic prey escaped digestion and remained within their hosts. These endosymbionts benefited by living within the protective environments of their hosts. The host benefited as the endosymbionts carried out chemical reactions that enabled the host to take better advantage of an atmosphere with increasing amounts of oxygen. This would be particularly true if oxygen were toxic to the host cell but could be siphoned off by the endosymbionts and used to create large amounts of energy by

┌─ FOSSILIZED BACTERIA ─┐

└─ PRESENT-DAY BACTERIA ─┘

**Figure 18.6** Fossilized bacteria. The fossilized bacteria were found in stromatolites located in South Africa. These stromatolites are about 3.4 billion years old. The present-day bacteria bear a striking resemblance to the fossilized bacteria.

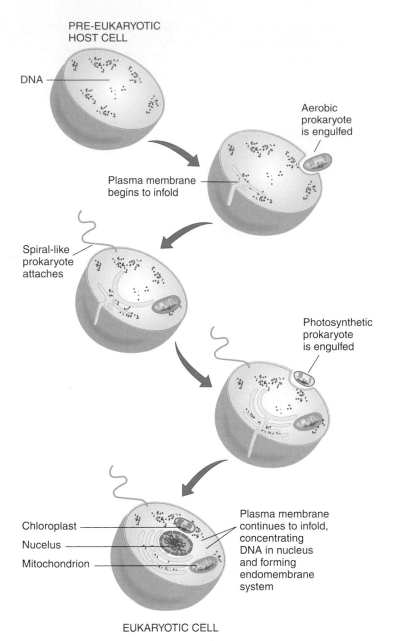

PRE-EUKARYOTIC HOST CELL

DNA

Aerobic prokaryote is engulfed

Plasma membrane begins to infold

Spiral-like prokaryote attaches

Photosynthetic prokaryote is engulfed

Chloroplast

Nucelus

Mitochondrion

Plasma membrane continues to infold, concentrating DNA in nucleus and forming endomembrane system

EUKARYOTIC CELL

**Figure 18.7 Endosymbiosis.** Pre-eukaryotes are thought to have entered into symbiotic relationships with prokaryotes, which likely gave rise to organelles such as mitochondria and chloroplasts as eukaryotic cells evolved. A separate theory addresses the infolding of the plasma membrane to form a nucleus and endomembrane system.

cellular respiration. Such endosymbionts could have been precursors of mitochondria (**Figure 18.7**). In addition, the host cell could have become more efficient at procuring nutrients if elongated spiral-like endosymbionts increased its potential for moving from one location to another. Such endosymbionts could have been precursors of cilia and flagella. In a similar manner, host cells may have engulfed photosynthetic prokaryotes, perhaps cyanobacteria. Such endosymbionts could have been precursors to chloroplasts. The pre-eukaryotic cells that engulfed these prokaryotes thus had a reproductive advantage over those cells not associated with endosymbionts, and therefore they flourished.

Evidence supporting the SET can be seen in cells today. Mitochondria and chloroplasts are each surrounded by a double membrane, corresponding to their own "prokaryotic" membranes as well as the pre-eukaryotic cell membrane that surrounded and engulfed them. These organelles also divide independently within the eukaryotic cell in a manner similar to that seen in bacteria. In addition, mitochondria and chloroplasts have their own DNA and can use it to make proteins. The organization of this DNA is more similar to prokaryotic DNA than it is to modern eukaryotic DNA.

The fungi, a kingdom of organisms that live on dead and decaying matter, also arose during the Proterozoic era, possibly more than 600 million years ago. The ancestry of fungi is not well understood. Scientists think fungi were successful because of their chitin-rich cell walls. Chitin is a stiff, hard substance, also found in the outer skeletons of insects and some other arthropods, that helps fungi remain "drought resistant." Fungi are also thought to be successful land organisms because of their roles, along with certain bacteria, as decomposers.

### CONCEPT CHECKPOINT

2. The mitochondria and chloroplasts have some of their own DNA, which is organized much like it is in bacteria. Explain how this observation supports the endosymbiotic origin of these organelles.

3. Explain how organelles like the nucleus and endoplasmic reticulum may have originated.

## When did multicellular organisms arise?

### 18.5 An "explosion" of multicellular life occurred in the Cambrian period (544–490 mya).

The Paleozoic era (PAY-lee-oh-ZOE-ik; 544 to 250 million years ago) comprises six periods. Although this era appears to span a long time

compared to the other eras shown on the time line in Figure 18.4, it is comparatively short, taking up only November 18 to December 11.

The Paleozoic era is marked by an abundance of easily visible, multicellular fossils. How did multicellular organisms arise? Current thinking is that they probably arose from several protist ancestors, not just one. The earliest multicellular fossils are algae approximately 1.2

billion years old. Fossils of multicellular animals date from much later in the Proterozoic, approximately 600 million years ago, and are found in rocks of southern Australia. For many years, however, the Paleozoic fossils were the oldest known, which led scientists to name this era after this "old" (*paleos*)"life" (*zoos*). Of the roughly 250,000 different kinds of fossils that have been identified, described, and named, only a few dozen are more than 600 million years old.

The Cambrian (KAM-bree-un) period, which ended roughly 490 million years ago, is the oldest period within the Paleozoic era. It represents an important point in the evolution of life; all of the main phyla and divisions of organisms that exist today—except for the land plants—evolved by the end of the Cambrian period. Because so many new kinds of organisms appeared in such a relatively short time span, paleontologists speak of a Cambrian "explosion" of living forms and often refer to geological strata older than Cambrian time as Precambrian.

The evolution of Cambrian organisms took place in the sea. **Figure 18.8** shows fossils of the organisms that lived on the sea floor

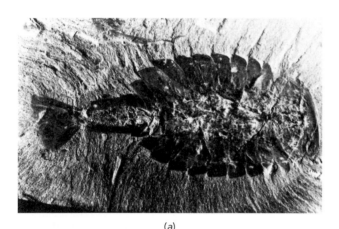

(a)

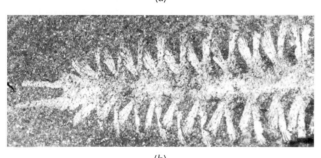

(b)

(c)

**Figure 18.8** Cambrian period fossils. These specimens are from the Burgess Shale, British Columbia, Canada, and are about 520 million years old. (a) A fossil of *Sidneyia inexpectans*, an arthropod (size: 2–5 inches long). (b) *Burgessochaeta setigera*, a segmented worm (size: about 5/8 inch long), showing a pair of tentacles on its front end. Footlike structures are attached to the worm in pairs along the sides of the body. (c) *Wiwaxia corrugata*. The body of this animal was covered with scales and also bore spines (size: about 1 1/4 inches across, excluding the spines). It may have been a distant relative of the mollusks.

(a)

(b)

**Figure 18.9** (a) Trilobite fossils (found in Ginetz, Czech Republic) and (b) horseshoe crabs. Both organisms appear to have arisen from the same evolutionary lineage.

during this period—quite different from those found there now! These fossils, together with well over 100 other species, have been found in the **Burgess Shale**, which are sedimentary rocks formed from fine-grained mud. During past movements of the Earth's crust, these geological strata, which were originally underwater, were uplifted and are now part of the Rocky Mountains of British Columbia, Canada.

Many kinds of multicellular organisms that thrived in the early Paleozoic era have no relatives living today, while others ultimately led to the contemporary phyla of organisms. The trilobites, for example (**Figure 18.9a**), appear to be derived from the same evolutionary line that gave rise to one living group of arthropods, the horseshoe crabs (Figure 18.9b).

## 18.6 Chordates likely evolved from early echinodermlike organisms in the Cambrian period.

During the Cambrian period, wormlike aquatic animals similar to lancelets and lampreys began to evolve from early echinodermlike

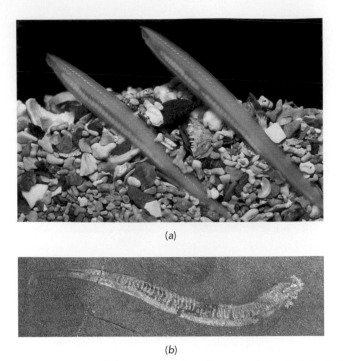

(a)

(b)

**Figure 18.10 Chordates.** (a) Two lancelets partly buried in shell gravel. Note the resemblance to *Pikaia gracilens* (b), the earliest known chordate. *Pikaia gracilens* is a small fishlike animal, and it is one of the first organisms with a notochord. This particular fossil was found in the Burgess Shale.

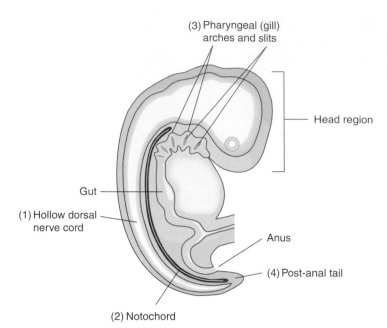

**Figure 18.11 The chordate embryo.** The four main chordate features are shown in the chordate embryo: (1) a single hollow nerve cord, (2) a rod-shaped notochord, (3) pharyngeal (gill) arches and slits and (4) a post-anal tail. All features may not be present in the adult but are always present in the embryo.

**Visual Thinking:** Which of these chordate characteristics do humans exhibit as adults?

organisms. Characteristic of these organisms was the stiff, internal rod that ran down the back, ventral to the central nerve. This stiffening rod is called a notochord, meaning "back cord." These organisms are therefore called **chordates** (KOR-dates). One of the earliest known chordate fossils is shown in **Figure 18.10**.

The approximately 42,500 species of chordates that exist today include fishes, amphibians, reptiles, birds, and mammals. Mammals are warm-blooded vertebrates that have hair and whose females secrete milk from mammary glands to feed their young. Chordates are distinguished by four principal features: (1) a single hollow *nerve cord* located along the back; (2) a rod-shaped *notochord*, which forms between the nerve cord and the developing gut (stomach and intestines); (3) *pharyngeal (gill) arches and slits*, which are located at the throat (pharynx) and (4) a muscular post-anal tail, which is a tail posterior to the anus. Nearly all other animals have a terminal anus. These four chordate features are shown in **Figure 18.11** as they appear in the chordate embryo of today. All of the distinctive features of chordates are evident in their embryos, even if they are not present in the adult form of the organism. The bony structures that support the pharyngeal arches evolved into jaws with teeth, as shown in **Figure 18.12**, allowing these organisms to feed differently than their ancestors.

With the exception of the tunicates and lancelets (see Figure 18.10a), all chordates are **vertebrates** (VER-teh-brits *or* VER-teh-braytes). Vertebrates, a subphylum of the chordates, differ from these two chordate groups in that the adult organisms have a vertebral column, or backbone, that develops around and replaces the embryological notochord. In addition, most vertebrates have a distinct head and a bony skeleton, although the living members of two classes of fishes, *Cyclostomata* (lampreys and hagfishes) (**Figure 18.13**) and *Chondrichthyes* (sharks, skates, and rays), have a cartilaginous skeleton.

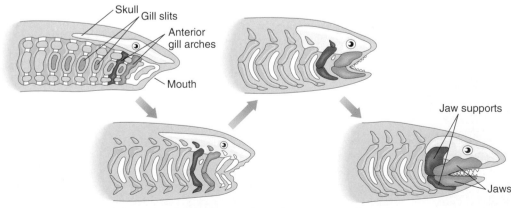

**Figure 18.12 Evolution of the jaw.** Jaws evolved from the anterior gill arches of the jawless fishes. With jaws, organisms fed differently than their jawless ancestors.

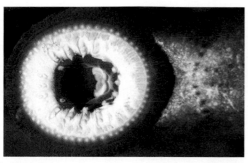

**Figure 18.13 A lamprey.** Lampreys attach themselves to fish by means of their suckerlike mouths, rasp a hole in the body cavity, and suck out blood and other fluids from within. Lampreys do not have a distinct jaw.

**Figure 18.15** Reconstruction of *Ichthyostega*, one of the early amphibians. *Ichthyostega* had efficient limbs for crawling on land, an improved olfactory sense associated with a lengthened snout, and a relatively advanced ear structure for picking up airborne sounds. Despite these features, *Ichthyostega*, which lived about 360 million years ago, was still quite fishlike in overall appearance.

## 18.7 Fishes with bony skeletons appeared in the Devonian period (417–360 mya).

The first vertebrates to evolve were jawless fishes, members of the class *Cyclostomata*, about 500 million years ago (see Figure 18.13). Although traces of their fossils are found in Cambrian strata, most date back to the Ordovician and Silurian periods, about 490 to 417 million years ago.

Jaws first developed among vertebrates that lived about 410 million years ago, near the beginning of the Devonian period. The first jawed fishes that evolved were placoderms, ancestors of today's bony fishes (such as salmon, trout, cod, and tuna) and cartilaginous fishes (such as sharks, skates, and rays). They were also one of the first groups of fishes to have paired fins. During the Devonian period, fishes having a bony skeleton appeared: the ray-finned fishes and the fleshy-finned fishes. The ray-finned fishes are the ancestors of the wide variety of bony fishes that exists today.

In the late Devonian period, the lungfishes and the lobe-finned fishes appear in the fossil record. Fleshy-finned fishes are ancestors to both. Only a few species of lungfishes exist today; one is shown in **Figure 18.14**. Lungfishes generally live in oxygen-poor stagnant

**Figure 18.14 A lungfish.** As the name suggests, lungfishes have lungs as well as gills. Although their lungs are supplementary to their gills, they allow the fish to come to the surface and breathe when oxygen levels are too low in their stagnant water environments. Four species of lungfishes exist today; they are found only in the waters of Australia, Africa, and South America.

ponds and swamps. Most of the Devonian period lobe-finned fishes also had primitive lungs similar to the lungfishes.

One group of the lobe-finned fishes is thought to be the ancestor of the land-living tetrapods (four-limbed vertebrates), a group that includes humans. The first land vertebrates were amphibians, animals that live in water and on land.

The early amphibians arose in the Devonian period and had fishlike bodies, short stubby legs, and lungs (**Figure 18.15**). They flourished during the next period of the Paleozoic era—the swampy Carboniferous period. Although most mature amphibians spend most of their time on land, the large majority of them must return to the water frequently. It is vital that they keep their thin skin moist because they use it for respiration in addition to gills and/or lungs. Amphibians also lay their eggs in water or in moist places and live in water during their early stages of development.

## 18.8 Land plants became abundant in the Carboniferous period (360–285 mya).

How did the land plants arise, such as the tall Sequoias, beautiful roses, or pesky crabgrass? There is evidence that plants independently invaded the land several times. If you look at Figure 20.17a, you'll see an organism similar to the ancient relatives of the land plants—single-celled green algae.

Until 2002, scientists thought that the first terrestrial plants evolved at the interface of the Silurian and Devonian periods, approximately 417 million years ago. However, more recent research has pushed back this date to about 430 million years ago, with the find of fossilized land plants from the Silurian period. Earlier plant fossils exist dating back to 475 to 460 million years ago, in the Ordovician, but they show plants having a simple tubular structure, with neither leaves nor roots. The new fossils show multicellular plants with mechanisms for transporting water within their bodies as well as for conserving it.

Within the next 100 million years, plants became abundant and diverse on the land and eventually formed extensive forests. The Carboniferous period (360 to 285 million years ago) is named for the great coal (carbon) deposits formed during this time. Much of the land was low and swampy due to a worldwide moist, warm climate—conditions that contributed to the fossil preservation of the still-prevalent forests. These coal deposits

**Figure 18.16** A forest during the Carboniferous period. This diorama of a reconstructed swamp forest of the Carboniferous period is displayed at the Field Museum of Natural History in Chicago. Notice the bushy, vinelike plants, horsetails, and ferns . . . and the huge dragonfly!

provide a relatively complete record of the horsetail plants, ferns, and primitive seed-bearing plants of the Carboniferous period. The conifers—a group of seed-bearing plants that is represented today by pines, spruces, firs, and similar trees and shrubs—also originated then. A reconstruction of a Carboniferous forest is shown in **Figure 18.16**.

## 18.9 Reptiles dominated the Triassic and Jurassic periods (250–145 mya).

If you check back to the geological time line in Figure 18.4, you'll see that the Mesozoic era, or "middle life" (250 to 65 million years ago), takes up most of the month of December. This era was a time of adaptive radiation of the terrestrial plants and animals that had been established during the mid-Paleozoic era, but that had not been wiped out by the mass extinctions of the Permian period (see *How Science Works*, p. 294). The radiation of these organisms led to the establishment of the major groups of organisms living today. The first reptile fossils appear in strata from the Carboniferous period. By the beginning of the Permian period, many new reptilian groups existed alongside a variety of amphibians.

The reptiles gradually replaced the amphibians, which had been the dominant land vertebrates for about 100 million years. Reptiles began to take over as the climate became drier during the Mesozoic era. Adaptations to dry climates, such as water-resistant skin and more efficient lungs, gave reptiles a survival advantage over the amphibians. In addition, reptiles developed a reproduc-

tive advantage—the amniotic egg. An amniotic egg has a thick shell that encloses the developing embryo and a nutrient source within a water sac, like a chicken's egg (see Chapter 22). This type of egg protects the embryo from drying out, nourishes it, and enables it to develop away from water.

The Mesozoic era (250 to 65 mya) is divided into three periods: the Triassic, the Jurassic, and the Cretaceous. During the Triassic period, small dinosaurs and primitive mammal-like reptiles appeared. During the Jurassic period, large dinosaurs dominated the Earth, and birds first appeared. The oldest undisputed fossil bird is *Archaeopteryx*, estimated to be about 145 million years old (**Figure 18.17a**).

### From reptiles to birds

Since *Archaeopteryx* represents a fairly advanced stage in bird evolution, and since no earlier well-accepted fossils showing bird evolution have been discovered, the origin of birds remains in dispute as alternate hypotheses are continually discussed. One hypothesis states that birds arose more than 200 million years ago from archosaurian reptiles which were direct ancestors to the dinosaurs. The archosauria are a large branch of reptiles that appeared early in the Triassic period. Their present-day relatives are crocodiles. The other major hypothesis is that birds originated from theropod dinosaurs in the later Cretaceous period (80 to 110 million years ago). Theropod means "beast-footed." This group includes the large carnivorous dinosaurs, such as the *T. rex*. This "dinosaur hypothesis" suggests that birds are the descendants of small theropods.

In 1998, a team of paleontologists from China, Canada, and the United States announced that they had discovered two new species of

(a)

(b)

**Figure 18.17** Links to the birds. (a) About the size of a crow, *Archaeopteryx* lived in the forests of Central Europe 145 million years ago. The teeth and long jointed tail are features not found in any modern birds. Discovered in 1862, *Archaeopteryx* was cited by Darwin in later editions of *On the Origin of Species* in support of evolution. (b) *Caudipteryx*. This feathered dinosaur appeared to provide evidence that birds might be descended from the dinosaurs. Not all scientists agree with this hypothesis, however, and suggest that birds and dinosaurs evolved from a common ancestor. They contend that *Caudipteryx* is a flightless bird that descended from more ancient birds.

small dinosaurs covered with feathers, *Protarchaeopteryx robusta* and *Caudipteryx zoui* (Figure 18.17b). These dinosaur fossils are not as old as *Archaeopteryx*, dating back to only 120 million years, but they initially seemed to add more evidence to the hypothesis that birds evolved from the dinosaurs. An analysis of *Caudipteryx* published in 2000, however, suggests that *Caudipteryx* is a bird that lost the ability to fly and is not a form transitional between birds and dinosaurs. Others, however, still contend that *Caudipteryx* is a transitional animal. Today, there are about 9000 species of birds.

The origin of flight is also uncertain. *Archaeopteryx* had feathers, but apparently not enough of them to fly very effectively. Feathers, thought to have evolved from reptilian scales, might have evolved first in dinosaurs as an adaptation that helped them control heat loss in their bodies. In addition, feathers on their forelimbs might have acted as traps for insects. Such feathered forelimbs and enlarged chest muscles might have served as **preadaptations** for flight. Preadaptations are inherited traits used for new functions for which they are later selected.

Early birds likely glided from tree to tree. Ultimately, feathers made possible the subsequent evolution of birds that could fly well. This ability allowed birds to populate unoccupied habitats and resulted in the evolution of a large and diverse class of organisms. Questions still remain, however, about whether feathers evolved in the manner just described or evolved primarily associated with flight and secondarily as insulators.

 CONCEPT CHECKPOINT

4. One concept that reappears throughout the evolution of life is preadaptation. Revisit the history of life as presented thus far and describe some examples of preadaptations.

## 18.10 Flowering plants appeared in the Cretaceous period (145–65 mya).

During the early Cretaceous period, about 140 million years ago, flowering plants (angiosperms) began to appear, becoming the dominant form of plant life about 100 million years ago. Seed-bearing plants with fernlike leaves, similar to the living cycads, were abundant at that time. Present-day cycads are shown in **Figure 18.18**. They are chiefly tropical plants that have thick stout trunks, palmlike fronds, and massive cones. Only about 100 species exist today. However, about 240,000 species of flowering plants exist, which makes them the dominant group in the plant kingdom, greatly outnumbering all other kinds of plants.

Until recently, all evidence suggested that insects and flowering plants coevolved. *Coevolution* is the reciprocal evolutionary influence between two species that are dependent on one another. Each exerts selective pressure on the other, so they evolve together. As flowering plants changed over time, insects did also because their feeding structures and habits were linked to the characteristics of the flowers. (An example of such linkage can be seen in the pollinators and flowers shown in Figure 17.10.) As the flowering plants became more diverse and plentiful, so did the insects.

Fossil evidence analyzed by American paleontologists Conrad Labandeira of the Smithsonian's National Museum of Natural History and Jack Sepkoski shows that all major insect functional–feeding

**Figure 18.18 Present-day cycads.** Ancient cycads that looked much like this were abundant about 100 million years ago.

groups predate the earliest fossil record of angiosperms. Their data suggest that insects (which arose in the Devonian period) experienced a major radiation in diversity during the Carboniferous period. Then, during the mass extinctions of the Permian period, about 30% of all insect orders died out. The adaptive radiation of the insects began again in the Triassic period, *before* the diversification of the angiosperms, and it continues today. The Labandeira and Sepkoski research suggests that the diversification of the insects actually *slowed down* as angiosperm diversity speeded up. Since the first publication of their data in 1993, further fossil evidence has bolstered their hypothesis.

Around 65 million years ago, as the Cretaceous period (and the Mesozoic era) ended, sudden shifts occurred in the kinds of marine organisms that existed. Some became extinct and others began to flourish. Scientists can infer these changes in the populations of marine organisms living at that time by studying marine fossils exposed in certain European geological strata. For example, many of the larger plankton (drifting or weakly swimming organisms suspended in water) disappeared about 65 million years ago, and a small number of the tinier ones took their place. The same rapid changes occurred in at least some nonplanktonic marine animal groups, such as the bivalve molluscs (clams and their relatives). The ammonites, a large and diverse group that are related to octopuses but that have shells, abruptly disappeared. A major extinction occurred on land: The dinosaurs disappeared, though at a pace much slower than that of the large plankton just mentioned. The *How Science Works* box discusses extinctions, in particular the extinction at the end of the Cretaceous period.

## 18.11 Mammals dominated the Cenozoic era beginning 65 mya.

The mass extinctions of the Cretaceous period marked the end of the Mesozoic era and heralded the Cenozoic era about 65 million

# How Science Works

## Mass extinctions: the development of hypotheses to explain these events

One of the most prominent features of the history of life on Earth has been periodic *mass extinctions*, the global death of whole groups of organisms. Five such events are noted in the Chapter 17 *Just Wondering* box. The fifth of these occurred at the close of the Cretaceous period, 65 million years ago. This was the famous event when dinosaurs became extinct.

Scientists have long discussed hypotheses regarding the extinction event of the late Cretaceous period. In 1980, a group of scientists headed by Luis Alvarez of the University of California, Berkeley, presented a dramatic hypothesis about the cause of this fifth great extinction. Alvarez and his associates observed that the usually rare element iridium was abundant in a thin layer in the geological strata that marked the end of the Cretaceous period. Alvarez and his colleagues proposed that if a large, iridium-rich meteorite or asteroid had struck the surface of the Earth, a dense particulate cloud would have been thrown up. This cloud would have darkened the Earth for a time, greatly slowing or temporarily halting photosynthesis and driving many kinds of organisms to extinction. By disrupting and killing plant life, other organisms dependent on the plants would die out as well. Then, as

its particles settled, the iridium in the cloud would have been incorporated into the layers of sedimentary rock that were being deposited at that time. Until recently, the "meteorite hypothesis" has been the most plausible explanation for the extinctions and

**Figure 18.A  A meteorite hit.** A crater on the Yucatan Peninsula in Mexico was thought to be the site of a meteorite impact that was implicated in a mass extinction at the end of the Cretaceous period. The illustration shows the location of that crater.

was widely accepted by the scientific community. An impact crater on the Yucatan Peninsula in Mexico (**Figure 18.A**) has been

considered the likely site of the key meteorite hit. This 110-mile-wide crater dates from the end of the Cretaceous period.

New evidence emerged in 2004 suggesting that the cause of the extinction event of the late Cretaceous might have been more complex than a single meteorite impact. Gerta Keller, a paleontologist at Princeton University, found evidence in core samples taken from the crater that the meteorite hit Mexico approximately 300,000 years *before* the dinosaurs and other organisms disappeared. This evidence lends support to the hypothesis that species were already endangered prior to the extinction event in the late Cretaceous. Over the past 10 years, Keller and her research team have studied foraminifera at the Mexican meteorite impact site and in many other countries. Foraminifera are microscopic eukaryotes that are extremely sensitive to environmental changes. By studying the shells of these organisms, Keller and her team were able to gather data on the temperature, salinity, and other environmental conditions of the time in which the organisms lived. Keller's data and that of others suggest that multiple meteorite impacts, intensive volcanic activity, and resulting greenhouse effects had negatively affected populations of organisms leading up to the Cretaceous extinction. Possibly a single impact after the Yucatan impact and at another site caused already stressed and diminished populations to become extinct. ●

---

years ago. Extending to the present, the Cenozoic era has two periods—the Tertiary and the Quaternary—which are subdivided into epochs. The epochs of the Tertiary period are listed in Figure 18.4. Notice also that most of the "year" has passed on the geological calendar. Although 65 million years may seem like an incredibly long time, in terms of the history of life, it is short.

At the beginning of the Cenozoic era during the Paleocene epoch, the small mammals that survived the extinctions of the Mesozoic era underwent adaptive radiation, quickly filling the habitats vacated by the dinosaurs. Organisms having *both* mammalian and reptilian characteristics, called transitional forms, first appear in the fossil record approximately 245 million years ago. Then, about 200 million years ago in the early Mesozoic era, the first known mammals appear. These early mammals resemble today's shrews, as seen in **Figure 18.19**. They were small, fed on insects, and were probably nocturnal (active at night). About 65 million years ago, following the extinction of the dinosaurs, mammals became abundant.

Various natural selection pressures resulted in the emergence of a number of significant anatomical and physiological changes in

the mammals as they evolved from their reptilian ancestors. These changes resulted in a class of organisms that not only survived but also flourished in a wide variety of habitats. Changes occurred in the reptilian arrangement of limbs as the mammals evolved, raising them high off the ground and allowing them to walk quickly and to run. The lower jaw became a single bone, and the hinge between it and the skull became repositioned. The teeth became flattened and opposed, allowing mammals to grab, tear, chew, and grind food.

The reptilian heart, having two ventricles with an incomplete separation, developed a complete wall in the mammals. Although very little mixing of oxygenated and deoxygenated blood occurs in the reptilian heart due to the coordinated actions of the heart valves and beating rhythm, a full wall between the ventricles completely prevents the mixing of oxygenated and deoxygenated blood (see Figure 22.9). This change was significant because mammals had also developed *warm-bloodedness*: a constant internal body temperature. Warm-blooded animals are also called *endotherms*, meaning "within" (*endo-*) "temperature" (*-therm*). Their reptilian ancestors were *cold-blooded*, (*ectotherms*, meaning "outside" [*ecto-*] "tem-

**Figure 18.19** The first mammals looked much like today's shrews. Shrews look mouselike, but have pointy faces.

perature" [-*therm*]), having internal body temperatures that followed the temperature of their environments within certain limits. Many scientists, however, now suggest that at least some dinosaurs were warm-blooded, which might help explain why they became so plentiful and dominant for so long. One major advantage of en-

dothermy is physiological independence, within limits, from environmental temperatures. Thus, warm-blooded dinosaurs could hunt at any time or in any season. Nevertheless, one disadvantage of endothermy is a higher rate of metabolism than ectotherms requiring more oxygen for the increase in the rate of cellular respiration. A four-chambered heart was more efficient at delivering oxygen.

Important changes also occurred in reproduction as the mammals evolved from the reptiles. The most "reptilian" type of reproduction is seen in the monotremes, one of the three subclasses of mammals living today. Monotremes lay eggs with leathery shells and incubate them in a nest. The underdeveloped young that hatch from these eggs feed on their mother's milk until they mature. The ability of a mother to feed her young with milk she produces in mammary glands is another characteristic that has been advantageous to the adaptive radiation of the mammals. The present-day monotremes are the platypuses and spiny anteaters (echidnas) of Australia. These organisms are thought to have arisen on Gondwana, which became today's South America, Africa, and Australia (see *How Science Works*, p. 296) and were then isolated in Australia as the continents drifted apart.

The marsupials, a second subclass of mammals, do not lay eggs but give birth to immature young. These blind, fetal-looking young crawl to the mother's pouch and nurse there until they are mature enough to venture out on their own. Like the monotremes, marsupials were present in Gondwana; today most marsupials, such as kangaroos, wombats, and koalas, are found in Australia. The marsupials of Australia resemble the placental mammals that are present on the other continents. **Figure 18.20** compares individual members of these two sets of mammals. The members of each pair have similar habitats and find their food in similar ways. These characteristics, along with their strikingly similar anatomy, suggest that the resemblances of these two subclasses of mammals are a product of convergent evolution.

In placental mammals, the young develop to maturity within the mother. They are named for an organ formed during the course of their embryonic development, the placenta. The placenta is located within the walls of the uterus, or womb. Composed of both maternal and fetal tissues, the placenta is connected to the fetus by the umbilical cord. At the placenta, fetal wastes pass into the bloodstream of the mother, and oxygen and nutrients in the mother's bloodstream pass into the bloodstream of the fetus, a mechanism that allows the fetus to develop within the mother until it reaches a certain age of maturity.

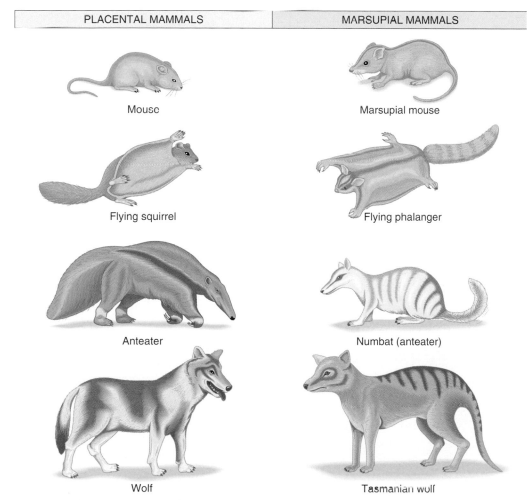

| PLACENTAL MAMMALS | MARSUPIAL MAMMALS |
|---|---|
| Mouse | Marsupial mouse |
| Flying squirrel | Flying phalanger |
| Anteater | Numbat (anteater) |
| Wolf | Tasmanian wolf |

**Figure 18.20** Convergent evolution. This illustration demonstrates the convergent evolution of marsupials in Australia and placental mammals in the rest of the world.

# How Science Works

## The merging of sciences: using the theory of plate tectonics to understand evolution

During the Mesozoic era, the continents did not exist as they do today. Instead, they were all joined in one giant continent called *Pangea*, meaning "all Earth." Pangea began to break up into smaller pieces during the Mesozoic era, with this movement continuing during the following era, the Cenozoic. These changes greatly affected evolution.

After the Earth coalesced approximately 4.6 billion years ago, its surface was hot and violent for about 600 to 700 million years. At first, these conditions were too inhospitable for living things or even for biochemical evolution to take place. Scientists hypothesize that geological activity eventually resulted in the formation of fragments of continents. Some geologists think that these land masses were formed as melted rock material rose to the surface of the Earth, pushed upward by other heavier, sinking, melted metals such as iron and nickel. Others suggest that the highly active volcanoes prevalent at that time spewed so much lava into the seas that it eventually accumulated, forming land masses.

Approximately 240 million years ago, small land masses united to form the single, large "supercontinent" Pangea. Its northern half was called *Laurasia* and consisted of present-day North America, Europe, and Asia. Its southern half was called *Gondwana* and was a combination of present-day South America, Africa, India, Antarctica, and Australia. Pangea remained as a supercontinent for approximately 100 million years. Forces were at work beneath it, however, that eventually divided it—once again—into smaller land masses.

Today, the outer shell of the Earth is made up of six large "plates" and several smaller ones (**Figure 18.B**). According to the *plate tectonics theory*, these plates are rigid pieces of the Earth's crust, which are from 75 to 150 kilometers (approximately 50 to 100 miles) thick. The rocks beneath these plates are less rigid than the plates they support. The surface plates move, or drift, over this underlying semiliquid rock. Scientists think that this "liquid" rock flows slowly, rising as it heats and sinking as it cools. At the mid-oceanic ridges—long, narrow mountain ranges under the sea—the plates move away from one another in a process called *sea floor spreading*. Molten rock wells up from deep within the Earth, pushing the plates apart and filling in the space between. At other places, the plates move toward each other. As they meet, one plate may sink below the other, or they may collide, pushing up mountain ranges in the process. The Himalayas, for example, have been thrust up to the highest elevations on Earth as a result of the prolonged collision of the Indian subcontinent with Asia. Other types of violent geological activity, such as earth-

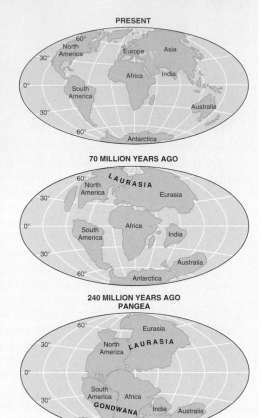

**Figure 18.C** Movement of the continents. The changes that occurred in the positions of all the continents as a result of shifting plates have played a major role in the distribution of organisms seen today.

quakes, also have been associated with the movements of the crustal plates.

**Figure 18.C** summarizes the movements of the continents as they gradually moved apart from their positions as parts of Pangea beginning about 200 million years ago. As land masses move away from one another, populations inhabiting those lands become separated from one another. Geographically isolated, populations of the same species often undergo speciation as differing pressures of natural selection act on them. Similarly, as land masses collide, species are thrust together and must compete with one another to survive. Shorelines disappear and with them go a variety of habitats and species. And as continents move, ocean currents change, causing climatic changes and yet more new selection pressures. ●

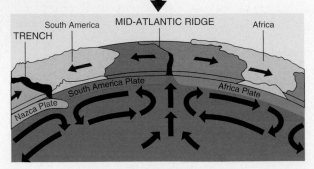

**Figure 18.B** Plate tectonics theory. Plate tectonics explains the movement of the continents. The semiliquid rock on which the plates rest rises as it heats and sinks as it cools, causing movement.

# What is the history of human evolution?

## 18.12 Primates appeared in the early Cenozoic era about 65 mya.

There are many orders of placental mammals, which include animals familiar to you, such as dogs, cats, horses, whales, squirrels, rabbits, bats, and a variety of others. The order of placental mammals that humans belong to is the **primates**. Primates are mammals that have characteristics reflecting an arboreal, or tree-dwelling, lifestyle. Among these characteristics are hands and feet that are able to grasp objects (such as tree branches), flexible limbs, and a flexible spine. **Table 18.1** lists primate characteristics.

Sometime in the early Cenozoic era, the primates may have developed from the shrewlike mammals. As thousands of years passed, primates' ancestors probably moved into the trees, eating insects that fed on the fruits and flowers growing on tree branches. Selection pressures must have been great for these arboreal creatures. Those that survived in this habitat developed excellent depth perception, the ability to see in more than one plane as they moved from tree to tree. The main feature of eyes that allows depth perception is the overlap of their visual fields. As primates evolved from their shrewlike ancestors, the eyes moved closer together from their placement on either side of the head, allowing stereoscopic vision.

The primates also developed long limbs with flexible hands and feet adapted to grasping and swinging from branch to branch. Primates have two bones in the lower part of a limb that enable the wrists and ankles to rotate. In addition, the hands and feet of primates have digits that can be spread apart from one another, helping them to balance themselves when walking or running or enabling them to grasp objects. An opposable thumb, one that can touch the tip of each finger, helps in grasping. Most primates also developed flattened nails at the end of the digits, replacing the claws of their mammalian relatives.

| TABLE 18.1 | Some Characteristics of Primates |
|---|---|

- Ability to spread apart toes and fingers
- Opposable thumb (thumb can touch the tip of each finger)
- Nails instead of claws
- Omnivorous diet (teeth and digestive tract adapted to eating both plant and animal food)
- A semierect to an erect posture
- Binocular vision (overlap of visual fields of both eyes) resulting in depth perception
- Well-developed eye–hand coordination
- Bony sockets protecting eyes
- A complex brain that is large in relation to body size

## 18.13 Anthropoids appeared about 50 mya.

Primates can be grouped into two broad types: the prosimians and the anthropoids (the type that includes humans). The **prosimians** (pro-SIM-ee-uns; meaning "before ape") are small animals such as lemurs, indris, aye-ayes, tarsiers, and lorises; they range in size from less than a pound to approximately 14 pounds—about the size of a cat or a small dog. Most are nocturnal and have large ears and eyes to help them see and hear at night. Prosimians have elongated snouts, reflecting their highly developed sense of smell. These characteristics, in addition to its elongated rear limbs, which help in leaping from tree to tree in a tropical rain forest habitat, are clearly seen in the prosimian pictured in **Figure 18.21**. By the end of the Eocene epoch (38 million years ago), prosimians were abundant in North America and Eurasia and were probably also present in Africa. Their descendants, however, now live only in the tropics of Asia, in tropical Africa, and on the island of Madagascar.

The **anthropoids** (AN-thruh-poyds; meaning "humanlike") include the monkeys, apes (gibbons, orangutans, gorillas, and chimpanzees), and humans. These primates differ from the prosimians in the structure of their teeth, brain, skull, and limbs. The prosimians have pointed molars and horizontal lower front teeth. These horizontal teeth are used to comb the coat or to get at food. The anthropoids have more rounded molars and no horizontal lower front teeth. The brain of the prosimian is much smaller in

**Figure 18.21 Prosimian.** The prosimian shown in the picture is a ringtail lemur, *Lemur catta*. Today, lemurs are found primarily on the island of Madagascar, and all are in danger of extinction as their rain forest habitats are destroyed. Notice its elongated snout, elongated rear limbs, large eyes, and large ears, all adaptations to an arboreal and nocturnal existence.

**Figure 18.22** Tarsier, *Tarsius syrichta*, tropical Asia. Note the large eyes of the tarsier, an adaptation to nocturnal living.

features in the prosimian in Figure 18.21 and the anthropoids (Figures 18.23 and 18.24). Both molecular and anatomical studies have shown that the tarsiers (**Figure 18.22**) are more closely related to the anthropoids than are the other prosimians; modern classification of tarsiers reflects this closeness.

In addition to their structural differences, prosimians and anthropoids also exhibit behavioral differences. Anthropoids are diurnal; that is, they are active during the day, whereas the prosimians are nocturnal. The anthropoids have evolved color vision, probably in relation to their diurnal existence. Also, the anthropoids live in groups with complex social interactions. In addition, they tend to care for their young for prolonged periods.

Anthropoids comprise three broad types: (1) the **hominids** (HAHM-uh-nidz; formerly called hominoids), which include all living and extinct ape and human species, (2) New World monkeys, and (3) Old World monkeys. Both New and Old World monkeys are shown in **Figure 18.23**. In general, New World monkeys are found in South and Central America; they have flat noses with nostrils that face outward. Old World monkeys have noses similar to those of humans, with nostrils next to each other that point downward; these monkeys are found in Africa, southern Asia, Japan, and Indonesia. Monkeys are considered to be different from apes for a variety of reasons. Most apes are bigger than monkeys, have larger brains than monkeys, and lack tails.

The lesser apes, the great apes, and the **hominins** are all hominids. Hominins include all living and extinct human species, including immediate nonhuman ancestors; this group was formerly called hominids. The *lesser apes* include the gibbons, which are the smallest hominids and closest in size to monkeys. They are about the size of a small dog, weighing 4 to 8 kilograms (9 to 18 pounds). They live in tropical rain forests where they move through the trees by brachiating, or hooking their long fingers over branches and swinging by their long arms (**Figure 18.24**). They often make long swings from branch to branch, like leaps, in which neither arm supports them.

relation to body size than the brain of an anthropoid. The face of an anthropoid is somewhat flat. In addition, the eyes of an anthropoid are closer together than those of a prosimian. Lastly, a prosimian's front limbs are short in relation to its long hind limbs, whereas both the front and hind limbs of an anthropoid are long. Compare these

The *great apes* include the orangutans, gorillas, and chimpanzees (**Figure 18.25**). These apes are much larger than the gibbons. The orangutans are about the size of humans, weighing between 50 and 100 kilograms (110 to 220 pounds). These apes exhibit sexual dimorphism; the females' weight is about half that of the males'. Like the gibbons, the orangutans have long arms, and smaller orangutans move through the trees by brachiating. However, because of their large size they use a leaning form of brachiation—they carefully shift their body weight to bend a supporting branch and then grab the next one before the first one breaks.

(a)

(b)

**Figure 18.23** Facial differences between Old World monkeys and New World monkeys. (a) Hamadryas baboon, an Old World monkey. (b) Geoffroy's marmoset, a New World monkey.

**Figure 18.24 Brachiation.** Brachiators such as the white-handed gibbon, *Hylobates lar*, locomote by swinging from branches with their arms and hooked fingers and reaching from hold to hold.

(a)                                        (b)

(c)

**Figure 18.25 The great apes.** (a) Gorilla, *Gorilla gorilla*. (b) Chimpanzee, *Pan troglodytes*. (c) Orangutan, *Pongo pygmaeus*.

Large orangutans cannot move from tree to tree by brachiating and must come down from the trees and walk. The gorillas are the largest of the apes. Males weigh about 160 kilograms, or 350 pounds; females are smaller, ranging in weight from 165 to 240 pounds. Gorillas spend most of their time on the ground. When alarmed, male gorillas beat on their chests in a behavioral display. The smallest of the great apes are the chimpanzees—humans' closest relatives. Males weigh between 40 and 50 kilograms, or 90 to 110 pounds; females are slightly lighter. Like the other apes (except the gorillas), they spend much of their time in the trees.

The hominins are the most intelligent of the hominids. They are distinguished from the other hominids in that they are *bipedal*; that is, they walk upright on two legs. In addition, hominins communicate by language and exhibit *culture*. Culture is a way of life—the behaviors, beliefs, values, and symbols accepted and passed along by communication and imitation from one generation to the next. Humans are the only living hominins.

Scientists have found jaw fragments suggesting that monkeys, apes, and humans began evolving from their common ancestor approximately 50 million years ago. The tarsioids may have begun their evolution even earlier—about 54 million years ago. Scientists do not know what adaptive pressures resulted in the appearance of these early anthropoid primates, nor do they know which prosimian was their ancestor.

Fossils and biochemical studies show that the New World monkeys branched from the line leading to the Old World monkeys and the hominins about 45 million years ago, in the mid-Eocene epoch. In addition, scientists have discovered many fossils in North Africa that date from 25 to 30 million years ago, in the Oligocene epoch. One of the earliest of these fossils, which scientists have named *Parapithecus*, probably led to the line of Old World monkeys, splitting from the hominids approximately 30 million years ago. The others, scientists think, are probably members of the evolutionary line that leads to the hominids.

The most well known of the Oligocene fossils found in North Africa is called *Aegyptopithecus*, thought to be the ancestor of the early Miocene hominids of Africa. Looking at the partially restored skull in **Figure 18.26**, you can see that *Aegyptopithecus* had some

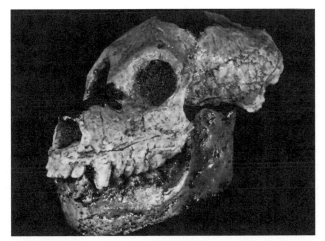

**Figure 18.26** *Aegyptopithecus zeuxis.* This primate fossil dates to the late Oligocene era and is thought to be the ancestor of the hominids that lived in Africa during the early Miocene era.

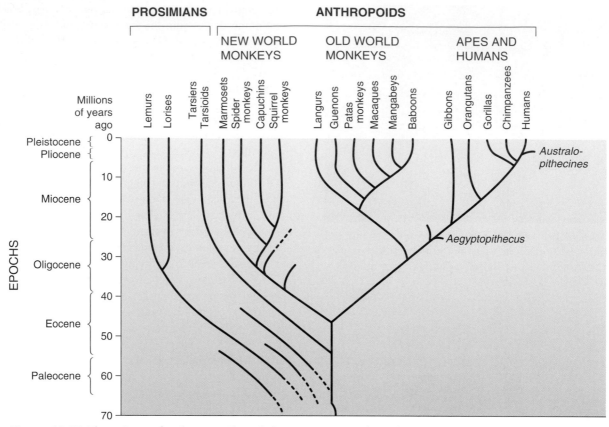

**Figure 18.27 The primate family tree.** This phylogenetic tree is based on DNA comparisons between living species. The dating of the branches is based on the ages of fossils.

**Visual Thinking:** Which organism on this primate family tree has the most recent shared ancestor with humans?

prosimian characteristics. It had a pronounced snout, leading scientists to think that its sense of smell was still highly specialized, much like that of the prosimians.

Although the only fossil evidence of the emergence of the anthropoids consists of pieces of jaw, biochemical studies complement the knowledge gained from the fossil record. Together, these techniques tell scientists a great deal about the evolution of the anthropoids and the relationships among humans, apes, and monkeys. The phylogenetic tree of the primates, including the anthropoids, is shown in **Figure 18.27**. (See Chapter 15 for a discussion of phylogenetic trees.)

Fossil evidence is still being accumulated that will help tell the story of hominid evolution. From mid-Miocene times on, the hominid fossil record consists primarily of skull and teeth fragments. Investigators have calculated that the evolutionary line leading to gibbons diverged from the line leading to the other apes about 18 to 22 million years ago; the line leading to orangutans split off roughly 13 to 16 million years ago; the line leading to gorillas diverged 8 to 10 million years ago; and the split between hominins and chimps occurred approximately 5 to 8 million years ago. This last statement suggests that chimpanzees and gorillas are humans' closest relatives, with a common relative alive 5 to 8 million years ago.

## 18.14 The oldest evidence of hominins is 6 to 7 million years old.

The two critical steps in the evolution of humans were the evolution of bipedalism and the enlargement of the brain. For many years, scientists have hypothesized how and when bipedalism arose. The current evidence suggests that bipedal humans evolved from ancestors adapted to knuckle walking and climbing. Knuckle walking is a mode of locomotion in which these apelike ancestors, just like present-day gorillas and chimps, walked on the soles of their feet and the second joints of their fingers. The skeletons and muscles of human ancestors were structured in a way that allowed these hominids to knuckle walk even though they lived in the trees. These structural preadaptations developed as a part of their arboreal life and the brachiation they exhibited in the trees (see Figure 18.24).

In their wooded habitats, large-bodied ancestral primates lived in environments in which trees were spaced widely apart. Thus, they had to move along the ground at least part of the time. More upright bipedalism likely evolved as an adaptation to some aspect of food acquisition. Those exhibiting upright bipedal standing could reach higher into short trees, thereby picking more fruit, and could continuously gather fruit with both hands. Being able to reach more food and use both hands to gather it confers an adaptive advantage.

The oldest evidence of the hominins was found in the Sahel Desert of Central Africa and reported in 2002. This "find" by French researcher Michel Brunet and his team consists of a 6–7-million-year-old cranium and jaw fragments of *Sahelanthropus tchadensis*, nicknamed Toumai, meaning "hope of life" (**Figure 18.28**). Toumai is very chimpanzee-like but has features typical of hominins rather than apes, such as a "taller" face, shorter snout, and smaller canine teeth.

In 2004, analysis of the femur (thigh bone) of *Orrorin tugenensis* provided strong evidence that this 6-million-year-old hominin often walked upright; it appears to be an intermediate between humans and apes. Twenty fossils representing at least five individuals have been discovered.

In 1994, 4.4-million-year-old skull fragments of *Ardipithecus ramidus* were found. Its teeth have characteristics intermediate between the apes and *Australopithecus* (see next paragraph). It is unclear to scientists whether *A. ramidus* was fully bipedal. It is considered likely that these hominins lived in a variety of habitats just as chimpanzees do today; how much time they spent in each is very difficult to know—it might have varied geographically and between species as in present-day chimps.

In 1995, skeletal remains of the hominin *Australopithecus anamensis* were found in Africa by an international research team led by Meave G. Leakey of the National Museums of Kenya in Nairobi. *A. anamensis* had apelike jaws, teeth, wrist bones, and a small brain. The structure of its arm and leg bones allowed it to be fully bipedal. *Anamensis* existed between 4.2 and 3.9 million years ago.

In 1974, Donald Johanson and an international team of scientists found and pieced together one of the most complete fossil skeletons of *Australopithecus afarensis*, or "southern ape of Afar." This 3.5 million-year-old fossil hominin skeleton has since become famous as Lucy (**Figure 18.29**), named after the Beatles song "Lucy in the Sky with Diamonds," which was playing on the researchers' tape machine at the time they made their fossil find. In 1976, Mary Leakey and an international team of scientists found fossil footprints of *A. afarensis*, which are older than Lucy and are dated at 3.65 million years old (**Figure 18.30**).

**Figure 18.28** Toumai, "hope of life." Cranium of *Sahelanthropus tchandensis*, nicknamed Toumai, the oldest hominin known at this time.

Although *Australopithecus* and *Ardipithecus* are hominins, they are not humans (members of the genus *Homo*). Their brains were still small in comparison to present-day human brains, about the size of the brains of great apes, and they had long, apelike arms. In addition, their faces were apelike, as shown in the photo of a reconstruction of Lucy in Figure 18.29.

*A. afarensis* may have evolved into two (and possibly more) lineages, including the species *A. africanus*, *A. robustus*, and *A. boisei*. These australopithecines lived on the ground, spending at least part of their time in the open savannah of eastern and southern Africa. Their diets consisted primarily of plants. At night, they probably slept in the few trees that existed in these grasslands, much like the savannah baboons do today to protect themselves from predators.

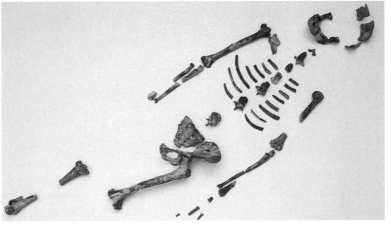

**Figure 18.29** Lucy (*Australopithecus afarensis*). Lucy, from Ethiopia, is the most complete skeleton of *Australopithecus* discovered so far. The reconstruction was made by a careful study of muscle attachments to the skull.

**Figure 18.30 Fossil footprints.** These fossil footprints, found in the Afar region of Ethiopia, have been preserved in volcanic ash for over 3.5 million years. They were made by our nonhuman ancestor *Australopithecus afarensis* and give scientists important clues to our heritage. The differing sizes of the footprints may reflect our ancestors' sexual dimorphism, which is size difference with gender. Also, the footprints show that these organisms walked erect on two legs.

The evolutionary relationships among these australopithecines is not clear; a phylogenetic tree that is widely accepted at this time is shown in **Figure 18.31**. Notice from the diagram that no australopithecines are alive today; the last ones disappeared about 1 million years ago.

---

CONCEPT CHECKPOINT

**5.** How does the structure of the human hand demonstrate the concept of preadaptation?

---

## 18.15 The oldest human fossils are about 2 million years old.

Climatic changes during the Pleistocene epoch (1.8 million to 12,000 years ago) may have contributed to the eventual disappearance of the australopithecines and the survival of a new, more intelligent genus of hominins: the human (genus *Homo*). To emphasize their intelligence, these early humans were named ***Homo habilis***, or "skillful human." *H. habilis* coexisted for at least 500,000 years with the smaller-brained *Australopithecus*. *Homo habilis* existed between 2.4 and 1.5 million years ago. The first *H. habilis* fossils were discovered in 1964 by Louis Leakey, Philip Tobias, and John Napier in the Olduvai Gorge of eastern Africa.

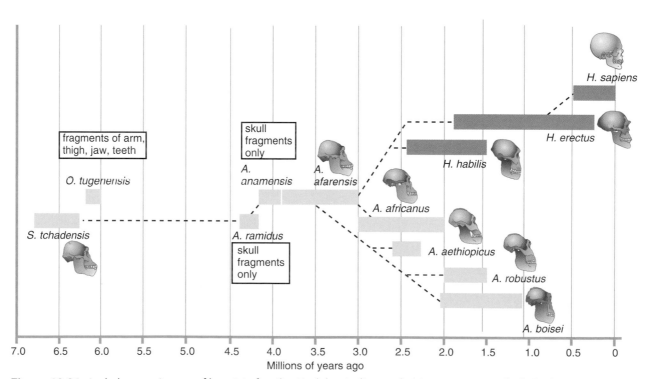

**Figure 18.31 A phylogenetic tree of hominin fossils.** Each bar indicates the time span over which the hominin lived.

**Visual Thinking:** Are any species on this phylogenetic tree considered apes or monkeys? Which species are ancestors of humans? Which species is/are living today?

Judging from the structure of the hands, *H. habilis* regularly climbed trees, as had their australopithecine ancestors, although *H. habilis* spent much time on the ground and walked erect on two legs. Skeletons found in 1987 indicate that *H. habilis* was small in stature like the australopithecines, but fossil skulls and teeth reveal that the diet of *H. habilis* was more diverse than that of the australopithecines, including meat as well as plants. The tools found with *H. habilis* were made from stones fashioned into implements for chopping, cutting, and pounding food. The use of stone tools by these early humans marks the beginning of the *Stone Age*, a time that spans approximately 2 million to 35,000 years ago.

## 18.16 A now-extinct species of humans arose 1.8 mya.

All of the early evolution of the genus *Homo* seems to have taken place in Africa. There, fossils belonging to the second, also extinct species of *Homo*—**Homo erectus**—are widespread and abundant from 1.8 million to about 300,000 years ago. *Homo habilis, Australopithecus,* and *Ardipithecus* fossils have been found only in Africa, but *H. erectus* fossils have been found in Africa, Asia, and Europe.

*H. erectus* was about the size of modern-day humans, was fully adapted to upright walking, and had a brain that was roughly twice as large as those of their ancestors. However, *H. erectus* still retained prominent brow ridges, rounded jaws, and large teeth. The tools of *H. erectus* were much more sophisticated than those of *H. habilis* and were used for hunting, skinning, and butchering animals. *H. erectus* was a lineage of hunter-gatherers, which means that they collected plants, small animals, and insects for food while occasionally hunting large mammals. Researchers may have found the first evidence of the use of fire by humans at 1.4-million-year-old campsites of *H. erectus* in the Rift Valley, in Kenya, in Africa. However, many dispute this evidence because it is difficult to distinguish hearths from natural fires. All of these activities (tool making, hunting, using fire, and building shelters) are signs of culture.

Inherent in the concept of culture is the concept of language. The ability to communicate enhances the ability of a species to survive, especially in harsh conditions such as during an ice age, by enabling members of the species to share survival tactics and to warn each other of danger. Anthropologists think that the development of language was probably one of the most important factors in the appearance of *Homo sapiens*. The results of recent studies indicate that language skills may have been present in primitive humans up to 300,000 years ago, earlier than the appearance of symbolic behavior such as cave paintings and symbols.

## 18.17 The earliest fossils of *Homo sapiens* are about 500,000 years old.

The earliest fossils of **Homo sapiens** (meaning "wise humans") are about 500,000 years old. They most likely evolved from *H. erectus* in Africa, although some scientists contend that they evolved simultaneously from *H. erectus* populations in Asia and Europe too. The oldest *H. sapiens* are not considered to be anatomically modern; that is, they do not have the same anatomical features as today's humans. They are therefore referred to as an early or archaic form. In general, early *H. sapiens* had larger brains, flatter heads, more sloping foreheads, and more protruding brow ridges and faces than today's humans.

The fossil record shows gradual change of the species *H. sapiens*, with the early form evolving over a 75,000-year span to a subspecies of *H. sapiens* called **Neanderthal** (also spelled Neandertal) (nee-AN-dur-thol or nee-AN-dur-tal). This subspecies was named after the Neander Valley in Germany, where their fossils were first found. The Neanderthals lived from about 230,000 to 30,000 years ago in Europe and the Middle East.

Compared with modern humans, the Neanderthals were powerfully built, short, and stocky. Their skulls were massive, with protruding faces, projecting noses, and rather heavy bony ridges over the brows. Their brains were somewhat larger than those of modern humans, but were differently shaped. The frontal lobe, the site of a large proportion of thinking, was flattened in the Neanderthal as it was in the earlier hominins. The Neanderthals made diverse tools, including scrapers, borers, spear heads, and hand axes. Some of these tools were used for scraping hides, which they used for clothing. They lived in hut-like structures or in caves. Neanderthals took care of their injured and sick and commonly buried their dead, often placing food and weapons and perhaps even flowers with the bodies. Such attention to the dead strongly suggests that they believed in a life after death. This is the first evidence of the kinds of thought processes—including symbolic thought—that are characteristic of modern *H. sapiens*.

Approximately 10,000 years before the Neanderthal subspecies died out, the "modern" subspecies of *H. sapiens* made its appearance. This modern subspecies (our subspecies) is called **Homo sapiens sapiens**. The early members of this subspecies are called the Cro-Magnons, after the cave in southwestern France where scientists found some of their fossils.

The **Cro-Magnons** (KRO-MAG-nuns) had a stocky build, much like the Neanderthals, but their heads, brow ridges, teeth, jaws, and faces were much smaller than the Neanderthals' and were more similar to those of today's humans. However, just as modern humans show variation among races, so too did the Cro-Magnons. The Cro-Magnons used sophisticated tools that were made not only from stone but also from bone, ivory, and antler—materials that were not used earlier. Hunting was an important activity for the Cro-Magnons, evidenced by the abundance of animal bones found with human bones and by elaborate cave paintings of animals and hunting scenes (**Figure 18.32**). The paintings appear to have been part of a ritual to ensure the success of the hunt. The subspecies of *H. sapiens* that preceded us showed a gradual development of culture and society, which was the foundation for the development of "modern" culture and society.

In 2004, an Indonesian–Australian team of archeologists discovered a near-complete skeleton of a three-foot-tall female belonging to a species of human that appears to have *co-existed* with modern humans (**Figure 18.33**). Called *Homo floresiensis* for the Indonesian island of Flores on which the skeleton was found, the

**Figure 18.32 Cave painting.** Cave paintings, almost always showing animals and sometimes hunters, were made by the Cro-Magnons, our immediate ancestors. These paintings are found primarily in Europe and were made for about 20,000 years, until 8000 to 10,000 years ago.

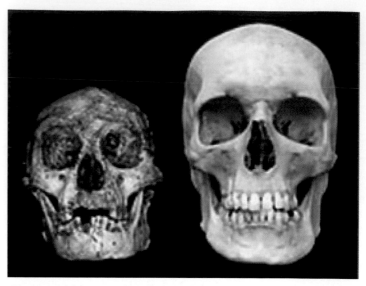

**Figure 18.33** *Homo floresiensis*, a human species that existed as recently as 13,000 years ago. This comparison of the skull of *H. floresiensis (left)* with the skull of a modern human *(right)* indicates how small these extinct humans were.

species seems to have existed during the period from 95,000 to 13,000 years ago. Its ancestors were probably a form of *H. erectus.* Although members of this species had a brain smaller than most chimpanzees, evidence indicates they used fire and made stone tools.

About 10,000 years ago, the last ice age came to a close, the global climate began to warm, and various groups of *H. sapiens*

*sapiens* began to cultivate crops and breed animals for food. Archeologists have uncovered the remains of small, ancient cities, such as those of Jericho, which give evidence that by 9000 years ago, humans had developed complex social structures. By 5000 years ago, the first large cities and great civilizations had appeared, such as those in Egypt and Mesopotamia.

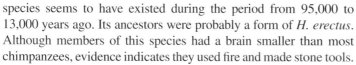

## Summary of Key Concepts and Key Terms

### How did life on Earth originate?

**18.1** Scientists theorize that approximately 15 billion years ago the condensed material of the universe exploded in an event called the **Big Bang** (p. 283), which formed an early generation of stars.

**18.1** Approximately 4.6 billion years ago, these stars exploded, forming Earth and the other planets of the solar system.

**18.2** In the 300 million years between the time when the Earth cooled and the first bacteria appeared, biochemical evolution became biological evolution.

**18.2** The long-held "primordial soup theory" states that life arose in the sea as elements and simple compounds of the primitive atmosphere reacted with one another to form simple organic molecules, such as amino acids and sugars.

**18.2** After the synthesis of complex organic molecules had taken place, self-replicating systems of RNA molecules may have begun the process of biological evolution.

**18.2** To be considered life, this "early" genetic material must have been organized within a cellular structure.

### When did cells arise?

**18.3** Earth's crust can be visualized as a geological time line in which five main divisions of rock strata are recognized as associated with major **eras** (p. 285) of geological time.

**18.3** Eras are subdivided into shorter time units called **periods** (p. 285); In addition, the periods of the Cenozoic era are subdivided into **epochs** (EP-uks or EE-poks, p. 285).

**18.3** Scientists speculate that the first cells to evolve fed on organic materials in the environment.

**18.3** The evolution of a pigment system that could capture the energy from sunlight and store it in chemical bonds most likely evolved next.

**18.3** After the evolution of an energy-capturing pigment system, oxygen-producing bacteria evolved, gradually oxygenating the atmosphere.

**18.4** The first unicellular eukaryotes appeared about 2.7 billion years ago during the Archean era.

**18.4** The most widely accepted hypothesis regarding the origin of eukaryotes, is the **serial endosymbiotic theory** (**SET**, p. 287); eukaryotic cells likely arose as "pre-eukaryotic" cells became hosts to endosymbiotic prokaryotes.

## When did multicellular organisms arise?

**18.5** Multicellular organisms first appeared about 1.5 to 1.2 billion years ago.

**18.5** Many of the phyla of organisms in existence today, except the land plants, appear to have evolved during the Cambrian period (543–490 million years ago) of the Paleozoic era.

**18.5** Cambrian fossils, representing well over 100 species, have been found in the **Burgess Shale** (p. 289), which are sedimentary rocks formed from fine-grained mud.

**18.6** **Chordates** (p. 290) and **vertebrates** (p. 290) evolved in the late Cambrian.

**18.7** Fishes became abundant and diverse during the Devonian period (417–360 million years ago).

**18.7** The amphibians, organisms that spend part of their time on land and part in the water, evolved from the fishes.

**18.8** Terrestrial plants appeared about 420 million years ago and flourished during the Carboniferous period (360–285 million years ago).

**18.9** The amphibians gave rise to the reptiles.

**18.9** A major extinction event occurred during the Permian period, wiping out many of the species living at that time.

**18.9** The Mesozoic era (250–65 million years ago) was a time of the adaptive radiation of the organisms surviving the Permian extinction, dominated by the evolution of the reptiles.

**18.9** The reptiles gave rise to the birds at least 150 million years ago.

**18.9** Feathered forelimbs and enlarged chest muscles might have served as **preadaptations** (p. 293) for flight.

**18.10** During the early Cretaceous period, about 140 million years ago, flowering plants (angiosperms) appeared, becoming the dominant form of plant life about 100 million years ago.

**18.11** The reptiles gave rise to the mammals about 200 million years ago.

**18.11** Mammals are warm-blooded (endotherms), able to maintain a constant internal body temperature.

**18.11** Most other living animals are cold-blooded (ectotherms); their body temperatures vary with the temperature of the environment.

**18.11** Changes in the structure and placement of the mammalian limbs, the structure of the heart, and reproductive strategies resulted in mammals' ability to survive in a variety of climates and to eventually become abundant.

## What is the history of human evolution?

**18.12** **Primates** (p. 297), one of the many orders of mammals, first appeared in the early Cenozoic era.

**18.12** Primates have large brains in proportion to their bodies, binocular vision, and five digits, including an opposable thumb; they exhibit complex social interactions.

**18.13** The primates are comprised of the **prosimians** (small animals such as lemurs, indris, and aye-ayes, p. 297) and the **anthropoids** (a group that includes monkeys, apes, and humans, p. 297).

**18.13** The **hominids** (HAHM-uh-nidz, meaning "humanlike" and including apes and humans, p. 298) are one of three broad types of anthropoids; the two other types of anthropoids are New World monkeys and Old World monkeys.

**18.13** Ancestors to the apes gave rise to the gibbons, orangutans, chimpanzees, gorillas, and **hominins** (p. 298), which include all living and extinct human species and immediate nonhuman ancestors.

**18.13** The New World monkeys branched from the evolutionary line leading to the Old World monkeys and the hominids about 45 million years ago; the Old World monkeys split from the hominids approximately 30 million years ago.

**18.14** The two critical steps in the evolution of humans were the evolution of bipedalism (walking on two feet) and the enlargement of the brain.

**18.14** The oldest hominin fossil belongs to the genus *Sahelanthropus* and is 6 to 7 million years old.

**18.14** The probable direct ancestors of humans belong to the genus *Australopithecus*.

**18.15** The first species of the human genus *Homo* is **H. habilis** (p. 302), appearing in Africa about 2.4 million years ago.

**18.15** Now extinct, *H. habilis* is considered human because this species exhibited an intelligence far greater than that of its ancestors by making and using tools.

**18.16** The second species of *Homo*, **H. erectus** (p. 302), appeared approximately 1.8 million years ago; *H. erectus* lived in Africa, Asia, and Europe.

**18.16** *H. erectus* used fire, built shelters, fashioned sophisticated tools, and exhibited culture.

**18.17** Early **H. sapiens** (p. 303) probably evolved from *H. erectus* about 500,000 years ago.

**18.17** The fossil record shows gradual change of the species *H. sapiens*, with the early form evolving over a 75,000-year span to a subspecies called *Neanderthal* (nee-AN-dur-thol or nee-AN-dur-tal, p. 303), which lived from about 230,000 to 30,000 years ago in Europe and the Middle East.

**18.17** Approximately 10,000 years before the Neanderthal subspecies died out, the "modern" subspecies of *H. sapiens*, called *Homo sapiens sapiens* (p. 303), made its appearance.

**18.18** The early members of *Homo sapiens sapiens* were the **Cro-Magnons** (p. 303).

---

## Level 1 | Learning Basic Facts and Terms

**Multiple Choice**

1. The serial endosymbiotic theory describes the origin of
   a. early eukaryotes.
   c. chloroplasts.
   b. mitochondria.
   d. all of the above.

2. When did life first arise on planet Earth?
   a. 6000 years ago
   d. 3 trillion years ago
   b. 3.5 million years ago
   e. Unknown
   c. 3.5 billion years ago

3. Which of the following groups of organisms arose during the Cambrian explosion?
   a. Hominins
   d. Bony fish
   b. Land plants
   e. Invertebrates
   c. Dinosaurs

4. Which group of organisms is directly responsible for the vast coal reserves on Earth?
   a. Dinosaurs
   b. Terrestrial plants that lived during the Carboniferous period
   c. The large diversity of invertebrate animals during the Cambrian period
   d. Planktonic algae

5. Which of the following adaptations made it possible for organisms to move from an aquatic environment to a terrestrial environment?
   a. Amniotic egg
   d. Roots
   b. Chitin
   e. All of the above
   c. Scales

6. The longest geologic period in the history of the Earth is the
   a. Cambrian.
   b. Mesozoic.
   c. Cenozoic.
   d. Proterozoic.
   e. Paleozoic.

**Matching**

Match the following fossils that are thought to be transitional with the groups between which they are intermediate in form.

7. _____ *Archaeopteryx*
8. _____ *Aegyptopithecus*
9. _____ *Sahelanthropus tchadensis*
10. _____ *Australopithecus afarensis* (Lucy)

a. early hominins and *Homo* species
b. reptiles and birds
c. apes and hominins
d. Old World monkeys and hominids

---

## Level 2 | Learning Concepts

1. What is the primordial soup theory, and what does it attempt to explain? Describe an alternative hypothesis that explains the same event.
2. What are cyanobacteria, and when did they first appear? Explain their evolutionary significance.
3. When was the Cambrian period, and why was it important?
4. Describe the major events of the Carboniferous period. What were the living conditions on Earth during this time?
5. Summarize the events of the Cenozoic era. When did it begin?

6. What are primates? What two characteristics have helped them to be successful?
7. Distinguish between the prosimians and the anthropoids. Give an example of each.
8. Which statement is true? (Or are they both false?) Explain your answer.
   a. All hominins are hominids.
   b. All hominids are hominins.
9. How did *Homo sapiens sapiens* differ from earlier *Homo sapiens*?

---

## Level 3 | Critical Thinking and Life Applications

1. **Misconception or Reality?** Consider the following statements about human evolution. Which are well supported by scientific evidence and which are misconceptions?
   a. Humans evolved from apes.
   b. Human evolutionary history is more like a multibranched bush rather than a single unbranched ladder, our species being the tip of one human branch that still exists.
   c. During the course of hominin history many species of humans (genus *Homo*) existed, but none of them ever coexisted at the same time.
   d. Ancient hominins hunted dinosaurs as a food source.
   e. Our understanding of human ancestry and evolution is complete.
2. Rocks from Mars land on Earth as often as every three days. By 1999, the hypothesis that life on Earth arose first on Mars and traveled to Earth during its early days was considered plausible by many NASA and university scientists. If life did arise on Mars and traveled to Earth on rocks, how would that information change current hypotheses regarding the origin and evolution of life?

3. The inner mitochondrial membrane has proteins that are more similar to prokaryotic plasma membrane proteins than to those found in eukaryotic plasma membranes. Explain how this observation supports the endosymbiotic origin of these organelles.
4. If invaginations of the outer plasma membrane led to the formation of the nucleus and ER, this presupposes that these invaginations were somehow initially adaptive to these early cells. Based on what you know about biological membranes, what kinds of advantages might these initially small invaginations have conferred on these early cells?
5. Scientists hypothesize that significant eukaryotic diversification did not occur until sexual reproduction evolved. Why did sexual reproduction lead to diversification?
6. Scientists hypothesize that significant eukaryotic diversification did not occur until there was significant concentration of oxygen in the oceans. What is the relationship between oxygen and this diversification?

---

## In The News | Critical Thinking

**WALKING WHALES**

Now that you understand more about the evolution of life on Earth, reread this chapter's opening story about Thewissen's research on whale ancestry. To better understand this information, it may help you to follow these steps:

1. Review your immediate reaction to this information about whale ancestry that you wrote when you began reading this chapter.
2. Based on your current understanding, again summarize the main point of the research in a sentence or two.
3. What questions do you now have about this information that this chapter's opening story does not answer?
4. Collect new information about the research. Visit the *In The News* section of this text's companion Web site at www.wiley.com/college/alters and

watch the "Walking Whales" video. Then use the "summary" link to read the accompanying story and access related links. Use this information, the links provided, and other online and library resources to answer your questions and find updates about this topic. State the sources of your information. Explain why you think the information is accurate. Also determine whether the information expresses a particular point of view or is biased in any way.
5. What in your view are the most significant aspects of this information? Give reasons for your opinion and for any changes in your ideas based on the additional information you have collected and the analysis you have done.

# VIRUSES AND BACTERIA

## In The News | Not So Harmless

Even after you step out of the shower, you are still home to billions of resident bacteria—what microbiologists call our normal microbiota. Although bacteria can at times make us sick, the species that normally inhabit our bodies provide services of which you might not be aware. For example, resident bacteria on the skin produce antimicrobial substances that help thwart infection from pathogenic (disease-causing) bacteria. Resident bacteria in the vagina produce acids that create an environment unfavorable to yeast, which would otherwise grow and cause an uncomfortable infection. Resident bacteria in the intestine produce vitamin K and other vitamins as they metabolize our undigested food—vitamins we absorb and use.

Results of recent research suggest that bacterial viruses, once thought as harmless to humans as the normal microbiota they sometimes invade, can sometimes turn our resident bacteria against us. Vincent Fischetti and Thomas Broudy from The Rockefeller University in New York City have been studying viruses that attack bacteria. A virus that infects bacteria is called a bacteriophage (bac-TEAR-ee-oh-fage), or simply a phage. They have discovered that phages capable of infecting disease-causing bacteria, such as the organism that causes scarlet fever, can transfer genes encoding toxins from the pathogen to a nonpathogenic organism, such as our normal microbiota. How does this occur?

The nose and throat, for example, normally contain low numbers of pathogenic bacteria, such as the causative agents of pneumonia, bacterial meningitis, and scarlet fever. The normal microbiota help keep these pathogens at low levels, and we stay healthy. Fischetti and Broudy have discovered, however, that bacteriophages can infect these pathogens and merge their hereditary material with that of the pathogen. When triggered to reproduce, the phage may take with it some bacterial genes and, in some cases, genes that encode disease-producing toxins. Viruses replicate within the pathogen, burst from these cells, and infect the usually harmless bacteria residing in the nose and throat of the same person or a different one. The virus then merges its hereditary material, now containing bacterial genes from the pathogen, with the genome of normal microbiota. As these bacteria replicate, they create a population of cells now capable of causing disease. In addition, the researchers note, the eukaryotic host (humans in our example) actively participates in the diversification and evolution of a bacterial population. You can learn more about this research and view the video "Not So Harmless" by visiting the *In The News* section of this text's companion Web site at www.wiley.com/college/alters.

---

*Write your immediate reaction to Broudy and Fischetti's research on the phage-mediated evolution of virulence in bacterial populations: first, summarize the main point of the research in a sentence or two; then suggest what you think its significance is. You will have an opportunity to reflect on your responses and gather more information on this topic in the* In The News *feature at the end of this chapter. In this chapter, you will learn more about bacteria and viruses.*

# CHAPTER GUIDE

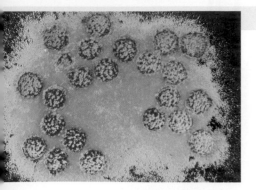

## What are viruses and how do they multiply?

**19.1** Viruses are nonliving infectious agents.

**19.2** Viruses vary in shape, but all consist of nucleic acid and protein.

**19.3** Viruses use cells' enzymes and ribosomes to make more viruses.

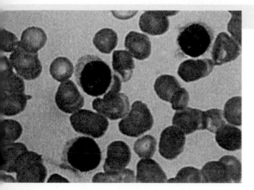

## How do viruses cause disease?

**19.4** Infection with certain viruses can initiate cancer.

**19.5** Viruses invade cells and take over cell functions.

**19.6** Prions and viroids are simpler than viruses and cause disease.

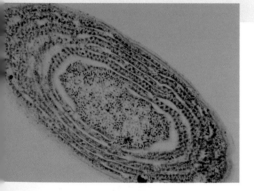

## What are bacteria and how do they reproduce?

**19.7** Bacteria are prokaryotic cells that are both helpful and harmful.

**19.8** Bacteria reproduce asexually by splitting.

## How do bacteria cause disease?

**19.9** Bacteria attach to cells and cause tissue damage.

# What are viruses and how do they multiply?

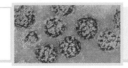

## 19.1 Viruses are nonliving infectious agents.

You may have had a sexual encounter with a virus and not even know it. Approximately 20 million people in the United States are currently infected with a sexually transmitted virus called human papillomavirus, or HPV. In the color-enhanced photo in **Figure 19.1a**, these virus particles appear as red spheres speckled with green; the human cell they are destroying is yellow. The odds of becoming infected with HPV are incredibly high—half to three-quarters of all sexually active men and women acquire this virus at some point in their lives. HPV is transmitted through sexual intercourse and causes a sexually transmitted infection (STI) commonly known as *genital warts* (Figure 19.1b). (STIs are also sometimes called sexually transmitted diseases [STDs] or venereal diseases.)

**Viruses** are *infectious agents*, which means that they enter living organisms, causing disease. Although they invade living things and cause cells to make more viruses, the viruses themselves are not living! They do not have a cellular structure, which is the basis of all life. They are nonliving *obligate parasites*, which means that viruses cannot reproduce outside of a living system. They must exist in association with and at the expense of living organisms. Unfortunately, that "living organism" may be you!

At the end of the nineteenth century, several groups of European scientists, working independently, first realized that viruses existed. As they filtered fluids derived from plants with tobacco mosaic disease and cattle with hoof-and-mouth disease, the scientists discovered that the infectious agents passed right through the fine-pored filters they used, which were designed to hold back bacteria. They concluded that the infectious agents associated with these diseases were *not* bacteria—they were too small. As they studied the filtrate containing these mysterious agents, the scientists also discovered that the disease-causing agents could multiply only within living cells. These infectious agents, they hypothesized, must lack some of the critical "machinery" cells use to reproduce.

For many years after their discovery, viruses were regarded as very primitive forms of life, perhaps the ancestors of bacteria. Today, scientists know that viruses show no living qualities when outside living cells. Also, viruses have no cellular structure. Therefore, these infectious agents are considered nonliving.

Because viruses are not living things, they are not included in classifications of living things. That is, they do not belong to any of the three domains of life (see Chapter 1). Scientists have, however, devised a classification scheme for viruses that is based on the host they infect. Viruses are first grouped according to whether they infect plants, animals, or bacteria. Further classification usually focuses on differences in morphology (shape and structure), type of nucleic acid, and manner of replication.

How did these nonliving infectious particles arise? The simple answer is that no one knows. Virologists—scientists who study viruses—hypothesize that they may have evolved from self-replicating strands of DNA or RNA prior to biological evolution, over 3.5 billion years ago. Others hypothesize that they evolved more recently from ancient parasitic cells that lost the ability to replicate on their own, becoming totally dependent on host cells.

## 19.2 Viruses vary in shape, but all consist of nucleic acid and protein.

Viruses infect primarily plants, animals, and bacteria, but they also infect archaea, protists, and fungi. Viruses that infect bacteria are discussed in this chapter's opening story. Many viruses can infect only a certain species: Your dog cannot catch your cold. Some viruses, however, can infect more than one species: Both dogs and people can be infected with the rabies virus.

It is hard to conceptualize how small viruses are. Some viruses, such as the poliovirus, are as small as the width of the plasma membrane on a human cell (**Figure 19.2**). Each virus has its own unique

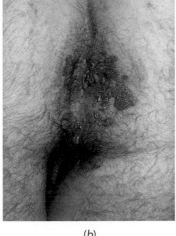

TEM 46,000×

(a)                    (b)

**Figure 19.1** Human papillomavirus. (*a*) Virus particles (red spheres speckled with green) attacking a human cell (yellow). Notice the difference in size between the cell and the virus particles. (*b*) Genital warts caused by HPV, growing around the anus of a man, nearly obscuring that opening.

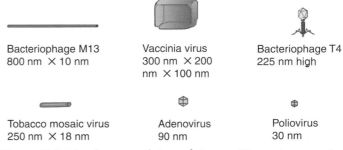

Bacteriophage M13
800 nm × 10 nm

Vaccinia virus
300 nm × 200 nm × 100 nm

Bacteriophage T4
225 nm high

Tobacco mosaic virus
250 nm × 18 nm

Adenovirus
90 nm

Poliovirus
30 nm

**Figure 19.2** The shapes and sizes of viruses. Dimensions are given in nanometers (nm) (1 nm = one billionth of a meter). A human red blood cell is 10,000 nm in diameter; an *Escherichia coli* bacterium is 3000 nm long by 1000 nm wide.

shape, but all contain the same basic parts: a nucleic acid **core** (either DNA or RNA, which can be single stranded, double stranded, circular, or linear) and a protein "overcoat" called a **capsid**.

The true nature of viruses became evident in the 1930s after the groundbreaking work of an American scientist, Wendell Stanley. Stanley prepared an extract of tobacco mosaic virus (TMV), purified it, and studied its chemical composition. He concluded that TMV was a protein, and he was partially right. Scientists later discovered that TMV also contains ribonucleic acid (RNA). In the late 1930s, with the development of the electron microscope, scientists were able to see the virus that Stanley purified (**Figure 19.3**). This virus is helical, with its single strand of RNA coiled like a spring, surrounded by a spiraling capsid of protein molecules.

Many viruses have another chemical layer over the capsid called the **envelope**, which is rich in proteins, lipids, and carbohydrates. **Figure 19.4** shows a typical enveloped virus, the causative agent of herpes. The spikes on the herpes virus are part of its envelope.

## 19.3 Viruses use cells' enzymes and ribosomes to make more viruses.

Viruses cannot multiply on their own. A virus must enter a cell and use the cell's enzymes and ribosomes to make more viruses. This process of viral multiplication within cells is called *replication*. The process of viral replication has been studied most extensively

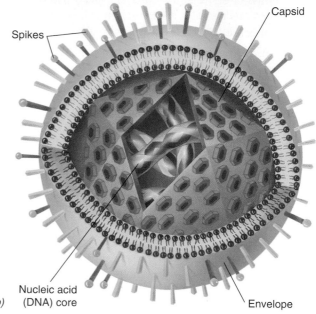

(a)

(b)

**Figure 19.4 Structure of an enveloped virus.** (a) Electron micrograph of the herpes virus. (b) Structure of the herpes virus.

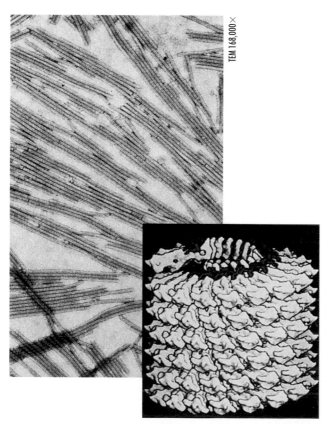

**Figure 19.3 Tobacco mosaic virus.** (a) Electron micrograph of purified tobacco mosaic virus. (b) Computer-generated model of a portion of tobacco mosaic virus. An entire virus consists of 2130 identical protein molecules—the yellow knobs—which form a cylindrical coat around a single strand of RNA (red).

in bacteria because bacteria are easier to grow in the laboratory and infect with viruses than are plant or animal cells.

Various patterns of viral replication exist. Some viruses enter a cell, replicate, and then cause the cell to burst, releasing the replicated viruses. This pattern of viral replication is called the **lytic cycle** (LIT-ic). Other types of viruses enter into a long-term relationship with the cells they infect, with their nucleic acid replicating as the cells multiply. This pattern of viral replication is called the **lysogenic cycle** (lye-suh-JEN-ik).

As shown in **Figure 19.5**, a bacteriophage first attaches to a receptor site on a bacterium **1**. The phage then injects its nucleic acid into the host cell, while its protein capsid is left outside the cell **2**. Next, the bacteriophage genes take over cellular processes and direct the bacterium to produce phage "parts" that will be used to assemble whole viruses **3**. After their manufacture, these strands of nucleic acid and proteins are assembled into mature phages **4** that *lyse* (LICE), or break open, the host cell **5**. Each bacterium releases many phage particles. Each "new" phage is capable of infecting another bacterial cell.

Some bacteriophages, instead of killing host cells, integrate their genetic material with that of the host (Figure 19.5 **2a**). Then, each time the host cell reproduces **2b**, the phage nucleic acid is replicated as if it were a part of the cell's genetic makeup **2c**. In this way, the phage is passed on from cell to cell. Bacteriophage infection of this sort is called a *latent infection*. The integrated phage genes may not cause any change in the host for a long time. Then, triggered by an appropriate stimulus, the phage may enter a lytic cycle and produce symptoms **2d** .

Some animal viruses infect cells in a manner similar to bacterial virus infection, but they enter animal cells by endocytosis (see Chapter 5) and must be uncoated before they can cause the cell to manufacture viruses. These cells may die as new virus particles are released, or they may survive if virus particles are slowly budded from the cell by a process similar to exocytosis (see Chapter 5). This "slow budding" causes a persistent infection.

Plant cells are somewhat protected from viral infection by their rigid cell walls and protective outer waxy cuticles. Viruses can enter plants only if the plant cells are damaged or if other organisms, such as sucking insects or fungi, assist them.

### CONCEPT CHECKPOINT

1. Develop an argument for why viruses are not considered to be alive. You may want to revisit Chapter 1 for a discussion of the characteristics of living organisms.
2. How is replication of phage genetic material different during the lytic cycle than it is during the lysogenic cycle?

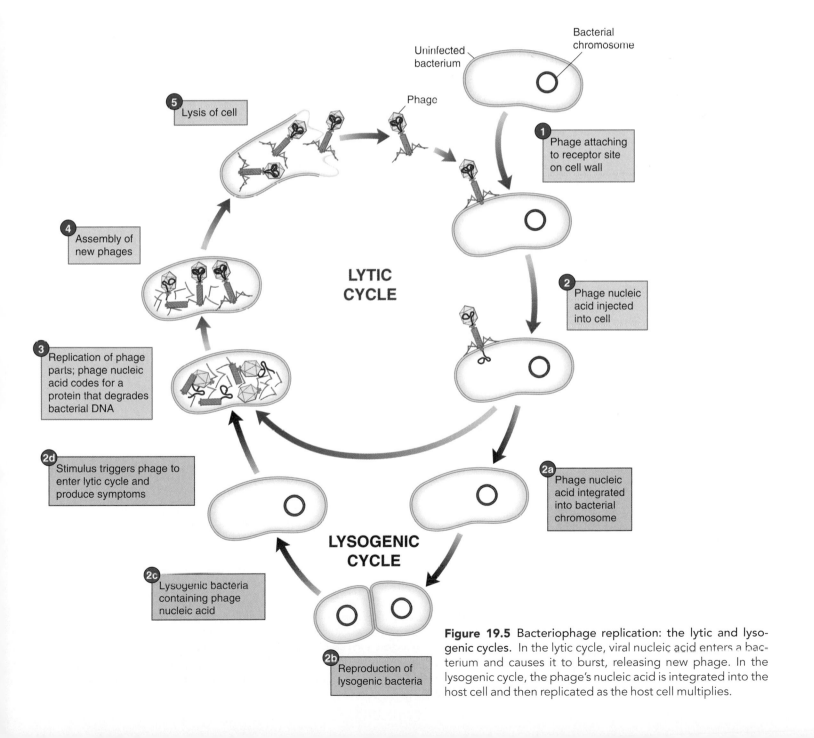

**Figure 19.5** Bacteriophage replication: the lytic and lysogenic cycles. In the lytic cycle, viral nucleic acid enters a bacterium and causes it to burst, releasing new phage. In the lysogenic cycle, the phage's nucleic acid is integrated into the host cell and then replicated as the host cell multiplies.

# How do viruses cause disease?

## 19.4 Infection with certain viruses can initiate cancer.

You can't "catch" cancer like you can catch a cold, but you can develop cancer as a result of infection with one of several viruses including: hepatitis B virus (HBV), human T-cell lymphotropic/leukemia virus (HTLV), human papillomavirus (HPV), human cytomegalovirus (CMV), and the Epstein-Barr virus (EBV). These viruses are initiators of human cancer. That means that they cause cell changes that can lead to cancer. For example, HTLV is associated with leukemia, a cancer of the red bone marrow. A person with leukemia manufactures a greater than normal number of white blood cells. Moreover, these cells are immature, abnormal, and unable to perform their infection-fighting roles (**Figure 19.6**). **Table 19.1** provides information about the cancers associated with these viruses. Chapter 12 discusses the development of cancer in more detail.

Except for EBV, these viruses may be transmitted by sexual contact, by contact with blood or blood products as may occur in health care workers or when intravenous drug users share needles, or by transfusion with contaminated blood. An infected mother may pass the virus to her offspring through the placenta or in breast milk. EBV is found in saliva and can be transmitted by mouth-to-mouth contact. You may be familiar with the noncancerous condition it causes—mononucleosis, sometimes called the "kissing disease." The virus can also be transmitted on contaminated cups, eating utensils, and similar objects.

## 19.5 Viruses invade cells and take over cell functions.

Viruses are infective agents; they enter cells and use them to make more viruses. In so doing, they often cause **infectious diseases**, those

in which *pathogens* (agents of infection) enter organisms and disrupt their normal functioning. Viruses upset normal cell functioning in many ways, such as by taking over protein synthesis, disrupting the cytoskeleton, changing the permeability of the cell membrane, and degrading cellular DNA. Different viruses have different cellular effects, so the signs and symptoms of viral diseases vary.

Eukaryotic viruses can cause acute infections and latent infections. *Acute infections* are active infections that often have a rapid onset, run a particular course, and then subside. You are probably familiar with many human acute viral diseases, such as the common cold, influenza (the flu), measles, and mumps. The common cold is the most frequent infectious human disease. To learn why

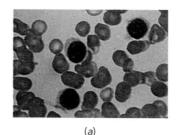  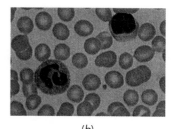

(a)                                   (b)

**Figure 19.6** A hairy cell leukemia blood smear contrasted with a normal blood smear. (*a*) This micrograph of a hairy cell leukemia blood smear shows three abnormal white blood cells among red blood cells. (*b*) This micrograph of a normal blood smear shows one neutrophil and one monocyte, both types of white blood cells, among red blood cells.

 **Visual Thinking:**   Describe the differences you see between the abnormal white blood cells in (*a*) and the normal ones in (*b*).

| TABLE 19.1 | Viruses and Cancer | |
|---|---|---|
| **Virus** | **Cancer Type** | **Comments** |
| Hepatitis B virus (HBV) | Primary liver cancer | HBV causes a serious liver infection that may lead to cirrhosis, a condition in which liver tissue is destroyed and replaced by scar tissue. Chronic infection increases the risk of liver cancer. |
| Human T cell lymphotropic/ leukemia virus (HTLV) | Leukemias and lymphomas | Cancers caused by HTLV are grouped as adult T-cell leukemia/lymphoma (ATLL); T cells are a type of white blood cell that is key in the immune response against disease. Leukemia is a disease of the bone marrow. Lymphoma is a malignant condition of the lymphatic system (see Chapters 30 and 31). |
| Cytomegalovirus (CMV) | Kaposi's sarcoma | Kaposi's sarcoma, often seen in AIDS patients, is a skin cancer characterized by red purplish lesions. Although HIV (human immunodeficiency virus—the cause of AIDS) is not a cancer-causing virus itself, it causes a breakdown of immunity that leaves its victims susceptible to other infections and cancers. |
| Human papillomavirus (HPV) | Vaginal/vulval/cervical cancer<br>Penile cancer | At least two types of HPV are strongly linked to the development of cancer, particularly cervical cancer. (The cervix is the tissue surrounding the opening to the uterus, or womb.) Development of cervical cancer may be due to an interaction of factors such as infection with other viruses or smoking. |
| Epstein-Barr virus (EBV) | Burkitt's lymphoma<br>Nasopharyngeal cancer | Cancers caused by EBV are rare in the United States but are one of the leading causes of cancer deaths in China. In Africa, EBV is related to Burkitt's lymphoma, a cancer of the lymph system. |

scientists haven't found a cure for the common cold, read the answer to the student question in the *Just Wondering* box.

Latent infections are not active infections; the infective agent (in this case the virus) is present, but it is not manifesting disease. For example, the herpes simplex virus causes latent infections of the skin. Herpes nucleic acid remains in nerve tissue (sensory ganglia) without damaging the host until a cold, a fever, or other factor such as ultraviolet radiation from the sun acts as a trigger, and the cycle of cell damage begins. This "damage" manifests as cold sores or fever blisters. The herpes zoster virus (the chickenpox virus) can also act in the same way. This virus may remain latent in the nerve tissue of a person who has had chickenpox, only to be triggered at a later time, causing the painful nerve disorder *shingles*.

Several sexually transmitted infections (STIs) are caused by viruses. Viral STIs are extremely serious because they cannot be cured. Certain medications may ease the discomfort they may cause, and some symptoms may disappear on their own. However, viral particles stay in the tissues, and symptoms may recur. Three sexually transmitted infections caused by viruses are genital herpes, genital warts, and human immunodeficiency virus (HIV) infection. Infection with HIV eventually leads to acquired immunodeficiency syndrome, or AIDS. This disease is fatal; it is discussed extensively in Chapter 30. To eliminate or reduce your chances of becoming infected with a sexually transmitted infection, follow the recommendations in **Table 19.2**.

## Genital herpes

More than one out of every five people aged 12 or older in the United States—approximately 45 million adolescents and adults—have genital herpes. Each year, more than one million people contract this incurable disease. As you can see from **Figure 19.7**, the incidence of herpes has been on the rise since the 1960s.

Genital herpes is caused most often by the *herpes simplex virus type 2 (HSV-2)*. However, HSV-1 is found to be the causative agent of genital herpes approximately 10 to 15% of the time. Otherwise, HSV-1 is associated primarily with the development of cold

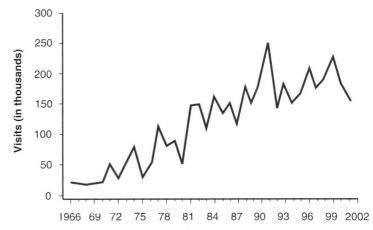

**Figure 19.7** The rise of genital herpes in the past 35 years. This graph charts initial visits to physicians' offices and shows how the incidence of this disease rose from 1966 through 2001.

 **Visual Thinking:** How does the incidence of genital herpes in 2001 compare with its incidence in prior years?

sores and fever blisters. Genital herpes caused by HSV-1 results from transmission through oral sex.

Genital herpes develops within a few days after sexual contact with an infected person. The first symptoms are an itching or throbbing in the genital area. Blisterlike, painful, itchy lesions and swollen lymph nodes in the groin area develop soon after. In women, lesions develop in the vagina and on the cervix, vulva, and thighs (**Figure 19.8**). In men, lesions develop primarily on the penis. The virus also causes an inflammation of the urethra in men, resulting in a watery discharge from the urethral opening.

The herpes viruses cause latent infections. In genital herpes, after the initial infection appears to subside, the virus particles travel via sensory nerve fibers from the genital area to ganglia in the sacral region of the spinal cord. There, the virus either remains dormant or slowly replicates. As is typical with a latent infection, the virus is reactivated by one of a variety of stimuli, such as stress, fever, and menstruation. Once reactivated, the virus moves along the nerve fibers back to the genital area, where it replicates, causing lesions once again. This cycle of dormancy followed by a recurrence of lesions continues, but successive recurrences usually become milder. Eventually, the recurrences may cease, but the viruses remain latent in the ganglia and can be reactivated by severe emotional or physical stress even a long time later.

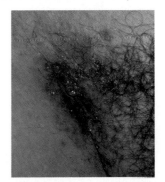

**Figure 19.8** Genital herpes. Lesions in the groin of a woman.

A person with genital herpes (or oral herpes) can pass the virus on to another person any time viruses are being shed, or cast off. Scientists know that shedding occurs when lesions are present and usually starts a few days before lesions appear, but shedding may *never* cease. There is no way to completely protect yourself from infection if you have sex with an infected person. Using latex condoms only reduces the chance of transmission of the virus.

| TABLE 19.2 | **Eliminating or Reducing Your Risk of Acquiring a Sexually Transmitted Infection (STI)** |
|---|---|

**To eliminate your risk of becoming infected with an STI:**
- Never share needles (HIV, HTLV, and HBV can be transmitted this way; see Table 19.1)
- Abstain from sex, or
- Do not have sex with infected individuals, or
- Engage in sex only in a mutually monogamous relationship in which you and your partner are both uninfected

**To reduce your risk of becoming infected with an STI:**
- Reduce your number of sexual partners
- Avoid having sex with persons who have casual sex with many partners and practice other risky behaviors such as sharing needles
- Avoid having sex while under the influence of drugs or alcohol
- Use a new latex condom during each act of sexual intercourse

**Note:** Persons in certain professions, such as health care workers and police officers, may become infected with HIV, HTLV, or HBV as they carry out their duties. Infection can occur if their blood mixes with the blood or bodily secretions of an infected individual. The above precautions may not eliminate their risk of infection.

### How come scientists haven't found a cure for the common cold?

Although we often refer to the causative agent of colds as "the cold virus," actually more than 200 viruses cause colds. Rhinoviruses and coronaviruses top the list, but within these viral groups alone there are well over 100 cold-causing viral types. Developing a vaccine against 200 agents of infection is not practical. However, in late 2003, a California-based drug company was finishing clinical trials on a drug developed to inactivate all types of rhinovirus, the most common cause of colds. This drug is similar to the protease inhibitors that help battle HIV infection. It inactivates a key enzyme any rhinovirus needs to replicate and should reduce the duration of a cold by a few days.

Another line of investigation to find a cold cure is to develop an antibody that would block viral receptor sites on human cells. Although more than 200 viruses cause colds, most of them use the same two receptor sites on cells of the upper respiratory tract mucosa. As yet, this cold-fighting antibody has not been developed.

Antibiotics, those "magic bullets" against bacterial diseases, do not work against viruses. Antibiotics attack bacterial structures and metabolic processes, which is the reason they cause little harm to human cells. To help keep antibiotic-resistant strains of bacteria from developing, it is important to use antibiotics prescriptions completely, even after symptoms go away. It is also important not to take antibiotics to treat viral infections such as colds.

Viruses are very different from bacteria and are *not* affected by antibiotics. Your immune system helps you fight each invasion of cold viruses you encounter. You develop immunity against colds as you age, because each time you contract a cold, you develop some immunity to the particular cold virus that infected you. Therefore, by the time you turn 60 years of age, you should be catching only one cold per year, on average, compared with the three or four you probably contracted when you were a child.

Some medications can reduce cold-related symptoms. Decongestants can lessen nasal congestion, and antihistamines may help dry a "runny" nose. Saline nasal sprays have been shown to have no effect on congestion. Health care providers also urge cold sufferers to rest, get sufficient sleep, and drink plenty of fluids, avoiding caffeine and alcohol. ●

*Are you wondering about a topic in biology and how it relates to your life? Submit your question by clicking the* Just Wondering *link in this text's companions Web site at* www.wiley.com/college/alters.

---

### CONCEPT CHECKPOINT

3. List four ways in which cancer-causing viruses may be transmitted from one person to another.
4. Why do AIDS patients often experience virally induced tumors?

---

## 19.6 Prions and viroids are simpler than viruses and cause disease.

An infectious agent simpler than a virus has almost reached celebrity status in the news. This pathogen causes "mad cow" disease, which was first diagnosed in cattle in Great Britain in 1985. Veterinarians expected to see about 500 cases per month through 1991. These reports prompted some European countries to ban the importation of British beef, and the U.S. Department of Agriculture watched the situation carefully.

Although the disease has declined dramatically in recent years, cows are still becoming infected. In December 2003, one "mad cow" was discovered in the United States, and as a result more than 30 countries stopped buying American beef. In a quick response, the U.S. secretary of agriculture imposed a broad set of new rules aimed at protecting the U.S. food supply from mad cow disease.

What is mad cow disease, and why is there such a fuss over a disease of cows? Its scientific name is bovine spongiform encephalopathy (en-SEF-uh-LOP-uh-thee). Bovine means "cattle," and spongiform encephalopathy is a brain disorder that causes the brain to become full of holes, like a sponge (**Figure 19.9**). This disease is caused by an abnormal prion that can be transmitted to humans.

**Prions** are proteins that cause several diseases of the brain in addition to mad cow disease, such as kuru and Creutzfeldt-Jacob disease (CJD) in humans and scrapie in sheep. Humans can contract a form of CJD called variant Creutzfeldt-Jakob disease (vCJD) by eating infected beef. Variant CJD predominantly affects younger persons (those in their 20s) and includes atypical symptoms such as psychiatric disorders. Similar forms of the disease can be inherited or can arise spontaneously, but generally arise in much older persons.

Prions exist in two forms—the normal form and the abnormal form. Normal prions (prionlike polypeptides) play important roles in brain functioning, such as helping nerves communicate with one another. Abnormal prions destroy the brain. Prions are the only infectious agent that has no genetic material.

How do organisms acquire abnormal prions? There are three ways: by infection with abnormal prions, by having inherited genes that give rise to abnormal prions, or by spontaneous genetic mutation that gives rise to abnormal prions. In the case of Mad Cow disease, cows are infected by eating contaminated protein supplements made from the remains of infected cows and other ruminants. The prion is then transmitted to humans when they eat infected beef, causing vCJD. After ingestion, the abnormal prions invade uninfected nerve tissue and convert normal prions to abnormal forms. Variant Creutzfeldt-Jacob disease has killed more than 100 people worldwide, but mostly residents of Britain. (Read the chapter opening story in Chapter 4 for more information on prions.)

**Viroids** are different from prions in that they are small strands of RNA rather than strands of protein. They also differ from viruses in that they are smaller than the strands of genetic material in viruses, and they have no protein coat. Nevertheless, viroids are replicated in host cells like viruses, using the host cell machinery. Viroids cause plant diseases, including potato spindle tuber, chrysanthemum stunt, and avocado sunblotch (**Figure 19.10**).

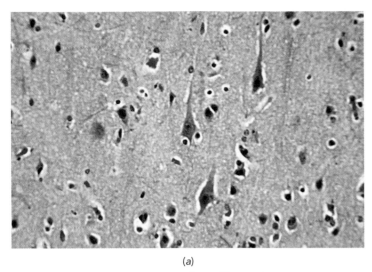
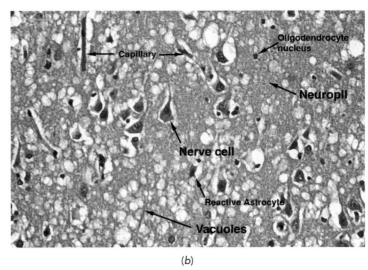

**Figure 19.9** Healthy and prion-infected human brain tissue. (*a*) Non-infected cerebral cortex. (*b*) Prion-infected (CJD) cerebral cortex. Note the labeled structures and compare in each view. The neuropil is an area of synapses (connections between neurons). The vacuoles seen within the neuropil are located within the nerve cell processes and result in a degeneration of synaptic connections. Astrocytes are small, star-shaped glial cells that surround and support nerve cells. Reactive astrocytes respond to injury by walling off dead brain tissue.

**Figure 19.10** Plants infected with viroids. (*a*) Malformed potatoes resulting from infection with the potato spindle tuber viroid. (*b*) Yellow blotches on a chrysanthemum leaf from a plant infected with chrysanthemum stunt viroid. (*c*) Avocados from a plant infected with avocado sunblotch viroid.

# What are bacteria and how do they reproduce?

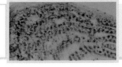

## 19.7 Bacteria are prokaryotic cells that are both helpful and harmful.

**Bacteria** and viruses both have "reputations" as being agents of infection. However, the role of bacteria in the world of living things is much broader than that of a pathogen. Bacteria are present on and in virtually everything you eat and touch. Bacteria make life on Earth possible because they perform essential functions as **decomposers** of organic material, and they are natural recyclers of nitrogen and other organic compounds in ecosystems (see Chapter 40). They are also used to produce certain foods, such as yogurt, sauerkraut, dill pickles, and olives. **Table 19.3** lists some of the ways in which bacteria are helpful.

| TABLE 19.3 | Some Beneficial Activities of Bacteria |
| --- | --- |

- Decompose and recycle organic materials that no other organisms can degrade
- Degrade environmental pollutants in the process of bioremediation (removing hazardous substances from an area)
- Digest cellulose in a stomach chamber (rumen) of cows
- Make available essential elements such as nitrogen to other organisms
- Produce oxygen
- Reside inside and outside our bodies, making it more difficult for pathogenic microorganisms to cause disease
- Produce vitamin K in the large intestine of humans
- Help produce foods such as yogurt and pickles
- Produce useful products such as ethanol and antibiotics
- Help produce useful proteins such as insulin or vaccines via genetic engineering
- Help produce useful characteristics in other organisms, such as pest resistance in crops, via genetic engineering
- Are inexpensive models used to study systems of higher organisms

Bacteria are prokaryotes and were abundant for well over 2 billion years before eukaryotes appeared in the world. The cellular organization of prokaryotes differs from eukaryotic cellular organization in two main ways: (1) the prokaryotic cell has no membrane-bounded nucleus and (2) the prokaryotic cell contains no membrane-bounded organelles that compartmentalize the cell. This structure places bacteria in a domain all their own. Bacterial structure is described briefly in Chapter 5 and shown in Figure 5.4. Prokaryotes were largely responsible for creating the properties of the atmosphere and the soil during the long ages in which they were the only form of life on Earth. Eukaryotes are thought to have arisen as pre-eukaryotic cells engulfed prokaryotes (see Chapter 18).

Bacteria occur in a broad range of habitats and play key ecological roles in each of them. Some thrive in hot springs in which the water temperature can be as high as 78°C (172°F). Others live more than a quarter of a mile beneath the surface of the ice in Antarctica. Still other bacteria, capable of dividing only under high pressures, exist around deep-sea vents formed by undersea volcanoes. Bacteria live practically everywhere, even in ground water, where they were once thought to be absent. They are able to play many ecological roles because they are extremely diverse in their physiology.

Bacteria are classified according to criteria listed in **Table 19.4**. For the level of genus and higher, rRNA sequencing provides the best information. Analyzing the rRNA sequences of bacteria provides insight into their phylogeny, or evolutionary history. Using this approach, the thousands of species of bacteria are placed into at least 12 different major groups. At the species level, however, the phenotypic characteristics listed in Table 19.4 are generally more useful to determine classification.

Reflecting their tremendous diversity, bacteria have three very different modes of nutrition. Some bacteria are photoautotrophic (photosynthetic). Like plants, photosynthetic bacteria use energy from the sun to produce "food" in the form of carbohydrates. They are therefore called **photoautotrophs** (FOTE-oh-AWE-toe-trofes), which means light (*photo*) self (*auto*) feeders (*trophs*).

Photosynthesis takes place in green bacteria, purple bacteria, and cyanobacteria (SIGH-eh-no back-TEAR-ee-ah). All these bacteria contain chlorophyll, which is found within a system of membranes that ring the interior periphery of the cell. Bacterial chlorophyll is chemically different from the chlorophyll in plants. In addition, in the green and purple bacteria, photosynthesis does not take place exactly as it does in plants, and oxygen is not a byproduct. In fact, since oxygen tends to break down their chlorophyll, these species usually live in polluted water that contains little oxygen. The cyanobacteria, however, release oxygen during photosynthesis as plants do and can be found living near the surface of lakes and ponds (**Figure 19.11**). The cyanobacteria were among the first cells to evolve and probably oxygenated Earth's primitive environment.

Other bacteria are chemoautotrophic. Chemical self-feeders, or **chemoautotrophs** (KEE-mo-AWE-toe-trofes), derive the energy they need from inorganic molecules such as ammonia ($NH_3$), methane ($CH_4$), and hydrogen sulfide ($H_2S$) gases. With this energy and carbon dioxide ($CO_2$) as a carbon source, they can manufacture all their own carbohydrates, fats, proteins, nucleic acids, and other growth factors. Bacteria called nitrogen-fixing bacteria, which play

| TABLE 19.4 | Some Criteria for Classifying Bacteria |
| --- | --- |

- Cell wall structure
- Shape of individual cells
- Arrangements of groups of cells
- Presence of flagella
- Staining characteristics
- Nutritional characteristics
- Temperature, pH, and oxygen requirements
- Biochemical nature of cellular components, such as RNA and ribosomes
- Genetic characteristics, such as percentages of DNA bases

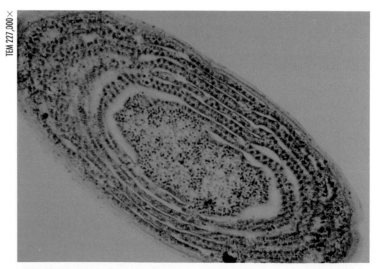

TEM 227,000×

**Figure 19.11 A cyanobacterium.** The outer regions of the cell are filled with photosynthetic membranes. The dark spots between the membranes are storage areas for the carbohydrates produced by photosynthesis.

**Figure 19.12 Soybean root nodules.** The nodules on the roots of this soybean plant contain nitrogen-fixing bacteria. These bacteria convert atmospheric nitrogen ($N_2$) to ammonia ($NH_3$), which can then be used as a nitrogen source by the plants.

an important role in the cycling of nitrogen between organisms and the environment, are an example of chemoautotrophs. Nitrogen-fixing bacteria live in nodules (spherical swellings) on the roots of legumes, such as beans, peas, and clover (**Figure 19.12**), and they convert atmospheric nitrogen ($N_2$) to ammonia ($NH_3$), a form of nitrogen used by plants. Other species of chemoautotrophs play a key role in the sulfur cycle, using $H_2S$ or elemental sulfur for energy and converting it in the process to sulfates, which are also plant nutrients.

Most bacteria are heterotrophic. **Heterotrophs** (HET-ur-oh-TRO-FES) feed on other organisms for their nutrition. They obtain their energy from organic material that enters their cells by diffusion and active transport. Humans are heterotrophs too, eating plants and animals—organic material that once lived. Heterotrophic bacteria play a key role in the carbon cycle by functioning as **decomposers**. That is, they break down large organic compounds, such as proteins and carbohydrates, into small compounds, such as $CO_2$, which is released into the atmosphere to be recycled as it is "fixed" by plants (used as a building block in synthesizing carbohydrates) during the process of photosynthesis.

### Archaea

The "other" prokaryotes are the **Archaea**. Archaeans are simple cells like bacteria, and scientists have only recently realized that they are not bacteria, classifying them in a domain separate from bacteria. A phylogenetic tree illustrating the evolutionary relationship of the three domains (Bacteria, Archaea, and Eukarya) is shown in **Figure 19.13**.

Archaeans do not cause common diseases like bacteria do. They differ chemically from bacteria in their cell walls, cell membranes,

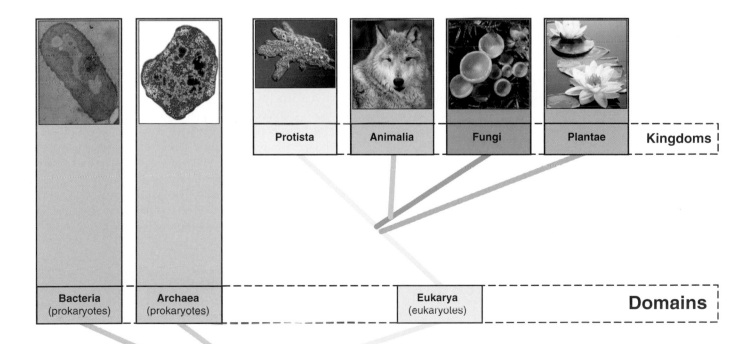

**Figure 19.13** Evolutionary relationships among the Bacteria, Archaea, and Eukarya. This "tree of life" shows that prokaryotes arose first, with eukaryotes evolving from prokaryotes. Protists were the first eukaryotes to develop, with plants, fungi, and animals arising from the protists.

and ribosomal RNA. Some of their physiological processes differ as well. In fact, archaea have such unusual physiologies that they often live in quite unusual places! Genera from one group of archaeans produce methane, or "marsh gas," from carbon dioxide and hydrogen. These prokaryotes can be found in places such as the guts of cattle and the depths of landfills. Unfortunately, the methane they produce adds to the blanket of greenhouse gases surrounding the Earth and therefore to the problem of global warming (see Chapter 42). Other genera of this group live only in areas that have high concentrations of salt, such as salt marshes in the intertidal zone, where fresh water meets seawater in stagnant, concentrated salt pools. Another group of archaeans lives in incredibly hot, acidic, sulfur-rich, and oxygen-depleted environments, such as geysers, sulfur hot springs, and erupting undersea volcanoes. Nevertheless, the Archaea are a diverse group and can be found in many environments that are not extreme, including fresh water at normal pH and common soils.

## 19.8 Bacteria reproduce asexually by splitting.

Reproduction is much simpler among bacteria than in eukaryotes, especially multicellular eukaryotes. Bacterial reproduction is asexual; one cell divides into two, with no exchange of genetic material among cells. This process is called **binary fission**. Before fission, or division of the cell, the genetic material replicates so it can divide along with the cell.

Bacterial DNA exists as a single double-stranded helix that forms a circle and is attached at one point to the interior surface of the cell membrane. As eukaryotic cells do, bacteria make a copy of their genetic material before cell division (**Figure 19.14 ①**). The bacterium also grows in size and manufactures sufficient ribosomes, membranes, and macromolecules for two cells before dividing. When the cell reaches an appropriate size and the synthesis of cellular components is complete, binary fission begins.

The first step of binary fission is the formation of a new cell membrane and cell wall (a septum) between the attachment sites of the two DNA molecules ②. As the new membrane and wall are added, the cell is progressively constricted in two. Eventually, the invaginating membrane and wall reach all the way into the cell center, forming two cells from one ③. The cells then separate ④.

Many bacteria reproduce every one to three hours, but some take a great deal longer. Fast-reproducing bacteria having conditions favoring their growth could produce a population of billions in little more than a day, as each cell divides into two, those two into four, and

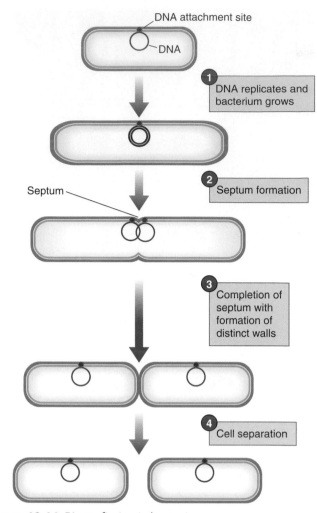

**Figure 19.14** Binary fission in bacteria.

so on. The rate of increase remains constant, but the actual increase in the number of cells accelerates rapidly as the size of the population grows. This type of mathematical progression found in the growth pattern of bacteria is termed *exponential* (see Chapter 15 for an explanation of exponential progressions). However, exponential growth can continue only as long as the conditions for growth are ideal. Eventually, resources become limited, the entire population cannot be maintained, and cells begin to die. Wastes also accumulate, poisoning some of the cells. The growth of the population "levels out." If the growth requirements of the bacterial population are no longer met, the population may begin to die as rapidly as it once grew.

## How do bacteria cause disease?

## 19.9 Bacteria attach to cells and cause tissue damage.

Bacteria that cause disease are heterotrophic. That is, they do not manufacture their own food but obtain food from the environment—and that environment might be you! Most plant diseases are not caused by bacteria but by fungi (see Chapter 20). However, a few genera of bacteria do infect plants, primarily causing types of

plant rot and wilt. Wilt occurs when bacteria block water from moving up the xylem (a type of vascular tissue) in the plant. In addition, some genera of bacteria cause tumorlike growths called *galls* in plants. **Figure 19.15** is a photograph of a plant with crown gall disease caused by bacteria of the genus *Agrobacterium*. Interestingly, scientists have learned how to use these disease-causing bacteria in productive ways in the genetic engineering of plants (see Chapter 11).

**Figure 19.15 Crown gall disease on a tobacco plant.** This disease is caused by bacteria of the genus *Agrobacterium*.

Many human diseases are caused by bacteria, including the diseases listed in **Table 19.5**. Most disease-producing bacteria use their hosts for food, but some poison their hosts. To cause disease in humans, bacteria or the poisons they produce must first get into the body. Usually, this happens if you eat contaminated food or drink contaminated water. You might also inhale bacteria present in the air after an infected person coughs or sneezes, touch a contaminated object, or have sexual intercourse with an infected partner. Sometimes bacteria enter the body through broken skin as the result of an injury or injection with a contaminated needle.

After entering the body, bacteria attach to body cells and cause various types of tissue damage. Bacteria produce chemicals that digest the tissues, and the breakdown products of digestion are taken into the cell and metabolized. In addition, certain bacteria produce *toxins*, or poisons. These bacteria include species that cause staphylococcal food poisoning, gas gangrene, tetanus, and cholera. Bacterial toxins can have powerful effects, such as high fevers, violent muscle spasms, vomiting, diarrhea, heart damage, and respiratory failure. All the while, your body combats these cells using its nonspecific and specific defense systems (see Chapter 30). The ability

| TABLE 19.5 | Some Diseases Caused by Bacteria |
|---|---|
| **Respiratory diseases**<br>• Bacterial pneumonia<br>• Strep throat<br>• Tuberculosis<br>• Legionnaires' disease<br><br>**Gastrointestinal diseases**<br>• Cholera<br>• Typhoid fever | **Nervous system diseases**<br>• Bacterial meningitis<br>• Botulism<br><br>**Sexually transmitted infections**<br>• Syphilis<br>• Gonorrhea<br>• Chlamydial infection |

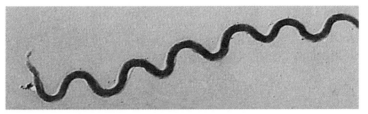

**Figure 19.16 Causative agent of syphilis.** This micrograph shows the spiral bacterium *Treponema pallidum*, which causes syphilis.

of your body to wage this war affects the degree and length of the illness and its eventual outcome.

The sexually transmitted infections caused by bacteria include syphilis, gonorrhea, and chlamydial infection. The bacteria that cause these diseases enter the body via the mucous membranes of the genitals. Persons harboring these organisms transmit them to other individuals during sexual acts. After entering the body, bacteria attach to body cells and cause various types of tissue damage.

### Syphilis

Syphilis (SIF-uh-lis) is caused by a spiral bacterium, *Treponema pallidum* (**Figure 19.16**). It is curable if treated with penicillin, but if left untreated, syphilis can result in death. The untreated disease progresses through a series of stages. Approximately three weeks after sexual contact with an infected partner, the newly infected person usually develops a lesion on either the cervix (in women) or the penis (in men) where the organism entered the body. This lesion, which is called a *hard chancre*, is the hallmark of the *primary stage* of syphilis. At first, the lesion is painless. It is raised above the skin and has a hard base. However, it soon develops into a painful, ulcerated sore. During this time, the bacterium enters the lymphatic system, and the lymph nodes closest to the lesion enlarge. Once in the lymph, the organisms quickly travel to the bloodstream and throughout the body, infecting other tissues. Within 4 to 12 weeks, the sore heals and the primary phase is over.

The *secondary stage* of syphilis begins from six weeks to several months after infection. Because the organisms are now dispersed throughout the body, lesions of the skin and mucous membranes develop away from the site of the original infection—usually on the trunk, arms, and legs. Sometimes, the lesions appear as discolored, flat spots on the skin. Other times, they are small, solid, raised bumps. A person with syphilis is highly contagious during the primary stage and through the secondary stage until these secondary lesions heal. **Figure 19.17** shows skin lesions typical of the secondary stage of syphilis.

During the next stage, an infected person has no symptoms and may think the disease has been successfully combated. Unfortunately, the disease is only in the *latent stage*. This stage may last a

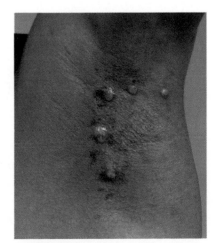

**Figure 19.17** Skin lesions of the secondary stage of syphilis shown in an armpit.

lifetime, but more often, it leads to the final, *tertiary stage*. Once again, lesions develop, but the disease is no longer communicable. The most typical tertiary lesions are *gummas* (GUM-ahz). These lesions are tumorlike masses that can invade the skin, tissues beneath the skin, mucous membranes, bones, and internal organs. When gummas develop in the cardiovascular and central nervous systems, paralysis and death often result.

### Gonorrhea

Gonorrhea (GON-uh-REE-uh) is caused by *Neisseria* (nye-SEAR-ee-ah) *gonorrhoeae*, a paired, coffee-bean-shaped bacterium (**Figure 19.18a**). The word *gonorrhea* is derived from two Greek words meaning "a flow of semen." This name was given to the disease in A.D. 130 by the Greek physician Galen when he mistook the genital discharge caused by gonorrhea for seminal fluid. This discharge, a fluid composed of mucus intermingled with pus, is a result of infection of tissues of the urogenital system. Figure 19.18*b* shows this typical symptom of gonorrheal infection in the male. If the infection remains untreated, other urogenital structures may become infected. Scar tissue may develop and block these passageways, resulting in sterility.

Women who become infected can develop an inflammation of the urethra, vagina (and nearby glands), cervix, and rectum. The initial symptoms of infection are abdominal or pelvic pain, vaginal discharge, and painful or difficult urination.

Both women and men may have no symptoms and so may not seek treatment, which puts them at risk for chronic infection and places their partners at risk for contracting this disease. Treatment with antibiotics is usually effective, although some strains of *Neisseria gonorrhoeae* have developed resistance to penicillin and tetracycline. Chronic infections in women can lead to pelvic inflammatory disease (PID) if the bacteria migrate up through the uterus and to the uterine (fallopian) tubes. PID can cause sterility if scar tissue develops and blocks the uterine tubes, which are passageways that lead from the ovaries to the uterus.

### Chlamydia

The most prevalent STI in the United States today is chlamydial infection (klah-MID-ee-ahl). The CDC estimates that 4 million cases

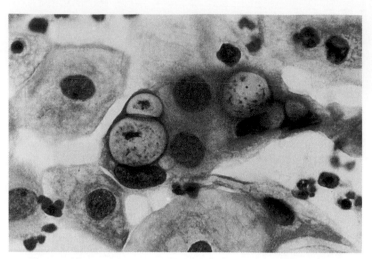

**Figure 19.19 Chlamydial and elementary bodies in a host cell.** This micrograph shows a Pap smear of cells taken from the cervix. The cytoplasm of the cells is stained blue, with nuclei stained dark pink. Normal cells surround a cell infected with *Chlamydia*. The infected cell is in the center of the photo. The chlamydial cells, which appear as various shades of pink, take up most of the infected cell. The reticular bodies are the dark pink cells. These cells are the metabolically active, reproductive form of *Chlamydia*. The light pink cells with red centers are in the process of condensing to form elementary bodies: the infective, dormant form of *Chlamydia*.

occur each year in the United States. It has become so common among teenage girls in some parts of the United States that physicians are recommending that sexually active girls be tested twice a year for this disease. As with gonorrhea, chronic infection with *Chlamydia* can lead to PID and sterility in women and to inflammation of the urogenital passageways and sterility in men.

Chlamydial infection is caused by an unusual bacterium, *Chlamydia trachomatis* (trah-koh-MAH-tiss). These organisms, among the smallest of the bacteria, grow only within other cells and obtain adenosine triphosphate (ATP) (see Chapter 6) from their host cells. For this reason, scientists call chlamydia an "energy parasite." Another unusual feature of chlamydia is that it exhibits two different forms depending on whether it is inside or outside host cells. The forms capable of infecting cells are metabolically inactive and small and have dense bodies with rigid cell walls. These infective forms of chlamydia are called *elementary bodies*. They attach to host cells and are taken into these cells by phagocytosis. Once inside their hosts, the elementary bodies transform into *reticulate bodies* (**Figure 19.19**): relatively large, metabolically active cells having flexible cell walls. The reticulate bodies then multiply, condense to form elementary bodies, and cause the host cell to burst, releasing the elementary bodies. Each new elementary body is capable of infecting a new host cell.

### CONCEPT CHECKPOINT

5. Like viruses, viroids, and prions, many bacteria infect living organisms, but bacteria are considered living while the others are not. What characteristics do bacteria have that these other nonliving entities lack, that place them in the category of "living"?

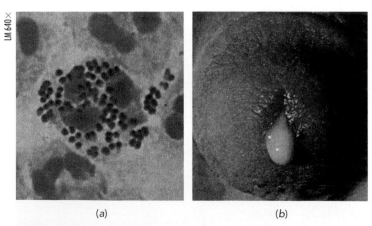

LM 640×

(a)                    (b)

**Figure 19.18 Gonorrhea.** (*a*) *Neisseria gonorrhoeae*, the causative agent of gonorrhea, appears as pairs of coffee-bean-shaped cells (the dark purple dots). Most have been engulfed by a white blood cell. (*b*) The mucuslike discharge of gonorrhea from the urethra of a male.

# CHAPTER REVIEW

## Summary of Key Concepts and Key Terms

### What are viruses and how do they multiply?

**19.1 Viruses** (p. 309) are nonliving infectious agents.

**19.1** Viruses are not living organisms (as they were once considered) because they are not cells.

**19.1** Viruses were discovered at the end of the nineteenth century.

**19.2** Viruses primarily infect plants, animals, and bacteria, replicating within their cells.

**19.2** Viruses are composed of a protein coat (**capsid**, p. 310) around a **core** (p. 310) of nucleic acids; some viruses have an outer layer called an **envelope** (p. 310).

**19.3** Various patterns of viral replication exist.

**19.3** In the bacteriophage **lytic cycle** (p. 310), the phage enters a cell and causes it to produce viral nucleic acid and protein coats. These parts are assembled into new phage particles, which may then burst from the host cell or may leave the host cell by budding.

**19.3** In the bacteriophage **lysogenic cycle** (p. 310), viruses enter into a long-term relationship with the cells they infect, their nucleic acid replicating as the cells multiply.

### How do viruses cause disease?

**19.4** Some viruses can cause cell changes that lead to cancer. These viruses include the hepatitis B virus, human T-cell lymphotropic/leukemia virus, human papillomavirus, cytomegalovirus, and Epstein-Barr virus.

**19.5** Viruses enter cells and upset normal cell functioning in many ways.

**19.5 Infectious diseases** (p. 312) are those in which pathogens, or agents of infection, enter organisms and disrupt their normal functioning.

**19.5** Three sexually transmitted infections (STIs) caused by viruses are genital herpes, genital warts, and human immunodeficiency virus (HIV) infection.

**19.5** Viral STIs are not curable but can be treated to reduce the severity of symptoms and the possibility of transmission.

**19.6 Viroids** (p. 314) and **prions** (p. 314) are simpler than viruses and cause disease.

**19.6** Prions are proteins that cause disorders of the brain.

**19.6** Viroids are small segments of RNA that cause certain plant diseases.

### What are bacteria and how do they reproduce?

**19.7 Bacteria** (p. 314) are prokaryotic cells.

**19.7** Bacteria play diverse ecological roles and are extremely important as **decomposers** (p. 314).

**19.7** Bacteria differ from eukaryotic cells in two main ways: they have no membrane-bounded nucleus and no membrane-bounded cellular organelles.

**19.7** Bacteria comprise thousands of genera and have three modes of nutrition: **photoautotrophs** (p. 316), which make their own food by photosynthesis using the energy of the sun; **chemoautotrophs** (p. 316), which make their own food by deriving energy from inorganic molecules; and **heterotrophs** (p. 317), which obtain energy from organic material.

**19.7 Archaea** (p. 317) are also prokaryotes but differ from bacteria in their cell structure and their physiology.

**19.7** Archaeans inhabit harsh environments, such as landfills, salt marshes, and deep-sea vents, as well as common environments, such as fresh water and common soils.

**19.8** Bacteria reproduce asexually by **binary fission** (p. 317); after replication of the genetic material, the bacterial cell splits in two.

**19.8** Populations of bacteria can increase rapidly when growth conditions are favorable.

### How do bacteria cause disease?

**19.9** Disease-causing bacteria are heterotrophs.

**19.9** A few genera of bacteria cause diseases in plants, and many genera cause diseases in animals, including humans.

**19.9** After entering the body, disease-causing bacteria attach to body cells and cause various types of tissue damage.

**19.9** Disease-causing bacteria produce chemicals that digest tissues, and the breakdown products of digestion are taken into the bacteria and metabolized.

**19.9** Disease-causing bacteria may also produce toxins, which poison the body.

**19.9** The sexually transmitted infections caused by bacteria include syphilis, gonorrhea, and chlamydial infection.

## Level 1    Learning Basic Facts and Terms

**Multiple Choice**

1. Viruses share which of the following in common with living organisms?
   a. Ability to synthesize proteins when outside of host cells
   b. Ability to replicate and transcribe their genetic material when outside of host cells
   c. Ability to harness the energy from their environment that they need to grow and reproduce when outside of host cells
   d. Have membrane-bounded organelles
   e. Have nucleic acid as genetic material

2. In order for a virus to be replicating via the lysogenic cycle, it must
   a. be a prion.
   b. be a viroid.
   c. have incorporated its genetic material into the host cell's chromosome.
   d. be encapsulated.
   e. cause cancer.

3. Which of the following diseases can potentially be treated with antibiotics?
   a. syphilis
   b. variant Creutzfeld-Jakob disease
   c. genital herpes
   d. cervical cancer
   e. HIV

4. Which of the following is *not* used to classify viruses?
   a. genetic makeup
   b. the organism they infect
   c. morphology
   d. mechanism of exchanging genetic information among each other
   e. All of the above are used.

**True–False**

5. _____ Some bacteria can synthesize organic compounds using the energy from the sun or oxidizing inorganic chemicals.

6. _____ Sterility, multiple infections, and pelvic inflammatory disease (PID) are all conditions associated with infection by the gonorrhea bacterium.

7. _____ Prions are infectious agents not composed of RNA or DNA.

8. _____ All bacteria cause disease.

9. _____ Most sexually transmitted infections can be treated successfully with antibiotics.

## Level 2    Learning Concepts

1. Many biologists argue that the most successful organisms on the planet are bacteria. Some of the measures of success they use include pervasiveness, diversity of environments they inhabit, metabolic diversity, adaptability, length of existence, and reproductive potential. For each of these measures explain why bacteria might be considered the most successful of all forms of life.

2. Distinguish between the lytic and the lysogenic viral replication cycles.

3. Many bacteria can cause dangerous diseases, but in general, bacteria make life on Earth possible. What critical functions do bacteria play that are integral to supporting life on Earth?

4. Are high standards of personal hygiene sufficient protection against sexually transmitted infections? Why or why not?

5. Many scientists think that bacteria are the most primitive biological organisms and probably evolved long before eukaryotes. Can you support this?

6. Diagram the structure of a generalized virus. Label the core, capsid, and envelope.

7. Explain how persons with gonorrhea or syphilis may unknowingly pass on these sexually transmitted infections.

8. Are condoms a completely adequate barrier to sexually transmitted infections? Is there still a risk factor if condoms are used correctly?

## Level 3    Critical Thinking and Life Applications

1. Your doctor prescribed an antibiotic for you when she diagnosed your strep throat. Antibiotics kill or inhibit the growth of bacteria. The instructions were to take the antibiotic for 10 days, but because you felt much better after 4 days, you stopped taking it. Why do you think the doctor told you to take the antibiotic for 10 days, and what could happen as a result of stopping the drug prematurely?

2. One of your friends usually develops a cold sore on his mouth when he gets a cold. What may cause this to occur? What pattern of viral replication is involved? How could this virus be transmitted to another person, resulting in a sexually transmitted infection?

3. Analyze your current behaviors regarding your risk for contracting STIs. What can you do to lower your risk? If you think you are not at risk, state reasons why.

4. Genetically modified forms of viruses are often used in human gene therapy (see Chapter 11). What characteristics of viral replication make them useful for inserting correct versions of genes into human cells?

## In The News | Critical Thinking

### NOT SO HARMLESS

Now that you understand more about bacteria and viruses, reread this chapter's opening story about research on the phage-mediated evolution of virulence in bacterial populations. To assist you in understanding this research better, it may help you to follow these steps:

1. Review your immediate reaction to Broudy and Fischetti's research on the phage-mediated evolution of virulence in bacterial populations that you wrote when you began reading this chapter.
2. Based on your current understanding, again summarize the main point of the research in a sentence or two.
3. What questions do you now have about this research that this chapter's opening story does not anwser?

4. Collect new information about the research. Visit the *In The News* section of this text's companion Web site at www.wiley.com/college/alters and watch the "Not So Harmless" video. Then use the "summary" link to read the accompanying story and access related links. Use this information, the links provided, and other online and library resources to answer your questions and find updates about this research topic. State the sources of your information. Explain why you think the information is accurate. Also determine whether the information expresses a particular point of view or is biased in any way.
5. What in your view are the most significant aspects of this research? Give reasons for your opinion and for any changes in your ideas based on the additional information you have collected and the analysis you have done.

# PROTISTS, FUNGI, AND PLANTS

## In The News  Aging and Cancer

It is common knowledge: the incidence of cancer increases as we age. Although some cancers are prevalent in children and young adults, most cancers occur as we pass through our golden years. What happens as we age to make us more prone to developing this disease? Scientists have found a way to use yeast to help them understand the link between aging and the increased incidence of cancer.

Daniel Gottschling and Michael McMurray, scientists at the Fred Hutchinson Cancer Research Center in Seattle, Washington, are using *Saccharomyces cerevisiae*—better known as baker's yeast—as a model biological system to develop an understanding of age-related chromosomal abnormalities, which are a type of genomic instability. Scientists have long known of the relationship between lifespan and one type of chromosomal abnormality: random mutations. The number and type of mutations that occur in our genomes accumulate with age. As described in Chapter 12, mutations can occur in the genes that regulate cell growth and cell division. The result can be cancer, a deregulation of the cell cycle and loss of control over mitosis.

Gottschling and McMurray have developed a system in yeast in which they can study the phenomenon of aging and chromosomal abnormalities. Although yeast cells do not get cancer, they do age, as measured by how many daughter cells each produces. After budding a few daughter cells, a yeast mother cell is still young. After producing 30 or so, it is old and ready to die. Budding is a method of asexual reproduction in yeast.

The researchers investigated whether genomic instability was related to the age of a yeast mother cell. The researchers created strains of yeast with marker genes, which, when lost, change the normal color of developing

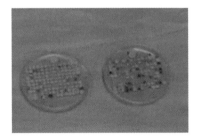

yeast colonies. (A colony is a group of yeast cells that developed from a single cell.) Gottschling and McMurray incubated yeast cells with marker genes under conditions favorable for them to bud. They took every daughter cell that budded from the yeast mother cell and allowed it to form a colony of its own. If marker genes had been lost in the mother, that would show up as a changed colony color of the daughter as shown in the photo.

The results showed that daughter cells derived from old mother cells were about 40 to 200 times more likely to reveal a lost marker gene than daughter cells derived from the same mother when she was young. The researchers suggest that the results point to the existence of a switch that is turned on during the aging process that results in increased genetic instability and consequent loss of marker genes. A deeper understanding of this phenomenon in yeast may help to solve the link between the development of cancer and aging in humans. You can learn more about Gottschling and McMurray's research on using yeast to gain insight into cancer and aging by viewing the video and reading the story "Aging and Cancer" in the *In The News* section of this text's companion Web site at www.wiley.com/college/alters.

*Write your immediate reaction to this research about using yeast systems to study genetic instability: first, summarize the main point of the research in a sentence or two; then suggest what you think its significance is. You will have an opportunity to reflect on your responses and gather more information on this topic in the* In The News *feature at the end of this chapter. In this chapter, you will learn more about yeasts and the kingdom Fungi of which they are a part.*

## CHAPTER GUIDE

### What are protists and how did they arise?

**20.1**   Protists are a diverse kingdom of eukaryotes that evolved from prokaryotes.

**20.2**   Some protists are heterotrophic and therefore animallike.

**20.3**   Some protists are photosynthetic autotrophs and therefore plantlike.

**20.4**   Some protists live on dead or decaying material and are therefore funguslike.

### What are fungi and how did they arise?

**20.5**   Fungi are eukaryotic heterotrophs that digest food outside their bodies.

**20.6**   The chytrids are aquatic fungi.

**20.7**   Zygote-forming fungi include the bread molds.

**20.8**   Sac fungi include the yeasts, cup fungi, and truffles.

**20.9**   Club fungi include the mushrooms and shelf fungi.

**20.10**  The imperfect fungi are a catchall group.

**20.11**  Lichens are associations of fungi and algae or cyanobacteria.

### What are plants and how did they arise?

**20.12**  Plants are multicellular, photosynthetic organisms that probably arose from the green algae.

# What are protists and how did they arise?

## 20.1 Protists are a diverse kingdom of eukaryotes that evolved from prokaryotes.

If you took a sample of water from a fresh water pond, possibly near the edge where plants grow, and looked at a drop of your sample under a microscope, it might look like **Figure 20.1**—teeming with protists. The **protists** (domain Eukarya, kingdom Protista) are a diverse group of eukaryotic organisms that live in moist or aquatic environments. They are found not only in fresh water, such as rivers, streams, ponds, and lakes, but also are abundant in the ocean and the soil, where they play important ecological roles. Many protists are single celled, as you can see in Figure 20.1, but some protists are multicellular or colonial (single cells that live together as a unit).

Protists represent an important step in early evolution. They were the first eukaryotes; they evolved from the prokaryotes about 2.1 billion years ago. Eventually, the protists gave rise to the other eukaryotes: plants, animals. and fungi (**Figure 20.2**). The serial endosymbiotic theory, as described in Chapter 18, states that pre-eukaryotic cells engulfed prokaryotes and lived with them in symbiotic relationships—relationships in which they both benefited. Scientists think that this engulfing process involved the infolding of pre-protist cell membranes, creating a membrane-bounded nucleus and other membranous organelles characteristic of eukaryotes. Mitochondria, cilia and flagella, and chloroplasts likely arose when prokaryotes came to reside within pre-protists (see Figure 18.7). They remain, playing important roles as eukaryotic cell organelles.

The classification of protists is in a state of flux because our knowledge of their evolutionary history is incomplete. The most recent classification schemes are based on similarity of DNA.

Nevertheless, biologists do not agree on how protists are interrelated or how they should be classified, so the phylum names we note in this chapter are not recognized universally. Many biologists, particularly microbiologists (those who study single-celled organisms), think that the kingdom Protista should be split into about 20 kingdoms. However, kingdom Protista is still used as a convenient way of referring to eukaryotes that are not plants, animals, or fungi. The Protista include animallike, plantlike, and funguslike organisms. These groupings are useful to help organize this array of organisms and discuss them, but they actually reflect protists' roles in nature and not their evolutionary relatedness. Therefore, our discussion of protists will be mainly in terms of phylum names, which *do* reflect evolutionary relatedness. Those with the same phylum name are the most closely related.

The animal-like protists are called **protozoans** (PRO-tuh-ZOE-unz) and are considered animallike because they are heterotrophs: They take in and use organic matter for energy. The plantlike protists are the **algae** (AL-jee) and are considered plantlike because they are eukaryotic photosynthetic autotrophs: They manufacture their own food using energy from sunlight. The funguslike protists consist of *slime molds* and *water molds*. Slime molds live on dead and decaying material. Water molds extend funguslike threads (hyphae) into organisms, release digestive enzymes, and absorb the "predigested" food.

Although many protists are single celled like prokaryotes, these organisms are very different from prokaryotes. The single-celled protists are much larger than prokaryotes, having approximately 1000 times the volume. Also, they contain typical eukaryotic, membrane-bounded cellular organelles, which means that protists are much more highly organized than prokaryotes. This higher level of organization allows these single-celled eukaryotic organisms to carry out the cell functions that support their larger cell volume.

## 20.2 Some protists are heterotrophic and therefore animallike.

The protozoans obtain their food in diverse ways, although all are heterotrophic. Protozoans' food-acquiring characteristics are linked to the ways that they move and provide a means by which they can be grouped. Five major groups of protozoans are amebas (uh-MEE-buhs), foraminiferans (feh-RAM-eh-NIF-er-ehnz), flagellates (FLADGE-uh-lates) , ciliates (SILL-ee-uts), and sporozoans (spore-oh-ZOE-unz). The amebas, foraminiferans, ciliates, and sporozoans each correspond to an individual phylum, so the organisms within each of these groups are closely related. In contrast, the flagellates we discuss comprise five phyla, which means that the flagellates vary more than the organisms in the other groups, even though the flagellates all have flagella.

LM 100×

**Figure 20.1** Protists in a sample of pond water.

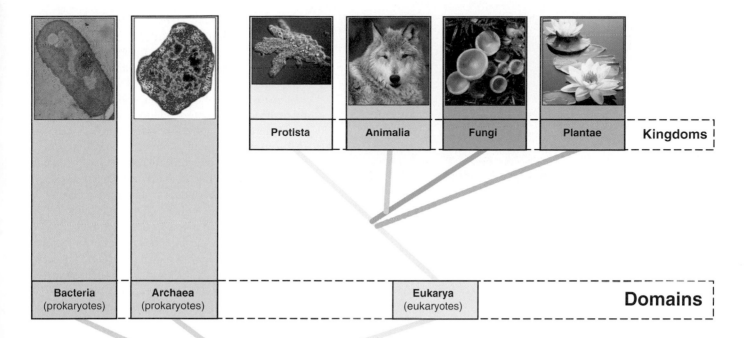

**Figure 20.2** Protists were the first eukaryotes to evolve from the prokaryotes. Plants, fungi, and animals arose from the protists.

## Amebas (phylum Rhizopoda)

Have you ever heard of amebic dysentery? If so, you might recall that this disease is found primarily in the tropics and is associated with poor sanitation. Nevertheless, medical researchers estimate that about 2 million Americans are infected with the causative agent, *Entamoeba histolytica* (EN-tah-ME-bah HISS-tow-LIH-tih-cah). When ingested, this organism infects the digestive system, producing diarrhea containing blood and mucus. Amebic dysentery is not the only disease caused by amebas. Many species are parasites of animals, including humans, and can cause disease. Amebas are abundant throughout the world in fresh water and saltwater, as well as in the soil.

If you were to look at **amebas** under a microscope (**Figure 20.3**), they would appear as soft, shapeless masses of cytoplasm. Within the cytoplasm lie a nucleus and other eukaryotic organelles. The ameba's cytoplasm continually flows, pushing out certain parts of the cell while retracting others. These cytoplasmic extensions are called *pseudopods* (SUE-doe-pods; from the Greek meaning "false feet") and are a means of both locomotion and food procurement. These cell extensions give the amebas their phylum name, Rhizopoda (rye-ZOH-poh-dah), which means "rootlike feet." The pseudopods simply stream around the ameba's prey, engulfing it within a *food vacuole*. Enzymes digest the contents of the vacuole, which are then absorbed into the cytoplasm to be further broken down for energy and biosynthesis. A second type of vacuole, the *contractile vacuole*, pumps excess liquid from the ameba. Amebas reproduce by binary fission: The nucleus reproduces by mitosis, and then the cell splits in two.

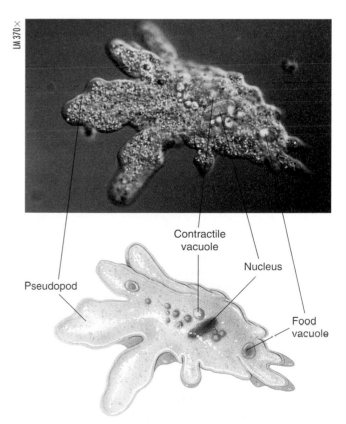

**Figure 20.3** Structure of *Amoeba proteus.* The pseudopods of this ameba are constantly flowing and changing as it moves and takes in food.

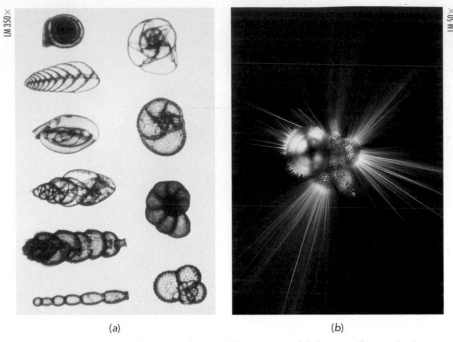

(a)                                    (b)

**Figure 20.4 Foraminiferan shells.** (a) These beautiful foraminiferan shells are made of calcium carbonate. (b) A live foraminaferan.

## Foraminiferans (phylum Granuloreticulosa)

The *foraminiferans* secrete shells that cover and protect their cells. These beautifully sculpted shells are made of calcium carbonate, $CaCO_3$ (**Figure 20.4a**). The name foraminiferans means "hole bearers," referring to the microscopic holes in their shells through which their pseudopods protrude. Figure 20.4b shows spinelike pseudopods protruding from a live foraminiferan's shell. Food particles stick to these cellular extensions and are then absorbed into the cell.

Foraminiferans are abundant in the sea—so abundant, that their shells litter the sea floor. When studying geological strata, scientists often use the foraminiferans as indicators of geological age by noting the types of foraminiferans present in ancient rock. Interestingly, the white cliffs of Dover (**Figure 20.5**) are actually masses of foraminiferan shells, uplifted millions of years ago with the sea floor in an ancient geological event.

## Flagellates (phyla Choanoflagellata, Parabasilida, Euglenida, Kinetoplastida, Diplomonadida)

**Flagellates** are an interesting group because they are so diverse (**Figure 20.6**). All flagellates have at least one *flagellum* (a long, whiplike organelle of motility), others have many flagella, and still others have thousands.

All flagellates have a relatively simple cell structure. They do not have cell walls or protective outer shells as some of the amebas or ciliates do. They also have no complex internal digestive system of organelles as the ciliates do. A flagellate simply absorbs food through its cell membrane, sometimes using its flagella to ensnare food particles.

The flagellates are generally found in lakes, ponds, or moist soil where they can absorb nutrients from their surroundings. The colonial flagellate *Codosiga* (phylum Choanoflagellata) shown in Figure 20.6a, for example, consists of groups of cells that are often found anchored to the bottom of a lake or pond by a cellular stalk. The flagella create currents in the water that draw food toward the

cells. The flagellate *Trichonympha* (TRICK-eh-NYM-fah; phylum Parabasilida), shown in Figure 20.6b, lives a protected life in the gut of termites, digesting cellulose from the wood particles the termite eats. Many flagellates are found living within other organisms; some of these relationships are not harmful to the hosts, but other relationships are.

Figure 20.6c shows the flagellate *Trypanosoma* (trih-pah-neh-SO-mah; phylum Kinetoplastida), which lives in a parasitic relationship with certain mammals, including humans, causing the disease African sleeping sickness. Initial symptoms of sleeping sickness include high fever, weakness, headache, and joint pains. Eventually, the brain and its coverings become inflamed, resulting in symptoms that include slurred speech, seizures, and sleeping for long periods during the day. If not treated, African sleeping sickness is generally fatal.

One group of flagellates, the **euglenoids** (yoo-GLEE-noyds; phylum Euglenida), generally have chloroplasts and make their own food by photosynthesis, so some biologists classify them with the green algae. However, they have certain pigments not present in the green algae, and many forms are heterotrophic, which places them with the other flagellates. In addition, many of the photosynthetic forms become heterotrophs in the absence of light.

Each euglenoid has two flagella, a short one and a long one, that are located on the anterior end of the cell. They move by whip-

**Figure 20.5 The white cliffs of Dover.** The picturesque white cliffs are composed of masses of foraminaferan shells, which were uplifted from the sea floor millions of years ago.

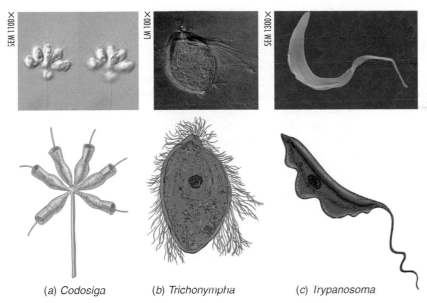

(a) *Codosiga*        (b) *Trichonympha*        (c) *Trypanosoma*

**Figure 20.6** The flagellates. (a) *Codosiga*, a colonial flagellate that remains attached to its substrate. (b) *Trichonympha*, one of the flagellates that inhabits the gut of termites and woodfeeding cockroaches. *Trichonympha* ingests wood cellulose, which it finds in abundance in the digestive tracts of its hosts. (c) *Trypanosoma*, which causes sleeping sickness in humans. It has a single flagellum.

ping the long flagellum (**Figure 20.7**). The euglenoids reproduce asexually by transverse fission, a process in which the parent cell divides across its short axis (see Figure 20.11a). No sexual reproduction is known among this group, which is named after its most well-known member: *Euglena*.

*Euglena* lives in ponds and lakes and can survive in stagnant water. It has a hard, yet flexible, covering called a *pellicle*, with ridges spiraling around its body. These ridges can be seen clearly in Figure 20.7. Two organelles, an *eyespot* and a *photoreceptor*, help *Euglena* stay near the light. (The name *Euglena* means "true eye.") The photoreceptor, located near the base of its longer flagellum, is shaded by the nearby eyespot. As light filters through the pigment of the eyespot, the receptor senses the direction and intensity of the light source. Information from the receptor assists the movement of *Euglena* toward the light, a behavior known as **positive phototaxis**.

The flagellate *Trichomonas vaginalis* (trih-KOM-oh-nas vaj-in-AY-lis; phylum Parabasilida) (**Figure 20.8**) causes genital infections called trichomoniasis (TRIK-oh-moh-NYE-ah-sis) in men as well as women, despite its "female" name. It feeds on bacteria and cell secretions in the urogenital structures it infects. In women, it infects the vagina, cervix, and vulva. However, it can cause infection only when the pH level of the vagina is elevated because it cannot survive in the vagina's normally acidic environment. The pH of the vagina may become elevated when a woman is taking antibiotics, because these drugs sometimes kill the resident vaginal bacteria that secrete acids.

If you go hiking or camping, you may be familiar with the flagellate *Giardia lamblia* (phylum Diplomonadida), which causes giardiasis, or "hiker's diarrhea." The symptoms are nausea, cramps, bloating, and diarrhea. *G. lamblia* is found throughout the world, including all parts of the United States and Canada. It occurs in water, including the clear water of mountain streams and the water supplies of some cities. In addition to humans, it infects at least 40 species of wild and domesticated animals. These animals can transmit *G. lamblia* to humans by contaminating water with their feces.

Flagella protrude from one end of *G. lamblia*, allowing it to move along the intestinal wall of its host (**Figure 20.9**). This motile form of the protist exists only while it is inside the body of its victim. Dormant, football-shaped cysts are expelled in the feces of a host animal. These cysts can survive for long periods of time outside

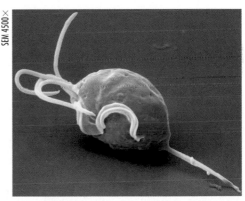

**Figure 20.8** *Trichomonas vaginalis.* This flagellate has five flagella in the front and one flagellum in the back.

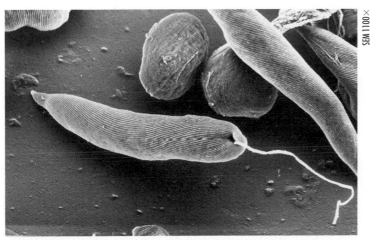

**Figure 20.7** *Euglena.* The long flagellum of the *Euglena* in the center of this colorized scanning electron micrograph is used for movement. This *Euglena*'s short flagellum is not visible in the micrograph.

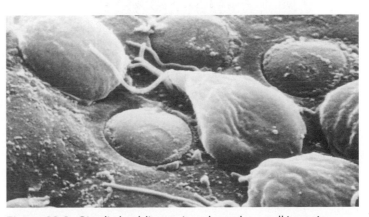

**Figure 20.9** *Giardia lamblia* moving along the small intestine.

their hosts—especially in the cool water of mountain streams. When ingested by other hosts, the cysts develop into their motile, feeding form.

What should you do to prevent infection by *G. lamblia* when hiking or camping? Never drink untreated water, no matter how clean it looks. *G. lamblia* is resistant to the usual water-treatment agents, including chlorine and iodine, so you should boil the water you drink for at least a minute. Better still, *do not* drink the water—bring your own water to be safe.

### Ciliates (phylum Ciliophora)

**Ciliates** get their name from a Latin word meaning "eyelash"—a name that is descriptive of the fact that all or parts of these cells are covered with hairlike extensions called *cilia* (**Figure 20.10**). These cilia beat in unison, moving the cell about or creating currents that move food particles toward the gullet of the cell.

Ciliates possess a wide array of cellular organelles that perform functions similar to the organs of multicellular organisms. The diagram of *Paramecium* in Figure 20.10*b* shows this interesting cellular organization. Cilia protrude through holes in the paramecium's outer covering, or pellicle. The micronuclei (there may be several) function in sexual reproduction; the single macronucleus controls cell metabolism and growth. The beating cilia of the paramecium sweep food into its *gullet*. From the gullet, food passes into *food vacuoles*, where enzymes and hydrochloric acid aid in digestion. After absorption of the digested material is complete, food vacuoles empty their waste contents into the *anal pore*, located in a special region of the pellicle. The waste then leaves the cell by a process similar to exocytosis. The contractile vacuoles expel excess water from the cell.

Paramecia reproduce asexually by *transverse fission* (**Figure 20.11a**). In addition, paramecia exchange genetic material by a process called *conjugation* (Figure 20.11*b*). Although conjugation does not produce offspring cells, it does promote genetic variability among cells that would otherwise produce generations of identical cells when they undergo fission. Genetic variability enhances the ability of the population to survive. Some algae, fungi, and bacteria also exchange genetic material in similar processes, which are also termed conjugation.

Most ciliates live in fresh water or saltwater and do not infect other organisms. However, the species *Balantidium coli* (BAL-an-TIH-dee-um KOH-lie) inhabits the intestinal tracts of pigs and rats. People who come in contact with this protozoan on farms or in slaughterhouses can become infected.

### Sporozoans (phylum Apicomplexa)

**Sporozoans** are nonmotile spore-forming parasites of vertebrates such as cows, chickens, and humans. Many cause serious and sometimes fatal diseases. Their old name, phylum Sporozoa, focused on their spore-forming capabilities. Their new name, apicomplexans (AY-pee-com-PLEX-ans), refers to the "apical complex," a group (a complex) of cell organelles at one end (the apex) of the cell specialized for penetrating host cells and tissues (**Figure 20.12**).

Apicomplexans have complex life cycles that involve both asexual and sexual phases and may require several species of host. Their spores are small, infective bodies that are transmitted from host to host by various species of insects. The sporozoan used classically to represent this phylum of protists is *Plasmodium*, the causative agent of malaria. Approximately 1 million people die of this disease each year; therefore, it is considered one of the most serious disease in the world. Read about the connection between sickle cell disease and malaria in Chapter 14.

---

**CONCEPT CHECKPOINT**

1. When protists are said to be animallike, plantlike, and funguslike, what does this mean? What aspect of their life is similar to animals, plants, and fungi?

---

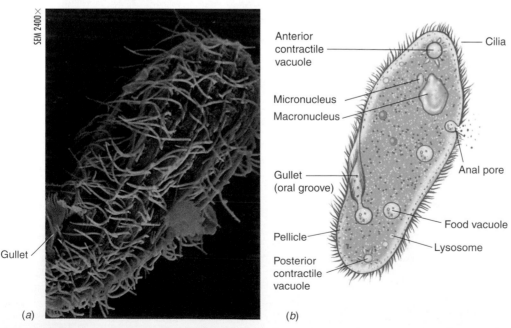

**Figure 20.10** *Paramecium*, a ciliate. (*a*) Cilia can be seen covering the paramecium shown in this photo. Note the concentration of cilia at the gullet. (*b*) Cell structures of a paramecium.

### 20.3 Some protists are photosynthetic autotrophs and therefore plantlike.

You may know algae from your swimming pool or the beach. However, algae are widely distributed in the oceans and lakes of the world, too, floating on or near the surface of the water, with their photosynthetic parts no lower than the sun's rays can reach. They are eukaryotic organisms that contain chlorophyll and carry out photosynthesis, so they can be thought of as plantlike. However, algae lack true roots, stems, leaves, and vascular tissue (an internal water-carrying system). In addition, many contain other pigments that mask the green chlorophyll.

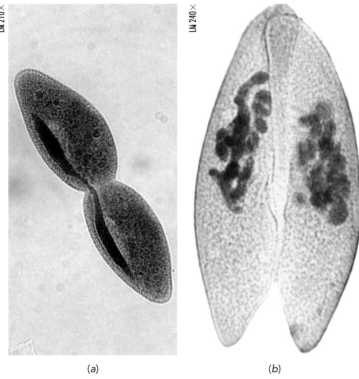

Figure 20.12 An apicomplexan attached at its apex to a red blood cell. Notice that many cell organelles occupy the attached end of the apicomplexan. Eventually, the entire apicomplexan will invade the red blood cell. Notice how the apicomplexan is much smaller than the red blood cell, of which only a part is visible.

(a)                    (b)

**Figure 20.11** *Paramecium caudatum.* (*a*) Reproduction by transverse fission. (*b*) Exchange of genetic material by conjugation.

## Dinoflagellates (phylum Dinoflagellata)

Although the **dinoflagellates** (DYE-no-FLAJ-uh-luts) have flagella, they look nothing like the flagellates. Many dinoflagellates have outer coverings of stiff cellulose plates, which give them very unusual appearances (**Figure 20.13**). Their flagella beat in two grooves, one encircling the cell like a belt and the other perpendicular to it. As they beat, the encircling flagellum causes the dinoflagellate to spin like a top; the perpendicular flagellum causes movement in a particular direction. In fact, the word "dinoflagellate" comes from Latin words meaning "whirling swimmer."

Most dinoflagellates live in the sea and carry on photosynthesis. Their photosynthetic pigments are usually golden brown, but some are green, blue, or red. The red dinoflagellates are also called fire algae. In coastal areas, these organisms often experience population explosions, or "blooms," causing the water to take on a reddish hue referred to as a red tide. Red tides can be harmful because many species of dinoflagellates produce powerful toxins that kill fishes, birds, and marine mammals. In addition, shellfish strain these dinoflagellates from the water and store them in their bodies. Although the shellfish are not harmed, they are poisonous to humans and other animals that eat them. Many species of dinoflagellates are bioluminescent and can be seen as twinkling light in the ocean at night.

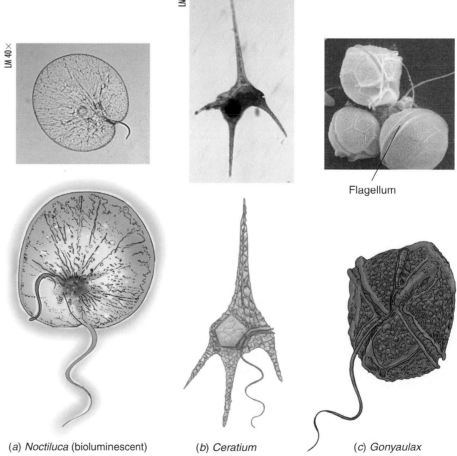

Flagellum

(a) *Noctiluca* (bioluminescent)    (b) *Ceratium*    (c) *Gonyaulax*

**Figure 20.13** The dinoflagellates. (*a*) *Noctiluca*, which lacks the heavy cellulose armor characteristic of most dinoflagellates. (*b*) *Ceratium*. (*c*) *Gonyaulax*. Notice the groove surrounding the organism. In the scanning electron micrograph, the flagellum that beats in this groove can be seen on the lower right.

**Figure 20.14** A red tide (harmful algal bloom) at Bountiful Islands, Gulf of Carpentaria, NW Queensland, Australia.

## Brown algae (phylum Phaeophyta)

The **brown algae** (Phaeophyta; literally, "brown plants") are the dominant algae of the rocky, northern ocean shores of the world, are all multicellular, and are the largest protists. The types of brown algae that grow attached to rocks at the shoreline are known as rockweed (**Figure 20.16a**). Their puffy air bladders keep the algae afloat during high tide. One type of rockweed, the so-called sargasso weed, gave the Sargasso Sea its name. The Sargasso Sea is an area of ocean in the mid-Atlantic, east of Bermuda, with unusual water currents that cause it to be quite calm. This calmness allows floating species of the sargasso weed to proliferate and dominate the area.

Kelp is a large brown alga with enormous leaflike structures (Figure 20.16b). These algae are an important source of food for fish and invertebrates, as well as for some marine mammals and

Dinoflagellates reproduce primarily by longitudinal cell division, but sexual reproduction also occurs in more than 10 genera of dinoflagellates.

**Figure 20.14** shows a red tide in Australia.

### Diatoms (phylum Chrysophyta)

**Diatoms** (DIE-uh-toms) look like fancy microscopic glass boxes of various shapes made up of top and bottom shells that fit together snugly (**Figure 20.15a**). These organisms reproduce asexually by separating their top from their bottom, each half then regenerating another top or bottom shell. The shells are composed of silica, so the diatoms look somewhat glasslike. Their shells are so characteristically striking and intricate (Figure 20.15b) that it would be hard to confuse them with any other group of protists.

The shells of fossil diatoms often form very thick deposits on the sea floor, which are sometimes mined commercially. The resulting "diatomaceous earth" is used in water filters and as an abrasive, and adds a sparkling quality to products such as the paint used on roads and frosted fingernail polish.

Air bladders

(a)

(b)

**Figure 20.16** The brown algae. (a) Rockweed. The air bladders help keep the plant afloat. (b) Giant kelp. Notice the stemlike stipe, leaflike blades, and air bladders. (Note: Brown algae do not always appear brown.)

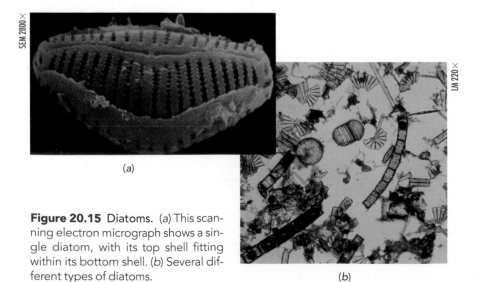

SEM 2800×

LM 220×

(a)

(b)

**Figure 20.15** Diatoms. (a) This scanning electron micrograph shows a single diatom, with its top shell fitting within its bottom shell. (b) Several different types of diatoms.

birds that live among these seaweeds. Some genera of the kelps are among the longest organisms in the world (rivaling the height of the giant sequoia trees), reaching lengths of up to 100 meters (328 feet)! These algae usually have a structure in which a root-like portion, descriptively termed a *holdfast*, anchors the seaweed to the ocean floor or to rocks. A stemlike *stipe* carries the leaflike *blades* of the seaweed, which float on or in the water, capturing the sun's rays.

Most brown algae have life cycles that parallel the generalized life cycle of plants (see Chapter 23 and Figure 23.1*b*). Other multicellular algae (green algae and red algae) also have plantlike life cycles. However, the life cycles of the red algae are quite complex.

## Green algae (phylum Chlorophyta)

The **green algae** (Chlorophyta; literally, "green plants") are an extremely varied phylum of protists. More than 16,000 species have been described. Of these species, most are aquatic, as are other algae, and make up a major component of phytoplankton, single-celled plantlike organisms that drift near the water surface and feed other organisms. Some green algae are semiterrestrial, living in moist places on land, such as on tree trunks, on snow, or in the soil. These algae are primarily unicellular microscopic forms, but some are multicellular.

Green algae show many similarities to land plants: They store food as starch, they have a similar chloroplast structure, many genera have cell walls composed of cellulose, and their chloroplasts contain chlorophylls *a* and *b*. For these reasons, scientists think the green algae were ancestors to the plant kingdom.

Well known among the unicellular green algae is the genus *Chlamydomonas* (**Figure 20.17a**). Individuals are microscopic, green, and rounded and have two flagella at their anterior ends. They are aquatic and move rapidly in the water as a result of the beating of their flagella in opposite directions. They have eyespots that are sensitive to sunlight. The life cycle of *Chlamydomonas* is very simple. This haploid organism reproduces asexually by cell division and sexually by the functioning of some of its cells as gametes. After gametes fuse, the diploid zygote divides by meiosis, restoring the haploid state.

Some genera of green algae live together in groups. *Volvox* is one of the most familiar of these colonial green algae. Each colony is a hollow sphere made up of a single layer of individual, biflagellated cells (Figure 20.17*b*). The flagella of all the cells beat in such a way as to rotate the colony in a clockwise direction as it moves through the water. In some colonies of *Volvox*, daughter colonies and even granddaughter colonies can be seen within the sphere. In some species of *Volvox*, there is a division of labor among the different types of cells, making them truly multicellular organisms.

The multicellular forms of green algae grow in either fresh water or saltwater, such as the sea lettuce *Ulva* shown in Figure 20.17*c*. Sea lettuce is extremely plentiful in the ocean and is often found clinging to rocks or pilings. This alga consists of sheets of tissue only two cells thick.

## Red algae (phylum Rhodophyta)

Almost all **red algae** (Rhodophyta; literally, "red plants") are multicellular, and most of their species are marine. Their color comes

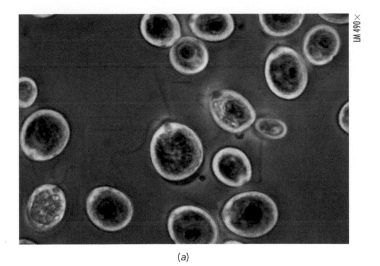
(a)

(b)

(c)

**Figure 20.17** The green algae. (a) *Chlamydomonas*. (b) *Volvox*. (c) *Ulva*, or sea lettuce.

 **Visual Thinking:** Which of these organisms lives in colonies? What do you see in the photo that leads you to this conclusion?

from the types and amount of photosynthetic pigments present in their chloroplasts. Many species have a predominance of red pigments in addition to chlorophyll and so are red, as their name suggests. Some red algae have a predominance of other pigments so that they look green, purple, or greenish black. The red pigment, however, is especially efficient in absorbing the green, violet, and blue light that penetrates into the deepest water. It *reflects* red light. Its absorption spectrum enables some of the red algae to grow at greater depths than other algae and to inhabit areas in which most algae cannot exist.

The red algae produce substances that make them interesting both ecologically and economically. The coralline algae, for example, deposit calcium carbonate (limestone) in their cell walls (**Figure 20.18**). Along with coral animals, these red algae play a major role in forming of coral reefs. Also, all red algae have gluelike substances in their cell walls: agar and carrageenan. Agar is used to make gelatin capsules, is an ingredient in dental impression material, and is used as a base for cosmetics. It is also a main component of laboratory media on which bacteria, fungi, and other organisms are often grown. Carrageenan is used mainly as a stabilizer and thickener in dairy products such as creamed soups, ice cream, puddings, and whipped cream, and as a stabilizer in paints and cosmetics. In certain parts of the world, such as in Japan, some red algae are used as food

---

CONCEPT CHECKPOINT

2. Some plantlike protists are unicellular, others are colonial, and still others are truly multicellular. Compare these categorizations.

---

(a)

(b)

**Figure 20.18 The red algae.** Red algae do not always appear red. Some algae show different colors depending on their daily exposure to light. (a) *Ahnfeltia plicata* growing on rocks. This type of alga is edible and is considered a delicacy in many Asian cultures. (b) *Bossiella*, a coralline red alga. It secretes the hard substance calcium carbonate.

## 20.4 Some protists live on dead or decaying material and are therefore funguslike.

The slime molds make up two unique and interesting phyla of protists, the cellular slime molds and the plasmodial slime molds. You might have seen a slime mold when hiking in the woods when the ground was moist. It might have looked like a flat, gooey mass lying on dead leaves or on the soil.

Both phyla of slime molds are called molds because they give rise to moldlike (funguslike) spore-bearing stalks during one stage of their life cycle. Both are heterotrophs and feed on bacteria, which they find in damp places rich in nutrients, such as rotting vegetation (especially rotting logs), damp soil, moist animal feces, and water.

### Cellular slime molds (phylum Acrasiomycota)

During most of their lives, the **cellular slime molds** (**Figure 20.19**) look and behave like amebas, moving along and capturing bacteria by means of pseudopods ❶. They have plantlike and funguslike, as well as animallike, characteristics and so their classification is continually argued.

At a certain phase of their life cycle, which is often triggered by a lack of food, the cellular form produces a chemical attractant, acrasin, which causes the cells to aggregate, or move toward one another ❷ and form a large, moving mass called a *slug* ❸. The slug eventually stops moving and begins to rise up in the middle ❹, transforming into a fruiting body ❺. The tips of the fruiting bodies release dormant cystlike forms of the amebalike cells, called *spores*. Some of these spores fuse and undergo a type of sexual reproduction before being released ❻; others do not. Each spore becomes a new "ameba" if it falls onto a suitably moist habitat such as damp soil, decaying plant material, dung, or fallen logs ❼. The "amebas" begin to feed and continue the life cycle.

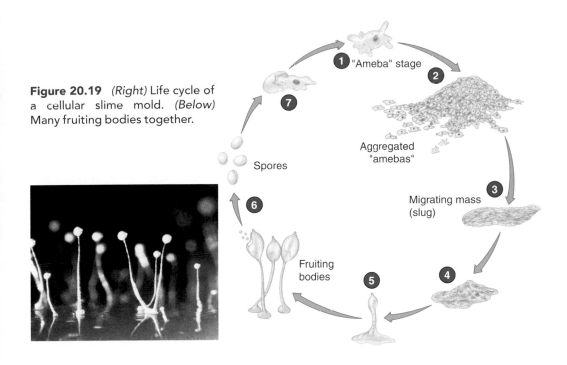

**Figure 20.19** *(Right)* Life cycle of a cellular slime mold. *(Below)* Many fruiting bodies together.

1 "Ameba" stage
2
Aggregated "amebas"
3 Migrating mass (slug)
4
5 Fruiting bodies
6
7
Spores

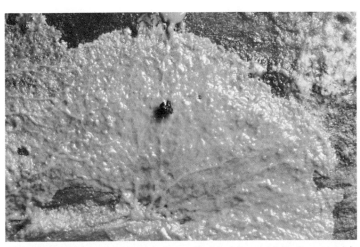

When food or moisture is in short supply, the plasmodium moves to a new area and forms spores. These spores are held in spore cases, which have a characteristic look for each genus of acellular slime mold (**Figure 20.21**). The spores are resistant to unfavorable environmental influences and may last for years if they remain dry. Meiosis occurs in the spores, and when conditions are favorable, the spore cases open and release the spores, which germinate into flagellated, haploid cells called *swarm cells*. The swarm cells can divide, producing more swarm cells, or can act as gametes. Gametes can fuse and form a new plasmodium by repeated mitotic divisions.

## Plasmodial (acellular) slime molds (phylum Myxomycota)

The **plasmodial slime molds**, or Myxomycota (MIK-sah-my-COH-tah; literally, "mucus mold"), have an ameboid stage in their life cycle like the cellular slime molds, but these "amebas" are quite unusual. These bizarre organisms stream along as a **plasmodium** (plaz-MOH-dee-um)—a nonwalled, multinucleate mass of cytoplasm—that resembles a moving mass of slime (**Figure 20.20**). The plasmodia engulf and digest bacteria, yeasts, and other small particles of organic matter as they move along. At this stage of its life cycle, a plasmodium may reproduce asexually; the nuclei undergo mitosis simultaneously, and the entire mass grows larger.

(a)

(b)

**Figure 20.20** A plasmodial slime mold. Plasmodial slime molds move about as a plasmodium, a multinucleated mass of cytoplasm.

**Figure 20.21** Spore cases of two types of plasmodial slime molds. (a) *Lycogala.* (b) *Physarium.*

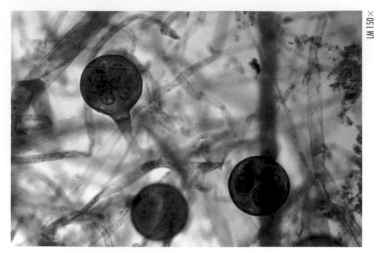

LM 150×

**Figure 20.22** *Saprolegnia*, a water mold. This water mold has filaments and spherical oogonia that encase eggs, which can be seen with the naked eye. *Saprolegnia* (sap-ro-LEG-knee-ah) is the main genus of water molds that infect fresh-water fish.

### Water molds (phylum Oomycota)

Because of their funguslike attributes and their common name "water molds," the Oomycota are grouped with the funguslike protists. If you have an aquarium and have seen white fuzz on any of your fish, you have been introduced to the **water molds** (**Figure 20.22**). They live not only in fresh water, in saltwater, and on aquatic animals, but also in moist soil and on plants. Among the plant diseases caused by this group are late blight and downy mildew. Although this group is named water molds because many species thrive in moisture; they are sometimes also called *egg fungi* because of the large egg cells present during their sexual reproduction ("oomycota" means "egg fungus").

CONCEPT CHECKPOINT

3. List a few similarities and differences between the life cycles of cellular and acellular slime molds.

---

# What are fungi and how did they arise?

## 20.5 Fungi are eukaryotic heterotrophs that digest food outside their bodies.

Have you ever heard of the Monterey pines? Perhaps not, because there are only five stands of Monterey pines in the world; that is, these trees are growing in only five locations. California is home to three of these stands.

A healthy Monterey pine is shown in **Figure 20.23a**. These trees are in jeopardy, however, from a parasitic fungus called pine pitch canker (*Fusarium subglutinans*). This fungus, which is transmitted by certain types of tree-inhabiting beetles, has been attacking various species of pine trees in California since 1986 and may destroy 85% of native Monterey pines within the next decade. The fungus infects the branches first, turning the needles brown. The infection also creates a flow of amber pitch that runs down the trunk. This area is called a canker and gives the fungus its name. Figure 20.23*b* shows a flow of pitch on the trunk of a Monterey pine.

Pine pitch canker is only one species of fungus, a kingdom that consists of eukaryotic heterotrophs that digest food outside their bodies. That is, **fungi** (FUN-jye or FUN-ghee) secrete enzymes

(a)

(b)

**Figure 20.23** The effects of pine pitch canker. (a) A healthy Monterey pine. (b) An infected, oozing Monterey pine.

onto a food source to break it down and then absorb the breakdown products. Although unicellular fungi exist, such as yeast, fungi are mostly multicellular.

Most fungi are **saprophytic** (SAP-roe-FIT-ik)—that is, they feed on dead or decaying organic material. Many fungi live in symbiotic relationships with other organisms. Some fungi are **parasites**, feeding off living organisms, as happens in ringworm and athlete's foot, pine pitch canker, and other plant and animal diseases. Unlike plants, fungi have no chloroplasts and do not produce their own food by photosynthesis. They are composed of slender filaments that may form cottony masses or that may pack together to form complex structures, such as mushrooms.

The slender filaments of fungi are barely visible to the naked eye. Termed *hyphae* (HI-fee; sing. *hypha* [HI-fah]), they may be divided into cells by cross walls called *septa* (sing. *septum*) or may have no septa at all. The septa rarely form a complete barrier, however; thus, cytoplasm streams freely throughout either type of hypha. Because of this streaming, proteins made throughout the hyphae may be carried to their actively growing tips. As a result, the growth of fungal hyphae may be very rapid when food and water are abundant and the temperature is optimal.

A mass of hyphae is called a *mycelium* (my-SEE-lee-um; pl. *mycelia*). Part of the mycelial mass grows above the food source, or substrate, and bears reproductive structures (**Figure 20.24**). The rest grows into the substrate. The part of the mycelium embedded in the food source secretes enzymes that digest the food. For this reason, many fungi are harmful because their mycelia grow into and decay, rot, and spoil foods and, sometimes, other organic substances such as leathers. In addition, some fungi cause serious diseases of plants and animals, including humans.

In their roles as decomposers, fungi may seem troublesome to humans, but they are essential to the cycling of materials in ecosystems (see Chapter 40). In addition, fungi produce many antibiotics, which are drugs that act against bacteria. Yeasts are single-celled fungi that are also useful in the production of foods such as bread, beer, wine, cheese, and soy sauce. All these foods depend on the biochemical activities of yeasts, which produce certain acids, alcohols, and gases as byproducts of their metabolism of sugar.

Most fungi reproduce both asexually and sexually. **Figure 20.25** shows a generalized life cycle shared by many fungi. In asexual reproduction, fungi may reproduce by budding, by growing new hyphae from fragments of parent hyphae, or by producing **spores** by mitosis. Spores are reproductive bodies formed by cell division in the parent organism. They are formed by meiosis in a diploid parent and by mitosis in a haploid parent; the spores are always haploid. During sexual reproduction, hyphae of two genetically different mating types (called + and − strains) come together; their haploid nuclei fuse and produce a diploid zygote. This zygote germinates into diploid hyphae in which meiosis occurs to form spores. The spores germinate into haploid hyphae after being released and finding appropriate growth conditions. The cycle comes full circle when hyphae of genetically different mating types fuse, forming zygotes.

How and when did the fungi arise? Fungi may have evolved from flagellated chemotropic protists—those that used sulfur or nitrogen for energy. Another hypothesis is that fungi evolved from obligate parasites, some of which became saprophytic. Fungi first

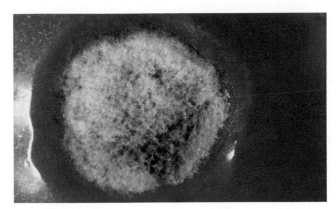

**Figure 20.24 The structure of mold.** A cottony mold growing on a tomato.

appear in the fossil record from 480 to 460 million years ago, in the Ordovician period. However, results of gene-based research published in 2001 by a team from The Pennsylvania State University suggest that fungi may have arisen as early as approximately 1300 million years ago. Scientists think that these early fungi likely grew in association with cyanobacteria, forming *lichens* (see Section 20.11).

There are four phyla of fungi that have a distinct sexual phase of reproduction: the *chytrids*, the *zygote-forming fungi*, the *sac fungi*, and the *club fungi*. Fungi in which a sexual stage of reproduction has never been observed are placed in a catchall group called the *imperfect fungi*.

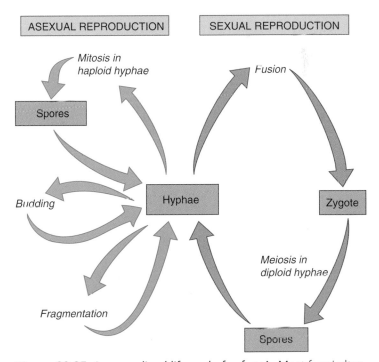

**Figure 20.25 A generalized life cycle for fungi.** Most fungi alternate between sexual and asexual reproductive stages.

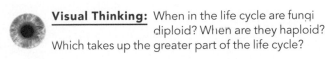

**Visual Thinking:** When in the life cycle are fungi diploid? When are they haploid? Which takes up the greater part of the life cycle?

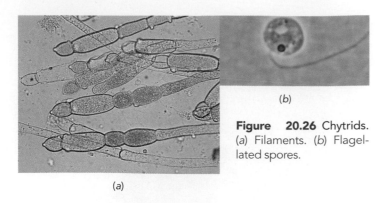

(b)

**Figure 20.26** Chytrids. (a) Filaments. (b) Flagellated spores.

(a)

## 20.6 The chytrids are aquatic fungi.

The chytrids (KIT-rids) are the aquatic fungi discussed in the *How Science Works* box as the causative agent of a deadly infection in frogs. In addition, several species of chytrids are known to parasitize plants and insects, although many live on decaying plant and insect parts.

The chytrids are very different from other fungi in many respects. First, they have flagellated spores and gametes, which is unique in the fungal kingdom. Second, they are aquatic. Some forms are fresh water and some are marine. Third, chytrid cells have centrioles like animal cells do. Other phyla of fungi do not have centrioles. Although some chytrids are unicellular, many produce a mycelium much like other fungi (**Figure 20.26**).

Chytrids are the oldest phylum of fungi and likely gave rise to the other phyla of fungi. They give scientists clues to fungal evolution, suggesting that fungi had their start in water, that they may be closely related to the animals, and that the other phyla lost their flagella during their evolution.

## 20.7 Zygote-forming fungi include the bread molds.

Black bread mold (*Rhizopus nigricans*; rye-ZOH-pus NYE-gricans) is an example of the **zygote-forming fungi** (phylum Zygomycota; ZYE-gah-my-COH-tah), which are generally found on decaying food and other organic material. They are characterized by their formation of sexual spores called **zygospores**, which gives this phylum its name. The life cycle of these fungi (**Figure 20.27**) parallels the generalized life cycle shown in Figure 20.25 quite closely.

**Figure 20.27** The life cycle of black bread mold.

 **Visual Thinking:** If you were to scrape the mold off a piece of food, would you have removed all the mold? Explain how the illustration shows you the answer to this question.

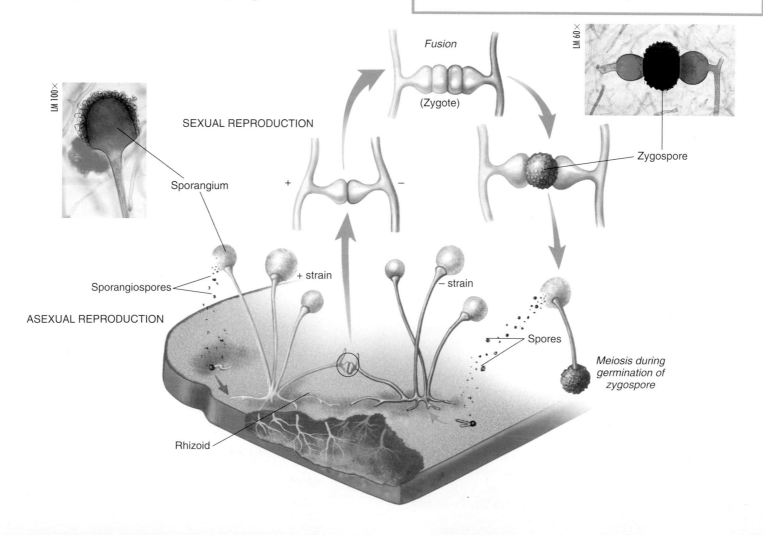

You may have seen black mold growing on bread or other food. It is a decomposer like other molds, but it also helps produce products useful to humans. It carries out fermentations during the manufacture of steroids such as cortisone that are used to control inflammation.

## 20.8 Sac fungi include the yeasts, cup fungi, and truffles.

The **sac fungi** are the largest phylum of fungi, with at least 30,000 named species. They live in a wide variety of places, such as in the soil, in saltwater and fresh water, on dead plants and animals, and on animal feces. They are called sac fungi because their sexual spores, which occur in groups of eight, are enclosed in saclike structures called **asci** (AS-kye; sing. ascus). Their life cycles are similar to the club fungi, which will be discussed shortly.

Among the sac fungi (phylum Ascomycota) are such familiar and economically important fungi as yeasts and cup fungi. This class of fungi also includes many of the most serious plant pathogens, including the causative agent of Dutch elm disease, which has killed millions of elms in North America and Europe. Elm and European bark beetles carry the spores of the mold that causes this disease from tree to tree. The beetles feed on the bark of the tree, which creates a wound through which the fungus can enter the tree and spread. If the mold finds its way to the vascular system of the tree, death can occur in as little as one month. The bark beetles were transported to the United States in shipments of wood from the Netherlands in the 1920s.

**Figure 20.28a** shows an orange cup fungus, and Figure 20.28*b* shows the life cycle of the sac fungi using a cup fungus as an example. During sexual reproduction, hyphae of two genetically different mating types (called + and − strains) come together ❶; the haploid nuclei from the + strain are transferred into the − strain, but the nuclei remain separate ❷. Hyphae with two nuclei in each cell emerge from this structure and form the cup ❸. The nuclei fuse, forming a brief diploid stage ❹, but then first divide by meiosis and second by mitosis. A protective wall forms around each of the resulting eight haploid nuclei, forming eight ascospores. The spores are released when the mature ascus ruptures ❺, and they germinate asexually into + and − hyphae.

Though single cells, the yeasts are also sac fungi and are one of the most economically important of the phylum because of their use in producing various foods. Most of the reproduction of the yeasts is asexual and takes place by binary fission or by budding (the formation of a smaller cell from a larger one). Sometimes, however, whole yeast cells may fuse, forming sacs with zygotes. These cells divide by meiosis and then mitosis, forming eight spores within each sac. When the spores are released, each functions as a new cell.

One species of yeast, *Candida albicans* (CAN-dih-dah AL-bih-cans), causes superficial infections of the mucous membranes of the vagina. (Other species may cause such infections, but *C. albicans* is most often the cause.) *Candida* organisms are normally

**Figure 20.28** Sac fungi (phylum Ascomycota). (*a*) An orange cup fungus, *Sowerbyella rhenana*, from the Willamete Valley, in Oregon. (*b*) The life cycle of a cup fungus.

(*a*)

ASEXUAL REPRODUCTION

Spores

+ strain

❷

Hyphae with two nuclei form the cup

− strain

❶

+

−

❸

Mitosis  Meiosis  Fusion

❹

❺

(*b*)  Release of spores

SEXUAL REPRODUCTION

found in small numbers in the vagina, but the acid environment of the vagina and the proliferation of normal vaginal bacteria limit its growth. When the environment of the vagina changes, this organism may flourish, producing raised gray or white patches on the vaginal walls, a scanty but thick whitish discharge, and itching. Pregnancy and the use of birth control pills can change the vaginal pH, allowing yeast to grow, as can the taking of broad-spectrum an-

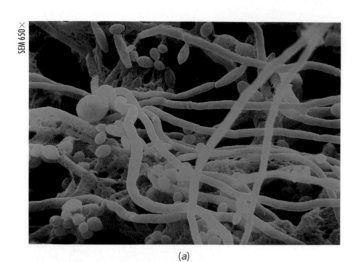

(a)

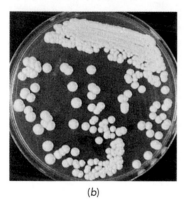

(b)

**Figure 20.29** *Candida.* (a) Although a yeast, *Candida* produces elongated cells called pseudohyphae. (b) The appearance of *Candida* on an agar plate. Colonies of *Candida* look similar when growing on a penis infected with the yeast.

**Figure 20.30** Truffles. You might think of fancy chocolates when you think of truffles, but the word also refers to a type of fungus, such as these French truffles.

tibiotics such as tetracycline, which kills the normal vaginal bacteria. In addition, patients with AIDS or uncontrolled diabetes often develop yeast infections. Candidiasis (CAN-dih-DIE-ah-sis) can also occur in other areas of the body, such as the mouth, hands, feet, skin, and nails. **Figure 20.29a** is a photomicrograph that shows the elongated structures termed *pseudohyphae* that *Candida* produces as it grows.

Candidiasis is not considered a sexually transmitted infection because it is usually contracted without sexual contact, but a vaginal yeast infection can spread to a partner during sexual activity. The infection in the male is characterized by the growth of small, elevated yeast colonies on the penis, similar to those shown in Figure 20.29b. In addition, like most sexually transmitted infections, a vaginal yeast infection can be transmitted to a newborn as it passes through the birth canal. A number of antifungal drugs are available that can be applied locally to treat infections caused by *Candida* organisms.

The sac fungi also include truffles (**Figure 20.30**). Fungal truffles are regarded as a delicacy. They grow underground in France and northern Italy and are detected by pigs or dogs trained to sniff them out.

## 20.9 Club fungi include the mushrooms and shelf fungi.

**Club fungi** include the mushrooms, toadstools, puffballs, jelly fungi, and shelf fungi (**Figure 20.31**). Some species of club fungi are commonly cultivated; for example, the button mushroom—often served in salad bars—is grown in over 70 countries, producing a crop with a value of over $15 billion annually. Other kinds of club fungi are the rusts and the smuts, which are devastating plant pathogens. Wheat rust, for example, causes huge economic losses to wheat farmers.

The club fungi are so-named because they have club-shaped structures from which unenclosed spores are produced. These structures are called **basidia** (buh-SID-ee-uh; sing. basidium) and give the phylum its name, Basidiomycota. In mushrooms, basidia line the gills found under the cap. **Figure 20.32** shows these gills and the life cycle of a toadstool, a poisonous mushroom. In rusts and smuts, basidia arise from hyphae at the surface of the plant.

As shown in Figure 20.32, a mushroom (or toadstool) is formed when pairs of hyphae—often of two different mating strains—fuse and intermingle their cytoplasm and haploid nuclei ❶. The fused hyphae develop into the mushroom, including the gills on the underside of its cap ❷. The basidia develop as single cells on the free edges of the gills. Within the cells that develop into basidia, the nuclei present from the fusion of hyphae now fuse to form zygotes ❸. Each zygote undergoes meiosis twice, producing four haploid spores in each basidium ❹. The zygotes within some basidia also undergo mitosis, producing basidia having eight haploid spores. Turgor pressure (water buildup within the basidia) bursts the basidia and hurls the spores from the mushroom ❺—at an average rate of 40 million per hour! These spores germinate to produce hyphae ❻, which fuse to begin a sexual cycle of reproduction once again. Asexual reproduction is rare in this class of fungi.

**Figure 20.31** Club fungi (phylum Basidiomycota). (*a*) A morel, a highlyprized mushroom. (*b*) The toadstool *Amanita muscaria*, a poisonous mushroom. (*c*) The puffball *Lycoperdon* dispersing spores into the air. (*d*) A yellow jelly fungus known commonly as witch's butter. (*e*) The shelf fungus *Polyporus squamosus*, commonly known as Dryad's saddle fungus. (*f*) *Gymnosporangium*, apple-cedar rust.

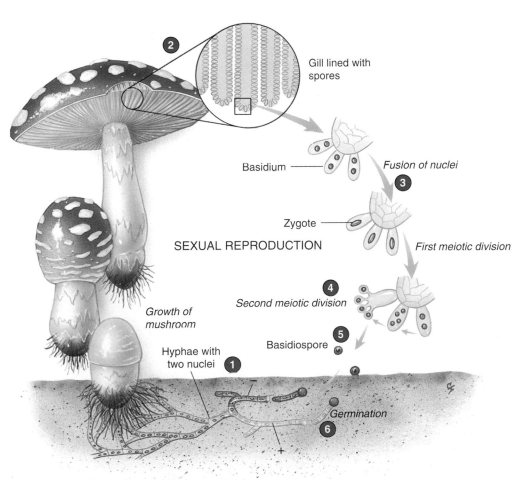

**Figure 20.32** Life cycle of a common club fungus—a mushroom. This class of fungi rarely undergoes an asexual stage of reproduction.

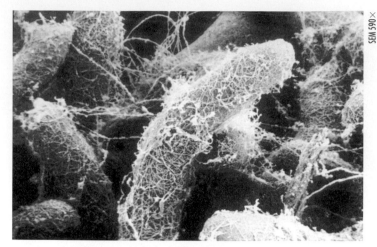

**Figure 20.33** Mycorrhizae. The fungal filaments can be seen growing in a tangled web surrounding plant roots. The fungus actually penetrates the root, which cannot be seen here.

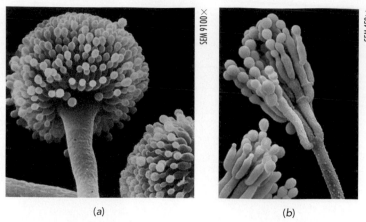

(a)                     (b)

**Figure 20.34** Examples of the Deuteromycota, the imperfect fungi. (a) Hyphae of *Aspergillis* bearing spores. (b) Hyphae of *Penicillium* bearing spores.

The club fungi play a predominant role in the formation of symbiotic relationships with the roots of plants, although other phyla are sometimes involved. These mutualistic associations involving plant roots and fungi are called **mycorrhizae** (MY-kuh-RYE-zuh). Approximately 80 to 95% of the plant species with roots have mycorrhizae. What do the plants and the fungi get out of these associations? The fungi benefit by feeding on plant fluids. The plants benefit because the fungi help make nutrients and water available to the roots, greatly increasing the growth potential of certain plant species. **Figure 20.33** shows stringlike fungal hyphae entwined around the roots of a plant.

### 20.10 The imperfect fungi are a catchall group.

Nearly everyone is familiar with the antibiotic penicillin, and we all have the fungi to thank for it. *Penicillium* is one member of the catchall

fungal group called the **imperfect fungi**, or more formally, the Deuteromycota. These fungi are not imperfect in any way; rather, the name comes from the fact that scientists understand them imperfectly. That is, their sexual stages of reproduction have not been observed, so classification into one of the phyla of fungi is impossible. These fungi may have even lost the ability to reproduce sexually. Nonetheless, the hyphae of different mating types often fuse in these fungi, providing some genetic recombination.

The imperfect fungi reproduce asexually by spores. The structure of the spore cases found at the tips of their hyphae is enormously diverse. These variations represent adaptations of the fungi to promote dispersal of their spores. For example, spores distributed by insects usually have sticky, slimy spore cases with odors attractive to insects, while those dispersed by the wind have dry spore cases.

Among the economically important genera of asexual sac fungi are *Penicillium* and *Aspergillus*. Their hyphae and spores are shown in **Figure 20.34**. Some species of *Penicillium* are sources

## just wondering . . .

*Questions students ask*

**I love to eat mushrooms and often see them growing in my yard or when I go hiking in the woods. How can I tell which are safe to eat?**

You should never eat mushrooms unless you are absolutely sure they are safe. For most people, that means eating only the ones sold at the grocery store.

The mushrooms most often sold in grocery stores are *Agaricus bisporus*, a fungus closely related to the common field mushroom. Other popular edible mushrooms are shiitake mushrooms, *Lentinus edodes*. Together, these two species make up about 86% of the world crop of mushrooms.

Within other genera of mushrooms, such as *Amanita*, there are both poisonous and edible species. An untrained person cannot

easily distinguish these closely related organisms; even mushrooms from different genera can be difficult to identify with certainty. Because the stakes are *so* high—one bite of a highly poisonous *Amanita* could kill you—it makes no sense to take a chance eating wild mushrooms.

Certain mushrooms produce powerful hallucinogenic compounds. Their "street names" include magic mushrooms, purple passion, and shrooms. Ingesting these hallucinogens can be extremely dangerous, possibly resulting in anxiety, paranoia, nausea, and chronic mental disorder. ●

*Are you wondering about a topic in biology and how it relates to your life? Submit your question by clicking the Just Wondering link in this text's companions Web site at www.wiley.com/college/alters.*

**Figure 20.35 Athlete's foot, *Tinea pedis*.** This itchy, painful condition is caused by members of the imperfect fungi, including the genus *Trichophyton*.

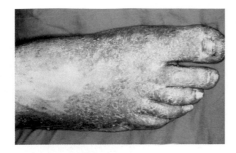

of penicillin, and other species of the genus give cheeses such as Roquefort and Camembert their characteristic flavors and aromas. Species of *Aspergillus* are used for fermenting soy sauce and soy paste, processes in which certain bacteria and yeasts also play important roles.

Other imperfect fungi cause diseases in humans, including infections of the skin, such as athlete's foot (**Figure 20.35**) and other forms of ringworm. The term *ringworm* refers to any fungal infection of the skin, including the scalp and nails. No worms are present in such an infection. The name comes from a pattern of infection on the skin that sometimes looks like a worm has burrowed beneath the epidermis.

## 20.11 Lichens are associations of fungi and algae or cyanobacteria.

**Lichens** (LIE-kins) (**Figure 20.36**) are associations between sac fungi or, occasionally, club fungi and either cyanobacteria or green algae (or sometimes both). Most of the visible body of a lichen consists of fungus, but the photosynthetic organism lives within the

fungal tissues in a type of fungal "sandwich" (**Figure 20.37**). The top and bottom layers of the lichen are formed by fungal hyphae, while the algae reside in the middle.

Lichens might provide an example of *mutualism*, a living arrangement in which both the fungus and the alga benefit. The alga provides nutrients for itself and the fungus. Specialized fungal hyphae penetrate or envelop the photosynthetic cells and transfer water and minerals to them, although this part of the mutualistic relationship has been questioned recently. Researchers now suspect that the fungus acts more like a parasite than a mutualistic partner.

Lichens are able to invade the harshest of habitats—at the tops of mountains, in the farthest northern and southern latitudes, and on dry, bare rock faces in the desert. In such harsh, exposed areas, lichens are often the first colonists, breaking down the rocks and setting the stage for the growth of other plants. These amazing alga-fungus or bacterium-fungus partnerships are able to dry or freeze and then recover quickly and resume their normal metabolic activities. The growth of lichens may be extremely slow in harsh environments—so slow that many small lichens appear to be thousands of years old and may be among the oldest living things on Earth.

> ### CONCEPT CHECKPOINT
>
> **4.** Describe the diploid and haploid stage of the sexual life cycle of these groups of fungi: Zygomycota, Ascomycota, and Basidiomycota. Which is the more conspicuous stage?

(a)

(b)

(c)

(d)

**Figure 20.36 Types of lichens.** (a) Foliose (leafy) lichens have a flat leaflike structure. This foliose lichen is growing on the bark of a tree. (b) Squamulos (scalelike) lichens consist of minute, scalelike structures. This British soldier lichen, *Cladonia cristatella*, has scalelike structures at its base but also produces erect fruiting bodies with red tops (spore-producing structures). Its common name refers to the red fruiting bodies that are reminiscent of the red headgear worn by British soldiers during the Revolutionary War. (c) Fruticose (shrubby) lichens have an erect, bushy structure. Fruticose lichens predominate in deserts because they are more efficient in capturing water from moist air than other types of lichens. (d) Crustose (encrusting) lichens produce a flat crust on rocks and trees. This crustose lichen is growing on a rock.

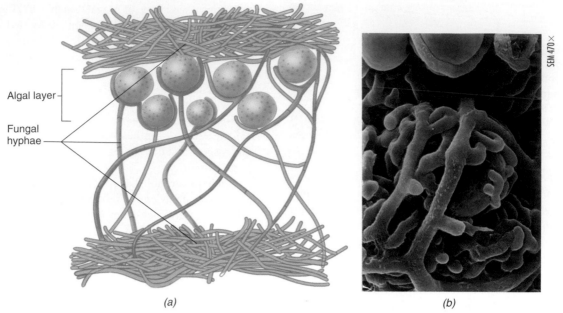

(a)

(b)

**Figure 20.37** Symbiosis within a lichen. (*a*) The bottom hyphae of the fungus anchor the lichen to a substrate, such as a rock or a tree. The fungus produces enzymes that digest the substrate, helping it gain a firm foothold. The alga is protected within the hy- phal structure. It produces sugars by photosynthesis, which the fungus consumes. (*b*) A false-color scanning electron micrograph showing the algal cells (green) being enveloped by the fungal hy- phae (brown).

# How Science Works

### DISCOVERIES

## *Amphibian deaths linked to killer fungus*

Their habitats are relatively undisturbed. Agriculture and urban sprawl have not en- croached on their living space. Foreign predators have not been in- troduced. So why are they dying?

"They" are the cloud- forest frogs in Costa Rica and Panama, Yosemite toads in California, and boreal toads in Colorado (**Figure 20.A**). Their populations are declin- ing dramatically owing to mass deaths. This problem was noticed in the 1990s, raising many questions. Were the amphibian deaths the result of normal climatic fluctuations? Was the problem pollution, acid rain, ultraviolet radiation, or global warming?

In the mid-1990s an international team of scientists set out to find the culprit, and by 1998 they had their answer—Chytrid- iomycosis (kit-RID-ee-oh-my-KOH-sis), a fungal disease caused by an organism in the

Chytrid phylum, an ancient group of fungi that probably gave rise to the other groups of fungi.

Most chytrids live in water, which ex- plains how these fungi might get close to amphibians. Amphibians, as you might re- call, depend on water in the early stages of

**Figure 20.A** Great barred frogs (*Mixophyes fasciolatus*). The frog on the right has chytridiomycosis. There is excessive shedding of skin over the body. The frog on the left is uninfected. The color differences are due to photography and variations within the species.

their development, and they reproduce in water. As adults, they are capable of living in both aquatic and terrestrial environments.

Although this disease was recognized only recently, it seems to have been killing amphibians since the 1970s. The fungus also appears to be widespread; infected frogs

have been found in Australia, Africa, Europe, and South, Central, and North America. In 2003, chytridiomycosis was still considered an emerging disease, meaning that it was just becoming known; scientists are still research- ing how it kills. They know that the fungus at- tacks keratin, a tough protein that forms a re- sistant layer in animal skin. Since frogs breathe and drink through their skin, sci- entists suspect that dam- age to the keratin may actu- ally kill the animals. Infected frogs shed a great deal of skin. Some scientists, how- ever, think that the fungus may produce a deadly toxin.

In 2003, Australian re- searchers discovered that they could clear frogs of chytridiomycosis if they heated them to human body temperature (98.6° F, 37°C) and kept them there for about 16 hours. Scientists are hopeful that the warming technique will be useful in helping rid captive populations of frogs of the disease before they are released into the wild. ●

# What are plants and how did they arise?

## 20.12 Plants are multicellular, photosynthetic organisms that probably arose from the green algae.

All of the photos in **Figure 20.38** are of plants. Although these organisms look quite different from one another, they all belong to the same kingdom: Plantae. What do they all have in common? What characteristics define them as plants?

**Plants** are multicellular, eukaryotic, photosynthetic autotrophs (producers). That is, they produce their own food, which most plants store as starch, by using energy from the sun and carbon dioxide from the atmosphere—a process described in Chapter

(a)

(b)

(c)

(d)

(e)

**Figure 20.38 All these organisms belong to the kingdom Plantae.** Plants are multicellular, eukaryotic, photosynthetic autotrophs. All of these plants except the moss are vascular plants (those having internal tubes to transport plant fluids). (a) Water lilies, an aquatic plant. (b) Hairycap moss *Polytrichum*, a bryophyte. (c) Giant sequoia next to a cabin in Yosemite National Park, California. (d) "Carefree Wonder" roses. (e) Purple pitcher plant, *Sarracenia purpurea*.

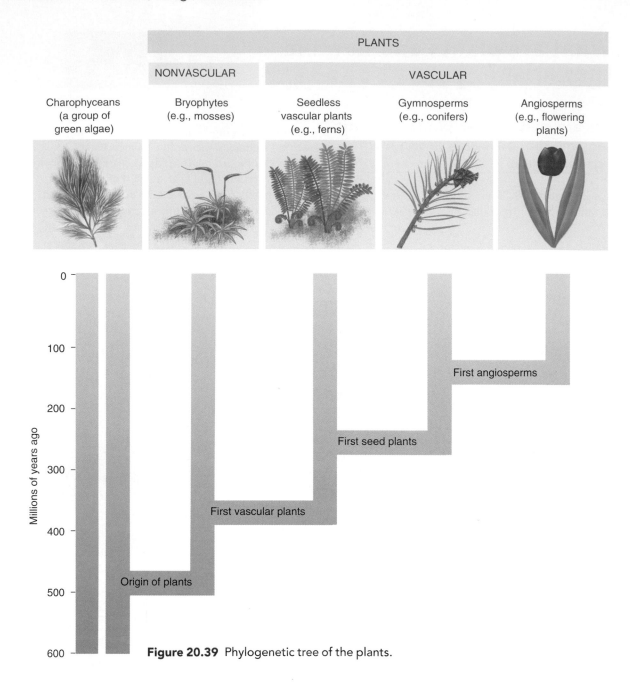

Figure 20.39 Phylogenetic tree of the plants.

7. The multicellular algae, at one time classified as plants and now classified by most scientists as protists, also fit this description, but other characteristics distinguish them from plants.

Unlike multicellular algae, which are aquatic, most plants are adapted for life on land. The few species of plants that live in the water evolved on land and later adapted to aquatic environments. Also, unlike multicellular algae, plants develop from embryos, which are multicellular structures enclosed in maternal tissue. The embryo is the young plant present in the seed. Given proper growth conditions, the embryo has the potential to grow into a new plant.

A third characteristic that differentiates the multicellular algae from the plants is their type of chlorophyll, the pigment that absorbs energy from the sun during the process of photosynthesis. Plants have the pigments chlorophyll *a* and chloro-

phyll *b*. The red and brown algae do not have chlorophyll *b* and so are thought not to have evolved from the same ancestors as the plants. The green algae do have chlorophylls *a* and *b*, but because they have many single-celled forms, they are classified as protists.

Most scientists think the multicellular green algae are the evolutionary predecessors of the land plants. **Figure 20.39** is a phylogenetic tree of the plants. Plants colonized the land second only to fungi, developing adaptations that made this colonization possible and that differentiated them from their algae ancestors. Roots not only anchored plants to the ground but also took in water and nutrients. Their cellulose-containing cell walls provided rigid support for those parts of the plant that grew above ground. A waxy cuticle covered their stem and leaf surfaces, preventing much water loss, while pores in the cuticle allowed oxygen, carbon dioxide, and water vapor to diffuse across. A vascular network—a system of fine tubes—distributed water carrying dissolved minerals and sugars.

Most plants—approximately 95% of all living plant species—are **vascular plants** (VAS-kyuh-ler), those having specialized tissues to transport fluids. The three major groups of vascular plants are seedless vascular plants such as ferns, vascular plants with naked seeds such as pine trees, and vascular plants with protected seeds such as flowering plants. Plants lacking these specialized transport tissues such as mosses and liverworts, are called **nonvascular plants**. The plant kingdom has 12 phyla (**Table 20.1**). Plants are described in more detail in Chapters 23–25.

| TABLE 20.1 | The Twelve Phyla of Plants | |
|---|---|---|
| **Group** | **Phylum** | **Examples** |
| Nonvascular plants | Bryophyta | Mosses |
| | Hepatophyta | Liverworts |
| | Anthocerophyta | Hornworts |
| Seedless vascular plants | Psilophyta | Whisk ferns |
| | Lycophyta | Club mosses |
| | Sphenophyta | Horsetails |
| | Pterophyta | Ferns |
| Vascular plants with naked seeds (gymnosperms) | Coniferophyta | Conifers |
| | Cycadophyta | Cycads |
| | Ginkgophyta | Ginkgos |
| | Gnetophyta | Gnetae |
| Vascular plants with protected seeds (angiosperms) | Anthophyta | Flowering plants |
| | Class Monocotyledons | Grasses, irises |
| | Class Dicotyledons | Flowering trees, shrubs, roses |

# CHAPTER REVIEW

## Summary of Key Concepts and Key Terms

### What are protists and how did they arise?

**20.1** The **protists** (kingdom Protista, p. 326) are a varied group of eukaryotic organisms that evolved from prokaryotes.

**20.1** Many protists are single celled, but some are multicellular or colonial.

**20.1** The Protista include animallike organisms (**protozoans**, p. 326), which take in organic matter for nutrition; plantlike organisms (**algae**, p. 326), which manufacture their own food; and funguslike organisms (slime molds and water molds), which live on dead or decaying organic matter.

**20.2** The protozoans (animallike protists) are heterotrophs and are classified according to the means by which they move and feed.

**20.2** The **amebas** (p. 327) move and eat by means of cell extensions, or pseudopods.

**20.2** The foraminiferans secrete shells that cover and protect their cells; they absorb food via pseudopods that protrude through holes in their shells.

**20.2** The **flagellates** (p. 328) move by means of long, whiplike cellular extensions (flagella); they absorb food through their cell membrane.

**20.2** One group of flagellates, the **euglenoids** (yoo-GLEE-noyds; p. 328), generally have chloroplasts and make their own food by photosynthesis, so some biologists classify them with the green algae; the euglenoid *Euglena* exhibits **positive phototaxis** (p. 329), a movement toward light; other flagellates can cause infections in humans.

**20.2** The **ciliates** (p. 330) move and sweep food toward themselves by means of short, hairlike cellular processes, their cilia; they also possess a wide array of cellular organelles.

**20.2** The **sporozoans** (apicomplexans, p. 330) are nonmotile parasites of vertebrate animals, carried from one host to another by insects.

**20.3** The algae (plantlike protists) contain chlorophyll and carry out photosynthesis, but they lack true roots, stems, leaves, and vascular tissue.

**20.3** The **dinoflagellates** (p. 331) and the **diatoms** (p. 332) are unicellular organisms, the green algae has multicellular and unicellular forms, and the brown algae and red algae are almost all multicellular.

**20.3** The **brown algae** (p. 332) are the dominant algae of the rocky, northern ocean shores of the world and are the largest protists.

**20.3** The **green algae** (p. 333) are an extremely varied phylum of protists; most are aquatic, but some are semiterrestrial.

**20.3** Almost all **red algae** (p. 333) are multicellular, and most of their species are marine.

**20.4** The slime molds have unique life cycles with funguslike spore-bearing stages and "slimy" amebalike stages.

**20.4** Most of their lives, the **cellular slime molds** (p. 334) look and behave like amebas, but they have characteristics that are plantlike and funguslike as well as animallike.

**20.4** Plasmodial slime molds (p. 335) have an ameboid stage to their life cycle but stream along as a **plasmodium** (plaz-MOH-dee-um, p. 335)—a nonwalled, multinucleate mass of cytoplasm—that resembles a moving mass of slime.

**20.4** The **water molds** (p. 336) grow in fresh water and saltwater, on aquatic animals, in moist soil, and on plants; water molds produce large egg cells during sexual reproduction.

### What are fungi and how did they arise?

**20.5 Fungi** (FUN-jye or FUN-ghee, p. 336) are mostly multicellular eukaryotic heterotrophs.

**20.5** Most fungi are **saprophytic** (p. 337), which means that they feed on dead or decaying organic material, but some are **parasites** (p. 337) and feed off living organisms.

**20.5** Fungi are important decomposers that are essential to the cycling of materials in ecosystems, are important in the production of certain foods, and also cause certain diseases.

**20.5** Most fungi are composed of slender microscopic filaments, or hyphae, that form cottony masses or compact plantlike structures.

**20.5** Most fungi reproduce both asexually via **spores** (p. 337) and sexually via fusion of haploid nuclei of hyphae.

**20.6** The chytrids are aquatic fungi that have flagellated spores and gametes, which is unique in the fungal kingdom.

**20.6** Chytrids are the oldest phylum of fungi and likely gave rise to the other phyla of fungi.

**20.7** The **zygote-forming fungi** (such as black bread mold, p. 338) are characterized by their formation of sexual spores called **zygospores** (p. 338).

**20.8** The **sac fungi** (such as cup fungi and yeasts, p. 339) have sexual spores that occur in groups of eight and are enclosed in saclike structures called **asci** (p. 339).

**20.9** The **club fungi** (such as mushrooms, puffballs, and shelf fungi, p. 340) are so-named because they have club-shaped structures (**basidia**, p. 340) from which unenclosed spores are produced.

**20.9** The club fungi play a predominant role in **mycorrhizae** (p. 342), mutualistic relationships with the roots of plants.

**20.10** The **imperfect fungi** (p. 342) are a catchall group in which a sexual stage of reproduction has not been observed.

**20.10** The imperfect fungi include organisms that produce antibiotics and are used in the production of certain foods but that also cause certain diseases in humans.

**20.11 Lichens** (p. 343) are associations of sac fungi with green algae or cyanobacteria.

**20.11** Lichens are able to exist in harsh environments and live for long periods of time.

### What are plants and how did they arise?

**20.12 Plants** (p. 345) are multicellular, eukaryotic, photosynthetic autotrophs that are primarily adapted for life on land.

**20.12** Plants contain chlorophyll *a* and chlorophyll *b*, and develop from embryos.

**20.12** Plants are both **nonvascular** (lacking specialized transport tissues for fluids, p. 347) and **vascular** (having specialized transport tissues for fluids, p. 347).

---

## Level 1    Learning Basic Facts and Terms

**Multiple Choice**

1. Which of the following groups of protozoans contain species that are parasitic?
   a. Amebas
   b. Flagellates
   c. Ciliates
   d. Sporozoans
   e. All of the above

2. Which group of plantlike protists is most closely related to plants?
   a. Rhodophyta
   b. Chlorophyta
   c. Stramenopila
   d. Dinoflagellata

3. Most fungi are saprophytic organisms, but _____ species have also evolved.
   a. parasitic
   b. pathogenic
   c. mutualistic
   d. All of the above

**Matching**

Match each type of protist or fungus with the correct statement.

4. _____ *Giardia*     a. Without them, termites would starve.

5. _____ *Plasmodium*     b. Campers beware . . . don't drink untreated water.

6. _____ Diatoms     c. Not so bad if you're a carrier for sickle cell anemia.

7. _____ Mycorrhizae     d. You may brush your teeth in the morning with their silica shells.

8. _____ *Trichonympha*     e. Without some members of this group, your bread wouldn't rise.

9. _____ Sac fungi     f. Without these, plants would have a hard time meeting their minimum daily nutritional requirements.

---

## Level 2    Learning Concepts

1. Explain how the following organisms are related: protists, protozoa, algae, diatoms, and slime molds.

2. Differentiate between a mutualistic and a parasitic relationship. Give an example of each among the protists and the fungi.

3. What do flagellates and ciliates have in common? How do they differ?

4. What are sporozoans (apicomplexans)? Briefly summarize a sporozoan's life cycle.

5. What do dinoflagellates and diatoms have in common? Describe each one.

6. Why are the algae classified as green, red, and brown? At the molecular level how do you think they are fundamentally different? How do these differences affect the type of environment these algae can inhabit?

7. Fungi are both helpful and harmful to humans. Give some examples of each.

## Level 3    Critical Thinking and Life Applications

1. Some single-celled protists like *Volvox* form spherical colonies of hundreds of cells. Within the colony, some of the cells become specialized for sexual reproduction. Would such a colony be considered a multicellular organism or a colonial organism? Why?

2. If the protist *Euglena* is "plantlike" because it photosynthesizes in the presence of light and "animallike" because it ingests food in the absence of light, in which protist group does it belong? What does this quandary say about the usefulness of such an approach to classifying protists?

3. Many fresh water lakes, when polluted by high-phosphate detergents, become overgrown with mats of photosynthetic green algae. If the pollution continues, all other life in the lake soon dies. Why? Why can you not avoid this lethal result by simply poisoning the algae?

4. Certain antibiotics kill certain types of bacteria. Many of these antibiotics are produced naturally by fungi. Penicillin, for example, is produced by the mold *Penicillium*. What would be the adaptive advantage for molds to produce antibiotics?

5. Although the protists are an extremely diverse kingdom of organisms, they all are aquatic or live in moist environments. What shared characteristics of this diverse group necessitate such an environment?

6. When the pesticide DDT was being used in countries around the world, the number of deaths from malaria dropped dramatically. What is the relationship between DDT, insects, and malaria?

## In The News    Critical Thinking

### AGING AND CANCER

Now that you understand more about yeasts, reread this chapter's opening story about research into using yeast systems to study genetic instability. To understand this research better, it may help you to follow these steps:

1. Review your immediate reaction to Gottschling and McMurray's research on using yeast systems to study genetic instability and to develop an understanding about aging and cancer that you wrote when you began reading this chapter.

2. Based on your current understanding, again summarize the main point of the research in a sentence or two.

3. What questions do you now have about this research that this chapter's opening story does not answer?

4. Collect new information about the research. Visit the *In The News* section of this text's companion Web site at www.wiley.com/college/alters and watch the "Aging and Cancer" video. Then use the "summary" link to read the accompanying story and to access related links. Use this information, the links provided, and other online and library resources to answer your questions and find updates about this research topic. State the sources of your information. Explain why you think the information is accurate. Also determine whether the information expresses a particular point of view or is biased in any way.

5. What in your view is the most significant aspect of this research? Give reasons for your opinion and for any changes in your ideas based on the additional information you have collected and the analysis you have done.

# ANIMALS: INVERTEBRATES

Experienced Spiders

Learning and remembering in spiders? These tasks may seem impossible for an animal with such a small brain. However, Eileen Hebets, a behavioral scientist at Cornell University, in Ithaca, New York, has found that young female wolf spiders (*Schizocosa uetze*) remember social encounters with males over a two- to three-week period and that these encounters influence the female's choice of mates as an adult.

As part of their normal courtship display, male wolf spiders wave their striped forelegs at females. These stripes range in color from brown to black. Sexually mature females react by either mating with or cannibalizing males that display. How does the female decide whether the male will be her mate or her meal? To answer this question, Hebets investigated the effects of the social experiences of sexually immature females (called subadults) on life decisions they make as adults.

Hebets took subadult female wolf spiders and exposed them to mature males whose forelegs she had painted with fingernail polish. She placed some subadult females with males having black polish on their forelegs and some with males having brown polish on their forelegs. Females were exposed to and interacted with these males for 30 minutes every other day until they experienced their final molt, which is the time of sexual maturity. The females were then kept isolated from the males for 11 to 20 days. Hebets next ran mate-choice trials, in which she paired the females, now sexually mature, with either familiar males (the type to which they had been exposed) or unfamiliar males (the type to which they had not been exposed).

The results showed that the females were more likely to cannibalize a male that was unfamiliar and to mate with a male that was familiar. Mature female wolf spiders that had not been exposed to males as subadults served as controls. Unexposed females showed no preference; their rates of mating and cannibalizing were the same with males with forelegs painted black or forelegs painted brown. The results suggest to Hebets that subadult experience in female wolf spiders influences their mate choice as adults. Thus, learning and memory appear to be factors in adult mate choice in these spiders. These factors have previously been shown to be present in vertebrates, but no invertebrate examples existed prior to Hebets' research. Hebets suggests that social experience and the memory of that experience influences adult behavior and possibly the evolution of traits. You can read more about Hebets' research by viewing the video and reading the story "Experienced Spiders" in the *In The News* section of this text's companion Web site at www.wiley.com/college/alters.

*Write your immediate reaction to this research on the influence of subadult experience on adult mate choice in wolf spiders: first, summarize the main point of the research in a sentence or two; then suggest what you think its significance is. You will have an opportunity to reflect on your responses and gather more information on this topic in the* In The News *feature at the end of this chapter. In this chapter, you will learn more about spiders and other invertebrates—animals without a backbone.*

## CHAPTER GUIDE

### What are the characteristics of animals?

**21.1**   Animals are multicellular, heterotrophic eukaryotes.

**21.2**   Animals arose from protist ancestors.

### What do developmental patterns among animals reveal?

**21.3**   Developmental patterns among animals reveal their evolutionary relationships.

### What are the basic organizational features of animals?

**21.4**   Animals have levels of organization, from cells to organ systems.

**21.5**   Almost all animals exhibit either bilateral or radial symmetry.

**21.6**   Animals have three types of body plans.

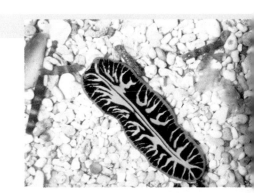

### What are the major invertebrate phyla and their distinguishing characteristics?

**21.7**   Sponges are asymmetrical and have no coelom.

**21.8**   Cnidarians are radially symmetrical and have no coelom.

**21.9**   Flatworms are bilaterally symmetrical and have no coelom.

**21.10**  Roundworms are bilaterally symmetrical and have a false coelom.

**21.11**  Mollusks, annelids, arthropods, and echinoderms are bilaterally symmetrical and have a true coelom.

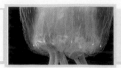

# What are the characteristics of animals?

## 21.1 Animals are multicellular, heterotrophic eukaryotes.

How are animals different from the other kingdoms of living things? Like plants, fungi, and protists, animals are eukaryotic organisms. Their cells have a distinct nucleus and a cellular structure different from the prokaryotic structure of bacteria. Also like plants, most fungi, and some protists, animals are multicellular. There are no single-celled animals. The gametes of animals are single celled, but these cells are not independently living organisms, and as soon as fertilization takes place, the development of a new multicellular individual begins. This pattern is shown in the sexual life cycle of animals (**Figure 21.1**).

Animals are heterotrophs, unable to make their own food. Therefore, animals must feed on other organisms. Some simple animals, such as the sponges, take organic matter directly into their cells. Most animals digest food within a body cavity. The resulting molecules are then taken into the body cells to be broken down further by the chemical reactions of cellular respiration (see Chapter 8). The end product of cellular respiration is energy, which is used to drive the activities of life, including growth, maintenance, reproduction, and response to the external environment. As part of this response, most animals are capable of movement to capture food or to protect themselves from injury.

Animals are extraordinarily diverse in their forms and in how they function. This diverse kingdom is often divided informally into two subgroups: the invertebrates (animals without a backbone) and the vertebrates (animals with a backbone). However, a more correct division would be between invertebrates and *chordates*. The chordates are animals with a special rod-shaped structure down their backs called a *notochord*. Vertebrates are a subphylum of chordates. In this chapter we study the characteristics of animals in general and then focus on the invertebrates. In Chapter 22, we study the chordates, focusing on vertebrates.

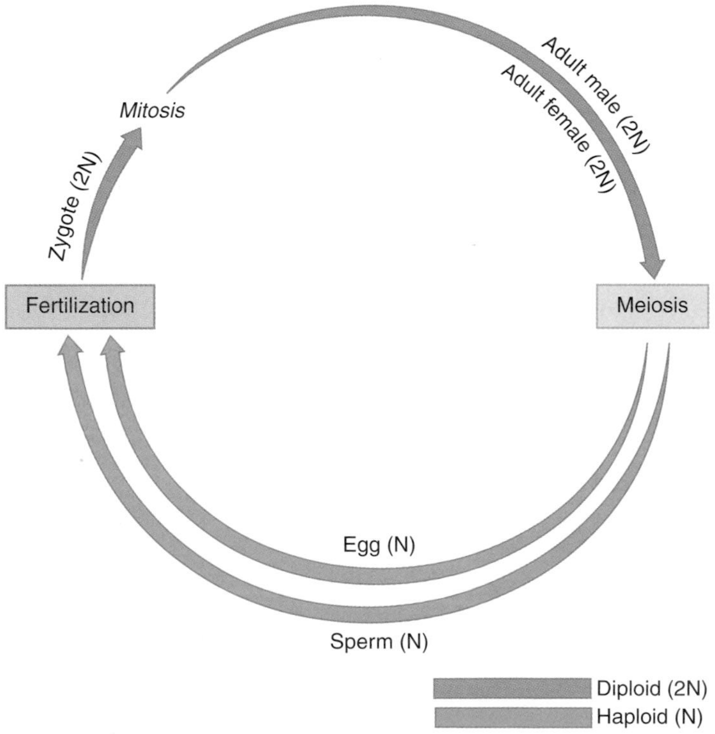

## 21.2 Animals arose from protist ancestors.

Most biologists agree that the animals arose from protist ancestors—specifically, from flagellated cells that lived and worked together in groups (colonies). These colonial organisms called choanoflagellates looked similar to the collar cells of sponges (see discussions in Sections 21.6 and 21.7).

The evolutionary relationships among the animals and their protist ancestors are shown in **Figure 21.2**. Each animal pictured in the phylogenetic tree represents a present-day phylum in the animal kingdom. The mammals, birds, reptiles, amphibians, and fishes are all classes of animals in the chordate phylum. We will discuss the chordates, including the vertebrates, in Chapter 22.

The remaining phyla of animals shown in the phylogenetic tree are invertebrates—animals without a backbone. They are the primary subjects of this chapter. The arthropods include many of the more commonly known invertebrates (e.g., crabs, insects, centipedes, spiders, and ticks). With respect to body plan, the sponges, cnidarians, flatworms, and roundworms are the present-day phyla of organisms without a true coelom (see Section 21.6). In contrast, the mollusks, annelids, arthropods, echinoderms, and chordates all have a true coelom. Each branch of the phylogenetic tree in Figure 21.2 represents an evolutionary pathway that diverged from an ancestral pathway. (Figure 21.2 does not include certain minor invertebrate phyla that are briefly discussed later in this chapter.)

**Figure 21.1 The animal sexual life cycle.** The animal zygote is diploid (2N), and it develops into a diploid, multicellular adult. Only the single-celled gametes are haploid (N).

Mitosis

Adult male (2N)

Adult female (2N)

Zygote (2N)

Fertilization

Meiosis

Egg (N)

Sperm (N)

Diploid (2N)

Haploid (N)

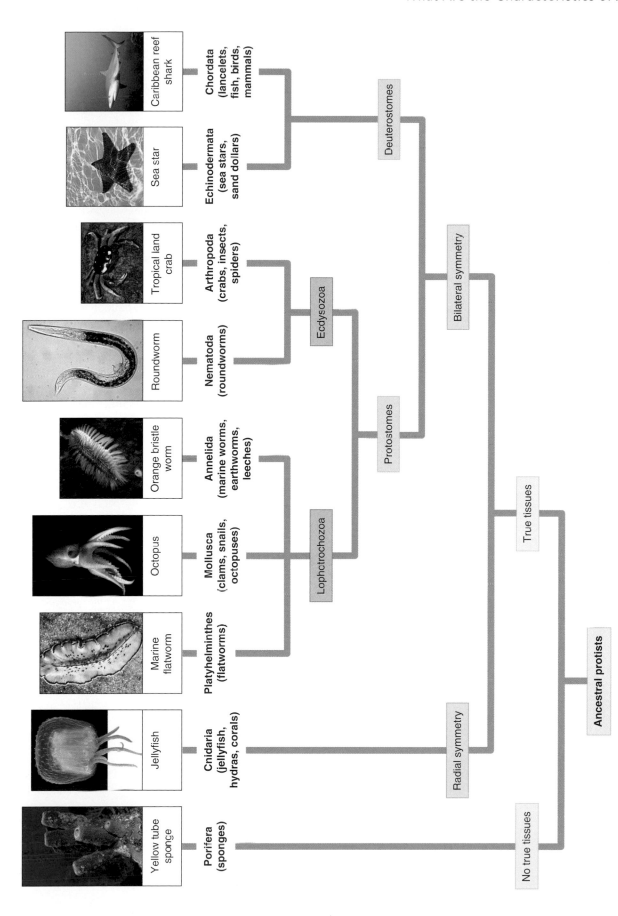

**Figure 21.2 The animal ancestral tree.** At the base of this phylogenetic tree are the protists that are ancestral to all the taxa on the tree. Each node, or branching point, represents where one population splits into two or more descendant populations. At the first node up from the base, the population splits into two branches. One population has no true tissues and most closely resembles the ancestral protists. The other population has true tissues and is ancestral to the rest of the phyla on the tree. This population then splits into two populations that differ in body symmetry. Continue reading the tree, identifying the fundamental characteristics that define the populations at each branching point.

# What do developmental patterns among animals reveal?

## 21.3 Developmental patterns among animals reveal their evolutionary relationships.

In Figure 21.2, look at the two distinct evolutionary lines represented by the **protostomes** (PRO-toe-stowmz) and the **deuterostomes** (DOOT-uh-row-stowmz). The protostomes include the flatworms, mollusks (clams), annelids (segmented worms), nematodes (roundworms), and arthropods (crabs, insects, spiders). In addition, the protostomes divide into two main groups (lophotrochozoa and ecdysozoa) based on the presence of certain genes. The deuterostomes include the echinoderms (sea stars, sea cucumbers) and chordates (amphibians, reptiles, birds, mammals).

Protostomes and deuterostomes differ from one another in their patterns of embryological development. These developmental differences suggest diverging evolutionary relationships between these groups and close evolutionary relationships among the organisms within each group.

During early development, all animals consist of a solid ball of cells usually called a morula. The morula is formed by cleavage of the zygote (the fertilized egg). The protostomes and the deuterostomes show different cleavage patterns during the morula stage of development.

The cells of the protostomes divide in a way that forms a spiral pattern (**Figure 21.3 ❶** ). The cells of the deuterostomes cleave in a radial pattern ❷ . As development proceeds, the cells of the morula secrete fluid that fills the interior of the ball and pushes the cells to the edges. This stage is called the blastula or, in mammals, the blastocyst. The fluid-filled ball of cells then forms an indentation, assuming the shape of a blown-up balloon with a fist pushing in one side ( ❸ and ❹ ). This stage of development, the *gastrula* (GAS-truh-lah), gives rise to a three-layered embryo and begins the formation of the gut. In the protostomes, this first indentation becomes the mouth of the organism ❸ and a second one becomes the anus (if present). In the deuterostomes, the first indentation becomes the anus ❹ and a second one becomes the mouth.

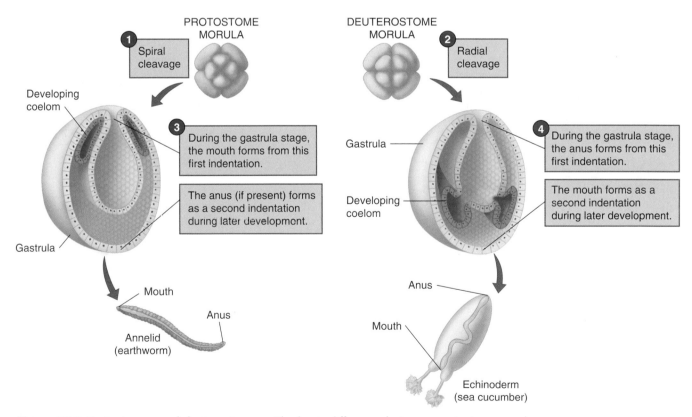

**Figure 21.3 Protostomes and deuterostomes.** The basic difference between protostomes and deuterostomes arises during embryonic development. In protostomes like the earthworm, the mouth develops from the first indentation of the gastrula, and the anus, if present, forms as a second indentation later. In deuterostomes like the sea cucumber, the anus develops from the first indentation, and the mouth develops from a second indentation.

# What are the basic organizational features of animals?

## 21.4 Animals have levels of organization, from cells to organ systems.

The bodies of animals are made up of many different types of cells, the smallest units of life (see Chapter 5). The human body, for example, contains more than 100 different kinds of cells. These cells are not distributed randomly but are organized to form the structures and perform the functions of the human body. This organization is hierarchical. Starting with cells, the next level up in the hierarchy is *tissues* (**Figure 21.4**). Tissues are groups of similar cells that work together to perform a function. Traditionally, tissues are divided into four basic types based on their function: epithelial, connective, muscle, and nervous (see Chapter 26). Epithelial tissue covers body surfaces and lines the body's cavities. The function of the other three types of tissues correspond to their names.

Two or more tissues grouped together to form a structural and functional unit are called an *organ*. Your heart is an organ. It contains cardiac muscle tissue wrapped in connective tissue and "wired" with nerves. All of these tissues work together to pump blood through your body. Other examples of organs are the stomach, skin, liver, and eyes.

An *organ system* is a group of organs that function together to carry out the principal activities of the animal body. For example, the digestive system is composed of individual organs concerned with the breaking up of food (teeth), the passage of food to the stomach (esophagus), the storage and partial digestion of food (stomach), the digestion and absorption of food and the absorption of water (intestines), and the expulsion of solid waste (rectum). The human body contains 11 principal organ systems, which are discussed in depth in Part Seven of this book.

With the possible exception of sponges (**Figure 21.5a**), most simple invertebrates have epithelial tissue and connective tissue. Some simple invertebrates have tissue in addition to these as well. For example, the jellyfish *Aurelia* (Figure 21.5b) has muscle tissue that lies below its epidermis, and it has longitudinal muscle fibers in its tentacles. It even has a stomach, some nervous tissue, and reproductive organs. The flatworms are more complex still. The zebra flatworm (Figure 21.5c) has a brain and nervous system, excretory system, reproductive system, and digestive system. It also has muscles and sense organs (the pigment cups that look like eyes, for example).

**Figure 21.4** Levels of organization. Cells form tissues, tissues form organs, and organs form organ systems. Complex organisms like human beings have all four of these levels of organization.

More complex animals have more diverse tissues. Within each animal phylum, organisms have the same basic tissue types that build their organs and organ systems. Organs and organ systems differ widely among animal phyla, but each phylum has specific patterns of organization that unite its members.

## 21.5 Almost all animals exhibit either bilateral or radial symmetry.

The word "symmetry" refers to the distribution pattern of the parts of an object or living thing. All animals except the sponges exhibit either radial symmetry or bilateral symmetry in their body plans. Sponges are asymmetrical, or without symmetry.

**Radial symmetry** means that body parts emerge, or radiate, from a central axis. The cnidarians (jellyfish, corals, sea anemones, and their relatives) are radially symmetrical. These organisms have an oral side (side containing the mouth) and an aboral side (literally, side without a mouth), but they have neither a dorsal (back) surface nor a ventral (belly) surface.

Animals exhibiting radial symmetry are all aquatic organisms. Many are sessile; that is, they spend most of their lives anchored in one place. Their radially symmetrical bodies allow them

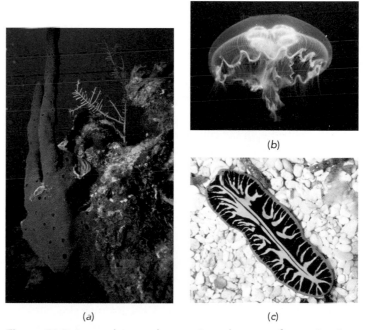

**Figure 21.5** Invertebrates show various degrees of organization. (a) Many scientists contend that sponges have no tissues because each specialized cell functions independently. The colorful sponge extending upward from an underwater rock is the finger sponge (*Heliciona rubena*). (b) This cnidarian, the moon jellyfish *Aurelia aurita*, has tissues and some simple organs. (c) The zebra flatworm (*Pseudoceros zebra*) has tissues, organs, and organ systems.

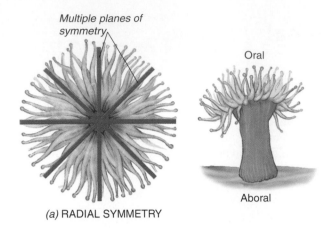

*Multiple planes of symmetry*

Oral

Aboral

*(a)* RADIAL SYMMETRY

*(a)*

LM 70×

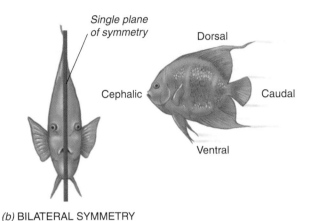

*Single plane of symmetry*

Dorsal

Cephalic

Caudal

Ventral

*(b)* BILATERAL SYMMETRY

**Figure 21.6** Patterns in symmetry. (*a*) In radial symmetry, body parts are arranged regularly around a central axis. Any plane through the center divides the animal into mirror-image halves. (*b*) In bilateral symmetry, only one single plane divides the animal into mirror-image halves.

**Figure 21.7** Sea star. (*a*) This adult echinoderm exhibits radial symmetry. (*b*) Its larval stage, however, is bilaterally symmetrical. *Peter Parks/ Imagequestmarine.com*

*(b)*

to interact with the watery environment in all directions. The echinoderms, such as sea stars, are also radially symmetrical but are not grouped with the cnidarians because their embryonic development and internal anatomy suggest that they are more closely related to organisms having bilateral symmetry. Echinoderm larvae, for example, are bilaterally symmetrical.

**Bilateral symmetry** means that the right side of an organism is a mirror image of the left side. Animals with bilateral symmetry have a dorsal and a ventral surface and a cephalic (head) end and a caudal (tail) end. All phyla of animals other than the sponges, cnidarians, and echinoderms exhibit bilateral symmetry.

Bilaterally symmetrical animals generally move headfirst. Sense organs and nervous tissue are concentrated in the head end, a characteristic called **cephalization**. Most internal structures are arranged symmetrically on either side of the body. In addition, bilaterally symmetrical animals all develop embryologically from three layers of tissue: an inner layer or **endoderm**, an outer layer of **ectoderm**, and a middle layer or **mesoderm**. (See Figure 36.23, which shows these body layers in the human.)

**Figure 21.6** shows the difference between radial and bilateral symmetry. **Figure 21.7** contrasts the radial symmetry of an adult echinoderm with the bilateral symmetry of its larval stage.

## 21.6 Animals have three types of body plans.

Single-celled organisms such as amebas, small multicellular organisms such as certain algae and the flatworms, and larger multicellular organisms having only a few layers of cells between themselves and the environment, such as sponges and cnidarians, are able to move substances easily to all the cells of their bodies. However, the phyla of animals with larger, more complex bodies, such as earthworms, clams, insects, fish, and humans, must have body plans that provide efficient means of moving substances within and between tissues—or they would not survive.

Most bilaterally symmetrical organisms accomplish this task with the *coelomate* body plan (**Figure 21.8a**). A **coelom** (SEE-lum) is a fluid-filled body cavity lined with connective tissue. It may function as a simple circulatory system in some organisms, transporting materials from one part of the body to another. Organs are located within this body cavity, somewhat protected from injury and compression, just as a fetus is protected as it develops in a fluid-filled sac within the uterus. Humans are **coelomates**. For example, you have an abdominal cavity lined with a

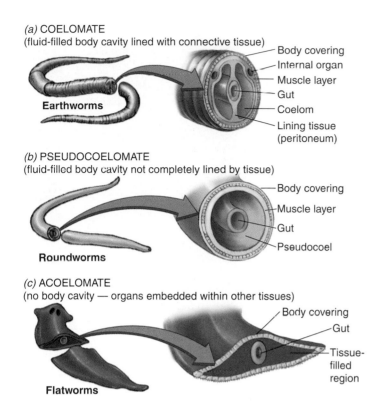

**(a) COELOMATE**
(fluid-filled body cavity lined with connective tissue)

Earthworms

- Body covering
- Internal organ
- Muscle layer
- Gut
- Coelom
- Lining tissue (peritoneum)

**(b) PSEUDOCOELOMATE**
(fluid-filled body cavity not completely lined by tissue)

Roundworms

- Body covering
- Muscle layer
- Gut
- Pseudocoel

**(c) ACOELOMATE**
(no body cavity — organs embedded within other tissues)

Flatworms

- Body covering
- Gut
- Tissue-filled region

**Figure 21.8 Animal body plans.** Animals can be (a) coelomate (having a coelom), (b) pseudocoelomate (having a false coelom), or (c) acoelomate (having no coelom or body cavity).

thin, nearly transparent sheet of connective tissue called the peritoneum. Suspended by thin sheets of connective tissue arising from the peritoneum, the stomach and intestines hang in this coelom.

Roundworms are **pseudocoelomates**. These organisms, along with tiny aquatic invertebrates called rotifers, have a so-called false coelom (pseudocoel): a fluid-filled cavity that houses the organs but is not lined completely by connective tissue. In addition, the organs are not suspended within the cavity by thin sheets of connective tissue. The pseudocoelomate body plan is illustrated in Figure 21.8b.

The **acoelomates** (the flatworms, cnidarians, and sponges) have no body cavity. Since their organs (if present) are embedded within the other tissues of the body, they are compressed as the animal moves. These compressions aid the movement of substances throughout the body. The acoelomate body plan is illustrated in Figure 21.8.

### CONCEPT CHECKPOINT

**1.** Biologists have used the following features to establish evolutionary relationships among animals: patterns of embryonic development, levels of organization, symmetry, and body plan. Describe in more detail what is meant by each of these features.

# What are the major invertebrate phyla and their distinguishing characteristics?

## 21.7 Sponges are asymmetrical and have no coelom.

You probably have some experience with natural sponges. Maybe you have used them while bathing or for washing your car. You may have also observed live sponges in a reef tank in a public aquarium. The **sponges** (phylum Porifera [poe-RIF-er-ah]) are aquatic organisms; most of the 9000 species live in the ocean rather than in fresh water. These sessile creatures are considered simpler than other animals in their organization because they have no tissues (although biologists debate this), no organs, and no coelom, or body cavity. The asymmetrical bodies of sponges consist of little more than masses of cells embedded in a gelatinous matrix.

Sponges reproduce both sexually and asexually. Sponges frequently reproduce asexually by fragmentation; groups of cells become separated from the body of the sponge and develop into new individuals. In addition, sponges may develop branches that grow over the rocks on the sea floor, much like a plant develops underground runners. Colonies of sponges grow along these branches.

During sexual reproduction, most species of sponges produce both female sex cells (eggs) and male sex cells (sperm),

which arise from cells in the matrix. Both types of sex cells are produced within the same organism. Such individuals are called **hermaphrodites** (hur-MAF-rah-dytes), after the Greek male god Hermes and female god Aphrodite. The sperm are released into the interior of the sponge and are carried out of the sponge with water currents and into neighboring sponges through their pores. Fertilization occurs in the gelatinous matrix where the eggs are held. There, the fertilized eggs develop into flagellated, free-swimming larvae that are released into the sponge's interior. After the larvae leave the interior of the sponge, they settle on rocks and develop into adults.

The body wall of the sponge is covered on the outside by a layer of flattened epithelial cells called the *epithelial wall*. Lining the inside cavity of the sponge are specialized, flagellated cells called *collar cells*. These collar cells are also called *choanocytes* [koh-ANN-oh-sites] and are very similar to ancient colonial choanoflagellates, which were likely the protist ancestors to all the animal phyla and to which the sponges are closely related. The matrix makes up the substance of the sponge, sandwiched between the outer epithelial layer and the inner layer of collar cells. Within the matrix are ameboid-type cells, which produce skeletal fibers consisting of either tough, flexible protein (as in the

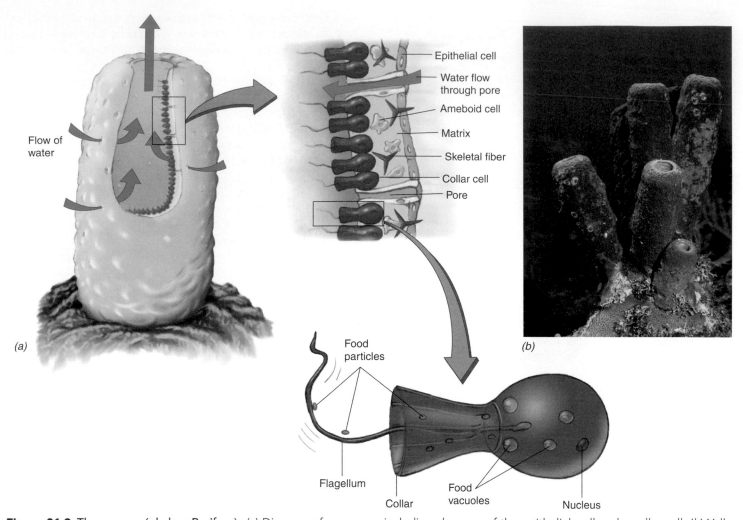

**Figure 21.9 The sponge (phylum Porifera).** (*a*) Diagram of a sponge, including closeups of the epithelial wall and a collar cell. (*b*) Yellow tube sponges off the coast of the Cayman Islands in the Caribbean Sea.

 **Visual Thinking:** The figure shows the flow of water through the body of the sponge. Why does the life of the sponge rely on this flow of water?

sponges we use to wash a car) or needlelike crystals of calcium carbonate or silica. Pores, channellike openings that span the matrix, are dispersed throughout the sponge. These pores are integral to the movement of water, dissolved substances, and particulate matter into the interior of the sponge. They give the sponges their phylum name Porifera, which means "pore bearers."

The flagella of the collar cells beat to create a current of water that flows from the outside of the sponge, through the pores, into the interior of the sponge, and then out again through the large opening at the top of the sponge. Scientists estimate that the "average" sponge filters about 27 liters of water per day—slightly over 7 gallons. This circulation of water brings particulate matter and bacteria to the collar cells. The food is trapped on a mucous membrane within the collar cells, which then engulf the food in vacuoles. The ameboid-type cells then pick up the packaged food from the collar cells, digest it, and distribute it throughout the body of the sponge, as shown in **Figure 21.9a**.

## 21.8 Cnidarians are radially symmetrical and have no coelom.

Nicknamed "the stingers," the **cnidarians** (neye-DARE-ee-uhnz; phylum Cnidaria), including jellyfish, hydras, sea anemones, and corals, have a definite shape and symmetry, as do all other animals except the sponges. However, the cnidarians are classified as radially symmetrical, unlike all other animals except the Ctenophora (teh-NOF-ah-rah), a minor phylum that includes the comb jellies. Two examples of cnidarians are shown in **Figure 21.10**, and a comb jelly is shown in **Figure 21.11**.

How did the cnidarians get their nickname? You may already know if you have had a close encounter with a jellyfish. The tentacles of a cnidarian help it capture prey, such as small fishes, shrimp, and aquatic worms. The tentacles bear stinging cells called *cnidocytes*, which give the phylum its name; the root word is from the Greek meaning "stinging nettle." You can see cnidocytes as tiny

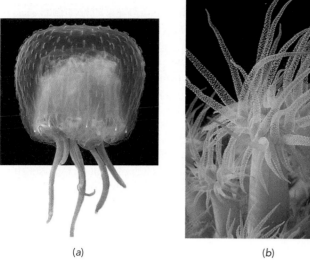

(a)                                              (b)

**Figure 21.10 Medusa and polyp: the two cnidarian body plans.**
(a) Medusa: In this photo of a jellyfish, many of the cnidarian tissues and structures are clearly visible. (b) Polyp: The tentacles of this orange cup coral are covered with stinging cells (cnidocytes).

dots in the tentacles of the orange cup coral shown in Figure 21.10b. These stinging cells work much like harpoons. Powered by water pressure, threadlike stingers shoot out of the cells, spearing and immobilizing the prey with what can be a powerful and painful sting. The tentacles then draw the prey back to the mouth.

There are thousands of cnidarian species, and most are marine. Some are builders of rich and complex ecosystems—the coral reefs. Others are important predators in the open ocean.

The cnidarians exhibit two basic body plans: **medusae** (meh-DOO-see) and **polyps** (POL-ups). The medusa form is exhibited by the jellyfish in Figure 21.10a, and the polyp form is exhibited by the coral in Figure 21.10b.

Medusae only reproduce sexually. Ovaries hang from the underside of the females, and testes hang from the males. Eggs and sperm are shed into the water, where fertilization takes place. In species that exist only as medusae, the fertilized egg develops into a larva that then develops directly into a medusa. In most species, however, medusae have a life cycle in which the larvae develop into sessile polyps, some of which then produce medusae. This type of alternating life cycle is shown in **Figure 21.12**.

Some cnidarians, such as the hydras, sea anemones, and corals, occur only as polyps. Simple polyps such as hydras usually reproduce asexually by budding; a

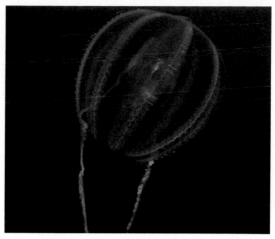

**Figure 21.11** A comb jelly (phylum Ctenophora). Note the comblike plates and two tentacles.

daughter organism simply grows from the parent organism. However, some of their tissue is organized into primitive ovaries that produce eggs and testes that produce sperm, as in the more complex polyps, such as sea anemones and corals. (Some species of these more complex organisms *never* reproduce by budding.) Like the sponges, some hydras, are hermaphrodites; others exist as separate sexes. In either case, eggs remain attached to the hydra but are exposed to the water. Sperm are discharged from the testes and swim to the egg. After fertilization, developing hydras grow while attached to the parent.

Polyps are aquatic, cylindrical animals with a tentacle-ringed opening at one end that serves as both a mouth and an anus. Most

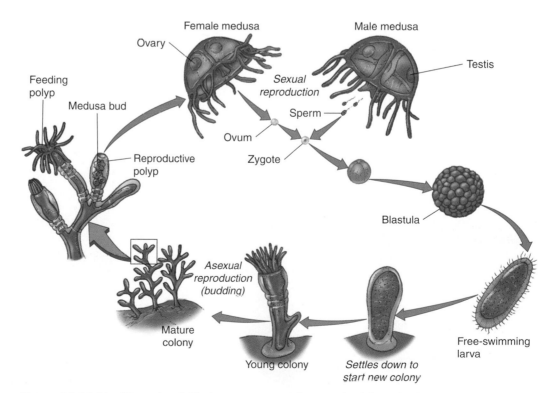

**Figure 21.12** The life cycle of Obelia, a marine cnidarian. This life cycle demonstrates the alternation of medusa and polyp forms found in cnidarians.

polyps, such as the sea anemones and corals, are sessile—they live attached to rocks. Like the corals, many polyps build up a hard outer shell, an internal skeleton, or both. Some polyps are free floating, such as the fresh water *Hydra*.

In contrast, most medusae are free floating. They are often umbrella shaped. Commonly known as jellyfish, medusae have a thick, gelatinous interior. The mouth of a medusa is usually located on the underside of its umbrella shape, with its tentacles hanging down around the umbrella's edge.

Structurally, epithelial tissue covers the outside of cnidarians; an inner tissue layer, the gastrodermis, lines the digestive cavity. The mesoglea (literally, "middle glue") lies between. This layer is quite thick within medusae and gives them their jellylike appearance. A network of nerve cells extends through cnidarians, but they have no brainlike controlling center.

## 21.9 Flatworms are bilaterally symmetrical and have no coelom.

The **flatworms** (phylum Platyhelminthes [PLAT-eh-hel-MIN-these]) have the simplest body plan among the bilaterally symmetrical animals. The phylum name comes from Greek words meaning "flat" (*platys*) "worm" (*helminthos*) and describes their flattened ribbon or leaflike shapes. Though simple in structure, the flatworms have organs and some organ systems, but because they have no coelom, the organs are embedded within the body tissues (see Figure 21.8). Flatworms are also the simplest animals that have a distinct head, a characteristic of many of the bilaterally symmetrical animals.

The three classes of flatworms are the turbellarians, flukes, and tapeworms. Examples of these invertebrates are shown in **Figure 21.13**. The flukes and tapeworms are parasites; they live on or in other animals (including humans), deriving nutrition from their hosts. Flukes feed on the tissues of the host by sucking in nutrients and digesting the material in their gastrovascular cavity. Tape-

worms, however, have neither mouth nor digestive tract; they simply absorb nutrients through their epidermis. The life cycles of tapeworms are discussed in the *Just Wondering* box.

The turbellarians are free living and are found in fresh water, saltwater, or damp soil. They move from place to place by means of ciliated epithelial cells that are concentrated on their ventral surfaces, feeding on a variety of small animals and bits of organic debris. The name "turbellarian" comes from a Latin word meaning "to bustle or stir" and refers to the water turbulence created by their movement. Sensory pits or tentacles along the sides of their heads detect food, chemicals, and movements of the fluid in which they are moving. They also have eyespots on their heads, which contain light-sensitive cells that enable the worms to distinguish light from dark. These organs are part of the turbellarian nervous system and connect to a ladderlike paired nerve cord that extends down the length of the animal. Tiny swellings at the cephalic, or head, end of the organism are considered a primitive brain. The anatomy of the fresh water turbellarian *Dugesia* is shown in **Figure 21.14**.

Turbellarians have a digestive system consisting of a digestive cavity, open only at one end (the pharynx—see Figure 21.14). The digestive system has a branching structure that reaches all parts of the body. Muscular contractions in the upper end of the digestive cavity of flatworms cause a strong sucking force by which the flatworms ingest their food through the pharynx and tear it into small bits. The cells making up the digestive cavity wall engulf these particles; most digestion takes place within these cells. Nutrients diffuse out of the digestive cavity and into nearby cells. Wastes from within the cells diffuse into the digestive tract and are expelled through the mouth, the opening to the pharynx. In addition, the flatworm excretes excess water and some wastes by means of a primitive excretory system, a network of fine tubules that runs along the length of the worm. Specialized bulblike cells called flame cells are located along these tubules. As cilia within them beat (looking like a flickering flame), they move the water and wastes into the tubules and out excretory pores.

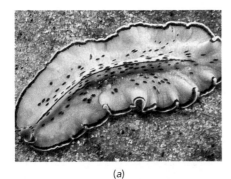

(a)

(b)

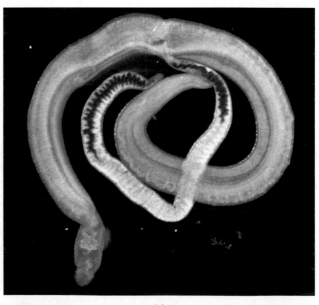

(c)

**Figure 21.13** Flatworms (phylum Platyhelminthes). (a) A marine turbellarian, a class of flatworms that is free living. (b) The human liver fluke, *Clonorchis sinensis*. (c) An intestinal tapeworm.

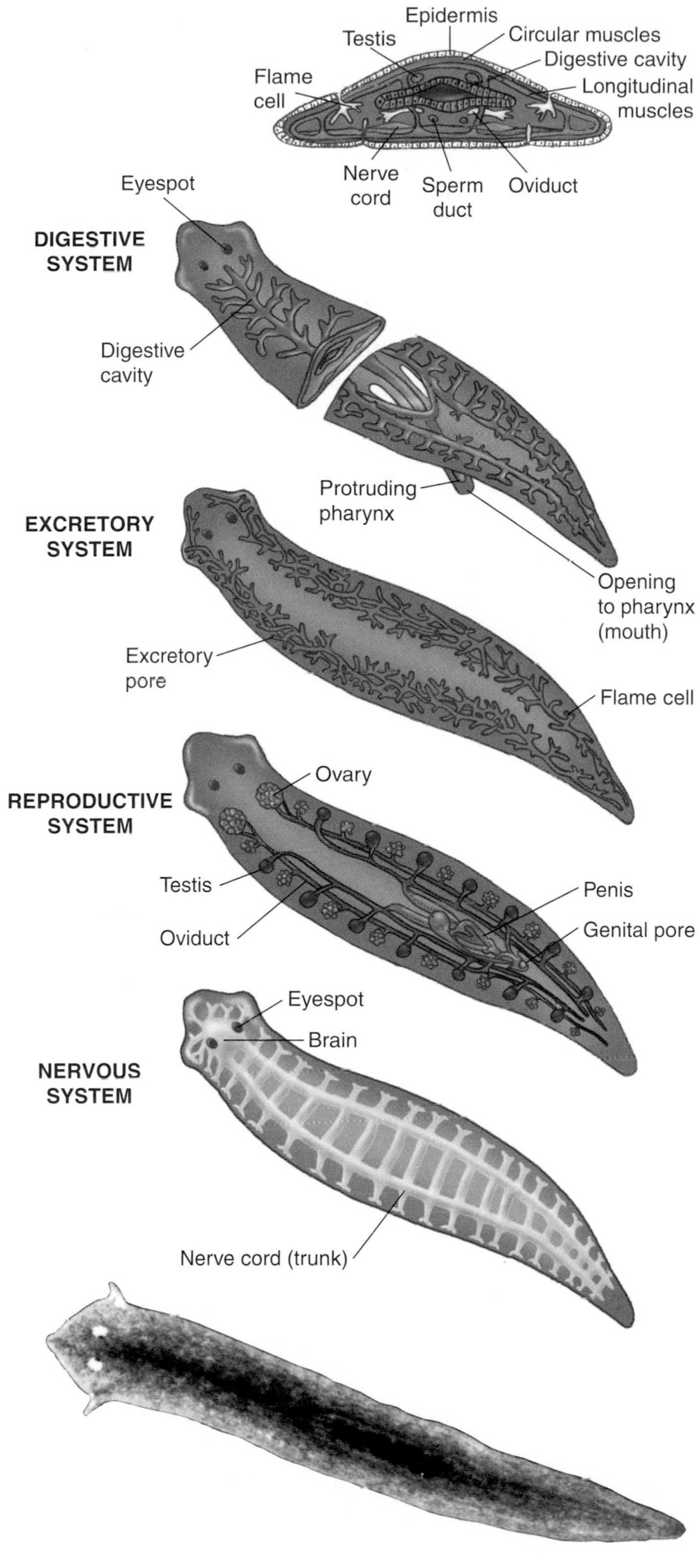

DIGESTIVE SYSTEM

Eyespot

Digestive cavity

Epidermis

Testis

Circular muscles

Digestive cavity

Longitudinal muscles

Flame cell

Nerve cord

Sperm duct

Oviduct

EXCRETORY SYSTEM

Protruding pharynx

Excretory pore

Opening to pharynx (mouth)

Flame cell

REPRODUCTIVE SYSTEM

Ovary

Testis

Oviduct

Penis

Genital pore

NERVOUS SYSTEM

Eyespot

Brain

Nerve cord (trunk)

**Figure 21.14** Turbellarian anatomy. The organism shown is *Dugesia*, the familiar fresh-water flatworm used in many biology laboratories.

**Visual Thinking:** Is the flatworm a coelomate, pseudocoelomate, or acoelomate? How can you tell from this illustration?

Reproduction is much more complicated in flatworms than it is in sponges or cnidarians. Although most flatworms are hermaphroditic, a characteristic shared by many sponges and cnidarians, the organs of reproduction are better developed. When flatworms mate, each partner deposits sperm in a genital pore of the other. The sperm travel along special tubes to reach the eggs. The fertilized eggs are then laid though oviducts. In free-living flatworms, the fertilized eggs are laid in cocoons and hatch into miniature adults. In some parasitic flatworms, there is a complex succession of distinct larval forms. Flatworms are also capable of asexual reproduction. In some genera, when a single individual is divided into two or more parts, each part can regenerate an entirely new flatworm.

## CONCEPT CHECKPOINT

2. Describe levels of organization, symmetry, feeding, and reproduction for each of the following phyla: Porifera, Cnidaria, and Platyhelminthes.

## 21.10 Roundworms are bilaterally symmetrical and have a false coelom.

A single spadeful of fertile soil may contain *millions* of **roundworms Figure 21.15**. Roundworms (phylum Nematoda) are one of seven phyla that have a pseudocoelomate (false coelom) body plan—a body cavity with no connective tissue lining (see Figure 21.8). Of these seven pseudocoelomate phyla, only the roundworms include a large number of species. The phylum name Nematoda comes from a Greek word meaning "thread." These cylindrical worms are diverse in size, but some of them are so small and slender that they look like fine threads. Although most are similar in form, they range in length from about 0.2 millimeters (about the width of a human hair) to about 6 millimeters (the diameter of the head of a tack). Some species are abundant in fresh water or saltwater.

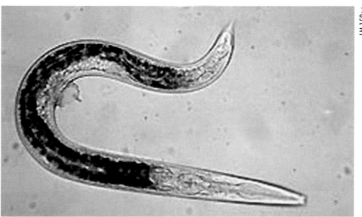

**Figure 21.15** Roundworms. Roundworms such as this one populate the soil. Its intestine can be seen as dark brown due to granules of stored food.

## *just wondering . . .*

### How do people get tapeworms? Are they common in the United States?

The most common tapeworm that affects humans is the beef tapeworm (*Taenia saginata*). The life cycle of this tapeworm begins when cattle (the secondary hosts) graze in areas in which tapeworm eggs contaminate the soil. After the eggs are eaten, they hatch within the intestine, and the larvae burrow through the intestinal wall until they reach muscle. Here, the larvae encyst—they become encapsulated and quiescent.

A person (the primary host) who eats raw or undercooked infected beef ingests the cysts along with the meat. Digestive enzymes break down the capsule surrounding the larvae, and they attach themselves to the intestinal wall with grasping hooks of their head (**Figure 21.A**). Feeding on the digested food of the host, the tapeworms mature within a few weeks. As adults, they consist of long chains of segments called *proglottids*. Each proglottid has both male and female sex organs and can produce up to 100,000 eggs, which are shed with proglottids in the feces. If the feces contaminate areas where cattle graze and they are ingested, the life cycle begins once again.

Parasites such as tapeworms are more prevalent in countries having poor sanitation systems (or no sanitation systems at all) than

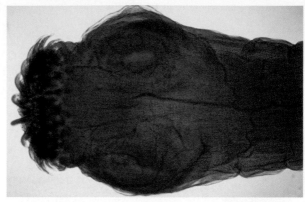

**Figure 21.A** Grasping hooks of the head of the beef tapeworm. By means of these hooks, tapeworms attach themselves to the intestinal wall. They then absorb nutrients from the intestine.

in countries with effective sewage treatment. In the United States, water is treated and purified in sewage treatment plants before it is returned to rivers, streams, or the ocean. In addition, beef is inspected before it can be sold. Still, infection is possible. To protect yourself, do not eat beef that is raw or undercooked.

Two other tapeworms, pork tapeworms and fish tapeworms, can also infect humans. Their life cycles are similar to that of the beef tapeworm; humans are their primary hosts. Therefore, avoid eating undercooked pork, raw fish, or undercooked fish.

Infection with tapeworms is serious. The organisms can live for years within the body and absorb nutrients essential for proper nutrition. Malnutrition is a common result of a tapeworm infestation. In addition, long tapeworms can block the movement of materials through the intestine. Tapeworm infections can be treated with certain medications, but treatment is difficult if the worms invade tissues beyond the intestines, as sometimes happens. Following the precautions mentioned here will help you avoid sharing your body with these unpleasant creatures! ●

*Are you wondering about a topic in biology and how it relates to your life? Submit your question by clicking the* Just Wondering *link in this text's companions Web site at* www.wiley.com/college/alters.

Many roundworms are parasites of vertebrates. About 50 species of roundworms parasitize humans as well as other animals. The people shown in **Figure 21.16** have elephantiasis, a condition caused by roundworms that live in the lymphatic passages and block the flow of lymph. As a result, fluids cannot drain, and swelling occurs. Roundworms that cause elephantiasis are transmitted to humans as larvae by the bite of an infected mosquito.

Roundworms also parasitize invertebrates and plants. For this reason, some nematodes are being investigated as agents for the biological control of insects and other agricultural pests.

Roundworms are covered by a flexible, tough, transparent multilayered tissue (cuticle) that is shed as they grow. A layer of muscle lies beneath this epidermal layer and extends lengthwise (**Figure 21.17**). These longitudinal muscles pull against both the cuticle and the firm, fluid-filled pseudocoel, similar to how your muscles pull against your bones. All this effort gets them nowhere in water, but they can move in mud or soil, which provide surfaces against which the worm's body pushes. Whether or not the movements result in locomotion, they compress the fluid-

**Figure 21.16** Elephantiasis. These people in southern India all have elephantiasis, a swelling of the feet, ankles, and legs caused by roundworms blocking the flow of lymph.

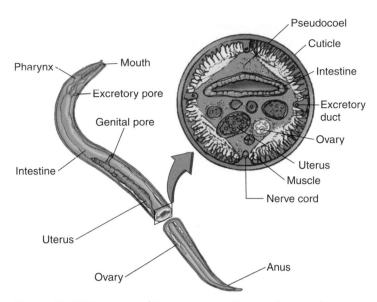

**Figure 21.17** Anatomy of *Ascaris*, a parasitic roundworm of humans.

filled pseudocoel and aid in the distribution of food and oxygen throughout the worm.

The roundworm digestive system has two openings—a mouth and an anus. Most roundworms have raised, hairlike sensory organs near their mouths. The mouth itself often has piercing organs, or stylets. Food passes through the mouth as a result of the sucking action of a muscular pharynx. After passing through these organs, food continues through the digestive tract, where it is broken down and then absorbed. The roundworms that parasitize animals take in digested food of the host; the cells lining the digestive system simply absorb these nutrients.

The roundworms also contain primitive excretory and nervous systems. The nervous system consists of a ring of tissue surrounding the pharynx and solid dorsal and ventral nerve cords, unlike the hollow nerve tube of a chordate (see Chapter 22). The excretory system consists of two lateral canals (excretory ducts) that unite near the anterior end to form a single tube ending in an excretory pore.

Unlike hermaphroditic flatworms, roundworms are male or female. Their reproductive systems are quite sophisticated. The female has ovaries in which eggs develop. From the ovaries, eggs pass through oviducts to the uterus where they are fertilized. The male produces sperm cells in the testis and stores them in the vas deferens. When it is time to reproduce, the sperm pass through penile spicules, which the male uses to attach to the female during copulation. The female empties fertilized eggs from the uterus into the vagina, and they then pass out through a genital pore. A female nematode may lay over 200,000 eggs a day.

## 21.11 Mollusks, annelids, arthropods, and echinoderms are bilaterally symmetrical and have a true coelom.

The remaining phyla of invertebrate animals all have a true coelom—they are coelomates. The mollusks (clams, snails, octopuses), the annelids (earthworms, leeches), and the arthropods (lobsters, crabs, insects, spiders) are all protostomes (see Section

21.3). That is, they show a spiral pattern of cleavage in early development. The pseudocoelomate and acoelomate animals already discussed are not classified as protostomes, but they exhibit developmental patterns similar to those of the protostomes. These similarities suggest an evolutionary closeness among these groups. The echinoderms, which are discussed last in this section, are deuterostomes. That is, they show a radial pattern of cleavage in early development.

Among the protostomes, the annelids and the arthropods have segmented bodies, whereas the mollusks do not. Segmentation underlies the organization of all of the more complex animals: the annelids (earthworms, leeches), arthropods (lobsters, insects, spiders), echinoderms (sea urchins, sea stars), and chordates (fishes, reptiles, mammals). Segmentation allows for flexibility and mobility, and it probably evolved as an adaptation for movement. Sometimes segmentation is apparent only in their embryological development. In the chordates, too, segmentation is revealed in embryological development. Because segmentation in animals is different among phyla, scientists think it arose independently in more than one line of evolution.

Most coelomate animals have a closed circulatory system, but some have an open system. Both types of circulatory system have a heart to pump blood. In a **closed circulatory system**, blood is enclosed within vessels as it travels throughout the body of the organism. In an **open circulatory system**, blood flows in vessels leading to and from the heart, but it flows through irregular channels called *blood sinuses* in many other parts of the body. (See Sections 29.2 and 29.3.)

The excretory system of coelomates varies, basically depending on whether the circulatory system is open or closed. In either case, wastes are removed from the organisms by tubular structures called *nephridia* (referred to as *nephron tubules* in vertebrates; see Chapter 31).

### Mollusks

The **mollusks** (phylum Mollusca) are a large phylum with which you may be very familiar, since the human diet often includes them—for example, clams, oysters, calamari (squid), and escargot (snails). These invertebrates have a muscular foot and a soft body that have a culinary appeal to many people. These soft body parts are covered by a soft epithelium called a *mantle*, and in most cases the entire organism is covered with a hard shell.

Mollusks are widespread and often abundant in marine and fresh water environments, and some, such as certain snails and the slugs, live on land. They range in size from near microscopic to huge: the largest giant squid on record measured approximately 18 meters long (about 59 feet). Most squid, however, are much smaller.

Mollusks exhibit four body plans: cephalopod, bivalve, gastropod, and chiton. The name of each group describes its prominent features. Although each group (with the exception of the bivalves) has a head end, the cephalopods (literally, "head–foot") have the most well-differentiated head and the most well-developed nervous system. Along with a large head, cephalopods have long, armlike tentacles that surround a mouth and a pair of large eyes. Examples of cephalopods are octopuses and squids.

The bivalves ("two-shelled") gastropods have the least well-developed nervous system of these groups. Bivalves are usually sedentary and may have a muscular foot with which they bury themselves. Examples of bivalves are clams, mussels, oysters, and scallops.

Gastropods (from Greek words meaning "stomach foot") have eyes and feelers on a distinct head. Most have a shell that is spiral or cone-shaped and a muscular foot for locomotion. Examples of gastropods are snails, slugs, and limpets.

Chitons are marine organisms that adhere to rocks. Their name comes from a Greek word meaning "tunic," which refers to their shell of eight overlapping plates.

**Figure 21.18** shows the four mollusk body plans.

All mollusks have a *visceral mass*, or group of organs, consisting of the digestive, excretory, and reproductive organs. The visceral mass is covered with the mantle, which arises from the dorsal body wall and encloses a cavity between itself and the visceral mass. This cavity is *not* the coelom; the coelom surrounds the heart only. Mollusks exhibit both open circulatory systems (in clams, for example) and closed circulatory systems (in squid, for example). The mollusk's gills, the organs of respiration, lie within the mantle cavity. Gills are a system of filamentous projections of the mantle tissue that is rich in blood vessels. These projections greatly increase the surface area available for oxygen-carbon dioxide exchange. In land-dwelling mollusks, a network of blood vessels within the mantle cavity serves as a primitive lung. All mollusks except bivalves have *radula*, a horny band or ribbon with tiny teeth, used to tear up food and draw it toward the mouth.

Mollusks were one of the earliest evolutionary lines to develop an efficient excretory system. In mollusks with open circulatory systems, wastes move from the coelom into the nephridia and are discharged into the mantle cavity. From there, they are expelled by the continuous pumping of the gills. In mollusks (and other animals with closed circulatory systems), the coiled tubule of a nephridium is surrounded by a network of capillaries. Wastes move from the circulatory system to the nephridium (referred to as nephron tubules in vertebrates) for removal from the body.

In most mollusks, the sexes are separate. Many of the gastropods and all of the bivalves are broadcast spawners, which means that each adult releases hundreds or thousands of gametes into the water, and fertilization and development occur in the water. Some cephalopods and gastropods copulate and lay eggs.

### Annelids

The **annelids** (AN-ul-idz; phylum Annelida) are worms characterized by a soft, elongated body composed of a series of ringlike segments (the word "annelid" means "tiny rings"). Internally, their segments are divided from one another by partitions called *septa* (sing. *septum*). Circulatory, excretory, and neural structures are repeated in each segment. Annelids are abundant in the soil and in both marine and fresh water environments throughout the world.

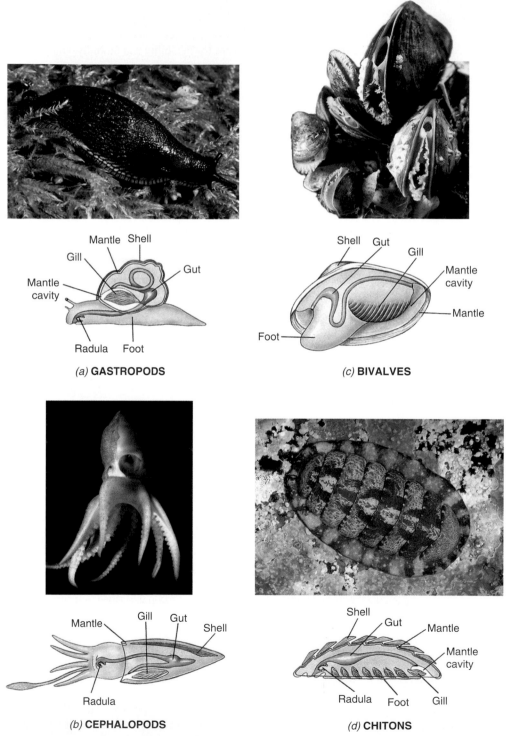

(a) **GASTROPODS**

(b) **CEPHALOPODS**

(c) **BIVALVES**

(d) **CHITONS**

**Figure 21.18** Body plans among the mollusks. (*a*) Gastropods—a slug. (*b*) Cephalopods—an octopus. (*c*) Bivalves—mussels. (*d*) Chiton.

(a)

(b)

**Figure 21.20** A medicinal leech and a mechanical leech. The medicinal leech in the beaker (right) bites the patient and introduces clot-dissolving chemicals while stimulating blood flow. The mechanical leech (left) does the same job in a more high-tech way. It is a glass vacuum chamber with tubes for maintaining suction and rinsing a patient's wound.

**Figure 21.19** Annelids. (a) The fine fiberglass-like bristles on this orange bristle worm from the Philippines can inflict a painful burning sensation when touched. (b) These feather duster plume worms from the Caribbean use their flowerlike gill plumes in respiration and feeding.

There are two commonly accepted classes of annelids: marine worms (class Polychaeta) and earthworms and leeches (class Clitellata). **Figure 21.19** shows two examples of marine worms. Exhibiting unusual forms and sometimes iridescent colors (like the orange bristle worm in Figure 21.19a), marine worms live in burrows, under rocks, inside shells, and in tubes of hardened mucus they manufacture.

Leeches occur mostly in fresh water. Most are predators or scavengers, and some suck blood from mammals, including humans. The best-known leech is the medicinal leech. In the late Middle Ages (1300–1400), the idea developed that some diseases were caused by an excess of body fluids, so bloodletting was used to cure the problem. Most often a vein was opened for the process, but sometimes bloodsucking leeches were used to accomplish the task. Bloodletting remained a popular form of treatment for centuries. Today, medicinal leeches are used in some plastic and reconstructive surgery procedures, especially in situations where severed body parts are reattached. The leeches are placed on the tissue, and they bite the patient and begin to remove blood that sometimes builds up in the area. They also secrete numerous compounds that widen blood vessels, promote blood flow, prevent the blood from clotting, and dissolve

clots that may have formed. In spite of their usefulness, most people are not anxious to have a leech sucking their blood; moreover, the leeches can introduce harmful bacteria into the circulatory system. In response to these concerns, researchers at the University of Wisconsin–Madison have developed a mechanical leech that does the same job without the negative factors (**Figure 21.20**).

Earthworms exhibit the generalized body plan of the annelid phylum: a tube within a tube (**Figure 21.21**). The digestive tract, a

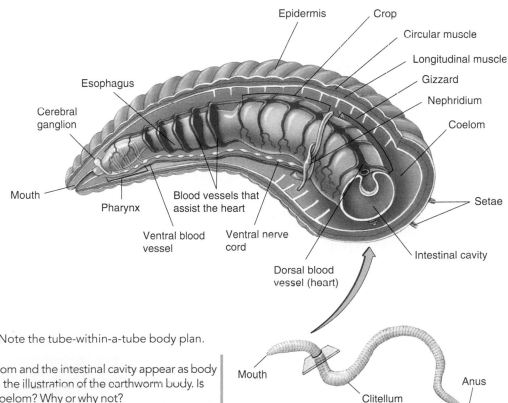

**Figure 21.21** Anatomy of an earthworm. Note the tube-within-a-tube body plan.

 **Visual Thinking:** Both the coelom and the intestinal cavity appear as body cavities within the illustration of the earthworm body. Is the intestinal cavity considered a coelom? Why or why not?

straight tube running from mouth to anus, is suspended within the coelom. An earthworm sucks in organic material by contracting its strong pharynx. It holds the food in its crop and then grinds this material little by little in its muscular gizzard, aided by the presence of soil particles it takes in with its food.

The anterior segments of an earthworm contain a well-developed *cerebral ganglion*, or brain, and a few muscular blood vessels that act like hearts, pumping the blood through the closed circulatory system. Sensory organs are also concentrated near the anterior end of the worm. Some of these organs are sensitive to light, and elaborate eyes with lenses and retinas have evolved in certain members of the phylum. Separate nerve centers, or ganglia, are located in each segment and are connected by nerve cords. Each segment also contains both circular and longitudinal muscles, which annelids use to crawl, burrow, and swim. *Setae*, or bristles, help anchor the worms during locomotion or when they are in their burrows.

Reproduction differs among annelids. In the marine worms, the sexes are usually separate, and fertilization is often external, occurring in the water and away from both parents. The earthworms and leeches, on the other hand, are hermaphroditic. When they mate, their anterior ends point in opposite directions and their ventral surfaces touch. The *clitellum*, a thickened band on an earthworm's body, secretes a mucus that holds the worms together as they exchange sperm. Ultimately, the worms release the fertilized eggs into cocoons also formed by mucous secretions of the clitellum.

## Arthropods

Crabs and lobsters (subphylum Crustacea; krus-TAY-sha); insects, centipedes, and millipedes (subphylum Uniramia); and spiders, horseshoe crabs, mites, and ticks (subphylum Chelicerata; KEH-lis-er-AH-tah) are representatives of the diverse phylum Arthropoda (ARE-thra-PO-dah or are-THROP-ih-dah). **Figure 21.22** shows various **arthropods** and points out the diversity of this phylum. The name *arthropod* comes from the two Greek words *arthros* (jointed) and *podes* (feet) and describes the characteristic jointed appendages of all arthropods.

The arthropods have a rigid external skeleton, or **exoskeleton**, which varies significantly in toughness and thickness among arthropods. Some arthropods have a tough exoskeleton, like the South American scarab beetle shown in **Figure 21.23a**. Others have a fragile exoskeleton, like the green darner dragonfly shown in Figure 21.23b. The exoskeleton provides places for muscle attachment, protects the animal from predators and injury, and most importantly, protects from water loss. As an individual outgrows its exoskeleton, that exoskeleton splits open and is shed. This process is called **molting**. A new, soft exoskeleton lies underneath, which subsequently hardens. The animal then grows into its new outer covering.

(a)

(b)

(c)

(d)

LM 12×

**Figure 21.22 Diversity of arthropods.** (a) The tropical land crab of Costa Rica (*Gecarcinus quadratus*) lives in half-meter-deep burrows by day and forages for plant material by night. (b) The house centipede (*Scutigera coleoptrata*) prefers to live in damp portions of basements, closets, and bathrooms. (c) The wolf spider (*Lycosidde sp.*), discussed in the opening story of this chapter, is unusual in that it does not spin a web but roams at night to hunt for food. When disturbed, it runs away. (d) The pubic louse (*Phthirus pubis*) causes "crabs," which is often spread by sexual contact. Females lay eggs in pubic hair. Infestation causes intense itching.

**Figure 21.23** Two exoskeletons. (a) The South American scarab beetle has a tough exoskeleton. (b) The green darner dragonfly has a fragile exoskeleton.

(a)

(b)

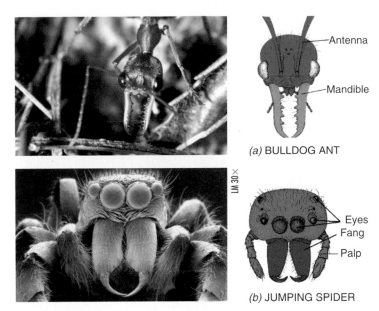

(a) BULLDOG ANT

(b) JUMPING SPIDER

**Figure 21.24 Arthropod mouthparts.** (a) The bulldog ant has mandibles (jaws) formed by the modification of one of the pairs of anterior appendages, as do all crustaceans and insects. (b) The jumping spider has chelicerae (fangs) formed by the modification of the appendages nearest the anterior end of the animal, as do the spiders, horseshoe crabs, mites, and ticks. The jumping spider's leglike palps are transformed into copulatory organs in the adult male.

All arthropods can be placed into one of two groups: those with jaws and antennae (the Crustacea and Uniramia) and those without jaws and antennae (the Chelicerata). The crustaceans, insects, centipedes, and millipedes have jaws, or *mandibles*. These chewing mouthparts are formed by the modification of one of the pairs of anterior appendages (but *not* the first pair). The appendages nearest the anterior end are sensory antennae. The mandibles and antennae of the bulldog ant are illustrated in **Figure 21.24a**. The remaining arthropods, which include the spiders, horseshoe crabs, mites, and ticks, lack jaws. Their mouthparts usually take the form of fangs (pincerlike mouthparts also called *chelicerae* [keh-LIS-er-ee], meaning claw), which evolved from the appendages nearest the anterior end of the animal. The fangs of the jumping spider are illustrated in Figure 21.24b. This spider uses these fangs to catch its prey.

**Figure 21.25** The compound eyes of the robberfly. Compound eyes are composed of many independent visual units, each containing a lens.

An important structure of many arthropods, such as bees, flies, moths, and grasshoppers, is the *compound eye*. The compound eyes of the robberfly are shown in **Figure 21.25**. Compound eyes are composed of many independent visual units, each containing a lens. *Simple eyes*, composed of a single visual unit having one lens, are found in many arthropods that also have compound eyes and function in distinguishing light and darkness. In some flying insects, such as locusts and dragonflies, simple eyes function as horizon detectors and help stabilize the insects during flight.

In the course of arthropod evolution, the coelom has become greatly reduced, consisting only of the cavities that house the reproductive organs and some glands. **Figure 21.26** illustrates the major structural features of a grasshopper as a representative of the arthropods. Like the annelids, the arthropods have a tubular gut that extends from the mouth to the anus. The circulatory system of arthropods is open; their blood flows through cavities between the organs. One longitudinal dorsal vessel functions as a heart, helping move the blood along.

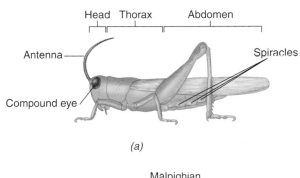

(a)

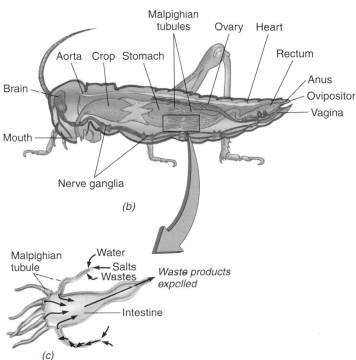

(b)

(c)

**Figure 21.26 Anatomy of a grasshopper.** (a) External anatomy. (b) Internal anatomy. (c) Detail showing how Malpighian tubules function in the excretory system.

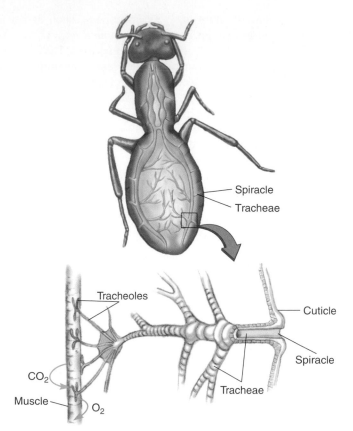

**Figure 21.27** The tracheal system of an ant. Sets of spiracles open into tracheae, which branch into tracheoles where gas exchange takes place.

Most aquatic arthropods breathe by means of gills. Their feathery structure provides a large surface area over which gas exchange takes place between the surrounding water and the animal's blood. The respiratory systems of terrestrial arthropods generally have internal surfaces over which gas exchange takes place. The respiratory systems of insects, for example, consist of small, branched air ducts called *tracheae* (**Figure 21.27**). These tracheae, which ultimately branch into very small *tracheoles*, are a series of tubes that transmit oxygen and take up carbon dioxide throughout the body. The tracheoles are in direct contact with the individual cells, and oxygen and carbon dioxide diffuse directly across the cell membranes. Air passes into the tracheae by way of specialized openings called *spiracles*, which in most insects can be closed and opened by valves.

Although different groups of arthropods possess various kinds of excretory systems, a unique excretory system evolved in terrestrial arthropods (notably, the insects) in relation to their open circulatory system. The principal structural elements are the *Malpighian tubules* (mal-PIG-ee-en), which are slender projections from the digestive tract (see Figure 21.26). Fluid is passed through the walls of the Malpighian tubules to and from the blood in which the tubules are bathed. The nitrogenous wastes in the blood are separated out as a solid (precipitated) and then emptied into the hindgut—the posterior part of the digestive tract—and eliminated. Most of the water and salts in the fluid are reabsorbed by the hindgut, thus conserving water.

With few exceptions, the sexes are separate in the arthropods. Males and females have paired sex organs, which connect directly to ducts that open onto the ventral surface of the trunk. In many species, males transfer free sperm to females, although in the majority of species males transfer sperm to females within scaled packets known as *spermatophores*. In this way, the sperm are not diluted by water in the aquatic arthropods, nor do they become dried out in the terrestrial arthropods.

Insects (class Insecta) are the major group of arthropods. This diverse class of organisms includes an incredible number of species. They are humans' greatest competitors for food; they damage crops, fabrics, and building materials; and they transmit diseases such as malaria and sleeping sickness. Nevertheless, they also play ecological roles beneficial to humans and other organisms, such as pollinating crops and flowers, controlling unwanted pests, and producing honey and silk.

Insect development involves **metamorphosis**. This term means "change," signifying that as insects develop and grow, they change form. There are two types of metamorphosis: complete and incomplete.

About 9 out of 10 species of insects undergo *complete metamorphosis*, which includes four stages of development: egg, larva, pupa, and adult (**Figure 21.28**). Larvae hatch from eggs. They do not look like adults, nor do they eat what adults eat, so they do not compete with adults for food. Instead, larvae look like worms and may be called caterpillars, maggots, or grubs. As they outgrow their "skin," they molt several times. After a period of growth and successive moltings, larvae spin silken cocoons around themselves. When finished, they stop eating, while their bodies, now called pupae (sing. pupa), develop into adults. When the metamorphosis is complete, the adult emerges from the cocoon.

About 1 out of 10 species of insects undergoes *incomplete metamorphosis*, with three stages: egg, nymph, and adult (**Figure 21.29**). The young hatch from eggs as nymphs, which look similar to the adult form but without wings. Nymphs grow into adults, molting from four to eight times during the growth process. At each *instar* (growing stage) after each molt, the nymph looks more like the adult, until wings are developed.

(a)  (b)  (c)  (d)

**Figure 21.28** Complete metamorphosis in the monarch butterfly. (a) Eggs. (b) Larvae. (c) Pupa. (d) Adult.

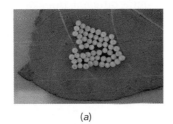
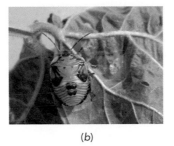

**Figure 21.29** Incomplete metamorphosis in the stinkbug. (a) Eggs. (b) Nymph. (c) Adult.

(a)             (b)             (c)

## Echinoderms

The term **echinoderm** means "spine skin," an appropriate name for many members of the phylum Echinodermata (ih-KEYE-neh-der-MA-tah). This phylum includes sea lilies, sea stars, brittle stars, sea urchins, sand dollars, and sea cucumbers. All are marine animals that live on the sea floor, with the exception of a few swimming sea cucumbers. **Figure 21.30** shows a variety of echinoderms.

The echinoderms are different from the other invertebrates in that they are deuterostomes, not protostomes, as discussed in Section 21.3. This connects the echinoderms with the chordates. The echinoderms and the chordates, along with two smaller phyla not discussed here, are thought to have evolved from a common ancestor.

Echinoderms are bilaterally symmetrical as larvae but radially symmetrical as adults (see Figure 21.7). They are closely related to and grouped with the bilaterally symmetrical animals, however. Adult echinoderms have a five-part body plan corresponding to the arms of a sea star or the design on the "shell" of a sand dollar.

The water vascular system (**Figure 21.31a**) is used for locomotion and is unique to echinoderms. Five *radial canals*, the positions of which are determined early in the development of the embryo, extend into each of the five parts of the body. Water enters the vascular system through small filters on the top, or aboral, surface of the sea star. The filters ensure that the fluid pressure stays the same inside the water vascular system as it is in the outside environment. Cilia move the water, which now contains cells, proteins, and ions, within the canals. The tubular feet, located on the bottom, or oral, surface of the sea star, are a part of the water vascular system. The sea star moves by a combination of hydraulic pressure and vacuum effects caused by muscle contraction, with the hundreds of tube feet acting as suction cups.

(a)

(b)

(c)

**Figure 21.30** Representatives of the phylum Echinodermata. (a) Sea stars off Belize, Central America. (b) A brittle star and a sea urchin. The sea urchin is attached to the left side of the red vase sponge, and the long-armed brittle star is on top of the sponge. (c) A sand dollar.

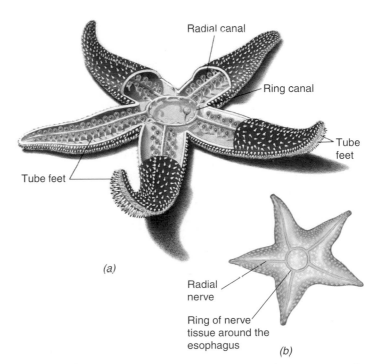

Radial canal

Ring canal

Tube feet

Tube feet

(a)

Radial nerve

Ring of nerve tissue around the esophagus

(b)

**Figure 21.31** Anatomy of an echinoderm. (a) The water vascular system of a sea star is shown in blue. (b) The sea star's ringlike nervous system.

As adults, these animals have no head or brain. Their nervous systems consist of central *nerve rings* from which branches arise (Figure 21.31*b*). The animals are capable of complex response patterns, but there is no centralization of function.

Echinoderms do not have a well-organized circulatory system. Food from the digestive tract is distributed to all the cells of the body in the fluid that lies within the coelom. In many echinoderms, respiration takes place by means of skin gills, which are small, fingerlike projections that occur near the spines. Waste removal also takes place through these skin gills. The digestive system is simple, consisting of a mouth, gut, and anus.

Echinoderms are well known for regenerating lost parts, but most reproduction is sexual and external. The sexes in most echinoderms are separate, although usually there is little external difference between them. The fertilized eggs of echinoderms usually de-

velop into free-swimming bilaterally symmetrical larvae that look quite different from the adults. The larvae eventually metamorphose (change through a series of stages) to adult forms.

## CONCEPT CHECKPOINT

**3.** Describe levels of organization, symmetry, and body plan in roundworms, mollusks, annelids, arthropods, and echinoderms.

**4.** There have been several distinctive anatomical trends in the evolution of animals. Review the diverse forms of invertebrate life. While you are reviewing, try to identify as many of these evolutionary trends as you can.

# CHAPTER REVIEW

## Summary of Key Concepts and Key Terms

### What are the characteristics of vertebrate and invertebrate animals?

**21.1** Both vertebrates and invertebrates are multicellular, heterotrophic eukaryotes.

**21.2** Animals arose from protist ancestors, specifically from flagellated cells that lived and worked together in groups.

### What do developmental patterns among animals reveal?

**21.3** The **protostomes** (p. 354) and the **deuterostomes** (p. 354) represent two distinct evolutionary lines.

**21.3** The protostomes (mollusks, annelids, and arthropods) differ from the deuterostomes (echinoderms and chordates) in cleavage patterns during early development.

**21.3** The similarities in developmental patterns within the protostomes and within the deuterostomes suggest close evolutionary relationships.

### What are the basic organizational features of animals?

**21.4** Animals have levels of organization, from cells to tissues to organs to organ systems.

**21.4** With the exception of sponges, most simple invertebrates have epithelial tissue and connective tissue, and some simple invertebrates have additional tissue types.

**21.5** Almost all animals exhibit either bilateral or radial symmetry.

**21.5** Animals exhibiting **radial symmetry** (p. 355) are aquatic organisms; many are sessile.

**21.5** All phyla of animals other than sponges, cnidarians, and echinoderms exhibit **bilateral symmetry** (p. 355); echinoderm larvae are bilaterally symmetrical.

**21.5** All animals other than the sponges and the cnidarians are bilaterally symmetrical or, in the case of the echinoderms, evolved from bilaterally symmetrical ancestors.

**21.5** A characteristic of bilaterally symmetrical animals is **cephalization** (p. 356), the concentration of sense organs and nervous tissue in the head end.

**21.5** Bilaterally symmetrical animals (including echinoderms) develop embryologically from three tissue layers: **endoderm** (p. 356), **ectoderm** (p. 356), and **mesoderm** (p. 356).

**21.6** Animals have three types of body plans: **coelomate** (p. 356), **pseudocoelomate** (p. 357), and **acoelomate** (p. 357).

**21.6** A **coelom** (p. 356) is a fluid-filled body cavity lined with connective tissue.

**21.6** A pseudocoel, or false coelom, is a fluid-filled cavity that houses organs, but they are not suspended within the cavity, and the cavity is not lined completely by connective tissue.

**21.6** In an acoelomate body plan, the organs are embedded within the other tissues of the body.

## What are the major invertebrate phyla and their distinguishing characteristics?

**21.7 Sponges** (phylum Porifera, p. 357) are acoelomate, asymmetrical, aquatic animals having no tissues or organs.

**21.7 Hermaphrodites** (hur-MAF-rah-dytes, p. 358) produce both male and female sex cells in the same organism; most species of sponges are hermaphrodites.

**21.8** The **cnidarians** (phylum Cnidaria, p. 358) are acoelomate, radially symmetrical, aquatic animals.

**21.8** Cnidarians have two layers of tissues and exist either as cylindrically shaped **polyps** (POL-ups, p. 359) or **medusae** (meh-DOO-see, p. 359).

**21.9** The **flatworms** (p. 360, phylum Platyhelminthes) are acoelomate, bilaterally symmetrical, ribbonlike worms that live in the soil or water or within other organisms.

**21.9** Having no coelom, flatworms are considered to have the simplest body plan of the bilaterally symmetrical animals.

**21.9** The flatworms have organs and some organ systems, including a primitive brain.

**21.10 Roundworms** (p. 361, phylum Nematoda ) are bilaterally symmetrical and have a false coelom.

**21.10** The roundworms are cylindrical worms that live in the soil, in water, or within other organisms.

**21.10** Roundworms have simple digestive, excretory, and nervous systems.

**21.11** Most coelomate animals have a **closed circulatory system** (p. 363), but some have an **open circulatory system** (p. 363).

**21.11** Mollusks, annelids, arthropods, and echinoderms are bilaterally symmetrical and have a true coelom.

**21.11** Mollusks, annelids, and arthropods are protostomes, echinoderms (and chordates) are deuterostomes.

**21.11** The **mollusks** (p. 363, phylum Mollusca) are a large phylum of coelomates that exhibit four body plans: cephalopod, gastropod, bivalve, and chiton.

**21.11** All mollusks have a visceral mass, or group of organs, consisting of the digestive, excretory, and reproductive organs.

**21.11** Mollusks were one of the first evolutionary lines to develop an efficient excretory system.

**21.11** The **annelids** (p. 364, phylum Annelida) are worms characterized by a soft, elongated body composed of a series of ringlike segments, each with its own circulatory, excretory, and neural structures.

**21.11** Annelids have a tube-within-a-tube body plan.

**21.11** The **arthropods** (p. 366, phylum Arthropoda) are an extremely diverse group that includes organisms such as lobsters, insects, and spiders.

**21.11** Arthropods are characterized by a rigid external skeleton (**exoskeleton**, p. 366) and jointed appendages.

**21.11** Arthropods periodically shed their exoskeleton in a process called **molting** (p. 366).

**21.11** Insect development involves **metamorphosis** (p. 368), in which immature insects change form as they grow into adults.

**21.11** The **echinoderms** (p. 368, phylum Echinodermata), including sea urchins and sea stars, are generally spiny-skinned marine animals that live on the sea floor.

**21.11** Echinoderms are the only deuterostome invertebrates, a characteristic that shows their relatedness to the chordates.

**21.11** Although adult echinoderms are radially symmetrical, their larvae are bilaterally symmetrical.

**21.11** Echinoderms have a water-filled vascular system used for locomotion.

---

## Level 1    Learning Basic Facts and Terms

### Matching

*Directions*: For questions 1–5, match the characteristics to the invertebrate phyla.

  a. Cnidaria        d. Mollusca
  b. Platyhelminthes    e. Porifera
  c. Nematoda

1. _____ During the life of these animals they may occur as a floating medusa stage or a sessile polyp stage. Many species produce stinging tentacles that aid in the capture of prey.

2. _____ Members of this phylum have a visceral mass consisting of the digestive, excretory, and reproductive organs.

3. _____ Many members of this group are parasitic acoelomate worms inhabiting the digestive tract of a host organism.

4. _____ These nonsymmetrical animals are sessile and feed on particulate organic matter in marine environments.

5. _____ These pseudocoelomate worms are very abundant and diverse.

### Multiple Choice

6. Which of the following invertebrate phyla is most closely related to vertebrates?
  a. Cnidaria        d. Arthropoda
  b. Annelida       e. Echinodermata
  c. Mollusca

7. Choanocytes (collar cells) are important for which of the following functions?
  a. immobilization of prey    d. waste excretion
  b. feeding             e. skeletal support
  c. reproduction

8. Which of the following cells/organs is correctly paired with its function?
  a. malpighian tubules—gas exchange
  b. flame cells—reproduction
  c. exoskeleton—muscle attachment
  d. cnidocytes—digestion
  e. nephridia—feeding

9. All animals share which of the following characteristics?
  a. hermaphroditic
  b. heterotrophic
  c. prokaryotic
  d. coelomate
  e. complex nervous system

10. Protostomes and deuterostomes differ mainly in
  a. their modes of reproduction.
  b. their ability to move.
  c. the complexity of their nervous system.
  d. their patterns of embryonic development.
  e. symmetry.

## Level 2 | Learning Concepts

1. What do you and jellyfish have in common? What is an important taxonomic difference between the two of you?
2. "Animals are a diverse group of eukaryotic, multicellular, heterotrophic organisms." Explain this statement.
3. Which organisms are considered the simplest animals? Why?
4. What two basic body plans are shown by cnidarians? Summarize the differences between these two plans.
5. Which organisms are the most primitive bilaterally symmetrical animals? Briefly describe the structure of these animals.
6. People can become very ill if infested by roundworms. Describe these organisms and why they can be a health risk to humans.
7. Flatworms do not have a circulatory system. How do they transport essential nutrients from food to the tissues of the body?
8. Insects are among the most active of all animals on Earth. How can insects sustain such high-activity levels when their circulatory system is a relatively inefficient open circulatory system?
9. In casual conversation, you refer to a spider on the wall as an insect. A friend who has studied biology informs you that spiders are not insects. Explain what she means.
10. How do echinoderms differ from all other invertebrates? What is the significance of this fact?

## Level 3 | Critical Thinking and Life Applications

1. Some invertebrates, such as most flatworms, are hermaphroditic. How can populations of flatworms maintain diversity if they are hermaphroditic?
2. Living things often have structures that are shaped in ways that result in large surface areas. For example, the human small intestine has villi and microvilli that result in an enormous surface area over which the absorption of nutrients takes place (see Chapter 8). Name one structure in an invertebrate animal that has a large surface area. Describe the structure and how a large surface area assists its function in this animal.
3. Echinoderms are grouped with the bilaterally symmetrical organisms, yet they exhibit radial symmetry. Why are they considered bilaterally symmetrical? Support the view that they evolved from a bilaterally symmetrical ancestor.
4. Insects are a highly diverse phylum of animals. Name one characteristic that distinguishes them from all other arthropods. State a hypothesis explaining why you think insects are so abundant on Earth. Give a rationale for your hypothesis.
5. What are the major differences between annelids (e.g., earthworms) and flatworms? What advantage does segmentation give earthworms over the flatworms, if any?
6. What are the advantages of each of the following invertebrate adaptations over more primitive invertebrate body designs?
   - bilateral symmetry and cephalization
   - body cavity
   - complete digestive tract

## In The News | Critical Thinking

**EXPERIENCED SPIDERS**

Now that you understand more about invertebrates, reread this chapter's opening story about Hebets' research on the influence of subadult experience on adult mate choice in wolf spiders. To better understand this information, it may help you to follow these steps:

1. Review your immediate reaction to this information about wolf spiders that you wrote when you began reading this chapter.
2. Based on your current understanding, summarize the main point of the research in a sentence or two.
3. What questions do you now have about this research that the opening story does not answer?
4. Collect new information about the research. Visit the *In The News* section of this text's companion Web site at www.wiley.com/college/alters and watch the "Experienced Spiders" video. Then use the "summary" link to read the accompanying story and access related links. Use this information, the links provided, and other online and library resources to answer your questions and find updates about this topic. State the sources of your information. Explain why you think the information is accurate. Also determine whether the information expresses a particular point of view or is biased in any way.
5. What in your view is the most significant aspect of this research? Give reasons for your opinion and for any changes in your ideas based on the additional information you have collected and the analysis you have done.

# ANIMALS: CHORDATES, INCLUDING THE VERTEBRATES

## In The News | Sticky Feet

Geckos, such as the one staring at you in the photo, are small, usually nocturnal lizards—and they have sticky feet! Not only can they climb walls and walk across ceilings, but they can also do it quite quickly. Scientists are studying gecko feet to determine what makes them stick so well, yet loosen in a flash.

Each foot of a gecko has nearly 500,000 stiff hairs, which are called *setae* (SEE-tah). Each of these hairs is only one-tenth the diameter of a human hair, yet each splits into hundreds of spatula-shaped projections at its end. This structure might lead one to think that the setae act like microscopic suction cups, but that is not the case. Possibly, you might think, their feet secrete a sticky substance, but their feet have no gland cells, so gecko glue is not the adhesive mechanism.

To answer the centuries-old question of how gecko feet stick to nearly every surface imaginable, Kellar Autumn of Lewis and Clark College in Portland, Oregon, and his colleagues have been studying gecko setae. They have determined that the gecko's sticky secret is Van der Waals forces, which are weak electrical attractions between atoms. The subdivided setae provide a tremendous surface area over which these forces occur. So, despite the weakness of the forces, the cumulative effect results in the adhesive properties of gecko feet. The ani-

mal releases its foot by peeling it off the surface to which it is stuck, much like one might peel adhesive tape off a surface. Researchers suggest that a possible design principle also underlies how gecko feet stick to surfaces, noting the repeated convergent evolution of dry adhesive microstructures in geckos, anoles and skinks (two other types of lizards), and insects.

Can scientists build adhesive microstructures like gecko setae? The answer appears to be yes! Autumn and his colleagues have built working artificial setal tips and hope their research will lead to the manufacture of the first dry adhesive microstructures. You can read more about Autumn and his colleagues' research by viewing the video and reading the story "Sticky Feet" in the *In The News* section of this text's companion Web site at www.wiley.com/college/alters.

*Write your immediate reaction to this research on the stickiness of gecko feet: first, summarize the main point of the research in a sentence or two; then suggest what you think its significance is. You will have an opportunity to reflect on your responses and gather more information on this topic in the In The News feature at the end of this chapter. In this chapter, you will learn more about lizards and other vertebrates—animals with a backbone.*

# CHAPTER GUIDE

## What are chordates?

**22.1** Chordates have a cartilaginous, rod-shaped notochord that forms during development.

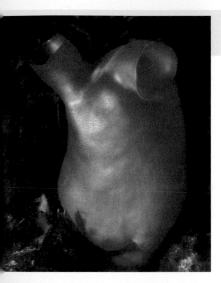

## What are the major chordate subphyla and their distinguishing characteristics?

**22.2** Tunicate larvae have a notochord in their tails.

**22.3** Lancelets have a notochord that extends into the head.

**22.4** In vertebrates, the notochord develops into the vertebral column.

## What are the major vertebrate classes and their distinguishing characteristics?

**22.5** Jawless fishes are tubelike aquatic animals with round mouths.

**22.6** Cartilaginous fishes are aquatic animals with flexible internal skeletons and toothlike scales.

**22.7** Bony fishes are aquatic animals with hard internal skeletons and thin, bony scales.

**22.8** Amphibians live in water and on land, and breathe both through their skin and by means of gills and/or lungs.

**22.9** Reptiles live primarily on land and have dry skins covered with scales.

**22.10** Birds have feathers, and their forelimbs are modified as wings.

**22.11** Mammals have hair, and females secrete milk from mammary glands.

# What are chordates?

## 22.1 Chordates have a cartilaginous, rod-shaped notochord that forms during development.

Sponges, worms, insects, and sea stars—these are some of the invertebrates discussed in the previous chapter. Now we move to the world of chordates, including vertebrates—animals with a stiffening rod down their backs. Vertebrates are generally much larger than invertebrates and may be more familiar to you. They include fishes, amphibians, reptiles, birds, and mammals.

**Chordates** are characterized by four principal features: (1) a single, hollow **nerve cord** located along the back, (2) a rod-shaped **notochord**, which forms between the nerve cord and the gut (stomach and intestines) during development, (3) **pharyngeal (gill) arches** and **slits**, which are located at the throat (pharynx) at some stage of life, and (4) a muscular **post-anal tail**, which is a tail posterior to the anus. These four features are present in the embryos of all chordates (**Figure 22.1**). In addition, lancelets exhibit all these features as adults, and tunicates exhibit them as larvae.

In vertebrates, the nerve cord differentiates into a brain and spinal cord. The notochord serves as a core around which the vertebral column develops, encasing the nerve cord in bone

and protecting it. The pharyngeal arches develop into gill structures in fishes and into ear, jaw, and throat structures in terrestrial (land) vertebrates. The presence of the gill arches in all vertebrate embryos indicates the aquatic ancestry of the vertebrates.

In addition to these four features, chordates have many other traits in common. All have a true coelom and bilateral symmetry. The embryos of chordates exhibit segmentation. The photo in Figure 22.1 shows segments of tissue called *somites* in a human embryo, which develop into the skeletal muscles. Most chordates have an internal skeleton to which their muscles are attached; the muscles work against the skeleton, providing movement. Only tunicates and lancelets do not have internal skeletons; muscles are attached to the notochord in larval tunicates and adult lancelets.

The chordate phylum includes three subphyla (**Figure 22.2** on page 376): tunicates (TOO-nih-kits), lancelets (LANS-lets), and vertebrates (VER-tuh-bruts). As you can see from the tree, the chordates are most closely related to the echinoderms, which are the only nonchordate invertebrates that are deuterostomes, suggesting evolutionary closeness between chordates and echinoderms. The evolution of the chordates is described in Section 18.6.

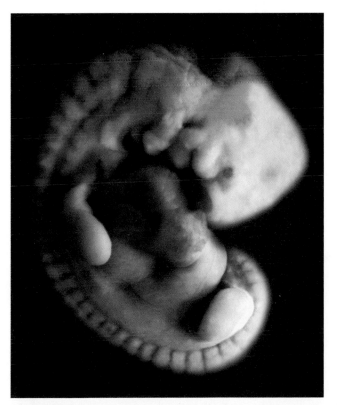

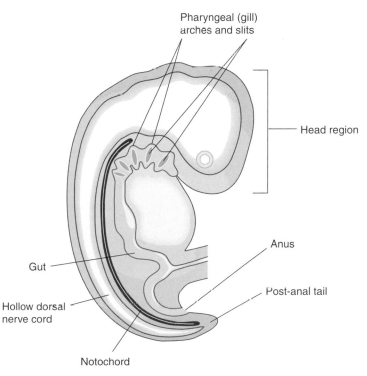

**Figure 22.1 A chordate embryo.** Chordates are characterized by four principal features: a single, hollow nerve cord located along the back, a rod-shaped notochord, pharyngeal (gill) arches and slits, and a post-anal tail. The photo is of a human embryo in the sixth week of development.

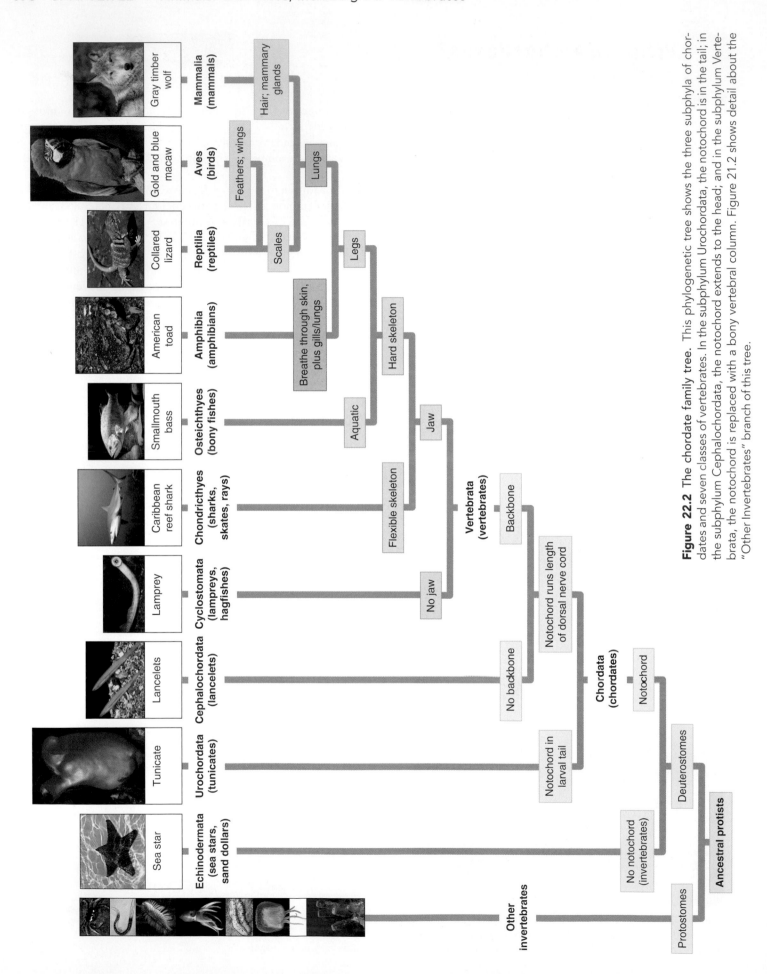

**Figure 22.2** The chordate family tree. This phylogenetic tree shows the three subphyla of chordates and seven classes of vertebrates. In the subphylum Urochordata, the notochord is in the tail; in the subphylum Cephalochordata, the notochord extends to the head; and in the subphylum Vertebrata, the notochord is replaced with a bony vertebral column. Figure 21.2 shows detail about the "Other Invertebrates" branch of this tree.

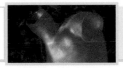

# What are the major chordate subphyla and their distinguishing characteristics?

## 22.2 Tunicate larvae have a notochord in their tails.

If you were ever to spot a tunicate while walking along the shore, you might mistake it for a sponge. You'd be in good company because Aristotle himself made this mistake. However, tunicates are not sponges; they are much more closely related to humans than to sponges.

The **tunicates** (subphylum Urochordata; YUR-eh-core-DA-tah) comprise a group of about 2500 species of marine animals, most of which look like living sacs attached to the floor of the ocean (**Figure 22.3a**). As shown in Figure 22.3b, a tunicate is not much more than a large pharynx covered with a protective tunic, a tough outer "skin." Colonial tunicates reproduce asexually by budding. Individual tunicates are hermaphrodites, with each organism having both male and female sex organs.

The pharynx of a tunicate is lined with numerous cilia. As these cilia beat, they draw a stream of water through the incurrent siphon into the pharynx, which is lined with sticky mucus. Plankton are trapped within the pharynx, and the filtered water flows

out of the animal through the excurrent siphon. (Plankton are small, even microscopic, organisms that float or drift in fresh water or saltwater and serve as food for fish and other larger organisms.) Because 90% of all species of tunicates have this structure and forcefully squirt water out their excurrent siphons when disturbed, they are also called *sea squirts*.

As adults, tunicates lack a notochord and a nerve cord. The gill slits, which develop from the gill arches, are the only clue that adult tunicates are chordates. Only the larvae, which look like tadpoles (Figure 22.3c), have a notochord and nerve cord. The notochord is in the tail, and the nerve cord runs dorsal to the notochord almost the entire length of the body. The subphylum name Urochordata comes from the placement of the larval notochord and literally means "tail chordate." In most species, the larvae remain free swimming for no more than a few days. Then they settle to the bottom and become sessile.

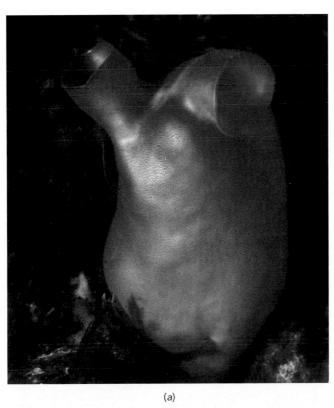

(a)

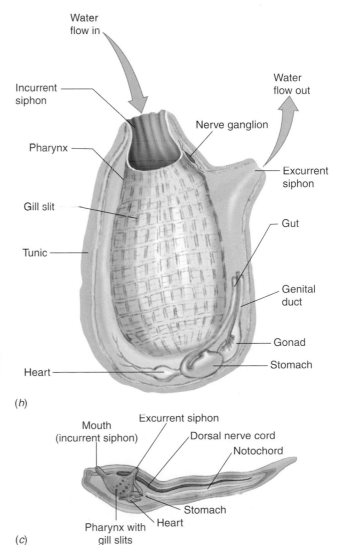

(b)

(c)

**Figure 22.3** Tunicates (subphylum Urochordata). (a) *Halocynthia auranthium*, a tunicate commonly called the sea peach. (b) The structure of an adult tunicate. (c) The structure of a larval tunicate.

## 22.3 Lancelets have a notochord that extends into the head.

The **lancelets** are tiny, scaleless, fishlike marine chordates that are just a few centimeters long and pointed at both ends (**Figure 22.4**). They look very much like tiny surgical blades called lancets, from which they get their name. You may have had a few drops of blood taken in the doctor's office from a "fingerstick" done with a lancet.

Lancelets have a segmented appearance because of blocks of muscle tissue that are easily seen through their thin, unpigmented skin. Although they have pigmented light receptors, lancelets have no real head, eyes, nose, or ears. However, unlike the tunicates, the lancelet's notochord runs the entire length of its dorsal nerve cord, so these organisms are called *cephalochordates* or *head chordates*, and their phylum name is Cephalochordata. The lancelet retains its notochord throughout its lifespan.

The 23 species of lancelets live in the shallow waters of oceans all over the world. They spend most of their time partly buried in the sandy or muddy bottom with only their anterior ends (oral hoods) protruding, feeding on plankton. In a manner similar to the tunicates, lancelets filter plankton from the water. Cilia line the anterior end of the alimentary canal (digestive

tract) and beat to create an incoming current of water. The filtered water exits at an excurrent siphon. An oral hood projects beyond the mouth, or incurrent siphon, and bears sensory tentacles. The sexes are separate, but no obvious external differences exist between them.

## 22.4 In vertebrates, the notochord develops into the vertebral column.

The **vertebrates** differ from the other chordates in that most have a vertebral column in place of a notochord. A vertebral column, or backbone, is a stack of bones, each with a hole in its center, which forms a cylinder surrounding and protecting the dorsal nerve cord. Each bone in the column is a vertebra (pl., vertebrae). Present-day representatives of one class of vertebrates have a notochord but no vertebral column: the jawless fishes. However, their ancestors had a bony skeleton and a vertebral column. Members of another class of vertebrates, the cartilaginous fishes, have a skeleton and vertebral column composed of cartilage (a tough, yet elastic, type of connective tissue) rather than bone. Although not all present-day vertebrates have a bony vertebral column, they all have a distinct head with a skull that encases the brain. For this reason, some current classification systems use the term *craniates* (denoting chordates with skulls) instead of the term *vertebrates*.

Vertebrates have a closed circulatory system—the blood flows within vessels—and a heart to pump the blood. Most vertebrates also have a liver, kidneys, and endocrine glands. Endocrine glands are ductless glands that secrete hormones, which play a critical role in controlling the functions of the vertebrate body. The human body plan is representative of the vertebrate body plan and is described extensively in Part Seven.

Oral hood with tentacles

Gill slits in pharynx

Excurrent siphon

Alimentary canal

Notochord

Dorsal nerve cord

Anus

Tail

(a)

(b)

**Figure 22.4** Lancelets (subphylum Cephalochordata). (a) The structure of an adult lancelet. (b) Lancelets stay partly buried, with only their anterior ends protruding.

 **Visual Thinking:** Compare the structure of the larval tunicate shown in Figure 22.3c with the structure of the adult lancelet shown in this illustration.

Vertebrates are a diverse group, consisting of animals adapted to life in the sea, on land, and in the air. There are eight classes of living vertebrates: Four classes are fishes and four classes are land-dwelling *tetrapods* (four-footed, or four-limbed, animals). Only three classes of fish are described in detail in this chapter. The fourth class consists of the lungfish, which live only in African fresh water lakes that regularly dry up. When this occurs, the lungfish burrow into the mud, leaving the mouth exposed to gulp air.

**CONCEPT CHECKPOINT**

1. What four characteristics do all chordates share?
2. You are a vertebrate, but as an adult you lack several of the characteristics shared by all chordates. Explain why you are still classified as a chordate.

## What are the major vertebrate classes and their distinguishing characteristics?

### 22.5 Jawless fishes are tubelike aquatic animals with round mouths.

You may be very familiar with fish in aquaria, on the end of a fishing line, or on your dinner plate, but you do know the scientific definition of this group of animals? A fish is an aquatic vertebrate that breathes by means of gills, is cold-blooded (cannot internally regulate its body temperature), has fins, and usually has a skin with scales. The jawless fishes are exceptions to this last characteristic. Their skin is scaleless and slimy.

Other than the lamprey eels and hagfishes, the major groups of the **jawless fishes** (class Cyclostomata, meaning "round mouth") have been extinct for hundreds of millions of years. Only about 20 to 30 species of each of these two groups are alive today. Both groups are long, tubelike aquatic animals that usually live in the sea or in brackish (somewhat salty) water where the fresh water of a river meets the ocean. Although they lack jaws, they do have a mouth, but it has no internal cartilaginous or bony supports. In addition, they have no paired fins to help them swim. They do have a notochord, however, and portions of a cartilaginous skeleton that are remnants from their extinct ancestors.

Lampreys parasitize other fishes. In fact, they are the only parasitic vertebrates. Their round mouth functions like a suction cup (**Figure 22.5a** and **b**), which they attach to their prey—the bony fishes. When a lamprey attaches to a fish, it uses its spine-covered tongue like a grater, rasping a hole through the skin of the fish and then sucking out body fluids. Sometimes, lampreys are so abundant that they become a serious economic threat, preying on salmon, trout, and other commercially valuable fishes. Entering the Great Lakes from the sea, they have become major pests there; millions of dollars are spent annually on their control. Although hagfishes (Figure 22.5c) are similar in size and shape to lampreys, they are not parasitic. Rather, they are scavengers, often feeding on the internal organs of dead or dying fishes or large invertebrates.

To reproduce, jawless fishes *spawn*, as do most fishes, amphibians, and shellfish. (Shellfish are aquatic animals that have shells or exoskeletons, such as clams and lobsters.) During spawning, the females and males deposit eggs and sperm directly into the water, where the eggs are fertilized. This type of fertilization, outside the body of the female, is called **external fertilization**. Lampreys swim upstream to spawn, as salmon do, and create nests in which

they deposit eggs and sperm. The fertilized eggs develop into larvae that feed on plankton. Over a period of years, the larvae mature and metamorphose (change) into parasitic adults. In contrast, the eggs of hagfishes do not develop into larvae. Completely formed hagfishes hatch directly from fertilized eggs.

### 22.6 Cartilaginous fishes are aquatic animals with flexible internal skeletons and toothlike scales.

The 1975 movie *Jaws* and its sequels immortalized the shark as predator. Sharks, skates, and rays are the **cartilaginous fishes** (class Chon-

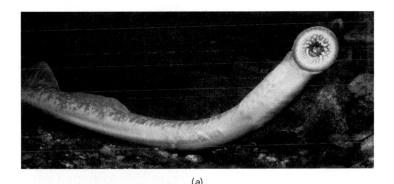

(a)

(b)

(c)

**Figure 22.5** Jawless fishes (class Cyclostomata). (a) A lamprey, showing the mouth and the elongated body. (b) A lamprey parasitizing a bony fish. (c) A Pacific hagfish.

(a)

(b)

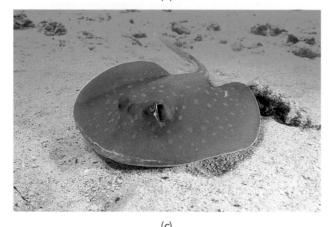

(c)

**Figure 22.6** Cartilaginous fishes (class Chondrichthyes). (a) Caribbean reef shark. (b) Starry skate (juvenile). (c) Bluespotted stingray.

 **Visual Thinking:** All three of these organisms display some amount of camouflage; that is, they are colored so as to blend in with their surroundings. Describe the type of camouflage you can see on each. What advantage does camouflage afford these cartilaginous fishes?

drichthyes [kon-DRIK-thee-eeze]; *chondri* meaning "cartilage," *ichthyes* meaning "fishes"). Cartilage is a type of connective tissue that is hard and strong but more flexible than bone. It makes up the skeleton of the Chondrichthyes. The skin of cartilaginous fishes is covered with small, pointed, toothlike scales called *denticles*, which give the skin a sandpaper texture. Sharks (**Figure 22.6a**) have

streamlined bodies and two pairs of fins: pectoral fins just behind the gills and pelvic fins just in front of the anal region. In addition, two dorsal (back) fins provide stability. Cartilaginous fishes are a very successful group of animals. Hundreds of extinct species have been identified in the fossil record, and more than 700 species still exist today.

Many sharks are predators that eat large fishes and marine mammals. Some feed on plankton rather than prey on other animals. These sharks swim with their mouths open, and the plankton are strained from the water by specialized denticles on the inner surfaces of the gill arches.

Skates and rays (Figure 22.6b and c) are generally smaller than sharks and have flattened bodies with enlarged pectoral fins that undulate (rise and fall regularly) when these fishes move. Their tails, which are not a principal means of locomotion as in sharks, are thin and whiplike. The tails are sometimes armed with poisonous spines that are used for defense rather than for predation. Their mouth is on their underside, and they feed mainly on invertebrates on or near the ocean floor.

The sensory systems of sharks, skates, and rays are quite sophisticated and diverse. Sharks have a lateral line system, as do many other fishes and some aquatic amphibians. A **lateral line system** is a complex system of mechanoreceptors that lie in a single row along the sides of the body and in patterns on the head. These receptors can detect mechanical stimuli such as sound, pressure, and movement. Moreover, sharks and some other fishes have electroreceptors in their lateral line systems, which can detect electrical fields emitted by other fishes. Chemoreception is another important sense in the Chondrichthyes. Sharks have been described as "swimming noses" because of their acute sense of smell. Vision is also important to the feeding behavior of sharks; they have well-developed mechanisms for vision in low light. These senses are also described in Chapter 33.

Another important characteristic of fishes is their ability to regulate their buoyancy, enabling them to float at various depths in the water. Cartilaginous fishes can adjust the size and oil content of their livers. Because oil is less dense than water and the liver has a high oil content, adjusting the oil content and size of the liver can regulate buoyancy to a certain degree. Most species of sharks, however, must swim continually to keep from sinking and to keep water flowing over their gills.

Aquatic organisms show various adaptations regarding **osmoregulation**, the control of water movement into and out of their bodies. Marine organisms tend to lose water to their surroundings because their body fluids usually have a solute concentration—concentration of dissolved substances—lower than that of seawater. The cartilaginous fishes maintain solute concentrations close to that of seawater, however, because they change potentially toxic nitrogen-containing wastes, such as ammonia, into a less toxic compound called urea and retain it in their bodies rather than excreting it (**Figure 22.7**). Shark meat, before it is eaten, must be soaked in fresh water to remove most of the urea. With similar solute concentrations both inside and outside the fish, water movement remains relatively equal in both directions.

Most aquatic animals reproduce by means of external fertilization, but the cartilaginous fishes have developed a method of **internal fertilization**. Male cartilaginous fishes have *pelvic claspers*, which are rodlike projections between their pelvic fins. During copulation (coupling of the male and female animals), the male in-

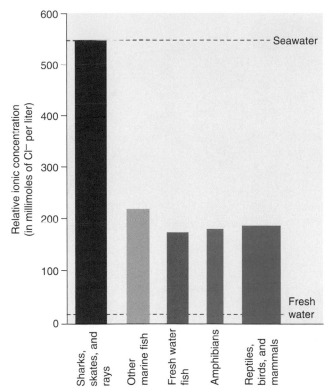

**Figure 22.7 Ion concentrations for different classes of vertebrates.** Ion concentrations in the bodies of different classes of vertebrates are roughly similar, with the exceptions of the cartilaginous fishes. Sharks, skates, and rays keep their ion concentrations close to that of seawater by adding urea to the bloodstream.

**Visual Thinking:** Use this graph to determine whether water will tend to move into or out of the cells of marine fish (other than sharks, skates, and rays) and fresh water fish when in seawater and fresh water, respectively. Explain your reasoning.

serts his clasper into the female's *cloaca* (kloe-AY-kuh). The cloaca is the terminal part of the gut into which ducts from the kidney and reproductive systems open. The male ejaculates into a groove in the clasper, which directs the sperm into the female's cloaca.

In addition to having a mechanism of internal fertilization, many cartilaginous fishes are ovoviviparous. **Ovoviviparous** (OH-vo-vye-VIP-uh-rus) organisms retain fertilized eggs within their oviducts until the young hatch (*ovum* meaning "egg," *vivus* meaning "alive," *pario* meaning "to bring forth"). However, the developing young within the eggs do not receive nutrition from the mother but from the yolk of the egg. The mother acts like an internal "nest" for the eggs until the young hatch. Certain fishes, some reptiles, and many insects are ovoviviparous.

In contrast, **oviparous** (oh-VIP-uh-rus) organisms lay eggs, and the young hatch from the eggs outside of the mother. Many skates and rays, for example, are oviparous and release eggs within protective egg cases. Birds and most reptiles are also oviparous.

Like ovoviviparous organisms, **viviparous** (vye-VIP-uh-rus) organisms such as mammals also bear live young, but the developing embryos derive nourishment primarily from the mother and not the egg. A few sharks, such as hammerhead sharks, blue sharks, and lemon sharks, are viviparous fishes.

## 22.7 Bony fishes are aquatic animals with hard internal skeletons and thin, bony scales.

The **bony fishes**, which include the vast majority of known species of fishes, inhabit almost every body of water, whether tropical, temperate, or polar. Approximately 58% of their species live in marine environments and 42% in fresh water. Some species migrate between fresh water and saline environments.

Bony fishes belong to the class Osteichthyes (OS-tee-IK-thee-eeze; *oste* meaning "bone"). The Osteichthyes get their name from their bony internal skeletons. However, the skin of Osteichthyes is also covered with thin, overlapping bony scales. Sometimes these scales have spiny edges, which provide some protection for the animal. Bony fishes also have a protective flap that extends posteriorly from the head and protects the gills. This flap is called the **operculum** (**Figure 22.8**). Along with protecting the gills, the movement of the operculum enhances the flow of water over the gills, bringing more oxygen in contact with the gas-exchanging surfaces and allowing the fish to breathe while stationary.

Bony fishes are divided into two subclasses: *lobe-finned fishes* and *ray-finned fishes*. Lobe-finned fishes have paired rounded fins supported by thick bones. Ray-finned fishes (Figure 22.8a) have flat fins supported by long, segmented, flexible, thin bones called rays. Only seven species of lobe-finned fishes exist today; most are lungfish (Figure 22.8b). Most bony fishes are ray-finned fishes and include fish that are probably familiar to you, such as perch, cod, trout, tuna, herring, and salmon.

Ancient lobe-finned fish had modified pouches in their digestive tracts, which evolved into the lungs of lungfish and *swim bladders* in most ray-finned fish. In the lungfish the modified swim

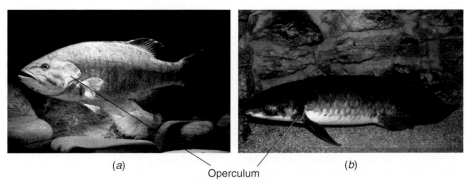

(a)          Operculum          (b)

**Figure 22.8** Bony fishes (class Osteichthyes) consist of two groups: ray-finned fishes and lobe-finned fishes. In both groups, the operculum covers the gills, protecting them and helping them move water across their surfaces. (a) A smallmouth bass, a ray-finned fish. (b) An Australian lungfish, a lobe-finned fish.

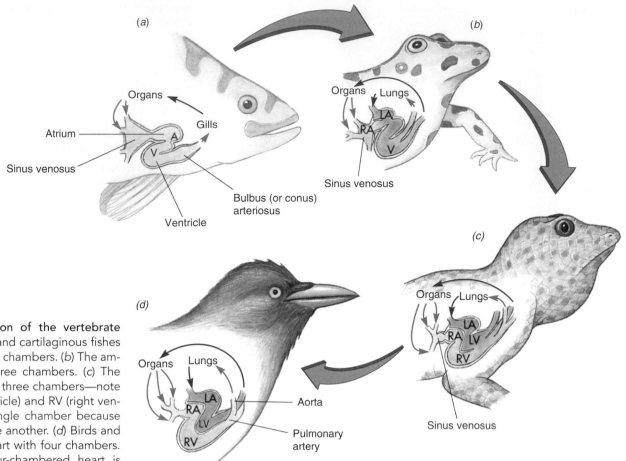

**Figure 22.9 Evolution of the vertebrate heart.** (*a*) The bony and cartilaginous fishes have a heart with two chambers. (*b*) The amphibian heart has three chambers. (*c*) The reptile heart also has three chambers—note that the LV (left ventricle) and RV (right ventricle) are really a single chamber because they are open to one another. (*d*) Birds and mammals have a heart with four chambers. The mammalian four-chambered heart is shown in Figure 29.7.

bladder (single or paired lungs) absorbs oxygen and removes wastes. Lungfish can survive by breathing air through their lungs instead of their gills and by lowering their metabolic rate. One group of the lobe-finned fishes is thought to have given rise to the amphibians, with their lobe fins and lungs being preadaptations for walking and breathing on land.

In ray-finned fishes, swim bladders regulate buoyancy. Gas exchange between the swim bladder and the blood regulates the density of the fish and allows it to remain suspended in water without sinking to the bottom. Unlike most of the cartilaginous fishes, fishes with swim bladders can remain motionless at various depths because of their ability to more finely regulate buoyancy and move water across their gills with their operculum.

The bony fishes, like the cartilaginous fishes and all other chordates, have a closed circulatory system consisting of blood vessels and a pump. (Figure 29.1 shows the differences between closed and open circulatory systems.) In the cartilaginous fishes and bony fishes, the pump is a tubelike heart with two chambers (an *atrium* and a *ventricle*) (**Figure 22.9a**). The atrium is preceded by an enlarged section of the vein called the *sinus venosus*, which collects blood returning from the body and supplies it to the atrium. The atrium pumps blood to the ventricle, which then pumps it to a thickened portion of the aorta, the large artery leaving the heart. This structure, called the *conus arteriosus* in cartilaginous fishes and *bulbus arteriosis* in bony fishes, has valves to prevent the backflow of blood into the ventricle. From there, blood moves to the gills and

then to the rest of the body. Because of the great resistance in the narrow passageways of the capillaries at the gills, the movement of blood to the rest of the body is sluggish. Figure 22.9 shows how the heart has evolved from fishes to amphibians to reptiles to birds (and mammals), as described further in later sections of this chapter.

Unlike the cartilaginous fishes, bony fishes do not maintain a solute concentration of their body fluids near that of seawater. Therefore, bony fishes must control the amount of water in their tissues, a process called *osmoregulation*. The body fluids of marine bony fishes are *hypoosmotic* with respect to seawater (see Section 5.13). That is, their body fluids contain a lower concentration of dissolved substances than seawater does (see Figure 22.7). Consequently, water tends to leave these fishes (at the gill epithelium) by osmosis. To regulate water balance, marine bony fishes drink seawater and then excrete the salt by means of active transport at the gill epithelium and the kidneys.

Marine and fresh-water bony fishes have opposite osmoregulation problems. In contrast to marine fishes, the body fluids of fresh-water fishes are *hyperosmotic* with respect to their fresh-water environment. That is, their body fluids contain a higher concentration of dissolved substances than fresh water does. Consequently, water tends to enter fresh-water fishes by osmosis. To regulate water balance, fresh-water fishes do not drink water and excrete large amounts of very dilute urine. They reclaim some of the ions they lose in their urine by the uptake of sodium and chlorine ions by the gills and in the food they eat.

(a)

(b)

(c)

**Figure 22.10** Amphibians (class Amphibia). The American toad (a) and the red-eyed tree frog (b) are amphibians without tails. Toads are generally more terrestrial than frogs and are shorter and squatter with weaker hind legs. Toads have rough, dry, warty skin, whereas frogs have smooth, moist skin. Both lay their eggs in the water. (c) The eastern newt is a salamander, an amphibian that has a tail.

Most Osteichthyes are oviparous, fertilizing their eggs externally. These eggs are often food for other marine organisms, however, and must survive other risks such as drying out. Most fishes, as well as other organisms that externally fertilize their eggs, lay large numbers of eggs, with some surviving these dangers. Many species of fishes build nests for their eggs and watch over them, which also enhances the chances of survival. Ovoviviparous and viviparous bony fishes bear live young.

### CONCEPT CHECKPOINT

3. Compare and contrast a lamprey, a shark, and a trout with respect to feeding, reproduction (fertilization, birthing process), and skeletal anatomy.
4. What adaptations do bony fishes possess (that sharks lack) that allow them to maintain their buoyancy and stay motionless in the water?

## 22.8 Amphibians live in water and on land, and breathe both through their skin and by means of gills and/or lungs.

You may have heard the expression "It's all in the name." That expression couldn't be truer for the **amphibians**, because their name (*amphi* meaning "both," *bios* meaning "life") describes their unique characteristic of having two lives: amphibians live in water and on land. Unlike other terrestrial animals such as reptiles, birds, and mammals, amphibians lay their eggs in water or in moist places and live in water during their early stages of development. Many amphibians live in moist places such as swamps and in tropical areas even when they are mature, which lessens the constant loss of water through their thin skin, which is an organ of respiration in addition to gills and/or lungs. The two most familiar orders of amphibians (class Amphibia) are those that do not have tails—the frogs and toads (**Figure 22.10a**)—and those that have tails—the salamanders, including mud puppies and newts (**Figure 22.10b** and **c**). In addition, adult amphibians, unlike fish, have legs.

Most frogs and toads fertilize their eggs externally. The male grasps the female and sheds sperm over the eggs as they are expelled from the female (see Figure 36.8). Most salamanders use internal fertilization but are still oviparous (lay their eggs). Because amphibian eggs have no shells or membranes to keep them from drying out, amphibians lay their eggs directly in water or in moist places (**Figure 22.11**).

The young of frogs and toads undergo *metamorphosis* (MET-uh-MORE-feh-sis), or change, during development from a larval to an adult form. The larvae are immature forms that do not look like the adult. The larvae of frogs and toads are tadpoles, which usually live in the water and have internal gills and a lateral line system like that of fishes. They feed on minute algae. These fishlike forms develop into carnivorous adults having legs and lungs; their gills and lateral line system disappear. The

**Figure 22.11** Nest of tropical tree frogs. Tropical tree frogs create "nests" of foam in tree branches to incubate their eggs. When the tadpoles develop, they drop from the tree branches into the water. Some amphibians protect their eggs by incubating them in their mouths, or their backs, or even in their stomachs!

lungs of the adults are inefficient, however. Much of the gas exchange takes place across the skin and on the surfaces of the mouth (see Section 28.2). The skin of amphibians, therefore, must remain damp to allow gases to diffuse in and out.

The adults of certain salamanders, such as mud puppies, live permanently in the water and retain gills and other larval features as adults. Other salamanders are terrestrial but return to water to breed; they do not retain their larval features. Like frogs and toads, they usually live in moist places, such as under stones or logs or among the leaves of certain tropical plants.

Along with having lungs rather than gills (in most cases), amphibians have a pattern of blood circulation different from that of the fishes. After the blood is pumped through the fine network of

capillaries in the amphibian lungs, it does not flow directly to the body as it does in fishes. Instead, it returns to the heart. It is then pumped out to the body at a much higher pressure than if it were not returned to the heart. Figure 22.9*b* shows this pathway of blood flow. The blood from the lungs and the blood from the body enter the right and left atria of the heart, respectively. The blood in both of these chambers flows into the single ventricle of the three-chambered amphibian heart and is pumped through two large vessels to the lungs and the body. To a large extent, oxygenated and deoxygenated blood are kept from mixing by the sequential contraction of the right and left atria and by the action of a valve that alternately blocks and unblocks the openings to the vessels leaving the heart.

# *just wondering . . .*

## *Questions students ask*

### I know that chameleons are "famous" for their ability to change color. How and why do they do that?

There are over 100 species of chameleons—some more colorful than others. The one shown in **Figure 22.A** (a *Chamaeleo pardalis* male) is a particularly colorful chameleon and is easily tamed, so you might have seen one like it in a pet store.

The common perception is that chameleons change their color to protect themselves from predators. However, the chameleon's camouflage consists primarily of its body shape and behavior; the ability to change color plays only a small role in its blending with the environment. Most chameleons have a body shape that is flattened vertically, making them look somewhat like a leaf. In addition, chameleons walk in a faltering or hesitating manner, looking a bit like the "leaf" is being moved by the wind. Many chameleons are shades of green or brown, which adds to the look of a live or dead leaf. Some species are camouflaged by looking like moss, and yet others look like wood chips on the forest floor.

If color change is not the chameleon's primary method of camouflage, then why does color change occur—and how does it occur? The chameleon's color change is primarily triggered by a combination of internal and external cues. Internal factors that affect a chameleon's color are its physical health, its level of sexual maturity, and its hormonal response to the presence of other chameleons (male or female) of the same species.

External conditions such as light intensity and the temperature of the environment also affect a chameleon's color. Chameleons are lizards and therefore ectothermic. When a chameleon is cold, certain cells called iridocytes allow more light to enter the skin. The animal flattens out and its colors darken, resulting in greater absorption of heat.

Chameleon skin is structured much like human skin (and like that of all vertebrates), with dead, hardened cells forming the topmost

layer of the epidermis and living cells lying just below. Beneath the epidermis is the dermis, a layer of dense connective tissue (see Figure 34.1). Three main types of skin cells are involved in chameleon color change: guanocytes, iridocytes, and chromatophores.

The living layer of the epidermis contains guanocytes. These cells contain nonpigmented crystals that diffract (bend) light rays. They appear yellow when light is reflected up through them by the iridocytes beneath and blue when light is not reflected up. The iridocytes lie between the epidermis and the dermis. These cells contain reflecting plates stacked like decks of cards. The plates scatter the light and, depending on their positions, may direct light back up through the epidermis (and the guanocytes) or down into the dermis where the chromatophores are located.

The chromatophores are pigmented cells with fingerlike cell processes that extend upward toward the epidermis. Depending on the species of chameleon, these cells may contain red, brown, yellow, orange, or blue-violet pigments. Chromatophores are under neuro-hormonal control. For example, if a chameleon is threatened by another animal, the chromatophores are stimulated to send pigment to the upward-reaching cell processes. This results in a coloration of the skin as pigment is "pumped" into upper cell layers.

**Figure 22.A Chameleon.** This colorful chameleon is a *Chamaeleo pardalis* male.

Chameleons are not the only animals that can change color. For example, certain fish change color as a part of their behavioral repertoire. Certain frogs can also darken, camouflaging themselves when threatened. Cephalopods (octopus and squid) change colors as well when mating or hunting or when threatened. The mechanisms of color change are similar among all animals capable of this "feat." However, chameleons are often so strikingly colorful and their color changes so dramatic that they have become the animals most noticed for this ability. ●

*Are you wondering about a topic in biology and how it relates to your life? Submit your question by clicking the* Just Wondering *link in this text's companions Web site at* www.wiley.com/college/alters.

**Figure 22.12** Reptiles (class Reptilia). (*a*) American alligator. (*b*) Sonora mountain kingsnake. (*c*) Collared lizard. (*d*) Eastern box turtle.

## 22.9 Reptiles live primarily on land and have dry skins covered with scales.

Steve Irwin, the Australian "Crocodile Hunter," has made crocodiles and other reptiles much more familiar to audiences around the world than they were previously. What are these animals that Steve is continually wrestling, chasing, or holding?

The three major orders of **reptiles** (class Reptilia) are (1) the crocodiles and alligators, (2) the turtles and tortoises, and (3) the lizards and snakes. Representatives of these orders are shown in **Figure 22.12**. Reptiles have dry skins covered with scales that help retard water loss. As a result, reptiles can live in a wider variety of environments on land than amphibians can, but the crocodiles, alligators, and turtles are aquatic organisms.

The hearts of reptiles differ from amphibian hearts in that a partial partition called a *septum* subdivides the ventricle, the pumping chamber of the heart. The septum reduces even further the mixing of oxygenated and deoxygenated blood in the heart. In most crocodiles, the separation is complete. Figure 22.9*c*, shows that the reptilian heart closely resembles the four-chambered heart of birds (Figure 22.9*d*), which evolved from the reptiles. The reptilian heart is considered a three-chambered heart, however, because the ventricles are only partially separated in most species.

One of the most critical adaptations of reptiles to life on land is the evolution of the shelled **amniotic egg** (AM-nee-OT-ik). Amniotic eggs are also characteristic of birds and egg-laying mammals

(monotremes—see Section 22.11). The amniotic egg protects the embryo from drying out, nourishes it, and enables it to develop outside of water. (See Section 18.9 for more details on the evolution of the reptiles.)

As shown in **Figure 22.13**, the amniotic egg contains a yolk and albumin (egg white). The yolk is the primary food supply for the embryo, and the albumin provides additional nutrients and water. The

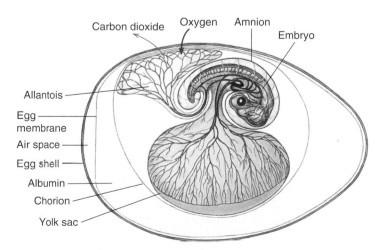

**Figure 22.13** The amniotic egg. The amniotic egg is an important adaptation that allows reptiles (and birds and monotremes) to live in a wide variety of terrestrial habitats.

embryo's nitrogenous wastes are excreted into the allantois (ah-LAN-toe-us), a sac that grows out of the embryonic gut. Blood vessels grow out of the embryo through the sac surrounding the yolk and through the allantois to the egg's surface, where gas exchange takes place. The amnion (AM-nee-on) surrounds the developing embryo, enclosing a liquid-filled space within which the embryo develops and is protected. A membrane called the *chorion* (KORE-ee-on) surrounds the embryo, amnion, yolk sac, and allantois and, along with the shell, controls the movement of gases into and out of the egg. In most reptiles, the eggshell is leathery, unlike the hard shell of bird eggs. Because of this difference, reptile eggs are somewhat permeable to water, whereas bird eggs are not.

Reptiles, like amphibians and fishes (but unlike birds and mammals), are **ectothermic** (*ectos* meaning "outside," *thermos* meaning "heat"). Ectothermic animals regulate body temperature by taking in heat from the environment. They are often called "cold blooded," but this is a misleading term, because they often maintain body temperatures much warmer than their surroundings. Nevertheless, ectotherms still may become quite cold and protect themselves with physiological and behavioral adaptations. For instance, fishes as well as certain lizards, invertebrates, and plants produce their own internal antifreeze—chemical compounds that lower the freezing temperature of the body fluids of an organism, allowing it to survive in freezing conditions. Some ectothermic animals protect themselves from high heat by burrowing under rocks or remaining in shady, somewhat cooler areas. Desert tortoises, for example, construct shallow burrows to stay in during the summer and deeper burrows for hibernation in winter. Reptiles often bask in the sun, which raises their body temperature and their metabolic rate. When cold-blooded animals are cold, their metabolic rate slows down and they are unable to hunt for food or to move about very quickly.

(a)

(b)

**Figure 22.14 Birds (class Aves).** The birds are a large and successful group of about 9000 species, more than any other class of vertebrates except the bony fishes. (a) Great blue herons in a courtship display. (b) Gold and blue macaw.

 CONCEPT CHECKPOINT

5. Most amphibians inhabit terrestrial environments, so why must they still maintain close ties with water (other than for drinking)?

6. What adaptations have evolved in reptiles that have allowed them to live a completely terrestrial life?

## 22.10 Birds have feathers, and their forelimbs are modified as wings.

The gold and blue macaw and the herons shown in **Figure 22.14** represent only two of the approximately 9000 species of **birds** (class Aves) living today. In birds, the wings are homologous to the forearms of other vertebrates. That is, they have the same evolutionary origin and the same basic anatomy, but they differ in function. (See

Section 18.9 for more information on the evolution of birds.) Birds have reptilianlike scales on their legs and lay amniotic eggs as reptiles do, but with hard shells. They have hard, horny extensions of the mouth called beaks that tear, chisel, or crush their food. They also have digestive organs called gizzards, often filled with grit, that grind food. Beaks are not limited to birds. Many reptiles have beaks, including turtles, and so do some fishes, such as the long, narrow needlefish. Birds, however, are the only animals that have feathers.

Feathers are flexible, light, waterproof epidermal structures. Several types of feathers form the body covering of birds, including contour feathers and down feathers. Contour feathers are flat (except for a fluffy, downy portion at the base) and are held together by tiny barbules as shown in **Figure 22.15**. These feathers provide a streamlined surface for flight, and some are modified to reduce drag on the wings or act like individual propeller blades. Feathers also provide birds with waterproof coats and play an important role in insulating birds against temperature changes.

Other characteristics also aid flight in birds. The light, hollow bones of birds that fly and the presence of beaks rather than teeth and jaws make birds lighter than jawed animals. Birds also have highly efficient lungs that supply large amounts of oxygen, which is necessary to sustain muscle contraction during prolonged flight. Unlike fishes, amphibians, and reptiles (except for the crocodiles), birds and mammals have a heart that acts as a double pump. The sides of the heart are completely separated with a septum (see Figure 22.9d). The right side of the heart pumps blood to the lungs, whereas the left side pumps blood to the body. Oxygenated blood and deoxygenated blood do not mix. As you compare the pathway of blood in the hearts of fishes, amphibians, reptiles, and birds in Figure 22.9, note that the four-chambered heart of birds and mammals evolved from the two chambers of the fish heart.

Birds and mammals are **endothermic** (*endo* means "within"); they regulate body temperature internally. The evolution of the four-chambered heart with separate pathways to the lungs and the body is

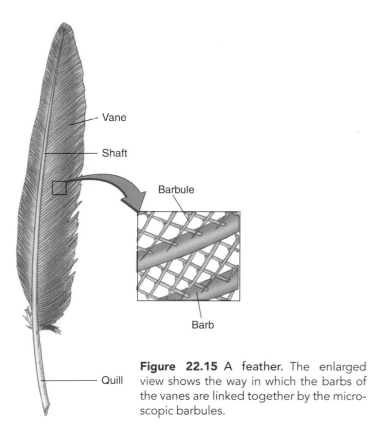

Figure 22.15 A feather. The enlarged view shows the way in which the barbs of the vanes are linked together by the microscopic barbules.

Labels on figure: Vane, Shaft, Barbule, Barb, Quill

thought to have been important in the evolution of endothermy in the birds and mammals. More efficient circulation is necessary to support the great increase in metabolic rate that is required to generate body heat internally. In addition, blood is the carrier of heat in the body, and an efficient circulatory system is required to distribute heat evenly throughout the body. Endothermic animals are sometimes called "warm-blooded," but their body temperatures may be cooler than the temperature of the surroundings (such as when you are in 100° F heat). Usually, however, because endotherms maintain a high internal body temperature (37° C [98.6° F] in humans, for example), their internal temperature is higher than that of the environment.

Endotherms maintain a constant high body temperature by adjusting heat production to equal heat loss from their bodies under various environmental conditions. The high metabolic rate of endotherms and the energy released during the chemical reactions of metabolism produce much of this body heat. Increasing the action of skeletal muscles increases the metabolic rate and the amount of heat produced. Shivering in the cold, for example, is an action that produces body heat. Animals that live in extremely cold temperatures, such as arctic birds and polar bears, are well insulated with either feathers or hair that traps air restricting heat loss. Raising or lowering the feathers or hair adjusts the insulating capacity. Getting "goosebumps" when you are cold is your body's way of raising your hairs and increasing your insulation. Humans no longer have substantial body hair, but this mechanism is important in reducing heat loss in other endothermic vertebrates.

### CONCEPT CHECKPOINT

**7.** Describe five adaptations for flight that have evolved in birds.

## 22.11 Mammals have hair, and females secrete milk from mammary glands.

There are about 4500 species of living **mammals**, including humans. Mammals (class Mammalia) are endothermic vertebrates that have hair, and their females secrete milk from mammary glands to feed their young. The evolution of mammals is described in detail in Sections 18.11–18.17. Mammals, like birds and most crocodiles, have a four-chambered heart with circulation to the lungs and separate circulation to the body. The locomotion of mammals is advanced over that of the reptiles, which in turn is advanced over that of the amphibians. The legs of mammals are positioned much farther under the body than those of reptiles and are suspended from limb girdles, which permit greater leg mobility.

Specialized teeth have evolved in mammals, different from the teeth of fishes, amphibians, and reptiles. The teeth of these other organisms are all essentially the same size and shape, but in mammals, evolutionary specialization has resulted in a variety of tooth types: incisors, which are chisel-like teeth used for cutting; canines, which are used for gripping and tearing; and molars, which are used for crushing and breaking (see Figure 27.5). There are three subclasses of mammals: monotremes (MON-oh-treems), marsupials (mar-SOO-pee-uhlz), and placental (pluh-SENT-uhl) mammals.

The duckbilled platypus (**Figure 22.16a**) and two genera of spiny anteaters (Figure 22.16b) are the only monotremes that exist today. **Monotremes** are mammals that lay eggs having

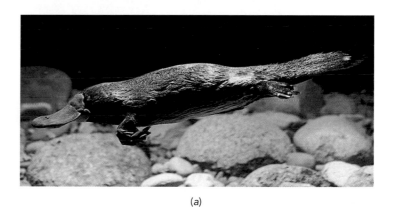

(a)

(b)

Figure 22.16 Monotremes (class Mammalia). (a) Duckbilled platypus. (b) Spiny anteater (echidna) from Australia.

leathery shells, similar to reptile eggs. The platypus generally lays one egg and incubates it in a nest. The spiny anteater generally lays two eggs and incubates them in a pouch. When the young hatch, they feed on milk produced by the mother's specialized sweat glands.

**Marsupials** are mammals in which the young are born early in their development and are retained in a pouch. After birth, the embryos crawl to the pouch and nurse there until they mature. The kangaroo and koala are familiar examples of marsupials (**Figure 22.17**).

In **placental mammals**, the young develop to maturity within the mother. They are named for the first organ to form during the course of their embryonic development, the placenta. (Human development as an example of the development of a placental mammal is described in detail in Chapter 36.) Placental mammals are extraordinarily diverse, as you can see from just the three examples shown in **Figure 22.18**.

(a)  (b)

**Figure 22.17** Marsupials (class Mammalia). (a) Eastern gray female kangaroo with a young kangaroo in her pouch. (b) Koala with young.

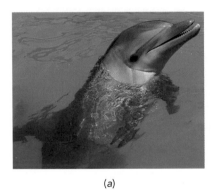
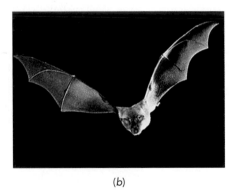

(a)  (b)  (c)

**Figure 22.18** Placental mammals (class Mammalia). (a) Bottle-nosed dolphin. (b) Greater horseshoe bat in flight. (c) Gray timber wolf.

# CHAPTER REVIEW

## Summary of Key Concepts and Key Terms

### What are chordates?

**22.1** The **chordates** (p. 375) are a phylum of animals characterized by four main features: (1) a single, hollow **nerve cord** (p. 375) located along the back; (2) a rod-shaped **notochord** (p. 375), which forms between the nerve cord and the gut during development; (3) **pharyngeal (gill) arches** and **slits** (p. 375), which are located at the throat (pharynx), and (4) a **post-anal tail** (p. 375). Some of these features may be present only in the embryo or fetus.

**22.1** There are three subphyla of chordates: the tunicates, the lancelets and the vertebrates.

### What are the major chordate subphyla and their distinguishing characteristics?

**22.2** The **tunicates** (subphylum Urochordata; p. 377) are sessile, saclike marine organisms that filter food from the surrounding water.

**22.2** Although adult tunicates have gill slits, only the larvae have a notochord and nerve cord.

**22.3** The **lancelets** (subphylum Cephalochordata; p. 378) are tiny, scaleless, fishlike marine chordates that are just a few centimeters long.

**22.4 Vertebrates** (subphylum Vertebrata; p. 378) are characterized by a vertebral column surrounding a dorsal nerve cord.

**22.4** Vertebrates have a distinct head with a skull that encases the brain, a closed circulatory system, and a heart to pump the blood.

**22.4** Most vertebrates also have a liver, kidneys, and endocrine glands.

**22.4** In spite of many similarities, the vertebrates are an extremely diverse group of organisms.

**22.4** Four classes of vertebrates are fishes, and four are tetrapods (animals with four limbs).

## What are the major vertebrate classes and their distinguishing characteristics?

**22.5** The **jawless fishes** (p. 379)—lampreys and hagfishes—are tubelike animals that lack paired fins and scales, and live in the sea or in brackish water.

**22.5** Lampreys are parasites of other fishes, and hagfishes are scavengers.

**22.5** To reproduce, jawless fishes deposit eggs and sperm directly into the water, where **external fertilization** (p. 379) takes place.

**22.6** The **cartilaginous fishes** (p. 379)—sharks, skates, and rays—have a cartilaginous skeleton, are covered with toothlike scales, and have sophisticated and diverse sensory systems.

**22.6** Sharks have a **lateral line system** (p. 380), a complex system of mechanoreceptors that lie in a single row along the sides of the body and in patterns on the head, as do other fishes and some aquatic amphibians.

**22.6** Aquatic organisms show various adaptations regarding **osmoregulation** (p. 380), the control of water movement into and out of their bodies; the cartilaginous fishes maintain solute concentrations close to that of seawater by retaining urea, which is their method of osmoregulation.

**22.6** The cartilaginous fishes have developed a method of **internal fertilization** (p. 380), in which the male inserts his clasper into the female's cloaca, the terminal part of the gut into which the reproductive system opens.

**22.6** Many cartilaginous fishes are **ovoviviparous** (OH-vo-vye-VIP-uh-rus, p. 381), retaining fertilized eggs within their oviducts until the young hatch.

**22.6** Many skates and rays are **oviparous** (oh-VIP-uh-rus, p. 381), meaning that they lay eggs, and the young hatch from the eggs outside of the mother.

**22.6** A few sharks are **viviparous** (vye-VIP-uh-rus, p. 381); they bear live young, as do ovoviviparous organisms, but the developing embryos derive nourishment primarily from the mother and not from the egg.

**22.7** The vast majority of fishes are the **bony fishes** (p. 381).

**22.7** Along with having a bony internal skeleton, the bony fishes have thin, bony, platelike scales.

**22.7** Bony fishes have a protective flap called the **operculum** (p. 381) that extends posteriorly from the head, protects the gills, and enhances the flow of water over the gills.

**22.7** Bony fish are divided into two classes: the ray-finned fishes and the lobe-finned fishes.

**22.7** Only seven species of lobe-finned fishes exist today, and most are lungfish; amphibians are thought to have evolved from a group of lobe-finned fishes.

**22.7** Ray-finned fishes have swim bladders to regulate buoyancy; this feature plus the ability to move water across their gills using their operculum, lets them remain motionless at various depths.

**22.7** Both the bony fishes and the cartilaginous fishes have a two-chambered heart that pumps blood to the gills; from there, the blood moves sluggishly around the body.

**22.7** Osmoregulation in bony fishes differs in marine (hypoosmotic) species and fresh water (hyperosmotic) species.

**22.7** Most bony fishes are oviparous, fertilizing their eggs externally.

**22.8** The **amphibians** (p. 383) live both in water and on land; they include salamanders, frogs, and toads.

**22.8** Because amphibian eggs have no shells or membranes to keep them from drying out, amphibians lay their eggs directly in water or moist places.

**22.8** The young of frogs and toads undergo change from larval to adult forms during development.

**22.8** The adults of certain salamanders live permanently in the water and retain gills and other larval features as adults.

**22.8** Although amphibians have a three-chambered heart, oxygenated and deoxygenated blood mix in the heart.

**22.9 Reptiles** (p. 385) include alligators and crocodiles, turtles and tortoises, and lizards and snakes.

**22.9** Reptiles are better adapted to life on land than the amphibians because of their dry, scaly skin that retards water loss and their shelled (amniotic) egg.

**22.9** In the reptile heart, the ventricle is partially subdivided.

**22.9** The **amniotic egg** (p. 385) retains a watery environment within the egg while protecting and nourishing the developing embryo.

**22.9** Fishes, amphibians, and reptiles are **ectothermic** (p. 386); that is, they regulate body temperature by taking in heat from the environment.

**22.9** Ectothermic animals protect themselves from the cold and high heat in behavioral ways.

**22.10 Birds** (p. 386) are winged vertebrates that are covered with feathers and are adapted to flight.

**22.10** Birds lay amniotic eggs like the reptiles but have a four-chambered heart like the mammals.

**22.10** Birds, like mammals and unlike reptiles, amphibians, and fishes, are **endothermic** (p. 386); that is, they regulate body temperature internally.

**22.11 Mammals** (p. 387) are vertebrates that have hair, and their females secrete milk from mammary glands to feed their young.

**22.11 Monotremes** (p. 388) are mammals that lay eggs having leathery shells.

**22.11 Marsupials** (p. 388) are mammals in which the young are born early in their development, are retained in a pouch, and crawl to the pouch after birth, nursing there until they mature.

**22.11** In **placental mammals** (p. 388), the young develop to maturity within the mother.

---

## Level 1    Learning Basic Facts and Terms

**Multiple Choice**

*Directions*: For questions 1–3, choose from the following classes of vertebrates:

   I. Mammalia     IV. Reptilia
   II. Aves        V. Osteichthyes
   III. Amphibia    VI. Chondrichthyes

1. On a hike in the outback of Australia you discover partial remains of a extremely decomposed animal. Its skeleton appears to be made mostly of sturdy bone. The animal appears to have died protecting a nest of shelled amniotic eggs. The animal had teeth and scales. This animal could be
   a. I only         c. I or IV         e. I, II, III, or IV
   b. II only        d. IV only

2. Your biology professor describes an animal with a two-chambered heart that gives birth to live young. This animal could be classified as
   a. III only
   b. III, V, or VI
   c. III, IV, or VI
   d. V or VI
   e. I, II, or III

3. Classify an animal that is ectothermic, has gills, and maintains an internal solute concentration that is roughly equal to that of its surroundings.
   a. III only
   b. III, V, or VI
   c. IV only
   d. VI only
   e. III, IV, V, or VI

4. Which of the following is an invertebrate chordate?
   a. shark
   b. lamprey
   c. lancelet
   d. octopus
   e. None of the above; all chordates are vertebrates.

5. Which of the following is considered an adaptation that supports a terrestrial existence?
   a. feathers
   b. being ovoviviparous
   c. shelled amniotic egg
   d. notochord
   e. lateral line

6. Which of the following characteristics is unique to birds?
   a. shelled amniotic eggs
   b. four-chambered heart
   c. feathers
   d. endothermy
   e. flight

7. The lateral line system of bony fishes and sharks functions in
   a. osmoregulation.
   b. hydrodynamics.
   c. insulation.
   d. gas exchange.
   e. sensory perception.

## Level 2 │ Learning Concepts

1. The skeleton of hagfish is cartilaginous just like that of a shark. Why aren't the hagfish classified with the Chondrichthyes?

2. Some species of sharks and most mammals are viviparous. How is the development of the embryo within the mother different between these two groups of vertebrates?

3. Why is "cold-blooded" not a very good term to use to describe regulation of body temperature in reptiles? Why is "warm-blooded" not a very good term to describe mammals?

4. Compare and contrast osmoregulation in sharks and bony fishes.

5. Distinguish among ovoviviparous, oviparous, and viviparous. In which groups of vertebrates would you find each?

6. The term *amphibian* means "both lives," referring to the notion that amphibians spend part of their life living in water and part living on land. However, the term *amphibian* does not accurately describe the lives of all species within this group. Explain why.

7. Describe two important adaptations of reptiles to life on land.

8. What characteristics differentiate mammals from other vertebrates?

9. Compare and contrast the structure of the heart in fishes, amphibians, reptiles, and birds. Which is the most efficient and why?

## Level 3 │ Critical Thinking and Life Applications

1. On your trip down the California coast you stop to do some tide pooling. In these marine tide pools you find several organisms. Put the following in order of their relatedness to humans: octopus, sea squirt, sea star, sponge, skate (you found the egg case in the tide pool), flatworm, rockfish (a bony fish), and jellyfish. The animal listed first should share a common ancestor with humans in the most recent past. The animal listed last should share a common ancestor with humans in the most distant past.

2. Most species of snakes inhabit tropical and subtropical countries. Based on information from this chapter relating to the physiology of reptiles, why do you think this is so? Why don't snakes live in the arctic?

3. Bony fishes are far more diverse than cartilaginous fishes. What adaptations do bony fishes possess that allow them to occupy a wider diversity of habitats than cartilaginous fishes?

4. *Archaeopteryx*, a 150-million-year old fossilized animal, had forelimbs with claws and scales, cone-shaped teeth, a long tail containing vertebrae, and feathers. This fossil is one of many that demonstrates the evolutionary relationship between which two groups of vertebrates?

## In The News │ Critical Thinking

**STICKY FEET**

Now that you understand more about chordates, including the vertebrates, reread this chapter's opening story about Autumn's research on the stickiness of gecko feet. To better understand this information, it may help you to follow these steps:

1. Review your immediate reaction to this research on the stickiness of gecko feet that you wrote when you began reading this chapter.

2. Based on your current understanding, again summarize the main point of the research in a sentence or two.

3. What questions do you now have about this information that this chapter's opening story does not answer?

4. Collect new information about the research. Visit the *In The News* section of this text's companion Web site at www.wiley.com/college/alters and watch the "Sticky Feet" video. Then use the "summary" link to read the accompanying story and access related links. Use this information, the links provided, and other online and library resources to answer your questions and find updates about this topic. State the sources of your information. Explain why you think the information is accurate. Also determine whether the information expresses a particular point of view or is biased in any way.

5. What in your view is the most significant aspect of this information? Give reasons for your opinions and for any changes in your ideas based on the additional information you have collected and the analysis you have done.

# PATTERNS OF REPRODUCTION IN PLANTS

## In The News | Purely Organic

Plants and insects exhibit a variety of interspecies relationships. Some insects, such as honeybees, are vital to sexual reproduction in plants. While the bees are collecting nectar or pollen from flowers to use as food, they are also transporting the male sex cell contained in the pollen from one plant to another. Other insects, however, eat more of the plant than just the nectar. They may eat the entire plant, or major portions of it.

Most farmers use synthetic pesticides to protect their crops and may use synthetic fertilizers to boost growth. Other farmers, however, cultivate "organic foods," those grown without the use of synthetic pesticides and fertilizers. Although organic foods usually cost more than their conventionally grown counterparts, many Americans purchase organic foods regularly. What factors influence their decision?

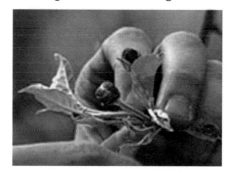

Advances in modern (conventional) farming technology have dramatically increased crop yields, but they also have produced some negative effects on the environment, pose health risks to farm workers who handle pesticides, and leave traces of chemical residues in food. The long-term health effects of ingesting pesticide residues are unknown. Such concerns have convinced a growing number of Americans to eat more organically grown foods.

In the past, many agricultural experts contended that organic farming systems were not viable alternatives to conventional farming systems because they were more costly to run and not productive enough to be sustainable.

To determine the sustainability of organic farming systems for apple production, John Reganold and his colleagues at Washington State University compared various factors, including fruit yield, taste, and profitability, in the farming of Golden Delicious apples. The scientists grew the apple trees under three experimental conditions that incorporated an organic, a conventional, and an integrated system of production. The integrated system relied on pest control methods and fertilizers that were used in both organic and conventional systems.

After six years of collecting data on the trees and on the fruit harvested from them, the investigators determined that fruit yield, pest infestation, and tree growth were similar among the three systems. Compared to the conventional system, however, the organic and integrated systems produced better soil quality. People who served as taste-testers rated apples grown in the organic system as sweeter than apples produced in the two other systems. Furthermore, the results indicated that organic farming was the most sustainable system for producing Golden Delicious apples, particularly when consumers were willing to pay a premium for the organically grown fruit. To learn more about organic farming and Reganold's research, visit the *In The News* section of this text's companion Web site at www.wiley.com/college/alters and view the video "Purely Organic."

*Write your immediate reaction to the research of Reganold and his colleagues on the sustainability of organic apple farming: first, summarize the main point of the information in a sentence or two; then suggest what you think its significance is. You will have an opportunity to reflect on your responses and gather more information on this topic in the* In The News *feature at the end of this chapter. In this chapter, you will learn more about plant reproduction, including the reproduction of angiosperms such as apples.*

# CHAPTER GUIDE

## What are the general patterns of reproduction in plants?

**23.1** Plants have two multicellular generations that alternate to create their life cycles.

**23.2** In nonvascular plants, the gametophyte generation is dominant.

**23.3** In vascular plants, the sporophyte generation is dominant.

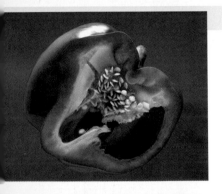

## How does pollination take place?

**23.4** Animals and wind help pollinate plants.

## How do seeds develop, become dispersed, and germinate?

**23.5** Seeds develop from ovules, and fruits develop from ovary walls.

**23.6** Animals, wind, and water help in seed dispersal.

**23.7** Germination begins when the seed takes up water and begins to sprout.

## What are types of vegetative propagation in plants?

**23.8** Underground runners and stems are two types of vegetative propagation.

**23.9** Using cell culture techniques, scientists can grow plants from individual cells.

# What are the general patterns of reproduction in plants?

## 23.1 Plants have two multicellular generations that alternate to create their life cycles.

The events that take place from one stage in the life of an organism, through a reproductive phase, to that same stage in the next generation is called a **life cycle**. Animals, plants, some protists, and some fungi have sexual life cycles, which are characterized by the alternation of meiosis and fertilization as described for the human life cycle. **Figure 23.1a** shows this type of sexual life cycle, which is common to most animals. Notice in the diagram that during the animal life cycle, the multicellular organism has a double set (2N) of hereditary material—the usual complement of chromosomes for that species; it is **diploid** (DIP-loyd). One set of chromosomes was contributed by a father and one set by a mother. This diploid, multicellular phase is the dominant and noticeable phase of the animal life cycle. This is also true in vascular plants. Gametes (eggs and sperm) represent the haploid phase of the life cycle; they are individual, haploid cells. **Haploid** (HAP-loyd) cells have one set (N) of hereditary material, or *half* the usual number of chromosomes for that species.

Plants and some species of algae have sexual life cycles unlike those of animals. They have both a multicellular haploid phase and a multicellular diploid phase. Because both phases are multicellular, this type of life cycle is called **alternation of generations** (Figure 23.1b). The alternating generations of plants are called the sporophyte (spore-plant) generation and the gametophyte (gamete-plant) generation.

**Gametophytes** form gametes. Because they are haploid, they form haploid gametes by the process of mitosis. Eggs are formed in female reproductive structures called **archegonia** (AR-kih-GO-nee-uh; sing. **archegonium**). The archegonium is flask-shaped and contains a single egg. Sperm are produced in structures called **antheridia** (AN-thuh-RID-ee-uh; sing. **antheridium**). The egg and sperm fuse during fertilization. The diploid plants produced from fertilized eggs are the **sporophytes**. They use the process of meiosis to produce haploid spores. In the nonvascular plants and several phyla of vascular plants, spores are borne in spore cases called **sporangia** (spoh-RAN-jee-uh; sing. **sporangium**). When dispersed, spores grow into haploid gametophyte plants.

The **gametophyte (haploid) generation** usually dominates the life cycles of the nonvascular plants such as mosses, whereas the **sporophyte (diploid) generation** dominates the life cycles of the vascular plants such as ferns, conifers, and flowering plants. The dominance of the sporophyte generation in vascular plants reflects their adaptation for life on land as they evolved from earlier, nonvascular forms (see Chapter 20). Other adaptations to land by vascular plants include the development of spores with protective walls able to tolerate dry conditions and efficient water and food-conducting systems (see Chapter 24), as well as gametophyte generations that became protected by and nutritionally dependent on the sporophyte generation.

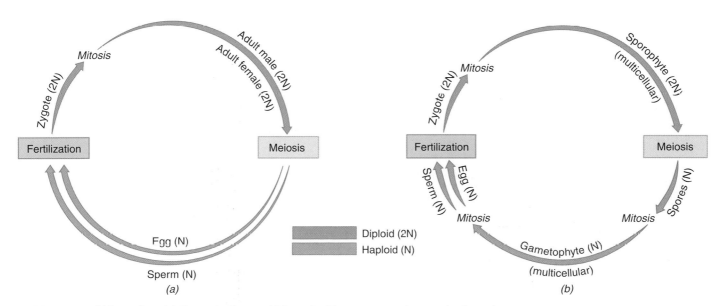

**Figure 23.1 Sexual life cycles.** (a) The animal sexual life cycle. The organism that results from the union of haploid gametes is diploid. Only the diploid phase is multicellular. (b) A generalized plant life cycle showing the alternation of generations. The sporophyte generation (diploid) alternates with the gametophyte generation (haploid). Both the haploid and diploid phases are multicellular.

## 23.2 In nonvascular plants, the gametophyte generation is dominant.

The nonvascular plants comprise three phyla of small, low-growing plants that are commonly found in moist places: mosses, liverworts, and hornworts. In all three, the gametophyte is the dominant generation, larger than the sporophyte and nutritionally independent of it. Conversely, the sporophyte is nutritionally dependent on and grows out of the gametophyte.

### Mosses

The largest phylum of nonvascular plants and probably the one most familiar to you includes the **mosses** (phylum Bryophyta). The gametophytes of most mosses have small, simple leaflike structures often arranged in a spiral around stemlike structures. Their "stems" and "leaves" are different from the stems and

(a)

(b)

**Figure 23.2** *Sphagnum,* peat moss. (a) The glistening black, round objects are spore cases. The spore cases of *Sphagnum* have lids that blow off explosively, releasing the spores. (b) Peat bog in Scotland. Cut sections have been laid out to dry.

leaves of vascular plants; these differences are described in Chapter 24. In **Figure 23.2a** you can see the green, leafy gametophyte generation. The sporophyte generation rises out of the gametophyte.

There are some 10,000 species of mosses, and they are found almost everywhere on Earth. One kind of moss that is important economically is *Sphagnum.* This moss grows in boggy places, forming dense and deep masses that are often dried and sold as peat moss (Figure 23.2b). Peat moss is used in gardening as mulch to protect the roots from temperature fluctuations and to retain moisture, control weeds, and enrich the soil. Its tissues have special water storage cells that allow the peat to absorb and retain up to 90% of its dry weight in water. Dried peat moss can also be used as fuel.

The life cycle of the hairy-cap moss (*Polytrichum*) is typical of the mosses (**Figure 23.3**). A slender sporophyte stalk grows out of the parent gametophyte **1**. This stalk is initially green, but its chlorophyll disintegrates as it matures, leaving the stalk yellow or brown. At this stage, the sporophyte stalk derives its nourishment from the gametophyte. Each sporophyte stalk bears a sporangium (spore capsule) near its tip. Haploid spores are produced by meiosis within this capsule. When the top of the sporangium pops off, the spores are freed **2**. Under the proper conditions, these spores **germinate** (begin to grow) into threadlike filaments; the characteristic leafy gametophytes arise from buds that form on these filaments **3**. The flask-shaped archegonia **4**, each with one egg, are found among the top "leaves" of the female gametophyte plants. Antheridia **5** are found in a similar place in the male gametophytes. Each antheridium produces many sperm. The flagellated sperm swim through drops of water from rain, dew, or other sources into the neck of the archegonium and then to the egg. After fertilization takes place, the zygote **6** develops into a young sporophyte embryo within the archegonium **7**. The cycle begins anew when the embryo then grows out of the archegonium and differentiates into a slender sporophyte stalk.

### Liverworts and hornworts

**Liverworts** (phylum Hepatophyta) were named in medieval times when people believed that plants resembling particular organs of the body could be used to treat diseases of those organs. Some liverworts are shaped like a liver and were thought to be useful in treating liver ailments. The ending *-wort* simply means "herb."

A well-known example of a liverwort is *Marchantia.* When conditions are moist, motile sperm from the antheridia reach eggs within archegonia. The zygotes develop into sporophytes. **Figure 23.4** shows a female gametophyte with archegonia elevated on stalks. Fertilized eggs within the archegonia have developed into sporophytes, which grow embedded within archegonial tissues. Their sporangia are dark spheres under the umbrellalike archegonia. Spores are freed from these structures and germinate into gametophyte plants.

The **hornwort** (phylum Anthocerophyta) is so named because it has elongated sporophytes that protrude like horns from the surface of the creeping gametophytes (**Figure 23.5**). The life cycle of the hornwort parallels quite closely the life cycles of the mosses and liverworts.

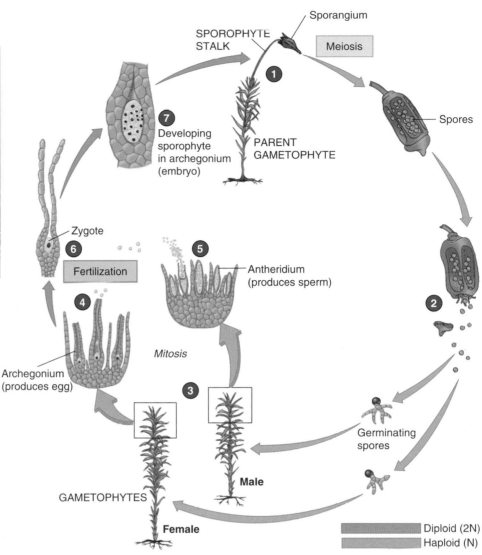

**Figure 23.3 Life cycle of a moss.** The sporophyte stalk on the parent gametophyte (top) produces spores, which are released and germinate into male and female gametophyte plants. These plants produce eggs and sperm, which, after fertilization, develop into sporophyte stalks.

**Visual Thinking:** In mosses, the sporophyte stage of the life cycle is nutritionally dependent on the gametophyte. The gametophyte is nutritionally independent of the sporophyte. How can you tell that these statements are true from this illustration?

**Figure 23.4 Liverwort, *Marchantia*.** The green, leafy part is the gametophyte. Antheridia and archegonia develop in the heads of the stalks that rise above the low-growing gametophyte.

**Figure 23.5 Hornwort, *Antheros*.** The gametophyte is on the rock. The sporophyte grows from the gametophyte and looks stalk-like. The long, thin sporangium splits open when mature, releasing spores that develop into gametophyte plants.

## 23.3 In vascular plants the sporophyte generation is dominant.

### Whisk ferns, club mosses, horsetails, and ferns do not form seeds

As listed in Table 20.1, the members of only four phyla of vascular plants do not form seeds. **Seeds** contain sporophyte embryos; they protect the embryonic plant from drying out or being eaten when it is at its most vulnerable stage. Seeds also contain stored food for the new plant. The seedless plants overcome these problems in ways explained next. **Figure 23.6** diagrams the life cycle of a fern, a representative member of the **seedless vascular plants**.

When a fern plant is mature ❶, it produces spores by meiosis. Each cluster of sporangia ❷ looks like a dot on the underside of the fern leaf, or *frond*. These clusters of sporangia in ferns are called *sori* (sing. *sorus*). Because it produces sporangia and spores, the fern is the sporophyte generation—the dominant form in the life cycle of seedless vascular plants. After its spores are dispersed ❸, those that settle in a moist environment will **germinate**, or sprout, into haploid plants that look very *unlike* familiar ferns. Each plant is a small, ground-hugging, heart-

shaped gametophyte called a *prothallus* (pl. *prothalli*), which is anchored to the ground by filaments called *rhizoids* ❹. Their antheridia and archegonia ❺ are protected somewhat because they are located on the underside of the plant. Sperm ❻, when released, swim through moisture collected on the underside of the gametophyte to the archegonia. Each fertilized egg (zygote) starts to grow within the protection of the archegonium and develops into an embryo ❼. After this initial protected phase of growth, the fern sporophyte is able to grow on its own and becomes much larger than the gametophyte ❽.

### Conifers, cycads, ginkgoes, and gnetophytes form naked seeds

Four phyla of vascular plants have naked (exposed) seeds: the conifers, cycads, ginkgoes, and gnetophytes. This group is called the **gymnosperms** (JIM-no-spurmz), a name derived from Greek words meaning "naked seed." The seeds of gymnosperms are not

**Figure 23.6** Life cycle of a fern. Mature ferns (shown at the top of the cycle and in the photo) are the dominant, sporophyte generation of the plant.

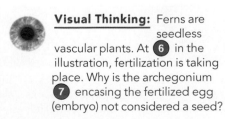

**Visual Thinking:** Ferns are seedless vascular plants. At ❻ in the illustration, fertilization is taking place. Why is the archegonium ❼ encasing the fertilized egg (embryo) not considered a seed?

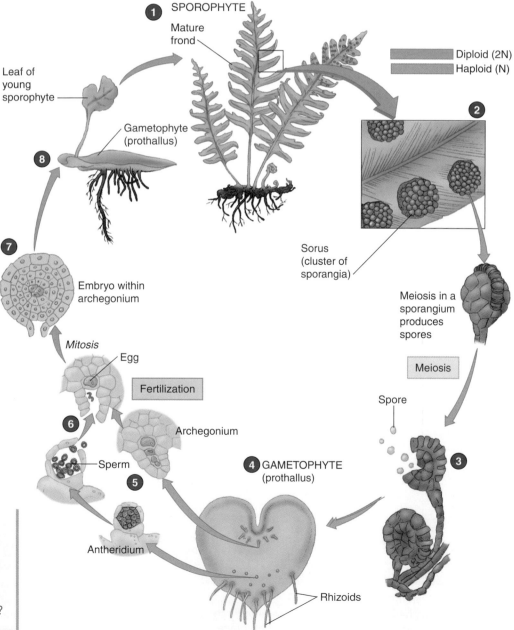

completely enclosed by the tissues of the parent, in contrast to angiosperms.

The most familiar phylum of gymnosperms includes the conifers, or cone-bearing trees, such as pines, spruces, firs, redwoods, and cedars. **Figure 23.7** diagrams the life cycle of a pine.

The pine tree is the sporophyte (diploid) generation. Pines and most other conifers bear both male cones and female cones ❶ on the same tree. You can tell them apart because the female cones are larger and often grow on the lower branches of the tree, with the male cones growing on higher branches. In some pines, the male cones are borne on the ends of the branches, while the female cones grow closer to the trunk.

Both types of cones produce spores. In the male cones, the microsporocytes ❷ undergo meiosis to form pollen ❸. Each pollen grain is a gametophyte that arose from a haploid spore. The male gametophytes produce sperm, which remain in the pollen grains. In the female cones, the megasporocytes ❷ undergo meiosis to form

the female gametophtyes containing egg cells ❸. The lives of gametophytes are comparatively short, and, unlike the gametophytes of nonvascular plants, they are dependent on the sporophyte.

Eggs develop within protective structures called *ovules* within the scales of the female pinecones. In the spring, the male cones release their pollen, which is blown about by the wind ❹. As some of this pollen passes by female cones, a sticky fluid produced by open female cones traps it there. As this fluid evaporates, the pollen is drawn further into the cone. When the pollen comes into contact with the outer portion of the ovule, it forms a pollen tube that slowly makes its way into the ovule—to the egg ❹.

After 15 months, the tube reaches its destination and discharges its sperm. Fertilization takes place, producing a zygote ❺. The development of the zygote into an embryo takes place within the ovule, which matures into a seed ❻. Eventually, the seed falls from the cone ❼ and germinates as the embryo resumes growing and becomes a new pine tree ❽.

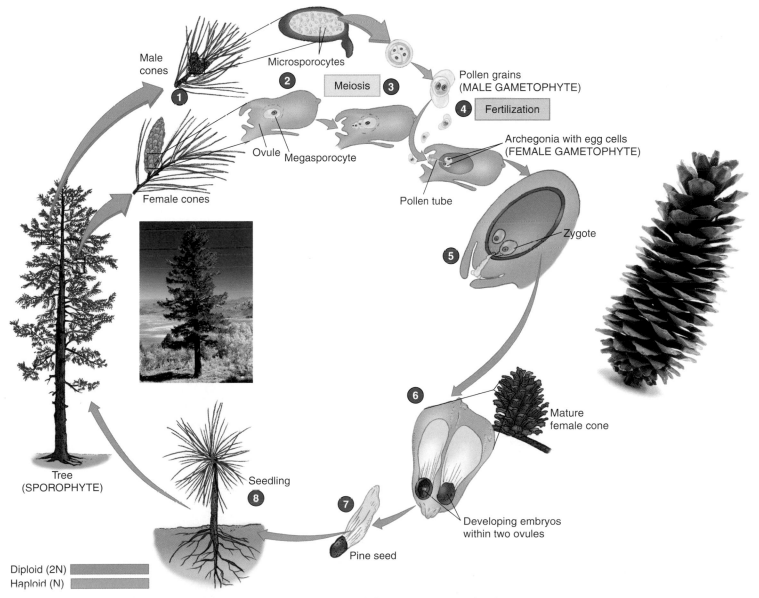

**Figure 23.7 Life cycle of a pine.** Mature pine trees (shown at the left in the cycle) are the dominant, sporophyte generation of the plant.

## Angiosperms form protected seeds

When you think of plants, you likely think of **angiosperms** (AN-jee-oh-spurmz), the flowering plants. Angiosperms are vascular plants with protected seeds, and they are the dominant plants in the world today. There are over 250,000 species of angiosperms. Many of these plants provide products that are critically important to the survival of humans, as well as some products that make life easier or more pleasant (**Table 23.1**).Angiosperms provide food, shelter, and nesting sites for other members of the animal kingdom as well.

Why are angiosperms the dominant plants on Earth today? Among the many traits contributing to their great success are (1) the evolution of the flower for sexual reproduction, (2) double fertilization, and (3) enclosure of the ovule within a fruit, characteristics that we will discuss in this section as well as in Section 23.5.

Flowers provide humans with visual and olfactory delight, but their colors, structure, and aroma promote their real function—not as people pleasers—but as organs of sexual reproduction. In the flower, gametes unite and develop into an embryo within the seed. These plants bear seeds within a fruit; therefore, the seeds are "protected."

Angiosperm flower parts are modified leaves. The outermost whorl or ring of modified leaves is comprised of sepals. Sepals enclose and protect the growing flower bud. In flowers such as roses, the sepals remain small, green, and somewhat leaflike. In other flowers, such as tulips, the sepals become colored and look like the petals, the next whorl of flower parts.

Prominent and colorful flower petals attract pollinating animals, especially insects (see Section 23.6). Angiosperms (such as grasses) that are primarily pollinated by the wind—rather than by insects and other animals—may lack sepals and petals, but they retain the sex organs of the flower. The sex organs are the innermost whorls of modified leaves—the male **stamens** and the female **pistil**.

In the life cycle of angiosperms, the conspicuous flowering plant is the sporophyte (**Figure 23.8** **1a**). Each stamen of the sporophyte consists of a supporting filament and an anther **2a**, a compartmentalized structure in which haploid microspores **3a** are produced by meiosis from diploid cells. Haploid pollen grains **4a** are produced from these microspores by mitosis. Each pollen grain is a male gametophyte enclosed within a protective outer covering. This gametophyte produces sperm by mitosis, which remain enclosed within the pollen grain.

The pistil consists of three parts: stigma, style, and ovary. At its tip is a sticky surface called the stigma to which pollen grains can adhere **1b**. The style is a narrow stalk arising from the top of the ovary that bears the stigma. *Pollination* in the flowering plants is the transfer of pollen from an anther to a stigma. At the base of the pistil is the ovary, a chamber that completely encloses and protects the ovules **2b**. Within each ovule, a single megaspore develops and then divides meiotically, producing four cells **3b**. One of these cells develops into a female gametophyte. When mature, the female gametophyte is called an embryo sac **4b**. Within this sac are typically eight cells, one of which is the egg. The ovule will become a seed when its egg is fertilized by the sperm from the male gametophyte.

After a pollen grain lands on the stigma, it produces a long pollen tube that grows from the pollen grain down the style and penetrates the ovary, entering an ovule **5**. One of two haploid sperm nuclei in the pollen tube fertilizes the haploid egg nucleus in the ovule, producing a diploid zygote **6**. The zygote will become a new plant embryo. The other sperm nucleus fuses with two other nuclei of the embryo sac, producing a triploid endosperm nucleus. The endosperm nucleus develops into tissue that will feed the embryo as it grows into a plant. The ovule becomes the seed within which the embryo develops, and the ovary ripens into a fruit **7**. The fruit is a food that attracts animals, which play a role in seed dispersal. The seed will eventually germinate under favorable conditions, and the embryo will grow into a new plant **8**.

## Dead plants formed coal

Many of the seedless vascular plants that lived about 270 million years ago were converted long ago to a fuel we use today—coal. Club mosses that grew on trees, horsetails, ferns, and tree ferns made up great swamp forests during the Carboniferous period (see Chapter 18). Dead plants did not completely decay in stagnant, swampy waters, and they accumulated. Ocean waters later covered these swamps, and marine sediments piled on top of the plant remains. Pressure and heat acted on the layers of dead plant material beneath the ocean floor and converted the remains to coal. When you burn fossil fuels such as coal, you are burning a resource that was formed under special conditions that have not been repeated in the last 270 million years. Coal is therefore called a *nonrenewable resource*.

| TABLE 23.1 | Products of Angiosperms |
|---|---|
| **Product** | **Example** |
| Food | rice, wheat, and corn |
| Fibers | cotton |
| Medicines | codeine and the cancer drugs vinblastine and vincristine |
| Materials | lumber, rubber |
| Beverages | coffee, tea |
| Oils | perfumes |

### CONCEPT CHECKPOINT

1. What are the similarities and differences between the life cycles of mosses and ferns with respect to the following:
   - Which generation is dominant (most conspicuous)?
   - Does the gametophyte generation exist as a free-living plant separate from the sporophyte?
   - How are the young dispersed?
   - How does the sperm get to the egg?
2. What adaptation in the life cycle of conifers allows them to reproduce without being partially submerged in water?
3. What is the function of the flower in angiosperms?

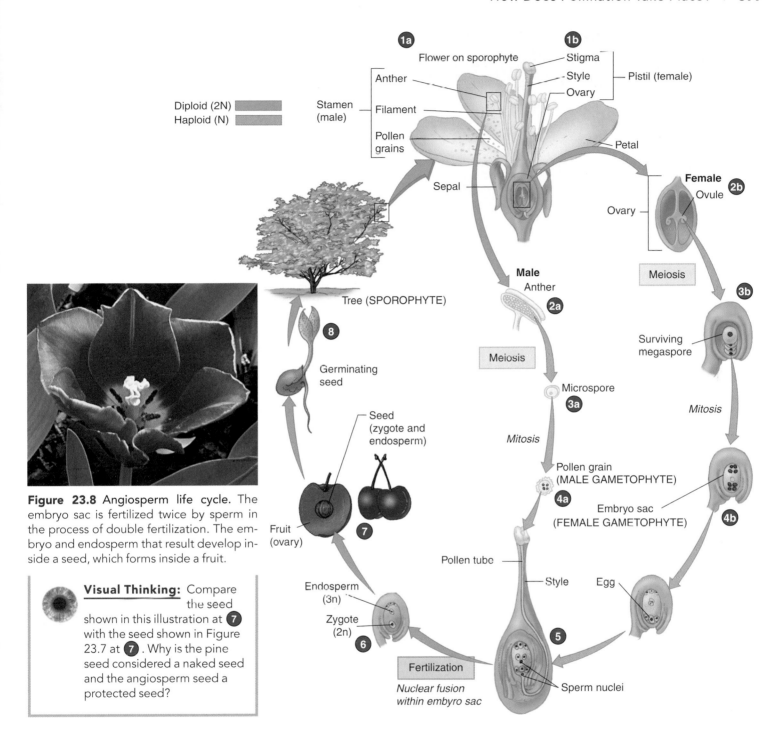

Diploid (2N) ▭
Haploid (N) ▭

**1a** Flower on sporophyte
**1b** Stigma — Style — Ovary — Pistil (female)
Anther
Stamen (male) — Filament
Pollen grains
Sepal
Petal
**Female** **2b** Ovule
Ovary
Meiosis
**3b**
**Male** Anther **2a**
Tree (SPOROPHYTE)
**8**
Meiosis
Surviving megaspore
Germinating seed
Microspore **3a**
Mitosis
Mitosis
Seed (zygote and endosperm)
Pollen grain (MALE GAMETOPHYTE) **4a**
Embryo sac (FEMALE GAMETOPHYTE) **4b**
Fruit (ovary)
**7**
Pollen tube
Style
Egg
Endosperm (3n)
Zygote (2n) **6**
**5**
Fertilization
Sperm nuclei
*Nuclear fusion within embyro sac*

**Figure 23.8 Angiosperm life cycle.** The embryo sac is fertilized twice by sperm in the process of double fertilization. The embryo and endosperm that result develop inside a seed, which forms inside a fruit.

**Visual Thinking:** Compare the seed shown in this illustration at **7** with the seed shown in Figure 23.7 at **7**. Why is the pine seed considered a naked seed and the angiosperm seed a protected seed?

## How does pollination take place?

### 23.4 Animals and wind help pollinate plants.

In certain angiosperms and in most gymnosperms, pollen is blown about by the wind and passively reaches stigmas or female cones. Compared with other vehicles of pollen transfer, however, such as insects or other animals, wind does not carry pollen far or precisely. Therefore, plants pollinated by the wind usually grow close together and produce large quantities of pollen.

Insects and other animals often visit the flowers of angiosperms for nectar, which is rich in sugars, amino acids, and other substances. While they are gathering nectar from a flower, pollen rubs off on their bodies, and when they visit other flowers, they unwittingly deposit the pollen on their stigmas. For plants to be effectively pollinated by animals, a particular insect or other animal must visit many plants of the same species. Flowers have evolved various colors and forms that attract certain pollinators, thereby

**Figure 23.9** Ponderosa pine, *Pinus ponderosa*, shedding pollen. Note the "puffs" of pollen being shed by the male cones borne on the ends of the branches.

(a)                                              (b)

**Figure 23.10** Different modes of pollination. (*a*) As this bumble-bee, *Bombus*, collects nectar from the flame azalea, the stigma contacts its back and picks up any pollen clinging there that the bee might have acquired during a visit to another flower. (*b*) This long-tailed hermit hummingbird is extracting nectar from the flowers of *Helinconia imbricata* in the forests of Costa Rica. There are pollen grains on the tip of the bird's beak.

promoting effective pollination. Yellow flowers, for example, are attractive to bees, whereas red flowers attract birds, but not insects. Insects in turn have evolved a number of special traits that enable them to obtain food efficiently from the flowers of the plants they visit. For example, the copper butterfly has a long, coiled, tongue-like organ that it uses to extract nectar from flowers. Hummingbirds have long curved beaks for the same purpose.

**Figure 23.9** and **Figure 23.10** illustrate these various modes of pollination.

In some angiosperms, the pollen does not reach other individuals at all. Instead, it is shed directly onto the stigma of the same flower. This process is termed *self-pollination*. Mendel did much of his work on garden peas (see Chapter 13) by taking advantage of the architecture of the flower, which traps pollen inside and leads to self-pollination.

# How do seeds develop, become dispersed, and germinate?

## 23.5 Seeds develop from ovules, and fruits develop from ovary walls.

Fertilization of an egg nucleus in the ovule, by a sperm within a pollen grain, produces a diploid zygote (see Figure 23.8 **6** ). This zygote will become a **plant embryo**, or undeveloped plant, housed within the seed. One end of the embryo will form the shoot when the seed germinates, or sprouts. The shoot will develop into the stems, leaves, and flowers of the plant. The other end of the embryo will give rise to the roots.

What makes up the rest of the seed? The other sperm nucleus fuses with two other nuclei of the embryo sac, producing a triploid endosperm nucleus (see Figure 23.8 **6** ). This cell divides quickly, and its daughter cells absorb nutrients from the parent plant. This nutritive **endosperm** feeds the embryo as it grows until the seed germinates and the seedling is able to produce its first true leaves and begin photosynthesis.

While in the seed, the embryo develops one or more **cotyledons** (COT-ul-LEED-uns, seed leaves). **Monocots** are plants with a single cotyledon (see Section 23.7). The monocot seed is packed with endosperm, and the cotyledon absorbs the nutrients and distributes them to the growing parts of the embryo. Examples of familiar monocots are corn and bread wheat. When you eat corn, you are eat-

ing corn seeds (actually, the fruits of the plant), which are packed with starchy endosperm. White flour is made from the endosperm of bread wheat. Bran is the outermost layer of the endosperm. Wheat germ is the embryo of bread wheat.

**Dicots** are plants with two cotyledons. In many dicots, the growing embryo absorbs the nutrients from the endosperm, developing fleshy, food-storing cotyledons. The cotyledons eventually take up most of the seed. Dicots have seeds with two halves, such as beans and peanuts.

A **fruit** is a mature ovary of an angiosperm, consisting of the seeds, the tissues connected with them, and their coverings (**Figure 23.11**). The seeds and the ovary develop at the same time. The ovule becomes the seed within which the embryo develops, and the outer covering of the ovule develops into the **seed coat**, a covering of the seed. The ovary ripens into a fruit. Many fruits are edible and may attract animals that play a role in seed dispersal.

CONCEPT CHECKPOINT

4. Describe three modifications in the life cycle of angiosperms that are lacking in the gymnosperms.

(a)

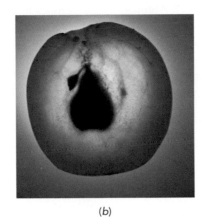

(b)

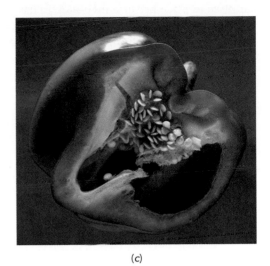

(c)

**Figure 23.11. Fruits.** A fruit is a mature ovary of an angiosperm, consisting of the seeds, the tissues connected with them, and their coverings. (a) In some angiosperms, such as oranges and lemons, many ovules, and therefore many ova, are present in the ovary. (b) In other angiosperms, such as grapes, only one ovule and one ovum is present. (c) You might call this pepper a vegetable, but it is really a fruit.

## 23.6 Animals, wind, and water help in seed dispersal.

Parallel to the evolution of the angiosperm flower, and nearly as spectacular, has been the evolution of the fruit. As fruits are dispersed, so are the seeds within them. Fruits that have fleshy coverings—often black, bright blue, or red—are normally dispersed by birds and other vertebrates. Just as red flowers attract birds, red fruits signal an abundant food supply. By feeding on these fruits, birds and other animals carry seeds from place to place and thus transfer the plants from one suitable habitat to another (**Figure 23.12a**).

(a)

(b)

(c)

(d)

**Figure 23.12 How seeds are dispersed.** (a) This cedar waxwing feeds berries to the waiting young. Birds that eat fruits digest them rapidly so that much of the seed is left intact. What is eliminated from the bird can grow into a mature plant. (b) Dog covered with spiny hedge parsley seeds. (c) The seeds of a coconut, *Cocos nucifera*. One of the most useful plants in the tropics, coconuts have become established even on the most distant islands by drifting in the waves. (d) The seeds of a dandelion, *Pyropappus caroliniana*, are dispersed by the wind. Dandelion fruits are slender and have a single seed; they are barely visible at the base of the stalks on which the umbrellalike hairs grow.

Other fruits, such as snakeroot, beggar ticks, and burdock (Figure 23.12*b*), have evolved hooked spines and are often spread from place to place because they stick to the fur of mammals or the clothes of humans. Others, such as the coconut and those that develop near rivers, ponds, lakes, or oceans, are regularly spread by water (Figure 23.12*c*). Some fruits have wings and are blown about by the wind. The dandelion is a familiar example (Figure 23.12*d*).

---

CONCEPT CHECKPOINT

**5.** Describe the mechanisms of seed dispersal that have evolved in angiosperms.

---

## 23.7 Germination begins when the seed takes up water and begins to sprout.

Did you ever wonder why seeds can remain packaged in a nursery—ready to plant—yet not "go bad?" You may also wonder why they don't sprout in the package. The answer to these questions is that seeds are a dehydrated, dormant stage in the angiosperm life cycle. They remain dormant because they have too little water to carry on the activities of life at more than a subsistence level. Seed germination depends on a variety of environmental factors, especially the presence of water. The availability of oxygen for aerobic respiration in the germinating seed, a suitable temperature, and sometimes the presence of light or disruption of the seed coat are also necessary.

The first step in seed germination occurs when it imbibes (takes up) water. Once this has taken place, metabolism within the embryo resumes. Germination and early seedling growth require the mobilization of food storage reserves within the seed. A major portion of almost every seed consists of food reserves. In dicots, most of the stored food is in the cotyledons, or seed leaves (see Section 23.5). In monocots, most of the food is stored in endosperm.

Monocots and dicots also differ from one another in a number of recognizable features (**Figure 23.13**). Monocots usually have parallel veins (fluid-carrying tissues) in their leaves, and their flower parts are often in threes. Among the monocots are the lilies, grasses, cattails, orchids, and irises (**Figure 23.14a** and **b**). Dicots usually have netlike veins in their leaves, and their flower parts are in fours or fives. The dicots include the great majority of familiar plants: almost all kinds of flowering trees and shrubs and most garden plants, such as snapdragons, chrysanthemums, roses, and sunflowers (Figure 23.14*c* and *d*).

**Figure 23.15** shows germination and seedling development in a monocot, using the example of corn (*Zea mays*), and in a dicot, using the example of a bean (*Phaseolus vulgaris*). Usually, the first portion of the embryo to emerge from the germinating seed is the **radicle** (young root), which anchors the seed and absorbs water and minerals from the soil. Then, the shoot of the young seedling elongates, breaking through the seed coat and emerging from the ground. In most monocots and some dicots, the **epicotyl**, which is the portion of the shoot above the cotyledons, undergoes extensive elongation. As the shoot emerges from the soil as in Figure 23.15*a*, the first true leaves are rolled up and, in grasses, are protected by a straight sheath called a **coleoptile** (KO-lee-OP-tle). In some di-

MONOCOTS

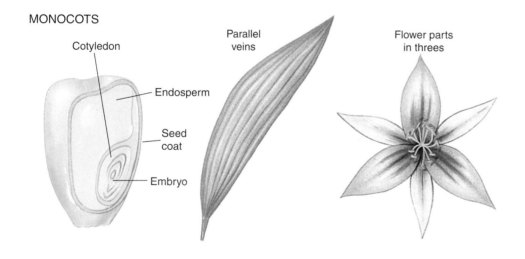

Cotyledon
Parallel veins
Flower parts in threes
Endosperm
Seed coat
Embryo

DICOTS

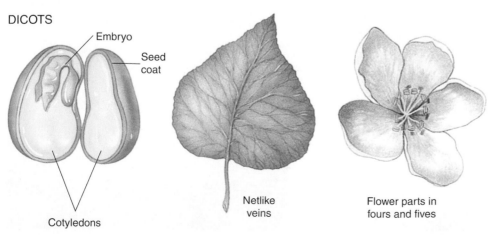

Embryo
Seed coat
Netlike veins
Flower parts in fours and fives
Cotyledons

**Figure 23.13 Monocots and dicots.** (*a*) In monocots, most of the food for the embryo is stored in endosperm. Other characteristics of monocots are the parallel veins in their leaves and the occurrence of their flower parts in threes. (*b*) Dicots store most of their food in cotyledons, or seed leaves. They have netlike veins in their leaves, and their flower parts occur in fours or fives.

**Figure 23.14 Examples of monocots and dicots.** Monocots include the cattail (a) and the crested dwarf iris (b). Both of these plants show the parallel leaf venation typical of monocots. Dicots include the rose (c) and the bloodroot (d), which show the netlike venation typical of dicots.

cots the **hypocotyl**, which is the shoot below the cotyledons, is the first to elongate. It pulls the cotyledons and the enclosed first true leaves up through the soil, protecting them by pushing through first as seen in Figure 23.15b.

During germination in corn, the endosperm is digested and absorbed by the cotyledon, and it is shuttled to the embryo as it grows. The cotyledon remains below ground within the seed during and

after germination. During germination in the bean, the cotyledons stay below the ground as the embryo absorbs their stored food. As the cotyledons emerge from the ground, they are somewhat shriveled, and as their nutrients are drained during the development of the shoot, they wither and fall away. Multiplication of the cells in the tips of the stem and roots, along with their elongation and differentiation, continues the growth of the young seedling.

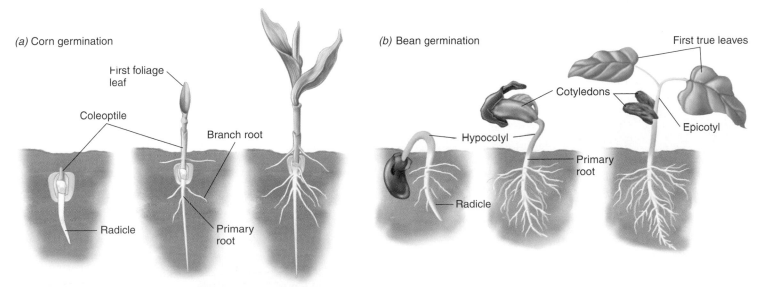

**Figure 23.15 Germination and seedling development.** (a) Stages in the germination of a bean. (b) Stages in the germination of corn.

# What are types of vegetative propagation in plants?

## 23.8 Underground runners and stems are two types of vegetative propagation.

In asexual reproduction, new individuals are generated without the union of gametes. **Vegetative propagation** is an asexual reproductive process in which a new plant develops from a portion of a parent plant. Some plants, such as irises and grasses, produce new plants along underground stems called **rhizomes** (RYE-zomz). Other plants, such as strawberries, have horizontal stems that grow above the ground called runners or **stolons** (STOW-lunz). These plants produce new roots and shoots at nodes (places where one or more leaves are attached) along these stems. New plants can also arise vegetatively from specialized underground storage stems called tubers. A white potato, for example, is a tuber and can grow a new plant from each of its eyes (which are nodes containing lateral buds).

In some plants, new shoots can arise from roots and grow up through the surface of the soil. For example, a group of aspen trees often consists of a single individual that has given rise to a colony of genetically identical trees by producing new shoots from its horizontal roots (**Figure 23.16**).

In a few species of plants, even the leaves can serve as reproductive structures. The plant in **Figure 23.17** is commonly called the maternity plant because small plants arise in the notches along the margins of the leaf. When mature, these small plants drop to the soil and form roots. Gardeners commonly propagate African violets from leaf cuttings and many other plants from stem cuttings.

**Figure 23.17** *Kalanchoe daigremontiana.* The small plants growing in the notches along the leaf margins will drop to the soil and form roots.

## 23.9 Using cell culture techniques, scientists can grow plants from individual cells.

A major breakthrough in the asexual propagation of plants has been the development of **cell culture** techniques (**Figure 23.18**). Using these techniques, scientists are able to remove individual cells from a parent plant and grow these cells into new individuals. These techniques are successful, however, only because individual plant cells (unlike animal cells) have the inherent capability to direct the growth and development of a new plant.

Using cell culture techniques, horticulturalists are able to produce virtually unlimited numbers of genetically identical offspring. These techniques have been particularly useful in propagating plants that are slow to multiply on their own, such as coconut palms and redwoods, and in cultivating varieties of individual plants with special characteristics, such as large flowers. Award-winning varieties of orchids, for example, are often produced in this way. Cell culture is also used for the commercial production of tremendous numbers of plants in a short period of time—such as the chrysanthemums you may buy at the grocery store. The timber industry uses cell culture techniques in developing rapidly growing conifers such as the Douglas fir.

The use of genetic engineering techniques may revolutionize worldwide agriculture. Scientists are able to insert genes that express favorable characteristics into the chromosomes of culture cells, producing plants with those characteristics. Genetic engineering and its applications are described in Chapter 11.

**Figure 23.16 Clones of aspen.** Notice that the aspens appear to be grouped by color. How does this phenomenon occur? Each group of aspens is comprised of clones of one tree. A single tree has given rise to a colony of genetically identical trees by producing new shoots from its horizontal roots. Because all the trees of one colony are genetically identical, their fall foliage is of the same color. The differences in coloration among colonies of trees reflect the genetic diversity among them.

**Figure 23.18** A plant growing in a cell culture. The small shoots in the petri dish are growing from a mass of undifferentiated tissue and are genetic clones of carrots. Researchers use clones and tissue cultures for a variety of purposes such as finding heartier and more insect-resistant varieties of plants.

## just wondering . . .

**I heard that one reason we need to save the tropical rain forests is so that we won't lose plants that could cure diseases. How do plants cure diseases? And how do scientists find these plants?**

Some plants produce chemicals that affect animals. These chemicals can be poisons, such as the blow-gun poison, curare, or the breast cancer drug, taxol. You may have used a medicine derived from plants yourself, because approximately 25% of all prescription drugs dispensed each year in North America were originally derived from flowering plants and ferns. Many of these drugs are now synthesized in the laboratory. Interestingly, the science of botany (the study of plants) was considered a branch of medicine until the early to mid-1800s.

*Ethnobotany* is the study of how people use plants for such purposes as medicine and shelter. Ethnobotanists study the therapeutic use of plants by indigenous peoples that have populated a particular area for many generations and pass their knowledge of "folk medicine" from generation to generation. Scientists also choose to study cultures living in areas where there is great diversity in plant life. Researchers then isolate and study the biologically active chemicals they extract from the targeted plants and determine whether they are indeed active against the type of condition for which they are used by folk healers. Many useful drugs have been developed this way, such as reserpine, which is derived from the Indian snake-

**Figure 23.A Useful drugs have been developed from plants.** (a) Reserpine, a drug used to treat hypertension, is derived from Indian snakeroot (*Rauwolfia serpentina*). (b) Extracts of rosy periwinkle (*Catharanthus roseus*) have provided two cancer drugs.

root (*Rauwolfia serpentina*) and is used to treat hypertension. Two cancer drugs have been extracted from rosy periwinkle (*Catharanthus roseus*). Both plants are shown in **Figure 23.A**.

Another approach researchers use is to study plants that are closely related to those shown to have medicinal value. Closely related plants often have similar biochemical traits. Researchers also observe plants in their natural surroundings and choose to study those plants that remain untouched by herbivores such as insects.

In their laboratories, scientists run automated tests to screen the chemicals they extract from plants. Plant extracts can be quickly screened to see whether they are active against certain types of cancer cells or whether they affect enzymes that are active in particular diseases. After initial screenings, drug companies often take over further analyses to evaluate whether the drug could be a promising addition to their pharmaceutical line.

It takes many years for a plant with suspected healing powers to reach your local pharmacy as a prescription drug. Many never make it. However, the drugs that prove to be useful are extremely important additions to our arsenal against disease. Thus, saving the rain forest, where about two-thirds of the world's plant species are found, not only will simply preserve biological diversity, but also will preserve plants that one day may cure many of our most devastating illnesses. ●

*Are you wondering about a topic in biology and how it relates to your life? Submit your question by clicking the Just Wondering link in this text's companion Web site at www.wiley.com/college/alters.*

## CHAPTER REVIEW

## Summary of Key Concepts and Key Terms

### What are the general patterns of reproduction in plants?

**23.1** Animals, plants, some protists, and some fungi have sexual **life cycles** (p. 393) that are characterized by the alternation of the processes of meiosis and fertilization.

**23.1** Plants and some species of algae have life cycles with both a multicellular **haploid** (p. 393) phase and a multicellular **diploid** (p. 393) phase. Because both phases of the life cycle are multicellular, this type of life cycle is called **alternation of generations** (p. 393).

**23.1** In plant life cycles, multicellular haploid plants called **gametophytes** (p. 393) produce gametes by mitosis.

**23.1** The gametes produced by haploid gametophytes fuse during fertilization and grow into multicellular, diploid spore-producing plants called **sporophytes** (p. 393).

**23.1** Sporophytes produce haploid spores by meiosis. The spores grow into gametophytes.

**23.1** In general, the **gametophyte generation** (p. 393) dominates the life cycles of the nonvascular plants, whereas the **sporophyte generation** (p. 393) dominates the life cycles of vascular plants, a difference that reflects an adaptation of the vascular plants for life on land.

**23.1** In the nonvascular plants and several phyla of vascular plants, spores are borne in spore cases called **sporangia** (spoh-RAN-jee-uh; sing. **sporangium**, p. 393); eggs are formed in female reproductive structures called **archegonia** (AR-kih-GO-nee-uh; sing. **archegonium**, p. 393); sperm are produced in structures called **antheridia** (AN-thuh-RID-ee-uh; sing **antheridium**, p. 393).

**23.2** The nonvascular plants include three phyla: the **mosses** (p. 394), **liverworts** (p. 394), and **hornworts** (p. 394).

**23.2** The life cycles of most nonvascular plants have two distinct phases; the gametophyte phase is dominant.

**23.2** In nonvascular plants, the sporophyte plants live on and derive nutrients from the gametophyte plants

**23.3** The sporophyte generation of the **seedless vascular plants** (such as ferns, p. 396) produces spores, which **germinate** (p. 396) to form gametophytes.

**23.3** Gametophytes of the seedless vascular plants produce motile sperm that swim through water to fertilize the eggs.

**23.3** The fertilized eggs of seedless vascular plants produce young sporophytes that begin to grow while protected within the tissues of the gametophyte, but eventually become free living.

**23.3** In pine trees, an example of plants with naked seeds (**gymnosperms**, JIM-no-spurmz, p. 396), both male cones and female cones of the sporophyte generation produce spores, which undergo meiosis to form male and female gametophytes.

**23.3** The male gametophytes (pollen) of pine trees produce sperm, which remain in the pollen grains.

**23.3** The female gametophytes of pine trees produce eggs, which develop in ovules within the scales of the female pinecones.

**23.3** In the spring, the male pinecones release their pollen grains into the wind. When a pollen grain comes into contact with the ovule, it grows a tube to the egg and discharges sperm.

**23.3** After a sperm reaches an egg in the pinecone ovule and fertilization takes place, the zygote develops into an embryo and the ovule matures into a **seed** (p. 396), enclosing the embryo.

**23.3** The organs of sexual reproduction in flowering plants (**angiosperms**, AN-jee-oh-spurmz, p. 398), are the male **stamens** (p. 398) and the female **pistil** (p. 398) in the flower.

**23.3** In angiosperms, the male gamete-producing cells are pollen grains, and the female gamete-producing cells are in the ovules in the ovary.

**23.3** Angiosperm seeds, formed within the ovary after the union of sperm and egg, protect the young sporophytes as they begin to develop into new plants; endosperm nourishes them during this initial development.

**23.3** The seeds of angiosperms remain within the ovary, which develops into a fruit.

## How does pollination take place?

**23.4** Pollen may be transferred by the wind.

**23.4** Insects (particularly bees), birds, and other animals, are attracted to certain flowers and have evolved special traits enabling them to efficiently gather food while transferring pollen from flower to flower.

## How do seeds develop, become dispersed, and germinate?

**23.5** Fertilization of an egg nucleus in the ovule, by a sperm within a pollen grain, produces a diploid zygote; this zygote will become a **plant embryo** (undeveloped plant, p. 400) housed within the seed.

**23.5** Seeds develop from ovules and fruits develop from ovary walls.

**23.5** Nutritive **endosperm** (p. 400) feeds the embryo as it grows until the seed germinates and the seedling is able to produce its first true leaves and begin photosynthesis.

**23.5** While in the seed, the embryo develops one or more **cotyledons** (COT-ul-LEED-uns, seed leaves, p. 400) that store or absorb nutrients and distribute them to growing parts of the embryo.

**23.5** **Monocots** (p. 400) are plants in which the embryo has a single cotyledon; **dicots** (p. 400) are plants with two cotyledons.

**23.5** A **fruit** (p. 400) is a mature ovary of an angiosperm, consisting of the seeds, the tissues connected with them, and their coverings.

**23.5** The ovule becomes the seed within which the embryo develops, and the outer covering of the ovule develops into the **seed coat** (p. 400), a covering of the seed.

**23.6** Seeds are dispersed by sticking to or being eaten and eliminated by animals, by being blown by the wind, or by floating to new environments across bodies of water.

**23.7** Seeds germinate only when they receive water and appropriate environmental cues.

**23.7** Usually, the first portion of the embryo to emerge from the germinating seed is the **radicle** (young root, p. 402), which anchors the seed and absorbs water and minerals from the soil.

**23.7** In most monocots and some dicots, the **epicotyl** (p. 402), which is the portion of the shoot above the cotyledons, undergoes extensive elongation; as the shoot emerges from the soil, the first true leaves are rolled up and, in grasses, are protected by a straight sheath called a **coleoptile** (KO-lee-OP-tle, p. 402).

**23.7** In some dicots, the **hypocotyl** (p. 403), which is the shoot below the cotyledons, is the first to elongate.

**23.7** Developing embryos in seeds use the food reserves stored in the cotyledons and in the endosperm during initial development.

## What are types of vegetative propagation in plants?

**23.8** Plants can reproduce asexually by **vegetative propagation** (p. 404); a new plant grows from a portion of another plant.

**23.8** Plants can reproduce vegetatively in many ways, including underground runners (**rhizomes**, RYE-zomz, p. 404) and stems (**stolons**, STOW-lunz, p. 404).

**23.9** A major breakthrough in the asexual propagation of plants has been the development of **cell culture** (p. 404) techniques, removing individual cells from a parent plant and growing these cells into new individuals.

---

| Level 1 | Learning Basic Facts and Terms |
|---|---|

**Matching**

Match each of the following characteristics with the appropriate group of plants:

1. _____ Seedless vascular plants
2. _____ Nonvascular plants
3. _____ Vascular plants with naked seeds
4. _____ Vascular plants with seeds enclosed in fruit

   a. Gymnosperms
   b. Angiosperms
   c. Mosses
   d. Ferns

**Fill in the Blank**

5. During the sexual life cycle of plants, gametes are produced through _____ cell division, while spores are produced through _____ cell division.

6. _____ are produced in archegonia; _____ are produced in antheridia.

7. _____ provides nourishment to the embryo in angiosperms.

8. Fruit is the ripened _____ in flowering plants.

9. The male gametophyte of angiosperms is produced in the _____ of the flower, while the female gametophyte is produced in the _____ of the flower.

**Multiple Choice**

10. Which of the following characteristics are *not* shared in common by monocots and dicots?
    a. the ability to propagate asexually
    b. production of endosperm
    c. presence of vascular tissue
    d. production of seeds
    e. pattern of seed germination

**Sequencing**

11. Place the following events in the correct sequence:
    a. The radicle emerges from the seed.
    b. Fertilization occurs.
    c. The shoot emerges from the ground.
    d. An embryo develops.
    e. The seed germinates.

## Level 2    Learning Concepts

1. Why is dispersal so important to plants? How do the dispersal mechanisms seen in primitive plants compare to those of more evolutionarily advanced plants?

2. Draw a generalized diagram of a flower from memory. Label the sepals, petals, stamens, pistil, anther, pollen grains, stigma, ovary, ovules, and style.

3. Ferns produce one type of spore, while conifers and flowering plants produce two different types of spores. What is the result of development of each of these three types of spores?

4. What do the life cycles of conifers and of flowering plants have in common?

## Level 3    Critical Thinking and Life Applications

1. What aspect(s) of the life cycles of nonvascular and seedless vascular plants has limited their ability to colonize many habitats that have been colonized by seed-bearing vascular plants?

**For questions 2–5:** You've become a world-famous naturalist who leads nature hikes in Rocky Mountain National Park. While hiking, some students make the following observations and ask you to give a biological explanation for them.

2. While hiking through a forest of aspens in the fall, a student notices that the trees seem to be growing in groves. All of the trees within each grove appear to be displaying the same fall foliage colors, but other nearby groves display different colors. Explain.

3. When you stop for a lunch break, one student comments that apple trees must expend enormous amounts of energy to encase their seeds in fruit. He wonders why they would do this. Explain.

4. During a hike through a dense pine forest early in the spring, a student brushes against the branches of a tree and notices that a cloud of yellowish dust is released, much of which ends up on her jacket. She wonders what this is. Explain.

5. During a hike in a mountain meadow, a student notices a boggy area carpeted with moss. Upon closer examination, he sees that a small stalk with a little capsule on the end is protruding from many of the mosses. He flicks a few of the capsules with his finger and claims to see tiny "puffs of smoke" released from the capsule. Explain.

6. The flowers of a species of plant attract only one or a few species of animals, and these pollinators may rely completely on the nectar of that species of plant as a food source. Explain why this has been adaptive both for the flowering plants and for their pollinators.

7. California is a state where large forest fires break out nearly every year. As a result, many of the trees that are native to California have adapted to forest fires and can live through a small to moderately intense fire. Some species of conifers, such as giant sequoias, actually use fire in reproduction. Some of their cones open and release their seeds only in response to the high temperatures experienced during a forest fire. Why might this be adaptive?

## In The News    Critical Thinking

**PURELY ORGANIC**

Now that you understand more about angiosperms and plant reproduction, reread this chapter's opening story about the sustainability of organic apple farming. To assist you in better understanding this information, it may help you to follow these steps:

1. Review your immediate reaction to the research regarding organic apple production that you wrote when you began reading this chapter.

2. Based on your current understanding, again summarize the main point of the research in a sentence or two.

3. What questions do you now have about this research that this chapter's opening story does not answer?

4. Collect new information about the research. Visit the *In The News* section of this text's companion Web site at www.wiley.com/college/alters and watch the "Purely Organic" video. Then use the "summary" link to read the accompanying story and access related links. Use this information, the links provided, and other online and library resources to answer your questions and find updates about this topic. State the sources of your information. Explain why you think the information is accurate. Also determine whether the information expresses a particular point of view or is biased in any way.

5. What in your view are the most significant aspects of this research? Give reasons for your opinion and for any changes in your ideas based on the additional information you have collected and the analysis you have done.

# PATTERNS OF STRUCTURE AND FUNCTION IN PLANTS

## In The News  Brain Viagra

Do you take herbal supplements? If you do, you are not alone. According to a recent report in the *Tufts University Health & Nutrition Letter*, Americans spend over $14 billion a year on dietary supplements that include herbals. Ginseng, echinacea, garlic, saw palmetto, and St. John's wort are among the most widely used herbal supplements in the United States.

Plants produce a wide array of chemicals, many of which have medicinal effects on humans. For thousands of years, people relied heavily on extracts made from the leaves, stems, bark, roots, and other parts of plants to treat illnesses, and some still do. Today, many Americans use botanical products in an attempt to prevent or alleviate a variety of common health problems, including colds, fatigue, depression, and memory loss.

A popular herbal product is ginkgo, which is derived from the fan-shaped leaves of *Ginkgo biloba*, a tree indigenous to Korea, China, and Japan. Many people think that taking ginkgo supplements is a natural way to improve cognitive functions, including memory, alertness, and mood. To determine whether the herb does produce such effects when taken as a dietary supplement, neuroscientist Paul Gold at the University of Illinois at Urbana-Champaign and his colleagues reviewed more than 40 studies on ginkgo. Overall, the scientists concluded that the studies did not provide enough information to determine whether ginkgo does or does not improve cognition in any substantive way. The investigators noted that, in general, the studies were poorly designed, and included insufficient numbers of subjects. Moreover, there was little or no comparison of cognitive function before and after treatment. You can learn more about Gold's findings and view the video "Brain Viagra—

Part 1" by visiting the *In The News* section of this text's companion Web site at www.wiley.com/college/alters.

Along with doubts about whether herbals are effective, many health care practitioners are concerned about the safety of herbal products. The Dietary Supplement Health and Education Act of 1994 exempts herbals from regulation by the U.S. Food and Drug Administration (FDA). As a result, manufacturers of herbal products do not have to provide evidence to the FDA that their products are pure, potent, and effective, as do manufacturers of prescription drugs.

Should the FDA regulate the testing and manufacture of herbals? The Center for Science in the Public Interest (CSPI) warns that not enough research has been conducted on the safety of herbal products. Some consumer rights advocates contend that many herbal products are useless, expensive, and possibly even dangerous. Manufacturers of herbal supplements, however, defend their products by pointing out that they are natural, have a long history of human use, and often are more reasonably priced than their prescription medication counterparts in the United States.

*Write your immediate reaction to the issue of FDA regulation of herbal supplements: first, summarize the issue in a sentence or two; then suggest how you think this issue could be or should be resolved. You will have an opportunity to reflect on your responses and gather more information on this issue in the* In The News *feature at the end of this chapter. In this chapter, you will learn more about vascular and nonvascular plants, including their structures and functions.*

# CHAPTER GUIDE

## What are vascular plants?

**24.1** Vascular plants have tissues that transport water and nutrients.

## What are the tissues of vascular plants, and what are their functions?

**24.2** Vascular plants have four types of tissues that carry out specialized functions.

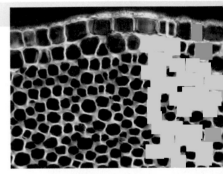

## What are the organs of vascular plants, and what are their functions?

**24.3** Vascular plants have three types of organs.

**24.4** Roots absorb water and minerals, and anchor the plant.

**24.5** Stems support the leaves and transport fluids throughout the plant.

**24.6** Leaves are the sites of photosynthesis.

## How do fluids move in plants?

**24.7** Fluids move in vascular plants by transpirational pull and bulk flow.

**24.8** Nonvascular plants have less sophisticated transport systems than those in vascular plants.

# What are vascular plants?

## 24.1 Vascular plants have tissues that transport water and nutrients.

The coastal redwoods of California (**Figure 24.1**) are the tallest trees in the world, with the tallest of the tall reaching a height of 367 feet—about 60 feet taller than the Statue of Liberty, including its granite pedestal and foundation. For any plant to grow as tall as a coastal redwood, it must have a means to transport substances to and from its various parts, including its highest branches.

Most plants (approximately 95% of all living plant species) have such a transport system—specialized vascular tissues that transport water and nutrients. Plants with these tissues are called **vascular plants**. Despite their reproductive diversity (see Chapter 23), all vascular plants have the same basic architecture.

A vascular plant is organized along a vertical axis as shown in **Figure 24.2**. The part below the ground is called the **root system**, and the part above the ground is called the **shoot system**. The root system penetrates the soil, absorbs water and ions crucial for plant nutrition, and anchors the plant. The shoot system consists of **stems** and **leaves**. Stems serve as a framework for positioning the leaves, the structures where most photosynthesis takes place. The arrangement, size, and other characteristics of the leaves are critically important in the plant's production of food. Flowers, and ultimately fruits with their enclosed seeds, are also formed on the shoots of flowering plants (angiosperms).

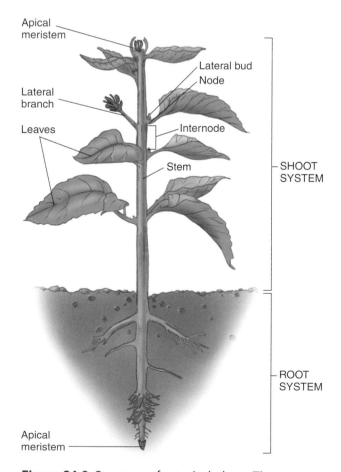

**Figure 24.2 Structure of a typical plant.** The root system is below the ground; the shoot system, consisting of stems and leaves, is above the ground. The purple areas indicate growth zones.

**Figure 24.1 Redwoods are the tallest trees in the world.** This one grows in Sequoia National Park in California.

# What are the tissues of vascular plants, and what are their functions?

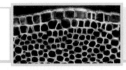

## 24.2 Vascular plants have four types of tissues that carry out specialized functions.

The organs of a vascular plant—the leaves, roots, and stems—are made up of different mixtures of tissues, groups of cells that work together to carry out specialized functions. Vascular plants have three types of differentiated tissues: *vascular tissue, ground tissue,* and *dermal tissue.* They also have one type of undifferentiated tissue: *meristematic tissue* (**Figure 24.3**).

The word "vascular" comes from a Latin word that means vessel. Thus, **vascular tissue** forms the "circulatory system" of a plant, conducting water and dissolved mineral nutrients (see Table 24.1) up from the root of the plant and conducting the products of photosynthesis throughout the plant. **Ground tissue** comprises most of the substance of the plant. It is the tissue in which the vascular tissue is embedded, and it stores carbohydrates that the plant produces. **Der-**

**mal tissue** covers the plant, protecting it—like skin—from water loss and injury to its internal structures. Plants also contain **meristematic tissue** (MER-uh-stuh-MAT-ik), often just called *meristem.* Cell division in undifferentiated meristematic tissue gives rise to cells of the other three tissue types, and results in growth of the plant.

### Vascular tissue

There are two principal vascular tissues: **xylem** (ZY-lem), which transports water and dissolved inorganic nutrients, and **phloem** (FLO-em), which conducts carbohydrates the plant uses as food along with needed minerals.

There are two types of conducting cells in xylem—the tracheids and the vessel elements (**Figure 24.4**). Both types of cells become functional only after they die and lose their cytoplasm, becoming hollow and thick-walled. Stacked end to end, these cells form pipelines that extend throughout the plant conducting water and dissolved inorganic nutrients. *Tracheids* (TRAY-kee-idz) have tapering ends with connections between them called pits. A layer of cell wall material covers the pits, through which water flows. In ad-

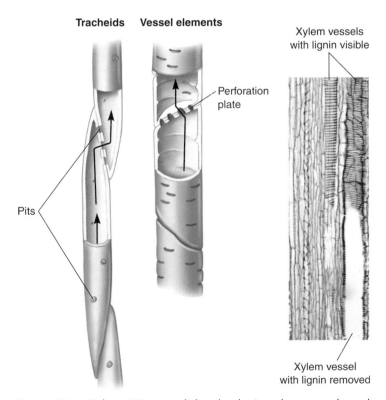

**Figure 24.3 Tissue types in plants.** Vascular tissue (orange) conducts materials throughout the plant. It is embedded in ground tissue (yellow). Dermal tissue (blue) forms the outer protective covering of the plant. Undifferentiated meristematic tissue is shown in purple.

**Figure 24.4 Xylem.** Water and dissolved minerals are conducted through stacks of tracheids with tapering ends containing pits, and through vessel elements that are joined end-to-end with openings between them called perforations. Xylem vessel elements are dead, hollow tubes. The walls are strengthened with spirals of lignin (shown as red in the photo). Lignin, a substance related to cellulose, imparts strength and rigidity.

dition to pits, *vessel elements* have openings between them called perforations through which water moves unimpeded. Pits along the lengths of both cell types allow water to move laterally to cells surrounding the xylem pipelines. They work much like soaking hoses used in gardens.

**Figure 24.5** shows two types of phloem cells: a *sieve-tube member* and a *companion cell*. The conducting cells (sieve-tube members) are alive and contain cytoplasm, but are not typical cells. They lack nuclei, for example. The ends of these elongated cells have pores, forming a sieve plate, that allow the easy passage of sugar-filled water from where it is produced in the leaves to where it is used—at the reproductive structures, for example. Companion cells, which do contain nuclei, secrete substances into and remove substances from the sieve-tube members.

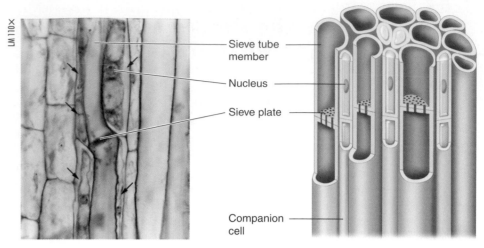

**Figure 24.5** Phloem. Sieve-tube members are joined at sieve plates where pores allow movement of sugars and other substances from cell to cell as they are conducted throughout the plant. Companion cells with nuclei nourish and maintain the sieve-tube members. In the photo, arrows point to companion cells that flank a sieve tube member.

## Ground tissue

Ground tissue is composed of three main cell types: *parenchyma* (pah-RENG-keh-mah), *collenchyma* (coh-LENG-keh-mah), and *sclerenchyma* (skli-RENG-keh-mah). Ground tissue parenchyma cells function in photosynthesis and storage. They are the most com-

mon of all plant cell types and make up the bulk of leaves, stems, and roots. Parenchyma cells are thin-walled, as you can see in **Figure 24.6a**, and usually contain large vacuoles. Parenchyma cells exposed to light are packed with chloroplasts. Those not exposed to light are generally packed with starch granules, which is the form of food storage in plants. Fleshy storage roots, such as carrots or sweet potatoes, contain a predominance of storage parenchyma, as do corn, rice, and wheat, the major sources of starch in the human diet. The flesh of most fruits is also made up of parenchyma cells.

Collenchyma tissue is support tissue. These cells have unevenly thickened cell walls and are elongated, commonly forming bundles of supportive "columns" inside the epidermis of the stem (Figure 24.6b). They are capable of stretching, so they elongate as the plant grows. Collenchyma is commonly encountered as the "strings" in celery stalks.

Sclerenchyma cells are dead, hollow cells with thick and strong walls. These cells help support and strengthen the ground tissue as well, but since they do not stretch, they support plant parts that have stopped growing. There are two types of ground tissue sclerenchyma cells: fibers and sclereids (Figure 24.6c).

Fibers are generally long, slender cells, which commonly occur in strands or bundles. They help support structures such as vascular bundles containing pipelines of xylem and phloem. In addition, sclerenchyma fibers have economic importance in that they provide the material for fabrics such as flax, hemp, jute, and ramie. Flax,

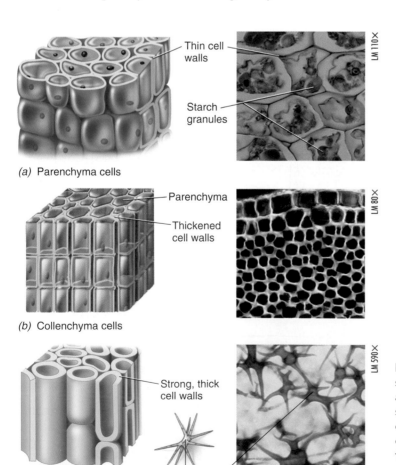

(a) Parenchyma cells

(b) Collenchyma cells

(c) Sclerenchyma cells

**Figure 24.6** Ground tissue. (*a*) Parenchyma cells are thin-walled, as shown by the dark line surrounding each cell in both the micrograph and the illustration. The cells in the micrograph are packed with starch granules. (*b*) The micrograph shows a cross section of celery collenchyma. The illustration shows a side view of these bundled, elongated cells. (*c*) The micrograph shows a bundle of sclerenchyma fibers, with their thickened cell walls, among the thin-walled parenchyma cells. Sclerenchyma cells, which help support a plant, occur as elongated fibers or as branched sclereids. The shells of nuts and the pits of fruits such as cherries are made up of sclereids.

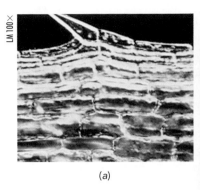

LM 100×

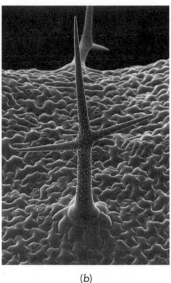

(a)

(b)

**Figure 24.7** Dermal tissue. (a) Epidermal cells cover and protect the outside of a plant. (b) Trichomes are outgrowths of the epidermis that may reflect sunlight or defend the plant.

hemp, and ramie are linen-like fabrics. Jute is a glossy fiber used most often in potato sacks, rope, twine, and backing on carpeting.

Sclereids are variable in shape and are often branched. Sclereids may occur singly or in groups throughout the ground tissue. They are usually found in seed coats and fruit. The gritty texture of a pear is due to schlerenchyma cells that are dispersed among the softer parenchyma cells (see Figure 24.6c).

### Dermal tissue

Dermal tissue is the "skin" of a plant. It covers and protects the outside of a plant just as skin protects an animal. Epidermal cells are the most abundant type of cell found in the dermal tissue of plants. These cells are often covered with a thick, waxy layer called a *cuticle* that protects the plant and provides an effective barrier against water loss (**Figure 24.7a**).

Other types of cells found in the dermis are guard cells, which surround openings in the leaves through which gases and water vapor pass (see Figure 24.20), and trichomes (TRY-kohmz; Figure 24.7b), which are outgrowths of the epidermis that have various functions. For example, on "air plants," trichomes help provide a large surface area for the absorption of water and inorganic nutrients. This is important because air plants obtain these substances through their leaves rather than through their roots as do other vascular plants. On

leaves or fruits, fuzzlike trichomes reflect sunlight, which helps control water loss. Some trichomes defend a plant against insects or larger animals with their sharpness or by secreting chemicals.

Woody shrubs and trees do not have dermal tissue, but have bark in its place. Bark is made up of other tissue types (see Section 24.5).

### Meristematic tissue

Plants contain *meristems*, areas of undifferentiated cells that are zones of plant growth. Every time one of these cells divides, one of the two resulting cells remains undifferentiated in the meristem while the other differentiates into vascular, ground, or dermal tissue, ultimately becoming part of the plant body. In this way, the meristem remains "forever young," capable of repeated cell division.

Plants can grow only in relationship to where their meristematic tissue is located. Growth tissue located at the tips of the roots and the tips of the shoots are called *apical meristems*. (Apex means "tip.") Locations of this type of tissue are purple in Figure 24.2. Apical meristems allow shoots to grow longer and their roots to grow deeper. This type of plant growth, which occurs mainly at the tips of the roots and shoots, is called **primary growth**.

Some plants grow not only taller but also thicker. This type of growth, called **secondary growth**, occurs in all woody trees and shrubs such as pines, oaks, and rhododendrons. These plants have a cylinder of meristematic tissue along the length of their stems and roots called *lateral meristem* (also called *cambium* in Figure 24.14).

Herbaceous (nonwoody) plants, such as tulips, have only apical meristems and therefore only primary growth. However, their stems do grow somewhat thicker as they develop. One reason for this growth is that plant development involves both cell division *and* cell growth; as cells enlarge, plant parts enlarge. In addition, some cell division occurs in the cells differentiating from meristematic tissue.

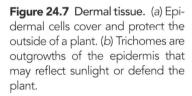

### CONCEPT CHECKPOINT

1. Which type of plant tissue gives a plant the ability to grow continuously throughout its life?
2. To which type of plant tissue and cell do each of the following refer? (a) hollow dead transport cells, (b) source of flax and hemp, (c) undifferentiated, (d) storage, (e) waxy exterior surface, and (f) growth in thickness.

---

## What are the organs of vascular plants, and what are their functions?

### 24.3 Vascular plants have three types of organs.

Roots, stems, and leaves are the organs of vascular plants. Together, the stems and leaves make up the shoot. The bulk of this chapter discusses the structure of the plants that dominate the plant world today: the flow-

ering plants (angiosperms). As was discussed in Chapter 23, the other types of vascular plants are seedless vascular plants (mosses, ferns) and vascular plants with naked seeds (gymnosperms). The principal differences between the structures of the roots, stems, and leaves in the seedless vascular plants and those with naked seeds lie primarily in the relative distribution of the vascular and ground tissue systems.

## 24.4 Roots absorb water and minerals, and anchor the plant.

The functions of a *root system* are to anchor a plant in the soil and to absorb water and mineral nutrients. Structurally, the roots of dicots (plants having netlike veins in their leaves; see Chapter 23) have a central column of xylem with radiating arms. In **Figure 24.8a** (left), the xylem looks like an "X." Between these arms are strands of phloem. Ringing this column of vascular tissue (often called the vascular cylinder) and forming its outer boundary is another cylinder of cells called the pericycle (Figure 24.8b). This tissue is made up of parenchyma cells able to undergo cell division to produce branch roots (roots that arise from other older roots; Figure 24.8c). Surrounding the pericycle is a mass of parenchyma called the *cortex*. These cells store food for the growth and metabolism of the root cells. The innermost layer of the cortex is called the endodermis, which consists of specialized cells that regulate the flow of water between the vascular tissues and the outer portion of the root. The outer layer of the root is the epidermis, which absorbs water and inorganic nutrients. Some epidermal cells have extensions

called *root hairs* that provide a large surface area over which absorption can take place.

Monocot roots (plants having parallel veins in their leaves; see Chapter 23) are similar to dicot roots, but often have centrally located parenchyma (storage) tissue called *pith*. The xylem and phloem are arranged in rings around the pith (see Figure 24.8a, right).

The end of a root grows in length due to cell division in the apical meristem shown in **Figure 24.9**. Outward cell division results in the formation of a thimblelike mass of relatively unorganized cells called the root cap, which secretes slime, easing root penetration into the soil. The root cap covers and protects the root's apical meristem, which regenerates cells that slough off as the root advances. Just behind its tip, the root cells elongate. Velvety root hairs, the tiny, delicate projections from epidermal cells, form above this area of elongation. Root hairs increase the surface area of the root for absorption of water and dissolved nutrients. Cells also differentiate in this portion of the root.

If you have ever pulled up plants and looked at their roots, you probably noticed that not all roots look the same. Differ-

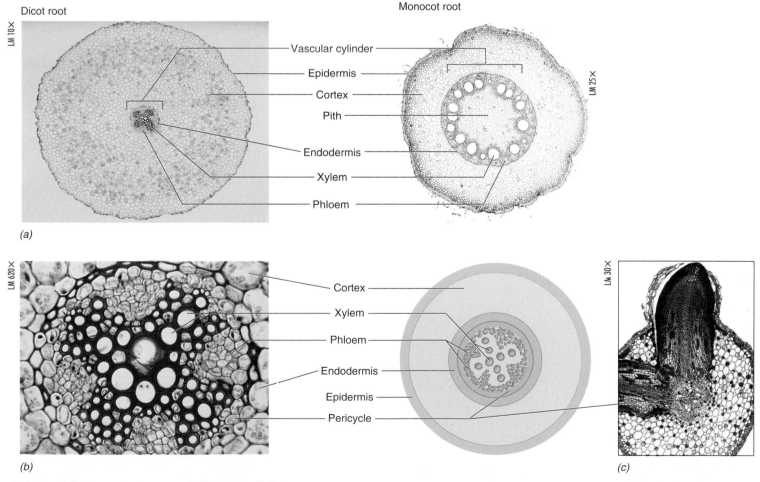

(a)

(b)                                                                                (c)

**Figure 24.8 Root structure.** (a) A dicot root (left) is compared to a monocot root (right). Dicot roots have a small vascular cylinder, which includes a central column of xylem with radiating arms in an X shape. Phloem tissue lies between the xylem arms. Monocot roots often have centrally located parenchyma (storage) tissue called *pith*. The xylem and phloem form a ring around the pith. (b) An enlargement of the vascular cylinder of the dicot root, as shown in a micrograph (left) and an illustration (right). The location of the pericycle is most clearly shown in the illustration. (c) Branch roots arise from the pericycle of the dicot root.

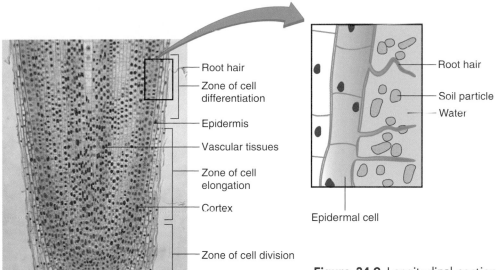

Root hair

Zone of cell differentiation

Epidermis

Vascular tissues

Zone of cell elongation

Cortex

Zone of cell division

Apical meristem

Root cap

Root hair

Soil particle

Water

Epidermal cell

**Figure 24.9 Longitudinal section of an onion root tip.** Apical meristem cells create a zone of cell division protected by the root cap. Inward cell division results in a zone of cell elongation. Above this zone, cells differentiate to produce specialized cell types. In addition, root hairs are found in this zone of cell differentiation.

**Visual Thinking:** Describe the structure of the root hair. How does this structure relate to the fact that no root hairs grow at the root cap?

taproots, such as carrots and radishes, are fleshy because they are modified for food storage. The plant draws on these food reserves when it flowers or produces fruit; thus, taproot crops are harvested before that time. Plants with taproots also have extensive secondary root systems.

Monocots do not usually have taproots. The primary root, which may look like a taproot, often dies during the early growth of the plant, and new roots develop from the lower part of the stem. These are called *adventitious roots* (AD-ven-TISH-us). Often, adventitious roots help anchor a plant, such as "prop" roots in corn (Figure 24.10*b*). Grass has *fibrous roots*, a type of root system typical in monocots that has no predominant root. Fibrous root systems work well to anchor the plant in the ground and absorb nutrients efficiently because of their large surface area (Figure 24.10*c*). They also help prevent soil erosion by holding soil particles together.

The essential nutrients that plants use are listed in **Table 24.1**. The nonmineral nutrients are carbon, hydrogen, and oxygen. Plants get carbon from carbon dioxide ($CO_2$) in the air and hydrogen and oxygen from water ($H_2O$). The remaining nutrients are minerals dissolved in water in the soil, which plants absorb through their roots. Primarily, water absorbed at the roots replaces the water released into the air by the leaves in a process called *transpiration* (see Section 24.7).

Why do plants need nutrients? Plants manufacture carbohydrates during photosynthesis using the non-mineral nutrients. These sugars are not the only substances that plants need in order

ences in appearance are linked to differences in function. Many dicots—dandelions, for example—have a single, large root called a *taproot* (see **Figure 24.10a**). Even a small section of a taproot can regenerate a new plant, which is one reason dandelions are so difficult to eliminate from lawns and gardens. Taproots grow deep into the soil, firmly anchoring the plant and often "tapping into" water sources far below the surface. Some

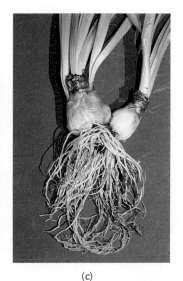

**Figure 24.10 Types of roots.** (*a*) Taproots in a dandelion, *Taraxacum officinale*. (*b*) Prop roots in corn, *Zea mays*, are adventitious—they arise from stem tissue and take over the function of the primary root. (*c*) Fibrous roots in a monocot, the daffodil, *Narcissus* hybrid.

(a)

(b)

(c)

| TABLE 24.1 | Plant Nutrients* |
|---|---|
| **Macronutrients** | **Micronutrients** |
| Carbon | Chlorine |
| Oxygen | Iron |
| Hydrogen | Manganese |
| Nitrogen | Boron |
| Potassium | Zinc |
| Calcium | Copper |
| Magnesium | Molybdenum |
| Phosphorus | |
| Sulfur | |

*Macronutrients are present in greater than 10,000 ppm (parts per million); nutrients of lesser concentrations are considered micronutrients. Parts per million equals units of an element by weight per million units of oven-dried plant material. The prefixes "macro-" and "micro-" relate only to the amounts plants require for normal growth and do not imply that one set of nutrients is more important to the needs of the plant than the other. The nutrients are listed in the order of their usual concentrations in healthy plants, from highest to lowest. Carbon, oxygen, and hydrogen are the non-mineral nutrients that plants obtain from the air and water. The remaining nutrients are minerals, inorganic elements that plants obtain from the soil.

to live, however. Plants need nucleic acids, proteins, fats, and vitamins. These substances are formed from the carbohydrates that plants manufacture and from the mineral nutrients that plants take in from the soil. For example, nitrogen is part of many molecules such as proteins, nucleic acids, certain hormones, and chlorophyll; phosphorus is part of nucleic acids and the energy molecule ATP. Without all their essential nutrients, plants exhibit nutritional deficiencies (**Figure 24.11**) just as humans exhibit nutritional deficiencies if they do not take in sufficient amounts and types of nutrients. Because plants incorporate and concentrate mineral nutrients in their bodies, many of them are important sources of these nutrients in the human diet. Broccoli and cabbage, for example, are excellent sources of calcium. Bananas provide potassium.

## 24.5 Stems support the leaves and transport fluids throughout the plant.

Although the organism in **Figure 24.12a** looks like the discarded shell of a huge turtle, it is actually a plant. In fact, one nickname for this plant is the turtleback. The turtleback's "shell" is really a modified stem that stores water for the plant. Turtlebacks (*Dioscorea elephantipes*) commonly grow in Africa, where native people call this fleshy, edible storage organ Hottentot's bread. The cactus, such as the Saguaro that grows in the American Southwest desert (Figure 24.12b), is another example of a plant with stems modified as giant water storage organs. In your study of plants and animals you will often notice that structures give you clues about function. In this case, the enlarged, fleshy stems

**Figure 24.12** Both of these plants have stems modified for water storage. (a) The turtleback grows in Africa. Its fleshy, edible stem has the texture of a sweet potato. (b) Saguaro cacti also with fleshy, modified stems grow in the American Southwest.

(a)

**Figure 24.11** Nutrient deficiency in tomato plants. The plant on the left is growing well in a complete nutrient solution. The other plants exhibit symptoms of nutrient deficiencies. From left to right, plants in the next four containers are growing minus phosphorus, minus calcium, minus iron, and minus nitrogen.

(b)

of these two plants that live on opposite sides of the Earth perform the same function—water storage.

Although these two plants have modified stems that perform an important function in hot, dry climates, stems in general function to transport fluids throughout the plant and to support the leaves, which carry out photosynthesis. Leaves are arranged on the stem so that light will fall on them. Stems conduct water and inorganic nutrients from the roots to all plant parts and bring the products of photosynthesis to where they are needed or stored.

Looking back at Figure 24.2, you can see that the leaves form on the stem at locations called **nodes**. The portions of the stem between the nodes are called the **internodes**. As the leaves grow, tiny undeveloped side shoots called **lateral buds** develop at the angles between the leaves and the stem. Given the proper environmental conditions, these buds, which contain their own embryonic leaves, may elongate and form lateral branches. These lateral buds can be thought of as "back-up" meristems, providing reserves of undifferentiated tissue to take over should something happen to the apical meristem, such as consumption by an herbivore. Nodes and internodes are characteristics that distinguish stems from roots. Some

modified stems grow underground and play a role in vegetative propagation of plants, such as rhizomes, tubers, and stolons (see Section 23.8).

The stem "transportation system" is made up of strands of xylem and phloem tissue that are positioned next to each other, forming cylinders of tissue called vascular bundles (often referred to as veins in leaves) (**Figure 24.13**). The xylem tissue characteristically forms the part of each bundle closer to the interior of the stem. The phloem lies closer to the epidermis. In herbaceous dicots (those with soft stems rather than woody, treelike stems), the vascular bundles are arranged in a ring near the periphery of the stem (Figure 24.13a). In monocots, the vascular bundles are also arranged in a ring but are scattered throughout the stem too (Figure 24.13b).

Because of the arrangement of their vascular bundles, dicot stems have a mass of ground tissue in the center of the stem called *pith* and a ring of ground tissue between the epidermis and the vascular bundles called *cortex*. Monocot stems also have ground tissue, but because it surrounds scattered vascular bundles, ground tissue does not form areas of pith or cortex. The epidermis of both

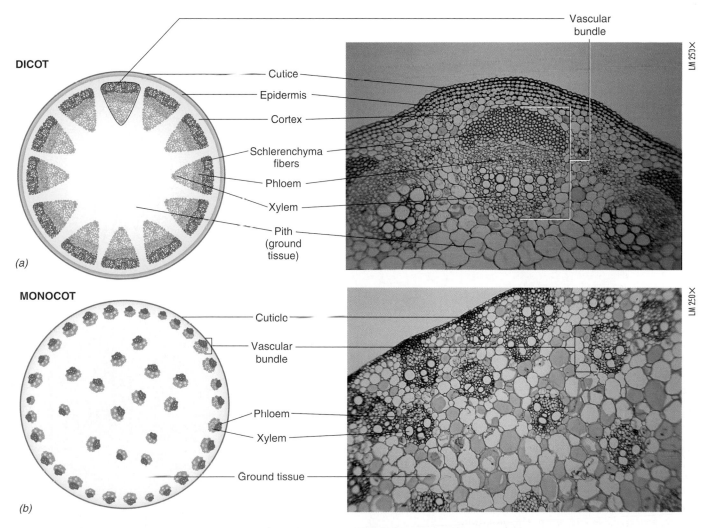

**Figure 24.13 Stem structure.** (a) A dicot stem from the common sunflower (*Helianthus*), in which the vascular bundles are arranged around the outside of the stem. Notice the sclerenchyma fibers that support the vascular bundle. (b) A monocot stem from corn (*Zea mays*), showing how the vascular bundles ring the periphery and are scattered throughout the interior of the stem.

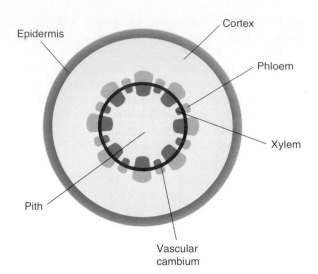

**Figure 24.14 The vascular cambium in a woody stem.** The vascular cambium (*red*) is the growth tissue in woody stems and lies between the xylem and the phloem.

monocot and herbaceous dicot stems is covered with a protective waxy coating called the cuticle.

As with roots, the growing end of a stem is tipped with apical meristematic tissue. Young leaves cluster around the apical meris-

tem, unfolding and growing as the stem itself elongates. Dicots with woody stems or trunks (such as flowering trees) and gymnosperms (such as pine trees) also have lateral meristems called *cambia*. One type of cambium in woody stems is called vascular cambium. As the name suggests, this growth tissue lies between the vascular tissue—the xylem and phloem—connecting the bundles to form a ring (**Figure 24.14**). As the cambial cells divide during secondary growth, one of the resulting daughter cells remains as a cambial cell, and the other differentiates into either a xylem or a phloem cell. This new xylem and phloem is called secondary xylem and secondary phloem (**Figure 24.15**).

The wood of trees is actually accumulated secondary xylem. The wood of dicot trees (such as cherry, hickory, oak, and walnut) is commonly referred to as hardwood, whereas the wood of conifers (such as fir, cedar, pine, and spruce) is called softwood. These names, however, are not accurate descriptions of each group. Each contains trees having woods of varying hardness. For example, soft hardwoods include red maple, paper birch, aspen, and American elm. Hard hardwoods include black walnut, oak, and hickory.

The hardness of wood relates to its density, which depends on its proportion of cell wall substance to the space bounded by the cell wall. The denser a wood, the more wall substance it has in relation to the space it bounds and the stronger it is. When used for building, denser woods are harder to nail and machine, but they generally shrink and swell less than less dense, softer woods. Denser woods are also better fuel woods.

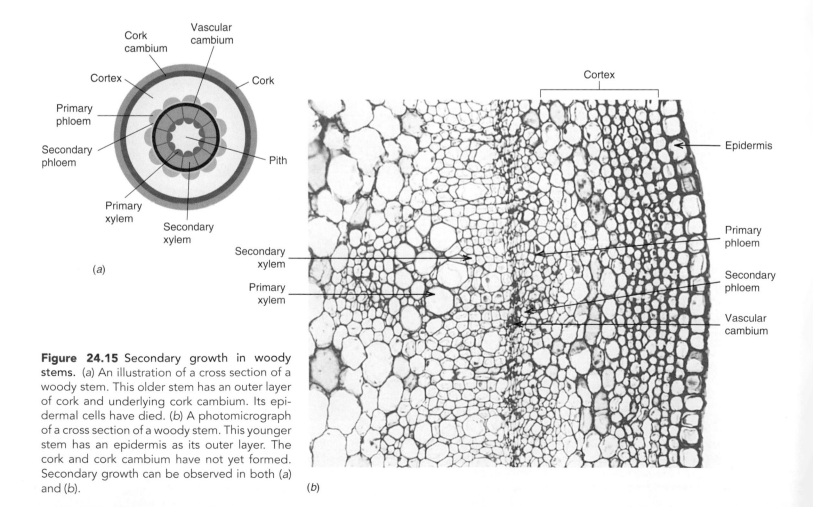

**Figure 24.15 Secondary growth in woody stems.** (*a*) An illustration of a cross section of a woody stem. This older stem has an outer layer of cork and underlying cork cambium. Its epidermal cells have died. (*b*) A photomicrograph of a cross section of a woody stem. This younger stem has an epidermis as its outer layer. The cork and cork cambium have not yet formed. Secondary growth can be observed in both (*a*) and (*b*).

Other lateral meristem tissue called cork cambium lies just under the epidermis in woody stems. The outermost cells of this growth tissue produce densely packed cork cells. Cork cells have thick cell walls that contain fatty substances, making these cells waterproof and resistant to decay. When mature, the cork cells lose their cytoplasm and become hardened. The cork, the cork cambium, and all the tissues outside the vascular cambium make up the outer protective covering of the plant called the **bark** (see Figure 24.16).

---

### CONCEPT CHECKPOINT

3. Imagine that you are a nutrient being absorbed by a dicot root. Trace your pathway through each of the different root tissues as you make your way to the vascular cylinder.

4. Use Figure 24.15 illustrating a two-year-old woody dicot to answer the following:
   a. Which tissues were produced by differentiation of apical meristem?
   b. Which tissues were produced by differentiation of vascular cambium?
   c. Which tissues were produced during the second year of growth?
   d. Which tissues transport water and inorganic nutrients?

---

## 24.6 Leaves are the sites of photosynthesis.

Lettuce, kale, cabbage, and spinach are some of the leaves that humans incorporate into their diets. Leaves, which are outgrowths of the shoot apex, are the light-capturing photosynthetic organs of most plants. A major exception is found in most cacti, in which the stems are green and have largely taken over the function of photosynthesis for the plants. Their leaves are modified to form water-conserving spines.

Most leaves have a flattened portion, the **blade**, and a slender stalk, the **petiole** (**Figure 24.17**). Veins, continuations of the vas-

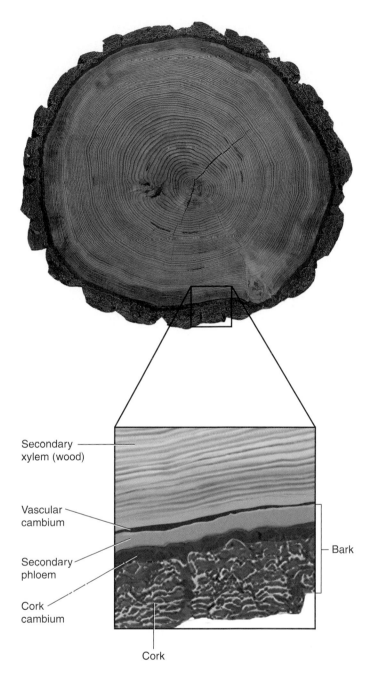

Secondary xylem (wood)

Vascular cambium

Secondary phloem

Cork cambium

Bark

Cork

**Figure 24.16 Annual rings in the cross section of a tree trunk.** Each two-part ring (lighter and darker) represents the growth produced during one year. Thus, rings can be used to calculate the age of a tree. The structure of bark is also shown.

**Visual Thinking:** About how old is this tree? Less than 10 years? Between 10 and 100 years? More than 100 years?

When growth conditions are favorable, as in the spring and early summer in most temperate regions, the cambium divides most actively, producing large, relatively thin-walled cells. During the rest of the year, the cambium divides more slowly, producing small, thick-walled cells. This pattern of growth results in the formation of rings in the wood. These rings are called annual rings and can be used to calculate the age of a tree (**Figure 24.16**).

Blade

Petiole

**Figure 24.17 Structure of a leaf.** This leaf is from an American linden tree, which is often used as a shade tree.

**Figure 24.18 Dicot versus monocot leaves.** (a) The leaves of most dicots, such as this ivy, have net venation; (b) those of most monocots, like this palm, have parallel venation.

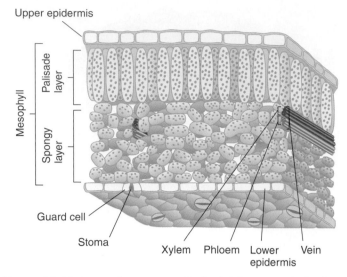

**Figure 24.19 Internal structure of a leaf.** The mesophyll of most leaves is divided into two layers: the palisade layer and the spongy layer.

cular bundle in the stem consisting of both xylem and phloem, run through the leaves. In monocots, veins are usually parallel, and in dicots, they are usually netlike (**Figure 24.18**). Many gymnosperms (vascular plants with naked seeds) have needlelike leaves suited for growth under dry and cold conditions.

Microscopically, a cross section of a typical leaf looks somewhat like a sandwich: parenchyma cells in the middle, bounded by epidermis. The vascular bundles, or veins, run through the parenchyma. The leaf parenchyma is appropriately called the *mesophyll*, or "middle leaf" (**Figure 24.19**).

The mesophyll of most leaves is divided into two layers: the palisade layer and the spongy layer. The palisade layer lies beneath the upper epidermis of the leaf and consists of one or more layers of loosely packed columnlike cells. These cells contain most of the organelles in which photosynthesis takes place in the leaf, the chloroplasts. The spongy layer lies beneath the palisade layer and is made up of loosely packed cells with many air spaces between them. These spaces are connected, directly or indirectly, with openings to the outside called **stomata** (STOW-muh-tuh; sing. **stoma**).

Each stoma is bracketed by two **guard cells** that regulate its opening and closing. Opening occurs when solutes (dissolved substances) are actively accumulated in the guard cells. This accumulation of solutes results in the movement of water into the guard cells by osmosis, creating water pressure commonly referred to as *turgor pressure*. As **Figure 24.20** shows, the two guard cells surrounding a stoma are long cells that are held together at their ends. Although invisible to the naked eye, guard cells are ringed by microfibrils of cellulose, much like hoops encircle a barrel. When the guard cells fill with water, they cannot increase in diameter because of the

restricting cellulose microfibrils. Therefore, they increase in length and bow out. This change in the shape of the guard cells opens the stoma.

During photosynthesis, guard cells actively transport potassium ions to their interior. Water follows the potassium ions by osmosis because of the osmotic gradient the ions create. This increases turgor pressure in the guard cells, causing them to elongate and bow out, opening the stoma. The oxygen produced by photosynthesis diffuses into the atmosphere through the stomata, whereas the carbon dioxide needed for photosynthesis to take place diffuses into the leaf. In addition, water exits the leaf through these openings in the form of water vapor—a process called **transpiration**.

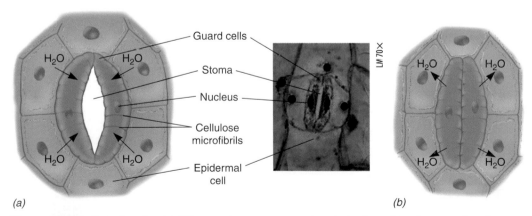

**Figure 24.20 Guard cells.** (a) When solute concentration is high within the guard cells, water moves into them. They elongate and bow outward, opening the stoma. (b) When solute concentration is low, the guard cells have low turgor pressure and the stoma is closed.

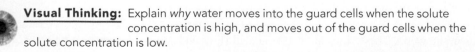

**Visual Thinking:** Explain *why* water moves into the guard cells when the solute concentration is high, and moves out of the guard cells when the solute concentration is low.

# How do fluids move in plants?

## 24.7 Fluids move in vascular plants by transpirational pull and bulk flow.

Did you ever wonder how trees like the redwoods in Figure 24.1 manage to move water and nutrients to their uppermost leaves? You may recall from Section 24.2 that two vascular tissues transport plant fluids: xylem, which transports water and dissolved mineral nutrients, and phloem, which conducts carbohydrates that the plant uses as food (commonly called sap) along with other needed substances. Water moves through xylem by transpirational pull. Fluid moves through phloem by bulk flow.

### Transpirational pull

The theory that explains water movement in plants is called the *adhesion-cohesion-tension theory*, or simply **transpirational pull (Figure 24.21)**. The word "tension" as used here means a "pull" on the water molecules within the plant as transpiration takes place, that is, as water evaporates from the leaves. This is how the pull occurs: Water evaporates from air spaces within the leaf and moves out of the leaf via the stomata ❶. Then, water within the mesophyll cells next to these spaces moves by osmosis into the spaces.

As water moves out of these cells, the concentration of water within them decreases. Water therefore moves from adjacent cells having a higher concentration of water into those cells having a lower concentration of water. A "pulling" or "sucking" effect is created, with water moving from cell to cell until the xylem is reached ❷, resulting in a pull on the water in the xylem. Notice the close placement of the xylem, the mesophyll cells, and the air spaces.

The column of water in the xylem of the plant "holds together" because of two other forces: cohesion and adhesion. *Cohesion* means that water molecules tend to stick together. (See Chapter 3 for a more complete discussion of the properties of water molecules.) Therefore, as molecules at the top of the xylem are pulled up ❸, this force is transmitted all the way down the xylem to the root, from one water molecule that sticks to the next and so forth. In addition, the water molecules adhere to the walls of the very narrow tracheids and vessel elements of the xylem, a property known as *adhesion*. These two forces link water molecules together and to the sides of the xylem with weak chemical bonds called hydrogen bonds.

Where does the water come from that is in the xylem of plants? Most of the water absorbed by a plant comes in through its root hairs ❹. Mineral nutrients also pass into the cells of the root hairs by means of special cellular "pumps." (Chapter 5 discusses

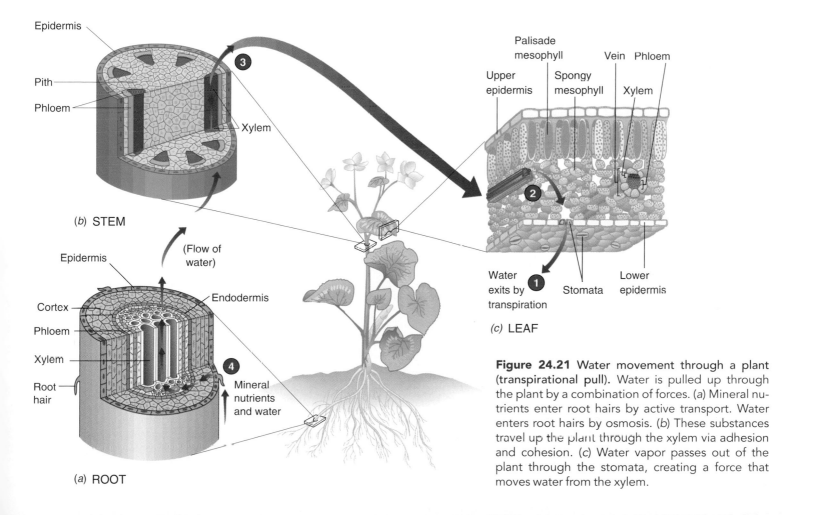

**Figure 24.21** Water movement through a plant (transpirational pull). Water is pulled up through the plant by a combination of forces. (a) Mineral nutrients enter root hairs by active transport. Water enters root hairs by osmosis. (b) These substances travel up the plant through the xylem via adhesion and cohesion. (c) Water vapor passes out of the plant through the stomata, creating a force that moves water from the xylem.

active transport and cell pumps.) In this way, root cells maintain a higher concentration of dissolved mineral nutrients than the concentration of mineral nutrients in the water of the soil. Therefore, water tends to steadily move into the root hair cells from an area of higher concentration of water (in the soil) to an area of lower concentration of water (in the root hairs) by osmosis, developing a force called root pressure. Once inside the roots, the water and dissolved mineral nutrients pass inward to the conducting elements of the xylem. Once inside the xylem, water molecules are pulled upward as other water molecules are lost through the stomata ❶ .

### Bulk flow

Fluid is also transported in plants by the phloem. As mentioned previously, these tissues transport the products of photosynthesis—sugars—dissolved in water. This sugary solution is commonly called sap. Phloem sap contains 10 to 25% sucrose in addition to inorganic nutrients, amino acids, and plant hormones, and may travel as fast as 1 meter (about 1 yard) per hour!

The forces of diffusion and osmosis alone cannot account for this rapid movement of phloem sap. Instead, a pressure-flow, or **bulk-flow** system performs this function. The bulk-flow system is shown in **Figure 24.22** and works in the following way. Sucrose is produced at a source, such as a photosynthesizing leaf, and is actively transported into sieve-tube members by companion cells. As the concentration of sucrose increases in the phloem, water follows by osmosis. In the roots below or at some other "sink" where sucrose is used, companion cells actively transport sucrose out of the phloem. Water again follows by osmosis. The high hydrostatic

(water) pressure in the phloem near the source and the low pressure near the sink cause the rapid flow of the sap. After the sap reaches its destination in the plant, the water can be recycled by moving back to the source through the xylem.

---

**CONCEPT CHECKPOINT**

5. Trace the pathway of a water molecule through the various tissues of a plant from the time it is absorbed from the soil to when it is transpired from a leaf into the atmosphere.
6. Trace the pathway of a sucrose molecule through the various tissues of a plant from the time it is synthesized in a plant leaf to when it is converted into starch in a root cortex cell.

---

## 24.8 Nonvascular plants have less sophisticated transport systems than those in vascular plants.

The nonvascular plants comprise three phyla: mosses, liverworts, and hornworts (see Chapter 23). None has true roots, stems, or leaves. Some nonvascular plants have distinct stems and leaflike structures, such as most mosses, whereas other nonvascular plants do not. These organs have an outer layer of epidermis made up of protective cells and growth cells. The cortex is made of parenchyma cells, much like those found in vascular plants. In addition, these stems and leaflike structures have a central strand of water-conducting cells.

---

## *just wondering . . .*

*Questions students ask*

### If all plants produce their own food through photosynthesis, why do Venus flytraps consume flies?

The Venus flytrap (*Dionaea muscipula*) is one of a few carnivorous, or animal-eating, plants. Carnivorous plants often grow in soils depleted of nutrients, especially nitrogen. Digesting insects provides these plants with nutrients that they are unable to get from the soil. Like other plants, however, carnivorous plants make their own carbohydrates during the process of photosynthesis.

All carnivorous plants must first trap their prey and then digest it to absorb nutrients. The Venus flytrap ensnares prey by means of its touch response, which works as follows. On the inner sides of each pair of the cusplike modified leaves of the plant are sensitive hairs. Attracted by nectar on the leaves' surfaces, insects touch the hairs as they walk onto the leaves. If two hairs are touched in succession or if one hair is touched twice (the plant's way of distinguishing between a living

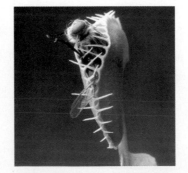

**Figure 24.A.** Venus flytrap.

and a nonliving stimulus), the leaves close on the insect, imprisoning it as shown in **Figure 24.A**. The leaf closure results from biochemical changes that occur within the epidermal cells of the leaves when the sensitive hairs are stimulated. These biochemical changes result in the expansion of the outer epidermal cells of each leaf, whereas the inner epidermal cells do not change. As the leaves change shape because of the cellular changes, they come together.

The Venus flytrap produces digestive enzymes, storing them in vacuoles within the cells of the leaves. As the insect becomes enclosed by the leaves, it gets pressed against the inner surfaces of the leaves. This stimulus results in the discharge of the digestive enzymes by the vacuoles onto the trapped prey. As the organism is digested, the nutrients are absorbed by the leaf cells. Seems like quite a gory story for the plant world, doesn't it? ●

---

*Are you wondering about a topic in biology and how it relates to your life? Submit your question by clicking the Just Wondering link in this text's companion Web site at www.wiley.com/college/alters.*

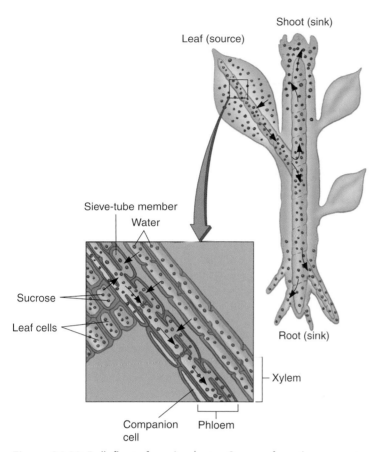

**Figure 24.22 Bulk flow of sap in plants.** Sucrose from the source in leaf cells (pink dots) is transported into sieve-tube members by companion cells, and water follows from the xylem by osmosis (blue dots). At a sink—a place where sucrose is used—sucrose is actively transported from the phloem, and water again follows by osmosis. The high water pressure in the phloem near the source and the low pressure near the sink cause this flow of sucrose, which can be very rapid.

Other than having water-conducting cells, the nonvascular plants have no specialized vascular tissues. Nonvascular plants, having less sophisticated transport systems than the vascular plants, do not grow very tall. Some appear to be creeping over their substrate, or food source.

Nonvascular plants have no roots, but some have slender, usually colorless projections called **rhizoids** (RYE-zoyds) that anchor these simple plants to their substrate (**Figure 24.23**). Unlike roots, however, rhizoids consist of only a few cells and do not play a major role in the absorption of water or inorganic nutrients. These substances often enter a nonvascular plant directly through its stems or leaves.

**Figure 24.23 Rhizoids.** Some nonvascular plants have rhizoids that anchor the plant to the substrate. They are the colorless filamentous projections seen in the photo.

# CHAPTER REVIEW

## Summary of Key Concepts and Key Terms

### What are vascular plants?

**24.1 Vascular plants** (p. 410) have tissues that transport water and nutrients.

**24.1** A vascular plant is organized along a vertical axis; the part below the ground is called the **root system** (p. 410), and the part above the ground is called the **shoot system** (p. 410).

**24.1** The root system penetrates the soil, absorbs water and ions crucial for plant nutrition, and anchors the plant; the shoot system consists of **stems** (p. 410) and **leaves** (p. 410).

### What are the tissues of vascular plants, and what are their functions?

**24.2** Vascular plants are made up of four types of tissue: **vascular tissue** (p. 411), **ground tissue** (p. 411), **dermal tissue** (p. 411), and **meristematic tissue** (MER-uh-stuh-MAT-ik, p. 411).

**24.2** Vascular tissue conducts water and dissolved mineral nutrients up from the root of the plant; ground tissue forms most of the substance of the plant and stores carbohydrates that the plant produces; dermal tissue covers the plant; and meristematic tissue provides for growth.

**24.2** Vascular tissue is of two types: **xylem** (ZY-lem, p. 411) and **phloem** (FLO-em, p. 411).

**24.2** Xylem conducts water and dissolved mineral nutrients through stacks of cells that form pipelines extending throughout the plant.

**24.2** Phloem conducts water and dissolved sugars throughout the plant.

**24.2** Ground tissue contains parenchyma cells, which function in photosynthesis and storage, and sclerenchyma cells, which help support and strengthen the ground tissue; collenchyma tissue is also support tissue.

**24.2** Dermal tissue covers the outside of a plant.

**24.2** Various types of cells can be found in dermal tissue serving a number of functions, such as protection, gas exchange, and prevention of water loss.

**24.2** Plants undergo **primary growth** (growth in length, p. 413) by cell division in apical meristems at the ends of the roots and the shoots.

**24.2** **Secondary growth** (growth in width, p. 413) in both shoots and roots takes place in woody trees and shrubs by means of cell division in lateral meristems along the length of their stems and branches.

## What are the organs of vascular plants, and what are their functions?

**24.3** Roots, stems, and leaves are the organs of vascular plants.

**24.4** The root system anchors a plant in the ground and absorbs water and inorganic nutrients from the soil.

**24.4** Most dicot roots have a central column of xylem with radiating arms and strands of phloem between these arms.

**24.4** Surrounding the vascular tissue in dicot roots is a layer of cells called the pericycle, which is capable of cell division, surrounded by parenchyma (storage) cells of the cortex.

**24.4** The entire dicot root is covered with a protective epidermis; root hairs are extensions of epidermal cells that increase their surface area for absorption.

**24.4** Monocot roots often have an additional storage tissue called pith, which is located in the center of the root; the xylem and phloem are arranged in rings around the pith.

**24.4** Dicots often have a taproot system with a single large root that may be used to store food as in carrots and radishes; monocots often have an extensive fibrous root system.

**24.5** Stems support the leaves, conduct water and inorganic nutrients from the roots to all plant parts, and bring the products of photosynthesis to where they are needed or stored.

**24.5** Leaves form on stems at locations called **nodes** (p. 417); portions of stems between nodes are called **internodes** (p. 417).

**24.5** As leaves grow, tiny undeveloped side shoots called **lateral buds** (p. 417) develop at the angles between leaves and stems.

**24.5** Together, the xylem and phloem make up the vascular bundles.

**24.5** Herbaceous dicot stems are characterized by an inner cylinder of ground tissue called pith surrounded by a ring of vascular bundles.

**24.5** Encircling the ring of vascular bundles in herbaceous dicot stems is additional ground tissue called cortex.

**24.5** Monocot stems are characterized by scattered vascular bundles embedded in ground tissue.

**24.5** Dicots with woody stems and gymnosperms have lateral meristem tissue, or cambium.

**24.5** The vascular cambium lies between the xylem and phloem; its dividing cells form xylem toward the interior (secondary xylem) and phloem (secondary phloem) toward the exterior of the stem.

**24.5** Wood is accumulated secondary xylem; it often displays rings because it exhibits different rates of growth during different seasons.

**24.5** Lateral meristem tissue called cork cambium lies just under the epidermis; the outermost cells of this growth tissue produce densely packed cork cells.

**24.5** The cork, the cork cambium, and all the tissues outside the vascular cambium make up the outer protective covering of the plant called the **bark** (p. 419).

**24.6** Leaves, the photosynthetic organs of most vascular plants, are made up of specialized ground tissue cells, or parenchyma, bounded by epidermis.

**24.6** Most leaves have a flattened portion, the **blade** (p. 419), and a slender stalk, the **petiole** (p. 419).

**24.6** Vascular bundles run through the parenchyma as veins, and parenchymal cells contain chloroplasts, the organelles in which photosynthesis takes place.

**24.6** Openings in the epidermis of the leaf (**stomata**, STOW-muh-tuh, sing. **stoma**, p. 420) allow the carbon dioxide needed for photosynthesis to enter and the oxygen produced by photosynthesis to escape.

**24.6** Each stoma is bracketed by two **guard cells** (p. 420) that regulate its opening and closing.

**24.6** Water vapor evaporates from a plant through its stomata, which is a process called **transpiration** (p. 420).

## How do fluids move in plants?

**24.7** Fluids move in vascular plants by **transpirational pull** (p. 421) and **bulk-flow** (p. 422).

**24.7** Water flows through plants in a continuous column, driven mainly by the evaporation of water vapor from the stomata.

**24.7** The cohesion of water molecules and their adhesion to the walls of the xylem through which they pass are important factors in maintaining the flow of water to the tops of plants.

**24.7** The process of bulk flow moves sucrose from where it is produced in a plant to where it is used.

**24.7** During bulk flow, sucrose is actively transported into phloem cells, where it is produced, and is actively transported out of the phloem cells, where it is used; these processes produce a sugar gradient and a water pressure gradient, which cause the movement of sugar and water.

**24.8** The organization and transport systems of nonvascular plants are much less complex than those of the vascular plants.

**24.8** Nonvascular plants have no roots, but some have slender, usually colorless projections called **rhizoids** (RYE-zoyds, p. 423) that anchor these simple plants to their substrate.

---

## Level 1 | Learning Basic Facts and Terms

**Matching**

Match each type of tissue to its function:

1. ____ Meristem
2. ____ Dermal tissue
3. ____ Ground tissue
4. ____ Vascular tissue

   a. Stores food manufactured by the plant.
   b. Protects the plant.
   c. Produces new plant cells during growth.
   d. Conducts water, inorganic nutrients, carbohydrates, and other substances throughout the plant.

5. Roots may do all of the following except
   a. anchor the plant.
   b. store food.
   c. absorb water and minerals.
   d. undergo secondary growth.
   e. Roots can do all of these things.

6. Which of the following is a plant nutrient but is not a macronutrient?
   a. O
   b. C
   c. Zn
   d. N
   e. K

7. Primary growth is responsible for
   a. elongation of roots and stems.
   b. transpiration.
   c. increase in thickness of a root or stem.
   d. opening and closing of stomata.
8. Transpiration refers to
   a. the way in which water is absorbed by roots.
   b. the evaporation of water through leaf stomata.
   c. gas exchange in plants.
   d. the bulk flow of phloem sap.
9. Unlike the flow of water in xylem, sap transport through the phloem
   a. is always down from the leaves to the roots.
   b. is always up from the roots to the leaves.

   c. is bidirectional but always from a place where sugars are being produced to a place where sugars are being stored or used.
   d. occurs laterally from the pith outward to the epidermis.
10. Water flows into the source end of a sieve-tube member because
   a. sucrose has been actively transported into the sieve-tube member.
   b. water pressure outside the sieve-tube member forces water in.
   c. the companion cell of the sieve-tube member actively pumps water in.
   d. sucrose has been transported out of the sieve-tube member by active transport.

## Level 2    Learning Concepts

1. Compare and contrast xylem and phloem transport with respect to: (a) direction of transport; (b) composition of fluid transported; (c) mechanism of flow.
2. Distinguish between primary and secondary growth.
3. What type of root system would you expect to find in a dandelion, an ivy plant, and a clump of grass? What are the advantages of each type of root system?

4. How can annual rings help you estimate the age of a tree? What type of tissue is involved? In which type of growth does this result?
5. Diagram the movement of phloem sap in a bulk-flow system. Label the source and sink, and show the direction of flow.
6. In general, how does the appearance of nonvascular plants differ from that of vascular plants? Why?

## Level 3    Critical Thinking and Life Applications

1. In tropical climates, many tall plants shut their stomata during the hot days and open them at night. If their stomata are closed during the day, why doesn't the water within the plant fall down the stem?
2. The roots of many plants have permanent mutualistic associations with fungi. What might be the advantage of this association to the plant?
3. Compare and contrast the structure and function of the "circulatory system" of a vascular plant with the circulatory system of a human. List and compare at least three major differences and two similarities.
4. When you were in middle/junior high school, you carved your initials about 4.5 feet up the trunk of a tree in the woods near your house. It's 10 years later, you've moved, and that tree has grown 10 feet taller. If you stopped to "visit" the tree, how high up the trunk would you look to find your initials, and why?

5. If you look at both the Food Guide Pyramid and the Healthy Eating Pyramid in Figures 27.A and 27.B, you can see that a large proportion of each pyramid—that is, a large proportion of a healthy diet—consists of "plant foods." Using your knowledge of plant structure and physiology from this chapter, describe two benefits of eating "plant food." (Hint: cellulose.)
6. During the winter, water in xylem vessels freezes and expands. The following spring the water melts and leaves behind air pockets that can disrupt the flow of water through this xylem vessel. What is it about the structure of xylem vessels and the growth of woody dicots that prevents this from completely disrupting transport through xylem over time?
7. Why does it make sense to put a Christmas tree or a bouquet of roses in water even though they no longer have roots to absorb the water?

## In The News    Critical Thinking

**BRAIN VIAGRA—PART 1**
Now that you understand more about plants, reread this chapter's opening story about the issue of FDA regulation of herbal supplements. To develop an informed and thoughtful stand on this issue, it may help you to follow these steps:
1. Review your immediate reaction to the issue that you wrote when you began reading this chapter.
2. Based on your current understanding, again identify the issue using either a statement or a question. An issue is a point on which various persons hold differing views.
3. Collect new information about the issue. Visit the *In The News* section of this text's companion Web site at www.wiley.com/college/alters and watch the "Brain Viagra—Part 1" video. Then use the "summary" link to read the accompanying story and access related links. Use this information, the

links provided, and other online and library resources to find updates about this issue. State the sources of that information. Explain why you think the information is accurate. Also determine whether the information expresses a particular point of view or is biased in any way.
4. Determine which individuals, groups, or organizations have a stake in the issue. What does each stand to gain or lose depending on the outcome?
5. List possible outcomes (resolutions) of the issue. List the pros and cons of each outcome.
6. Which outcome do you think would be best? Why? Note whether your opinion differs from or is the same as what you wrote when you began reading this chapter. Give reasons for your views and for any changes in them based on the additional information you have collected and the analysis you have done.

# CONTROL SYSTEMS IN PLANTS

Plant Wake-up Call

"Time-out!" "Time to get up!" "Timing is everything." Timing is a crucial aspect of life in modern societies. You must get to class and work *on time*. You must register for classes *on time*. Even your bills must be paid *on time*. By paying attention to the time of day, the month, or the semester, you can manage your time more effectively.

The internal processes of your body also respond to time and other environmental cues. The biological clock that is programmed within you runs on approximately 24-hour cycles. As a result of these circadian rhythms, you produce certain hormones at night and your body temperature fluctuates depending upon the time of day.

Like animals, plants also respond to environmental stimuli such as light and dark, sensing the time of day and season. In addition, plants respond to temperature, and many do not flower unless first exposed to cold, a process called *vernalization*. University of Wisconsin-Madison molecular biologist Richard Amasino and graduate student Sibum Sung studied this phenomenon in the small flowering biennial weed, *Arabidopsis thaliana*, which belongs to the mustard family. Biennials, such as onions, lettuce, cabbage, carrots, radishes, and beets, and flowers like Hollyhocks, Sweet William, and Forget-Me-Nots, are plants that are started from seeds one year and flower and die the next year. Biennials must go through a cold winter before they flower.

Amasino and Sung located a gene (VIN3) that plays a role in the vernalization of *Arabidopsis*. The researchers determined that VIN3 is switched on by a long period of cold. When turned on, VIN3 encodes a protein that blocks the action of a gene called FLC, which normally prevents the plant from flowering. When its action is blocked by the action of VIN3, flowering occurs. More research is needed to understand the complex process of vernalization completely, but information such as this will help scientists control the spring flowering and reproduction of certain weeds that show up in your yard every spring or that remove nutrients from farmland before summer crops can emerge. In addition, the findings could help repress flowering in crops like cabbage and carrots that yield more if they don't bloom. You can learn more about vernalization research and view the video "Plant Wake-up Call" by visiting the *In The News* section of this text's companion Web site at www.wiley.com/college/alters.

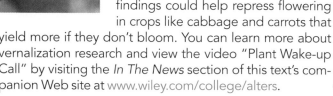

*Write your immediate reaction to Amasino and Sung's research on vernalization: first, summarize the research in a sentence or two; then suggest what you think its significance is. You will have an opportunity to reflect on your responses and gather more information on this topic in the* In The News *feature at the end of this chapter. In this chapter, you will learn more about factors that influence and control plant responses.*

## CHAPTER GUIDE

### How do plants control their activities?

**25.1** Plant hormones and other regulatory chemicals control plant physiology.

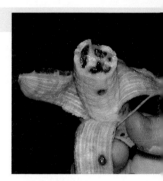

### How is development controlled in plants?

**25.2** In plants, development is a lifelong process.

**25.3** External environmental stimuli are the primary controls of seed germination.

**25.4** Programmed sequences of gene expression, along with hormones and environmental signals, control plant development.

### How do plants respond to the environment?

**25.5** The interactions of environmental stimuli and hormones govern growth responses.

**25.6** Plant responses to stress may be physical, chemical, or biological.

**25.7** Plants measure time by photoperiod and endogenous (internal) clocks.

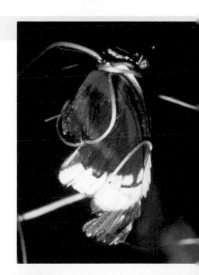

### What happens to plants as they die?

**25.8** Nutrient recycling occurs as plants redistribute substances from dying tissues to actively growing tissues.

# How do plants control their activities?

## 25.1 Plant hormones and other regulatory chemicals control plant physiology.

You might wonder how plants "know" to turn toward the light or to grow up and not down. They have no brains—no nervous tissue at all. Nevertheless, plant activities are well coordinated. Shoots grow up and roots grow down. Trees drop their leaves in the fall, and bulbs sprout in the spring. How do plants "know" what to do?

As plants grow, they are influenced by environmental factors such as the amount of water and light they receive. In addition, chemicals called **hormones** affect plant growth, differentiation, maturation, flowering, and many other activities. Hormones are chemical substances produced in small, often minute quantities in one part of an organism and then transported to another part of the organism where they bring about physiological responses.

Plant hormones and their actions are complex. Each type of hormone produces multiple effects, interacting with one another and with environmental signals as they control developmental and other phenomena in plants. There are at least six major kinds of hormones in plants: auxins (AWK-sinz), gibberellins (JIB-eh-REL-enz), cytokinins (SIGH-toe-KYE-ninz), abscisic acid (ab-SIS-ik), ethylene (ETH-ih-lean), and brassinosteroids (BRASS-in-oh-STARE-oydz).

**Auxins** induce cell elongation in stems and usually affect other plant processes as well, such as the development of roots; the differentiation of phloem and xylem (vascular tissue); the develop-

ment of buds, flowers, and fruits; and responses of plants to stimuli such as gravity and light. Auxins (often referred to in the singular as auxin) are primarily found in actively growing regions of the plant, such as new leaves and the tips of shoots and roots. If you have ever dipped plant cuttings in rooting compound to propagate them (**Figure 25.1**), then you likely have used auxin—the active ingredient in most rooting compounds. In general, you can think of auxins as a group of hormones that promote and regulate growth. In fact, the word "auxin" comes from a Greek word meaning "to grow."

Like auxins, **gibberellins** promote growth through cell elongation. Gibberellins stimulate hyper-elongation of stems in plants treated with the hormone as shown in **Figure 25.2**. Plants that naturally produce insufficient gibberellin, or cannot respond to the hormone, are developmental dwarfs. Many such dwarfs are known, some of which are used in agriculture and horticulture since they are more manageable than taller species and are higher yielding (**Figure 25.3**); short plants put less energy into growing stalks and more energy into growing grain. Gibberellins also promote the germination of seeds, ready endosperm reserves for use during early embryo growth, and support flower and fruit development. As you might expect, gibberellins are primarily found in developing seeds and fruits, the young leaves of developing apical buds, elongating shoots, and the apical regions of roots.

**Cytokinins** are plant hormones that stimulate cell division. Their name comes from the word for cell division: cytokinesis. Plants manufacture cytokinins primarily in their roots, especially in the xylem and at the root tip where mitosis is active. They then transport these hormones via the xylem to other parts of the plant, such as developing fruits.

**Abscisic acid (ABA)** is a plant hormone that may improve chances of survival in a plant. It induces and maintains dormancy as it inhibits shoot growth and induces seeds to synthesize storage proteins. ABA also regulates the opening and closing of stomata, which are openings in the epidermis of leaves. The movement of water into

**Figure 25.1 Auxin.** The plant on the left was dipped in a rooting compound containing auxin, which stimulates the development of roots. The plant on the right was not treated.

**Figure 25.2 Gibberellins.** The California poppy (*Eschscholtzia californica*) on the left is normal. The California poppy on the right was treated with gibberellins, resulting in elongation of the stem.

**Figure 25.3** Two closely related lines of "super dwarf" spring wheat (*Triticum aestivum*). These plants have gene mutations that result in their producing only small amounts of gibberellin. They were developed at Utah State University for NASA to fit in the small growth chambers on the space station. The plants on the left are 40 cm (about 16 inches) tall, and the ones on the right are 25 cm (about 10 inches) tall, considerably shorter than the non-mutant average height for spring wheat of about 3 feet.

guard cells opens stomata, while movement of water out of guard cells closes stomata. The manufacture of ABA is stepped up in response to water loss. This stimulates closure of stomata, thus conserving water. This plant hormone is found predominantly in mature, green leaves.

**Ethylene** is the one plant hormone that is a gas. You might have learned that fruit ripens quickly when you put an unripe fruit into a paper bag with a ripe fruit. The ethylene given off by the ripe

fruit ripens the other. Putting fruit together in a bowl has the same effect. Ethylene is used commercially to enhance ripening in bananas and other fruits that are picked for shipment while they are still green. Ethylene also affects flower blossoming and can shorten the life of blooms. To retard ripening and blossoming, fruit and plants are often refrigerated, which reduces internal ethylene production and activity. Ethylene does not appear to be necessary for normal plant growth, although it stimulates shoot and root growth and differentiation, flower opening, flower and leaf senescence (aging), and leaf and fruit abscission (detachment). It is produced in all plant organs, such as roots, stems, leaves, fruits, and seeds.

As their name suggests, **brassinosteroids** are steroid hormones. Their chemical structure is similar to the steroid hormones in animals, such as the male and female sex hormones, in that they are comprised of four carbon rings. Brassinosteroids cause a wide array of responses in plants, such as stimulating stem and pollen tube elongation, stimulating ethylene production, and inhibiting root growth and development. Due to their potential agricultural uses related to these functions, brassinosteroids are currently a research focus.

Plant regulatory chemicals also influence seed germination, plant growth, and plant development by affecting a plant's response to the external environment. One important group of such regulatory chemicals includes the **phytochromes**, plant photoreceptor pigments that absorb red and far-red light. To understand what that means, look back to Figure 7.2. Notice that red light consists of wavelengths from about 625 nm to about 700 nm. The term *far-red light* refers to wavelengths of more than 680 nm within the red range. The term *red light* refers to wavelengths of less than 680 nm within the red range. Phytochromes monitor light conditions in the environment; this information helps regulate virtually all aspects of plant development from seed germination to flowering.

# How is development controlled in plants?

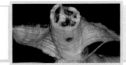

## 25.2 In plants, development is a lifelong process.

**Development** refers to the changes in an organism throughout its life cycle. In plants, development is never completed, and plants lack the fixed developmental blueprint of animals. The process is continuous, spanning the lifetime of the plant. Moreover, external influences directly affect development in plants, whereas they usually play a less important role in the development of animals. Why? The production of plant hormones, crucial in plant development, is sensitive to changes in the environment. This does not apply to most animal hormones, because animals' internal environment generally influences hormone production. Hormone production in animals is usually regulated by feedback loops that are controlled by the nervous system and by the concentration of specific substances in the bloodstream, such as glucose or calcium ions. Conversely, external environmental conditions, such as the intensity of the sun, the amount of available water, and the length of days influence hormone production in plants. The ability of plants to respond to environmental cues is an evolutionary adaptation

that helps produce individuals well suited to their environments, for plants cannot move from unsuitable environments as animals can.

Another characteristic that distinguishes plant development from animal development is that almost all differentiation in plants is reversible. Developmental reversibility is rarely possible in animals. The only truly *totipotent* cells in mammals are the fertilized egg and cells that arise from cleavage of the fertilized egg during the first four days after fertilization. It is these cells that sometimes separate from one another and develop on their own, resulting in identical twins, triplets, or even quadruplets. (A totipotent cell has the potential to develop into any other kind of cell in the organism.) Animal embryos at the stage of implantation, however, do contain *pluripotent stem cells* that have the potential to develop into almost any cell in the body but cannot develop into a new organism. Adult animals also have *multipotent stem cells* in every organ of the body. These cells can differentiate into a limited number of cell types and do so when they replace dead or damaged cells in an organ. For example, multipotent stem cells of the bone marrow can differentiate into any type of blood cell but no other

(a)                                    (b)                                    (c)

**Figure 25.4 Growing new plants from plant tissue.** (a) A small piece of plant tissue, called an *explant*, is taken from dividing plant cells such as root or shoot tips, dipped in antiseptic to kill any bacteria present, and placed on an agar medium that is high in nutrients and low in hormones. (b) The callus (undifferentiated tissue) develops from the explant and is transferred to an agar medium having the proper balance of nutrients and growth regulators (auxins and cytokinins) for that species of plant to produce roots, shoots, or both. (c) Plantlets develop form the callus.

cell types. The *In The News* opening story for Chapter 12 discusses stem cell research and its potential use in treating human diseases.

In contrast to animals, all plant cells are totipotent. For example, mature cells within plants can be removed from the plant, cultured on a nutrient medium, and stimulated to divide by using plant hormones such as auxins, cytokinins, and gibberellins. The shapeless mass of undifferentiated cells that results is called a *callus*. Roots and shoots form, which will eventually develop into a new plant, as shown in **Figure 25.4**.

### CONCEPT CHECKPOINT

1. One cannot predict how many branches most plants will have when they die, but we know exactly how many legs most animals will have. What explains this difference?

2. Which plant hormone(s) do you think might elicit each of the following plant responses? (a) tree branches tend to grow toward openings in the forest canopy; (b) trees lose their leaves in the fall; (c) the roots of a plant growing from the side of a cliff will bend downward and the stem will bend upward; and (d) pine pollen tubes only germinate when they land on the female cone.

## 25.3 External environmental stimuli are the primary controls of seed germination.

We often think of the life cycle of plants as beginning with the **seed**, which is comprised of an embryo, stored food, and a protective seed coat. **Seed germination** occurs when the embryo begins to grow and the radicle (young root) emerges through the seed coat (see Section 23.7).

What triggers and controls seed germination? Seeds need water, oxygen, and the appropriate temperature for germination to occur. Most small seeds with thin seed coats, such as lettuce seeds, require light for

germination. Moreover, the seeds of most plants, especially those that grow in areas that have seasonal temperature changes, require a period of dormancy and cold temperatures before they will germinate. **Seed dormancy** is an evolutionary adaptation that allows seeds to germinate only when growing conditions are good. You may be familiar with this phenomenon if you have ever planted bulbs in the fall that sprouted in the spring after being dormant during a cold winter.

Requirements for breaking seed dormancy include abrasion of the seed coat, or being dried out and then rehydrated, or soaking in water or fat solvents to leach out germination inhibitors, such as auxins, gibberellins, or ABA that may have accumulated in seed coats. Some seeds cannot have exposure to light before germination, while others must have light exposure. Some seeds germinate only at certain wavelengths of red light. This evolutionary adaptation assures that the seed is located in an area free of a thick canopy of leaves, which would shade the germinating seedling too much for optimal growth. Moreover, these surrounding plants would compete with the seedling for nutrients and moisture in the soil. When phytochromes in the seed detect certain wavelengths of red light, it is likely that the seed has been dispersed away from a dense population of trees or shrubs and heavy competition. The change in the phytochromes due to absorption of red light triggers seed germination.

When the requirements for breaking seed dormancy have been met, changes in the seed allow it to imbibe (take up) water. Water uptake by the seed is necessary for metabolic activities to occur. It helps activate enzymes already in seeds and provides a fluid medium in which chemical reactions can take place, such as the synthesis of new enzymes to help digest and use the food stored in the seed. With the intake of water, the embryo and stored food swell, rupturing the seed coat and allowing oxygen to enter. Anaerobic metabolism becomes aerobic. Cell growth and division begin, and if the seed continues to be supplied with water, germination will continue.

## 25.4 Programmed sequences of gene expression, along with hormones and environmental signals, control plant development.

Control of plant development does not lie in a single source—there is no one "captain at the helm." Instead, control mechanisms come from three sources: intracellular (within cell) activities, intercellular (between-cell, cell-to-cell) activities, and environmental stimuli. These three "levels" of control interact and overlap, determining the overall development of a plant.

The primary intracellular control mechanism in the development of any organism, including plants, is gene expression. Genes express themselves through the production of particular polypeptides. Polypeptides form proteins that comprise structures and drive metabolism via enzymatic action. Genes are turned on and off as plant development proceeds—as meristematic tissue differentiates into leaf tissue in one spot and bud tissue in another, for example. Thus, a programmed sequence of gene activations and deactivations produces required gene products at appropriate times, resulting in the orderly development of a plant. Other intracellular factors may modify development. These factors include level of turgor, which is the rigidity of cells due to the uptake of water; presence of necessary minerals; presence of pathogens; and action of internal clocks, which are discussed in Section 25.7.

The primary intercellular control mechanism in plant development is the action of hormones. Auxins, gibberellins, cytokinins, and brassinosteroids all play roles in plant growth and development, coordinating activities of groups of cells, either singly or in combination.

Extrinsic controls in plant development are varied and may include light, temperature, gravity, soil composition, and soil moisture. These controls originate outside the plant and convey information about the environment. Signals from the environment may modulate development.

## *just wondering . . .*

### Questions students ask

### I know that some fruits are seedless, such as some watermelons and grapes. How can that happen?

Based on your question, we suspect that you already know that fruit usually develops when flowers are pollinated and fertilization occurs, producing a plant embryo housed within a seed. Fruits are mature ovaries consisting of seeds, the tissues connected with them, and their coverings. So how can fruit develop without seeds? There are a few ways this can happen.

One method of producing seedless fruit is to treat flowers with auxin or gibberellin before their pollen matures. This hormone application triggers some angiosperm species to produce fruit without fertilization. Such fruit is called parthenocarpic (PAR-theh-no-CAR-pick), a term that comes from two Greek words meaning virgin (*parthenos*) fruit (*karpos*). How does this work? Researchers have determined that parent tissues, which form the fruit, can act independently from embryonic and other tissues that make up the seed. Therefore, when hormonally stimulated, some species develop fleshy ovary tissue (fruit) without seed formation. Some seedless fruits that develop in this way are apples, currants, cucumbers, tomatoes, and eggplants.

Seedless grapes have arisen through to two gene mutations that cause seeds to initially develop after normal fertilization, but then stop development when the seeds are extremely small. Therefore, grapes

are not really seedless, but have such small underdeveloped seeds that consumers do not notice them most of the time. Grape breeders generally propagate seedless grapes by removing the embryos before their development is arrested and nurturing them into mature plants using tissue culture techniques. Many varieties of seedless citrus fruit, such as seedless oranges, are the result of similar gene mutations.

Some plants that produce seedless fruit have separate male and female individuals. If a female plant is not near a male plant and cannot be pollinated, the female plant may form seedless fruit. Persimmon and pineapple produce seedless fruit in this way.

Seedless bananas and seedless watermelons arise in yet another way. Both seedless varieties are triploid (they have three sets of chromosomes) rather than diploid, so chromosomes cannot pair normally during meiosis to form sperm and eggs. How are the triploid plants formed? First, a diploid seedling is exposed to the chemical mutagen colchicine. The result is a tetraploid plant—one that has four sets of chromosomes. These plants produce diploid eggs, and breeders pollinate them with haploid sperm (pollen) produced by normal diploid plants. The result is triploid seeds that develop into triploid plants. And yes, *wild* bananas do have seeds, which can be seen in **Figure 25.A**. ●

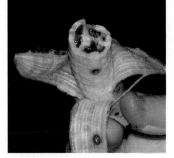

**Figure 25.A** A wild banana has seeds!

*Are you wondering about a topic in biology and how it relates to your life? Submit your question by clicking the* Just Wondering *link in this text's companion Web site at www.wiley.com/college/alters.*

## How do plants respond to the environment?

### 25.5 The interactions of environmental stimuli and hormones govern growth responses.

Environmental stimuli and hormones govern many growth responses of plants, including responses to light and gravity. Such plant responses, which involve plant parts bending or curving to-

ward or away from an external stimulus, are called **tropisms**. When a plant bends toward the stimulus, the tropism is positive; when a plant bends away from the stimulus, the tropism is negative.

### Phototropism

A plant growth response to light is called **phototropism** (foe-TOE-trah-PIZ-um), and plant parts exhibit both positive and negative

**Figure 25.5 Positive phototropism.** Plants grow toward the light because auxin, a hormone that stimulates growth, is in a higher concentration on the darker side of the stem. Thus, the side of the plant away from the light grows faster, causing the plant to bend toward the light. The stems of this flowering shamrock are oriented toward the window, which lets in plenty of sunlight.

 **Visual Thinking:** Draw the stems of this plant showing the difference between the growth of cells on the sunny side of the stem and those on the dark side of the stem.

phototropisms. For example, stems generally exhibit positive phototropism as they lean toward the light (**Figure 25.5**). Some tendrils (modified stems or leaves) of climbing plants exhibit negative phototropism as they cling to shadowed crevices while climbing a wall. The wrapping and clinging of tendrils as they contact objects is called *thigmotropism* (**Figure 25.6**). Leaves generally exhibit *plagiotropism* (PLAY-gee-OT-rah-PIZ-um) as they orient at angles intermediate to light, maximizing their photosynthetic output by sens-

ing and responding to changes in the amount and direction of light. Plant roots are nonphototropic or even negatively phototropic.

What causes plants to bend toward the light? Today scientists know that the plant hormone auxin is redistributed in the plant stem based on light exposure. Darkened parts of the stem receive more of this growth hormone, which causes elongation of the cells in the stem. As cells on the darkened side of the stem elongate and cells on the lighted side do not, the plant stem bends toward the light. The discovery of this fact and of the hormone auxin is one of myriad interesting stories of discovery in science. The *How Science Works* box chronicles this tale.

### Gravitropism

What would happen to a growing plant if you turned its pot on its side? Many school children have performed this simple experiment to find that the shoot will curve up away from the center of the Earth, while the roots will grow "down" toward the center of the Earth. Both grow parallel with the direction of gravitational pull. This plant growth response to gravity is called **gravitropism** and is shown in corn seedlings in **Figure 25.7**.

Do other plant organs such as leaves and stems exhibit gravitropism? Plant structures that grow horizontally, such as some lateral branches, stolons, and rhizomes are said to exhibit *diagravitropism*. (Stolons and rhizomes are stems that run parallel to the ground, giving rise to roots and shoots along their lengths.) Those plant organs oriented between the horizontal and the vertical, such as many branches and roots, are said to exhibit *plagiogravitropism*. **Figure 25.8** illustrates these differences.

How do plants sense gravity? The current explanation is the **starch-statolith hypothesis**. Statoliths are granules that are pulled downward by gravity. Invertebrates have statoliths of sand or calcium carbonate for gravity sensing; vertebrates have otoliths or "ear stones" of calcium carbonate (see Section 33.12). In plants, statoliths are starch granules and are grouped into larger, membrane-bound structures called *amyloplasts* (AM-uh-low-plasts). Amyloplasts are found within cells called *statocytes*. Gravity sensing involves the movement of the amyloplasts within the statocytes in response to gravity (**Figure 25.9**).

When a plant senses a change in gravity, such as when its pot is laid on its side, how do the plant shoot and roots curve in response,

**Figure 25.6 Thigmotropism.** The tendrils of a passion flower plant (*Passiflora* sp.) wrap around a butterfly that set down on it for the night.

**Figure 25.7 Gravitropism.** After beginning its growth, these corn seedlings were turned horizontally. The roots exhibit positive gravitropism, growing toward gravity (the Earth), while the shoots exhibit negative gravitropism, growing away from gravity.

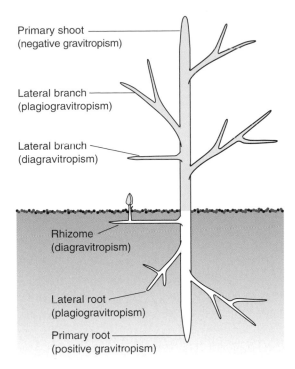

Primary shoot
(negative gravitropism)

Lateral branch
(plagiogravitropism)

Lateral branch
(diagravitropism)

Rhizome
(diagravitropism)

Lateral root
(plagiogravitropism)

Primary root
(positive gravitropism)

**Figure 25.8** The range of gravitropic responses in shoots and roots. These responses ensure that the shoot, branches, and leaves will grow up through the soil and catch the sun's rays, the roots will grow down into the soil where nutrients and moisture are found, and plant parts will efficiently fill the surrounding space.

as shown in Figure 25.7? As in phototropism, a plant bends toward or away from gravity because of auxin. In negatively gravitropic organs such as shoots, auxin diffuses to the lower side of the organ so that it is in a higher concentration than on the upper side, causing greater elongation of the cells on the side of the shoot closer to the ground and resulting in the shoot curving upward. As you can see, the Cholodny-Went hypothesis of asymmetric auxin distribution

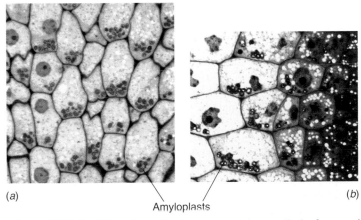

(a)                                                                (b)
Amyloplasts

**Figure 25.9** Position of amyloplasts in root cap cells before and after turning a plant on its side. (a) The position of amyloplasts in root cap statocyte cells in a normally growing plant. (b) The position of amyloplasts in the same root cells about 10 minutes after the plant was placed on its side. Since the amyloplasts are pulled downward by gravity, they now occupy a different position in the cell, providing the plant with data as to "up" and "down" with respect to the gravitational pull of the Earth.

**Figure 25.10** Nastic movements in plants. Thigmonasty is a plant response to touch. In this two-image sequence, the compound leaves of the mimosa, or "sensitive plant" (*Mimosa pudica*), fold up when the plant is touched even slightly. This response may help prevent leaf damage as predators walk by or when herbivores nibble the plant. After 15 to 30 minutes, the leaves open again.

(see the *How Science Works* box) applies here just as it does to phototropism. In positively gravitropic organs such as roots, auxin diffuses to the lower side of the organ as well; the higher auxin concentration on the lower side of the root is thought to inhibit cell elongation relative to the upper side, resulting in the roots curving downward.

## Nastic movements

Have you ever seen time-lapse photography of plants in which they appear to sway, or in which their buds swell, open, and eventually fall off? These plant movements are called nastic (NAS-tick) movements. **Nastic movements** differ from tropisms in that the direction of the response is not related to the direction of the stimulus as it is in tropisms. In addition, stimuli may be external, coming from one or many directions, or they may be internal. Stimuli may affect the whole plant or just one part.

Nastic movements in plants include the opening and closing of flower petals in response to changes in temperature or light and dark, and the closing of the leaves in carnivorous plants in response to insects stimulating sensitive hairs on these organs (see the *Just Wondering* box in Chapter 24). Another example of a nastic movements is shown in **Figure 25.10**.

Nastic movements generally result from changes in the sizes of special cells called *motor cells*. These cells change size as water moves into or out of them. The movement of water follows the movement of ions (see Chapter 5), which pass into or out of cells as ion channels open or close in response to the stimulus. In many plants, shrinkage of the motor cells due to water loss causes the overall movement of the plant. Nastic movement may also result from differing rates of growth in parts of a plant.

### CONCEPT CHECKPOINT

3. For each of the following, describe (1) the extrinsic control signal and (2) the intercellular control mechanism that regulates the response of the plant: (a) seed germination, (b) growth of plant stems toward a window, and (c) growth of plant roots down into the soil.

# How Science Works

## What causes plants to bend toward the light?

Charles Darwin and his son Francis asked that question in 1880 and determined through a series of experiments that it is the tip of the plant that detects a light stimulus. They covered the tips of seedlings with opaque caps, and the seedlings no longer responded to the light stimulus. Yet, when they buried the shoot in sand and left the tip uncovered, they observed that the seedling bent toward the light as usual although the shoot was in the dark (**Figure 25.B**). Darwin hypothesized that the tip of the plant detected the light and somehow signaled the stem, causing it to bend.

In 1913, Danish plant physiologist Peter Boysen-Jensen showed that this "something" was a chemical passing from the seedling tip to the stem. Boysen-Jensen separated the tip from the shoot by a block of gelatin (**Figure 25.C**). Although the shoot and the tip were separated, chemicals could diffuse from the tip to the shoot through the gelatin. The results were that the seedling behaved normally in response to light. However, when he placed a thin piece of mica, which is impervious to chemical flow, between the cut tip and the shoot, no bending occurred in the presence of directional light. Boysen-Jensen concluded that the tip of the seedling was sending chemicals to the

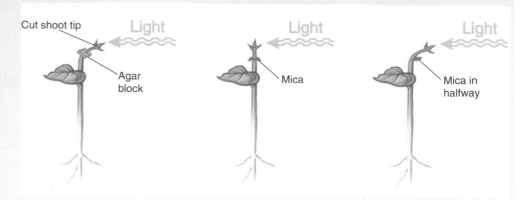

**Figure 25.C** Peter Boysen-Jensen's experiment. In the seedling on the left, the tip was cut from the shoot, and an agar (gelatin) block was placed between the cut tip and the shoot. The shoot *did* respond to the light. In the seedling in the middle, an impervious square of mica was placed between the cut tip and the shoot. The shoot *did not* respond to the light. In the seedling to the right, an impervious square of mica was placed between the cut tip and the shoot only on the side receiving direct light. The shoot *did* respond to the light.

shoot, which diffused through the gelatin but could not flow through the mica. He further showed that, when the mica was placed between the tip and the shoot on the side receiving direct light, the shoot still bent toward the light. Boysen-Jensen concluded from these results that the chemicals sent from the tip stimulated the darkened side of the shoot to grow while not stimulating the side of the shoot exposed to the direct light.

In 1926, Dutch graduate student Fritz Went isolated the substance responsible for the bending and named it auxin. How did he do that? First he exposed seedlings to light from one side. Then he cut the tips from the shoots and split them longitudinally into two halves: the half that had been directly exposed to the light and the half that had been shaded. He placed each half on an agar block to collect the chemical that diffused out of the bases of the tip halves. By doing so, he was able to characterize this chemical (auxin) and

determine how much auxin was collected on each block. Went determined that the shaded half had more auxin than the lighted half, leading him to conclude that the portion of the shoot tip that was exposed to the direct light allowed less auxin to migrate down its side of the shoot than the shaded portion of the shoot tip.

In 1937, Went and N. Cholodny, an investigator working independently in Germany, pulled their ideas together in the **Cholodny-Went hypothesis** of phototropism. The hypothesis states that the growth curvatures seen as a plant bends toward the light are due to an unequal distribution of auxin between the two sides of the curving organ. Auxin is responsible for cell growth and elongation. Therefore, the cells on the shaded side of the shoot grow longer than on the illuminated side, bending the stem toward the light.

Although the Cholodny-Went hypothesis has been challenged and re-researched in recent years, the evidence supporting the hypothesis seems strong. Scientists recently discovered that the photoreceptor for light involved in phototropic responses is a blue light receptor called *phototropin*. However, the mechanism by which blue light establishes lateral redistribution of auxin is not yet known. ●

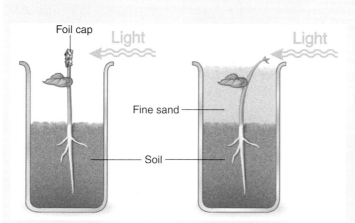

**Figure 25.B** Charles and Francis Darwin's experiment. In the seedling on the left, the tip was covered with foil. The shoot *did not* respond to the light. In the seedling on the right, the shoot was buried in sand with the tip exposed. The shoot *did* respond to the light.

## 25.6 Plant responses to stress may be physical, chemical, or biological.

**Plants respond to environmental stress in three ways: avoidance, resistance, or susceptibility. Figure 25.11** shows the relationships between environmental stress, plant responses, and plant survival or death.

Some plants avoid stress because their life cycles allow them to grow, produce seeds, and die during favorable environmental conditions. The seeds survive during periods of adverse conditions, and when favorable conditions return, the seeds germinate. For example, during the spring in the American Southwest the desert is in bloom (**Figure 25.12a**). At other times of the year, the desert landscape offers dead grass, some bushes, and cacti (**Figure 25.12b**).

Some plants effectively resist stress. In spite of the stress, such plants maintain homeostasis (internal equilibrium). They have the ability to acclimate to the stress. **Acclimation** (ACK-lih-MAY-shun) is the ability of a plant to become accustomed to new conditions. How does this ability develop? As the plant is gradually exposed to the stress, physiological changes occur that enable the plant to function and ultimately to survive and reproduce in the stressful environment. When the stress is removed, resistant plants continue normal growth and development. For example, the resurrection fern is capable of losing much of its water content during dry periods—so much so that it looks dried up and dead, as shown in **Figure 25.13a**. Nevertheless, when moisture returns, the plant rehydrates. Its fronds unfurl and are soft and green, as shown in Figure 25.13b.

Conversely, plants that are susceptible to stress stop flowering and forming seeds. Stress eventually causes senescence in these plants. **Senescence** (sih-NEH-sense) simply means "growing old" and is the final stage in the development of a plant. During senescence, parts of the plant begin to die.

You know what causes stress in your life: serious illness or death of family members, personal or work-related relationship dif-

(a)  (b)

**Figure 25.12 Stress avoidance.** (a) Flowering desert plants avoid the stress of heat and drought because they grow, bloom, produce seeds, and die during the wet period of spring. (b) During the rest of the year, their seeds survive and will germinate when environmental conditions become favorable once again.

ficulties, too little sleep, too many demands on your time, and long commutes on crowded roads. The list could go on and differs from person to person. But what if you had to make up a "stress list" for plants? What would you include?

Plant stress falls into two categories: abiotic stress (physical or chemical) and biotic stress (biological). The abiotic and biotic causes of plant stress are listed in **Table 25.1**.

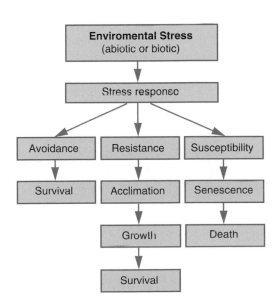

**Figure 25.11 Plant responses to stress.** Plants capable of avoiding or resisting stress generally survive, but plants susceptible to stress generally die.

(a)  (b)

**Figure 25.13 The resurrection fern is resistant to water stress.** Resurrection ferns grow on tree trunks and limbs, rocks, and dry ground. (a) The fern looks dried up and dead, but it is not. It has simply lost much of its water content due to drought conditions. (b) The resurrection fern "comes back to life" when moisture returns to its environment. Its soft, green fronds unfurl, with no permanent damage having been done to the plant.

| TABLE 25.1 | Causes of Plant Stress |
|---|---|
| **Abiotic Causes** | **Biotic Causes** |
| • Too much or too little *water*<br><br>• *Temperature* too high or too low<br><br>• Concentration of *inorganic ions*, such as sodium ions ($Na^+$) or chloride ions ($Cl^-$), too high or too low<br><br>• *Heavy metals* such as lead, cadmium, or mercury<br><br>• *Airborne pollutants* such as nitrogen and sulfur oxides | • Disease-causing *insects*<br><br>• Disease-causing *microorganisms*<br><br>• *Soil compaction* (soil that is pressed together and hard)<br><br>• Plant-to-plant *competition*<br><br>• *Herbivore* damage |

## 25.7 Plants measure time by photoperiod and endogenous (internal) clocks.

You may have seen "flower clocks" in your travels—they exist in many cities of the world, such as Geneva, Switzerland; Valparaíso, Chile; Alexandria, Egypt; and Niagara, Ontario, Canada. Such clocks are decorated with flowers but have the usual clock mechanism with hands to tell the time. Two hundred and fifty years ago, Swedish botanist Carolus Linnaeus, noted in Chapter 1 for establishing our system of binomial nomenclature, proposed a different type of flower clock. Linnaeus knew that some flowers have fixed times for opening and closing. **Figure 25.14** shows a representation of the flower clock proposed by

**Figure 25.14 A representation of the Linnaean flower clock.** The flowers of some plants open and close every day at regular times. Linnaeus proposed a clock that would tell the time by the opening and closing of flowers.

Linnaeus, which would tell the time without clock hands or a clock mechanism, but simply by the opening and closing of flowers.

How is this daily rhythm of opening and closing of flowers controlled? Plants, like animals, protists, fungi, and bacteria, have internal **biological clocks**, or timing mechanisms, that regulate 24-hour cycles of physiological activity called **circadian rhythms** (sir-KAY-dee-un). Biological clocks often interact with environmental cues, which help keep the internal clock timed to the outside world. The basis of biological clocks is genetic; that is, genes regulate the cyclical patterns of physiology. Circadian rhythms and biological clocks in animals are discussed in Chapter 37.

Along with plants that open and close every day at certain times, there are plants that bloom at certain predictable times during the year. You might know that crocuses bloom in the spring, roses bloom in the summer, and mums bloom in the fall. How do plants regulate the process and timing of developing flowers?

Flowers are the reproductive structures of angiosperms (see Chapter 23). The change that plants make from nonflowering to flowering is a major transition in their life cycles, moving from production of strictly vegetative (nonreproductive) structures to production of both vegetative and reproductive structures. The proper timing of this transformation is critical for the reproductive success of plants. One factor in regulating this transition is the developmental stage of a plant; a seedling, for example, is developmentally too young to produce flowers. Other factors are environmental. Daylength is one important environmental control factor.

Flowering plants have the ability to measure the proportion of daylength in a 24-hour period. This ability is called **photoperiodism** (FOE-toe-PEER-ee-oh-dizm). There are three basic categories of plants based on their photoperiodic responses: short-day plants, long-day plants, and day-neutral plants. Short-day plants flower in response to daylengths that are shorter than a certain value within a 24-hour cycle; long-day plants flower in response to daylengths longer than a certain value within a 24-hour cycle; and day-neutral plants flower regardless of daylength. Other variations on this theme exist as well. For example, some plants will flower only if a certain number of short days precede a certain number of long days. **Table 25.2** lists some short-day, long-day, and day-neutral plants.

How does photoperiodism work? The name suggests that plants measure the light (photo-) period, but this is not the case. Plants measure the period of darkness in between periods of light. Experiments by American scientists Karl C. Hamner and James F. Bonner in 1938

| TABLE 25.2 | Examples of Short-Day, Long-Day, and Day-Neutral Plants | | |
|---|---|---|---|
| | **Short-Day Plants** | **Long-Day Plants** | **Day-Neutral Plants** |
| | • chrysanthemum<br>• poinsettia<br>• soybean<br>• Japanese morning glory<br>• cocklebur | • dill<br>• Swiss chard<br>• rye grass<br>• radish<br>• spinach<br>• spring wheat | • cucumber<br>• sunflower<br>• common bean<br>• garden pea<br>• corn |

(a) Most seeds germinate in the spring.

(b) The vegetative rosettes require an overwintering to flower the next spring.

(c) In spring, if large enough, the stems lengthen and the plant flowers in May.

(d) By late July, the seeds have matured and the plant dies.

**Figure 25.15** Vernalization: life cycle of the wild parsnip. The wild parsnip (*Pastinaca sativa*) is a biennial plant and requires a prolonged period of cold in order to flower during its second spring. (a) Wild parsnip seeds germinate in the spring. (b) The vegetative plants grow during the spring and summer, and require a cold winter to (c) flower the next spring, usually in May and June. (d) By the end of July, the seeds are mature and the plant dies. These seeds will germinate in the spring (a).

revealed this fact. When they interrupted the day periods with periods of darkness, there was no effect on flowering. When they interrupted the dark periods with bursts of light, however, flowering did not occur. Thus, short-day plants actually require long dark periods, and long-day plants require short dark periods.

Scientists have also determined that leaves are the parts of the plant that sense the photoperiodic signal, with phytochromes (see Section 25.1) acting as the likely light sensor. Since the response to the light-on/light-off signal is flowering, scientists know that somehow the signal is transmitted from plant leaves to the apical meristems that will produce buds and consequently flowers.

Temperature can modify photoperiodic behavior. In some plants, flowering is enhanced at certain temperatures. In others, high temperatures elicit one response, while low temperatures elicit another. One well-studied interaction between photoperiod and temperature is the low-temperature requirement of biennial

flowering plants that is discussed in this chapter's opening story. Biennials, such as the wild parsnip shown in **Figure 25.15**, must go through a cold winter before they flower. This requirement for prolonged exposure to cold temperatures prior to flowering is called **vernalization**.

The opening story tells how scientists have recently come to understand the molecular mechanism of vernalization. In the species of plants that they have studied, Amasino and Sung, along with colleague S.D. Michaels, have determined that flowering is controlled by the regulation of a floral inhibitor gene (FLC). The process works like this: When biennials first germinate, they have high levels of a protein transcribed from FLC that represses flowering. After vernalization, very little of the protein is produced, since vernalization switches on the gene VIN3, which encodes a protein that blocks the action of FLC. Flowering is no longer repressed, and the plants flower rapidly in the spring.

## What happens to plants as they die?

### 25.8 Nutrient recycling occurs as plants redistribute substances from dying tissues to actively growing tissues.

Plant death is different from animal death in many ways. A plant has neither a heart that stops beating nor a nervous system that stops functioning. In addition, plant cells are totipotent as described in Section 25.2. Most of a plant's cells can die, yet those

cells still living could give rise to a new plant if cultured appropriately. The only certain time of plant death is when none of a plant's cells produce energy and its cell membranes are no longer intact.

Plants often die from environmental stress, as noted in Section 25.6, or they may die at a genetically programmed time. Annuals, for example, generally live for one growing season, biennials for two, and perennials for a few.

Senescence is the final stage in the development of plant cells, tissues, and organs, prior to cell death (**Figure 25.16**). A great deal of activity occurs in dying plant parts: respiration increases, photosynthesis decreases, and large molecules are broken down into smaller ones. Nutrient recycling occurs as plants redistribute substances from senescing tissues to actively growing tissues. For example, older leaves may die as a plant continues to grow and flower petals may die as seeds form; their catabolic (breakdown) products are then transported to other parts of the growing plant. It is clear—plants die productively.

**Figure 25.16** Senescence in sunflowers.

## CONCEPT CHECKPOINT

4. Choose among avoidance, acclimation, or senescence to explain each of the following:
   a. Features that distance a plant from the impact of stress even though the source of the stress remains in the environment
   b. Nonheritable physiological or structural modifications that occur over a plant's lifetime enabling it to live and reproduce in the stressful environment
   c. The plants succumbing to the environmental stress
5. Winter rye, when planted in the spring, will take four to five months to flower after the seed germinates. When planted in the fall, however, winter rye will flower about two months after seed germination, unless the plant is artificially exposed to very short daylengths during that period. What aspects of the coordination of timing of flowering are exhibited in winter rye?

## CHAPTER REVIEW

## Summary of Key Concepts and Key Terms

### How do plants control their activities?

**25.1** Plant growth is influenced by environmental factors and plant **hormones** (p. 428).

**25.1** Plant hormones produce multiple effects, interacting with one another and with environmental signals as they control developmental and other phenomena in plants.

**25.1** There are at least six major kinds of hormones in plants: **auxins** (p. 428), **gibberellins** (p. 428), **cytokinins** (p. 428), **abscisic acid** (p. 428), **ethylene** (p. 429), and **brassinosteroids** (p. 429).

**25.1** Auxins and gibberellins promote growth through cell elongation; cytokinins stimulate cell division; abscisic acid regulates seed

dormancy and the opening and closing of stomata; ethylene stimulates ripening in fruit; and brassinosteroids cause a wide array of responses.

**25.1** **Phytochromes** (p. 429) are photoreceptor pigments that regulate plant development from seed germination to flowering.

### How is development controlled in plants?

**25.2** In animals, **development** (p. 429) is completed with the mature organism, but in plants, development is a lifelong process.

**25.2** Almost all differentiation in plants is reversible, and each living plant cell possesses the potential to develop into any other kind of plant cell.

**25.3** Various environmental stimuli and processes are the primary controls of **seed germination** (p. 430).

**25.3** The life cycle of plants begins with **seeds** (p. 430), which need water, oxygen, and the appropriate temperature for germination.

**25.3** The seeds of many plants, especially those that grow in areas with seasonal temperature changes, require a period of **seed dormancy** (p. 430) and cold temperatures before they will germinate.

**25.3** Various conditions must be met before some seeds will germinate, such as abrading the seed coat, drying and rehydration, soaking in

water or fat solvents, having no light exposure, or having some light exposure.

**25.3** Water uptake up by seeds is necessary for metabolic activities.

**25.4** The control of plant development comes from three sources: intracellular activities, intercellular activities, and environmental stimuli.

**25.4** The primary intracellular control mechanism in the development of any organism, including plants, is gene expression.

**25.4** The primary intercellular (cell-to-cell) control mechanism in plant development is the action of hormones.

**25.4** Signals from the environment are used to modulate development and include light, temperature, gravity, soil composition, and soil moisture.

## How do plants respond to the environment?

**25.5** Environmental stimuli and hormones govern many growth responses of plants, including responses to light and gravity.

**25.5** Plant responses that involve plant parts bending or curving toward or away from an external stimulus are called **tropisms** (p. 431).

**25.5** Plant growth responses to light are called **phototropisms** (p. 431).

**25.5** Phototropic responses are caused by the redistribution of auxin in a plant organ, such as a stem.

**25.5** Plant growth responses to gravity are called **gravitropisms** (p. 432).

**25.5** The **starch-statolith hypothesis** (p. 432) explains gravity sensing in plants, which involves the movement of starch granules (amyloplasts) within cells called statocytes.

**25.5** The **Cholodny-Went hypothesis** (p. 434) explains gravitropic and phototropic responses that are caused by the redistribution of auxin in a plant organ.

**25.5** **Nastic movements** (p. 433) are plant movements in which the direction of the response is not related to the direction of the stimulus.

**25.5** Nastic movements generally result from changes in the sizes of motor cells, caused by movement of water into or out of them.

**25.6** Plant stress has either abiotic (physical or chemical) or biotic (biological) sources.

**25.6** Plants respond to environmental stress in three ways: avoidance, resistance, or susceptibility.

**25.6** **Acclimation** (ACK-lih-MAY-shun, p. 435) is the ability of a plant to resist stress by becoming accustomed to new conditions.

**25.6** Plants that are susceptible to stress stop flowering and forming seeds; **senescence** (sih-NEH-sense, p. 435) begins.

**25.7** Plants have internal **biological clocks** (p. 436), or timing mechanisms, that regulate 24-hour cycles of physiological activity called **circadian rhythms** (p. 436).

**25.7** Flowering plants have the ability to measure the proportion of daylength in a 24-hour period by measuring the period of darkness in between periods of light; this ability is called **photoperiodism** (p. 436).

**25.7** Leaves sense the photoperiodic signal, with phytochromes the likely light sensor.

**25.7** Currently, scientists hypothesize that many factors are involved in flowering, including genetic promoters and inhibitors as well as the presence and appropriate concentrations of certain hormones and nutrients.

**25.7** Temperature can modify photoperiodic behavior.

**25.7** The requirement for prolonged exposure to cold temperatures prior to flowering is called **vernalization** (p. 436).

## What happens to plants as they die?

**25.8** During senescence, a time when plants or plant parts grow old, plants redistribute substances from dying tissues to actively growing tissues.

## Level 1    Learning Basic Facts and Terms

**Multiple Choice**

1. Which of the following responses to the environment are controlled by plant hormones?
   a. growth response to gravity
   b. growth response to light
   c. dropping of leaves in the fall
   d. seed germination
   e. all of the above

2. Circadian rhythms in plants probably play a role in which of the following?
   a. flowering
   b. opening and closing of leaf stomata
   c. chemical defenses against herbivores
   d. resistance to pathogens
   e. Both a and b are correct.

3. Which of the following are paired correctly?
   a. auxin // seed germination
   b. phytochrome // photoperiodic control of flowering
   c. abscisic acid // fruit ripening
   d. ethylene // phototropism

4. At the molecular level, plant hormones induce physiological or growth responses by
   a. regulating the expression of genes.
   b. controlling the production of transcription factors.
   c. mutating the DNA.
   d. inducing the plant to synthesize new genes.
   e. Both a and b are correct.

**True–False**

5. _____ Nastic movements are strongly related to the direction from which the stimulus is coming.

6. _____ Phytochromes are photoreceptor pigments that regulate plant development from seed germination to flowering.

7. _____ A plant's only response to environmental stresses is avoidance.

8. _____ Photoperiodism could be classified as a type of circadian rhythm or biological clock.

9. _____ Internal environmental conditions are the primary influences on plant hormone production, while the external environment is the primary influence on animal hormone production.

10. _____ Plant cells can best be characterized as pluripotent.

## Level 2 — Learning Concepts

1. For each of the following environmental stimuli, determine which plant hormones may be involved in initiating the appropriate plant response: gravity, light, and availability of water. Explain how the action of the hormone leads to a physiological or growth response.
2. How is adaptation to an environmental stress different from acclimation to stress?
3. The meristems of plants are comprised of totipotent cells. Why is this characteristic essential for the survival of a plant?
4. Describe some of the environmental signals that stimulate plants to break seed dormancy. Why is seed dormancy adaptive? How do phytochromes trigger the seed to break dormancy?
5. What prevents seeds from using up their energy reserves (i.e., endosperm) prior to germination?
6. Plants are capable of a variety of movements that are not the result of growth responses to stimuli. Provide two examples of these types of movements and explain how this movement is generally induced.

## Level 3 — Critical Thinking and Life Applications

1. Classify each of the following plant responses as either acclimation or avoidance. Explain your classification of each response.
   a. Giant sequoia trees produce cones that only open and drop their seeds in response to the heat of a forest fire, which clears away underbrush on the forest floor.
   b. Some desert species of plants germinate, grow, flower, and die very quickly during a brief rainy season, and survive over the dry season as dormant seeds.
   c. Gradual exposure of some plants to low temperatures makes them frost-hardy and makes them better able to withstand the cold stress of winter.
2. What is the biological explanation for the saying, "One bad apple spoils the lot?"
3. Prior to the spring, people living in the southern United States dig up their tulip bulbs and put the bulbs in the refrigerator for several weeks so they will flower. Explain the biology behind this behavior.
4. Poinsettias are short-day plants that flower around Christmas. Develop a strategy for stimulating poinsettias to flower in the summer.

## In The News — Critical Thinking

### PLANT WAKE-UP CALL

Now that you understand more about control systems in plants, reread this chapter's opening story about research into certain factors that influence flowering. To better understand this research, it may help you to follow these steps:

1. Review your immediate reaction to Amasino and Sung's research on *Arabidopsis* that you wrote when you began reading this chapter.
2. Based on your current understanding, again summarize the main point of the research in a sentence or two.
3. What questions do you now have about this research that this chapter's opening story does not answer?
4. Collect new information about the research. Visit the *In The News* section of this text's companion Web site at www.wiley.com/college/alters and watch the "Plant Wake-Up Call" video. Then use the "summary" link to read the accompanying story and access related links. Use this information, the links provided, and other online and library resources to answer your questions and find updates about this topic. State the sources of your information. Explain why you think the information is accurate. Also determine whether the information expresses a particular point of view or is biased in any way.
5. What in your view are the most significant aspects of this research? Give reasons for your opinion and for any changes in your ideas based on the additional information you have collected and the analysis you have done.

# TISSUES

## In The News | Spare Parts

In the early nineteenth century, British author Mary Shelley wrote a now classic horror story with a biological twist. In her novel, a scientist fabricates a hideous monster from various body parts of dead people. Does this story sound familiar? Shelley's book *Frankenstein, or the Modern Prometheus*, was published in 1818, and since then a variety of movies, academic essays, and Halloween costumes have made Frankenstein a household name.

Although the idea of constructing a living person from human tissues remains purely science fiction, biomedical engineers have been able to culture (grow) certain animal tissues in the laboratory as a prelude to culturing human tissue. Bioengineer Tony Keaveny at the University of California–Berkeley and his colleagues at Tulane University in Louisiana are growing rat bone in their laboratories to learn more about the bio-

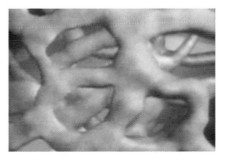

chemical and mechanical processes necessary for culturing human bone. Keaveny's eventual hope is to use a person's own bone cells to grow new or stronger bone at particular sites in the body, treating diseases such as thinning of the bones (osteoporosis) in the spine. Rena D'Souza, a professor and research director in the Department of Orthodontics at the University of Texas–Houston, and her colleagues are studying the genetics of tooth development and have been able to culture mice teeth in the laboratory. Based on what they have learned thus far, D'Souza expects that one day researchers will be able to grow human teeth suitable for transplantation. To learn more about Keaveny's and D'Souza's research, visit the *In The News* section of this text's

companion Web site at www.wiley.com/college/alters and view the video "Spare Parts."

Such pioneering efforts provide hope for thousands of people who need to replace diseased, damaged, or worn out tissues or organs. According to the Organ Procurement and Transplant Network (OPTN), over 86,000 Americans await organ transplantation. Many of these persons need a kidney, liver, lung, or heart, but the demand for these and other organs suitable for donation far exceeds the supply. Each year, approximately 6000 people die before they can receive an organ donation. Even after receiving donated tissues or organs, most recipients are not able to live normally; they usually take medication to prevent their bodies from rejecting the foreign tissues. Thus, the ability to "grow" body tissues or organs in the laboratory could make organ donation and the need for anti-rejection medication as obsolete as traveling by stagecoach, which was probably how Shelley herself traveled.

*Write your immediate reaction to scientists' research efforts to grow tissues in the laboratory: first, summarize the main point of the research in a sentence or two; then suggest what you think the significance of this research is. You will have an opportunity to reflect on your responses and gather more information on this topic in the* In The News *feature at the end of this chapter. In this chapter, you will learn more about tissues and the cells that compose them.*

## CHAPTER GUIDE

### How are multicellular organisms organized?

**26.1** Multicellular organisms are organized hierarchically, from cells to organ systems.

**26.2** The human body has the same general architecture as all vertebrates.

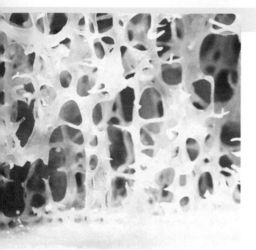

### What are the functions of the different types of animal tissues?

**26.3** Epithelial tissue covers and lines.

**26.4** Connective tissues and cells support, bind, defend, and isolate.

**26.5** Muscle tissue contracts.

**26.6** Nervous tissue transmits.

### How are tissues organized and their activities coordinated?

**26.7** Tissues are organized into organs, and organs into organ systems.

**26.8** Cells, tissues, organs, and organ systems work together to maintain internal equilibrium.

# How are multicellular organisms organized?

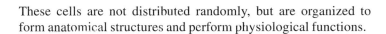

## 26.1 Multicellular organisms are organized hierarchically, from cells to organ systems.

The word "autopsy" comes from a Greek word meaning "eyewitness." That is what pathologists do when they autopsy a body—they act as an eyewitness and determine via observation and testing why a person died. A pathologist is a physician who identifies diseases by studying cells and tissues under a microscope. When pathologists perform autopsies, they look not only at cells and tissues but also at all the components of the body's hierarchy of organization to give their scientific and medical opinions on cause of death.

First, the pathologist looks at the outside of the body, examining the whole **organism**, which is the highest level of organization of multicellular animals (**Figure 26.1**). Then the pathologist cuts into the body to look at the organ systems, the next level "down" in organization. An **organ system** is a group of organs that function together to carry out one of the principal activities of the body, such as the digestive system and the circulatory system. The human body has 11 principal organ systems, which are listed in Section 26.7 and discussed in depth in later chapters. During an autopsy, entire organ systems may be removed from the body for close examination.

The pathologist also focuses on examining the individual organs that make up the organ systems. An **organ** consists of two or more tissues grouped together to form a structural and functional unit. The heart, lungs, brain, and intestines are organs. During an autopsy, the pathologist weighs and examines these organs; the intestines and stomach are opened using special scissors.

The pathologist also takes samples of body fluids, such as urine and blood (a tissue) to have them tested for pathogens and drugs. A **tissue** is a group of cells that work together to perform a function. Tissues are divided into four basic types based on their function: **epithelial**, **connective**, **muscle**, and **nervous**. For example, the cells making up the walls of the air sacs of your lungs are a type of epithelial tissue. Epithelial tissue covers body surfaces and lines its cavities. All four tissue types are described in this chapter. The pathologist examines many body tissues under the microscope in addition to the blood (connective tissue), such as the heart muscle tissue, or brain (nervous tissue). Under the microscope, the pathologist can see the cells that make up the various tissues.

**Cells** are the smallest living units of structure and function of organisms. The bodies of all multicellular organisms are made up of many different types of cells. The human body, for example, contains more than 200 different kinds of cells!

**Figure 26.1 Levels of organization.** Cells form tissues, tissues form organs, organs form organ systems, and organ systems form the organism.

These cells are not distributed randomly, but are organized to form anatomical structures and perform physiological functions.

## 26.2 The human body has the same general architecture as all vertebrates.

This book uses the human organism as the vertebrate example of how living things work, so that you can gain insight into the animal world while learning more about yourself. The human body has the same general body architecture that all vertebrates have, including a coelom. A coelom is a fluid-filled body cavity lined with connective tissue (see Chapter 21).

In humans the coelom is divided into two main parts: (1) the thoracic cavity, which contains the heart and lungs, and (2) the abdominal cavity, which contains organs such as the stomach, intestines, and liver (**Figure 26.2**). The diaphragm forms the floor of the thoracic cavity and the roof of the abdominal cavity. The lower portion of the abdominal cavity is often referred to as the pelvic cavity. The body is supported by an internal scaffold, or skeleton, made up of bones that grow as the body grows. A bony skull surrounds the brain, which is located in a cavity separate from the coelom, called the cranial cavity. In addition, a column of bones called the vertebral column forms the backbone, or spine. These bones surround a nerve cord, the spinal cord, that relays messages between the brain and other parts of the body.

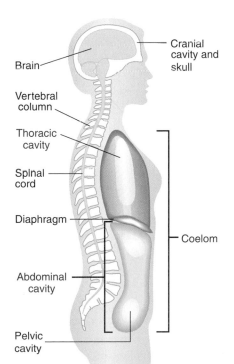

**Figure 26.2 Architecture of the human body.** Humans, like all vertebrates, have a spinal cord and brain enclosed in the vertebral column and skull. In mammals, a muscular diaphragm divides the coelom into the thoracic cavity and the abdominal cavity.

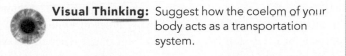

**Visual Thinking:** Suggest how the coelom of your body acts as a transportation system.

# What are the functions of the different types of animal tissues?

## 26.3 Epithelial tissue covers and lines.

*Epithelial tissues* guard and protect your body; they cover and line the surfaces of the body, both internal and external. You can think of your body as a tube, with your skin forming the outside of the tube and your digestive system the inside of the tube. The surfaces of your inner tube, as well as your skin, are made up of epithelial tissues. Because epithelial tissues cover the surfaces of the body, they determine which substances enter and which do not.

All epithelial tissues, collectively called the epithelium, are broadly similar in form and function and occur as sheets of cells packed together. The epithelial tissues of the body function in six different ways:

- *Protection.* They protect the tissues beneath them from drying out, sustaining chemical and mechanical injury, and being invaded by microorganisms.
- *Absorption.* Epithelium lines the digestive tract. During absorption, the products of digestion enter the epithelial cells of the stomach and the small intestine, and from there these products pass into the blood or lymph.
- *Sensation.* Many sensory nerves end close to epithelial tissues. The epithelium provides a surface through which stimuli reach nerves.
- *Secretion.* Certain epithelial tissues (glands) are specialized to produce and discharge substances. Glands may be single cells, such as the mucus-secreting cells lining the intestine, or multicellular structures, such as the thyroid and sweat glands. Although many glands lie deep within your body, they originally formed as infoldings of epithelial cell layers during embryological growth and development.
- *Excretion.* Specialized epithelial tissue in the kidney excretes waste products in urine. Also, the epithelium forming the air sacs of the lungs excretes the waste gas carbon dioxide.
- *Surface transport.* Some epithelial tissues have hairlike projections called cilia. These cilia beat in unison, causing a wavelike movement in the thin film of mucus that bathes the cells' surfaces, sweeping particles along in the process. The ciliated cells lining your respiratory passageways keep foreign particles from entering the lungs. Cigarette smoking damages these cilia, resulting in accumulations of mucus in the throat and lungs.

Structurally, epithelial tissues share some common characteristics. Epithelial tissues are usually only one or a few cells thick (with the exception of the skin), are packed together and stacked very tightly, and have very few blood vessels running through them. The circulation of nutrients, gases, and wastes in epithelial tissue occurs by diffusion from the capillaries of neighboring tissue. In addition, although the chemical reactions of many types of epithelial cells take place at a very slow rate, many have amazing regenerative pow-

ers. Your skin cells, for example, are continually being replaced throughout your lifetime.

Epithelial cells and tissues are illustrated in **Figure 26.3**. There are three main shapes of epithelial cells—*squamous*, *cuboidal*, and *columnar*. *Squamous cells* (SKWAY-muss) are thin and flat, allowing for the rapid movement of substances through them. They are found in places such as the air sacs of the lungs, the lining of blood vessels, and the skin. *Cuboidal cells* have complex shapes but look like cubes when the tissue is cut at right angles to the surface. These cells function in secretion and absorption; they are found lining tubules in the kidney and the ducts of glands. *Columnar cells* look like tiny columns when they are viewed from the side. Columnar cells often function in secretion and absorption as well. Much of the digestive tract is lined with columnar epithelium.

Epithelial tissues are usually attached to, yet compartmentalized from, underlying tissue by a basement membrane ❶. This membrane is made up of a substance secreted by cells and contains proteins such as collagen for strength (see Section 26.4).

Epithelial tissues include *simple*, *stratified*, *pseudostratified*, and *transitional* tissues. *Simple epithelium* (Figure 26.3 ❶ – ❸) is epithelial tissue that is only one cell thick. It is usually found in areas where substances diffuse through the tissue and where substances are secreted, excreted, or absorbed. For example, tissue lining your blood vessels is simple squamous epithelium ❶, as is the tissue forming the air sacs of your lungs.

*Stratified epithelium* is made up of two or more layers of cells. These layers are usually protective—for example, the surface of your skin is *stratified squamous epithelium* ❹. The ducts of your sweat glands are lined with *stratified cuboidal epithelium*, and part of your urethra (the tube through which you urinate) is lined with *stratified columnar epithelium*. *Pseudostratified epithelium* ❺ looks as though it is layered but is actually made up of only one layer of cells—some tall, some short. The cells that reach the surface are either ciliated or secrete mucus. Parts of your airways are lined with pseudostratified epithelium. *Transitional epithelium* ❻ is tissue that can stretch. The cells of transitional epithelium in the bladder, for example, are cubelike when the bladder has little or no urine. However, when the bladder fills with urine, the cells become thin and flat as the bladder expands like a water balloon to hold the urine.

### CONCEPT CHECKPOINT

1. For each type of epithelial tissue, state its primary function(s) and the type(s) of cell(s) (squamous, cuboidal, columnar) that can compose it. Also give an example of an organ or structure in which each type of epithelial tissue is found.

# Visual Summary     Figure 26.3 Epithelial tissues.

## Simple squamous epithelium

LOCATION   Lining of blood vessels, air sacs of lungs, skin, and lining of body cavities
DESCRIPTION   Single layer of thin, flat cells
FUNCTION   Diffusion, filtration, and passage of materials where little protection is needed

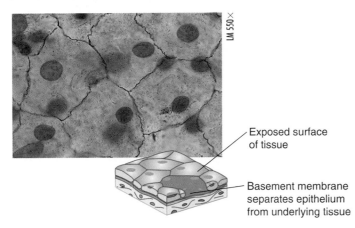

Exposed surface of tissue

Basement membrane separates epithelium from underlying tissue

## Simple cuboidal epithelium

LOCATION   Kidney tubules, glands and their ducts, terminal bronchioles of lungs, and surface of ovaries and retina
DESCRIPTION   Single layer of cube-shaped cells; some have microscopic microvilli, and some have cilia
FUNCTION   Secretion, absorption, and movement of substances

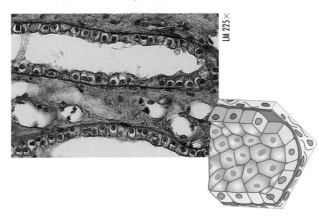

## Simple columnar epithelium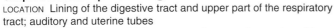

LOCATION   Lining of the digestive tract and upper part of the respiratory tract; auditory and uterine tubes
DESCRIPTION   Single layer of tall, narrow cells; some have microvilli or cilia
FUNCTION   Secretion of mucus

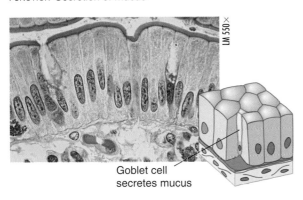

Goblet cell secretes mucus

## Stratified squamous epithelium

LOCATION   Skin, mouth, and throat lining; vaginal lining; anal lining; and cornea
DESCRIPTION   Several layers of cells: the lower layers are columnar and active; the upper layers are flattened at the surface
FUNCTION   Protection, hard outer layer being continuously removed by friction and replaced from below

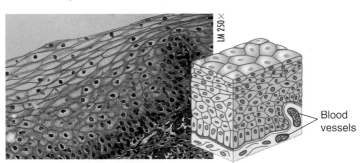

Blood vessels

## Pseudostratified epithelium

LOCATION   Nasal cavity and sinuses, ducts of some glands, and some ducts of the male reproductive system
DESCRIPTION   Single layer of cells similar to columnar epithelium, except of varying heights: some reach the surface and others do not; some have cilia, and some may have microvilli
FUNCTION   Protection and secretion of mucus

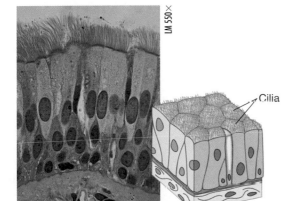

Cilia

## Transitional epithelium

LOCATION   Lining of urinary bladder and ureters
DESCRIPTION   Several layers of cuboidal cells beneath layers of surface cells; cells are flattened when tissue is stretched
FUNCTION   Accommodation of fluid fluctuations in an organ or tube by stretching easily

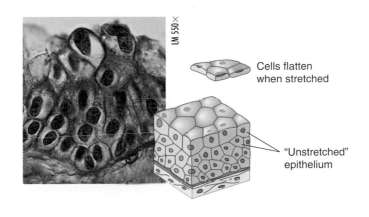

Cells flatten when stretched

"Unstretched" epithelium

## 26.4 Connective tissues and cells support, bind, defend, and isolate.

When you are trying to lose weight, you probably aren't thinking about connective tissue . . . but you are losing fat (triglyceride) that is stored in your adipose tissue! There are a variety of types of connective tissue that perform an assortment of important jobs for the body; adipose tissue is just one of those types, and it performs a few of those jobs—insulation, cushioning, and long-term nutrient storage. What other jobs does connective tissue perform? The cells of connective tissue provide the body with structural building blocks (such as bones) and potent defenses (such as white blood cells). In addition, connective tissue joins the other tissues of the body.

The varied composition of connective tissues reflects the diversity of this category of tissue. Nevertheless, these diverse types of connective tissue have, in general, a common structural theme. Connective tissue is generally made up of cells that are usually spaced well apart from one another and are embedded in a nonliving substance called a *matrix*. Connective tissue is usually made up of a great deal more matrix than cells. This matrix varies in consistency among the different types of connective tissue—from a fluid to a gel to crystals—and is usually secreted by the connective tissue cells.

The different types of connective tissues and cells can be categorized in many different ways. One way to group them is by function: *structural*, *defensive*, and *isolating connective tissue*. *Structural connective tissue cells*, such as bone and cartilage cells, stay in one place, secreting protein fibers into the empty spaces between them. These proteins provide structural connective tissue with a fibrous matrix, giving it the strength it needs to support the body and provide connections among tissues. Thus, structural connective tissue supports and binds. *Defensive connective tissue cells*, which protect the body from attack, float in a matrix of blood plasma or reside in certain other tissues. They roam the circulatory system, hunting invading bacteria and foreign substances. An example of this type of cell is the lymphocyte, a special type of white blood cell. *Isolating connective tissue cells* act as storehouses, accumulating specific substances such as fat (in adipose cells) and hemoglobin (in red blood cells).

### Cells and tissues that support and bind

The three principal types of cells found in structural connective tissue are fibroblasts, cartilage cells (chondrocytes), and bone cells (osteocytes) (**Figure 26.4**). These cells produce substances that cause the tissues of which they are a part to have distinctive characteristics.

**Fibroblasts** Of all the connective tissue cells, *fibroblasts* are the most numerous. They are flat, irregular, branching cells that secrete fibers into the matrix between them. These fibers are of three basic types: collagen, reticular, and elastic.

Both the collagen and reticular fibers are made up of the protein collagen, the most abundant protein in the human body. *Dense fibrous connective tissue* (Figure 26.4 **1**) is made up primarily of

collagen fibers—the figure shows bundles of collagen fibers with widely spaced rows of fibroblasts. This type of connective tissue is very strong; it makes up tendons connecting muscles to bones. It also composes the lower layer of the skin and makes strong attachments between organs. Reticulin is a fine branching fiber that forms the framework of many glands, such as the spleen and the lymph nodes. It also makes up the junctions between many tissues. The tissue formed by fibroblasts and reticulin alone is called *reticular connective tissue* **2** .

Elastic fibers, as the name suggests, act much like rubber bands. They are not made of collagen, but of a "stretchy" protein called elastin. *Elastic connective tissue* **3** ), made up of branching elastic fibers with fibroblasts interspersed throughout, is found in structures that must expand and then return to their original shape, such as the lungs and large arteries.

*Loose connective tissue* **4** contains cells and fibers within a semifluid matrix. Fibroblasts and macrophages are the most common cells in loose connective tissue. This tissue also contains loosely packed elastic and collagen fibers; it is therefore a somewhat strong but very flexible tissue. Loose connective tissue is distributed widely throughout the body and is found wrapping nerves, blood vessels, and tissues; filling spaces between body parts; and attaching the skin to the tissues beneath it.

**Cartilage cells** Chondrocytes are the cells that make up cartilage, a specialized connective tissue that is firm and flexible and does not stretch. The chondrocytes secrete a matrix consisting of a semisolid gel and fibers. As they secrete the matrix, they wall themselves off from it and eventually come to lie in tiny chambers called lacunae. Although there are three different types of cartilage—hyaline cartilage (HI-uh-lynn), elastic cartilage, and fibrocartilage—they all have cartilage cells within lacunae. Their differences lie in the matrix.

In its matrix, *hyaline cartilage* **5** has very fine collagen fibers that are almost impossible to see under the light microscope. During your development before birth, most of your skeleton was composed of hyaline cartilage. As an adult, you have hyaline cartilage on the ends of your long bones, cushioning the places where these bones meet. Hyaline cartilage also forms C-shaped rings around the windpipe, keeping this airway propped open. *Elastic cartilage* **6** has elastic fibers embedded in its matrix. It is found where support with flexibility is needed, such as in the external ear. *Fibrocartilage* **7** has collagen fibers embedded in its matrix. It is therefore a very tough substance and is used in places of the body where shock absorbers are needed. It is found, for example, as disks between the vertebrae and in the knee joint.

**Bone cells** Osteocytes are mature bone cells **8** . As in cartilage, these cells are isolated in lacunae. They lay down a matrix of collagen fibers that become coated with small, needle-shaped crystals of calcium salts. The calcium salts make the bone rigid, and the fibers keep the bone from being brittle. Bone makes up the adult skeleton, which supports the body and protects many of the organs. Bone is also a storehouse for calcium.

# Visual Summary    **Figure 26.4** Connective tissues that support and bind.

### Dense fibrous connective tissue (1)
LOCATION Tendons, ligaments, and attachments between organs and dermis of the skin

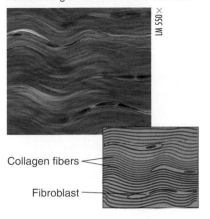

LM 550×

Collagen fibers

Fibroblast

### Reticular connective tissue (2)
LOCATION Liver, lymph nodes, spleen, and bone marrow

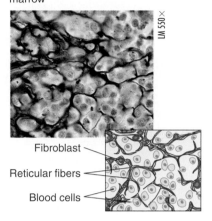

LM 550×

Fibroblast

Reticular fibers

Blood cells

### Elastic connective tissue (3)
LOCATION Lung tissue, arteries

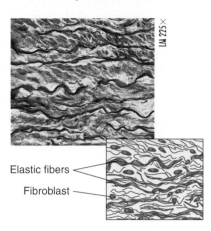

LM 225×

Elastic fibers

Fibroblast

### Loose connective tissue (4)
LOCATION Packing between glands, muscles, and nerves; attachments between skin and underlying tissue

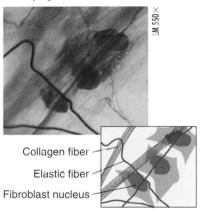

LM 550×

Collagen fiber

Elastic fiber

Fibroblast nucleus

### Hyaline cartilage (5)
LOCATION Ends of long bones, joints, respiratory tubes, costal cartilage of ribs, nasal cartilage, and embryonic skeleton

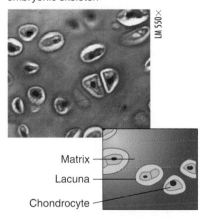

LM 550×

Matrix

Lacuna

Chondrocyte

### Elastic cartilage (6)
LOCATION Auditory tube, external ear, epiglottis

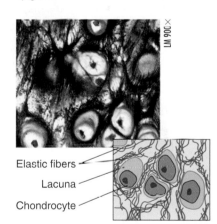

LM 900×

Elastic fibers

Lacuna

Chondrocyte

### Fibrocartilage (7)
LOCATION Connection between pubic bones, intervertebral disks

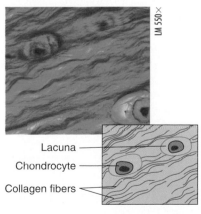

LM 550×

Lacuna

Chondrocyte

Collagen fibers

### Bone (8)

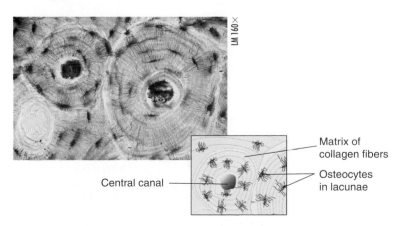

LM 160×

Matrix of collagen fibers

Osteocytes in lacunae

Central canal

The bones of your skeleton have two types of internal structure: compact bone and spongy bone. *Compact bone* is denser than spongy bone and gives the bone the strength to withstand mechanical stress. It forms the outer surface of all bones. In compact bone, the cells lay down matrix in thin concentric rings, forming tubes of bone around narrow channels or canals that contain blood vessels and nerves. *Spongy bone* is composed of an open lattice of bone that fills the ends of long bones and the interiors of flat bones and irregular bones. It lines the interiors of long bones. Spongy bone

supports the bone just as beams support a building. It also helps keep bones somewhat lightweight. The spaces within the latticework of bone are filled with red bone marrow, the substance that produces the body's blood cells. **Figure 26.5** illustrates the structure of a long bone.

## Cells that defend

Leukocytes are white blood cells that defend the body from foreign cells and substances. The two major groups of leukocytes are *granulocytes* (GRAN-yuh-low-sites) and *agranulocytes*. Granulocytes (**Figure 26.6** ❶ ) are circulating leukocytes with lobed nuclei; they get their name from the tiny granules in their cytoplasm. Granulocytes play various roles in protecting the body, including engulfing and digesting microorganisms and other foreign particles, and releasing chemicals to enhance the body's responses to injury and infection. Agranulocytes ❷ have no cytoplasmic granules, nor are their nuclei lobed. The three types of agranulocytes are lymphocytes, macrophages (MACK-row-fayge-ez), and mast cells.

*Lymphocytes* circulate in the blood or reside in the organs, vessels, and nodes of the lymphatic system; they play key roles in the body's defense against infection. Your body has an amazing trillion or so lymphocytes ready to attack foreign cells or viruses that enter the body or to produce antibodies that can act against specific substances (see Chapter 30).

*Macrophages*, which are abundant in the bloodstream as well as in the lungs, spleen, and lymph nodes, develop, or differentiate, from white blood cells called *monocytes*. Macrophages may be thought of as the janitors of the body, cleaning up cellular debris and invading bacteria by a process known as *phagocytosis* (FAH-guh-sigh-TOE-sis)—an engulfing and digesting of particles.

*Mast cells* (**Figure 26.7**) produce substances that are involved in the body's inflammatory response. Generally, they are found alongside blood vessels that supply connective tissue. One important substance produced by mast cells is histamine, which causes blood vessels to dilate (widen). As more blood then flows through the vessels, it brings added oxygen and nutrients and dilutes any toxins, or poisons. The increased blood flow also aids the movement of defensive leukocytes (white blood cells) coming to the area.

## Cells and tissues that isolate

Isolating connective tissue is composed of cells that specialize in accumulating and transporting particular

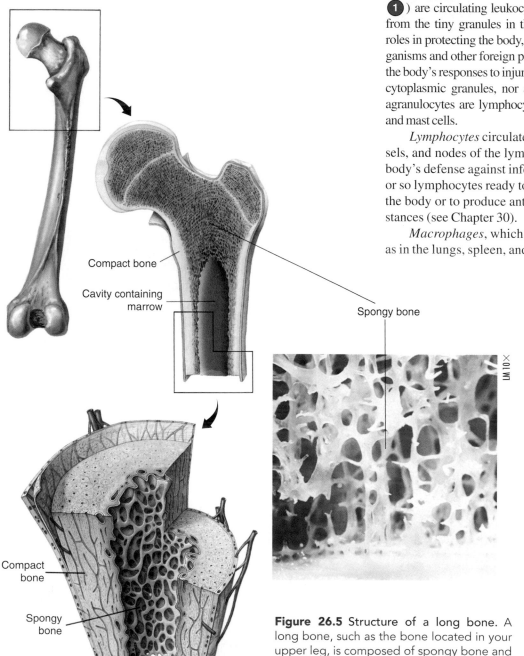

Compact bone

Cavity containing marrow

Spongy bone

LM 10×

Compact bone

Spongy bone

Marrow cavity

**Figure 26.5** Structure of a long bone. A long bone, such as the bone located in your upper leg, is composed of spongy bone and compact bone. Spongy bone (photo) is composed of a delicate latticework and adds support. Along with the hollowed-out shaft of compact bone, it keeps the bone somewhat lightweight. Compact bone adds strength.

**Visual Summary**   **Figure 26.6** Connective tissues and cells that defend and isolate.

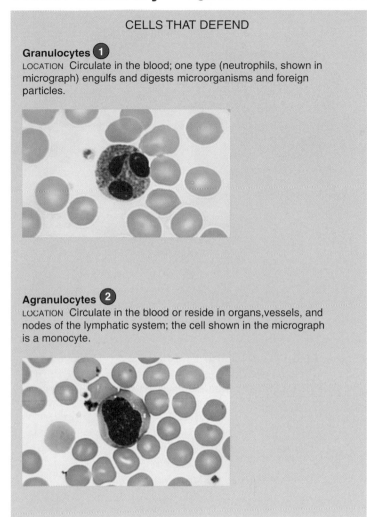

## CELLS THAT DEFEND

**Granulocytes** ①
LOCATION  Circulate in the blood; one type (neutrophils, shown in micrograph) engulfs and digests microorganisms and foreign particles.

**Agranulocytes** ②
LOCATION  Circulate in the blood or reside in organs, vessels, and nodes of the lymphatic system; the cell shown in the micrograph is a monocyte.

## CELLS AND TISSUES THAT ISOLATE

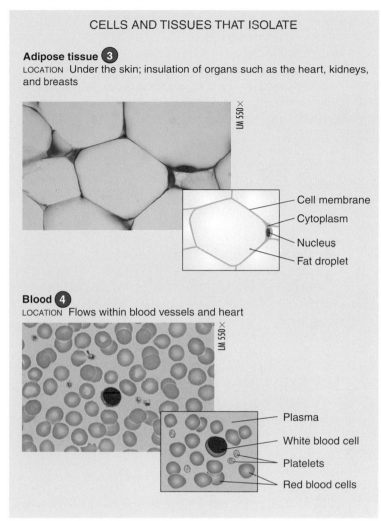

**Adipose tissue** ③
LOCATION  Under the skin; insulation of organs such as the heart, kidneys, and breasts

LM 550×

- Cell membrane
- Cytoplasm
- Nucleus
- Fat droplet

**Blood** ④
LOCATION  Flows within blood vessels and heart

LM 550×

- Plasma
- White blood cell
- Platelets
- Red blood cells

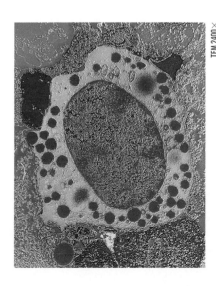

TEM 2400×

**Figure 26.7** A mast cell, colorized to better visualize structures, has produced histamine (red granules). Mast cells, which are found in the blood and other tissues, are involved in the body's inflammatory response to physical injury or trauma. They also play a role in allergic reactions.

molecules. These cells include the fat cells of adipose tissue, as well as pigment-containing cells.

Many people think that they have more fat cells than they need for fuel emergencies, but fat serves other purposes too. *Adipose tissue* (Figure 26.6 ③ ) helps shape and pad the body and insulates against heat loss. The cytoplasm and nucleus of each cell is pushed against the cell membrane by a large fat droplet. The fat takes up much of the cell and is released when the body needs it for fuel.

Possibly the most important isolating cells are red blood cells (erythrocytes), one of the solids that float in the fluid connective tissue called *blood* ④ . Red blood cells are the most common of the blood cells. There are about 5 million in every milliliter of blood. Red blood cells act as mobile transport units, picking up and delivering the gases carbon dioxide and oxygen.

Oxygen—as well as some carbon dioxide—is carried by the iron-containing pigment hemoglobin, which imparts the color to red blood cells. An amazing 300 million molecules of hemoglobin become isolated within each red blood cell.

Blood cells float in a fluid intercellular matrix, or plasma. This fluid is both the banquet table and the refuse heap of your body because practically every substance used and discarded by cells is found in the plasma. The plasma also contains minerals such as calcium used to form bone; proteins such as fibrinogen that help the blood to clot; albumin, which gives the blood its viscosity; and antibodies produced by lymphocytes. Cell fragments called *platelets* are also present in the blood. These cell pieces play an important role in the clotting of blood.

### CONCEPT CHECKPOINT

2. All connective tissues are made up of living cells interspersed within a nonliving extracellular matrix. Describe how the cellular and extracellular matrix structure of the tissue relates to its general functions for the following: blood, bone, and tendons.

## 26.5 Muscle tissue contracts

Muscle cells are the workhorses of your body. The distinguishing characteristic of muscle cells, which are also called muscle fibers, is the abundance of special thick and thin protein filaments. These filaments are highly organized to form strands called *myofibrils*. Each muscle cell is packed with many thousands of these myofibril strands. Their arrangement is shown in skeletal muscle tissue in **Figure 26.8**. The myofibrils shorten when the filaments slide past each other, causing the muscle to contract.

### Smooth muscle

Smooth muscle is found in organs such as the stomach and intestines. Smooth muscle cells are long, with bulging middles, tapered ends, and a single nucleus. The cells are organized into sheets, forming smooth muscle tissue (**Figure 26.9** ). This tissue contracts involuntarily—you cannot consciously control it.

Some smooth muscle contracts when a nerve or hormone stimulates it. Examples of smooth muscles that contract in this way are the muscles that line your blood vessels and those that make up the iris of your eye. In many cases, impulses may pass directly from one smooth muscle cell to another, resulting in a self-propagating wave of contraction throughout a layer of the muscle. In other smooth muscle tissue, such as that found in the wall of the intestines, individual cells might contract sponta-

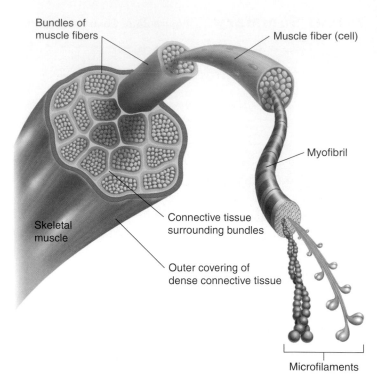

**Figure 26.8** Skeletal muscle. Muscle fibers (cells) are comprised of many myofibrils. Myofibrils, in turn, are made up of microfilaments, which are responsible for muscle contraction. The arrangement of these structures is shown here in a skeletal muscle.

**Visual Thinking:** What characteristics of a tissue do you see present in this skeletal muscle? How many types of tissue are present?

neously or when they are stretched, leading to a slow, steady squeezing of the tissue.

### Skeletal muscle

*Skeletal muscles* ❷ are attached to your bones and allow you to move your body. These muscles are called voluntary muscles because you have conscious control over their action. They are also called striated muscles because the tissue has "stripes"—microscopically visible bands, or striations. These striations result from the organization of thick and thin microfilaments within the myofibrils and the alignment of the myofibrils with one another. The myofibrils are organized so that all within a single cell contract at the same time when the muscle cell is stimulated by a nerve.

Striated muscle cells are extremely long. A single muscle cell, or *fiber*, may run the entire length of a muscle. Each fiber has many nuclei that are pushed to the edge of the cell and lie just under the cell membrane. Bundles of these muscle cells are wrapped with connective tissue and joined with other bundles to form the muscle.

## Visual Summary    **Figure 26.9** Muscle tissue.

### Smooth muscle ①
LOCATION  Walls of hollow organs, iris of eye, skin (attached to hair), and glands
DESCRIPTION  Tissue is not striated; spindle-shaped cells have a single, centrally located nucleus
FUNCTION  Regulation of size of organs, forcing of fluid through tubes, control of amount of light entering eye, production of "gooseflesh" in skin; under involuntary control

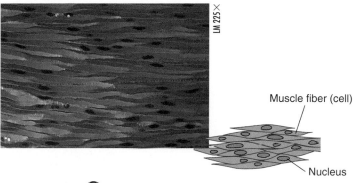

Muscle fiber (cell)

Nucleus

### Skeletal muscle ②
LOCATION  Attachment to bone
DESCRIPTION  Tissue is striated; cells are large, long, and cylindrical with several nuclei
FUNCTION  Movement of the body; under voluntary control

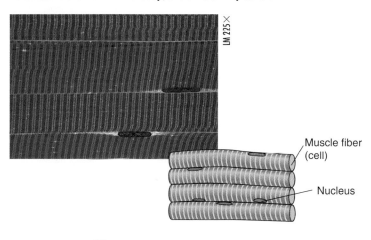

Muscle fiber (cell)

Nucleus

### Cardiac muscle ③
LOCATION  Heart
DESCRIPTION  Tissue is striated; cells are cylindrical and branching with a single centrally located nucleus
FUNCTION  Pumping of blood; under involuntary control

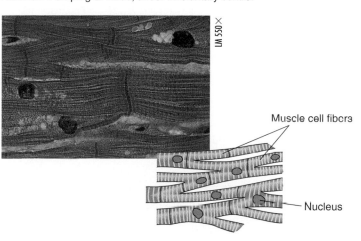

Muscle cell fibers

Nucleus

### Cardiac muscle

The heart, which acts as a pump for the circulatory system, is made up of striated muscle fibers. These fibers are arranged differently from the way they are arranged in skeletal muscle. Instead of very long cells running the length of the muscle, *cardiac muscle* ③ is composed of chains of single cells. These chains of cells are organized into fibers that branch and interconnect, forming a latticework. This lattice structure is critical because it allows an entire portion of the heart to contract at one time. Cardiac muscle is like smooth muscle in that it is involuntary and the muscle fibers usually have only one centrally located nucleus.

## 26.6 Nervous tissue transmits.

The fourth major class of tissue in humans is nervous tissue. It is made up of two kinds of cells: (1) *neurons*, which conduct nerve impulses, and (2) *supporting cells*, which surround neurons, nourishing, protecting, insulating, and holding them in place.

Neurons are cells specialized to conduct an electrochemical impulse (**Figure 26.10**). The cell body of a neuron contains the nucleus of the cell. Two different types of projections extend from the cell body. Multiple projections called *dendrites* act as antennae for the reception of nerve impulses. A single long projection called an *axon* transmits impulses toward another neuron or another tissue. An axon may give out branches, which usually split off at right

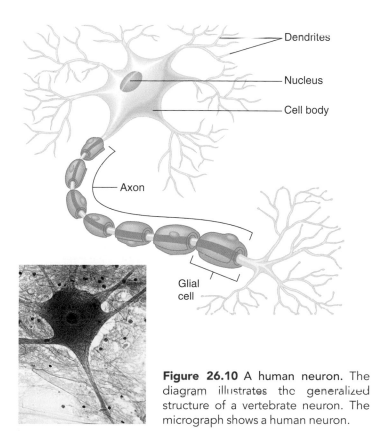

Dendrites

Nucleus

Cell body

Axon

Glial cell

**Figure 26.10** A human neuron. The diagram illustrates the generalized structure of a vertebrate neuron. The micrograph shows a human neuron.

angles, and is usually covered with myelin insulation. When axons or dendrites are long, they are referred to as *nerve fibers*. Some nerve fibers are so long, in fact, that they can extend from your spinal cord all the way to your fingers or toes. Single neurons over a meter in length are common.

The *nerves*, which appear as fine white threads when they are viewed with the naked eye, are actually composed of clusters of axons and dendrites. Like a telephone cable, they include large numbers of independent communication channels—bundles of hundreds of axons and dendrites, each connecting a different nerve cell with a different muscle fiber or sensory receptor. In addition,

each nerve contains numerous supporting cells called glial (GLEE-uhl) cells. One type of glial cell produces the myelin insulation that covers most axons.

CONCEPT CHECKPOINT

3. Compare and contrast stimulation and contraction in the three types of muscle tissue.
4. How does the structure of neurons relate to their function?

# How are tissues organized and their activities coordinated?

## 26.7 Tissues are organized into organs, and organs into organ systems.

The four major classes of tissues that we have discussed in this chapter are the building blocks of the human (vertebrate) body. These tissues form the *organs* of the body. Each organ contains several different types of tissue coordinated to form the structure of the organ and to perform its function. The heart, for example, is composed of muscle cells that together make up the cardiac

muscle tissue. Bundles of this tissue are wrapped in connective tissue and wired with nervous tissue. Muscles can help you walk, pump your blood, and digest your food. Different combinations of tissues are found in different organs that perform different functions.

An *organ system* is a group of organs that function together to carry out the principal activities of the body. **Table 26.1** lists the functions and components of the 11 major organ systems of the human body. **Figure 26.11** illustrates each of these systems.

| TABLE 26.1 | The Major Human Organ Systems | |
| --- | --- | --- |
| **System** | **Functions** | **Components** |
| Skeletal | Protects the body and provides support for locomotion and movement | Bones, cartilage, and ligaments |
| Muscular | Produces body movement | Skeletal, cardiac, and smooth muscles |
| Nervous | Receives and helps body respond to stimuli | Nerves, sense organs, brain, and spinal cord |
| Endocrine | Coordinates and regulates body processes and functions | Pituitary, adrenal, thyroid, and other ductless glands |
| Circulatory | Brings nutrients and oxygen to cells and removes waste products | Heart, blood vessels, blood, lymph, and lymph structures |
| Immune | Helps defend the body against infection and disease | Lymphocytes, macrophages, and antibodies |
| Integumentary | Covers and protects the body | Skin, hair, nails, oil and sweat glands |
| Respiratory | Supplies blood with oxygen and rids it of carbon dioxide | Trachea, lungs, and other air passageways |
| Digestive | Breaks down food and absorbs breakdown products | Mouth, esophagus, stomach, intestines, liver, and pancreas |
| Urinary | Removes wastes from the bloodstream | Kidney, bladder, and associated ducts |
| Reproductive | Produces sex cells and carries out other reproductive functions | Testes, ovaries, and other associated reproductive structures |

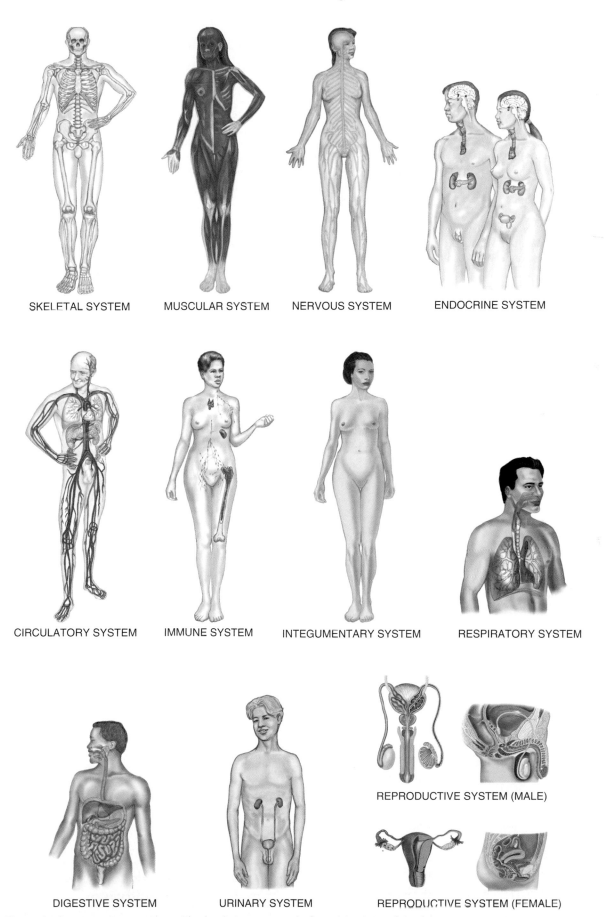

SKELETAL SYSTEM    MUSCULAR SYSTEM    NERVOUS SYSTEM    ENDOCRINE SYSTEM

CIRCULATORY SYSTEM    IMMUNE SYSTEM    INTEGUMENTARY SYSTEM    RESPIRATORY SYSTEM

REPRODUCTIVE SYSTEM (MALE)

DIGESTIVE SYSTEM    URINARY SYSTEM    REPRODUCTIVE SYSTEM (FEMALE)

**Figure 26.11 The major human organ systems.** The body is composed of combinations of the four types of tissue, assembled in various ways. The many organs working together to carry out the principal activities of the body are traditionally grouped together as organ systems.

## *just wondering . . .*

### Why can some animals regenerate certain tissues and body parts and we can't?

The idea of humans regenerating limbs is, of course, science fiction. But as you suggested, many kinds of animals can regenerate parts. Sponges, for example, are capable of extensively regenerating and replacing lost parts. Sometimes such regeneration results in a new organism in addition to the original one and is considered a form of asexual reproduction. This type of asexual reproduction also occurs in echinoderms such as sea stars. When one arm is removed from a sea star, the animal will grow a new arm; however, if the cut arm has a piece of the central disk attached, it will grow into a new individual. Vertebrates show some capacity for tissue regeneration. The best examples are the amphibians; salamanders have been extensively studied to try to answer just the question you raised.

Developmental biologists have been intrigued by the regeneration question for years. They have found that the regenerated limbs of salamanders arise from epidermal tissue that surrounds a mass of rapidly growing cells capable of differentiating into connective tissue, bone, cartilage, vascular tissue, and lymphatic vessels. These cells apparently arise from the stump and eventually grow and develop into a new limb.

But what controls the growth and differentiation of these cells? And what controls the ability of the proper type of cell to develop in the appropriate place, forming a fully functional and anatomically correct limb? We are a long way from having complete answers to these questions. Researchers know that hormones play a role in influencing the normal course of growth during regeneration. They are also studying the role of one class of regulatory genes called homeotic genes, which scientists know control patterns of development. However, the exact role of these genes remains unclear. Key immune system cells called T cells likely play a role in blocking regeneration in vertebrates similar to the way in which they block the overproliferation of cells in cancer.

Most serious injuries of the human body—and some not so serious—result in the development of scar tissue. This tissue is formed by fibroblasts, which divide rapidly in the injured area and secrete large quantities of collagen. This substance "fills in" the area of lost cells—certainly not a process nearly as complex or useful as that of salamander limb regeneration! ●

*Are you wondering about a topic in biology and how it relates to your life? Submit your question by clicking the* Just Wondering *link in this text's companion Web site at www.wiley.com/college/alters.*

---

## FOCUS

## *on Plants*

**Figure 26.A**   Vascular plant.

**Figure 26.B**   Mosses are nonvascular plants.

### Patterns and Levels of Organization in Plants

Plants fall into one of two groups with respect to patterns and levels of organization: vascular plants and nonvascular plants. Vascular plants have specialized tissues to transport fluids and have the plant organs—roots, stems, and leaves—shown in **Figure 26.A**. More than 95% of all living plant species are vascular plants. Nonvascular plants do not have true roots, stems, or leaves, so they usually grow close to the ground, like the mosses in **Figure 26.B**. You can learn more about patterns and levels of organization in plants in Chapter 24.

The skeletal system supports and protects your body. It is moved by the large, voluntary muscles of the muscular system. Other muscles in this system help move internal fluids throughout your body. The nervous system regulates most of the organ systems. It can sense conditions in both your internal and external environments and help your body respond to those conditions. The organs of your endocrine system secrete chemicals called hormones that also regulate body processes and functions. The circulatory system is the main transportation system of the body. It brings nutrients and oxygen to your cells and removes the waste products of metabolism. Along with the immune and integumentary (skin) systems, it also helps defend the body against infection and disease. The respiratory system works hand in hand with the circulatory system, supplying the blood with oxygen and ridding it of the waste gas carbon dioxide. The food you eat is broken down by the digestive system and is absorbed through the intestinal walls into the bloodstream. Solid wastes are also eliminated from the body by this organ system. The urinary system collects waste materials and excess water from the bloodstream and excretes them as liquid wastes. The reproductive system produces gametes, or sex cells, that can join in the process of fertilization to produce the first cell of a new individual.

### 26.8 Cells, tissues, organs, and organ systems work together to maintain internal equilibrium.

As you can see from their descriptions, the organ systems interact with one another to keep the organism—you—alive and well. This state of "wellness" is called **homeostasis**. Put another way, homeostasis is the maintenance of a stable internal environment despite what may be a variable external environment.

Your body maintains a steady state by means of **feedback loops**. These are mechanisms by which information regarding the status of a physiological situation or system is fed back to the body's control system. The thermostat in your home works by means of a feedback loop. For example, if you set the thermostat at 68°F, a sensor in the thermostat monitors the temperature in the room and switches off the furnace when that temperature is reached. Likewise, when the temperature drops below 68°F, the sensor in the thermostat switches on the furnace. This ensures that the temperature will fluctuate only within a small range. This is called a **negative feedback loop** because the change that takes place is opposite to the condition that prompted the change.

Most regulatory mechanisms of your body work by means of negative feedback loops. For instance, sufficient levels of specific hormones in the blood trigger mechanisms that result in the shutdown of their secretion. Similarly, your body temperature is maintained by means of negative feedback (**Figure 26.12**), much as the temperature of your home is maintained. When your body temperature rises too high ❶, your thermostat (a special portion of the brain called the *hypothalamus*) senses this rise and

counters it. Messages race along your nerve fibers ❷ to your sweat glands, triggering the release of sweat, which evaporates and cools the body ❸. Blood vessels near the surface of the skin dilate, or widen, bringing blood near the surface, where the heat is dissipated ❹ until a normal body temperature is achieved. When your body temperature falls ❺, the hypothalamus triggers tiny muscles under the skin to contract ❻, causing a shiver that generates heat ❼. Blood vessels near the surface of the skin constrict ❽, lessening the flow of blood there and the amount of heat that radiates from the body until a normal body temperature is restored.

In all negative feedback loops, the change that takes place is opposite to whatever triggers the change. In some cases, the change may be an increase rather than a decrease. When blood glucose levels fall, the body responds by releasing glucose into the bloodstream from the liver, thereby raising blood glucose levels.

Very few body mechanisms are regulated by means of **positive feedback loops**. In positive feedback loops the response of the regulating mechanism moves in the same direction as whatever triggers the change. You can see the problems inherent with this type of regulatory mechanism—a situation that continues to intensify rather than moving in an opposite direction. Positive feedback loops most often disrupt the steady state

**Figure 26.12 Body temperature is controlled by a negative feedback loop.** In a negative feedback loop, the response of the regulating mechanism is opposite to whatever triggers the change.

**Visual Thinking:** Describe why the feedback in each of these loops is considered negative feedback.

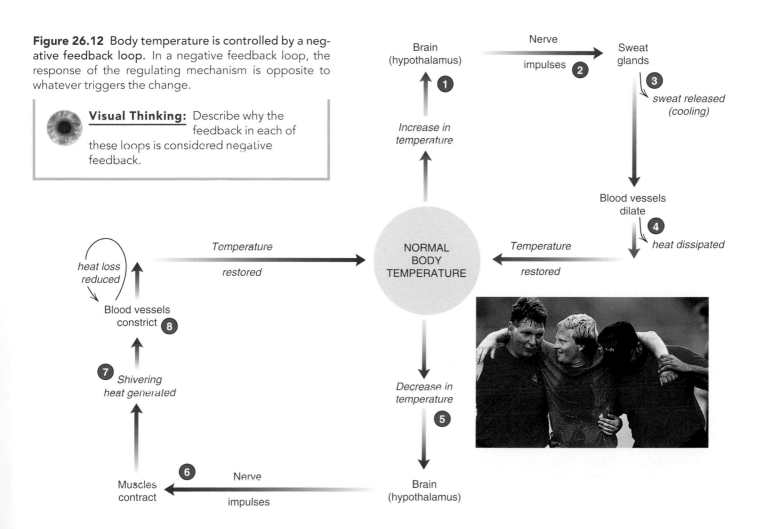

and can even lead to death. In circulatory shock, for example, a severe loss of blood can result in such a decreased blood volume that homeostasis cannot be maintained. The blood pressure drops and the blood flow is reduced so much that the heart and brain do not receive enough oxygen. As the heart weakens, the blood flow decreases more, weakening the heart even further. As the body becomes damaged from the loss of oxygen and nutrients, the shock worsens and can result in death. There are some good uses for positive feedback in the body, including oxytocin release during childbirth, which stimulates contraction of the uterus.

---

**CONCEPT CHECKPOINT**

**5.** Classify each of the following as a positive or negative feedback loop.
- An increase in blood glucose that stimulates the release of the hormone insulin, which stimulates the cellular uptake of glucose from the blood.
- The release of a hormone in the mother, stimulated by a nursing infant that increases the release of milk from the breast.
- An increase in blood calcium that causes a release of a hormone that stimulates the removal of calcium from the blood and increases the storage of calcium in the bones.

---

# CHAPTER REVIEW

## Summary of Key Concepts and Key Terms

### How are multicellular organisms organized?

**26.1** Multicellular **organisms** (p. 443) are organized hierarchically, from the lowest level of organization to the highest: **cells** (p. 443), **tissues** (p. 443), **organs** (p. 443) and **organ systems** (p. 443).

**26.1** Tissues are divided into four basic types based on their function: **epithelial** (p. 443), **connective** (p. 443), **muscle** (p. 443), and **nervous** (p. 443).

**26.2** The human body has the same general architecture as all vertebrates.

**26.2** In humans the coelom is divided into the thoracic cavity and the abdominal cavity.

### What are the functions of the different types of animal tissues?

**26.3** Epithelium covers and lines the surfaces of the body and composes the glands. These cells are of three main shapes and are arranged in ways that best suit their functions.

**26.4** Connective tissue and its cells provide a framework for the body, join its tissues, help defend it from foreign invaders, and act as storage sites for specific substances.

**26.5** There are three kinds of muscle tissue: smooth muscle, which is found in internal organs; skeletal muscle, which moves the body parts; and cardiac muscle, which makes up the heart.

**26.6** Nervous tissue is made up of two kinds of nerve cells: neurons and supporting cells. Neurons are specialized to conduct electrochemical impulses.

### How are tissues organized and their activities coordinated?

**26.7** Organs are body structures composed of several different tissues grouped into a structural and functional unit.

**26.7** An organ system is a group of organs that function together to carry out the principal activities of the body.

**26.8** The human body is organized so that its parts form an integrated whole. These parts work together at each level of organization to

maintain **homeostasis** (p. 455)—a stable internal environment within the human organism.

**26.8** Your body maintains a steady state by means of **feedback loops** (p. 455).

---

## Level 1 — Learning Basic Facts and Terms

**Multiple Choice**

1. Which of the following is the best example of negative feedback?
   a. An increase in body temperature resulting from shivering due to a decrease in body temperature
   b. An increase in body temperature resulting from exercise
   c. An increase in body temperature resulting from exposure to sun
   d. An increase in body temperature resulting from fever
   e. All of the above are equally good examples of negative feedback.

2. Which of the following is not considered a tissue?
   a. Cartilage
   b. Blood
   c. Cardiac muscle
   d. Tendons
   e. Mucus

**Matching**

3. ____ Elastic connective tissue
4. ____ Cartilage
5. ____ Bone
6. ____ Blood
7. ____ Dense fibrous connective tissue

a. The cells in its fluid matrix include erythrocytes and leukocytes.
b. The extracellular matrix is a semisolid gel secreted by chondrocytes.
c. The extracellular matrix of this tissue is mineralized with calcium salts laid down in concentric rings.
d. This type of tissue is found in structures that expand and then return to their original shape, such as the lungs and large arteries.
e. Comprises the lower layer of the skin and makes strong attachments between organs.

8. ____ Smooth muscle
9. ____ Epithelial tissue
10. ____ Skeletal muscle
11. ____ Nervous tissue

a. The type of tissue that lines the surface of the digestive tract and respiratory surfaces and functions mainly in absorbing and secreting chemicals.
b. The cells of this tissue are striated and contract under the control of the voluntary nervous system.
c. This type of tissue has a variety of supporting cells that nourish, protect, and insulate the cells that transmit impulses.
d. Waves of contraction of this tissue help move food through the digestive tract.

## Level 2    Learning Concepts

1. Distinguish among cells, tissues, organs, and organ systems. Give an example of each.
2. What are the four basic types of tissue in the human body? Give an example of each.
3. Draw and label the three main shapes of epithelial cells. Give an example of where each is found in your body.
4. What do lymphocytes, erythrocytes, and fibroblasts have in common? How do they differ?
5. Red blood cells provide the life-sustaining role of carrying oxygen to your body cells. For what cellular process is this oxygen necessary?

6. Do fat cells serve a purpose, and if so, what is it?
7. Which type of muscle tissue is under voluntary control and may be specifically strengthened by such exercises as lifting weights and doing pushups? Which type is not under voluntary control and may be strengthened by aerobic exercises such as running and cross-country skiing?
8. Create a table that lists the major function(s) and at least two components of each of the principal organ systems in your body.
9. What is homeostasis? How is it achieved?

## Level 3    Critical Thinking and Life Applications

1. Our bones generally have reached their full growth and mature length by the time we are 20 years old. How then is the femur, a leg bone, of a 30-year-old man able to heal fully after being broken in an accident?
2. Your cat scratches your finger, and the skin surrounding the scratch turns reddish and feels warm. Explain why.
3. The text states that connective tissues are generally made up of cells spaced well apart in a nonliving matrix. Using your know-

ledge of cells and tissues, suggest what the functions of a matrix might be.
4. Certain tissues in your body do not repair themselves if damaged (for example, nervous tissue within the brain and spinal cord), whereas other tissues such as skin rapidly repair themselves. Based on what you learned in this chapter, suggest a reason why nerve tissue in your central nervous system does not repair itself as readily as skin.

## In The News    Critical Thinking

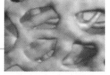

**SPARE PARTS**

Now that you understand more about animal tissues, reread this chapter's opening story about biotechnological advances in growing bone and teeth in the laboratory. To understand this research better, it may help you to follow these steps:

1. Review your immediate reaction to Keaveny and D'Souza's research on growing bone and teeth that you wrote when you began reading this chapter.
2. Based on your current understanding, again summarize the main point of the research in a sentence or two.
3. What questions do you now have about this research that are not answered by this chapter's opening story?

4. Collect new information about the research. Visit the *In The News* section of this text's companion Web site at www.wiley.com/college/alters and watch the "Spare Parts" video. Then use the "summary" link to read the accompanying story and access related links. Use this information, the links provided, and other online and library resources to answer your questions and find updates about this research topic. State the sources of your information. Explain why you think the information is accurate. Also determine whether the information expresses a particular point of view or is biased in any way.
5. What in your view is the most significant aspect of this research? Give reasons for your opinions and for any changes in your ideas based on the additional information you have collected and the analysis you have done.

# NUTRITION AND DIGESTION

## Vitamin A and Learning

"Carrots have vitamin A—the vitamin that's good for your eyes." You may have learned this in the third grade and still associate the vitamin with vision. The vitamin does have a role in the biochemical processes that occur in the retina of the eye and enable you to see, especially in dim light. Like most nutrients, however, vitamin A multitasks—it is also required for growth, development, reproduction, immunity to infectious disease, and maintenance of nervous tissue.

Ronald M. Evans, a professor at the Salk Institute for Biological Studies in Southern California, and his colleagues recently discovered yet another function of vitamin A. It appears to play a role in memory and learning. In recent experiments, the researchers studied two groups of mice: those with normal diets and those with diets lacking vitamin A. The young rats were each placed in a large tub of water containing a slightly submerged platform on which they could stand. Motivated by their dislike of water, the rats on normal diets quickly found the platforms. However, the rats deprived of vitamin A became progressively unable to perform this task. When vitamin A was reintroduced to their diets, their ability to find the platform returned. The results of these experiments indicate that the nutrient may have a role in recalling information and learning new tasks. Evans suggests that future research on the cognitive effects of vitamin A may lead to therapies for memory-robbing illnesses, such as Alzheimer's disease. To learn more about Evans's research, visit the *In The News* section of this text's companion Web site at www.wiley.com/college/alters and view the video "Vitamin A and Learning."

Should you worry about not having enough vitamin A in your diet? In the United States, vitamin A deficiency rarely occurs because the population has access to a variety of foods that contain the vitamin or its precursor *beta carotene*, a substance in many plant foods that the body converts to vitamin A. In countries where people often lack sufficient food to eat, however, vitamin A deficiency disease is a major cause of blindness and death.

Could taking extra amounts of vitamin A make it easier for you to read science textbooks or recall information during tests? More research is needed to determine whether vitamin A affects cognitive functioning in humans. It's not wise to take supplements of the vitamin in an attempt to boost your brain power. According to the National Academy of Sciences, the daily recommended dietary allowance (RDA) of vitamin A is 1000 micrograms for adult males and 800 micrograms for adult females. If a person consumes excess amounts of the vitamin, it accumulates in the body and becomes toxic. Signs and symptoms of vitamin A poisoning include scaly skin, headache, vomiting, and liver problems. A person can die as a result.

*Write your immediate reaction to Evan's research into the effects of vitamin A deficiency on learning in mice: first, summarize the main point of the research in a sentence or two; then suggest what you think its significance is. You will have an opportunity to reflect on your responses and gather more information on this topic in the* In The News *feature at the end of this chapter. In this chapter, you will learn more about the roles of nutrients, including vitamins, in the body.*

## CHAPTER GUIDE

### What are nutrients and how are they used?

**27.1** Nutrients provide building blocks and energy for the body.

**27.2** A healthy diet includes a variety of foods with a large proportion of fruits, vegetables, and whole grains.

### What are the patterns of digestion among organisms?

**27.3** Digestion can be intracellular or extracellular.

### How do humans digest food?

**27.4** Humans digest food mechanically and chemically.

**27.5** Digestion begins in the mouth.

**27.6** After swallowing, peristalsis moves food from the mouth to the stomach.

**27.7** Protein digestion begins in the stomach.

**27.8** The liver and pancreas help digestion take place in the small intestine.

**27.9** Digestion is completed in the small intestine.

**27.10** Absorption of nutrients occurs in the small intestine.

**27.11** Hormones control digestive gland secretions.

**27.12** The large intestine concentrates solids.

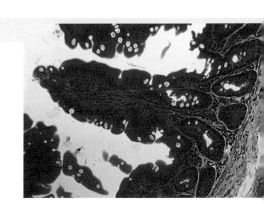

### Why is it unhealthy to be overweight or obese?

**27.13** Overweight and obesity are risk factors for many medical conditions.

# What are nutrients and how are they used?

## 27.1 Nutrients provide building blocks and energy for the body.

**Nutrients** are molecules in the diet that provide (1) raw materials for growth, repair, maintenance, or reproduction and (2) energy for all body activities. There are six classes of nutrients: carbohydrates, lipids (fats and oils), proteins, vitamins, minerals, and water. These substances make up most of your body and the bodies of all living things. Assuming that you are a proper weight for your height, your body is made up of about 60% water and about 20% lipids. The other nearly 20% is mostly protein, carbohydrate, combinations of these two substances, and two major minerals found in your bones: calcium and phosphorus. Other minerals and vitamins make up less than 1% of your body.

As you may recall from Chapter 4, carbohydrates, lipids, and proteins are carbon-containing (organic) compounds that organisms can use as a source of energy. They are often referred to as the energy nutrients. These nutrients provide more than energy, however; they also provide the raw materials to build the substances the body needs. Because the body needs substantial amounts of these nutrients for energy and raw material, they are called *macronutrients*.

Organisms do not obtain energy or organic building blocks from vitamins and minerals. Without them, however, your body could not unlock the energy in the macronutrients or perform a number of other important functions. Vitamins are organic molecules, and minerals are inorganic substances that the body needs in very small, but necessary, amounts. They are called *micronutrients*. Among other functions, micronutrients regulate vital chemical reactions.

Water is an essential nutrient that provides the medium in which all the body's chemical reactions take place. Without water, other nutrients could not be broken down in the body, absorbed into the bloodstream, or transported to your cells. It travels through your arteries and veins, lubricates your joints, and cushions organs such as the brain and spinal cord.

## 27.2 A healthy diet includes a variety of foods with a large proportion of fruits, vegetables, and whole grains.

### Macronutrients

How much protein, fat, and carbohydrate should you have in your diet? The Food and Nutrition Board of the National Academy of Sciences' Institute of Medicine report *Dietary References Intakes for Energy, Carbohydrates, Fiber, Fat, Protein and Amino Acids (Macronutrients)* recommends that 45 to 65% of your daily energy come from carbohydrates, including fruits, vegetables, and whole grains. Fat should supply 20 to 35% of your daily energy, and the remaining 10 to 35% should come from protein. **Figure 27.1** shows typical foods that are high in proteins, fats, and carbohydrates.

Of the 20 different amino acids that make up proteins, nine are usually cited as being indispensable or **essential amino acids** that must be included in the human diet. That is, humans cannot synthesize them and must obtain them from food. However, some "indispensable" amino acids can be synthesized from precursors that are structurally very similar. From this perspective, only three amino acids are indispensable: lysine, threonine, and tryptophan. Two others are dispensable or *nonessential*: glutamate and serine. That is, humans can manufacture them from raw materials in food. The rest of the amino acids are *conditionally essential*. Humans can manufacture conditionally essential amino acids, but their synthesis may be limited by a variety of factors, such as the dietary supply of the materials needed to make them and the health of the individual.

Protein foods that contain the essential amino acids in amounts proportional to the body's need for them are called high-quality, or complete, proteins. However, many high-quality protein foods such as meat, cheese, and eggs are high in animal (saturated) fat as

(a)

(b)

(c)

**Figure 27.1** Representative proteins (*a*), fats (*b*), and carbohydrates (*c*). Although eggs and cheese are shown in both the protein group and the fat group, they yield more calories from fat than from protein.

well. To reduce saturated fat intake, yet still get the essential amino acids, you could combine any legume (dried peas, beans, peanuts, or soy-based food) with any grain, nut, or seed.

Research supports the idea that the type of fat in the diet has a link with disease. The "good" fats are monosaturated and polyunsaturated fats (see Chapter 4). The "bad" fats are **saturated fats** and **trans fats**. Trans fats are processed fats, created by adding hydrogen to oil to harden it, as is done to produce margarine. Saturated and trans fats tend to raise blood cholesterol levels, greatly increasing the risk for heart disease. A healthy diet includes more foods containing "good" fats and fewer foods containing "bad" fats. **Table 27.1** categorizes foods based on the predominant type of fat that they contain.

Just as nutritional researchers have determined that all fats are not the same with respect to their effects on our bodies, they have also determined that all carbohydrates may not be the same. Some carbohydrates are **high-glycemic index foods** and cause higher rises in blood sugar levels when eaten than do low-glycemic index foods. They have been linked to an increased risk for both diabetes and heart disease. Some nutrition experts think the metabolic significance of the glycemic index is minimal, particularly when healthy people eat meals with foods that have a wide range of glycemic indices. In general, less processed, whole grain foods, including fresh fruits and vegetables, have a low glycemic index. Healthy diets contain more minimally processed whole-grain foods and fewer highly processed grains, cereals, and sugars, which are high-glycemic index foods.

## Vitamins and minerals

Although required only in small amounts, each of the 13 vitamins plays a vital role in the human body (**Table 27.2**). Many, such as the B vitamins, are used for the synthesis of coenzymes—small molecules that help enzymes work. For example, niacin and riboflavin, two of the B vitamins, are used to make the coenzymes NAD and FAD, which function in cellular respiration (see Chapter 8).

Vitamin C, vitamin E, and beta carotene, a yellow pigment in vegetables that the body converts to vitamin A, are *antioxidants*. Antioxidants prevent molecules in the body that lack an electron (many are called free radicals) from oxidizing (taking an electron from) other molecules, thus damaging them. Such damage is thought to contribute to the development of cancer, heart disease, and aging. Moreover, antioxidants have been shown to lower the risk of heart disease as well as the risk of breast, lung, stomach, colon, uterine, and cervical cancers.

Vitamins fall into two general categories: fat soluble (vitamins A, D, E, and K) and water soluble (the other nine vitamins). The water-soluble vitamins enter the cells of the intestinal lining—the mucosa—and move into the bloodstream. Excesses tend to be eliminated from the body. The fat-soluble vitamins are carried across the membranes of the intestinal cells associated with the fatty acids and glycerols, which are the breakdown products of triglycerides. They are transported to the bloodstream by means of the lymphatic system. Excesses tend to accumulate in the body and may become toxic.

Minerals are inorganic substances and are absorbed into the bloodstream and transported around the body as ions dissolved in the blood and other bodily fluids. Your body uses a variety of minerals that perform a variety of functions (see Table 27.2). Calcium, for example, does many jobs, including making up a part of the structure of your bones and teeth, helping your blood to clot, and helping your muscles contract. Phosphorus performs many jobs as well, including the maintenance of your blood pH. Magnesium is an important player in the process of releasing energy from carbohydrates, lipids, and proteins.

Most people who eat a sufficient amount and variety of foods get the vitamins and minerals they need in the food they eat. Many nutritionists, however, suggest that adults take a multiple vitamin/mineral supplement every day.

## Water

Healthy diets also include water, but there is no set amount that individuals should take in each day. Normal hydration can be maintained over a wide range of water intakes; the water needs of individuals depend on their age, weight, health, level of physical activity, and the temperature of their environment. The National Academy of Sciences, in its 2004 recommendations, suggests that the amount of fluid a person takes in should be driven by thirst and desire for beverages. The food and beverages individuals eat and drink during the day generally provide adequate hydration.

| **TABLE 27.1** | **Primary Types of Fats in Various Foods**[1] |
|---|---|

**Monounsaturated fats ("best" fats)**
- Olive, canola, and peanut oils
- Most nuts, such as peanuts, cashews, and almonds
- Olives
- Avocados

**Polyunsaturated fats ("good" fats)**
- Corn, soybean, safflower, and cottonseed oils
- Fish
- Chicken

**Saturated fats ("bad" fats)**
- Whole milk
- Butter
- Cheese
- Ice cream
- Red meat
- Chocolate
- Coconuts and coconut milk
- Palm and palm kernel oil

**Trans fats ("worst" fats)**
- Margarine unless stated otherwise on the package
- Vegetable shortening
- Deep-fried chips
- Many fast foods, such as deep-fried French fries, onion rings
- Most commercial baked goods, snack foods, and processed foods
- Any product that lists partially hydrogenated vegetable oil on the label

[1]Foods contain mixtures of monounsaturated, polyunsaturated, and saturated fatty acids. One type usually predominates, which is the categorization used here. Although margarine contains predominantly monounsaturated fatty acids, it is a hydrogenated product and therefore contains trans fats.

CONCEPT CHECKPOINT

**1.** Contrast each of the following sets of terms: (a) mineral : vitamin, (b) micronutrient : macronutrient, (c) essential amino acid : conditionally essential amino acid : nonessential amino acid, and (d) unsaturated fats : saturated fats : trans fats.

| TABLE 27.2 | Major Vitamins and Minerals | | | | |

| Vitamins Minerals | Major Functions | Rich Food Sources | RDA* (mg) | Signs and Symptoms of Deficiency | Toxicity |
|---|---|---|---|---|---|
| **Vitamins** | | | | | |
| A | Formation of visual pigments, maintenance of epithelial cells | Green or yellow fruits and vegetables, milk products, liver | 0.7–0.9 | Night blindness, dry skin, growth failure | Headache, nausea, birth defects |
| Thiamin | Coenzyme in $CO_2$ removal during cellular respiration | Whole grains, breads, cereals | 1.1–1.2 | Mental confusion, loss of muscular coordination | None from food |
| Riboflavin | Part of coenzymes FAD and FMN | Liver, leafy greens, dairy products | 1.1–1.3 | Dry skin, cracked lips | None from food |
| Niacin | Part of coenzymes NAD and NADH | Wheat bran, tuna, chicken, beef | 14–16 | Skin problems, diarrhea, depression, death | Skin flushing |
| Pantothenic acid | Part of coenzyme A, energy metabolism | Widespread in foods | 5** | Rare | None from food |
| $B_6$ | Protein metabolism | Animal proteins, spinach, broccoli, bananas | 1.3–1.7 | Anemia, headaches, convulsions | None from food |
| $B_{12}$ | Coenzyme for amino acids and nucleic acid metabolism | Meat, fish, poultry, fortified cereals | 0.0024 | Pernicious anemia | None from food |
| Biotin | Coenzyme in carbohydrate and fat metabolism | Liver, peanuts, cheese, egg yolk | 0.03** | Rare | Rare |
| Folate | Nucleic acid and amino acid synthesis | Green leafy vegetables, whole grain breads | 0.40 | Anemia, embryonic neural tube defects | None from food |
| C | Collagen synthesis antioxidant | Citrus fruits, broccoli, greens, tomatoes, strawberries | 75–90 | Scurvy, poor wound healing, bruises | Diarrhea, kidney stones |
| D | Absorption of calcium and phosphorous, bone formation | Vitamin D fortified milk, fish | 0.005–0.015** | Rickets (bone deformities) | Calcium deposits in soft tissues |
| E | Antioxidant | Vegetable oils, nuts, fruits, vegetables, meats | 15 | Rare | None from food |
| K | Synthesis of blood clotting substances | Green leafy vegetables, liver, plant oils | 0.09–0.12** | Hemorrhage | None from food |
| **Major minerals** | | | | | |
| Calcium (Ca) | Component of bone and teeth, blood clotting, nerve transmission, muscle action | Dairy products, canned fish, broccoli, kale | 1,000–1,200 | Osteoporosis | Kidney stones |
| Phosphorus (P) | Components of bone and teeth, energy transfer (ATP) | Dairy products, meat, peas, eggs | 700 | None | Bone loss if low calcium intake |
| Magnesium (Mg) | Bone formation, muscle and nerve function | Wheat bran, green vegetables, nuts, milk | 310–420 | Weakness | Rare |
| Iron (Fe) | Hemoglobin synthesis, oxygen transport | Liver, meat, fortified breads and cereals | 8–18 | Anemia | Constipation |
| Zinc (Zn) | Component of many enzymes | Seafood, liver, meats, whole grains, fortified cereals | 8–11 | Skin rash, poor growth, hair loss | Diarrhea, depressed immune function |
| Iodine (I) | Thyroid hormone production | Iodized salt, seafood | 0.15 | Goiter | Interference with thyroid function |
| Fluorine (F) | Strengthener of teeth | Fluoridated water, tea, marine fish | 0.003–0.004** | Increased risk of dental caries | Stained teeth during development |
| Copper (Cu) | Iron metabolism, component of many enzymes | Liver, cocoa, whole grains, nuts | 0.9 | Anemia | Rare: vomiting, liver damage |

*Recommended dietary allowances (2003) values are for adults (men and nonpregnant women age 19 and up).
**Estimated adequate intake for adults (no RDA established).

# How Science Works

## The evolution of food guides

Americans have used a variety of food guides over the past six decades. In 1943, in an effort to promote nutrition education, the U.S. Department of Agriculture (USDA) introduced the *Basic Seven Food Guide*. The basic seven guide divided foods into seven groups: (1) green and yellow vegetables; (2) oranges, tomatoes, and raw salads; (3) potatoes; (4) milk and cheese; (5) meat, poultry, fish, and eggs; (6) bread, flour, and cereals; and (7) butter and margarine.

The basic seven plan remained in use until 1956, when it was replaced by the *Basic Four Food Guide*. The basic four guide combined the basic seven's three fruit and vegetable groupings into one group and eliminated the butter and margarine group. By the late 1970s, critics charged that the basic four guide did not reflect current scientific findings about the role of such nutrients as fats in the development of heart disease and some cancers. Responding to these concerns, the USDA presented the *Hassle-Free Guide to a Better Diet* in 1979. This guide, sometimes referred to as the "basic five," added a fifth food group, fats, sweets, and alcoholic beverages, and recommended limiting intake of foods rich in these substances. However, critics still were not satisfied that these recommendations went far enough.

Again, the USDA responded with a revised food guide. *The Food Wheel: A Daily Pattern for Food Choices* recommended eating more fiber-rich fruits, vegetables, and whole grains and limiting dairy products and meats—foods that are high in fat—in the diet. By this time, the development and use of food guides had become a political and economic issue in addition to one of health. Meat and dairy producers, for example, were not pleased with a food guide that recommended the limitation of their products in the American diet.

The latest food guide, which is under review by a USDA panel of experts at the time of this writing, was introduced in 1992. Called the *Food Guide Pyramid*, this guide continues to reflect health concerns (**Figure 27.A**). Grouped together at the base of the pyramid are cereals and other grain products. These foods have the highest number of recommended servings. Fruits and vegetables, grouped separately, form the next tier. Fewer servings of these foods are considered to be needed for a healthy diet, so their tier is smaller. Meats, meat alternatives, and dairy products form the next, still smaller tier, because they include dietary fat. At the top are high-sugar or fat foods and alcoholic beverages, which offer few vitamins and minerals in relation to calories.

Like the food guides of the past, the *Food Guide Pyramid* is not perfect. Researchers at the Harvard School of Public Health suggest that the current food pyramid was based on shaky scientific evidence and that it has not changed to incorporate recent research findings on diet and health. Therefore, they have developed a food guide of their own, called the *Healthy Eating Pyramid*.

The *Healthy Eating Pyramid* is shown in **Figure 27.B**. Unlike any other food guide, it has a foundation of daily exercise and weight control. The nutritional base of the pyramid is composed of whole-grain foods and plant oils, which shows a major change in understanding the role of fats in the diet. These fats are "good" fats and can improve blood cholesterol levels when eaten in place of highly processed carbohydrates. They also lessen the risk of potentially deadly rhythm problems of the heart. Another major change is that the *Healthy Eating Pyramid* separates red meat from chicken and fish, recognizing that red meat is high in saturated ("bad") fats, while the fats in chicken and fish are much healthier. What other differences do you see? How should you change your diet to reflect the healthy eating patterns shown in the *Healthy Eating Pyramid*? ●

### FOOD GUIDE PYRAMID
#### A guide to daily food choices

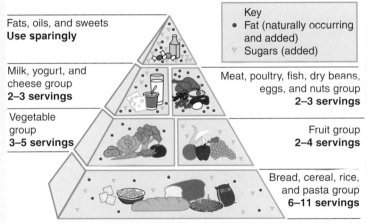

Fats, oils, and sweets
**Use sparingly**

Key
● Fat (naturally occurring and added)
▽ Sugars (added)

Milk, yogurt, and cheese group
**2–3 servings**

Meat, poultry, fish, dry beans, eggs, and nuts group
**2–3 servings**

Vegetable group
**3–5 servings**

Fruit group
**2–4 servings**

Bread, cereal, rice, and pasta group
**6–11 servings**

**Figure 27.A** The *Food Guide Pyramid*. Developed in 1992 and under review at the time of this writing, the *Food Guide Pyramid* was produced by the U.S. Department of Agriculture to show Americans a healthy way of eating.

### HEALTHY EATING PYRAMID

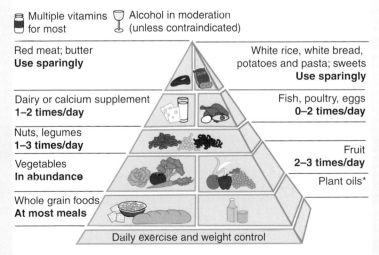

Multiple vitamins for most · Alcohol in moderation (unless contraindicated)

Red meat; butter
**Use sparingly**

White rice, white bread, potatoes and pasta; sweets
**Use sparingly**

Dairy or calcium supplement
**1–2 times/day**

Fish, poultry, eggs
**0–2 times/day**

Nuts, legumes
**1–3 times/day**

Vegetables
**In abundance**

Fruit
**2–3 times/day**

Plant oils*

Whole grain foods
**At most meals**

Daily exercise and weight control

*Including olive, canola, soy, corn, sunflower, peanut, and other vegetable oils.

**Figure 27.B** The *Healthy Eating Pyramid* in 2003. The developers of the guide, from the Harvard School of Public Health, state that this guide will be updated as research reveals healthier ways to eat.

# What are the patterns of digestion among organisms?

## 27.3 Digestion can be intracellular or extracellular.

**Digestion** is a process in which food particles are broken down into small molecules that can be absorbed by the body. **Autotrophs**, or self-feeders, make food within their bodies, and digestion is unnecessary. Plants, algae, some protists, and some bacteria are autotrophs. **Heterotrophs** digest their food either within their cells (intracellularly), outside of their cells (extracellularly), or both.

Intracellular digestion usually takes place in single-celled protists and the sponges. **Figure 27.2** shows the process of intracellular digestion in the paramecium, a single-celled protist. The beating cilia of the paramecium sweep food into its gullet **1**. From the gullet, food passes into a developing food vacuole. Lysosomes fuse with the food vacuole, bringing enzymes that aid in digestion. Wastes are expelled from the anal pore **2**.

Although sponges are multicellular organisms, they process food in a similar way. Water containing food particles passes through the sponge's body by means of its pores **3**. Collar cells, located in the interior of the vase-shaped sponge, trap food particles as their flagella generate currents in the water. Food trapped in the collars is ingested by phagocytosis and digested by lysosomal enzymes **4**.

Intracellular digestion also occurs in conjunction with extracellular digestion in some *cnidarians* (hydra, jellyfish, corals, and sea anemones), *ctenophores* (comb jellies), and *turbellarians* (free-living flatworms). When these organisms take in food, digestion begins in the digestive cavity **5**. Some food, however, may remain only partially digested. These particles are taken into the cells of the digestive cavity, where they are broken down completely **6**.

Extracellular digestion can take place with or without a **digestive system**, a series of organs specialized for breaking down food into nutrients that can be absorbed and ridding the body of undigested wastes. In fungi and bacteria, extracellular digestion takes place without a digestive system. Fungi secrete enzymes directly onto their food source,—usually organic matter such as a dead tree or a slice of bread. After the enzymes digest the food, the fungi absorb the breakdown products **7**. Bacteria often take in food directly by phagocytosis or pinocytosis. However, if food particles are too large to ingest this way, some bacteria produce enzymes that digest food outside of the cell much like the fungi.

In animals, extracellular digestion usually takes place in a digestive system. Enzymes are secreted in various parts of the system, digesting the food into small molecules that can be absorbed by the body **8** and **9**. The food you digest is really *outside* of your cells and tissues. Digestive systems may provide mechanical breakdown of food as well as chemical. Your teeth, for example, mechanically break down food, as does the gizzard of a bird or an earthworm. The bird's gizzard usually contains bits of sand or gravel that physically break food apart as it is churned in this "second stomach." In addition to digesting and absorbing food (and perhaps storing food—in a crop, for example), digestive systems also eliminate indigestible material, or wastes.

Digestive systems vary depending on the complexity of the animal and its diet. Animals with relatively simple anatomy, such as jellyfish and flatworms, have only one opening to the digestive system. Food is taken into a digestive cavity through this opening, where it is digested and nutrients are absorbed. Wastes are expelled through the same opening. More complex animals, such as earthworms, birds, and humans, have digestive systems with two openings—a mouth and an anus.

> ### CONCEPT CHECKPOINT
>
> 2. What general evolutionary trends do you observe in how animals digest their food?
> 3. Tapeworms have no digestive system. How do you explain this? (You may find it helpful to review the *Just Wondering* box in Chapter 21, which discusses tapeworms.)

# How do humans digest food?

## 27.4 Humans digest food mechanically and chemically.

The digestion of food directly yields no usable energy, but changes a diverse array of complex molecules into a small number of simpler molecules. These molecules are absorbed by the body and are used by your cells as fuel for cellular respiration. During cellular respiration, adenosine triphosphate (ATP) molecules are produced—the energy currency of your body. That is how you derive energy from the food you eat—by the process of cellular respiration.

In the human body, digestion is carried out mechanically in the oral cavity as teeth crush food, and in the stomach as its muscular walls churn food. In addition, digestion is carried out chemically in three ways:

- Hydrochloric acid (HCl) denatures or unfolds, protein molecules and disrupts the protein glue that holds cells together.
- Bile salts emulsify, or separate, large lipid droplets into much smaller lipid droplets.
- A variety of highly specific enzymes cleave certain chemical bonds.

# Visual Summary

**Figure 27.2** Patterns of digestion.

**NO DIGESTION** (autotrophs)

**Plants**

**INTRACELLULAR DIGESTION**
(single-celled protists and sponges)

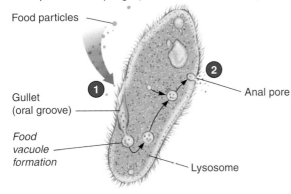

Food particles

Gullet
(oral groove)

*Food
vacuole
formation*

**1**

**2**
Anal pore

Lysosome

**Paramecium**

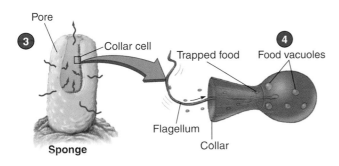

Pore

**3**

Collar cell

Trapped food

Food vacuoles

**4**

Flagellum

Collar

**Sponge**

**INTRACELLULAR AND EXTRACELLULAR DIGESTION**
(hydra, jellyfish, corals, sea anemones, comb jellies, flatworms)

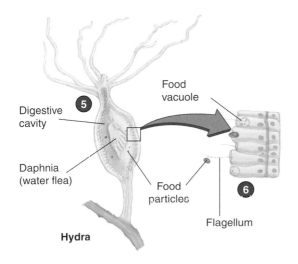

Digestive
cavity

**5**

Food
vacuole

Daphnia
(water flea)

Food
particles

**6**

Flagellum

**Hydra**

**EXTRACELLULAR DIGESTION**
No digestive system (fungi and some bacteria)

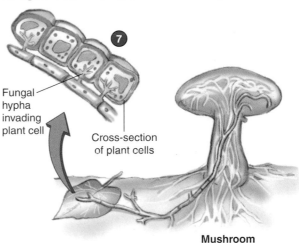

Fungal
hypha
invading
plant cell

**7**

Cross-section
of plant cells

**Mushroom**

**EXTRACELLULAR DIGESTION**
Digestive system (most animals)

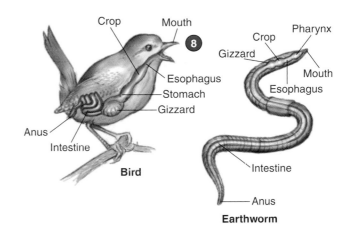

Crop

Mouth

**8**

Esophagus

Stomach

Gizzard

Anus

Intestine

**Bird**

Crop

Pharynx

Gizzard

Mouth

Esophagus

Intestine

Anus

**Earthworm**

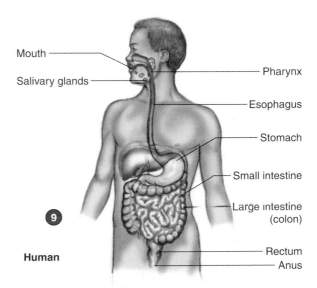

Mouth

Salivary glands

Pharynx

Esophagus

Stomach

Small intestine

**9**

Large intestine
(colon)

Rectum

Anus

**Human**

Enzymes are proteins that speed up the rate of chemical reactions in living things (see Chapter 6). **Table 27.3** shows the enzymes that help digest the energy nutrients—proteins, carbohydrates, and lipids. These enzymes are of three basic types:

- **Proteases** (PRO-tee-ace-es) first break down proteins to smaller polypeptides and then to amino acids.
- **Amylases** (AM-uh-lace-es) break down starches and glycogen to sugars.
- **Lipases** (LYE-pays-es) break down the triglycerides in lipids to fatty acids and glycerol. Most dietary fat is in the form of triglycerides, an abundant type of lipid.

| TABLE 27.3 | Digestive Enzymes | | |
|---|---|---|---|
| **Source** | **Enzyme** | **Substrate** | **Digestion Product** |
| Salivary gland | Amylase | Starch, glycogen | Disaccharides |
| Stomach | Pepsin | Proteins | Short polypeptides |
| Small intestine | Peptidases | Short peptides | Amino acids |
| Pancreas | Lactase, maltase, sucrase | Disaccharides | Glucose, monosaccharides |
| | Pancreatic amylase | Starch, glycogen | Disaccharides |
| | Trypsin, chymotrypsin, carboxypeptidase | Proteins | Polypeptides |
| | Lipase | Triglycerides | Fatty acids, glycerol |

## 27.5 Digestion begins in the mouth.

Does your mouth ever water when you think about your favorite food? Does the smell of some of your favorite foods also evoke this reaction? Do you know why this reaction occurs?

The "water" in your mouth is really a secretion from a set of glands called the **salivary glands**. The locations and names of each of these glands are shown in **Figure 27.3**. Their secretion, saliva, moistens and lubricates food so that it is swallowed easily and does not scratch the throat. Saliva consists primarily of water, mucus, and the digestive enzyme **salivary amylase**. Other substances can be found in smaller amounts, such as antibodies and a bacteria-killing enzyme.

Salivary amylase is a hydrolyzing enzyme that breaks down starch (carbohydrate) into molecules of the disaccharide maltose. Hydrolyzing enzymes break down substances with the addition of water—a process called hydrolysis. The nervous system maintains a continual secretion of saliva but when stimulated by the presence (or sometimes the sight, smell, or thought) of food, secretion of saliva speeds up. When food is in the mouth, nerve endings called chemoreceptors, which are sensitive to the presence of certain chemicals, send a signal to the brain, which responds by stimulating the salivary glands. Have you ever sucked on a slice of lemon? If you have, then you know that the most potent stimuli for secretion of saliva are acid solutions; lemon juice can increase the rate of salivation eightfold.

Mechanical digestion begins when the teeth tear food apart into tiny pieces. In this way, the surface area of the food is increased, allowing the digestive enzymes to mix with the food and break it down more quickly and completely. The teeth of humans, as well as those of other organisms, are specialized depending upon the type of food an organism eats and how it obtains its food. Humans are omnivores. An omnivore eats both plant and animal foods. As a result, human teeth are structurally intermediate between the pointed, cutting teeth characteristic of carnivores (meat eaters), and the flat, grinding teeth characteristic of herbivores (plant eaters) (**Figure 27.4**).

The four front teeth in the upper and lower jaws of humans are incisors (**Figure 27.5**). These teeth are sharp and chisel shaped and are used for biting. On each side of the incisors are pointed teeth called canines, which are used in tearing food. Behind each canine, on each side of the mouth and along both top and bottom jaws, are two premolars and three molars, all of which have flattened ridged surfaces for grinding and crushing food.

## 27.6 After swallowing, peristalsis moves food from the mouth to the stomach.

As food is chewed and moistened, the tongue forms it into a ball-like mass called a *bolus* and pushes it into the upper part of the **pharynx** (FAIR-inks), known commonly as the throat. As this happens, the soft palate raises up, sealing off the nasal cavity and preventing any food from entering this chamber (**Figure 27.6 ❶** ). The soft palate is the tissue at the back of the roof of the mouth. The pressure of the food in the pharynx stimulates nerves in its walls that begin the swallowing reflex, an involuntary action. As part of this reflex action, the **larynx** (LAIR-inks), which is commonly known as the voice box,

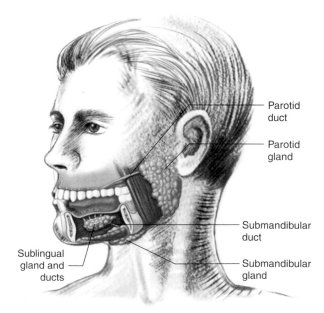

Parotid duct

Parotid gland

Submandibular duct

Submandibular gland

Sublingual gland and ducts

**Figure 27.3** The salivary glands.

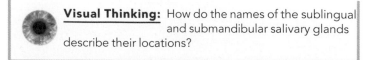

**Visual Thinking:** How do the names of the sublingual and submandibular salivary glands describe their locations?

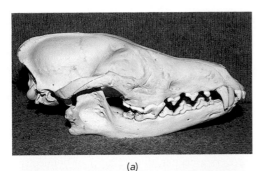

(a)

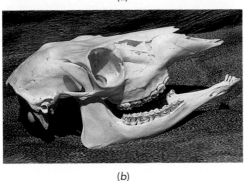

(b)

**Figure 27.4 Carnivore teeth vs. herbivore teeth.** (a) The skull of a coyote, a carnivore. Notice the pointed tooth surfaces, front and back, that can rip and tear the flesh of prey. (b) The skull of a mule deer, an herbivore. Notice the flattened tooth surfaces that crush and grind plant material, and the cutting teeth in the front that snip bits from plants.

rises to meet the **epiglottis** (ep-ih-GLOT-iss), a flap of tissue that folds back over the opening to the larynx **2**. With this action the epiglottis acts much like a trapdoor, closing over the **glottis**, the opening to the larynx and trachea (your windpipe), so that food will not go down the wrong way. If you place your hand over your larynx (Adam's apple), you can feel it move up when you swallow.

After passing through the pharynx and bypassing the windpipe, the food enters the **esophagus** (ih-SOF-uh gus) **3**, a food tube that connects the pharynx to the stomach. The esophagus does not take part in digestion but instead acts like an escalator, moving food down toward the stomach. The circular muscles in its walls produce successive waves of contractions called **peristalsis**

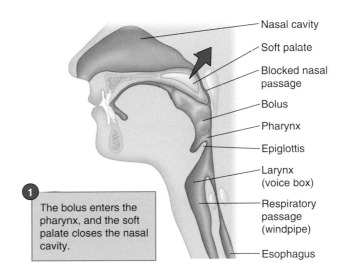

Nasal cavity

Soft palate

Blocked nasal passage

Bolus

Pharynx

Epiglottis

Larynx (voice box)

Respiratory passage (windpipe)

Esophagus

**1** The bolus enters the pharynx, and the soft palate closes the nasal cavity.

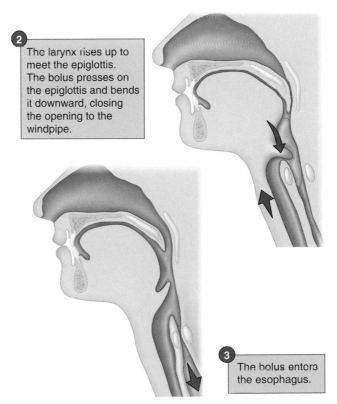

Molars

Premolars

Canine

Incisors

**Figure 27.5 Teeth in humans.** The four front teeth are incisors, used for biting. Canines are used for tearing food. The premolars and molars are used for grinding and crushing.

**2** The larynx rises up to meet the epiglottis. The bolus presses on the epiglottis and bends it downward, closing the opening to the windpipe.

**3** The bolus enters the esophagus.

**Figure 27.6** The process of swallowing.

**Visual Thinking:** How does the pattern of human teeth compare to the pattern of teeth in the carnivore and the herbivore in Figure 27.4?

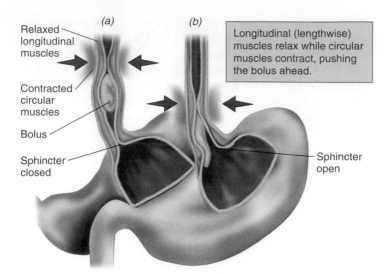

Longitudinal (lengthwise) muscles relax while circular muscles contract, pushing the bolus ahead.

**Figure 27.7** Peristalsis. Food does not simply slide down the esophagus; it is pushed along by waves of muscular contraction.

(PAIR-ih-STALL-sis) that push the bolus along. The movement of food to the stomach is therefore not dependent on gravity, so even astronauts in zero gravity—or you standing on your head—can swallow without difficulty.

The esophagus, which is about 25 centimeters long (a bit less than a foot), ends at the lower esophageal sphincter, a circular muscle that regulates entry into the stomach (**Figure 27.7a**). When this sphincter relaxes, food can enter the stomach (Figure 27.7b).

## 27.7 Protein digestion begins in the stomach.

The lower esophageal sphincter relaxes when food reaches it, allowing food to enter the **stomach**. The stomach is a muscular sac in which the food is collected and partially digested by hydrochloric acid and proteases. The stomach then "feeds" this food, little by little, to the primary organ of digestion, the small intestine. The stomach and small intestine have the same basic structural plan (**Figure 27.8**). Their interiors are lined with a layer of tissue called the mucosa, which covers a deeper, thicker layer of connective tissue, the submucosa, which is rich in blood vessels and nerves. Surrounding the submucosa are layers of smooth muscle tissue that churn and continue to push food along by peristalsis. An envelope of tough connective tissue called serosa (seh-ROH-sa) serves as the outer covering of the digestive tract (as well as the other abdominal organs). Thin sheets of connective tissue called mesentery (MEH-zen-TAR-ee) are attached to the serosa along most of the intestinal tract, holding it in place and serving as a highway for blood vessels and nerves.

The inner surface of the stomach is dotted with **gastric glands** that are part of the epithelium that dips deeply into the mucosa (**Figure 27.9**). Two different kinds of cells in these glands secrete a gastric juice made up of hydrochloric acid and the protein pepsinogen. After the secretion of both, the acid chemically interacts with the pepsinogen, converting it to the protein-hydrolyzing enzyme **pepsin**. The hydrochloric acid also softens the connective tissue in foods; denatures, or unfolds, large protein molecules; and kills most bacteria that may have been ingested with the food. The pepsin digests only proteins, breaking them down into short polypeptides. Other epithelial cells are specialized for the secretion of mucus. This mucus, produced in large quantities, helps to protect the stomach from digesting itself.

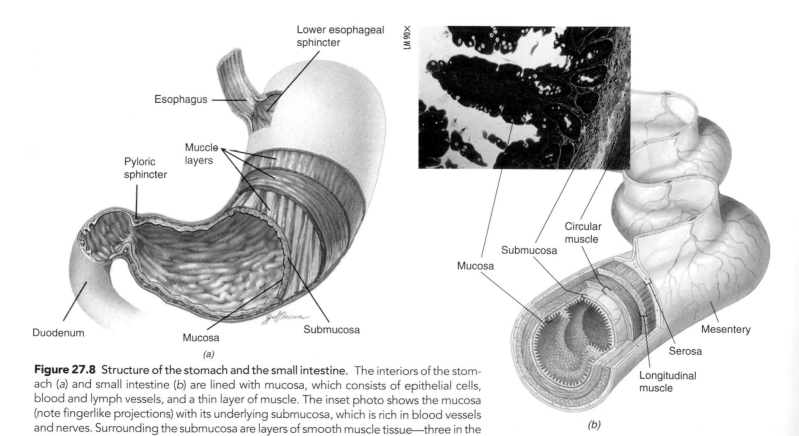

**Figure 27.8** Structure of the stomach and the small intestine. The interiors of the stomach (a) and small intestine (b) are lined with mucosa, which consists of epithelial cells, blood and lymph vessels, and a thin layer of muscle. The inset photo shows the mucosa (note fingerlike projections) with its underlying submucosa, which is rich in blood vessels and nerves. Surrounding the submucosa are layers of smooth muscle tissue—three in the stomach and two in the intestines.

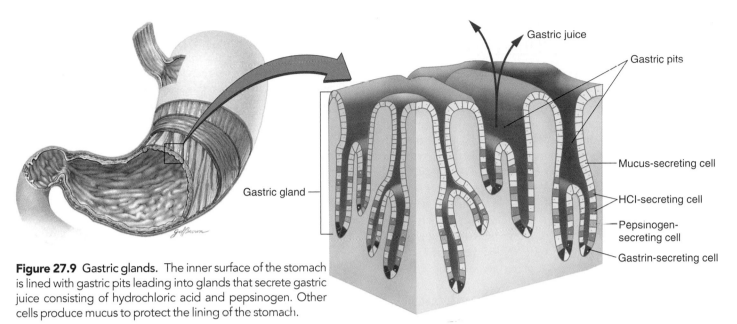

**Figure 27.9 Gastric glands.** The inner surface of the stomach is lined with gastric pits leading into glands that secrete gastric juice consisting of hydrochloric acid and pepsinogen. Other cells produce mucus to protect the lining of the stomach.

Gastric juice could damage the stomach itself if it were constantly secreted. Instead, gastric glands are stimulated by the nervous system in response to the sight, smell, or taste of food. Once food has been ingested, the stomach controls the production of gastric juice by means of a hormone called **gastrin**. Gastrin is produced by cells scattered throughout the epithelium of the stomach and travels through the bloodstream back to the gastric glands.

Food stays in the stomach for approximately two to six hours. During this time, the contractions of the muscular wall of the stomach churn the food, mixing it with the gastric juice and mucus. By the time the food is ready to leave the stomach as a substance called chyme (KIME), it has the consistency of pea soup. The gate to the small intestine, the pyloric sphincter, opens to allow just a bit of the chyme to pass. When the acid in this chyme is neutralized and the food is digested, the pyloric sphincter is signaled by the nervous system to open again, allowing the next bit of chyme to pass.

Acid from the stomach may attack the walls of the first portion of the small intestine, or duodenum, burning holes through the wall. These holes are called duodenal ulcers. Because the contents of the small intestine are not normally acidic, this organ is much less able to withstand the disruptive actions of stomach acid than the wall of the stomach is. For this reason, the majority of ulcers are duodenal. The bacterium *H. pylori* is responsible for the majority of ulcers; it weakens the protective mucus coating of the stomach and duodenum, which allows acids to attack the tissues beneath.

### CONCEPT CHECKPOINT

4. Trace the pathway of a morsel of food from the time it enters your mouth until it just enters the small intestine. At each point along the way, briefly describe the processes that have contributed to the initial digestion of that food.

## *just wondering . . .*

### *Questions students ask*

### I get heartburn a lot. What causes it?

Heartburn results when the acid contents of the stomach splash back into the esophagus. The acid burns the esophagus—but it feels like your heart is on fire! You're not alone in your pain: 10% of adults report experiencing heartburn about once a week, and more than 30% experience it once a month.

Occasional heartburn is not a serious condition. Frequent heartburn, however, can irritate the delicate lining of the esophagus. In severe cases, stomach acid injures the esophagus so much that it bleeds. If your heartburn is severe and frequent, you should consult a physician for treatment and evaluation, since this condition can be a risk factor for esophageal cancer.

There are a few things you can do to avoid or relieve heartburn. If you smoke, understand that cigarette smoke relaxes the lower esophageal sphincter that closes the opening to the stomach. Therefore, quitting smoking will reduce your heartburn, or get

rid of it altogether. In addition, try not to overeat, which stuffs your stomach too full of food and promotes reflux (movement of stomach contents into the esophagus). Avoid eating close to bedtime because lying down with food in your stomach also promotes reflux. You might also try elevating the head end of your bed about six inches. Tight clothing or obesity puts extra pressure on the abdominal organs and may force food into the esophagus. Last, try to avoid the following foods and medications because they all relax the sphincter between the esophagus and the stomach: alcohol, chocolate, fats, peppermints, birth control pills, antihistamines, antispasmodics, some heart medications, and some asthma medications. As a last resort, many persons find relief by taking over-the-counter antacids. ●

*Are you wondering about a topic in biology and how it relates to your life? Submit your question by clicking the Just Wondering link in this text's companion Web site at www.wiley.com/college/alters.*

## 27.8 The liver and pancreas help digestion take place in the small intestine.

Most of the digestion of food takes place in the **small intestine**, the tubelike portion of the digestive tract that begins at the pyloric sphincter and ends at its T-shaped junction with the large intestine. The small intestine is approximately 6 meters long—long enough to stretch from the ground to the top of a two-story building. The initial portion of the small intestine (the first 25 centimeters [8 inches]) is called the **duodenum** (DOO-oh-DEE-num *or* doo-ODD-un-um). Within the duodenum, starches and glycogen are broken down to sugars, proteins to amino acids, and triglycerides to fatty acids and glycerol. These products of digestion then pass through the cells of the intestinal mucosa and diffuse into the blood in underlying blood vessels. Fatty acids and glycerol move into the lymph in neighboring lymphatic vessels.

Some of the enzymes necessary for digestion are secreted by the salivary glands, epithelial cells of the stomach, and epithelial cells of the duodenum. The others are secreted by the **pancreas**, a long gland that lies beneath the stomach and is surrounded on one side by the curve of the duodenum (**Figure 27.10**). A tiny duct runs from the pancreas to the small intestine and serves as the passageway for the pancreatic juice. The pancreatic enzymes are secreted in an inactive form and become active in the presence of a particular enzyme secreted by cells in the intestinal epithelium. This way, they do not digest their way down the pancreatic duct to the small intestine. As you can see from Table 27.3, this secretion of

the pancreas includes a number of digestive enzymes. The pancreas also secretes an alkaline solution that helps neutralize the acidic chyme. This helps to provide a suitable environment in which pancreatic and intestinal enzymes can work.

The **liver** is another organ that works with the duodenum to digest food. This organ, which weighs over 3 pounds, lies just under the diaphragm (see Figure 27.10). It is one of the most complex organs of the body and performs more than 500 functions. Although the liver produces no digestive enzymes, it does help in the digestion of lipids by secreting **bile**. Among the many components of bile are bile salts, which are substances that act much like detergents, breaking lipids into minute droplets, similar to droplets of cream suspended in milk. The liver manufactures bile salts from cholesterol. Excess bile is stored and concentrated in the **gallbladder** on the underside of the liver. A bile duct brings bile from the liver and gallbladder to the small intestine.

## 27.9 Digestion is completed in the small intestine.

Some of the food that enters the small intestine has already been partially digested. Salivary amylase has broken some of the bonds in the starches and glycogen, producing the disaccharide maltose. However, much starch and glycogen remain undigested. In the small intestine, **pancreatic amylase** breaks down these carbohydrates to maltose. Maltose, sucrose (table sugar), and lactose (milk sugar) are digested to the monosaccharides glucose, fructose, and galactose by enzymes called **disaccharidases** that are produced and secreted by specialized epithelial cells of the small intestine, called brush border cells (see Section 27.11). Some people do not produce the enzyme to digest lactose, however, and are therefore unable to completely digest milk. The undigested lactose retains fluid in the feces, and bacteria in the intestines ferment the lactose, producing gas. Therefore, individuals who fail to synthesize enough lactase cannot drink milk or eat other dairy products containing lactose without experiencing cramps, bloating, and, in some cases, diarrhea. This condition is known as lactose intolerance.

Some of the proteins in the food entering the small intestine have also been partially digested. The hydrochloric acid of the stomach has unfolded these proteins, and pepsin has cleaved some of them to shorter polypeptides. Three other enzymes produced by the pancreas complete the digestion of proteins: **trypsin** (TRIP-sin), **chymotrypsin** (KYE-moe-TRIP-sin), and **carboxypeptidase** (kar-BOK-see-PEP-ti-dace). These enzymes work as a team with **peptidases** produced by the brush border cells of the intestinal epithelium, breaking down polypeptides into shorter chains and then to amino acids (see Table 27.3).

Virtually all lipid digestion takes place in the small intestine. Because they are insoluble in water, lipids tend to enter the small intestine as globules. Before these globules can be digested, they are emulsified, or separated into small droplets, by bile salts. Emulsification increases the surface area available for **pancreatic lipase** to act on the individual triglyceride molecules, which make up the lipids. Fatty acids and glycerol are the digestion products of triglycerides.

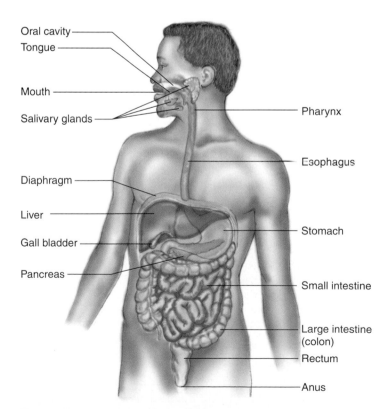

Oral cavity
Tongue
Mouth
Salivary glands
Pharynx
Esophagus
Diaphragm
Liver
Stomach
Gall bladder
Pancreas
Small intestine
Large intestine (colon)
Rectum
Anus

**Figure 27.10** The pancreas and liver work with the small intestine to digest food.

**5.** Describe the enzymatic digestion of carbohydrates and proteins from the time they enter the stomach until the breakdown products are absorbed by the small intestine. Be sure to list the enzymes involved and the digestive organ that secretes each of them.

## 27.10 Absorption of nutrients occurs in the small intestine.

The internal surface area of the small intestine is tremendously large, which aids in the absorption of nutrients. How does a large surface area exist in such a relatively small space? First, the mucosa and submucosa of the small intestine are thrown into folds; they do not have a smooth inner surface like a garden hose. This folded surface, in turn, is covered by fine, fingerlike projections of the epithelium. These projections are called **villi** (singular *villus*) and are so small that a microscope must be used to see them. In addition, the epithelial cells of the villi are covered on their exposed surfaces by cytoplasmic projections called microvilli (**Figure 27.11**). This brushlike structure has given these epithelial cells the name "brush border" cells, and the digestive enzymes secreted by these cells (see Table 27.3) are called "brush border" enzymes. The infoldings, the villi, and the microvilli provide the small intestine with a surface area of about 300 square meters, or 2700 square feet, an area greater than the floor space in many homes.

Within each villus is a network of capillaries and a lymphatic vessel called a lacteal. Each monosaccharide and amino acid must catch a ride on a carrier molecule to get into an epithelial cell of a villus. Energy is needed to ferry many of these molecules across cell membranes—a process called active transport (see Chapter 5). Others are taken up by facilitated diffusion. Once the monosaccharides and amino acids are in the epithelial cells, however, they accumulate and eventually move by facilitated diffusion through the base of the cell and into the blood. When in the bloodstream, they are quickly swept away to the liver for processing and storage. When blood levels of glucose are sufficient to supply cells with fuel for cellular respiration (see Chapter 6), the liver stores excess glucose as glycogen. When more glucose is needed, such as between meals, the liver readily converts the glycogen back to glucose. In this way, the liver is your metabolic bank, accepting deposits and withdrawals in the currency of glucose molecules.

The absorption of lipids takes place somewhat differently. After triglycerides are broken down into fatty acids and glycerol, they become surrounded by bile salts. Packaged in this way, they are called *micelles* (my-SELZ) and move to the cell membranes of the villi. There, they discard their shell of bile salts and easily move across the membrane and into the cell. Short-chained fatty acids are absorbed directly into the bloodstream. Longer-chained fatty acids are reassembled into triglycerides by the endoplasmic reticulum and are then encased in protein. These large spherical masses are called *chylomicrons* (KYE-low-MY-chronz). After this processing, they pass out of the epithelial cells and into the lacteal. These protein-coated triglycerides are then transported in the lymphatic fluid through a system of vessels that drains the lymph into the blood at the left subclavian vein, a major blood vessel at the base of the neck. As the chylomicrons pass through the capillaries of the liver and adipose tissue, an enzyme breaks down the tryglycerides and other lipoproteins to fatty acids and glycerol. Eventually, these

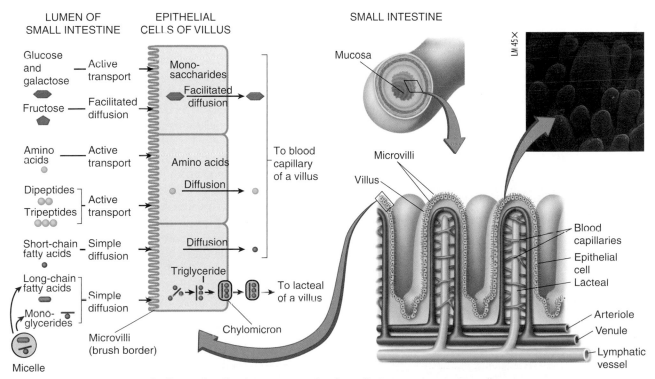

**Figure 27.11** Structure of villi. Each villus has a network of capillaries serving it, as well as a lacteal, a fingerlike projection of a lymphatic vessel. The microvilli of the brush border cells, which are the epithelial cells covering the villi, are seen in the electron micrograph.

breakdown products diffuse from the bloodstream into liver and adipose (fat) cells, where they are then resynthesized into triglycerides and stored for later use when energy is needed. •

## 27.11 Hormones control digestive gland secretions.

As you can see, the digestion of food involves so many players that the digestive team could use a manager. The key players in digestion—the liver, gallbladder, pancreas, stomach, and small intestine—have more than one manager. These managers of digestion are hormones (**Figure 27.12**). Earlier in this chapter you read how the hormone gastrin controls the release of gastric juice in the stomach **1**. Two other hormones, **secretin** and **cholecystokinin** (KOL-uh-SIS-tuh-KINE-un, CCK), control digestion in the small intestine.

When chyme enters the small intestine, the acid in it stimulates cells in the intestinal mucosa to produce secretin **2**. This hormone does two things. First, it stimulates the release of an alkaline fluid called sodium bicarbonate from the pancreas **3**. This solution neutralizes the acid in the chyme so that it will not damage the wall of the small intestine and produces the proper pH in which the pancreatic and intestinal enzymes will work **4**. Second, secretin increases the rate of bile secretion in the liver **5**.

The presence of fatty acids and partially digested proteins in the chyme stimulates the mucosa to produce CCK **2**. This enzyme is a stimulus to the pancreas to release its digestive enzymes **3**. It also signals the gallbladder to contract and pour its contents into the small intestine **6**.

CONCEPT CHECKPOINT

6. How is the digestion of lipids handled differently from that of proteins or carbohydrates?
7. Describe two ways in which the structure of the small intestine relates to its function.

## 27.12 The large intestine concentrates solids.

The **colon** (large intestine) is much shorter than the small intestine—only about a meter and a half, or 5 feet, long. It is wide, however, having a diameter slightly less than the width of your hand. (In contrast, the small intestine has a diameter only slightly larger than the width of two of your fingers.) The small intestine joins the large intestine about 7 centimeters from its end, creating a blind pouch called the cecum (SEE-kum) at the beginning of the large intestine (**Figure 27.13**). Hanging from this pouch is the appendix, a structure that serves no digestive function in humans, but does contain lymphatic tissue that is part of the immune system (see Chapter 30). An infection of the appendix is called appendicitis and can be quite serious and painful.

The junction of the small and large intestines is in the lower right side of the abdomen. From there, the large intestine goes up the right side of the abdomen to the liver. It then turns left, crossing the abdominal cavity just under the diaphragm. On the left side of the abdominal cavity, it turns downward, ending at a short portion of the colon called the **rectum** (see Figure 27.10). The rectum terminates at the **anus**, the opening for the elimination of the feces.

The large intestine is not an organ of digestion. Its role is to absorb sodium ions and water, to eliminate wastes, and to provide a home for friendly bacteria. These bacteria help keep out disease-causing microbes and produce certain vitamins, especially vitamin K. In many plant-eating vertebrates, the cecum houses bacteria and other microorganisms that digest the cellulose of plant cell walls. Humans cannot digest cellulose, and lack significant numbers of microorganisms to do so. Cellulose therefore becomes a digestive waste, but is nonetheless important to health. Also called dietary fiber, undigested plant material provides bulk against which the muscles of the large intestine can push. Dietary fiber is important to the regular movement of the feces through the large intestine and the avoidance of constipation.

Research results over the past decade have provided conflicting findings concerning fiber intake and risk of colon and rectal cancer. Research results published in 2003 and 2004 from studies involving over 550,000 Europeans and North Americans indicated that high-fiber diets protect against colon and rectal cancer. Although more research is still needed to clarify this correlation, sufficient dietary fiber does appear to reduce the risk of developing certain other medical conditions, including heart disease, diabetes, and diverticulosis. Diverticulosis refers to outpouchings of the colon called diverticula. These may be caused by pressure within the colon thought to occur when stools are small and slow-moving.

**Visual Summary** **Figure 27.12** Digestive hormones manage many digestive enzymes.

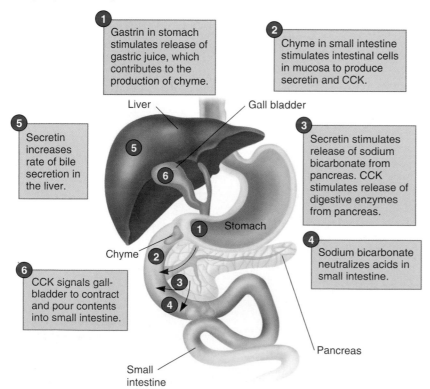

**1** Gastrin in stomach stimulates release of gastric juice, which contributes to the production of chyme.

**2** Chyme in small intestine stimulates intestinal cells in mucosa to produce secretin and CCK.

**5** Secretin increases rate of bile secretion in the liver.

**3** Secretin stimulates release of sodium bicarbonate from pancreas. CCK stimulates release of digestive enzymes from pancreas.

**4** Sodium bicarbonate neutralizes acids in small intestine.

**6** CCK signals gallbladder to contract and pour contents into small intestine.

Liver

Gall bladder

Stomach

Chyme

Pancreas

Small intestine

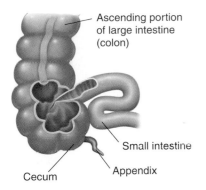

Ascending portion of large intestine (colon)

Small intestine

Cecum

Appendix

**Figure 27.13** The junction between the small and large intestines. A one-way valve prevents waste material in the large intestine from reentering the small intestine, yet allows waste from the small intestine to enter the colon.

**Visual Thinking:** Locate the junction of the small and large intestine in the illustration. How do you think the one-way valve operates?

How do you increase the fiber in your diet? Eat plenty of whole fruits and vegetables (rather than juices), whole grains, nuts, seeds, and legumes, such as dried beans and peas. The current recommendation from the National Academy of Sciences is an intake of between 19 grams and 38 grams of dietary fiber per day depending on age and gender. The average American consumes only 14 grams of dietary fiber per day.

Waste materials move slowly along the smooth interior of the large intestine as water and sodium are slowly reabsorbed.

(Water is absorbed by osmosis, and sodium by diffusion and active transport; see Chapter 5.) As they move along, the wastes become more compacted. If the wastes move too slowly through the colon, perhaps because the diet is low in fiber, too much water may be reabsorbed, leading to a difficulty in elimination called constipation. Conversely, if the wastes move too quickly, as happens with certain intestinal illnesses, not enough water may be removed, resulting in diarrhea. Eventually, the solids within the colon pass into the rectum as a result of the peristaltic contractions of the muscles encasing the large intestine. From the rectum, the solid material passes out of the anus through two anal sphincters. The first of these is composed of smooth muscle. It opens involuntarily in response to a pressure-generated nerve signal from the rectum. The second sphincter, in contrast, is composed of skeletal muscle. It is subject to voluntary control from the brain, thus permitting a conscious decision to delay defecation, or the elimination of waste.

# Why is it unhealthy to be overweight or obese?

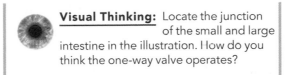

## 27.13 Overweight and obesity are risk factors for many medical conditions.

Any intake of food in excess of that required to maintain the blood sugar (glucose) level and the glycogen reserve in the liver results in one of two consequences. Either the excess glucose is metabolized by the muscles and other cells of the body, or it is converted to lipid and stored within adipose tissue (fat cells). Only when all the body's energy needs have been met—including the energy needed to run chemical reactions, move muscles, and digest food—will calories be stored as fat. Think of your body as a giant scorecard, keeping track of calories eaten and calories used. If you eat more than you use, you will gain weight. (Calories are units of heat energy.)

The National Heart, Lung, and Blood Institute (NHLBI), a division of the National Institutes of Health (NIH), released the first federal guidelines on the identification, evaluation, and treatment of overweight and obesity in adults on June 17, 1998. The final report was completed in 2000. In 2004, the World Health Organization published a report titled *Global Strategy on Diet, Physical Activity, and Health*.

Are you overweight or obese? The NHLBI report uses the concept of the Body Mass Index (BMI) to determine overweight and obesity. The BMI is the weight of a person divided by his or her height and is a measure of fatness of both males and females. A BMI of 25 to 29.9 is considered overweight, and 30 or above is considered obese (**Figure 27.14**). You can determine your BMI

(a)

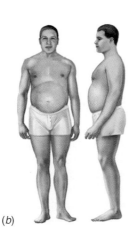

(b)

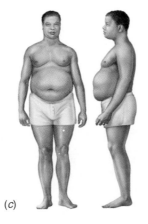

(c)

**Figure 27.14 The BMI varies based on weight and height.** All these men are six feet tall. (a) Weight = 170 pounds, BMI = 23; this man is a healthy weight for his height. (b) Weight = 200 pounds, BMI = 27; this man is overweight. (c) Weight = 230 pounds, BMI = 31; this man is obese.

by using **Table 27.4**. The report notes that the most successful strategies for weight loss include reducing the number of calories consumed, increasing physical activity, and engaging in behavior therapy designed to improve eating and exercise habits. (Of course, it is also possible to be underweight, defined as a BMI of 18.5 or less.)

The NHLBI report states that overweight and obesity affect 97 million American adults—55% of the population. In 2002, U.S. Surgeon General David Satcher announced that obesity was reaching "epidemic proportions" in the nation, soon could cause as much preventable disease as cigarette smoking, and caused about 300,000 premature deaths each year. Overweight and obese people are at increased risk of illness from chronic high blood pressure, blood lipid disorders, type 2 diabetes, heart disease, stroke, gallbladder disease, osteoarthritis, certain sleep disorders and respiratory problems, and certain cancers.

| TABLE 27.4 | Body Mass Index (BMI) |
|---|---|

| Height | Body Weight (pounds) |
|---|---|

| Height | | | | | | | | | | | | | | | | | | | | | | | | | | |
|---|---|---|---|---|---|---|---|---|---|---|---|---|---|---|---|---|---|---|---|---|---|---|---|---|---|---|
| 4'10" | 91 | 96 | 100 | 105 | 110 | 115 | 119 | 124 | 129 | 134 | 138 | 143 | 148 | 153 | 158 | 162 | 167 | 172 | 177 | 181 | 186 | 191 | 196 | 201 | 205 | 210 |
| 4'11" | 94 | 99 | 104 | 109 | 114 | 119 | 124 | 128 | 133 | 138 | 143 | 148 | 153 | 158 | 163 | 168 | 173 | 178 | 183 | 188 | 193 | 198 | 203 | 208 | 212 | 217 |
| 5' | 97 | 102 | 107 | 112 | 118 | 123 | 128 | 133 | 138 | 143 | 148 | 153 | 158 | 163 | 168 | 174 | 179 | 184 | 189 | 194 | 199 | 204 | 209 | 215 | 220 | 225 |
| 5'1" | 100 | 106 | 111 | 116 | 122 | 127 | 132 | 137 | 143 | 148 | 153 | 158 | 164 | 169 | 174 | 180 | 185 | 190 | 195 | 201 | 206 | 211 | 217 | 222 | 227 | 232 |
| 5'2" | 104 | 109 | 115 | 120 | 126 | 131 | 136 | 142 | 147 | 153 | 158 | 164 | 169 | 175 | 180 | 186 | 191 | 196 | 202 | 207 | 213 | 218 | 224 | 229 | 235 | 240 |
| 5'3" | 107 | 113 | 118 | 124 | 130 | 135 | 141 | 146 | 152 | 158 | 163 | 169 | 175 | 180 | 186 | 191 | 197 | 203 | 208 | 214 | 220 | 225 | 231 | 237 | 242 | 248 |
| 5'4" | 110 | 116 | 122 | 128 | 134 | 140 | 145 | 151 | 157 | 163 | 169 | 174 | 180 | 186 | 192 | 197 | 204 | 209 | 215 | 221 | 227 | 232 | 238 | 244 | 250 | 256 |
| 5'5" | 114 | 120 | 126 | 132 | 138 | 144 | 150 | 156 | 162 | 168 | 174 | 180 | 186 | 192 | 198 | 204 | 210 | 216 | 222 | 228 | 234 | 240 | 246 | 252 | 258 | 264 |
| 5'6" | 118 | 124 | 130 | 136 | 142 | 148 | 155 | 161 | 167 | 173 | 179 | 186 | 192 | 198 | 204 | 210 | 216 | 223 | 229 | 235 | 241 | 247 | 253 | 260 | 266 | 272 |
| 5'7" | 121 | 127 | 134 | 140 | 146 | 153 | 159 | 166 | 172 | 178 | 185 | 191 | 198 | 204 | 211 | 217 | 223 | 230 | 236 | 242 | 249 | 255 | 261 | 268 | 274 | 280 |
| 5'8" | 125 | 131 | 138 | 144 | 151 | 158 | 164 | 171 | 177 | 184 | 190 | 197 | 203 | 210 | 216 | 223 | 230 | 236 | 243 | 249 | 256 | 262 | 269 | 276 | 282 | 289 |
| 5'9" | 128 | 135 | 142 | 149 | 155 | 162 | 169 | 176 | 182 | 189 | 196 | 203 | 209 | 216 | 223 | 230 | 236 | 243 | 250 | 257 | 263 | 270 | 277 | 284 | 291 | 297 |
| 5'10" | 132 | 139 | 146 | 153 | 160 | 167 | 174 | 181 | 188 | 195 | 202 | 209 | 216 | 222 | 229 | 236 | 243 | 250 | 257 | 264 | 271 | 278 | 285 | 292 | 299 | 306 |
| 5'11" | 136 | 143 | 150 | 157 | 165 | 172 | 179 | 186 | 193 | 200 | 208 | 215 | 222 | 229 | 236 | 243 | 250 | 257 | 265 | 272 | 279 | 286 | 293 | 301 | 308 | 315 |
| 6' | 140 | 147 | 154 | 162 | 169 | 177 | 184 | 191 | 199 | 206 | 213 | 221 | 228 | 235 | 242 | 250 | 258 | 265 | 272 | 279 | 287 | 294 | 302 | 309 | 316 | 324 |
| 6'1" | 144 | 151 | 159 | 166 | 174 | 182 | 189 | 197 | 204 | 212 | 219 | 227 | 235 | 242 | 250 | 257 | 265 | 272 | 280 | 288 | 295 | 302 | 310 | 318 | 325 | 333 |
| 6'2" | 148 | 155 | 163 | 171 | 179 | 186 | 194 | 202 | 210 | 218 | 225 | 233 | 241 | 249 | 256 | 264 | 272 | 280 | 287 | 295 | 303 | 311 | 319 | 326 | 334 | 342 |
| 6'3" | 152 | 160 | 168 | 176 | 184 | 192 | 200 | 208 | 216 | 224 | 232 | 240 | 248 | 256 | 264 | 272 | 279 | 287 | 295 | 303 | 311 | 319 | 327 | 335 | 343 | 351 |
| 6'4" | 156 | 164 | 172 | 180 | 189 | 197 | 205 | 213 | 221 | 230 | 238 | 246 | 254 | 263 | 271 | 279 | 287 | 295 | 304 | 312 | 320 | 328 | 336 | 344 | 353 | 361 |
| BMI | 19 | 20 | 21 | 22 | 23 | 24 | 25 | 26 | 27 | 28 | 29 | 30 | 31 | 32 | 33 | 34 | 35 | 36 | 37 | 38 | 39 | 40 | 41 | 42 | 43 | 44 |

To use the table, find the appropriate height in the left-hand column. Move across to a given weight. The number at the bottom of the column is the BMI at that height and weight. Pounds have been rounded off. *Source*: The National Heart, Lung, and Blood Institute.

# CHAPTER REVIEW

## Summary of Key Concepts and Key Terms

### What are nutrients and how are they used?

**27.1 Nutrients** (p. 460) are molecules in the diet that provide building blocks and energy for the body and provide substances needed for proper metabolism.

**27.1** Carbohydrates, lipids (fats and oils), and proteins are macronutrients that provide energy and building blocks.

**27.1** Vitamins, minerals, and water help various chemical reactions and bodily processes take place.

**27.2** The Food and Nutrition Board of the National Academy of Sciences' Institute of Medicine recommends that 45 to 65% of your daily energy come from carbohydrates, 20 to 35% from fat, and 10 to 35% from protein.

**27.2** Three amino acids, the building blocks of proteins, are indispensable, or **essential amino acids** (p. 460) in the human diet, which means that humans cannot synthesize them and must obtain them from food.

**27.2 Saturated fats** (p. 461) and **trans fats** (p. 461) tend to raise blood cholesterol levels, and high blood cholesterol levels greatly increase the risk for heart disease.

**27.2** When eaten, **high-glycemic index foods** (p. 461) may cause higher increases in blood sugar levels than low-glycemic index foods and have been linked to an increased risk for both diabetes and heart disease.

**27.2** A healthy diet includes a variety of foods with a large proportion of fruits, vegetables, and whole grains.

## What are the patterns of digestion among organisms?

**27.3** Intracellular **digestion** (p. 464) usually takes place in single-celled protists and the sponges.

**27.3** **Autotrophs** (p. 464), or self-feeders, make food within their bodies, and digestion is unnecessary.

**27.3** **Heterotrophs** (p. 464) digest their food either within their cells (intracellularly), outside of their cells (extracellularly), or both.

**27.3** In fungi, extracellular digestion takes place without a **digestive system** (p. 464).

**27.3** In animals, extracellular digestion usually takes place within a digestive system.

## How do humans digest food?

**27.4** Human digestion is both mechanical and chemical.

**27.4** Food is digested mechanically in the mouth and stomach, and chemically by three basic types of enzymes: **proteases** (PRO-tee-ACE-es, p. 466), **amylases** (AM-uh-lace-es, p. 466), and **lipases** (LYE-pays-es, p. 466).

**27.5** Teeth shred and grind food in mechanical digestion, while the chemical digestion of carbohydrates begins in the mouth with the action of **salivary amylase** (p. 466), secreted by **salivary glands** (p. 466).

**27.6** As food is chewed and moistened, the tongue forms it into a ball-like mass called a bolus and pushes it into the upper part of the **pharynx** (FAIR-inks, p. 466).

**27.6** The **larynx** (LAIR-inks, p. 466) rises to meet the **epiglottis** (ep-ih-GLOT-iss, p. 467), a flap of tissue that folds back over the **glottis** (p. 467), the opening to the larynx and trachea.

**27.6** When food is swallowed, it moves via **peristalsis** (p. 467) along the **esophagus** (p. 467) to the stomach.

**27.7** The **gastric glands** (p. 468) of the **stomach** (p. 468) secrete hydrochloric acid and pepsinogen; the acid unfolds proteins and converts pepsinogen to its active protein-hydrolyzing form, **pepsin** (p. 468).

**27.7** The stomach controls the production of gastric juice by means of a digestive hormone called **gastrin** (p. 469).

**27.8** Most of the digestion of food takes place in the **small intestine** (p. 470) with the help of the liver and pancreas, accessory organs to digestion.

**27.8** The first portion of the small intestine is the **duodenum** (DOO-oh-DEE-num *or* doo-ODD-un-um, p. 470).

**27.8** The **pancreas** (p. 470), an accessory organ to the digestive system, secretes digestive enzymes into the small intestine, and an alkaline solution to help neutralize acidic chyme.

**27.8** The **liver** (p. 470) is another accessory organ to the digestive system and secretes **bile** (p. 470), which is stored in the **gallbladder** (p.

470); bile helps emulsify fats so they can be digested more easily.

**27.9** In the small intestine, **pancreatic amylase** (p. 470) breaks down starches and glycogen to maltose.

**27.9** **Disaccharidases** (p. 470) digest the disaccharides maltose, sucrose, and lactose to the monosaccharides glucose, fructose, and galactose.

**27.9** Three other enzymes produced by the pancreas complete the digestion of proteins: **trypsin** (TRIP-sin, p.470), **chymotrypsin** (KYE-moe-TRIP-sin, p. 470), and **carboxypeptidase** (kar-BOK-sce-PEP-ti-dace, p. 470).

**27.9** The pancreatic enzymes work with **peptidases** (p. 470) produced by the brush border cells of the intestinal epithelium, breaking down polypeptides into shorter chains and then to amino acids.

**27.9** Emulsification of lipids increases the surface area available for **pancreatic lipase** (p. 472) to act.

**27.10** The internal surface area of the small intestine is tremendously large, having a folded surface covered by microscopic projections called **villi** (p. 471).

**27.10** During absorption across the walls of the small intestine, amino acids and monosaccharides pass into the bloodstream through capillaries in the villi and are carried to the liver.

**27.10** Fatty acids and glycerols pass into the lymphatic system through the lacteal, and are then transported to the bloodstream.

**27.11** **Secretin** (p. 472) and **cholecystokinin** (KOL-uh-SIS-tuh-KINE-un, p. 472) control digestion in the small intestine.

**27.12** The **colon** (p. 472) functions principally to absorb water and sodium from, and to eliminate digestive waste.

**27.12** The final, short portion of the colon is the **rectum** (p. 472), and it terminates at the **anus** (p. 472), the opening for the elimination of the feces.

## Why is it unhealthy to be overweight or obese?

**27.13** Overweight and obesity affect from 55 to 80% of the adult population in the United States.

**27.13** Overweight and obese persons are at higher risk than others for certain diseases such as type 2 diabetes, heart disease, and stroke.

---

| **Level 1** | **Learning Basic Facts and Terms** |

**Multiple Choice**

1. Bile is secreted into the _____, where it functions in _____.
   a. large intestine—solidifying feces
   b. duodenum—emulsifying fat
   c. stomach—digesting proteins
   d. stomach—absorbing fat
   e. colon—facilitating the absorption of water

2. Which structure of humans is not correctly matched with one of its functions?
   a. small intestine—enzymatic digestion
   b. large intestine—absorption of most nutrients
   c. stomach—digestion of proteins
   d. oral cavity—mechanical breakdown of food
   e. gallbladder—storage of bile

3. To enter the body, a substance must cross a cell membrane. During which stage of food processing does this happen?
   a. Ingestion
   b. Digestion
   c. Hydrolysis
   d. Absorption
   e. Elimination

4. Sponges are limited to feeding on small food particles because
   a. their digestion is entirely intracellular.
   b. they lack a mechanism for bringing food into their bodies.
   c. they have a very small mouth.
   d. they have an incomplete digestive tract.
   e. their cell membranes are highly selective.

**Matching**

Match the following digestive enzymes with their substrate AND the location in the digestive tract where they operate.

| | | | | |
|---|---|---|---|---|
| 5. ____ trypsin | a. carbohydrates | uu. oral cavity |
| 6. ____ lipase | b. triglycerides | vv. stomach |
| 7. ____ carboxypeptidase | c. proteins | ww. duodenum |
| 8. ____ pancreatic amylase | d. nucleic acids | xx. colon |
| 9. ____ pepsin | | yy. rectum |
| 10. ____ salivary amylase | | zz. esophagus |

## Level 2  Learning Concepts

1. Name the six classes of nutrients, and explain why you need each one in your diet.
2. Describe the digestive fates of proteins, lipids, and carbohydrates. What three types of enzymes are involved?
3. Where does the process of human digestion begin?
4. Why do you think that the liver and pancreas are not called organs of the digestive system even though they perform important roles in digestion?
5. For years, baking soda (sodium bicarbonate) has been a popular home remedy for indigestion and heartburn. Explain why.
6. What important tasks does hydrochloric acid accomplish in the human stomach? What protects the stomach wall from potential damage by this strong acid?

7. What is the duodenum? Summarize the digestive activities that take place there.
8. Identify and give a function for each of the following: gastrin, secretin, and cholecystokinin. What do they have in common?
9. Discuss the features of the small intestine that increase its internal surface area. Why is this important?
10. Summarize how your body produces ATP from glucose, amino acids, triglycerides, and fatty acids.

## Level 3  Critical Thinking and Life Applications

1. Many people are vegetarians. Discuss how a vegetarian could create a balanced diet while eliminating meat and animal products.
2. While on a hike, a student finds the dry skull of a small mammal. How could the student determine whether that animal ate only plant products?
3. Draw a graph of what you might expect the relationship to be between enzyme activity levels (Y-axis) and pH (X-axis) for pepsin and for trypsin. Explain your graphs.

4. Pepsin is first secreted into the stomach as an inactive enzyme called pepsinogen, and becomes activated upon exposure to the stomach's acidic environment. Speculate on why this occurs.
5. Keep a record of everything you eat and drink for three days. Then (this could be painful!) analyze your diet. Compare it with the recommendations in this chapter. How could you improve your nutritional intake?
6. Explain the biology behind why someone might choke on a piece of food. What prevents us from choking during normal swallowing?

## In The News  Critical Thinking

**VITAMIN A AND LEARNING**

Now that you understand more about nutrients, reread this chapter's opening story about vitamin A and learning in mice. To understand this research better, it may help you to follow these steps:

1. Review your immediate reaction to Evans's research on vitamin A-deficient mice that you wrote when you began reading this chapter.
2. Based on your current understanding, again summarize the main point of the research in a sentence or two.
3. What questions do you now have about this research that this chapter's opening story does not answer?
4. Collect new information about the research. Visit the *In The News* section of this text's companion Web site at www.wiley.com/college/alters and

watch the "Vitamin A and Learning" video. Then use the "summary" link to read the accompanying story and access related links. Use this information, the links provided, and other online and library resources to answer your questions and find updates about this research topic. State the sources of your information. Explain why you think the information is accurate. Also determine whether the information expresses a particular point of view or is biased in any way.

5. What in your view are the most significant aspects of this research? Give reasons for your opinion and for any changes in your ideas based on the additional information you have collected and the analysis you have done.

# RESPIRATION

Asthma Treatment

Doug, a 22-year-old student attending a small Midwestern college, is an avid paintball enthusiast who also enjoys playing the drums. "I've had asthma since I was born," he reports. "When I was little, it was a constant struggle . . . cold weather and pet dander were my triggers." *Triggers* are factors that precipitate episodes referred to as asthma attacks.

Asthma is a complex inflammatory lung disease caused by both environmental and genetic factors. During an asthma attack, the tissues lining the bronchi and bronchioles (airways of the respiratory tract) swell, muscles encircling the airways constrict, and goblet cells that line the airways secrete excess mucus. Symptoms include wheezing, coughing, and difficulty in breathing. According to Doug, "It's like trying to breathe through a soda straw . . . you can inhale but not enough to feel comfortable." In acute attacks, breathing is impossible—death results if the attack is not treated immediately. To treat his asthma, Doug

inhales albuterol, a medication that relaxes bronchial muscles, dilating the airways and increasing airflow.

Doug and the millions of other people who have asthma may soon be able to breathe a lot easier. Kenneth Adler, professor of molecular biomedical sciences at North Carolina State University, and a team of scientists have synthesized a peptide called MANS that may become a new treatment for asthma and other respiratory diseases, such as chronic bronchitis and cystic fibrosis, in which excess mucus production interferes with breathing. MANS is a fragment of the MARCKS protein, which is necessary for mucus production. MARCKS binds to the contractile cytoskeleton of goblet cells that produce mucus and to the mucin granules within them, resulting in the secretion of the thick, sticky fluid. Mucin is a component of mucus.

To simulate the allergic inflammation of the airways that characterizes asthma, Adler and his team treated mice with a protein that made the rodents allergic to certain environmental triggers. After sensitizing the mice, the scientists exposed them to an inhaled chemical that triggered an immune response within the bronchial cells of the animals' airways, and the cells secreted excess mucus as a result. By treating a group of the affected mice with MANS, the scientists were able to prevent the airways of the mice from becoming filled with mucus. In test tube studies, the researchers also discovered that MANS inhibited mucus production in human bronchial cells. Adler and his team think that the MANS peptide binds at sites the MARCKS protein normally would, thus preventing the binding of the MARCKS protein and preventing mucus secretion. Such findings are likely to pave the way for the development of new medications for human asthma and other conditions in which excess mucus secretion occurs. To learn more about Adler's research, visit the *In The News* section of this text's companion Web site at www.wiley.com/college/alters and view the video "Asthma Treatment."

*Write your immediate reaction to this information about asthma and the research into the MANS peptide that prevents excess mucus secretion: first, summarize the research in a sentence or two; then suggest what you think its significance is. You will have an opportunity to reflect on your responses and gather more information on this topic in the* In The News *feature at the end of this chapter. In this chapter, you will learn about the respiratory system and some common diseases that affect it. The* How Science Works *box in this chapter discusses asthma in more detail.*

## CHAPTER GUIDE

### What is respiration?

**28.1** Respiration is the uptake of oxygen and the release of carbon dioxide.

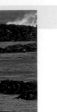

### What patterns of respiration exist among organisms?

**28.2** Some organisms respire over cell and skin surfaces.

**28.3** Some organisms respire using specialized organs.

### How do humans respire?

**28.4** Human respiration takes place in the lungs.

**28.5** During breathing, air moves into the lungs via the nose, throat, windpipe, and bronchi.

**28.6** Low pressure causes air to flow into the lungs, and high pressure forces air out of the lungs.

**28.7** The circulatory system transports gases throughout the body.

### What are some diseases and potential problems of the airways?

**28.8** Chronic obstructive pulmonary disease is a group of disorders that impair breathing.

**28.9** Abdominal thrusts can dislodge substances blocking the airway.

# What is respiration?

## 28.1 Respiration is the uptake of oxygen and the release of carbon dioxide.

Awkward in movement out of the water, the northern elephant seals shown in **Figure 28.1** are master divers. Taking a mile-deep dive and staying down for about two hours is a normal occurrence for these seals and the other master divers of the ocean, the sperm whales. The diving abilities of these animals are far superior to those of humans; the world diving record for humans while breath holding is 133 meters, which is approximately 435 feet. Both humans and seals breathe in much the same way. So what confers this diving advantage upon the northern elephant seal? Put simply—myoglobin.

Found in the muscles, myoglobin is an oxygen-carrying protein that is similar to hemoglobin, the oxygen-carrying protein of the blood. Myoglobin stores the oxygen needed for the muscles to burn fuel. Human muscle cells have small amounts of myoglobin. In contrast, elephant seals (and the other adept divers) have high concentrations of myoglobin in their muscles. Because their muscles can store so much oxygen, seals, whales, sea turtles, and penguins can stay submerged for long periods. In addition, many of their body processes slow down, allowing their muscles to have a large share of the oxygen supply. Amazingly, the heart rate of elephant seals can fall to a low of 3 beats per minute during a dive.

Although the diving abilities of elephant seals may be amazing and unusual in the animal world, their use of oxygen is not. Seals, whales, penguins, and humans are only a few of the myriad organisms that **respire**; they take up oxygen from the environment and release the waste gas, carbon dioxide. Oxygen is used within cells during the breakdown of fuel, primarily glucose. The chemical process by which cells break down fuel molecules using oxygen, producing carbon dioxide, and releasing energy is called *aerobic cellular respiration*. (This process is described in Chapter 8.)

Respiring organisms obtain oxygen from their environments—either the air or water. They also rid themselves of the waste gas carbon dioxide, expelling it to the environment. In order for an organism to move oxygen and carbon dioxide into and out of its body, it must have a thin, moist, and relatively large respiratory surface. Body fluids move past the inside of this surface, and air or water moves past the outer, environmental side of the surface.

*Diffusion* is the mechanism by which oxygen and carbon dioxide move across a respiratory surface. During this process, molecules of oxygen and carbon dioxide each move from areas of high concentration and pressure to areas of low concentration and pressure. (Because the pressure of each individual gas makes up part of the total, it is called partial pressure.) They are said to move down a concentration and pressure gradient. (This process is described in detail in Chapter 5.) The oxygen supply within a respiring organism is continually being depleted because it is used during cellular respiration. Oxygen therefore diffuses from the environment (the area of higher concentration) and into the organism (the area of lower concentration). Similarly, carbon dioxide continually increases in concentration within a respiring organism because it is a waste product of cellular respiration. Thus, carbon dioxide diffuses from the organism and into the environment.

### FOCUS *on Plants*

Plants undergo cellular respiration, *and* they respire. Plants use cellular respiration to break down carbohydrates that they produce during photosynthesis, extracting energy from these molecules as other organisms do. This process requires oxygen. When light levels are high, plants produce oxygen as a by-product of photosynthesis, so they do not need to get it from the environment. When light levels are low, however, plants take in oxygen because they cannot photosynthesize in the dark. The processes of photosynthesis and cellular respiration are described in more detail in Chapters 7 and 8.

**Figure 28.1 Northern elephant seals can hold their breath under water for two hours.** A secret behind this ability is the oxygen-carrying molecule myoglobin, which stores large quantities of oxygen in the muscles.

# Visual Summary    **Figure 28.2** Patterns of respiration.

## RESPIRATION OVER CELL AND SKIN SURFACES

### Cell membrane
In single-celled organisms, gas exchange takes place across the cell membrane.

### Outer layer of cells
In certain small, multicellular organisms, gas exchange takes place across the outer layer of cells.

### Skin
In some large, multicellular organisms, gas exchange takes place across the skin.

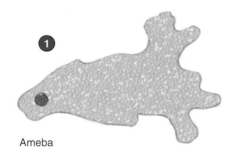

Ameba

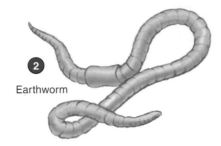

Earthworm

Amphibian

## RESPIRATION USING SPECIAL ORGANS

### Tracheal system
In some spiders and most insects, gas exchange takes place within the organism over cell surfaces in contact with the tracheal system, a branching network of microscopic air tubes.

### Gills
In most large, multicellular, aquatic animals that are not mammals, gas axchange takes place over evaginations of the body surface called gills.

### Lungs
In most large, multicellular, land-dwelling animals and in marine mammals, including most animals that also respire through the skin, gas exchange takes place over invaginations of the body surface called lungs.

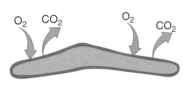

Grasshopper

Salmon

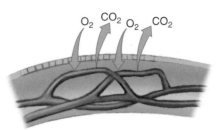

Grizzly bear

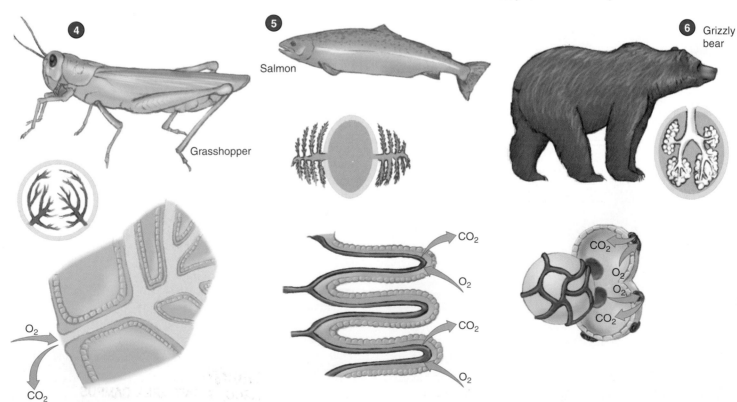

# What patterns of respiration exist among organisms?

## 28.2 Some organisms respire over cell and skin surfaces.

In organisms that are single cells, such as bacteria and single-celled protists, the cell membrane serves as the respiratory surface (**Figure 28.2 ①**). Oxygen and carbon dioxide simply diffuse across this thin membrane. These organisms, which live in aquatic or moist environments, are small and their cell membranes provide adequate surface area over which enough oxygen and carbon dioxide can diffuse for their survival. (See Chapter 5 for a description of surface-to-volume ratio.)

The outer layer of cells of certain multicelled animals (some sponges, corals, and jellyfish; many flatworms, tapeworms, and roundworms; and a few earthworms) also act as respiratory surfaces ②. All of these organisms live in watery environments with the exception of the roundworms and earthworms, which live in soil. These soil-dwelling organisms must keep their outermost cells (the cuticle) moist or respiration will cease.

Some of these organisms are small, such as the flatworms. Others, though much larger, have great surface-to-volume ratios because of their shapes. Corals, for example, are usually branching, fanlike, or fringed, each shape providing a large surface area over which respiration can take place. Earthworms, roundworms, and tapeworms have large surface areas as a result their elongated shapes. Sponges, vase-shaped masses of cells embedded in a jelly-like matrix, also have large surface areas.

Some multicellular organisms can breathe through their skin. The most familiar are amphibians such as salamanders and frogs, but certain fishes, snakes, turtles, and lizards rely on the skin as a respiratory organ to one degree or another ③. These animals have thin, moist skins with an abundant supply of blood vessels just beneath the skin surface. In addition, most of these animals have lungs or gills that also provide a surface over which oxygen and carbon dioxide can diffuse, supplementing their "skin breathing."

## 28.3 Some organisms respire using specialized organs.

Many animals are large, and their bodies have small surface-to-volume ratios. However, they have special organs with large surface areas, which provide large surface-to-volume ratios for efficient gas exchange. Although there are many types of respiratory systems, the three primary types are the tracheal systems of insects and some spiders, the gills of fishes and other aquatic animals, and the lungs of reptiles and mammals.

### Tracheal systems

A **trachea** (TRAY-kee-uh) is an air-filled tube. This name comes from a Latin word meaning "windpipe" and is also used to denote the windpipe in mammalian respiratory systems, such as in humans. A **tracheal system** is a branching network of microscopic air tubes. Some spiders and most insects have tracheal systems (Figure 28.2 ④). A variety of types of these systems exist. The tracheal systems of these two groups appear to have evolved independently from one another, and are a good example of convergent evolution.

In general, the tracheal system opens to the environment by means of small holes in the cuticle (exoskeleton). These openings are called **spiracles**, a term that comes from a Latin word meaning "to breathe." In many tracheal systems, spiracles are muscular openings that can open and close under the direction of the nervous system, triggered by carbon dioxide levels. Spiracles open and close at different times, regulating the flow of air through the system.

In insects, tracheas originate at the spiracles and usually interconnect to form a pair of longitudinal trunks that extend the length of the thorax (midsection) and abdomen (last section) of the organism. **Figure 28.3** shows how tracheas branch into narrower and narrower passageways called tracheoles. The tracheoles end in microscopic branches that are lined with moist epithelium and extend to all parts of the body. Some extend into individual cells. Oxygen uptake and carbon dioxide release takes place at the moist ends of the tracheoles.

In small insects and those that use little energy, gases move through the tracheal system by diffusion. In large insects and ones that fly, diffusion is not enough. The body movements of large, active insects compress parts of the tracheal system, thereby changing its volume and consequently the pressure within the tracheas. These movements help pump air through the system.

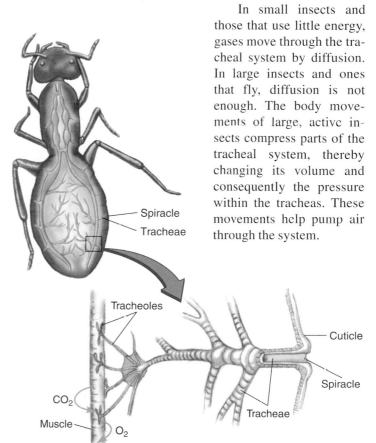

**Figure 28.3** The tracheal system of an ant. Sets of spiracles open into tracheae, which branch into tracheoles where gas exchange takes place.

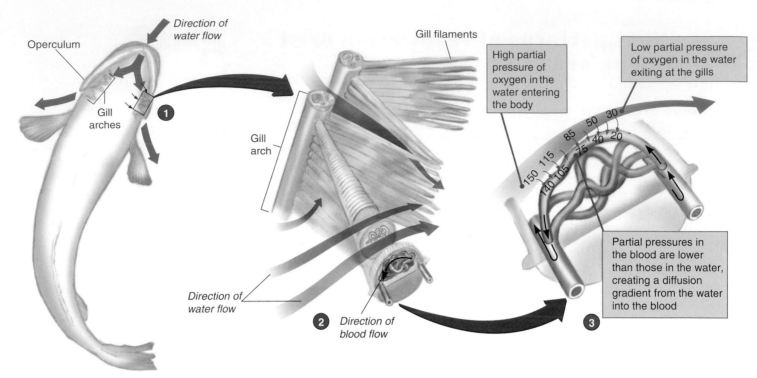

**Figure 28.4 Countercurrent flow in gills.** During countercurrent flow, the partial pressure of oxygen in the water is always higher than the partial pressure of oxygen in the blood. Oxygen diffuses from the water and into the blood as they flow past one another.

## Gills

**Gills** are organs by which many large aquatic animals respire. Crustaceans (such as lobsters and crabs), most mollusks (such as clams), aquatic worms, fishes, and some amphibians have gills (see Figure 28.2 **5**).

Just as tracheas and lungs are invaginations (a turning inward) of the body surface, gills are evaginations of the body surface (a turning outward). Often, however, gills have protective coverings because they are feathery and delicate, and are therefore easily damaged. Bony fishes, for example, have a protective flap covering the gills called an *operculum* (see Figure 22.8). Gills can be distributed throughout an organism, as in the case of aquatic worms, or can be in a particular location, as in bony fishes.

Just as humans inhale and exhale to ventilate, or pass air through, the lungs so do gill breathers ventilate the gills. **Figure 28.4** shows how water passes over the gills of fishes. Water enters the fish's mouth, passes through slits in the throat (pharynx), flows over the gills, and exits at the back of the gill cover (operculum) **1**. Notice in the enlarged view of the gill that the water flows in a direction opposite to the flow of blood **2**. This movement is called *countercurrent flow*—one fluid flows counter to, or against the flow of the other. Countercurrent flow enhances the diffusion of oxygen into the blood of the fish and the diffusion of carbon dioxide into the water **3**.

## Lungs

**Lungs** are organs by which most amphibians, aquatic mammals, such as whales, seals, and dolphins, and large land-dwelling animals respire (with the exception of the insects, spiders, roundworms, and earthworms mentioned previously). Lungs are invaginations of the body surface, but are not located throughout the body as are tracheal systems. Lungs are confined to one part of the organism. Therefore, lung-breathers, like gill-breathers, have a circulatory system to distribute the oxygen to other parts of the body (see Figure 28.2 **6**).

Lungs are made up of microscopic air sacs that are covered with blood vessels and coated with a thin film of moisture. It is at the air sacs that oxygen diffuses into the blood and carbon dioxide diffuses out. Narrow passageways connect the lungs to the outside, their thinness minimizing the loss of moisture to the environment.

As organisms evolved on land, variations in lung development occurred. Snails, for example, lack the gills of their aquatic mollusk relatives: clams, octopus, and squid. Instead, their mantle cavity acts like a lung. Most spiders have structures called book lungs, which are stacks of spaced membranes within their abdomens, connected to the outside by a tracheal system. Other organisms, such as most amphibians, reptiles, and birds, have lungs that are closer in their anatomy and physiology to human lungs (and those of other mammals), but are ventilated differently. Frogs, for example, gulp air and pump it into their simple lungs. Birds have air sacs in addition to lungs, which help ventilate the lungs.

### CONCEPT CHECKPOINT

1. What adaptations in body structure allow for efficient uptake of oxygen and release of carbon dioxide in organisms without gills, lungs, or tracheal systems?

2. How does the structure of the gills of fishes fit (or relate to) their function?

# How do humans respire?

## 28.4 Human respiration takes place in the lungs.

In humans and other mammals, air moves into the lungs through the nose, mouth, and airways of the **respiratory system**. The movement of air into and out of the lungs is called **breathing**. A breath consists of taking air into the lungs, or **inspiration**, and expelling air from the lungs, or **expiration**. Try to breathe normally while counting the number of breaths you take in 15 seconds. Multiply the number of your breaths by 4. Is your breathing rate within the average of 14 to 20 breaths per minute?

**Figure 28.5** shows the general plan of the human respiratory system. As blood moves within microscopic blood vessels around the body and past its cells, carbon dioxide diffuses from the tissue fluid into the blood. *Tissue fluid*—also called interstitial fluid or intercellular fluid—is a waterlike fluid derived from the blood that bathes all the cells of the body. Oxygen moves from the blood into the tissue fluid. This process is called *internal respiration*, the exchange of oxygen and carbon dioxide between the blood and the tissue fluid. *External respiration* is the exchange of gases at the lungs.

## 28.5 During breathing, air moves into the lungs via the nose, throat, windpipe, and bronchi.

### The nasal cavities, pharynx, and larynx

As you breathe in, air first enters your body through the nostrils. The nostrils are lined with hairs that filter out dust and other particles from the air. The air is warmed and moistened as it swirls around in the **nasal cavities**. These cavities, located above the roof of your mouth, are bordered by projections of bone covered with epithelial tissue. This tissue stays moist with mucus secreted by its many mucous glands. This sticky fluid helps trap dirt and dust that you breathe in. The epithelium is also covered with tiny, hairlike projections called *cilia*. The word "cilia" comes from a Latin word meaning "eyelashes." These "cell eyelashes" beat in unison, creating a current in the mucus that carries the trapped particles toward the back of the nasal cavity. From here, the mucus drips into the throat and is swallowed—at a rate of over a pint per day!

After passing through the nasal cavities, air enters the **pharynx** (FAIR-inks), or throat. The pharynx extends from behind the nasal cavities to the openings of the esophagus and larynx. The esophagus, as you may recall from Chapter 27, is the food tube, a passageway for food to the stomach. The **larynx** (LAIR-inks), or voice box, lies at the beginning of the trachea, the passageway that brings air to the lungs.

The larynx is a cartilaginous box shaped somewhat like a triangle. Stretched across the upper end of the larynx are the **vocal cords**. The vocal cords are two folds of elastic tissue that produce the sounds you make as air rushes by and causes them to vibrate. Your mouth, lips, and tongue form the sounds into words. In addition, the size and shape of the passageways in your chest, head, nose, throat, and mouth affect the quality of the sound. Your lungs add a power supply and volume control to your voice.

### The trachea and bronchi

Put your hand on your larynx, or Adam's apple, and then picture about 4 or 5 inches of garden hose attached to its bottom end. A garden hose is about the diameter of your **trachea** (windpipe), which extends downward from your larynx toward your lungs. The trachea has thin walls, similar to the thickness of those in the hose. Garden hoses are reinforced with materials such as rubber or vinyl to keep them from collapsing; your trachea is reinforced with rings of cartilage. The cartilage wraps part of the way around the trachea, forming C shapes that begin and end where the windpipe lies next to the esophagus. Run your fingers gently up and down the front of your neck to feel some of these cartilaginous rings.

If you could look down your trachea, as in the upper photograph in **Figure 28.6**, the two black holes you would see would be your **primary bronchi** (BRON-keye). These airways are structured much like the trachea but are smaller in diameter. One bronchus goes to each lung, branching into three right and two left **secondary bronchi** serving the three right and two left lobes of the lungs. The heart is nestled into the left side of the lungs, taking up some of the space that a third left lobe might occupy.

The secondary bronchi divide into smaller and smaller branches, looking much like an upside-down tree until they end in thousands of passageways called **respiratory bronchioles** (BRON-kee-olz). These airways have a diameter less than that of a

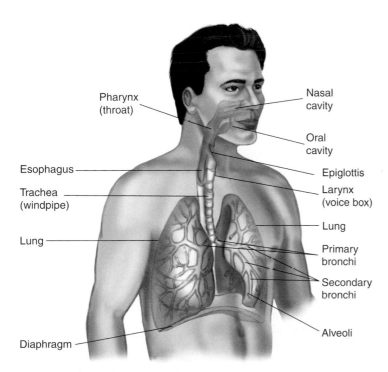

Pharynx (throat)
Nasal cavity
Oral cavity
Esophagus
Epiglottis
Trachea (windpipe)
Larynx (voice box)
Lung
Lung
Primary bronchi
Secondary bronchi
Diaphragm
Alveoli

**Figure 28.5** The human respiratory system.

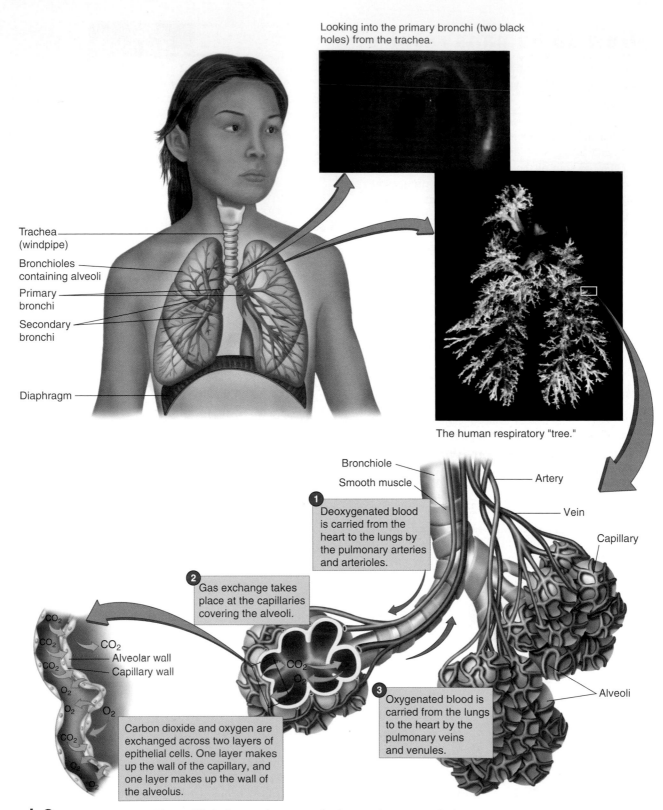

Looking into the primary bronchi (two black holes) from the trachea.

Trachea (windpipe)

Bronchioles containing alveoli

Primary bronchi

Secondary bronchi

Diaphragm

The human respiratory "tree."

Bronchiole

Smooth muscle

Artery

Vein

Capillary

**1** Deoxygenated blood is carried from the heart to the lungs by the pulmonary arteries and arterioles.

**2** Gas exchange takes place at the capillaries covering the alveoli.

$CO_2$
$CO_2$
$CO_2$
$CO_2$
$CO_2$
$CO_2$
$O_2$
$O_2$
$O_2$
$O_2$
$O_2$
$O_2$

Alveolar wall
Capillary wall

$CO_2$
$O_2$

Alveoli

**3** Oxygenated blood is carried from the lungs to the heart by the pulmonary veins and venules.

Carbon dioxide and oxygen are exchanged across two layers of epithelial cells. One layer makes up the wall of the capillary, and one layer makes up the wall of the alveolus.

**Visual Summary** **Figure 28.6 Gas exchange at the lungs.** A section of a lung is enlarged to show the diffusion of the respiratory gases, oxygen ($O_2$) and carbon dioxide ($CO_2$), between the blood and alveoli.

 **Visual Thinking:** Blue traditionally denotes oxygen-poor blood that is carried in veins. In the illustration, what type of vessel carries oxygen-poor blood to the lungs? What type of vessel carries oxygen-rich blood? To which organ of the body is blood returned after picking up oxygen at the lungs?

pencil lead. Their walls have clusters of tiny pouches that, along with the respiratory bronchioles, are the sites of gas exchange. These pouches, or air sacs, are called **alveoli** (al-VEE-uh-lye).

How do the airways help keep foreign particles from entering the alveoli? The inner walls of the trachea are lined with ciliated epithelium (**Figure 28.7**). Certain cells in the epithelium secrete mucus. Together, the cilia and the mucus provide your windpipe with an "up escalator" for any particles or microbes you may have inhaled. This escalator brings substances up to the pharynx, where they are swallowed and eliminated through the digestive tract. In the trachea of a cigarette smoker, however, action of the cilia is impaired, causing mucus to build up in the airway. The result is that the tars in cigarettes are not caught and expelled with the action of cilia. They move easily into the lungs and settle there. A chronic cough, often called smoker's cough, is triggered by accumulations of mucus below the larynx.

### The alveoli: where gas exchange takes place

The alveoli provide a perfect place for carbon dioxide and oxygen to diffuse between the air in the lungs and the blood. Membranes made up of a single layer of epithelial cells bound these clusters of microscopic air sacs, and a network of capillaries tightly clasps each alveolar sac. The capillary walls are also only one cell thick and press against the alveolar epithelium. These two adjacent membranes provide the thinnest possible barrier between the blood in the capillaries and the air in the alveoli.

The alveoli also provide another important component of efficient gas exchange: a large surface area. If the epithelial membrane of all your alveoli were spread out flat, it would cover a tennis court! The capillaries cover this enormous surface, creating patterns much like tightly woven spider webs, providing nearly a continuous sheet of blood over the alveolar surface.

In the magnified view of the bronchioles and alveoli in Figure 28.6, the blue vessels denote arteries that carry deoxygenated blood from the right side of the heart to the lungs ❶. After the blood picks up oxygen at the lungs and gets rid of carbon dioxide ❷, it travels to the heart via veins (shown in red ❸) and gets a push out to the rest of the body.

---

### CONCEPT CHECKPOINT

**3.** Trace the pathway of a molecule of oxygen through the human respiratory tract from the time it enters the nose until it is absorbed into the bloodstream.

---

## 28.6 Low pressure causes air to flow into the lungs, and high pressure forces air out of the lungs.

Air moves into and out of your lungs as the volume of your thoracic cavity is made larger and smaller by the action of certain muscles. The **thoracic cavity** (thu-RASS-ick, chest cavity) is within the trunk of your body above your diaphragm and below your neck. The **diaphragm** (DYE-uh-fram) is a sheet of muscle that forms the horizontal partition between the thoracic cavity and the abdominal cavity. Various blood vessels and the esophagus penetrate it as they

SEM 8600×

**Figure 28.7** Colorized scanning electron micrograph of cilia. Cilia such as these line the trachea and sweep trapped particles out of the respiratory tract.

traverse these two body cavities. The position of the diaphragm is shown in Figure 28.6 and **Figure 28.8**.

Other muscles of breathing assist the diaphragm. These muscles extend from rib to rib—from the lower border of each rib to the upper border of the rib below—and are called **intercostal muscles**. The word "intercostal" literally means "between" (*inter*) "the ribs" (*costal*). You have two sets of intercostals: the internal intercostals and the external intercostals. The *internal intercostals* are those that lie closer to the interior of the body (as

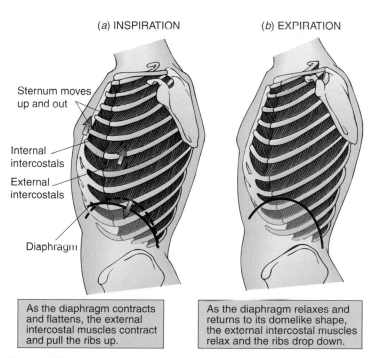

*(a)* INSPIRATION  *(b)* EXPIRATION

Sternum moves up and out

Internal intercostals

External intercostals

Diaphragm

| As the diaphragm contracts and flattens, the external intercostal muscles contract and pull the ribs up. | As the diaphragm relaxes and returns to its domelike shape, the external intercostal muscles relax and the ribs drop down. |

**Figure 28.8 How a human breathes.** The volume of the thoracic cavity increases during inspiration as the diaphragm contracts and lowers, and the ribs are raised by the external intercostal muscles. Volume decreases during expiration as the ribs lower and the diaphragm relaxes, assuming a domelike shape.

their name suggests) and that have fibers extending obliquely downward and backward (from front to back). The *external intercostals* extend from back to front, having fibers that are directed downward and forward.

## Inspiration

When you are breathing quietly—not exerting yourself physically—your diaphragm and external intercostals alone are responsible for the change in the size of your thoracic cavity. This change in size results in the movement of air into and out of your lungs. During inspiration the dome-shaped diaphragm contracts, flattening somewhat and thereby lowering the floor of the thoracic cavity. The external intercostals contract, raising the rib cage (Figure 28.8*a*). Notice in Figure 28.8 how these two actions increase the size of the thoracic cavity.

The interior walls of the thoracic cavity are lined with a thin, delicate, sheetlike membrane called the *pleura*. The pleura folds back on itself to cover each lung. These two layers of the membrane are quite close to one another, separated only by a thin film of fluid. As the volume of the thoracic cavity increases during inspiration, the lungs also expand, held to the wall of the thoracic cavity by cohesion of the water molecules between the two membranes. Put simply, the lungs "stick to" the thoracic wall and move with it.

As the lungs expand in volume, the air pressure within the lungs decreases because there are fewer air molecules per unit of volume. As a result, air from the environment outside the body flows into the lungs, equalizing the pressure inside and outside the thoracic cavity. By means of this process, you breathe in 13,638 liters (more than 3000 gallons) of air every day.

## Expiration

The lungs contain special nerves called *stretch receptors*. When the lungs are stretched to their normal inspiratory capacity, these receptors send a message to a respiratory center located in parts of the brainstem called the medulla and the pons (see Chapter 32). In response, the respiratory center stops sending "contract" messages to the muscles of breathing, which causes them to relax—a passive process in contrast to the active process of inspiration. As the diaphragm relaxes, it assumes its domelike shape, reducing the volume of the thoracic cavity. Similarly, as the external intercostals relax, the rib cage drops (Figure 28.8*b*), further reducing the volume of the thoracic cavity. The volume of the lungs, in turn, decreases, aided by the recoil action of the lungs' elastic tissue. The reduced volume of the lungs results in an increase in the air pressure within them. Air is forced out of the lungs until the pressure outside and inside the thoracic cavity is equalized. The cycle of inspiration and expiration is shown in **Figure 28.9**.

The brainstem also contains areas that control the rate of the cycle of inspiration and expiration. These areas are sensitive to changes in the level of carbon dioxide in the blood. When you have a high level of carbon dioxide in your blood, the carbon dioxide diffuses into the fluid that surrounds the brain. Hydrogen ions ($H^+$) are produced from a series of chemical reactions involving carbon dioxide (see the discussion of internal respiration in Section 28.7) and stimulate the activity of the respiratory center. Rapid breathing allows faster diffusion of carbon dioxide out of the bloodstream and into the lungs where it can be expelled.

When you breathe deeply, such as during physical exercise, the internal intercostals as well as other muscles in the chest and abdomen help the muscles of respiration. During deep inspiration, certain muscles attached to the sternum—the breastbone in the center of your chest to which your ribs are attached—pull up on it. In addition, other muscles pull up on the first two ribs. This action increases the volume of the chest cavity more than during quiet inspiration, so more air flows into the lungs. By deep breathing and increasing the rate of breathing, world champion runners have been shown to increase their air intake fifteenfold.

Expiration, a passive process during quiet breathing, becomes an active process during deep breathing. The internal intercostals contract and pull down the rib cage. Abdominal muscles also contract, pulling down on the lower ribs and compressing the abdominal organs, causing them to push up on the diaphragm. The volume of the thoracic cavity is actively decreased, causing more air to be expelled than during quiet expiration.

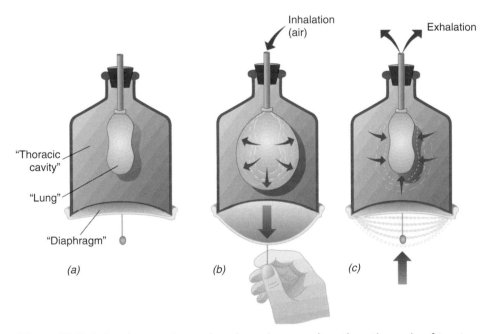

**Figure 28.9** A simple experiment that shows how you breathe—the cycle of inspiration and expiration. In the jar is a balloon (*a*). When the diaphragm is pulled down, as shown in (*b*), the balloon expands. When it is relaxed (*c*), the balloon contracts. In the same way, air is taken into your lungs when your diaphragm contracts and flattens, increasing the volume of your chest cavity. When your diaphragm relaxes and resumes its dome shape, the volume of your chest cavity decreases and air is expelled.

 **Visual Thinking:** Why does air fill the balloon when the diaphragm is pulled down?

## Lung volumes

How much air do you move into and out of your lungs during inspiration and expiration? Fill a large jar with water and invert it in a pan of water. Try not to let much water spill out of the jar. Mark the level of the water in the jar with tape or a wax pencil. Take a piece of rubber tubing about a foot long and slip one end up into the inverted jar. Put the other end into your mouth and exhale normally. The air you breathe out will displace the water. Mark the new level of the water. Remove the jar from the water and fill it to this line with water. Now measure the volume of the water between the two lines with a measuring cup. Its volume equals the volume of air you breathed out. **Figure 28.10** shows a woman having her lung volume measured in a more sophisticated way, via machine, in the hospital. If her breathing is average, she will expel about 500 milliliters of air during quiet breathing, or slightly more than 1 pint. This volume of air—the amount inspired or expired with each breath—is called the **tidal volume**.

Of the 500 milliliters of air a person normally breathes in, only about 350 milliliters reach the alveoli. The other 150 milliliters is on its way either into or out of the lungs, occupying space in the nose, pharynx, larynx, trachea, and bronchial tree. This space is called dead air space because it serves no useful purpose in gas exchange. Some of this air, in fact, will never reach the lungs. And some air, called residual air, remains in the lungs—even during deep breathing.

 CONCEPT CHECKPOINT

4. Organ systems in the body such as the respiratory system are regulated or coordinated by the nervous system. Describe one way in which the nervous system controls the respiratory system.

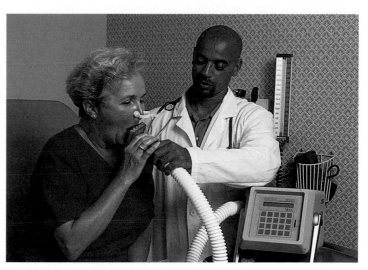

**Figure 28.10 Measuring lung volume.** This machine will measure the volume of air this woman breathes out. If volumes are abnormal, it may be an indication of respiratory disease.

## 28.7 The circulatory system transports gases throughout the body.

The transport of gases throughout the body is assisted by the circulatory system. Without the help of a "highway" of blood, scientists estimate that it would take a molecule of oxygen three years to diffuse from your lung to your toe! A person dies within a few minutes if this transport system breaks down.

Although the pathway of blood throughout the body will be described in detail in the next chapter, it is helpful to understand the basic routing of blood when discussing gas transport. Notice in

## *just wondering . . .*

*Questions students ask*

### I'm a smoker thinking about quitting. Does my smoking really affect the health of my children?

Cigarette smoking damages the lungs of children in more ways than people realize, and the damage can start in the womb. Reports from the Harvard Six Cities Study, a 20-year investigation into the effects of air pollution and lung hazards, conclude that smoking during pregnancy can detrimentally affect a child's lung function for the rest of the child's life! Lung function refers to the *vital capacity* of the lungs—the maximum volume of air that can be pushed out of the lungs during one exhalation after a single, deep inhalation. The reason for reduced lung function in the children of smokers is unclear. Many investigators hypothesize that they develop smaller lungs producing less elastic tissue that helps the lungs recoil during exhalation. There is also a positive correlation between mothers who smoke and the incidence of asthma in their children. In addition, questions have been raised about the incidence of maternal smoking and hearing defects in their children, as well as the occurrence of sudden infant death syndrome (SIDS).

Results from the Six Cities Study showed that children exposed to tobacco smoke before the age of 6 had a 2 to 3% decrease in lung function, typical of a person in his or her late 20s or early 30s. Children growing up in households with smokers also develop more lower respiratory conditions such as shortness of breath, wheezing, coughing, and bronchitis. Statistics from the American Heart Association state that secondhand smoke causes up to 300,000 lower respiratory tract infections such as pneumonia and bronchitis in children younger than the age of 1 1/2. In 5% of these cases, the infections are so severe that the child must be hospitalized. The choice to quit smoking will benefit not only your health, but the health of your children as well. ●

*Are you wondering about a topic in biology and how it relates to your life? Submit your question by clicking the* Just Wondering *link in this text's companion Web site at* www.wiley.com/college/alters.

**Figure 28.11** that blood from the upper and lower body first enters the upper right chamber of the heart ( 1 on the left as you face the illustration). This blood has given up some of its supply of oxygen to the tissues, so it is considered oxygen poor. Along its route, however, it collected the waste product of cellular respiration, carbon dioxide. Deoxygenated blood such as this (more carbon dioxide than oxygen) is shown as blue in the diagram. The deoxygenated blood passes from the upper to the lower right chamber of the heart 2 , where it is pumped to the lungs. Some of this blood goes to the right lung, and some goes to the left lung 3 .

### External respiration

At the lungs, carbon dioxide within the blood of the capillaries surrounding the alveoli and the oxygen in the air of the alveoli are exchanged 4 , which is a process called **external respiration**. This exchange of gases at the lungs works by the process of diffusion, down a concentration and pressure gradient as described in Section 28.1. Carbon dioxide, which has a higher concentration and partial pressure in the blood and lower partial pressure and concentration in the air of the alveoli, moves from the blood into the alveoli. Oxygen, which has a higher concentration and partial pressure in the alveoli and lower con-

centration and partial pressure in the blood, moves from the alveoli into the blood. Oxygenated blood (more oxygen and less carbon dioxide) is shown as red in Figure 28.11. The oxygenated blood then flows from the lungs to the left side of the heart 5 , where the lower left chamber 6 pumps it out to other parts of the body 7 .

How do oxygen and carbon dioxide travel within the blood? Most of the carbon dioxide in the blood (69%) travels as bicarbonate ions ($HCO_3^-$) in the plasma, the fluid portion of the blood (**Figure 28.12**). Carbon dioxide combines chemically with water to form molecules of carbonic acid ($H_2CO_3$), but quickly breaks apart, or dissociates, into bicarbonate ions ($HCO_3^-$) and hydrogen ions ($H^+$).

$$CO_2 + H_2O \rightarrow H_2CO_3 \rightarrow HCO_3^- + H^+$$

One-fourth (25%) of the carbon dioxide in the blood is carried in the red blood cells, bound to the oxygen-carrying molecule *hemoglobin (Hb)*. It is carried, however, by a different portion of the molecule than oxygen is. A small amount (6%) of the carbon dioxide is simply dissolved in the blood. At the alveoli, the dissolved carbon dioxide moves out of the blood first. This decrease in carbon dioxide concentration triggers a reversal of the chemical reactions just described:

$$H^+ + HCO_3^- \rightarrow H_2CO_3 \rightarrow H_2O + CO_2$$

Carbon dioxide also dissociates from hemoglobin, and the released carbon dioxide molecules diffuse into the alveoli.

As oxygen diffuses from the alveoli, very little of it (1.5%) dissolves in the plasma. Instead, it combines with hemoglobin within the red blood cells. When the pressure gradient is high, as in the capillaries of the lung, hemoglobin binds large amounts of oxygen. When oxygenated, hemoglobin turns bright red, which makes blood look red. You will notice, however, that the blood vessels on the underside of your wrists look blue. These blood vessels are veins carrying deoxygenated blood. Deoxygenated blood is dark red but appears blue through layers of skin.

At high altitudes, such as in mountainous regions, the air is thinner and the pressure of the oxygen molecules within the air is lower than at sea level. Therefore, less of a pressure gradient is created at the alveoli. As a consequence, less oxygen diffuses into the blood; shortness of breath, nausea, and dizziness may result in those unaccustomed to the altitude. This condition is referred to as *high-altitude sickness*. Mountain climbers, athletes working out or playing at high altitudes, such as Mile High Stadium in Denver, Colorado, and tourists visiting an area of high altitude, such as Mexico City, Mexico, need to slowly work up to their normal levels of activity to give their bodies time to adjust to the lower oxygen pressure.

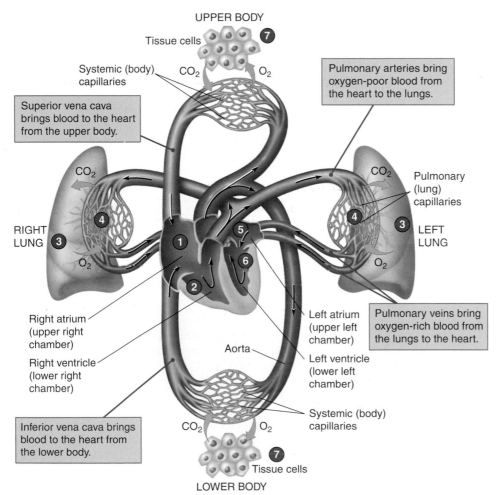

**Figure 28.11 The route of blood showing gas transport.** Oxygen diffuses into the bloodstream within the lungs and is delivered to all cells of the body. Carbon dioxide, a product of aerobic cellular respiration, diffuses into the bloodstream from the tissues and is carried to the lungs where it diffuses out of the bloodstream and is then expelled.

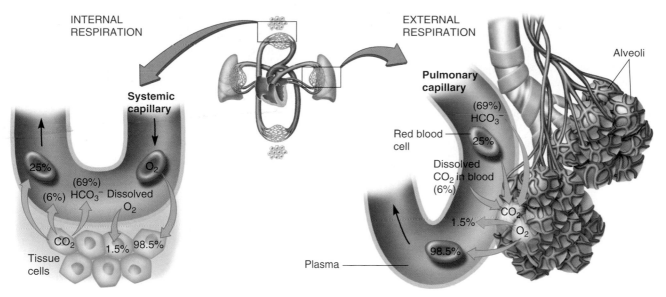

**Figure 28.12** External and internal respiration. During external respiration, carbon dioxide moves from the bloodstream into the alveoli, while oxygen moves from the alveoli into the bloodstream. During internal respiration, oxygen moves from the bloodstream into the cells of the body, while the waste carbon dioxide moves from the cells of the body into the bloodstream.

The body adjusts by increasing the number of red blood cells and thus, the amount of hemoglobin available to carry oxygen.

### Internal respiration

Oxygenated blood is pumped out to the body by the left ventricle, or lower left chamber, of the heart (see Figure 28.11). It travels through the aorta to other large arteries that soon branch into smaller and smaller arteries. Eventually, the blood vessels become so small that only one red blood cell at a time can pass. These vessels are the capillaries and are the sites of internal respiration.

**Internal respiration** is the uptake of oxygen and release of carbon dioxide by the body tissues (see Figure 28.12). As capillaries pass through tissues and movement of respiratory gases takes place, oxygenated blood becomes progressively deoxygenated. In essence, the processes of internal respiration are the reverse of the processes of external respiration. As the blood begins its journey from the lungs

around the body, it is oxygen-rich. The oxygen molecules in the blood exert a higher pressure than those in the cells because the cells (in the process of cellular respiration) are continuously using oxygen. Consequently, oxygen molecules begin to dissociate from the hemoglobin, diffuse into the tissue (interstitial) fluid, and from there diffuse into the cells. Conversely, the levels of carbon dioxide are higher in the cells than in the blood because cellular respiration continuously produces carbon dioxide. Consequently, carbon dioxide diffuses from the cells into the tissue fluid and then into the blood.

As the diffusion of respiratory gases at the capillaries continues, an interesting thing happens. The partial pressure of oxygen in the blood plasma decreases. Diffusion of oxygen into the tissue fluid does not slow down, however, because the pH at the capillaries lowers as carbon dioxide diffuses into the blood. Remember that carbon dioxide is carried in the blood primarily as bicarbonate ions. As these ions are formed, hydrogen ions are also produced. The buildup of these hydrogen ions makes the blood increasingly acidic as more carbon dioxide diffuses in. (See Chapter 3 for a discussion of the relationship between hydrogen ions and pH.) This acidic environment helps split more oxygen from hemoglobin, thereby enhancing oxygen's diffusion into the tissue fluid.

# What are some diseases and potential problems of the airways?

## 28.8 Chronic obstructive pulmonary disease is a group of disorders that impair breathing.

The term **chronic obstructive pulmonary disease (COPD)** is used to refer to disorders that block the airways and impair breathing. COPD is the fourth leading cause of death and is responsible for at least 50,000 deaths per year in the United States and 2.9 million worldwide. Two disorders commonly included in COPD are **chronic bronchitis** and **emphysema**.

*Chronic bronchitis* is an inflammation of the bronchi and bronchioles that lasts for at least three months of two consecutive years, with no accompanying disease as an alternate cause. The primary cause is cigarette smoking. Air pollution and occupational exposure to industrial dust are other, much less frequent causes. Cigarette smoking paralyzes ciliated epithelial cells so that they can no longer effectively remove incoming particles and microbes, and also causes increased mucus production by cells lining the trachea. A continued buildup of mucus provides food for bacteria, and infection can result.

Normally, bronchioles widen during inspiration; they narrow during expiration. If mucus is plugging various bronchioles, some air may be able to get to the alveoli beyond the plugged bronchioles, but may not be able to get out. Coughing produces pressure within these continuously inflated alveoli that ruptures their walls, decreasing the surface area over which gas exchange takes place. In addition, the lungs lose their elasticity and the ability to recoil during exhalation thus remaining filled with air. The result is a disorder aptly called *emphysema*, meaning "full of air." A person with emphysema has to work voluntarily to exhale.

People with COPD find that they have more respiratory infections than they did before the disorder and that these infections last longer. They have a morning cough or may cough all day. They may also tire easily and become short of breath with minimal physical exertion. Some people with COPD feel as though they cannot breathe at times. As the disorder progresses, some find it difficult to do a day's work or to accomplish the daily activities of living. Some persons suffer bouts of respiratory failure. COPD is serious and deadly—but in most cases, it is avoidable with wise health choices.

## 28.9 Abdominal thrusts can dislodge substances blocking the airway.

Have you ever been eating with someone who started to choke? Did you know what to do? Choking is caused when food or a foreign object becomes lodged in the windpipe. When someone is choking, first notice whether the person can talk, breathe, or cough. If so, stay with the person until coughing clears the airway. Do not try to slap the person on the back. The slapping may only cause the food to become more deeply lodged in the windpipe.

If a person cannot talk or cough and appears not to be breathing, administer several short, quick *abdominal thrusts*, as shown in **Figure 28.13**. These thrusts push in on the diaphragm and the thoracic cavity, suddenly decreasing its volume. This sudden decrease creates a surge in air pressure below the obstruction. This usually dislodges the item blocking the respiratory passage and projects it forcefully from the windpipe (Figure 28.13*c*).

# How Science Works

**DISCOVERIES**

### Asthma—new findings about an "old" ailment

Asthma is a chronic inflammatory disorder of the airways that currently affects about 6 million children and adolescents and about 14 million adults in the United States. Chronic means that a disease persists for a long time. New drugs and preventive measures, however, lessen the severity of asthma attacks and allow sufferers to lead normal lives. In addition, scientists have recently discovered that some persons with hard-to-treat asthma may have underlying bacterial or viral infections of the lungs. This new discovery may lead to new treatments for these sufferers.

The airways (bronchi and bronchioles) of a person with asthma are much more sensitive to certain stimuli than are the airways of other people. Triggers that are known to cause asthma attacks include allergens such as pollen, dust, tobacco smoke, dust mites (shown in **Figure 28.A**), animal dander, or certain foods; physical factors such as coughing, sneezing, rigorous exercise, or cold temperatures; certain viral or bacterial infections; and chemical irritants such as cigarette smoke. In an asthma attack or episode, the lining of the airways swells, the bronchial muscles contract, and mucus production increases. This narrows the airways. The asthma sufferer has trouble bringing air into the lungs, and also has trouble getting air out. Forcing air through the narrowed airways can produce a whistling or wheezing sound.

Scientists know that there is a genetic link to asthma and that asthma is more likely to be passed to children by the mother than the father. The mechanisms involved in the higher rate of maternal transfer, such as prenatal influences, are unknown.

An important element in the treatment of asthma is teaching the asthma patient to recognize the triggers that bring on attacks and to prevent attacks before they can occur. Treatment can also involve several types of drugs. Maintenance drugs control

**Figure 28.A** A dust mite, shown magnified 500 times, is a common trigger of asthma attacks.

and prevent symptoms. Rescue drugs relieve acute symptoms. Inhaled bronchodilators are rescue drugs used during an attack—these fast-acting drugs work to open constricted airways.

Recent studies have shown that overuse of bronchodilator inhalers can lead to a worsening of asthma and even to death because the asthma patient comes to rely on the fast-acting nature of the drug and does not notice

that his or her asthma might be worsening. A severe attack of asthma can take such a person by surprise, leading to serious complications and death. That is why asthma patients must participate as much as possible in monitoring and treating their condition.

The discovery that the lungs of some young asthma patients are chronically infected with a virus that causes bronchitis led to the understanding that antiviral therapy might be a useful treatment for some children. In addition, research has revealed that a virus or the bacterium *Chlamydia pneumoniae* (not the same bacterium implicated in sexually transmitted infections) is sometimes present in adult-onset asthma. Treatment for these pathogens cured or improved the asthma of over half the infected adults. Research results published in 2003 noted that mild repeated viral infections early in life could confer protection from asthma.

In 1998, researchers reported that obesity was also a risk factor in the development of adult-onset asthma. Researchers were unsure, however, whether obesity was a cause or a result of asthma. Some hypotheses suggested that those with asthma were more likely to avoid exercise, adopt a sedentary lifestyle, and subsequently gain weight. A study published in 2003, however, revealed evidence that obesity and a diet high in fats contributed to the symptoms of asthma. Researchers suggested that asthma sufferers lower their intake of fats to reduce asthma symptoms and enhance their overall health. ●

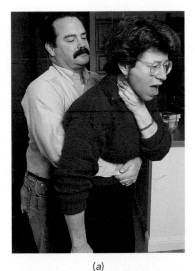

*(a)*

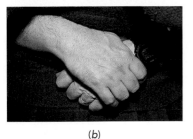

*(b)*

*(c)*

**Figure 28.13** How to unblock the windpipe when someone is choking. (a) Stand behind the person with your arms encircling the lower chest. (b) Place one hand over the fist of the other positioned just below the breastbone. (c) Administer a series of quick, sharp, upward and inward thrusts.

### CONCEPT CHECKPOINT

**5.** Based on the partial pressures of $O_2$ and $CO_2$ in the blood in the following table, propose one place in the body this capillary bed might be found.

| | Beginning of capillary bed | End of capillary bed |
|---|---|---|
| Partial pressure $O_2$ | 100 mmHg | 42 mmHg |
| Partial pressure $CO_2$ | 40 mmHg | 46 mmHg |

**6.** Based on the partial pressures of $O_2$ and $CO_2$ in the blood in the following table, propose one place in the body this next capillary bed might be found.

| | Beginning of capillary bed | End of capillary bed |
|---|---|---|
| Partial pressure $O_2$ | 42 mmHg | 100 mmHg |
| Partial pressure $CO_2$ | 46 mmHg | 40 mmHg |

## CHAPTER REVIEW

## Summary of Key Concepts and Key Terms

### What is respiration?

**28.1** Respiration (**respire**, p. 479) is the uptake of oxygen and the release of carbon dioxide by an organism.

### What patterns of respiration exist among organisms?

**28.2** Respiration occurs across the cell membranes of single-celled organisms and the outer layer of cells of certain multicelled animals.

**28.2** Respiration occurs across the skin of amphibians and certain fishes, snakes, and turtles.

**28.3** A **trachea** (TRAY-kee-uh, p. 481) is an air-filled tube.

**28.3** Respiration occurs in the **tracheal systems** (p. 481) of most spiders and insects, the **gills** (p. 482) of most large multicellular aquatic animals, and the **lungs** (p. 482) of most large land-dwelling animals.

**28.3** In general, the tracheal system is open to the environment by means of small holes (**spiracles**, p. 481) in the cuticle (exoskeleton) of the insect or spider.

**28.3** Water flows in the opposite direction of blood flow through the gills of fishes; this countercurrent flow enhances diffusion of oxygen into the blood of the fish and the diffusion of carbon dioxide into the water.

**28.3** Lungs are invaginations of the body surface and are served by a circulatory system that delivers oxygen to other parts of the body.

### How do humans respire?

**28.4** Human respiration uses special organs and relies on pressure changes within the thoracic cavity for ventilation.

**28.4** In humans and other mammals, air moves into the lungs through the nose, mouth, and airways of the **respiratory system** (p. 483); the movement of air into and out of the lungs is called **breathing** (p. 483).

**28.4** A breath consists of taking air into the lungs, or **inspiration** (p. 483), and expelling air from the lungs, or **expiration** (p. 483).

**28.4** Human respiration includes the processes of internal respiration, external respiration, and breathing.

**28.5** Air passes through many respiratory structures on its way to the lungs: the **nasal cavities** (p. 483), the **pharynx** (throat, p. 483), the **larynx** (voice box containing **vocal cords**, p. 483); the **trachea** (windpipe, p. 483); two **primary bronchi** (p. 483), three right and two left **secondary bronchi** (p. 483), **respiratory bronchioles** (p. 483), and microscopic air sacs called **alveoli** (al-VEE-uh-lye, p. 485).

**28.5** The alveoli provide a thin, large, moist, surface area over which gas exchange takes place.

**28.6** Inspiration occurs when the volume of the **thoracic cavity** (thu-RASS-ick, p. 485) is increased and the resulting lowered pressure causes air to flow into the lungs.

**28.6** The muscles of breathing are the **diaphragm** (DYE-uh-fram, p. 485), which forms the floor of the thoracic cavity, and the **intercostal muscles**, (p. 485) which extend from rib to rib.

**28.6** Expiration occurs when the volume of the thoracic cavity is decreased and the resulting raised pressure forces air out of the lungs.

**28.6** The volume of air inspired or expired with each breath is called the **tidal volume** (p. 487).

**28.7** The circulatory system aids in respiration by transporting gases throughout the body.

**28.7** Oxygen in the air diffuses into the blood, and carbon dioxide diffuses out of the blood at the alveoli; this process is called **external respiration** (p. 488).

**28.7** At the capillaries in the body tissues, oxygen in the blood diffuses into the cells and carbon dioxide diffuses out of the cells and into the blood; this process is called **internal respiration** (p. 489).

---

### What are some diseases and potential problems of the airways?

**28.8** **Chronic obstructive pulmonary disease** (COPD, p. 489) refers to disorders that impair movement of the air in the respiratory system.

**28.8** **Chronic bronchitis** (p. 489) and **emphysema** (p. 489) are two common COPD disorders that are most frequently caused by cigarette smoking.

**28.9** To assist a person who is choking and cannot talk or breathe, administer several short, quick, abdominal thrusts with your arms around the victim and your fist just below the breastbone.

---

## Level 1    Learning Basic Facts and Terms

**Multiple Choice**

1. The specific sites of gas exchange between the air and the bloodstream in the lungs of mammals are
   a. alveoli.
   b. bronchi.
   c. larynx.
   d. tracheas.
   e. book gills.

2. In humans the _____ transport(s) respiratory gases to and from the tissues, while in insects; a system of _____ is responsible for oxygen and carbon dioxide transport.
   a. bronchi // blood vessels
   b. blood // tracheas
   c. the diaphragm // lymph vessels
   d. skin // gills
   e. lungs // gills

3. $CO_2$ produced by a muscle cell during cellular respiration first diffuses into _____ before it enters the muscle capillaries.
   a. lymph vessels
   b. blood plasma
   c. interstitial fluid
   d. alveoli
   e. tracheas

4. The epiglottis covers the _____ during swallowing.
   a. esophagus
   b. pharynx
   c. glottis
   d. stomach
   e. alveoli

5. Which of the following represents the pathway of a $CO_2$ molecule produced in the body tissues through the respiratory tract of a mammal until it is exhaled?
   a. alveolar capillaries → alveoli → bronchioles → bronchi → trachea → glottis → larynx → pharynx → nasal cavities
   b. nasal cavities → pharynx → alveolar capillaries → alveoli → bronchioles → bronchi → trachea → glottis → larynx
   c. nasal cavities → pharynx → larynx → glottis → trachea → bronchi → bronchioles → alveoli → alveolar capillaries
   d. trachea → alveoli → alveolar capillaries → bronchi → bronchioles → pharynx → glottis → larynx → nasal cavities

6. Capillary beds are the sites of
   a. transfer of waste $CO_2$ from blood into lungs in mammals.
   b. diffusion of $O_2$ from water into blood within the gills of fish.
   c. exchange of $CO_2$ between the blood and the water within the skin of amphibians.
   d. transfer of metabolic waste products into bloodstream from muscle.
   e. All of the above.

**True–False**

7. _____ Most of the $CO_2$ transported in the blood is bound to hemoglobin.

8. _____ Hemoglobin has little capacity to alter the amount of oxygen it releases to the tissues.

9. _____ Contraction of the diaphragm enlarges the chest cavity, which draws air into the lungs.

10. _____ Countercurrent exchange in the gills of fish is an adaptation that maximizes the diffusion of oxygen into the bloodstream.

---

## Level 2    Learning Concepts

1. Describe the ways in which organisms without special organs of respiration can exchange oxygen and carbon dioxide with the environment.

2. Distinguish among respiration, cellular respiration, internal respiration, and external respiration.

3. Your biology instructor poses a problem in class, and you suggest a solution. Explain how you produce the necessary sounds.

4. Explain how differences in air pressure help you to breathe.

5. Explain what happens during gas exchange at the alveoli. What gases are exchanged, and what forces "drive" this exchange?

6. A normal hemoglobin level is an essential circulatory function. What does the hemoglobin carry?

7. The accompanying diagram represents changes in the volume of air in the lungs during inspiration, expiration, and deep breathing. Which letter represents tidal volume? Which represents the maximum amount of air that you can breathe in and then breathe out during deep breathing? Which represents residual air that is retained in the lungs even during deep breathing?

8. Explain the process of gas exchange at the capillaries. What gases are exchanged, and why does an exchange of gases occur?

9. While on vacation in the mountains, you find that you feel dizzy and short of breath. Explain why.

10. Athletes often choose to train at high altitude because the body responds by increasing its production of hemoglobin. Explain why this would be beneficial to an athlete.

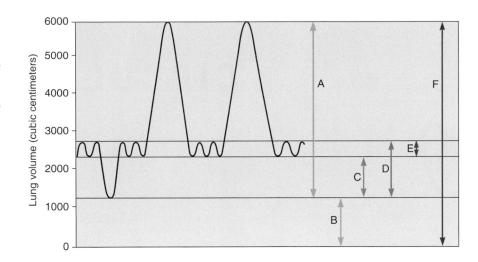

## Level 3  Critical Thinking and Life Applications

1. During the winter in cold regions, frogs often hibernate in the mud at the bottom of ponds. How do they get enough oxygen to survive?

2. Hiccups occur when the diaphragm contracts spasmodically, causing a sudden inhalation. What do you hypothesize causes the sound effect that results from hiccups?

3. How does the circulatory system assist the process of respiration? Why is this important?

4. Why do you think that people with COPD and emphysema tire easily?

5. Study Figure 28.4 regarding countercurrent exchange. Describe in detail how this process works to aid the diffusion of oxygen into the blood of organisms that use gills for respiration.

6. Using the diagram showing changes in lung volume during breathing in question 7 above, explain what would happen to the magnitude of A, B, C, D, and E in a person with emphysema.

## In The News  Critical Thinking

### ASTHMA TREATMENT

Now that you understand more about respiration, reread this chapter's opening story about asthma and the peptide that blocks excess mucus secretion in mice. To better understand this research, it may help you to follow these steps:

1. Review your immediate reaction to research on blocking excess mucus secretion by the respiratory tract that you wrote when you began reading this chapter.

2. Based on your current understanding, again summarize the main point of the research in a sentence or two.

3. What questions do you now have about this research regarding a possible treatment for asthma that this chapter's opening story does not answer?

4. Collect new information about the research. Visit the *In The News* section of this text's companion Web site at www.wiley.com/college/alters and watch the "Asthma Treatment" video. Then use the "summary" link to read the accompanying story and access related links. Use this information, the links provided, and other online and library resources to answer your questions and find updates about this research topic. State the sources of your information. Explain why you think the information is accurate. Also determine whether the information expresses a particular point of view or is biased in any way.

5. What in your view are the most significant aspects of this research? Give reasons for your opinion and for any changes in your ideas based on the additional information you have collected and the analysis you have done.

# CIRCULATION

## In The News | Bridge to Recovery

Can only love mend a broken heart? A new love interest may help heal emotional heartache, but a heart that's damaged physically and fails to pump effectively needs far more than love.

According to the American Heart Association, approximately 5 million Americans currently suffer from *heart failure*, a chronic condition characterized by the organ's ineffective pumping. Damage to the lower left chamber of the heart, the left ventricle, often results in heart failure. The left ventricle pumps oxygen-rich blood out of the heart and throughout the body.

Patients suffering from heart failure may be able to slow the illness's progression by following their physician's recommendations concerning medication and lifestyle practices. As their hearts grow weaker, however, those with heart failure have increasing difficulty performing basic physical activities. Such patients may be candidates for heart transplantation surgery, but they usually have to wait for a new heart. In the meantime, they may die of heart failure.

Just as you can benefit from a relaxing vacation from school or work, a diseased heart may benefit from some rest too. But how can a living heart take a break from beating? A special biomechanical pump, the *left ventricular assist device (LVAD)* may provide an answer. The LVAD (see photo) takes over the pumping chores of the left ventricle and allows the heart muscle to rest. This device may enable some patients with heart failure to survive until heart transplantation surgery. Moreover, during LVAD treatment, heart tissue may undergo structural and functional modifications called remodeling. As a result, some patients recover enough cardiac function that they do not need heart transplantation after the device is removed.

Burns Blaxall, a professor of medicine at the University of Rochester Medical Center, and his colleagues used DNA microarrays (see Chapter 11) to perform genetic analyses on muscle tissue taken from the left ventricles of a small group of patients suffering from heart failure. They measured patterns of gene expression both before and after LVAD support treatment. DNA microarrays are used to study the expression of entire genomes or large sets of genes. They found the LVAD support treatment produced significant changes in the levels of gene expression in over 500 genes. These changes occurred in a specific pattern that enabled the scientists to identify patients who did not need to undergo heart transplantation after the LVAD was removed. Dr. Blaxall thinks physicians may someday rely on DNA microarrays to help determine which heart failure patients are likely to recover after LVAD support and which will need to receive donor hearts. To learn more about Dr. Blaxall's research, visit the *In The News* section of this text's companion Web site at www.wiley.com/college/alters and view the video "Bridge to Recovery." Also, the How Science Works box in this chapter chronicles the development of heart assist devices, including the LVAD.

*Write your immediate reaction to the research concerning the effects of LVAD support in some people with heart failure: first, summarize the main point in a sentence or two; then suggest what you think its significance is. You will have an opportunity to reflect on your responses and gather more information on this topic in the* In The News *feature at the end of this chapter. In this chapter, you will learn about the circulatory system.*

## CHAPTER GUIDE

### What patterns of circulation exist among organisms?

**29.1**  Simple organisms in moist environments have no circulatory systems.

**29.2**  Most arthropods, certain mollusks, and tunicates have open circulatory systems.

**29.3**  Cephalopod mollusks, annelids, echinoderms, and vertebrates have closed circulatory systems.

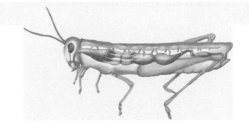

### How does human circulation work?

**29.4**  Human circulation has a variety of functions.

**29.5**  Blood circulates within arteries, capillaries, and veins.

**29.6**  The heart is the pump of human circulation.

**29.7**  Physicians can monitor the heartbeat in various ways.

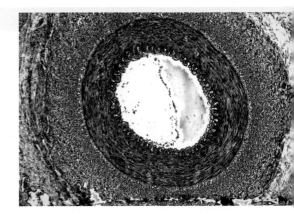

### What makes up the blood?

**29.8**  Blood plasma is the fluid portion of the blood.

**29.9**  Blood cells and platelets are suspended in the plasma.

**29.10**  Some fluid that leaves the blood returns via the lymphatic system.

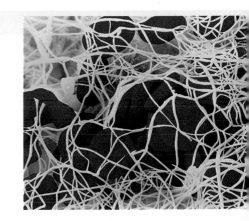

### What are some common cardiovascular diseases and conditions?

**29.11**  Cardiovascular diseases and conditions include atherosclerosis, heart attacks, and strokes.

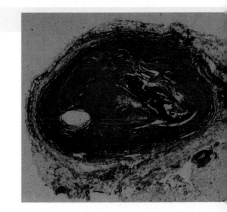

# What patterns of circulation exist among organisms?

## 29.1 Simple organisms in moist environments have no circulatory systems.

Not all organisms have hearts as humans do. In fact, not all organisms have circulatory systems. A **circulatory system** is a transport system that uses a fluid to move substances such as nutrients, wastes, and gases throughout an organism. **Circulation** is the movement of a fluid through an organism in a regular manner.

Single-celled organisms such as bacteria, yeasts, and most protists (see the paramecium in **Figure 29.1 ❶**) do not have circulatory systems. Gases and nutrients simply diffuse into the cell or are moved into the cell by means of active transport or phagocytosis (see Chapter 5). Within the cell, nutrients, wastes, and gases diffuse or are actively transported from one part of the cell to another.

Even fungi, though multicellular, have no circulatory system. Most fungi live in moist places and are composed of nearly microscopic filaments called *hyphae*. They are able to absorb nutrients, gases, and water directly from the soil, air, and decaying organisms. Many species have no cell partitions, which aids the movement of dissolved substances within the organism.

Some small animals have a body architecture that allows substances to move directly between their cells and the outside environment. As described in Chapter 27, aquatic animals with relatively simple anatomy, such as hydra, jellyfish, corals, sea anemones, and flatworms, have digestive systems with a single opening into a digestive cavity called the *gastrovascular cavity*. This a dead-end sac that functions not only in digestion (*gastro-*) but also in circulation (*-vascular*). The gastrovascular cavity of the hydra ❷ extends into the tentacles of the animal, allowing the complete interior of the hydra to be bathed in fluid, as is the exterior of the animal. No cells are far from its watery environment.

### FOCUS

*on Plants*

More than 95% of all living species of plants have a system of specialized tissues to transport substances to and from their various parts. Plants with these tissues are called vascular plants, and these specialized tissues are collectively called vascular tissue. Vascular tissue forms the "circulatory system" of a plant, conducting water and dissolved inorganic nutrients up the plant and the products of photosynthesis throughout the plant. Chapter 24 describes the structure of vascular plants in more detail.

## 29.2 Most arthropods, certain mollusks, and tunicates have open circulatory systems.

Larger animals and those that live on land have circulatory systems. Even most land plants have their own version of a circulatory system. (See the *Focus on Plants* box.) Terrestrial (land) animals have a wet internal environment, although their external environment is dry. Their cells are bathed in *tissue fluid*, a waterlike substance. The circulatory system helps maintain the composition of this fluid.

Some animals have an **open circulatory system**, in which the circulating fluid bathes internal organs directly. It may or may not have a pumping heart connected to open-ended vessels. Open circulatory systems are found in most arthropods such as spiders, lobsters, crabs, and insects; certain mollusks such as clams and snails (but not squid and octopus); and tunicates, which are saclike marine organisms (see Chapter 22). In an open circulatory system ❸, a fluid composed of blood mixed with tissue fluid bathes the internal tissues of an animal, oozing through spaces or cavities that surround the organs. This mixture is usually referred to as hemolymph.

The term *hemolymph* refers to tissue fluid, which is also called *lymph*, and to hemoglobin or hemocyanin, two oxygen-carrying pigments of blood. When combined with oxygen, hemoglobin becomes bright red. Hemocyanin becomes blue as seen, for example, in the horseshoe crab. Some organisms, such as the tunicates and the insects, have no oxygen-carrying pigment, so their hemolymph is colorless.

The fluid in an open circulatory system moves sluggishly, pushed by the movements of the animal or by a simple heart connected to open-ended vessels. The circulatory system of the tunicate ❹ consists of a heart with a large vessel at each end. The heart pushes blood first in one direction and then the other. This unusual reversal of blood flow is unique to the tunicates. In highly active insects such as the grasshopper ❺, diffusion of oxygen within the open circulatory system is not enough to serve all the cells of the body efficiently. Instead, oxygen is transported by means of a tracheal system (see Chapter 28), a set of tubes that connects to the outside and provides the transport mechanism for this essential gas.

## 29.3 Cephalopod mollusks, annelids, echinoderms, and vertebrates have closed circulatory systems.

Many organisms, such as you, have a **closed circulatory system**, in which blood is enclosed in vessels. Closed circulatory systems are found in cephalopod mollusks (that is, mollusks with a head, such as squid and octopus), annelids (earthworms and leeches), echinoderms (sea stars, sea urchins, sand dollars, sea cucumbers), and vertebrates (all animals with a backbone such as fish, amphibians, reptiles, birds, and mammals). In a closed circulatory system, blood is confined within blood vessels as it circulates throughout the body. This fluid is pumped either by a heart or by contractions of specialized blood vessels.

## NO CIRCULATORY SYSTEM
(single-celled organisms, fungi, and small animals having a gastrovascular cavity)

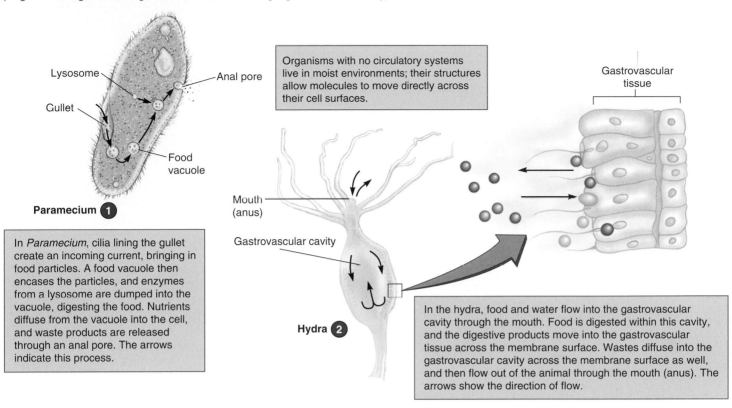

Organisms with no circulatory systems live in moist environments; their structures allow molecules to move directly across their cell surfaces.

**Paramecium 1**

In *Paramecium*, cilia lining the gullet create an incoming current, bringing in food particles. A food vacuole then encases the particles, and enzymes from a lysosome are dumped into the vacuole, digesting the food. Nutrients diffuse from the vacuole into the cell, and waste products are released through an anal pore. The arrows indicate this process.

**Hydra 2**

In the hydra, food and water flow into the gastrovascular cavity through the mouth. Food is digested within this cavity, and the digestive products move into the gastrovascular tissue across the membrane surface. Wastes diffuse into the gastrovascular cavity across the membrane surface as well, and then flow out of the animal through the mouth (anus). The arrows show the direction of flow.

## OPEN CIRCULATORY SYSTEM
(most arthropods, such as spiders, lobsters, crabs, and insects; certain mollusks, such as clams and snails; tunicates)

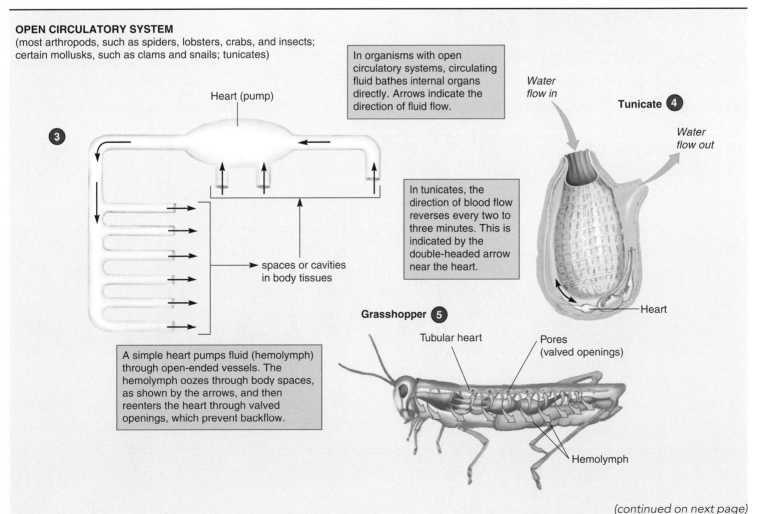

In organisms with open circulatory systems, circulating fluid bathes internal organs directly. Arrows indicate the direction of fluid flow.

**Tunicate 4**

In tunicates, the direction of blood flow reverses every two to three minutes. This is indicated by the double-headed arrow near the heart.

A simple heart pumps fluid (hemolymph) through open-ended vessels. The hemolymph oozes through body spaces, as shown by the arrows, and then reenters the heart through valved openings, which prevent backflow.

**Grasshopper 5**

(continued on next page)

**CLOSED CIRCULATORY SYSTEM**
(cephalopod mollusks such as squid and octopuses; annelids such as earthworms and leeches; echinoderms such as sea stars and sea urchins; vertebrates)

**Earthworm** 6

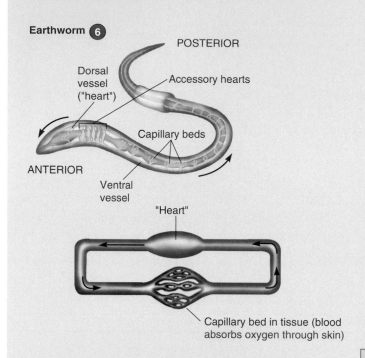

POSTERIOR

Dorsal vessel ("heart")

Accessory hearts

Capillary beds

ANTERIOR

Ventral vessel

"Heart"

Capillary bed in tissue (blood absorbs oxygen through skin)

**Fish** 7

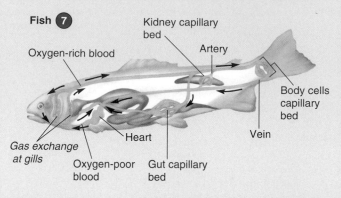

Kidney capillary bed

Oxygen-rich blood

Artery

Body cells capillary bed

Vein

*Gas exchange at gills*

Oxygen-poor blood

Heart

Gut capillary bed

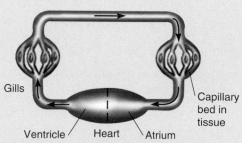

Gills

Ventricle

Heart

Atrium

Capillary bed in tissue

In organisms with closed circulatory systems, the blood is pumped by a heart and remains enclosed in vessels throughout its journey.

**Human** 8

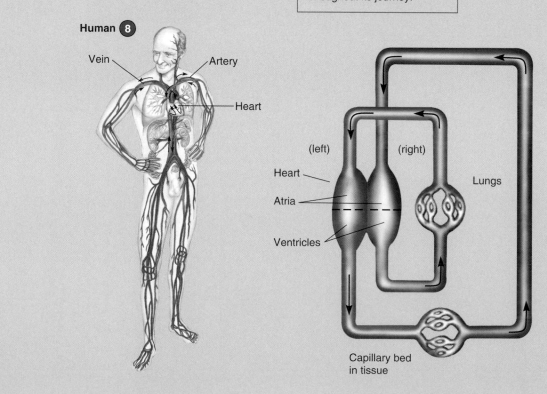

Vein

Artery

Heart

(left)

(right)

Heart

Atria

Ventricles

Lungs

Capillary bed in tissue

One of the simplest examples of a closed circulatory system is that of the earthworm  . One major vessel runs along the dorsal (top) side of the worm and carries blood anteriorly (toward the front). The other major vessel runs along the ventral (bottom) side of the worm and carries blood posteriorly (toward the back). Capillaries, which are microscopic blood vessels, connect the dorsal and ventral vessels through a network of capillary beds that bring blood to the tissues. The dorsal vessel is the pump of the system, pushing the blood forward by waves of contractions. Five pairs of pulsating anterior blood vessels connect the dorsal and ventral vessels. These vessels aid the movement of the blood around the worm's body.

Fish **7** have a two-chambered heart (see also Figure 22.9). The atrium receives the blood, and the ventricle pumps the blood. The circulatory "loop" is from the heart, through the gills, around the body, and back to the heart. Amphibians have a three-chambered heart (see Figure 22.9) and a double loop of circulation. A single ventricle pumps blood to both the lungs and the body. The blood from each circulatory loop returns to the heart, flowing into separate atria. Alternate contraction of the atria and coordinated movements of heart valves result in little mixing of oxygenated and deoxygenated blood. Some reptiles, and all birds and mammals have a four-chambered heart (Figure 29.1 **8** and Figure 22.9) with two atria and two ventricles. A double loop of circulation is maintained in which oxygenated blood from the lungs and deoxygenated blood from the body maintain separate pathways.

## CONCEPT CHECKPOINT

1. Briefly describe how each of the following organisms transports nutrients, wastes, and respiratory gases throughout its body: ameba, flatworm, oyster, housefly, octopus, salmon, salamander, hawk, wolf.

# How does human circulation work?

## 29.4 Human circulation has a variety of functions.

What comprises the human circulatory system? It is made up of three components: (1) the heart, a muscular pump; (2) the blood vessels, a network of tubelike vessels that permeate the body; and (3) the blood, which circulates within these vessels. The heart and blood vessels are known as the **cardiovascular system** (**Figure 29.2**). The prefix *cardio* refers to the heart, and the suffix *vascular* refers to blood vessels. The terms *cardiovascular system* and *circulatory system* are often used interchangeably.

The circulatory system connects the various muscles and organs of the body with one another via a network of vessels enclosing a fluid highway of blood. It serves four principal functions: nutrient and waste transport, oxygen and carbon dioxide transport, temperature maintenance, and hormone circulation.

### Nutrient and waste transport

Transportation is a major function of any circulatory system, and nutrients are major cargo. The blood takes absorbed nutrients to the liver for processing. There, some molecules are converted to glucose, which is released into the bloodstream. The blood also picks up molecules such as amino acids, which are used as building blocks to produce proteins. Some molecules, such as essential amino acids and vitamins, pass through the liver unchanged. Excess energy molecules are used for the synthesis of glycogen in the liver or body fat, and are stored for later use.

The cells, in turn, release the waste products of metabolism into the bloodstream. The blood carries most of these wastes, such as nitrogenous wastes from the deamination of amino acids, to the kidney, which captures and concentrates them for excretion in the urine (see Chapter 31).

### Oxygen and carbon dioxide transport

In all organisms with a closed circulatory system, respiratory gases are another major cargo. The cells of the body require oxygen to carry out

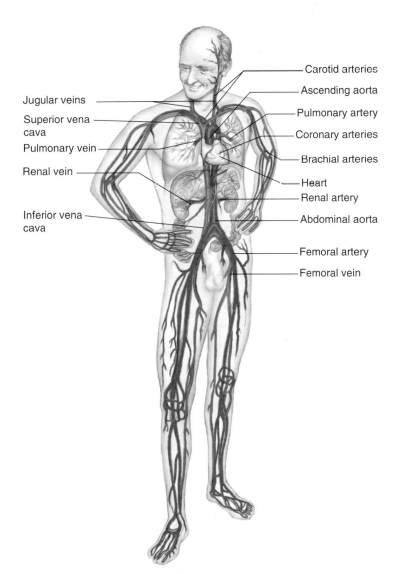

Figure 29.2 The human cardiovascular system.

Carotid arteries
Ascending aorta
Jugular veins
Pulmonary artery
Superior vena cava
Coronary arteries
Pulmonary vein
Brachial arteries
Renal vein
Heart
Renal artery
Inferior vena cava
Abdominal aorta
Femoral artery
Femoral vein

aerobic cellular respiration, which in turn produces carbon dioxide. Within the lungs, oxygen molecules diffuse into the circulating blood through the walls of capillaries (see Figure 28.12). From the lungs, the blood enters the left side of the heart which pumps oxygenated blood to all cells of the body (see Figure 29.1 **8** ). As the blood delivers oxygen to the tissues of the body, carbon dioxide molecules diffuse into the bloodstream, again through the walls of capillaries. Blood then returns to the right side of the heart which pumps it back to the lungs, where the carbon dioxide is released and a fresh supply of oxygen is captured.

### Temperature maintenance

As you read in Chapter 8, energy is continuously being lost as heat as a result of the chemical reactions that take place in the cells of your body. The blood distributes this heat, helping to maintain a stable body temperature. As the blood circulates, it passes through networks of blood vessels that lie under your skin. The blood passing through these vessels gives up heat because the environment is usually cooler than the body's temperature of 37°C (98.6°F).

To maintain stable body temperature, a regulatory center in a part of the brain called the hypothalamus acts like your own personal thermostat, constantly monitoring body temperature and stimulating regulatory processes. If your body temperature drops, signals from this center cause surface blood vessels to narrow, or constrict. This limits blood flow to the surface of the skin and lessens heat loss. Conversely, if your body temperature rises, signals from this center cause surface blood vessels to widen, or dilate. Dilation increases blood flow to the surface of the skin and increases heat loss (**Figure 29.3**).

### Hormone circulation

Nerve signals and hormones coordinate the chemical reactions and other activities of the body. Hormones are chemical messengers; they are produced in one place in the body and affect cells or tissues in another. The circulatory system is the highway within which hormones travel throughout the body, from their site of production to the target tissues that are capable of responding to them.

## 29.5 Blood circulates within arteries, capillaries, and veins.

Blood leaves the heart within vessels known as arteries (see Figure 29.2). From the arteries, the blood passes into a network of arterioles, or small arteries. From these, it eventually is forced through the capillaries, a fine latticework of very narrow tubes that get their name from the Latin word *capillus*, meaning "a hair." As blood passes through these capillaries, gases are exchanged between the blood and the tissues, nutrients are delivered to the tissues, and wastes are picked up from the tissues. After its journey through the capillaries, the blood passes into the venules (VAYN-yooles), or small veins. A network of venules and larger veins collects the blood and carries it back to the heart.

### Arteries and arterioles

Contrary to what many people think, arteries and veins are not defined by the type of blood they carry—oxygenated or deoxygenated—but by the direction in which they are carrying blood. **Arteries** are vessels that carry blood away from the heart. **Veins** are vessels that carry blood to the heart.

The largest artery in your body is the aorta, which is the main vessel that takes blood from the heart to the body. The vessels leading from the heart expand slightly in diameter during a heartbeat and then recoil before the next heartbeat. You can feel this expansion and recoil of an artery as you take your pulse at your wrist or neck.

The walls of the arteries are made up of three layers of tissue surrounding a hollow core called the *lumen* (**Figure 29.4a**). *En-*

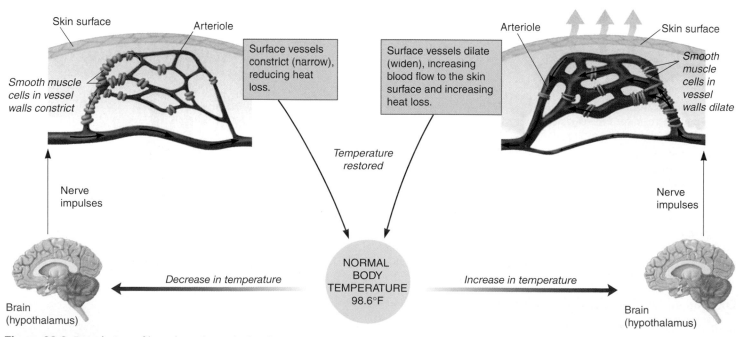

**Figure 29.3** Regulation of heat loss through the skin.

*dothelial cells* (the inner epithelium of blood vessels and the heart) line arteries and are in contact with the blood. Surrounding these cells is a thick layer of elastic fibers and smooth muscle. The elastic tissue allows the artery to expand and recoil in response to the pulses of blood. The steady contraction of the muscle strengthens the wall of the vessel against overexpansion. This layer of muscle and elastic tissue is encased within an envelope of protective connective tissue.

**Arterioles** are smaller in diameter than arteries. Their muscles tighten or relax in response to messages from nerves and hormones, which controls blood flow and blood pressure. As the arterioles dilate, the resistance to blood flow decreases, and blood pressure falls. Conversely, as arterioles constrict, resistance to blood flow increases, and blood pressure rises. The arterioles are the main site of regulation of blood pressure. As arterioles near the capillaries, their diameter decreases until they consist of only a layer of endothelium wrapped with a few scattered smooth muscle cells.

### Capillaries

**Capillaries** are microscopic blood vessels that connect arterioles with venules. They are little more than simple tubes with walls one cell thick and are barely long enough to stretch across the head of a pin. The internal diameter of a capillary is about the same as that of a red blood cell, causing these cells to squeeze through the capillaries in single file (Figure 29.4*b*). The closeness between the walls of the capillaries and the membranes of the red blood cells facilitates the diffusion of gases, nutrients, and wastes.

Your entire body is permeated with a fine mesh of capillaries, networks that amount to several thousand kilometers in overall length. If all the capillaries in your body were laid end to end, they would extend across the United States! These networks of capillaries are called *capillary beds*. In a capillary bed, some of the capillaries are called *thoroughfare channels* (**Figure 29.5**). From these channels, loops of true capillaries—those not on the direct flow route from arterioles to venules—leave and return. Almost all exchanges between the blood and the cells of the body occur through these loops. A ring of muscle called a precapillary sphincter guards the entry to each loop, and when contracted, it blocks flow through the capillary.

Restricting blood flow in capillary beds near the surface of the skin and constricting surface arterioles provide powerful means by which the body limits heat loss (see Figure 29.3). The body also

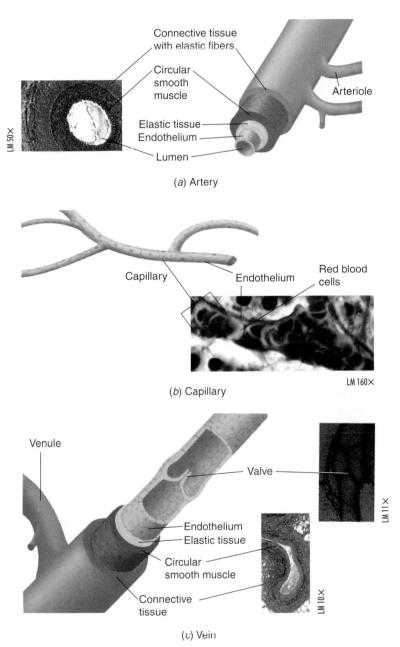

(a) Artery

(b) Capillary

(c) Vein

**Figure 29.4** The structure of blood vessels.

**Visual Thinking:** Study the structure of the valve in the vein. The function of this valve is to prevent backflow of blood in the vein, such as when blood is moving up the legs. How does the structure of the valve relate to its function?

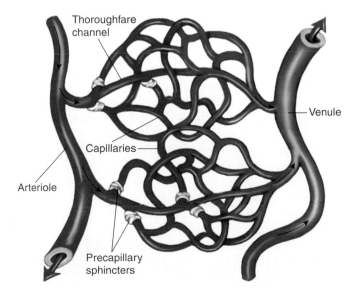

**Figure 29.5** Arterioles and venules are connected by capillary beds. Precapillary sphincters regulate the flow of blood through capillary beds.

cuts down on the flow within a capillary bed when temperature and carbon dioxide levels fall in the muscles, for example, when you are resting. In like manner, the body increases the flow within a capillary bed when temperature and carbon dioxide levels rise in the muscles during exercise. Interestingly, you do not have enough blood to fill all your capillary beds at the same time. If they were all open at the same time, you would faint because of insufficient blood to the brain.

### Veins and venules

**Venules** are small veins that collect blood from the capillary beds and bring it to larger veins that carry it back to the heart. The walls of veins, although similar in structure to those of the arteries, have a much thinner layer of muscle and elastic fiber (Figure 29.4*c*). The force of the heartbeat is greatly diminished by the time the blood reaches the veins, so these vessels do not experience pulsing pressures. This lack of pressure means that the pathway of blood back to the heart from much of the body is an uphill struggle.

How is blood moved against gravity from your foot back to your heart for example? As skeletal muscles in your leg contract, they press on veins and help move blood along. Larger veins contain one-way valves that prevent backflow and keep blood moving toward the heart (Figure 29.4*c*). If some of the valves in a vein are weak, however, gravity can force blood back through these valves, overloading a portion of a vein and pushing its walls outward. These "stretched out" veins, called varicose veins, are often seen in the legs at the surface of the skin (**Figure 29.6**).

## 29.6 The heart is the pump of human circulation.

The pump of the human circulatory system is the **heart**, but it is two pumps in one (**Figure 29.7**). The left side of the heart has two connected chambers, the left atrium and the left ventricle. The right side of the heart also has two connected chambers, the right atrium and the right ventricle. The two sides, or pumps, of the heart are not directly connected with one another.

### Circulatory pathways

Let's trace the journey of blood around the body starting with the entry of oxygenated blood into the heart from the lungs. Oxygenated blood from the lungs enters the left side of the heart, emptying directly into the upper left chamber of the heart, the **left atrium 1**, through large vessels called the **pulmonary veins**. These are unusual in that they carry oxygenated blood; other veins, because they carry blood back to the heart from the body tissues, carry deoxygenated blood. The word "pulmonary" refers to the lungs. The circulation of blood to and from the lungs is therefore called **pulmonary circulation**.

From the left atrium of the heart, blood flows through a one-way valve, the *left atrioventricular valve* (*bicuspid valve*), into the lower, adjoining chamber, the **left ventricle 2**. Most of this flow—roughly 70%—occurs while the heart is relaxed. The atrium then contracts, filling the remaining 30% of the ventricle with its blood. After a slight delay, the ventricle contracts. The walls of the ventricle are far more muscular than those of the atrium; note the relative thickness in Figure 29.7. As a result of its

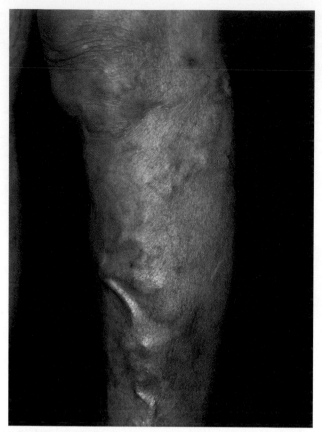

**Figure 29.6** Varicose veins. When skeletal muscles are inactive, such as during prolonged periods of sitting or standing, veins become dilated and valves may not operate properly. Defective valves allow blood to pool, causing varicose veins.

thick muscular wall, the ventricle contracts much more strongly than the atrium. The blood is prevented from going back into the atrium by the atrioventricular valve, whose flaps are pushed shut after the ventricle fills. Strong fibers are attached to the edges of the valve and prevent the flaps from moving too far when closing. If the flaps moved too far, they would project into the atrium and allow backflow. Prevented from reentering the atrium, the blood takes the only other way out of the contracting left ventricle: an opening that leads into the largest artery in the body—the **aorta 3**. The aorta is closed off from the left ventricle by a one-way valve, the *aortic semilunar valve*, shown in **Figure 29.8**. It is oriented to permit the flow of the blood out of the ventricle, but it snaps shut in response to this backflow.

Many arteries branch from the aorta, carrying oxygen-rich blood to all parts of the body. The pathway of blood vessels to the body regions and organs other than the lungs is called the **systemic circulation** (Figure 29.7 **4**). The first arteries to branch off the aorta are the coronary arteries, which carry freshly oxygenated blood to the heart itself. From the arch of the aorta, the *carotid arteries* branch off and bring blood to networks of vessels in the neck and head. The *subclavian arteries* bring blood to the shoulders and arms. The aorta then descends down the trunk of the body, with arteries branching off to supply various organs such as the kidneys, liver, and intestines. The aorta divides into two major vessels at the lower back, one traveling to each leg.

## Visual Summary

**Figure 29.7** Blood flow through the human heart.

**Visual Thinking:** The wall of the left ventricle has a thicker musculature than that of the right ventricle. How does this structure relate to the function of the left ventricle?

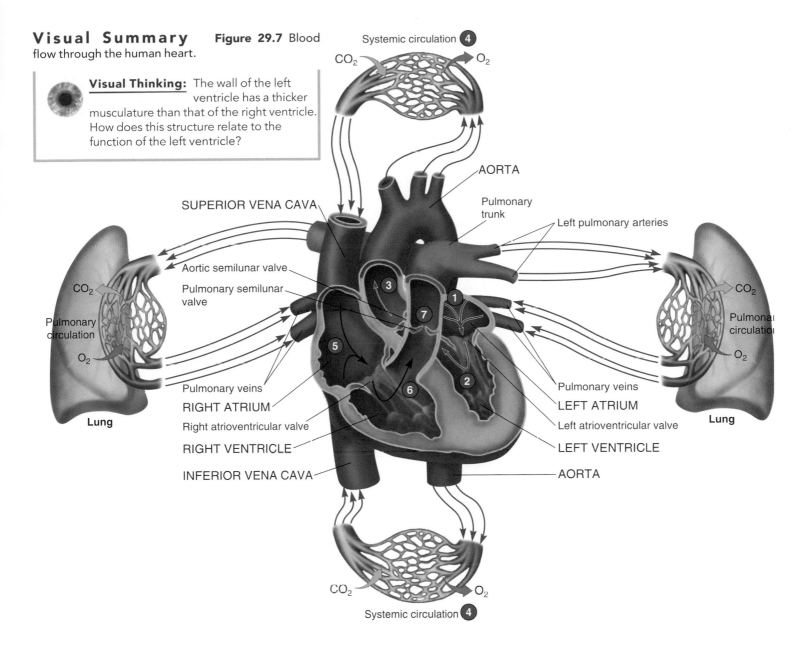

The blood that flows into the arterial system eventually returns to the heart after flowing through the capillaries. As it returns, blood passes through a series of veins, eventually entering the right side of the heart. Two large veins collect blood from the systemic circulation. The **superior vena cava** (VEE-nuh KAY-vuh *or* VAY-nuh KAH-vuh) drains the upper body, and the **inferior vena cava** drains the lower body. These veins dump deoxygenated blood into the **right atrium** ⑤.

The right side of the heart is similar in organization to the left side. However, the muscular walls of the right ventricle are not as thick as those of the left ventricle. Blood passes from the right atrium into the **right ventricle** ⑥ through a one-way valve, the *right atrioventricular valve* (*tricuspid valve*). It passes out of the contracting right ventricle through a second valve, the *pulmonary semilunar valve*, into a single pulmonary artery ⑦, sometimes called the **pulmonary trunk**, which subsequently branches into arteries that carry deoxygenated blood to the lungs. The blood then

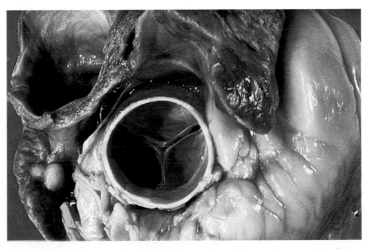

**Figure 29.8.** The aortic semilunar valve. This valve is at the base of the lumen of the aorta, which has been cut away.

returns from the lungs to the left side of the heart, replenished with oxygen and cleared of much of its load of carbon dioxide.

### How the heart contracts

Contraction of the heart depends on a small cluster of specialized cardiac muscle cells embedded in the upper wall of the right atrium (**Figure 29.9**). This cluster of cells, called the **sinoatrial (SA) node** (sye-no-AY-tree-uhl), automatically and rhythmically sends out impulses that initiate each heartbeat. The SA node is therefore nicknamed the "pacemaker" of the heart, but this pacemaker is modulated by both nerve signals and hormones.

The impulse initiated by the SA node causes the left and right atria to contract simultaneously and also excites a bundle of cardiac muscle cells located at the base of the atria. These cells are known as the **atrioventricular (AV) node** (AY-tree-oh-ven-TRIK-yuh-lur). The AV node conducts the impulse to a strand of specialized muscle in the **septum**, the tissue that separates the two sides of the heart. This strand of impulse-conducting muscle, known as the *atrioventricular (AV) bundle*, branches to the right and left. On reaching the lower tip of the heart, each branch further divides into conducting fibers called *Purkinje fibers* (per-KIN-gee), which initiate the almost simultaneous contraction of all the cells of the right and left ventricles.

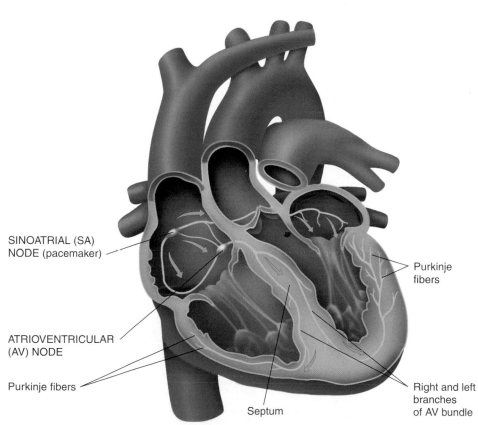

SINOATRIAL (SA) NODE (pacemaker)

ATRIOVENTRICULAR (AV) NODE

Purkinje fibers

Septum

Purkinje fibers

Right and left branches of AV bundle

**Figure 29.9 The conduction system of the heart.** Four structures comprise the conduction system of the heart—the sinoatrial (SA) node, the atrioventricular (AV) node, the AV bundle, and Purkinje fibers. These specialized groups of cardiac muscle cells initiate and conduct an electrical impulse throughout the heart. The impulse begins in the SA node resulting in simultaneous contraction of the atria, and spreads to the AV node. The AV node then initiates a signal that is conducted through the ventricles by way of the AV bundle and Purkinje fibers causing them to contract.

---

> ### CONCEPT CHECKPOINT
>
> **2.** Beginning with the vena cava, list the following structures in the order they would be encountered by a red blood cell traveling through your body: right atrium; pulmonary artery; left atrium; pulmonary vein; right atrioventricular valve; left atrioventricular valve; right ventricle; aortic semilunar valve; pulmonary semilunar valve; left ventricle; lungs; vena cava; aorta
>
> **3.** In which chambers/blood vessels noted in question 2 is the blood deoxygenated? In which vessel would you expect blood pressure to be highest?

## 29.7 Physicians can monitor the heartbeat in various ways.

The heartbeat is really a cardiac cycle of contraction and relaxation—a series of events that occurs in a predictable order. A physician can gain information about the health of the heart and events occurring during the heartbeat in several ways. The simplest is to listen to the heart at work by placing a stethoscope on the skin over the heart.

The first sound heard in a heartbeat is a low-pitched *lubb* caused by the turbulence in blood flow created by the closing of the atrioventricular valves at the start of ventricular contraction. A little later, a higher-pitched *dupp* can be heard, signaling the closing of the pulmonary and aortic semilunar valves at the end of ventricular contraction. If the valves do not operate properly, a sloshing sound known as a heart murmur can be heard.

A second way to examine the events of the heartbeat is to monitor blood pressure, the force exerted by the blood on vessel walls. While the atria are filling and contracting, the pressure in the arteries leading from the left side of the heart decreases slightly because they are not receiving blood from the left ventricle. This period of relaxation in the ventricles is referred to as ventricular **diastole** (dye-AS-tl-ee). The force the blood exerts on the arteries at this time is the diastolic pressure. During contraction of the ventricles, a pulse of blood is forced into the systemic arterial system by the left ventricle, immediately raising the blood pressure within these vessels. This period of contraction of the ventricles ends with the closing of the aortic semilunar valve and is referred to as ventricular **systole** (SIS-tl-ee). The force the blood exerts on the arteries at this time is the systolic pressure.

Blood pressure is measured in units called millimeters (mm) of mercury (Hg). These units refer to the height to which a column of mercury is raised in a narrow tube by the force of a beating heart. Blood pressure is expressed as the systolic pressure over the diastolic pressure, such as 110 mm Hg over 70 mm Hg. According to the American Heart Association, a normal blood pressure is less than 120 mm Hg over 80 mm Hg.

When the inner walls of arteries accumulate fats, as they do in the condition known as atherosclerosis (ATH-uh-ROW-skluh-ROW-sis), the diameters of the passageways are narrowed. Such narrowing is one cause of *high blood pressure* or **hypertension**. Hypertension describes people with blood pressures of 140/90mm Hg or above, or of 130/80 mm Hg or above when coupled with diabetes or kidney disease. Prehypertension describes people with blood pressures between 120 and 139 mm Hg systolic or 80 to 89 mm Hg diastolic. Why is hypertension a dangerous condition? The higher the blood pressure, the greater the risk of heart attack, heart failure, stroke, and kidney disease. Section 29.11 discusses prevention of this condition.

Another way to monitor the events of a heartbeat is to measure the electrical changes that take place as the heart's chambers both contract and relax. Because so much of the human body is made up of water carrying dissolved ions, it conducts electrical currents rather well. Therefore, as the impulses initiated at the SA node pass throughout the heart as an electrical current, this current also passes in a wave throughout the body. Although the magnitude of this electrical pulse is small, it can be detected with sensors placed on the skin. A recording made of these impulses (**Figure 29.10**) is called an **electrocardiogram (ECG)**.

Three electrical events are recorded in a normal heartbeat. The first event occurs just prior to atrial contraction; this is called the *P wave* on the ECG **1**. There is a much stronger electrical event called the *QRS complex* 0.2 seconds later **2**, which is the stimulus for contraction of the ventricles. Finally, a third event called the *T wave* occurs **3**, caused by the electrical changes that precede the relaxation of the ventricles. The electrical events that precede relaxation of the atria are masked by the much stronger QRS complex.

An ECG may identify erratic heartbeats or arrhythmias. Maybe you have experienced feelings of your heart racing, "skipped beats," or "fluttering" in your chest. Such things as exercise, emotional upsets, stress, and caffeine can bring on such heartbeat changes. Arrhythmias are extremely common, even among young adults, and usually are not cause for alarm. However, arrhythmias that occur frequently or cause severe symptoms, such as dizziness, chest pain, passing out, or shortness of breath should be discussed with a physician. An ECG can also detect damage to the heart caused by a heart attack (see Section 29.11).

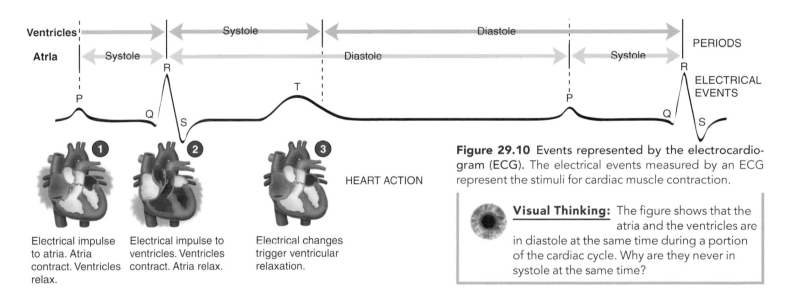

Electrical impulse to atria. Atria contract. Ventricles relax.

Electrical impulse to ventricles. Ventricles contract. Atria relax.

Electrical changes trigger ventricular relaxation.

HEART ACTION

**Figure 29.10** Events represented by the electrocardiogram (ECG). The electrical events measured by an ECG represent the stimuli for cardiac muscle contraction.

 **Visual Thinking:** The figure shows that the atria and the ventricles are in diastole at the same time during a portion of the cardiac cycle. Why are they never in systole at the same time?

# What makes up the blood?

## 29.8 Blood plasma is the fluid portion of the blood.

Blood is comprised of plasma, cells, and platelets. Blood **plasma** is a straw-colored liquid made up of water and dissolved substances. The dissolved substances can be grouped into three categories:

• *Nutrients, hormones, respiratory gases, and wastes.* Dissolved within the plasma are substances that are used or produced by the metabolism of cells. These substances include glucose, certain lipids, amino acids, vitamins, hormones, and the respiratory gases.

• *Salts and ions.* Plasma is a dilute salt solution. Chemically, the word "salt" is applied to any substance composed of positively and negatively charged ions (see Chapter 3). In water, salts dissociate into their component ions. The chief plasma ions are sodium ($Na^+$), chloride ($Cl^-$), and bicarbonate ($HCO_3^-$). In addition, there are trace amounts of other ions, such as calcium ($Ca^{2+}$), magnesium ($Mg^{2+}$), zinc ($Zn^{2+}$), and potassium ($K^+$). In living systems, these ions are called *electrolytes*.

# How Science Works

## The development of heart assist devices

The heart is an amazing living pump. It moves about 5 liters (5.3 quarts) of blood through each of its two sides per minute. The heart beats, on average, 72 times per minute, or over 100,000 times per day. So if you are about 20 years old, your heart has beat about 750 million times!

Physicians have not only constructed replacement parts for the vessels and valves of the heart, but they also have designed replacements for the heart itself. The Jarvik-7 was the first long-term-use total artificial heart (TAH), developed in 1982. The Jarvik-7 helped seriously ill heart patients live for months or even years with the mechanical device pumping their blood. After its first use, the Jarvik-7 was implanted in approximately 180 patients, but because of problems ranging from infection to strokes induced by blood clots, the Jarvik-7 was taken off the market in 1990.

In 2001, the U.S. Food and Drug Administration (FDA) approved for clinical trials a new artificial heart called AbioCor. It is the first artificial heart to be totally contained within the chest, different from the Jarvik-7, which was connected to a power source by tubes that exited and reentered the body. The AbioCor is intended for critically ill heart patients who are ineligible for human heart transplants because of other medical problems such as diabetes. Weighing 3 pounds, the artificial heart replaces the two ventricles and is the size of two fists. Its internal battery is con-stantly recharged wirelessly by an external battery pack worn on a belt. In late 2003, the FDA granted approval of the heart as a Humanitarian Use Device, a first step in making the artificial heart widely available.

Another pumping device, called the Left Ventricular Assist Device (LVAD), has

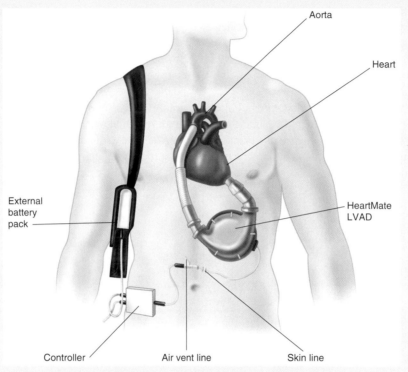

**Figure 29.A** The HeartMate left ventricular assist device. The pump, implanted within the body and connected to battery packs outside the body, performs the job of the left ventricle. (The left-side battery pack has been omitted from the illustration for clarity.)

also been helping seriously ill heart patients. The first human use of an LVAD occurred in 1963. Early LVADs were not implanted and were used temporarily to help a patient's heart recover from surgery. LVADs take over the work of the left ventricle, which pumps blood to the body. Early LVADs worked well, and in the early 1970s, medical researchers developed implantable LVADs.

During the 1990s, LVADs became widely used. A few types exist, but, in general, an LVAD is about the size of a man's fist and sits in the upper part of a patient's abdomen. A tube shunts blood from the left ventricle of the heart to the pump, and another tube sends blood from the pump to the aorta. The pump is connected to a battery pack and computerized control system worn outside the body. Some patients live for more than three years with these temporary devices.

Recently, advances have been made in the development of LVADs that allow them to be considered permanent assist devices. In 2001, the FDA approved for clinical trials the first fully implantable LVAD powered by wireless electric transmission. No wires, tubes, or other connections protrude through the skin to the battery pack, which lessens the risk of infection and makes the device more comfortable. Named the LionHeart, this LVAD is intended to help persons who need a human heart transplant, but are not eligible for one. Clinical trials were ongoing in the United States in 2003. In late November 2003, the device was approved for use in Europe.

In November 2002, the FDA approved a LVAD called the HeartMate for permanent implantation in terminally ill patients who are not eligible for heart transplant. **Figure 29.A** shows where the HeartMate is implanted and how it is connected to the external battery source.

Progress in the development and successful use of artificial hearts and assist devices has been significant over the past few decades. Although "high-tech" and amazing, the artificial heart and LVADs are still not as amazing as the body's own remarkable pump—the heart. ●

---

Electrolytes serve three general functions in the body. First, many are essential for various physiological processes such as the proper functioning of muscles and nerves, and the proper formation of bone. Second, they play a role in the movement of water—osmosis—between various compartments within the body. Third, they help establish acid-base balance, maintaining the proper pH required for normal cellular activities.

- *Proteins.* Blood plasma is approximately 90% water, but contains a concentration of proteins that helps balance osmotic pressure between the cells and the blood. The most abundant protein in blood is serum albumin, which acts as the primary osmotic counterforce. Other proteins in the plasma include antibodies and other proteins that are active in immunity, and fibrinogen and prothrombin, key players in blood clotting.

## 29.9 Blood cells and platelets are suspended in the plasma.

Although blood is liquid, 45% of its volume is actually occupied by cells and pieces of cells, collectively called **formed elements**. There are three principal types of formed elements in the blood: erythrocytes (red blood cells), leukocytes (white blood cells), and platelets (**Figure 29.11**).

### Erythrocytes

In only one teaspoonful of your blood, there are about 25 billion **erythrocytes** (ih-RITH-row-sites), or red blood cells. Almost the entire interior of each red blood cell is packed with the oxygen-car-

rying molecule hemoglobin. One estimate claims that each erythrocyte can carry 280 million molecules of hemoglobin!

Mature erythrocytes do not have nuclei or the ability to manufacture proteins. Red blood cells are therefore unable to repair themselves and consequently have a rather short lifespan—only about four months. New erythrocytes are constantly synthesized and released into the blood by cells within the soft interior marrow of bones at the amazing rate of 2 million per second!

### Leukocytes

Less than 1% of the cells in human blood are **leukocytes** (LOO-ko-sites), or white blood cells. Leukocytes are larger than red blood cells; they have nuclei, contain no hemoglobin, and are essentially

**Visual Summary**     Figure 29.11 Types of blood cells.

| | BLOOD CELL TYPE | DESCRIPTION | FUNCTION | LIFE SPAN |
|---|---|---|---|---|
| RED BLOOD CELLS | Erythrocyte | Flat disk with a central depression; no nucleus; contains hemoglobin. | Transports oxygen ($O_2$) and carbon dioxide ($CO_2$). | About 120 days. |
| WHITE BLOOD CELLS (LEUKOCYTES) — GRANULOCYTES | Neutrophil | Spherical; many-lobed nucleus; no hemoglobin; pink-purple staining cytoplasmic granules. | Cellular defense—phagocytosis of small microorganisms. | Hours to 3 days. |
| | Eosinophil | Spherical; two-lobed nucleus; no hemoglobin; orange-red staining cytoplasmic granules. | Cellular defense—phagocytosis of large microorganisms such as parasitic worms; releases anti-inflammatory substances in allergic reactions. | 8 to 12 days. |
| | Basophil | Spherical; generally two-lobed nucleus; no hemoglobin; large purple staining cytoplasmic granules. | Inflammatory response—contains granules that rupture and release chemicals enhancing inflammatory response. | Hours to 3 days. |
| WHITE BLOOD CELLS (LEUKOCYTES) — AGRANULOCYTES | Monocyte | Spherical; single nucleus shaped like kidney bean; no cytoplasmic granules; cytoplasm often blue in color. | Converted to macrophages, which are large cells that entrap microorganisms and other foreign matter. | Days to months. |
| | B-lymphocyte | Spherical; round single nucleus; no cytoplasmic granules. | Immune system response and regulation; antibody production; sometimes causes allergic response. | Days to years. |
| | T-lymphocyte | Spherical; round single nucleus; no cytoplasmic granules. | Immune system response and regulation; cellular immune response. | Days to years. |
| PLATELETS | Platelets | Irregularly shaped fragments; very small pink staining cytoplasmic granules. | Control blood clotting or coagulation. | 7 to 10 days. |

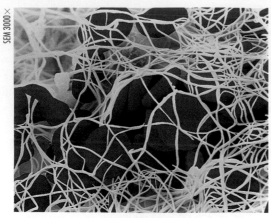

SEM 3000×

**Figure 29.12 Formation of a clot.** Many red blood cells will become caught in this net of insoluble protein forming a clot.

colorless. There are several kinds of leukocytes. Each has a different function, but all are related to the defense of the body against invading microorganisms and other foreign substances (see Chapter 30).

There are two major groups of leukocytes: **granulocytes** (GRAN-yuh-low-sites) and **agranulocytes**. Granulocytes are circulating leukocytes and get their name from the tiny granules in their cytoplasm. In addition, they have lobed nuclei (see Figure 29.11). The granulocytes are classified into three groups by their staining properties. About 50% to 70% of them are **neutrophils**. These cells migrate to the site of an injury and stick to the interior walls of the blood vessels. They then form projections that enable them to push their way into the infected tissues, where they engulf, or *phagocytize*, microorganisms and other foreign particles in a process called phagocytosis. **Basophils** (BAY-soh-filz), a second kind of granulocyte, contain granules that rupture and release chemicals that enhance the body's response to injury or infection. They play a role in causing allergic responses. The third kind of granulocyte, the **eosinophils** (EE-oh-SIN-oh-filz), are involved in allergic reactions and act against certain parasitic worms.

Agranulocytes, the second major group of leukocytes, have no cytoplasmic granules, nor are their nuclei lobed. One group of agranulocytes, the **monocytes**, circulate as the granulocytes do. Monocytes are attracted to sites of injury or infection, where they are converted into **macrophages** (MAK-row-FAY-djus), enlarged amebalike cells that entrap microorganisms and particles of foreign matter by phagocytosis. They usually arrive after the neutrophils and clean up any remaining bacteria and dead cells. **Lymphocytes**, the other type of agranulocyte, recognize and react to substances that are foreign to the body, sometimes producing a protective immunity to disease.

### Platelets

Certain large cells within the bone marrow called *megakaryocytes* regularly pinch off bits of their cytoplasm. These cell fragments, called **platelets**, enter the bloodstream and play an important role in blood clotting, or coagulation. The clotting of blood is a complicated process initiated by damage to blood vessels or tissues.

When an injury occurs, platelets clump at the damaged area, temporarily blocking blood loss. The damaged tissues and platelets release a complex of substances called *thromboplastin*, which interacts with calcium ions, vitamin K, and other clotting factors to form an enzyme called *prothrombin activator*. Prothrombin activator brings about the conversion of prothrombin to thrombin. Thrombin

converts fibrinogen, a soluble protein, to fibrin, an insoluble, thread-like protein. Fibrin threads, along with trapped red blood cells (**Figure 29.12**), form the clot—a plug at the damaged area of a blood vessel so that blood cannot escape.

## 29.10 Some fluid that leaves the blood returns via the lymphatic system.

Although the blood proteins and electrolytes help maintain an osmotic balance between the blood and the tissues, the blood loses more fluid to the tissues than it reabsorbs from them as it passes through the capillaries. Of the total volume of fluid that moves from the blood into and around the tissues, about 90% reenters the cardiovascular system. Where does the other 10% go? The answer is to the body's one-way, passive circulatory system, the **lymphatic system** (lim-FAT-ik) (**Figure 29.13**). This system counteracts the effects of net fluid loss from the blood.

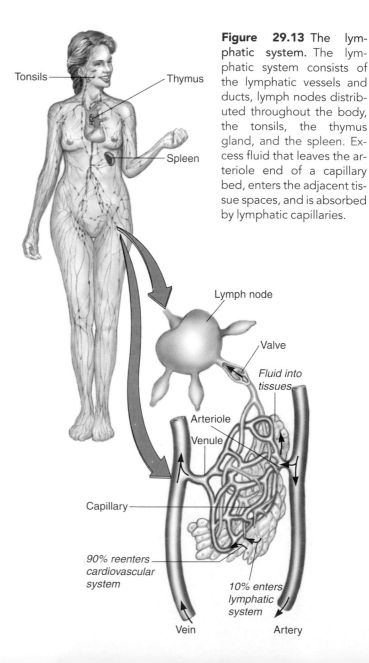

**Figure 29.13 The lymphatic system.** The lymphatic system consists of the lymphatic vessels and ducts, lymph nodes distributed throughout the body, the tonsils, the thymus gland, and the spleen. Excess fluid that leaves the arteriole end of a capillary bed, enters the adjacent tissue spaces, and is absorbed by lymphatic capillaries.

Tonsils

Thymus

Spleen

Lymph node

Valve

*Fluid into tissues*

Arteriole

Venule

Capillary

*90% reenters cardiovascular system*

*10% enters lymphatic system*

Vein

Artery

By osmosis and diffusion, blind-ended lymphatic capillaries fill with tissue fluid, including small proteins that have diffused out of the blood. From these capillaries, the tissue fluid—now called **lymph** (LIMF)—flows through a series of progressively larger vessels to two large lymphatic vessels, which structurally resemble veins. These vessels drain into veins near the base of the neck. Although the lymphatic system has no heart to pump lymph through its vessels, the movement of the body's muscles squeezes the fluid through them. Like veins, the lymphatic vessels contain a series of one-way valves (see Figure 29.13) that permit movement only in the direction of the neck.

Small, ovoid, spongy structures called **lymph nodes** are located in various regions of the body along the routes of lymphatic vessels. They are clustered in areas such as the groin, armpits, and neck, filtering the lymph as it passes through. Some lymphocytes, the cells that activate the immune response, reside in the lymph nodes. These cells are discussed in more detail in Chapter 30. In addition, the lymphatic system has two organs: the spleen and the thy-

mus. The *spleen* stores an emergency blood supply and also contains white blood cells. Specific types of white blood cells in the spleen destroy old red blood cells, filter microorganisms out of the blood as it passes through, and initiate an immune response against foreign microorganisms. The *thymus* plays an important role in the maturation of certain lymphocytes called *T cells*, which are an essential part of the immune system.

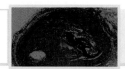

### CONCEPT CHECKPOINT

4. Oxygen is not very soluble in salty warm water, which is the main constituent of blood plasma. What is responsible for the relatively large amount of oxygen that can be carried in the blood?

5. Generally describe the structure and function of the formed elements of blood. Organize this information as an outline.

# What are some common cardiovascular diseases and conditions?

## 29.11 Cardiovascular diseases and conditions include atherosclerosis, heart attacks, and strokes.

Nearly 40% of all deaths in the United States are due to cardiovascular disease. More than 61 million people in this country—about one person in five—have some form of cardiovascular disease.

Atherosclerosis is a major cause of cardiovascular disease. **Atherosclerosis** is a thickening and hardening of the walls of the arteries in which masses of cholesterol and other lipids build up within the walls of large and medium-sized arteries (**Figure 29.14a**). These masses are referred to as **plaques** (PLAKS). The accumulation of plaques impairs the arteries' proper functioning. When this condition is severe, the arteries can no longer dilate and constrict properly, and the blood moves through them with difficulty. If a plaque ruptures, it can trigger a blood clot, or thrombus, to form, which can block blood flow (Figure 29.14b).

Atherosclerosis begins with damage to the endothelium of arteries. How does this inner layer of the artery wall become damaged? Damage can result from elevated levels of cholesterol and triglycerides in the blood, cigarette smoke, or high blood pressure. Subsequent inflammation (see Chapter 30), possibly from infectious bacteria, fuels the accumulation of fats, cholesterol, cellular debris, calcium, and other substances within vessel walls at the site. Physicians can now monitor inflammation in vessel walls by measuring the amount of C reactive protein in the blood. High levels of this protein as seen in a high sensitivity assay (hs-CRP) indicates an increased risk of heart attack or stroke.

*Heart attacks*, the most common cause of death in the United States, result from an insufficient supply of blood to an area of heart muscle. As a result, this area of heart muscle dies. A heart attack may result if a blood

clot forms in one of the coronary arteries that supplies the heart with blood, thereby blocking its blood supply. Alternatively, a floating blood clot, or *embolus*, may travel to the heart from another location and block the supply of blood to the heart muscle. In addition, if a vessel supplying the heart with blood is blocked sufficiently by fatty deposits, especially cholesterol and triglycerides, blood flow to the heart may be reduced or stopped, even without the complication of a blood clot. As heart muscle dies, it may cause the ventricles to beat irregularly or cause the heart to stop beating—a condition called cardiac arrest. This will result in death unless the person receives immediate medical care.

Signs of a heart attack are a pressure, squeezing, or pain in the center of the chest. This pain may extend to the neck, shoulder, or arms and is often accompanied by lightheadedness, nausea, sweating, and shortness of breath. *Those experiencing heart attack symptoms should be rushed to a hospital emergency room immediately or an*

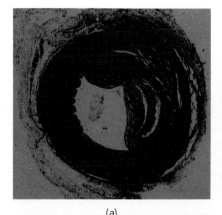
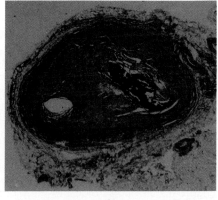

(a)  (b)

**Figure 29.14** Atherosclerosis. (a) Much of the passage of blood in the coronary artery is blocked by buildup of cholesterol and other lipids on the interior walls of the artery, forming masses or plaques. (b) The coronary artery is almost completely blocked. A dark red blood clot blocks blood flow on the right side of the artery.

*ambulance should be called.* Swift action can save the life of a heart attack victim. If a person reaches the hospital emergency room alive, he or she has a 95% chance of surviving their attack.

*Angina pectoris*, which literally means "chest pain," occurs for reasons similar to those causing heart attacks, but angina pain is not as severe as heart attack pain. In angina, a reduced blood flow to the heart muscle weakens the cells but does not kill them.

The amount of heart damage associated with a small heart attack may be relatively slight and thus difficult to detect. It is important that such damage be detected, however, so that the overall condition of the heart can be evaluated properly. Electrocardiograms are very useful for this purpose because they reveal abnormalities in the timing of heart contractions—abnormalities that are associated with the presence of damaged heart tissue. Damage to the atrioventricular (AV) node, for example, may delay as well as reduce the QRS wave in an ECG recording, indicating an abnormal contraction of the ventricles. Unusual conduction routes around dead tissue may lead to continuous disorganized contractions (arrhythmias) called *fibrillations*. In many fatal heart attacks, ventricular fibrillation is the immediate cause of death.

What can you do to lessen your risk of heart disease and heart attacks? **Table 29.1** lists risk factors for cardiovascular disease, and **Table 29.2** lists ways to reduce your risk of developing this disease. Following heart-healthy behaviors throughout your life, even when you are young (see the *Just Wondering* box), will help you avoid heart disease and its debilitating effects when you are older. Scien-

| TABLE 29.1 | Risk Factors for Cardiovascular Disease |
|---|---|

- Cigarette smoking and exposure to cigarette smoke
- Heavy alcohol use
- Physical inactivity
- Obesity (20% or more above the desirable weight for your height)
- Total serum cholesterol over 200 mg/dl
- High levels of "bad" (LDL) cholesterol (>100 mg/dl)
- Low levels of "good" (HDL) cholesterol (<40 mg/dl)
- High levels of C-reactive protein
- Family history—having a male first-degree relative who had a heart attack before age 55 or a female first-degree relative who had a heart attack before age 65
- Chronic high blood pressure (hypertension) 140/90 or higher
- Gender—women are at lower risk than men until menopause
- Increased age
- Elevated blood glucose levels present in diabetes mellitus
- Anxiety disorders

## *just wondering . . .*

### Questions students ask

**I often hear about young people who die suddenly of a heart attack even though they exercised and ate right. Why is that?**

As you have discovered, there is no guarantee that even if you follow health recommendations regarding cardiovascular disease, you will be free of this disease and the threat of heart attacks and strokes. Nevertheless, following health recommendations helps you reduce your risk of all types of cardiovascular disease, the number one killer of Americans. The more of the risk factors listed in Table 29.1 you have, the more likely you are to have a heart attack or stroke. Risk reduction behaviors are noted in Table 29.2.

So why does someone die from a heart attack when their risk factors are low? In some cases, persons are discovered, during autopsy perhaps, to have congenital heart abnormalities—problems of heart structure or function with which they were born and of which they were unaware. Suddenly, under some physical or emotional time of stress, their heart fails. Researchers have discovered that the vast majority of sudden deaths among athletes under the age of 35 are due to one of several congenital heart abnormalities.

One inherited heart disorder that can cause sudden death is arrhythmogenic right ventricular dysplasia, or ARVD. In persons with ARVD, muscle cells die in the wall of the right ventricle—the chamber that pumps blood to the lungs. Sudden death can then occur from arrhythmias in this weakened heart. Even if a person with ARVD does not die suddenly, his or her heart eventually weakens so much that it cannot deliver sufficient oxygen to the organs.

When the heart begins to fail in this way, death occurs within a few years.

Another inherited heart disorder that can cause sudden death is left ventricular hypertrophy (LVH), which simply means an enlarged heart. The gene mutation for this condition is found in about 20% of the population. It is the reason that some people develop a dangerously enlarged heart after heavy exercise or as a side effect of hypertension. An enlarged heart can lead to ineffective pumping of the heart, or congestive heart failure. LVH can also cause *sudden cardiac death*—death resulting from an abrupt loss of heart function.

You may know persons who have had heart attacks at young ages—before age 45 in men and age 50 in women. Researchers have discovered that inherited mutant genes are usually the cause, producing substances or conditions that place individuals at risk.

Nevertheless, it is rare that young persons die suddenly of heart attacks. It may simply seem frequent because these occurrences are notable and are often reported in the news media, especially when the person is a sports figure or other celebrity. Some physicians contend that aspiring athletes should have more than routine physical examinations prior to participation in sports, since such exams are unlikely to uncover previously unseen problems in the heart or major blood vessels. To increase your probability of living a long and healthy life, follow recommendations you *can* control for cardiovascular disease as well as for other "big killers" such as cancer. ●

*Are you wondering about a topic in biology and how it relates to your life? Submit your question by clicking the Just Wondering link in this text's companion Web site at www.wiley.com/college/alters.*

| **TABLE 29.2** | **How to Reduce Your Risk of Cardiovascular Disease** |
|---|---|

- Get regular medical checkups.
- Do not smoke cigarettes or breathe "second-hand" smoke.
- Restrict alcohol consumption to no more than 1 drink (1 oz. ethanol) per day for women and 2 for men.
- Exercise regularly.
- Maintain an appropriate weight for your height.
- Eat a diet low in saturated and trans fats (see Chapter 27).
- Eat foods rich in soluble fiber such as fruits, beans, and oats.
- Maintain salt intake at less than one teaspoon (2400 mg) per day.
- Manage diabetes mellitus properly.

tists now know that fatty deposits can begin to develop in the arteries of children. Heart disease is not just a disease of older persons—it is a life-long process affected by diet and lifestyle. Heart-healthy eating patterns and other behaviors will help you maintain a healthy cardiovascular system throughout your life.

*Strokes* ("brain attacks") are caused by an interference with the blood supply to the brain. A stroke may also be caused by a thrombus, embolus, and/or fatty deposits blocking blood flow as in heart attacks. The effects of strokes depend on the severity of the damage and the specific location of the stroke. Some of the signs of a stroke are: weakness, numbness, or paralysis on one side of the body; visual dimming, particularly in one eye; difficulty speaking; sudden, severe headache; and sudden dizziness or falling. As with heart attack victims, *those experiencing stroke symptoms should seek medical attention immediately* to help reduce brain damage and increase the possibility of survival.

# CHAPTER REVIEW

## Summary of Key Concepts and Key Terms

### What patterns of circulation exist among organisms?

**29.1** Single-celled organisms do not have **circulatory systems** (**circulation**, p. 496); gases and nutrients move into the cell directly from the environment.

**29.2** In an **open circulatory system** (p. 496), blood mixed with tissue fluid bathes the internal tissues; blood is not enclosed in vessels; organisms such as insects, spiders, clams, and snails have open circulatory systems.

**29.3** In a **closed circulatory system** (p. 496), the blood is located within blood vessels as it is pumped throughout the body; closed circulatory systems are found in cephalopod mollusks, annelids, echinoderms, and vertebrates.

### How does human circulation work?

**29.4** Human circulation is a closed circulatory system, also called a **cardiovascular system** (p. 499).

**29.4** The cardiovascular system is made up of three components: the heart, the blood vessels, and the blood.

**29.4** The circulatory system transports nutrients, wastes, respiratory gases, and hormones, and plays an important role in temperature maintenance of the body.

**29.5** **Arteries** (p. 500) and **veins** (p. 500) are the major vessels of the circulatory system. Arteries have a thick layer of muscle and elastic tissue; veins have thinner walls and may contain one-way valves.

**29.5** **Arterioles** (p. 501) are smaller in diameter than arteries; their muscles tighten or relax in response to messages from nerves and hormones, controlling blood flow and blood pressure.

**29.5** **Capillaries** (p. 501) are microscopic blood vessels through which the exchange of substances between the cells and the blood takes place.

**29.5** **Venules** (p. 502) are small veins that collect blood from the capillary beds and bring it to larger veins that carry it back to the heart.

**29.6** The **heart** (p. 502) is a double pump, pushing both **pulmonary circulation** (p. 502) between the heart and the lungs, and **systemic circulation** (p. 502) between the heart and the rest of body.

**29.6** Oxygenated blood from the lungs enters the left side of the heart, emptying directly into the **left atrium** (p. 502), through large vessels called the **pulmonary veins** (p. 502).

**29.6** The **left ventricle** (p. 502) pumps blood into the **aorta** (p. 502), the largest artery in the body; from there it will travel to smaller arteries, then to capillaries, then to small veins.

**29.6** The **superior vena cava** (VEE-nuh KAY-vuh *or* VAY-nuh KAH-vuh, p. 503) drains the upper body, and the **inferior vena cava** (p. 503) drains the lower body; these veins dump deoxygenated blood into the **right atrium** (p. 503).

**29.6** The **pulmonary trunk**, (p. 503) branches into arteries that carry deoxygenated blood to the lungs from the **right ventricle** (p. 503).

**29.6** The contraction of the heart depends on a small cluster of specialized cardiac muscle cells in the upper wall of the right atrium, called the **sinoatrial (SA) node** (sye-no-AY-tree-uhl, p. 504), that initiate the heartbeat and cause cells of the right atrium and the left atrium to contract simultaneously.

**29.6** The impulse initiated by the SA node excites a bundle of cardiac muscle cells at the base of the atria called the **atrioventricular (AV) node** (AY-tree-oh-ven-TRIK-yuh-lur, p. 504), which conducts the impulse to a strand of specialized muscle in the **septum** (p. 000), the tissue that separates the two sides of the heart.

**29.6** Impulses from the atrioventricular (AV) bundle branch into conducting fibers that cause cells in the right ventricle and the left ventricle to contract simultaneously.

**29.7** A physician can gather data about the health of the heart by listening to its beats, by determining the blood pressure in arteries leaving the heart, or by conducting an **electrocardiogram** (p. 505).

**29.7** **Diastole** (dye-AS-tl-ee, p. 504) is a period of relaxation of the chambers of the heart; **systole** (p. 504) is a period of contraction.

**29.7** Chronic high blood pressure is called **hypertension** (p. 504).

### What makes up the blood?

**29.8** The liquid portion, or **plasma** (p. 505), of the circulating blood contains the proteins and ions that are necessary to maintain the blood's osmotic equilibrium with the surrounding tissues.

**29.9** The **formed elements** (p. 507) of the blood are the red blood cells (**erythrocytes**, ih-RITH-row-sites, p. 507), white blood cells (**leukocytes**, LOO-ko-sites, p. 507), and platelets.

**29.9** There are two major groups of leukocytes: **granulocytes** (GRAN-yuh-low-sites, p. 508) and **agranulocytes** (p. 508).

**29.9** The granulocytes are classified into three groups by their staining properties: **neutrophils** (p. 508), **basophils** (BAY-soh-filz, p. 508), and **eosinophils** (EE-oh-SIN-oh-filz, p. 508).

**29.9** Agranulocytes are of two types: **monocytes** (which are converted into **macrophages**, MAK-row-FAY-djus, p. 508) and **lymphocytes** (p. 508).

**29.9** The red blood cells transport oxygen, the white blood cells defend the body against disease, and the **platelets** (p. 508) are essential to the process of blood clotting.

**29.10** The **lymphatic system** (lim-FAT-ik, p. 508) gathers fluid from the body that has been lost from the circulatory system by osmosis and returns it as **lymph** (LIMF, p. 509) via a system of lymphatic capillaries, lymphatic vessels, and two large lymphatic ducts to veins in the lower part of the neck.

**29.10** **Lymph nodes** (p. 509) are clustered in areas such as the groin, armpits, and neck, filtering the lymph as it passes through.

### What are some common cardiovascular diseases and conditions?

**29.11** Cardiovascular diseases, diseases of the heart and blood vessels, are the leading cause of death in the United States.

**29.11** Major cardiovascular diseases and conditions are atherosclerosis, heart attacks, and strokes.

**29.11** **Atherosclerosis** (p. 509) is a thickening and hardening of the walls of the arteries in which masses of cholesterol and other lipids called **plaques** (PLAKS, p. 509) build up within the walls of large and medium-sized arteries; it is a major cause of cardiovascular disease.

---

## Level 1 — Learning Basic Facts and Terms

**Matching**

1. Which of the following is not considered a function of the vertebrate cardiovascular system?
   a. transport of respiratory gases
   b. temperature regulation
   c. regulation of osmotic balance of internal body fluids
   d. transporting hormones
   e. absorption of nutrients

2. Arteries are different from veins in which one of the following ways?
   a. Arteries carry oxygenated blood; veins carry deoxygenated blood.
   b. Arteries carry blood away from the heart to capillary beds; veins carry blood toward the heart from capillary beds.
   c. Arteries are thin-walled; veins are thick-walled.
   d. Arteries have one-way valves to prevent backflow of blood; veins have two-way valves to allow for backflow of blood.

3. The P wave of the electrocardiogram (ECG) represents
   a. the closing of the semilunar valves.
   b. the electrical event that stimulates the contraction of the atria.
   c. the electrical event that stimulates the contraction of the ventricles.
   d. the inappropriate backflow of blood into the atria.

4. The function of precapillary sphincters is to
   a. prevent arterioles from collapsing so that all capillary beds stay open and filled with blood.
   b. regulate the flow of blood to capillary beds.
   c. prevent blood from flowing backward in veins.
   d. return fluid back to the blood lost during capillary exchange.

5. The *lub-dupp* sound that the heart makes is most directly caused by
   a. the contraction of heart muscle.
   b. blood alternately filling the atria and ventricles of the heart.
   c. the electrical excitation of the heart.
   d. the alternating turbulence in blood flow during the closing of the atrioventricular valves and the semilunar valves.

**True–False**

6. _____ In a closed circulatory system, blood and tissues that bathe the internal organs mix to form a fluid known as hemolymph.

7. _____ The lymphatic system returns some of the fluid lost during capillary exchange to the circulatory system.

8. _____ Arrhythmias are nearly always an indication that a heart attack has occurred.

9. _____ A healthy diet, exercise, and not smoking are the best ways to prevent future coronary artery disease.

10. _____ The primary role of platelets in the blood is to fight off infections by microorganisms such as bacteria.

---

## Level 2 — Learning Concepts

1. You have just finished a great lunch of a chicken sandwich and a tossed salad with oil and vinegar dressing. Discuss what happens to the molecules of the food after they pass into your bloodstream.

2. What significant change occurs in the blood from the time it enters the capillaries until it enters the venules?

3. Mystery novels sometimes describe people who "turn pale with fear." Using your knowledge of the circulatory system, explain the reason for this.

4. Compare the structure of arteries, capillaries, and veins. Relate any differences to their respective functions.

5. The text states that the heart is actually two pumps in one; explain this duality and its importance.
6. Why are valves necessary in heart function?
7. What are the SA node and the AV node? Explain their significance.
8. List the names of all the formed elements of the blood and describe their functions. What is the function of blood plasma?

9. Two of the statements below describe events that occur during the diastolic phase of the heart cycle. Identify each statement and explain why you think it is true. Correct the other two statements to make them true as well.
   a. Arterial blood pressure is at its lowest.
   b. The right ventricle is contracting, and the left ventricle is relaxing.
   c. Both of the semilunar valves are closed.
   d. The AV node stimulates the simultaneous contraction of the ventricles.

## Level 3  Critical Thinking and Life Applications

1. Analyze your diet, lifestyle, and family history to determine your risk factors for cardiovascular disease. Develop a list of these factors.
2. Which risk factors on your list can you change? Which cannot be changed? What changes can you make in your diet and lifestyle to lower your risk of heart disease?
3. Figure 29.10 shows an electrocardiogram, the electrical events of the heart. Redraw the ECG and show where the heart sounds would occur. Support your answer with an explanation.
4. Insects are among the most active of all animals on Earth, yet they have a very inefficient open circulatory system. How can they survive?

**For questions 5–8, you are the physician.**
5. When a woman is diagnosed with breast cancer you remove some of her underarm lymph nodes. Explain to her why this procedure is useful in determining whether the cancer has spread.
6. While listening to the heart of one of your patients you hear the following sound: . . . lub . . . hiss . . . dupp, lub . . . hiss . . . dupp, etc. What heart disorder does this patient have? How would your diagnosis change if you heard: lub . . . dupp . . . hiss, lub . . . dupp . . . hiss, etc?
7. After receiving a bone marrow transplant, your patient must be quarantined in a germ-free room. Why is this necessary?
8. Elephantiasis is a tropical disease caused by a nematode worm that invades lymph vessels and blocks lymphatic drainage. What symptoms would a person experience if infected with this parasite?

## In The News  Critical Thinking

### BRIDGE TO RECOVERY

Now that you understand more about the circulatory system, reread this chapter's opening story about research concerning the effects of LVAD support in some people with heart failure. To better understand this research, it may help you to follow these steps:
1. Review your immediate reaction to research on predicting how heart tissue will respond to the device that you wrote when you began reading this chapter.
2. Based on your current understanding, again summarize the main point of the research in a sentence or two.
3. What questions do you now have about this research that are unanswered by this chapter's opening story?

4. Collect new information about the research. Visit the *In The News* section of this text's companion Web site at www.wiley.com/college/alters and watch the "Bridge to Recovery" video. Then use the "summary" link to read the accompanying story and access related links. Use this information, the links provided, and other online and library resources to answer your questions and find updates about this research topic. State the sources of your information. Explain why you think the information is accurate. Also determine whether the information expresses a particular point of view or is biased in any way.
5. What in your view are the most significant aspects of this research? Give reasons for your opinion and for any changes in your ideas based on the additional information you have collected and the analysis you have done.

# DEFENSE AGAINST DISEASE

## In The News | Cell Scouts

Communication among the cells of your body is vital for your health and well being. Why? One reason is that communication among cells triggers responses against invading *pathogens*, agents of infection that include certain bacteria, viruses, and fungi. Relaying such information is essential for your body to mount an effective defense.

Hidde Ploegh, a professor of pathology at Harvard Medical School, and a team of scientists used time-lapse video to record communication between dendritic cells and T cells, both of which are types of immune system cells. Dendritic cells are in the skin, intestinal tract, and other tissues in contact with the environment, where they serve as scouts poised to engulf and destroy bacteria and viruses almost as soon as they enter the body. After destroying the pathogen, dendritic cells travel to lymph nodes carrying molecules that contain molecular fragments from the invaders to T cells that have had previous contact with the same invasion force. These particular T cells respond to the dendritic cell's call to arms and spread throughout the body, locating invading pathogens that may have gotten through the first line of defense and preventing them from gaining a foothold in the body.

To observe how dendritic cells use molecular fragments to "tell" T cells about the presence of pathogens, Ploegh and his colleagues took the gene that codes for a

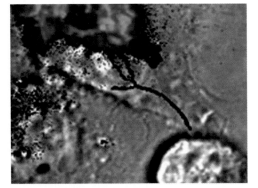

green fluorescent protein in certain jellyfish and inserted it into the genome of mice dendritic cells. When the genetically engineered live dendritic cells were examined under a special microscope, they glowed green; the researchers could easily identify them. The scientists were able to observe the dendritic cells transport the molecular fragments of the pathogens to nearby T cells via cell extensions called tubular endosomes (see photo). The scientists think the purpose of this transport is to convey information about the identity of specific pathogens to certain T cells. Ploegh calls this basic research into how the immune system works "essential" in finding new ways to combat disease. To learn more about this research and observe the dendritic cell's method of alerting T cells, view the video "Cell Scouts" by visiting the *In The News* section of this text's companion Web site at www.wiley.com/college/alters.

*Write your immediate reaction to this research on the communication between immune system cells: first, summarize the main point in a sentence or two; then suggest what you think its significance is. You will have an opportunity to reflect on your responses and gather more information on this topic in the* In The News *feature at the end of this chapter. In this chapter, you will learn more about the body's defense against pathogens.*

# CHAPTER GUIDE

## How do organisms defend themselves against disease?

**30.1** Humans and other vertebrates have both specific and nonspecific immunity.

**30.2** Invertebrates, protists, and plants have only nonspecific immunity.

**30.3** Nonspecific immunity includes mechanical, cellular, and molecular processes that combat any foreign invader.

**30.4** Specific immunity includes body structures, chemicals, cells, and processes that act against specific pathogens.

**30.5** Specific immunity includes cell-mediated and antibody-mediated responses.

**30.6** Antibody cell surface receptors recognize specific antigens.

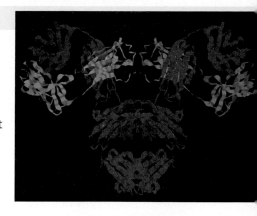

## How do vaccines work?

**30.7** Vaccines cause the body to develop specific immunity.

## Why is AIDS so deadly?

**30.8** Infection with HIV weakens the immune system.

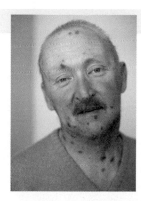

## What are allergies?

**30.9** Allergies are immune responses to usually harmless antigens.

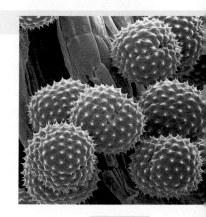

# How do organisms defend themselves against disease?

## 30.1 Humans and other vertebrates have both specific and nonspecific immunity.

Your body works in many ways to keep you healthy and free of infection. An *infection* results when microorganisms or viruses enter the tissues, multiply, and cause damage. **Immunity** is protection from disease, particularly infectious disease. You, and all other vertebrates, have two types of immunity: nonspecific and specific.

**Nonspecific immunity** works to keep out *any* foreign invader, just as the walls and roof of your home are barriers that protect you from any of the elements of the environment—rain, hail, wind, the sun's rays, insects, or the neighbor's dog, for example. Nonspecific immunity acts in the same way against any invader. In addition, it does not require previous exposure to an invader to be effective. Likewise, your roof does not have to have been rained upon in order for it to acquire the properties of being effective against rain.

Nonspecific immunity is also called *innate immunity* because all individuals are usually born with these nonspecific defenses. These defenses consist of mechanical and chemical barriers as well as cells that attack invaders in general. They will be discussed shortly in more detail (see Section 30.3).

The other set of defenses against disease in the human body is **specific immunity**. This specific immune response works against each particular type of microbe that may invade your body. A specific defense in your home, for example, may be a certain chemical you use to rid your kitchen of ants in the summer.

Specific immunity is also called *adaptive immunity* because it does not become active against a particular invader until the individual is first exposed to it. In other words, specific immunity is activated during a first infection and adapts to protecting you the next time you contact that same pathogen (disease-causing microbe). Put simply, the specific immune response has a memory. Although it works to protect you on first contact, the response is delayed; upon a subsequent contact with that same pathogen, specific immunity acts more quickly and is therefore more effective in thwarting disease. Specific immunity consists of cellular and molecular responses to particular foreign invaders, which will be discussed shortly (see Sections 30.4–30.6).

## 30.2 Invertebrates, protists, and plants have only nonspecific immunity.

Certain elements of immunity are found in almost all living things. Only vertebrates, however, have immune systems. Although there are differences among vertebrate immune systems, their mechanisms of specific immunity are remarkably similar. This chapter presents the human immune system and uses it as a model to explain how the vertebrate system works.

Invertebrates, on the other hand, have only nonspecific immunity. The cells of invertebrates do not cooperate during defensive reactions like the cells described in this chapter's opening story. In addition, they cannot recognize foreign molecules with specificity as can vertebrate immune systems. Their nonspecific defense systems work quite well for them, however, as defenses against infection.

One of the nonspecific defenses found in animals and protozoans is **phagocytosis** (fag-oh-si-TOE-sis). During this process, a host cell surrounds an invader, engulfing it (**Figure 30.1**). After being enclosed within the cell, the pathogen is digested. This same process

**Figure 30.1** Phagocytosis. Ameba engulfs *Pandorina morum.*

## FOCUS

### on Plants

**Figure 30.A** A tree's response to an infection.

Plants do not have immune systems, but they exhibit immunity against disease. Plants recognize pathogens, which is a key to immunity. They have a variety of cellular mechanisms that defend against these foreign invaders.

At the site of an infection, plants respond with the death of plant cells in the immediate area, the production of chemicals to kill the invader, the deposition of a substance to wall off the area, and the strengthening of cell walls. Notice how the tree in **Figure 30.A** has "sealed off" its injured tissues. In addition, certain types of pathogens induce defense responses in parts of a host plant far from the site of infection. They enhance the resistance of the plant against subsequent infection by a broad range of pathogens.

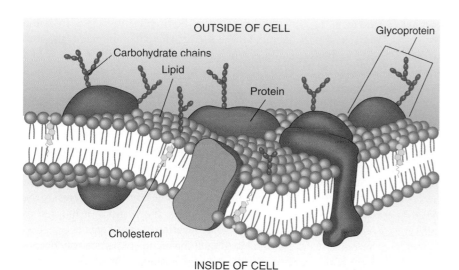

OUTSIDE OF CELL

Carbohydrate chains
Lipid
Protein
Glycoprotein
Cholesterol

INSIDE OF CELL

**Figure 30.2** The outer surface of a cell is studded with proteins, carbohydrates, and lipids. Glycoproteins (carbohydrate-protein complexes) often serve as specific cell surface markers that identify particular cell types and also identify the cell as "self."

| TABLE 30.1 | **The Nonspecific Defenses of the Human Body** |
|---|---|

- Mechanical barriers: skin and mucous membranes
- Lysozyme in tears and saliva
- Chemical barriers: acid environment of the stomach, urinary tract, and genital tract
- Normal microbiota
- Inflammatory response
- Phagocytic cells
- Natural killer cells
- Complement system
- Interferon

occurs in your body when white blood cells attack, engulf, and digest bacteria. This phagocytic process was observed as far back as 1882, when Elie Metchnikoff, the "father" of cellular immunology, observed phagocytes surrounding a rose thorn inserted into the transparent larva of a sea star.

Key to the process of phagocytosis—and to immunity—is the ability of an organism to distinguish its own cells from cells that are foreign—self from nonself. The human immune system distinguishes self from nonself by means of complex molecules present on the surface of foreign substances that are different from molecules present on host cells (**Figure 30.2**). These complex molecules, usually proteins or carbohydrates, elicit immune responses and are called **antigens**. How does your immune system recognize your own complex molecules as self? These molecules are continually present while your immune system cells mature. During this time, immature immune system cells that attack self molecules are destroyed, while immune cells that do not attack the body's own molecules are retained.

## 30.3 Nonspecific immunity includes mechanical, cellular, and molecular processes that combat any foreign invader.

The human body has a wide variety of nonspecific defenses, as shown in **Table 30.1**. Together, they slow the growth of a pathogen until activation of the specific immune response.

Mechanical barriers keep foreign substances and disease-causing microbes from entering the body. Unbroken skin is an effective mechanical barrier to any foreign substance. The mucous membranes of the body, such as those lining your mouth, nose, throat, eyelids, urinary tract, and genital tract, also defend the body against invaders. These membranes are sticky and trap microbes and foreign particles like flypaper traps flies.

Some of the mucous membranes of the body also have other physical or chemical aids for combating infection (**Figure 30.3**).

For example, the membranes covering the eyes and eyelids are constantly washed in tears, a fluid that contains a chemical called lysozyme, which is deadly to most bacteria. The ciliated mucous membranes of the upper respiratory system have a comparatively thick coating of mucus that not only traps invaders, but also efficiently transports them from these respiratory passageways to the upper part of the throat, where they are swallowed.

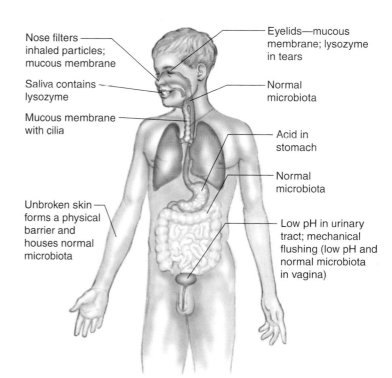

Nose filters inhaled particles; mucous membrane

Saliva contains lysozyme

Mucous membrane with cilia

Unbroken skin forms a physical barrier and houses normal microbiota

Eyelids—mucous membrane; lysozyme in tears

Normal microbiota

Acid in stomach

Normal microbiota

Low pH in urinary tract; mechanical flushing (low pH and normal microbiota in vagina)

**Figure 30.3** The first line of defense against infection. Mechanical barriers, chemical barriers, and normal microbiota work together to prevent entry of microorganisms into the human body.

 **Visual Thinking:** The title of this illustration refers to the body's "first line of defense." Explain why this is an accurate characterization.

In the digestive system, lysozyme in the saliva and the acid environment of the stomach kill microbes possibly eaten with food or taken in with liquids. The environments of the urinary and genital tracts are somewhat acidic too—an unfavorable situation for many foreign bacteria.

Foreign bacteria are also kept from invading the body by bacteria that normally inhabit it. These bacteria are called **normal microbiota**, and they colonize such areas as the mouth, throat, colon, vagina, and skin. Their presence makes it difficult for other organisms to flourish. Normal microbiota use the substances that invading bacteria need to live, and may produce chemicals that are toxic to other bacteria.

Physical barriers such as skin or mucous membranes, secretions such as lysozyme, and normal microbiota are together considered the first line of defense against disease (see Figure 30.3). Pathogens that get through these barriers are faced with additional defense mechanisms such as phagocytosis or the more complex response of inflammation.

An *inflammation* is a nonspecific response of the body to damage by microbes, chemicals, or physical injuries. Characterized by redness, pain, heat, and swelling, an inflammation consists of a series of events that remove the cause of the irritation, repair the damage that was done, and protect the body from further invasion and infection. During this nonspecific response, blood vessels dilate, or widen, bringing an increased supply of blood to the injured area as shown in **Figure 30.4 ①** . This increased flow brings defensive substances to the site of injury as fluid and phagocytic white blood cells move out of the capillaries ② . Lymphatic drainage dilutes and removes dissolved poisonous substances that may accumulate there ③ . Blood clots wall off the area, preventing the spread of microbes or other injurious substances to other parts of the body. Phagocytes, both neutrophils and monocytes (see Figure 29.11), migrate to the area and ingest microbes and other foreign substances ④ . Nutrients stored in the body are also released to the area to support these defensive cells.

Specialized white blood cells called *natural killer cells (NK cells)* are also part of the body's repertoire of nonspecific defenses. NK cells react to cell surface changes that occur on cancer cells or virally-infected cells and attack them, breaking them apart. Rather than ingesting infected or diseased cells as macrophages do, NK cells secrete proteins that create holes in the membranes of the cells they attack. In addition, NK cells secrete toxins that poison these cells. The cells under attack break apart and die.

The **complement** system is a nonspecific defense that acts both alone and in concert with the specific immune response. Its name comes from the fact that these proteins help (complement) the specific response. Complement proteins are produced in the liver and circulate in the blood in an inactive form. The presence of foreign antigens activates these proteins. The ultimate result is that the activated proteins form channels, or holes, in the cell membranes of the invading pathogen. The cell contents leak out and the organism dies.

*Interferons* are another nonspecific defense. Interferons, a family of proteins produced by virus-infected cells, circulate in the bloodstream and travel to both nearby and distant cells. These proteins bind with receptors on the cell membranes of healthy cells, triggering the cell to produce virus-blocking enzymes. Preparations of interferons are also used as medications to treat certain types of cancers, such as leukemia (cancer of the blood) and melanoma (a particularly deadly form of skin cancer). The mechanism of action of interferons in combating certain cancers is not entirely understood, but it seems to prevent the growth of tumor cells.

## CONCEPT CHECKPOINT

1. Differentiate between specific and nonspecific immunity. Explain how inflammation works as a nonspecific defense mechanism.

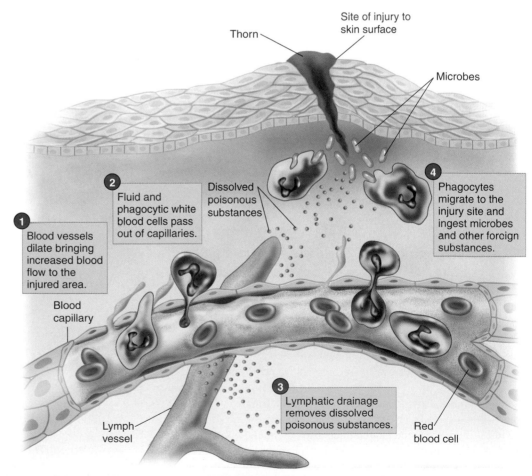

**Figure 30.4 The inflammatory process.** This nonspecific response disposes of foreign material and microbes at the injury site and prevents the spread of microbes.

**Figure 30.5** "Ah-choo!" Just a simple sneeze can propel 10,000 to 100,000 pathogens into the air traveling at 200 miles per hour.

## 30.4 Specific immunity includes body structures, chemicals, cells, and processes that act against specific pathogens.

What happens when your nonspecific defenses are not capable of stopping an infection? Nonspecific defenses sometimes fail, and for various reasons. For example, you may become inoculated with a "heavy dose" of a pathogen, such as when a person who has a bad cold comes into close contact with you while they are sneezing, coughing, or shaking your hand (**Figure 30.5**). You will likely become infected and develop a cold. Pathogens may also overcome your nonspecific defenses if they are extremely virulent, that is, if they are good at establishing an infection and damaging the body. *Yersinia pestis*, the bacterium that causes bubonic plague and killed about one-third of the world's population in the Middle Ages, and the Ebola virus, the pathogen that has caused outbreaks of Ebola hemorrhagic fever in Africa, are both considered virulent organisms. Even if people were infected with few organisms, they would likely develop the disease.

While the nonspecific immune response is at work, another line of defense is being summoned: the specific immune response. In a healthy person, this specific defense system staves off most infections and usually overcomes those that take hold. The body structures that perform specific defense responses constitute the **immune system**. The immune system is different from the other body systems in that it is not a connected system of organs. Instead, the immune system is made up of cells residing in organs and tissues that are scattered throughout the body (**Figure 30.6**). However, the cells of the immune system have a common function: reacting to specific foreign molecules.

The cells of the immune system are white blood cells called *leukocytes*. Two types of white blood cells are involved in the immune response: phagocytes and **lymphocytes** (LIM-foe-sites). Two classes of lymphocytes play roles in specific resistance: *T cells* and *B cells*. These cells arise in the bone marrow, circulate in the blood and lymph, and reside in the lymph nodes, spleen, and thymus. Although not bound together, these white blood cells exchange information and act in concert as a functional, integrated

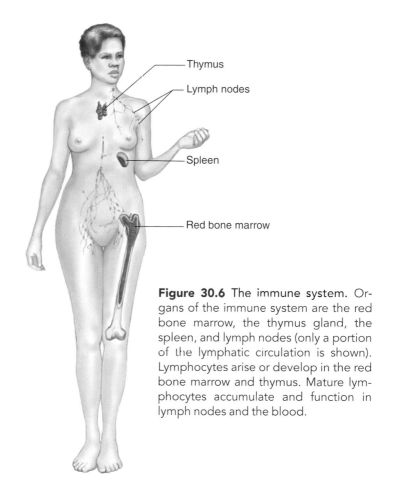

**Figure 30.6** The immune system. Organs of the immune system are the red bone marrow, the thymus gland, the spleen, and lymph nodes (only a portion of the lymphatic circulation is shown). Lymphocytes arise or develop in the red bone marrow and thymus. Mature lymphocytes accumulate and function in lymph nodes and the blood.

system—your "army" of 200 billion defenders—that is called into action when antigens invade your body.

## 30.5 Specific immunity includes cell-mediated and antibody-mediated responses.

The specific immune response works in several ways to defend the body against foreign invaders. Some of the cells of the immune system react immediately to invasion by foreign antigens. Other cells work to protect your body against future attack. The specific immune response is a complex yet coordinated effort, and it involves several types of protection.

### Sounding the alarm

Armies of white blood cells constantly guard the body. These white blood cells are the neutrophils, dendritic cells, natural killer cells, and macrophages. All play a role in nonspecific immunity by engulfing and destroying invaders. Macrophages and dendritic cells also play a role in the specific immune response. Lymphocytes, which include T cells and B cells, are white blood cells that are also active in the specific immune response.

Large, irregularly shaped cells, the macrophages act as the body's scavengers. Macrophages engulf anything that is not normal, including cell debris, dust particles in the lungs, and invading microbes. When the body is not under attack, only a

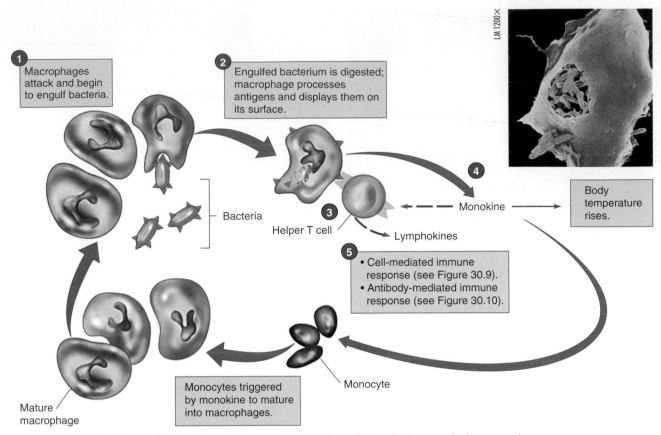

**Figure 30.7 Activation of the specific immune response.** The colorized micrograph shows a yellow macrophage phagocytizing green bacteria.

**Visual Thinking:** Using this illustration as a guide, explain how the immune reaction is triggered.

small number of macrophages circulate in the bloodstream and lymphatic system. In response to infection, precursors of macrophages called *monocytes* develop into mature macrophages in large numbers. Dendritic cells, named for their long spiky arms called dendrites (as in nerve cells), also form from monocytes.

When macrophages or dendritic cells encounter foreign microbes in the body, they attack (**Figure 30.7 ①**) and engulf them and then display parts of these microbes on their surfaces **②**. Helper T cells, which initiate the specific immune response, recognize the cells that have foreign antigens displayed on their surfaces and bind to them **③**. This binding stimulates the macrophage or dendritic cell to secrete a hormonelike protein called a *monokine* (from the word "monocyte," the precursor of the macrophage) **④**. The monokine (interleukin-1) allows communication between cells of the immune system, triggers the maturation of monocytes into macrophages (thereby increasing their numbers), and signals the brain to raise the body temperature, producing a fever. The higher temperature aids the immune response by increasing reaction rates of immune cells and inhibits the growth of invading bacteria by denaturing their enzymes. The monokine also stimulates the helper T cell to produce other hormonelike proteins called *lymphokines* **⑤**, which trigger the spe-

cific immune response by stimulating both B cells and T cells to multiply. The *In The News* feature at the beginning of this chapter presents recent research into how dendritic cells communicate with T cells.

## The two branches of the specific immune response

**T cells**, also called **T lymphocytes**, develop in the bone marrow but then migrate through the circulatory system to the thymus. The thymus, a small gland located in the upper chest and extending upward toward the neck (see Figure 30.6), is the place where T (thymus) cells mature. This organ is large and active in children; it dwindles in size throughout childhood until, by puberty, it is quite small. In the thymus, T cells develop the ability to identify invading bacteria and viruses by the foreign molecules—antigens—on their surfaces. Tens of millions of different T cells are produced, each specializing in recognizing one particular foreign antigen. The five principal kinds of T cells are listed and described in **Table 30.2**.

When they are triggered by the monokine interleukin-1 and the presentation of antigens, the helper T cells stimulate both branches of the specific immune response: the *cell-mediated immune response* and the *antibody-mediated immune response*.

| TABLE 30.2 | Principal Types of T Cells |
|---|---|
| **Type of Cells** | **What They Do** |
| Helper T cells | Initiate the specific immune response. |
| Cytotoxic T cells | Break apart cells that have been infected by viruses and break apart foreign cells such as incompatible organ transplants. |
| Inducer T cells | Oversee the development of T cells in the thymus. |
| Suppressor T cells | Limit the specific immune response. |
| Memory T cells | Respond quickly and vigorously if the same foreign antigen reappears. |

**The cell-mediated immune response** The activation of helper T cells by monokine and the binding of antigen to these activated helper T cells unleash a chain of events. One of these events is the **cell-mediated immune response**. The primary response of this arm of specific immunity is that cytotoxic T cells ("cell-poisoning" cells) recognize and destroy intracellular pathogens—those infecting body cells—such as viruses (**Figure 30.8**).

When a helper T cell has been activated, it produces chemical substances collectively called *lymphokines* (**Figure 30.9** ①). One type of lymphokine attracts macrophages to the site of infection ②, and another inhibits their migration away from it. Another of the lymphokines stimulates T cells that are bound to foreign antigens to undergo cell division many times. This cell division produces enormous quantities of T cells capable of recognizing the antigens specific to the invader ③. Each type of activated T cell does a specific job.

The activated inducer T cells trigger the maturation of immature lymphocytes in the thymus into mature T cells. Helper T cells ④ continue to stimulate the specific immune response. Activated cytotoxic T cells ⑤ kill the body's own cells that have been infected with fungi, some viruses, and bacteria that produce slowly developing diseases such as tuberculosis. In addition, they act to kill cells that have become cancerous. Cytotoxic T cells bind to these infected or abnormal cells by means of molecules on their surfaces that specifically fit antigens, much as a key fits a lock. Because the entire cell binds to the abnormal cells by means of specific cell-surface proteins, this response is called *cell-mediated*. The cytotoxic T cells then secrete a chemical that breaks apart the foreign cell (see Figure 30.8).

Unfortunately for patients undergoing organ transplantation, cytotoxic T cells also recognize foreign body cells. As a result, cytotoxic T cells attack transplanted tissue, leading to the rejection of transplanted organs. Although tissue donors are matched

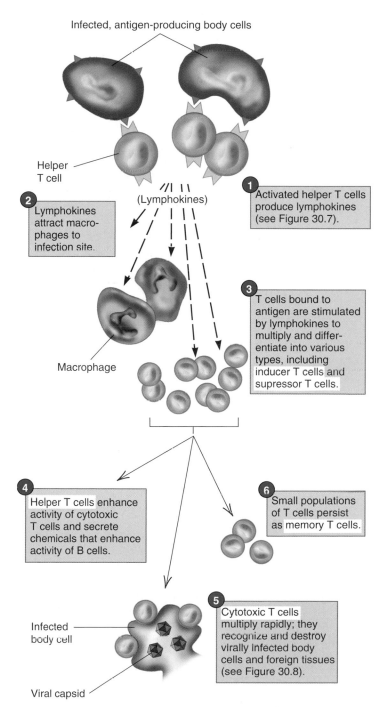

**Figure 30.9 The cell-mediated immune response.** This specific defense is directed against body cells infected with intracellular pathogens, such as viruses. It also works against foreign tissues such as some cancer cells and tissue transplants.

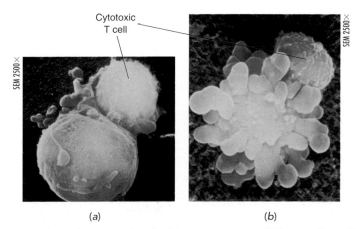

(a)                              (b)

**Figure 30.8 Cytotoxic T cell.** (a) A cytotoxic T cell (the smaller cell) binding with a tumor cell (the larger cell). (b) The cytoplasm of the tumor cell is leaking from holes in its membrane, which were created by the T cell.

closely with recipients to minimize rejection, transplant patients must take immunosuppressive drugs that selectively subdue cell-mediated immunity while leaving antibody-mediated immunity active. At this time, scientists are researching the use of custom-built antibodies that would block specific facets of the rejection process so that donors and recipients would not have to be close matches.

After the cytotoxic cells and macrophages do their jobs, the cell-mediated immune response begins to shut down. The cells in charge of shutdown are the suppressor T cells. The number of suppressor T cells slowly begins to rise after activation by the helper T cells. However, it takes about one to two weeks for their numbers to increase to a point where they are able to suppress the cytotoxic T cell response.

After suppression, or shutdown, a population of T cells persists, probably for the life of the individual. Referred to as *memory*

*T cells* **6** , these cells provide an accelerated and larger response to any later encounter with the same antigens.

**The antibody-mediated immune response** When helper T cells are stimulated to respond to foreign antigens, they activate the cell-mediated immune response as described and activate a second, more long-range defense called the **antibody-mediated immune response**. Depending on the types of antigens present, the helper T cells may stimulate either or both of these branches of the specific immune response. The primary antibody-mediated immune response is directed toward extracellular pathogens, such as bacteria and some viruses, recognizing and destroying them before they infect cells.

The key players in antibody-mediated immunity are **B cells**, also called **B lymphocytes**. B cells mature in the bone marrow. Their role in the antibody response is to secrete antigen-specific chemicals into the bloodstream.

## just wondering . . .

*Questions students ask*

### How do antibiotics work? Why do I have to take them for so long after I feel better? How come they don't work on a cold or the flu?

Antibiotics are a group of chemicals that kill bacteria or inhibit their growth. Many of these chemicals are actually produced by bacteria themselves, which helps them compete with other types of bacteria in nature (**Figure 30.B**). They cannot help rid your body of a cold or the flu because viruses, not bacteria, cause these infections. Sometimes, however, a physician will prescribe an antibiotic when you have a cold or the flu because of a secondary bacterial infection.

Antibiotics work by disrupting essential functions necessary for bacterial survival, such as interfering with the synthesis of the bacterial cell wall. Because antibiotics target bacterial structures that human cells do not have, you can take an antibiotic and it will not damage or kill your body cells. Antibiotics, however, can kill the "friendly" bacteria that inhabit your body—your normal microbiota (see Section 30.3). Normal microbiota help keep foreign bacteria from invading because they are well established and use the substances invading bacteria need to live. Therefore, one risk of taking antibiotics is that your normal microbiota will be affected, possibly causing intestinal upset, diarrhea, or infection. Women are particularly prone to developing vaginal yeast infections when the normal microbiota of the vagina are killed. The normal bacteria inhabiting the vagina produce an acid environment. When the acid-producing bacteria die, the pH of the vagina changes and the yeast thrive.

When taking an antibiotic, you'll usually feel better within 24 to 48 hours. Nevertheless, you must continue the full course of treatment. Although many of the bacteria have been killed or their reproduction inhibited, some will have survived. Stopping the antibiotic early will allow these more resistant bacteria to multiply and the infection will return.

The resistance of bacteria to antibiotic treatment is a problem we are now facing worldwide. Tuberculosis, gonorrhea, malaria, and childhood ear infections are only a few of the diseases whose causative agents are antibiotic resistant. How did this problem arise? Some bacteria are naturally resistant to antibiotics, and others develop resistance through mutation. Genes conferring resistance to antibiotics can be

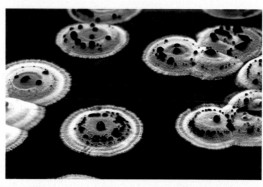

**Figure 30.B One source of antibiotics.** The large circles are bacterial colonies of *Streptomyces coelicolor*, a soil bacterium that, along with its close relatives, naturally produces many of our antibiotics, such as tetracycline and erythromycin. The blue globules are droplets of water containing the blue-pigmented antibiotic actinorhodin produced by this strain.

passed from bacterium to bacterium as bacteria exchange genetic material (see Chapter 11). Taking antibiotics when they are not needed, or not finishing the full course of treatment of an antibiotic, sets up ideal conditions for the emergence of drug-resistant bacteria. Those situations provide opportunities for resistant bacteria to thrive because competitors are killed. In hospitals, where antibiotics are used frequently, resistant strains of bacteria can become a serious problem and can be passed quickly and easily from sick patient to sick patient.

How can you help thwart the rise of antibiotic resistant strains? First, take an antibiotic only when you have a bacterial infection. Do not pressure your health care provider to prescribe one for a cold, the flu, or other viral infection. When you do take an antibiotic, take it exactly as it is prescribed and until it is gone. Never save an antibiotic to take at another time or to give to others. ●

*Are you wondering about a topic in biology and how it relates to your life? Submit your question by clicking the* Just Wondering *link in this text's companion Web site at* www.wiley.com/college/alters.

On their surfaces, B cells each have about 100,000 identical copies of a protein receptor that binds to a specific antigen. Because different B cells bear different receptors, each recognizes a different, specific antigen. At the onset of a bacterial infection, for example, the receptors of one or more B cells bind to bacterial antigens (**Figure 30.10** **1a**), while macrophages engulf and digest bacteria, process bacterial antigens, and display the antigens on their surface **1b**. The B cells also process and display bacterial anitgens **2a**. Helper T cells detect and bind to the antigen-displaying macrophges **2b** and then bind to the antigen-displaying B cells **3**. After this "double" binding, the helper T cells release lymphokines that trigger cell division in the B cell **4**.

After about five days and numerous cell divisions, a large clone of cells is produced from each B cell that was stimulated to divide. A *clone* is a group of identical cells that arise by repeated mitotic divisions from one original cell. After these clones are formed, the B cells begin secret-

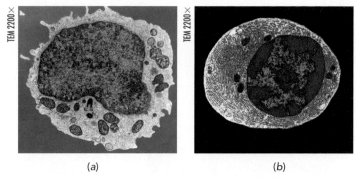

(a)          (b)

**Figure 30.11** Comparison of a B cell and a plasma cell. (*a*) Colorized micrograph of a B cell. (*b*) Colorized micrograph of a plasma cell. The large amount of endoplasmic reticulum (ER) on which antibodies are made can be seen as orange in the cytoplasm. The B cell shown in (*a*) does not have large quantities of ER.

ing copies of the receptor protein that responded to the antigen **5**. These receptor proteins are called **antibodies**, or *immunoglobulins*. The secreting B cells are called *plasma cells*. After B cells become plasma cells, they live for only a few days, but secrete a great deal of antibody during that time (**Figure 30.11**). One plasma cell will typically secrete more than 2000 antibody molecules per second!

Antibodies do not destroy a virus or bacterium directly, but rather inactivate it and mark it for destruction by either the complement system or macrophages. The binding of both antibodies and complement to the antigen enhances the phagocytic activity of macrophages. Types of inactivation of antigens by antibodies are discussed in the next section.

The B cell clones that did not become plasma cells live on as circulating lymphocytes called *memory B cells* (Figure 30.10 **6**). These cells provide an accelerated response to any later encounter with the stimulating antigen (**Figure 30.12**). As in the case of the cell-mediated immune response, the antibody response is shut down after several weeks by suppressor T cells.

**Figure 30.13** shows an overview of both nonspecific and specific immunity.

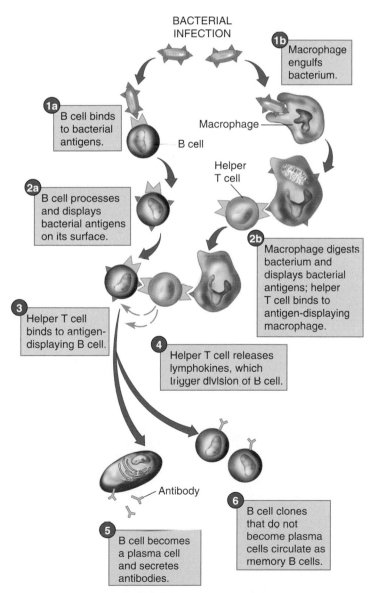

**Figure 30.10** The antibody-mediated immune response. This specific defense is directed against extracellular pathogens such as bacteria.

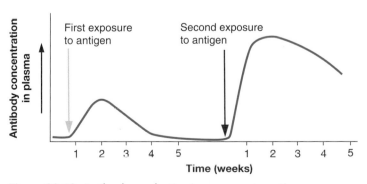

**Figure 30.12** Antibody production in response to antigen exposure. The first arrow shows the initial exposure to the antigen. After the exposure, the body responds by producing antibodies. It takes about two weeks for the antibody level to peak, and as the foreign antigen is removed from the body, antibody levels fall. However, memory cells persist. The next time the body is exposed to that same antigen, the memory cells respond rapidly and produce even higher levels of antibodies.

**INFECTION
(ANTIGEN)**

**NONSPECIFIC IMMUNITY**

- Mechanical barriers
- Lysozyme
- Chemical barriers
- Normal microbiota
- Inflammatory response
- Phagocytic cells
- Natural killer cells
- Complement system
- Interferon

**SPECIFIC IMMUNITY AND THE IMMUNE SYSTEM**

*Initial contact with antigen*

*Macrophages and dendritic cells
migrate to site of infection*

*Macrophages and dendritic cells
engulf and display antigen*

| CELL-MEDIATED IMMUNE RESPONSE | HELPER T CELLS | ANTIBODY-MEDIATED IMMUNE RESPONSE |

Lymphokines

*Additional macrophages
and dendritic cells
attracted to infection site*

**T CELLS**

*Multiply and differentiate*

Inducer T cells

Helper T cells

Cytotoxic T cells

Supressor T cells

Memory T cells

*Stimulated on
future contacts with
the same antigen*

**B CELLS**

*Multiply and differentiate*

Plasma cells

Memory B cells

*Produce
antibodies*

*Stimulated on
future contacts with
the same antigen*

**Visual Summary**   **Figure 30.13** An overview of nonspecific immunity and specific immunity of the human body. Notice the central role of the helper T cells in both the antibody-mediated and cell-mediated immune responses.

**Visual Thinking:** Using this illustration as a guide, explain why helper T cells, more than any other, are critical to the body's specific resistance to infection.

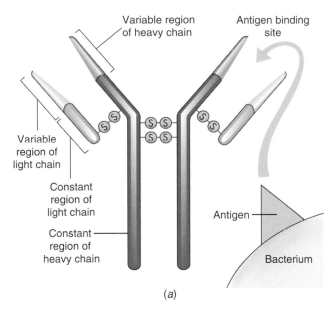

(a)

(b)

**Figure 30.14** The structure of an antibody molecule. (a) Each antibody molecule is composed of two identical light chains and two identical heavy chains. Disulfide bonds hold the chains together. The antibody binds the antigen to which it is specific at the antigen binding site (the variable region of the molecule). The arrow shows where the antigen binds. (b) A computer-generated image of an antibody molecule.

## 30.6 Antibody cell surface receptors recognize specific antigens.

The cell surface receptors of lymphocytes, T cells, and B cells, can recognize specific antigens with great precision. Even single amino acid differences between proteins can often be discriminated, with a receptor recognizing one form and not the other. This high degree of precision is a necessary property of specific immunity because without it, the identification of foreign antigens would not be possible in many cases.

A typical antibody, or immunoglobulin molecule, consists of four polypeptide chains. A polypeptide is a chain of amino acids linked end-to-end by peptide bonds. There are two identical short strands, called *light chains*, and two identical long strands, called *heavy chains*. Disulfide (-S-S-) bonds hold the four chains together, forming a Y-shaped molecule (**Figure 30.14**). The two "arms" of the Y determine which antigen will bind to the antibody. Antibodies recognize, or lock onto, small parts of antigen molecules called *epitopes* by means of special binding sites in the variable regions of their arms. These binding sites are made up of specific sequences of amino acids that determine the shape of the site. The specificity of the antibody molecule for an antigen depends on this shape. An antigen fits into the binding site on the antibody like a key into a lock.

Changes in the amino acid sequence of an antibody in the variable regions can alter their shape so that a different antigen can bind to that antibody, just as changing a lock will alter which key can fit into it.

What happens when an antibody binds with an antigen? Three processes can occur: *neutralization*, *opsonization*, and *activation of complement*. In the process of neutralization, antibodies immobilize, coat, and agglutinate (clump) pathogens (**Figure 30.15a**).

ANTIBODIES PROTECT AGAINST DISEASE BY

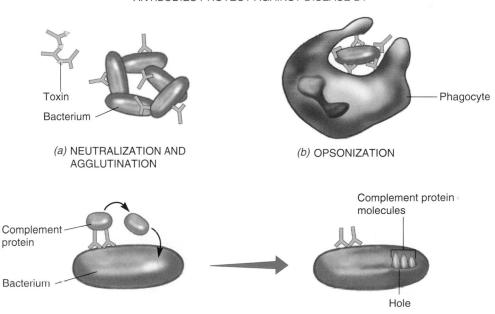

(a) NEUTRALIZATION AND AGGLUTINATION

(b) OPSONIZATION

(c) ACTIVATION OF COMPLEMENT

**Figure 30.15** What happens when an antibody binds with an antigen? (a) Agglutination may occur in which each binding site on the antibody binds to antigens on a different cell, causing cells to clump. Toxins may be neutralized. (b) Opsonization may occur, resulting in phagocytosis of antibody-bound antigens. (c) The complement system may be activated, resulting in holes being punched in cell walls and membranes.

This prevents them from entering cells or attaching to sites of infection. Antibodies also bind to toxins secreted by pathogens, coating them and rendering them harmless. The binding of antibodies to antigens also results in opsonization. The bound antibodies act like tags—scientists call them opsonins—marking the antigens for destruction. Opsonins bind to receptors on phagocytic cells, such as macrophages and neutrophils, which is a signal to these cells to engulf and destroy the bacterium or virally infected cell (Figure 30.15b). The binding of antibody to antigen also triggers the complement system. The interacting proteins of the complement system punch holes in cell walls and membranes, causing the death of these tagged cells (Figure 30.15c).

---

### CONCEPT CHECKPOINT

2. Your study partner was confused about the roles of the two branches of the specific immune response. Explain, as concisely as possible, the basic workings of these two immune responses.
3. The following is an essay written by a student who is attempting to use an analogy to demonstrate how the immune system responds when the body is infected by a pathogen. Explain, how this analogy could be improved.

> The immune system can be compared to purchasing a suit from a custom tailor. The tailor designs the suit to fit your measurements and specifications. This is similar to specific immunity in the sense that when a pathogen invades the body, the immune system responds by designing and constructing antibodies that are tailored to the antigens on the pathogen.

---

# How do vaccines work?

## 30.7 Vaccines cause the body to develop specific immunity.

Did you ever wonder what you were receiving when you got a flu shot or other vaccine? A vaccine is made up of disease-causing microbes or toxins (poisons) that have been killed or changed in some way so as not to produce disease. For example, both influenza and polio vaccines are comprised of whole viruses that have been inactivated so that they can no longer cause disease, yet their presence triggers an immune response (**Figure 30.16**).

Injection with antigens in vaccines causes B cells to launch an antibody response, with the production of memory cells. Vaccination also induces memory T cell formation. A booster shot induces these memory cells to differentiate into antibody-producing plasma cells and more memory cells (see the second response in Figure 30.12). In summary, **vaccination**, or immunization, causes the body to build up antibodies against a particular disease without causing the disease. Since the body actually produces antibodies and generates memory cells, vaccination results in **active immunity**.

Scientists are now applying the techniques of genetic engineering to the development of vaccines to overcome problems with "traditional" vaccines. Some of these problems include vaccine preparations that do not confer immunity well enough, or others that produce unwanted—and sometimes dangerous—side effects. Section 11.6 describes genetically engineered vaccines, including DNA vaccines.

Vaccination to produce active immunity works well for many diseases, but it is a somewhat slow process. After the injection of the antigen, it takes weeks for the body to develop sufficient antibodies and memory cells to combat the disease. Another method of implementing specific immunity, called **passive immunity**, can be used when protection is needed quickly. Passive immunity involves the transfer of antibodies or antitoxins directly to an individual. Since the recipient does not produce antibodies or generate memory cells, passive immunity is temporary.

One example of passive immunity is the use of antitoxin when a person has been bitten by a snake. An antitoxin is a preparation of antibodies specific for a toxin. (Antitoxins to snake venoms are also called *antivenins.*) When a poisonous snake bites a person or other animal, it injects a toxic venom into the body. Certain toxins damage nerve cells and may result in death. Antibodies to the toxin are injected into the victim and they bind with the toxin, preventing it from causing damage. Though effective, this type of "borrowed" specific immunity lasts only a short time. The body soon uses or eliminates these antibodies, so that they will not be available if another snakebite occurs.

Newborns have a type of passive immunity borrowed from their mothers. Antibodies that pass from mother to fetus through the placenta provide babies with a short-term specific immunity that subsides by the time the baby has produced its own system of antibodies. In addition, if a mother nurses her baby, antibodies in the mother's milk provide the baby with some protection.

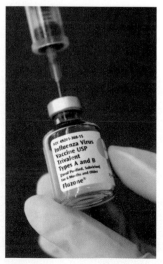

**Figure 30.16** A flu vaccination being drawn from a bottle. Health officials urge people at highest risk of the flu, such as children aged 6–23 months, adults aged 65 years and older, and persons with chronic medical conditions, to get a preventive vaccine during the flu season.

---

### CONCEPT CHECKPOINT

4. Why don't people who are vaccinated with a "live" or active version of a virus contract the disease in the course of being vaccinated?

# Why is AIDS so deadly?

## 30.8 Infection with HIV weakens the immune system.

Helper T cells are the key to specific immunity because they initiate both the cell-mediated and the antibody-mediated immune responses. Without helper T cells, the immune system is unable to mount a response. The **human immunodeficiency virus (HIV)** is deadly because it mounts a direct attack on T lymphocytes, mainly targeting the helper T cells. HIV kills helper T cells by entering them, using the cells' own machinery to produce more virus particles, and then bursting open the cells to release the new viruses (see Chapter 19). These newly formed viruses then infect other helper T cells.

Infection with HIV does not mean that a person has **acquired immunodeficiency syndrome (AIDS)**. During the first years of the disease process, the HIV-infected individual is usually asymptomatic, which means that no symptoms are apparent because the immune system has not been weakened sufficiently. This phase of HIV disease usually lasts about 7 to 10 years.

The asymptomatic phase of HIV can be extended with multidrug highly active antiretroviral therapy, or HAART. Antiretroviral therapy means that these drugs work against retroviruses, such as HIV. What are retroviruses? They are virus particles that contain RNA as the hereditary material instead of DNA. Retroviruses also contain the enzyme reverse transcriptase, which causes synthesis of a complementary DNA molecule (cDNA) using virus RNA as a template. (See Chapter 11.) When a retrovirus infects a cell, it injects its RNA into the cell along with the reverse transcriptase enzyme. The cDNA produced from the RNA template contains the genetic instructions for continuing the infection of the host cell.

HAART appears to work well in holding HIV to low levels in the bloodstream and retarding its debilitating effects on the immune system. Physicians also combine other treatments with HAART, such as reverse transcriptase inhibitors. However, during treatment the virus replicates at low levels and cannot be eliminated. There is no cure for HIV infection at this time and it usually progresses to AIDS.

Fusion inhibitors are the newest class of anti-HIV drugs under study. They are designed to prevent HIV from fusing with host cells so that the virus cannot inject these cells with its genetic material. Another area of research is gene therapy, in which a patient's HIV-infected cells are replaced with cells engineered to resist viral replication. Clinical trials for HIV gene therapy have not yet reported positive results, nor have clinical trials for AIDS vaccines (see Chapter 11).

During the symptomatic phase of HIV infection, before it can be considered AIDS, the helper T cell count has usually declined from the normal level of 800 to 1200 helper T cells/mm³ to about 200 helper T cells/mm³. Signs and symptoms of this phase of HIV infection include chronic diarrhea, minor infections of the mouth, fever, night sweats, headache, and fatigue. (Signs are observable, measurable features of an illness such as a fever, whereas symptoms are subjective complaints of an ill person, such as reporting a headache.)

AIDS is a syndrome, a set of signs and symptoms that occur together. A person with HIV infection is usually diagnosed as having AIDS when the helper T cell count falls below 200 cells/mm³ and certain diseases typical of this stage are present. Such diseases include *Pneumocystis carinii* pneumonia (a fungal infection of the lungs) and Kaposi's sarcoma (**Figure 30.17**), a formerly rare type of cancer. In a normal individual, helper T cells make up 60 to 80% of circulating T cells; in patients with AIDS, helper T cells often become too rare to detect. Because the specific immune response cannot be initiated, any one of a variety of infections proves fatal. This is the primary reason that AIDS is a devastating disease.

HIV is *not* highly infectious; that is, it is not transmitted from person to person by casual contact. In addition, HIV is *not* transmitted by insects. It is transmitted only by the direct transfer of body fluids, typically in semen or vaginal fluid during sexual activity, in blood during transfusions, or by contaminated hypodermic needles. It also crosses the placenta to infect babies within the womb of an infected mother.

To *eliminate* your risk of becoming infected with HIV, do not have sex with HIV-infected individuals and abstain from using illegal injected drugs. These precautions will not eliminate the risk

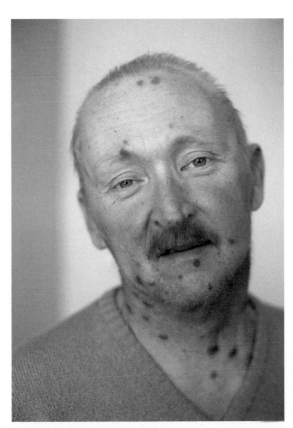

**Figure 30.17** A man with AIDS has purple marks on his face from Kaposi's sarcoma.

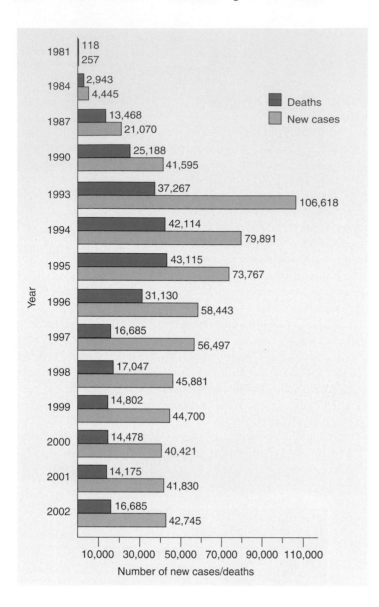

of infection for persons in certain professions, such as health care workers and police officers, who may become infected by HIV-contaminated blood during the course of their work.

To help *reduce* your risk of becoming infected with HIV, remember these key points: Use a new latex condom during each act of sexual intercourse, never share needles and syringes, avoid having sex while under the influence of drugs (including alcohol), reduce your number of sexual partners, and avoid having sex with persons who practice risky behaviors with regard to HIV infection.

**Figure 30.18** shows the decline of AIDS-related deaths in the United States first noted in 1996. In 1997, HIV infection fell from the eighth to the fourteenth leading cause of death in the United States. In 1998, it was no longer one of the top 15 causes of death and has remained off this list since then. This decline is thought to be the result of increasingly effective drug therapies. Nevertheless, AIDS is the leading cause of death in other regions of the world, such as sub-Saharan Africa and Thailand. In the United States, the number of diagnosed AIDS cases rose slightly in 2001, the first increase since 1993. The number also increased in 2002.

CONCEPT CHECKPOINT

5. How does the elimination of helper T cells affect the specific immune response, and how does this relate to the development of AIDS?

**Figure 30.18** Reported cases and known deaths from AIDS, United States, 1981–2002. The extreme rise in reported cases in 1993 is the result of a change in the definition of AIDS, which broadened the list of conditions reportable as AIDS. *Source:* National Center for Health Statistics, National Vital Statistics System.

# What are allergies?

## 30.9 Allergies are immune responses to usually harmless antigens.

Although the human immune system provides effective protection against viruses, bacteria, parasites, and other microorganisms in healthy people, sometimes it does its job too well, mounting a major defense against a harmless antigen called an *allergen*. Such specific immune responses are called **allergic reactions**. Hay fever, the sensitivity that many people exhibit to proteins released from plant pollen, is a familiar example of an allergy (**Figure 30.19**). Many people are also sensitive to proteins released from the feces of a tiny dust mite. This microscopic arthropod lives in the house dust present on mattresses and pillows and eats the dead epithelial tissue that everyone sheds from their skin daily.

**Figure 30.19** Ragweed plant. The inset shows a scanning electron micrograph of ragweed pollen, the cause of many allergic reactions.

An allergic reaction is uncomfortable and sometimes dangerous due to the involvement of class E (IgE) antibodies. The binding of antigens to these antibodies initiates an inflammatory response; powerful chemicals such as histamines cause the dilation of blood vessels and a host of other physiological changes. Sneezing, runny nose, and fever often result. In some instances, allergic reactions can produce anaphylactic shock, a severe and life-threatening response of the immune system during which chemicals that dilate blood vessels are released suddenly into the bloodstream, causing many vessels to widen abruptly. This causes the blood pressure to fall dangerously low. In addition, the muscles of the trachea, or windpipe, may contract, making breathing difficult.

## CHAPTER REVIEW

## Summary of Key Concepts and Key Terms

### How do organisms defend themselves against disease?

**30.1** All living things have some elements of **immunity** (p. 516).

**30.1** Some defenses of the human body provide **nonspecific (innate) immunity** (p. 516), acting against any foreign invaders; other defenses are specific, acting against particular invaders.

**30.1** Only the vertebrates have **specific (adaptive) immunity** (p. 516).

**30.2** Invertebrates, protists, and plants have only nonspecific immunity.

**30.2 Phagocytosis** (fag-oh-si-TOE-sis, p. 516) is one important nonspecific defense in invertebrates and protozoans; this mechanism is also used extensively in vertebrates.

**30.2** The human immune system distinguishes self from nonself by means of complex molecules called **antigens** (p. 517), usually proteins or carbohydrates, present on the surface of foreign substances, which are different from molecules present on host cells.

**30.3** The nonspecific defenses, those acting against any foreign invader, include the skin and mucous membranes, **normal microbiota** (p. 518), acidic secretions of the body, chemicals (including **complement**, p. 518), cells that kill bacteria, and the inflammatory process.

**30.4** The body structures that perform specific defense responses constitute the **immune system** (p. 519).

**30.4** Two types of **lymphocytes** (p. 519) play roles in specific immunity: T cells and B cells, which are capable of recognizing foreign substances by means of receptor proteins.

**30.5 T cells** (**T lymphocytes**, p. 520) develop in the bone marrow but then migrate to the thymus where they mature; there, T cells develop the ability to identify invading bacteria and viruses by the antigens on their surfaces.

**30.5 B cells** (**B lymphocytes,** p. 522) mature in the bone marrow and secrete antigen-specific chemicals called **antibodies** (p. 523).

**30.5** In the **cell-mediated immune response** (p. 521), cytotoxic T cells attack infected body cells.

**30.5** In the **antibody-mediated immune response** (p. 522), B cells are converted to plasma cells that secrete proteins called antibodies that specifically bind circulating antigen and mark cells or viruses bearing antigens for destruction.

**30.6** The high level of specificity of an antibody for a particular antigen is caused by the three-dimensional shape of the ends of the arms, or antigen-binding sites, of the molecule.

### How do vaccines work?

**30.7 Vaccination** (p. 526), which is an injection with disease-causing microbes or toxins that have been killed or altered in some way that renders them harmless, causes the body to build up antibodies against a particular pathogen; this type of specific immunity is called **active immunity** (p. 526).

**30.7 Passive immunity** (p. 526) is the transfer of antibodies directly to an individual.

**30.7** Scientists can now use laboratory-made antigenic proteins as vaccines.

### Why is AIDS so deadly?

**30.8** The **human immunodeficiency virus** (**HIV**, p. 527) is deadly because it primarily attacks helper T cells, the key to the entire specific immune response.

**30.8** Infection with HIV does not mean that a person has **acquired immunodeficiency syndrome** (**AIDS**, p. 527).

**30.8** Because helper T cells are no longer present in AIDS patients to initiate the specific immune response, any one of a variety of otherwise commonplace infections proves fatal.

### What are allergies?

**30.9** An **allergic reaction** (p. 528) is a specific immune response against a harmless antigen.

**30.9** Allergic reactions are produced by the combination of a particular kind of antigen and a high level of IgE antibody.

## Level 1 — Learning Basic Facts and Terms

**Multiple Choice**

1. All of the following are correct about nonspecific defenses except:
   a. They include inflammatory responses.
   b. They include physical and chemical barriers such as lysozymes and mucous membranes.
   c. They are initiated by B cells and T cells.
   d. They may involve the formation of proteins by nonspecific cells such as natural killer cells that attack the membranes of pathogens.
   e. They are not directed at one particular strain or species of pathogen; rather they work against any foreign invading organism.

2. Which one of the following statements about antibody-mediated immunity is correct?
   a. It is responsible for defending the body against larger pathogens like fungi and protozoans.
   b. It protects the body against cells that become cancerous.
   c. It is mounted by lymphocytes that have matured in the thymus gland.
   d. It primarily defends against extracellular pathogens.
   e. It acts as a barrier to prevent entry of pathogens.

3. An antigen may
   a. be found on or secreted by a pathogen.
   b. be displayed on the surface of a macrophage or dendritic cell.
   c. be used in distinguishing self cells from foreign cells.
   d. All of the above are true.

**Matching**

Pair the descriptions in questions 4–9 with the words listed in a–h.

4. _____ The human immunodeficiency virus (HIV) compromises the immune system by primarily attacking this type of cell.

5. _____ These cells produce antibodies against invading bacteria.

6. _____ These cell are produced late in the immune response and are responsible for shutting down the immune response.

7. _____ These cells engulf pathogens and display their foreign antigens on their own surface.

8. _____ These cells differentiate into macrophages.

9. _____ These cells are responsible for attacking your own body's cells that have been infected by a virus.

   a. B lymphocytes
   b. Helper T cells
   c. Cytotoxic T cells
   d. Suppressor T cells
   e. Memory cells
   f. Monocytes
   g. Macrophages
   h. Plasma cells

## Level 2 — Learning Concepts

1. Name your body's nonspecific defenses, and summarize how each protects you against infection or injury.

2. Read each of the following statements concerning the specific components of the immune response. Which one is correctly stated? Modify the other two statements to make them correct as well.
   a. When a virus invades the body for the first time, there are usually memory T cells in the body prepared to destroy virally-infected body cells.
   b. When a virus invades the body and infects cells, generally only one component of specific immunity is armed: the antibody-mediated immune response.
   c. When a virus re-invades the body, the immune system is generally prepared for that attack because the body already has small populations of memory T cells that will recognize those virally-infected body cells.

3. A cell membrane has proteins and carbohydrates on its surface that act as "cell markers." How do you think these markers would be interpreted if this cell were placed into the tissues of another organism?

4. What are the five principal types of T cells? Summarize the function of each.

5. Summarize the events of the cell-mediated immune response. What cells are involved?

6. Summarize the events of the antibody-mediated immune response. What cells are involved?

7. How is an antibody molecule able to discriminate between millions of cells in a tissue and bind to only one specific antigen?

8. A nurse vaccinates a child against the tetanus bacterium. Later, the child receives a booster shot. Describe how these two vaccinations affect the child's specific immune response during subsequent exposures to tetanus.

9. Which virus is responsible for AIDS? Explain why this virus is so deadly.

10. Many people—including you, perhaps—are allergic to something. Explain what an allergic reaction is and why it occurs. What causes its uncomfortable symptoms?

##  Critical Thinking and Life Applications

1. The prevention of disease through vaccination was discovered by Edward Jenner in the 1800s, when he noticed that women who milked cows ("milkmaids") rarely contracted smallpox. Jenner also noticed that cows develop a disease called cowpox, whose symptoms are similar to those of smallpox. Jenner deliberately infected a child with the pus from a cowpox blister, hoping that it would give the child immunity to smallpox—and it worked! Based on what you know about the immune system, explain why the child developed this immunity.

2. You had measles as a child. Now your younger brother has measles, but you don't catch it from him. Explain why.

3. Pediatricians often encourage mothers to breastfeed infants because maternal antibodies circulate in milk and provide the baby with some protection against infection. What sort of immune response is involved here? Why is this immunity only temporary?

4. If you have allergies, you've probably taken over-the-counter medications containing antihistamines. Using your understanding of allergic reactions, what do you think is the physiological role of antihistamines? Why do they make you feel better?

5. In 1918 the world experienced a strain of flu virus that killed over 200 million people worldwide. Based on what you know about the immune response, develop a hypothesis as to why this strain of flu was so deadly.

6. When Europeans arrived in the New World in the 1500s, millions of native Indians died from diseases the Europeans brought with them, even though these diseases were rarely fatal to Europeans who contracted them. Why do you suppose these diseases were so deadly to the native Indians but not so to Europeans?

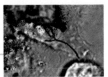

## In The News | **Critical Thinking**

### CELL SCOUTS

Now that you understand more about the immune system, reread this chapter's opening story about research into communication between dendritic cells and T cells. To better understand this research, it may help you to follow these steps:

1. Review your immediate reaction to research into the communication process that occurs between cells of the immune system that you wrote when you began reading this chapter.

2. Based on your current understanding, again summarize the main point of the research in a sentence or two.

3. What questions do you now have about this research that this chapter's opening story does not answer?

4. Collect new information about the research. Visit the *In The News* section of this text's companion Web site at www.wiley.com/college/alters and watch the "Cell Scouts" video. Then use the "summary" link to read the accompanying story and access related links. Use this information, the links provided, and other online and library resources to answer your questions and find updates about this research topic. State the sources of your information. Explain why you think the information is accurate. Also determine whether the information expresses a particular point of view or is biased in any way.

5. What in your view are the most significant aspects of this research? Give reasons for your opinion and for any changes in your ideas based on the additional information you have collected and the analysis you have done.

# EXCRETION

## 31

Kidney Bones

What do bones and kidneys have in common? At first glance, not much. Bones make up the skeletal system; kidneys are part of the excretory system. Bones are hard tissues; kidneys are soft tissues. Although there doesn't seem to be any obvious connection between bones and kidneys, they work together to maintain homeostasis. How? Healthy bones remove calcium, phosphorus, and sodium from the blood and store these minerals in the bones, while healthy kidneys remove excess amounts of these minerals from the blood and excrete them. Furthermore, kidneys synthesize *erythropoetin*, a hormone that stimulates red blood cell production in bone marrow, and *bone morphogenetic protein-7* (*BMP-7*), a molecule that stimulates skeletal growth.

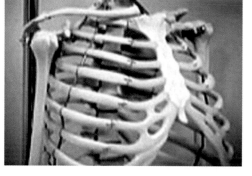

As you can see, the physiological roles of bones and kidneys are intimately connected. As a result, people with chronic renal (kidney) failure—a slowing of filtration of the blood at the kidneys—usually suffer from a condition called adynamic bone disorder (ABD). ABD is characterized by severely weakened bones that become deformed and fracture easily. No effective treatments for ABD are currently available, but Keith Hruska and his colleagues at Washington University in St. Louis, Missouri, are studying the effects of chronic renal failure on bone health in an effort to develop a possible remedy.

The team of scientists tested the hypothesis that suppression of BMP-7 production in patients with chronic renal failure results in ABD. The researchers used mice to test their hypothesis because mouse kidneys also synthesize BMP-7. By damaging one kidney and removing the other kidney in each individual in a group of mice, the scientists simulated chronic renal failure in the animals. One group of the surgically altered mice received a special diet and a vitamin D supplement. Another group of the mice, the treatment group, received a special diet, the vitamin D supplement, and BMP-7 injections once a week for 12 weeks. After the mice were sacrificed, the scientists removed the animals' femurs (thigh bones), examined the bone tissue microscopically, and compared the tissue of the mice treated with BMP-7 to the tissue of the mice that had not received the protein. Hruska and his team found that the mice treated with BMP-7 had normal numbers of osteoblasts (bone-building cells) and normal bone formation, whereas the mice that did not receive the protein had indications of ABD. The results supported their hypothesis, and the researchers think BMP-7 injections may be an effective treatment for human cases of ABD. Additional research, however, is needed to provide evidence that the protein is useful in treating ABD in human patients. To learn more about Hruska's research, visit the *In The News* section of this text's companion Web site at www.wiley.com/college/alters and view the video "Kidney Bones."

*Write your immediate reaction to the research involving BMP-7 treatment for mice with damaged kidneys: first, summarize the main point in a sentence or two; then suggest what you think its significance is. You will have an opportunity to reflect on your responses and gather more information on this topic in the* In The News *feature at the end of this chapter. In this chapter, you'll learn more about the excretory system, how the kidneys work, and chronic renal failure. Bone structure and function are discussed in Chapter 34.*

## CHAPTER GUIDE

### What are the functions of excretion?

**31.1** Excretion removes metabolic wastes and balances water and salt.

**31.2** Excretory organs are varied among organisms.

### What substances does the human body excrete?

**31.3** The human body excretes carbon dioxide, water, salts, and nitrogenous molecules.

**31.4** The primary organs of excretion in humans are the lungs and the kidneys.

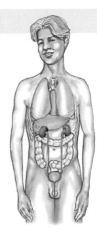

### How do kidneys work?

**31.5** Kidneys are comprised of microscopic filtering units, collecting ducts, and blood vessels.

**31.6** Kidneys regulate the balance of water and dissolved substances in the blood.

**31.7** The urinary system is made up of the kidneys, ureters, urinary bladder, and urethra.

**31.8** The principal function of the kidney is homeostasis.

### What health concerns relate to the kidneys?

**31.9** Kidney stones, which are crystals of salts, may develop in the kidney.

**31.10** A reduction in the filtration rate in the glomerulus leads to renal failure.

# What are the functions of excretion?

## 31.1 Excretion removes metabolic wastes and balances water and salt.

The story of excretion is, to a large extent, a story about water and salt. That story explains why the sea turtle in **Figure 31.1a** appears to be crying. In animal bodies, water is contained within compartments—any spaces bounded by membranes. Cells themselves are compartments, as are structures such as the interiors of the blood vessels and the heart. Water moves among the compartments within an organism, pumped by muscles surrounding those compartments, and by osmosis, moving from areas of high concentration of water to areas of lower concentration of water. The concentration of salts—compounds that form ions in water—affects the concentration of water in the various body compartments and therefore water's movement. To regulate the water in the various body compartments, animals must therefore regulate the amount of salt as well—a balancing act of sorts.

Animals have evolved various mechanisms that regulate their water and salt levels. The production of the turtle's saltwater "tears" is only one mechanism. Because the sea turtle lives in a salty environment, water tends to leave its less salty body. To compensate, the turtle drinks a lot of seawater but gets rid of sea salt via excretory organs located in the corners of its eyes. So, shedding salty tears is one way the sea turtle accomplishes its balancing act and maintains a steady internal environment.

Although excretion ends the salt/water balance story for one animal, it may begin the story for another kind of animal. Butter-

(a)

(b)

**Figure 31.1 Reptiles balance salt and water.** (*a*) Water tends to leave the sea turtle's body and move into the saltier environment. The sea turtle drinks seawater to replenish the water its body loses, but it must dispose of the salt. Excretory organs located in the corners of its eyes accomplish this task. (*b*) This side-necked turtle sheds "tears" containing salts and minerals on which butterflies feed.

## FOCUS

## *on Plants*

Plants have no excretory organs, so how do they regulate the concentration of water in their cells? How do they rid themselves of oxygen, the end product of photosynthesis? Plants take in water via their roots and lose water as water vapor, principally from openings in their leaves called stomata (**Figure 31.A**, and see the discussion of transpiration in Chapter 24.) Oxygen passes from the plant through the stomata also, providing oxygen to the environment.

**Figure 31.A** Stomata on a rose leaf.

LM 330×

flies primarily feed on the nectar of flowers, a rich energy source packed with sugars. Still, like other animals, butterflies need salts and minerals, too. They sip the moisture containing these substances from the eyes and nose of the side-necked turtle in Figure 31.1*b*. These secretions are not as salty as the tears of the sea turtle, but they are saltier than the water in their Amazon River habitat.

Although **excretion** is about water and salt balance, this process also regulates the concentrations of other substances in the body and removes metabolic wastes. **Table 31.1** lists the functions of excretion.

Do not confuse metabolic wastes removed by excretion with digestive wastes removed during elimination. Digestive wastes arise from the process of digestion—not from cellular metabolism, which comprises all the chemical reactions that take place within the cells of organisms. Even in unicellular organisms, food is really digested outside the organism because it is trapped within a vacuole, a sac whose contents are separated from the cell by a membrane. Food is digested within the vacuole, and wastes remain in the vacuole and leave the cell by exocytosis (see Figure 5.28). Metabolic wastes, on the other hand, are produced within cells themselves—within the cytoplasm or cellular oganelles—and are removed from the body by excretion.

| TABLE 31.1 | Functions of Excretion |
|---|---|

- Maintain proper concentrations of dissolved substances throughout an organism's cells and fluids
- Maintain a proper concentration of water throughout an organism's cells and fluids
- Remove metabolic wastes, such as carbon dioxide and nitrogenous wastes, from an organism
- Remove foreign substances, such as drugs and/or their breakdown products, from an organism

## 31.2 Excretory organs are varied among organisms.

As you can see from the turtle story in the previous section, organisms have more than one way to balance water and salt. Methods vary, and specialized excretory organs vary as well. Some organisms have no specialized excretory organs at all. This section, along with **Figure 31.2**, provides an overview of excretion among organisms.

### Cell membrane

In simple single-celled organisms such as bacteria ❶, the processes of diffusion, osmosis, and active transport across the cell membrane complete the job of excretion (see Chapter 5). Other living things that have no specialized excretory organs, but move substances across their cell membranes to help regulate their balance of water, salts, and other substances are plants (see a Focus on Plants), fungi, the plantlike and funguslike protists (such as algae and slime molds), the cnidarians (such as hydra, jellyfish, and corals), and the echinoderms (such as sea stars and sea urchins).

### Contractile vacuoles

More complex unicellular organisms, such as protozoans, and simple multicellular organisms, such as sponges, use organelles called **contractile vacuoles** for excretion ❷. Fresh water organisms are always hypertonic to the surrounding water. That is, their body fluids contain a higher concentration of dissolved substances than fresh water, so water flows into their body. Contractile vacuoles collect this excess water. When they reach a critical size, they contract, squeezing their contents out of the organism.

### Nephridia

Platyhelminthes, such as flatworms and tapeworms, and annelids, such as earthworms and leeches, have excretory organs called **nephridia** (neh-FRID-ee-uh, ❸). The simplest nephridia—*protonephridia*—are blind-ended tubes that open to the outside of the organism. A good example is the excretory system of the planarian, or flatworm.

Although questions remain regarding how the nephridia of flatworms work, research indicates that they exhibit one of the basic processes of excretion: *filtration*. During filtration, the body fluid of an organism is forced through a semipermeable membrane, which allows the fluid and small, dissolved molecules such as salts, sugars, and amino acids to pass, but holds back large molecules such as proteins ❹. The fluid and small molecules that pass through the flame cell, which together are now called the filtrate, become the excreted material ❺.

The nephridia of earthworms are more complex than those of flatworms and open into the body cavity (coelom) at one end and to the outside on the other. In these *metanephridia*, some substances are reabsorbed by active transport. Such *selective reabsorption* is another of the basic processes of excretion. By this means, certain substances are taken from the filtrate and returned to the body, helping maintain the right balance of dissolved substances.

## Visual Summary   Figure 31.2 Patterns of excretion.

**CELL MEMBRANE ❶**
(simple single-celled organisms including bacteria; fungi; plants; plantlike and funguslike protists [algae, slime molds]; cnidarians [hydra, jellyfish, corals]; and echinoderms [sea stars, sea urchins])

Processes at work: diffusion, osmosis, and active transport

**CONTRACTILE VACUOLES ❷**
(complex single-celled organisms [protozoans] and simple multicellular organisms [sponges])

**Paramecium**

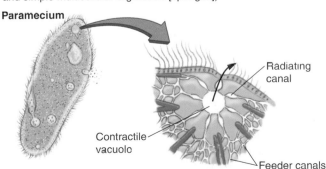

Radiating canal

Contractile vacuole

Feeder canals

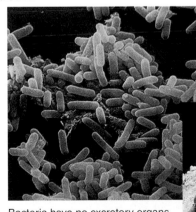

SEM 25,400×

Bacteria have no excretory organs.

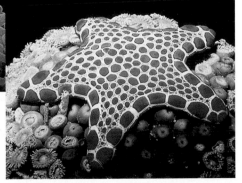

Echinoderms, such as this sea star, get rid of wastes by diffusion over thin surfaces or by exocytosis.

(*continued on next page*)

## Visual Summary    Figure 31.2 Patterns of excretion. (*continued*)

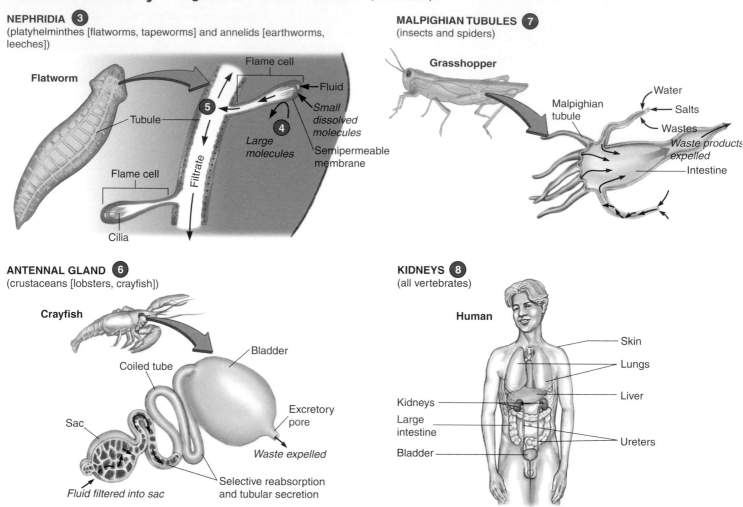

**NEPHRIDIA** ③
(platyhelminthes [flatworms, tapeworms] and annelids [earthworms, leeches])

Flatworm

Flame cell

Tubule

Fluid

⑤

④

*Small dissolved molecules*

*Large molecules*

Semipermeable membrane

Flame cell

Filtrate

Cilia

**MALPIGHIAN TUBULES** ⑦
(insects and spiders)

Grasshopper

Malpighian tubule

Water

Salts

Wastes

*Waste products expelled*

Intestine

**ANTENNAL GLAND** ⑥
(crustaceans [lobsters, crayfish])

Crayfish

Coiled tube

Bladder

Sac

Excretory pore

*Waste expelled*

*Fluid filtered into sac*

Selective reabsorption and tubular secretion

**KIDNEYS** ⑧
(all vertebrates)

Human

Skin

Lungs

Liver

Kidneys

Large intestine

Ureters

Bladder

Mollusks have complex metanephridial systems called kidneys, although they differ from vertebrate kidneys. The mollusks secrete wastes into the nephridial tubes by active transport. This process, called *tubular secretion*, is yet another basic process of excretion. These excretory processes of filtration, selective reabsorption, and tubular secretion are described again in more detail in Section 31.6.

### Antennal gland

If you've ever eaten lobster or crayfish, you may remember the green material located in the head. This is the excretory organ of crustaceans and is called the **antennal gland** ⑥ because its excretory pore opens near the base of the antennae. It is also called the *green gland* because of its color. These paired glands each consist of a sac, a long coiled tube, and a bladder.

In crustaceans, nitrogenous wastes, which are nitrogen-containing wastes from protein metabolism, are removed at the gills and the antennal gland. The green gland, however, is primarily involved with the excretion of water and salts. This is how it works: Fluid from the blood is filtered into the sac. The composition of this filtrate is adjusted by selective reabsorption and tubular secretion as it moves along the coiled tube. The bladder then stores the waste fluid until it is expelled through the pore.

### Malpighian tubules

Insects and certain other arthropods such as spiders have excretory organs called **Malpighian tubules** (mal-PIG-ee-en, named after the seventeenth century Italian anatomist who discovered them, Marcello Malpighi). The blind ends of these tubes lie within the blood-filled body cavity and empty into the intestine ⑦. Salts and wastes move into the tubules by either diffusion or active transport. The composition of this material is adjusted by selective reabsorption and tubular secretion as it moves along the tubule. Insects and spiders that live on fresh vegetation take in large amounts of water with their food and excrete liquid urine that is mixed with digestive wastes from the intestinal tract. Insects and spiders that live on dry food excrete very little water; their waste products are expelled as *uric acid*, which is a paste.

### Kidneys

Vertebrates have excretory/regulatory organs called **kidneys** ⑧. Vertebrate kidneys are made up of microscopic units called **nephrons** that carry out filtration, selective reabsorption, and tubular secretion to produce liquid waste called *urine*. The kidney is quite a versatile excretory and regulatory organ, balancing water and salt concentrations of marine, fresh water, and terrestrial animals.

# What substances does the human body excrete?

## 31.3 The human body excretes carbon dioxide, water, salts, and nitrogenous molecules.

The major products of excretion are water, salts, carbon dioxide, and nitrogen-containing molecules. The primary salt excreted is sodium chloride (NaCl)—table salt. Carbon dioxide ($CO_2$) is a gas produced during cellular respiration and is excreted primarily by the lungs. The lungs also excrete water in the form of water vapor—which is the reason you can see your breath on a cold day as the water vapor condenses to droplets. A small amount of salt and often a large amount of water also leave the body by means of glands in the skin (see Chapter 34).

The liver produces **bile pigments** from the breakdown of old, worn-out red blood cells. These pigment molecules are combined with other substances to form bile, which is carried by a duct from the liver, to the gallbladder where it is stored, and then to the small intestine. You may recall from Chapter 27 that bile plays a role in the emulsification of fat globules during the digestion of food. The bile pigments leave the body mixed with the digestive wastes and are the cause of the characteristic brown color of the feces, the digestive excrement. Because of its role in the excretion of the bile pigments, the liver is considered an organ of excretion. The large intestine is also considered an organ of excretion because cells lining its walls excrete the salts of some minerals such as calcium and iron (**Figure 31.3**).

Nitrogen-containing molecules, or **nitrogenous wastes**, are produced from the breakdown of proteins and nucleic acids (DNA and RNA). As discussed in Chapter 27, humans need to take in protein as a source of amino acids. These amino acids are used to construct body-building proteins. The liver breaks down any extra amino acids you may eat—those not needed for building new body parts or repairing old ones. Enzymes remove the amino groups ($-NH_2$), a process called *deamination*. The molecules that result can supply the body with energy. The amino group is a leftover that cannot be used.

In the liver, amino groups ($-NH_2$) are chemically converted to ammonia ($NH_3$). Ammonia is quite toxic, and the body must get rid of it quickly. A bit of chemical reshuffling meets this challenge, as the ammonia combines with carbon dioxide to form **urea**, the primary excretion product from the deamination of amino acids. Two other nitrogenous wastes found in small amounts in the urine are uric acid and **creatinine** (kree-AT-uh-neen). **Uric acid** is formed from the breakdown of nucleic acids found in the cells of the food you eat and from the metabolic turnover of your nucleic acids and adenosine triphosphate (ATP). Creatinine is derived primarily from a nitrogen-containing molecule called *creatine* (KREE-ah-teen) found in muscle cells.

### CONCEPT CHECKPOINT

1. List the mechanisms that have evolved for waste excretion and regulation of water/salt balance. List the type of excretory organ, briefly describe how it functions, and give an example of an organism that utilizes that organ.

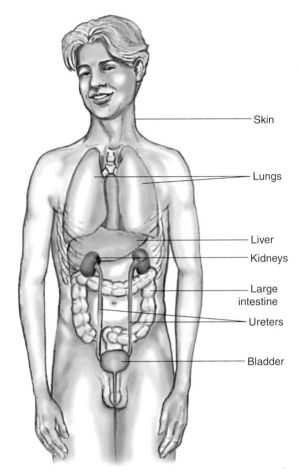

**Figure 31.3 Organs that carry out excretion in humans.** In addition to the kidneys, the skin, lungs, liver, and large intestine all play a role in the excretion of metabolic wastes from the body.

 **Visual Thinking:** List the excretory products for each of the organs of excretion shown here.

## 31.4 The primary organs of excretion in humans are the lungs and the kidneys.

The skin, liver, and large intestine play a minor role in the excretion of metabolic wastes from the body. The primary organs of excretion are the lungs and the kidneys. As mentioned previously, the lungs excrete carbon dioxide and water vapor. The kidneys excrete the ions of salts, such as $Na^+$ (sodium), $K^+$ (potassium), $Cl^-$ (chloride), $Mg^{2+}$ (magnesium), and $Ca^{2+}$ (calcium). In addition, they excrete the nitrogenous wastes urea, creatinine, and uric acid, along with small amounts of other substances that may vary depending on diet and general health. These substances, together with water, form the excretion product called **urine**.

The kidneys produce urine by first filtering out most of the molecules dissolved in the blood, then selectively reabsorbing useful components, and finally secreting a few other waste products into this remaining *filtrate*, which becomes urine. As the kidneys process the blood in this way, they regulate its chemical composition and water content and, in turn, the chemical and fluid environment of the body. For example, almost no proteins are excreted by the kidneys; nearly all proteins that are filtered from the blood are reabsorbed. However, more than half of the urea filtered from the blood is excreted. In addition, the kidneys maintain the concentrations of all ions within narrow boundaries. This strict maintenance of specific levels of ion concentrations keeps the blood's pH at a constant value, maintains the proper ion balances for nerve conduction and muscle contraction, and affects the amount of water reabsorbed into the bloodstream from the filtrate.

# How do kidneys work?

## 31.5 Kidneys are comprised of microscopic filtering units, collecting ducts, and blood vessels.

The kidneys look like what their name implies: two gigantic, reddish-brown kidney beans. Your body has two kidneys, each about the size of a small fist, located in the lower back region and partially protected by the lower ribs (see Figure 31.3 and **Figure 31.4**). Substances enter and leave the kidneys through blood vessels that pierce the kidney near the center of its concave border. *Collecting ducts* empty into an open area called the *renal pelvis*, carrying the urine produced by the nephrons. From here, urine exits the kidney via a tube called a **ureter** (YER-ih-ter). The word "renal" means kidney.

Covered with a fibrous connective tissue capsule and cushioned with fat, the kidney is made up of outer, reddish tissue called the *cortex* and inner, reddish-brown tissue called the *medulla*. Lined up within the medulla are triangles of tissue called the *renal pyramids*. The cortex and pyramids of each kidney are made up of approximately 1 million nephrons (Figure 31.4b), their collecting ducts, and the blood vessels that surround them (Figure 31.4c). Blood enters each kidney via the right and left *renal arteries*. Branches split off these major arteries, travel up the sides of the pyramids, and meet at the interface of the cortex and medulla. Small branches extend into the cortex, giving rise to the arterioles that enter each individual nephron. The total volume of blood in your body passes through this network of blood vessels every 5 minutes!

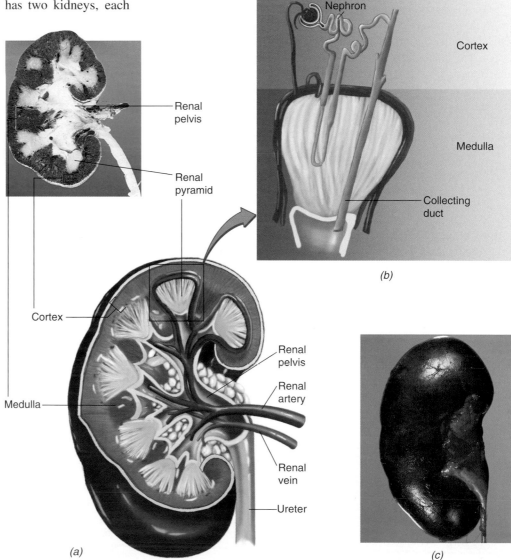

**Figure 31.4 Structure of the human kidney.** (a) This section shows some of the kidney's major internal structures. (b) This section of a renal pyramid shows the location of a nephron relative to the medulla and cortex of the kidney. This view of the nephron is greatly enlarged. Each renal pyramid contains tens of thousands of nephrons. (c) Each kidney is about 4 inches long, 2 1/2 inches wide, and 1 inch thick.

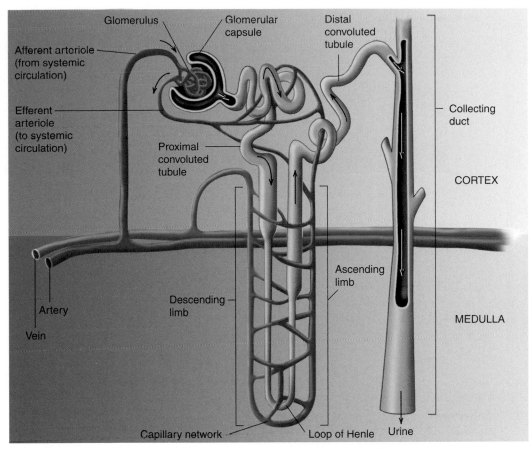

**Figure 31.5** Structure of a nephron. Part of the nephron is located in the cortex, and part is located in the medulla of the kidney. Nephrons filter the blood, reabsorb substances into the blood from the filtrate, and secrete certain ions and potentially harmful substances into the filtrate.

## 31.6 Kidneys regulate the balance of water and dissolved substances in the blood.

Each nephron (**Figure 31.5**) is structured to accomplish three tasks: filtration, selective reabsorption, and tubular secretion.

- *Filtration.* During filtration, the blood is passed through membranes that separate blood cells and most large proteins from much of the water and molecules dissolved in the blood. Together, the water and dissolved substances are referred to as the *filtrate.*
- *Selective reabsorption.* During reabsorption, desirable ions and metabolites and most of the water from blood plasma are recaptured from the filtrate, leaving nitrogenous wastes, excess water, and excess salts behind for later excretion. The filtrate is now called *urine.*
- *Tubular secretion.* During secretion, the kidney adds materials such as potassium and hydrogen ions, ammonia, and potentially harmful drugs to the filtrate from the blood. These secretions rid the body of certain materials and help control the blood pH.

### Filtration

At the front end of each nephron tube is a filtration apparatus called the *glomerular capsule*. The capsule is shaped much like a caved-in tennis ball and surrounds a tuft of capillaries called the *glomerulus* (pl. *glomeruli*). The glomerular capillaries branch off an entering

arteriole termed the *afferent arteriole* (named after the Latin verb *affero*, which means "going toward"). The diameter of the *efferent arteriole* leaving the capsule (*effero*, "going away from") is smaller than that of the afferent arteriole. As a consequence, the pressure of the incoming blood forces the fluid within the blood through the capillary walls and through spaces between specialized capsular cells that surround the capillaries. Capillary walls do not allow molecules such as large proteins and the formed elements of the blood to pass through. Water and smaller molecules such as glucose, ions, and nitrogenous wastes pass through easily.

This filtrate passes into the tubule of the nephron at the glomerular capsule. From there, it passes into a coiled portion of the nephron called the *proximal convoluted tubule*. This name describes the coiled (convoluted) tubule that is closest (proximal) to the glomerular capsule. Both of these structures lie in the cortex of the kidney. As the filtrate passes through the proximal tubule, selective reabsorption begins.

### Selective reabsorption

Epithelial cells throughout the length of the nephron tubule carry out reabsorption. During this highly selective process, specific amounts of certain substances are reabsorbed depending on the body's needs at the time. In the proximal convoluted tubule, glucose and small proteins are brought back into the blood by active

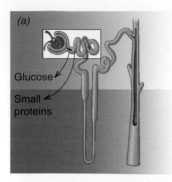

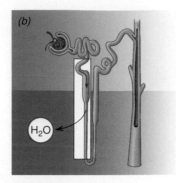

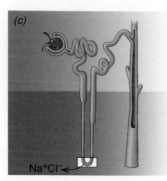

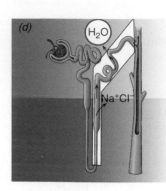

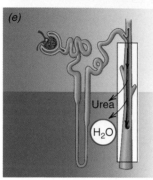

**Figure 31.6 How selective reabsorption works.** For ease of viewing the nephron, surrounding capillaries are not shown. (*a*) After filtration, in which fluid from the blood is forced into the glomerular capsule, selective reabsorption begins. First, glucose and small proteins in the filtrate move from the proximal convoluted tubule back into the blood by active transport. (*b*) Water moves from the descending limb of the loop of Henle back to the blood by osmosis. (*c*) Salt diffuses from the turn of the loop of Henle into the surrounding tissue. (*d*) Salt moves from the ascending limb of the loop of Henle into surrounding tissues by active transport. Water moves out of the distal convoluted tubule by osmosis. (*e*) Some urea diffuses from the collecting duct into surrounding tissues. Water moves from the collecting duct back into the blood by osmosis.

transport. These substances move out of the tubule and into the blood within surrounding capillaries (**Figure 31.6a.**)

The filtrate moves from the proximal convoluted tubule to the descending limb of the loop of Henle. The *loop of Henle* is the part of the tubule that extends into the medulla of the kidney. The *descending limb* of the loop of Henle dips downward, and the *ascending limb* extends in the opposite direction, back up into the cortex. The cells making up the walls of the descending limb do not permit salt and urea to pass out of the tubule; however, they are freely permeable to water. Water moves out of the tubule by osmosis, passing into the surrounding tissue fluid and then into the blood in surrounding capillaries (Figure 31.6b). As water passes out of the descending limb, it leaves behind a more concentrated filtrate.

At the turn of the loop, the walls of the tubule become permeable to salt but much less permeable to water. Salt passes out of the tubule and into the surrounding tissue fluid by diffusion (Figure 31.6c). This movement of salt produces a high concentration of salt in the tissue fluid surrounding the bottom of the loop.

In the ascending limb, the walls of the tubule contain active transport channels that pump even more salt out of the filtrate. This active removal of salt from the ascending limb increases the salt concentration within the renal medulla, causing water to diffuse outward from the filtrate as it passes through the descending limb (see Figure 31.6b). The walls of the ascending limb are impermeable to water. The filtrate passes from the ascending limb into the *distal convoluted tubule*. The walls of this portion of the kidney tubule are again permeable to water (Figure 31.6d). The filtrate entering the distal convoluted tubule contains some water and a relatively high concentration of urea.

Finally, the filtrate empties into a collecting duct that passes back into the medulla. The collecting ducts of all the nephrons bring the urine to the renal pelvis, an open area in which the urine collects before it flows out of the kidney through the ureter. The lower part of the collecting duct is permeable to urea. During this final passage, some of the concentrated urea in the filtrate diffuses into the surrounding tissue fluid, which has a lower urea concentration than the filtrate (Figure

31.6e). A high urea concentration in the tissue fluid surrounding the loop of Henle results. This high concentration of urea helps produce the osmotic gradient that causes water to move out of the filtrate as it passes down the descending limb (see Figure 31.6b). Water also passes out of the filtrate as it moves down the collecting duct. This movement occurs as a result of the osmotic gradient created by both the movement of urea out of the collecting duct and the movement of salt out of the ascending limb. The water is then collected by blood vessels in the kidney, which carry it back into the systemic circulation.

## Tubular secretion

A major function of the kidney is the elimination of a variety of potentially harmful substances that you may eat, drink, or inhale or that may be produced during metabolism. The human kidney has evolved the ability to detoxify the blood through the process of tubular secretion. During this process, the cells making up the walls of the distal convoluted tubule take substances from the blood within surrounding capillaries and from within the surrounding tissue fluid and put them into the filtrate to be excreted in the urine (**Figure 31.7**). (The proximal convoluted tubule plays a minor role in this process as well.)

Ammonia ($NH_3$) that has not been converted to urea is removed from the blood by tubular secretion. Certain prescription drugs such as penicillin are also removed by this process. In addition, the body rids itself of harmful drugs such as marijuana, cocaine, heroin, and morphine by tubular secretion. These drugs are processed by the liver, and their breakdown products are secreted into the filtrate. Specific tests for the breakdown products of these drugs can be performed on urine samples to determine whether a person has particular drugs in his or her body.

By removing specific amounts of hydrogen ions ($H^+$) from the blood and secreting them into the filtrate, the kidney can keep the pH of the blood at a constant level (7.35 to 7.45). (See Chapter 3 for an explanation of pH.) Likewise, the potassium ion ($K^+$) concentration of the blood is fine tuned by tubular secretion. The proper concentration of potassium ions is important to the proper functioning of muscles, including the heart.

## Summary: the process of urine formation

The roughly 2 million nephrons that form the bulk of the two human kidneys receive a flow of approximately 2000 liters of blood per day—enough to fill about 20 bathtubs! The nephron first filters the blood, removing most of its water and all but its largest molecules and its cells. The nephron then selectively reabsorbs substances back into the blood. This process is driven by two factors: (1) the varying permeability of the membranes of the cells lining the kidney tubule and (2) the development of a high osmostic gradient surrounding the loop of Henle.

Because of the varying permeability of kidney tubule cells, substances move from the filtrate back into the blood by diffusion and active transport mechanisms at specific places along the length of the tubule. Each substance will be reabsorbed only to a particular threshold level, however, with the rest remaining in the filtrate. For example, glucose is reabsorbed at the proximal convoluted tubule to a threshold level of 150 milligrams per 100 milliliters of blood. Any glucose above this threshold will be excreted in the urine. The amount of glucose in the blood of a healthy person does not normally exceed this limit. Persons with diabetes mellitus, however, excrete glucose because they fail to produce a hormone called *insulin* that promotes glucose uptake by the cells and the glucose level in the bloodstream rises. Urea, on the other hand, has a very low threshold for reabsorption that is reached quickly. Therefore, most urea stays in the filtrate until a portion of it diffuses from the collecting duct, adding to the osmotic gradient surrounding the loop of Henle.

The high osmotic gradient surrounding the loop of Henle causes most of the water filtered from the blood to be reabsorbed, or conserved, as the filtrate passes through the descending limb of the loop. Water conservation is the job of this loop. Of the 2000 liters of filtrate in humans, only about 1.5 liters leave the body as urine.

In contrast, fresh-water fish have no loop of Henle (**Figure 31.8a**). Their bodies take on water by osmosis; therefore, they rid their bodies of excess water rather than conserve it. Fresh-water fish drink little water and produce large amounts of urine. These mechanisms work to maintain their body fluids hypertonic to their fresh-water surroundings. That is, their body fluids contain more dissolved substances and have a lower concentration of water than

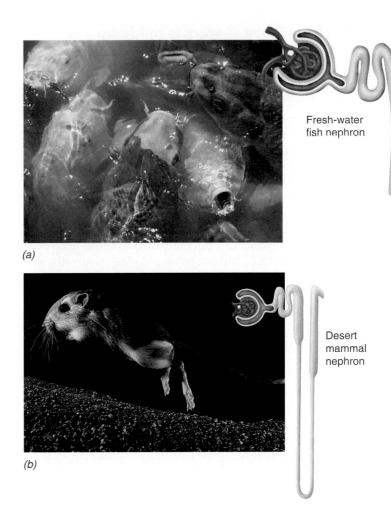

(a)

(b)

**Figure 31.8** The job of the loop of Henle is to conserve water. (a) The nephrons of these fresh-water fish have no loop of Henle. Since their bodies take on water by osmosis, they must rid themselves of excess water rather than conserve it. (b) The nephrons of desert mammals have long loops of Henle, allowing them to conserve a great deal of water, producing highly concentrated urine.

**Visual Thinking:** How does a long loop of Henle allow desert mammals to conserve a great deal of water?

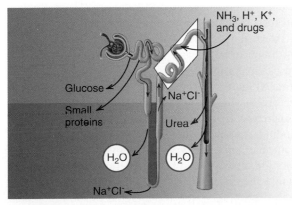

**Visual Summary**  **Figure 31.7** Tubular secretion and a summary of the process of selective reabsorption.

the water in which they live. Marine fish, on the other hand, lose water by osmosis to their environment. To compensate, they drink seawater and secrete the excess salt from their gills.

Only birds and mammals have a loop of Henle and are able to produce urine that is more concentrated than their body fluids. Among these animals, the longer the loop, the greater the ability to concentrate the urine and conserve water. Animals such as desert rodents that have highly concentrated urine have exceptionally long loops of Henle (Figure 31.8b).

The nephron also helps to control blood pH, fine tunes the concentrations of certain ions and molecules, and removes potentially harmful drugs from the blood by the process of tubular secretion. In these ways, the kidney contributes to the chemical and water balance of the body, thus functioning as a major organ of internal equilibrium, or homeostasis. The resulting fluid is urine, a waste that is excreted from the body.

## 31.7 The urinary system is made up of the kidneys, ureters, urinary bladder, and urethra.

The kidneys are only one part of the **urinary system**—a set of interconnected organs (**Figure 31.9**) that not only remove wastes, excess water, and excess ions from the blood but also store this fluid, or urine, until it can be expelled from the body. The urine exits each nephron by means of the collecting duct. From there, it flows into the renal pelvis (see Figure 31.4). This area narrows into a tube called the *ureter* that leaves the kidney on its concave border. By means of the rhythmic contractions of peristalsis, these muscular tubes—one from each kidney—bring the urine to a storage bag called the **urinary bladder**.

The urinary bladder is a hollow muscular organ that, when empty, looks much like a deflated balloon. As the bladder fills with urine, it assumes a pear shape. On average, the bladder can hold 700 to 800 milliliters of urine, or almost a quart. When less than half this amount is in the bladder, however, special nerve endings in the walls of the bladder, called *stretch receptors*, send a message to the brain that results in the desire to urinate.

The urinary bladder empties into a tube called the **urethra**. This tube leads from the underside of the bladder to the outside of the body. The urethra in men plays a role in the reproductive system, carrying semen to the outside of the body during ejaculation. The urinary and reproductive systems have no connection in women. In addition, the urethra in women is much shorter than the urethra in men and is closer to the rectum and its associated bacterial population. For these reasons, women contract bladder or urinary tract infections more easily than men.

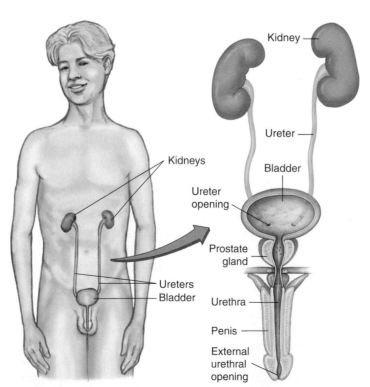

**Figure 31.9 The human urinary system.** The urinary system in males and females includes the kidneys, ureters, urinary bladder, and urethra. This diagram shows details of the male urinary system.

CONCEPT CHECKPOINT

2. Describe the pathway of a glucose molecule and of a urea molecule from the time it enters the kidney in the blood until it is excreted from the body, or reabsorbed back into the blood.

3. For each of the molecules (1–5) in the first column, indicate which processes (a–c) in the second column regulate its concentration as it passes through the kidney:
   1. glucose       a. filtration
   2. urea          b. tubular secretion
   3. $H^+$ ions     c. reabsorption
   4. $Na^+$ ions
   5. water

## 31.8 The principal function of the kidney is homeostasis.

The kidney is an excellent example of an organ whose principal function is homeostasis, the maintenance of constant physiological conditions within the body. It is concerned with both water balance and ion balance. To maintain homeostasis, however, the urinary system depends in part on the endocrine system and its hormones.

A *hormone* is a chemical messenger secreted by a gland into the bloodstream. It regulates the activity of cells in another part of the body. Although several hormones influence kidney function, among the most important are antidiuretic hormone and aldosterone. **Antidiuretic hormone (ADH)** regulates the rate at which water is lost or retained by the body. **Aldosterone** (al-DOS-terown) regulates the level of sodium ions ($Na^+$) and potassium ions ($K^+$) in the blood.

Antidiuretic hormone (ADH) is secreted by the pituitary gland found at the base of the brain. It works by controlling the permeability of the distal convoluted tubules and collecting ducts to water (**Figure 31.10**, left). When ADH levels increase, distal tubules and collecting ducts become more permeable to water. Water therefore moves by osmosis ❶ back into the blood within surrounding capillaries. When ADH levels decrease, the permeability of the collecting ducts and distal tubules to water decreases. Therefore, less water is reabsorbed from the filtrate ❷ and is excreted from the body in urine. Alcohol inhibits the release of ADH, resulting in decreased reabsorption of water and therefore increased urination, which can cause dehydration.

Aldosterone (Figure 31.10, right) is secreted by the adrenal glands, which are located on top of both kidneys. When aldosterone levels increase, the kidney tubule cells increase their reabsorption of sodium ions from the filtrate ❸ and decrease their reabsorption of potassium ions ❹. Put simply, aldosterone promotes the retention of sodium and the excretion of potassium. In addition, because the concentration of sodium ions in the blood affects the reabsorption of water, the increase in sodium ions in the blood causes water to move by osmosis from the filtrate into the blood ❺.

| ADH LEVEL | EFFECT ON KIDNEY |
|---|---|
| Increased ADH Levels | The distal convoluted tubules and collecting ducts become permeable to water; water moves out of ducts and back into blood. |
| Decreased ADH Levels | The distal convoluted tubules and collecting ducts are impermeable to water; water is not reabsorbed from the filtrate and is excreted in the urine. |

| ALDOSTERONE LEVEL | EFFECT ON KIDNEY |
|---|---|
| Increased Aldosterone Levels | Tubules increase reabsorption of sodium from the filtrate and decrease reabsorption of potassium; water and sodium thus move from filtrate into the blood, and excess potassium is excreted. |
| Decreased Aldosterone Levels | Tubule absorption of sodium and potassium is normal; water is not reabsorbed from the filtrate and is excreted. |

**Figure 31.10** Hormones regulate kidney function.

**Visual Thinking:** Caffeine is a diuretic. When you drink a caffeinated beverage, which of the hormones described in this illustration does it affect? Does it result in the increase or decrease of that hormone? How is water balance in your body affected?

## *just wondering . . .*

### Questions students ask

**I play sports and we take random drug tests. Are these tests accurate? I worry about being accused of something I didn't do.**

In 1986, the National Collegiate Athletic Association (NCAA) instituted mandatory drug screening of student athletes. Therefore, if you attend an NCAA school and are involved in sports, you are subject to the NCAA regulations prohibiting drug use. In addition, many colleges and universities have their own drug use and drug-testing policies.

In addition to its use in collegiate sports, drug testing is conducted in professional and Olympic sports as well as in the workplace. It was first used in the United States by the Department of Defense in the 1960s and early 1970s to screen servicemen and women returning from the war in Vietnam. In the 1980s, drug-testing methods became more reliable, and private companies, especially those concerned about public safety, such as some in the transportation industry, began testing employees. Many laws were passed in the late 1980s that established drug-free workplaces within the federal government and random drug-screening programs in transportation industries such as aviation, rail, mass transit, and trucking.

Major improvements have come with the introduction of tests that use recombinant DNA techniques to amplify trace chemicals in urine (see Chapter 11). The enzyme-multiplied immunoassay technique (EMIT) has become the standard approach for screening large numbers of urine samples. EMIT can be used in a robot tester that employs a light sensor to read urine samples and prints out a value for each of five or six drugs present; it can process 18,000 samples an hour. The detection period depends on the type and dose of the drug. EMIT is more than 98% accurate, with error biased toward nondetection.

Because EMIT responds to a broad range of opiate and amphetamine compounds, it sometimes produces a positive result for harmless prescription drugs such as ibuprofen and decongestants, as well as for foods such as herbal teas and poppy seeds. For this reason, samples that show a positive result on EMIT are then retested by more cumbersome but 100% accurate procedures such as gas chromatography–mass spectrometry. For example, eating poppy seeds may trigger a positive signal on EMIT, but the confirmation test tells that a key heroin breakdown product (6-0-acetyl-morphine) is not present. So don't worry; the tests are highly reliable, and you shouldn't be wrongly accused. ●

*Are you wondering about a topic in biology and how it relates to your life? Submit your question by clicking the Just Wondering link in this text's companion Web site at www.wiley.com/college/alters.*

## What health concerns relate to the kidneys?

### 31.9 Kidney stones, which are crystals of salts, may develop in the kidney.

Do you know someone who has had a kidney stone? If so, they probably told you how painful it was to "pass" the stone. Kidney stones are not uncommon. About 10% of Americans will develop a kidney stone sometime during their lives.

What are kidney stones, and how do they form? During the formation of urine, sometimes certain salts do not stay dissolved in the filtrate—most notably calcium salts or uric acid—but instead form crystals called **kidney stones**. These stones form for a variety of reasons, including too little water intake or use of certain medications.

Kidney stone attacks are excruciatingly painful; the stone, which may be quite jagged, moves through the kidney and then blocks the flow of urine. **Figure 31.11** shows a kidney stone on a fingertip.

Since 1984, a treatment for kidney stones called *shock-wave lithotripsy* (LITH-oh-TRIP-see) has been available. In this procedure, the patient is immersed up to the neck in a tank of water. Guided by X-ray monitors, intense sound waves are directed at the stone, shattering it. Lithotripsy is a Greek word meaning "stone crushing." Once the stone is broken apart, the pieces are excreted in the urine. Surgery is still used to remove stones larger than 2 centimeters in diameter or stones that are causing infection.

### 31.10 A reduction in the filtration rate in the glomerulus leads to renal failure.

**Renal failure** is a condition in which the kidneys stop working properly or at all. It occurs when the filtration of the blood at the glomeruli either slows or stops.

**Figure 31.11 A kidney stone.** Shown on a fingertip, this kidney stone is small. Others may be as large as a pearl or even, rarely, a golf ball.

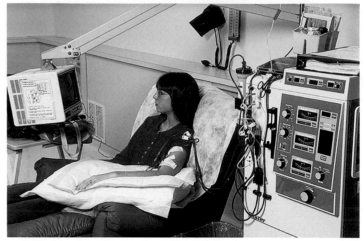

**Figure 31.12 This patient is undergoing hemodialysis.** The artificial kidney is the machine at right. The patient's blood is visible coursing through the tubes from the patient to the machine and back again.

In acute renal failure, filtration stops suddenly. Acute renal failure can have many causes, such as decreased blood flow through the kidneys as a result of heart problems or blockage of a blood vessel, damage to the kidney by disease, the presence of a kidney stone blocking urine flow, or severe dehydration as occurs occasionally in marathoners and other endurance athletes.

In chronic renal failure, the filtration of the blood at the glomeruli slows gradually. This condition is usually irreversible because it is most commonly caused by injury to the glomeruli. These injuries have many causes, such as the deposit of toxins, bacterial cell walls, or molecules produced by the immune system within the glomeruli; the coagulation of the blood within the glomeruli; or the presence of a disease such as diabetes mellitus.

If the kidneys become unable to excrete nitrogenous wastes, regulate the pH of the blood, or regulate the ion concentration of the blood, the individual will die unless the blood is filtered in another manner. A machine called an artificial kidney accomplishes this job (**Figure 31.12**). The process of filtering blood to replace the normal function of the kidney is called **dialysis**.

During *hemodialysis*, a patient's blood is pumped through tubes to one side of a selectively permeable membrane. On the other side of the membrane is a fluid called the dialysate, which contains the same concentration of ions as is normally found in the bloodstream. Because small molecules can pass across the membrane, any extra ions in the patient's blood move by diffusion into the dialysate until their concentrations on both sides of the membrane are equal. In addition, the dialysate contains no wastes and is changed frequently, so the wastes in the patient's blood also diffuse into the dialysate. Hemodialysis is a slow process, usually taking four hours, three times a week. The results of research studies show that some patients with kidney failure can undergo hemodialysis successfully at home, while they sleep.

Another type of dialysis is *periotoneal dialysis*. This process uses the thin lining of the abdominal cavity, the peritoneum, as the dialyzing membrane. The patient has a thin tube called a catheter implanted in the abdominal wall. The dialysate flows into the abdomen through the catheter from a bag outside the body. Waste materials and excess water pass from blood vessels in the peritoneum into the dialysate. After several hours, the dialysate is drained from the abdomen.

Only 10% of persons with kidney failure survive for at least 10 years on dialysis due to the stress that toxic waste products place on the body organs. Many more persons with kidney failure survive for at least 10 years with a kidney transplant: 56% with a transplant from a deceased donor and 76% with a transplant from a living donor.

Each year in the United States, over 13,000 patients with renal failure receive kidney transplants. It is the second most common transplant operation in this country, with corneal transplants being the first.

## CONCEPT CHECKPOINT

4. You did not drink enough water during a 10-K run. What effect will this have on: (a) the production of ADH, (b) the concentration of water in the urine, (c) the concentration of urea in the urine, (d) the permeability of the collecting duct to water, and (e) the production of aldosterone?

# CHAPTER REVIEW

## Summary of Key Concepts and Key Terms

### What are the functions of excretion?

**31.1 Excretion** (p. 534) is a process whereby metabolic wastes, excess water, and excess salts are removed from the blood and passed out of the body.

**31.2** Some organisms have no specialized excretory organs and rely on diffusion, osmosis, and active transport to rid themselves of metabolic wastes and excess water.

**31.2** The simplest excretory organ is the **contractile vacuole** (p. 536) and is found in protozoans and sponges.

**31.2** Other types of excretory organs are **nephridia** (neh-FRID-ee-uh, p. 536), **antennal glands** (p. 536), **Malpighian tubules** (mal-PIG-ee-en, p. 536), and **kidneys** (p. 536).

**31.2** Vertebrate kidneys are made up of microscopic units called **nephrons** (p. 536) that carry out filtration, selective reabsorption, and tubular secretion to produce liquid waste called urine.

### What substances does the human body excrete?

**31.3** The excretory products of the body are **bile pigments** (p. 537), nitrogen-containing molecules (**nitrogenous wastes**, p. 537), carbon dioxide, water, and salts.

**31.3** Nitrogenous wastes can be in the form of ammonia, **urea** (p. 537), **creatinine** (p. 537), and **uric acid** (p. 537).

**31.4** The primary organs of excretion in humans are the lungs and the kidneys.

**31.4** Bile pigments are excreted by the liver and passed out of the body by means of the digestive system.

**31.4** Most carbon dioxide is excreted by the lungs.

**31.4** The kidneys excrete most of the nitrogenous wastes along with excess water and salts in an excretion product called **urine** (p. 537).

### How do kidneys work?

**31.5** Via tubes called **ureters** (YER-ih-ter; p. 538), the kidneys excrete wastes as urine, which is formed within microscopic tubular units of the kidney called nephrons.

**31.5** The kidney is made up of an outer region called the cortex and an inner region called the medulla; these tissues are composed of nephrons, their collecting ducts, and blood vessels.

**31.6** During the formation of urine, most of the water and other small molecules are first filtered out of the blood; this fluid is called the filtrate.

**31.6** As the filtrate flows through the nephron tubule, various substances are reabsorbed because the permeability of its walls varies along its length.

**31.6** By means of tubular secretion, the kidney also excretes the breakdown products of a variety of potentially harmful substances from the blood, such as marijuana, cocaine, heroin, morphine, and prescription drugs.

**31.6** Most of the water removed from the blood is reabsorbed from the filtrate because of a high osmotic gradient that surrounds certain sections of the kidney tubule.

**31.6** Different groups of vertebrates can function in a wide variety of environments because of adaptations in their nephrons that allow selective reabsorption of molecules valuable to their particular habitats.

**31.7** The **urinary system** (p. 541) is a set of interconnected organs that not only remove wastes, excess water, and excess ions from the blood but also store this fluid, or urine, until it can be expelled from the body.

**31.7** Urine leaves the kidneys by means of muscular tubes called ureters and is conveyed to a storage pouch called the **urinary bladder** (p. 542).

**31.7** A tube called the **urethra** (p. 542) brings urine to the outside.

**31.8** Two hormones, **antidiuretic hormone** (**ADH**, p. 543) and **aldosterone** (p. 543), help the kidneys control the balance of water and ions in the blood, thus helping it perform its primary function, the maintenance of homeostasis.

## What health concerns relate to the kidneys?

**31.9** Crystals of salts that sometimes form within the kidney are called **kidney stones** (p. 544); surgery and shock-wave lithotripsy are two ways in which kidney stones are removed.

**31.10** When the filtration of the blood at the glomerulus is seriously impaired, the blood must be filtered by means of **dialysis** (p. 545).

**31.10** In severe cases of **renal failure** (p. 544), a patient may require kidney transplantation.

---

### Level 1 | Learning Basic Facts and Terms

**Matching**

Match the phrases in items 1–5 with the structures in a–e.

1. _____ The structure that passes urine into the renal pelvis.
2. _____ The part of the nephron that is permeable only to water.
3. _____ The walls of this structure contain active transport channels that pump salt out of the filtrate.
4. _____ The structure from which small proteins are first reabsorbed by the blood.
5. _____ The structures that ADH affects to regulate water content of urine.

a. Proximal convoluted tubule
b. Ascending limb of the loop of Henle
c. Descending limb of the loop of Henle
d. Distal convoluted tubule
e. Collecting duct

**Multiple Choice**

6. All of the following are functions of the mammalian kidney except:
   a. regulation of water balance in the blood.
   b. filtration of the blood.
   c. excretion of metabolic wastes like urea.
   d. regulation of salt balance in the blood.
   e. detoxification of harmful compounds.

7. Which of the following organs in mammals could be considered responsible for excretion?
   a. Lungs
   b. Kidneys
   c. Liver
   d. Skin
   e. All of the above

8. The function of reabsorption by metanephridia and nephrons is to
   a. return molecules to the blood that were initially filtered out.
   b. regulate blood pH.
   c. excrete urea.
   d. excrete $CO_2$.
   e. return waste from the blood to the filtrate.

9. A function of tubular secretion in nephrons is to
   a. return molecules to the blood that were initially filtered out.
   b. regulate blood pH.
   c. produce bile.
   d. produce urea from ammonia.
   e. return waste from the blood to the filtrate.

10. Which of the following excretory systems is correctly matched with the animal within which it is found?
    a. Flame cells—earthworms
    b. Metanephridia—flatworms
    c. Kidneys—crabs
    d. Gills—reptiles
    e. Malpighian tubules—terrestrial arthropods

---

### Level 2 | Learning Concepts

1. What evolutionary trends do you observe in how animals excrete waste and regulate water balance?
2. Some organisms require no specialized excretory organs to rid themselves of toxic waste materials. How can these organisms survive when the excretion of wastes is so important for survival?
3. What are the primary metabolic waste products that you excrete? What metabolic process results in their production? What organ is responsible for their excretion?
4. What do ammonia, urea, and uric acid have in common? How do they differ? Why do animals expend the energy to create urea and uric acid?

5. Draw a diagram of a human kidney from memory. Show the location of the renal pelvis, cortex and medulla, and ureter. Now draw an enlarged nephron overlaid on the kidney diagram from memory. Show the location of the glomerulus, glomerular capsule, nephron tubules, and collecting duct with respect to their positions in different sections of the kidney.
6. Summarize how your kidneys help your body conserve water.
7. How does the endocrine system work together with your kidneys to maintain homeostasis?

---

### Level 3 | Critical Thinking and Life Applications

1. People who have high blood pressure often take diuretics that inhibit the secretion or action of ADH. Explain how this would treat high blood pressure, or hypertension.
2. The primary nitrogenous waste product produced by birds and reptiles is uric acid, which is insoluble in water and can be excreted as crystals. It is more energy expensive, however, to synthesize than urea. Is it advantageous to these organisms to produce uric acid rather than urea?

3. Normally, glucose is not present in the urine. Persons with diabetes mellitus often show glucose in their urine upon testing. Why? Using the information in this chapter, propose a reasonable hypothesis as to why persons with poorly controlled diabetes urinate frequently.
4. In organisms with contractile vacuoles, a layer of mitochondria surrounds the vacuole. What role might you propose for this high concentration of mitochondria?
5. Explain why a urine test can reveal the use of certain drugs.

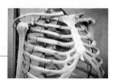

## In The News | **Critical Thinking**

**KIDNEY BONES**

Now that you have a greater understanding of the excretory system, reread this chapter's opening story about research into kidney disease and bone loss. To better understand this research, it may help you to follow these steps:

1. Review your immediate reaction to research concerning kidney failure and BMP-7 treatment of mice that you wrote when you began reading this chapter.

2. Based on your current understanding, again summarize the main point of the research in a sentence or two.

3. What questions do you now have about this research that this chapter's opening story does not answer?

4. Collect new information about the research. Visit the *In The News* section of this text's companion Web site at www.wiley.com/college/alters and watch the "Kidney Bones" video. Then use the "summary" link to read the accompanying story and access related links. Use this information, the links provided, and other online and library resources to answer your questions and find updates about this research topic. State the sources of your information. Explain why you think the information is accurate. Also determine whether the information expresses a particular point of view or is biased in any way.

5. What in your view are the most significant aspects of this research? Give reasons for your opinion and for any changes in your ideas based on the additional information you have collected and the analysis you have done.

# NERVOUS SYSTEM COMMUNICATION

Kid Concussions

Being sidelined, not able to participate with the rest of the team, is the last thing a competitive athlete wants. Nevertheless, an athlete who sustains a head injury needs to be removed immediately from the activity and evaluated medically.

Although the skull protects the brain, a blow to the head not only injures the brain at the place where the blow strikes, but it also causes the brain to move and hit against the opposite side of the skull. Such forces damage brain cells and result in temporary loss of cognitive functioning, including the capacity for thought or memory, and sometimes, permanent neurological problems. The person experiencing this type of head injury, referred to as a *concussion*, may lose consciousness immediately. If conscious or after regaining consciousness, the injured person is often confused, disoriented, and unable to answer simple questions. In addition,

he or she may have memory loss and report having a headache and a ringing sensation in the ears. Many people who suffer a concussion seem to recover within a few minutes, but subtle long-term deficits in cognitive abilities and changes in personality may result. Furthermore, persons who experience a concussion have a high risk of future concussions following head injuries, and the consequences can be catastrophic.

Second impact syndrome (SIS) can occur when a person who has had a concussion sustains another head injury before recovering fully from the initial injury. In response to the additional trauma, the brain tissue swells rapidly and applies pressure on the brain stem, the location of control centers for vital body functions such as breathing and heart rate. When this happens, the person lapses into a coma and may die within minutes. Therefore, it's essential for coaches and members of the athletic health care team to

know when it's safe to permit an athlete who has had a concussion to participate in sports again.

Michael (Micky) Collins, assistant director of the Concussion Program at the University of Pittsburgh's Center for Sports Medicine, and his colleagues developed the Immediate Post-concussion Assessment and Cognitive Testing (ImPACT) system to evaluate cognitive functioning of athletes who have suffered concussions. The computerized testing program uses an array of stimuli, including designs, words, symbols, and numbers, and evaluates multiple aspects of cognitive functioning, such as memory, attention span, reaction time, and problem solving. Prior to their sport's season, athletes undergo ImPACT testing to obtain baseline values concerning their cognitive functioning. If an athlete sustains a concussion, he or she is retested using the ImPACT program, and the results are compared with those obtained earlier, before the athlete is allowed to return to play. Such objective testing enables coaches and parents to feel more confident about the decision to allow the athlete to rejoin the action with his or her teammates. To learn more about the ImPACT computer program, visit the *In The News* section of this text's companion Web site at www.wiley.com/college/alters and view the video "Kid Concussions."

*Write your immediate reaction to this information about concussions and the ImPACT system: first, summarize the main point in a sentence or two; then suggest what you think its significance is. You will have an opportunity to reflect on your responses and gather more information on this topic in the* In The News *feature at the end of this chapter. In this chapter, you will learn more about the structure and functioning of the brain and other components of the nervous system.*

# CHAPTER GUIDE

## Do all animals have nervous systems?

**32.1**    Almost all animals have nerve cells and nervous systems or networks.

**32.2**    Nervous systems exhibit great diversity throughout the animal kingdom.

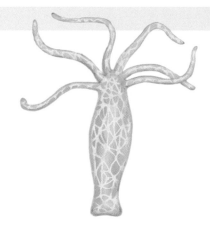

## How do nerve cells work?

**32.3**    Nerve cells are structured to receive and send impulses.

**32.4**    Nerve impulses are electrical signals transmitted along membranes of neurons.

**32.5**    Insulated neurons carry impulses more quickly than uninsulated neurons.

**32.6**    Neurotransmitters send signals across junctions between neurons or between neurons and muscle cells.

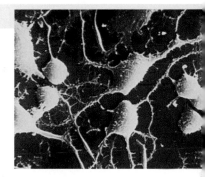

## How does the human nervous system work?

**32.7**    The human nervous system is highly organized.

**32.8**    The human central nervous system integrates sensory and motor impulses.

**32.9**    The spinal cord connects the peripheral nervous system with the brain.

**32.10** The human peripheral nervous system carries sensory and motor impulses.

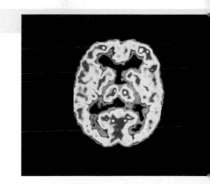

## What are psychoactive drugs and how do they work?

**32.11** Psychoactive drugs affect the action of neurotransmitters in specific parts of the brain.

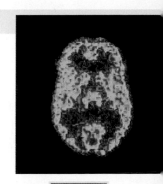

# Do all animals have nervous systems?

## 32.1 Almost all animals have nerve cells and nervous systems or networks.

Humans are not the only members of the animal kingdom to have brains and nervous systems. With the exception of the sponges, both invertebrates and vertebrates have nerve cells, or **neurons** (NER-ons), and nervous systems or nerve networks. Neurons are specialized cells with cell bodies containing a nucleus and long cell extensions, or *fibers*, (**Figure 32.1**) that transmit signals within an animal, forming networks of communication. Bundled in groups, the long cell extensions of neurons make up *nerves*. Their message signals are called **nerve impulses**. Like the dots and dashes of Morse code, the signals that neurons transmit are the same, differing only in their frequencies, their points of origin, and their destinations.

## 32.2 Nervous systems exhibit great diversity throughout the animal kingdom.

A complex web of neurons is a nervous system. More specifically, a **nervous system** is a network of neurons specialized for transmitting information from sensory receptors to neurons, between neurons, and from neurons to effectors, such as muscles and glands.

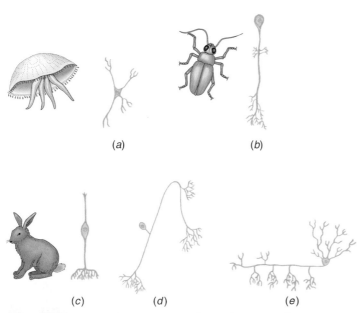

(a)   (b)

(c)   (d)   (e)

**Figure 32.1 Invertebrate neurons have the same characteristics as vertebrate neurons.** Neurons typically have a cell body and long cell extensions that transmit impulses. (a) A neuron from the nerve net of a cnidarian. (b) An arthropod motor neuron. (c) A bipolar neuron from the vertebrate retina. (d) A mammalian spinal sensory neuron. (e) A basket cell from the mammalian cerebellum.

The job of a nervous system is to gather information about an organism's internal and external environments, and then to process and respond to the information it has gathered. These nervous system responses are nerve impulses—electrochemical commands sent out to muscles, glands, or other nerves, directing them to react in an appropriate way.

Although invertebrate neurons look and act very much like vertebrate neurons (see Figure 32.1), the nervous systems of invertebrates vary in their organization among invertebrate phyla. Sponges are the only multicellular animals with no nerve cells. The simplest nervous system organization is the nerve net, found in the cnidarians (hydra, jellyfish, sea anemones—**Figure 32.2 ❶** ).

A **nerve net** is a system of interconnecting nerve cells with no central controlling area, or **brain**. Impulses are transmitted in all directions, resulting in a response in the stimulated region of the organism. Responses are simple, such as the withdrawing of tentacles. The stronger the stimulus, the greater the area of response.

Echinoderms (sea stars, sea urchins, sea cucumbers) are another phylum of invertebrates with no brain. The nervous system of echinoderms consists of a ring around the esophagus with five nerves radiating outward ❷ . In species such as sea stars, these nerves extend down each arm to the tube feet. Some species of echinoderms have a second nerve ring system that controls motor function.

The flatworms (*Platyhelminthes*), have a variety of nervous system plans. Some have a simple nerve net. The most complex have a distinct brain and one to three pairs of nerve cords, also called trunks, that run the length of the animal. The flatworm shown in Figure 32.2 ❸ has a small brain that sends nerve impulses down its single pair of nerve cords, resulting in a ladderlike nervous system. Such simple brains are aggregations of nervous tissue often referred to as *ganglia* (GANG-glee-uh), which serve as a central processing station for incoming and outgoing impulses. Ganglia are not capable of conscious thought or emotion. The nematodes, or roundworms, also have simple brains and nerve cords.

Annelids (earthworms and leeches ❹ ) also have an anterior brain at the "head end" attached to one or two nerve cords. In the case of annelids, however, the nerve cords run along the ventral, or "stomach" side of the animal. These segmented worms also have segmental ganglia. That is, at each segment, the nerve cord has a swelling of nervous tissue, making this annelid nervous system a bit more complex than those of the flatworms and roundworms. Arthropods (insects, spiders, lobsters; see ❹ ) have a similar plan: a well-defined ventral nerve cord with segmental ganglia and a prominent brain at the anterior end.

The mollusks—clams, squid, octopus—have varying nervous system structures that correspond to the level of activity of the organism. For example, the slow-moving clam has a ring of ganglia surrounding its esophagus with nerve cords running to its major body parts. In contrast, the fast-moving cephalopods, such as squid and octopus, have large, complex brains. Results of research studies show that cephalopods are capable of memory and learn-

# Visual Summary    **Figure 32.2** Patterns of nervous systems.

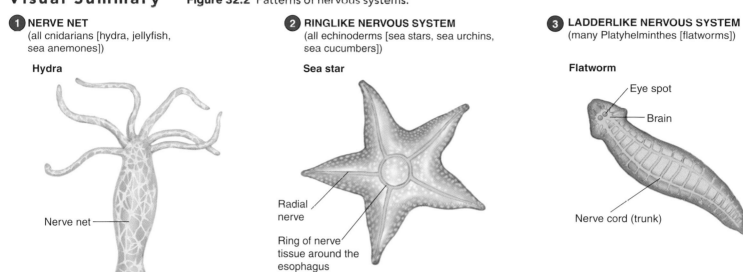

**1** **NERVE NET**
(all cnidarians [hydra, jellyfish, sea anemones])

**Hydra**

Nerve net

**2** **RINGLIKE NERVOUS SYSTEM**
(all echinoderms [sea stars, sea urchins, sea cucumbers])

**Sea star**

Radial nerve

Ring of nerve tissue around the esophagus

**3** **LADDERLIKE NERVOUS SYSTEM**
(many Platyhelminthes [flatworms])

**Flatworm**

Eye spot

Brain

Nerve cord (trunk)

**4** **NERVOUS SYSTEM WITH SEGMENTAL GANGLIA, VENTRAL NERVE CORD, AND A BRAIN**
(annelids [earthworms, leeches], arthropods [insects, spiders, lobsters])

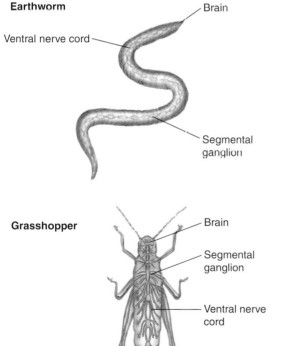

**Earthworm**

Brain

Ventral nerve cord

Segmental ganglion

**Grasshopper**

Brain

Segmental ganglion

Ventral nerve cord

**5** **CENTRAL (BRAIN AND SPINAL CORD) AND PERIPHERAL NERVOUS SYSTEM**
(all vertebrates)

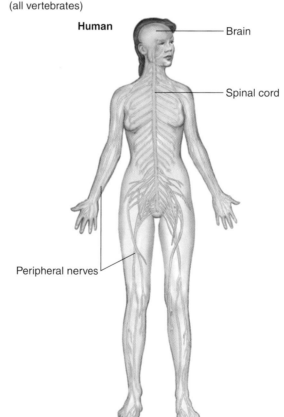

**Human**

Brain

Spinal cord

Peripheral nerves

ing. In addition, cephalopods have neurons specialized for rapid impulse conduction, allowing them to move very quickly. The processes of these neurons, called giant fibers, connect the brain with the muscles. Giant fibers have also been important to humans; our present knowledge of how nerve impulses are generated and transmitted comes largely from studies of cephalopod giant fibers.

Humans and all vertebrates have the most sophisticated nervous systems **5**: a central nervous system (CNS) consisting of a brain and a spinal cord, and a peripheral nervous system (PNS) consisting of nerves that run to and from the CNS. First we will discuss how neurons work, and then we will describe the vertebrate nervous system, using humans as the example.

# How do nerve cells work?

## 32.3 Nerve cells are structured to receive and send impulses.

Even though individual neurons vary widely in size and shape within an individual organism and among organisms, all neurons have the same basic parts: a cell body and cell extensions called dendrites and axons. The **cell body** is the region of the neuron around the nucleus that is responsible for producing substances that are necessary for the nerve cell to live and carry out its functions. The main organelles of the cytoplasm are in this portion of the cell. An **axon** is a single, long cell extension that often makes distant connections. It may give out branches, which often split off at right angles, and usually has a myelin covering, a type of insulation. Axons function to conduct and pass on information. A **dendrite** is a cell extension that is usually much shorter than an axon, has no myelin covering, and is specialized to receive impulses from sensory cells or from axons of other neurons. Most neurons contain multiple dendrites.

**Sensory neurons** receive information and transmit it to the central nervous system (**Figure 32.3**). Most of these neurons have one long axon bringing messages from particular receptors. Certain specialized sensory neurons, such as some in the eye, do not fit this description and may have multiple dendrites bringing in messages. The cell bodies of sensory neurons lie near the central nervous system.

**Motor neurons** transmit commands away from the central nervous system to the muscles and glands. Each of these neurons has one long axon, usually branched at the end, bringing messages to a muscle or a gland. The axons that control the muscular activity in the legs and feet can be more than a meter long, depending on height. The cell bodies of most motor neurons lie in or near the central nervous system.

**Interneurons**, which are located within the brain or spinal cord, integrate incoming information with outgoing messages (see Figure 32.3). These neurons usually have a highly branched system of dendrites able to receive input from many different neurons converging on a single interneuron. The axons of interneurons may not be myelinated and are usually much shorter than the axons of sensory and motor neurons.

The cell body of a neuron has the structures typical of a eukaryotic cell, such as a membrane-bounded nucleus. Surrounding the nucleus is cytoplasm, which contains the various cell organelles.

Most neurons have companion cells nearby. These companion cells are called *neuroglia* (nuh-ROG-lee-uh) and provide support, protection, and nutritional stability to neurons. Special neuroglial cells in the peripheral nervous system called **Schwann cells** are often wrapped around individual axons of sensory and motor neurons. The fatty wrapping of axons created by the cell membrane of Schwann cells forms a **myelin sheath** (MY-eh-len). The myelin sheath insulates the axon; however, the Schwann cells are wrapped around an axon in such a way that uninsulated spots occur at regular intervals (**Figure 32.4**). These uncovered spots are called **nodes of Ranvier** (ron-VYAY). These nodes and the myelin sheath (discussed in more detail later in this chapter) create conditions that speed the nerve impulse as it is conducted along the surface of the axon. *Oligodendrocytes* in the CNS also produce a myelin sheath around nerve cells, but operate somewhat differently than Schwann cells do. Oligodendrocytes spiral broad, flat processes around several nearby CNS axons. Many invertebrates, such as insects, earthworms, and crabs, have insulation covering their neurons that is quite similar to the myelination of vertebrate neurons.

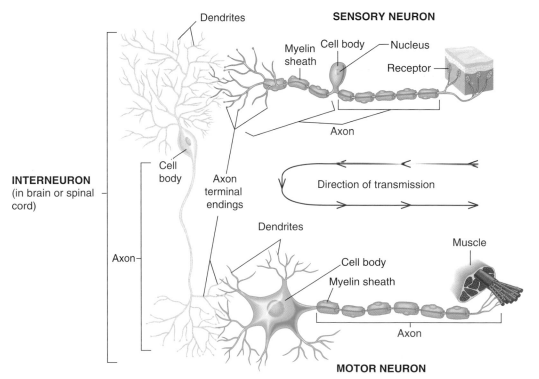

**Figure 32.3 Types of vertebrate neurons.** An incoming message travels along sensory neurons from a sense organ, such as specialized nerve endings in the skin, to the spinal cord and/or brain. There, an interneuron perceives the message and "decides" on an appropriate outgoing message, if any. A motor neuron carries the outgoing message to a muscle or gland.

CONCEPT
CHECKPOINT

1. Compare and contrast the structure and function of sensory, motor, and interneurons.

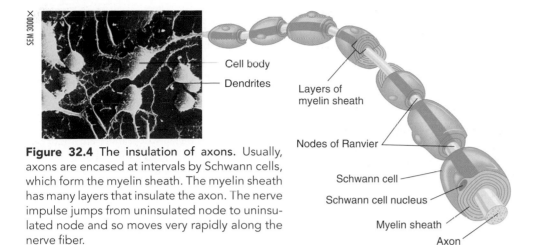

**Figure 32.4 The insulation of axons.** Usually, axons are encased at intervals by Schwann cells, which form the myelin sheath. The myelin sheath has many layers that insulate the axon. The nerve impulse jumps from uninsulated node to uninsulated node and so moves very rapidly along the nerve fiber.

## 32.4 Nerve impulses are electrical signals transmitted along membranes of neurons.

The job of neurons is to transmit information from the environment to the spinal cord and brain, from one cell to another within the CNS, and from the brain and spinal cord to other parts of the body. Nerve cells transmit this information in the form of electrical signals. The stimulation of specialized receptor cells, such as the rods and cones in your eyes or pressure receptors in your skin, causes electrical signals to be generated in these cells. Once an electrical signal is generated in a receptor, it can travel in the nervous system.

### The neuron at rest: the resting potential

The story of how a neuron conducts an impulse begins with the neuron at rest—a neuron that is not conducting an impulse. While at rest, a neuron is electrically charged. This electrical charge can be mea-

sured in the laboratory by means of microelectrodes and is approximately −70 millivolts (mV) (**Figure 32.5**, left). The negative charge means that the inside of the cell is negatively charged relative to extracellular fluid along the outside of the membrane. A millivolt is one thousandth of a volt, the unit measure of electrical potential.

The term *electrical potential* refers to the amount of potential energy created by a separation of positive and negative charges—in this case, along the inside and outside of the cell membrane of the neuron. The electrical potential of the nerve cell membrane is called the *membrane potential*. The charges that are separated from one another along the nerve cell membrane are carried as ions—atoms with unequal numbers of protons and electrons. The ions that play the principal role in the development of the electrical potential along the membrane of the neuron are sodium ions ($Na^+$) and potassium ions ($K^+$).

How are charges separated from one another along the nerve cell membrane? Proteins embedded within the cell membranes of neurons include enzymes called **sodium–potassium pumps**, which actively transport sodium ions ($Na^+$) and potassium ions ($K^+$) across the cell membrane (see Chapter 5 and Figure 32.5). These transmembrane proteins use the energy stored in molecules of adenosine triphosphate (ATP) to move $K^+$ into the neuron at the same time that they move $Na^+$ out of the neuron (Figure 32.5 **1**). Potassium ions diffuse back out of the cell through voltage-gated channels open to them at –70 mV **2**. Sodium voltage-gated channels, however, are closed at −70mV **3**. Sodium ions ($Na^+$) move back into the cell by other means, but only very slowly. These mechanisms place most of the $Na^+$ outside the cell, and most of the $K^+$ inside the cell, but as more potassium ions leak out of the cell

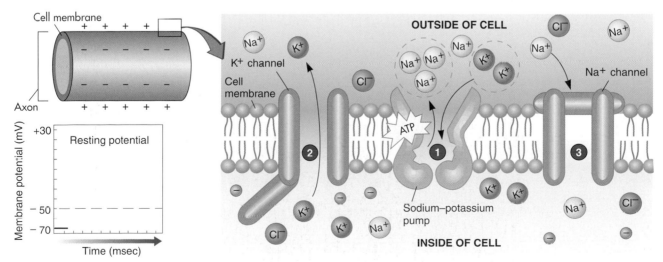

**Figure 32.5 The resting potential in a neuron.** The sodium–potassium pump moves sodium ions ($Na^+$) to the outside of the cell and potassium ($K^+$) ions to the inside of the cell. $K^+$ ions move back out of the cell through open voltage-gated channels, while $Na^+$ voltage-gated channels are closed. More positively charged ions build up outside the cell while larger molecules, such as negatively charged proteins, remain inside the cell. Thus, an electrical gradient is established at the membrane of a resting neuron.

than sodium ions move into the cell, the fluid along the inner surface of the plasma membrane becomes increasingly more negative. Moreover, proteins, which have a negative charge, are most abundant inside the cell. Thus, an electrical gradient is established: the inside of a nerve cell is more negatively charged than the outside of the cell along the membrane. The membrane of the neuron is *polarized*, which means there is a difference in charge on the two sides of the membrane. The difference in charge is called the **resting potential**. This electrical potential difference across the membrane is the basis for the transmission of signals by nerves.

## Conducting an impulse: the action potential

A neuron transmits a nerve impulse when it is excited by an internal or external environmental change called a *stimulus* (plural, *stimuli*). Examples of stimuli are pressure, chemical activity, sound, and light. In receptors, the energy of a stimulus is converted to a nerve impulse. Stimuli affect ion channels, resulting in the movement of ions through membrane channels or closing

channels to movement. This ion movement typically *depolarizes* the receptor membrane. This depolarization, which is a change in voltage across the nerve cell membrane, increases the movement of sodium ions ($Na^+$) through the $Na^+$ voltage-gated channels. Put simply, it opens the sodium channels. At the same time, it closes the $K^+$ channels.

As the sodium channels become increasingly permeable, a few sodium ions move rapidly into the cell, diffusing from where there are more sodium ions (outside the cell) to where there are fewer (inside the cell). This permeability for sodium ions across the cell membrane further depolarizes the nerve cell membrane. When a neuron is sufficiently stimulated, depolarizing the membrane to a level called the **threshold potential** (**Figure 32.6a**), it initiates an action potential in the sensory neurons with which the receptor communicates. The **action potential** (nerve impulse) is a rapid reversal in the membrane's electrical potential across a portion of the membrane. The interior of the cell develops a positive charge of approximately 30 mV relative to the outside (Figure 32.6b), a 100 mV electrical difference from the −70 mV resting potential.

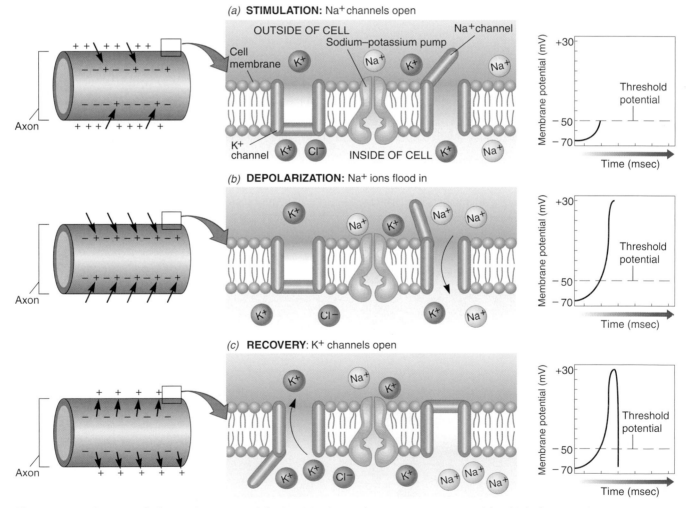

**Figure 32.6 Stages of the action potential.** An initial stimulus and rise in the membrane potential opens $Na^+$ channels and closes $K^+$ channels. $Na^+$ begins to diffuse across the membrane through these channels, depolarizing the membrane further. When the neuron is depolarized to the threshold potential (a), it initiates an action potential in the sensory neurons with which the receptor communicates. The action potential (b) is a rapid reversal in the membrane's electrical potential. The $Na^+$ channels then quickly close (c), the $K^+$ channels quickly reopen, $K^+$ ions move out of the cell, and the resting membrane potential is restored.

The depolarization of the cell membrane (the action potential) lasts only a *few thousandths of a second* (milliseconds) because the sodium channels close quickly. They cannot reopen until after the resting potential is reestablished and another depolarization occurs, triggering them again. During this inactive state of the sodium channels, a nerve impulse cannot be conducted. This period, which is only milliseconds long, is called the *refractory period*.

When the sodium channels close the potassium channels open, allowing potassium ions to move outward. Many neurons contain voltage-sensitive potassium channels that open as the membrane depolarizes and the sodium channels shut down. It is this event—the movement of potassium ions out of the cell—that repolarizes the membrane (Figure 32.6c). The sodium–potassium pump works to maintain the resting potential on a more long-term basis.

An action potential at one point on the nerve cell membrane is a stimulus to neighboring regions of the cell membrane. The change in membrane potential causes sodium channels to open, depolarizing the adjacent section of membrane. In this way, the initial depolarization passes outward over the membrane, spreading out in all directions from the site of stimulation. Like a burning fuse, the signal is usually initiated at one end of a nerve fiber and travels in one direction, as shown in **Figure 32.7**, but it would travel out from both directions if it were lit in the middle. The self-propagating wave of depolarization that travels along the nerve cell membrane is the nerve impulse.

## 32.5 Insulated neurons carry impulses more quickly than uninsulated neurons.

How do insulated neurons work? Schwann cells wrap around the length of an axon, one Schwann cell after the other, with spaces separating one from the next (see Figure 32.4). These spaces, the nodes of Ranvier, are critical to the propagation of the nerve impulse in myelinated cells. As an insulator, the myelin sheath prevents the transport of ions across the neuron membrane beneath it. However, within the small gaps between Schwann cells, the surface of the axon is exposed to the extracellular fluid. An action potential can be generated only at these gaps.

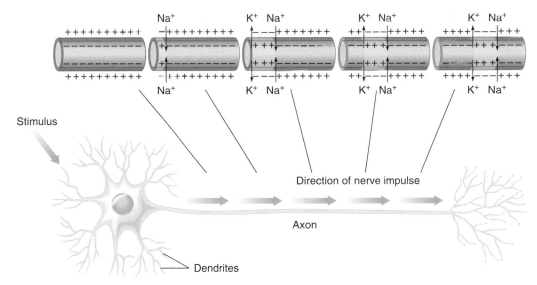

**Figure 32.7** Transmission of a nerve impulse: propagation of the action potential. Depolarization moves along a neuron in a self-propagating wave.

The action potential moves along the nerve cell membrane much faster in myelinated cells than in unmyelinated cells. In unmyelinated neurons, the wave of membrane depolarization simply travels down the axon (**Figure 32.8a**). In myelinated neurons, however, the action potential jumps from one node to the next, causing a depolarization only at these specific points (see Figure 32.8b). This very fast form of nerve impulse conduction is known as saltatory conduction, from the Latin word *saltare*, meaning "to jump." Impulses conducted in large-diameter myelinated neurons

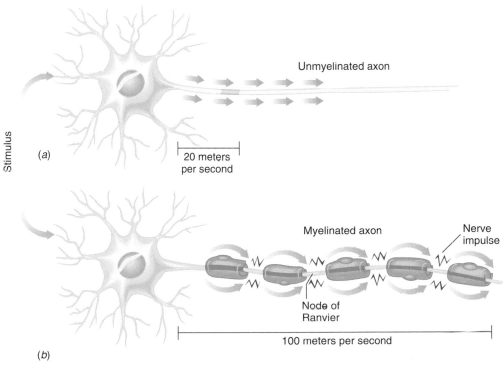

**Figure 32.8** Saltatory conduction. (a) In a fiber without Schwann cells, each portion of the membrane becomes depolarized in turn, like a row of falling dominoes. (b) In fibers of the same diameter, as in (a), the nerve impulse moves faster along a myelinated fiber because the wave of depolarization jumps from node to node.

travel up to 270 miles per hour (120 meters per second). These myelinated neurons can transmit a signal from your toes to your brain in a fraction of a second!

**2.** Describe the changes in membrane ion channels and the movement of ions across the membrane that occur at each of the labeled points on the graph during an action potential.

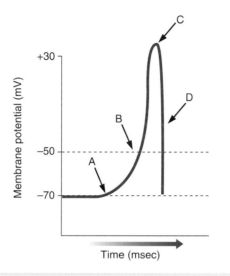

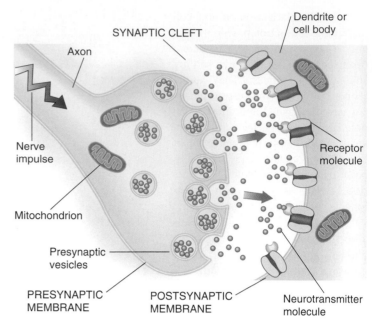

**Figure 32.9** A synaptic cleft between neurons. Neurotransmitters are released from the presynaptic membrane of an axon into the synaptic cleft. These molecules diffuse across the cleft and reach receptor molecules in the postsynaptic membrane of the dendrite of another neuron, triggering the nerve impulse to continue.

## 32.6 Neurotransmitters send signals across junctions between neurons or between neurons and muscle cells.

When the nerve impulse reaches the end of an axon, it must be transmitted to another neuron or to muscle or glandular tissue. Muscles and glands are called *effectors* because they effect (or cause) responses when stimulated by nerves. The place where a neuron communicates with another neuron or an effector cell is called the **synapse** (SIN-aps).

Most neurons do not touch other neurons or cells with which they communicate. Instead, there is a minute space (*billionths* of a meter across) separating these cells called the **synaptic cleft** (**Figure 32.9**). The nerve impulse must cross this gap, and it does so by changing an electrical signal to a chemical signal. Chemical synapses are the prevalent type of synapse in humans and all vertebrates. Neurons that communicate with one another by means of electrical signals *do* touch one another; they have no synaptic cleft.

The membrane on the axon side of the synaptic cleft is called the *presynaptic membrane*. In chemical synapses, when a wave of depolarization reaches the presynaptic membrane, it stimulates a flow of calcium into the cell. The sudden rise in the cytoplasmic concentration of calcium triggers the release of organic molecules called **neurotransmitters** into the cleft (see Figure 32.9). These molecules, released from thousands of small vesicles located at the tips of axons, diffuse to the other side of the gap. Once there, they combine with receptor molecules in the *postsynaptic membrane*, usually associated with either a dendrite or a cell body of the target cell. When they do, they cause ion channels to open.

Chemical junctions between neurons and other neurons or effector cells have a distinct advantage over direct electrical connections—flexibility. The chemical transmitters can vary in different junctions. More than 60 different chemicals, such as norepinephrine, serotonin, and dopamine, have been identified that act as neurotransmitters or that act to modify the activity of neurotransmitters.

### Neuron-to-muscle cell connections

Synapses between neurons and skeletal muscle cells are called **neuromuscular junctions**, as shown in **Figure 32.10**. The neurotransmitter found at neuromuscular junctions is *acetylcholine* (uh-SEE-tl-KOH-leen). The nerve impulse travels down the axon, reaches the axon tip ❶, and stimulates a flow of calcium into the cell, which triggers the release of acetylcholine (ACh) from vesicles ❷. Passing across the gap, the acetylcholine molecules bind to receptors in the postsynaptic (muscle cell) membrane, opening sodium channels ❸. This influx of sodium ions depolarizes the muscle cell membrane, which initiates a wave of depolarization that passes down the muscle cell ❹.

After an impulse has been transmitted across the synaptic cleft, the neurotransmitter must be broken down. In general, some or all of the breakdown products are transported back to the presynaptic cell to be reused. For example, the neurotransmitter acetylcholine is broken down to acetic acid and choline by an enzyme called *acetylcholinesterase* (uh-SEE-tl-koh-lin-ESS-ter-ase) **5**. Choline is transported back to the presynaptic cell, where it is used to make more molecules of acetylcholine **6**.

Nerve gases and the agricultural insecticide parathion work by blocking the action of acetylcholinesterase. These chemicals can cause death because they produce continuous neuromuscular transmission, which results in a continuous muscular contraction of vital muscles such as those involved in breathing and the circulation of blood. Many drugs also work by affecting synapses. For example, cocaine, local anesthetics, and some tranquilizers work by altering nerve impulse frequency.

### Neuron-to-neuron connections

Impulses are transmitted from one neuron to another by a variety of neurotransmitters. Some neurotransmitters depolarize the postsynaptic membrane, which results in continuation of the nerve impulse. This type of synapse is called an *excitatory synapse*. Other neurotransmitters have the reverse effect, reducing the ability of the postsynaptic membrane to depolarize. This type of synapse is called an *inhibitory synapse*. A single nerve cell can have *both* kinds of synaptic connections to other nerve cells. As you might expect, exci-

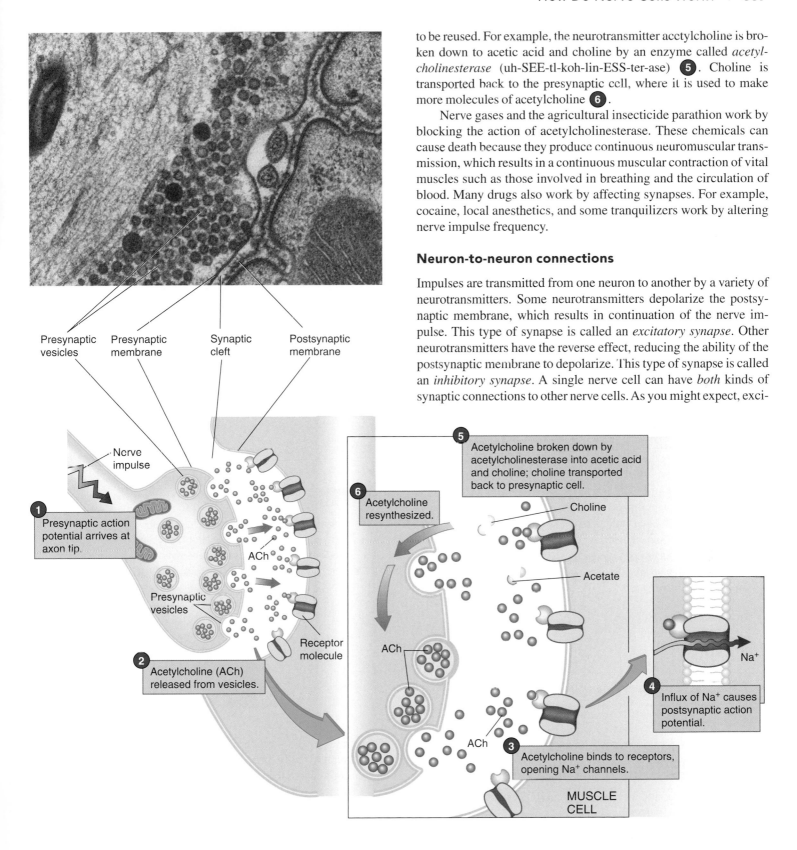

Presynaptic vesicles   Presynaptic membrane   Synaptic cleft   Postsynaptic membrane

Nerve impulse

**1** Presynaptic action potential arrives at axon tip.

ACh

Presynaptic vesicles

**2** Acetylcholine (ACh) released from vesicles.

Receptor molecule

**5** Acetylcholine broken down by acetylcholinesterase into acetic acid and choline; choline transported back to presynaptic cell.

**6** Acetylcholine resynthesized.

Choline

Acetate

ACh

ACh

ACh

**3** Acetylcholine binds to receptors, opening Na⁺ channels.

**4** Influx of Na⁺ causes postsynaptic action potential.

$Na^+$

MUSCLE CELL

**Visual Summary**   **Figure 32.10** Synaptic transmission at a neuromuscular junction.

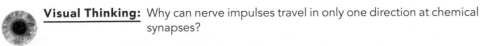 **Visual Thinking:** Why can nerve impulses travel in only one direction at chemical synapses?

**Figure 32.11 Integration of nerve impulses.** The synapses made by some axons are inhibitory, tending to counteract depolarization of the postsynaptic membrane; these synapses are indicated in *reddish-brown*. The synapses made by other axons are stimulatory, tending to depolarize the postsynaptic membrane; these synapses are indicated in *green*. The summed influences of these inputs determine whether the postsynaptic neuron will be sufficiently depolarized to initiate a nerve impulse.

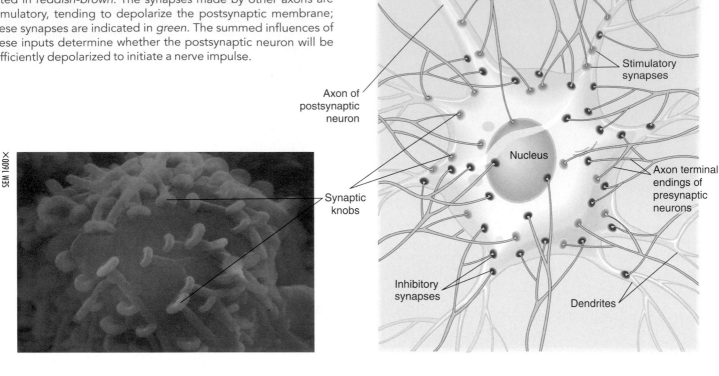

tatory signals cancel out inhibitory signals, modifying each other's effects. The postsynaptic neuron keeps score as the impulses reach its dendrites and cell body, and it responds accordingly. For this reason, the postsynaptic neuron is called an *integrator* (**Figure 32.11**). Within your body, synapses are organized into functional units with definite patterns, similar to electrical circuits.

### CONCEPT CHECKPOINT

**3.** What is the mechanism that determines whether an interneuron receiving impulses from thousands of sensory neurons, generates or does not generate an impulse of its own?

## How does the human nervous system work?

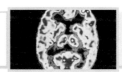

### 32.7 The human nervous system is highly organized.

Structurally, human and all vertebrate nervous systems can be divided into two main parts (**Figure 32.12**):

- The **central nervous system**, made up of the brain and spinal cord, is the site of integration (information processing) within the nervous system.

- The **peripheral nervous system** (puh-RIF-uh-rul) is an information highway made up of nerves that bring messages to and from the brain and spinal cord.

The nerves of the peripheral nervous system are made up nerve fibers, support cells, blood vessels, and connective tissue. These nerves contain the fibers of sensory neurons and motor neurons.

One group of motor neurons controls voluntary responses, such as coordinating the movement of muscles in your legs. Motor neurons that control voluntary responses make up the **somatic nervous system** (soe-MAT-ik). The word "somatic" means "body" and refers to the fact that these neurons carry messages to your skeletal muscles, which move the parts of your body. Certain voluntary activities that may seem somewhat out of your control, such as blinking and breathing, are also directed by the somatic nervous system. Such activities are actually *reflexes*, automatic responses that are mediated by the spinal cord or lower portions of the brain—those closest to the spinal cord.

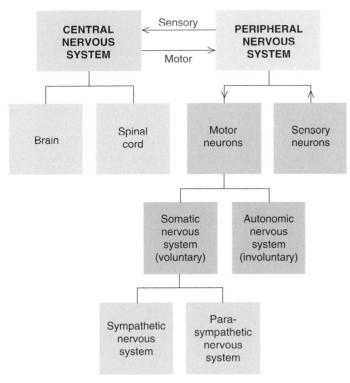

**Figure 32.12 The vertebrate nervous system.** The nervous system consists of the central nervous system (the brain and spinal cord) and the peripheral nervous system, a highway of nerves throughout the body.

Another group of motor neurons carries messages that control *involuntary responses* such as mixing the food in your stomach or pumping adrenaline into your bloodstream during times of stress. Motor neurons that carry messages about involuntary activities make up the **autonomic nervous system** (awe-tuh-NOM-ik). The word "autonomic" comes from Greek words meaning "self" (*auto*) "governing" (*nomos*). This portion of the nervous system literally takes care of you by itself! In general, it works to promote homeostasis—a "steady state"—within your body. The autonomic nervous system accomplishes this feat by carrying opposing messages that either speed up or slow down the activities of your glands, heart muscle, and smooth muscles. These opposing messages are carried on separate neurons, dividing the autonomic nervous system functionally into two parts: the sympathetic and parasympathetic systems, which are discussed in Section 32.10.

## 32.8 The human central nervous system integrates sensory and motor impulses.

The most complex or highest functions of the nervous system, such as thinking, remembering, and feeling, are all integrated in the brain, in a portion called the *cerebrum* (seh-REE-brum). In addition to the cerebrum, the human brain is made up of the cerebellum (sere-eh-BEL-um), the **diencephalon** (DIE-en-SEF-eh-lon)—including the thalamus and hypothalamus—and the brain stem. These parts of the brain and a list of their functions are shown in **Figure 32.13**.

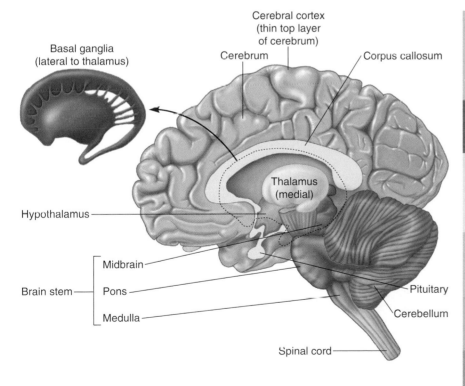

**Cerebral cortex**
- Receives sensory information
- Sends messages to move skeletal muscles
- Integrates incoming and outgoing nerve impulses
- Performs activities such as thinking, learning, and remembering

**Basal ganglia**
- Helps coordinate slow, sustained movements
- Suppresses useless patterns of movement

**Thalamus**
- Relays most sensory information from the spinal cord and certain parts of the brain to the cerebral cortex
- Interprets certain sensory messages such as those of pain, temperature, and pressure

**Hypothalamus**
- Controls various homeostatic functions such as body temperature, respiration, and heartbeat
- Directs hormone secretions of the pituitary

**Cerebellum**
- Coordinates subconscious movements
- Contributes to muscle tone, posture, and balance

**Brain stem**
- Origin of many cranial nerves
- Reflex center for movements of eyeballs, head, and trunk
- Regulates heartbeat and breathing
- Plays a role in consciousness
- Transmits impulses between brain and spinal cord

**Visual Summary**   **Figure 32.13** Major structures of the brain and their functions.

The **cerebrum** is the dominant part of the human brain. In the brains of humans and other mammals the cerebrum is split into two halves, or hemispheres, the right and left sides of the brain. (Figure 32.13 shows the brain split in half, and, therefore, you see only one hemisphere.) These two sides of the cerebrum are connected by a thick bundle of nerve fibers called the *corpus callosum* (KOR-pus kuh-LOW-sum) (see Figure 32.13) that allows information to pass from one side of the brain to the other, so each side "knows" what the other is thinking and doing.

Most of the activity of the cerebrum takes place within the **cerebral cortex**, a thin layer of tissue that forms its outer surface. Your cortex cap is densely packed with unmyelinated neuron cell bodies, so the cerebral cortex appears gray (although it shown as beige in the illustration) and is referred to as gray matter. Three major types of activities take place within the cerebral cortex: motor, sensory, and association activity. Association areas are the sites of higher cognitive activities, such as planning and contemplation, and are clustered in different areas of the cortex. Many of the neurons in the cerebral cortex send out myelinated axons that form the central core of white matter of the cerebrum. Within the central nervous system, bundles of nerve fibers are called *tracts*. In the peripheral nervous system, they are called *nerves* (see Section 32.7).

In addition to the gray matter of the cerebral cortex, other masses of gray matter are located deep within the cerebrum. These islands of gray matter, shown in Figure 32.13, are often collectively referred to as the **basal ganglia**. These ganglia play important roles in the coordination of slow sustained movements, such as maintaining posture, and in the suppression of useless patterns of movement. Injury to the basal ganglia can result in various types of uncontrolled muscular activities. Parkinson's disease (PD) is associated with the degeneration of parts of the basal ganglia, including the substantia nigra. (See the *How Science Works* box on recent research into the link between head trauma and increased risk of PD.)

The **cerebellum** is a relatively large part of the brain, weighing about a half pound. This part of the brain coordinates subconscious movements of the skeletal muscles. Sensory nerves bring information to the cerebellum about the position of body parts relative to one another, the state of relaxation or contraction of the skeletal muscles, and the general position of the body in relation to the outside world. After synthesizing these data, the cerebellum issues orders to motor neurons that result in smooth, well-coordinated muscular movements, contributing to overall muscle tone, posture, balance, and equilibrium.

At the base of the cerebrum are paired oval masses of gray matter called the **thalamus** (THAL-uh-muss). (Figure 32.13 shows only one-half of the brain and, therefore, only one of these oval masses.) The thalamus acts as a relay station for most sensory information. This information comes to the thalamus from the spinal cord and certain parts of the brain. The thalamus then sends these sensory signals to appropriate areas of the cerebral cortex. In addition, the thalamus interprets certain sensory messages such as pain, temperature, and pressure.

The **hypothalamus**, located beneath the thalamus (hypo means "under"), controls the activities of various body organs. The hypothalamus is a major regulator of homeostasis, a steady state within the body, by means of its various activities such monitoring body temperature, heartbeat rate, osmotic pressure, and glucose level. It also directs the hormone secretions of the pituitary, which is shown as a light beige knoblike area hanging beneath the hypothalamus in Figure 32.13.

If you think of the brain as being shaped somewhat like a mushroom, the cerebrum, cerebellum, thalamus, and hypothalamus would be its cap. The **brain stem** would be the mushroom's stalk. Its 3 inches of length consists of three parts—the midbrain, pons (PONZ), and medulla (mih-DULL-uh). Each part contains tracts of nerve fibers that bring messages to and from the spinal cord. In addition, each portion of the brain stem contains nuclei that govern important reflex activities of the body. In this context, the term *nuclei* means groups of nerve cell bodies having a similar function. With the exception of the basal ganglia, this term is used to denote groups of nerve cell bodies within the brain, and the term *ganglia* is used to denote groups of nerve cell bodies within the peripheral nervous system.

Many of the cranial nerves enter at the brain stem. These cranial nerves bring messages to and from the regulatory centers of the brain stem or use the brain stem as a relay station.

The **midbrain** sits at the top of the brain stem. Its white matter consists of nerve tracts that connect the upper parts of the brain (cerebrum, thalamus, and hypothalamus) with lower parts of the brain (pons and medulla). In addition, the midbrain contains reflex centers for movements of the eyeballs, head, and trunk in response to sights, sounds, and various other stimuli. For example, if a plate falls off the counter behind you, you probably turn around quickly and automatically. That is your midbrain at work.

The term **pons** means "bridge," and it is actually two bridges. One bridge consists of horizontal tracts that extend to the cerebellum, connecting this part of the brain to other parts and to the spinal cord. The other bridge consists of longitudinal tracts that connect the midbrain and structures above to the medulla and spinal cord below. In addition, the gray matter of the pons contains nuclei that work with other nuclei in the medulla to help control respiration.

The **medulla**, the lowest portion of the brain stem, is continuous with the spinal cord below. Because of its location, a large portion of the medulla consists of tracts of neurons that bring messages to and from the spinal cord. Most of these tracts cross over one another within the medulla. Therefore, sensory information from the right side of the body is perceived in the left side of the brain and vice versa. Similarly, the right side of the brain sends messages to the left side of the body, and the left side of the brain controls the right side. In addition to these tracts, the medulla contains reflex centers that regulate heartbeat, control the diameter of blood vessels, and adjust the rhythm of breathing. Centers there also control less vital functions such as coughing, sneezing, and vomiting.

## 32.9 The spinal cord connects the peripheral nervous system with the brain.

How do nerve impulses get to and from the brain, and to and from various body parts? The spinal cord extends from the brain stem and runs down the neck and back within the bony casing of the vertebral column. It receives information from the body by means of spinal nerves (**Figure 32.14**) and carries this information to and from the brain along tracts of white matter. In addition, the gray matter of the spinal cord itself integrates responses to certain kinds of stimuli. These integrative pathways are called reflex arcs and are

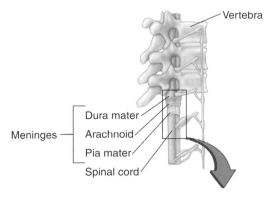

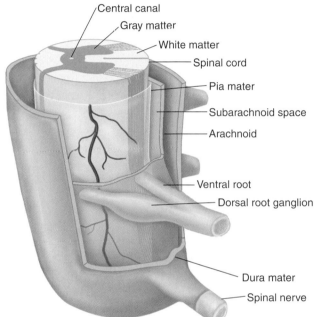

**Figure 32.14 Section through the spinal cord.** The spinal cord is protected by three layers of membranes called the meninges, by circulating fluid, and by the bones of the vertebral column. The spinal nerves carry information between the spinal cord and the rest of the body.

discussed in more detail later in this chapter. The spinal nerves then carry commands back to the body.

The spinal cord has a tiny central canal that pierces its length. This tubelike space is filled with cerebrospinal fluid and is continuous with fluid-filled spaces, or ventricles, in the brain. This fluid acts as a shock absorber, and brings nutrients, hormones, and white blood cells to different parts of the brain.

Along with bone and fluid, three layers of membranes called the **meninges** (muh-NIN-jeez) protect both the brain and spinal cord (see Figure 32.14). The outermost of these layers is called the *dura mater*. Between it and the bones of the vertebrae is a space filled with fat, connective tissue, and blood vessels called the *epidural* (outside the dura) *space*. Another space exists between the middle (arachnoid layer) and inner (*pia mater*) membranes of the meninges. The cerebrospinal fluid circulates within this space, called the *subarachnoid space*. The subarachnoid space is continuous with the central canal of the vertebral column and the ventricles of the brain.

## 32.10 The human peripheral nervous system carries sensory and motor impulses.

The peripheral nervous system connects the central nervous system with receptors, which are structures that detect stimuli, and effectors, which are muscles and glands that respond to that stimuli. The information that the brain derives from sensory input is based solely on the source of the nerve impulses and their frequency. If the auditory nerve is artificially stimulated, for example, the central nervous system perceives the stimulation as sound. Sensory receptors react to specific stimuli by producing nerve impulses because they have low thresholds for these particular types of stimuli and high thresholds for others. Receptors in the retina of your eye, for example, have a low threshold for light.

**Cranial nerves** communicate directly with the brain and are shown in bright yellow in **Figure 32.15**. Some cranial nerves conduct only sensory information from specialized receptors to the

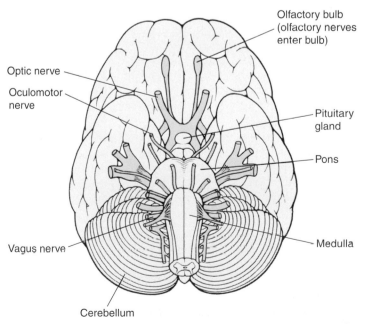

**Figure 32.15 A view of the brain from its underside showing the origins of the cranial nerves.** The cranial nerves are shown in bright yellow, and a few of the more familiar ones are labeled.

# How Science Works

## Solving the mystery of Alzheimer's disease

Alzheimer's disease was first described nearly a century ago in 1907 by Swiss psychiatrist Alois Alzheimer. One of Alzheimer's patients, aged 51, was having substantial memory difficulties. She continued to deteriorate, forgetting words and losing her ability to reason. She died four years later. At the autopsy, the woman's brain was found to be much smaller than average for her age and gender, and the ventricles (cavities) of her brain were much larger. Her brain was also riddled with waxy-looking patches where nerve cells had once been.

Just what happens in Alzheimer's disease to cause such devastating mental deterioration? In those suffering from Alzheimer's, the synthesis of a component of nerve cell membranes called amyloid protein goes awry. Pieces of amyloid pile up in needlelike masses within the brain, punching holes in brain cells and killing them. These masses are usually referred to as *amyloid beta plaques*. Beta refers to amyloid's structure of accordionlike sheets.

In 2002, scientists developed a blood test for Alzheimer's, and in 2003 it was successfully used on living mice. The test uses a chemical called PIB, which is able to bind to the amyloid plaques in the brains of mice with Alzheimer's. Researchers were able to determine the amount of amyloid plaque in the brains by detecting the amount of PIB. Human trials will be conducted next. In addition, positron emission tomography (PET) scans of the brain show characteristic changes in glucose metabolism in the brain of Alzheimer's patients (**Figure 32.A**).

Currently affecting about 4 million Americans, Alzheimer's disease is characterized by a gradual loss of memory and reasoning ability. Affected individuals cannot remember things that they heard or saw just a few minutes previously. They have trouble finding their way around and eventually forget how to talk, feed themselves, and even swallow. There is no cure; however, the drug galantamine (Reminyl), approved for use by

the FDA in 2001, appears to improve memory and learning along with slowing disease progression in Alzheimer's patients. In addition, the drug metrifonate, which belongs to the class of drugs called acetylcholinesterase (ACE) inhibitors, appears to slow mental decline and to control behavior in persons with mild or moderate Alzheimer's. The drug appears to work by slowing the breakdown of the chemical messenger acetylcholine, which is found at lower than normal levels in the brains of Alzheimer's patients.

As the population ages, Alzheimer's disease is becoming more common. By 2047, 8.6 million people in the United States are expected to have Alzheimer's. In response to this coming national health crisis, researchers are now focusing on finding ways to block the formation of amyloid fragments or prevent their attachment to the surface of brain cells to thwart the disease. Researchers at New York University Medical Center have found protein frag-

ments called beta sheet breakers that appear to prevent amyloid plaques from forming and appear to dissolve existing plaques. Other researchers from Cornell University in Ithaca, New York, are experimenting with implanting plastic pellets that release a growth factor that may keep nerve cells from degenerating. Yet others at the University of Washington and a Washington-based biotechnology company are working with a plant extract that may be useful in disrupting the amyloid.

What can you do to help prevent this disease as you age? Smoking cigarettes increases the risk of Alzheimer's, so do not smoke. Researchers have also discovered that persons less likely to develop Alzheimer's are those who exercise regularly, have intellectually demanding occupations, have a high level of education, and stay socially active. Researchers hypothesize that the brain is like other organs of the body—it ages better if it is used. ●

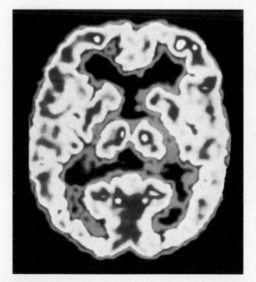
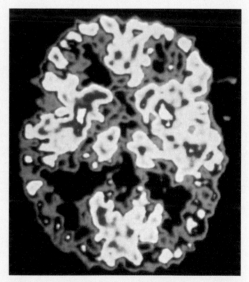

**Figure 32.A** Positron emission tomography (PET) scan of the brain of a normal subject (left) and of a person with Alzheimer's disease. This scan shows how the brain uses glucose. Red areas show the highest level of glucose use, yellow less, blue still less, and black the least. The scan of the normal subject shows a higher level of glucose use, while the scan of the Alzheimer's patient shows where certain regions of the brain have decreased glucose use. This pattern is typical of Alzheimer's and can be used as a diagnostic tool to differentiate Alzheimer's from other types of dementia or depression.

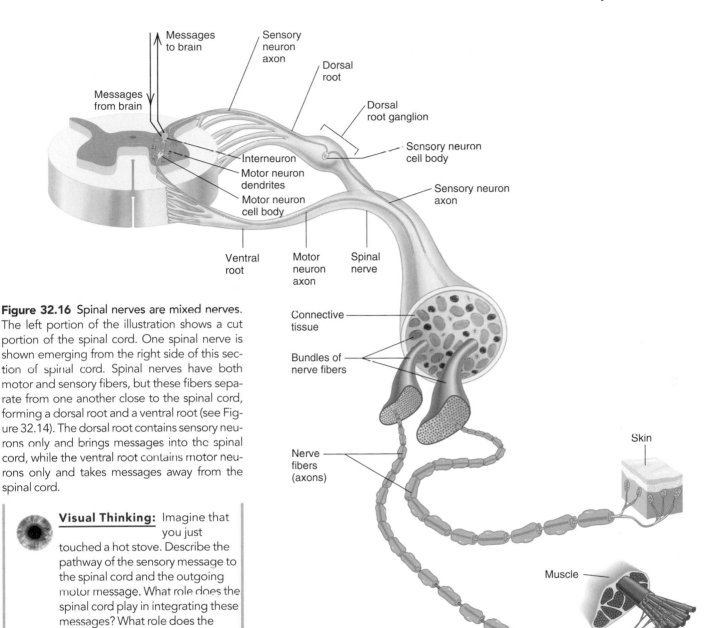

**Figure 32.16** Spinal nerves are mixed nerves. The left portion of the illustration shows a cut portion of the spinal cord. One spinal nerve is shown emerging from the right side of this section of spinal cord. Spinal nerves have both motor and sensory fibers, but these fibers separate from one another close to the spinal cord, forming a dorsal root and a ventral root (see Figure 32.14). The dorsal root contains sensory neurons only and brings messages into the spinal cord, while the ventral root contains motor neurons only and takes messages away from the spinal cord.

**Visual Thinking:** Imagine that you just touched a hot stove. Describe the pathway of the sensory message to the spinal cord and the outgoing motor message. What role does the spinal cord play in integrating these messages? What role does the brain play?

brain and are associated with sight, sound, smell, and equilibrium. They arise in your eyes, nose, and ears. Other cranial nerves called *mixed nerves* are made up of both sensory and motor nerve fibers, such as the vagus nerve, which regulates the function of your heart rate, respiration rate, and digestive activities. Other cranial nerves conduct only motor information such as the oculomotor nerve that helps to control eye movement.

The **spinal nerves** are made up of both sensory and motor nerve fibers that travel directly to and from the spinal cord. Nerve fibers that serve the same general area of the body, whether they are sensory or motor neurons, are bundled together to form these mixed nerves (**Figure 32.16**). Close to the spinal cord, motor and sensory fibers separate from one another. The sensory fibers are myelinated axons that extend from the source of stimulation—your fingers or toes, for example—to a swelling near their entrance to the spinal cord. This swelling is the dorsal root ganglion (see Figure 32.16), which contains the cell bodies of these sensory neurons.

Shorter portions of the axons of these neurons extend from this ganglion to the gray matter of the spinal cord. Here, the axon ends of these sensory neurons synapse with interneurons that play a role in integrating incoming messages with outgoing responses. Interneurons, may extend from the spinal cord to the brain. They may make up part of the brain tissue itself, or in the case of certain reflex pathways, may be located within the gray matter of the spinal cord.

Interneurons synapse with the short dendrites of spinal motor neurons, located in anterior portions of the gray matter of the spinal cord. These dendrites then conduct the impulses to their cell bodies, also within the gray matter of the spinal cord (see Figure 32.16). From there, the nerve impulse sweeps along the membrane of the axon of each motor neuron to effectors: muscles or glands that produce a response. The motor pathway to a skeletal muscle contains a single motor neuron that traverses the distance from the spinal cord to the muscle. The motor pathways to smooth and cardiac muscles and to glands are made up of a series of two motor neurons.

# How Science Works

## DISCOVERIES

### Head injuries linked to Parkinson's risk

Sixty-one boxing bouts over 21 years, with 56 wins, 5 losses, and 37 knockouts. That's quite a record, especially when you add in his titles of 1960 Olympic Light Heavyweight Gold Medalist and Heavyweight Champion 1964–1967, 1974–1978, and 1978–1979. Now, however, "the greatest," Muhammad Ali, has Parkinson's disease (PD) and has had the disease for over 20 years.

Over the decades he has been afflicted, Ali and others have wondered about the relationship between the punches he took and his disease. The relationship between head trauma and the development of PD has been a research topic for neurologists ever since James Parkinson described the disease in 1817. Is head trauma a risk factor for PD?

PD affects regions of the brain essential for creating smooth, coordinated movement. In PD, some of the cells in a part of the brain called the *substantia nigra* are dam-aged. They no longer produce the neurotransmitter dopamine, and the result is trembling muscles, muscle stiffness, and slowed movement.

The link between head trauma and the development of PD was tenuous until 2003. Some studies showed that a history of severe head trauma occurred significantly more often in patients with PD than in those without the disease. Other studies show no relationship at all. Researchers explained the discrepancy by citing recall bias; that is, the research was based on people recalling whether they had experienced head trauma decades earlier. Sometimes, physicians suggest, PD patients overreport previous life events to explain their disease. In addition, the recall of non-PD patients may or may not have been accurate.

In 2003, researchers from the Mayo Foundation in Rochester, Minnesota, published the results of a study in which they were able to eliminate recall bias (Bower et al., 2003, *Neurology*, 60: 1610–1614). The researchers compared the medical charts of 196 patients with PD with those of 196 persons without PD. They found that overall, persons having experienced head trauma were about four times more likely to develop PD than those not experiencing head trauma. They then separated the "head trauma" study participants into two groups: (1) those who experienced mild head trauma with mild memory loss and (2) those who experienced a mild head trauma with loss of consciousness or a more severe trauma. They found that those in the first group had no increased risk for PD, whereas those in the second group were 11 times more likely to develop PD than persons with no head trauma. The researchers concluded that there is an association between head trauma and development of PD, and that the risk rises with the severity of the injury.

Although the cause of PD is still unknown, researchers have found that the inheritance of certain genes is linked to development of the disease. Now, experiencing head trauma may be associated with this disease as well. ●

---

The bundles of axons leaving the spinal cord join with the axons of sensory neurons entering the same level of the cord, forming spinal nerves. Figure 32.5 ❺ shows the approximate location of the 31 pairs of spinal nerves and, in general, the areas of the body that they serve.

### The somatic and autonomic nervous systems

In the vertebrate nervous system, motor responses are of two types: voluntary and involuntary. Movements of the skeletal muscles are voluntary and are controlled by messages from the brain and spinal cord via pathways composed of single motor neurons. These motor pathways are referred to as the *somatic nervous system* (see Figure 32.12). Although skeletal muscle movements are primarily subject to conscious control by the associative cortex of the cerebrum, not all movements of the skeletal muscles are conscious. Blinking your eyes, putting one foot in front of the other when you walk, and breathing can be consciously controlled, but most often take place without conscious thought; they are controlled by reflex activity.

A *reflex* is an automatic response to nerve stimulation. Very little, if any, integration—and certainly no thinking—takes place during reflex activity. The knee jerk is one of the simplest types of reflexes in the human body; a sensory neuron synapses directly with a motor neuron within the spinal cord. This pathway—the pathway an impulse follows during reflex activity—is called a **reflex arc**.

Simple reflex arcs such as the knee jerk are called monosynaptic reflex arcs (literally, "one synapse"). A monosynaptic reflex arc works without integration by the central nervous system. Most voluntary muscles within your body possess these monosynaptic reflex arcs, such as those that maintain posture, although they usually operate in conjunction with other more complex reflex pathways. It is through these more complex paths that voluntary control is established.

The motor pathways that control involuntary and automatic responses of the glands and nonskeletal muscles of the body are known as the *autonomic nervous system*. The autonomic nervous system is made up of two divisions—the **parasympathetic nervous system** and the **sympathetic nervous system**—which act in opposition to each other, speeding up or slowing down certain bodily processes (**Figure 32.17**). Together, the two divisions fine-tune internal physiological parameters such as body temperature or blood pressure to maintain homeostasis, a relatively stable internal state. Each gland (except the inner portion of the adrenal gland), smooth muscle, and cardiac muscle is "wired" to both systems, which use different neurotransmitters that elicit the opposing responses. The autonomic nervous system is a key player in the "fight-or-flight" response of stress (see Chapter 35). In general, the sympathetic nervous system prepares the body for action, whereas the parasympathetic system prevails during periods of inactivity. The "decision" to stimulate or inhibit a muscle, organ, or gland is "made" by integrating centers in the spinal cord and lower regions of the brain.

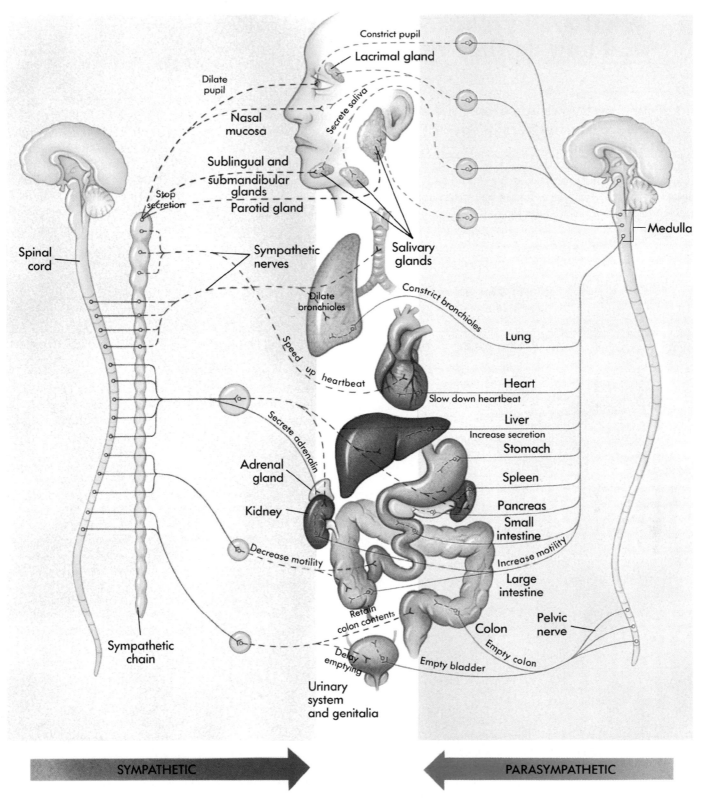

**Figure 32.17 Structure and function of the autonomic nervous system.** The ganglia of sympathetic nerves are located near the spine, and the ganglia of parasympathetic nerves are located far from the spine, near the organs they affect.

# What are psychoactive drugs and how do they work?

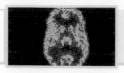

## 32.11 Psychoactive drugs affect the action of neurotransmitters in specific parts of the brain.

Drugs are chemical substances that affect normal body functions. **Psychoactive drugs** affect the action of neurotransmitters in the brain. Specific drugs interfere with the normal activity of neurotransmitters to which they are chemically similar, and affect parts of the brain or spinal cord that have receptors for those neurotransmitters. In this way, psychoactive drugs affect communication between neurons or between neurons and muscles or glands. **Table 32.1** lists the major classes of psychoactive drugs.

One way in which drugs work is to decrease the amount of neurotransmitter that is released from a presynaptic neuron. Reserpine, a tranquilizer that also lowers blood pressure, is such a drug. It interferes with the storage of the neurotransmitter norepinephrine (nor-EP-ih-NEF-rin), also called noradrenaline

| TABLE 32.1 | Major Classes of Psychoactive Drugs | | |
|---|---|---|---|
| **Depressants** | | | |
| **Drugs** | **Nervous System Action** | **Medical Uses** | **Possible Side Effects** |
| **Alcohol** <br><br> **Sedative-hypnotics** <br> • Barbiturates (Seconal®, Nembutal®, Phenobarbital) <br> • Benzodiazepines (Valium®) <br> *Street names:* yellow jackets, reds, blues, Amy's, rainbows | • Slow down activity of CNS <br> • Interact with GABA, an inhibitory brain neurotransmitter <br> • In higher doses, depression of inhibitory brain circuits <br> • Produce sedation and reduction of anxiety | *Sedative-hypnotics* <br> • Treat sleep disorders, epileptic convulsions, anxiety <br> • Sometimes used as anesthetics for dental surgery | *Alcohol* <br> • Increased heart rate, loss of alertness, blurred vision, decreased coordination <br> • Long-term use can result in liver damage, ulcers, inflammation of the pancreas, nutritional disorders, heart disease, fetal alcohol syndrome <br> • Taking sedative-hypnotics with alcohol can be fatal <br><br> *Sedative-hypnotics* <br> • Muddled thinking, childlike personality, emotional instability and irritability, unsteadiness when walking <br> • Overdose can be fatal |
| **Opiates** | | | |
| **Drugs** | **Nervous System Action** | **Medical Uses** | **Possible Side Effects** |
| Opium, morphine, codeine, heroin, paregoric, Dextromethorphan, Dilaudid®, Percodan®, Vicodin®, Demerol®, Darvon® <br> *Street names:* smack, horse, junk, Mexican brown, China white, dolls | • Mimic naturally occurring morphinelike neurotransmitters called endorphins, which help people cope with pain and help modulate their response to emotional trauma <br> • Produce sedation and euphoria | *Dextromethorphan* <br> • Cough control <br> *Paregoric* <br> • Controls diarrhea and cramps in infants and small children <br> *Codeine and morphine* <br> • Pain control | • Sleepiness, depressed breathing, nausea, slowed reflexes, cold skin <br> • Long-term use destroys the ability to reason |

(nor-ah-DREN-ul-in). This neurotransmitter is one of two neurotransmitters active in the sympathetic nervous system. One of norepinephrine's functions within the sympathetic nervous system is the constriction of blood vessels, which increases blood pressure. Decreasing the amount of norepinephrine decreases the constriction of the blood vessels and lowers the blood pressure.

Some drugs increase the amount of a neurotransmitter or its effects at the synapse. These types of drugs may enhance the release of neurotransmitter molecules, inhibit the action of enzymes that degrade neurotransmitter molecules at the postsynaptic neuron, or chemically resemble the neurotransmitter and mimic its effects at the postsynaptic neuron. In all cases, the postsynaptic neuron becomes or remains stimulated by the neurotransmitter or its mimic. Amphetamines, drugs that stimulate the brain, are of this type.

Some drugs that chemically resemble specific neurotransmitters act in still another way. These mimics occupy receptor sites but do not stimulate postsynaptic neurons. They simply block the neurotransmitter molecules from the sites. Therefore, these drugs block the effects of the neurotransmitter.

Some psychoactive drugs are used medically. For example, the drug diazepam (die-AZ-eh-pam), commonly known as Valium®, is used to control anxiety. Imipramine (ih-MIP-rah-meen) is a psychoactive drug that works as an antidepressant. Morphine is sometimes used to control pain after surgery or in terminally ill cancer patients. However, many psychoactive drugs are abused. That is, they are used for nonmedical reasons, are taken in doses that may cause damage to the body (**Figure 32.18**), and often result in personally destructive, antisocial, and crime-related behaviors.

The chronic use (usually abuse) of psychoactive drugs results in **drug addiction**: a compulsive urge to continue using the drug, physical and/or psychological dependence on the drug, and a tendency to increase the amount of the drug taken. Persons physically dependent on a drug show symptoms of this dependence when they

## Stimulants

| Drugs | Nervous System Action | Medical Uses | Possible Side Effects |
|---|---|---|---|
| **Mild stimulants**<br>• Nicotine, caffeine<br>**Powerful stimulants**<br>• Amphetamines, cocaine<br>*Street names:* speed, dex, zoom | • Enhance the activity of the neurotransmitters norepinephrine and dopamine<br>• Function in brain pathways that regulate emotions, sleep, attention, and learning<br>• Produce alertness and euphoria | *Ritalin® (amphetamine-related drug)*<br>• Treatment of attention deficit disorder<br>*Synthetic cocaine derivatives*<br>• Local anesthetic in dental and eye surgery | *Normal use*<br>• Increased pulse rate and blood pressure, sleeplessness, lack of appetite, restlessness, anxiety<br>*Heavy use of powerful stimulants*<br>• Schizophrenialike disorder characterized by paranoia, the hearing of voices, and irrational thought<br>• Long-lasting and severe physical changes in the brain<br>• Severe damage to the tissues of the nose and the lungs from snorting<br>• Heart disease, epileptic seizures, respiratory failure |

## Hallucinogens

| Drugs | Nervous System Action | Medical Uses | Possible Side Effects |
|---|---|---|---|
| LSD, marijuana (cannabis), PCP (phencyclidine), psylocibin/ mushrooms, peyote, ecstasy/MDMA, ketamine<br>*Street names:* acid, blotter, pot, MaryJane, 'shrooms, special K, vitamin E, candy | • Bear a close chemical resemblance to the neurotransmitters norepinephrine, dopamine, and serotonin, a transmitter involved with mood, anxiety, and sleep induction<br>• May disrupt the sorting process and allow a surge of sensory data to overload the brain<br>• Produce euphoria, hallucinations, and poor time perception | *Marijuana*<br>• Relieves nausea in some patients undergoing chemotherapy treatment for cancer<br>• Appears to reduce pain and suffering for persons with AIDS<br>• Reduces pressure within the eye in persons suffering from glaucoma | *All*<br>• Psychotic behavior (derangement of the personality and loss of contact with reality)<br>*Marijuana*<br>• Impaired eye–hand coordination, increased heart rate, panic attacks, anxiety, paranoia, depression, immune system impairment, upper respiratory system damage, and decreased levels of sex hormones |

**What are psychoactive drugs and how do they work?**

**32.11 Psychoactive drugs** (p. 566) affect the action of neurotransmitters in specific parts of the brain.

**32.11** Some psychoactive drugs are used medically to alter the mood or to treat diseases or disorders.

**32.11** Many psychoactive drugs are used for nonmedical reasons, are taken in doses that may cause damage to the body, and often result in personally destructive, antisocial, and crime-related behaviors.

**32.11** The chronic use of psychoactive drugs results in **drug addiction** (p. 567), which is a compulsive urge to continue using the drug, physical and/or psychological dependence on the drug, and a tendency to increase the strength (dosage) of the drug.

**32.11** Major classes of psychoactive drugs are depressants (sedative-hypnotics, alcohol), opiates (opium, morphine), stimulants (amphetamines, cocaine), and hallucinogens (LSD, marijuana).

## Level 1 — Learning Basic Facts and Terms

**Multiple Choice**

1. The myelin sheath
   a. insulates axons.
   b. increases the speed of impulses along axons.
   c. insulates dendrites.
   d. releases neurotransmitter.
   e. Both a and b are correct.

2. Which of the following represents the path a nerve impulse would take through the human nervous system?
   a. sensory neuron → interneuron → motor neuron → muscle cell
   b. interneuron → motor neuron → sensory neuron → muscle cell
   c. muscle cell → motor neuron → interneuron → sensory neuron
   d. sensory neuron → motor neuron → interneuron → muscle cell

3. Which of the following is true of a neuron at its resting membrane potential?
   a. The concentration of $Na^+$ ions is greater outside the cell than inside.
   b. The concentration of $K^+$ ions is greater inside the cell than outside.
   c. $Na^+$ voltage-gated channels are closed.
   d. The inside of the cell has a negative charge.
   e. All of the above are true.

4. Put the steps below (1–5) in the correct sequence for transmission of a nerve impulse across a chemical synapse.
   1. Neurotransmitter binds to the postsynaptic cell receptor.
   2. An action potential depolarizes the presynaptic membrane.
   3. Ion channels open allowing $Na^+$ to enter the postsynaptic cell.
   4. Neurotransmitter is released from presynaptic vesicles.
   5. Enzymes degrade the neurotransmitter in the synaptic cleft.
      a. 2,4,1,3,5          d. 4,3,1,2,5
      b. 2,3,5,4,1          e. 5,1,2,4,3
      c. 3,2,5,1,4

5. Which of the following statements is true regarding an action potential?
   a. It results in a long depolarization of the membrane that lasts several minutes.
   b. The magnitude of the depolarization changes with the strength of the stimulus.
   c. It travels more slowly in myelinated axons.
   d. It skips or jumps from one node to the next in myelinated axons.
   e. All of the above are true.

**True–False**

6. _____ In most sensory neurons, the impulse travels directly to an interneuron that they are touching, whereas in most interneurons the impulse is converted to a chemical signal before traveling to a motor neuron.

7. _____ Sensory neurons are the main integrators of the nervous system.

8. _____ An interneuron can receive both inhibitory and excitatory impulses from different sensory neurons at the same time.

9. _____ Heart rate is controlled exclusively by the sympathetic nervous system.

## Level 2 — Learning Concepts

1. Explain the term *resting potential*. What creates it, and why is it important?

2. Explain how a nerve impulse travels along an axon.

3. You decide to move your finger, and it moves. Explain how your nervous system communicated your decision to the muscles in your finger.

4. Why are neurons of the central nervous system called *integrators*? What do they integrate?

5. Some invertebrate brains are more accurately called ganglia. What is the definition of a ganglion, and why is it more descriptive than the term *brain* for these invertebrates?

6. Distinguish between the central nervous system and peripheral nervous system. What are the functions and components of each?

7. What is the brain stem? Name its three major components and summarize its functions.

8. You're studying in a quiet library. Suddenly someone drops an armful of books; you hear the loud crash and turn your head toward the noise. Explain how your sensory and motor pathways allowed you to detect and respond to this stimulus.

9. Two divisions of the autonomic nervous system, the sympathetic and parasympathetic, control the organs of your body. Each organ is affected by neurons from each division. Why do you think the body uses two neurons to control an organ rather than just one?

## Level 3 | Critical Thinking and Life Applications

1. Is the nerve impulse (action potential) a chemical reaction? Why or why not?
2. The nerve impulse is an all-or-nothing response, and for any one neuron the action potential is always the same. How, then, do you explain your ability to experience differing intensities of feeling such as a mild pain or a severe pain?
3. Doctors warn parents to avoid feeding their toddlers a low-fat diet because fat is essential for the body's production of myelin. Why is this important in prenatal care and in the diet of infants and toddlers?

4. If the spinal cord is severed near the neck in an injury, the resulting disability is more severe than that seen if the spinal cord is severed lower in the back. Why do you think is so?
5. Although you may enjoy indulging in strawberry sundaes from time to time, you will not become addicted to them. Why not?

## In The News | Critical Thinking

### KID CONCUSSIONS

Now that you understand more about the nervous system, reread this chapter's opening story about the ImPACT method of evaluating athletes who have had a concussion. To better understand this information, it may help you to follow these steps:

1. Review your immediate reaction to the testing of brain functioning after a concussion that you wrote when you began reading this chapter.
2. Based on your current understanding, again summarize the main point of the research in a sentence or two.
3. What questions do you now have about this information that this chapter's opening story does not answer?
4. Collect new information about the research. Visit the *In The News* section of this text's companion Web site at www.wiley.com/college/alters and

watch the "Kid Concussions" video. Then use the "summary" link to read the accompanying story and access related links. Use this information, the links provided, and other online and library resources to answer your questions and find updates about this topic. State the sources of your information. Explain why you think the information is accurate. Also determine whether the information expresses a particular point of view or is biased in any way.
5. What in your view are the most significant aspects of this information? Give reasons for your opinion and for any changes in your ideas based on the additional information you have collected and the analysis you have done.

# THE SENSES

## In The News | Super Snout

Before the Airbus A320 could completely back out of the gate and head to the runway, it stopped briefly and then returned to its original position. The passengers looked at each other and wondered what had happened. Moments later the pilot announced, "I'm sorry . . . there's a problem. You'll have to deplane and return to the terminal. Leave your bags in the overhead compartments." Six hours later, the tired passengers were allowed to board again. After they were seated, the pilot explained the reason for the delay. "An anonymous caller reported that an explosive device was on board this aircraft. Everything that was stowed in the baggage compartment was removed, and two bomb-sniffing dogs checked everything that was removed from the plane and then sniffed this aircraft 'up and down.' No explosives were found. If you'd rather take another flight, you may leave now." None of the passengers left.

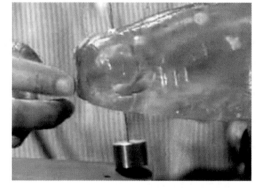

Certain dogs can be trained to use their sense of smell to locate not only explosive materials, but also illegal drugs, lost children, and even people buried beneath the rubble caused by an earthquake. It takes considerable time and money to train dogs to perform these tasks, thus a limited number of these animals are available. With the threat of terrorism, there is a pressing need for technology that can detect explosives in public places. An ideal solution would be a portable device able to detect certain odors as well as a dog's nose can.

Efforts to develop such a device are underway. David Walt, a chemistry professor at Tufts University in Massa-chusetts, and his research team used computerized tomography (CT) scans of a canine's nose to build a plastic model of the snout, which is shown in the photo. To simulate dog olfaction, Walt and his team made sensors out of optical fibers, each tipped with a fluorescent-dyed bead able to bind with odor molecules. The researchers arranged the sensors inside the snout model. When a sensor's bead bound to a molecule in the odor, its dye changed color, with different colors indicating different molecules. The different colors' wavelengths of light passed along the optical fibers to a computer that interpreted the overall pattern of wavelengths collected from the various sensors and identified the smell.

Although Walt's artificial nose is not as sensitive as a dog's, he and other scientists are continuing to refine their olfactory devices. In the future, mechanical noses are likely to find work "sniffing" the air in a variety of public places. To learn more about the artificial nose, visit the *In The News* section of this text's companion Web site at www.wiley.com/college/alters and view the video "Super Snout."

*Write your immediate reaction to this information about the development of an odor-detecting device: first, summarize the main point in a sentence or two; then suggest what you think its significance is. You will have an opportunity to reflect on your responses and gather more information on this topic in the* In The News *feature at the end of this chapter. In this chapter, you will learn more about the senses.*

# CHAPTER GUIDE

## What are sensory receptors?

**33.1** Sensory receptors change environmental stimuli into nerve impulses.

## How do animals sense their internal and external environments and their position in space?

**33.2** Receptors within blood vessels and organs sense the internal environment.

**33.3** Receptors within skeletal muscles, tendons, and the inner ear sense position in space.

**33.4** Mechanoreceptors detect stimuli such as touch, movement, and pressure.

**33.5** Vertebrates have nerve endings in the skin that are sensitive to temperature.

## Can some animals detect stimuli humans cannot?

**33.6** Some animals can detect infrared radiation, and others can detect UV light.

**33.7** Some animals can detect electrical fields.

**33.8** Some migratory animals and those with homing instincts detect magnetic fields.

## How do the senses of taste, smell, sight, hearing, and balance work?

**33.9** Taste buds and olfactory receptors sense chemicals.

**33.10** Humans and other organisms sense light.

**33.11** Humans and other organisms sense sound.

**33.12** The vestibule and semicircular canals of the inner ear detect position and direction of movement.

# What are sensory receptors?

## 33.1 Sensory receptors change environmental stimuli into nerve impulses.

Sensing the external environment is important for all animals, not just humans. How else can animals find food and mates, and escape from environmental dangers such as predators? Animals can sense a variety of environmental stimuli. Their sensory receptors change environmental stimuli such as sound, light, and pressure into nerve impulses. Some receptors are composed of nervous tissue; others are not but are capable of initiating a nerve impulse in an adjacent neuron. Many kinds of receptors have evolved among animals, with each type of receptor sensitive to a different aspect of the environment (**Table 33.1**).

For an organism to be aware of, or to sense, its internal or external environment, certain events must take place (**Figure 33.1**). First, a change in the environment acting as a stimulus ❶ must occur in the presence of a sense organ ❷. It must be of sufficient magnitude to open ion channels within the membranes of receptor cells ❸, thus depolarizing them (see Chapter 32). This depolarization is called a *generator potential.* When the generator potential reaches the threshold level, it initiates a nerve impulse, which is an action potential ❹ in the sensory neurons with which the receptor synapses. In some sensory neurons, the dendrites themselves act as simple receptors. In this case, the generator potential initiates an action potential along that same neuron. The impulse is conducted by nerve fibers to either the spinal cord or the brain ❺.

Impulses conducted to specific sensory areas of the cerebral cortex (see Chapter 32) in vertebrates produce conscious sensations. Only the cerebral cortex can "see" a flower or "feel" a paper cut. **Figure 33.2** shows areas of the human cerebral cortex that perform various tasks, including the special senses of sight, hearing, taste, and smell. **Special senses** are those that are restricted to particular locations of the body. Figure 33.2 also shows the primary sensory area, in which the general senses are perceived. **General (somatic) senses**

| TABLE 33.1 | | Types of Environmental Stimuli | |
|---|---|---|---|
| **Mechanical** | **Thermal** | **Chemical** | **Electromagnetic** |
| • Pressure<br>• Touch<br>• Motion<br>• Sound<br>• Vibration<br>• Gravity | • Heat<br>• Cold<br>• Infrared radiation | • Individual types of molecules | • Visible light<br>• Electricity<br>• Magnetism |

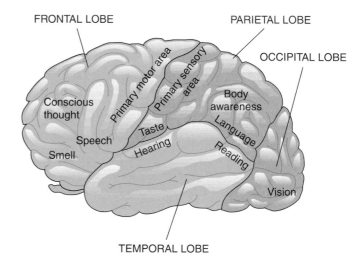

**Figure 33.2** Areas of the human cerebral cortex that sense external stimuli. Find the areas that perceive smell, hearing, taste, and vision. Also note the primary sensory area.

are those that have more than one location in the body, such as sense receptors distributed throughout the body for heat, cold, pain, and pressure.

Receptors provide animals with information about the organism's internal environment, position in space, and the external environment. Receptors that sense the internal environment are located deep in the body within the walls of blood vessels and organs. Receptors that provide information about body position and movement are located in the muscles, tendons, joints, and inner ear. Receptors sensitive to stimuli outside the body are located at or near the body surface.

**GENERALIZED SENSORY PATHWAY (SIGHT)**

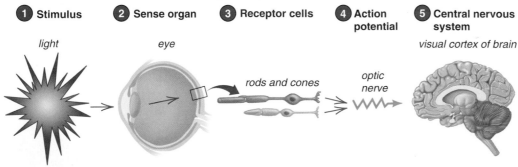

**Figure 33.1** Generalized sensory pathway. A stimulus of sufficient strength causes a generator potential in receptor cells that initiates action potentials in the sensory nerve carrying information to the central nervous system for processing.

# How do animals sense their internal and external environments and their position in space?

## 33.2 Receptors within blood vessels and organs sense the internal environment.

Are all sense receptors complex? No, many are simply nerve endings that depolarize in response to direct physical stimulation—to temperature, to chemicals such as carbon dioxide, or to a bending or stretching of the neuron cell membrane. Among the simplest of these neurons are those involved in regulatory mechanisms that maintain the body temperature and blood chemistry within a narrow range; these mechanisms are most well developed in mammals.

Temperature-sensitive neurons in the hypothalamus, highlighted in beige in **Figure 33.3**, act as your body's thermostat. These neurons constantly take your temperature by monitoring the temperature of your blood. If the temperature of the blood rises, such as when you are sick or vigorously exercising, neurons in the hypothalamus trigger your body's heat-loss mechanisms. Such mechanisms include dilating, or widening, the blood vessels closest to the skin so that excess heat can be lost to the environment (see Figure 29.3). Similarly, if the temperature of the blood drops, such as when you are in a cold environment, neurons in the hypothalamus trigger your body's heat production mechanisms. Such mechanisms include shivering, a cycle of contraction and relaxation of your skeletal muscles producing body heat as a waste product of the cellular respiration that generates adenosine triphosphate (ATP) for muscle contraction. Once heat is generated, blood vessels closest to the skin constrict, conserving the heat.

Other receptors are sensitive to the levels of carbon dioxide and oxygen in your blood as well as its pH. These receptors are embedded within the walls of your arteries at several locations in the circulatory system. Bathed by the blood that flows through the arteries, these chemical receptors provide input to respiratory centers in the medulla and pons, which are highlighted in red in Figure 33.3. The medulla and pons use this information to regulate the rate of respiration. When carbon dioxide and pH levels in the blood rise and the oxygen level falls, the respiratory centers respond by increasing the respiration rate.

Various other receptors that sense the environment within your body have membranes whose ion channels open in response to mechanical force. Put simply, twisting, bending, or stretching these nerve endings results in depolarization of their membranes, causing these neurons to fire. These receptors differ from one another primarily in their location and in their orientation to the stimulus. Pain receptors, for example, are widely distributed throughout the body. They respond to many different types of stimuli when these stimuli reach a level that can endanger the body. Pain receptors deep within the body detect such internal environmental stresses as inadequate blood flow to an organ, excessive stretching of a structure, and spasms of muscle tissue.

## 33.3 Receptors within skeletal muscles, tendons, and the inner ear sense position in space.

If you close your eyes, you still know how your arms and legs are positioned; if you're lying, sitting, or standing; or if you're moving or still. You, like all animals, have receptors called **proprioceptors** (PRO-pree-oh-SEP-turz) that sense the position of the body in space. It is your proprioceptors that tell you how much your arms and legs are bent, where they are in relationship to your body, and where your head is relative to the ground.

The muscles of vertebrate animals have proprioceptors buried deep within them, keeping track of the degree to which the muscles are contracted. These receptors are actually specialized muscle cells called *muscle spindles*. Wrapped around each spindle is the end of a sensory neuron called a **stretch receptor** (**Figure 33.4a**). When a muscle is stretched, the muscle spindle gets longer and stretches the nerve ending, repeatedly stimulating it to fire. Conversely, when the muscle contracts, the tension on the fiber lessens and the stretch receptor ceases to fire.

Figure 33.4b (top left) shows a student carrying books. The books are a constant load on the biceps muscle, which is balanced by a constant force. Equal motor and sensory impulses are sent to and from the spinal cord (bottom left). A second student places his books on those of the first student (top middle). This sudden increase in load stretches the muscle. The muscle spindles send signals back to the spinal cord (bottom middle). There, additional motor neurons are stimulated and send impulses back in the motor nerve (top right), causing increased contraction to balance the load (bottom right).

Other proprioceptors are found in vertebrate tendons, the connective tissue that joins muscles to bones, and in the tissue surrounding joints. These receptors are also stimulated when they are stretched and help protect muscles, tendons, and joints from excessive tension and pulling.

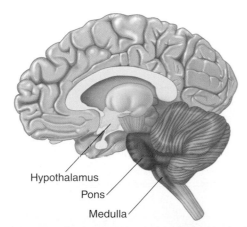

**Figure 33.3** Centers for homeostasis in the brain. The hypothalamus, shown in beige, is the body's thermostat. The medulla and pons, highlighted in red, contain respiratory centers.

Hypothalamus
Pons
Medulla

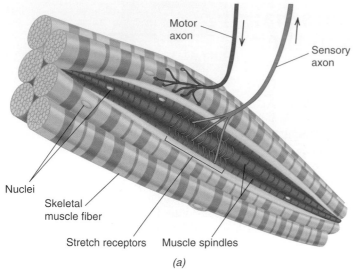

Nuclei

Skeletal muscle fiber

Stretch receptors    Muscle spindles

*(a)*

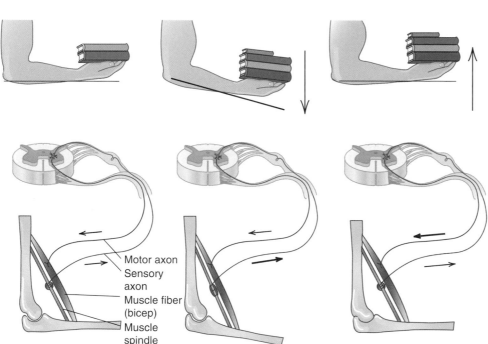

Motor axon
Sensory axon
Muscle fiber (bicep)
Muscle spindle

*(b)*

**Figure 33.4 Stretch receptors in skeletal muscles.**
(*a*) Stretch-sensitive neurons wrap around muscle spindles, forming stretch receptors that detect and control muscle contraction. (*b*) When the force of the load of books equals the force of the contraction of the biceps muscle, the arm carrying the books is held steady (left). When the load is greater than the force exerted by the biceps muscle, the arm drops (center). Stretch receptors in the biceps then send a signal to the spinal cord, which sends back a signal to increase the force of contraction (right).

 **Visual Thinking:** Why is the sensory neuron shown arising at the muscle spindle and the motor neuron shown ending at the muscle fiber?

Gravity and motion receptors help an animal maintain equilibrium, or balance. In humans, these receptors are located in the inner ear. Balance is discussed in Section 33.12.

---

### CONCEPT CHECKPOINT

1. How is the initiation of a generator potential in a sensory receptor cell different from how an action potential is initiated in a neuron?
2. Describe some of the internal stimuli that elicit generator potentials in sensory receptor cells.

---

## 33.4 Mechanoreceptors detect stimuli such as touch, movement, and pressure.

Animals have a variety of *mechanoreceptors*, which are stimulated when they are physically deformed by mechanical energy such as pressure. Fish and amphibians have a complex system of mechanoreceptors called a *lateral line system*. This system consists of a single row of mechanoreceptors along the sides of the body (**Figure 33.5**); other mechanoreceptors form various patterns on the head. The lateral line system detects mechanical stimuli such as sound vibrations and bodily movement. It also senses pressure changes produced by water currents.

Invertebrates as well as vertebrates have mechanoreceptors. For example, flatworms (see Figure 21.13) have a variety of cells that sense pressure changes and mechanical stimuli. Many species of roundworms (see Figure 21.15) have external setae (SEE-tee), which are thin, stiff, bristlelike structures at various loca-

Lateral line

**Figure 33.5** The lateral line system in a fish.

tions on the body. These are thought to be mechanical receptors. Earthworms also have setae, but they function in locomotion. In addition, they have nerve endings along their length that act as touch receptors or vibration receptors. The tentacles of most cephalopods—octopus and squid—are studded with receptors sensitive to touch.

**Figure 33.6** shows a cross section of skin with specialized receptors in humans. The simplest of these are free nerve endings, which are distributed throughout most of the body. In addition to detecting pain, free nerve endings detect temperature, itch, and movement. For example, free nerve endings wrap around the roots of hairs and detect any stimulus, such as the wind or an insect that moves body hair.

Disk-shaped dendrite endings called *Merkel's disks* (MER-kelz) and egg-shaped receptors called *Meissner's corpuscles* (MYZ-nerz) are two types of touch receptors. (The word "corpuscle" means little body.) Both are widely distributed in the skin but are most numerous in the hands, feet, eyelids, tip of the tongue, lips, nipples, clitoris, and tip of the penis.

Pressure receptors called *Pacinian corpuscles* (pah-SIH-nee-an) are located deeper within the skin. Pacinian corpuscles look layered, much like an onion. The layers are made up of connective tissue, with dendrites sandwiched between. These receptors are most numerous in the nipples and external genitals of both sexes. Pacinian corpuscles are also a part of the body's internal sensing system because of their locations around joints and tendons, in muscles, and in certain organs.

**Figure 33.6** Mechanoreceptors in the human skin. The layers of the skin are described and discussed in Chapter 34.

## 33.5 Vertebrates have nerve endings in the skin that are sensitive to temperature.

In humans and other vertebrates, two populations of nerve endings in the skin called *thermoreceptors* are sensitive to coldness and warmth. The temperature of the environment is sensed only indirectly. The temperature of the skin near these receptors is the true stimulus. The environmental temperature, the body's internal temperature, and other factors that cool the body, such as the evaporation of sweat, all affect skin temperature.

# Can some animals detect stimuli humans cannot?

## 33.6 Some animals can detect infrared radiation, and others can detect UV light.

Heat radiated from animals or machines is called *infrared radiation*. Have you ever seen police or military personnel on TV using infrared sensors from helicopters to detect humans and warm engines in vehicles on the ground? A few animals can detect infrared radiation without the equipment!

Some snakes, such as rattlesnakes and other pit vipers, can detect heat radiated from warm-blooded animals. Pit vipers detect infrared radiation by means of *pit organs*, which are located on either side of the head between the nostrils and the eyes (**Figure 33.7**). The term *pit viper*, in fact, refers to poisonous snakes that possess pit organs. Heat guides the direction at which the rattlesnake strikes. It moves its head back and forth; when the intensity of detected heat is equal on both sides of the snake's head, it perceives that its prey is straight ahead.

**Figure 33.7** Pit organs of a rattlesnake. These organs detect heat radiated by warm-blooded animals.

(a)                    (b)

**Figure 33.8** Evening primrose in (a) normal light and (b) UV light. Notice the pattern differences. Many insects, such as bees, can see UV light.

Whereas pit vipers can detect infrared radiation, which consists of wavelengths longer than visible light, insects can detect ultraviolet (UV) radiation, which consists of wavelengths shorter than visible light. Many flowers emit patterns of UV light that are visual cues to insects (**Figure 33.8**). Many flowering plants depend on insects for pollination, and these visual cues aid the process.

## 33.7 Some animals can detect electrical fields.

No human can detect electric fields emitted by other organisms. However, some fishes, some salamanders, and the duck-billed platypus have receptors that can. *Electroreceptors* help these organisms find prey and navigate waters when the conditions are dark and/or murky. Sharks (see Figure 22.6a) can use their electroreceptors to detect electrical fields emitted by other fishes that may become "dinner."

"Electric fish" generate electric fields as well as detect them. Some of these fish, such as electric catfish and electric eels, generate a strong electrical pulse that may stun potential prey. Electricity is generated by means of electric organs, which consist of stacks of modified muscle tissue. Other electric fish generate a much weaker electric field around themselves by which they can detect objects or organisms in the environment when the objects' conductivity differs from the surrounding water and they distort the fish's electric field. The lines of flow converge on an object of higher conductivity, such as another fish, and diverge around a poorer conductor, such as a rock (**Figure 33.9**). These weakly electric fish typically have electroreceptors scattered over most of their body.

The electroreceptors of the duck-billed platypus (see Figure 22.16) are located in its bill. This egg-laying mammal finds its food in muddy streams, diving with its eyes, ears, and nose closed. Its ability to home in on weak electric fields with its bill helps it search out prey.

## 33.8 Some migratory animals and those with homing instincts detect magnetic fields.

A wide variety of animals are affected by magnetic fields, including salmon, salamanders, turtles, hornets, honeybees, and homing pigeons. Scientists have found that most animals affected by magnetic

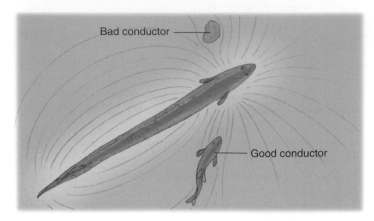

Bad conductor

Good conductor

**Figure 33.9** Weakly electric fish can detect distortions in a self-generated electric field. Electric fish detect objects or organisms in the environment when they differ in conductivity from the surrounding water and they distort the fish's electric field, stimulating electroreceptors that are scattered over its body.

fields have magnetic granules containing iron called magnetite in their bodies. There is evidence that some migratory animals and those with homing instincts use the Earth's magnetic lines of force to help orient themselves (**Figure 33.10**). Certain bacteria are also affected by magnetic fields; they also contain similar magnetic particles. In their natural environment, these bacteria are attracted by the magnetic forces of the Earth and orient themselves deep in mud and away from oxygen, which is a toxin to them.

**Figure 33.10** Animal migration. After reaching sexual maturity, female loggerhead turtles migrate long distances from their feeding sites to lay their eggs where they were born. This loggerhead sea turtle, *Caretta caretta*, is laying her eggs on a Florida beach at night. Upon hatching, young turtles become imprinted to the Earth's magnetic field and possibly to the smell of the water near their nesting beaches. Imprinting is a rapid and irreversible type of learning that takes place during an early developmental stage of some animals.

# How do the senses of taste, smell, sight, hearing, and balance work?

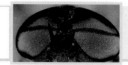

## 33.9 Taste buds and olfactory receptors sense chemicals.

The ability to sense chemicals, or *chemoreception*, is widespread in the animal kingdom. Many flatworm species have a variety of cells that sense chemicals, such as those in potential foods. Roundworms are more complex in that they have *chemoreceptors* located on their anterior end, and some species have additional chemoreceptors on their posterior end. Earthworms have chemoreceptors distributed along the length of the body, while cephalopods (squid and octopus) have tentacles covered with chemoreceptors. Insects have their six legs studded with chemoreceptors; such receptors are also found on their mouthparts and elsewhere on the body, as can be seen dispersed over the head of the acrobat ant shown in **Figure 33.11**.

Used for finding food or mates, locating enemies, or warning an animal of danger, chemoreception is most highly developed in vertebrates and arthropods (spiders, lobsters, and insects). Even the single-celled protists, however, respond to chemical stimuli, although they have no sense organs. Like the white blood cells that roam the human body fighting infection, protists detect chemicals when these stimuli reach their cell membranes. They respond with directed movements either toward or away from the stimuli. Such directed movements are called *taxes* (sing. *taxis*). A directed movement toward or away from a particular chemical is a positive or negative chemotaxis.

Taste and smell are the two types of chemoreception in humans. Taste usually means sensing chemicals in solution; smell usually means sensing chemicals in the air. These terms are applied to chemoreception in other animals, too. The catfish, for example, has taste buds distributed all over its body. Imagine going for a swim and using your back or chest for tasting!

### Taste

Taste receptors—**taste buds**—detect chemicals in the foods you eat. Humans detect many taste qualities, far more than the four primary tastes: salty, sweet, sour, and bitter. Some sensory physiologists have suggested a potential fifth primary taste, monosodium-glutamate-induced savory taste, or umami.

Each of the 10,000 taste buds on the upper surface of the tongue, the lips, and the throat responds in varying degrees to the four primary tastes. The rich and diverse array of tastes humans perceive are composed of different combinations of impulses from chemoreceptors distributed all over the human tongue. In addition, the sense of taste interacts with the sense of smell to produce the taste sensation, as you may have noticed when you had a "stuffy nose" and lost much of your ability to taste.

Taste buds are packed within short projections known as *papillae*. **Figure 33.12** shows one type of papilla on the tongue and points out the location of taste buds deep within the papilla. Taste buds are microscopic structures shaped like tiny onions. Each is made up of 30 to 80 receptor cells bound together by support cells. Hairlike projections of receptor cells called *microvilli* poke through an opening in the taste bud called a *taste pore*. The various chemicals in food stimulate the receptor cells as they dissolve in the saliva and come into contact with the microvilli. Generator potentials within the receptor cells stimulate neighboring sensory nerve fibers that travel from the tongue and throat to the cerebrum, medulla, and thalamus.

### Smell

Although smell receptors are the simplest of the special receptors in humans, they can distinguish several thousand odors. Smell receptors are also called **olfactory receptors** (ole-FAK-tuh-ree), a

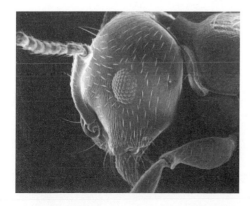

**Figure 33.11** Insect chemoreceptors. The short hairlike structures distributed over the head of this acrobat ant are chemoreceptors.

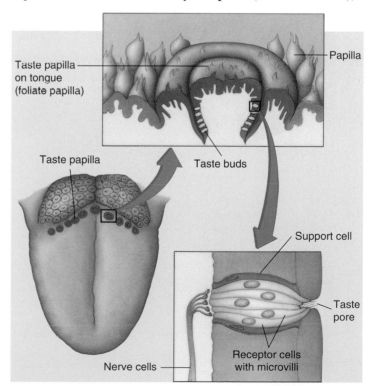

**Figure 33.12 Structure of taste buds.** Papillae are the projections on the tongue that give it its velvety appearance. The human tongue has four types of papillae. One type has no taste buds but senses the texture of food. The other three types contain taste buds, including the type shown in this illustration, foliate papillae.

name derived from the Latin words *oleo*, meaning "a smell," and *facio*, meaning "to make." Olfactory receptors consist of neurons whose cell bodies are embedded in the nasal epithelium, the tissue that lines the nasal cavity (**Figure 33.13**). Dendrites extend from these cell bodies. Cilia, microscopic projections of the dendrites, poke out of the epithelium like minute tufts of hair and are bathed by the mucus that covers this tissue. Molecules in the air dissolve in the mucus and come into contact with the cilia.

Evidence suggests that the olfactory receptors detect different smells because specific airborne molecules bind with receptor proteins located within the cilia. This interaction opens ion channels within the membrane of the receptor so that a generator potential is developed, firing the neuron when it reaches a particular threshold. The nerve impulse then travels to the olfactory area of the cerebral cortex. (This area is labeled "smell" in Figure 33.2.) On its way, it travels through the brain's limbic system, an area of the brain responsible for many drives and emotions. Does the aroma of baking cookies please you? Does the odor of rotting garbage cause you to turn your head in disgust? That is your limbic system at work in conjunction with your cerebrum and your sense of smell.

---

### CONCEPT CHECKPOINT

3. List the different categories of external sensory receptors discussed thus far. For each category, describe the following: (1) the type of stimuli sensed, (2) the location in organisms, and (3) how the organism uses or interprets information in the stimuli.

---

## 33.10 Humans and other organisms sense light.

Photoreception is the sensing of visible light, which consists of certain wavelengths of electromagnetic energy. You can visualize these disturbances as waves, much like the repeating disturbances or tiny waves caused by a stone thrown into a still pond.

Visible light is only a part of the full range of the electromagnetic waves coming from the sun called the *electromagnetic spectrum* (see Figure 7.4). The waves in this spectrum vary in length. The shortest wavelengths are gamma rays, and the longest are radio waves. Visible light has wavelengths in between these two extremes, but it, too, is made up of many different wavelengths. Remember the last rainbow you saw? Its array of colors was caused by the separation of the various wavelengths of visible light as they passed through tiny droplets of water in the air.

### Photoreception in organisms other than humans

Sensitivity to light is extremely widespread. Organisms other than animals, such as plants (see *Focus on Plants*) and certain protists, can detect light. The red, cup-shaped, light-sensing organelle of certain flagellated protists, called the *stigma* or *eyespot*, is one of the few true sensory organelles known among protists. *Euglena* (**Figure 33.14**) is the best known in this photoreceptive group.

Although many invertebrates have "eyes," these sensory organs are not always image-forming. For example, various species of flatworms have eyes on their anterior ends, and earthworms have photoreceptor cells in their epidermis along the body. Like the protist eyespot, however, the flatworm and earthworm "eyes" are only sensitive to light and changes in its intensity. These receptors contain light-absorbing pigments that send impulses to the animal's brain when stimulated.

Although flatworms and earthworms have only light-sensitive eyes, many arthropod species (crabs, insects, and spiders), especially many insect species, have multifaceted image-forming eyes called *compound eyes*. Horror movies have led us all to believe that animals with compound eyes, such as flies, see thousands of identical, complete images. This is not the case. Each unit of the multifaceted compound eye "sees" only a part of the complete image. The brain puts the

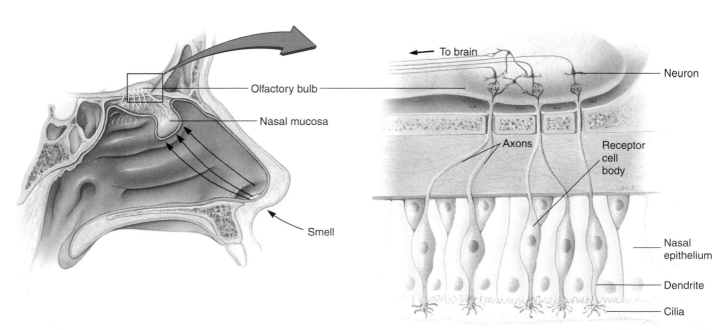

**Figure 33.13 The sense of smell in humans.** (*a*) Olfactory receptor cells are located in the lining of the nasal cavity. (*b*) Airborne molecules bind with receptor proteins located in the cilia. This triggers a nerve impulse, which is sent to the olfactory area of the brain.

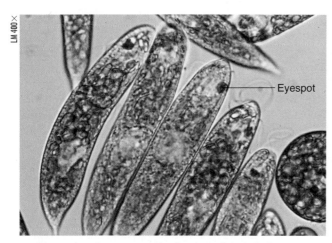

LM 400×

**Figure 33.14** *Euglena*, a protist that can detect light.

**Visual Thinking:** What is the adaptive advantage to *Euglena* of having an eyespot? A clue is in its green color.

"pieces" together to form a single image. **Figure 33.15** shows the compound eyes of a horsefly. Most cephalopods (squid and octopus) have image-forming eyes similar to the eyes of vertebrates.

The detection of light can affect an organism's behavior. In organisms displaying *positive phototaxis*, a behavioral adaptation in which an organism moves toward the light, the animal moves until the sensations coming from the eyes or eyespots are equal and strong. Many species of flying insects exhibit positive phototaxis. In organisms displaying *negative phototaxis*, the animal moves until the sensations coming from the eyes are equal and weak. The common cockroach, which thrives in dark areas with high humidity, exhibits negative phototaxis.

## Rods and cones in the retina of the human eye sense light

Image-forming eyes are either of the multifaceted type as in insects, or the single-lens, cameralike type as in cephalopods and vertebrates. The cameralike eye in humans works basically as follows: receptors sensitive to the various wavelengths of visible light are located in the back of the eye and act somewhat like film in a camera. Light that falls on the eye is focused by a lens onto these receptors, just as the lens of a camera focuses light on film.

The structure of the human eye is shown in **Figure 33.16**. Human eyes are each about 1 inch in diameter and are covered and protected by a tough outer layer of connective tissue called the **sclera** (SKLEAR-uh). The front of the eye is transparent, allowing light to enter the eye. This portion of the eye's outer layer is

called the **cornea**. Because the cornea is rounded, it not only allows light to enter the eye, but bends the light as well. This bending, or refraction, of light occurs when the light waves slow down as they move from the air and pass through the tissue of the cornea. The cornea is the first structure of the eye that begins to focus incoming light onto the rear of the eye where the receptors are located.

A compartment behind the cornea is filled with a watery fluid called the **aqueous humor**. The word "humor" refers to "fluid within the body," and the word "aqueous" means "water." Fresh aqueous humor is continually produced as old fluid is drained into the bloodstream. This fluid nourishes both the cornea and the lens because neither structure is supplied with blood vessels. In addition, the aqueous humor, along with fluid in the main compartment of the eye called the *vitreous* (Latin for "glass") *humor*, creates a pressure within the eyeball that maintains the eyeball's shape. It also keeps the **retina** (RET-un-nuh), the sensory portion of the eye, pressed properly against the back of the eyeball.

The **lens** lies just behind the aqueous humor and plays a major role in focusing the light that enters the eye onto the retina. It looks much like a lemon drop candy or a somewhat flattened balloon. If you were to cut a lens in half, it would look similar to an onion because it is made up of layer upon layer of protein fibers. Ligaments encircle the lens and suspend it within the eye. These ligaments are attached to a tiny circular muscle called the **ciliary muscle** (SILL-ee-err-ee), which, by contracting or relaxing, slightly changes the shape of the lens. The greater the curve of the lens, the more sharply it bends light rays toward one another.

**Figure 33.15** Compound eyes in a horsefly.

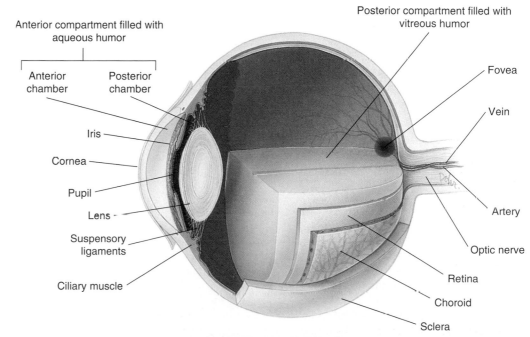

**Figure 33.16** Structure of the human eye.

Anterior compartment filled with aqueous humor

Anterior chamber

Posterior chamber

Iris

Cornea

Pupil

Lens

Suspensory ligaments

Ciliary muscle

Posterior compartment filled with vitreous humor

Fovea

Vein

Artery

Optic nerve

Retina

Choroid

Sclera

In individuals with normal vision (**Figure 33.17**, left), light is focused on the retina. When focusing on distant objects, the lens is flattened **1**. When focusing on near objects, the lens is more rounded **2**. Nearsighted people (those with *myopia*), however, have an elongated eyeball **3** or a thickened lens. In either case, the light from distant objects is focused in front of the retina. (Compare with **1**.) The near object can be brought into focus, but distant objects cannot. Nearsightedness can be corrected with a concave lens that spreads the light rays entering the eye from distant objects so that they focus on the retina **4**. Some people are farsighted (they have *hyperopia*), having either a shortened eyeball **5** or a thin lens. Light from near objects would be in focus behind the retina (compare with **1** and **3**), whereas light from distant objects can be focused on the retina. Farsightedness can be corrected with a convex lens that converges entering light rays so that they focus on the retina **6**. *Presbyopia* is farsightedness, which generally occurs in people over the age of 45, caused by a loss of elasticity of the lens as a person ages.

Some vision problems can also be corrected with photorefractive keratectomy (PRK) and laser-assisted in-situ keratomileusis (LASIK) surgery. PRK and LASIK are similar surgical techniques that use a laser to reshape the cornea so that incoming light is focused on the retina.

The **iris**, which lies between the cornea and the lens, controls the amount of light entering the eye. It works much like the aperture setting on a film camera, controlling the size of the **pupil** of the eye—what you see as a black dot—through which the light passes. Together, the ciliary muscle and the iris make up two of the structures of the middle layer of the eye. The third structure is the **choroid** (KOR-oyd). As shown in Figure 33.16, the choroid extends toward the back of the eye as a thin, dark-brown membrane that lines the sclera. It contains vessels carrying blood that nourishes the retina and a dark pigment that absorbs light rays so that they will not be reflected within the eyeball.

The inner layer of the eye is the retina. The retina lines the back of the eye and contains sensory receptors called **rods** and **cones** (**Figure 33.18**). The tips of these cells contain a pigment that undergoes a chemical alteration when activated by light. Activated *rhodopsin* found in rod cells, for example, triggers a series of events that eventually reach the brain and are interpreted as patterns of light and dark.

Rod cells are stimulated by low levels of light and detect only the presence of light and not its color; thus, vision with rod cells alone is black, white, and shades of gray. Did you ever try to see color in extremely dim light? Your rods alone are at work, and they cannot see color. Color vision is achieved by your cone cells, which function in bright light.

Although they all look the same under the electron microscope, there are three kinds of cone cells. Each type possesses rhodopsinlike molecules that have slightly different shapes from one another. These differences determine which wavelengths of light the cone containing that pigment will absorb. One type absorbs wavelengths of light in the 455-nanometer (blue-absorbing) range. Another type absorbs in the 530-nm (green-absorbing) range, and the third in the 625-nm (red-absorbing) range. The color you perceive depends on how strongly each group of cones is stimulated by a light source. The pigment in rod cells, on the other hand, absorbs in the 500-nm range. This range encompasses the wavelengths of light (blue, green, and red) that together make up white light.

The human retina contains about 7 million cones, most of which are located at a central region of the retina called the **fovea**

**NORMAL VISION**

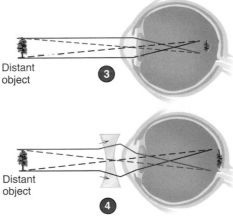

**NEARSIGHTED** (elongated eyeball)

**FARSIGHTED** (shortened eyeball)

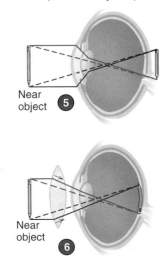

**Figure 33.17 Focusing the human eye.** Although light rays are reflected to the eye from all points on an object, not all are shown. In normal vision, images are in focus on the retina. With nearsightedness (myopia), images are in focus in front of the retina. A concave lens spreads the light rays entering the eye so that images are focused on the retina. With farsightedness (hyperopia), objects would hypothetically be in focus behind the retina. A convex lens converges light rays entering the eye so that images are focused on the retina.

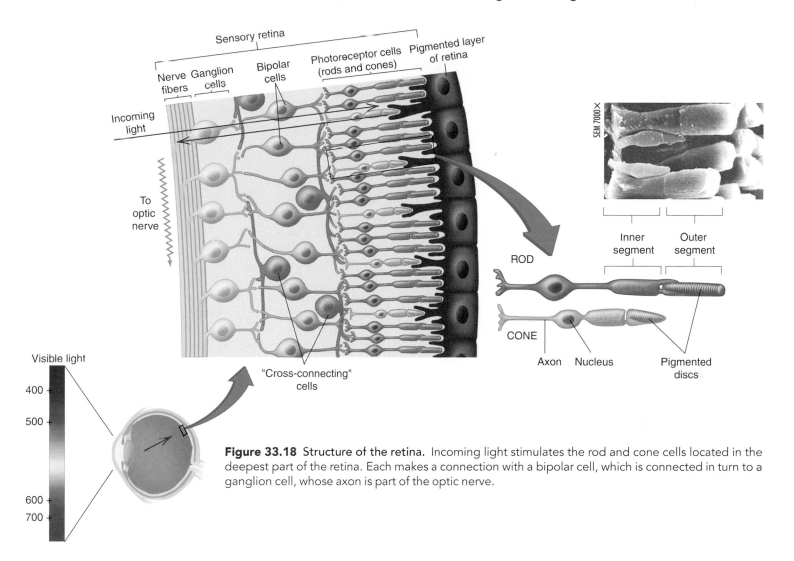

**Figure 33.18 Structure of the retina.** Incoming light stimulates the rod and cone cells located in the deepest part of the retina. Each makes a connection with a bipolar cell, which is connected in turn to a ganglion cell, whose axon is part of the optic nerve.

(FOE-vee-uh). The lens focuses incoming light on this spot, which is the area of sharpest vision. The blind spot, which is where the **optic nerve** leaves the eye carrying sensory information to the visual cortex of the brain. Sometimes called the optic disc, the blind spot is devoid of photoreceptors, but the brain "fills in" the missing spot with information from each eye, since the blind spot for each eye receives light from a different point in visual space. In addition, the eyes move a great deal, so each eye gets a full view of the surrounding space. The brain integrates this information to provide a "spotless" view of our surroundings.

Each foveal cone cell makes a one-to-one connection with a special kind of neuron called a *bipolar cell* (see Figure 33.18). Each of the bipolar cells is connected in turn to a single ganglion cell, whose axon is part of the optic nerve. The frequency of impulses transmitted by any one receptor provides information about light intensity. The pattern of firing among the foveal axons provides a point-to-point image. The different cone cells provide information about the color of the image.

Outside the fovea the output of many cones and rods is channeled to one bipolar cell. Many bipolar cells, in turn, synapse with one ganglion cell. In the outer edge of the retina, more than 125 re-

ceptor cells, mostly rods, feed stimuli to each ganglion cell in the optic nerve. This portion of the retina does not transmit a point-to-point image as the fovea does, but instead transmits a processed version of the visual input that may be interpreted simply as movement. Think of the periphery of your retina as a detector, whereas your fovea is an inspector.

Your two eyes are in two different positions, so they send two slightly different sets of information to the brain about what you are seeing. Each set of information is slightly different because each eye views an object from a slightly different angle. This slight displacement of images is one of the many cues that gives *depth perception*—a three-dimensional quality—to your sight.

## 33.11 Human and other organisms sense sound.

Sound is a type of mechanical energy resulting from the vibration of an object. This vibration disturbs the air around the object, rhythmically pushing the molecules in the air closer to one an-

### FOCUS

*on Plants*

#### Plants and light detection

Plants bend toward the light—this is a well-known phenomenon (**Figure 33.A**). However, plants have no nervous system and no sense organs, so how does this happen? When a plant is exposed to light on one side only, a shadow or darker side results. Under such conditions, the plant growth hormone *auxin* redistributes, resulting in a higher concentration of this hormone on the darker side of the stem. This makes the side of the plant away from the light grow faster, causing the plant to bend toward the light. This response of plants to light is known as *phototropism*.

**Figure 33.A** The flowering shamrock is bending toward the light.

other, compressing them. Areas of molecules that are not compressed border these areas of compression. Vibrating objects can cause sound waves in substances other than air, too, such as in water or wood.

#### Sound reception in organisms other than humans

In invertebrates, the organs responsible for perception of vibration are very different from the vertebrate ear. Scientists have been unable to identify vibration or sound receptors in many phyla, but they do know that earthworms have vibration receptors located along the length of their bodies. They also know that insects perceive sounds, which many generate to attract mates.

Many animals perceive sounds that humans cannot hear, such as the infrasound, or low-frequency sound, heard by elephants, homing pigeons, and probably hippos. Infrasound is that deep rumbling you might feel, rather than hear, when loud bass sounds are played, or during thunder and earthquakes.

Elephants not only can hear infrasound, but can also make these low-frequency sounds to communicate with distant members of their herd, especially during mating. Using specialized elephant-sized radio collars, researchers discovered that when a female elephant is ready to mate, she sends out a call that informs males for miles around of her condition. The female elephant sings for only half an hour, but within a day she is surrounded by bull elephants.

Elephants use infrasound for purposes other than mating as well. Warnings are issued by infrasound, and the locations of groups of elephants are transmitted to other groups by infrasound. Observers have

## *just wondering . . .*

*Questions students ask*

### How does exposure to loud noise levels contribute to a loss of hearing?

Listening to loud sounds contributes to hearing loss by damaging hair cells in the inner ear. As described in Section 33.11, the hair cells of the inner ear form part of the organ of Corti, which is the organ of hearing. These hair cells are embedded in a "roof" called the tectorial membrane, and the entire organ of Corti sits on the basilar membrane below. When the basilar membrane vibrates, the hairs move up and down, bending when they are pushed upward. Very loud sounds can cause violent vibrations of the basilar membrane, which may re-

sult in the hairs breaking or becoming permanently deformed. Such damage leads to partial hearing loss. The scanning electron micrograph shows portions of the organ of Corti with rows of hair cells. The photo on the left in **Figure 33.B** shows hair cells from a normal guinea pig. The photo on the right shows hair cells from a guinea pig after it experienced 24 hours of 120 decibel noise—about the level of sound at a rock concert. This damage is permanent. Do you allow such damage to your hearing? ●

**Figure 33.B** Normal hair cells (*left*) and damaged hair cells (*right*) in the ears of a guinea pig. The damage occurred as a result of exposure to 120 decibel noise for 24 hours.

*Are you wondering about a topic in biology and how it relates to your life? Submit your question by clicking the* Just Wondering *link in this text's companions Web site at www.wiley.com/college/alters.*

noted for years that entire groups of elephants will suddenly freeze, raising and spreading their ears. Researchers surmise that these elephants are listening to faint, distant calls and are remaining as silent as possible to do so. Moments later, the elephants will come out of their trance and rush off together to follow the signal.

Are there other sounds that animals can hear but that humans cannot detect? Dogs can hear high-pitched sounds that our ears cannot. Moreover, whales, dolphins, shrews, some birds, and bats can hear extremely high-pitched sounds. Most well developed in bats, these animals emit extremely high-frequency (short wavelength) sounds that can be used in *echolocation* or *animal sonar*. Echolocation allows an animal to gather information about the environment from hearing the "echos" of sound waves it has emitted as they bounce off living and nonliving things.

### Receptor cells in the inner ear of humans sense sound

How is the human ear structured and how does it work? The human ear has three parts: the outer ear, middle ear, and inner ear. The outer ear and middle ear work together to transmit sound waves to the inner ear, where sound stimuli are changed into nerve impulses (**Figure 33.19**).

The flaps of skin on the outside of the head that we call ears are only one part of the outer ear. Each flap, or **pinna** (PIN-uh), funnels sound waves into an auditory canal. The **external auditory canal** leads directly to the eardrum, or **tympanic membrane**.

The tympanic membrane is a thin piece of fibrous connective tissue that is stretched over the opening to the **middle ear**. The eardrum is connected to one of the smallest bones in the body—the **malleus** (MAL-ee-us, hammer). The malleus is connected to another tiny bone called the **incus** (ING-kus, anvil). The incus, in turn, is connected to a third bone called the **stapes** (STAY-peez, stirrup). These three bones pick up sound vibrations from the outer ear and transfer them to the inner ear.

Sound waves entering the outer ear beat against the tympanic membrane, causing it to vibrate. Vibrations of this membrane cause the malleus to move with a rocking motion because it is attached to the internal surface of the tympanic membrane. This rocking is transferred, in turn, to the incus and stapes. These three bones are hinged to one another in a way that produces a lever system, a mechanism that causes them to act like an amplifier and increase the force of the vibrations.

Also located in the middle ear is an opening to the **eustachian tube** (you-STAY-kee-en *or* you-STAY-shun), a structure named after an Italian physician who lived in the 1500s. The eustachian tube connects the middle ear with the nasopharynx—the upper region of the throat. Its function is to equalize air pressure on both sides of the eardrum. You have probably experienced "popping" in your ears while driving up or down a mountain, taking off and landing in an airplane, or riding in an elevator. It is the result of the pressure equalization between these two sides of the eardrum.

The stapes, the third in the series of middle ear bones, is attached to a membrane that separates the middle ear from the inner

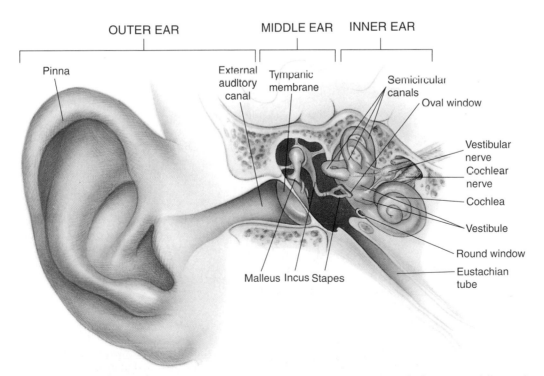

OUTER EAR     MIDDLE EAR    INNER EAR

Pinna

External auditory canal

Tympanic membrane

Semicircular canals

Oval window

Vestibular nerve

Cochlear nerve

Cochlea

Vestibule

Round window

Eustachian tube

Malleus Incus Stapes

**Figure 33.19 Structure of the human ear.** The human ear is composed of outer, middle, and inner sections (not to scale). The outer ear extends from the pinna to the tympanic membrane. The middle ear contains bones that transmit sound vibrations from the tympanic membrane to the cochlea. The cochlea contains the organ of hearing and makes up part of the inner ear.

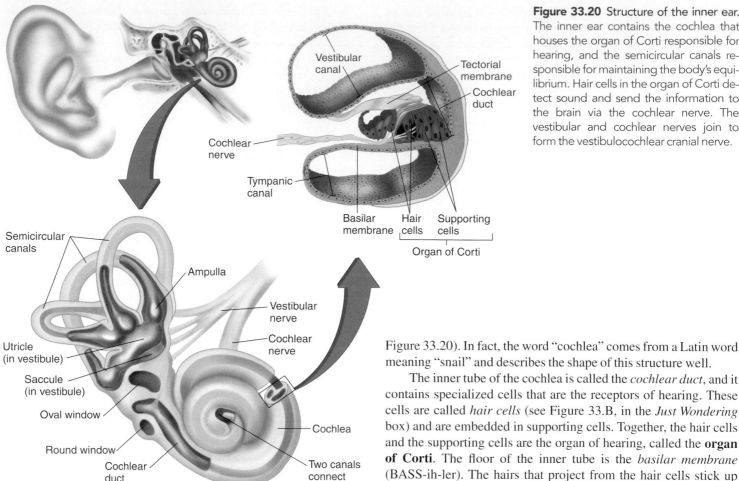

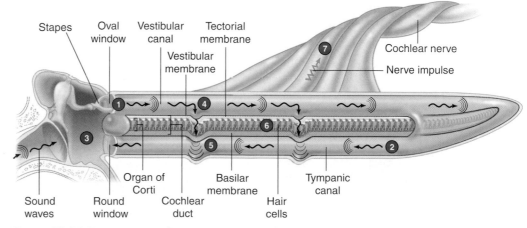

**Figure 33.20 Structure of the inner ear.** The inner ear contains the cochlea that houses the organ of Corti responsible for hearing, and the semicircular canals responsible for maintaining the body's equilibrium. Hair cells in the organ of Corti detect sound and send the information to the brain via the cochlear nerve. The vestibular and cochlear nerves join to form the vestibulocochlear cranial nerve.

Figure 33.20). In fact, the word "cochlea" comes from a Latin word meaning "snail" and describes the shape of this structure well.

The inner tube of the cochlea is called the *cochlear duct*, and it contains specialized cells that are the receptors of hearing. These cells are called *hair cells* (see Figure 33.B, in the *Just Wondering* box) and are embedded in supporting cells. Together, the hair cells and the supporting cells are the organ of hearing, called the **organ of Corti**. The floor of the inner tube is the *basilar membrane* (BASS-ih-ler). The hairs that project from the hair cells stick up into the cochlear duct and are embedded in a covering called the *tectorial membrane* (tek-TORE-ee-ul) (see Figure 33.20).

How does the organ of hearing detect sound? As the stapes rocks in the oval window, it sets the fluid within the vestibular canal in motion, as shown in **Figure 33.21 ①** . As the oval win-

ear. This membrane covers the **oval window**, the entrance to the fluid-filled **cochlea** (KOCK-lee-uh)—the part of the **inner** ear that contains the organ of hearing (**Figure 33.20**). To understand the cochlea's structure and function, imagine a tapering, blind-ended tube having two membrane-covered holes at its wider end. The upper hole is the oval window, and the lower hole is the *round window*. Also imagine that another, smaller membranous tube runs down its center, serving as a partition between the oval and round windows and bisecting the large tube into upper and lower channels, or canals. The upper channel is the *vestibular canal*; the lower channel is the *tympanic canal*. The inner tube stops short of the blind, tapered end of the outer tube, however, so the two canals connect with one another there (see Figure 33.21). Imagine this tube rolled up like a jellyroll, and you have the basic structure of the cochlea (see

**Figure 33.21 Transmission of pressure waves in the inner ear.** Waves in the fluid of the inner ear push on the vestibular membrane, which in turn causes the basilar membrane to vibrate. As hair cells embedded within the basilar membrane move up and down with the vibration, they are bent, which stimulates action potentials along neurons that are part of the auditory nerve.

**Visual Thinking:** Explain how the waves of pressure created in the fluid of the inner ear allow humans to hear. Where is sound perceived?

dow membrane rocks inward, it pushes on the fluid in the vestibular canal. This fluid moves forward in waves that pass into the tympanic canal **2**. When the waves reach the membrane-covered round window, it bulges outward **3**, compensating for the pressure increase. However, none of this movement results in the perception of sound. What you hear is the result of the following events: The waves in the fluid of the vestibular canal also push on the thin vestibular membrane **4**. This "push" transfers through the cochlear duct and causes the basilar membrane to vibrate **5**. The organ of Corti sits on the basilar membrane, so the hair cells move up and down when the basilar membrane vibrates. The hairs, embedded in the stiff and stationary tectorial membrane, are therefore bent back and forth with the vibration **6** (see also Figure 33.20). The bending of the hairs are the stimuli that generate action potentials along neurons that are all part of the auditory (cochlear) nerve **7**. The brain interprets these impulses as sound.

The brain is able to interpret the highness or lowness of a sound—its pitch—because sounds of different pitches produce sound waves of different frequencies. Different frequencies of sound set up differing wave patterns within the fluid of the inner ear that cause specific regions of the basilar membrane to vibrate more intensely than others. In addition, louder sound waves increase the magnitude of vibrations of the basilar membrane. Repeated exposure to extremely loud noises such as gunshots, jet engines, and loud music can actually damage the hairs of the receptor cells embedded within the membrane and cause partial but permanent hearing loss (see the *Just Wondering* box).

## 33.12 The vestibule and semicircular canals of the inner ear detect position and direction of movement.

Sensitivity to the force of gravity is extremely widespread in the animal kingdom and allows animals to distinguish up from down. In most animals, hearing and balance are related, and the mechanisms involved in balance are strikingly similar.

In humans and other vertebrates, the inner ear has two parts that are each structured as a fluid-filled tube within a tube like the cochlea. As shown in Figure 33.19, the cochlea makes up one side of the inner ear. The bulge in the midsection of the inner ear is called the **vestibule**, and it contains structures that sense whether you are upside down or right side up. In other words, it detects the effects of gravity on the body and helps you to maintain balance.

Inside this bulge are the **utricle** (YOO-trih kul) and the **saccule** (SAK-yool) (**Figure 33.22, 1** and **2**). Both words mean "a little bag or sac." Within each of these sacs is a flat area composed of both ciliated receptor cells **3** and nonciliated supporting cells. A layer of jellylike material is spread over the surfaces of these cells. Their long cilia and thin cell extensions are embedded within the jelly. Also embedded within the jelly are small pebbles of calcium carbonate called *otoliths* (OH-teh-liths; literally, "ear stones") **4**. When the head is moved, the otoliths pull on the jelly, which pulls on the cilia, bending them **5**. Any shift in the position of the otoliths results in different cilia being bent to specific degrees and in specific directions. These stimuli initiate generator potentials that are transmitted to neurons that make up a branch of vestibulo-

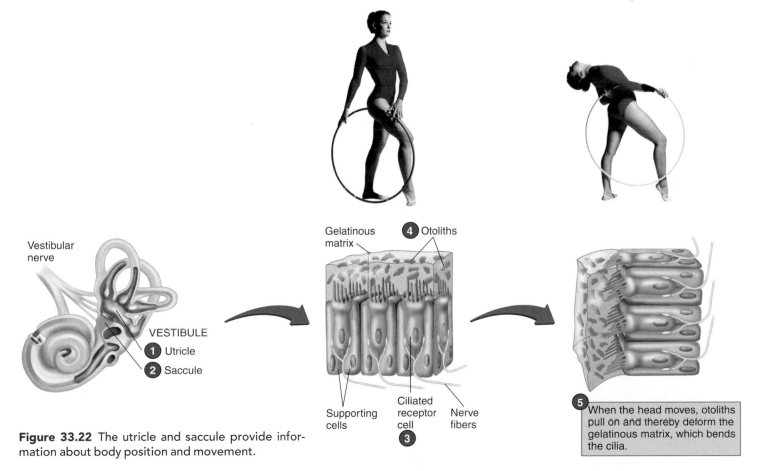

**Figure 33.22** The utricle and saccule provide information about body position and movement.

cochlear cranial nerve. The brain interprets these messages, resulting in your perception of "up."

Positioned above the saccule and utricle are three fluid-filled **semicircular canals** that help to provide sensory information about movement of the body or the head (see Figure 33.20). These loops are oriented at right angles to one another in three planes. At the base of each loop is a group of ciliated sensory cells that are connected to neurons. Lying above these cells is a mass of jellylike material. When the head moves, the fluid within the semicircular canals moves and pushes the jelly and the cilia in a direction opposite to that of the motion. You have experienced this phenomenon when you accelerate suddenly in your car and your head is thrown backward. This movement initiates the depolarization of the cell membrane, which triggers a nerve impulse. Because the three canals are each oriented differently, movement in any plane is sensed by at least one of them. Complex movements are analyzed in the brain as it compares the sensory input from each canal.

In invertebrates, gravity receptors are called *statocysts*. They have a similar plan to the organ of balance in vertebrates: a stonelike body (*statolith*) rests on sensory hairs. Displacement of the stone or stones relative to the force of gravity causes the statolith to stimulate sensory cells, which then send information to the central nervous system.

## CONCEPT CHECKPOINT

4. Continue with the list you developed in Concept Checkpoint 3 by adding other external sensory receptors discussed in Sections 33.10 and 33.12.
5. Using only your peripheral vision, look at something that has color. Do not look directly at the object. Can you determine what color it is? Using your knowledge of the anatomy of the human eye, explain why it is so difficult to see colors that are in the periphery of your vision.
6. Describe the effect each of the following would have on hearing and explain why: (a) ruptured eardrum (tympanic membrane), (b) exposure to a very loud noise, and (c) ruptured vestibule.

# CHAPTER REVIEW

## Summary of Key Concepts and Key Terms

### What are sensory receptors?

**33.1** Animals possess an interesting array of sense organs, specialized structures that trigger nerve impulses, which ultimately provide information about the surroundings.

**33.1** For an organism to sense its environment, a stimulus must open ion channels within receptor cells, depolarizing them and initiating a nerve impulse that is conducted by nerve fibers to the central nervous system.

**33.1** **Special senses** (p. 574) are those that are restricted to particular locations of the body; **general (somatic) senses** (p. 574) are those that have more than one location in the body, such as sense receptors distributed throughout the body for heat, cold, pain, and pressure.

### How do animals sense their internal and external environments, and position in space?

**33.2** Receptors within blood vessels and organs sense internal body temperature, blood chemistry, and physical stimuli.

**33.3** Receptors called **proprioceptors** (PRO-pree-oh SEP turz, p. 575) that sense the body's internal environment are located within the walls of blood vessels and organs.

**33.3** Proprioceptors called **stretch receptors** (p. 575) are located within the skeletal muscles; gravity and motion receptors are other types of proprioceptors.

**33.4** The body has a variety of mechanoreceptors, which are stimulated when they are physically deformed by mechanical energy such as pressure.

**33.4** Fishes and amphibians have a complex system of mechanoreceptors along their bodies called a lateral line system, which detects mechanical stimuli such as sound vibrations, bodily movement, and pressure changes produced by water currents.

**33.4** Invertebrates as well as vertebrates have mechanoreceptors.

**33.5** In humans and other vertebrates, two populations of nerve endings buried in the skin are sensitive to changes in temperature.

### Can some animals detect stimuli humans cannot?

**33.6** Pit vipers can detect heat in the form of infrared radiation from warm-blooded animals.

**33.6** Insects, especially pollinating insects, can see the ultraviolet radiation emitted by some flowering plants.

**33.7** Humans have no sensory organs to detect electric currents, but some fishes and the duck-billed platypus have receptors that can.

**33.8** There is evidence that some migratory animals and those with a homing instinct detect magnetic fields and use the Earth's magnetic lines of force to help orient themselves.

## How do the senses of taste, smell, sight, hearing, and balance work?

**33.9** The ability to sense chemicals, or chemoreception, is widespread in the animal kingdom.

**33.9** In humans, taste receptors called **taste buds** (p. 579) are located within the short projections on the tongue known as papillae.

**33.9** The **olfactory receptors** (p. 579), which are located in the nasal epithelium, detect various smells.

**33.10** Image-forming eyes are either of the multifaceted type as in insects, or the single-lens, cameralike type as in cephalopods and vertebrates.

**33.10** Human eyes are covered and protected by a tough outer layer of connective tissue called the **sclera** (SKLEAR-uh, p. 581).

**33.10** The front of the eye, the **cornea** (p. 581), is transparent, allowing light to enter the eye; a compartment behind the cornea is filled with a watery fluid called the **aqueous humor** (p. 581).

**33.10** The **lens** (p. 581) lies just behind the aqueous humor and plays a major role in focusing the light that enters the eye onto the **retina** (RET-un-nuh, p. 581), the sensory portion of the eye; the **ciliary muscles** (SILL-ee-err-ee, p. 581) contract or relax, slightly changing the shape of the lens.

**33.10** The **iris** (p. 581) controls the amount of light entering the eye, reducing the size of the **pupil** (p. 582) through which the light passes.

**33.10** The **choroid** (KOR-oyd, p. 582) contains vessels carrying blood that nourishes the retina and a dark pigment that absorbs light rays so that they will not be reflected within the eyeball.

**33.10** The tips of the **rods** (p. 582) and **cones** (p. 582) in the retina contain a pigment that undergoes a chemical alteration when activated by light; in rods this pigment is rhodopsin.

**33.10** Most of the cones are located at a central region of the retina called the **fovea** (FOE-vee-uh, p. 582).

**33.10** Impulses initiated by the rods and cones are transmitted by the **optic nerve** (p. 583) to the visual cortex of the brain.

**33.11** Sound waves are disturbances of molecules in the air—and in certain other substances—that result from the vibration of an object.

**33.11** Scientists have been unable to identify vibration or sound receptors in many phyla of invertebrates, but they do know that earthworms and insects perceive sounds.

**33.11** Each **pinna** (PIN-uh, p. 585) funnels sound waves into the **external auditory canal** (p. 585), which leads directly to the eardrum.

**33.11** The **tympanic membrane** (p. 585) is a thin piece of fibrous connective tissue that is stretched over the opening to the **middle ear** (p. 585).

**33.11** The **malleus** (MAL-ee-us, p. 585), **incus** (ING-kus, p. 585), and **stapes** (STAY-peez, p. 585) are the three bones of the middle ear; they pick up sound vibrations from the outer ear and transfer them to the inner ear.

**33.11** The **eustachian tube** (you-STAY-kee-en *or* you-STAY-shun, p. 585) connects the middle ear with the nasopharynx—the upper region of the throat.

**33.11** The **oval window** (p. 586) is the entrance to the fluid-filled **cochlea** (KOCK-lee-uh, p. 586)—the part of the **inner ear** (p. 586) that contains the **organ of Corti** (p. 586), which is the organ of hearing.

**33.11** The ear collects sound waves and changes them into other forms of mechanical energy that stimulate the organ of hearing within the inner ear; impulses are carried from there to the brain.

**33.12** Gravity is detected by the **utricle** (YOO-trih kul, p. 587) and the **saccule** (SAK-yool, p. 587), located in the **vestibule** (p. 587) of the inner ear.

**33.12** Motion is sensed by the **semicircular canals** (p. 588), also located in the inner ear.

---

## Level 1 — Learning Basic Facts and Terms

### Matching

Match each type of sensory receptor below to the stimulus to which it responds, *and* to its location in the human body or another organism.

1. proprioceptor
2. thermoreceptors
3. mechanoreceptor
4. photoreceptors
5. chemoreceptors
6. electroreceptors

a. stretching
b. pressure changes
c. temperature
d. electrical field
e. chemicals
f. light

uu. pit organs in snakes
vv. muscle spindles
ww. lateral line in fish
xx. retina
yy. taste buds
zz. ear

### Multiple Choice

7. Which of the following parts of the vertebrate eye is responsible for sensing light?
   a. Choroid
   b. Retina
   c. Iris
   d. Lens
   e. Optic nerve

8. These structure(s) transmit and amplify sound-wave–induced vibrations of the tympanic membrane to the inner ear.
   a. Incus, stapes, and malleus
   b. Eustachian tube
   c. Oval window
   d. Cochlea
   c. Round window

9. Which of the following provides the *direct* stimulus that results in the generation of impulses in the auditory nerve?
   a. bending of hair cells
   b. vibration of the tympanic membrane
   c. bending of cilia in the utricle
   d. changes in the structure of rhodopsin

10. Focusing light on the retina is the job of the
    a. fovea.
    b. choroid.
    c. lens and ciliary muscle.
    d. cornea.
    e. c and d are correct

---

## Level 2 — Learning Concepts

1. Explain the events occurring at the cellular level that allow you to "sense" something.

2. Which receptors would be most highly stimulated when riding on a roller coaster at an amusement park? Why?

3. Researchers have found that the hypothalamus is involved in the sense of satiation or "feeling full" after a meal, and thus integrally involved in weight control. What type of information do you think is being sent to the hypothalamus?

4. Explain how your sense of smell can distinguish between your favorite perfume and your least favorite food.
5. You may have noticed that certain smells are able to evoke specific memories and emotions—both good and bad. Explain why.
6. What are taste buds? Explain their role in allowing you to taste the flavors in your favorite food. What other sense is involved in your ability to taste?
7. Draw a diagram of the human eye. Label the sclera, cornea, aqueous humor, lens, iris, pupil, and fovea.

8. How is the manner in which you see in a well-lit environment different from your vision in a darkened room? What type of cell is involved in each situation?
9. Explain how your ears enable you to hear the music on your favorite CD.
10. Even with your eyes closed, you can tell when your head tilts to one side. Explain how.

Level 3 **Critical Thinking and Life Applications**

1. In a very interesting series of experiments, migrating songbirds were placed in a cage covered with dark paper so that they could not see out. The birds would peck at the wall of the cage, attempting to get out, but not at random—by far the greatest density of pecks was in the direction they were migrating when captured! When a large magnet was placed in the room, they pecked in the direction of the magnet, even though they could not see it. What sort of sensory system do you imagine the birds possess that orients them in this way? Why does the magnet affect the birds? Explain the rationale of your answer.
2. Some children are born with the inability to feel pain. Do you think this is an advantage or a disadvantage? Explain the reasons for your answer.

3. Mr. Garcia went for an eye examination. The eye doctor placed drops in his eyes that dilated (widened) his pupils so that he could view Mr. Garcia's retina. What structures were affected by the drops? What happened within the eye that caused the pupil to enlarge?
4. Pick a sense in an animal that has been described in this chapter and suggest how this sense helps the animal survive and/or reproduce.
5. In their preschool years, many children experience frequent middle ear infections. Which structures of the ear are affected by such infections? How do these infections affect hearing and why?

**In The News** | **Critical Thinking**

**SUPER SNOUT**

Now that you understand more about the senses, reread this chapter's opening story about the development of an artificial nose. To better understand this information, it may help you to follow these steps:

1. Review your immediate reaction to the development of an odor-detecting device that you wrote when you began reading this chapter.
2. Based on your current understanding, again summarize the main point of the research in a sentence or two.
3. What questions do you now have about this information that this chapter's opening story does not answer?
4. Collect new information about the research. Visit the *In The News* section of this text's companion Web site at www.wiley.com/college/alters and

watch the "Super Snout" video. Then use the "summary" link to read the accompanying story and access related links. Use this information, the links provided, and other online and library resources to answer your questions and find updates about this topic. State the sources of your information. Explain why you think the information is accurate. Also determine whether the information expresses a particular point of view or is biased in any way.
5. What in your view are the most significant aspects of this information? Give reasons for your opinion and for any changes in your ideas based on the additional information you have collected and the analysis you have done.

# PROTECTION, SUPPORT, AND MOVEMENT

## In The News | Jeweled Joints

"Diamonds are forever." You've probably heard that phrase before and know that in many cultures diamonds are considered a symbol of love and fidelity. Why? Many would say this symbolism arises from the fact that diamonds are the hardest mineral, and therefore, relatively indestructible, like the relationship symbolically bound by the diamond ring.

Although you are most likely familiar with the use of diamonds in jewelry, over 80% of all diamonds mined annually are used for industrial purposes. They can be used to cut, grind, and polish other hard substances, and can be used in tools that are able to slice through hard metal or bore into solid rock.

Although natural diamonds are created by geologic processes that incorporate high temperature and pressure deep within the Earth, synthetic diamonds can be made in the laboratory. In the 1950s, researchers determined how to synthesize diamonds from graphite. Such synthetic diamonds are used for a variety of applications.

Yogesh Vohra, a physics professor at the University of Alabama-Birmingham, and his colleagues have been researching ways to use artificial diamonds to extend the life of artificial joints. Joints are places in the body that may experience wear and tear because they are where bones come together. Vohra notes that the crystalline structure of diamonds—the repeating arrangements of the carbon atoms—produces surfaces that may require expensive polishing because they could scratch metal implants and not adhere to their surfaces. While experimenting with ways to synthesize smooth diamond films for use with implants, Vohra and his team produced an unusual batch of crystals that were hard, but only a few

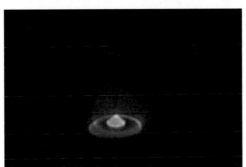

nanometers thick. A nanometer is *one-billionth* of a meter; it would take 2500 of these crystals to equal the thickness of just one red blood cell! Along with their extreme thinness, the diamond film was smooth and stuck to metal. The scientists investigated why this particular batch had these unusual, yet desirable, properties. They found a small leak in the chamber in which they synthesized the diamond film. The leak had allowed air to enter the chamber, as shown in the photo. The researchers determined that nitrogen in the air prevented normal diamond formation and enabled the crystals to stick to metal.

If this diamond film is compatible with living tissues, it can be used to coat the surfaces of surgical implants, creating more durable artificial joints and reducing patients' need for future replacement surgeries. Vohra and his team have applied the nanodiamond film to a metallic simulated jaw at the mandibular joint, and they plan to test the durability of the joint. In the future, nanodiamond-coated joints may become widely used as implants for humans. To learn more about the nanoscale diamond-coated film, visit the *In The News* section of this text's companion Web site at www.wiley.com/college/alters and view the video "Jeweled Joints."

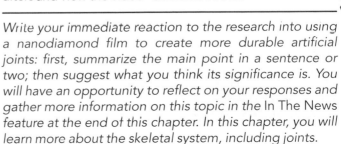

*Write your immediate reaction to the research into using a nanodiamond film to create more durable artificial joints: first, summarize the main point in a sentence or two; then suggest what you think its significance is. You will have an opportunity to reflect on your responses and gather more information on this topic in the In The News feature at the end of this chapter. In this chapter, you will learn more about the skeletal system, including joints.*

## CHAPTER GUIDE

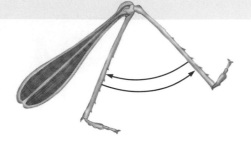

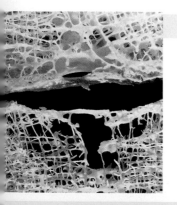

# How do organisms move?

## 34.1 Skin, bones, and muscles together provide movement.

In humans, the skin is a soft and flexible covering, stretching to accommodate movements of the human body. That movement is determined largely by muscles, which are attached to bones.

Bones not only give shape to the human body and protect its delicate inner structures, but they also act as levers in movement, pulling on one bone while anchored to another. The bones increase the strength of a movement, similar to the way a claw hammer increases the force exerted on a nail to pull it out of a piece of wood. Rigid yet flexible, bones are able to bear a considerable amount of weight.

To supply the force needed for movement, your body uses the chemical energy of ATP to move certain filaments within your muscle cells, resulting in the shortening of those cells (see Section 34.10). When a large group of muscle cells shortens all at once, the muscle cells exert a great deal of "pull." For a muscle to use this force to produce movement, it must direct its force against an ob-

ject. In humans, muscles pull on bones, so, for example, muscle contraction results in the lifting of an arm or the bending of a finger.

## 34.2 Skeletons provide support and a framework against which muscles pull.

What do you think you would look like without your skeleton? Most animal tissues are soft; the skeleton provides support for a soft body. On land, this support keeps the animal body from collapsing under its own weight. In aquatic environments, skeletal support keeps the shape of the animal body even though water pressure acts on it. Muscles attach to the rigid, skeletal portion of a body; the skeleton acts as a framework against which muscles pull.

Look down the column labeled "skeleton" in **Table 34.1**. Aside from the sponge, which has hard, spiny pieces of tissue embedded in its gel-like middle layer to support its body, three pat-

| **TABLE 34.1** | **The Skeleton, Musculature, and Movement of Representative Animals** | | |
|---|---|---|---|
| **Representative animal (phylum)** | **Skeleton** | **Musculature** | **Movement** |
| Sponge (Porifera) | Gelatinlike middle layer supported by hard, sharply pointed skeletal pieces | No true musculature | Sessile |
| Jellyfish (Cnidaria) | Absent, but water in bell acts as hydroskeleton | Circular muscles in swimming bell | Propulsion: circular muscles in bell contract, pushing water out and moving the jellyfish forward |
| Flatworm (Platyhelminthes) | Hydroskeleton | Circular and longitudinal muscles underneath the epidermis | Waves of contraction that generate thrusting and pulling forces |
| Roundworm (Nematoda) | Hydroskeleton | Longitudinal muscles underneath the epidermis and cuticle | Movement by alternate contraction of muscles on either side of body |
| Earthworm (Annelida) | Hydroskeleton | Circular and longitudinal muscles underneath the epidermis | Alternate contraction of circular and longitudinal muscles results in peristaltic movement |
| Clam (Mollusca) | Shell (exoskeleton) | Muscular foot | Muscles in foot allow clam to burrow into sand or mud when shell is open |
| Insect (Arthropoda) | Exoskeleton (integument is skeleton) | Primarily striated muscle that attaches to exoskeleton | Ground movement by means of limbs; flight by movement of wings |
| Sea star (Echinodermata) | Hydroskeleton and endoskeleton of hard calcium carbonate plates | Retractor muscles in tube feet | Protraction and retraction of tube feet connected to the water vascular system |
| Human (Chordata) | Bony endoskeleton | Striated muscles attached to skeleton for locomotion; smooth muscle in organs; cardiac muscle in heart | Movement by lever action of limbs |

terns of skeleton emerge: the hydroskeleton, the exoskeleton, and the endoskeleton.

A **hydroskeleton** is a support system that uses a fluid under pressure as scaffolding. For example, erection in the mammalian penis takes place this way: Blood rushes into the spongy erectile tissue and the penis becomes rigid. When the blood flows out of this tissue, the penis becomes limp.

In a true hydroskeleton, an organism has a fluid-filled body cavity with flexible body walls. Muscles are attached to the body wall, and body fluids are used to transmit force. Movement results when the muscles pull in on one part of the body, compressing the fluid and causing it to push out another part.

The earthworm is an excellent example of an organism with a true hydrostatic skeleton. Its body consists of enclosed segments, each housing a fluid-filled body cavity (coelom). Two sets of muscles are attached to the body wall: circular muscles and longitudinal (lengthwise) muscles. When the earthworm contracts the circular muscles in some of its segments, it elongates those segments. That is, when the segments get smaller in diameter as the circular muscles tighten, the pressure exerted on the body fluid inside each segment forces each segment to get longer. This process is how the earthworm pushes its front end forward, as shown in **Figure 34.1**. This wave of circular and then longitudinal contraction that travels down the worm's body is called peristaltic locomotion and is similar to peristalsis in humans.

Jellyfish do not have a true hydroskeleton. They use the water of their environment, rather than their own body fluids, as a type of hydroskeleton. Water is pushed out of the jellyfish as the circular muscle at the bottom of its bell contracts. As water is forced out of the bell, it pushes the animal forward.

Flatworms are acoelomates (see Figure 21.5c), which means that they have no body cavity. Therefore, they have no true hydroskeleton, but their body contents do act in this fashion. Like earthworms, they

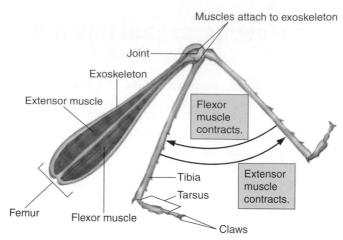

Figure 34.2 How muscles move appendages in an animal with an exoskeleton. In this example of an insect leg, you can see the extensor and flexor muscles inside the exoskeleton. At the joint, the extensor muscle is attached to the topside of the tibia. When it contracts, it lifts this leg. At the joint, the flexor muscle is attached to the underside of the tibia. When it contracts, it pulls the leg inward toward the femur.

have circular and longitudinal muscles attached to their integument. By alternately contracting these sets of muscles, flatworms generate thrusting and pulling forces as their body contents become alternately compressed and decompressed, moving them along a surface.

Roundworms are pseudocoelomates, with turgid bodies and rigid cuticles. Unlike earthworms and flatworms, they have only longitudinal muscles. To move, these worms alternately contract the longitudinal muscles on either side of the body. As the longitudinal muscles contract on one side of the body, the longitudinal muscles on the other side relax. Since the cuticle cannot expand to relieve the pressure, the nematode body forms S-shaped curves as the worm thrashes forward.

Even though the sea star has an internal skeleton called an **endoskeleton** comprised of a system of hard plates, it moves by using a water vascular system, a unique system of water-filled canals connected to its tube feet on its underside. By a complex set of muscular actions, suction is created or released on the tube feet. The sea star crawls along by extending, attaching, and releasing the grip of its tube feet on a surface.

Arthropods like insects and spiders have striated muscle as vertebrates do (see Section 34.10), which makes rapid movement and flight possible. Other invertebrates have slower-acting smooth muscle. Insects and spiders also have lightweight **exoskeletons**—rigid support systems that cover their bodies. Their muscles are attached to the inside of the exoskeleton, which is paper thin at the joints, allowing arthropods to bend their appendages at joints (**Figure 34.2**) using them as levers to increase force as vertebrates do. The arthropod exoskeleton does not grow, however, so insects and spiders periodically shed their skin in a process called molting. A new exoskeleton is present under the old one. The arthropod "puffs up" and expands this exoskeleton, and it hardens after a few hours.

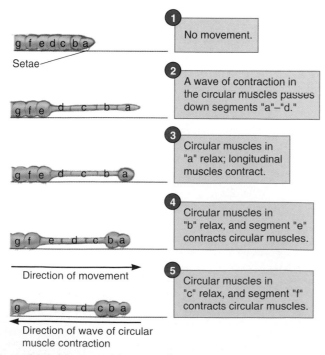

1. No movement.

2. A wave of contraction in the circular muscles passes down segments "a"–"d."

3. Circular muscles in "a" relax; longitudinal muscles contract.

4. Circular muscles in "b" relax, and segment "e" contracts circular muscles.

Direction of movement

5. Circular muscles in "c" relax, and segment "f" contracts circular muscles.

Direction of wave of circular muscle contraction

**Figure 34.1** Movement in an earthworm.

CONCEPT CHECKPOINT

**1.** Define the term "hydroskeleton." How does a hydroskeleton provide support and allow movement?

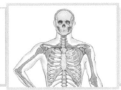

# What is bone, what are its functions, and how do bones make up the skeleton?

## 34.3 Bone is connective tissue containing collagen and minerals.

**Bone** is a type of connective tissue (see Chapter 26) consisting of widely separated bone cells embedded in a matrix of collagen fibers and mineral salts in a semisolid gel. The bones' mineral salts are needle-shaped crystals of calcium phosphate and calcium carbonate. The collagen fibers are coated and surrounded by these mineral salts within the gel, producing bones that are rigid yet flexible. Have you ever performed the "classic" experiment of soaking chicken bones overnight in vinegar? The results show this interplay of flexibility and hardness. The acetic acid of the vinegar dissolves the mineral salts in the bone. Without the hardness of the mineral salts, the collagen fibers leave the bone so flexible that you can tie it in a knot!

Living cells are as important a component of bone as are the collagen fibers and mineral salts. Cells called *osteoblasts* form new bone. These cells secrete the collagen fibers and semisold gel in which the body later deposits mineral salts. The osteoblasts lay down bone in thin, concentric layers called *lamellae* (**Figure 34.3**), like layers of insulation wrapped around an old pipe. You can think of the osteoblasts as being embedded within the insulation and the open tube of the pipe as a narrow channel called the *central canal*. *Osteocytes* are mature bone cells that maintain the bone. Central canals run parallel to the length of the bone and contain blood vessels and nerves. Central canals and the surrounding lamellae are together called *osteons*.

The bones of the human skeleton are composed of compact bone and spongy bone. *Compact bone* has the concentric ring structure described previously. As shown in Figure 34.3, it runs the length of long bones, such as those in your arms and legs. However, if bones were completely made up of compact bone, your body would be very heavy, and your arms and legs would be nearly impossible to move. Instead, the central core of long bones is a hollow cylinder lined with spongy bone. This cavity is filled with a soft, fatty connective tissue called *yellow bone marrow*.

*Spongy bone* also makes up most of the ends of long bones and most of the bone tissue of short bones (like your wrist and ankle bones), flat bones (like your ribs), and irregularly shaped bones (like some of your facial bones). Spongy bone is an open latticework of thin plates, or bars, of bone. Microscopically, its structure does not show a regular concentric ring structure like compact bone; it has a somewhat

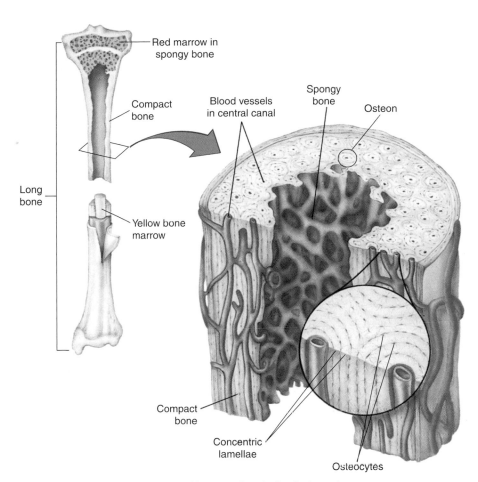

**Figure 34.3 The organization of bone.** The shaft of a long bone is compact, giving the bone strength. The ends of the bone and the lining of the cavity within the long bone are spongy, with a more open lattice. Blood cells are formed within red bone marrow that fills the lattice at the ends of the bone. Yellow bone marrow, a fat storage tissue, fills the cavity in the shaft of the long bone.

more irregular organization. The spaces within its bony latticework are filled with *red bone marrow* where the body's blood cells are formed. Surrounding the spongy bone tissue are layers of compact bone. The compact bone gives bones the strength to withstand mechanical stress, and spongy bone provides some support and a storage place for the red bone marrow, while helping to lighten bones.

## 34.4 Bones support the body, protect internal structures, store minerals, produce blood cells, and aid in movement.

The 206 bones of the adult human body make up a working whole called the **skeleton**, or **skeletal system**. The bones of the skeleton perform many important functions, some of which have already been mentioned. To summarize, the skeletal system:

- Provides support for the body.
- Provides for movement, with individual bones serving as points of attachment for the skeletal muscles and acting as levers against which muscles can pull.
- Protects delicate internal structures such as the brain, heart, and lungs.
- Gives rise to new red and white blood cells in its marrow.
- Acts as a storehouse for minerals such as calcium and phosphorus.

Bone tissue is not the inert tissue it may seem to be. In fact, it is continually broken down and re-formed, its minerals being transported to other parts of the body on demand. Bone is laid down in areas exposed to physical stress, which is why weight-bearing exercise results in stronger bones. Conversely, loss of bone mass occurs when physical stress is removed, which is why people confined to bed or in outer space lose bone density. Under "normal" conditions, your body completely replaces its skeleton over a period of seven years.

## 34.5 The axial skeleton includes the skull, vertebral column, rib cage, and hyoid.

The skeleton is made up of two divisions: the axial skeleton and the appendicular skeleton (**Figure 34.4**). The 80 bones of the **axial skeleton** include the bones of the skull, vertebral column, and rib cage.

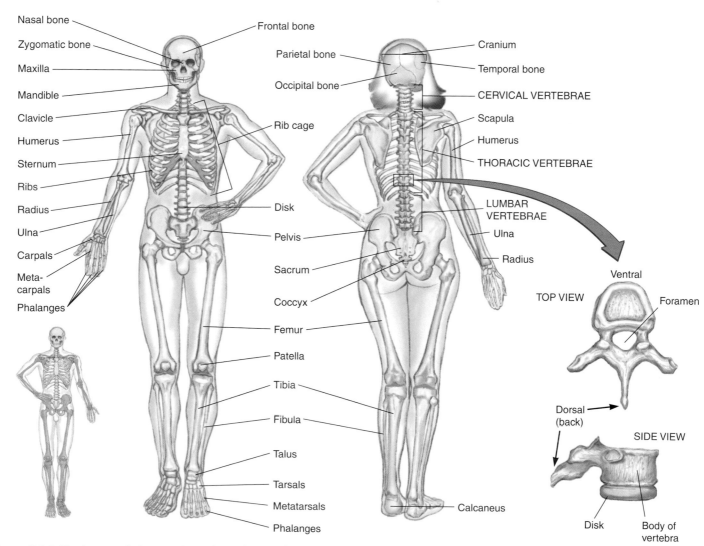

**Figure 34.4** The human skeleton. The enlarged view of the top of a vertebra shows where the foramen is located. The foramen is the hole in each vertebra. Between each vertebra and the next, there is a disk composed of an outer layer of fibrocartilage and an inner, more elastic layer. The small figure in the upper right shows the axial skeleton (green) and the appendicular skeleton (orange). The axial skeleton comprises the central axis of the body, while the appendicular skeleton comprises the appendages and their attachments to the axial skeleton.

 **Visual Thinking:** The top view of a vertebra shows a foramen (hole), which is also mentioned in the caption. What is the function of this hole?

The **skull** contains 8 cranial bones and 14 facial bones. The cranial bones enclose and protect the brain. These include the frontal bone that forms the forehead and the roof of each orbit (eye socket). A "black eye" results from damage to the ridge of this bone or the skin over it at the top of the eye socket.

Some of the 14 facial bones include the mandible (lower jawbone) and the upper jawbone formed by two maxillae (max-ILL-ee). Two zygomatic (zeye-go-MAT-ik) bones (cheekbones) give definition to the cheeks, while two nasal bones form the bridge of the nose. Below and behind the nasal bones is the nasal cavity. Bones and cartilage act as a septum (divider) of the nasal cavity into right and left halves. A person can be born with a deviated septum—the septum deflects to the left or right—or develop one due to trauma to the nose. The deviation partially or fully blocks the nasal passageway and can lead to inflammation or infection, causing congestion, sinusitis, headache, and nosebleeds. Usually, surgery can correct this condition.

The **vertebral column** is made up of 26 individual bones called *vertebrae* (see Figure 34.4). Stacked one on top of the other, these bones act like a strong yet very flexible rod that supports the head. In addition, some of the vertebrae serve as points of attachment for the ribs. Each vertebra has a central hole, or foramen (fo-RAY-men). Lined up, the vertebrae form a bony canal protecting the spinal cord, which runs down much of its length.

The seven vertebrae closest to the head are the cervical (neck) vertebrae. Next are twelve thoracic (chest) vertebrae. Following these are five lumbar (lower back) vertebrae. The sacrum (SAY-crum) and the coccyx (COCK-sicks), or tailbone, are the last vertebrae in the column. Positioned between the cervical, thoracic, and lumbar vertebrae are disks of fibrocartilage called *intervertebral disks*. These disks act as shock absorbers, provide the means of attachment between one vertebra and the next, and permit movement of the vertebral column.

Attached to the 12 thoracic vertebrae are 12 pairs of ribs surrounding the chest. The ribs curve around to the front of the thoracic cavity, producing a bony cage that protects the heart and lungs. The upper 7 pairs of ribs directly connect to a flat bone that lies at the midline of the chest called the breastbone, or sternum.

The remaining bone of the axial skeleton is the hyoid bone, a name that means "U-shaped." Ligaments in the neck support the hyoid, and the tongue attaches to it. (Figure 34.4 does not show the hyoid bone.) **Ligaments** (LIG-uh-munts) are bundles, or strips, of dense connective tissue that usually hold bones to bones, but in this case are supportive structures. The hyoid is the only bone in the body that does not form a joint with other bones. Because this bone is often broken when a person is strangled, the hyoid can provide important evidence in certain murder cases.

## 34.6 The appendicular skeleton includes the appendages and their attachments to the axial skeleton.

The **appendicular skeleton** is made up of the bones of the appendages, the *pectoral (shoulder) girdle*, and the *pelvic (hip) girdle*. Appendages include the arms, or upper extremities, and the legs, or lower extremities. The **pectoral girdle** is made up of two pairs of bones: the clavicles (CLAH-vih-kuhlz), or collarbones, and the scapulae (SKAP-you-lee), or shoulder blades. Find these bones in **Figure 34.5** and then locate the edges of your own clavicles and scapulae. The bones of the pectoral girdle support and articulate with, or form joints with, the arms. Together, your arms contain 60 bones.

As shown in Figure 34.5, the bone of the upper arm is called the humerus. This bone articulates with the scapula. The other end of this long bone articulates with the two bones of the forearm: the radius and the ulna. The wrist is made up of eight short bones called carpals (KAR-pulz), lined up in two rows of four bones. These bones are held together by ligaments and articulate with the bones of the hand, or metacarpals. The five metacarpals articulate with the bones of the fingers, or phalanges (fuh-LAN-geez) (singular: phalanx [FAY-lanks]). Each finger has three phalanges; the thumb has only two phalanges.

The **pelvic girdle** is made up of two bones that you know as pelvic bones or hipbones. Find these bones in Figure 34.5. Have you ever watched crime-related television programs or read such novels in which the pelvic bones are used to distinguish male and female

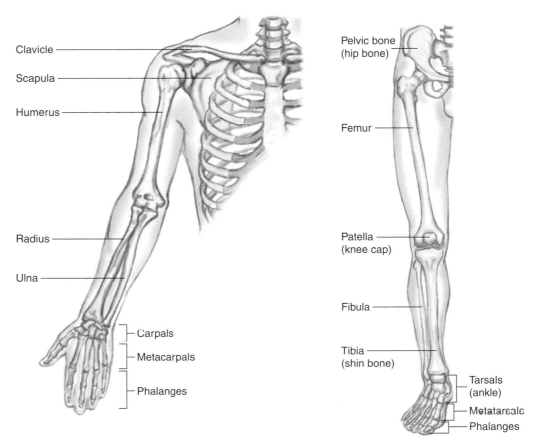

**Figure 34.5** The bones of the appendicular skeleton.

Clavicle
Scapula
Humerus
Radius
Ulna
Carpals
Metacarpals
Phalanges

Pelvic bone (hip bone)
Femur
Patella (knee cap)
Fibula
Tibia (shin bone)
Tarsals (ankle)
Metatarsals
Phalanges

skeletons? Females have a wider pelvis with a large, oval opening in the center for giving birth. The male pelvic girdle, in contrast, is narrower and has a smaller, heart-shaped opening in the center.

The bones of the pelvis support and articulate with the legs. Together, both legs contain 60 bones as the arms do. As shown in Figure 34.5, the bone of the upper leg, or thigh, is called the femur (FEE-mur). This bone has a rounded, ball-shaped head that articulates with a depression, or socket, in the hipbone. The femur, the longest and heaviest bone in the body, articulates at its lower end with the tibia (TIB-ee-ah), one of the two bones of the lower leg. The fibula (FIB-you-lah) is the other bone of the lower leg. The tibia is the bone commonly referred to as the shinbone and can be felt at the front of the lower leg. Have you ever experienced shin splints? If so, you are well aware of this bone. Shin splints are usually an inflammation of the outer covering of the tibia caused by repeated tugging by muscles and tendons. The tendons themselves

may also be inflamed. **Tendons** are cords of dense connective tissue that attach muscles to bones. Vigorous walking, running, or other types of exercise can sometimes result in shin splints, which can be very painful.

The patella (pah-TELL-uh), or kneecap, is a small, triangular bone that sits in front of the joint formed by the femur, tibia, and fibula. The patella is a sesamoid (SESS-uh-moid) bone, one that is formed within tendons where pressure develops.

Each ankle is made up of seven short bones called tarsals (TAR-suhlz). One of these bones (the calcaneus, shown in Figure 34.4) forms the heel and is the largest and strongest ankle bone. The tarsals are held together by ligaments and articulate with the bones of the foot, or metatarsals. The five metatarsals articulate with the bones of the toes, or phalanges. Like the fingers, each toe has three phalanges; like the thumb, the "big toe" has only two phalanges.

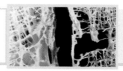

# What are joints, and how does exercise affect them?

## 34.7 Joints are places where bones come together.

Have you ever broken a leg or an arm and had a cast covering a joint? If so, you learned firsthand just how important joints are to movement. A **joint**, or **articulation**, is a place where bones, bone and cartilage, or bone and teeth come together. Not all joints are alike. The amount of movement allowed by a joint is a direct result of how tightly the bones are held together at that location.

The bones of the skull articulate with one another in a type of firm or tight joint called a **suture** (**Figure 34.6** ①), which is usually classified as immovable. A layer of dense connective tissue helps keep the bones from separating (see Chapter 26). Another example of immovable joints is the articulation of your teeth with the mandible and maxillae. A ligament lies between each tooth and its socket, holding the tooth in place. A third example is the articulation between the first pair of ribs and the breastbone. Each of these two ribs is connected to the sternum by a strip of hyaline cartilage, connections that change to bone during adult life.

As you twist your foot to the right and left, imagine the movement taking place between the lower ends of the tibia and fibula ②. A ligament (dense connective tissue) holds the ends of these bones together while permitting some flexibility. This articulation is an example of a slightly movable joint. The articulation between one vertebra and the next is also an example of a slightly movable joint. Here, a broad, flat sheet of fibrocartilage covers the top and bottom of each intervertebral disk, which is sandwiched between adjacent vertebrae, creating a somewhat flexible connection.

In immovable and slightly movable joints, there is no space between the articulating bones. The bones forming these joints are held together by dense connective tissue or cartilage. In freely movable joints, there is a space between the articulating bones. Such joints are common in your body and are called **synovial joints** (suh-NO-vee-uhl).

Although synovial joints permit a wide range of movement, this movement is limited by several factors, including the tension, or tightness, of the ligaments and muscles surrounding the joint and the structure of the articulating bones. You can see the various types of movement of synovial joints on your own body. For example, move your hand at the wrist; your radius forms an *ellipsoidal joint* (Figure 34.6 ③) with the closest of the wrist bones, or carpals. Another joint, located in the thumb, is called a *saddle joint* ④, so named because one bone of the joint is shaped like a saddle and the other like the back of a horse. Some of your carpals and tarsals glide over one another when you move your ankles and wrists, as seen in the *plane* or *gliding joint* ⑤.

Other joints allow angular movements, which increase or decrease the angle between two bones. When body builders flex their biceps, for example, they bring their forearms closer to their upper arms, decreasing the angle between the radius/ulna and the humerus. The elbow joint permits this movement and is called a **hinge joint** ⑥. A hinge joint allows movement in one plane only, similar to how a door opens and closes by means of its hinges.

The knee is a hinge joint that often sustains injuries to the ligaments holding the femur (thighbone) and the tibia (shinbone) together. Two common injuries are to the anterior cruciate ligament (ACL) and the medial collateral ligament (MCL). Both of these ligaments extend from the tibia to the femur and help keep the knee stable. Tears or complete rupture of these ligaments occur most often in sports such as basketball, soccer, volleyball, and football, when the foot is planted and the knee is twisted.

Another type of movement is rotation. You see this type of movement in the articulation between the thigh bone and the hip bone; this is an example of a **ball-and-socket joint** ⑦. You see the same type of movement in the shoulder joint when a pitcher winds up and throws a baseball. In the shoulder joint, the rounded head or

# Visual Summary

**Figure 34.6** Joints. The arrows in the schematic diagrams show the types of movements possible at freely movable (synovial) joints.

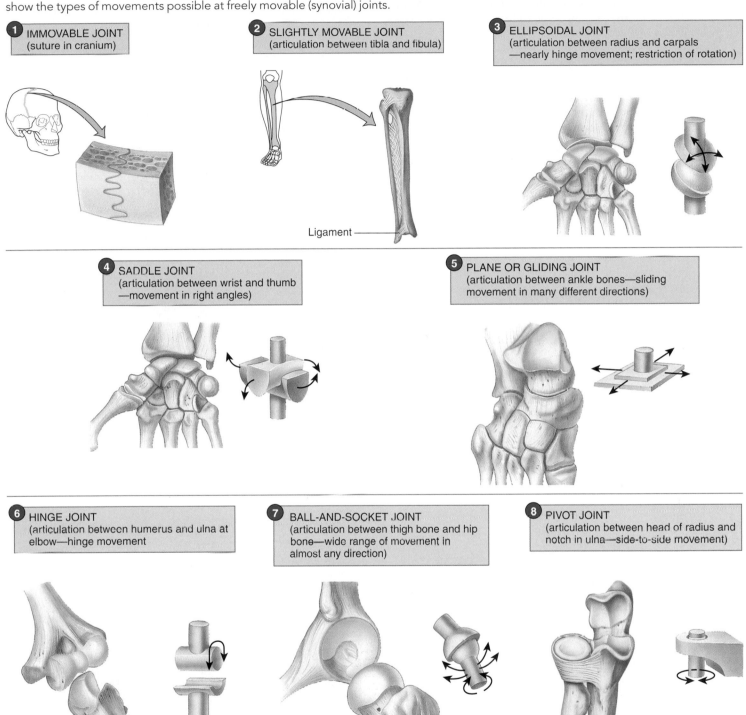

**1** IMMOVABLE JOINT
(suture in cranium)

**2** SLIGHTLY MOVABLE JOINT
(articulation between tibia and fibula)

Ligament

**3** ELLIPSOIDAL JOINT
(articulation between radius and carpals
—nearly hinge movement; restriction of rotation)

**4** SADDLE JOINT
(articulation between wrist and thumb
—movement in right angles)

**5** PLANE OR GLIDING JOINT
(articulation between ankle bones—sliding
movement in many different directions)

**6** HINGE JOINT
(articulation between humerus and ulna at
elbow—hinge movement

**7** BALL-AND-SOCKET JOINT
(articulation between thigh bone and hip
bone—wide range of movement in
almost any direction)

**8** PIVOT JOINT
(articulation between head of radius and
notch in ulna—side-to-side movement)

ball of the humerus articulates with the concavity or socket formed by the ends of the clavicle and scapula. Unlike the knee, the shoulder is stabilized not only by ligaments, but also by four muscles and their tendons that join the scapula (shoulderblade) to the humerus (upper arm bone). The tendons of these muscles fuse to form a complete circle, or cuff, around the shoulder joint. Rotator cuff in-

juries, a common injury among baseball pitchers and volleyball players, are tears to these tendons.

When you twist your arm from side to side, the type of joint you are using is a **pivot joint** **8**, formed by the by the head of the radius and the radial notch of the ulna. Similarly, when you twist your head from side to side, you are using a pivot joint formed by the first two vertebrae, called the atlas and the axis. Your skull rests on the ringlike atlas—named after the mythological Greek god who could lift the Earth—and pivots on a projection of the axis rising from below.

**Figure 34.7** shows the components of a synovial joint. Hyaline cartilage covers the ends of the articulating bones, which are separated by a film of fluid continuous with the joint cavity. The joint cavity is the space between a joint capsule composed of connective tissue that surrounds the bones, and the bones themselves. This cavity is filled with fluid.

The joint capsule is composed of a double layer of tissue. The outer layer is made up of dense connective tissue that holds the bones of the joint together. Some of the connective tissue fibers are arranged in bundles forming strips of tissue, or ligaments, as shown in **Figure 34.8**. The inner layer of the capsule is the synovial membrane, which secretes the fluid of the joint cavity, called synovial fluid. The synovial joint gets its name from this "egg white–like" fluid; the word "synovial" comes from a Greek word and a Latin word meaning "with" (*syn*) "an egg" (*ovum*). Synovial fluid lubricates the joint, provides nourishment to the cartilage covering the bone, and contains white blood cells that battle infection.

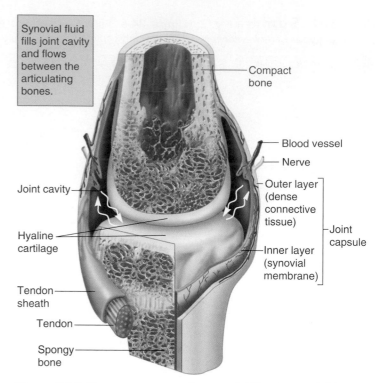

**Figure 34.7** The components of a synovial (freely movable) joint. In synovial joints, there is a space between the articulating bones. Synovial fluid fills the joint cavity and flows between the articulating bones (arrows).

 CONCEPT CHECKPOINT

2. Describe the difference in the internal structure of the bone that comprises most of the length of the femur to that of your ankle bones.

3. Starting with your thumb and moving up to your shoulder, name the bones forming joints and describe the articulation and movement of each joint.

## 34.8 Moderate exercise enhances bone and joint health.

Exercise can help your bones, but it can also hurt your joints. Researchers have known for many years that weight-bearing exercise helps maintain bone density and helps avoid *osteoporosis*, or bone loss. Osteoporosis is a condition in which the bones gradually lose mineral density, becoming weak and brittle (**Figure 34.9**). Although some bone loss begins in both men and women at about age 40, postmenopausal women are at greatest risk for developing osteoporosis. At menopause, women stop producing estrogen, a hormone that stimulates bones to retain calcium. For this same reason, young women who stop menstruating due to eating disorders, and young women athletes who train so strenuously that they no longer menstruate, are also at risk. Researchers have found that the bone density of the vertebrae in some female athletes in their twenties is similar to that of women in their seventies and eighties.

To help maintain your bone density throughout life, engage regularly in moderate physical activities that force you to work against gravity, such as walking, jogging, racquet sports, and weight lifting. Also, be sure that you take in enough calcium—approximately 1200 mg to 1500 mg per day, the amount in about

4 cups of milk. Lastly, do not smoke; cigarette smoking tends to increase bone loss. A variety of medications that counter bone loss in postmenopausal women are available for women at risk for osteoporosis.

Although moderate exercise has a beneficial effect on bones—and even on cartilage—excessive exercising can damage the cartilage in joints. Evidence suggests that a high level of practice and training in certain sports, such as soccer, racquet sports, and track,

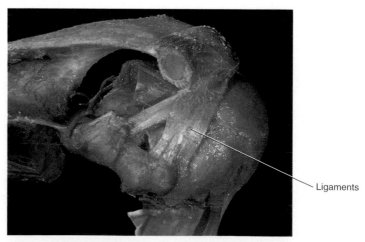

**Figure 34.8** Human left shoulder joint. Dense connective tissue fibers are arranged in bundles to form ligaments that hold the bones of a joint together.

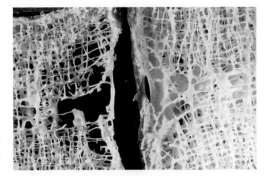

**Figure 34.9** Bone affected by osteoporosis (left) is less dense than normal bone (right).

or doing heavy manual work over a long time are risk factors for cartilage degeneration and osteoarthritis. *Osteoarthritis* is a condition in which the cartilage that covers the articulating ends of bones degenerates. With the cartilage cushion missing, the bones move against one another noisily and painfully. This condition develops particularly in the hip and knee joints.

The bottom line regarding bone and joint health is that exercise is good—in moderation. The old adage that "more is better" is not true when it comes to exercise and its effects on your bones, joints, and general health.

# How do muscles attach to bones and move body parts, and how are muscles structured?

## 34.9 Muscles attach to bones by tendons and move body parts by contracting.

Bones and joints would be of no use in movement without the muscles attached to them, providing the power for movement. Humans have three different kinds of muscle tissue: smooth muscle, cardiac muscle, and skeletal muscle (see Figure 26.9). Smooth muscle tissue is made up of sheets of cells (muscle fibers) found in many organs, doing jobs like mixing food in your stomach or narrowing the interior of your arteries to restrict blood flow. Cardiac muscle tissue makes up the heart—the pump of your circulatory system. Skeletal muscle tissue makes up the muscles that are attached to bones, allowing you to move your body; these muscles are the nearly 700 muscles of the muscular system (**Figure 34.10**).

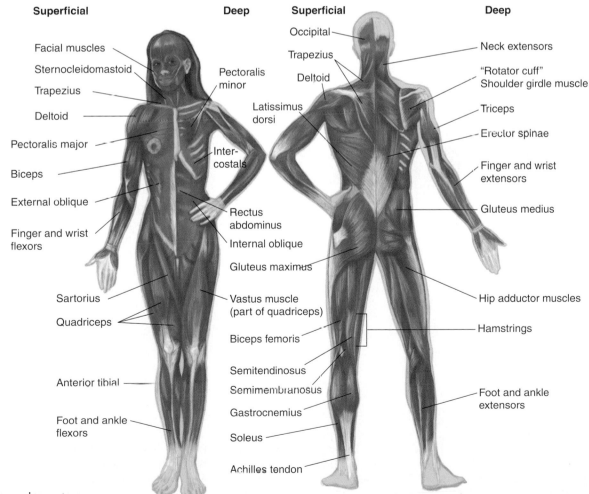

**Figure 34.10** The human muscular system.

The end of a skeletal muscle is attached by tendons to a stationary bone; this attachment is called the *origin*. The other end of the muscle is attached to a bone that will move; this attachment is called the *insertion*. The origin serves to anchor the muscle as it pulls at the insertion. As an example, **Figure 34.11** shows the biceps and triceps muscles and their origins and insertions. The biceps brachii is located on the front of the upper arm. In fact, the word "brachii" means "arm." This muscle has two upper ends, or heads, which is how it got its name—biceps, meaning "double-headed." These two heads have origins on the edges of the scapula near the front side of the humerus. Their tendons pass over the upper end of the humerus and then blend into the fattened midsection, or belly, of the biceps. The insertion of the biceps is by means of flattened tendons into the radius. When the biceps shortens, the insertion is brought closer to the origin of the muscle, an action that pulls the forearm to the upper arm.

To lower the forearm, the triceps brachii goes into action. Skeletal muscles oppose each other in this way; such opposing muscle pairs are called *antagonists*. In this case, the action of the triceps opposes the action of the biceps. The triceps sits on the back of the arm with one of its three heads originating on edges of the scapula near the back side of the humerus. Another head originates along the upper half of the back side of the humerus and the third along the lower half of the back side of the humerus. The triceps inserts on the ulna near the elbow. When the triceps shortens, its insertion is brought closer to its origin and the forearm is brought downward.

## 34.10 Muscles are made up of muscle fibers containing filaments that slide past one another.

Skeletal muscles are called *striated muscles* because their cells appear marked with striations, or lines, when viewed under the light microscope (see micrograph in Figure 34.12*b*). These striations are caused by an orderly arrangement of filaments called myofilaments within skeletal muscle cells; *myo-* means "muscle." Special groupings of these myofilaments are the contractile units of muscle cells.

Skeletal muscle cells are extremely long cells formed by the end-to-end fusion of shorter cells during embryonic development. These long muscle cells are called *muscle fibers*. Each muscle fiber contains all the nuclei of the fused cells pushed out to the periphery of the cytoplasm (**Figure 34.12*a***). Notice that the names of certain parts of muscle cells shown in the figure have the prefix sarco-, which comes from a Greek word meaning "flesh."

Each muscle fiber is packed with *myofibrils*, which are cylindrical, organized arrangements of special thick and thin myofilaments capable of shortening the muscle fiber. The so-called thin myofilaments are made up of the protein **actin** (AK-tin). The thick myofilaments are made up of a much larger protein, **myosin** (MY-uh-sin). All the myofilaments within a muscle fiber are lined up in such a way that the cells appear to have interior bands of light and dark. The diagram in Figure 34.12*b* shows how the actin and myosin filaments are arranged, forming this banding pattern.

The thin actin filaments are attached to plates of protein that appear as dark lines called *Z lines*. The actin filaments extend from the Z lines equally in two directions, perpendicular to the plates. In a resting muscle, the actin filaments that extend from two sequential Z lines are not long enough to reach each other. Instead, they are joined to one another by interdigitating myosin filaments (see the M line), similar to the way your fingers interlock when you fold your hands. This arrangement of protein plates and myofilaments produces the banding patterns in muscle fibers.

The I band in Figure 34.12*b* contains the Z line and the thin actin filaments. The A band contains a portion of the thin actin filaments and the thick myosin filaments. The H zone appears as a light zone running down the center of the A band; it contains only thick filaments held in place by a series of fine threads called the *M line*. The part of the myofibrils lying between adjacent Z lines is called the **sarcomere** (SAR-koe-mere), the contractile unit of muscles.

Molecularly, each thin actin filament consists of two strings of proteins wrapped around one another, like two strands of loosely wound pearls (**Figure 34.13**). The result is a long, thin, helical filament. Myosin has an unusual shape: One end of the molecule is a coil of two chains that forms a very long rod, whereas the other end

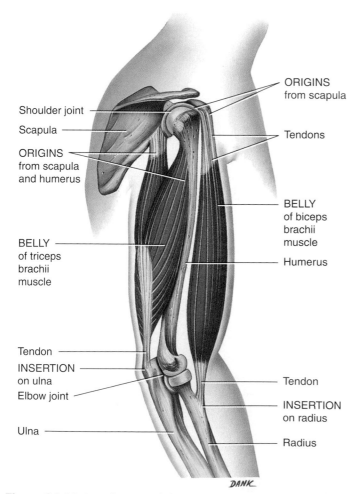

Shoulder joint
Scapula
ORIGINS from scapula and humerus
BELLY of triceps brachii muscle
Tendon
INSERTION on ulna
Elbow joint
Ulna

ORIGINS from scapula
Tendons
BELLY of biceps brachii muscle
Humerus
Tendon
INSERTION on radius
Radius

DANK

**Figure 34.11 Attachment of skeletal muscles to bones.** The origin of a skeletal muscle is attached to a stationary bone. The insertion is attached to a bone that will move. The origin serves to anchor the muscle as it pulls at the insertion. The biceps and triceps are antagonistic in their actions. That is, when one contracts, the other relaxes.

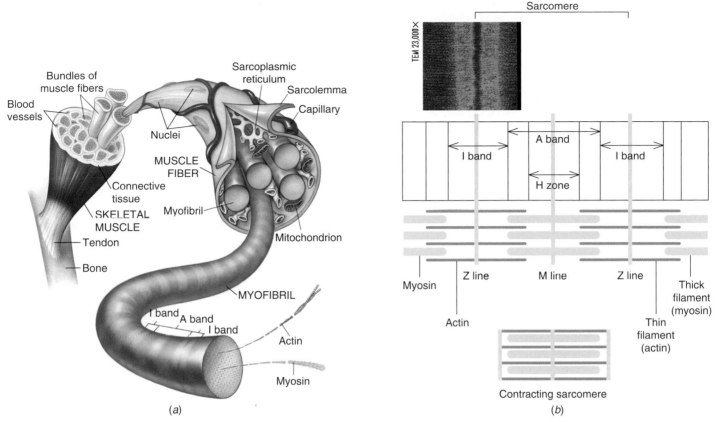

**Figure 34.12 Structure of a skeletal muscle and a sarcomere.** (a) Progressively magnified views of a skeletal muscle. (b) Banding patterns of the thick and thin filaments in a sarcomere.

**Visual Thinking:** Notice in (a) that mitochondria are located near myofibrils. What is the function of the mitochondria?

consists of a double-headed globular region. In electron micrographs, a myosin molecule looks like a two-headed snake or golf club. The contraction of myofilaments occurs when the heads of the myosin filaments change shape. Their globular heads "walk" step by step along the actin, causing the actin filaments to slide inward, toward the M Line. Because actin myofilaments are anchored to the Z lines, the Z lines are pulled closer to one another, contracting the sarcomere as shown in Figure 34.13.

All the sarcomeres of a myofibril contract simultaneously, shortening the myofibril. In addition, all the myofibrils of a muscle fiber usually contract at the same time. However, not all the muscle fibers within a muscle contract simultaneously. The forcefulness, speed, and degree of a muscle contraction depend on the number of muscle fibers that contract, their positions in relation to one another, and the frequency of nerve stimuli.

**CONCEPT CHECKPOINT**

4. When muscles contract, the contraction results in a relatively small change in the length of the muscle. Why, then, do you think the origin of muscles is generally so close to the joint? What would result if the origin were much farther away?

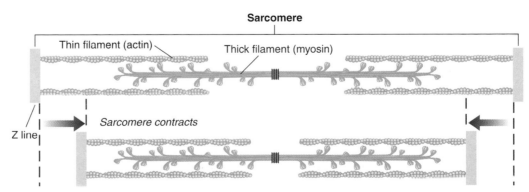

**Figure 34.13 How the myofilaments move during muscle contraction.** Muscle cells contract as actin filaments slide past myosin filaments, producing contraction of the myofilament.

# What functions does skin perform, and how is it structured?

## 34.11 Skin protects underlying structures, provides a barrier to infection, and helps conserve water.

Animals have all sorts of skins! **Figure 34.14** lists examples of animals from the major animal phyla and their types of skin. A general term for the outer covering of living things is **integument** (in-TEG-you-ment).

Many of the organisms listed in Figure 34.14 live in watery environments (sponge, jellyfish, flatworm, scallop, sea star) or in the soil (earthworm and roundworm). All these organisms, with the exception of the scallop and the sea star, breathe through their body surfaces. Notice in Figure 34.14 that their integuments are adapted to this function, being simply a layer of epidermal cells. The earthworm and the roundworm, which live in the soil, must keep their integuments moist or they will be unable to respire and will lose valuable internal moisture. The multilayered but porous cuticle that covers the roundworm does not protect the worm from dehydration.

Bivalve mollusks, such as scallops and clams, have a shell (exoskeleton) that also functions as integument. The shell protects the animal's soft body from being eaten by most predators; mollusk means "soft body." The sea star, however, a major predator of the clam, can pull on a closed clamshell for several hours, eventually opening this hard, protective integument.

Echinoderm means "spiny-skinned." The integument of the sea star serves as a flexible, protective covering for the skeleton and internal organs. Respiration takes place at the tube feet, which cover the underside of the sea star and also play a role in movement.

The integument of insects, which live on land, is structured to reduce the evaporation of water from the body. They have a hard outer covering with a waxy surface. If its wax covering is scratched, an insect will die from dehydration. For this reason, abrasives are mixed with stored grain to kill insect pests, both of which are later removed from the grain. The hard, celluloselike chitin covering of insects is even more important as an exoskeleton than as a structure for water conservation (see Section 34.2).

All land animals have integuments that not only protect them but also conserve water. Vertebrate skin provides an effective but partial barrier to water loss. In humans and all vertebrates, the integument consists of keratinized skin. *Keratin* (KARE-ah-tin) is a hardened, protective, fairly waterproof protein in integumentary structures such as vertebrate skin, nails, hair, horns, and feathers. Although the skins of vertebrates have certain structural differences, their integuments are all instrumental in helping to balance the gain and loss of water, which is crucial to life. This keratized surface also helps prevent the penetration of microorganisms to deeper tissues, thus being an effective barrier to infection.

Surface differences among cold-blooded (ectothermic) vertebrate skins range from thick scales in animals like alligators to mucus coverings in amphibians. Ectothermic organisms absorb heat from their surroundings rather than generate their own body heat. In reptiles, epidermal scales help prevent water loss as these animals bake in the sun to raise their body temperature and metabolic rate. In skin-breathers (see Section 28.2), such as frogs and salamanders, the naked skin is covered with mucus to help keep that respiratory surface moist and yet reduce water loss.

## *just wondering . . .*

### Questions students ask

### Is body piercing safe? What are the risks?

Even a simple puncture can compromise the skin barrier and open it to disease-causing microorganisms. *Pseudomonas aeruginosa* and *Staphylococcus aureus* bacteria, which are widely distributed in nature and commonly infect wounds of all types, cause most post-piercing infections. Both of these organisms can cause serious problems that can be difficult to treat successfully because many strains are resistant to some antibiotics commonly used to treat infections of these organisms. In addition, there is a high rate of infection from piercing in the ears and nose. Nevertheless, if you follow a few simple rules, you will reduce your risk of infection from body piercing:

- Use a reputable piercing salon.
- Ask about their health and safety guidelines, such as methods for hygienically inserting jewelry and recommendations for aftercare.
- See a physician immediately if you experience redness, swelling, or pain in the pierced area.
- If you have oral piercing, visit your dentist for instructions on how to avoid infection and damage of the oral area. ●

*Are you wondering about a topic in biology and how it relates to your life? Submit your question by clicking the* Just Wondering *link in this text's companion Web site at* www.wiley.com/college/alters.

## Visual Summary

**Figure 34.14** The integument of representative animals.

**REPRESENTATIVE ANIMAL**
(Phylum)

**INTEGUMENT**

**Sponge** (Porifera)

Layer of epidermal cells

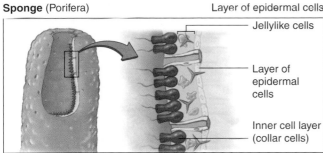

Jellylike cells

Layer of epidermal cells

Inner cell layer (collar cells)

**Jellyfish** (Cnidaria)

Layer of epidermal cells

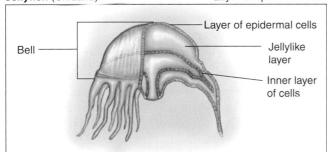

Layer of epidermal cells

Bell

Jellylike layer

Inner layer of cells

**Flatworm** (Platyhelminthes)

Layer of epidermal cells

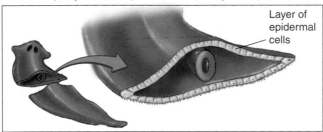

Layer of epidermal cells

**Roundworm** (Nematoda)

Thick cuticle covering layer of epidermal cells

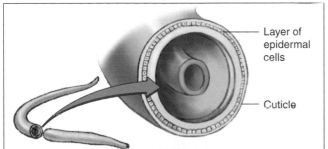

Layer of epidermal cells

Cuticle

**Earthworm** (Annelida)

Thin cuticle covering layer of epidermal cells

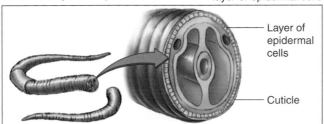

Layer of epidermal cells

Cuticle

**REPRESENTATIVE ANIMAL**
(Phylum)

**INTEGUMENT**

**Scallop** (Mollusca)

Calcium carbonate shell

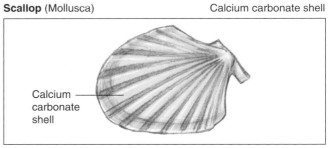

Calcium carbonate shell

**Insect** (Arthropoda)

Chitin cuticle with a waxy surface and underlying epidermis

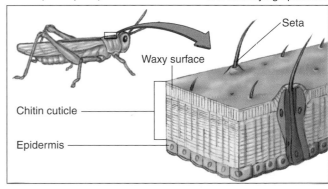

Seta

Waxy surface

Chitin cuticle

Epidermis

**Sea star** (Echinodermata)

Thin, prickly epidermis

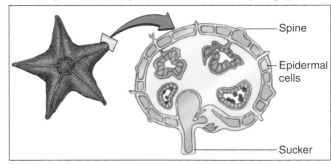

Spine

Epidermal cells

Sucker

**Human** (Chordata)

Keratinized skin

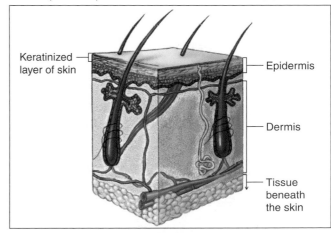

Keratinized layer of skin

Epidermis

Dermis

Tissue beneath the skin

## FOCUS

### on Plants

Plants, like animals, have integuments. Dermal tissue is the "skin" of a plant. It covers the plant, protecting it from water loss and injury to its internal structures.

As in animals, the most abundant type of cell found in the dermal tissue of plants is epidermal cells. These cells are often covered with a thick, waxy layer called a cuticle that protects the plant and provides an effective barrier against water loss. Plant "skin" also has outgrowths. For example, on leaves or fruits, fuzzlike outgrowths reflect sunlight, which helps control water loss, as in the lamb's ear (**Figure 34.A**). Some outgrowths, such as the thorns of a rosebush, protect a plant against animal predators.

Woody shrubs and trees have protective bark in place of dermal tissue. Bark is made up of other tissue types. Chapter 24 discusses the structure and function of plants—including their dermal tissues—in more detail.

**Figure 34.A.** Lamb's ear is a gray, fuzzy, low-growing foliage plant.

Aquatic mammals and other aquatic vertebrates also have a variety of integument adaptations that help them survive in a watery environment. For example, fresh water fishes are covered by scales and mucous, which retard the movement of water *into* their bodies.

## 34.12 Human skin has three layers and performs a variety of functions.

Human skin is far more than simply an elastic covering of epithelial cells encasing your body's muscles, blood, and bones. It consists of an inner layer called the dermis and an outer layer called the epidermis. The subcutaneous layer, or hypodermis, lies just beneath the skin and consists primarily of fat tissue. Human skin is a dynamic organ that performs many functions (**Figure 34.15**):

- *Skin is a protective barrier*. It keeps out microorganisms that would otherwise infect the body. Because skin is waterproof, it keeps the fluids of the body in and other fluids out. Skin cells also contain a pigment called melanin, which absorbs potentially damaging ultraviolet radiation from the sun.
- *Skin provides a sensory surface*. Nerve endings in skin act as your body's pressure gauge, telling you how gently to caress a loved one and how firmly to hold a pencil. Other sensors em-

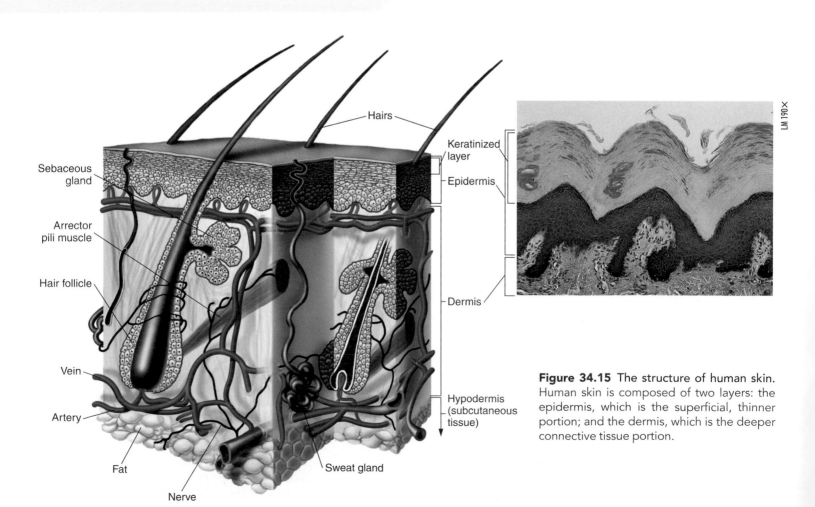

**Figure 34.15** The structure of human skin. Human skin is composed of two layers: the epidermis, which is the superficial, thinner portion; and the dermis, which is the deeper connective tissue portion.

bedded in the skin detect pain, heat, and cold (see Figure 34.15). Skin is the body's point of contact with the outside world.

- *Skin compensates for body movement.* Skin stretches when you reach for something and retracts quickly when you stop reaching. It also grows as you grow.
- *Skin helps control the body's internal temperature.* When the environmental temperature is cold, the blood vessels in the skin constrict, so less of the body's heat is lost to the surrounding air. When it is hot, these same vessels dilate, giving off heat. In addition, glands in the skin release sweat, which then absorbs body heat and evaporates, cooling the body surface.
- *Skin manufactures vitamin D in the presence of sunlight.* This vitamin helps the body absorb calcium from the digestive tract. It is important for bone formation. Because you wear clothes and may not be in the sun for lengthy periods, milk is typically vitamin-D fortified.

Your skin is the largest organ of your body. In an adult human, 15% of the total body weight is skin. Much of the multifunctional role of skin reflects the fact that its tissues are made up of a variety of specialized cells. One typical square centimeter of human skin contains 200 nerve endings, 10 hairs with accompanying microscopic muscles, 100 sweat glands, 15 oil glands, 3 blood vessels, 12 heat receptors, 2 cold receptors, and 25 pressure-sensing receptors. Together with the hair and nails, human skin is called the **integumentary system**.

Keratinized human skin develops as follows: As cells produced deep within the skin rise to the surface, pushed by the continual formation of new cells, they produce the strengthening protein keratin and change in shape and structure. By the time they reach the topmost layer of the skin, they are dead, flattened, and hardened, providing a surface that is resistant to abrasion and impermeable to most pathogens and noncorrosive substances. You lose these dead surface cells on a regular basis.

### CONCEPT CHECKPOINT

5. Match the following functions in living organisms with skin, skeleton, or muscle tissue: infection barrier, body support, movement, gas exchange, protection from predators, thermoregulation, and water conservation.

# CHAPTER REVIEW

## Summary of Key Concepts and Key Terms

### How do organisms move?

**34.1** In humans, movement results from the contraction of skeletal muscles anchored to bones, which are used like levers to direct force against an object.

**34.2** The skeleton provides support for the animal body; an animal may have a **hydroskeleton** (p. 594), an **endoskeleton** (p. 594), or an **exoskeleton** (p. 594).

**34.2** Muscles attach to the rigid, skeletal portion of a body; the skeleton acts as a framework against which muscles pull.

### What is bone, what are its functions, and how do bones make up the skeleton?

**34.3** **Bone** (p. 595) contains living cells embedded in a matrix of collagen fibers, which provide flexibility, and mineral salts, which provide hardness.

**34.3** Compact bone is found along the length of long bones, and spongy bone is found within the ends of long bones and within short, flat, and irregularly shaped bones.

**34.4** The **skeletal system** ( **skeleton**, p. 595) provides support for the body and attachment sites for skeletal muscles, protects delicate internal structures, produces blood cells in its red marrow, and stores minerals.

**34.5** The **axial skeleton** (p. 596) is the central column of bones and consists of the **skull** (p. 597), **vertebral column** (p. 597), and rib cage.

**34.5** **Ligaments** (LIG-uh-munts, p. 597) are bundles, or strips, of dense connective tissue that usually hold bones to bones, or may act as supportive structures.

**34.6** The **appendicular skeleton** (p. 597) consists of the bones of the appendages and the bones off which the appendages hang: the **pectoral girdle** (p. 597) and the **pelvic girdle** (p. 597).

**34.6** **Tendons** (p. 598) are cords of dense connective tissue that attach muscles to bone.

### What are joints, and how does exercise affect them?

**34.7** A **joint** (**articulation**, p. 598), is a place where bones, or bone and cartilage, or bone and teeth come together.

**34.7** The immovable joints of the skull are called **sutures** (p. 598), and freely movable joints are called **synovial joints** (p. 598).

**34.7** Joints that allow angular movements are **hinge joints** (p. 598), joints that allow rotation are **ball-and-socket joints** (p. 598), and joints that allow side-to-side movements are **pivot joints** (p. 599).

**34.8** Bone and joint health is enhanced by moderate physical activity but can be hurt by excessively strenuous, long-term physical activity.

## How do muscles attach to bones and move body parts, and how are muscles structured?

**34.9** There are three kinds of muscle tissue: smooth muscle tissue, which is found in many body organs; cardiac muscle tissue, which makes up the heart; and skeletal muscle tissue, which is attached to bones and move the body.

**34.10** Skeletal muscle cells contract as a result of thick **myosin** (p. 602) and thin **actin** (p. 602) filaments, which are packed within the muscle cells, sliding past one another.

**34.10** The contractile unit of muscles is the **sarcomere** (SAR-koe-mere, p. 602).

## What functions does skin perform, and how is it structured?

**34.11** In animals, the outer covering (**integument**, p. 604) protects soft body parts beneath, provides a barrier to infection, and helps conserve water in land animals.

**34.11** In many aquatic animals, the integument also acts as the respiratory surface.

**34.12** Human skin, the **integumentary system** (p. 607) provides an elastic covering for the body, serves as a protective barrier against water loss and invading microbes, provides a sensory surface, compensates for body movement, and helps control the body's internal temperature.

## Level 1 | Learning Basic Facts and Terms

**Multiple Choice**

1. A hydroskeleton can be found in all of the following organisms except:
   a. Sea stars
   b. Flatworms
   c. Earthworms
   d. Nematodes
   e. Insects

2. The human integumentary system functions in
   a. thermoregulation.
   b. posture.
   c. production of blood cells.
   d. storage of minerals.

3. Which of the following is a difference between spongy bone and compact bone?
   a. Spongy bone is more dense than compact bone.
   b. Spongy bone houses red bone marrow and compact bone does not.
   c. Spongy bone is found throughout long bones, while compact bone is found only in irregularly shaped bones.
   d. Spongy bone generally surrounds compact bone to make it lighter.

4. Which of the following bones does not belong with the others?
   a. Temporal
   b. Occipital
   c. Parietal
   d. Frontal
   e. Sacrum

5. Which of the following bones does not belong with the others?
   a. Sternum
   b. Femur
   c. Tibia
   d. Radius
   e. Ulna

6. How are ligaments different from tendons?
   a. Ligaments are usually shorter than tendons.
   b. Ligaments attach muscle to skin; tendons attach bone to muscle.
   c. Ligaments attach bone to bone; tendons attach muscle to bone.
   d. The point of insertion of a ligament is closer to the joint than that of a tendon.

**Matching**

For questions 7–11, match the following joints with their location in the human body:

7. _____ ball and socket joint
8. _____ pivot joint
9. _____ hinge joint
10. _____ ellipsoidal joint
11. _____ plane/gliding joint

a. between atlas and axis vertebrae
b. shoulder joint
c. elbow
d. tarsals
e. radius/carpals

## Level 2 | Learning Concepts

1. You raise your hand to answer a question in class. Explain the roles played by your bones, skin, and skeletal muscles in this movement.

2. Human babies learn extensively by touching and manipulating toys and objects. What physiological role does skin serve in this process?

3. What is bone, and what are its functions? Explain its importance.

4. Leukemia patients sometimes have their red bone marrow irradiated, which kills the cancerous cells as well as healthy blood cells, and then subsequently receive a bone marrow transplant as treatment for their disease. From what kind of tissue would this marrow be obtained?

Why are these patients kept in sterile rooms until their bone marrow is restored?

5. What type of bone tissue makes up most of the bones of the axial skeleton?

6. What do you think might happen to excess calcium that is taken in by the body, but not needed at the moment?

7. Mrs. Gorman has injured her back; her doctor tells her she has a "slipped disk." What type of "disk" is involved, and what is its function?

8. What do hinge, pivot, and ball-and-socket joints have in common? How are they different?

## Level 3    Critical Thinking and Life Applications

1. From your observations, describe the changes that occur to skin and bone as people age.
2. Tremendous progress has been made in the development and use of metallic artificial joints and bone replacement. What activities and bone functions would be easily served by such artificial structures? What functions of the bone could not be filled by such replacements?
3. The trunk of an elephant has no endoskeleton to which muscles can attach, yet the elephant is able to move its trunk quite skillfully. What type of skeleton must the elephant trunk have? Describe how you think the elephant moves its trunk.

4. Many bacteria colonize human skin; these organisms are called the normal microbiota. Based on your knowledge of the structure of human skin, why do you think that these bacteria do not usually cause skin infections, yet can quickly cause an infection when the skin surface is broken, as in a wound?
5. Analyze and evaluate your patterns of exercise (or nonexercise) to determine whether you are exercising in a manner that will help strengthen your bones and joints, or in a manner that may cause damage to your bones and joints. Give reasons that support your evaluation as you describe your exercise or training routine.

## In The News    Critical Thinking

### JEWELED JOINTS

Now that you understand more about the skeletal system, reread this chapter's opening story about the development of a nanodiamond coating for artificial joints. To better understand this information, it may help you to follow these steps:

1. Review your immediate reaction to the nanodiamond technology that you wrote when you began reading this chapter.
2. Based on your current understanding, again summarize the main point of the research in a sentence or two.
3. What questions do you now have about this information that this chapter's opening story does not answer?
4. Collect new information about the research. Visit the *In The News* section of this text's companion Web site at www.wiley.com/college/alters and

watch the "Jeweled Joints" video. Then use the "summary" link to read the accompanying story and access related links. Use this information, the links provided, and other online and library resources to answer your questions and find updates about this topic. State the sources of your information. Explain why you think the information is accurate. Also determine whether the information expresses a particular point of view or is biased in any way.

5. What in your view are the most significant aspects of this information? Give reasons for your opinion and for any changes in your ideas based on the additional information you have collected and the analysis you have done.

# HORMONES

## In The News | Teen Steroids

Anabolic steroids have been the downfall of many winners. These controversial drugs are really synthetic hormones, chemicals that affect the activity of specific organs or tissues. Anabolic steroids affect the body in ways similar to the male sex hormone testosterone and stimulate the buildup of muscles. Along with building a championship body, however, anabolic steroids strikingly change the body's metabolism.

Female athletes on steroids experience side effects such as shrinking breasts, a deepening voice, and an increase in body hair. Male athletes find that their testicles shrink. Some users also experience life-threatening heart, kidney, and liver damage. Youngsters who take these drugs risk stunting their growth because anabolic steroids cause bones to stop growing prematurely. Furthermore, some scientific evidence suggests high doses of anabolic steroids promote aggressive behavior in humans.

Surveys conducted by the National Institute on Drug Abuse (NIDA) indicate that steroid use among adolescents is increasing, despite the health risks associated with the practice. Richard Melloni, a professor of psychology and behavioral neuroscience, and his colleagues at Northeastern University in Boston are extremely interested in the effects of anabolic steroids on the teen brain. To learn more about these effects, the researchers investigated the effects of anabolic steroids on the behavior and brains of young male golden hamsters (*Mesocricetus auratus*).

As in human teenagers, hamster adolescence is a critical period for the maturation of neural systems in the brain. For 30 days during the animals' adolescent period, the sci-

entists administered daily anabolic steroid injections with sesame oil to one group of hamsters, mimicking a chronic heavy use regimen of the drug. Daily sesame oil injections without the steroids were administered to a control group of hamsters. After this period, the researchers tested each animal for signs of aggression by introducing an "intruder" hamster into its cage. Animals in the treatment group exhibited significantly more signs of aggression, such as offensive attacks and bites, than hamsters in the control group. Microscopic examination of the hamsters' brains indicated that the treatment altered levels of neurotransmitters implicated in this behavioral response.

Preliminary research suggests that the changes observed by the researchers are long lasting. Although Melloni suggests that steroid use by teenagers might result in behavioral problems in adulthood, he notes that more research is needed to determine the long-term effects of steroid use on the human brain. To learn more about this research, visit the *In The News* section of this text's companion Web site at www.wiley.com/college/alters and view the video "Teen Steroids."

*Write your immediate reaction to Melloni's research into the effects of anabolic steroids on behavior and the brain: first, summarize the main point in a sentence or two; then suggest what you think the significance of this research is. You will have an opportunity to reflect on your responses and gather more information on this topic in the In The News feature at the end of this chapter. In this chapter, you will learn more about hormones.*

### What are hormones, and how do they work?

**35.1** Hormones are chemical messengers.

**35.2** Hormones are control mechanisms that exist in a wide variety of organisms.

**35.3** Hormones help regulate the internal environment of the body.

**35.4** Hormones affect target tissues by binding to receptor molecules.

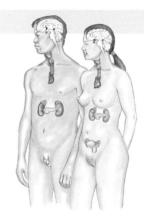

### Where are human hormones produced, and what do they do?

**35.5** The secretion of hormones from the pituitary is regulated by the hypothalamus.

**35.6** The pineal gland secretes melatonin, a sleep/wake cycle hormone.

**35.7** The thyroid secretes hormones that regulate the body's metabolism.

**35.8** The parathyroids secrete hormones that regulate blood levels of ions such as calcium.

**35.9** The adrenals secrete stress hormones and those that regulate the level of water in the blood.

**35.10** The pancreas secretes hormones that regulate the level of glucose in the bloodstream.

**35.11** Other endocrine glands include the thymus, ovaries, testes, and digestive hormone–secreting cells.

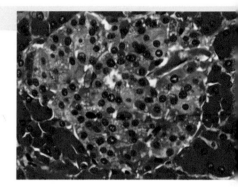

# What are hormones, and how do they work?

## 35.1 Hormones are chemical messengers.

Have you ever heard preteens referred to as "containers of raging hormones?" Have you ever felt like you could almost feel hormones coursing through your body when you are frightened or otherwise stressed, or (if you are female) when you are experiencing monthly hormonal swings? What are these powerful chemicals that flow throughout our bodies?

**Hormones** are chemical messengers that are produced in one part of the body but have an effect on another, perhaps distant, part of the body. In vertebrates, hormones are released into the bloodstream where they are carried to all of the tissues. Hormones, however, only affect particular tissues called target tissues. Such selective interaction takes place because hormones affect only those tissues bearing hormone-specific receptors.

In vertebrates, such as humans, the hormonal system is referred to as the **endocrine system** (ENN-doe-krin). This name comes from the **endocrine glands**, which secrete hormones. Glands are individual cells or groups of cells that secrete substances. Endocrine glands are ductless glands; they secrete hormones directly into the bloodstream. In contrast, **exocrine glands** (EK-so-krin) secrete substances through ducts to the interior of the body, as with digestive enzymes, for example, or to the outside of the body. Sweat glands are an example of this type of gland.

In concert with the nervous system, the endocrine system integrates and coordinates the multitude of diverse physiological processes occurring in the body at all times. This control ensures that the body can maintain homeostasis, or a stable state, in a variety of conditions. In general, a nerve cell exerts control in its imme-diate vicinity by releasing neurotransmitters at the synapse, affecting another nerve cell, a muscle, or a gland. An endocrine gland, however, can exert widespread control by releasing its chemical messenger, a hormone, into the bloodstream. A specialized nerve cell called a *neurosecretory cell* (NEW-row-SEE-creh-tor-ee) has characteristics of both an endocrine cell and a neuron, or nerve cell. It conducts nervous impulses, but also releases hormones into the bloodstream (**Figure 35.1**).

A hormone may affect parts of the body that are distant from the gland that secreted it. Moreover, a hormone may affect numerous parts of the body at the same time. For example, release of adrenaline into the bloodstream by the adrenal glands stimulates heart rate, breathing rate, blood flow to the muscles, and a number of other physiological events that prepare the body for "fight or flight." This response is a good example of how the nervous and endocrine systems work together. The adrenal gland is stimulated to release its hormone by input from the sympathetic nervous system.

The effect of nervous stimulation is immediate. The effect of hormonal stimulation takes a bit longer. First the hormone must be released into the bloodstream, and then it must reach a target cell. Once the hormone reaches the target cell, it may initiate reactions with enzymes already present in the cell, or it may direct synthesis of new enzymes. These events take more time than the electrical shifts across the cell membrane of a nerve cell. Moreover, the effects of hormones may be relatively long-lasting, such as effects on growth, metabolism, or reproductive structures.

Not all hormones travel great distances to reach their target cells. And not all hormones are produced by organs that do nothing else. For example, the stomach produces a hormone called gastrin

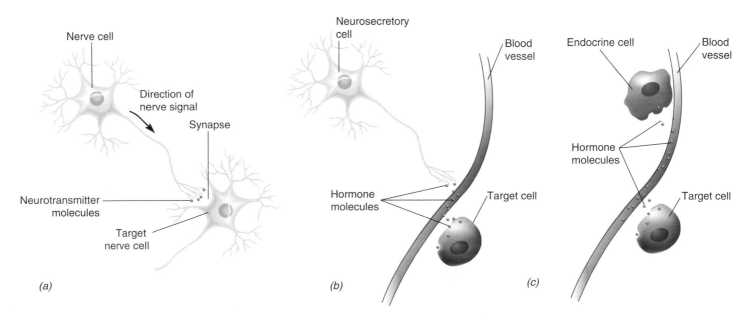

**Figure 35.1 Chemical signals.** (*a*) Nerve cells release neurotransmitters into synapses. (*b*) Neurosecretory cells are specialized nerve cells that release hormones directly into the blood stream. (*c*) Endocrine cells also release hormones into the bloodstream.

that signals the presence of food in the stomach and stimulates the release of a digestive enzyme by other cells in the stomach. You may not think of the kidney or the heart as endocrine glands, but they, too, produce chemical signals that are important in the function of those glands. They are considered hormones because they are chemicals released into the bloodstream and affect cells with a specific receptor for that chemical.

## 35.2 Hormones are control mechanisms that exist in a wide variety of organisms.

The function of chemical messengers—that is, hormone function—has been studied not only in vertebrates, but also in a wide variety of invertebrates such as squid, earthworms, lobsters, and insects, and in other organisms such as protozoans and plants. (See Chapter 25 for discussion of plant hormones.) Researchers have discovered that these organisms also have hormones. Most of their hormones differ from vertebrate hormones in their chemical structures and modes of action, and, of course, not all of the organisms have a bloodstream in which hormones can flow. A few vertebrate hormones and cell receptor molecules, however, have also been found in nonvertebrates. For example, adrenaline has been found in protists. Insulin, the hormone in humans that stimulates the uptake of glucose from the blood, has been found in insects, annelids, mollusks, and protists. Scientists deduce from these discoveries that hormones are ancient molecules that first appeared in simple organisms evolving hundreds of millions of years ago.

Hormones play a major role in metamorphosis and growth in insects (the most extensively studied invertebrates with regard to hormones). *Metamorphosis* refers to the changes that most insects go through as they develop from egg to adult. It includes such processes as molting, the shedding of skin to allow further growth, and pupation, the development of a "cocoon" stage.

## 35.3 Hormones help regulate the internal environment of the body.

There are 10 major endocrine glands in the human body that make up the endocrine system (**Figure 35.2**). These glands are scattered throughout the body and in some cases are parts of other glands. Together, the glands of the endocrine system produce over 50 different hormones. The messages of the endocrine hormones are varied but can be grouped into four categories:

- *Regulation*: Hormones control the internal environment of the body by regulating the secretion and excretion of various chemicals in the blood, such as salts and acids.
- *Response*: Hormones help the body respond to changes in the environment and cope with physical and psychological stress.
- *Reproduction*: Hormones control the female reproductive cycle and other reproductive processes essential to conception and birth. In both sexes, hormones control the development of sex cells, the reproductive organs, and secondary sexual characteristics that make men and women different.
- *Growth and development*: Hormones are essential to the proper growth and development of the body from conception to adulthood.

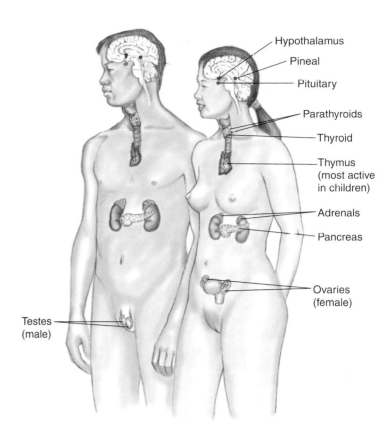

**Figure 35.2** The human endocrine system. There are 10 major endocrine glands scattered throughout the human body.

CONCEPT CHECKPOINT

1. The nervous system and the endocrine system function together to coordinate and control the body. How is control by the endocrine system different from control by the nervous system?
2. Pair each of the following events with a category of hormone discussed in Section 35.3: (a) the menstrual cycle, (b) differentiation of cells during embryonic development, (c) increased heart rate when you are frightened, and (d) control of water concentration in the urine.

## 35.4 Hormones affect target tissues by binding to receptor molecules.

Once molecules of a hormone are released into the bloodstream, they travel throughout the body. Although hormone molecules may pass billions of cells, specific hormones affect only target cells. How does this happen? Hormones recognize target cells because they bind to receptor molecules that are embedded within the cell membrane or that are located within the cytoplasm of the cell. The binding of a hormone molecule to a receptor molecule activates a chain of events in the target cell that results in the effect of the hormone being expressed.

An easy way to think about how hormones affect target cells—and only target cells—is to compare this interaction to the reception of radio waves by a radio. Radio waves (hormones) are sent by the antenna of a radio station (endocrine gland) and are transmitted in all directions (throughout the body via the bloodstream). The receiver, or radio (the target cells) must be tuned to the frequency of the radio waves (the receptor molecules) to receive the transmission. Any radio not tuned to that frequency (cells without receptor molecules for that hormone) will not receive the signal.

Endocrine hormones in humans are divided into two major classes, peptide hormones and steroid hormones, based upon how and where they interact with their receptors. See Figures 10.8 and 10.9 and related text for an explanation of each.

### Hormonal feedback loops

The production of peptide and steroid hormones is regulated by a mechanism called a *feedback loop*. In general, hormonal feedback loops work in the following way (**Figure 35.3**): Endocrine glands are initially stimulated ❶ to synthesize and secrete hormones. Stimulation of an endocrine gland occurs in one of three ways:

- *Direct stimulation by the nervous system:* The sensation of fear, for example, can cause the autonomic nervous system to trigger the release of the hormone adrenaline from the adrenal medulla.
- *Indirect stimulation by the nervous system by means of releasing hormones:* The hypothalamus is a specialized portion of the brain that produces and secretes releasing hormones. Some releasing hormones stimulate the release of other hormones; some prevent their release.
- *The concentration of specific substances in the bloodstream:* The blood level of a substance such as glucose or calcium ions, for example, may signal a gland to turn on or turn off.

After an endocrine gland secretes its hormone into the bloodstream ❷, the hormone travels throughout the body via

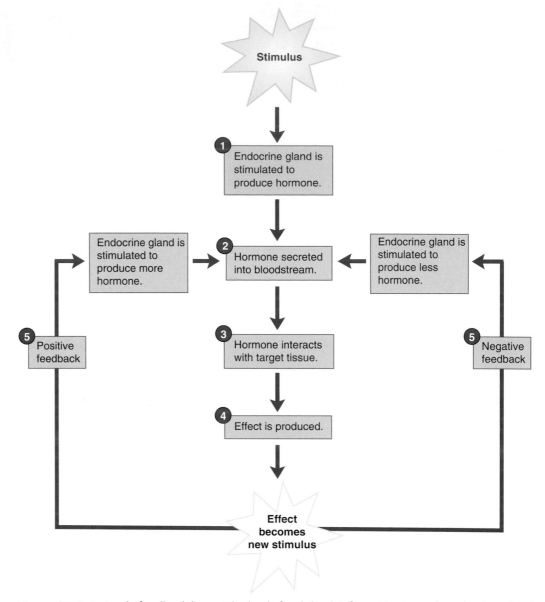

**Figure 35.3** A simple feedback loop. The body feeds back information to each endocrine gland after it releases its hormone.

**Visual Thinking:** In the negative feedback loop, what happens between step 5 and step 2?

the circulatory system and interacts with target tissues ❸. The target tissues produce the desired effect ❹. This effect acts as a new stimulus to the endocrine gland ❺. Put simply, the body feeds back information to each endocrine gland after it releases its hormone. In a positive feedback loop, the information that is fed back causes the gland to produce more of its hormone. In a *negative feedback loop*, the feedback causes the gland to slow down or to stop the production of its hormone. Most hormones work by means of negative feedback loops. Specific examples of feedback mechanisms and interactions are discussed throughout this chapter.

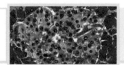

# Where are human hormones produced, and what do they do?

## 35.5 The secretion of hormones from the pituitary is regulated by the hypothalamus.

The **pituitary** (puh-TOO-ih-tare-ee) is a powerful gland that secretes nine major hormones. Even though it secretes so many hormones, it is amazingly tiny—slightly smaller than a marble. The pituitary gland hangs from the underside of the brain, supported and cradled within a bony depression.

The pituitary secretes seven major hormones from its larger front lobe, the *anterior pituitary*, and two from its rear lobe, the *posterior pituitary*. The secretion of these hormones is regulated by a mass of nerve cells that lies directly above the pituitary, making up a small part of the floor of the brain. This regulatory nervous tissue, the **hypothalamus** (HYE-poe-THAL-uh-muss), is connected to the pituitary by a stalk (**Figure 35.4**).

The hypothalamus uses information it gathers from other parts of the brain and its own receptors to stimulate or inhibit the secretion of hormones from the pituitary. In this way, the hypothalamus acts like a production manager, receiving information about the needs of the company's customers and regulating the manufacturing of products to satisfy those needs. However, the hypothalamus manages the anterior and posterior lobes of the pituitary differently.

The hypothalamus regulates the secretion of hormones from the anterior pituitary by producing *releasing hormones*. These seven hormones flow directly from the hypothalamus to the anterior pituitary via a network of blood vessels. Five releasing hormones are stimulatory, each causing the release of one or more pituitary hormones. For example, gonadotropin-releasing hormone (GnRH) stimulates the release of gonadotropins (see the next subsection) in both males and females. Two releasing hormones are inhibitory. One inhibits the release of growth hormone and thyroid-stimulating hormone. The other inhibits the release of prolactin.

The hypothalamus also regulates the release of two hormones

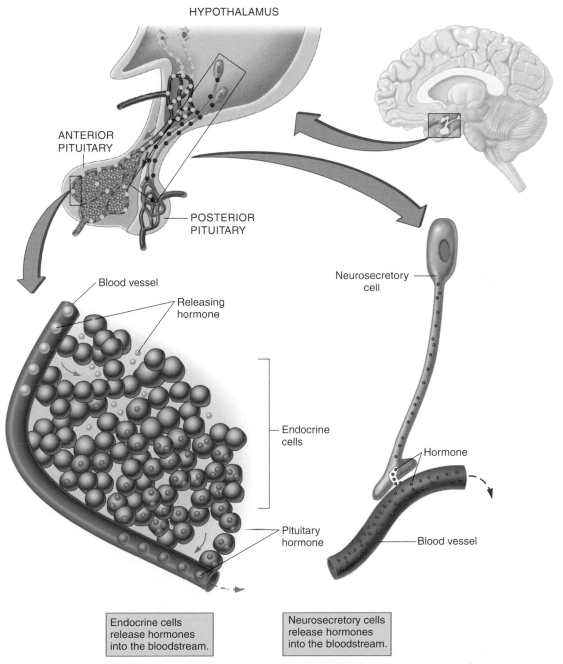

HYPOTHALAMUS

ANTERIOR PITUITARY

POSTERIOR PITUITARY

Blood vessel

Releasing hormone

Neurosecretory cell

Endocrine cells

Hormone

Pituitary hormone

Blood vessel

| Endocrine cells release hormones into the bloodstream. | Neurosecretory cells release hormones into the bloodstream. |

**Figure 35.4** Connection between the hypothalamus and the pituitary. The pituitary shown in the enlarged view is larger than actual size. Endocrine cells release hormones from the anterior pituitary, while neurosecretory cells originating in the hypothalamus release hormones from the posterior pituitary.

from the posterior pituitary. This regulation is neural, not hormonal. Neurons extend from the hypothalamus to the posterior pituitary. The hormones released from the posterior pituitary are produced in the cell bodies of neurosecretory cells in the hypothalamus. After production, the hormones travel down the axons of these neurons to the posterior pituitary. They are held there until the hypothalamic neurons signal their release into the bloodstream.

**Table 35.1** shows the hormones secreted by the central endocrine glands.

### The anterior pituitary

The hormones produced by the anterior pituitary regulate a wide range of bodily functions (**Figure 35.5**). Four of these hormones are called **tropic hormones** (TROW-pick). The word "tropic" comes from a Greek word meaning "turning" and refers to the ability of tropic hormones to turn on or stimulate other endocrine glands. Of the four tropic hormones, two are **gonadotropins** (GON-ah-duh-TROP-inz).

The gonads are the male and female sex organs, the testes and the ovaries. The gonadotropins are hormones that affect these sex organs, which are also considered endocrine glands because they too secrete hormones. The two gonadotropins are *follicle-stimulating hormone (FSH)* and *luteinizing hormone (LH)* (LOO-tea-in-EYE-zing). In females, FSH targets the ovaries and triggers the maturation of one egg each month. In addition, it stimulates cells in the ovaries to secrete female sex hormones called *estrogens*. In men, FSH targets the testes and triggers the production of sperm. LH stimulates cells in the testes to produce the male sex hormone *testosterone*. In females, a surge of LH near the middle of the menstrual cycle stimulates the release of an egg. In addition, LH triggers the development of cells within the ovaries that produce another female sex hormone— *progesterone*. (See Chapter 36 for a description of the organs and processes of the reproductive system.)

The two other tropic hormones produced by the anterior pituitary are *adrenocorticotropic hormone (ACTH)* (ah-DREE-no-KOR-tih-coh-TROW-pick) and *thyroid-stimulating hormone (TSH)*. ACTH triggers the adrenal cortex to produce certain steroid hormones. The adrenal glands are located on top of the kidneys (see Figure 35.2) and have two distinct parts: an outer cortex and an inner medulla. ACTH stimulates the adrenal cortex to produce hormones that regulate the production of glucose from noncarbohydrates such as fats and proteins. Others regulate the balance of sodium and potassium ions in the blood. Still others contribute to the development of the male secondary sexual characteristics. TSH triggers the thyroid gland to produce two thyroid hormones. This endocrine gland is located on the front of the neck, just below the voice box (see Figure 35.2). Its hormones control normal growth and development and are essential to proper metabolism. (Further discussion of ACTH in Section 35.9, and further discussion of TSH is in Section 35.7.)

*Growth hormone (GH)*, also known as somatotropin, is produced by the anterior pituitary and works with the thyroid hormones to control normal growth. GH increases the rate of growth of the skeleton by causing cartilage cells and bone cells to reproduce and lay down their intercellular matrix. In addition, GH stimulates the deposition of minerals within this matrix. GH also stimulates the skeletal muscles to grow in

| TABLE 35.1 | Central (Brain) Endocrine Glands and Their Hormones | |
|---|---|---|
| **Endocrine Gland and Hormone** | **Target Tissue** | **Principal Actions** |
| **Hypothalamus** | | |
| Releasing hormones | Anterior pituitary | Stimulate the release of hormones by the anterior pituitary |
| **Anterior pituitary** | | |
| Follicle-stimulating hormone (FSH) (gonadotropic hormone) | Sex organs | Stimulates ovarian follicle, spermatogenesis |
| Lutenizing hormone (LH) (gonadotropic hormone) | Sex organs | Stimulates ovulation and corpus luteum formation in females, and testosterone production in males |
| Adrenocorticotropic hormone (ACTH) | Adrenal cortex | Stimulates secretion of adrenal cortical hormones |
| Thyroid-stimulating hormone (TSH) | Thyroid | Stimulates secretion of thyroid hormones |
| Growth hormone (GH) | Cartilage and bone cells, skeletal muscle cells | Stimulates division of cartilage and bone cells, growth of muscle cells, and deposition of minerals |
| Prolactin | Mammary glands | Stimulates milk production |
| Menalocyte-stimulating hormone (MSH) | Melanocytes | Stimulates production of melanin |
| **Posterior pituitary** | | |
| Oxytocin | Uterus Mammary glands | Stimulates contraction of uterus and milk release |
| Antidiuretic hormone (ADH) | Kidneys and smooth muscle | Stimulates reabsorption of water by the kidneys |
| **Pineal gland** | | |
| Melatonin | Hypothalamus and/or reproductive organs | Possible stimulation of immune system; inhibits secretion of GnRH; other specific actions unknown |

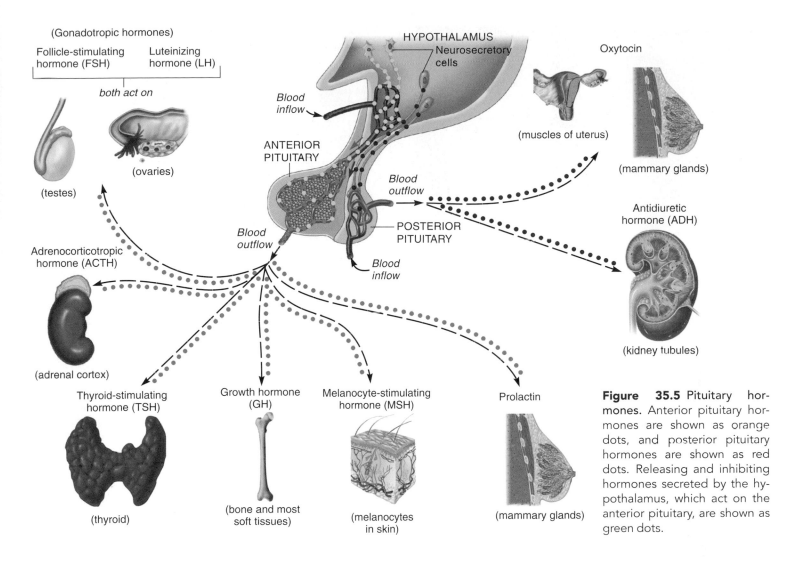

**Figure 35.5 Pituitary hormones.** Anterior pituitary hormones are shown as orange dots, and posterior pituitary hormones are shown as red dots. Releasing and inhibiting hormones secreted by the hypothalamus, which act on the anterior pituitary, are shown as green dots.

both size and number. In the past, children who did not produce enough GH did not grow to an average height range; this condition is called hypopituitary dwarfism. However, in the past decade, scientists have been able to use the techniques of genetic engineering to insert the human GH gene into bacteria to produce human GH. Currently, this laboratory-made hormone is being used successfully for treating growth disorders caused by hyposecretion (underproduction) of GH in children. The opposite problem may also occur: During the growth years, some children produce too much GH. This hypersecretion (overproduction) can cause the long bones to grow unusually long and result in a condition known as giantism (**Figure 35.6**). In adults, hypersecretion of GH causes the bones of the hands and face to thicken, resulting in a condition known as acromegaly (AK-row-MEG-ah-lee), as shown in **Figure 35.7**.

*Prolactin* is another hormone secreted by the anterior pituitary. Prolactin works with estrogens, progesterone, and other hormones to stimulate the mammary glands in the breasts to produce milk after a woman has given birth to a child. During the menstrual cycle, milk is not produced and secreted because prolactin levels in the bloodstream are very low. Midcycle, however, as the level of and estrogens falls and progesterone is low, the pituitary is stimulated by the hypothalamus to secrete some prolactin. This rise in prolactin, though not sufficient to cause milk production, does cause the breasts of some women to feel sore.

*Melanocyte-stimulating hormone (MSH)* acts on cells in the skin called melanocytes, which synthesize a pigment called melanin. Epidermal cells in the skin contain this pigment, which produces genetically determined skin colorations from pale yellow (in combi-

**Figure 35.6** Giantism.

**Figure 35.7 Acromegaly.** These photographs of a woman with acromegaly, at ages 16 (*left*) and 33 (*right*), show the thickening of the facial bones that results from the oversecretion of growth hormones in an adult.

nation with another pigment called carotene) to black, depending on the amount of melanin in the skin. Ultraviolet (UV) light triggers additional MSH production. In response to MSH, melanocytes produce more melanin, which helps protect the skin from UV damage. This increased melanin production is visible as a suntan in light-skinned people and as a darkening of the skin in dark-skinned individuals, and is a visible sign that UV light is damaging the skin.

### The posterior pituitary

The posterior lobe of the pituitary stores and releases two hormones that are produced by the hypothalamus: *antidiuretic hormone (ADH)* (AN-tih-die-yuh-RET-ick) and *oxytocin* (OK-sih-TOE-sin). ADH helps control the volume of the blood by regulating the amount of water reabsorbed by the kidneys. For example, receptors in the hypothalamus can detect a low blood volume by sensing when the solute concentration of the blood is high. When the hypothalamus detects such a situation, it triggers neurosecretory cells to make ADH. This hormone is transported within axons to the posterior pituitary, which releases the hormone into the bloodstream. ADH binds to target cells in the collecting ducts of the nephrons of the kidneys, increasing their permeability to water. An increased amount of water moves out of these ducts and back into the blood, resulting in more concentrated urine and a lower solute concentration in the blood. ADH also acts on the smooth muscle surrounding arterioles. As these muscles tighten, they constrict the arterioles, an action that helps raise the blood pressure despite fluid loss.

Oxytocin is the other hormone of the posterior pituitary. In women, oxytocin is secreted during the birth process, triggered by a stretching of the cervix as the baby descends toward the birth canal. Oxytocin binds to target cells of the uterus, enhancing contractions. The mechanism of oxytocin secretion is an example of a *positive feedback loop* in which the effect produced by the hormone enhances the secretion of the hormone. For this reason, physicians use synthetic oxytocin (pitocin) to induce uterine contractions when labor must be brought on by external means. Oxytocin also targets muscle cells around the ducts of the mammary glands, allowing a

new mother to nurse her child. The suckling of the infant triggers the production of more oxytocin, which aids in the nursing process and also helps contract the uterus to its normal size.

> ### CONCEPT CHECKPOINT
>
> **3.** How does a tropic hormone differ from a releasing hormone?

## 35.6 The pineal gland secretes melatonin, a sleep/wake cycle hormone.

The **pineal gland** (PIN-ee-uhl) gets its name from its shape: It looks like a tiny pinecone embedded deep within the brain between the two cerebral hemispheres (see Figure 35.2 and Table 35.1). The major endocrine product of the pineal gland is *melatonin*. Signaled by nerve impulses arising at the eyes, the pineal gland increases its secretion of melatonin by nearly 10-fold at nighttime. Similarly, during the daytime, nervous stimulation from the eyes suppresses secretion of this hormone.

Endogenous rhythms, also called biological clocks, are body activities that change throughout the day and night or with the seasons. One of the most widely accepted roles of melatonin is the synchronizing of the body's circadian rhythms (Latin *circa* = about and *dies* = a day) with the light/dark cycle. Many animals show seasonal rhythms such as patterns of hibernation, migration, or mating as their melatonin production fluctuates with increasing or decreasing numbers of daylight hours. These seasonal biological patterns of behavior are discussed in more detail in Chapter 37.

## 35.7 The thyroid secretes hormones that regulate the body's metabolism.

Sitting like a large butterfly just below the level of the voice box, the **thyroid** gland can be thought of as your metabolic switch. This gland secretes hormones that determine the rate of the chemical reactions of your body's cells, thus affecting growth and development in children. Put simply, thyroid hormones determine how fast bodily processes take place. Most body cells have receptors for thyroid hormones, and these hormones exert their effects throughout the body.

The hypothalamus and the thyroid gland work together to keep the proper level of thyroid hormones circulating in the bloodstream. This level is detected by the hypothalamus (**Figure 35.8 ❶**). A low level of thyroid hormones stimulates the hypothalamus to secrete a releasing hormone to the anterior pituitary ❷. This hormone message tells the pituitary to release more thyroid-stimulating hormone (TSH) ❸. The thyroid responds to stimulation by TSH, secreting more of the thyroid hormones, thereby raising their levels in the blood ❹. Shutdown occurs when thyroid hormone levels are sufficient ❺. This mechanism of action is an example of a negative feedback loop, in which the ultimate effect produced by stimulation of a gland (inhibition of TSH secretion) is opposite to the stimulus (TSH secretion).

In certain disease conditions, the amount of thyroid hormones in the bloodstream cannot be regulated properly. If the thyroid produces

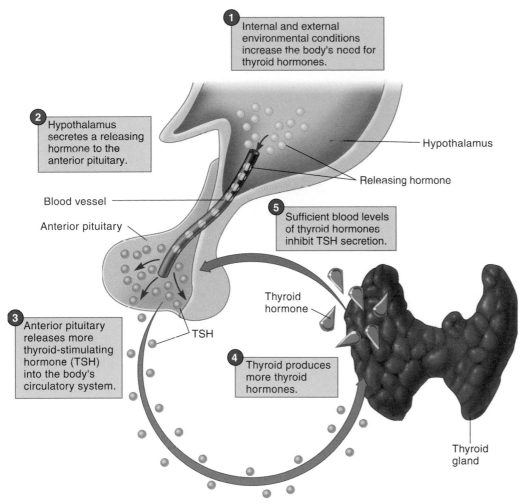

1. Internal and external environmental conditions increase the body's need for thyroid hormones.

2. Hypothalamus secretes a releasing hormone to the anterior pituitary.

Hypothalamus

Releasing hormone

Blood vessel

Anterior pituitary

5. Sufficient blood levels of thyroid hormones inhibit TSH secretion.

3. Anterior pituitary releases more thyroid-stimulating hormone (TSH) into the body's circulatory system.

TSH

Thyroid hormone

4. Thyroid produces more thyroid hormones.

Thyroid gland

**Figure 35.8** A negative feedback loop regulates release of thyroid hormone.

**Figure 35.10** A goiter. Iodide is used in the production of the thyroid hormones; when not enough iodide is available, the thyroid cannot produce thyroid hormones, and the thyroid enlarges from overstimulation by the anterior pituitary.

**Figure 35.9** Graves' disease. Cells behind the eye produce too much connective and adipose tissue, in which push the eye out of the socket.

too much of the thyroid hormones, a person may feel as though the "engine is racing," with such symptoms as a rapid heartbeat, nervousness, weight loss, and protrusion of the eyes. This condition is called *hyperthyroidism*. There are many causes of hyperthyroidism; one that has become well known is Graves' disease (**Figure 35.9**), a condition that both former President George H. W. Bush and former first lady Barbara Bush developed within two years of each other in the early 1990s.

The thyroid can also produce too little of the thyroid hormones, causing a person to feel run down and develop symptoms such as weight gain and slow growth of the hair and fingernails. This condition is called *hypothyroidism*. If the diet contains an insufficient amount of iodide, the thyroid gland enlarges forming a *hypothyroid goiter* (**Figure 35.10**). This condition is due to overstimulation by the anterior pituitary in response to low blood levels of thyroid hormone. Foods such as seafood and iodized salt are good sources of dietary iodide. Various factors can be the underly-

ing causes of both hyper- and hypothyroidism; often medication or surgery can correct the situation.

The thyroid gland secretes a hormone called *calcitonin*, or *CT*. This hormone works to balance the effect of another hormone called *parathyroid hormone*, or *PTH*. PTH and CT regulate the concentration of calcium in the bloodstream, as described in Section 35.8. Calcium is an important structural component in bones and teeth and aids in the proper functioning of nerves and muscles.

See Table 35.2, at the end of this chapter, which shows the hormones produced by the thyroid and the other major peripheral (nonbrain) endocrine glands discussed in Sections 35.7–35.11.

## 35.8 The parathyroids secrete hormones that regulate blood levels of ions such as calcium.

Embedded in the posterior side of the thyroid are the **parathyroid glands**. These glands secrete PTH, which works antagonistically to CT to help maintain the proper blood levels of various ions, primarily calcium.

PTH and CT work in the following way to keep calcium at an optimum level in the blood: If the calcium level is too low (**Figure 35.11**

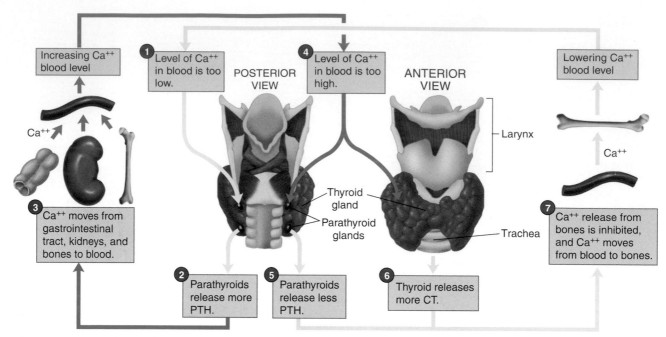

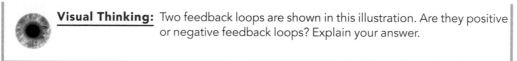

**Figure 35.11** Parathyroid hormone and calcitonin maintain proper calcium levels in the blood.

**Visual Thinking:** Two feedback loops are shown in this illustration. Are they positive or negative feedback loops? Explain your answer.

), the parathyroids are stimulated to release more PTH ❷. PTH stimulates the activity of osteoclasts, or bone-destroying cells. These cells liberate calcium from the bones and put it into the bloodstream. PTH also stimulates the kidneys to reabsorb calcium from urine that is being formed and stimulates cells in the intestines to absorb an increased amount of calcium from digested food ❸. CT acts in opposition to PTH. When the level of calcium in the blood is too high ❹, the parathyroids secrete less PTH ❺ and the thyroid secretes more CT ❻. The CT inhibits the release of calcium from bone and speeds up its absorption ❼, decreasing the levels of calcium in the blood. These interactions of PTH and CT are an example of a negative feedback loop that does not involve the hypothalamus or pituitary gland. The level of calcium in the blood directly stimulates the thyroid and parathyroid glands.

Two of the many problems related to abnormal calcium levels in the blood are kidney stones and osteoporosis. If calcium levels in the blood remain high, tiny masses of calcium may develop in the kidneys. These masses called kidney stones can partially block the flow of the urine from a kidney (see Figure 31.11). If calcium levels in the blood remain low, calcium may be removed from the bones (see Figure 34.9), a disorder known as osteoporosis. Osteoporosis is most common in middle-aged and elderly women, who have stopped secreting estrogens at menopause (see Chapters 34 and 36).

## 35.9 The adrenals secrete stress hormones and those that regulate the level of water in the blood.

The two **adrenal glands** (uh-DREE-nul) are named for their position in the body: above (*ad*, meaning "near") the kidneys (*renal*,

meaning "kidney"). Each of these triangular glands has two parts (**Figure 35.12**). The **adrenal cortex** is the outer, yellowish portion of each adrenal gland. The word "cortex" comes from a Latin word meaning "bark" and is often used to refer to the outer covering of a tissue, organ, or gland. The **adrenal medulla** (muh-DULL-uh) is the inner, reddish portion of the gland and is surrounded by the cortex. Not surprisingly, the word "medulla" comes from a Latin word meaning "marrow" or "middle."

### The adrenal cortex

As you may recall, the anterior pituitary gland secretes the hormone ACTH, adrenocorticotropic hormone. This hormone, as its name implies, stimulates the adrenal cortex to secrete a group of hormones known as **corticosteroids** (KORT-ik-oh-STARE-oydz). These steroid hormones act on the nucleus of target cells, triggering the cell's DNA to direct the manufacture of certain proteins. The two main types of corticosteroids produced by the adrenal cortex are the *mineralocorticoids* (MIN-er-al-oh-KOR-tih-koidz) and the *glucocorticoids* (GLOO-ko-KOR-tih-koidz).

The mineralocorticoids are involved in regulating the levels of certain ions within the body fluids. The most important of this group of hormones is *aldosterone* (al-DOS-ter-own). It affects tubules within the kidneys, stimulating them to reabsorb sodium ions and water from the urine that is being produced, putting these substances back into the bloodstream. The secretion of aldosterone is triggered when the volume of the blood is too low, such as during dehydration or blood loss. Special cells in the kidneys monitor the blood pressure. When the blood pressure drops, these cells secrete an enzyme that begins a chain of reactions ending with the secretion of aldosterone. Conversely, when the blood

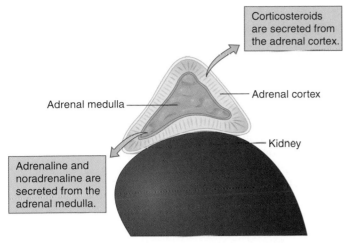

Corticosteroids are secreted from the adrenal cortex.

Adrenal medulla

Adrenal cortex

Kidney

Adrenaline and noradrenaline are secreted from the adrenal medulla.

**Figure 35.12** The adrenal glands are located atop the kidneys. The adrenal cortex and the adrenal medulla secrete hormones in response to stress.

pressure is within a normal range, the cellular detectors in the kidneys are not stimulated, the release of aldosterone is not triggered, and the kidney tubules are not stimulated to conserve sodium and water.

The glucocorticoids affect glucose metabolism, causing molecules of glucose to be manufactured in the body from noncarbohydrates such as proteins. This glucose enters the bloodstream, is transported to the cells, and is used for energy as part of the body's reaction to stress.

Almost everyone is familiar with the term *stress* and can give examples of stressful situations: their boss "chewing them out" in front of co-workers, their kids fighting constantly with one another, or their sustaining a physical injury. The stress reaction was first described in 1935 by Hans Selye, an authority on stress. Selye explained how the body typically reacts to stress—any disturbance that affects the body—and called this reaction the *general adaptation syndrome*. Over a prolonged period of stress, the body reacts in three stages: (1) the alarm reaction, (2) resistance, and (3) exhaustion. Contrary to maintaining homeostasis within the body, the general adaptation syndrome works to help the body gear up to meet an emergency.

During the alarm reaction, the body goes into quick action. Imagine waking up late one morning during final exam week and realizing you've missed your biology final. Your body reacts with a quickening pulse, increased blood flow, and an increased rate of chemical reactions within your body. Why does your body react in this way? Although the adrenal cortex is involved in the stress reaction by manufacturing glucose, the adrenal medulla, the inner section of the adrenal glands, plays a central role.

## The adrenal medulla

The adrenal medulla is different from most other endocrine tissue in that its cells are derived from cells of the peripheral nervous system, similar to the neurosecretory cells of the posterior pituitary. The sympathetic division of the autonomic nervous system, which controls involuntary or automatic responses, stimulates the adrenal medulla. Tropic hormones do not trigger these cells.

The principal hormone made by the adrenal medulla is *adrenaline*, also called *epinephrine*. This hormone is primarily responsible for the alarm reaction, along with a similar hormone secreted by the adrenal medulla called *noradrenaline*. The hypothalamus is responsible for sending the alarm signal via the autonomic nervous system to the adrenal medulla. The hypothalamus picks up the alarm signal as it monitors changes in the emotions via its neural connections with the emotional centers in the cerebral cortex. It can therefore sense when the body perceives an emotional stress. It can also sense physical stress, such as cold, bleeding, and poisons in the body.

The hypothalamus initially reacts to stress by readying the body for fight or flight; it first triggers the adrenal medulla to secrete adrenaline. Adrenaline and noradrenaline cause the heart rate and breathing to quicken, the rate of chemical reactions to increase, and glucose, which is stored in the liver, to be released into the bloodstream.

A summary of the first stage of the stress reaction—the alarm reaction—is shown in **Figure 35.13**. For example, when a hiker sees a bear, the fear of bears is detected by the hypothalamus. The hypothalamus triggers the anterior pituitary to secrete ACTH **1**, which stimulates the adrenal cortex to secrete glucocorticoids **2**. The glucocorticoids trigger the breakdown of glycogen to glucose **3**. The hypothalamus also sends out a nerve signal **4** that triggers the adrenal medulla to secrete adrenaline and noradrenaline **5**. Adrenaline and noradrenaline increase the heart rate, breathing rate, and blood sugar level by releasing glucose from the liver **6**.

During the second stage of the stress reaction, the resistance stage, the hypothalamus triggers continuing responses by releasing regulating factors. These factors stimulate the pituitary to release

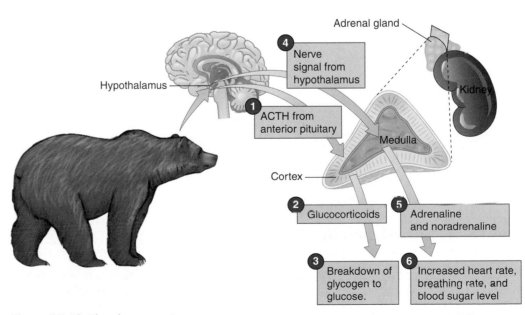

**Figure 35.13** The alarm reaction to stress.

ACTH, TSH, and growth hormone (GH). TSH stimulates the thyroid to secrete thyroid hormones, which stimulate the liver to break down stored glycogen to glucose. GH also stimulates the liver to produce glucose from glycogen, providing the body with an abundant energy source. ACTH stimulates the adrenal cortex to secrete both mineralo-corticoids and glucocorticoids. The mineralocorticoids cause the body to retain sodium ions and water, raising the blood pressure and providing more blood volume in the case of blood loss. The gluco-corticoids also promote the production of glucose.

If a person continues to be highly stressed over a long time, the body may lose the fight and enter the third stage of the stress reaction: exhaustion. This stage is serious and can result in death. One cause of exhaustion is the loss of potassium ions, which are excreted when sodium ions are retained under the influence of mineralocorticoids. This loss severely affects the body's ability to function properly. Another cause is depletion of the glucocorticoids, resulting in a sharp drop in the blood glucose level. The organs also become weak and may cease to function.

## 35.10 The pancreas secretes hormones that regulate the level of glucose in the bloodstream.

The pancreas, located alongside the stomach, is two glands in one: an exocrine gland and an endocrine gland. As an exocrine gland, it secretes the digestive enzymes discussed in Chapter 27. As an endocrine gland, it secretes the hormones *insulin* and *glucagon*.

The endocrine portion of the pancreas consists of clusters of cells that lie among the exocrine cells. For this reason, these cells are called islets—the **islets of Langerhans** (**Figure 35.14a**). Sep-

arate types of cells within the islets produce insulin and glucagon. These hormones act antagonistically to one another to regulate the level of glucose in the bloodstream. Glucagon increases the blood glucose level by triggering the liver to convert stored carbohydrates (glycogen) into glucose and to convert other nutrients such as amino acids into glucose. Insulin decreases the blood glucose level by helping body cells transport glucose across their membranes. In addition, insulin acts on the liver to convert glucose into glycogen and fat for storage, as shown in Figure 35.14b.

Diabetes mellitus is a set of disorders in which a person tends to have a high level of glucose in the blood. The underlying cause is the lack or partial lack of insulin or the inability of tissues to respond to insulin. The high levels of glucose in the blood result in water moving out of the body's tissues by osmosis. The kidneys remove this excess water through increased urine production, which causes excessive thirst and dehydration.

The primary types of diabetes mellitus are termed type 1 and type 2 diabetes. Insulin-dependent (type 1) diabetes was formerly known as juvenile diabetes because it strikes chiefly in childhood. Non-insulin-dependent (type 2) diabetes was known as adult onset diabetes because it strikes primarily in adulthood, although recently the disease is appearing in some obese children and adolescents as well. This disease is described in the *Just Wondering* box.

Type 1 diabetes mellitus is a disorder in which the pancreas does not produce adequate amounts of the hormone insulin or the body becomes insensitive to it. Affecting about 5% of diabetics, type 1 diabetes is thought to be a hereditary disease in which the body's immune system attacks the pancreas, destroying its ability to produce insulin. As a result, muscle and fat cells are prevented from absorbing sugar from the blood; the sugar is excreted in urine while the undernourished cells literally starve. The high levels of

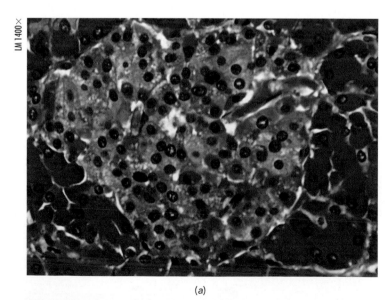

(a)

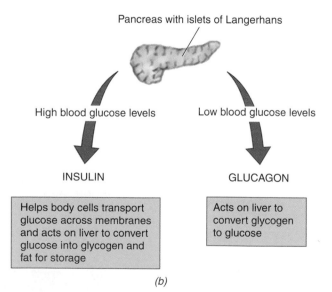

(b)

**Figure 35.14 Functions of insulin and glucagon.** (a) The islets of Langerhans, which secrete insulin and glucagon, are the lighter-stained cells. Pancreatic cells that secrete digestive enzymes are the darker-stained cells. (b) This illustration shows the opposing functions of insulin and glucagon, hormones that work together to maintain steady blood glucose levels.

## *just wondering . . .*

### What is adult onset diabetes, and why do people suddenly develop it?

Adult onset or maturity onset diabetes is called type 2 diabetes. Type 2 diabetes is the most common form of diabetes mellitus. Approximately 90% of all cases are of this type, and it affects about 17 million Americans.

Some persons with type 2 diabetes make insulin but not enough to meet their needs. Others have sufficient insulin, but the cells lack enough insulin receptors, so they cannot use the insulin they make. Without enough insulin, the body cannot move blood sugar (glucose) into the cells. Therefore, the level of blood glucose rises and can cause problems with the kidneys, legs and feet, eyes, heart, nerves, and blood flow. If left untreated, these problems can lead to kidney failure, gangrene and amputation, blindness, and stroke.

Type 2 diabetes only seems to develop suddenly because its symptoms are not always apparent. Many people have no symptoms, or they have such mild symptoms that they do not notice them. Older persons often ignore their symptoms, thinking they are just part of the aging process. Persons at highest risk for devel-

oping type 2 diabetes are those who are over 45, overweight, and have a family history of the disease. Researchers know that a person can inherit a tendency to get type 2 diabetes. However, it usually takes another factor such as obesity to bring on the disease.

Common symptoms of type 2 diabetes are an increased thirst and increased urination. Many persons also feel edgy, tired, and nauseous. Some people have an increased appetite, but they lose weight. Other symptoms of type 2 diabetes are repeated or hard-to-heal infections of the skin, gums, vagina, or bladder; blurred vision; tingling or loss of feeling in the hands or feet; and dry, itchy skin.

Individuals with type 2 diabetes must control their blood sugar levels by following their physician's instructions regarding diet and exercise. Persons who cannot control their blood sugar level sufficiently with these measures may have to take diabetes pills or insulin injections to help prevent problems. For more information on this disease, contact the branch of the American Diabetes Association in your area. ●

*Are you wondering about a topic in biology and how it relates to your life? Submit your question by clicking the* Just Wondering *link in this text's companion Web site at www.wiley.com/college/alters.*

---

blood glucose cause the thickening of capillary and artery walls, constricting blood flow and damaging critical organs.

Diabetes is treated by insulin injections, usually twice a day. As a protein, insulin would be destroyed in the stomach if administered as a pill. However, taking insulin injections is only part of the treatment for type 1 diabetes. Personal management plans help avoid the complications of diabetes. Such a plan usually includes nutrition and exercise recommendations, glucose monitoring instructions, and a medical plan to monitor organs and body systems that may be affected by type 1 diabetes.

Research on transplanting islets of Langerhans for persons who do not produce enough insulin, especially those with type 1 diabetes mellitus, holds much promise as a lasting treatment for this disease. Researchers are currently experimenting with ways to encapsulate islet cells collected from rats so that they will not trigger an immune response. Transplantation from cadavers and fetal cell transplantation holds promise as well (see the opening story to Chapter 12).

### 35.11 Other endocrine glands include the thymus, ovaries, testes, and digestive hormone–secreting cells.

The thymus gland is a small gland located in the neck a few inches below the thyroid (see Figure 35.2). In the thymus, certain immune system cells called *T lymphocytes* develop the ability to identify invading bacteria and viruses (see Chapter 30). The thymus produces a variety of hormones called thymosins to promote the maturation of

these cells. This gland is quite active during childhood, but is replaced by fat and connective tissue by the time a person reaches adulthood.

The ovaries produce female sex cells, or eggs, and the testes produce male sex cells, or sperm. In addition, the ovaries produce estrogens and progesterone, and the testes produce testosterone. These organs become active during puberty, the time of sexual maturation. These endocrine glands and the roles of their hormones are discussed in Chapter 36.

Certain cells in the walls of the stomach and small intestine secrete hormones that regulate the release of digestive juices by various cells. In addition, they regulate other processes related to digestion such as the emptying of the stomach, contracting of the gallbladder, and inducing mass movements in the large intestine. **Table 35.2** reviews the peripheral (nonbrain) endocrine glands and their hormones described in Sections 35.7–35.11.

### CONCEPT CHECKPOINT

4. How does the body's alarm response to stress demonstrate the structural and functional relationship between the nervous system and the endocrine system?

5. What is the hormonal response to the rise in the concentration of glucose in the blood following a large meal?
   a. Both glucagon and insulin are released.
   b. Glucagon is released, but not insulin.
   c. Insulin is released, but not glucagon.
   d. Neither glucagon nor insulin is released.

| TABLE 35.2 | | Major Peripheral (Nonbrain) Endocrine Glands and Their Hormones |
|---|---|---|
| **Endocrine Gland and Hormone** | **Target Tissue** | **Principal Actions** |
| **Thyroid** | | |
| Thyroid hormones | General | Regulate metabolism |
| Calcitonin (CT) | Bone | Regulates calcium levels in blood |
| **Parathyroid** | | |
| Parathyroid hormone (PTH) | Bone, kidney, small intestine | Regulates calcium levels in blood |
| **Adrenal Cortex** | | |
| Aldosterone | Kidney | Increases sodium and water reabsorption and potassium excretion |
| Glucocorticoids | General | Stimulate breakdown of glycogen to glucose |
| **Adrenal Medulla** | | |
| Adrenaline and noradrenaline | Heart, blood vessels, liver, fat cells | Regulate fight-or-flight response: increase cardiac output, blood flow to muscles and heart, conversion of glycogen to glucose |
| **Pancreas (Islets of Langerhans)** | | |
| Insulin | Liver, skeletal muscle, fat | Decreases blood glucose levels by stimulating movement of glucose into cells |
| Glucagon | Liver | Increases blood glucose levels by converting glycogen to glucose |
| **Ovary** | | |
| Estrogens | General, female reproductive organs | Stimulate development of secondary sex characteristics in females; control monthly preparation of uterus for pregnancy |
| Progesterone | Uterus | Completes preparation of uterus for pregnancy |
| | Breasts | Stimulates development |
| **Testis** | | |
| Testosterone | General | Stimulates development of secondary sex characteristics in males and growth spurt at puberty |
| | Male reproductive structures | Stimulates development of sex organs, spermatogenesis |
| **Thymus** | | |
| Thymosins | T lymphocytes | Promotes the maturation of T lymphocytes |

# CHAPTER REVIEW

## Summary of Key Concepts and Key Terms

### What are hormones, and how do they work?

**35.1 Hormones** (p. 612) are the chemical products of cells, secreted into the bloodstream, and used as messengers that affect other cells within the body.

**35.1** In vertebrates such as humans, the hormonal system is referred to as the **endocrine system** (ENN-doe-krin, p. 612).

**35.1 Endocrine glands** (p. 612) are ductless glands that secrete hormones; **exocrine glands** (EK-so-krin, p. 612) secrete substances via ducts.

**35.1** The endocrine system generally exerts slower control than the nervous system.

**35.2** A wide variety of organisms have hormonal control mechanisms, including invertebrates, protists, and plants.

**35.2** Hormones are ancient molecules that first appeared in simple organisms evolving hundreds of millions of years ago.

**35.3** The 10 major endocrine glands make over 50 different hormones.

**35.3** Hormones affect cells in one of four ways: by regulating their secretions and excretions, by helping them to respond to changes in the environment, by controlling activities related to reproductive processes, and by influencing their proper growth and development.

**35.4** The two primary types of endocrine hormones are peptide hormones and steroid hormones.

**35.4** The production of hormones is regulated by feedback mechanisms in the following way: an endocrine gland is initially stimulated by the nervous system or by the concentration of various substances in the bloodstream. The gland secretes a hormone that interacts with target cells, and the effect produced by the target cells acts as a new stimulus to increase or decrease the amount of hormone produced.

## Where are human hormones produced, and what do they do?

**35.5** The **pituitary gland** (puh-TOO-ih-tare-ee, p. 615), which hangs from the underside of the brain, secretes nine major hormones: seven from its anterior lobe and two produced by the hypothalamus that are released from its posterior lobe.

**35.5** The **hypothalamus** (HYE-poe-THAL-uh-muss, p. 615) is a small mass of brain tissue lying above the pituitary; the hypothalamus controls the anterior pituitary by means of releasing hormones.

**35.5** Four of the hormones produced by the anterior pituitary control other endocrine glands; they are called **tropic hormones** (p. 616); of the four tropic hormones, two are **gonadotropins** (GON-ah-duh-TROP-inz, p. 616), hormones that affect the sex organs.

**35.5** In addition to tropic hormones, the anterior pituitary secretes growth hormone, which works with the thyroid hormones to control normal growth; prolactin, which works with other female sex hormones to stimulate the mammary glands to secrete milk after childbirth; and melanocyte-stimulating hormone, which affects certain pigment-producing cells of the body.

**35.5** The posterior pituitary secretes two hormones produced by the hypothalamus: Antidiuretic hormone helps control the volume of the blood by regulating the amount of water reabsorbed by the kidneys, and oxytocin is secreted during the birth process and enhances uterine contractions.

**35.6** The primary endocrine secretion of the **pineal gland** (PIN-ee-uhl, p. 618), located at the base of the brain, is melatonin, a hormone that regulates the sleep/wake cycle in humans and certain seasonal behavior patterns in other animals.

**35.7** The **thyroid** gland (p. 618), located in the neck near the voice box, produces hormones that regulate the body's metabolism.

**35.8** The **parathyroid glands** (p. 619), located on the underside of the thyroid, secrete parathyroid hormone, which works in an antagonistic manner with one of the thyroid hormones, calcitonin, and helps maintain the proper blood level of calcium.

**35.9** The **adrenal glands** (uh-DREE-nul, p. 620), located on top of the kidneys, have two different secretory portions: the outer cortex and the inner medulla.

**35.9** The **adrenal cortex** (p. 620) secretes hormones known as **corticosteroids** (KORT-ik-oh-STARE-oydz, p. 620) that regulate the levels of certain ions, water, and glucose in the bloodstream.

**35.9** The **adrenal medulla** (muh-DULL-uh, p. 620) secretes adrenaline and nonadrenaline, which ready the body for action during times of stress.

**35.9** The general adaptation syndrome is a stress reaction with three stages: the alarm reaction, resistance, and exhaustion.

**35.10** The endocrine portion of the pancreas, the **islets of Langerhans** (p. 622), secretes two hormones, insulin and glucagon, that act antagonistically to one another, regulating the level of glucose in the bloodstream.

**35.11** The thymus, located in the neck and most active in childhood, produces a variety of hormones to promote the maturation of certain immune system cells.

**35.11** The ovaries and testes, which become active during puberty, produce the sex hormones.

**35.11** Cells in the walls of the stomach and small intestine secrete digestive hormones.

---

## Level 1     Learning Basic Facts and Terms

**Multiple Choice**

1. Hormones are used for all of the following except:
   a. Stress management.
   b. Regulation of metabolism.
   c. Regulation of the level of calcium in the blood.
   d. Stimulation of growth.
   e. Thermoregulation.

2. All of the following statements about hormones are correct except:
   a. They are produced by endocrine glands.
   b. They travel to different areas of the body.
   c. They are carried in the circulatory system.
   d. They elicit specific biological responses from target cells.
   e. They are used to communicate between different organisms.

3. All of the following statements about the hypothalamus are correct except:
   a. It functions as an endocrine gland.
   b. It is part of the central nervous system.
   c. It is subject to feedback inhibition by certain hormones.
   d. It secretes tropic hormones that act directly on the gonads.
   e. Its neurosecretory cells terminate in the posterior pituitary.

4. Which endocrine gland is part of the brain?
   a. Ovary           d. Thyroid
   b. Adrenal        e. Pancreas
   c. Pituitary

5. Which hormone is correctly paired with its action?
   a. Oxytocin—stimulates uterine contractions during childbirth.
   b. Calcitonin—stimulates conversion of glycogen to glucose in liver cells.
   c. Glucagon—stimulates synthesis of glycogen from glucose.
   d. ACTH—stimulates release of adrenaline from adrenal cortex.
   e. Parathyroid hormone (PTH)—stimulates production of red blood cells in bone marrow.

**True–False**

6. _____ Most of the hormones produced by the anterior pituitary are tropic hormones.

7. _____ Parathyroid hormone and calcitonin are antagonistic hormones much in the same way that insulin and glucagon are antagonists.

8. _____ There are two types of neurosecretory cells of the hypothalamus, one produces releasing hormones and the other produces tropic hormones.

9. _____ Hormones that play a role in regulating blood glucose levels include insulin, glucagon, glucocorticoids, adrenaline, noradrenaline, and thyroid hormones.

10. _____ The target organs for the gonadotropins are the mammary glands.

## Level 2 | Learning Concepts

1. Why do only certain cells in the body respond to insulin?
2. How do the "messages" sent by the endocrine system differ from those carried by the nervous system?
3. Describe how a feedback loop regulates the production of a hormone. What is the difference between a negative feedback loop and a positive feedback loop?
4. If the fluid content of the blood is low, where would this information initially be received and processed? What hormone would be released, and from which endocrine gland would it be released?

5. What do tropic hormones and releasing hormones share in common? How are they different?
6. Which two hormones regulate calcium ion levels in your blood? Describe how they do it.
7. Describe the three stages of the general adaptation syndrome. What hormones are involved?
8. Which hormones regulate the level of glucose in your blood? Describe how they do this. Where are they produced?

## Level 3 | Critical Thinking and Life Applications

1. In the past, people living in remote inland areas often suffered from goiter. Assess the reasons for increased goiter problems in such places.
2. a. Chronic (long-term) emotional stress has been linked to a greater than normal risk of hypertension (high blood pressure). People with high blood pressure are often given a class of drugs called diuretics. One of these drugs inhibits the secretion of mineralocorticoids by the adrenal cortex. Why would this drug treat high blood pressure? What effect would you predict the drugs would have on urine volume?

b. What is another hormone on which diuretics might act to reduce blood pressure? Would diuretics inhibit or enhance the secretion of this hormone?
3. Sometimes parents refer to their middle-schooled-aged children as being "containers of raging hormones." To what hormones are the parents referring? Why would parents consider these hormones to be "raging" in their children? What effects would these hormones have on adolescents?

## In The News | Critical Thinking

**TEEN STEROIDS**

Now that you understand more about hormones, reread this chapter's opening story about research into the effects of anabolic steroid use on behavior. To better understand this research, it may help you to follow these steps:

1. Review your immediate reaction to Melloni's research on the effects of anabolic steroids on the behavior and brain chemistry of hamsters that you wrote when you began reading this chapter.
2. Based on your current understanding, again summarize the main point of the research in a sentence or two.
3. What questions do you now have about this research that this chapter's opening story does not answer?
4. Collect new information about the research. Visit the *In The News* section of this text's companion Web site at www.wiley.com/college/alters

and watch the "Teen Steroids" video. Then use the "summary" link to read the accompanying story and access related links. Use this information, the links provided, and other online and library resources to answer your questions and find updates about this research topic. State the sources of your information. Explain why you think the information is accurate. Also determine whether the information expresses a particular point of view or is biased in any way.

5. What in your view are the most significant aspects of this research? Give reasons for your opinion and for any changes in your ideas based on the additional information you have collected and the analysis you have done.

# SEX, REPRODUCTION, AND DEVELOPMENT

## In The News | Smoking and Pregnancy

Among Americans, smoking cigarettes is not as widespread or socially acceptable as it was in the past. Nevertheless, about 23% of adult Americans still smoke cigarettes regularly, and this practice results in serious health concerns not only for those who smoke but for others as well. One major public health concern is smoking during pregnancy and its effects on the fetus.

A recent U.S. survey indicated that 12% of women who gave birth during the previous year had smoked while they were pregnant. Pregnant women who smoke are 1.5 to 3.5 times more likely to deliver low-birth-weight babies than nonsmoking pregnant women. A low-birth-weight newborn weighs 5.5 pounds or less and has a higher risk of having serious or even fatal health problems than a newborn whose birth weight is within the normal range. Moreover, women who smoke during pregnancy have a higher risk of developing *placenta previa*, a condition in which the placenta is located abnormally within the uterus, which can result in life-threatening problems for the mother during delivery. The CDC notes that smoking during pregnancy is the "single most preventable cause of illness and death among mothers and infants."

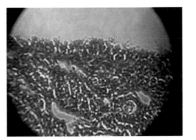

Smoking also reduces fertility in the female offspring of mothers who smoke. Scientists think they know why this occurs. Normally, about 70% of the 1 to 2 million oocytes (potential eggs) in a female newborn's ovaries undergo *apoptosis* (cellular death) before she reaches puberty. Exposure to polycyclic aromatic hydrocarbons (PAHs) during prenatal development hastens oocyte apoptosis in the female progeny. PAHs are a group of chemicals, including DMBA, that form during the incomplete combustion of organic substances such as tobacco.

Jonathan Tilly, a biology professor at Harvard Medical School, and his research team conducted experiments to learn more about how cigarette smoking reduces fertility in offspring of mothers who smoke. In one series of experiments, they treated a group of pregnant mice with an injection of DMBA in corn oil and a control group with just corn oil. After the animals gave birth, the scientists removed the ovaries of the female offspring and analyzed the tissues seen in the photo. They found significantly reduced oocyte numbers and an accumulation of a protein called Bax in ovaries of offspring whose mothers received the DMBA treatment but not in offspring of the control group. To investigate this phenomenon further, the scientists treated pregnant mice that lacked the Bax gene (so were unable to synthesize the Bax protein) with DMBA. They found no reduction in oocytes in the mutant female offspring's ovaries. The scientists suspect that Bax is involved in PAH-induced oocyte aptosis, but more research is needed to clarify its role. To learn more about Tilly's research into the effects of smoking on the fertility of female offspring, visit the *In The News* section of this text's companion Web site at www.wiley.com/college/alters and view the "Smoking and Pregnancy" video.

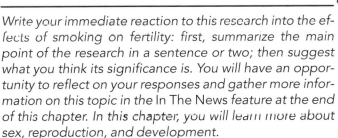

*Write your immediate reaction to this research into the effects of smoking on fertility: first, summarize the main point of the research in a sentence or two; then suggest what you think its significance is. You will have an opportunity to reflect on your responses and gather more information on this topic in the* In The News *feature at the end of this chapter. In this chapter, you will learn more about sex, reproduction, and development.*

## CHAPTER GUIDE

### What are the various ways in which organisms reproduce?

**36.1** Reproduction takes place both sexually and asexually.

**36.2** Sexual reproduction is prevalent in animals.

**36.3** Human reproduction is sexual, and fertilization is internal.

**36.4** Human males produce sperm and transport the sperm to the female reproductive tract.

**36.5** Human females produce eggs, receive sperm, and nourish the developing fetus.

**36.6** The human female reproductive tract prepares monthly for reproduction.

**36.7** The human sexual response can be described as having four phases.

### How can pregnancies be limited and sexually transmitted infections be avoided?

**36.8** The use of birth control methods can limit pregnancies.

**36.9** Some birth control methods can also protect against sexually transmitted infections.

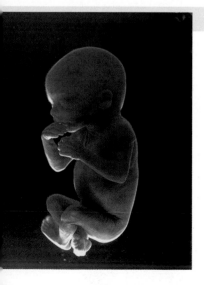

### What happens during human development and birth?

**36.10** Human development proceeds from zygote to fetus.

**36.11** Hormonal changes in the fetus trigger the birth process.

**36.12** The newborn makes many physiological adjustments.

# What are the various ways in which organisms reproduce?

## 36.1 Reproduction takes place both sexually and asexually.

Organisms reproduce in a variety of ways, and many of these ways do not include "having sex." Similarly, "having sex" may have nothing to do with reproduction. Let's first define some terms. **Sexual reproduction** usually refers to a process that involves the union of a male and a female sex cell, forming the first cell of a new individual. However, some organisms capable of sexual reproduction, such as the protist *Paramecium*, do not produce **gametes**, (GAM-eets; sex cells); instead, they transfer genetic material between organisms or undergo a fusion of nuclei of different mating types as part of the process of producing new individuals. Furthermore, not all organisms reproduce sexually. Many organisms reproduce by **asexual reproduction**, the generation of new individuals without the union of gametes or transfer of genetic material. With the exception of mutations that arise spontaneously, asexual reproduction is a cloning process—a method of replication of a parental organism.

What are the advantages and disadvantages of each process for the survival of populations and species? A disadvantage of asexual reproduction is genetic sameness. If environmental conditions become harmful for individuals of the cloned genetic makeup, or *genotype*, then the entire population could be at risk. Sexual reproduction, in contrast, is advantageous because it provides genetic variability among organisms of the same species. When environmental conditions change for populations with greater genetic variability, there is a greater chance that some individuals will survive. The capacity of some organisms to survive and reproduce in environments in which others of their species cannot survive and reproduce is, in turn, the basis of natural selection. Natural selection is a process in which organisms with adaptive traits survive in greater numbers than organisms without such traits. The concepts of adaptation and natural selection are explained in more detail in Chapters 15–18.

An advantage of asexual reproduction is that individual organisms can produce new organisms, adding to the population. This can increase the population size quite rapidly in many cases. For example, bacteria can generate a population of billions in little more than a day from one original cell. Producing a large population from few individuals can result in excluding potential competitors and flooding the population with a particularly successful genotype. A disadvantage of sexual reproduction is the inability to increase population size quickly and dramatically in this way.

Asexual reproduction occurs in all three domains of life and across the four kingdoms in the domain Eukarya. (See Chapter 1 for a discussion of domains and kingdoms.) In the Archaea and Bacteria—the prokaryotic domains—all reproduction is asexual. Most bacterial cells reproduce by a process called *binary fission* (literally, "two splitting") (**Figure 36.1**). Before fission, or division of the cell, the genetic material replicates and divides. Some bacteria transfer genetic material from one cell to another, but this is not a form of reproduction because additional cells do not result.

In the eukaryotic domain, asexual reproduction is common among the protists. Many of the protozoans (animallike protists) reproduce sexually as well. They have no gametes but exchange genetic material through a tube connecting the cytoplasm of two individuals before fission occurs. Many of the plantlike protists (the algae) reproduce primarily by fission, but most of the brown, green, and red algae have life cycles much like plants, which include both sexual and asexual stages. Similarly, the funguslike protists (slime molds and water molds) have life cycles with sexual stages.

In the kingdom of fungi, which includes yeasts, molds, and mushrooms, most organisms have a life cycle in which they reproduce both asexually and sexually. Asexual reproduction in fungi usually involves budding, growing new hyphae (slender filaments of cells) from fragments of parent hyphae, or producing *spores* that are contained within spore cases (**Figure 36.2**). Spores are reproductive bodies formed either by mitosis in a haploid parent or by meiosis in a diploid parent. (Haploid means an

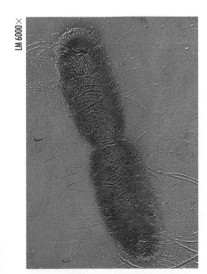

LM 6000×

**Figure 36.1 Binary fission.** Colorized electron micrograph of the bacterium *E. coli*, a normal inhabitant of the human intestine.

**Figure 36.2 Spore cases atop hyphae in fungi.** Hundreds of spores are in each spore case; spores are a means of asexual reproduction in fungi.

organism has only half the full amount of hereditary material of the original parent cell—one of each chromosome. Diploid means an organism has the full complement of its genetic material—two of each chromosome. See Chapter 12.) In fungal sexual reproduction, the haploid nuclei of two genetically different mating types fuse, forming a diploid zygote. A **zygote** (ZYE-goat) is a diploid cell formed by the fusion of two haploid cells or nuclei. Fungi do not form gametes.

How do organisms in the plant kingdom reproduce? The plant life cycle includes both asexual and sexual phases. A generalized life cycle of plants is illustrated in Figure 12.2. The sexual reproductive phase in plants' life cycles occurs when gametes are produced by haploid plants (gametophytes) by mitosis. Male and female gametes join during fertilization, forming diploid zygotes that grow into diploid plants (sporophytes). The asexual reproductive phase occurs when sporophytes produce spores by meiosis. The spores then germinate into haploid gametophytes. Chapter 23 focuses on the reproductive diversity of plants.

In the animal kingdom, asexual reproduction is quite common among the invertebrates. Some invertebrates, such as sponges, jellyfish, and sea anemones can reproduce asexually by budding, as shown in **Figure 36.3**. Other invertebrates, such as certain flatworms and sea stars, can reproduce by a process called **fragmentation**. If a part of the organism's body becomes detached, each of the two pieces can grow into a complete organism. The sea star shown in **Figure 36.4** grew from a detached arm—the long arm in the photo. Some flatworms, such as liver flukes and blood flukes, can reproduce asexually by fission when in the larval stage—an immature developmental form. The larvae simply split into two organisms, increasing the size of the population, which increases the probability that some of the parasites will find hosts in which to live.

Another type of asexual reproduction called **parthenogenesis** (PAR-theh-no-JEN-eh-sis) occurs in a variety of invertebrates, such as earthworms, mites, ticks, and beetles. During this process the eggs of the female develop into adults without being fertilized. Male bees called drones are produced this way, while female bees—both workers and queen bees—develop from fertilized eggs. Parthenogenesis even occurs among vertebrates. Certain

**Figure 36.4** Reproduction by fragmentation. Certain invertebrates can grow new individuals from detached body parts, like the sea star in this photo, which grew from a detached arm.

fishes, amphibians, birds, and lizards reproduce this way. Many species of whiptail lizards, for example, reproduce exclusively by parthenogenesis but engage in behaviors that simulate mating, as shown in **Figure 36.5**. The lizard on the top is simulating male behavior during mating in a closely related species of lizard that reproduces sexually.

---

## CONCEPT CHECKPOINT

1. For each of the following groups of organisms list the type(s) of asexual and sexual reproduction exhibited by members of that group: bacteria, protozoans, funguslike protists, plants, animals. You may find it helpful to organize this information in the form of a table.

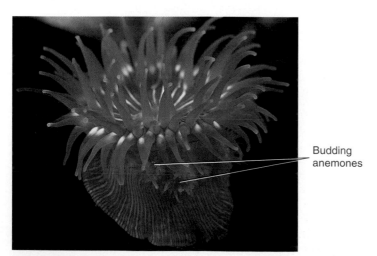

**Figure 36.3** Sea anemones can reproduce asexually by budding.

Budding anemones

**Figure 36.5** Many species of whiptail lizards reproduce asexually but engage in behaviors that simulate mating.

## 36.2 Sexual reproduction is prevalent in animals.

You probably know that sexual reproduction is prevalent among vertebrates—animals with a backbone, such as humans. Among invertebrates, sexual and asexual reproduction are equally common, and there are many variations. For example, among many invertebrates, a single individual, called a **hermaphrodite** (her-MAF-reh-dite), may be both male and female. In some cases, a hermaphroditic individual can reproduce sexually by self-fertilization. Usually, however, two individual organisms are involved.

Some species of invertebrates are *simultaneous hermaphrodites*; that is, an individual is both male and female at the same time. Earthworms, for example, are simultaneous hermaphrodites. **Figure 36.6** shows earthworms mating. They have aligned themselves so that they can exchange sperm, with each individual storing the other's sperm. The *clitellum* of each worm secretes a mucous covering that slips along the worm's body as it moves through the soil. The clitellum picks up eggs and stored sperm. These sacs then slip off each worm and remain in the soil, protecting the developing embryos.

*Sequential hermaphrodites* change their sex as they age. The American oyster, for example, first matures as a male and produces only sperm. The next year, it becomes female and produces eggs, and it continues to alternate its sex year by year. Certain fish change sex depending on dominance hierarchies (see Chapter 37) (**Figure 36.7**).

When two individuals are involved in sexual reproduction, which is most often the case in animals, the male gamete (sperm) must reach the female gamete (egg) in order for **fertilization** to take place, the union of a sperm and an egg. In **external fertilization**, the female sheds eggs and the male fertilizes them after they leave her body. In **internal fertilization**, the male deposits sperm in or near the female reproductive tract, and fertilization takes place within the body of the female.

External fertilization takes place in moist environments in which the eggs will not dry out and the sperm can swim. Aquatic invertebrates such as sponges, for example, simply shed their eggs and sperm into their watery surroundings at the same time. The individual organisms do not touch one another during this process. Aquatic vertebrates, such as fish and amphibians that use

**Figure 36.7** Hermaphrodites. Here, the green and black wrasse on the right is a dominant male, and the smaller brown and white fish on the left is a female. If the male dies, the female may become a dominant male.

external fertilization, show specific cooperative mating behaviors that result in one male fertilizing the eggs of one female. External fertilization in Southwestern toads is shown in **Figure 36.8**; the courtship behavior of stickleback fish, which leads to external fertilization, is illustrated in Figure 37.16. Internal fertilization requires cooperative behavior as well. Courtship behavior of the sage grouse, which leads to internal fertilization, is shown in Figure 37.15.

### CONCEPT CHECKPOINT

**2.** Compare and contrast external and internal fertilization.

Clitellum

**Figure 36.6** Earthworms mating.

**Figure 36.8** External fertilization in Southwestern toads. The female expels eggs into the surrounding water and the male then fertilizes them.

## 36.3 Human reproduction is sexual, and fertilization is internal.

Owing to the nature of sexual reproduction, the reproductive organs of sexually reproducing males and females are usually quite different. This distinction is apparent in humans, although there are some similarities because many male and female reproductive organs are derived from the same embryological (developmentally early) tissues.

The function of the human male reproductive system is to produce sperm cells and transport them to the female reproductive tract. The function of the human female reproductive system is to produce eggs, to receive sperm, and to facilitate their access to the eggs. Fertilized eggs are then housed and nourished within the female reproductive organs during development until birth.

The male and female reproductive organs, called **gonads**, produce the gametes (sex cells; eggs and sperm). Other organs and ducts store, transport, and receive the sex cells. Accessory glands help nourish the gametes and often produce hormones that facilitate gonad development, influence behavior, and affect other bodily processes.

## 36.4 Human males produce sperm and transport the sperm to the female reproductive tract

### The production of sperm: the testes

Sperm are produced in the male gonads, the **testes**. The testes (testicles) are located outside the lower pelvic area of the male, housed within a sac of skin called the **scrotum** (**Figure 36.9**). The place-

ment of the testes allows the sperm to successfully complete their development at a temperature slightly lower than the 37°C (98.6°F) internal temperature of the human body.

Each testis is packed with approximately 750 feet of tightly coiled tubes called **seminiferous tubules** (**Figure 36.10a**). Male gametes, or sperm cells, develop within the tissue lining these tubules during a process of meiosis and development called *spermatogenesis* (literally, "the making of sperm"). Spermatogenesis usually begins during the teenage years when a boy reaches sexual maturity, or *puberty*, and continues throughout his life. Sperm production is triggered by follicle-stimulating hormone (FSH).

Luteinizing hormone (LH) regulates the testes' secretion of testosterone, a sex hormone that plays a role in spermatogenesis and is responsible for the development and maintenance of male secondary sexual characteristics. These characteristics, such as a deepening of the voice and the growth of facial hair, begin to develop at puberty and are signs that sexual maturation is taking place. Both FSH and LH are secreted by the pituitary, a tiny gland located on the underside of the brain (see Chapter 35).

Packed within the walls of the seminiferous tubules are cells called *spermatogonia*, which give rise to sperm (Figure 36.10b). These cells have a diploid complement of hereditary material and constantly produce new daughter cells by the process of mitosis. Some of these daughter cells move toward the interior of the tubule and develop into primary spermatocytes (**Figure 36.11**). As part of this development process, the genetic material in these diploid cells duplicates. The primary spermatocytes then undergo meiosis I, producing secondary spermatocytes. The secondary spermatocytes undergo meiosis II, resulting in haploid spermatids, each with half the full amount of genetic material. Thus, each diploid original cell has produced four haploid cells. Each spermatid contains one member of each pair of the 23 chromosome pairs present in the cells of the human body.

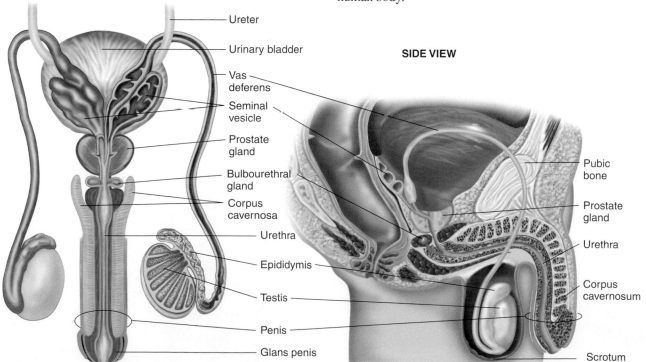

**FRONT VIEW**

- Ureter
- Urinary bladder
- Vas deferens
- Seminal vesicle
- Prostate gland
- Bulbourethral gland
- Corpus cavernosa
- Urethra
- Epididymis
- Testis
- Penis
- Glans penis

**SIDE VIEW**

- Pubic bone
- Prostate gland
- Urethra
- Corpus cavernosum
- Scrotum

**Figure 36.9** The human male reproductive system.

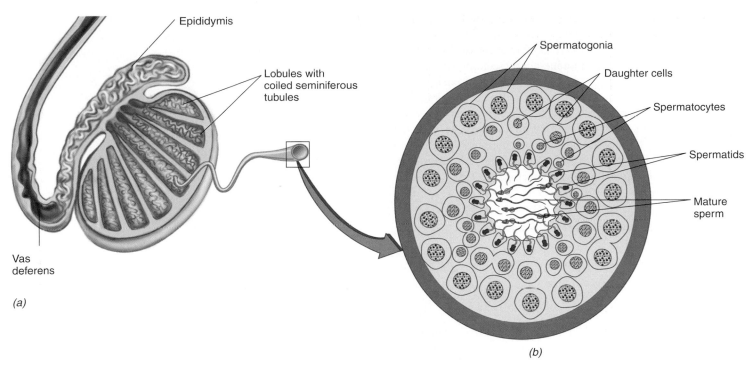

**Figure 36.10 The interior of the testis, site of spermatogenesis.** (*a*) Seminiferous tubules. (*b*) A cross section of a seminiferous tubule, in which diploid cells (spermatogonia) develop into haploid cells (sperm) (see Figure 36.11).

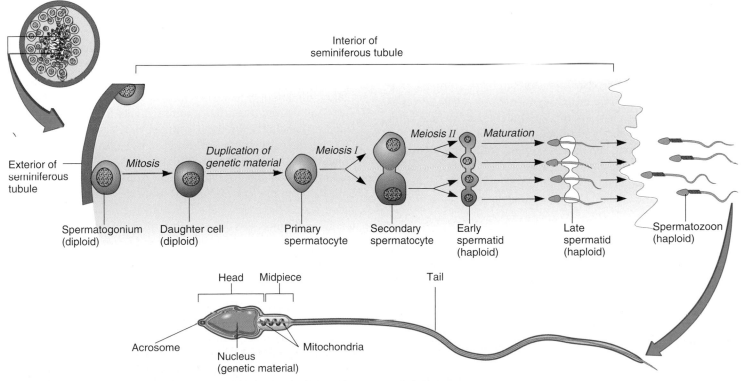

**Figure 36.11 Spermatogenesis.** Spermatogonia undergo mitosis, producing more spermatogonia (daughter cells). Some of these move to the interior of the seminiferous tubule and undergo meiosis. Meiotic cell division produces four haploid spermatids that mature into spermatozoa, or sperm.

The spermatids then undergo a process of development, producing sperm cells, or spermatozoa. Each has an anterior portion, or head, that consists primarily of a cell membrane encasing hereditary material. The *acrosome* is a membrane-bound section of the head of a sperm, located at its tip (see Figure 36.11). Derived from the Golgi body, the acrosome contains enzymes that aid in the penetration of the protective layers surrounding the egg. In addition, each **spermatozoon** has a whiplike tail, or flagellum, that propels the cell, and mitochondria, which produce adenosine triphosphate (ATP), from which a sperm derives energy that powers its flagellum. Sperm development takes about two months. On average, 300 million mature sperm are produced per day.

## Maturation and storage: the epididymis and vas deferens

After formation within the testes, the sperm move through its tubules to a nearby long, coiled tube called the **epididymis** (ep-ih-DID-uh-mis) (see Figure 36.10a), located on the back side of each testis. Here the sperm are stored for about 18 hours to 10 days, undergoing further development. The sperm then move toward the **vas deferens**, a long tube that ascends from the epididymis into the pelvic cavity, looping over the side of the urinary bladder (see Figure 36.9), and leading into the penis. Sperm retain their ability to fertilize an oocyte for a short while. If they are not ejaculated, "old" sperm are reabsorbed by the body to make room for "new" sperm.

## Nourishment of the sperm: the accessory glands

When a male ejaculates, the sperm are propelled through the vas deferens to the *urethra*, where the reproductive and urinary tracts join. On the way, accessory glands add fluid to the sperm (see Figure 36.9), producing an ejaculate called semen. The semen leaves the body through the urethral opening at the tip of the **penis**.

The first accessory glands to add fluid to the traveling sperm are the **seminal vesicles**. The fluid secreted by the seminal vesicles is primarily composed of the sugar fructose, which serves as a source of energy for the sperm, and bicarbonate, which helps buffer the acidity of the urethra and the vagina. Near its beginning, the urethra is surrounded by a gland called the **prostate**. About the size of a chestnut, the prostate adds a milky alkaline fluid to the semen. This fluid also helps neutralize the acidity of the urethra and the vagina. Sperm are unable to swim in an acid environment.

Just beneath the prostate lies a set of tiny, round accessory glands called the **bulbourethral glands** (BULL-bo-yoo-REE-thrul). These glands secrete a mucuslike alkaline substance that precedes the ejaculate. It serves as a lubricant for sexual intercourse and neutralizes the acidity of the urethra as well. This secretion can carry sperm to the outside before ejaculation and is one of the reasons that the withdrawal method of birth control is highly unreliable.

## The penis

The scrotum (which encloses the testes and epididymides) and the penis are the two male **external genitals** (external sexual organs). The penis is a cylindrical organ that transfers sperm from the male

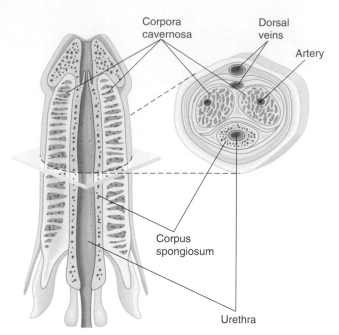

**Figure 36.12** A penis in cross section and longitudinal section. The three cylinders of spongy tissue that make up the penis are the corpus spongiosum (middle) and the corpora cavernosa (one cylinder on each side).

 **Visual Thinking:** The drug Viagra® works by increasing blood flow to the penis. Using this illustration and information from the text, explain how this mode of action enhances erections or makes erections possible in some men.

reproductive tract to the female reproductive tract. This organ is made up of three cylinders of spongy erectile tissue. Two veins run along the top (dorsal) surface of the penis. Beneath these veins, two of the cylinders of erectile tissue, the corpora cavernosa, sit side by side. And beneath them lies a third cylinder, the corpus spongiosum, surrounding the urethra (**Figure 36.12**).

Erectile tissue has spaces between its cells that normally contain a small amount of blood. During sexual stimulation, arterioles leading into this tissue dilate (widen), causing additional blood to collect within the spaces. Pressure from the increased volume of blood within the spaces causes the erectile tissue to become distended and compresses the veins that normally drain blood from the penis, causing it to become erect and rigid. Continued stimulation by the central nervous system is required for this erection to be maintained.

Erection can occur without any physical stimulation of the penis. However, physical stimulation usually is required for the ejaculation of semen, which corresponds to the culmination of sexual excitement, or orgasm (see Section 36.7). Prolonged stimulation of the penis leads first to the mobilization of the sperm. In this process, muscles encircling the vas deferens contract, moving the sperm into the urethra. Then ejaculation takes place: muscles at the base of the penis and within the walls of the urethra contract repeatedly, ejecting approximately a teaspoon of semen containing about 200 million sperm out of the penis.

**FRONT VIEW**  **SIDE VIEW**

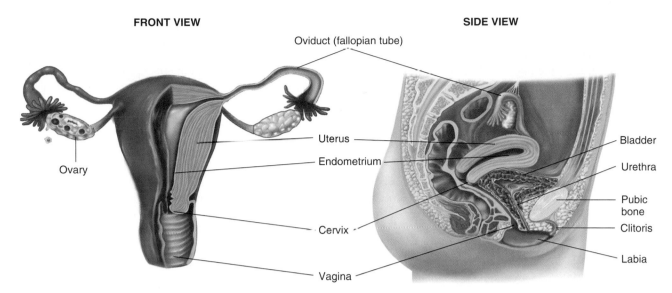

**Figure 36.13** The human female reproductive system.

 **Visual Thinking:** If a fertilized egg were to get "stuck" in the fallopian tube on its way to the uterus, and implanted in the tube, would the tube provide an appropriate place for the embryo/fetus to develop? Observe differences between the structure of the uterus and the structure of the fallopian tube as you formulate your answer.

## 36.5 Human females produce eggs, receive sperm, and nourish the developing fetus

### The vagina

The part of the female reproductive tract (**Figure 36.13**) that accepts the penis during sexual intercourse and receives the sperm is called the **vagina**, a muscular tube 9 to 10 centimeters (about 3 1/2 inches) long. The vagina is also the lower portion of the birth canal during childbirth and provides a passageway for the exit of the menstrual flow. The vagina can stretch considerably during intercourse and childbirth because its walls are composed of elasticlike smooth muscle tissue. The mucous membrane that lines the inside of the vagina produces a viscous fluid that keeps the vagina moist and provides lubrication during intercourse.

### The external genitals

The external genitals of a female surround the vaginal opening and are collectively called the **vulva** (**Figure 36.14**). The most anterior structure is the **mons pubis** (monz PYOO-bis). The word *mons* comes from a Latin word meaning "prominence" and refers to the mound of fatty tissue that lies over the place of attachment of the two pubic bones. Two longitudinal folds of skin called the **labia majora** (LAY-bee-uh muh-JORE-uh; *labia* means "lips") run posteriorly from the mons. These folds are homologous to the scrotum in the male; that is, they develop from the same embryonic tissue. Additional folds of skin called the **labia minora** are covered by the labia majora. Both sets of labia protect the vaginal and urethral openings beneath. Figure 36.14 shows that the urethra opens to the outside slightly in front of the vagina and is not connected to the reproductive system as it is in

males. In addition to the vaginal and urethral openings, the labia minora also cover the openings of two tiny ducts that produce a mucus-like secretion during sexual stimulation. This secretion helps lubricate the vagina and penis during sexual intercourse. Slightly anterior to the convergence of the labia minora is the **clitoris** (KLIT-uh-ris), a small mass of erectile and nervous tissue that responds to sexual stimulation. The clitoris is homologous to the penis in males.

### The production of eggs: the role of the ovaries

The female reproductive system usually produces one secondary oocyte (potential egg) each month. Secondary oocytes are produced in the female gonads, or **ovaries** (see Figure 36.13). These two almond-shaped organs are each about 3 centimeters long, approximately equal to the length of your thumb from the tip to where it bends. They are located in the lower portion of the abdominal cavity, called the *pelvic cavity*. The ovaries contain *primary oocytes*, cells that have the potential to develop into eggs. Supporting cells called *follicular cells* surround them. Together

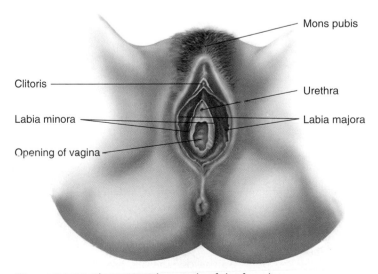

**Figure 36.14** The external genitals of the female.

follicular cells and an oocyte are called a follicle (see Figure 36.17).

The process that produces mature female sex cells, or eggs, is called *oogenesis* (literally, "the making of eggs"). At the time of this writing, understanding about some aspects of primary oocyte development was in flux. Groundbreaking research conducted by Jonathan Tilly and the team of researchers mentioned in this chapter's opening story suggests that our traditional understanding is incorrect. It has been widely accepted that (1) females at birth had all of the primary oocytes they would ever produce—about 1 to 2 million of them; and (2) as the woman aged, her potential eggs aged as well, until, by the age of 50 or so, no more viable primary oocytes existed. Tilly's research suggests that primary oocytes die relatively quickly—over three to five years—but are replenished with new primary oocytes. However, the production rate of new primary oocytes lags behind the death rate of older primary oocytes, so the pool of primary oocytes decreases gradually as a woman ages until she reaches approximately 50 years.

Primary oocytes are produced from diploid cells called *oogonia* (sing. *oogonium*). In this process, the genetic material duplicates, as in the process of spermatogenesis in males (see Section 36.4). About every 28 days after the onset of puberty, generally only one primary oocyte completes the first meiotic division and survives. As a result of this division, two cells of unequal size are produced (**Figure 36.15**). One is called the *secondary oocyte* and contains most of the cytoplasm of the primary oocyte. The other cell, called a *polar body*, contains little cytoplasm. The secondary oocyte is released from the follicle and breaks away from the ovary during **ovulation** and continues with the second meiotic division *only if it is penetrated by a sperm*. This second meiotic division produces a mature, haploid egg, or *ovum*, and a haploid polar body. The haploid nuclei of the sperm and the egg then fuse in the process of fertilization, producing a zygote, the first cell of a new individual. The polar body produced by the first meiotic division may undergo a second meiotic division or may degenerate. In any case, the polar bodies remaining after the second meiotic division soon degenerate. The follicle remnants left be-

hind in the ovary secrete hormones that support the development of the uterine lining in preparation for embryo implantation, should fertilization occur.

If the follicle containing the egg matures but the egg is not released, a fluid-filled sac called a *cyst* may form on the ovary. Most ovarian cysts are benign.

### Passage to the uterus: the oviducts

At ovulation, when the secondary oocyte is released from the ovary, it enters a passageway called an **oviduct**. In humans, the oviducts commonly are called the *fallopian tubes* (see Figure 36.13). The oviducts lead to the uterus, the organ in which a fertilized egg can develop. The uterus is shaped somewhat like an upside-down pear: wider at its top and narrower at its bottom. The narrower portion, called the **cervix** (or neck), opens into the vagina.

The journey from the ovary to the uterus is a slow one, taking about three days to complete. Because oocytes can be fertilized for only about 16 hours after ovulation, they must be fertilized while in the oviduct. In addition, sperm live for approximately six days. Therefore, fertilization can take place if sexual intercourse occurs within six days before to less than one day after ovulation. However, knowing exactly when ovulation will occur is usually a difficult task unless specific testing procedures are used, such as basal body temperature readings or cervical mucus changes.

If fertilization occurs, the mitotic division of the zygote begins as it moves along the oviduct. By the time the new embryo reaches the uterus, it is a tiny ball of cells. After a few days, it implants in the lining of the uterus to receive nourishment. Further development of the embryo is described later in this chapter.

### The site of prenatal development: the uterus

The uterus, the site of prenatal (pre-birth) development, sits above the urinary bladder and in front of the rectum and connects directly with the vagina (see Figure 36.13). Its walls are made up of thick

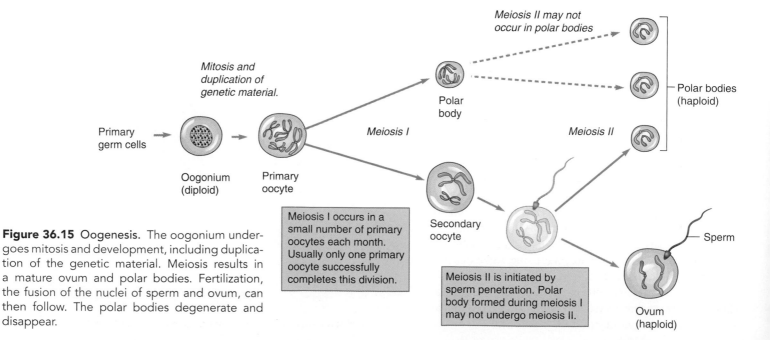

**Figure 36.15 Oogenesis.** The oogonium undergoes mitosis and development, including duplication of the genetic material. Meiosis results in a mature ovum and polar bodies. Fertilization, the fusion of the nuclei of sperm and ovum, can then follow. The polar bodies degenerate and disappear.

Meiosis II may not occur in polar bodies

Polar body

Polar bodies (haploid)

Mitosis and duplication of genetic material.

Primary germ cells

Oogonium (diploid)

Primary oocyte

*Meiosis I*

*Meiosis II*

Meiosis I occurs in a small number of primary oocytes each month. Usually only one primary oocyte successfully completes this division.

Secondary oocyte

Meiosis II is initiated by sperm penetration. Polar body formed during meiosis I may not undergo meiosis II.

Sperm

Ovum (haploid)

layers of smooth muscle that contract during the birth process and during menstruation. The interior of the uterus is hollow and provides a cavity for the development of a fertilized egg.

The inner lining of the uterus is called the **endometrium**. This lining has two layers: an underlying permanent layer and a functional transient layer that is in contact with the uterine cavity. The functional layer grows and thickens each month, readying itself for the implantation of an embryo. If an embryo is not present, the functional layer is shed. This monthly development and shedding of the functional layer of the endometrium (called the *menstrual cycle*) and the monthly maturation of an egg and its release (called the *ovarian cycle*) are both governed by levels of various female sex hormones. Together, these two cycles are commonly referred to as the *reproductive cycle*.

## 36.6 The human female reproductive tract prepares monthly for reproduction.

A small number of primary oocytes continue their development within an ovary each month (**Figure 36.16 ❶**). This development is triggered by follicle-stimulating hormone (FSH), released by the pituitary gland in the brain (see Chapter 32). After they are stimulated, these follicles start producing very low levels of a group of hormones called **estrogens**. Estrogens develop and maintain the female reproductive structures such as the ovarian follicles, as well as the lining of the uterus and the breasts. As the follicles secrete estrogens ❷, the lining of the uterus begins to thicken ❸, preparing for the implantation of a fertilized egg.

As the developing follicles mature, a membrane forms around each potential egg, fluid builds up within each follicle, and each follicle's cells produce increasing amounts of estrogens. **Figure 36.17** shows a developing follicle. During the second week of the reproductive cycle, most of the developing follicles die; usually only one follicle continues to mature. The hypothalamus senses the rise in blood level of estrogens and secretes gonadotropic-releasing hormone. This hormone triggers the pituitary to secrete a surge of luteinizing hormone, or LH (Figure 36.16 ❹), which triggers ovulation. A smaller surge of FSH also occurs ❺.

During ovulation, the follicle breaks open and releases its immature egg (secondary oocyte) into the abdominal cavity ❻, close to the opening of the oviduct. Ovulation usually (but not always) occurs at the midpoint of the reproductive (menstrual) cycle, around day 14 in a 28-day cycle. The lengths of the phases of the reproductive cycle vary with age and among women.

After ovulation, the movement of the ciliated cells at the fringes of the nearby oviduct sweeps the egg into the oviduct. The egg then begins its three- to four-day journey to the uterus. The ruptured follicle collapses ❼, and then, under the continued influence of LH, it begins to enlarge, forming a yellowish structure called the *corpus luteum*, or "yellow body" ❽. The corpus luteum plays an important role in preparing the endometrium for the implantation of the fertilized egg by secreting estrogens ❾ and **progesterone** ❿. The progesterone steps up the thickening of the lining of the uterus and dramatically increases its blood supply ⓫. The rise in progesterone and estrogens also acts as a negative feedback mechanism to the hypothalamus, ultimately inhibiting the secretion of FSH and LH from the pituitary ⓬, thus preventing additional ovulation near the end of the menstrual cycle.

If fertilization does not occur, the egg dies and cannot implant in the uterus and develop. Therefore, no placenta develops and no hCG is produced. The corpus luteum begins degenerating on approximately day 24 of our "example" 28-day cycle ⓭. As it degenerates, it stops producing estrogens ⓮ and progesterone ⓯. As the blood level of these hormones falls, the endometrial lining sloughs off ⓰ in a process known as **menstruation**, or menses.

If fertilization and implantation occur, the **placenta** (pluh-SEN-tuh) begins to form. The placenta is a flat disk of tissue that

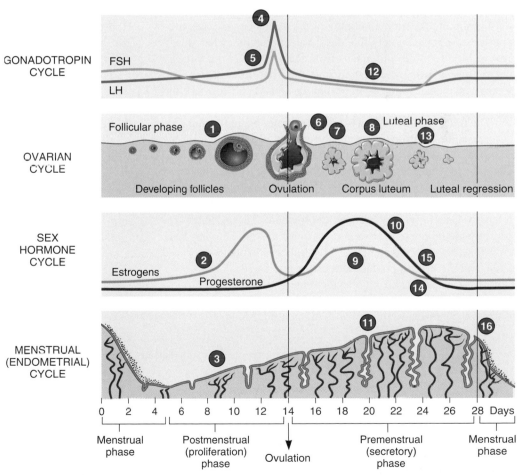

**Visual Summary**  **Figure 36.16** The human reproductive cycle. The growth and thickening of the uterine lining is governed by increasing levels of estrogens and progesterone. Menstruation, the sloughing off of the blood-rich tissue, is initiated by decreasing levels of these hormones.

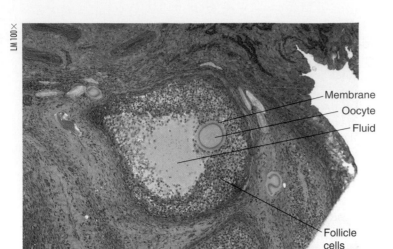

LM 100×

**Figure 36.17** A primary oocyte within an ovarian follicle.

provides an exchange of nutrients and wastes between the embryo and the mother. It secretes a hormone called *human chorionic gonadotropin (hCG)*, which maintains the corpus luteum so that it will continue to secrete progesterone and estrogens. Progesterone is needed for the maintenance of the uterine lining. Secretion of hCG occurs about eight days after fertilization. Because this hormone is excreted in the urine, its detection is the basis for both laboratory and home pregnancy tests.

During menstration, the lining of the uterus degenerates over a period of approximately five days, causing a somewhat steady flow of blood, tissue, and mucus from the uterus and out through the vagina, usually heaviest during the first three days. The process of menstruation is traditionally considered to mark the *beginning* of the reproductive cycle (days 1–5).

The reproductive cycle of women generally begins during adolescence and usually ends between the ages of 50 and 55 years. This cessation of menses is termed **menopause**. As a woman ages, changes take place in the reproductive cycle that leads to menopause. These changes usually begin between the ages of 40 and 50 years. The menstrual cycles may become irregualr, varying in length. Generally, they become less frequent. These changes result from the failure of the ovaries to respond to stimulation by FSH and LH secreted by the pituitary. Thus, follicles do not develop, and the ovaries produce less estrogens. The drop in estrogens causes the two major symptoms associated with menopause: flushes, the reddening to the face and a feeling of warmth, and flashes, short periods of intense flushing.

## CONCEPT CHECKPOINT

3. Compare and contrast spermatogenesis and oogenesis with respect to (a) time frame of gamete maturation, (b) site of production of gamete, (c) ploidy of gamete, and (d) splitting of the cytoplasm during meiosis. You may find it helpful to organize this information as a table.

4. Describe the hormonal control of the female reproductive cycle prior to menstruation.

## 36.7 The human sexual response can be described as having four phases.

The sexual act is referred to by a variety of names, including *intercourse, copulation,* and *coitus* (KOH-ih-tus), as well as a host of more informal ones. The physiological events associated with sexual intercourse can be described as having four phases: excitement, plateau, orgasm, and resolution (**Figure 36.18**).

The *excitement phase* precedes intercourse. During excitement, the heartbeat, blood pressure, and rate of breathing increase. The blood vessels of the face, breasts, and genitals widen, the nipples harden, and increased circulation leads to an erection in the male and a swelling and parting of the labia in females.

The act of sexual intercourse signals the beginning of the *plateau phase*, a period of intensification of the physiological changes initiated in the excitement phase. During sexual intercourse, the penetration of the vagina by the thrusting penis results in the repeated stimulation of nerve endings both in the head of the penis (the glans penis) and in the clitoris.

For both sexes, the climax of intercourse, *orgasm*, is reached when the stimulation is sufficient to initiate a series of reflexive muscular contractions associated with sexual pleasure. In females, there may be one or several intense peaks of contractions (orgasms), or the peaks may be more numerous but less intense (see Figure 36.18). In males, the analogous contractions are initiated by nerve signals from the brain. These signals first cause emission, in which rhythmic peristaltic contractions of the vas deferens and the prostate gland cause the sperm and seminal fluid to move to the base of the penis. Shortly after, nerve signals from the brain induce contractions of the muscles at the base of the penis, resulting in the ejaculation of the semen from the urethral opening.

The *resolution phase* begins after orgasm, as the bodies of both men and women return to their normal physiological states over the course of several minutes. After ejaculation, males rapidly lose their erection and enter a period lasting 20 minutes or longer in which sexual arousal is difficult to achieve and ejaculation is almost impossible. Women can be aroused again almost immediately.

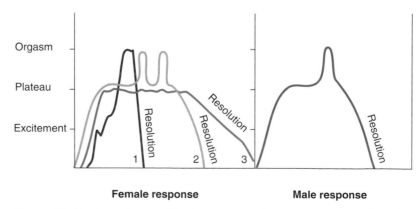

**Figure 36.18** Four phases of the human sexual response. Among females, the response is highly variable, usually resembling one of the three patterns illustrated here. Among males, the response is not as variable.

# How can pregnancies be limited and sexually transmitted infections be avoided?

## 36.8 The use of birth control methods can limit pregnancies.

In most vertebrates, sexual intercourse is associated solely with reproduction. In humans, however, sexual behavior serves a second important function—the reinforcement of pair bonding, the emotional relationship between two individuals. Couples who wish to limit the number of pregnancies resulting from sexual intercourse may choose from a variety of methods of family planning, or birth control (see **Table 36.1** on pp. 640–641 and **Figure 36.19** on p. 642). These methods differ from one another in their effectiveness and in their acceptability to different couples and individuals. In addition, some methods of birth control protect against sexually transmitted infections (STIs).

As methods of birth control, fertility awareness and withdrawal are not highly reliable. Condoms, diaphragms, and the contraceptive sponge are fairly effective when used correctly, but mistakes are common, as is also the case with foams, creams, jellies, and vaginal suppositories. Birth control pills, the contraceptive ring, implants, hormone shots, and intrauterine devices (IUDs) are very effective but some of these methods have serious disadvantages (see Table 36.1). Vasectomies are 99.9% effective and tubal ligations are 99.5% effective. Microsurgical techniques make reversals of these procedures possible in some instances. The success rates of tubal ligation reversals depend on factors of age and the type of tubal ligation that was performed. The success rate of vasectomy reversal is about 50%.

Emergency contraception can prevent pregnancy *after* unprotected intercourse. There are two types available by prescription only in the United States: inserting a copper T IUD or using hormonal contraceptive pills. In May 2004, the U.S. Food and Drug Administration rejected over-the-counter sales of a "morning after" contraceptive pill.

Insertion of the copper T IUD for emergency contraception works by interfering with the implantation of a fertilized egg. For this reason, it can be inserted to prevent pregnancy up to five days after unprotected intercourse. This method is highly effective, with only a 0.8% failure rate.

Emergency contraceptive pills (ECPs) act by delaying or inhibiting ovulation, altering the movement of the sperm, or inhibiting implantation. Two types of ECPs are available in the United States: pills that combine estrogen and certain progestins (hormones that have progesteronelike effects) and pills containing progestins only. The earlier ECPs are taken after unprotected intercourse, the more effective they are. The closer a woman is to ovulation at the time of unprotected intercourse, the less likely the method will succeed. Combined, ECPs have about a 25% failure rate; that is, the pills reduce the number of women who would become pregnant through unprotected intercourse by 75%. Progestin-only ECPs reduce the chance of pregnancy from unprotected intercourse by 89%.

## 36.9 Some birth control methods can also protect against sexually transmitted infections.

Sexually transmitted infections (STIs), formerly called sexually transmitted diseases (STDs), are caused by the transmission of certain agents of infection from one person to another during sexual activity. The idea that STIs are caused by poor personal hygiene and the idea that STIs are transmitted only by persons of low socioeconomic status are misconceptions. Statistics show that the incidence of STIs crosses all boundaries of race, gender, ethnicity, social class, and economic status. Although the incidence of STIs crosses boundaries of age too, young people are hardest hit. Today, 12 million new sexually transmitted infections occur each year, with two-thirds of these cases in persons younger than 25.

Currently, one out of every four Americans has an STI. All STIs are communicable, or contagious, meaning that they can be spread from one person to another. The most common STIs in the United States are chlamydial infection and gonorrhea, which are both caused by bacteria, and genital herpes, which is caused by a virus. Taken together, these three diseases make up about 10 million of the 12 million annual cases of STIs. Other organisms such as certain fungi, protists, mites, and lice can also be transferred from person to person by sexual contact.

The only way to totally protect yourself against contracting an STI is to abstain from sexual activity. If you do engage in sexual activity, your risk of contracting an STI increases as your number of partners increases. A monogamous relationship with an uninfected person is the safest situation for protecting against STIs.

How else can you protect yourself from STIs? The best protection is the proper and consistent use of latex condoms. The condom puts a barrier between yourself and any disease-producing organisms, but *condoms are no guarantee against contracting disease.* Some spermicides, particularly nonoxynol-9, have been found to kill HIV in the laboratory and appear to reduce the risk of gonorrhea and chlamydia transmission, although this effect has not been proven in actual practice. However, spermicides may cause irritation in some women, which increases the chance of being infected.

If you observe any of the signs or symptoms of an STI in yourself, see a physician immediately. **Table 36.2** on p. 643 lists common sexually transmitted infections and infestations, brief descriptions of these diseases, and treatments.

### CONCEPT CHECKPOINT

5. Which birth control methods prevent pregnancy by preventing implantation?
6. Besides abstinence, which of the methods of birth control will completely prevent the spread of infections that are transmitted through sexual intercourse?

| TABLE 36.1 | Methods of Birth Control |

## Abstinence, Fertility Awareness, and Withdrawal

| Method | Action | Failure Rate | | Advantages | Disadvantages |
|---|---|---|---|---|---|
| | | Actual* | Theoretical* | | |
| Abstinence | Refraining from vaginal intercourse | — | — | Protects against STIs; excellent alternative for young persons who are not yet ready to have sex or for others who choose to abstain for various reasons | Lack of this particular facet of sexual intimacy in one's life |
| Fertility awareness | Avoidance of vaginal intercourse during woman's fertile period | 20% | 9% | Contraceptive devices unnecessary; agreeable to some religious groups | High failure rate due to difficulty of predicting ovulation; no protection against STIs |
| Withdrawal | Removal of penis from vagina before ejaculation | 19% | 4% | Contraceptive devices unnecessary | Failure attributed to the secretion of sperm before ejaculation; self-control necessary; no protection against STIs |

## Barrier Methods (Sperm Blockage)

| Method | Action | Failure Rate | | Advantages | Disadvantages |
|---|---|---|---|---|---|
| | | Actual* | Theoretical* | | |
| Male latex condom | Thin rubber sheath for penis that collects semen | 14% | 3% | Easy to use, effective, and inexpensive; the most reliable protection against some STIs with the exception of abstinence | Possible decrease in spontaneity; possible deterioration on the shelf; lambskin condoms do not protect against STIs |
| Female condom | Polyurethane sheath with a ring at each end that lines the vagina | 21% | 5% | May protect against STIs | High failure rate |
| Diaphragm | Soft rubber cup that covers entrance to uterus, prevents sperm from reaching egg, and holds spermicide | 20% | 6% | No dangerous side effects; some protection against cervical cancer | Careful fitting required; some inconvenience associated with insertion and removal; possible dislodgment during sex; relationship to STI prevention not yet clear |
| Contraceptive sponge | Polyurethane foam cup that covers entrance to uterus, kills sperm with nonoxynol-9 present in the polyurethane, and absorbs sperm into the foam | *No previous births* 13% *Previous births* 16% | 9% 11% | Convenient, no prescription needed, no fitting required; can be inserted prior to sex and used for multiple acts of intercourse within a 24-hour period | May cause irritation in some women |
| Cervical cap | Miniature diaphragm that covers cervix closely, prevents sperm from reaching egg; and holds spermicide | *No previous births* 20% *Previous births* 40% | 9% 26% | No dangerous side effects; able to remain in place longer than diaphragm | Problems with fitting and insertion; limited number of sizes; not available in the United States; relationship to STI prevention not yet clear |

## Sperm Destruction

| Method | Action | Failure Rate | | Advantages | Disadvantages |
|---|---|---|---|---|---|
| | | Actual* | Theoretical* | | |
| Foams, creams, jellies, vaginal suppositories | Chemical spermicides inserted in vagina before intercourse that also prevent sperm from entering uterus | 26% | 6% | Possible use by anyone who is not allergic, no known side effects; some vaginal spermicides containing nonoxynol-9 may reduce the risk of gonorrhea and chlamydia transmission | Sometimes messy; necessary to use 5 to 10 minutes before each act of intercourse; may cause irritation, raising the possibility of an increased susceptibility to some STIs |

| | | **Prevention of Egg Maturation** | | | |
| | | **Failure Rate** | | | |
| **Method** | **Action** | **Actual*** | **Theoretical*** | **Advantages** | **Disadvantages** |
|---|---|---|---|---|---|
| Oral contraceptives (birth control pills) | Hormones, either in combination or progestin only, that primarily prevent release of egg; one product allows a woman to take the pill continuously for 91 days, with menstruation occurring only four times a year | *Estrogen/progestin* 3% *Progestin only* 3% | 0.1% 0.5% | Convenient and highly effective; may reduce risk of ovarian and endometrial cancers, but may increase risk of cervical and breast cancers | Necessary to take regularly; possible minor side effects, which new formulations have reduced; not for women with cardiovascular risks, mostly those over 35 who smoke |
| Contraceptive ring | Soft plastic ring inserted into the vagina and left in place for 3 weeks; releases hormones that prevent the release of eggs | 3% | 0.1% | Hormones enter the body through the vaginal wall, allowing lower dose than with oral contraceptives; convenient and highly effective; possible nonreproductive cancer benefits and risks like those of oral contraceptives | Not for women with cardiovascular risks, mostly those over 35 who smoke |
| Implant | Single matchstick-sized rod inserted just under the skin; releases a progestin slowly into the bloodstream that blocks the release of eggs | 0.9% | 0.9% | Very safe, convenient, and effective; very long lasting (3 years); possible nonreproductive health benefits like those of oral contraceptives | Irregular or absent periods, necessity of minor surgical procedure to insert and remove, no protection against STIs |
| Hormone shot | Injection of progestin that inhibits ovulation for 3 months | 0.3% | 0.3% | Very safe and effective; possible nonreproductive cancer benefits and risks like those of oral contraceptives | No protection against STIs; changes the menstrual cycle; must have injection every 3 months |
| | | **Intrauterine Devices (IUDs)** | | | |
| Intrauterine device | Small plastic device placed in the uterus that prevents fertilization or implantation | *Copper T* 0.8% *Progesterone T* 2% | 0.6% 1.5% | Convenient, highly effective, infrequent replacement | No protection against STIs; possible excess menstrual bleeding and pain; danger of perforation, infection, and expulsion; not recommended for those who are childless or not monogamous; risk of pelvic inflammatory disease or infertility; dangerous in pregnancy |
| | | **Surgical Intervention** | | | |
| Male sterilization | Vasectomy: cutting and tying off the vas deferens so that sperm cannot enter the ejaculate | 0.1% | 0.1% | Permanency; effectiveness | Involves surgical procedure; reversal is not always successful |
| Female sterilization | Tubal ligation: cutting and tying off the oviducts to prevent fertilization and implantation | 0.5% | 0.5% | Permanency; effectiveness | Involves a somewhat complicated and costly surgical procedure |

*Actual failure rate refers to the number of women who will become pregnant out of 100 couples using a method under usual conditions for 1 year. This number includes improper use of a method, forgetfulness, etc. *Theoretical failure rate* refers to the number of women who will become pregnant out of 100 couples using a method consistently and properly as their only method of birth control for 1 year.

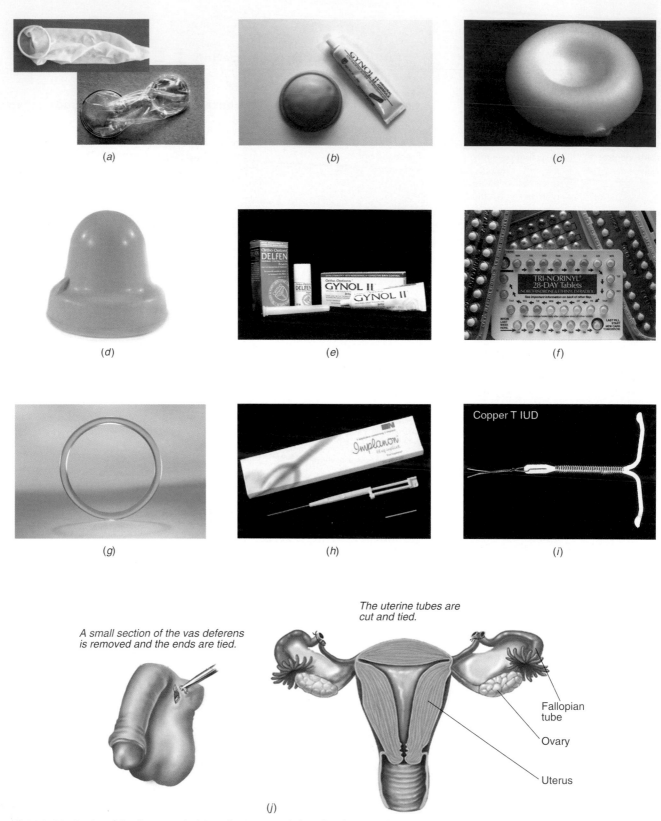

**Figure 36.19** Methods of birth control. (*a*) Male (*top*) and female (*bottom*) latex condoms. (*b*) Diaphragm and spermicide. (*c*) Contraceptive sponge. (*d*) Cervical cap. (*e*) Contraceptive foam, contraceptive jelly, and applicator. (*f*) Oral contraceptives (birth control pills). (*g*) Contraceptive ring. (*h*) Contraceptive implant. (*i*) Copper T IUD. (*j*) Surgical means of birth control: vasectomy (*left*) and tubal ligation (*right*).

| TABLE 36.2 | Sexually Transmitted Infections and Infestations | | | |
|---|---|---|---|---|
| **Disease** | **Causative Agent** | **Type of Agent** | **Brief Description of Disease** | **Treatment** |
| Genital herpes | Herpes simplex virus (HSV) | Virus | Disease causes painful sores in genital area. Virus causes latent infection so recurrences are common. | Acyclovir® and other antiviral drugs (no cure) |
| Genital warts | Human papillomavirus (HPV) | Virus | Disease causes soft, pink, flat, or raised growths on external genitals, rectum, vagina, cervix. Some of the viruses that cause genital warts are associated with the development of cervical cancer. | Electrocauterization, cryotherapy, surgery (no cure) |
| HIV infection | Human immunodeficiency virus (HIV) | Virus | Virus attacks and destroys T cells, a key component in the body's immune system. Infection with HIV eventually leads to acquired immune deficiency syndrome, or AIDS. On average, people diagnosed with AIDS live for about 2 years. | Various antiviral drugs (no cure) |
| Syphilis | *Treponema pallidum* | Bacterium | Disease progresses through stages of localized infection to widespread infection. If untreated, it can result in death. | Penicillin |
| Gonorrhea | *Neisseria gonorrhoeae* | Bacterium | Primary infection is of urethra (in men and women) and vagina and cervix (in women). Other urogenital structures maybecome infected. Disease often causes puslike discharge. Disease can result insterility and/or damage to other organs. | Penicillin and other antibiotics |
| Chlamydial infection | *Chlamydia trachomatis* | Bacterium | Infection is similar to gonorrhea but usually has milder symptoms. | Tetracycline and sulfa drugs |
| Candidiasis (yeast infection) | *Candida albicans* | Yeast | Vaginal infection results in raised gray or white patches on the vaginal walls, a thick whitish discharge, and itching. Penile infection results in the growth of small, raised yeast colonies on the penis. | Topical and oral antifungal drugs |
| Trichomoniasis | *Trichomonas vaginalis* | Protozoan | Infection in women resulting in vaginal itching, burning, and a profuse discharge that may be bloody or frothy. Infection in men results in a slight discharge from the urethra, painful urination, and increased urination. | Antiparasitic drugs |
| Scabies | *Sarcoptes scabiei* | Itch mite | Common sites of infestations are the base of the fingers, the wrists, the armpits, the skin around the nipples, and the skin around belt line. Female mites bore into the top layers of skin of an infected person to lay their eggs. Symptoms are dark, wavy lines in the skin and itching. | Topical antiparasitic medications |
| Pubic lice (crabs) | *Phthirius pubis* | Pubic louse | Females lay eggs in pubic hairs. Disease causes intense itching. | Topical antiparasitic medications |

# What happens during human development and birth?

## 36.10 Human development proceeds from zygote to fetus.

Following sexual activity, if fertilization takes place a zygote results—a new cell that contains intermingled genetic material from both the sperm and egg cells (**Figure 36.20**).

### Morula

The one-celled zygote begins to divide rapidly, in a process of cell division without cell growth called *cleavage*. The dividing cell mass is referred to as a *morula* (MORE-yuh-luh), from a Latin word

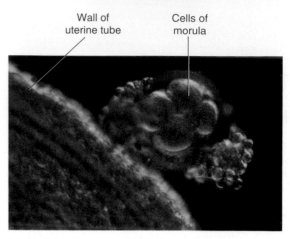

**Figure 36.21** The morula moving along the oviduct.

meaning "mulberry" (**Figure 36.21**). As the morula divides, it journeys along the remaining two-thirds of the oviduct. This trip takes approximately three days. When the morula reaches the uterus, it is still the same size as a newly fertilized ovum but consists of about 16 densely clustered cells.

### Blastocyst

After two days in the uterus, the morula has developed into a *blastocyst* (BLAS-tuh-sist), a ball of cells (**Figure 36.22**). One portion of the interior of the blastocyst consists of a concentrated mass of cells destined to differentiate (change) into the various body tissues of the new individual, a process called **tissue differentiation**. This group of cells is referred to as the *inner cell mass*. Each cell in this mass has the ability to begin the development of a complete individual. (These cells are called stem cells and may be used to treat certain diseases. See the *In The News* chapter-opening story for Chapter 12 for more discussion of stem cells and their use in medical therapy.) The rest of the interior is filled with fluid from the uterine cavity. The outer layer of cells, called the *trophoblast* (TRO-fuh-blast), will give rise to most of the extraembryonic membranes, including much of the placenta, an organ that helps maintain the developing embryo. The term **embryo** is a general term for the developing organism from the first division of the zygote until body structures begin to appear at about the ninth week of gestation. Thereafter, the developing organism is termed a *fetus*.

The term *blastula* is used to describe the saclike blastocyst of non-mammalian animals, the stage that develops a similar fluid-filled cavity. This stage of development is significant for all vertebrates; it is the first time that cells begin to migrate to shape the new individual in a process called **morphogenesis** (more-foe-JEN-uh-sis). However, the major morphogenetic events occur during the third to eighth weeks.

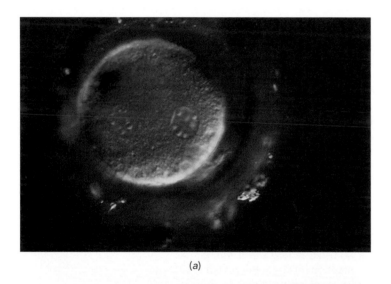

(a)

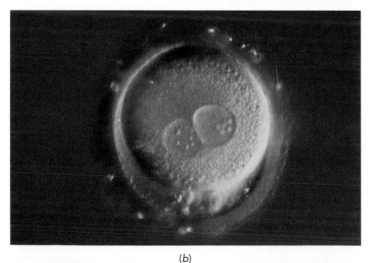

(b)

**Figure 36.20** Fertilization. (*a*) The nuclei of the sperm and ovum are clearly separate. (*b*) The nuclei are fusing to form the nucleus of the first cell of a new individual.

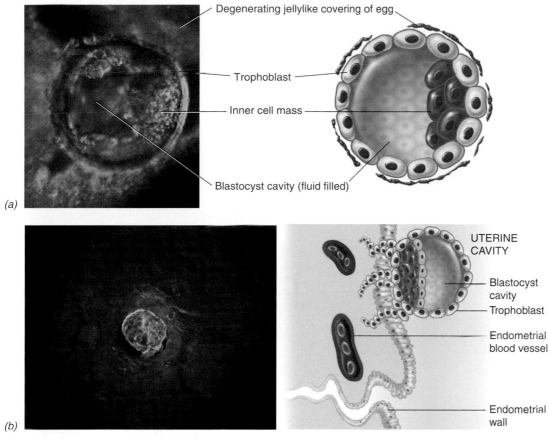

*(a)*

*(b)*

**Figure 36.22** Blastocyst and implantation. (*a*) The early blastocyst, before implantation. (*b*) About a week after fertilization, the blastocyst begins to embed itself into the wall of the uterus.

## Implantation

Approximately one week after fertilization, the blastocyst secretes enzymes that digest a microscopic portion of the endometrial wall of the uterus. It then nestles into this site, nourished by the digested uterine cells, in a process called **implantation**. The blastocyst secretes human chorionic gonadotropin (hCG), which acts on the cor-

pus luteum in the ovary. The corpus luteum responds by continuing to produce estrogens and progesterone, hormones that sustain the implanted blastocyst as the placenta develops. **Figure 36.23** summarizes the events that take place from ovulation to implantation.

During its second week of development, the embryo completes its implantation within the uterine wall, two of its three *primary germ layers* develop, and the extraembryonic membranes begin to form. The primary germ layers are three layers of cells that develop from the inner cell mass of the blastocyst and from which all the organs and tissues of the body develop. These three layers are called the *ectoderm* ("outside skin"), *endoderm* ("inside skin"), and *mesoderm* ("middle skin"). The ectoderm forms the outer layer of skin, the nervous system, and portions of the sense organs. The endoderm gives rise to the lining of the digestive tract, the digestive organs, the respiratory tract, and the lungs; the urinary bladder; and the urethra. The mesoderm differentiates into the skeleton (bones), muscles, blood, reproductive organs, connective tissue, and the innermost layer of the skin. **Figure 36.24** shows the body systems and tissues into which the germ layers develop. At the beginning of the second week, only the endoderm and the ectoderm have formed.

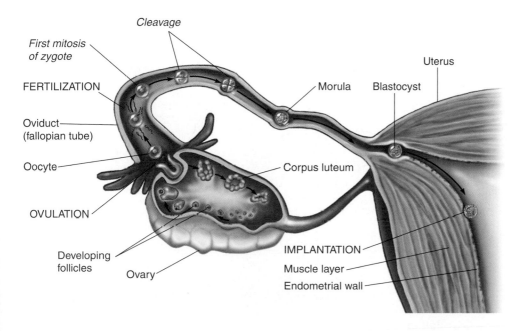

**Visual Summary** Figure 36.23 Ovulation, fertilization, and the journey of the zygote from the oviduct to implantation in the uterus. The first event in human development is fertilization, which takes place in the oviduct. The zygote begins to divide by mitosis as it moves toward the uterus, and is now called a morula. At 4 days after fertilization, the developing ball of cells begins to fill with fluid, a stage of embryonic development termed the *blastocyst*. Approximately 6 days after fertilization, the blastocyst implants in the endometrial wall of the uterus. During *implantation*, the embryo attaches firmly to this inner lining of the uterus. Occasionally, implantation takes place outside of the uterus, a condition known as *ectopic pregnancy*. Embryos may implant on an ovary, on the intestine, or in the fallopian tube. All ectopic pregnancies endanger the mother's life.

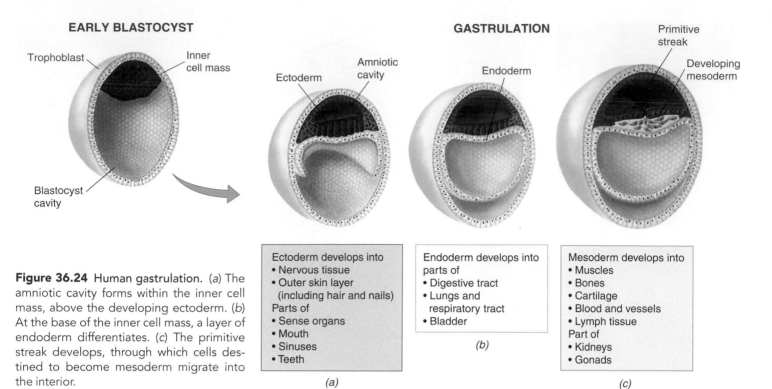

**EARLY BLASTOCYST**

Trophoblast

Inner cell mass

Blastocyst cavity

**GASTRULATION**

Ectoderm

Amniotic cavity

Endoderm

Primitive streak

Developing mesoderm

Ectoderm develops into
• Nervous tissue
• Outer skin layer (including hair and nails)
Parts of
• Sense organs
• Mouth
• Sinuses
• Teeth

*(a)*

Endoderm develops into parts of
• Digestive tract
• Lungs and respiratory tract
• Bladder

*(b)*

Mesoderm develops into
• Muscles
• Bones
• Cartilage
• Blood and vessels
• Lymph tissue
Part of
• Kidneys
• Gonads

*(c)*

**Figure 36.24** Human gastrulation. (*a*) The amniotic cavity forms within the inner cell mass, above the developing ectoderm. (*b*) At the base of the inner cell mass, a layer of endoderm differentiates. (*c*) The primitive streak develops, through which cells destined to become mesoderm migrate into the interior.

## Gastrulation

At the end of the second week and continuing into the third week, various cell groups of the inner cell mass begin to divide, move, and differentiate, forming the *primary germ layers*. These are three layers of cells called the *ectoderm* ("outside skin"), *endoderm* ("inside skin"), and *mesoderm* ("middle skin"). All the organs and tissues of the body develop from these three primary germ layers. This process is called **gastrulation** (**Figure 36.24**). This word is derived from a Greek word *gastros* meaning "belly." In fact, the prefix *gastr-* is found in many words denoting parts of the human body, such as gastric, referring to the stomach. This term is descriptive of the fact that at this stage in many animals, a primitive gut is formed by the infolding of the blastula, but the intestines develop differently in humans and many animals.

At the beginning of the third week, the embryo is an elongated mass of cells barely one tenth of an inch long (2.5 mm) A streak (called the *primitive streak*) runs down the midline of what will be the back side of the embryo. Cells at the streak migrate inward, producing the mesoderm. Cells at the head end of the streak grow forward to form the beginnings of the *notochord* (NO-toe-kord). The notochord is a structure that forms the midline axis along which the vertebral column (backbone) develops. An embryonic notochord forms in all vertebrate animals (see Chapter 22). Later in development it becomes the intervertebral disks. Gastrulation ends with the completion of the notochord midway through the third week. By the end of gastrulation, a layer of ectoderm covers the notochord tissue.

## Neurulation

In the photo of the embryo at 4 weeks (see Figure 36.26), its head can be seen to have the beginnings of eyes, a sign that **neurulation**

(NOOR-oo-LAY-shun) has taken place. Neurulation is the development of a hollow nerve cord, which later develops into the brain, spinal cord, and related structures such as the eyes.

Neurulation begins in the third week with the folding of the ectoderm lying above the notochord, forming an indentation along the back of the embryo. This indentation is called the *neural groove*. On either side of the groove are areas of tissue called *neural folds*. In **Figure 36.25**, the neural folds of the three-week embryo have come together at one spot and have fused. This spot is destined to be the neck region of the developing individual. The tissue above the fused region, looking somewhat like a pair of lips, will develop into the

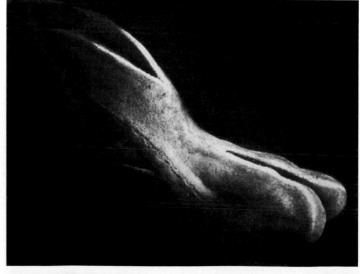

**Figure 36.25** Neurulation.

brain. The less broad area of tissue below the fused region will develop into the spinal cord. Eventually, the edges of the groove will fuse along its length, forming a neural tube, the precursor to the structures noted above. Neurulation results in an embryo called a *neurula* and signals the developmental process of tissue differentiation.

During the third week, the embryo's heart also begins its development as a pair of microscopic tubes. The cardiovascular system is the first system to become functional in the embryo.

Paired segments of tissue are also prominent in the embryo, as shown in **Figure 36.26**. These chunks of mesoderm are called *somites* (from a Greek word meaning "a body"). They will give rise to most of the axial skeleton (see Chapter 34) with its associated skeletal muscles and most of the dermis of the body—tissues that underlie the epidermis of the skin. The first somites appear during the third week of development and are added as the embryo grows—42 to 44 pairs develop by the middle of the fifth week.

**Figures 36.27–36.29** continue the story of human development to full term.

---

CONCEPT CHECKPOINT

**7.** What hormonal event occurs during pregnancy that prevents the degradation of the corpus luteum and endometrium?

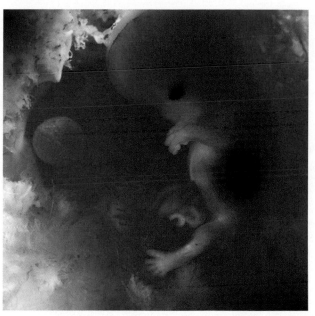

**Figure 36.27** Human fetus at 10–11 weeks of development. During the seventh week, eyelids begin to cover the eyes, and the face begins to look somewhat human. By the eighth week, the last week in the embryonic period, the embryo has grown to about an inch. Most of the body systems are functional by this time. The fetal period begins at week 9, lasts throughout the rest of the pregnancy, and is characterized by growth and functional maturation of the organs. During weeks 9 through 13 (the third month), facial features become increasingly well developed, the genitals begin to develop, and the heart is now a four-chambered structure. The fetus in the photo measures 3–3 1/2 cm (about 1 1/3 inches) from crown to rump Actual length.

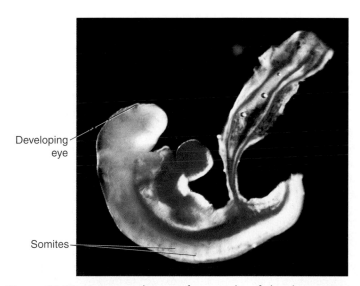

Developing eye

Somites

**Figure 36.26** Human embryo at four weeks of development. As the pregnancy moves into the third week, the embryo is only one-tenth of an inch long Actual length. The flattened embryo develops into a cylindrical embryo. Some of the organs, such as the heart, begin to develop. This four-week-old embryo is about 6 mm (about 1/4 inch) long Actual length and has established the beginnings of most of the major organ systems. Its C-shaped body has a relatively featureless head, a middle with the heart and liver bulging from the body, and a tail. During the fifth week, the embryo grows to 8 mm (3/8 inch Actual length). The brain grows rapidly. Wrists, fingers, and ears begin to form during the sixth week of development. Although development has been rapid, growth has not. By the end of the sixth week, the embryo is a mere half-inch long.

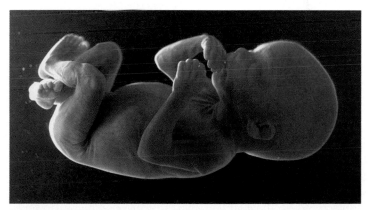

**Figure 36.28** Human fetus at five months (20 weeks) of development. The second trimester comprises the fourth, fifth, and sixth months of development, or weeks 14 through 26. The organs that developed during the first trimester mature and grow. As the fourth month passes, the genitals become fully formed, and the sensory organs nearly finish their development. The mother becomes aware of fetal movements around the fifth month of pregnancy, the stage of development of the fetus in the photo. By the end of the fifth month, as fat deposits are laid down, the fetus begins to look more like a baby. Only 12 inches long and weighing 1 pound, it probably could not survive on its own. However, if born by the end of the sixth month, the fetus has a chance of surviving with special medical care.

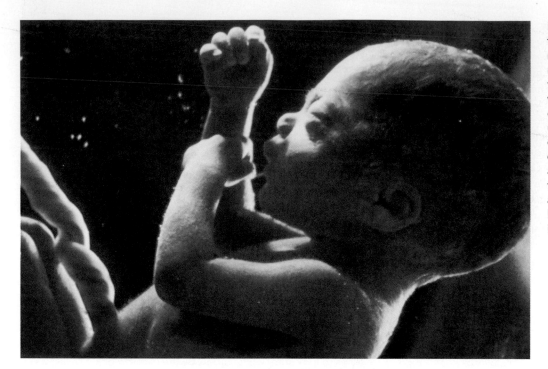

**Figure 36.29** Human fetus at nearly full term—8 to 9 months of development. During the third trimester, months seven through nine or weeks 27 through 38, the fetus primarily gains weight. Its lungs develop more fully, eyelids open, and the nervous system undergoes further development. However, the nervous system does not fully develop, and its maturation continues after birth. At the end of the third trimester, the average fetus weighs about 7 1/2 pounds and is 20 inches long (about the width of this book when it is open).

# just wondering . . .

### Questions students ask

**Will it one day be possible for men to take pills or injections that would cause temporary sterility and allow them to have sex without the worry of unplanned pregnancy?**

Yes, it is likely. At this time, scientists are working primarily on two categories of drugs for male contraception: (1) drugs that interfere with the hormonal mechanisms involved in the development of sperm and (2) drugs that render sperm incapable of fertilizing an egg.

In the first category, scientists are testing the effects of injections of high doses of testosterone or of testosterone combined with other compounds. In research studies on humans, weekly injections of high doses of testosterone caused 50 to 70% of male subjects to become infertile within three to four months. However, if male subjects also took daily pills of progestogen, which is similar to the "female" hormone progesterone, over 90% became infertile in under three months. Scientists are still testing these methods of male birth control along with various delivery methods, including injections and implants.

Scientists are also testing the effectiveness of combining testosterone with GnRH (gonadotropic-releasing hormone) antagonists to achieve the same results. GnRH antagonists suppress the release of LH and FSH, two hormones necessary for spermatogenesis. Scientists are currently testing a plan in which testosterone injections alone could maintain infertility after it is achieved with the GnRH antagonists–testosterone combination.

In the second category of drugs, scientists are testing a compound called triptolide, which has been isolated from the plant *Tripterygium wilfordii*, used in traditional Chinese medicine to treat inflammation. This compound acts primarily on sperm when they are in the epididymis, interfering with their maturation and thereby impairing sperm function.

Another discovery in this second category of drugs was accidental. While studying medications called calcium channel blockers to treat high blood pressure, scientists noticed that men taking the drugs were rendered sterile. It was determined that the drugs coated the sperm with cholesterol, blocking their ability to fertilize eggs. Researchers are working to develop testes-specific compounds, since the blood pressure drug has too many wide-reaching effects on the body to use it safely for contraception.

Most scientists would agree that a new form of male contraception will not be available until at least 2010, but two drug companies have stated that they expect to market a birth control pill for men by 2007. These German and Dutch pharmaceutical companies have not divulged the mode of action of their male pill but are pooling their resources to get the drug on the market. If they accomplish their goal. The drug would not be available in the United States until approved by the U.S. Food and Drug Administration. ●

---

*Are you wondering about a topic in biology and how it relates to your life? Submit your question by clicking the* Just Wondering *link in this text's companion Web site at* www.wiley.com/college/alters.

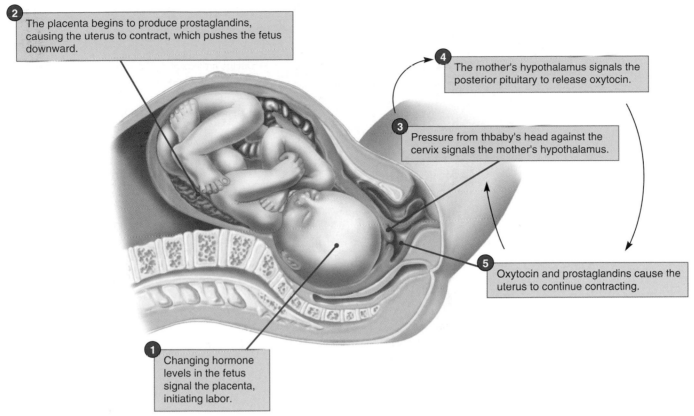

**2** The placenta begins to produce prostaglandins, causing the uterus to contract, which pushes the fetus downward.

**4** The mother's hypothalamus signals the posterior pituitary to release oxytocin.

**3** Pressure from thbaby's head against the cervix signals the mother's hypothalamus.

**5** Oxytocin and prostaglandins cause the uterus to continue contracting.

**1** Changing hormone levels in the fetus signal the placenta, initiating labor.

**Figure 36.30** Birth is initiated and sustained by hormones in a positive feedback loop. This feedback loop (**3** to **5**), in which the pressure of the fetus's head causes the release of oxytocin, continues until the baby is born.

## 36.11 Hormonal changes in the fetus trigger the birth process.

If a woman becomes pregnant, she usually will give birth to a newborn 38 weeks after fertilization, or conception. The delivery date, or birth date, is calculated to be 40 weeks from a woman's last menstrual period because fertilization usually occurs in the middle of the cycle.

Although the exact mechanism of the onset of *labor*, the sequence of events that leads to birth, is not well understood, scientists know that changing hormone levels in the developing fetus initiate this process (**Figure 36.30 1**). These hormones induce placental cells of the mother to manufacture *prostaglandins*, hormonelike substances that cause the smooth muscle of the uterine wall to contract **2**. In addition, the pressure of the fetus's head against the cervix sends nerve impulses to the mother's brain that trigger the hypothalamus to release the hormone oxytocin from her pituitary **3** and **4**. Working together, oxytocin and prostaglandins stimulate waves of contractions in the walls of the uterus **5**, forcing the fetus downward. Eventually, strong contractions, aided by the mother's pushing, expel the fetus through the vagina, or birth canal. The fetus is now a newborn.

After birth, uterine contractions continue and expel the placenta and associated membranes, collectively called the *afterbirth*. The umbilical cord is still attached to the baby, and to free the newborn, the cord is cut. Blood clots in the uterus, and contraction of its muscles prevent excessive bleeding.

During pregnancy, the anterior pituitary secretes the hormone prolactin, which is primarily responsible for stimulating the production of milk by the **mammary glands** within the breasts. These glands lie over the chest muscles. Each mammary gland is made up of 15 to 20 lobes that are separated by fat. The amount of fat determines the size of the breast and is not an indicator of a woman's ability to produce milk.

## 36.12 The newborn makes many physiological adjustments.

At birth, the baby's lungs are not filled with air. Its first breath is therefore unusually deep. Because the baby is now obtaining oxygen from the lungs rather than from the placenta, several major changes must take place in the circulation of the blood.

Until birth, the placenta was the source of nutrients and oxygen for the fetus; in addition, it was the site for the removal of waste products from the fetal circulation. The lungs were not functional as organs of gas exchange. The fetal body had two major adaptations to limit the flow of blood to the lungs. First, a hole between the two atria called the foramen ovale shunted most right atrial blood directly into the left atrium, thus avoiding the right ventricle and the pulmonary

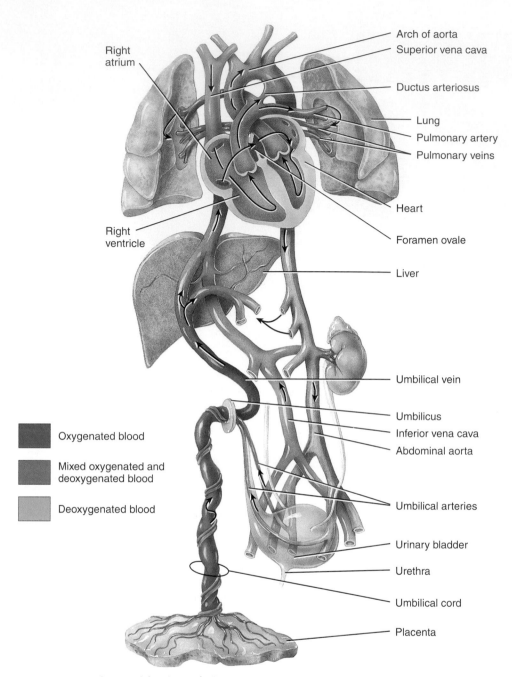

Right atrium

Arch of aorta

Superior vena cava

Ductus arteriosus

Lung

Pulmonary artery

Pulmonary veins

Heart

Right ventricle

Foramen ovale

Liver

Umbilical vein

Oxygenated blood

Mixed oxygenated and deoxygenated blood

Deoxygenated blood

Umbilicus

Inferior vena cava

Abdominal aorta

Umbilical arteries

Urinary bladder

Urethra

Umbilical cord

Placenta

**Figure 36.31 Placental-fetal circulation.** During its development, the fetus receives all its nutrients from its mother via the umbilical vein. Wastes are removed from the fetus through the two umbilical arteries and are excreted by the mother.

 **Visual Thinking:** What do you think might be the consequences if the foramen ovale did not close after birth? What symptoms of this condition might be seen in the newborn?

circulation (**Figure 36.31**). In addition, the right ventricular blood that, after birth, is pumped into the pulmonary artery and then into the lungs was mostly shunted into the aorta by the *ductus arteriosus*, a direct connection between the pulmonary artery and the aorta. At birth, the foramen ovale is closed by two flaps of heart tissue that fold together and fuse. The ductus arteriosus is shut off by contractions of muscles in its walls. Complete closure may take several months.

 CONCEPT CHECKPOINT

**8.** Why does blood circulation to the lungs for oxygenation not begin until after birth? Anatomically, how is blood shunted away from the lungs during embryonic development?

# CHAPTER REVIEW

## Summary of Key Concepts and Key Terms

### What are the various ways in which organisms reproduce?

**36.1** Usually during **sexual reproduction** (p. 629), a male **gamete** (GAM-eet, p. 629) and a female gamete combine to form the first cell of a new individual, a **zygote** (ZYE-goat, p. 630).

**36.1** During **asexual reproduction** (p. 629), new individuals are generated without the union of gametes.

**36.1** Sexual reproduction results in genetic variability in a population, while asexual reproduction does not.

**36.1** Types of asexual reproduction include binary fission, budding, **fragmentation** (p. 630) and **parthenogenesis** (PAR-theh-no-JEN-eh-sis, p. 630).

**36.2** Sexual reproduction in the invertebrates is common, as is hermaphrodism (**hermaphrodite**, her-MAF-reh-dite, p. 631), in which single individuals may be both male and female.

**36.2** During sexual reproduction, **fertilization** (p. 631) may be one of two types: **external fertilization** (p. 631) or **internal fertilization** (p. 631).

**36.3** Human reproduction is sexual and fertilization is internal.

**36.3** The male and female reproductive organs are called **gonads** (p. 632).

**36.4** The male reproductive system produces male sex cells called sperm that develop in the **testes** (p. 632) and mature in the **epididymis** (p. 634).

**36.4** The testes, housed within a sac of skin called the **scrotum** (p. 632), are packed with coiled **seminiferous tubules** (p. 632), in which **spermatozoa** (p. 634) develop.

**36.4** During ejaculation, sperm are propelled through the **vas deferens** (p. 634) to the urethra and out the **penis** (p. 634).

**36.4** The accessory glands of the male reproductive system—the **seminal vesicles** (p. 634), the **prostate** (p. 634), and the **bulbourethral glands** (BULL-bo-yoo-REE-thrul, p. 634)—secrete fluids that combine with the sperm during ejaculation.

**36.4** The scrotum and the penis are the two male **external genitals** (p. 634).

**36.5** Human females produce eggs, receive sperm, and nourish the developing fetus.

**36.5** The part of the female reproductive tract that accepts the penis during sexual intercourse and receives the sperm is called the **vagina** (p. 635).

**36.5** The external genitals of a female surround the vaginal opening and are collectively called the **vulva** (p. 635).

**36.5** The most anterior structure is the **mons pubis** (monz PYOO-bis, p. 635).

**36.5** Two longitudinal folds of skin called the **labia majora** (LAY-bee-uh muh-JORE-uh, p. 635) run posteriorly from the mons and cover additional folds of skin called the **labia minora** (p. 635).

**36.5** Slightly anterior to the convergence of the labia minora is the **clitoris** (KLIT-uh-ris, p. 636), a small mass of erectile and nervous tissue that responds to sexual stimulation.

**36.5** In the female reproductive cycle, each month primary oocytes in the **ovaries** (p. 636) continue their development; their surrounding cells start producing estrogens, which promote the thickening of the endometrial lining of the uterus.

**36.5** Of the developing oocytes, one matures into a secondary oocyte that bursts from the ovary during **ovulation** (p. 636) as a surge of luteinizing hormone (LH) occurs midcycle.

**36.5** The oocyte travels along the **oviduct** (p. 636) to the uterus.

**36.5** The inner lining of the uterus is called the **endometrium** (p. 636); the narrow portion of the uterus opens into the vagina at the **cervix** (p. 636).

**36.6** The ruptured follicular cells that surrounded the oocyte secrete **estrogens** (p. 637) and **progesterone** (p. 638).

**36.6** If fertilization and implantation occur, the **placenta** (pluh-SEN-tuh, p. 638) begins to form; it is a flat disk of tissue that provides an exchange of nutrients and wastes between the embryo and the mother.

**36.6** If fertilization does not take place, the egg dies; progesterone and estrogens maintain the endometrial lining until diminished levels of these hormones cause it to degenerate.

**36.6** A sloughing of the endometrial lining (**menstruation**, p. 638), follows.

**36.6** The human female reproductive tract prepares monthly for reproduction until about age 50 years, when the menses cease at **menopause** (p. 638).

**36.7** The human sexual response can be described as having four phases: excitement, plateau, orgasm, and resolution.

### How can pregnancies be limited and sexually transmitted infections be avoided?

**36.8** Methods of birth control differ in their reliability and in other advantages and disadvantages.

**36.9** Sexually transmitted infections (STIs) are caused by the transmission of certain agents of infection from one person to another during sexual activity.

**36.9** The best protection against STIs, other than abstention from sexual activity or sex within a monogamous relationship between two uninfected partners, is the consistent and proper use of latex condoms.

### What happens during human development and birth?

**36.10** If fertilization takes place, the fertilized egg, or zygote, begins to divide and to travel down the oviduct to the uterus; at this point, the developing embryo is termed a *morula* (MORE-yuh-luh).

**36.10** The developing organism is termed an **embryo** (p. 644) from the first division of the zygote until body structures begin to appear at about the ninth week of gestation; thereafter, it is termed a fetus.

**36.10** Cells begin to migrate to shape the new individual in a process called **morphogenesis** (more-foe-JEN-uh-sis; p. 644); the major morphogenetic events occur during the third to eighth weeks.

**36.10** After two days, the morula has developed into a blastocyst (BLAS-tuh-sist), a ball of cells consisting of an outer layer called the *trophoblast* (TRO-fuh-blast) and an interior consisting of an inner cell mass and a fluid-filled cavity.

**36.10** The cells of the inner cell mass differentiate (change) into the various body tissues of the new individual, a process called **tissue differentiation** (p. 644).

**36.10** At approximately one week after fertilization, **implantation** begins (p. 645)—the blastocyst embeds itself in the lining of the uterine wall.

**36.10** The implanted blastocyst secretes human chorionic gonadotropin (hCG), which acts on the corpus luteum in the ovary, stimulating the body to produce estrogens and progesterone to maintain the uterine lining.

**36.10** The cells of the inner cell mass form the three primary germ layers by the end of the third week, in a process called **gastrulation** (p. 646).

**36.10** The three primary germ layers are the ectoderm, the endoderm, and the mesoderm.

**36.10** **Neurulation** (p. 646) begins in the third week—an indentation called the neural groove develops, and tissue differentiation begins. This stage includes the development of a hollow nerve cord, which later develops into the brain, spinal cord, and related structures such as the eyes.

**36.11** After 38 weeks of development and growth, changing hormone levels in the fetus trigger the onset of labor, the sequence of events that leads to birth.

**36.11** Waves of contractions in the walls of the uterus force the fetus downward and out the birth canal.

**36.11** During pregnancy, the anterior pituitary secretes the hormone prolactin, which is primarily responsible for stimulating the production of milk by the mother for the newborn; the **mammary glands** (p. 649) produce and secrete milk.

**36.12** At birth, the circulation of the newborn changes as the lungs rather than the placenta become the organs of gas exchange.

## Level 1 | Learning Basic Facts and Terms

**Matching**

Pair the following human male reproductive structures with their functions.

1. ____ Seminal vesicles
2. ____ Vas deferens
3. ____ Bulbourethral glands
4. ____ Prostate
5. ____ Testes
6. ____ Epididymis

a. secrete a nutrient rich fluid that nourishes sperm during ejaculation
b. adds a fluid to semen that neutralizes the acidic environment of the vagina
c. secrete a mucus that acts as a lubricant during sexual intercourse
d. produce sperm cells
e. structure where sperm become capable of propelling themselves and penetrating an oocyte
f. a long tube that ascends from the epididymis into the pelvic cavity

Pair the following female reproductive structures with their functions.

7. ____ Uterus
8. ____ Endometrium

a. organs where oocytes are produced
b. structures that connect the ovaries to the uterus

9. ____ Fallopian tubes
10. ____ Ovaries
11. ____ Corpus luteum
12. ____ Cervix

c. secretes estrogens and progesterone after ovulation
d. opening of the uterus to the vagina
e. the uterine lining
f. an organ with thick muscular walls that contract during childbirth

**Multiple Choice**

13. Which of the following methods of birth control is most likely to help prevent the spread of sexually transmitted infections?
   a. IUD
   b. Condom
   c. Birth control pills
   d. Emergency contraceptive pills
   e. Rhythm method

14. Which mode of reproduction is exhibited within all three domains of life?
   a. Parthenogenesis
   b. Asexual
   c. Hermaphroditism
   d. Sexual
   e. Fragmentation

## Level 2 | Learning Concepts

1. Distinguish between asexual reproduction and sexual reproduction. Give an example of each.
2. Relate the structure of a spermatozoon to its function.
3. Summarize the hormonal control of the reproductive cycle of a human female when fertilization does not occur.
4. When pregnancy occurs, what prevents the endometrium (and the implanted embryo) from being shed from the body as it is during a woman's regular menstrual cycle?

5. Unicellular algal protists such as *Chlamydomonas* primarily reproduce asexually when conditions are optimal in the water where they live. However, during times of drought, they will switch to sexual reproduction. Explain the advantages of this strategy.
6. John and Mary have had difficulty conceiving a child, so their doctor recommends that they use in an in-home ovulation test to determine the day of Mary's ovulation. Given this information, during what window of time could fertilization take place?

## Level 3 | Critical Thinking and Life Applications

1. Table 36.1 gives actual and theoretical failure rates of a variety of birth control methods. Which type of data should carry more weight in your decision making regarding the type of birth control method you would choose to use? Why?
2. John was an oyster farmer, and sea stars were ruining his business by eating his oysters. John tried to rid his oyster beds of sea stars by collecting them, chopping them in half, and throwing them back into the water for fish to eat. His problem only worsened. Why wasn't John's solution working?

3. Many human pregnancies result in spontaneous miscarriage because early morphogenesis of cells during gastrulation occurs improperly. Why do you think proper gastrulation is so critical to the success of the early embryo?
4. Based on what you know about the hormonal control of a woman's reproductive cycle, why can't a woman get pregnant again while she is pregnant?
5. Birth control pills are composed of estrogens and synthetic progesterone. Often it is said that this combination of hormones "fools

the body into thinking it is pregnant." How do you think birth control pills prevent pregnancy?

6. When a woman is on the pill, the last few pills she takes during each cycle contain no hormones. What is the purpose of discontinuing these hormones at this time? What would happen if a woman continued to take hormones during this period?

7. Most animals that are simultaneous hermaphrodites are sessile (nonmoving) as adults. What do you think is the advantage of the hermaphroditic mating strategy for these animals?

8. Home pregnancy tests detect the presence of hCG in the urine. What is hCG, and why is it present in the bloodstream of a pregnant woman? Using your understanding from Chapter 31 of how nephrons in the kidney work, explain why hCG is also found in the urine.

9. Why don't condoms, even if used correctly, completely protect against sexually transmitted infections?

## In The News | **Critical Thinking**

**SMOKING AND PREGNANCY**

Now that you understand more about reproduction, reread this chapter's opening story about the effects of smoking on fertility. To better understand this information, it may help you to follow these steps:

1. Review your immediate reaction to the research on the effects of smoking on fertility that you wrote when you began reading this chapter.

2. Based on your current understanding, again summarize the main point of the research in a sentence or two.

3. What questions do you now have about this information that the opening story does not answer?

4. Collect new information about the research. Visit the *In The News* section of this text's companion Web site at www.wiley.com/college/alters and watch the "Smoking and Pregnancy" video. Then use the "summary" link to read the accompanying story and access related links. Use this information, the links provided, and other online and library resources to answer your questions and find updates about this topic. State the sources of your information. Explain why you think the information is accurate. Also determine whether the information expresses a particular point of view or is biased in any way.

5. What in your view is the most significant aspect of this information? Give reasons for your opinion and for any changes in your ideas based on the additional information you have collected and the analysis you have done.

# ANIMAL BEHAVIOR

## In The News | Child's Play

When you were a child, was recess or gym at the top of your list at school? Did you look forward to coming home and going out to play? Like human children, many young birds and mammals (such as the ape in the photo) spend much time engaging in play activities. Why do children and some young animals engage in so much play?

John Byers, an animal behaviorist and professor of zoology at the University of Idaho, investigates play behavior. He observed play behavior in many species of Australian marsupials because he is interested in the frequently cited hypothesis that the function of play is motor training. Playful behaviors such as the boxing, kicking, and sparring of juvenile kangaroos and the nipping, chasing, wrestling, and tumbling of young wombats involve motor skills the animals need to practice and master. Mastery of these skills is likely to help the animals find mates, obtain food, and defend themselves as adults. Byers wondered whether play improved the animals' motor skills as a result of muscle growth, increased aerobic capacity, changes in brain structure, or some combination of the three.

To gain insight into this question, Byers compared the brain sizes and rates of play of various Australian marsupial species. He noted that marsupials with a relatively large brain, such as the wombat and the Tasmanian devil, appeared to play more than marsupials having a relatively small brain, such as the koala. Byers noted no such correlation between rate of play and animals' body mass or metabolic rate, measured by the rate at which the body uses oxygen in the breakdown of glucose. These results supported the hypothesis that play acts to modify postnatal brain development, rather than muscle growth or increased aerobic capacity.

Byers then reviewed research that determined the timing and rates of play behavior in various mammalian species, as well as research about the brain development of these animals. He noted a correlation between the timing of play and the development of neural connections in the cerebellum, the portion of the brain that controls motor movement. After birth, an excessive number of connections between neurons form in the cerebellum. Some of these connections are maintained and some are removed, in a process called *synaptogenesis*. Byers determined that the rate of synaptogenesis rises, peaks, and falls in a pattern that matches the pattern of the animals' rate of play, supporting the hypothesis that motor activities, including playful ones, influence neural connections in the cerebellum during development.

Psychologist Steven Siviy at Gettysburg College in Pennsylvania also studies play and brain development. Siviy's studies of rats indicated that play behavior resulted in the production of proteins associated with learning and memory in the brain. Rats not allowed to play did not produce these proteins. Both Byers's and Siviy's studies provide evidence that play changes the brain. To learn more about the role of play in brain development, visit the *In The News* section of this text's companion Web site at www.wiley.com/college/alters and view the video "Child's Play."

*Write your immediate reaction to this research about play and brain development: first, summarize the main point of the research in a sentence or two; then suggest what you think its significance is. You will have an opportunity to reflect on your responses and gather more information on this topic in the* In The News *feature at the end of this chapter. In this chapter, you will learn more about animal behavior.*

## CHAPTER GUIDE

### What is the focus of those who study animal behavior?

**37.1** Animal behaviorists study behavioral mechanisms and how they evolved.

### What are the roles of innate and learned behaviors?

**37.2** Innate behaviors are directed by genes.

**37.3** Simple innate behaviors include orientation to gravity and light.

**37.4** Complex innate behaviors are unchanging sequences of simpler innate behaviors.

**37.5** Learning helps animals change behaviors in response to environmental changes.

**37.6** Animals are capable of various types of learning.

### How are regularly occurring behaviors controlled in animals?

**37.7** Organisms have internal clocks that regulate cycles of physiology and behavior.

**37.8** Migratory behavior results from environmental cues, physiological changes, and hormonal changes.

### How do social behaviors aid the reproductive fitness of individuals?

**37.9** Social behaviors are a form of communication.

**37.10** Competitive behaviors help animals vie for limited resources.

**37.11** Reproductive behaviors help males and females mate.

**37.12** Parenting behaviors help individuals survive to reproductive age.

**37.13** Members of social groups aid one another.

**37.14** Human behavioral capacity is genetically determined but can be modified by learning.

# What is the focus of those who study animal behavior?

## 37.1 Animal behaviorists study behavioral mechanisms and how they evolved.

What type of behavior do you think the "kissing" fish are exhibiting in **Figure 37.1**? You might be surprised to learn that they are exhibiting aggression, not affection. Engaged in a territorial dispute, each is trying to win the rights to its own "space." Their fight begins as they encounter one another at a territorial boundary. First, each tries to intimidate the other, puffing up its body to appear as big and threatening as possible. Then they swim side-by-side, pushing currents of water past one another. These currents provide clues to each contender regarding the size of the opponent. If neither retreats, the pushing match begins, mouths locked in battle. Soon, one fish emerges the victor. The loser signals submission, and the "tournament" is over.

Animals, including humans, continually exhibit a wide range of behaviors. The behaviors exhibited by the kissing fish are simply one example among countless others. **Behaviors** include the patterns of movements, sounds (vocalizations), and body positions (postures) exhibited by an animal. In addition, behaviors include changes in an animal, such as a change in coloration or the release of a scent that can trigger certain behaviors in another animal.

Some behaviors are simple responses to environmental stimuli. A bacterium "behaves" when it moves toward higher concentrations of sugar. Other types of behaviors, such as the territorial behaviors of the kissing fish or the eating behavior of the sea otter are quite complex. In **Figure 37.2**, a sea otter is shown having dinner while swimming on its back. It is using the rock as a hard surface against which to hit the clam and break it open. Often a sea otter will keep a favorite rock for a long time, which suggests that it understands the manner in which it

will use the rock. The sea otter may learn this pattern of eating behavior from others while young, but the capacity to use tools and consciously foresee their future use certainly depends on inherited abilities. Complex behaviors such as these are limited to multicellular animals with a neural network that can sense stimuli, process these stimuli in a central nervous system (brain), and send out appropriate motor impulses.

Before the late 1950s, the study of animal behavior was dominated by **ethology** (ee-THOL-uh-jee). Using a physiological perspective, ethologists observed and interpreted the behavior of animals in their natural environments. They "dissected" behavior patterns into recognizable units, named these patterns, and categorized them. They often focused on the activity of the nervous system and also sought to understand the connection between an animal's behavior and its genetic makeup.

Austrian scientist Konrad Z. Lorenz (see Section 37.6, on imprinting), is one of the best-known ethologists and is referred to by some as the father of modern ethology. In 1973, Lorenz, Dutch ethologist Nikolaas Tinbergen, and Austrian zoologist Karl von Frisch won the Nobel Prize for their contributions to the study of animal behavior. Lorenz based his work on the premise that behaviors change through a process of evolutionary adaptations, just as physical characteristics do. He referred to behaviors as being part of an animal's "equipment for survival." This concept is still fundamental in the study of animal behavior. Today, behaviors that enhance the ability of members of a population to live to reproductive age and that tend to occur at an increased frequency in successive generations are called **adaptive behaviors**.

Another group of animal behavior researchers prominent before the late 1950s was the *behaviorists*. These psychologists focused on behavior alone, studying specific behaviors in the laboratory without regard to any accompanying mental (cognitive) events. In the late 1950s and the early 1960s, the fields of ethology and behavioral psychology merged to form the discipline of **ani-**

**Figure 37.1** Are these "kissing fish" making love or war? This seemingly affectionate behavior is really a display of intimidation and aggression, typical of territorial displays.

**Figure 37.2** Clam bash! The clam appears as a blur as this sea otter tries to break it open.

**mal behavior**. This approach includes many features derived from both behaviorism and ethology.

The field of animal behavior expanded in the late 1970s as the science of *behavioral ecology* emerged. This biological discipline has an evolutionary focus. The central assumption of behavioral ecology is that an animal will behave in ways that will benefit it in the short term and that will maximize its Darwinian fitness, or ability to achieve reproductive success.

Today, the field of animal behavior is composed of researchers from a variety of disciplines: psychology, sociobiology/behavioral ecology (see Section 37.14), and neurobiology. A main difference between the work of animal behaviorists today and ethologists of the past has to do with explanation. Ethologists described, characterized, and cataloged behavior; animal behaviorists today pose hypotheses and test their predictions in an effort to describe how behavioral mechanisms work and explain how they evolved.

## What are the roles of innate and learned behaviors?

### 37.2 Innate behaviors are directed by genes.

Genes play a role in the development of behaviors because they direct the development of the nervous and endocrine systems upon which behaviors depend. In addition, automatic responses depend on specific nerve pathways within the central nervous system of an organism. These pathways are genetically determined.

Genetically determined neural programs are part of the nervous system at the time of birth or develop at an appropriate point in maturation, resulting in **innate behaviors**. These instinctive (inborn) behaviors are performed in a reasonably complete form the first time they are exhibited. A human newborn, for example, will turn to suckle when touched on the cheek near the mouth. Innate behaviors are important to the survival of an animal because they help it stay alive in certain situations and provide adaptive advantages that contribute to its fitness, or ability to achieve reproductive success.

Ethologists such as Lorenz and animal behaviorists of today would argue about the role of environment and learning on innate behaviors. **Learning** is an alteration in behavior based on experience. Today, the rigid distinction between innate and learned behavior that was once held by ethologists no longer exists. Evidence suggests that both genetic factors *and* the experience of the individual influence many aspects of behavior.

### 37.3 Simple innate behaviors include orientation to gravity and light.

Survival is linked to how an organism responds to its environment. Certain types of responses are characteristic of all members of a species and thus are considered innate. Reflexes, kineses, and taxes are simple types of innate behaviors.

## *just wondering . . .*

### Questions students ask

#### Why do people yawn?

Your question could be broadened to ask why most animals yawn—there is nothing particularly human about yawning. Most carnivores yawn. Few herbivores seem to yawn, although there are exceptions—hippos have enormous yawns. Frogs yawn. Even fish appear to yawn. Despite this widespread occurrence of yawning, there may be multiple explanations for it—different species may yawn for different reasons.

It used to be hypothesized that a human yawn was a silent scream for oxygen that usually occurs when people are tired or bored—a type of deep breath to increase oxygen in the blood or to get rid of excess carbon dioxide. Not so. When a group of freshman psychology students at the University of Maryland inhaled air containing different mixtures of oxygen and carbon dioxide and counted their yawns, only breathing rates went up or down to compensate for changing levels of oxygen and carbon dioxide. Yawning rates did not change.

Some researchers think that yawning is the body's way of promoting arousal in situations where you have to stay awake. Most humans yawn when stimulation is lacking. When a team of yawn counters observed people engaged in various activities, their results supported this "lack of stimulation" idea. For example, they found that people riding subway cars yawned far more often when the cars were empty. The arousal hypothesis helps explain why people driving late at night on the highway yawn a lot and why very few people yawn when they are actually in bed—they don't need to stimulate themselves with a yawn because it's okay to go to sleep.

Whether this hypothesis is plausible, one thing is certain: Yawning is highly contagious. Seeing another person yawn creates a powerful urge to yawn yourself. This trait seems to be restricted to humans; no other animal responds in this way. Although yawning is not a hotbed of research, there is still much to learn about this behavior because yawning seems to be associated with many diseases in ways not yet understood—brain lesions and epilepsy often lead to excessive yawning, but people with schizophrenia yawn very little. ●

*Are you wondering about a topic in biology and how it relates to your life? Submit your question by clicking the Just Wondering link in this text's companion Web site at www.wiley.com/college/alters.*

A **reflex** is an automatic response to nerve stimulation. In complex organisms, reflexes play a role in survival, such as when you pull your hand away from a hot stove before you consciously realize that your hand hurts. In animals with extremely simple nervous systems, such as the cnidarians (hydra, jellyfish, sea anemones, and coral, see Figure 32.2), most behaviors are the result of reflexes. In these animals, a stimulus is detected by sensory neurons, and the impulse is passed on to other neurons in the animal's nerve net, eventually reaching the body muscles and causing them to contract. There is no associative activity in which other neurons can influence the outcome, no control of complex actions, and little coordination.

A **kinesis** (pl. kineses) is a change in the speed of the movements of an animal with respect to changes in environmental stimuli. Put simply, movement slows down in an environment favorable to the animal's survival and speeds up in an unfavorable one. For example, pillbugs living under a rotting log or clump of leaves in a wooded area move slowly in this favorable environment. But if you pick up the log or clump of leaves, the pillbugs scurry about rapidly. Kineses are nondirected types of movements. In other words, an animal is not attracted to a favorable environment; it tends to "blunder" there as it moves about quickly, and then, once there, it moves about slowly.

A **taxis** (pl. taxes), unlike a kinesis, is a directed movement toward or away from a stimulus, such as light, chemicals, or heat. Animals having innate taxes also have receptors that sense the particular stimuli to which the animal can orient.

Female mosquitoes and ticks, for example, have sensory receptors that detect warmth, moisture, and certain chemicals emitted by mammals. Sensing these stimuli helps these organisms orient to their victims. Some mosquito repellents work by "blocking" the insect's receptors so that it cannot sense and then locate its victim. The crowding of flying insects around outdoor lights is a familiar example of a taxis called positive *phototaxis*. Other organisms, such as the common cockroach, avoid light; they are negatively phototactic.

Fish swim upright by orienting their ventral side to gravity and their dorsal side to light, as shown in **Figure 37.3** ①. In ②, when the light is moved to the bottom of the tank, the fish are unable to orient to both stimuli; the orientation to gravity is the stronger response. If the gravity-detecting organ is removed, the fish orient only to light, as shown in ③ and ④. Certain species of fishes, such as trout and salmon, also automatically orient against a current and therefore face and swim upstream, as shown in **Figure 37.4**. Orientation to water currents is called *rheotaxis*.

Even organisms without a nervous system, such as protists and bacteria, exhibit taxes. These organisms have cellular organelles and other cellular components that act as receptors or that react to specific environmental stimuli.

## 37.4 Complex innate behaviors are unchanging sequences of simpler innate behaviors.

**Fixed action patterns**—behavior patterns elicited by specific stimuli, apparently innate and specific to a particular species—are more complex than other innate behaviors. These patterns are made up of an unchanging order of muscular movements, such as the movements of a mother cardinal popping food into the mouths of her young. Such sequences are recognizable "units" of behavior. An environmental cue called a **sign stimulus** triggers the behavior. In the

**BEHAVIOR WITH RESPECT TO LIGHT AND GRAVITY**

UNALTERED FISH      FISH WITH GRAVITY-DETECTING ORGAN REMOVED

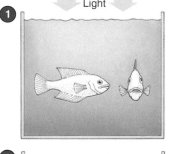
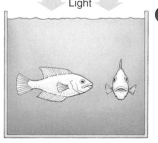
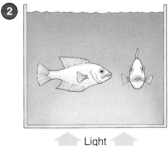
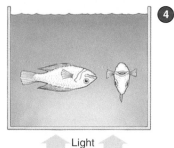

**Figure 37.3 Phototaxis and gravitaxis in fish.** Unaltered fish orient to both light and gravity. Fish that have had their gravity-detecting organ removed orient to light only and have difficulty staying upright. Orientation to gravity overrides orientation to light, as shown by ②.

 **Visual Thinking:** If the source of light were at the right of the container, how would the fish with a gravity-detecting organ orient themselves? How would the fish without a gravity-detecting organ orient themselves?

**Figure 37.4 Positive rheotaxis in salmon.** Salmon leap up a small waterfall in Alaska—swimming against the current—while a hungry grizzly bear watches. Positive rheotaxis means facing upstream.

case of the mother cardinal, the sign stimulus is the gaping mouths of the young birds. After the stimulus, the sequence of activity begins and is carried through to completion. Fixed action patterns are often seen in body maintenance behaviors, such as a cat washing its face.

The classic example used to illustrate a fixed action pattern is the behavior of a graylag goose retrieving an egg that has rolled out of its nest. As shown in **Figure 37.5**, if a goose notices that an egg has been knocked out of the nest (the sign stimulus), it will extend its neck toward the egg and roll the egg back into the nest with its bill. Because this behavior seems so logical, it is tempting to believe that

the goose saw the problem and figured out what to do. The entire behavior, however, is instinctive, as indicated by the fact that any rounded object, regardless of size or color, acts as a sign stimulus and triggers the response. A beer bottle, for example, is an effective sign stimulus. Also, the goose completes the action pattern even if the egg rolls away from its retrieving bill. Instead of stopping in the middle of the behavior, the goose completes the sequence and then begins it again, until the egg is brought back to the nest. This shows that the action pattern is truly a fixed "unit" of behavior.

### CONCEPT CHECKPOINT

1. European cuckoos lay their eggs in the nests of other species of birds and then abandon them. Soon after a cuckoo chick hatches, it uses its head and beak to eject any other eggs from the nest. When the "foster" parent returns to the nest, the cuckoo chick innately raises its head, opens its mouth, and begins chirping. Classify the cuckoo chick's behavior as fixed action pattern, taxis, kinesis, or reflex.

**Figure 37.5** Fixed action pattern in a graylag goose. An egg or other similarly shaped object near the nest triggers a response in the goose to extend its neck toward the object and roll it back into the nest with its bill.

## 37.5 Learning helps animals change behaviors in response to environmental changes.

Although important in many respects to the survival and fitness of animals, innate behaviors can become a liability if environmental conditions change, since innate behaviors cannot change. For this reason, behavior patterns that can change in response to experience have adaptive advantages over the set programs of instinct. In most animals, only some behaviors are innate and permanent; many behaviors can be changed or modified as an individual organism learns from experience. The more scientists learn about the nature of animal behavior the more the line between learned behaviors and innate behaviors is blurring.

Learned behaviors that help an animal become better suited to a particular environment or set of conditions can be grouped into five categories: imprinting, habituation, classical conditioning, trial-and-error learning, and insight. As you read about these types of learning, notice that the first four categories regard learning as automatic and machinelike, in a stimulus–response fashion. Learning portrayed in this way, possibly reinforced by some type of reward that may or may not be readily apparent, reflects a school of thought called **behaviorism**. Behaviorism, which flourished from 1900 to 1960, was the theoretical basis for many of the rote, drill-and-practice approaches to learning that were widely used in schools during that time. Behaviorism is still popular today; behavior analysis has been extensively applied not only in educational settings, but also in clinical and industrial settings.

More recently, another school of thought regarding how we learn has emerged. **Cognitivism**, the prominent view in psychology today, suggests that individuals acquire and then store information in memory. Learning takes place as new information builds and merges with the old, leading to new types of behavior. Cognitivism is the theoretical basis for the constructivist approach to learning, which has gained wide acceptance in education.

## 37.6 Animals are capable of various types of learning.

### Imprinting

Innate behavior and learning interact very closely in a time-dependent form of learning known as **imprinting**. Imprinting is a rapid and irreversible type of learning that takes place during an early developmental stage of some animals. Various types of imprinting exist. One type, object imprinting, has been observed in birds such as ducks, geese, and chickens. During a short time early in the bird's life, the young animal forms a learned attachment to a moving object. Usually, this object is its mother. However, animals can imprint on other types of objects regardless of size or color, such as balloons, clocks, and people. From an evolutionary viewpoint, parent–offspring imprinting enhances reproductive fitness by allowing parents and offspring to "recognize" one another; this enables parents to care for their offspring. Researchers have found that imprinting tends to occur in species that have a social organization in which attachment to parents, to the family group, or to a member of the opposite gender is important.

Konrad Lorenz performed classic experiments regarding object imprinting during the 1930s. In one of his most famous experiments, Lorenz divided a clutch of graylag goose eggs in half. He left one half of the eggs with the mother goose and put the other half in an incubator. The half that hatched with their mother displayed normal behavior, following her as she moved. As adults, these geese also exhibited normal behavior and mated with other graylag geese. The other half of the clutch, however, hatched in the incubator and then spent time with Lorenz. These goslings did not behave in the same manner as their siblings; they followed Lorenz around as if he were their mother, as shown in **Figure 37.6**. After their initial time with him, Lorenz introduced the geese to their mother, but they still preferred to follow Lorenz. As adults, these geese tried to "court" adult humans!

Many migrating birds and fishes learn to recognize their birthplace by another type of imprinting, called locality imprinting. Pacific salmon, for example, are imprinted with the odor of the stream or lake in which they were born. Amazingly, two to five years later, when they return from the sea to spawn, they are able to find their birthplace by its odor.

### Habituation

**Habituation** is the ability of animals to "get used to" certain types of stimuli. People get used to stimuli all the time. A gunshot would startle most people, but this response would lessen at the end of a day at a firing range. Not only would you get used to the noise level (and wear ear protection), but you would perceive the gunshots as nonthreatening. In the same way, you quickly get used to the feel of your clothing after dressing in the morning. Animals also stop responding to stimuli that they learn are neither harmful nor helpful. Learning to ignore unimportant stimuli is a critical ability in an animal confronting a barrage of stimuli in a complex environment and can help an animal conserve its energy. You may have observed that pigeons living in the city are undisturbed by usual city noises, while those living in quiet settings fly away when they hear noise. Similarly, hydra stop contracting if they are disturbed often enough by water currents, and sea anemones stop withdrawing if they are

**Figure 37.6** Konrad Lorenz with some of his imprinted goslings. These goslings hatched in an incubator and then spent time with Lorenz. They followed Lorenz as if he were their mother.

 **Visual Thinking:** Explain the behavioral mechanism that results in the goslings following Lorenz.

touched repeatedly. In another example, young black-headed gull chicks will initially crouch if any object flies overhead. Gradually, the chick stops crouching in response to familiar objects overhead—the chick has gotten used to them and has learned that they pose no threat. However, the chick will still crouch when an unfamiliar object appears perceiving it as a threat.

### Classical conditioning

**Classical conditioning** is a form of learning in which an animal is taught to associate a new stimulus with a natural stimulus that normally evokes a response in the animal. Repeatedly presenting an animal with the new stimulus in association with the natural stimulus can cause the animal's brain to form an association between the two stimuli. Eventually, the animal will respond to the new stimulus alone; that is, the new stimulus will act as a substitute for the natural stimulus. The connection between the new stimulus and the natural stimulus is the result of a learning process, but the connection must be reinforced periodically by the presence of the natural stimulus or the animal will stop responding to the substitute.

In his famous study of classical conditioning, the Russian psychologist Ivan Pavlov conditioned dogs to salivate in response to a stimulus normally unrelated to salivation. Interestingly, Pavlov's discovery of conditioning was a side effect of his work on the physiology of the digestive system. In 1904, he was awarded the Nobel Prize for his work on digestion. As part of his research, Pavlov routinely collected saliva from his laboratory dogs. He stimulated saliva flow by placing meat powder in the dog's mouth. He soon noticed that the dog began salivating as he approached it with the meat powder. He then paired other, unrelated stimuli with the meat powder. For example, he shone a light on the dog at the same time that meat powder was blown into its mouth. As expected, the dog salivated. After repeated trials, the dog eventually salivated in response to the light alone. The dog had learned to associate the unrelated light stimulus with the meat stimulus.

An animal's tendency to form stimulus–response connections is an inborn reflex, or an *unconditioned response*. Such innate responses are important to animals; many of them are protective, such as blinking, sneezing, vomiting, and coughing. The work of researchers such as Pavlov contributed to the understanding that animal behavior depends on innate neural circuitry but that most of these neural programs can be modified and directed by learning.

### Trial-and-error learning: operant conditioning

More complex than imprinting or habituation is **operant conditioning** (OP-uh-runt), a process in which the frequency of occurrence of a behavior is modified by the consequences of the behavior. (The word *operant* means "having the power to produce an effect"; it is a form of the verb "to operate.") In operant conditioning, an animal must make the proper association between its response (such as pressing a lever) and a positive stimulus (the appearance of a food pellet). It may also learn to avoid a behavior when associated with a negative stimulus. Animal trainers use the techniques of operant conditioning.

The American psychologist B. F. Skinner studied such conditioning in rats by placing each in a specially designed box (today called a *Skinner box*) fitted with levers and other experimental devices (**Figure 37.7**). Once inside, the rat would explore the box feverishly, running this way and that. Occasionally, it would accidentally press a lever, and then a pellet of food would appear. After this happened a few times, the rat would start purposely pressing the lever to obtain food. It had learned to associate this behavior with getting food.

This sort of trial-and-error learning is of major importance to most vertebrates in nature. For example, when a toad gobbles a bumblebee, it gets a sharp sting on its tongue and spits out the bee. The toad learns quickly not to eat a bumblebee. The toad does not use reasoning to determine that the bumblebee is not good to eat—it merely becomes conditioned by experience to avoid a response that causes it pain.

### Insight

Best developed in primates such as chimpanzees and humans, **insight**, or **reasoning**, is the most complex form of learning. An animal capable of insight can recognize a problem and solve it mentally before trying out the solution. This enables the animal to

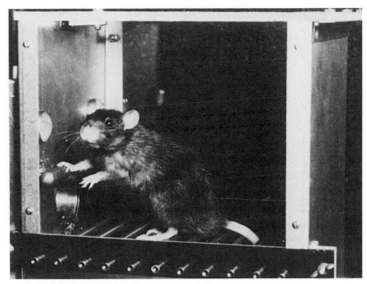

**Figure 37.7** Operant conditioning. This rat is in a Skinner box fitted with a lever that, when pressed, provides food. Such setups were used in early conditioning experiments.

perform a correct or appropriate behavior the first time it tries, without ever before having been exposed to the specific situation. However, previous experiences, such as situations in which conditioning or other types of learning took place, contribute to insight.

German psychologist Wolfgang Kohler was the first to describe learning by insight. Performing extensive experiments on chimpanzees in the 1920s, Kohler showed that an animal must perceive relationships and manipulate concepts in its mind to solve a problem on the first try. In his classic experiments, he placed a chimpanzee in a room with a few crates, poles that could be joined together, and a bunch of bananas hung high out of reach of the animal. After assessing the situation, the chimpanzee would stack the crates, join the poles, and get the bananas. Unlike chimpanzees, most animals are unable to use insight to solve problems.

> **CONCEPT CHECKPOINT**
>
> **2.** For each of the following examples of animal behavior, indicate whether the learning involves imprinting, habituation, classical conditioning, trial-and-error learning, or insight.
>
>   a. White crowned sparrows exposed to their species' song once during the first 50 days of life are able to sing their species' song as adults. However, sparrows raised in isolation without hearing their species' song during this 50-day period are not able to sing their species' song as adults.
>
>   b. Ravens exposed to food hanging from a long piece of string tied to their horizontal perch exhibit a variety of novel behaviors as they use different strategies to acquire the food. One raven, upon first exposure to the situation, was able to get the food by using its beak to pull up the string while intermittently using its foot to secure the string and prevent the food from dropping.

# How are regularly occurring behaviors controlled in animals?

## 37.7 Organisms have internal clocks that regulate cycles of physiology and behavior.

Animals, plants, protists, fungi, and even bacteria have internal clocks that regulate many of their activities. These regular cycles of physiological activity and behavior are called **circadian rhythms** (sir-KAY-dee-un). Circadian rhythms are regulated by organisms' **biological clocks**. Often, environmental cues such as light help keep these internal clocks set to the outside world. Recently, scientists have gained some insight into exactly what a biological clock is and how it governs rhythmic patterns such as sleep and wake cycles, feeding patterns, photosynthetic activity, and animal movements such as migration.

Circadian rhythms have a genetic basis; that is, genes regulate these cyclical patterns of behavior and physiology. Scientists have discovered ways in which these regulatory genes are turned on and off.

Studies in mice demonstrated that a protein known as CLOCK regulates circadian rhythms in mammals. In both fruit flies and mice, CLOCK binds with another protein called BMAL1. The CLOCK–BMAL1 complex binds to a specific sequence of DNA, activating a gene that drives the transcription of CLOCK proteins. The CLOCK proteins accumulate over the course of the day. A certain level of CLOCK proteins is a trigger that turns the gene off. Gene inactivation (after a lag) is, in turn, a trigger for CLOCK–BMAL1 to reactivate the gene.

This type of regulatory mechanism is known as a *negative feedback loop* (see Chapters 26 and 35). Scientists suspect that this negative feedback loop is only one factor among other, yet unknown factors that govern circadian rhythms. In addition, the same biological clock genes appear to govern circadian rhythms in fruit flies, mice, and people. This suggests that all animals depend on the same molecular mechanisms to keep time.

## 37.8 Migratory behavior results from environmental cues, physiological changes, and hormonal changes.

**Migrations** are movements of animals from one region to another. Seasonal migrations occur with a change of seasons. Diurnal migrations occur daily. In many animals, migrations occur once a year and result from interactions among various environmental factors, such as daylength, with the animals' physiological changes, including hormonal changes. Ducks, geese, and shorebirds for example, migrate down flyways from Canada across the United States each fall and return each spring. A good example of a long migration is that of the American golden plover, a bird that flies from Arctic breeding grounds to wintering areas in southeastern South America, a distance of approximately 13,000 kilometers, or about 8000 miles (**Figure 37.8**).

Monarch butterflies, which are thought to be the only migrating insects, travel long distances too, from the eastern United States to Mexico and back, a journey of more than 3000 kilometers, or approximately 2000 miles, each way. The monarchs that make the trip to Mexico are the result of breeding that took place in late August. The shorter days and colder temperatures of late August cause the emerging monarchs to postpone reproductive maturity. These butterflies will live for eight or nine months—much longer than monarchs that develop earlier in the summer—and can endure the flight to Mexico. In Mexico in the spring, the eight- or nine-month-old monarchs reach sexual maturity and begin migrating back to the

**Figure 37.8 Migration routes.** The American golden plover and its migration route from Arctic breeding grounds to wintering areas in southeastern South America.

 **Visual Thinking:** The American golden plover is a seasonal migrant with a long migration route, as you can see on this map. How does the plover "know" when to fly south and when to fly north?

eastern United States. They mate all along the migratory route. These monarchs do not complete the return trip, but their offspring or the next generations of offspring do.

Migrating animals also have the ability to navigate in order to reach their destinations, by orienting themselves in relation to environmental cues such as landmarks, the position of the sun or stars, the Earth's magnetic field. Animals that migrate by day, such as some birds, ants, and bees, primarily use the sun's position to chart a course. Night migrants, a group that includes birds such as the indigo bunting, primarily use the stars to chart a course. The biological clocks of all these animals interact with information from the environment to help migrant animals find their way.

How do animals actually use environmental cues to navigate? One good example is the *star compass*, used by many species of bird migrants that travel at night. Scientists first gained insight into how birds used the stars to aid their travel in planetarium studies in the late 1950s and early 1960s. Much of this work was conducted by behavioral ecologist Stephen Emlen in the 1960s on the indigo bunting, a bird that breeds in the eastern United States in the spring and migrates southward in the fall.

Emlen caged the birds during their normal periods of migration. During this time, the caged birds were restless, hopping and fluttering their wings. He outfitted the cages with an inked floor and sloping white paper on the sides, leaving the top uncovered so that the birds could see the night sky. From the markings the birds' inked feet made on the paper, Emlen determined that the birds' activity was oriented in the direction they would normally migrate at that time—northward in the spring and southward in the fall. If Emlen blocked the birds' view of the night sky, however, their movements became random.

Emlen also conducted planetarium studies in which he systematically covered part of the night sky in the planetarium dome and observed the direction in which the birds would fly. He deter-

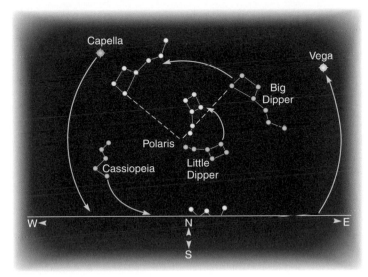

**Figure 37.9** Birds that migrate at night use the stars as a compass. The constellations seem to rotate around the North Star (Polaris) in the northern hemisphere. This illustration shows the positions of stars in the northern sky during the spring. The arrows indicate the direction of the apparent rotation and the change in star positions from early evening to about six hours later.

mined that the birds gauged direction using the pattern of stars nearest the North Star. The North Star provides a reliable reference point for due north in the Northern Hemisphere because Earth's axis points almost directly at it, which makes all other stars appear to rotate around it (**Figure 37.9**). Emlen determined that indigo buntings learn how to "read" the night sky during their first summer of life.

# How do social behaviors aid the reproductive fitness of individuals?

## 37.9 Social behaviors are a form of communication.

If you've ever observed chimpanzees at the zoo or in a nature preserve, you've probably seen the grooming behavior exhibited in **Figure 37.10**. Occasionally chimps groom themselves, but social grooming—one chimp grooming another—is more common. Not only does social grooming help chimps keep clean and relatively free of parasites, it strengthens the social bonds among them.

Grooming among chimps is only one among many **social behaviors**, which help members of the same species communicate and interact with one another. Each member responds to stimuli from others. Social animals seek out each other's company and interact with one another. They do not passively group together like pillbugs under a rotting log or like moths drawn to a light. The biology of social behavior is called **sociobiology**, or **behavioral ecol-**

**Figure 37.10** Social grooming among chimpanzees. One chimpanzee cleans dirt, debris, and parasitic insects from another's body, which strengthens the social bond between them.

ogy. This science applies the concepts of evolutionary biology to social behavior.

Living in social groups has numerous advantages and disadvantages, which are usually referred to as costs and benefits. The costs and benefits of group living differ from species to species. In general, however, the costs of sociality to individuals includes increased competition for the basics of life: mates, nesting sites, or food. Being with others also increases the probability of being infested with parasites or infected with disease. Among animals that care for their young, there may be an increased probability of caring for other individuals' young or losing their own young in the group. As you can see in **Figure 37.11**, it can be difficult to find your own offspring when the group is large and members live closely together!

On the other hand, there are many benefits of sociality. Predators have more difficulty attacking groups than attacking individuals. Groups are better able than individuals to detect predators, and one group member can warn the entire group. Groups also have a better ability to confuse and repel predators, thus defending their food, offspring, and territories. In addition, animals that hunt in groups have advantages over animals that hunt alone, including the ability to overcome prey larger than themselves.

Social interactions imply **communication**—information that is transferred from one individual to another and that is a stimulus for the other's reaction. Animals communicate with one another in a variety of ways and convey many types of messages, which serve many functions. For example, sea lions greet one another by rubbing noses (**Figure 37.12**). This social behavior communicates a message of nonaggression. That is, one sea lion communicates to another that he is friend, not foe. The animal being greeted, then, knows that a nonaggressive reaction is appropriate at the time. Both the sender of the message and the receiver benefit in that no fighting takes place and neither animal is injured.

**Figure 37.12** Sea lions greet one another by rubbing noses. This behavior between these two Australian sea lions communicates nonaggression.

How did animal communication evolve? Animal behaviorists think that communication began as animals associated preliminary actions, such as opening the mouth, with subsequent actions, such as biting. That is, the preliminary action would serve as a cue to the subsequent action. Animals noticing this cue would benefit by reacting defensively or fleeing before being bitten. Animals giving the cue might benefit as well. If animals could communicate to opponents that a fight was coming by opening their mouths and baring their teeth, and if that cue caused opponents to back down, then the senders of the cue would benefit by winning without fighting. Animals communicating well regarding this cue would have an adaptive advantage over others who did not. Fewer of the "good" communicators would be killed or injured in fights than would the "bad" communicators. Therefore, more of the good communicators would live to reproduce and would pass on the genes for effectively sending or receiving this cue. Thus, evolution would favor an increase in the amount of information sent and received related to the cue. The cue would thus become a sign stimulus, triggering a behavior. Sign stimuli communicated by members of the same species are called social releasers, or simply **releasers**.

## 37.10 Competitive behaviors help animals vie for limited resources.

Social interactions include conflicts within groups, because individuals seek to maximize their benefits and minimize their costs. Conflicts within a group can be resolved through a variety of aggressive behaviors such as threats, chasing, and fighting, as well as a variety of submissive behaviors such as appeasement or avoidance. Conflicts in stable groups are often minimized by territoriality and dominance hierarchies.

When two or more individuals strive to obtain the same needed resource, such as food, water, nesting sites, or mates, they are exhibiting **competitive behavior**. This type of behavior occurs when resources are scarce.

**Figure 37.11** A "daycare center" for Mexican freetailed bat young. Mothers leave the young at night to forage for food. On returning, it is difficult for them to tell their own offspring from others' in this large, closely packed group. In this species, one cost of group living is possibly nursing others' young and not your own.

## Threat displays and submissive behavior

How do animals compete? Many first engage in **threat displays**, or **intimidation displays**, a form of aggressive behavior. The purpose of these displays is to do as the name suggests: Scare other animals away or cause them to back down before fighting takes place. Threat displays vary widely among species but usually involve such behaviors as showing fangs or claws, making noises such as growls or roars, and making the body appear larger by standing upright, making the fur or hair stand on end, or inflating a throat sac. The common toad in **Figure 37.13** is exhibiting a threat display; it inflates its body and sways from side to side in an attempt to appear larger.

Threat displays are important social signals that communicate the intent to fight. Interestingly, some of the movements and body postures (body language) of threat displays that repel competitors also attract members of the opposite gender. Biologically and evolutionarily this makes sense, because the competition may be over a mate.

Animals also use submissive behavior to avoid fighting. **Submissive behavior** includes making the body appear smaller, "putting away weapons" by concealing fangs or claws or turning a vulnerable part of the body to an opponent. As you can tell, submissive behavior usually looks like the opposite of a threat display. If an animal is losing a fight, it might display submissive behaviors to stop the fight. Contrary to popular belief, animals of the same species rarely fight to the death. One reason is that most animals do not have the means to do so—they do not have sufficiently dangerous fangs, claws, or horns, for example. Species with dangerous weapons usually have defenses against those weapons too, such as

**Figure 37.14** The common green darner dragonfly (*Anax junius*). The male patrols its territory aggressively at speeds of up to 35 m.p.h. This pond dragonfly, also called the mosquito hawk and the lord of June, patrols about eight feet up and eight feet in from the pond's edge, hovering occasionally.

a strong hide, long hair, or a thick layer of body fat. Scientists hypothesize that aggression within a species has evolved as a means of chasing off rather than killing rivals.

## Territorial behavior

A territory is an area that an animal marks off as its own, defending it against the same-gender members of its species. These behaviors are called **territorial behaviors**. Members of the opposite gender are sometimes allowed into the territory, often for mating purposes.

Territoriality is very common in all classes of vertebrates and is even found in some invertebrates, such as crickets, wasps, and praying mantises. Some animals, such as dragonflies (**Figure 37.14**), mark their territories by conspicuously patrolling their borders, resting on and then moving from prominent landmarks. As an animal demonstrates the borders of its territory, it may exhibit specific movements or body postures as signals. Some animals, such as songbirds, monkeys, apes, frogs, and lizards, mark their territory by producing sounds that announce their ownership. Animals with a well-developed sense of smell, such as wolves, hippopotamuses, rhinoceroses, some rodents, and even domestic dogs, mark their territories with substances that have an odor, such as urine. Many of these species, in fact, possess special scent glands that secrete substances just for this purpose. The odor marking helps keep out members of the same gender and helps owners of large territories orient themselves to its borders.

Territorial behavior has several adaptively important consequences. In many mammalian societies, territoriality often has an important influence on the establishment of new populations, with the dominant male defeating the lesser males and leaving them to found their own populations. Individuals surviving in the surrounding marginal areas repopulate any vacant territories.

Although territoriality evolved because it benefits individuals, it also has an effect on the resources available to all members of a

**Figure 37.13** The common toad exhibiting a threat display. The toad inflates its body, attempting to appear larger to the snake. If the snake is intimidated, it may back down and not strike.

population. When resources such as nesting sites and food are limited, each member of a population may be endangered if resources are spread too thin and no member gets an adequate amount. In addition, breeding pairs of animals will not reproduce unless they have adequate resources. Territorial behavior ensures that at least some members of the population will have their own space, adequate food, and nesting sites, enabling them to survive and reproduce. Territorial behaviors also enhance reproductive capabilities by placing the peak time of competition and aggressive behavior at the time of the marking of the territory, *before* the time of reproduction and raising of young.

### Dominance hierarchies

Groups of fishes, reptiles, birds, and mammals exhibit a **dominance hierarchy**, or *rank order*. In chickens, the rank order is often called a *pecking order* because these animals establish the order by pecking at one another. In many species of animals, the rank order is linear, with the highest-ranking individual dominant over all the others and the lowest-ranking individual submissive to all others. Some rank orders show certain complexities within the ranking, such as a rank order in which animal "A," for example, may be dominant over animal "B" and "B" dominant over "C," but "C" may be dominant over "A." In some rankings, the sex of the animal may play no role. In other rankings, males and females are ranked separately, and in still other vertebrate groups, the female takes on the rank of the male with which she is associated.

Dominance hierarchies help reduce the costs of aggression and fighting within social groups, restricting aggressive behavior mainly to the period when the ranking is developed. By reducing the chance that individuals will be killed or injured in confrontations within the group, both the dominant and subordinate, or submissive, animals benefit. The dominant animal does not need to expend energy to win the fight, and the submissive animal does not become injured or killed in a fight it would lose anyway.

## 37.11 Reproductive behaviors help males and females mate.

You've likely heard the phrase "sex sells" and know that sex is a powerful force not only in marketing but also in many other aspects of human life. Without it, our species would be extinct. Sex is also a powerful force in other animals.

For sexual reproduction to take place, animals of most species must communicate and cooperate with one another. Individuals within many species of animals form pair bonds for the purposes of reproduction and parenting. An important outcome of these processes is the retention of genes responsible for the relevant social behavior. For natural selection to favor a particular gene within a population, the advantages of the particular social trait it encodes must outweigh the disadvantages—for the individual or for the individual and its *kin* (close relatives).

Behaviors that promote successful sexual reproduction, then, are highly adaptive. Organisms that reproduce sexually must exhibit patterns of male–female interactions that lead to fertilization—the penetration of an egg by a sperm. Fertilization can be preceded by copulation, the joining of male and female reproductive

organs. However, some organisms such as many fish and frogs do not copulate; instead, the male releases sperm over the eggs after they have been deposited by the female. The term **mating** refers to male–female behaviors that result in fertilization, regardless of whether copulation occurs. The term **courtship** refers to the behavior patterns that lead to mating.

For courtship to take place, males and females must first find each other. In some animal species, males and females live together in pairs or groups, but in many species they do not. Most often the job of attracting a mate goes to the male of the species. The male peacock has a brilliantly colored, ostentatious fan of tail feathers, but not the female peahen. The male lion has a beautiful mane, but not the female. Many species of male scarab beetles have impressive horns, while females have none.

Why do males display more than females? Females are choosy. They have more to lose than do males when selecting a mate. Why? Females invest more energy into developing eggs than do males in developing sperm. After fertilization, females generally incubate or care for developing embryos. Finally, females generally nurture the young. Males compete for "choosy" females in a variety of ways.

Charles Darwin originally proposed that a process called sexual selection provides the adaptive benefit that outweighs the cost of developing colorful and ostentatious structures needed to attract and compete for females. The cost of being conspicuously colored or horned, for example, is that it attracts predators as well as mates. Developing such structures also requires an investment of energy. The benefit, however, is being chosen by a female for mating, thereby passing on genes that are retained in the gene pool.

**Sexual selection** is a special kind of natural selection that depends on a struggle between the individuals of one sex, generally the males, for mates of the other sex (see *Just Wondering*, Chapter 16). There are two types of sexual selection. In one kind, individuals compete for opportunities to mate. The winners reproduce more than the losers. Selection acts on the traits that play a role in winning, such as having large horns, antlers, or teeth. In the other kind of sexual selection, members of one sex prefer some individuals over others as mates. Again, selection acts on the inherited traits that play a role in the preference, such as bright plumage or other characteristics that are signals to females of sufficient sperm, useful resources, good parenting skills, or simply "good genes."

Males often attract a mate by marking a territory and defending it against intrusion by other males. Aggressive behaviors against other males are often the same behaviors that attract females. In addition, males can attract females by displaying body colors or markings, like the peacock. Many birds sing songs to attract a mate, and certain frogs and many insects use other sounds, or mating calls, for this purpose. Many male mammals and insects produce odors that are attractive to females. In some species, the males congregate and perform dances or songs as a group to attract mates (**Figure 37.15**). See Chapter 33 for a discussion of how male and female elephants use infrasound to find each other for mating.

After the formation of a male–female pair, a **courtship ritual**, unique to each species, leads to mating. Courtship behaviors usually consist of a series of fixed action patterns of movement, each triggered by some action of the partner that acts as a releaser and, in

**Figure 37.15 Courtship dance of the sage grouse.** At one point in this series of behaviors, a male displays his fanned-out tail feathers and puffed-up throat sac to a female.

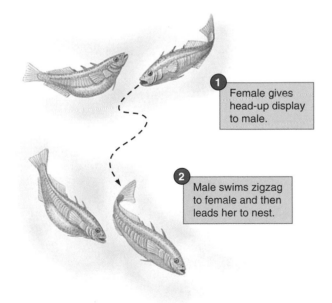

1. Female gives head-up display to male.

2. Male swims zigzag to female and then leads her to nest.

turn, triggers the next movement pattern by the potential mate. The courtship behaviors of the stickleback fish exemplify this repeating releaser–trigger (**Figure 37.16**). Some species, such as various types of geese, exhibit postcopulatory behaviors: Both birds rise out of the water, wings extended, facing one another. Such behaviors appear to strengthen the pair bond, which is necessary when both parents care for the young.

## 37.12 Parenting behaviors help individuals survive to reproductive age.

For offspring to survive, grow, and eventually reproduce, parents must either make preparations for the care of their young or care for their young themselves. Parenting behaviors are a type of **altruistic behavior**. Such behaviors benefit one organism at the cost of another. In animal behavior, this term is different from the everyday use of the word. When we speak of altruistic behaviors of humans, we imply that the person performing the altruistic act is knowingly and willingly risking or giving up something for another. For example, people who risk their lives to save others are performing altruistic acts. With animals other than humans, however, the focus of altruism is evolutionary: An altruistic act is one that increases the fitness of the recipient while decreasing the fitness of the donor. Natural selection favors genes that promote altruistic behavior among kin such as parents and offspring.

Parenting behaviors are found in almost all animal groups, but the participation of the parents varies. In many species of birds and fishes, both parents care for the young, and a division of labor usually exists, with each parent carrying out a specific role. Among mammals, females produce milk to nourish the young, so the mother most often assumes the parenting role. In only a few species, such as certain fishes (including sea horses) and birds, does the male alone take care of the offspring.

To prepare for their young, a food supply is usually gathered, and the animals build protective structures such as nests or cocoons in areas that lie near a ready source of additional food. Marsupials such as kangaroos—mammals in which the young are born developmentally early—keep their young safe and nourish them as they complete their development by storing them in a pouch, conveniently situated at the mammary glands.

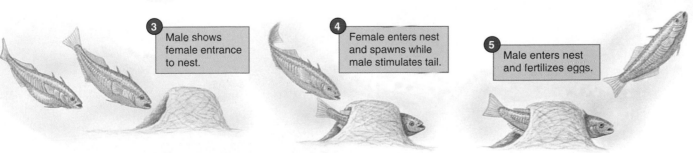

3. Male shows female entrance to nest.

4. Female enters nest and spawns while male stimulates tail.

5. Male enters nest and fertilizes eggs.

**Figure 37.16 Stickleback courtship.** Each set of movements in the courtship ritual is a fixed action pattern that triggers the next set of movements by the partner.

## 37.13 Members of social groups aid one another.

In addition to pair-bond relationships, the individuals of many species establish temporary or permanent associations with many other members of their species, forming social groups. Animals within social groups work together for common purposes and derive the benefits of sociality, as noted in Section 37.9. Individuals within a group, for example, have more protection against predators than do single individuals or pairs. They can warn each other of danger by using calls or chemical signals, or they can cluster together when predators appear, forcing predators to either separate individuals from the group or fight the entire group. Some animals, such as musk oxen and bison, place their young at the center of the group, forming a circle around them much like the early American settlers did with their covered wagons.

Many animals living within social groups search for food together and are more successful as group hunters than as single individuals. Food-gathering strategies become group strategies; often animals such as those that prey on fish will drive the prey to one location, encircle them, and feast. The eastern white pelicans shown in **Figure 37.17** hunt for fish in this way.

In social groups, animals can also ration the "work load" so that each animal does not have to perform an array of daily tasks but may focus on one task. Many species of insects, including ants, bees, wasps, and termites, are organized into social groups having a division of labor. These social groups are called **insect societies**.

### Insect societies

In **insect societies**, individuals are organized into highly integrated groups in which each member of the society performs one special task or a series of tasks that contributes to the survival of the group (**Figure 37.18**). In some insect societies, individual insects perform different roles during the course of their lives. In addition, a common feature among social insects is a queen that outlives other members of the society. The role of the queen is to produce offspring and to promote cooperation among members of the group.

A honeybee colony, for example, is made up of three different types of bees, called *castes:* the queen, the workers, and the drones.

**Figure 37.18** Ant society. This worker ant, a member of the genus *Polyergus*, is responsible for capturing slaves for the nest. The slave, a member of the ant genus *Formica*, is in its pupa stage. When it emerges as an adult, it will be a nonreproductive worker slave.

The queen is the focus of the colony; she lays the eggs from which all the other bees develop (**Figure 37.19**). The drones are the male bees that fertilize the eggs of the queen. They develop from unfertilized eggs and so have no male parent themselves—and only one set of chromosomes. The drones play only minor roles in the hive other than to mate with the queen. After mating, the drones die. The workers are females that develop from fertilized eggs. Genetically, a queen and a worker are no different; any fertilized egg can develop into either a queen or a worker depending on how the egg is housed and fed.

In one hive, there may be up to 50,000 workers and 5000 drones, and, of course, there is one queen. The workers perform nearly all the tasks of the hive except for mating and egg laying. In the winter months, the average length of a worker bee's life is 6 months. In the summer, workers live only 38 days! During the first 20 days of a worker's life, she performs various duties within the hive sequentially as she matures: feeding larvae, producing wax, building the honeycomb, passing out food, and guarding the hive.

**Figure 37.17** Hunting behaviors of pelicans. Eastern white pelicans hunt in groups, driving fish to one location and then circling their prey.

**Figure 37.19** A bee colony. The queen bee has a red spot on her back painted on by a researcher.

**Figure 37.20 The waggle dance.** This dance is performed by a foraging worker bee who, on spotting a food supply (a flower), goes back to the hive and communicates to the other bees through dance the exact location of the flower. The dance tells how far away the flower is and also the angle of the flower to the sun. The part of the dance that yields this information is the straight run (*zigzag line between two circles*). The forager shakes its abdomen back and forth and buzzes with its wings while moving in the straight run. Then it makes a semicircle and returns to the starting point of the straight run. After completing the next straight run, it circles in the opposite direction to land again at the beginning of the straight run, and so on. How fast the bee performs the straight run determines how far away the flower is—a distance of 500 meters equals a straight run duration of about 1 second. The direction of the flower is indicated by the angle that the straight run deviates from the vertical. This angle equals the angle of the flower to the sun.

Although a worker may go on short play flights around the hive during this time, she does not venture very far. It is only after the first 20 days of her life that the worker takes long flights away from the hive to forage for food.

Communication among the workers has evolved that enables them to pass on information about the location of flowers at which they can collect pollen and nectar. This communication is a form of body language called the *waggle dance*, a behavior discovered by Austrian zoologist Karl von Frisch. The manner in which a worker shakes, or waggles, her abdomen communicates the distance to the flowers. The angle at which the worker dances indicates the direction her co-workers must fly. **Figure 37.20** illustrates and describes how this dance works.

The queen bee controls the worker bees with a chemical she produces known as *queen substance*. This substance is a **pheromone** (FARE-uh-moan), a chemical produced by one individual that alters the physiology or behavior of other individuals of the same species. The bees ingest the queen substance by licking the queen and then pass the substance around from one to the other as they pass food. Queen substance renders the worker bees sterile

and also inhibits the workers from making queen cells, compartments in which fertilized eggs can develop into queens. If the hive becomes too congested, however, the queen may begin producing less queen substance, resulting in the removal of the inhibition to produce queen cells. Workers make a half dozen or more new queen cells in which replacement queens begin to develop. The old queen and a swarm of females and male drones leave to establish a new hive. The first new queen to emerge may kill the other candidate queens and assume rule or may create another swarm and leave to establish another hive.

The lifestyles of social insects can be so unusual as to be bizarre, none more so than that of the little reddish ants called *leafcutters* (**Figure 37.21**). Leafcutters live in the tropics, organized into colonies of up to several million individuals. These ants are farmers, growing crops of fungi beneath the ground. Their mound-like nests look like tiny underground cities covering more than 100 square yards, with hundreds of entrances and with chambers as deep as 16 feet underground. Long lines of leafcutters march daily from the mound to a tree or bush, hack its leaves into small pieces, and carry the pieces back to the mound. These ants are nearly blind; each ant finds it way by following a trail of secretions left by those who came before it. They follow the scent by holding their antennae close to the ground. At the underground site, worker ants chew the leaf fragments into a mulch that they spread like a carpet in underground chambers. They wet the leaf mulch with their saliva and fertilize it with their feces. Soon a luxuriant lawn of fungi is growing, which serves as the sole food for all the ants, no matter what their age. Nurse ants even feed this fungus to the larvae, carrying them around to browse on choice spots.

The complex caste system of bees and ants apparently evolved a long time ago. Ant fossils 80 million years old exhibit three castes

(a)

(b)

**Figure 37.21 Leafcutter ants.** (*a*) These leafcutter ants are carrying their day's harvest back to the nest. The ants do not live on the leaf material—they use it as a substrate on which to grow the fungi that they cultivate in underground farms. (*b*) Leafcutter ants in a "fungus garden."

(males, queens, and workers), indicating that their complex social system had already evolved.

We have described only two examples of social systems in insects, but many other variations exist. For example, some social insects have multiple queens per nest, not the single queen of the bumblebee. Other insect social systems do not consist of both haploid and diploid organisms, such as haploid drones and a diploid queen and workers. In termites, for example, all organisms are diploid. We have described general patterns within insect social systems, but not their rich diversity.

---

### CONCEPT CHECKPOINT

**3.** Many social interactions appear to decrease the fitness of individuals for the benefit of the social group. For these social behaviors to have evolved and be maintained, however, they must to some extent increase the fitness of all individuals. Speculate on how dominance hierarchies and altruism might increase individual fitness.

---

## 37.14 Human behavioral capacity is genetically determined but can be modified by learning.

In 1975, Harvard biologist E. O. Wilson published a groundbreaking work titled *Sociobiology: The New Synthesis*. In this book, Wilson argued that human social behavior could be explained in evolutionary terms. Some scientists question the social and political implications of human sociobiology, concerned that its ideas might lead to forms of racism, sexism, or other types of group stereotyping. Others argue that great insight into human behavior can be gained by viewing it from an evolutionary perspective.

The study of human sociobiology has inherent limitations: Scientists cannot experimentally manipulate the genes of humans or vary their environments in potentially harmful ways for scientific studies. It is therefore sometimes difficult to learn whether observed behaviors are a product of heredity or a person's environment. This "nature or nurture" dichotomy continually raises questions that scientists have a difficult time researching. Interesting observations have been made, however, in the study of identical twins raised by different families in different environments. From these studies and others, most biologists hold that much of the behavioral *capacity* of humans is genetically determined, but that the types of behaviors specified by genes can be shaped and molded—within certain limits—by learning.

At the end of the 1980s, a new science emerged from sociobiology: **evolutionary psychology**. The difference between sociobiology and evolutionary biology is this: Sociobiologists attempt to explain *human behavior of today* as a direct result of adaptive evolution. Evolutionary biologists attempt to explain *the behavior of our prehistoric hunter–gatherer ancestors* as a direct result of adaptive evolution. Our present-day societies developed only relatively recently from the hunter–gatherer heritage that dominated most of the two million years of human existence. Evolutionary biologists contend that human behavior of today is *not* like human behavior that was important in evolution, because the environment of today is so different from the environment in which our ancestors evolved. Therefore, evolutionary psychologists use evolutionary theory to predict which of the behaviors of our ancestors would have been selected at the time and in the environments in which they lived. Thus, our behavior today is seen in the context of our ancestors' behaviors. How does this work? Evolutionary psychologists first try to reconstruct the problems and the problem-solving mechanisms of our ancestors. Then they try to determine the roots of our ancestral behavior and how those behavioral roots are manifested today in our myriad cultures.

For example, results of extensive survey research across 37 cultures conducted by University of Texas evolutionary psychologist David Buss reveal that, in general, women tend to prefer older men and men tend to prefer younger women. In addition, women tend to highly value good earning potential in men, while men tend to highly value physical attractiveness in women. Buss suggests that these universal patterns in human preferences are actually adaptations from our Pleistocene ancestors. Women were attracted to men who could provide for their children, while men were attracted to women who were healthy, fertile, and able to bear children. Thus, these "conflicts of interest" made our male and female ancestors behave differently from one another, and these behavioral differences carry on in men and women today.

---

## CHAPTER REVIEW

### Summary of Key Concepts and Key Terms

#### What is the focus of those who study animal behavior?

**37.1** The study of animal behavior (**behaviors**, p. 656) encompasses patterns of change in animals, including movement, sounds, and body positions.

**37.1** Ethologists (**ethology**, ee-THOL-uh-jee, p. 656), the dominant animal behavior researchers before the late 1950s, study the behavior of animals in their natural environments, observing and interpreting their behavior in the context of physiology while attempting to uncover the biological significance of the behavior.

**37.1** **Adaptive behaviors** (p. 656) are those that enhance the ability of members of a population to live to reproductive age and that tend to occur at an increased frequency in successive generations.

**37.1** Today, the field of **animal behavior** (p. 656) is composed of researchers from a variety of disciplines: psychology, sociobiology/behavioral ecology, and neurobiology.

## What are the roles of innate and learned behaviors?

**37.2** Some patterns of behavior are inborn, **innate behaviors** (p. 657), resulting from neural pathways developed before birth.

**37.2** Such instinctive behaviors are important to the survival of animals, enabling them to elude predators, to mate, and to care for young.

**37.2** **Learning** (p. 657) is an alteration in behavior based on experience; the rigid distinction between innate and learned behavior that was once held by ethologists no longer exists.

**37.3** Reflexes (**reflex**, p. 658), kineses (**kinesis**, p. 658), and taxes (**taxis**, p. 658) are simple types of innate behaviors.

**37.3** A reflex is an automatic response to nerve stimulation.

**37.3** Kineses cause animals to slow down in suitable environments and speed up in unsuitable ones.

**37.3** Taxes cause some animals to move toward or away from certain stimuli.

**37.4** A **fixed action pattern** (p. 658) is a more complex innate behavior in which an organism performs sequences of movements triggered by a **sign stimulus** (p. 658), an environmental cue for the behavior.

**37.5** Behaviors can be changed within certain limits by means of various types of learning: imprinting, habituation, trial and error, and insight.

**37.5** Behaviorists (**behaviorism**, p. 659) portrayed learning as taking place in an automatic fashion, possibly reinforced by some type of reward that may or may not be readily apparent.

**37.5** **Cognitivism** (p. 659), the prominent view in psychology today, suggests that individuals acquire and then store information in memory; learning takes place as new information builds and merges with the old, leading to new types of behavior.

**37.6** Certain animals learn to recognize their kin or birthplace by means of **imprinting** (p. 660), a rapid and irreversible type of learning that takes place during an early developmental stage.

**37.6** **Habituation** (p. 660) helps animals adapt to certain types of stimuli.

**37.6** In **classical conditioning** (p. 660), animals learn to associate a new stimulus with a natural stimulus that normally evokes a response in the animal.

**37.6** Some animals learn to associate a behavior with a reward or punishment by means of trial-and-error learning, called **operant conditioning** (OP-uh-runt, p. 661).

**37.6** An animal capable of **insight**, or **reasoning** (p. 661), can recognize a problem and solve it mentally by using experience, thereby performing an appropriate behavior the first time it tries.

## How are regularly occurring behaviors controlled in animals?

**37.7** Organisms have internal clocks called **biological clocks** (p. 662), that regulate regular cycles of physiological activity and behavior called **circadian rhythms** (p. 662).

**37.7** Circadian rhythms are genetic mechanisms governed by negative feedback loops and other, as yet unknown, factors.

**37.8** Migratory behavior (**migrations**, p. 662) results from interactions among various environmental factors (such as daylength) with animals' cyclical physiological and hormonal changes (circadian rhythms).

## How do social behaviors aid the reproductive fitness of individuals?

**37.9** The biology of social behavior is called **sociobiology**, or **behavioral ecology** (p. 663).

**37.9** **Social behaviors** (p. 663) help members of the same species communicate (**communication**, p. 664) and interact with one another.

**37.9** Sign stimuli communicated by members of the same species are called social releasers, or simply **releasers** (p. 664).

**37.10** Animals of the same species naturally compete (**competitive behavior**, p. 664) for resources they need to survive.

**37.10** Within animals' repertoires of behaviors are aggressive behaviors that threaten and intimidate rivals (**threat displays**, **intimidation displays**, p. 665) and **submissive behaviors** (p. 665)—signals of "backing down."

**37.10** Animals also use **territorial behavior** (p. 665) to limit aggression within their species.

**37.10** A territory is an area that an animal marks off as its own, defending it against same-gender members of its species.

**37.10** Behaviors relating to rank order (the **dominance hierarchy**, p. 666) help contribute to the stability of the social relationships among animals and within groups.

**37.11** Reproductive behaviors include patterns of male-female interactions that lead to successful fertilization—the penetration of an egg by a sperm.

**37.11** The term **mating** (p. 666) refers to male–female behaviors that result in fertilization, regardless of whether copulation occurs. The term **courtship** (p. 666) refers to the behavior patterns that lead to mating.

**37.11** **Sexual selection** (p. 666) is a special kind of natural selection that depends on a struggle between the individuals of one sex, generally the males, for the possession of mates of the other sex.

**37.11** After the formation of a male–female pair, a **courtship ritual** (p. 666), unique to each species, leads to mating.

**37.12** Parenting behaviors include preparations for the care of young and their direct care to aid their survival and ultimate reproduction.

**37.12** Parenting behaviors are an example of **altruistic behaviors** (p. 667): those that are advantageous to the recipient, increasing its fitness, while disadvantageous to the donor, decreasing its fitness.

**37.13** Animals within social groups work together for a variety of common purposes.

**37.13** Some species of insects, particularly various species of bees and ants, are organized into social groups (**insect societies**, p. 669) having a division of labor.

**37.13** A **pheromone** (FARE-uh-moan, p. 669) is a chemical produced by one individual that alters the physiology or behavior of other individuals of the same species.

**37.14** Much of the behavioral capacity of humans is genetically determined, but behaviors can be shaped and molded by learning.

**37.14** Evolutionary psychologists (**evolutionary psychology**, p. 670) attempt to explain the behavior of our prehistoric hunter–gatherer ancestors as a direct result of adaptive evolution and see our behavior today in the evolutionary context of our ancestors' behavior.

## Level 1 Learning Basic Facts and Terms

**Multiple Choice**

1. How does ethology differ from behavioral ecology?
   a. Ethology focuses on the connection between behavior and an animal's genetic makeup, whereas behavioral ecologists attempt to understand how behaviors enhance survival and reproductive success.
   b. Ethologists are more concerned with describing behaviors, whereas behavioral ecologists are more concerned with the how, why, and context of the behavior.
   c. both of the above

2. Innate behaviors differ from learned behaviors in that
   a. Innate behaviors require practice, learned behaviors do not.
   b. Innate behaviors are performed the same way every time, whereas learned behaviors can be shaped by experience.
   c. Innate behaviors are controlled exclusively by genes, whereas learned behaviors are controlled exclusively by environment.
   d. Innate behaviors are instinctive, whereas learned behaviors are genetically programmed actions.

3. Sign stimuli elicit
   a. kineses.        d. imprinting.
   b. taxes.          e. learning.
   c. fixed action patterns.

4. Which of the following would not be characterized as a type of learning?
   a. Imprinting          d. Insight
   b. Habituation         e. All of the above are types of
   c. Operant conditioning     learning.

5. The main thrust of the "nature vs. nurture" dichotomy is
   a. the importance of genes vs. environmental influences on behavior.
   b. the degree to which behavior is innate or learned.
   c. the role of mating rituals in human behavior.
   d. the importance of insight vs. instinct in behavior.
   e. a and b are correct.

**True–False**

6. ____ Only humans have been shown to be capable of insight.
7. ____ Examples of social communication include courtship rituals, submissive behavior, and communication through pheromones.
8. ____ Male boat-tail grackles have very long tail feathers that are very attractive to females. This trait probably evolved through dominance hierarchies.
9. ____ Altruism will only occur between two individuals that are closely related.
10. ____ Elaborate courtship rituals strengthen the pair bond in species where both parents must care for the young in order to increase their survival chances.

## Level 2 Learning Concepts

1. Explain this statement: Behaviors are part of an animal's equipment for survival.
2. Explain the importance of innate and learned behaviors. How do they complement each other?
3. Identify the specific type of behavior involved in each of the following situations:
   a. Fireflies are attracted to the flashing luminescence of other fireflies.
   b. While studying your biology text, you hear a screen door slam. It startles you at first, but then you barely notice it when you realize the wind is occasionally opening and closing it.
   c. You praise your puppy every time it sits down when you say "sit." Soon you have trained it to sit on command.
4. Distinguish between operant conditioning and classical conditioning.
5. What is insight? Explain its significance.
6. Identify the specific type of behavior involved in each of the following situations:
   a. Every year, purple martins fly to Brazil and return to the United States around April.

   b. When a bright light flashes near your eyes, you automatically blink.
   c. Your cat frequently grooms itself by licking its fur and rubbing its paws over its face.
7. What is competitive behavior?
8. Identify the type of behavior shown in each of the following situations:
   a. The hair on a dog's back stands on end as the dog growls at an approaching stranger.
   b. When you scold the dog for growling, it rolls over on its back and exposes its belly.
   c. A robin builds its nest.
   d. A male peacock extends its tail feathers into a colorful fan.
   e. A cat chases another cat away from its food bowl.
9. Summarize the social structure of a honeybee colony.
10. Explain the term *rank order*. What is its significance, and in what animal(s) does it appear?

## Level 3 Critical Thinking and Life Applications

1. Very few mammals exhibit the complex societies seen among bees and ants—but a few do. Naked mole rats of central Africa, for example, maintain large colonies with queens and special worker castes, organized remarkably like the colonies of bees. Why do you think such societies are much rarer among mammals?
2. Pacific salmon are born at the headwaters of rivers, then swim downstream hundreds of miles to the sea where they spend their adult lives. Years later, when it is time to spawn (that is, to lay and

fertilize the eggs that will be the next generation), the adults swim up the same rivers to the precise location where they were born. Using information from this chapter, propose a hypothesis regarding how the fish know which way to go when they come to a fork in the river.

3. Do you think the principles of sociobiology can be applied to human behavior? Explain your answer using information from this chapter and from other sources. Cite your sources.

4. Researchers at the University of Chicago published a study in 1998 in which they contend that human pheromones exist. In the study, women between the ages of 20 and 35 wore a pad in their armpit for at least 8 hours during an early stage of their menstrual cycles. The researchers then placed these pads under the noses of 20 women in the same age group. The menstrual cycles of the second group of women shortened after this treatment. Why might these results be evidence of the existence of human pheromones? (Be sure to define "pheromone" in your answer.)

5. Why are genes that encode social behaviors retained in the gene pool (the total of all the genes of the breeding individuals in a population at a particular time)?

6. Some potential prey animals that are poisonous, somewhat harmful, or bad tasting also have very conspicuous coloration. These prey animals are taking advantage of what type(s) of learning exhibited by their potential predators?

---

## In The News | Critical Thinking

**CHILD'S PLAY**

Now that you understand more about behavior, reread this chapter's opening story about play behavior and brain development. To assist you in better understanding this information, it may help you to follow these steps:

1. Review your immediate reaction to the research on the effects of play on the brain that you wrote when you began reading this chapter.

2. Based on your current understanding, again summarize the main point of the research in a sentence or two.

3. What questions do you now have about this information that this chapter's opening story does not answer?

4. Collect new information about the research. Visit the *In The News* section of this text's companion Web site at www.wiley.com/college/alters and watch the "Child's Play" video. Then use the "summary" link to read the accompanying story and access related links. Use this information, the links provided, and other online and library resources to answer your questions and find updates about this topic. State the sources of your information. Explain why you think the information is accurate. Also determine whether the information expresses a particular point of view or is biased in any way.

5. What in your view is the most significant aspect of this information? Give reasons for your opinion and for any changes in your ideas based on the additional information you have collected and the analysis you have done.

# POPULATION ECOLOGY

## In The News  Tuna Tracker

The same satellite technology that monitors global oceanic and climatic conditions is also used to provide information about the migratory habits of populations of diverse species of wild animals, including green sea turtles, polar bears, and Siberian cranes. The Advanced Research and Global Observation Satellite (ARGOS) is a system that receives, processes, and distributes environmental information obtained from around the world for a variety of purposes. ARGOS is a joint venture of the U.S. National Aeronautics and Space Administration (NASA), U.S. National Oceanic and Atmospheric Administration (NOAA), and Centre National d'Etudes Spatiales (CNES), the French counterpart of NASA. According to ARGOS, animal tracking is one of the system's fastest growing applications. At any one time, ARGOS-equipped satellites as well as ground-based resources track global movements of several hundred animals. They are well suited for the study of populations, the individuals of a given species that occur together at one place and at one time.

The tracking devices that store electronic data about the behavior, physiology, and seasonal movement of populations of endangered marine species need to withstand exposure to saltwater. Some devices are surgically implanted as shown in the photo; other data storage devices are referred to as "popup" tags. A popup tag is attached to the outside of an animal's body where it remains for a period of time, then detaches and floats to the surface of the ocean where it transmits data to the satellite receiver. Scientists can access these data from the Internet. Accurate monitoring of the behavior and movements of a particular animal population to identify its habitat is helpful when managing species that are threatened with extinction.

Barbara Block, professor of marine biology at Stanford University, relies on ARGOS to track populations of endangered bluefin tuna. The bluefin is the largest living species of tuna; the fish can reach up to 10 feet in length and weigh over 1400 pounds. Bluefins are not as common as smaller species of tuna such as the skipjack or yellowfin tuna. In the United States, skipjack and yellowfin tuna usually are processed into canned tuna. If you can find it, bluefish tuna is one of the higher priced fish items on a seafood menu. In Japan, bluefin sushi is an expensive delicacy. The fish is so valuable that one bluefin can sell for $45,000. The demand for bluefin tuna makes it a prized catch for recreational and commercial ocean fishing enterprises. As a result, overfishing has seriously depleted bluefin populations around the world. According to the International Council for the Conservation of Atlantic Tuna (ICCAT), the bluefin tuna population has declined by 70 to 80% over the past 20 years. The monitoring of bluefin tuna populations by Dr. Block and other marine biologists will be helpful in efforts to keep the species from becoming extinct. To learn more about the use of satellite technology to monitor animal populations, visit the *In The News* section of this text's companion Web site at www.wiley.com/college/alters and view the video "Tuna Tracker."

*Write your immediate reaction to this information about using satellites to collect data about bluefish tuna: first, summarize the main point in a sentence or two; then suggest what you think its significance is. You will have an opportunity to reflect on your responses and gather more information on this topic in the* In The News *feature at the end of this chapter. In this chapter, you will learn more about population ecology.*

# CHAPTER GUIDE

## How do populations grow?

**38.1** Populations generally grow exponentially, then level off.

## What are some important characteristics of populations?

**38.2** Organisms within a population may be found in uniform, random, or clumped distributions.

**38.3** Density-dependent factors regulate population size.

**38.4** In animal populations, mortality characteristics are closely linked to parental care for offspring.

## What is the status of the world population?

**38.5** Developing countries have higher population growth rates than developed countries.

**38.6** The world population rose sharply after the Industrial Revolution and is still growing.

## How do populations grow?

### 38.1 Populations generally grow exponentially, then level off.

Looking like homes for oversized mud wasps, the cliff-swallow nests shown in **Figure 38.1** hang from the face of a rock outcropping. Any rough, vertical surface will serve as a nesting site for these birds. Cliff swallows have recently discovered that the sides of bridges work well for this purpose . . . and come complete with protective overhangs! As a result, cliff swallows, once found mainly in the western part of the United States, can now be found inhabiting the prairie, nesting on the sides of bridges that span the major prairie rivers.

Scientists study the interactions among organisms like these cliff swallows and study interactions between organisms and their environments both in the laboratory and in nature. This specialized field of biology is called **ecology** (eh-KOL-uh-gee); the scientists who work within this field are ecologists. Ecologists study the physical and biological variables governing the distribution and growth of living things. Ecologists also study the theoretical bases of these interactions. Some ecologists use computers to develop mathematical models of ecological systems. The knowledge gained by ecologists is essential to the basic understanding of the world and provides a foundation for finding solutions to the many environmental problems created by humans.

Scientists usually classify the study of ecological interactions into four levels: populations, communities, ecosystems, and the biosphere (see Figure 1.4). This chapter discusses the first level of this hierarchy: populations.

Cliff swallows that inhabit the nests shown in Figure 38.1 are a population of organisms. A **population** consists of the individuals of a given species that occur together at one place and at one time. This flexible definition allows the use of this term in many contexts, such as the world's human population, the population of protozoans in the gut of an individual termite, or the population of bloodsucking bugs living in the feathers of a cliff swallow.

**Population ecologists** study how populations grow and interact. Most populations will grow rapidly if optimal conditions for growth and reproduction of its individuals exist. Why, then, is the Earth not completely covered in bacteria, cockroaches, or houseflies? Why do some populations change from season to season or year to year?

### Population size and growth rate

To answer these questions, you need first to understand how the size of a population is determined. The size of a population at any given time is the result of (1) additions to the population from births and from **immigration** (the movement of organisms into a population) and (2) deletions from the population due to deaths and **emigration** (movement of organisms out of a population). Put simply:

(Births + immigrants) − (Deaths + emigrants) = Population change

Births and deaths are often expressed as *rates*: numbers of individuals per thousand per year. For example, the population of the United States at the beginning of 2002 was approximately 285 million people. During 2002, the following occurred:

- 4,021,726 live births: The birth rate was 4,021,726 per 285,000,000 people, or 14.1 births per 1000.
- 2,418,000 deaths: The death rate was 2,418,000 per 285,000,000 people, or 8.5 deaths per 1000.
- 1,063,732 (legal) immigrations: The immigration rate was 1,063,732 per 285,000,000 people, or 3.7 legal immigrants per 1000.
- 222,000 emigrations*: The emigration rate was 222,000 per 285,000,000 people, or 0.8 emigrants per 1000.

---

*The collection of statistics on emigration from the United States was discontinued in 1957; no direct measure of emigration has been available since then. The U.S. Bureau of the Census currently uses an annual emigration figure of 222,000.

**Figure 38.1** A population of cliff-swallows in their nests under a rock overhang in Montana. The birds are about six inches long, and difficult to see when in their nests. In the close-up view, you can see some birds peeking out from the small nest openings.

The population change in the United States in 2002 can be calculated as follows:

(14.1 births/1000/year + 3.7 legal immigrants/1000/year)
− (8.5 deaths/1000/year + 0.8 emigrants/1000/year)
= 8.5 people/1000/year.

This figure can also be expressed as a population change of 0.85%—an increase of slightly less than 1%.

In natural populations of plants and animals, immigration and emigration are often minimal. Therefore, a determination of the **growth rate** of a population does not include these two factors. Growth rate ($r$) is determined by subtracting the death rate ($d$) from the birth rate ($b$):

$$r = b - d$$

Using the figures from our previous example:

$$r = 14.1 \text{ births/1000/year} - 8.5 \text{ deaths/1000/year}$$
$$r = 5.6 \text{ people/1000/year or } 0.0056 \text{ or } 0.56\%$$

To figure out the number of individuals added to a population of a specific size ($N$) in a given time *without* regard to immigration and emigration, $r$ is multiplied by $N$:

$$\text{Population growth} = rN$$

Therefore, the population growth in the United States in 2002 solely from births and deaths was $0.0056 \times 285{,}000{,}000$ people = 1,596,000 people.

### Exponential growth

Even though the rate of increase in a population may stay the same, the actual size of the population—the number of individuals—grows. This sort of growth pattern is similar to the growth pattern of money in the bank as interest is earned and compounded. If you put $1000 in the bank at 4% per year, the first year you will earn $40. The second year you will earn 4% interest on $1040, or $41.60. Although your interest rate has stayed the same, the amount of money you earn grows as your money grows.

**Figure 38.2** illustrates this principle with a population of bacteria in which each individual divides into two every half hour. The rate of increase remains constant, but the actual increase in the number of cells accelerates rapidly as the size of the population grows. This type of mathematical progression found in the growth pattern of bacteria is termed **exponential growth**. For example, two cells split to form $2^2$ or 4, 4 become $2^3$ or 8, 8 become $2^4$ or 16, and so on. The number, or power, to which 2 is raised is called an

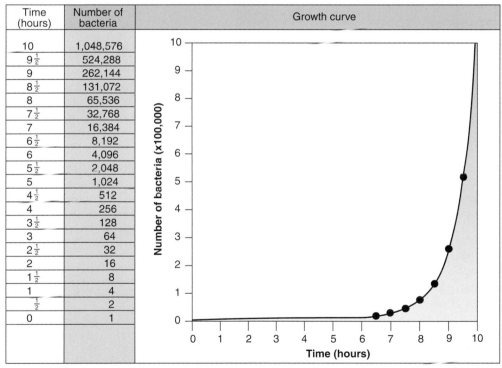

| Time (hours) | Number of bacteria |
|---|---|
| 10 | 1,048,576 |
| $9\frac{1}{2}$ | 524,288 |
| 9 | 262,144 |
| $8\frac{1}{2}$ | 131,072 |
| 8 | 65,536 |
| $7\frac{1}{2}$ | 32,768 |
| 7 | 16,384 |
| $6\frac{1}{2}$ | 8,192 |
| 6 | 4,096 |
| $5\frac{1}{2}$ | 2,048 |
| 5 | 1,024 |
| $4\frac{1}{2}$ | 512 |
| 4 | 256 |
| $3\frac{1}{2}$ | 128 |
| 3 | 64 |
| $2\frac{1}{2}$ | 32 |
| 2 | 16 |
| $1\frac{1}{2}$ | 8 |
| 1 | 4 |
| $\frac{1}{2}$ | 2 |
| 0 | 1 |

**Figure 38.2 Exponential growth in a population of bacteria.** The columns to the left show how the population doubles every half hour. The rate of increase remains constant, but the number of cells accelerates rapidly as the size of the population grows. This type of rapid acceleration in growth can be seen in the graph, which plots the number of bacteria versus time using the data in the left columns.

**Visual Thinking:** The time prior to the dramatic increases of exponential growth is often called the lag phase. How is the lag phase depicted on this graph? Why is the term *lag phase* an appropriate name for this period of growth?

(a)                                                        (b)

**Figure 38.3** A cactus takes over Australia. (a) An infestation of prickly pear cacti in scrub in Queensland, Australia, in October 1926. (b) The same view in October 1929, after the introduction of the cactus-feeding moth.

mathematical model that no natural population fits perfectly.

Sometimes populations do not level off at or below the carrying capacity. Instead, a sudden, massive die-off occurs. This pattern of exponential growth with a sudden die off of the population is called a **boom and bust cycle**. Such cycles occur for a variety of reasons and with a variety of organisms. Many short-lived species have seasonal boom and bust cycles that are linked to regularly occurring environmental changes, such as temperature shifts, food availability, and differences in levels of precipitation. If you live in a climate with seasons that differ greatly from one another, you regularly see boom and bust cycles in the growth of insect populations in the spring and summer, and their die-off in the fall and winter. Boom and bust cycles are described further in Section 38.3.

exponent. Exponential growth can refer to the rapid growth in numbers of a population of any species of organism.

A period of exponential growth can occur only when growth conditions are ideal. In nature, exponential growth often takes place when a population of individuals begins to grow in a new location having abundant resources. Such a situation occurred when the prickly pear cactus was introduced into Australia from Latin America. The species flourished, overrunning the ranges. The cactus became so abundant that cattle were unable to graze (**Figure 38.3a**). Scientists regulated the population by introducing a cactus-eating moth to the area. The larvae of the moth fed on the pads of the cactus and rapidly destroyed the plants. Within relatively few years, the moth had reduced the population; the prickly pear cactus became rare in many regions where it was formerly abundant (Figure 38.3b).

### Carrying capacity

No matter how rapidly a population may grow under ideal conditions, it cannot grow at an exponential rate indefinitely. As a population grows, each individual takes up space, uses resources such as food and water, and produces wastes. Eventually, shortages of important growth factors will limit the size of the population. In some populations such as bacteria, a buildup of poisonous wastes may also limit population growth. Ultimately, a population stabilizes at a certain size, called the **carrying capacity** of the particular place where it lives. The carrying capacity is the maximum number of individuals within a population that can be supported within a particular environment for an indefinite period. Populations often stabilize at levels below their carrying capacities. A population actually rises and falls in numbers at the stabilized level, but tends to be maintained at an average number of individuals (**Figure 38.4**). The exponential growth of a population and its subsequent stabilization at the level of the carrying capacity (or below) is represented by an S-shaped **sigmoid (logistic) growth curve**, named with reference to the Greek letter sigma. Keep in mind, however, that the sigmoid growth curve is an idealized

### Population size and ability to survive

The size of a population has a direct bearing on its ability to survive. Very small populations are less able to survive than large populations and are more likely to become extinct. Random events or natural disturbances can wipe out a small population, whereas a large population—simply due to its larger numbers and wider geographical distribution—is more likely to have survivors. Inbreeding—reproduction between closely related individuals—is also a negative factor in the survival of small populations. Inbreeding tends to produce many homozygous offspring, which results in the expression of many recessive deleterious traits that are usually masked by dominant genes. In addition, inbreeding reduces the level of variability in the gene pool, which includes all of the genes of breeding individuals of the population, detracting from the population's ability to adjust to changing conditions. Loss of genetic diversity therefore increases the probability of extinction of that species. Inbreeding, reduced genetic variability, and extinction are described in more detail in Chapter 17.

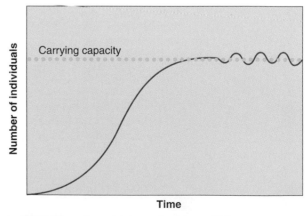

**Figure 38.4 The sigmoid growth curve.** This curve represents how populations generally grow: first exponentially, then stabilizing around the carrying capacity.

# What are some important characteristics of populations?

## 38.2 Organisms within a population may be found in uniform, random, or clumped distributions.

In addition to a population's size, its **density**—the number of organisms per unit of area—influences its survival. For example, if the individuals of a population are spaced far from one another, they may rarely come into contact. Sexually reproducing animals cannot produce offspring if they do not mate. Therefore, the future of such a population may be limited even if the absolute numbers of individuals over a wide area are relatively high.

A factor related to population density is **dispersion**, the way in which the individuals of a population are arranged. In nature, organisms within a population may be distributed in one of three different patterns: uniform, random, or clumped (**Figure 38.5**). Each of these patterns reflects the interactions between a given population and its environment, including the other species that are present.

*Uniform* distributions, those in which individuals are evenly spaced, are rare in nature and generally are indications of competition or interference. For example, populations of plants exhibiting allelopathy (AL-eh-LOP-eh-thee), the secretion of toxic chemicals that harm other plants, often show a uniform distribution. The creosote bush, often the dominant vegetation covering wide areas of deserts of Mexico and the southwestern United States, grows evenly dispersed, probably due to chemicals secreted by the bush that retard the establishment of other individuals near established ones. Uniform dispersion patterns in animal populations often result from territoriality (see Chapter 37). In both animals and plants, a uniform pattern of distribution allows members of a population similar access to resources.

*Random* distributions, those lacking any definite order, are also rare in natural populations. Random distributions occur if individuals within a population do not influence each other's growth and if environmental conditions are sufficient and uniform—that is, if the resources necessary for growth are distributed equally throughout the area. Random distributions are sometimes seen in plants as the result of certain types of seed dispersal, such as scattering by the wind. However, most seeds generally fall to the ground and take root near the adult plants that produced them, so most plants show clumped distribution patterns.

*Clumped* distributions, those in which individuals form clusters or groupings, are by far the most frequent in nature. Organisms that show a clumped distribution are close to some members of the population, but far from others. Clumping occurs as a result of the interactions among animals, plants, microorganisms, and unevenly distributed resources in an environment. Organisms are found grouped in areas of the environment that have resources they need. Furthermore, animals often congregate for a variety of other reasons, such as for hunting, mating, and caring for their young, which are described in more detail in Chapter 37.

**CONCEPT CHECKPOINT**

1. Territorial species would probably exhibit what pattern of dispersion?

(a)

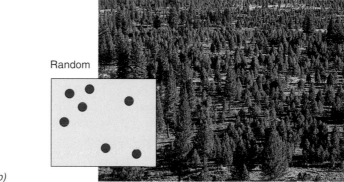

(b)

(c)

**Figure 38.5 Distribution patterns within populations.** (a) Creosote bushes in Death Valley, California, show a uniform distribution. (b) Pine trees in Tahoe National Forest, California, show a random distribution. (c) Molting African (jackass) penguins in Namibia in southern Africa, show a clumped distribution.

**Visual Thinking:** Describe each type of distribution. Which type is most prevalent in natural populations? Why?

## 38.3 Density-dependent factors regulate population size.

**Density-dependent factors** are environmental conditions that affect a population only if its density changes. For example, a supply of berries may be adequate to support a small population of gray catbirds, but it may not be adequate to support a large catbird population. The berries are one density-dependent factor that regulates the size of the catbird population. Conversely, density-independent factors are environmental conditions that affect a population regardless of its density. For example, a tornado may pass through an area and kill an entire population of catbirds. It would not matter if 5 or 50 catbirds comprised the population; all catbirds in the population would likely have been killed by this occurrence.

### Density-dependent factors

Density-dependent factors generally regulate the size of populations. For example, as a population grows and its density increases, competition among organisms for resources such as food, shelter, light, and mating sites increases and poisonous waste products accumulate. Factors such as these, which result from the growth of a population, regulate its subsequent growth. Of such factors, competition among organisms is a significant factor in the regulation of population size. **Competition** is the interaction among organisms that vie for the same resources, such as food and living space.

Individuals within a species and individuals of differing species may compete for the same limited resources. Competition among members of the same species is called *intraspecific competition*; it is a major density-dependent limiting factor. Each member of the same species requires access to the same or similar resources. Individuals within the species compete for these resources unless predation or other limiting factors reduce the population to levels in which the available resources are sufficient for the entire population. As population density rises, intraspecific competition increases and limits population size.

Competition among members of different species is called *interspecific competition*; it limits the size of at least one of the competing populations. Chapter 39 discusses interspecific competition in detail and points out that the species able to use a particular resource most efficiently will eventually eliminate the other species in that location.

Charles Darwin described competition among populations of the same and differing species as resulting in natural selection and survival of the most well adapted organisms under the conditions at that time. Competition, therefore, not only regulates the sizes of populations but is also one of the driving forces of evolutionary change.

**Predation** is another factor that may regulate the size of populations and works most effectively as the density of a population increases. Predators are organisms of one species that kill and eat organisms of another—their prey. Predators include animals that feed on plants, such as cows grazing on grass, and plants that feed on insects, such as the Venus flytrap (see the *Just Wondering* box in Chapter 24). The intricate interactions between predators and prey are an essential factor in the maintenance of diverse species living in the same area. By controlling the levels of some species, the predators make the continued existence of other species in that same community possible.

For example, populations of sea stars, barnacles, mussels, and nine other invertebrate species inhabit the rocky shoreline in one area of the Pacific Northwest. When the sea stars, which prey on the other eleven species, are removed from this community the acorn barnacles and mussels outcompete the remaining species. Eventually they are the only species left. Thus, sea star predation of strong competitors allows the weak competitors to survive.

This is an excellent example of how density-dependent factors may not only limit population size, but may also enhance population size and, in this case, the actual existence of multiple populations. Moreover, a given predator may very often feed on two or more kinds of plants or animals, switching from one to the other as their relative abundance changes. Similarly, a given prey species may be a primary source of food for an increasing number of predator species as it becomes more abundant, a factor that will regulate the size of its population automatically.

**Parasitism** also regulates the size of populations by weakening or killing host organisms. Parasites live on or in larger species of organisms and derive nourishment from them. As a population increases in density, parasites such as bacteria, viruses, and a variety of invertebrates can more easily move from one organism to another, infecting an increasing proportion of a population. Once again, this regulating factor of population size acts in negative feedback fashion, becoming more effective as the density of the population increases.

Density-dependent factors can also contribute to *boom and bust cycles*. For example, lemming populations experience striking increases in their populations followed by dramatic declines (crashes) every three to four years (**Figure 38.6**). Lemmings are small mouselike herbivores that live in the Arctic tundra. Although the causes of population oscillations of this rodent are not well understood, many hypotheses suggest density-dependent factors are key. One hypothesis suggests that as the lemming population increases ❶, populations of lemming predators increase. Lemming predators include weasels, arctic foxes, and certain arctic birds. As the predator populations increase, the lemming prey population then decreases ❷. The predator populations decrease as well because there is less food for them, which allows the lemming population to recover. As the lemming population increases dramatically again ❸, the populations of lemming predators increase again, and so on. Some hypotheses also suggest that the lemmings overgraze the arctic tundra in which they live, contributing to the crash in their populations. A recovery of the vegetation when lemming populations are small contributes to the next boom in lemming population growth.

### Density-independent factors

Factors such as weather, availability of nutrients, and physical disruptions such as volcanoes or earthquakes can limit the growth of a population. Because these factors operate regardless of the density of a population, they are called **density-independent factors**.

A variety of environmental conditions can limit populations. For example, freak summer snowstorms in the Rocky Mountains of Colorado can kill butterfly populations. The size of insect populations that feed on pollen and flower tissues varies seasonally with the blooming of flowering plants. Humans, too, can affect the sizes of populations. Poachers have killed so many African elephants for their ivory, for example, that the species may become extinct.

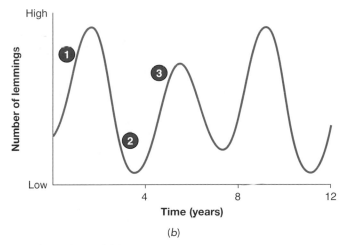

(a)　　　　　　　　　　　　　　　　　　(b)

**Figure 38.6** Boom and bust cycles in lemming populations. (a) The brown lemming, which makes its home in the Arctic tundra. (b) Although the causes of lemming population cycles over three- to four-year periods are not fully understood, oscillations in predator populations and in availability of food are likely important contributing causes.

## 38.4 In animal populations, mortality characteristics are closely linked to parental care for offspring.

Do you know what the human life expectancy is? Have you ever thought about life expectancy for other organisms? If you are a pet owner, you might have wondered how long certain breeds of dogs or cats live. Different species of organisms have different life expectancies. In addition, members of the same species may have different life expectancies depending on their living conditions. A baby born in the United States in 2001 may be expected to live to age 77, but one born in a developing country in which infection, starvation, and poverty are commonplace may not be so fortunate. Insurance companies use life expectancy data to construct life tables, which show how much longer people can expect to live when they reach certain ages. For example, if you were an American between 20 and 21 years of age in 2000, then you could expect to live an additional 55.2 years. Insurance companies use life tables for such purposes as determining the probability that a client will live long enough to pay the premiums on a life insurance policy.

Population ecologists construct life tables as well, which show the probability of survival for organisms at different times during their lifespans. When a population lives in a constant environment for a few generations, its **age distribution**—the proportion of individuals that survive to the different age categories—becomes stable. As with life expectancy, this distribution differs greatly from species to species and even to some extent within a given species from place to place.

Scientists express the mortality characteristics of a population by means of a survivorship curve. **Mortality** is the death rate. **Survivorship** refers to the proportion of an original population that survives to a certain age. The curve is developed by graphing the number of individuals within a population that survive through various stages of the lifespan. Population ecologists recognize three main types of survivorship curves (**Figure 38.7**).

First look at the human survivorship curve. The upper left corner of the graph shows 1000 people at birth. As you follow the blue line to the right, notice the number of people who survive throughout the various percentages of the lifespan. Most humans survive for about 75% of the maximum human lifespan. Using 100 years as the maximum human lifespan for our example, the graph tells us that most humans live to age 75 or so, about the life expectancy mentioned previously in this section. Some live past that age, of course, but deaths increase dramatically between 75 and 100% of the maximum human lifespan.

The survivorship curve for humans is characteristic for large vertebrates and is very different from those for hydra or oysters. Large vertebrates such as humans produce few offspring but protect

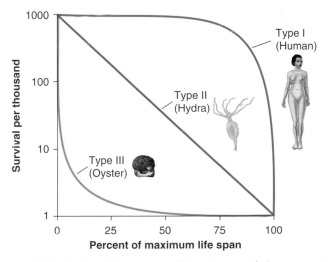

**Figure 38.7** Survivorship curves. The shapes of the respective curves are determined by the percentages of individuals in populations that are alive at different ages. In type I, organisms have few young and care for them until they can survive independently. Most organisms survive to old age. Type II organisms generally reproduce asexually and "young" do not exist. Mortality is constant over the lifespan. In type III, parents have abundant offspring but do not care for them. Many young die, but the ones that survive are likely to live out their maximum lifespans.

and nurture them; therefore, most humans, except in areas of great poverty, hunger, and disease, survive past their reproductive years. The human survivorship curve is called a type I survivorship curve.

Next, look at the hydra survivorship curve in Figure 38.7. In the hydra, mortality is more constant over the lifespan, as indicated by the straight type II survivorship curve. The probability of an organism dying does not increase or decrease with its age. This type of survivorship curve is characteristic of organisms that reproduce asexually, such as hydra, bacteria, and asexually reproducing protists. Care of the young is not a factor in survival because no "young" exist.

Oysters, on the other hand, produce vast numbers of young, but few of these offspring live to reproduce. The death rate of organisms that survive and reach reproductive age is extremely low. Those individuals that survive, however, are highly likely to live out their maximum lifespan. The result is a type III survivorship curve. This type of survivorship curve is characteristic of organisms producing offspring that must survive on their own and therefore die in large numbers when young because of predation or their inability to acquire the resources they need.

Many animal and protist populations have survivorship curves that lie somewhere between those characteristic of type II and type

III. Many plant populations, with high mortality at the seed and seedling stages, have survivorship curves close to type III. Humans have probably approached type I more and more closely through the years, with the lifespan being extended because of better health care and new medical technology.

## CONCEPT CHECKPOINT

2. American dune grass grows on shifting and transient sand dunes on ocean beaches. Grasses tend to survive better when other dune grasses are nearby to stabilize the sand dune soil. Even a well-established population of dune grasses, however, can be wiped out by wave action from a strong storm. How would these characteristics of dune grass and their environment affect the population's growth curve?
3. What factors probably play a role in regulating population growth in American dune grass? density dependent, density independent, or both?
4. Characterize the probable survivorship curve for American dune grass.

# What is the status of the world population?

## 38.5 Developing countries have higher population growth rates than developed countries.

**Demography** is the statistical study of human populations. Demographers predict the ways in which the sizes of populations will change in the future, taking into account the age distribution of the population and its changing size through time.

A population whose size remains the same through time is called a **stable population**. In such a population, births plus immigration exactly balance deaths plus emigration. In addition, the number of females of each age group within the population is similar. If this were not the case, the population would not remain stable. For example, if there were many more females entering their reproductive years than older females leaving the population, the population would grow.

The age distribution of males and females in human populations of Kenya, the United States, and Austria in 2003, and the projected populations in 2025, are shown as population pyramids in **Figure 38.8**. A **population pyramid** is a bar graph that shows the composition of a population by age and gender. Males are conventionally enumerated to the left of the vertical age axis and females to the right.

By using population pyramids, scientists can predict the future size of a population as shown in Figure 38.9b. First, the number of females in each age group is multiplied by the average number of female babies that women in that age group bear. These numbers

are added for each age group to see whether the new number will exceed, equal, or be less than the number of females in the population being studied. By such means, the future growth trends of the human population as a whole and of individual countries and regions can be determined.

The population pyramids show the differences in the patterns of a rapidly growing population seen in Kenya in sub-Saharan Africa, a slowly growing population in the United States, and a country experiencing negative growth seen in Austria in central Europe. The 2003 population pyramid of Kenya is characteristic of developing countries—those that have not yet become industrialized—such as countries in Africa, Asia, and Latin America. Each of these countries has a population pyramid with a broad base reflecting the large numbers of individuals yet to enter their reproductive years. Demographers consider the reproductive years to be ages 15–44. In Kenya, for example, about 50% of the population is younger than 15 years of age. These children will reach reproductive age in the near future. The fertility rate, which is number of children a woman will have in her lifetime, is dropping in Kenya and other developing countries due to illness from HIV/AIDS and increased use of contraception. It is still high at 5.0. The AIDS/HIV epidemic in sub-Saharan Africa affects infant and adult mortality more than it does the fertility rate. The base of the population pyramid for Kenya will remain wide in 2025, as shown in Figure 38.8b due to relatively high fertility rates and large numbers of women in their reproductive years, in spite of deaths from HIV/AIDS.

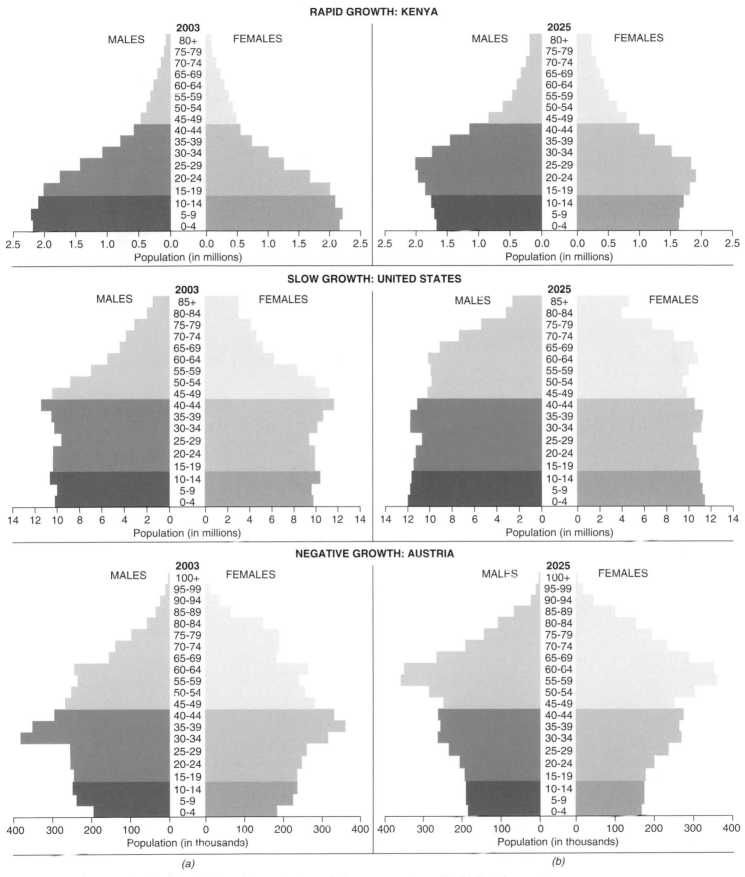

**Figure 38.8** Three patterns of population change. (*a*) Population patterns in 2003. (*b*) Population patterns estimated for 2025. (Dark color = people under 15 years old. Medium color = people 15–44 years old [the reproductive years]. Light color = people 45+ years old.) Source: U.S. Census Bureau, International Data Base.

## just wondering . . .

**On the news I often see photos of starving children. Why are their stomachs bloated? Also, is the world running out of enough food to feed everybody?**

Starving children often have swollen bellies, a condition known as kwashiorkor (KWASH-ee-ORE-kore) (**Figure 38.A**). This disease is a sign of protein deficiency in the child. Normally, a certain level of protein is present in the blood. Blood protein affects the movement of water into and out of the blood. When a child does not take in enough protein, blood protein become diminished. Fluid leaks out of the blood and into the belly and legs, building up there.

Starvation and malnutrition are issues of grave concern as the world faces such astonishing growth. There are no easy answers, and scientists themselves are polarized on these issues.

One group within the scientific community contends that new technologies will allow humans to expand the world food production and feed all of its people. Research assessing the climate and soil conditions in 93 developing countries suggests that three times as much land as is currently cultivated could be put to agricultural use. In areas without additional land to farm, the number of crops grown each year on land presently in use could be increased—an agricultural practice called *multicropping*. In addition, the use of high-yield crop varieties, fertilizer, and irrigation in areas where they are not currently employed could increase crop yields.

An opposing view is held by many scientists who suggest that intensifying agricultural practices as suggested will cause serious ecological damage to our world, such as extensive deforestation, loss of species diversity, erosion of the soil, and the pollution of aquifers, streams, and rivers from pesticides and fertilizers. In addition, they assert that our natural resources will be unable to support this future demand, suggesting that crop yields would have to rise by 112% to feed all the people in 2050. To also raise the standards of their presently inadequate diets would translate into each acre of land increasing its yield more than six times. Many scientists think that these goals are impossible to achieve.

Unfortunately, in countries torn apart by war, food production is only one of the many serious problems facing its citizens. War-

**Figure 38.A** Children with the swollen bellies of Kwashiorkor.

torn countries and those with inadequate natural resources and technologies will have to rely on food aid, placing even more pressure on the resources available in the rest of the world. There are no easy answers to your question, and feeding the people of the world will continue to be a critical issue. Possibly researchers in biotechnology, agriculture, and related areas will make contributions that will help solve this problem for future generations. Even if we solve our immediate problems of food production, if the world population continues to grow at the same rate as it does today, what will happen beyond 2050? Will we reach a final limit? ●

*Are you wondering about a topic in biology and how it relates to your life? Submit your question by clicking the* Just Wondering *link in this text's companion Web site at* www.wiley.com/college/alters.

In the United States, birth rates are higher than death rates at present, producing a growth rate of approximately 5.6 people/1000/year, or a population change of approximately 0.56%. The high birth rate is not due to couples having large families—the fertility rate is approximately 2.0, which is about the replacement rate. Rather, the growth rate in the United States is due to the large size of the "baby boom" generation that is just passing the peak of its reproductive years. The baby boom generation can be seen as the slight bulge from approximately ages 35–54 in the U.S. population pyramid. Individuals in this age group were born within the 20 years or so after World War II. The large number of women in this group causes the births to still out-

number the deaths. Notice in the population pyramid for the United States for 2025 that the pyramid will become more "squared off" as the baby boomers move past their reproductive years.

Austria and the United States are both experiencing a decline in fertility and mortality. Austria's population, however, does not include as high a percentage of women in their childbearing years as the United States, so deaths outnumber births. In addition, Austria's fertility rate is lower than that of the United States. It is approximately 1.4, well under the replacement rate. For these reasons, the population pyramid has a narrow base, which continues to narrow by the year 2025.

## 38.6 The world population rose sharply after the Industrial Revolution and is still growing.

Although some countries have populations that are no longer growing, such as Denmark, Germany, Hungary, and Italy, and some countries such as Austria are declining in numbers, the population of the world as a whole is growing 1.4% a year. This growth may sound low, but with the world population numbering about 6.3 billion, it adds over 88 *million* people to the population each year. The United Nations and the Population Reference Bureau estimate that the world population could reach about 9 billion by the year 2050.

How did the human population reach its present size? With the development of agriculture 11,000 years ago, human populations began to grow steadily. Villages and towns were first organized about 5000 years ago; the specialization of professions such as metallurgy (the science and technology of metals) became possible, and technology advanced. The Renaissance in Europe, with its renewed interest in science, ultimately led to the establishment of industry in the seventeenth century and to the Industrial Revolution of the late eighteenth and early nineteenth centuries. **Figure 38.9** shows the slow growth of the human population until it began an exponential increase around the beginning of the nineteenth century.

By the mid-nineteenth century, Louis Pasteur had arrived at the germ theory of disease, the understanding that microbes cause infection. In the 1920s, Alexander Fleming discovered penicillin and opened the door to antibiotic therapy—medicine's "magic bullets" against bacterial infection. These medical advancements decreased the death rate by increasing the number of individuals surviving infection. The advent of the Industrial Revolution also heralded new farming and transportation technology, which helped

provide better nutrition for many people. With better nutrition and increased medical understanding and technology, the death rate in developed countries has fallen steadily and dramatically since the mid-nineteenth century (**Figure 38.10a**). In developing countries, international foreign aid imported this new technology along with food aid after World War II. The mortality rate plunged in a matter of years (Figure 38.10*b*).

Birth rates fall, in both developed and developing countries, with higher rates of literacy and education, especially among

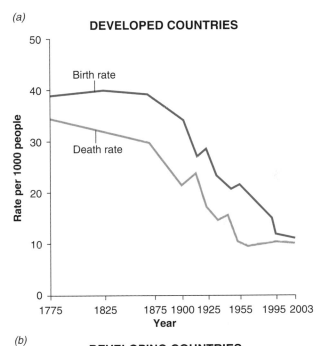

*(a)* **DEVELOPED COUNTRIES**

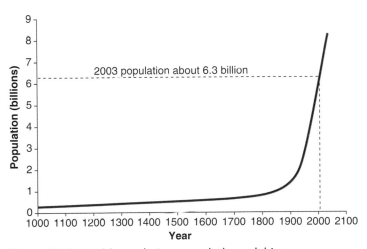

**Figure 38.9** World population growth through history.

**Visual Thinking:** This graph of world population growth does not look like a sigmoid (logistic) growth curve up to the present. Why not?

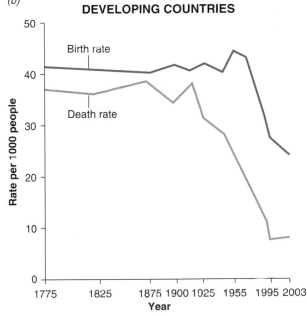

*(b)* **DEVELOPING COUNTRIES**

**Figure 38.10** Birth and death rates in developed and developing countries from 1775 to 2003. The increase in population in any given year is the vertical distance between the birth rate and the death rate. (*a*) In developed countries, the birth rate barely exceeds the death rate. (*b*) In developing countries, the birth rate substantially exceeds the death rate.

women. Not only do educated women have a greater understanding of how to use contraceptives properly, but they are also generally healthier and tend to produce healthier babies and smaller families.

Although birth rates have declined in developing countries, they are still high. Presently, about 80% of the people in the world are living in less developed countries; about 60% are living in countries that are at least partly tropical or subtropical, and about 20% are living in China. Approximately 20% are living in the more developed countries of Europe, the Commonwealth of Independent States (formerly the Soviet Union), Japan, the United States, Canada, Australia, and New Zealand (**Figure 38.11**). **Figure 38.12** graphically depicts the share of the world population living in less developed countries from 1970 through today, with projections through 2020.

Of the estimated 3.8 billion people living in tropical areas in 2003, a large proportion live in poverty. These people cannot reasonably expect to be able to consistently provide adequate food for themselves and their children. Even though some experts estimate that enough food is produced in the world to provide an adequate diet for everyone in it (see the *Just Wondering* box), the distribution is so unequal that large numbers of people live in hunger. The United Nations International Children's Emergency Fund (UNICEF) estimates that in the developing world today, about 6 million children younger than 5 years of age—about 16,500 per day—die each year, mainly of malnutrition and the complications associated with it. Fifty-four percent of the deaths of young children in developing countries are associated with malnutrition.

The size of human populations, like that of other organisms, is or will be controlled by the environment. Early in its history, human populations were regulated by both density-dependent and density-independent limiting factors, including food supply, disease, and predators. There was also ample room on Earth for migration to new areas to relieve overcrowding in specific regions. In the past century, humans have been able to expand the carrying capacity of the Earth through technological innovations. Gradually, changes in

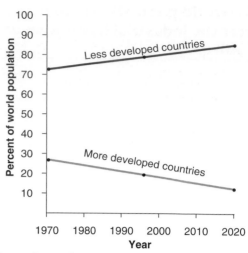

**Figure 38.12 Share of the world population.** The percentage of the human population in more developed countries is projected to continue to decline, shown here through 2020.

technology have given humans more control over their food supply, enabled them to develop superior weapons to ward off predators, and provided the means to cure diseases. Improvements in transportation and housing have increased the efficiency of migration. At the same time, improvements in shelter and storage capabilities have made humans less vulnerable to climatic uncertainties.

As a result of the ability to manipulate these factors, the human population has been able to grow explosively to its present level of about 6.3 billion people. The most effective means of dealing with the population explosion has been the support of governments to encourage small families, the establishment of family planning clinics, improvement in education, and socioeconomic development. The developed countries of Western Europe and North America, Japan, and Australia have very low birth rates at this time. China, Indonesia, Thailand, South Korea, Hong Kong, and Singapore have had considerable success with lowering their birth rates. Countries such as Mexico and India have had some success in reducing their birth rates, but are still striving toward this goal. The countries in sub-Sahara Africa have the highest birth rate of any countries in the world. Family planning programs are now being implemented across the continent.

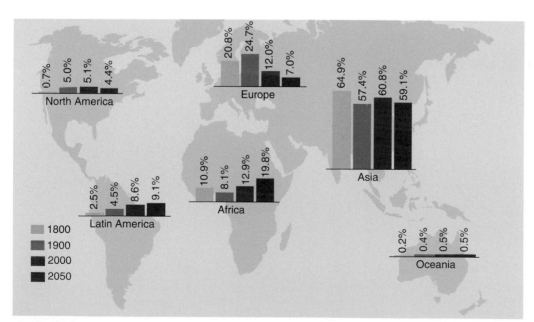

**Figure 38.11 World population distribution by region, 1800 to 2050.** If current trends continue, Asia will have 59% of the total world population in 2050, Africa nearly 20%, and the combined share of Europe and North America will drop to 11.4%.

CONCEPT CHECKPOINT

**5.** Besides simply having fewer children, propose some measures that people could take to reduce the rate of human population growth.

# CHAPTER REVIEW

## Summary of Key Concepts and Key Terms

### How do populations grow?

**38.1** The study of interactions among organisms and between organisms and their environments in the laboratory or in nature is called **ecology** (eh-KOL-uh-gee, p. 676).

**38.1** A **population** (p. 676) consists of the individuals of a given species that occur together at one place and at one time.

**38.1** **Population ecologists** (p. 676) study how populations grow and interact.

**38.1** The **growth rate** (p. 677) of any population is the difference between the birth rate and the death rate.

**38.1** The actual change in a population may also be affected by **emigration** (p. 676) from the population and **immigration** (p. 676) into it, but these factors are only considered in human populations.

**38.1** Most populations exhibit a **sigmoid (logistic) growth curve** (p. 678), which implies a relatively slow growth, a rapid increase (**exponential growth**, p. 678), and then a leveling off when the **carrying capacity** (p. 677) of the species' environment is reached.

**38.1** A pattern of exponential growth with a sudden die-off of the population is called a **boom and bust cycle** (p. 678).

### What are some important characteristics of populations?

**38.2** In addition to a population's size, its **density** (p. 678) —the number of organisms per unit of area—influences its survival.

**38.2** Individuals in a population may be dispersed (**dispersion**, p. 679) in a uniform, clumped, or random manner.

**38.2** Clumped dispersion patterns are the most frequent in nature.

**38.3** Each population grows in size until it eventually reaches the limits of its environment to support it; resources are always limiting.

**38.3** Some of the limits to the growth of a population are related to the density of that population, but others are not.

**38.3** Factors that result from the growth of a population and regulate its subsequent growth are **density-dependent factors** (p. 679).

**38.3** Three density-dependent limiting factors are **competition** (p. 680), **predation** (p. 680), and **parasitism** (p. 680).

**38.3** Factors such as weather conditions that regulate the growth of a population and that operate regardless of its density are **density-independent factors** (p. 681).

**38.4** When a population lives in a constant environment for a few generations, its **age distribution** (p. 681)—the proportion of individuals that survive to the different age categories—becomes stable.

**38.4** **Mortality** (p. 681) is the death rate; **survivorship** (p. 681) refers to the proportion of an original population that survives to a certain age.

**38.4** Survivorship curves are used to describe the characteristics of growth in different kinds of populations.

**38.4** Type I populations are those in which a large proportion of the individuals approach their physiologically determined limits of age.

**38.4** Type II populations have a constant mortality throughout their lives.

**38.4** Type III populations have very high mortality in their early stages of growth, but an individual surviving beyond that point is likely to live its maximum lifespan.

### What is the status of the world population?

**38.5** **Demography** (p. 682) is the statistical study of human populations.

**38.5** A population whose size remains the same through time is called a **stable population** (p. 682).

**38.5** A **population pyramid** (p. 682) is a bar graph that shows the composition of a population by age and gender.

**38.5** Developing countries—those that have not yet become industrialized—have disproportionately young populations.

**38.5** Although fertility is declining and mortality is rising in many developing countries, they are experiencing rapid growth due to still-high fertility rates and large numbers of women entering their reproductive years.

**38.5** Developed countries have populations with similar proportions of their populations in each age group and are growing very slowly or not at all.

**38.6** The world population rose sharply and dramatically after the Industrial Revolution because of new technology in agriculture, transportation, industry, and medicine.

**38.6** Today, the global population is about 6.3 billion people and is growing about 1.4% per year.

**38.6** Growth rates are higher in developing countries than in developed countries.

**38.6** Approximately 80% of the world's people live in developing countries, while only 20% live in developed countries.

## Level 1 | Learning Basic Facts and Terms

**Multiple Choice**

1. Ecologists study life at all of the following hierarchical levels except:
   - a. Cellular.
   - b. Populations.
   - c. Communities.
   - d. Ecosystems.
   - e. Biosphere.

2. Which of the following is a question that a population ecologist might ask?
   - a. How does the composition of species in a community change over time?
   - b. How does carbon flow through the ecosystem?

    c. Do members of two different species compete for food?

    d. What factors determine the distribution and abundance of organisms?

    e. How does the age structure of a population affect its growth rate?

    f. d and e

3. A calculation of population growth rate (*r*) directly utilizes which of the following parameters?

    a. Birth rate            d. Frequency of mating

    b. Death rate           e. Immigration

    c. The population's age structure      f. a and b

4. Density-independent factors

    a. affect population growth only when population size is small.

    b. affect mortality regardless of the population size.

    c. alter the survivorship curve for a population.

    d. affect population growth only when population size is large.

    e. cause death rate to be greater than birth rate even when the population is small.

5. Which of the following will affect population growth?

    a. Death rate            d. Immigration

    b. Birth rate           e. All of the above

    c. Age structure

**True–False**

6. _____ Human populations are stable or even declining in developed countries because they have reached the carrying capacity.

7. _____ Uniform distribution of individuals in a population is common in nature because most species are territorial.

8. _____ Organisms that invest a great deal of energy in parental care of their offspring typically exhibit survivorship curves that are close to type I.

9. _____ The human population could be characterized as a mosaic of age structures that differs from country to country and is influenced by a variety of social and environmental factors.

10. _____ We could solve the problems associated with human population growth if we could only find a way to adequately feed everyone in the world.

## Level 2    Learning Concepts

1. In the early twentieth century, the United States experienced a large influx of immigrants from many European nations. Does immigration affect the size or growth rate of the population in the United States?

2. How is the work of population ecologists similar to that of demographers? How does it differ?

3. Distinguish between density and dispersion. How does each affect a population's chances for survival?

4. Identify the three patterns of population dispersion found in nature. Into which pattern do human populations fall?

5. Distinguish between density-dependent and density-independent limiting factors. Give an example of each.

6. The following graph represents the changes in population sizes of a prey species and one of its predators. Develop a hypothesis as to why the populations fluctuate as they do. Would you classify the interactions that limit or alter population growth of each population as density dependent or density independent? Explain.

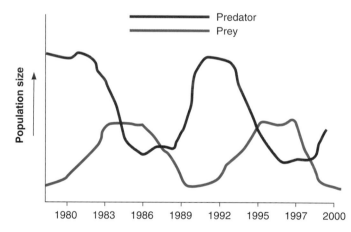

7. What do predation and parasitism have in common? How do they differ?

8. Draw type I, type II, and type III survivorship curves. Summarize the types of organisms that are characteristic of each curve, and give an example of each.

9. Compare the typical population pyramids of a developing country and an industrialized country.

## Level 3    Critical Thinking and Life Applications

1. Both the current human population level and the projected rate of growth worldwide have potential consequences for the future that are extremely grave. Explain why.

2. Suppose that you were given the political power to deal with the world's population explosion. What steps would you take?

3. Draw a hypothetical population pyramid of a stable population. Describe your pyramid and explain how it represents a stable population.

4. What conclusions can you draw about a population that has a low proportion of its members under reproductive age?

5. If the most successful parasite does not kill its host, then why are parasites considered to be density-dependent limiting factors?

6. Lobsters show a basic type III survivorship curve but have reduced chances of survival following molting (shedding of the exoskeleton) because they are more susceptible to predation. This occurs each

season. How would this characteristic affect the general shape of their survivorship curve?

7. Consider the following table based on population data for a group of a newly discovered species of monkey. How would you characterize the survivorship of this population? Do you think that these monkeys invest much energy to care for their young?

| Age | Number Alive | Deaths | % Mortality | Number of Offspring |
|-----|-----|-----|-----|-----|
| 1 | 1000 | 30 | 3.0 | 0 |
| 2 | 970 | 70 | 7.2 | 20 |
| 3 | 900 | 50 | 5.9 | 250 |
| 4 | 850 | 50 | 6.3 | 270 |
| 5 | 800 | 300 | 37.5 | 56 |
| 6 | 500 | 400 | 80.0 | 4 |
| 7 | 100 | 97 | 97 | 1 |

## In The News | Critical Thinking

### TUNA TRACKER

Now that you understand more about population ecology, reread this chapter's opening story about tracking endangered bluefin tuna. To better understand this information, it may help you to follow these steps:

1. Review your immediate reaction to using satellite technology to track tuna that you wrote when you began reading this chapter.
2. Based on your current understanding, again summarize the main point of the research in a sentence or two.
3. What questions do you now have about this information that this chapter's opening story does not answer?
4. Collect new information about the research. Visit the *In The News* section of this text's companion Web site at www.wiley.com/college/alters and

watch the "Tuna Tracker" video. Then use the "summary" link to read the accompanying story and access related links. Use this information, the links provided, and other online and library resources to answer your questions and find updates about this topic. State the sources of your information. Explain why you think the information is accurate. Also determine whether the information expresses a particular point of view or is biased in any way.

5. What in your view are the most significant aspects of this information? Give reasons for your opinion and for any changes in your ideas based on the additional information you have collected and the analysis you have done.

# INTERACTIONS WITHIN COMMUNITIES

## In The News | Parasites Lost

"Exotic Invader." Although the phrase may sound like a film title, an exotic invader is simply a nonnative species that is introduced into an ecosystem and becomes resident, often with devastating effects on native communities. The invading species may successfully compete against indigenous species for food, shelter, and territory; spread disease; and even drive native organisms to extinction.

Nonnative species enter new habitats in various ways. Humans may introduce organisms that are not native to an ecosystem either accidentally or intentionally. Exotic pets, for example, become invasive species when they escape from or are released by their owners. People may deliberately introduce nonnative species for practical reasons such as to provide food for grazing animals. In the 1800s, city pigeons, house sparrows, starlings, and kudzu were all introduced into the United States for useful purposes. Within a few decades, however, these species became pervasive pests. The feces of city pigeons can contain the fungus that causes histoplasmosis, a lung infection. House sparrows and starlings have taken over nesting sites of Eastern bluebirds, reducing the numbers of these native songbirds. Kudzu's vines grow so rapidly that they can break or crush small trees with their weight, and they can kill larger trees by preventing sunlight from reaching leaves.

One hypothesis for why nonnative species often experience dramatic growth at both organism and population levels is their reduced parasite load from that in their native habitat. Parasitism can hinder a species' success by reducing the size and vitality of the host organism. A species that has few kinds of parasites could have a competitive advantage over a similar species affected by a larger variety of parasites.

To test this hypothesis, Mark Torchin, a research biologist at the University of California, Santa Barbara, and his

colleagues compared numbers of different parasites hosted by 26 animal species in their former and new habitats. According to the scientists' findings, the animals had an average of 16 different parasites in their native habitats but averaged 7 kinds of parasites after invading new habitats. In Europe, for example, European green crabs are small and uncommon in areas inhabited by a species of barnacle that castrates the crabs. When green crabs invade areas where the barnacle does not live, such as waters off the coast of the United States, they successfully compete against native crabs for food, multiply relatively unchecked, and grow much larger than green crabs in Europe. The large "U.S. version" of the European green crab is shown in the photo.

In the United States, the green crabs' appetite for Dungeness crabs and softshell clams has severely affected native shellfish communities along the Eastern seaboard. The green crabs are now invading the west coast of the United States as well and are threatening native shellfish communities in the Northwest and California. Although scientists are exploring biological ways to control invasive green crab populations, they must be careful to avoid introducing yet other unwelcome organisms into the environment. To learn more about this research, view the video "Parasites Lost" by visiting the *In The News* section of this text's companion Web site at www.wiley.com/college/alters.

*Write your immediate reaction to this research on the role of parasites in communities: first, summarize the main point in a sentence or two; then suggest what you think the significance of this research is. You will have an opportunity to reflect on your responses and gather more information on this topic in the* In The News *feature at the end of this chapter. In this chapter you will learn more about parasites and other factors that affect the structure of communities.*

## CHAPTER GUIDE

### What are communities?

**39.1** Populations live together forming communities.

### What types of interspecies relationships occur in communities?

**39.2** Species often compete for limited resources.

**39.3** Predator populations have complex interactions with prey populations.

**39.4** In commensalism, one species benefits while the other neither benefits nor is harmed.

**39.5** In mutualism, both species benefit.

**39.6** In parasitism, one species benefits while the other is harmed.

### How do communities change over time?

**39.7** Communities change over time by succession.

# What are communities?

## 39.1 Populations live together forming communities.

The beautiful coral reef shown in **Figure 39.1** is home to a variety of organisms. Each organism contributes to the array of colors evident here. The reef provides a shallow-water environment favorable to many organisms. Nutrients are abundant, as are surfaces for attachment and hiding places into which animals can burrow. Although a number of species characteristic of coral reef populations are visible in the photograph, more than 3000 species may coexist in a large reef. Its populations interact in a variety of ways. The coral reef organisms compete with one another for space and food. Sponges, for example, may outcompete the corals for space on crowded reefs. In addition, reefs provide food and shelter to schools of fish that swim within and around the reef. Together, the interacting populations of a coral reef form a vibrant, colorful community. A **community** is a grouping of populations of different species living together in a particular area at a particular time.

The magnificent redwood forest that extends along the coast of central and northern California and into the southwestern corner of Oregon is another example of a community. Within it, the most obvious organisms are redwood trees. The huge trunk of a redwood can be seen in **Figure 39.2**. Populations of other organisms live in the filtered light beneath the towering redwoods, such as the rhododendrons visible in the photo, as well as sword ferns, ground beetles, and deer. The coexistence of these various populations is made possible in part because of the special conditions that are created by the redwoods: shade, water dripping from the branches, and relatively cool temperatures. For this reason and because the redwoods

**Figure 39.2** The redwood community.

**Figure 39.1 A coral reef.** This reef is not only beautiful, but is home to many populations of species that live together in complex associations, forming a community.

visually dominate the area, this distinctive group of populations is known as the redwood community.

Scientists have long known that the nonliving or **abiotic factors** (AYE-bye-OT-ik) within the environment—such as air, water, and even rocks—affect an organism's survival, as do the living or **biotic factors** (by-OT-ik)—such as surrounding plants, animals, and microorganisms. All the biotic and abiotic factors together within a certain area are called an **ecosystem** (EH-koe-SIS-tem). A rock is as important to the makeup of an ecosystem as are barnacles. Thus, an ecosystem can be thought of as a community of organisms, along with the abiotic factors with which the members of the community interact.

Within an ecosystem, each living thing has a home, an area in which it resides. This space, including the factors within it, is an organism's **habitat**. Organisms not only reside in their habitats, but they also interact with the biotic and abiotic factors within them. In addition, each organism plays a special role within an ecosystem; this role is called a **niche** (NICH). (You might liken a habitat to an office and a niche to a person's occupation—the role he or she plays—in that office.) In biological communities, the term *niche* refers to the organism's use of the biotic and abiotic resources in its environment. Thus, it may be described with reference to space, food, temperature, appropriate conditions for mating, and require-

ments for moisture, for example. A full portrait of an organism's niche also includes the organism's behavior and the ways in which this behavior changes at different seasons and at different times of the day. These concepts are important to the understanding of communities.

# What types of interspecies relationships occur in communities?

## 39.2 Species often compete for limited resources.

In a biological context, **competition** is a situation in which organisms that live near one another strive to obtain the same limited resources. When organisms use the same limited resource for survival, they must compete for that resource. Complex animals such as vertebrates compete by using innate behaviors such as threat displays and territorial behavior (see Chapter 37).

Competition among organisms has been observed by scientists for a long time. About 70 years ago, the Russian scientist G. F. Gause formulated the **competitive exclusion principle**, which was based partly on a mathematical model created by American biophysicist Alfred Lotka and Italian mathematician Vito Volterra in the 1920s. Gause conducted laboratory experiments to test the principle in the 1930s, as did Thomas Park in the 1940s to 1960s. The outcomes of these early experiments were sometimes dependent on the starting populations and other experimental factors, which helps demonstrate the complexity of competition. Most scientists today think that the competitive exclusion principle makes sound theoretical sense.

The competitive exclusion principle states that if two species are competing with one another for the same limited resource in a specific location, the species able to use that resource most efficiently will eventually eliminate the competing species in that location. An implication of the competitive exclusion principle is that if two species are found to coexist, then they must be using the environment or resource in different ways. In other words, they must be occupying different niches.

### The study of competition in the laboratory

Scientists sometimes study competition between species in the laboratory so that they can control the environmental conditions. John Harper and his colleagues at the University College of North Wales performed competition experiments with two species of clover: white clover and strawberry clover. Each species was sown with the other at one of two densities: 36 or 64 plants per square foot. Various plots of the two species were planted, using all the possible combinations of the two densities of plants. One plot contained 36 white clover plants and 36 strawberry clover plants per square foot. Another contained 64 white and 36 strawberry per square foot and so forth. The white clover initially formed a dense canopy of leaves in each experimental plot. The slower-growing strawberry clover, however, whose leaf stalks are taller, eventually produced leaves that grew above the white clover leaves. In competing more effec-

tively for light, the strawberry clover overcame the white clover, causing it to die out. The outcome was the same, regardless of the initial densities at which the seeds of the plants were sown. Results of studies such as these support the competitive exclusion principle.

### The study of competition in nature

Plants compete for soil nutrients as well as for sunlight. The roots of one species, for example, may outcompete another species by using up minerals in the soil essential to both species. In addition, one species may secrete poisonous substances that depress the growth of other species. Sage plants, for example, inhibit the establishment of other plant species nearby, producing bare zones around populations of these plants. In the aerial photograph of sage plants taken in the mountains above Santa Barbara, California (**Figure 39.3**), you can see these bare zones around the colonies of plants. These are zones in which chemicals secreted by the sage has killed the grass. Eventually, sage will grow in these bare zones, and the new plants will secrete chemicals that will kill more grass. In this way, the sage will continue to take over the now-grassy areas.

Acorn barnacles demonstrate another interesting example of competition in nature. Highly adapted to their environment, acorn barnacles are typically found in the intertidal zone of rocky ocean shores—the narrow strip of land exposed during low tide and covered during high tide. When submerged, an acorn barnacle feeds by extending appendages from the hole in its shell. Spread out, these appendages act like a net, sweeping the water and collecting food

**Figure 39.3** Sage outcompetes grass by secreting toxic chemicals. The bare zones around colonies of sage plants are areas of killed grass. Eventually the sage will grow in these bare areas, secreting more toxins and killing more grass.

**Figure 39.4** *Chthamalus* (smaller) and *Balanus* (larger) barnacles growing together on a rock. Both are well adapted for life in the intertidal zone and compete with one another.

that it then brings into its shell and eats. When exposed to the air, an acorn barnacle pulls in its feeding appendages and shuts down, actually using much less oxygen than when underwater. Interestingly, barnacles of the genus *Balanus* have been kept out of the water as long as six weeks without detectable ill effects. A relative of *Balanus* organisms, however, barnacles of the genus *Chthamalus* (pronounced with the first two letters silent), have been kept out of water for three years, being submerged only one or two days a month—and they survived! **Figure 39.4** shows both types of barnacles.

Although both organisms have adaptations that make them well suited to the intertidal environment, their differences play an important role in determining where each genus lives. Of the two, *Chthamalus* barnacles live in shallower water, where they are often exposed to air as the tide rolls in and out. *Balanus* barnacles live deeper in the intertidal zone and are covered by water most of the time.

In studying these two genera of barnacles, J. H. Connell of the University of California, Santa Barbara, found that in the deeper zones, *Balanus* barnacles could always outcompete *Chthamalus* barnacles. *Balanus* organisms would crowd *Chthamalus* barnacles off the rocks, replacing them even where they had begun to grow. When Connell removed *Balanus* barnacles from the area, however, *Chthamalus* organisms were easily able to occupy the deeper zone, indicating that no physiological or other general obstacles prevented it from becoming established there. *Balanus* barnacles, however, must use the resources of the deeper zone more efficiently than *Chthamalus* organisms do, even though *Chthamalus* barnacles are able to survive there in the absence of its competitor. In contrast, *Balanus* barnacles cannot survive in the shallow water where *Chthamalus* organisms normally occur. *Balanus* barnacles evidently do not have the special physiological and morphological adaptations that allow *Chthamalus* barnacles to occupy this zone.

Along with illustrating the competitive exclusion principle, these experiments with *Balanus* and *Chthamalus* barnacles illustrate that the role an organism plays in an ecosystem—its niche—can vary depending on the biotic and abiotic factors in the ecosystem. In this example, the niche occupied by *Chthamalus* barnacles is its **realized niche**—the role it actually plays in the ecosystem (**Figure 39.5**, pink solid line). It is distinguished from its **fundamental niche**—the niche that it might occupy if competitors were not present. Thus, the fundamental niche of the barnacle *Chthamalus* (pink dotted line) in

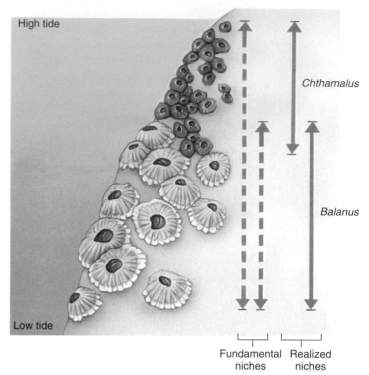

**Figure 39.5 Competition can limit niche use.** The fundamental niche is that which an organism might occupy if predators were not present. The realized niche is that which an organism actually occupies. *Chthamalus* has a fundamental niche that overlaps that of *Balanus*, but *Balanus* outcompetes *Chthamalus* for this portion of its niche.

 **Visual Thinking:** Which organism's realized niche is more restricted (narrower) than its fundamental niche? Would you usually expect an organism's realized niche to be more restricted than its fundamental niche? Why or why not?

Connell's experiments included the fundamental niche of *Balanus* barnacles (purple dotted line), but its realized niche was much narrower because *Chthamalus* organisms were outcompeted by *Balanus* organisms. The realized and fundamental niches of *Balanus* barnacles, however, are the same (purple solid and dotted lines).

Gause's competitive exclusion principle can be restated in terms of niches as follows: No two species can occupy exactly the same niche indefinitely. Certainly, species can and do coexist while competing for the same resources. Nevertheless, Gause's theory predicts that when two species coexist on a long-term basis, one or more features of their niches will always differ; otherwise, the extinction of one species will inevitably result. The factors that are important in defining a niche are often difficult to determine, however; thus, Gause's theory can sometimes be difficult to apply or investigate.

CONCEPT CHECKPOINT

1. State one way in which the scientific definition of the term "competition" is the same as the everyday use of the term; state one way in which it is different.

## 39.3 Predator populations have complex interactions with prey populations.

If you are a sci-fi action genre fan, when you hear the word "predator," you might think of Arnold Schwarzenegger and his 1987 movie of the same name. In biology, predation has the same definition as in the movie: **Predation** is a relationship in which an organism of one species—the **predator**—kills and eats an organism of another species—the **prey**. Predation includes one species of animal capturing and eating another, an animal feeding on plants, and even a plant, such as the Venus flytrap shown in the *Just Wondering* box in Chapter 24, capturing and eating insects.

How do predator populations affect prey populations? When experimental populations are set up under very simple conditions in the laboratory, the predator often exterminates its prey and then becomes extinct itself because it has nothing to eat. This fact was illustrated nicely in experiments performed by Gause using populations of the two protozoans shown in **Figure 39.6**: *Didinium* and its prey *Paramecium*. As shown in Figure 39.6a, when didinia

are introduced into a growing population of paramecia, the population of paramecia instantly begins to decline and quickly dies out. The didinia population lives on for a short while, then itself dies out.

If refuges are provided for the prey, however, even though its population may be driven to low levels, it can recover. In another experiment, Gause provided sediment in the bottom of the test tubes in which he was growing *Didinium* and *Paramecium*. Interestingly, as *Didinium* began to prey on *Paramecium*, only those organisms in the clear fluid of the test tubes were killed. Those in the sediment were not eaten. Eventually, *Didinium* protozoans died from lack of food; meanwhile, the *Paramecium* prey multiplied in the sediment (Figure 39.6b) and overtook the culture!

In another series of experiments, Gause discovered that when he introduced new prey at successive intervals (Figure 39.6c), the decline and rise in the numbers of the predator-prey populations followed a cyclical pattern. As the number of prey increased, the number of predators increased. As the numbers of the prey were lowered by predation, the large predator population did not have enough food to eat; some died and the predator population declined. As this decline occurred, the prey recovered—aided by the addition of new organisms—and again became abundant, starting the cycle once again.

At one time, scientists thought that predator–prey populations always cycled. They have come to realize, however, that in nature, conditions for survival are complex and do not always lead to such a cycling of populations. Predators cannot survive when the prey population is low. Immigration of prey, which is movement of new prey into the community, may be necessary to sustain the predator population. Changes in predator–prey populations also depend on how prey are dispersed in an area and the manner in which the predator searches for the prey. Factors other than the relationship between a single predator population and a single prey population also influence the survival and abundance of both

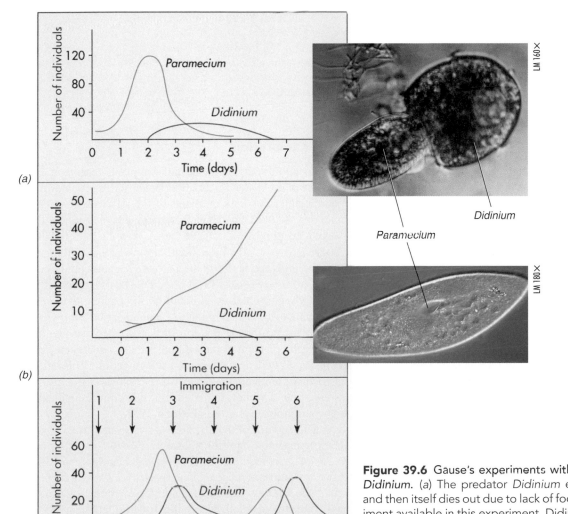

(a)

(b)

(c)

**Figure 39.6** Gause's experiments with the protozoans *Paramecium* and *Didinium*. (*a*) The predator *Didinium* exterminates the prey *Paramecium* and then itself dies out due to lack of food. (*b*) Some paramecia hide in sediment available in this experiment. Didinia eat those paramecia in the clear fluid, then die from lack of food. The paramecia in the sediment reproduce and its population grows. (*c*) When new prey is introduced at successive intervals, the predator-prey populations follow a cyclical pattern. The arrows show introduction of new prey.

predator and prey. For example, adverse weather conditions may result in the death of the predator and/or prey species; the prey may be eaten by more than one predator; or fluctuations may occur in the food source of the prey, limiting the survival of this population.

The intricate interactions between predators and prey often affect the populations of other organisms in a community. By controlling the levels of some species, for example, predators help species survive that may compete with their prey. In other words, predators sometimes prevent or greatly reduce competitive exclusion by limiting the population of one of the competing species. Such interactions among organisms involving predator-prey relationships are key factors in determining the balance among populations of organisms in natural communities.

## Plant–herbivore coevolution

Plants, animals, protists, fungi, and bacteria that live together in communities have changed and adjusted to one another continually over millions of years. Such interactions, which involve the long-term, mutual evolutionary adjustment of the characteristics of the members of biological communities in relation to one another, are examples of coevolution.

Plants and plant-eating predators called herbivores are organisms that change and adjust to one another over time. Natural selection favors plants that have developed some means of protection against herbivores. In the dynamic equation of coevolution, however, natural selection also favors adaptations that enable animals to prey on plants in spite of their protective mechanisms.

To avoid being eaten, for example, some plants have developed hard parts that are difficult to eat or are unpalatable. Certain grasses defend themselves by incorporating silica (a component of glass) into their structure, making them too tough to eat. Some groups of herbivores, however, have developed strong, grinding teeth and powerful jaws or adaptations of their digestive systems. One such adaptation allows them to store the grass they have eaten in a digestive pouch called a rumen. Bacteria that live in the rumen attack the grass chemically, aiding in the digestive process. This stored food is then regurgitated and rechewed at a later time, providing a better breakdown of the cell walls within the grass.

Some plants have developed chemical defenses against herbivores. The best-known plants with toxic effects are poison ivy, poison oak, and poison sumac. Other plants produce toxins that inhibit the growth of bacteria, fungi, and roundworms. Still others produce chemicals whose odors act as a warning or as a repellent to a predator. Today, using the techniques of genetic engineering, scientists have been able to grow plants containing genes for these toxins that chemically repel certain predators, thereby reducing the need for artificial pesticides (**Figure 39.7**).

Certain groups of herbivores are adapted to feed on plants that are usually protected against predators by chemical compounds. For example, the larvae of cabbage butterflies feed almost exclusively on plants of the mustard and caper families, which are characterized by the presence of protective chemicals—the mustard oils. Although these plants are protected against most potential herbivores, the cabbage butterfly caterpillars have developed the ability to break down the mustard oils, rendering them harmless. In a similar example of coevolution, the larvae of monarch butterflies are able to harmlessly feed on the toxic plants of the milkweed and dogbane families.

**Figure 39.7 Defense through genetic engineering.** Scientists have developed plants that, through genetic engineering, chemically repel predators. The tobacco plant on the right is a nonengineered plant and shows the effects of insect predation. The tobacco plant on the left, however, has been engineered to produce an insect toxin that deters insects and protects the plant from them.

## Protective coloration

Some groups of animals that feed on toxic plants receive an extra benefit—one of great ecological importance. When the caterpillars of monarch butterflies feed on plants of the milkweed family, for example, they do not break down the chemicals that protect these plants from most herbivores. Instead, they store them in fat within their bodies. As a result, the caterpillars and all developmental stages of the monarch butterfly are protected against predators by this "plant" poison (**Figure 39.8**). A bird that eats a monarch butterfly quickly regurgitates it. Although this is no help to the eaten insect, the bird will soon learn not to eat another butterfly with the bright orange and black pattern that characterizes the adult monarch. Such conspicuous coloration, which "advertises" an insect's toxicity, is called warning coloration. Warning coloration is characteristic of animals that have effective defense systems, such as poisons, stings, or bites. Other examples of animals that exhibit warning coloration are shown in **Figure 39.9**.

During the course of their evolution, many unprotected species have come to resemble harmful or distasteful ones that exhibit warning coloration. Provided that the unprotected, harmless animals are present in low numbers relative to the species they resemble, predators will avoid them as well. If the unprotected animals are too numerous, of course, predators will not learn to avoid these individuals because too few will have the harmful or distasteful characteristic. Such a pattern of resemblance is called Batesian mimicry, after the British naturalist H. W. Bates, who first described this concept in the 1860s. Many Batesian mimics occur among butterflies and moths.

**Figure 39.8 Monarch butterflies make themselves poisonous.** The monarch butterfly is protected from predators by the poisonous chemicals that occur in the milkweeds and dogbanes on which they feed as larvae. Both caterpillars and adult butterflies advertise their poisonous nature with warning coloration.

Another example is a grasshopper from Borneo (*Condylodera tricondyloides*) that resembles a tiger beetle. The tiger beetle is an aggressive and dangerous insect to its predators. The grasshopper not only looks like the beetle but runs like it as well, benefiting from the tiger beetle's reputation for aggression.

Another kind of mimicry, Müllerian mimicry, was named for the German biologist Fritz Müller, a contemporary of Bates. In Müllerian mimicry, the protective colorations of different animal species come to resemble one another as in Batesian mimicry. Unlike Batesian mimicry, however, the organism and its mimic possess similar defenses. The Monarch and Viceroy butterflies, both of which are toxic, are Müllerian mimics (**Figure 39.10**).

(a)

(b)

**Figure 39.10 Müllerian mimicry.** The viceroy butterfly (a) and the monarch butterfly (b) are Müllerian mimics. Both are distasteful and toxic.

(a)

**Figure 39.9 Warning coloration.** The coloring of these animals warns other animals to stay away. (a) The red and black African grasshopper feeds on highly poisonous *Euphorbia* plants. (b) This tropical frog is so poisonous that Indians in western Colombia use its venom to poison their blow darts.

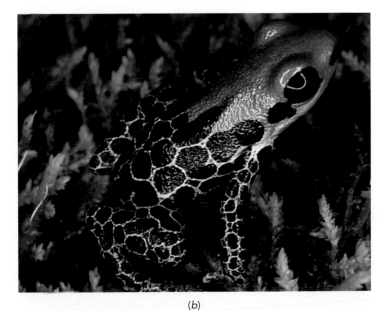

(b)

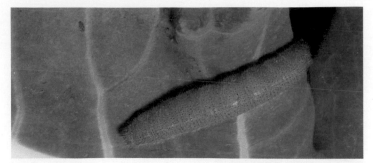

**Figure 39.11** Insect herbivores are well suited to their hosts. The green caterpillars of the cabbage butterfly are camouflaged on the leaves of cabbage and other plants upon which they feed. These caterpillars are able to break down the toxic mustard oils that prevent most insects from eating cabbage.

The appearance of some organisms allows them to blend in with their surroundings—a protective coloration called camouflage. Both cabbage caterpillars (**Figure 39.11**) and cabbage butterflies have evolved a green coloration, allowing them to hide while feeding. Although cabbage contains compounds that play a chemical defense role against most herbivores, plant pathogens, and weeds, cabbage caterpillars can feed on the toxic plants without harm and break down the toxin into harmless chemicals in their bodies. They do not store the toxin in their tissues. Insects that either eat plants lacking specific chemical defenses, or break down toxic plant chemicals to harmless compounds, seldom have warning coloration. Instead, their coloration helps them become "invisible" to predators. **Figure 39.12** shows examples of other animals that are camouflaged from their predators.

### CONCEPT CHECKPOINT

2. How would the loss of an important predator influence the diversity of species in a community?

**Figure 39.12** Two striking examples of camouflage. (a) A leaf insect displaying effective camouflage on a plant in Malaysia. (b) A spotted scorpionfish in the waters off Fiji.

**Figure 39.13** Commensalism: epiphytes and trees. The epiphytes use trees as supporting structures for their growth, but do not derive nourishment from them. The epiphytes benefit but the trees are neither helped nor harmed.

## 39.4 In commensalism, one species benefits while the other neither benefits nor is harmed.

**Symbiosis** is a general term referring to types of interactions in which one organism lives with another in an intimate association. There are three types of relationships within communities that are types of symbiosis: *commensalism*, in which one species benefits and the other is unharmed; *mutualism*, in which both species benefit; and *parasitism*, in which one species benefits and the other is harmed. Although symbiotic relationships in biological contexts occur between organisms of different species, you might have fun applying these ideas in a more general "everyday" way to characteristics of relationships in your life.

In **commensalism**, an organism of one species benefits from its interactions with another, whereas the other species neither benefits nor is harmed. Many "one-sided" relationships exist in nature. Often, the individuals deriving benefit are physically attached to the other species in the relationship. As shown in **Figure 39.13** for example, plants called epiphytes (EP-ih-fites) grow on the branches of other plants. The epiphytes derive their nourishment from the air and the rain—not from the plants to which they attach for support. Epiphytes do not harm the plant supporting them, but they benefit from this support by reaching positions where the light is better or where there is less competition for light. Similarly, various marine animals such as barnacles grow on other, often actively moving sea animals (**Figure 39.14**). These "hitchhikers" gain more protection from predation than if they were fixed in one place, and they continually reach new sources of food. They do not, however, harm the organisms to which they are attached.

(a)

(b)

**Figure 39.14 Commensalism: barnacles and whales.** (a) This gray whale displays the barnacles on its skin. (b) A close-up of a gray whale's skin reveals hitchhikers—lice and barnacles. The lice are parasites, whereas the barnacles cause no harm to the whale.

## 39.5 In mutualism, both species benefit.

**Mutualism** is a relationship in which two species live together in close association, both benefiting from the relationship. A particularly striking example of mutualism involves one genus of stinging ants and a Latin American plant of the genus *Acacia*. The modified leaves of acacia plants appear as paired, hollowed thorns. These thorns provide a home for the ants, protecting them and their larvae. In addition, the plants produce nectar that the ants eat. In turn, the ants attack any herbivore that lands on the branches or leaves of an acacia and clear away vegetation that comes in contact with their host shrub, increasing the plant's ability to survive.

Many other interesting examples of mutualism exist in nature. Certain birds, for example, spend most of their time clinging to grazing animals such as cattle, picking insects from their hides (**Figure 39.15**). The birds are provided with food, and the cattle benefit by having their parasites removed. In another similar mutualistic relationship, ants use the tiny insect aphids, or greenflies, as a provider of food. The aphids suck fluids from the phloem of plants (see Chapter 24), extracting a certain amount of

**Figure 39.15 Mutualism: red billed oxpeckers on impala in South Africa.** The oxpeckers eat parasites from the hide of the impala, and both organisms benefit.

sucrose and other nutrients. Many of these nutrients however, are not absorbed within the digestive tract of the aphid. A substantial portion runs out—somewhat altered—through the anus. The ants use this nutritional excrement as a food source and, in turn, actually carry the aphids to new plants so that they can continue eating!

Possibly one of the best-known examples of mutualism involves the relationship between certain small tropical fish called clownfish and sea anemones, marine animals that have stinging tentacles. The fish have developed an adaptation that allows them to live among the deadly tentacles of the anemones (**Figure 39.16**). These tentacles quickly paralyze other species of fishes, protecting the clownfish against predators. Although anemones can survive without clownfish, the clownfish appear to aerate the anemone tentacles, get rid of parasites, and lure other fish as prey.

**Figure 39.16 Mutualism: clownfish and sea anemones.** Two clownfish peer out from the tentacles of a large red sea anemone off the coast of Australia. Both species benefit from their close association.

## 39.6 In parasitism, one species benefits while the other is harmed.

**Parasitism** is a relationship in which an organism of one species—the parasite—lives in or on another—the host. Parasites include viruses, many bacteria, fungi, and an array of invertebrates. A different species of organism, usually larger than the parasite itself, is "home" for a parasite. During the intimate relationship between parasite and host, the parasite derives nourishment and benefits from the relationship; the host is harmed. Parasitism is sometimes considered a form of predation. Unlike a true predator, however, the successful parasite does not kill its host.

Many instances of parasitism are well known. Intestinal hookworms, for example, are parasites (**Figure 39.17**). A person is infected when walking barefoot in soil containing hookworm larvae. These larvae are able to penetrate the skin, entering the bloodstream. The blood carries the larvae to the lungs. From there, they are able to migrate up the windpipe to the esophagus. The larvae are then swallowed and reach the intestines. After growing into adult worms, they attach to the inner lining of the intestines. They remain attached there, feeding on the blood of the host.

Some parasites do not live within an organism as hookworms do, but attach to the outer surface of a plant or an animal. The attachment may be fleeting, as with the bite of a mosquito, or it may take place over a longer period, such as the burrowing of mites.

Many fungi and some flowering plants are parasitic. The dodder plant, for example, has lost its chlorophyll and leaves in the course of its evolution and is unable to manufacture food. Instead, it obtains food from the host plants on which it grows (**Figure 39.18**).

The more closely the life of a parasite is linked with that of its host, the more its morphology and behavior are likely to have been modified during the course of its evolution. The human flea, for example, is flattened from side to side and slips easily through hair. The ancestors of this species of flea were brightly colored, large, winged insects. The structural and behavioral modifications of the human flea have come about in relation to a parasitic way of life.

---

**CONCEPT CHECKPOINT**

3. Propose a hypothesis as to how Müllerian mimicry would evolve. In other words, why would it be individually adaptive for two toxic species to evolve similar coloration?

4. Although both species must be receiving a benefit from this co-mimicry in Müllerian mimicry, why is this not an example of mutualism?

**Figure 39.17 Parasitism: hookworms and humans.** These parasites of humans live in the intestine, feeding on the blood of the host.

LM 30×

**Figure 39.18 Parasitism: dodder.** The dodder plant is entwined around a host plant, deriving nutrients from it. It is yellow and without leaves because it does not obtain food via photosynthesis, but derives its nutrients from the plants it parasitizes.

## *just wondering . . .*

**Why were killer bees given that name? Is their sting so serious that you could die of it?**

African honeybees were dubbed "killer bees" by the news media, not scientists. The sting of a so-called killer bee will not kill a person, unless, of course, he or she is allergic to bee stings. There are aspects of the behavior of African honeybees however, that make them potentially life threatening.

When a bee stings a victim, glands attached to the stinger release an alarm pheromone. Pheromones are chemicals produced by one individual that alter the physiology or behavior of other individuals of the same species. African honeybees release more alarm pheromone when they sting than do other bees, and they are more sensitive to this chemical trigger than are other bees. Instead of a dozen bees pursuing a victim for 100 yards as a reaction to alarm pheromone, an entire colony of "killer bees," easily numbering 30,000 or more, may pursue a victim for a mile. Luckily, however, most people can outrun African honeybees. The problem arises if the vic-

tim is an animal that is tied down and cannot flee or a person who falls while fleeing, allowing the bees to catch up and inflict numerous stings. One death occurred in 1986 when a University of Miami graduate student on a field trip in Costa Rica stepped in a crack inside a cave, which disturbed a killer bee colony. The student caught his foot in the crack and was unable to run. He died of 8000 bee stings.

Anita Collins, a U.S. Department of Agriculture research leader specializing in honeybees, thinks that African honeybees are a minimal threat to the public. The range of these bees is shown in **Figure 39.A**. Steps are currently being taken to inject European honeybee queens in the United States with semen from African honeybees to produce a hybrid that will be as high a honey producer as the African variety and, it is hoped, that will be less sensitive to alarm pheromone. If scientists are successful, interactions within the bee community and between the bee community and human populations may become more productive and less threatening! ●

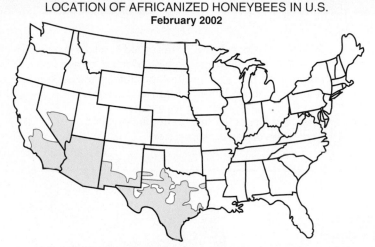

LOCATION OF AFRICANIZED HONEYBEES IN U.S.
**February 2002**

**Figure 39.A** The range of Africanized honeybees in the continental United States.

*Are you wondering about a topic in biology and how it relates to your life? Submit your question by clicking the Just Wondering link in this text's companion Web site at www.wiley.com/college/alters.*

# How do communities change over time?

## 39.7 Communities change over time by succession.

Communities change over time. Even when the climate of a given area remains stable year after year, the composition of the species making up a community, as well as the interactions within the community, show a dynamic process of directional, nonseasonal, cumulative change in the types of plant species in an area. This process of change is known as **succession**. Succession is generally linked to disturbances that set succession "in motion." This process is familiar to anyone who has seen a vacant lot or cleared woods slowly become occupied with plants and animals, or has seen a pond become filled with vegetation.

**Primary succession** takes place on surfaces not previously supporting organisms and on which there is no soil. It occurs in lakes formed from the retreat of glaciers, for example, or on volcanic islands that may rise above the sea. Primary succession also takes place

after volcanic activity kills the organisms in an area, depositing lava or rock over the soil. This process may begin as lichens take hold on bare rock. Lichens are made up of an alga and a fungus living together in a symbiotic relationship (see Chapter 20). As lichens grow, they produce acids that can break down rock, forming small pockets of soil. When enough soil has accumulated, mosses may begin to grow. This first community, a *pioneer community*, consists of plants that are able to grow under harsh conditions. The pioneer community paves the way for the growth and development of vegetation native to that climate. Over many thousands of years or even longer, the rocks may be completely broken down, and vegetation may cover a once-rocky area. As plants take hold, the area becomes able to sustain other forms of life (**Figure 39.19**).

As the plant community changes during succession, so too do the other living things. Eventually, the mix of plants and animals becomes somewhat stable, forming what is termed a **climax community**—the

(a)

(b)

**Figure 39.19 Primary succession.** Mount St. Helens, a previously dormant volcano in the state of Washington, erupted violently on May 18, 1980. The lateral blast devastated more than 600 square kilometers of forest and recreation lands within 15 minutes. (a) An area called Pumice Plain is shown four years after the blast. Pumice is a type of stone formed from cooled, hardened volcanic "froth." Notice the crater in the background. (b) In 1999, nineteen years after the blast, succession is underway at the same spot in Pumice Plain.

 **Visual Thinking:** How does succession in Mount St. Helens after the volcanic eruption fit the definition of primary succession?

last stage of succession. With an increasing realization, however, that (1) climates may change, (2) the process of succession is often very slow, and (3) the nature of a region's vegetation is determined to a great extent by human activities, ecologists do not consider the concept of a climax community as useful as they once did.

Succession occurs not only within terrestrial communities but in aquatic communities as well. A lake poor in nutrients, for example, may gradually become rich in nutrients as organic materials accumulate (**Figure 39.20**). Plants growing along the edges of the lake, such as cattails and rushes, and those growing submerged, such as pondweeds, may contribute to the formation of a rich organic soil as they die and are decomposed by bacteria. As this process of soil formation continues, the pond may become filled in with terrestrial vegetation. Eventually, the area where the pond once stood may become an indistinguishable part of the surrounding vegetation.

**Secondary succession** occurs in areas that have been disturbed, have soil, and were originally occupied by organisms. Humans are often responsible for initiating secondary succession throughout portions of the world that they inhabit. Abandoned farm fields, for example, undergo secondary succession as they revert to forest. Secondary succession may also take place after natural disasters such as a forest fire or a volcanic eruption that produces ash rather than lava flows.

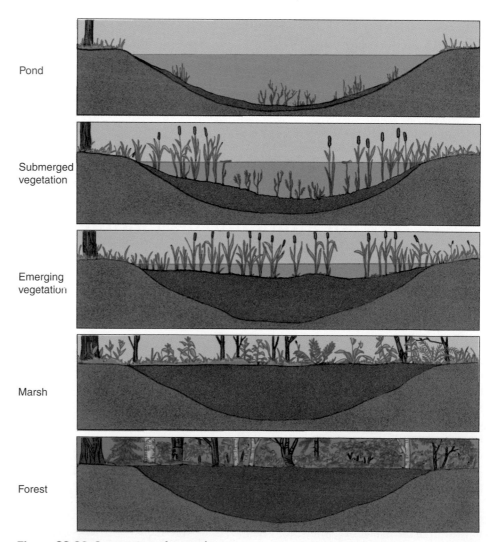

Pond

Submerged vegetation

Emerging vegetation

Marsh

Forest

**Figure 39.20** Succession of a pond.

# CHAPTER REVIEW

## Summary of Key Concepts and Key Terms

### What are communities?

**39.1** Various populations of organisms living together in a particular area at a particular time make up a **community** (p. 692).

**39.1** All of the living (**biotic**, by-OT-ik, p. 692) factors and nonliving (**abiotic**, AYE-bye-OT-ik, p. 692) factors within a certain area are an **ecosystem** (EH-koe-SIS-tem, p. 692).

**39.1** An ecosystem can be thought of as a community of organisms, along with the abiotic factors with which the community interacts.

**39.1** Within an ecosystem, each living thing has a **habitat** (p. 692), an area in which it lives that is characterized by its physical (abiotic) properties.

**39.1** Each organism plays a special role within an ecosystem, which is its **niche** (NICH, p. 692).

### What types of interspecies relationships occur in communities?

**39.2** **Competition** (p. 693) involves organisms striving to obtain the same needed resource.

**39.2** The **competitive exclusion principle** (p. 693) states that if two species are competing with one another for the same limited resource in a specific location, the species able to use that resource most efficiently will eventually eliminate the other species in that location.

**39.2** An organism's **realized niche** (p. 694) is the role it plays in an ecosystem, while its **fundamental niche** (p. 694) is the niche it might occupy if competitors were not present.

**39.3** During **predation** (p. 695), one species—the **prey**, (p. 695) becomes a resource, being killed and eaten by another species—the **predator**, (p. 695).

**39.3** Interactions that involve the long-term, mutual evolutionary adjustment of the characteristics of the members of biological communities in relation to one another are forms of coevolution.

**39.3** Some organisms within communities exhibit conspicuous coloration, or warning coloration, that advertises their ability to poison, sting, or bite.

**39.3** Organisms without specific defenses are often colored, or camouflaged, so as to blend in with their surroundings.

**39.4** **Symbiosis** (p. 698) is a general term referring to types of interactions in which one organism lives with another in an intimate association.

**39.4** **Commensalism** (p. 698) is a one-sided symbiotic relationship: One species in a relationship between organisms of different species benefits, whereas the other species neither benefits nor is harmed.

**39.5** In **mutualism** (p. 699), both species in a symbiotic relationship benefit.

**39.6** In **parasitism** (p. 700), one species (the parasite) benefits, but the other (the host) is harmed.

### How do communities change over time?

**34.7** The dynamic process of directional, nonseasonal, cumulative change in the types of plant species in an area, is **succession** (p. 701).

**34.7** **Primary succession** (p. 701) takes place on surfaces not previously supporting organisms and on which there is no soil.

**34.7** **Secondary succession** (p. 702) takes place in areas where the communities of organisms that existed initially have been disturbed and where there is soil.

**34.7** A **climax community** (p. 701) is one in which the mix of plants and animals becomes somewhat stable; it is the last stage of succession.

## Level 1 — Learning Basic Facts and Terms

**Multiple Choice**

1. Within an ecosystem, each organism has a _____, an area in which it resides. Each organism plays a role within the ecosystem, which is its _____.
   a. community // habitat
   b. habitat // niche
   c. fundamental niche // realized niche
   d. climax community // biotic community

2. Communities
   a. can change over time.
   b. can have stable compositions of species.
   c. can undergo succession.
   d. include more than one population.
   e. All of the above.

3. Which of the following is a community-level ecological question?
   a. How do the abundance and distribution of a species within a habitat change over time?
   b. How do nutrients cycle through the living and nonliving components of an ecosystem?
   c. How does predation affect the diversity of species in a community?
   d. How does the age structure of a population affect birth rate?

4. Which of the following would be considered a community-level ecological interaction?
   a. Symbiosis
   b. Succession
   c. Predation
   d. Coevolution
   e. Competition
   f. All of the above

5. Primary succession differs from secondary succession
   a. in the speed of successionary change.
   b. in the abiotic and biotic composition of the initial community.
   c. in the influence of predation on the process.
   d. by whether it occurs in an aquatic or a terrestrial community.

**True–False**

6. \_\_\_\_ The concept of a climax community is very useful to ecologists because once formed, a climax community remains stable and unchanging.

7. \_\_\_\_ An organism's fundamental niche can be larger than its realized niche but never the opposite.

8. \_\_\_\_ Coevolution is fundamentally a community level interaction because it involves interactions between more than one species.

## Level 2 | Learning Concepts

1. Distinguish among these terms: population, community, and ecosystem.
2. How and why might an organism's realized niche be different from its fundamental niche?
3. Summarize the pattern of the predator–prey relationships as shown in Gause's experiments with the protozoans, *Didinium* and *Paramecium*.
4. What other factors, in addition to those revealed in Gause's work, can affect the balance between predators and prey?
5. What is coevolution? Give an example of coevolution involving a herbivore.
6. What type of adaptation is shown in each of the following:
   a. The bright yellow and black stripes of a bumblebee's body.
   b. The unobtrusive color of some lizards that makes them difficult to see against surrounding rocks.
7. What is primary succession? Summarize the process.

## Level 3 | Critical Thinking and Life Applications

1. Some biologists argue that symbiotic relationships that are classified as commensalistic are really parasitic or mutualistic. Speculate on why biologists might think that this is the case.
2. Many people in America want to live in smaller rural towns with little traffic congestion, sprawl, strip malls, and other qualities that they find undesirable in more urban areas. As more people move to these initially attractive communities, however, they find that they become much like the community that they left behind. Describe how urbanization of a rural area is analogous to succession in natural communities. What does the "lesson" of ecological succession teach us about how to structure our communities so that the undesirable effects of urbanization can be avoided?
3. A famous ecologist, Joseph Connell, said that the structure of a community is often governed by "the ghost of competition past," because the structure we observe in communities is the result of evolution resulting from competition that occurred in the past. If this is true, and we cannot "wind back the clock to observe this competition," how do we know it occurred? What evidence is there that competition really plays a role in structuring communities?
4. Would the most successful predator feed on the very young of a population, the adults of a population, or the very old? Give a rationale for your answer.
5. What factors determine the type of organisms that grow in a climax community?
6. Herbivory is a situation in which grazing animals feed on plants. Under what conditions would herbivory be considered predation? Under what conditions would herbivory be considered parasitism?

## In The News | Critical Thinking

**PARASITES LOST**

Now that you understand more about interactions within communities, reread this chapter's opening story about research into parasites and invasive species. To better understand this research, it may help you to follow these steps:

1. Review your immediate reaction to research on the European green crab you wrote when you began reading this chapter.
2. Based on your current understanding, again summarize the main point of the research in a sentence or two.
3. What questions do you now have about this research that this chapter's opening story does not answer?

4. Collect new information about the research. Visit the *In The News* section of this text's companion Web site at www.wiley.com/college/alters and watch the "Parasites Lost" video. Then use the "summary" link to read the accompanying story and access related links. Use this information, the links provided, and other online and library resources to answer your questions and find updates about this research topic. State the sources of your information. Explain why you think the information is accurate. Also determine whether the information expresses a particular point of view or is biased in any way.
5. What in your view are the most significant aspects of this research? Give reasons for your opinions and for any changes in your ideas based on the additional information you have collected and the analysis you have done.

# ECOSYSTEMS

## In The News | Food of the Sea

Blue whales are the largest animals on Earth. Despite their weight of 160 tons or more, blue whales derive all of their nourishment from plankton, some of the tiniest organisms in the ocean.

Plankton are animals called *zooplankton* and plants called *phytoplankton* that live in bodies of water such as ponds and oceans. Plankton drift with the movement of the current or tides because some species cannot swim, and others swim poorly. Many species of plankton are single-celled organisms and others are multicellular animals, such as the larvae of fish and marine invertebrates.

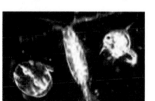

Blue whales capture plankton by opening their mouths and straining huge amounts of plankton from the water through baleen, long platelike structures that hang down from their upper jaw. A blue whale's stomach can hold over 2000 pounds of food, primarily krill and copepods, types of zooplankton.

Along with comprising part of the diet of the blue whale, phytoplankton form the base of aquatic food chains. Food chains are food transfers from producers, such as phytoplankton, to various levels of consumers, such as smaller fish, larger fish, and humans. As producers, phytoplankton contain chlorophyll and use solar energy to manufacture carbohydrates during photosynthesis. Since phytoplankton cannot survive without sunlight, they tend to float on or near the surface of water, where they are eaten by zooplankton. Many larger animal species that live in water, including shellfish, sponges, various fish, and massive blue whales, feed on the zooplankton and phytoplankton. Humans consume the fish, shellfish, and, in some cultures, marine mammals.

Along with being the foundation of aquatic food chains, phytoplankton remove $CO_2$ that enters the ocean from the atmosphere, and these organisms produce a considerable portion of the planet's supply of $O_2$. In addi-tion, scientists think that plankton play a role in reducing the buildup of certain gases in the atmosphere, including $CO_2$. These gases trap heat at the Earth's surface, thus contributing to the greenhouse effect and resulting in a rise in the Earth's average temperature (global warming).

Scientists can learn more about plankton's effects on ecosystems by monitoring biomass trends, including species variations, size, distribution, and density. Scott Gallager, a plankton ecologist at Woods Hole Oceano-graphic Institution in Massachusetts, and his team developed five different devices, collec-tively called video plankton recorders (VPRs), that gather data about plankton without re-moving the organisms from their ecosystems. One type of VPR is an underwater microscope with a video camera that takes 60 pictures per second, such as the one shown here, while being towed through the water. A computer analyzes the images and provides a count and classification of the plankton species "captured" in the photographs. As a result, scientists can generate maps of various plankton populations. In addi-tion, VPRs enable scientists to study the organisms' behav-iors and determine conditions that create boundaries be-tween plankton biomasses. To learn more about these new methods of collecting data about plankton populations, visit the *In The News* section of this text's companion Web site at www.wiley.com/college/alters.

*Write your immediate reaction to this information about plankton and plankton monitoring systems: first, summa-rize the main point in a sentence or two; then suggest what you think its significance is. You will have an oppor-tunity to reflect on your responses and gather more infor-mation on this topic in the* In The News *feature at the end of this chapter. In this chapter you will learn more about food chains, food webs, and ecosystems.*

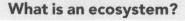

## What is an ecosystem?

**40.1** An ecosystem is a community of organisms and their environment.

## How do nutrients, energy, and organisms interact within ecosystems?

**40.2** Organisms feed on one another, transferring nutrients and energy.

**40.3** Energy flows through ecosystems.

## How do organisms obtain substances essential to life?

**40.4** Atmospheric water condenses in clouds and eventually falls to Earth.

**40.5** Carbon and oxygen in the atmosphere are usable by organisms.

**40.6** Nitrogen in the atmosphere is made available to organisms by soil bacteria.

**40.7** Phosphorus is held in the soil and taken up by plants.

# What is an ecosystem?

## 40.1 An ecosystem is a community of organisms and their environment.

A wriggling beetle may not be your idea of a gourmet meal, but to the collared lizard in **Figure 40.1** it is a five-star dinner. Beetles and other small animals, including smaller lizards, are on the daily menu of the collared lizard, an inhabitant of dry, rocky regions in southwestern North America. Minutes before this photograph was taken, the beetle was feeding on the fresh seedlings of desert plants just starting to emerge after a spring rain. Unknown to the lizard, a hawk soars overhead, waiting for the right moment to swoop down and pick up the foot-long lizard in its talons. The hawk will add yet another link to this chain of relationships among the plants and animals of this ecosystem.

Individuals of a species, such as lizards and beetles, are each part of a population. Together, interacting populations form communities (see Chapters 38 and 39). The living organisms in a community interact not only with each other but also with the nonliving substances in their environment, such as the soil, water, and air, to form an ecological system, or ecosystem. An **ecosystem** therefore includes the abiotic environment as well as the community of plants, animals, and microorganisms that interact with it and within it.

The living, or biotic, components of an ecosystem are made up of two types of organisms: those that can make their own food, or **producers**, and those that eat other organisms for food, or **consumers**. Producers, such as plants and photosynthetic protists and

some bacteria, are also called *autotrophs*. Consumers, such as animals, are also called *heterotrophs*. The Greek word *trophos*, which means "feeder" is the root of the word "heterotroph," or "other feeder," and of "autotroph," or "self-feeder."

Many consumers kill and eat their food. A special group of consumers, called **decomposers**, obtains nourishment from dead matter such as fallen leaves or the bodies of dead animals. You know the decomposers as fungi and certain bacteria.

**Figure 40.2** is a diagram of an ecosystem. Notice that the abiotic, or nonliving, components of the environment, such as decomposed material in the soil, contribute nutrients needed for the ecosystem to function. In addition, notice that the exchanges of nutrients and other chemical substances among the organisms within the ecosystem form a cyclical pattern. Energy, however, eventually flows *through* the ecosystem, first captured from the sun by the producers, then used by consumers that eat the producers, or decomposers that break them down, and ultimately used by

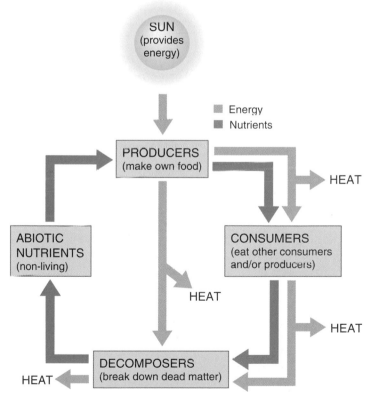

**Figure 40.2 An ecosystem.** Nutrients (green arrows) flow within an ecosystem in a cyclic pattern, whereas energy (orange arrows) flows *through* an ecosystem. Some energy is "lost" as heat.

**Visual Thinking:** Where would the lizard in Figure 40.1 fit in this scheme? Where do you fit?

**Figure 40.1 Collared lizard.** This lizard eats beetles and other small animals, making up part of a food chain in the dry, rocky regions of the western United States.

all the organisms living in the ecosystem. Much of the sun's energy originally captured by producers, however, is lost to the environment as heat. In summary, nutrients *cycle within* an ecosystem; energy *flows through* an ecosystem. Ecosystems display a regulated transfer of energy and an orderly, controlled cycling of nutrients.

Where does one ecosystem begin and another end? Some ecosystems have clearly recognizable boundaries, such as that of a pond or those found within the coastal ranges of California (**Figure 40.3**). Sometimes humans produce artificial ecosystems with human-made boundaries, such as the glass walls of an aquarium or the fencing surrounding a cultivated field. The boundaries of many natural ecosystems blend with one another, sometimes almost imperceptibly.

**Figure 40.3** **Two distinct ecosystems.** In the coastal ranges of California, the boundary between the evergreen shrub ecosystem known as chaparral and the grassland ecosystem is often relatively sharp, as shown in this photograph taken along the western edge of the Santa Clara Valley near Morgan Hill. These two ecosystems also include characteristic sets of nonliving factors.

# How do nutrients, energy, and organisms interact within ecosystems?

## 40.2 Organisms feed on one another, transferring nutrients and energy.

What is the ultimate source of most energy on Earth? Why, the sun, of course! Green plants, one of the primary producers of terrestrial ecosystems, are able to capture some of the sun's radiant energy that falls on their leaves and convert it to chemical energy during the process of photosynthesis (see Chapter 7). Green plants use energy from the sun to build organic molecules from carbon dioxide in the air, and water and nutrients in the soil. Producers, then, are key to life on Earth because no other organisms can capture this energy and use it to synthesize organic molecules for use in living systems. **Primary consumers** are herbivores (plant eaters). Herbivores feed directly on green plants, incorporating some of this energy and these nutrients into molecules that make up their own bodies, and using the rest to perform the activities of life. **Secondary consumers** are carnivores (meat eaters). Carnivores feed, in turn, on the herbivores. The chain continues, with one organism feeding on another, passing along nutrients as well as energy that was once captured from the sun.

Living things produce waste and eventually they die, but the Earth is not piled high with waste material or dead organisms. Why? Organisms that are decomposers break down the refuse, waste material, and dead organisms of an ecosystem, which is collectively known as **detritus** (dih-TRITE-us). Decomposers, sometimes called detritivores, recycle the organic materials in detritus to inorganic nutrients that can be reused by plants (**Figure 40.4**). The decomposers use some of the energy still held in the tissues of once-living things, but they are the last link in this transfer of energy and nutrients among organisms.

**Figure 40.4** **A decomposer doing its job.** The shelf fungi (phylum Basidiomycota) growing on this hardwood stump are decomposing it, converting the organic materials contained within it to nutrients usable by plants.

## *just wondering . . .*

### What happens if one species in a food chain dies off? Do all the other species in the chain die too?

A food chain is normally part of an intricate food web within an ecosystem. The organisms that make up the chain enter into a variety of relationships, in addition to predator–prey relationships, with other organisms. As species die off or move out of the ecosystem, new species move in and relationships among organisms change. The number of species lost to an ecosystem is critical. As more species are lost, the functioning of the ecosystem is increasingly degraded. In other words, diversity contributes to the health of the ecosystem—it survives better with more species.

Species that have the greatest impact on ecosystems are *keystone species*—those species that influence the structure and function of an ecosystem to a much greater degree than would be expected simply from the sizes of their populations. When a keystone is removed from an ecosystem—and therefore from a food chain and web—changes are certain to take place, and those changes depend on the role of the keystone species. In the forests of Peru, located on the Pacific coast of South America, for example, only a dozen or so species of fig and palm trees support an entire community of fruit-eating birds and mammals during the time of year when fruits are least available (**Figure 40.A**). Loss of these few tree species would probably result in the loss of most of the fruit-eating animal species even if hundreds of other tree species remained. Conversely, if a keystone predator species dies off, the populations of the species it kept in check through predation may soar. These prey may now overtake their competitors as predators of other species, or they may become prey for species that move into the area.

As you can see, the loss of a keystone species may have a great impact on a food chain and the ecosystem of which it is an integral

**Figure 40.A** Blue-and-yellow macaw eating palm fruit.

part. Because of the importance of keystone species, scientists are anxious to identify them because their extinction *could* result in the extinction of many other species in food chains, food webs, communities, and ecosystems.

Therefore, in answer to your question, the effects of the loss of a single species in a food chain vary. They depend on the adjustments made among species as relationships change in response to the loss and on whether the species was a keystone species. ●

*Are you wondering about a topic in biology and how it relates to your life? Submit your question by clicking the* Just Wondering *link in this text's companion Web site at* www.wiley.com/college/alters.

All the feeding levels described and additional levels such as tertiary consumers—carnivores that eat other carnivores—are represented in any fairly complicated ecosystem. These feeding levels are called **trophic levels** (TROW-fik). Organisms from each trophic level, feeding on one another in a linear series from consumer through sequential levels of consumers and decomposers, make up a

**food chain**. An example of a food chain can be seen in a pond ecosystem in which water fleas (primary consumers) feed on green algae (producers). Sunfish (secondary consumers) eat the water fleas, but in turn are eaten by green herons (tertiary consumers; **Figure 40.5**). The length and complexity of food chains vary greatly.

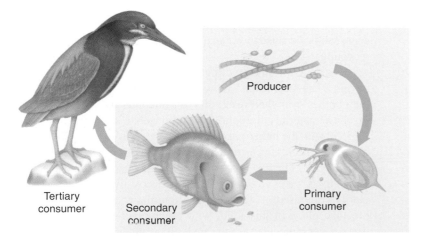

Tertiary consumer

Secondary consumer

Producer

Primary consumer

**Figure 40.5** A typical pond food chain. A food chain is a series of organisms from each trophic level that feed on one another.

 **Visual Thinking:** Where does energy enter the chain? What is the source of this energy?

In reality, rarely does any species of organism feed on only one other species. Organisms feed on many different species and types of organisms and are, in turn, food for two or more other kinds. Shown in a diagram such as **Figure 40.6**, these relationships appear as a series of branching and overlapping lines rather than as one straight line. The organisms in an ecosystem that have such interconnected and interwoven feeding relationships make up a **food web**. Figure 40.6 shows a food web in a salt marsh ecosystem.

## 40.3 Energy flows through ecosystems

In any ecosystem, the number of organisms, their total mass (biomass), and the amount of energy making up each successive trophic level are often less than the level that precedes it. Lamont Cole of Cornell University illustrated this concept in his study of the energy flow in a fresh water ecosystem in Cayuga Lake, New York. He calculated that approximately 150 calories of each 1000 calories of energy "fixed" by producers during photosynthesis was transferred into the bodies of small heterotrophs that feed on these plants and bacteria (**Figure 40.7**). Smelt, which are tiny fish, eat the heterotrophs; these secondary consumers obtain about 30 calories of each original 1000. If trout eat the smelt, they gain about 6 calories from each 1000 calories that originally entered the system. If humans eat the trout, humans gain only about 1.2 calories from each original 1000.

These types of calculations show that, on average, only a fraction of plants' accumulated energy is actually converted into the biomass of bodies of the organisms that consume them. What happens to the rest? A certain amount of the energy that is ingested by organisms goes toward heat production. It is actually "lost" as heat, consistent with the second law of thermodynamics, which states that no transformation of energy is 100% efficient. A great deal of energy is used by consumers for digestion and work; usually 40% or less goes toward growth and reproduction. An invertebrate typically only uses about a quarter of this 40% for growth. In other words, about 10% of the food that an invertebrate eats is turned into new body tissue. This figure varies from approximately 5% for carnivores to nearly 20% for herbivores, but 10% is an average value for the amount of energy, or organic matter, that organisms incorporate into their bodies from the energy available in the previous trophic level.

The relationships between trophic levels can be diagrammed as pyramids (**Figure 40.8**). *Pyramids of numbers* show, as the name suggests, the total number of organisms at each feeding level (Figure 40.8*a*). Diagrams that show the total weight of organisms supported at each trophic level in an ecosystem are referred to as *pyramids of biomass* (Figure 40.8*b*). Those that show the energy flow through an ecosystem are called *pyramids of energy* (Figure 40.8*c*).

Occasionally, pyramids of biomass and/or pyramids of number can be inverted. For example, in an ocean community, the photosynthesizing organisms (usually bacteria and protists) reproduce so rapidly that a small but constant biomass can feed a much larger biomass of herbivores. An inverted pyramid of biomass would depict this relationship. In another example, within a forest community, a single tree can support many herbivorous insects, resulting in an inverted pyramid of numbers.

### CONCEPT CHECKPOINT

1. If energy does indeed flow, rather than cycle, through ecosystems, will ecosystems eventually run out of energy? Explain.

# How do organisms obtain substances essential to life?

## 40.4 Atmospheric water condenses in clouds and eventually falls to Earth.

Although energy flows through ecosystems and most is lost at each successive transfer, the matter making up the organisms at each trophic level is not lost. All of the substances essential to life are recycled and are used only temporarily by living things. What are these substances? Hydrogen, carbon, nitrogen, and oxygen—the principal *elements* that make up all living things—are held primarily in the atmosphere in molecules of water, carbon dioxide, nitrogen gas, and oxygen gas. Other recycled elements necessary for life, such as phosphorus, potassium, sulfur, magnesium, calcium, sodium, iron, and cobalt, are held in rocks and, after weathering, enter the soil. The atmosphere, soil, and rocks are therefore referred to as the *reservoirs* (sinks) of inorganic substances that cycle within ecosystems, because they can always accept and hold elements for recycling.

The description of cycling of materials in ecosystems usually begins at the reservoirs. Living things incorporate substances into their bodies from their reservoirs or from other living things, passing these materials throughout the food web. Ultimately these substances, with the help of decomposers, move from the living world back to the nonliving world, becoming part of the soil or the atmosphere again. The first cycle we will discuss is the water cycle.

Why is water so important to life? Most living things are primarily composed of water. In addition, the chemical reactions of life take place in this fluid, with organic substances losing and using water molecules as they are built up and broken down.

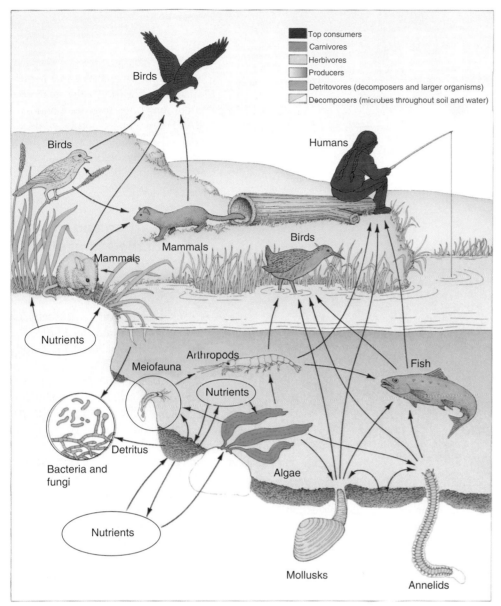

Top consumers
Carnivores
Herbivores
Producers
Detritovores (decomposers and larger organisms)
Decomposers (microbes throughout soil and water)

Birds

Birds

Humans

Mammals

Mammals

Birds

Nutrients

Meiofauna

Arthropods

Nutrients

Fish

Detritus

Bacteria and fungi

Algae

Nutrients

Mollusks

Annelids

**Figure 40.6** The food web in a salt marsh. A food web is a group of interwoven food chains within an ecosystem. In this diagram, each color represents a trophic level. Each trophic level feeds on, or gains energy from, the level below. (The term *meiofauna* refers to a group of animals that live in the spaces between grains of sand.)

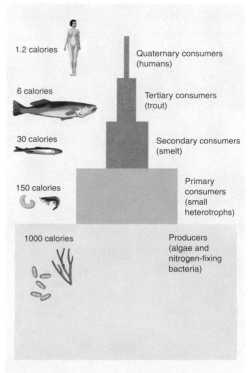

1.2 calories

Quaternary consumers (humans)

6 calories

Tertiary consumers (trout)

30 calories

Secondary consumers (smelt)

150 calories

Primary consumers (small heterotrophs)

1000 calories

Producers (algae and nitrogen-fixing bacteria)

**Figure 40.7** The flow of energy in Lake Cayuga. Results of experiments in Lake Cayuga demonstrated that only 150 calories of each 1000 calories of energy "fixed" by producers during photosynthesis was transferred into the bodies of small heterotrophs that feed on these plants and bacteria. Smelt, which are tiny fish and secondary consumers, obtain about 30 calories of each original 1000. If trout eat the smelt and humans eat the trout, humans gain only about 1.2 calories from each original 1000.

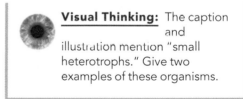

**Visual Thinking:** The caption and illustration mention "small heterotrophs." Give two examples of these organisms.

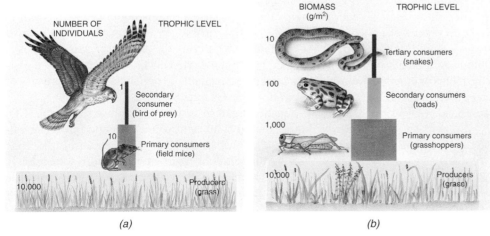

NUMBER OF INDIVIDUALS        TROPHIC LEVEL

1        Secondary consumer (bird of prey)

10        Primary consumers (field mice)

10,000        Producers (grass)

(a)

BIOMASS (g/m²)        TROPHIC LEVEL

10        Tertiary consumers (snakes)

100        Secondary consumers (toads)

1,000        Primary consumers (grasshoppers)

10,000        Producers (grass)

(b)

ENERGY (Kcal/m²/year)        TROPHIC LEVEL

48        Secondary consumers (first-level carnivores)

596        Primary consumers (herbivores)

36,380        Producers

(c)

**Figure 40.8** Pyramids of numbers, biomass, and energy. (a) The number of organisms supported at each trophic level in a hypothetical temperate grasslands ecosystem, with decomposers not shown. (b) The total weight (biomass) of organisms at each trophic level in that ecosystem. (c) The energy flow through a generalized ecosystem, indicating how much energy is present at each trophic level and how much is transferred to the next trophic level.

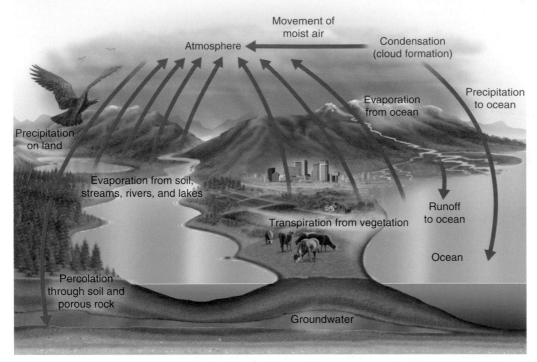

**Figure 40.9** The water, or hydrologic, cycle. In terrestrial ecosystems, 90% of the water that reaches the atmosphere comes from plant transpiration. In general, however, evaporation from oceans contributes the most water to the Earth's atmosphere.

Where do we get water? Water constantly circulates on the Earth in a pattern referred to as the **water cycle** (**hydrologic cycle**), shown in **Figure 40.9**. Heated by the sun, water evaporates into the atmosphere from the surfaces of oceans, lakes, and streams. Because oceans cover three-fourths of the Earth's surface, evaporation from these bodies contribute most of the water to the atmosphere worldwide. On land, as much as 90% of the water that reaches the atmosphere comes from plants as they release water vapor into the air during the process of transpiration (see Chapter 24).

Atmospheric water condenses in clouds and eventually falls back to the Earth as precipitation. Most of it falls directly into the oceans. Some falls onto the land, flowing into surface bodies of water or trickling through layers of soil and rock to form subsurface bodies of fresh water called *groundwater*. Plants take up water as it trickles through the soil. Crop plants, for example, use about 1000 kilograms of water just to produce 1 kilogram of biomass. Animals obtain water directly from surface water or from the plants or other animals they eat. In the United States, groundwater provides about a quarter of the water used by humans for all purposes and provides about half of the population with drinking water.

About 2% of the groundwater in the United States is polluted, and the situation is worsening. Pesticides are one source, being carried to *aquifers*, the underground reservoirs in which the groundwater lies within porous rock. Pesticides reach the groundwater in aquifers as rain washes the chemicals from the surfaces of leaves and the topsoil. Chemical wastes, stored in surface pits, ponds, and lagoons, are another key source of groundwater pollution.

Although technologies are available to clean water in underground aquifers, they are costly and time-consuming. Research has focused on developing less expensive methods. In one approach, scientists have identified a strain of bacteria known as BAV1 that metabolizes toxic waste in the anaerobic conditions of underground aquifers. BAV1 uses vinyl chloride as a food source, releasing harmless waste products. Vinyl chloride is one of the most common hazardous industrial chemicals in polluted aquifers.

## 40.5 Carbon and oxygen in the atmosphere are usable by organisms.

Like water, carbon cycles between the atmosphere and the Earth. The **carbon cycle** is based on carbon dioxide ($CO_2$), which makes up about 0.03% of the atmosphere and is found dissolved in surface bodies of water as shown in **Figure 40.10**.

Why do organisms need carbon? Terrestrial as well as marine producers use $CO_2$—along with energy from the sun—to build carbon compounds such as glucose during the process of photosynthesis (see Chapter 7). Without carbon, green plants could not transform the energy from the sun into chemical energy usable by the plants themselves as well as other organisms. Producers and the consumers that eat producers break down these carbon compounds during cellular respiration and use the energy locked in their chemical bonds to carry on the metabolic processes of life. Consumers use some of the carbon atoms and compounds from the food they eat to produce needed substances. Most of this carbon waste, however, is released to the atmosphere or to the oceans as $CO_2$. Carbon, then, forms a critical interface between the living organic world and nonliving inorganic world, being employed as an integral part of energy storage and energy use mechanisms. Intimately linked to the cycling of carbon is the cycling of oxygen ($O_2$), as shown in **Figure 40.11**. As plants use $CO_2$ during the process of photosynthesis, they produce $O_2$ as a byproduct. This $O_2$ is released into the atmosphere and becomes available to organisms for the process of cellular respiration (see Chapter 8).

Some aquatic organisms such as mollusks use the $CO_2$ dissolved in water and combine it with calcium to form their calcium carbonate ($CaCO_3$) shells. When these organisms die, their shells collect on the sea floor. Years of exposure to water slowly dissolves the $CaCO_3$, releasing the $CO_2$ and making its carbon once again available to aquatic producers for use in the process of photosynthesis.

When organisms die, decomposers break down the carbon compounds making up their bodies. Some carbon-containing compounds, such as the cellulose found in the cell walls of plants, are more resistant to breakdown than others, but certain bacteria, fungi, and protozoans are able to accomplish this feat. Some cellulose,

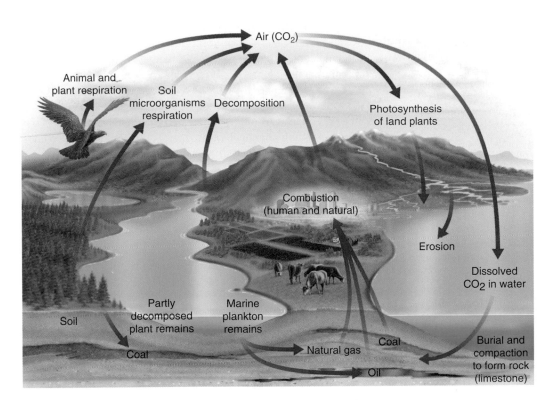

**Figure 40.10** The carbon cycle. Carbon dioxide is found in the oceans and in the atmosphere. Producers and some consumers incorporate this carbon into substances necessary for life.

however, accumulates as undecomposed organic matter. Over time and with heat and pressure, this undecomposed matter results in the formation of fossil fuels such as oil and coal. When these fuels are burned, as when wood is burned, the $CO_2$ is returned to the atmosphere. The release of this carbon as $CO_2$, a process that is proceeding rapidly as a result of human activities, may change global climates. Section 42.9 describes the role of $CO_2$ and global climate change in more detail.

### CONCEPT CHECKPOINT

2. Global climate change due to increasing atmospheric $CO_2$ has caused us to look more closely at those components of the global carbon cycle that add $CO_2$ to the atmosphere, acting as sources, or that take away $CO_2$ from the atmosphere, acting as sinks. Based on your knowledge of the carbon cycle, classify each of the following as a net $CO_2$ source or sink: Calvin cycle, Krebs cycle, burning of fossil fuels, plants, animals, phytoplankton, decomposers, forest fires.

## 40.6 Nitrogen in the atmosphere is made available to organisms by soil bacteria.

If you like gardening or raising houseplants, then you know that nitrogen is an important plant nutrient. Nitrogen is important for all living things, representing a critical part of the chemical structure of proteins. Where do living things obtain this essential element?

Nitrogen gas ($N_2$) makes up 78% of the Earth's atmosphere. However, it is useless to most organisms in that form. Only a few

**Figure 40.11** How carbon dioxide is linked to oxygen cycling. Plants use carbon dioxide in photosynthesis and give off oxygen. This oxygen is available to animals for the process of cellular respiration. Animals breathe out carbon dioxide; some carbon dioxide is liberated from the decomposition of dead organisms.

kinds of bacteria can convert $N_2$ into a form that can be used for biological processes, thus playing a crucial role in the cycling of nitrogen. These bacteria are called **nitrogen-fixing bacteria**, and they convert $N_2$ to ammonia ($NH_3$). Living things depend on this process of nitrogen fixation. Without it, they would ultimately be unable to continue to synthesize proteins, nucleic acids, and other necessary nitrogen-containing compounds.

**Figure 40.12** Nitrogen-fixing bacteria live in root nodules visible as swellings on this pea plant. In this mutualistic relationship, bacteria of the genus *Rhizobium* convert nitrogen in the atmosphere (N$_2$) to ammonia (NH$_3$) or ammonium ions (NH$_4^+$), a form of nitrogen the plant can use. Some of this "fixed" nitrogen leaks into the soil, enriching it for subsequent crops. The plant provides the bacteria with food—carbohydrates and other organic compounds.

Certain of the nitrogen-fixing bacteria are free-living in the soil. Others form mutualistic relationships with plants by living within swellings, or nodules, of plant roots as shown in **Figure 40.12**. Some of these plants are legumes (LEG-yoomz *or* lih-GYOOMZ)—plants such as soybeans, alfalfa, peas, beans, and clover. Plants having mutualistic associations with nitrogen-fixing bacteria can grow in soils having such low amounts of available nitrogen that they are unsuitable for most other plants. Growth of a leguminous crop can enrich the nitrate level of poor soil enough to benefit the next year's nonleguminous crop. This is the basis for crop rotation in which, for example, a field may be planted with soybeans—a legume—and corn (a nonlegume) in alternating years. Nitrogen-fixing bacteria of the genus *Anabaena* contribute large amounts of nitrogen to the rice paddies of China and Southeast Asia (**Figure 40.13**).

Other bacteria play roles in the nitrogen cycle as well as the nitrogen-fixing bacteria. Ammonifying bacteria convert biological compounds containing nitrogen in detritus to ammonia (NH$_3$) or ammonium ions (NH$_4^+$). Nitrifying bacteria convert NH$_3$ or NH$_4^+$ to nitrates (NO$_3^-$), another form of nitrogen that plants can use. NO$_3^-$ is also produced by lightning, which causes nitrogen gas (N$_2$) to react with oxygen gas (O$_2$) in the atmosphere. Humans add NO$_3^-$ to the soil by spreading chemical fertilizers. The nitrogen cycle comes full circle as nitrogen is continuously returned to the environment by denitrifying bacteria that break down NO$_3^-$, liberating N$_2$ to the atmosphere. The **nitrogen cycle** is illustrated in **Figure 40.14**.

Many commercial plant fertilizers include nitrogen. The amount of nitrogen in the fertilizer is indicated by the first of three numbers on the fertilizer package. These numbers, called N–P–K numbers, stand for nitrogen, phosphorus, and potassium, and indicate the percent by weight of ammonium nitrate, phosphorus oxide, and potassium oxide in the fertilizer. If the N–P–K rating is 15–30–15, then 15% by weight of the fertilizer is the nitrogen (N) component, ammonium nitrate.

## 40.7 Phosphorus is held in the soil and taken up by plants.

Do you take a vitamin and mineral supplement? If so, the minerals you are likely to see listed on the container include the elements phosphorus, potassium, magnesium, calcium, and iron. These elements are some of the nutrients, in addition to water, carbon, and nitrogen, that organisms need to form structural and functional components. Where do these elements come from? Other than from mineral supplements, we get them by eating plants, or by eating organisms that ate plants. Plants obtain these elements from the soil. These elements are cycled as shown in the **phosphorous cycle** in **Figure 40.15**.

Phosphorus, more than any of the other required plant nutrients except nitrogen, is apt to be so scarce that it limits plant

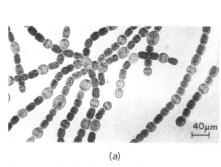

(a)        (b)        (c)

**Figure 40.13** The use of nitrogen-fixing bacteria in agriculture. The nitrogen-fixing bacteria *Anabaena azollae* (a) lives in the spaces between the leaves of the floating water fern *Azolla* (b), which is deliberately introduced into the rice paddies of the warmer parts of Asia. By means of this mutualistic relationship, the *Anabaena* thrives and not only fixes nitrogen for the water fern but also enriches the nitrate level of the rice paddy. Rice, here cultivated in Bali (c), is the major food for well over one-fourth of the human race.

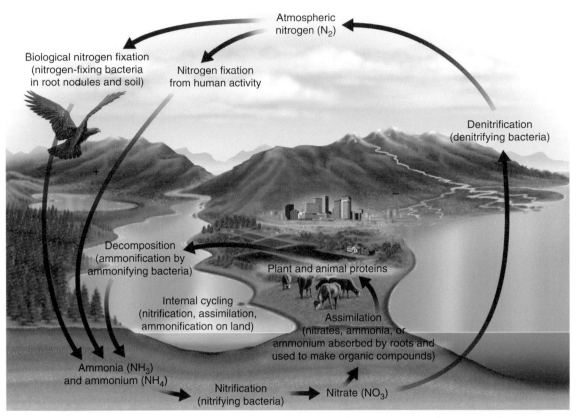

**Figure 40.14 The nitrogen cycle.** Bacteria that can convert nitrogen into usable forms by plants are extremely important in the nitrogen cycle. Most nitrogen is held in the atmosphere as a gas, which is unusable by living things. Bacteria "fix" the nitrogen by incorporating it into compounds plants can use. Ultimately, it is the bacteria that also return nitrogen to the atmosphere.

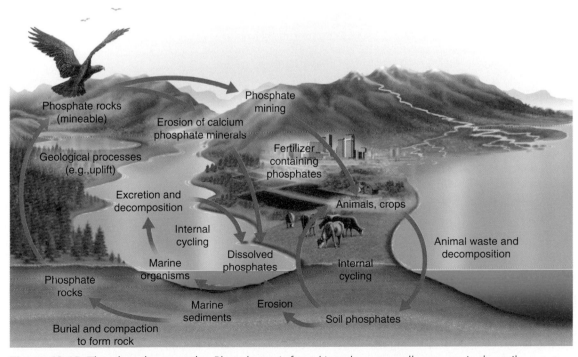

**Figure 40.15 The phosphorus cycle.** Phosphorus is found in only very small amounts in the soil, to make up for this lack of phosphorus, humans add phosphates to the soil. Plants take up phosphate from the soil, and animals obtain the necessary levels of phosphate by eating plants.

growth. In the soil, phosphorus is found as relatively insoluble phosphate ($PO_4^{-3}$) compounds, present only in certain kinds of rocks. For this reason, phosphates exist in the soil only in small amounts. Therefore, humans add millions of tons of phosphates to agricultural lands every year. Plants take up the phosphates from the soil, and animals obtain the phosphorus they need by eating plants or other plant-eating animals. When these organisms die, de-

composers release the phosphorus incorporated in molecules such as DNA and ATP, making it again available for plant use.

Rivers and streams carry any phosphate runoff to the oceans. Some of these phosphates become incorporated into the bodies of fishes and other marine animals. The seabirds that eat these animals deposit enormous amounts of guano (feces) rich in phosphorus along certain coastlines (**Figure 40.16**). These deposits have

# How Science Works

### Discoveries

## *The Hubbard Brook experiment: what have we learned about the cycling of nutrients?*

The Hubbard Brook Ecosystem Study (HBES), initiated over forty years ago, is the longest-running ecosystem study in the United States. The site of the study is the Hubbard Brook Experimental Forest, an 8000-acre site located in the White Mountain National Forest, near North Woodstock, New Hampshire. One of the areas of study in this research is the effects of the cycling of inorganic nutrients within ecosystems. The HBES has yielded much of the information that is now known about nutrient cycles, and the ingenious experimental designs used in the HBES have provided the basis for experimental methods used in the study of other ecosystems.

Hubbard Brook is the central stream located in the Hubbard Brook Forest. For measurement of the flow of water and nutrients within the Hubbard Brook ecosystem, concrete weirs with V-shaped notches (**Figure 40.B**) were built across six tributary streams that were selected for study. A weir is a low dam built across a stream. In this case, the dams were built so that all of the water that flowed out of the six valleys—the areas surrounding the six streams—had to pass through the notches in the dams so that samples could be collected and flow could be determined. The precipitation that fell in the six streams was measured, and the amounts of nutrients present in the water flowing in the six streams were also determined. By these methods, the scientists demonstrated that the undisturbed forests in this area were very efficient in retaining nutrients. The small amounts of nutrients that fell from the atmosphere with the rain and snow were approximately equal to the amounts of nutrients that ran out of the six valleys. For example, there was only a small net loss of calcium—about 0.3% of the total calcium in the system per year—and small net gains of nitrogen and potassium.

Then the researchers disturbed the ecosystem. In 1965, the investigators felled all of the trees and shrubs in one of the six valleys and prevented their regrowth by spraying the area with herbicides. The effects of these activities were dramatic. The amount of water running out of the valley increased by 40%, representing the water normally taken up by the trees and shrubs—and evaporated into the atmosphere from their leaves. That water was now running off. The amounts of nutrients running out of the system also increased. The loss of calcium was 10 times higher than it had been previously. The change in the status of nitrogen was particularly striking. The undisturbed ecosystem in this valley had been accumulating nitrogen at a rate of 2 kilograms per hectare (about 2.4 acres) per year, but the cut-down ecosystem lost it at a rate of about 120 kilograms per hectare per year! The nitrate level of the water rapidly increased to a

level exceeding that judged safe for human consumption, and the stream that drained the area carried these nutrients, which acted as a fertilizer for cyanobacteria and algae, resulting in massive blooms of these organisms.

The Hubbard Brook experiment demonstrated that nutrient cycling depends, among other things, on the vegetation present in the ecosystem. The fertility of the deforested valley decreased rapidly, and at the same time the danger of flooding greatly increased. The Hubbard Brook experiment is particularly instructive today because large areas of tropical rain forest are being and have been destroyed to make way for cropland. Many of the insights gleaned from the Hubbard Brook experiment can help scientists understand some of the consequences of rain forest destruction that may not have been readily apparent otherwise. Rain forest destruction is discussed in more detail in Chapter 42. ●

**Figure 40.B A notched weir.** This weir (dam) is representative of those built across the six tributary streams of the Hubbard Brook. Water could be easily sampled and its flow determined as it passed through the notches in the dams.

traditionally been used for fertilizer. Phosphates not incorporated into the bodies of animals precipitate out of the water and become part of the bottom sediment.

**Figure 40.16** A guano coast. Marine animals incorporate phosphates from runoff into their bodies by eating aquatic plants that have incorporated the phosphates into their tissues. The seabirds that feast on these marine animals deposit feces rich in phosphorus on certain coasts.

---

### CONCEPT CHECKPOINT

3. A limiting nutrient is one that restricts or limits an organism's growth when it is in low quantities in the environment. Based on what you know about the cycling of oxygen, nitrogen, phosphorus, and carbon, which of these nutrients would you hypothesize is most limiting to terrestrial plant growth? Explain your answer.

---

## CHAPTER REVIEW

## Summary of Key Concepts and Key Terms

### What is an ecosystem?

**40.1** **Ecosystems** (p. 707) are communities of organisms and the nonliving factors of their environments through which energy flows and within which nutrients cycle.

**40.1** The living, or biotic, components of an ecosystem are made up of two types of organisms: autotrophs that can make their own food, or

**producers** (p. 707), and heterotrophs, or **consumers** (p. 707) that eat other organisms for food.

**40.1** A special group of consumers, called **decomposers** (p. 707), obtains nourishment from dead matter such as fallen leaves or the bodies of dead animals.

### How do nutrients, energy, and organisms interact within ecosystems?

**40.2** Through photosynthesis, plants growing under favorable circumstances capture and lock up some of the sun's energy that falls on their green parts.

**40.2** Plants may be eaten by herbivores (**primary consumers**, p. 708), which in turn may be eaten by carnivores (**secondary consumers**, p. 708).

**40.2** Decomposers break down the remains of all organisms as well as refuse and waste material, which is collectively called **detritus** (dih-TRITE-us, p. 708).

**40.2** This sequence of organisms, one feeding on another, constitutes a food chain.

**40.2** Each link in a **food chain** (p. 709) is a **trophic level** (feeding level, p. 709).

**40.2** The organisms in an ecosystem that have such interconnected and interwoven feeding relationships make up a **food web** (p. 709).

**40.3** Each trophic level includes organisms that can transfer about 10% of the energy that exists at their level to the next level.

**40.3** From one level to the next, 90% of the energy in a food chain is lost as heat or is used for various metabolic activities.

### How do organisms obtain substances essential to life?

**40.4** Elements essential to life are cycled through the atmosphere, the soil, and the ocean.

**40.4** Carbon dioxide, nitrogen gas, oxygen gas, and water are the atmospheric reservoirs of the carbon, nitrogen, oxygen, and hydrogen used in biological processes; all of the other elements that organisms incorporate into their bodies come from the Earth's rocks and soil.

**40.4** In the **water cycle** (**hydrologic cycle**, p. 712), water in the atmosphere condenses in clouds and falls to the Earth as precipitation.

**40.4** Plants take up water from the soil, and animals obtain water from surface water or from the plants or other animals they eat.

**40.4** Water returns to the atmosphere through the evaporation of surface water and transpiration by plants.

**40.5** In the **carbon cycle** (p. 713), carbon is held in the atmosphere in the form of carbon dioxide.

**40.5** Carbon dioxide is taken in by photosynthetic organisms and is used to build the carbon compounds that plants manufacture during the process of photosynthesis.

**40.5** Both producers and consumers use the carbon compounds that plants manufacture as an energy source, metabolizing them during cellular respiration and releasing carbon dioxide to the atmosphere.

**40.6** In the **nitrogen cycle** (p. 715), atmospheric nitrogen is converted to ammonia by several genera of symbiotic and free-living bacteria called **nitrogen-fixing bacteria** (p. 714).

**40.6** The ammonia, in turn, is assimilated into amino groups in proteins of cells or is converted to nitrites and then to nitrates by other bacteria.

**40.6** The breakdown of nitrogen-containing molecules either converts them to recyclable forms or results in the release of atmospheric nitrogen.

**40.7** Phosphorus is a key component of many biological molecules; in the **phosphorus cycle** (p. 716) it weathers out of soils and is transported to the world's oceans, where it tends to be lost.

**40.7** Phosphorus is relatively scarce in rocks; this scarcity often limits or excludes the growth of certain kinds of plants.

**40.7** Humans add phosphate compounds to the soil for plants to take up, and animals obtain the phosphates they need by eating plants or plant-eating animals.

**40.7** When organisms die, decomposers release phosphorus back to the soil.

## Level 1 — Learning Basic Facts and Terms

**Multiple Choice**

1. Which of the following does not describe an organism in terms of its mode of nutrient acquisition?
   a. Heterotroph
   b. Autotroph
   c. Decomposer
   d. Detritus
   e. Tertiary consumer

2. The organisms that harness the energy from sunlight and thereby provide all of the energy to other trophic levels are called
   a. heterotrophs.
   b. producers.
   c. primary consumers.
   d. detritus feeders.
   e. chemotrophs.

3. Which of the three trophic level pyramids can sometimes be inverted (broad at the top and narrow at the base)?
   a. Energy
   b. Biomass
   c. Numbers
   d. Both a and b
   e. Both b and c

**Matching**

4. _____ To make this nutrient available for use in living processes, it must first be chemically altered through the action of symbiotic or soil-dwelling bacteria.

5. _____ The shells of mollusks are involved in the natural cycling of this nutrient.

6. _____ Because this element is only present in certain types of rocks, it is often lacking in many soils.

7. _____ Bird guano is used to fertilize soils that are depleted of this element.

8. _____ This substance is the primary constituent of the hydrologic cycle.

9. _____ Planting of legumes like soybeans helps replenish this nutrient in the soil.

10. _____ The burning of fossil fuels is affecting the natural balance of this nutrient's cycle.

   a. Carbon
   b. Water
   c. Nitrogen
   d. Phosphorus

## Level 2 — Learning Concepts

1. What is an ecosystem? Contrast the sources of nutrition within an ecosystem used by producers, consumers, and decomposers.

2. For most ecosystems, energy pyramids are similarly broad at the base and narrow at the top. What is the primary reason for this pattern?

3. Distinguish among trophic level, food chain, and food web.

4. Describe one way in which human activities can impact or affect each of these nutrient cycles: carbon, water, nitrogen, and phosphorus.

5. Why can crop rotation make soil more fertile? To what chemical cycle does this relate?

## Level 3 — Critical Thinking and Life Applications

1. Imagine you had a hollow glass sphere the size of a basketball and you wanted to create within it a self-sustaining stable ecosystem that needed only sunlight and moderate temperature to persist indefinitely. What would you put into this ecosphere?

2. Extensive cutting and burning of tropical rain forests often result in a drastic and permanent lowering of rainfall in the cleared area, an effect noted by Alexander von Humboldt more than 100 years ago. Why?

3. Some have argued that better technologies in finding and extracting fossil fuels from the Earth will provide the human population with a limitless supply of coal and oil to meet our ever-increasing energy demands. What do the lessons of ecological energy flow teach us about the validity of this statement?

4. Develop an argument to support the statement that the global carbon cycle is "out of balance."

5. In the 1990s thousands of acres of northwestern forests were protected from logging because of concern over the potential loss of the endangered spotted owl, an owl that makes its home in this forest ecosystem. Imagine that you are an environmentalist advocating the protection of the spotted owl habitat. Develop an ecologically-focused argument to convince the voting public that the spotted owl should be protected.

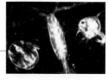

## In The News — Critical Thinking

**FOOD OF THE SEA**

Now that you understand more about food chains and the importance of producers as the base of food chains, reread this chapter's opening story about plankton and plankton monitoring systems. To better understand this information, it may help you to follow these steps:

1. Review your immediate reaction to new methods of collecting data about plankton that you wrote when you began reading this chapter.

2. Based on your current understanding, again summarize the main point of the research in a sentence or two.

3. What questions do you now have about this research that this chapter's opening story does not answer?

4. Collect new information about research into methods of plankton data collection. Visit the *In The News* section of this text's companion Web site at www.wiley.com/college/alters and watch the "Food of the Sea" video. Then use the "summary" link to read the accompanying story and access related links. Use this information, the links provided, and other online and library resources to answer your questions and find updates about this topic. State the sources of your information. Explain why you think the information is accurate. Also determine whether the information expresses a particular point of view or is biased in any way.

5. What in your view are the most significant aspects of this research? Give reasons for your opinion and for any changes in your ideas based on the additional information you have collected and the analysis you have done.

# BIOMES AND LIFE ZONES OF THE EARTH

## In The News  Food for Thought

Since the 1940s, dramatic increases in crop yields have enabled farmers in the United States to grow more food for consumption around the world. Farm mechanization, technological advances in plant breeding, pesticide and synthetic fertilizer application, and more efficient use of resources have contributed to the high productivity of many American farms. Although farmers can control their crop production methods, including the types of equipment they use and the kinds and amounts of pesticides they apply, they cannot control one of the most important factors that influences crop yields—the weather, or in general terms, the climate.

In the late 1980s, an increasing number of scientists became concerned about the possible consequences of the greenhouse effect on farm productivity. The greenhouse effect is the blocking of outward heat radiation from the Earth's surface by high levels of certain gases in the atmosphere, most notably carbon dioxide ($CO_2$). As a result of the greenhouse effect, the warming of the atmosphere could alter average temperatures and average amounts of precipitation throughout the world. In many food-producing regions, such climate changes may modify the distribution of farmland and reduce farmers' ability to grow certain crops. The effects on local economies, food availability, human health, and various ecosystems could be significant. Although scientists are still concerned today about the effects of climate changes on agriculture, they are uncertain about the rate and extent of the changes as well as their overall impact on global agricultural production.

David Lobell and Gregory Asner, graduate students in the Department of Global Ecology at Stanford University, studied the relationship between climate change and U.S. corn and soybean crop yields during the period from 1982 to 1998. The scientists examined crop produc-

tion and climatic data on a subset of midwestern counties that had exhibited higher yields of corn and soybeans during slightly cooler summers. The researchers found that the cooler climate during the study period was responsible for approximately 20 percent of the gains in crop yield, while other factors such as crop management accounted for the rest.

From these data, Lobell and Asner predicted that if the growing seasons warmed, rather than cooled, then for every degree Celsius (or nearly every two degrees Farenheit) that the growing season temperature increased; corn and soybean crop yields could decrease by 17%. Researchers caution, however, that other factors also influence crop yields as mentioned previously, including extreme events such as floods. In addition, the data from these counties may not be representative of the United States as a whole. More research is needed to determine whether these projections are valid for other parts of the country and other situations. Nevertheless, scientists agree that small, gradual shifts in climate can play an important role in crop yields and the food supply. You can learn more about this research and view the video "Food for Thought" by visiting the *In The News* section of this text's companion Web site at www.wiley.com/college/alters.

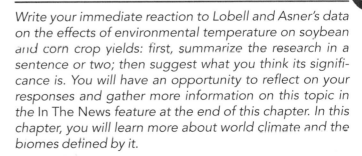

*Write your immediate reaction to Lobell and Asner's data on the effects of environmental temperature on soybean and corn crop yields: first, summarize the research in a sentence or two; then suggest what you think its significance is. You will have an opportunity to reflect on your responses and gather more information on this topic in the* In The News *feature at the end of this chapter. In this chapter, you will learn more about world climate and the biomes defined by it.*

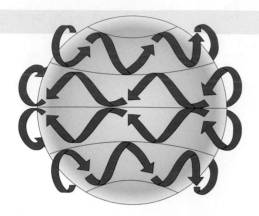

## CHAPTER GUIDE

### What are biomes, and how does climate define them?

**41.1** Biomes are large terrestrial ecosystems.

**41.2** Temperature, rainfall, and altitude determine an area's characteristic vegetation.

**41.3** Biomes are often classified into seven climatic regions.

### What are the features of the Earth's major biomes?

**41.4** Tropical rain forests are regions of high temperature and rainfall.

**41.5** Savannas are regions of high temperature with less rainfall than tropical rain forests.

**41.6** Deserts are regions of little rainfall.

**41.7** Temperate deciduous forests are regions having warm summers, cold winters, and moderate amounts of precipitation.

**41.8** Temperate grasslands are regions with greater rainfall than deserts but less than savannas.

**41.9** Taiga are regions having long, cold winters with little precipitation.

**41.10** Tundra encircles the top of the world.

### How are the Earth's aquatic life zones distributed?

**41.11** Fresh-water ecosystems lie near and are intertwined with terrestrial ecosystems.

**41.12** Estuaries occur where fresh water meets saltwater.

**41.13** The oceans provide a variety of habitats.

# What are biomes, and how does climate define them?

## 41.1 Biomes are large terrestrial ecosystems.

What are the biological characteristics of the region in which you live? What types of trees dominate the landscape, or are there no trees at all? Maybe grasses are the dominant vegetation in your area or desert cacti. Whatever its characteristics, the region in which you live is one of the Earth's major ecosystems—a biome.

**Biomes** are large ecosystems that occur over wide areas of land within specific climatic regions and that are easily recognized by their overall appearance. Each biome is similar in its structure and appearance wherever it occurs on Earth and differs significantly from other biomes. Biomes support communities of characteristic organisms, and are sometimes named for their stable plant communities, such as the tropical rain forest.

The characteristics of biomes are a direct result of the climate of the region, especially temperature and rainfall patterns. These patterns result from the interaction of:

- The amount of *solar radiation (energy)* that reaches different parts of the Earth, and the seasonal variations in that radiation.
- *Global atmospheric circulation* and the resulting patterns of oceanic circulation.
- The *features of the Earth* itself such as the presence of mountains and valleys.

## 41.2 Temperature, rainfall, and altitude determine an area's characteristic vegetation.

Because the Earth is a sphere, some parts receive more energy from the sun than others do. The tropics, which are near the equator, are warmer than the temperate regions farther north or south. As shown in **Figure 41.1**, the sun's rays arrive almost perpendicular to regions at and near the equator. These rays (let's call them units of solar energy) fall on a smaller area of the Earth's surface than do rays falling farther north or south, so their warmth is concentrated in a smaller area. Distance north or south of the equator is measured in degrees called latitude. The greater the latitude, the colder the climate.

The northern and southern hemispheres also experience a change of seasons as well as the gradations in temperature that vary with latitude. Seasons occur because the Earth is tilted on its axis, and as it takes its year-long journey around the sun, the northern and southern hemispheres receive unequal amounts of sunlight at various times. When the sun's rays strike the northern hemisphere more directly, summer occurs there; at the same time, winter occurs in the southern hemisphere where the sun's rays strike the earth at a greater angle (see Figure 41.1). The situation is reversed when the sun's rays strike the southern hemisphere more directly. The angle of the sun's rays striking the areas near the

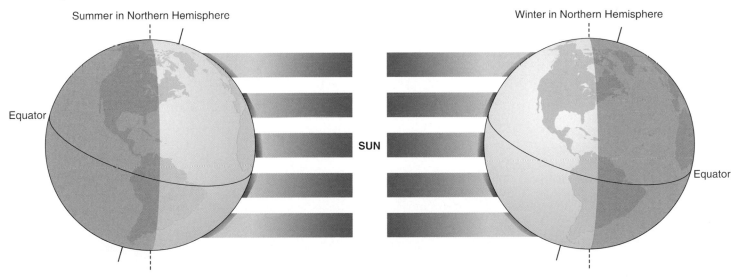

**Figure 41.1** The angle of solar energy striking the Earth affects climate. As latitude increases from the equator, the angle at which the sun's ray strike the Earth increases, resulting in a given unit of solar energy being spread over a larger area than at the equator. There-fore, the greater the latitude, the colder the climate. Seasons occur because the Earth is tilted on its axis, and as it moves around the sun, the northern and southern hemispheres receive unequal amounts of sunlight at various times.

**Visual Thinking:** 1. Which part of the Earth receives the most solar energy? Explain your answer.
2. In many areas of the United States, it is cold in the winter and warm in the summer. Why is this so?
3. In the summer in the United States a person standing in the sun for a half hour generally has greater exposure to the ultraviolet rays from the sun than in the winter. Why?

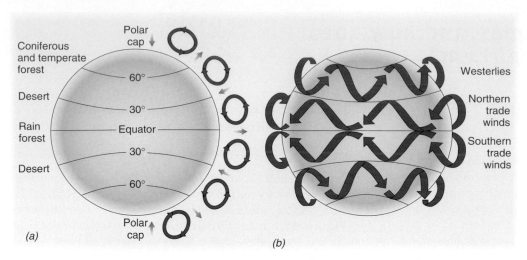

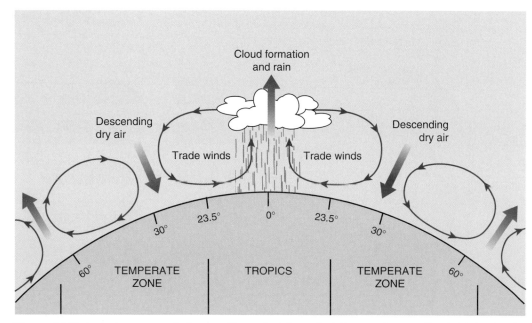

**Figure 41.2** The Earth's atmospheric circulation. (*a*) Differential warming of the Earth's surface creates six coils of rising and falling air. (*b*) Rotation of the Earth deflects winds, creating the trade winds near the equator and the westerlies in the middle latitudes.

equator does not vary a great deal. Those tropical regions show little seasonal variation.

Because of the unequal distribution of solar energy from the equator to the poles, the warmer air at the equator rises and flows toward the poles. By the time the air masses reach about 30° north or south of the equator, they have cooled and become more dense, so they sink back to the Earth's surface. This sinking air drives temperate air masses that rise at about 30° and fall at about 60° north or south of the equator. These in turn drive the polar air masses (**Figure 41.2a**). With increasing latitude, each air mass is cooler than the one before but warmer than the next.

The rotation of the Earth deflects the winds from the vertical paths illustrated in Figure 41.2a to the patterns of coils or belts shown in Figure 41.2b. Because the surface of the Earth near the equator moves faster than that at the poles, it causes the coils of air at the equator to blow from east to west, creating the Northern and Southern *trade winds*. It also causes the coils of air in the temperate zones north and south of the equator to blow from west to east, creating the *westerlies*.

These moving masses of air have a profound effect on precipitation. The moisture-holding capacity of air increases when it is warmed and decreases when it is cooled (**Figure 41.3**). Therefore, precipitation is relatively high where air is rising and being cooled. Conversely, precipitation is generally low near 30 degrees north and south latitude, where air is falling and being

warmed. Partly as a result of these factors, all the great deserts of the world such as the Sahara in Africa lie near 30 degrees north or 30 degrees south latitude, and some of the great temperate forests are near 60 degrees north or south latitude (see Figure 41.2a). Tropical rain forests lie near the equator.

Within these broad air masses, local features of the Earth may also affect climate. As air travels up a mountain (its windward side), it is cooled, and precipitation forms (**Figure 41.4**). As the air descends the other side of the mountain (its leeward side), it is warmed; its moisture-holding capacity increases. For these reasons, the windward sides of mountains are much wetter than the more arid leeward sides, and the vegetation is often very different. This phenomenon is known as the rain shadow effect. Seattle, Washington, for example, lies on the windward side of the Cascade Mountain range in the northwestern United States. This city receives 99 centimeters (39 inches) of rainfall per year. Yakima, Washington, on the leeward side of this mountain range, receives only 20 centimeters (8 inches) of rain per year.

**Figure 41.3** As hot or warm air rises and is cooled, it can hold less moisture. Precipitation is relatively abundant near the equator as rising air cools and loses moisture. Near 30 degrees north and south latitude, air falls and warms, absorbing available moisture and creating belts of dry or desert regions.

 **Visual Thinking:** Would you expect regions of high precipitation or low precipitation at about 60° latitude? Why?

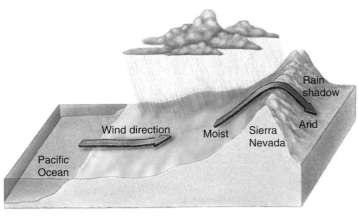

**Figure 41.4** **The rain shadow effect**. As air travels up a mountain, it is cooled, its moisture-carrying capacity decreases, and precipitation forms. As air descends the other side of the mountain, it is warmed and its moisture-holding capacity increases. The windward side of mountains tends to be moist; the leeward side tends to be arid.

British Isles. Because of the Gulf Stream, western Europe is much warmer and thus more temperate than eastern North America at similar latitudes. As a general principle, the western sides of continents in the temperate zones of the northern hemisphere are warmer than their eastern sides; the opposite is true in the southern hemisphere.

### CONCEPT CHECKPOINT

1. Imagine that a large comet struck the Earth causing its axis to become perpendicular to the line between the Earth and the sun. What effect would this event have on the Earth's climate?
2. Many millions of years ago, continental drift resulted in the convergence of most of the Earth's landmasses into one large "supercontinent" called Pangea. Predict what aspects of the Earth's climate might have been affected by this event. What aspects probably would not have been affected?

The patterns of atmospheric circulation influence patterns of circulation in the ocean, modified by the location of the land masses around and against which the ocean currents must flow. Oceanic circulation is dominated by huge surface currents (gyres) (**Figure 41.5**), which move around the zones of high pressure between approximately 30 degrees north and 30 degrees south latitude. These gyres move clockwise in the northern hemisphere and counterclockwise in the southern hemisphere. They profoundly affect life not only in the oceans, but also on coastal lands because they redistribute heat. For example, the Gulf Stream in the North Atlantic swings away from North America near Cape Hatteras, North Carolina, and reaches Europe near the southern

## 41.3 Biomes are often classified into seven climatic regions.

If you ask ecologists how many biomes exist, you will likely get a variety of answers. Scientists hold differing opinions on the exact number, or different types of biomes. One well-recognized approach is to classify biomes into seven categories: (1) tropical rain forests, (2) savannas, (3) deserts, (4) temperate grasslands, (5) temperate deciduous forests, (6) taiga, and (7) tundra. This list is arranged by distance from the equator, but the biomes do not encircle the Earth in neat bands.

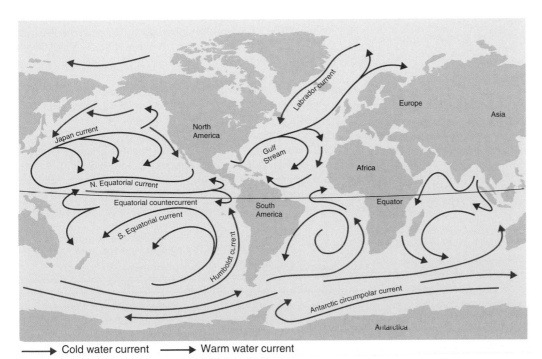

**Figure 41.5** Ocean circulation. The circulation in the oceans moves in great spiral patterns or loops, which are called surface currents (gyres), and affects the climate on adjacent continents by redistributing heat.

→ Cold water current    → Warm water current

The distribution of biomes around the world is shown in **Figure 41.6**. Their distribution is greatly affected by the climatic effects caused by the presence of mountains, the irregular outlines of the continents, and the temperature of the surrounding sea. In addition, climate and vegetation change with increas-ing elevation of land similar to the manner in which they change with increasing latitude. Thus, as shown in **Figure 41.7**, tundra-like vegetation occurs near the top of a mountain in the tropics, as well as on land surfaces approaching the North and South Poles.

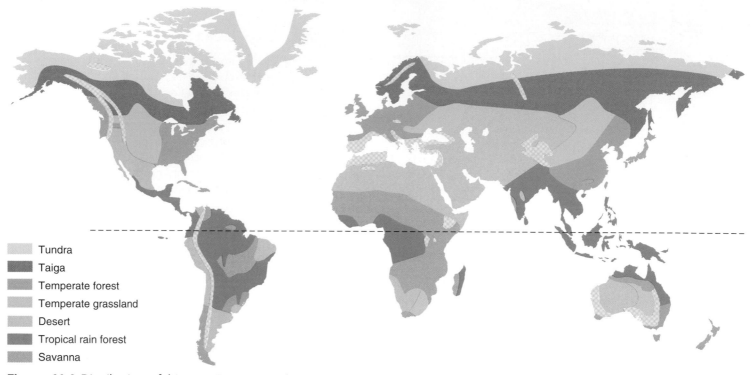

Tundra
Taiga
Temperate forest
Temperate grassland
Desert
Tropical rain forest
Savanna

**Figure 41.6 Distribution of biomes.** Biomes are large terrestrial ecosystems supporting characteristic plant communities and defined largely by climatic regions.

# *just wondering . . .*

*Questions students ask*

## What is the difference between El Niño and La Niña?

El Niño and La Niña are both climatic phenomena. The story of El Niño starts with the Humboldt current, which sweeps up the west coast of South America from the south (see Figure 41.5). Simultane-ously, southern trade winds push the warm, nutrient-poor surface water away from the coast, resulting in an upwelling of deeper water. An upwelling brings up cool, nutrient-rich waters from the depths, replacing the surface water as it is pushed westward.

Once every three to seven years, the southern trade winds di-minish due to changes in atmospheric pressures in certain parts of the world. Therefore, the upwelling of the Humboldt current diminishes, and warm water flows south, down the coast to southern Peru and northern Chile. This warm water flow has been named El Niño (mean-ing "The Boy Child") by local fisherman because the warm current oc-curs around the Christmas season; their reference is to the Christ child. Fishermen are familiar with this event because the warm, nutri-ent-poor water causes massive fish kills. These events can be devas-tating to South American fisheries, especially in Chile and Peru.

Conversely, La Niña ("The Girl Child") refers to colder than av-erage sea surface temperatures in the central or eastern equatorial Pacific region. Many scientists call this a *cold event* rather than La Niña. A cold event is a more widespread cooling of the eastern tropical Pacific than is the warming of the area off the cost of South America during El Niño.

El Niño and La Niña affect the weather near to and far from the eastern tropical Pacific. Both affect the winds that blow over these wa-ters, changing the weather those winds carry. Rain patterns shift and temperatures shift. Both El Niño and La Niña have different local weather effects in different parts of the world. For example, Galapagos El Niño events are wet, whereas Isthmus of Panama El Niño events are dry. La Niña events generally have opposite effects to El Niño events.

Scientists are still working to understand these unusual weather patterns and cannot predict them with accuracy. Never-theless, scientists have used the knowledge they have gained to recognize such events in their early stages. Further study may hold the key to understanding the complex interplay of atmospheric and water currents around the globe and both the positive and nega-tive effects of El Niño and La Niña on weather patterns. ●

*Are you wondering about a topic in biology and how it relates to your life? Submit your question by clicking the* Just Wondering *link in this text's companion Web site at* www.wiley.com/college/alters.

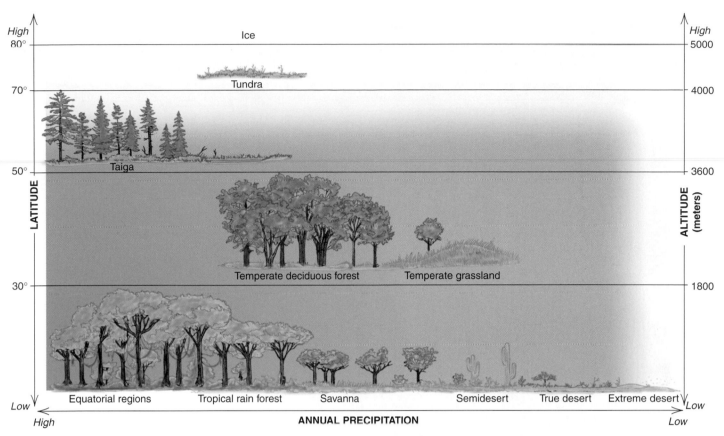

**Figure 41.7 Elevation and biomes.** Biomes that normally occur far north and far south of the equator occur also at high mountain elevations. Thus, as you climb a tall mountain in southern Mexico or Guatemala, you might pass through a sequence of biomes like those you might encounter if you traveled farther from the equator.

# What are the features of the Earth's major biomes?

## 41.4 Tropical rain forests are regions of high temperature and rainfall.

You've likely read or heard about the deforestation of tropical rain forests. That topic is discussed in Chapter 42. Here, we focus on the characterization and locations of these forests.

**Tropical rain forests** occur in Central America; in parts of South America, particularly in and around the Amazon Basin; in Africa, particularly in central and west Africa; and in southeast Asia.

As the name suggests, the tropical rain forests occur in regions of high temperature and rainfall, generally 200 to 450 centimeters (80 to 175 inches) per year, with little seasonal variation. The temperature averages 25°C (77°F). As a comparison, Houston, Texas—one of the hottest and wettest cities in the United States—receives an average of 121 cen-

timeters (48 inches) of rainfall per year; the average temperature is 20.5°C (69°F). Houston almost seems dry and cool in comparison to the tropical forest.

You may have the idea that the tropical rain forest is thick with lush vegetation and creeping vines, creating a network too dense to penetrate without a machete. These forests *are* thick and lush—but not at the forest floor. Little can grow on the ground beneath the canopy of trees whose branches and leaves form an overlapping roof. Only 2% of the light shining on the forest canopy reaches its floor. Plants that do grow there have large, dark green leaves adapted to conducting photosynthesis at low light levels. Other types of vegetation have interesting adaptations that enable them to compete successfully with the large trees for sunlight. Vines, for example, have their roots anchored in the soil, but they climb up the trees with their long stems. They reach the canopy where leaves grow to capture sunlight. Other interesting plants are the epiphytes, or "air plants." *Epiphytes*, such as those shown in Figure 39.13, grow on the trees or other plants for support, but draw their nourishment mostly from rainwater. Some epiphytes catch moisture with modified leaves or flower parts; others have roots that hang free in the air and absorb water (with its dissolved minerals) from the rain.

**Figure 41.8** The colorful diversity of the tropical rain forest: an orchid from Peru, a chestnut-mandibled toucan from Costa Rica, and a flag-legged insect from Costa Rica, with leaflike extensions of its hind legs.

The giant trees of the tropical forest support a rich and diverse community of animals on their branches. **Figure 41.8** provides examples of just a few of these interesting organisms. The roots of the trees are interesting also, spreading out from thickened trunks into a thin layer of soil, often no more than a few centimeters deep. These roots transfer the nutrients from fallen leaves and other organic debris quickly and efficiently back to the trees after bacteria and fungi break down the debris. Very few nutrients remain in the soil. Therefore, when humans cut down and then burn these trees to clear the land for agriculture, they are "burning away" the nutrients held in the trees, as well as breaking down organic matter to carbon dioxide. The small amount of ash that is left provides few nutrients for the crops farmers try to grow. In two to three years of farming, these few remaining nutrients are depleted from the soil, and the land becomes barren.

## 41.5 Savannas are regions of high temperature with less rainfall than tropical rain forests.

Not all areas near the equator are wet; some areas have a lower annual rainfall than the tropical forests do (generally, about 90 to 150 centimeters per year, or 35 to 60 inches) and may experience prolonged dry seasons. The heat, periodic dryness, and poor soils cannot support a forest, but have led to the evolution of **savannas**: open grasslands with scattered shrubs and trees, as shown in **Figure 41.9**. These areas, situated between the tropical rain forests and deserts, cover much of central and southern Africa, western India, northern Australia, large areas of northern and east-central South America, and some of Malaysia.

The vegetation of the savanna supports large grazing herbivores such as buffalo, wildebeests, and zebras. These animals, in turn, are food for carnivores such as lions. The savanna also supports a large number of plant-eating invertebrates, such as mites, grasshoppers, ants, beetles, and termites. The termites are one of the most important soil organisms in the savanna (**Figure 41.10**).

**Figure 41.9 Savannas: dry, tropical grasslands.** Marabou storks, elephants, and zebras visit the water on an African savanna in Tsavo Park, Kenya.

**Figure 41.10 Termite mound in Australia.** The termites of the savanna break down dried twigs, leaves, and grass to usable nutrients, thereby improving the nutrient status of the soil in and around their mounds. In addition, these mounds provide passageways for rainwater to deeply penetrate the ground rather than just running off or evaporating from the surface. The mounds are built of termite excrement or of soil held together with saliva. The mound is connected to underground chambers and passageways.

## 41.6 Deserts are regions of little rainfall.

Most everyone knows that deserts are dry—but how dry are they? **Deserts** are biomes that have 25 centimeters (10 inches) or less of precipitation annually. For this reason, the vegetation in deserts is characteristically sparse. The higher the annual rainfall a desert has, however, the greater the amount of vegetation it will be able to support.

Ecologists classify deserts based on their annual rainfall: Semideserts receive about 25 centimeters (10 inches) per year (Phoenix, Arizona, and San Diego, California, for example); true deserts receive less than 12 centimeters (4.7 inches) per year (Las Vegas, Nevada); and extreme deserts average below 7 centimeters (2.8 inches) per year (Namib Desert in southwestern Africa). The photos of these three types of deserts in **Figure 41.11** show the differences in their patterns of vegetation.

Major deserts occur around 20 to 30 degrees north and south latitude, where the warm air that rises from the equator falls. The air at the equator rises, cooling and releasing its moisture, which falls on the tropical forests. The dry air then falls over desert regions, resulting in little precipitation. Deserts also occur in the interiors of continents far from the moist sea air, especially in Africa, Eurasia (Europe and Asia), and Australia. Some deserts, such as the Baja region of California, are near the ocean and yet are dry; the winds blow from the north, carrying little moisture because they are cool. In addition, some deserts form on the leeward side of mountain ranges, such as in the Great Basin of Nevada and Utah in the United States on the leeward side of the Sierra Nevada Mountains.

Because desert vegetation is sparse and the skies are usually clear, deserts radiate heat rapidly at night. This situation results in substantial daily changes in temperature, sometimes more than 30° Centigrade (approximately 55° Fahrenheit) between day and night. Hot deserts like the Sahara are hot all year, whereas cool deserts like the Great Basin of North America are hot only in the summer. Summer daytime temperatures in all deserts, however, are extremely high, frequently exceeding 40°C (104°F). Temperatures of 58°C (136.4°F) have been recorded both in Libya and in San Luis Potosi, Mexico—the highest that have been recorded on Earth.

Plants have developed a wide variety of adaptations in this difficult environment. Annual plants—those that complete their life cycles and die within a single growing season—are often abundant in deserts and simply bypass the unfavorable dry season in the form of seeds. After sufficient rainfall, many germinate and grow rapidly, sometimes forming spectacular natural displays. Characteristic of deserts are the many species of succulent plants, those with tissues adapted to store water, such as cacti (see Figure 41.11a). The trees and shrubs that live in deserts often have deep roots that reach sources of water far below the surface of the ground. The woody plants that grow in deserts either lose their leaves during the hot, dry seasons of the year, or they are evergreen, with hard, reduced leaves. The creosote bush of the deserts of North and South America is an example of an evergreen desert shrub. Near the coasts in areas where there are cold waters offshore, deserts may be foggy, and the water that the plants obtain from the fog may allow them to grow luxuriantly. Examples are the Chilean-Peruvian and the outer Namib deserts.

Desert animals, too, have fascinating adaptations that enable them to cope with the limited water of the deserts. Many limit their activity to a relatively short period of the year when water is available or even plentiful. Many desert vertebrates live in deep, cool, and sometimes moist burrows. Organisms that are active for much of the year emerge from their burrows only at night, when temperatures are relatively cool. Other organisms, such as camels, can drink large quantities of water when it is available and can survive for long periods without water. A few animals simply migrate to or through the desert and exploit food that may be abundant seasonally.

(a)

(b)

(c)

**Figure 41.11 Deserts.** (a) The Sonoran Desert in Arizona, a semi-desert. (b) Death Valley, California, a true desert. (c) Namib Desert, Namibia in western Africa, an extreme desert.

CONCEPT CHECKPOINT

**3.** Organize your knowledge of the Earth's biomes discussed so far for each biome, describing its geographic location, climatological characteristics, and a few adaptations or characteristics of organisms inhabiting the biome.

## 41.7 Temperate deciduous forests are regions having warm summers, cold winters, and moderate amounts of precipitation.

The colors of the fall foliage in a temperate deciduous forest can be fabulous. Tourists come from all over the world to forests in the northeastern United States, particularly to the mountains of Vermont and New Hampshire, to see the autumn colors. The climate in areas of the northern hemisphere such as  the eastern United States and Canada and an extensive region in Eurasia supports the growth of trees whose leaves turn shades of red, orange, and yellow, and subsequently drop from the trees, during the autumn. Such trees are called *deciduous* (dih-SIJ-you-us), from a Latin word meaning "to fall." These vast areas of trees are therefore **temperate deciduous forests** (**Figure 41.12**) and thrive in climates where summers are warm, winters are cold, and the precipitation is moderate, generally from 75 to 150 centimeters (30 to 60 inches) annually. Precipitation is well distributed throughout the year, but water is generally unavailable during the winter because it is frozen.

The temperate forest has an upper canopy of dominant trees such as beech, oak, birch, hickory, and maple; often a lower tree canopy of eastern dogwood; and a layer of shrubs beneath dominated by blueberry, mountain laurel, or rhododendron. On the ground, herbs, ferns, and mosses abound, although temperate forests vary greatly in the amount of vegetation that grows near the forest floor. In addition, animal life in the temperate forest is abundant on the ground as well as in the trees.

## 41.8 Temperate grasslands are regions with greater rainfall than deserts, but less than savannas.

**Temperate grasslands** have various names in different parts of the world: the prairies of North America, steppes (STEPS)

**Figure 41.12 Temperate deciduous forests.** This photograph was taken in the fall at Baxter State Park, Maine. The inset images show an Eastern grey squirrel, Whitetail deer, and a cicada.

**Figure 41.13** A herd of bison on temperate grasslands.

of Russia, pusztas (PUZ-taz) of Hungary, veldt (VELT) of South Africa, and pampas (PAM-pahz) of South America. All temperate grasslands have 25 to 75 centimeters (10 to 30 inches) of rainfall annually, much less than that of savannas but more than that of deserts. Temperate grasslands also occur at higher latitudes than savannas.

Temperate grasslands are characterized by large quantities of perennial grasses—those living for three seasons or more. The rainfall is insufficient to support forests or shrublands. Grasslands are often populated by burrowing rodents, such as prairie dogs and other small mammals, and herds of grazing mammals, such as the North American bison (**Figure 41.13**). The grazing of herbivores

contributes to the maintenance of this biome by eating the few woody plants that might arise.

The soil in temperate grasslands is rich; much of the temperate grasslands are farmed for this reason. Grasslands are often highly productive when they are converted to agriculture, and many of the rich agricultural lands in the United States and southern Canada were originally occupied by prairies.

## 41.9 Taiga are regions having long, cold winters with little precipitation.

In latitudes north of the temperate deciduous forest, the trees that dominate the forests change to various species of evergreen (**Figure 41.14**). The northern coniferous forest is also called **taiga** (TIE-gah). The cone-bearing (coniferous) evergreen trees of this forest are primarily spruce, hemlock, and fir and extend across vast areas of Eurasia and North America.

The taiga is characterized by long, cold winters with little precipitation. Most of the precipitation falls in the summer. Because of the latitude where taiga occurs, the days are short in winter with as little as six hours of daylight, and correspondingly long in summer with as little as six hours of darkness. The light, warmth, and rainfall of the summer allows plants to grow rapidly, and crops often attain a large size in a surprisingly short time.

The trees of the taiga occur in dense stands of one or a few species of cone-bearing trees. Alders, a common species, harbor nitrogen-fixing bacteria in nodules on their roots; for this reason, they are able to colonize the infertile soils of the taiga. Lichens, which are made up of algae and fungi, can photosynthesize in low temperature and low light. In addition, their tissues are not easily damaged by frost and the cold, making lichens successful taiga and tundra inhabitants, and a plentiful food source. Marshes, lakes, and ponds also characterize the taiga; they are often fringed by willows or birches. Many large mammals can also be found there, such as the caribou (**Figure 41.15**). Other herbivores, including moose, elk, and deer, are stalked by carnivores such as wolves, bear, and lynx.

**Figure 41.14** Taiga: great evergreen forests of the north. This photograph was taken in Alaska.

**Figure 41.15** Caribou in the taiga in winter.

 **Visual Thinking:** What might serve as food for these animals during this challenging season?

## 41.10 Tundra encircles the top of the world.

Farthest north in Eurasia, North America, and their associated islands—between the taiga and the permanent ice—is the open, often boggy community known as the **tundra** (**Figure 41.16**). Dotted with lakes and streams, this enormous biome encircles the top of the world, covering one-fifth of the Earth's land surface. A well-developed tundra does not occur in the Antarctic because there is no land at the right latitude. The tundra is amazingly uniform in appearance, dominated by scattered patches of grasses and sedges (grasslike plants), heathers, and lichens. Some small trees do grow, but they are primarily confined to the margins of streams and lakes.

Annual precipitation in the tundra is very low, similar to desertlike precipitation of less than 25 centimeters (10 inches) annually. In addition, the precipitation that falls remains unavailable to plants for most of the year because it freezes. During the brief Arc-

tic summers, some of the ice melts. The permafrost, or permanent ice found about a meter down from the surface, never melts, however, and is impenetrable to both water and roots. When the surface ice melts in the summer, it has nowhere to go and forms puddles on the land. In contrast, the alpine tundra found at high elevations in temperate or tropical regions does not have this layer of permafrost.

The tundra teems with life during its short summers. As in the taiga, perennial herbs grow rapidly then, along with various grasses and sedges. Large grazing mammals, including musk ox and caribou, migrate from the taiga. Many species of birds nest in the tundra in the summer and then return to warmer climates for the winter. Populations of lemmings, small rodents that breed throughout the year beneath the snow, rise rapidly and then crash on a three- to four-year cycle, influencing the populations of the carnivores that prey on them (see Figure 38.6), such as snowy owls and arctic foxes.

### CONCEPT CHECKPOINT

4. Add to the list that you developed for the previous Concept Checkpoint including for the remaining biomes their geographic location, climatological characteristics, and adaptations of organisms inhabiting the biome.

**Figure 41.16** Tundra: cold boggy plains of the north. Mount McKinley National Park in Alaska. The inset photos show a snowy owl and arctic foxes.

# How are the Earth's aquatic life zones distributed?

## 41.11 Fresh-water ecosystems lie near and are intertwined with terrestrial ecosystems.

Only 2% of the Earth is covered by fresh water, found standing in lakes and ponds or moving in rivers and streams. Fresh-water ecosystems are intimately connected to terrestrial ecosystems. For example, some organisms such as amphibians may move from one ecosystem to another. In addition, organic and inorganic material continuously enters bodies of fresh water from terrestrial communities. Often, the wet, spongy land of marshes and swamps provides habitats intermediate between the two.

Where does life occur in a pond or a lake? Ponds and lakes have three life zones, or regions, in which organisms live: the shore zone, the open-water zone, and the deep zone (**Figure 41.17**). The **shore zone** is the shallow water near edges of a lake or pond in which plants with roots, such as cattails and water lilies, may grow. Consumers such as frogs, snails, dragonflies, and tiny shrimplike organisms live among these producers. The **open-water zone** is the main body of water through which light penetrates. Floating and drifting algae and plantlike organisms, or phytoplankton, grow here. Floating consumers such as protozoans comprise the zooplankton that feed on the phytoplankton in this aquatic food chain. They, in turn, are eaten by small fish, which are eaten by larger fish. The **deep zone**, the water into which light does not penetrate, is devoid of producers. This dark zone is inhabited mainly by decomposers and other organisms such as clams that feed on the organic material that filters down to them. Ponds differ from lakes in that they are smaller and shallower. Therefore, light usually reaches to the bottom of all levels of a pond; it has no deep zone.

Rivers and streams differ from ponds and lakes primarily in that their water flows rather than remains stationary. The nature of this ecosystem is therefore different from that of a pond or lake. One difference is that the level of dissolved oxygen is usually much higher in a river or stream than in a standing body of water because moving water mixes with the air as it churns and bubbles along. A high level of dissolved oxygen allows an abundance of fish and invertebrates to survive. In addition, only a few types of producers inhabit rivers and streams: various species of algae that grow on rocks and a few types of rooted plants such as water moss.

A river or stream is characterized as an open ecosystem. That is, it derives most of its organic material from outside sources. Detritus, composed of debris or decomposing material, flows from upstream or enters from the land. Leaves and woody material drop into the stream from vegetation bordering its banks. Rainwater washes organic material from overhanging leaves. In addition, water seeps into a river or stream from below the surface of adjoining land, carrying with it organic materials and, in some cases, fertilizers and other chemicals. These nutrients feed the producers and small consumers. The river/stream ecosystem is largely heterotrophic and is strongly tied to terrestrial ecosystems that surround it.

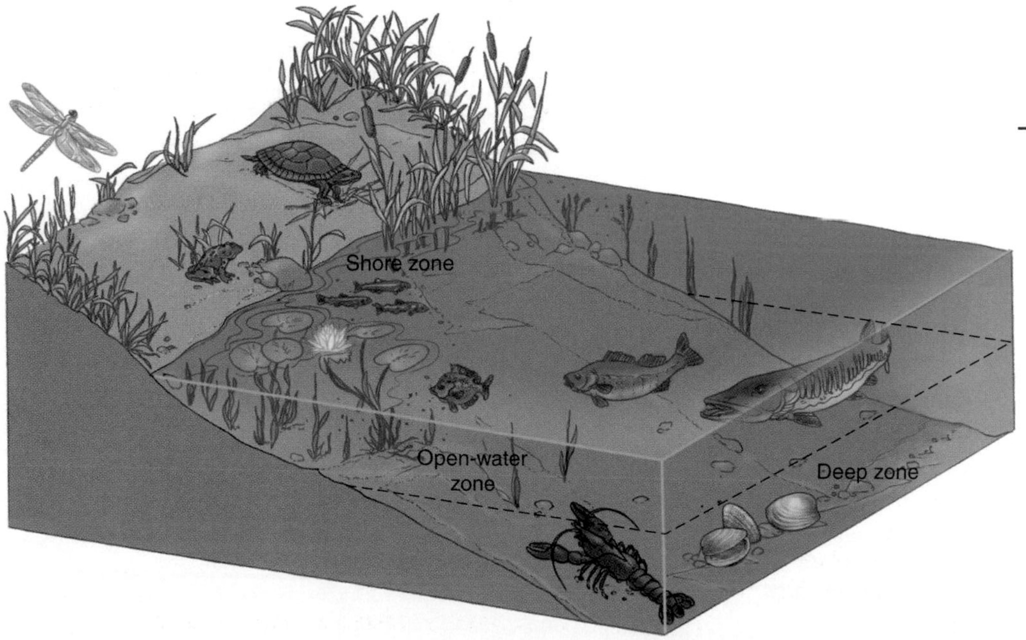

## 41.12 Estuaries occur where fresh water meets saltwater.

As rivers and streams flow into the sea, an environment called an **estuary** (ESS-choo-ER-ee) is created where fresh water joins saltwater. In the shallow water of the estuary, rooted grasses often grow. Other producers of the estuary are various types of algae and phytoplankton. Consumers are primarily mollusks, crustaceans, fish, and various zooplankton. All organisms inhabiting estuaries, however, have adaptations that allow them to survive in an area of moving water and changing salinity. *Salinity* refers to the concentration of dissolved salts in the water. For example, many estuarine organisms are bottom dwellers and attach themselves to bottom material or burrow in the mud. Each

**Figure 41.17** The three life zones of a "typical" lake. Rooted plants grow in the shallow shore zone. Small consumers live among these producers. Light penetrates through the open-water zone, supporting floating plants and animals, and fish. Light does not penetrate the deep zone. It supports decomposers and other organisms that feed on organic material that filters down to them.

Shore zone

Open-water zone

Deep zone

species of organism is found living in a region of salinity optimal for its survival.

Oysters are one of the most important bottom dwellers in estuaries, providing a habitat for many other species of organisms. These mollusks either bury themselves in the mud, forming oyster beds, or cement themselves in clusters to the partially buried shells of dead oysters, forming oyster reefs. Many other invertebrates, such as sponges and barnacles, attach themselves to the oysters and feed on plankton. Other species, such as crabs, snails, and worms, live on, beneath, and between the oysters, feeding on the oysters themselves or on detritus trapped in the oyster reef. One researcher has documented more than 300 species of organisms living in association with a single oyster reef.

The motile organisms of the estuary are primarily crustaceans such as crabs, lobsters, shrimp, and various species of fish. Fish exhibit interesting reproductive adaptations to the varying salinities of the estuary. Some species, such as the striped bass, spawn upstream from the estuary where the salinity of the water is low. The young fish move downstream through increasing concentrations of salt as they develop, moving into the ocean in adulthood.

Nutrients are more abundant in estuaries than in the open ocean because estuaries are close to terrestrial ecosystems and derive much of their nutrients from them, as rivers and streams do. Unfortunately, estuaries are also easily polluted from these sources. In Chesapeake Bay (**Figure 41.18a**), for example, complex systems of rivers enter the Atlantic Ocean, forming one of the most biologically productive bodies of water in the world. In the 1960s, the bay yielded an annual average of about 275,000 kilograms (600,000 pounds) of fish. As the human population in this area increased, however, along with oil transport and commercial shipping, pollution also increased. More than 290 oil spills were reported in the bay in 1983 alone. In addition, uncontrolled erosion from certain agricultural practices increased the level of nutrients—primarily nitrogen and phosphorus—in the water. These nutrients resulted in the growth of algae, which clouds the water and blocks the sunlight needed by the bay grasses. This can be seen in some of the rivers that empty into the bay in the aerial view shown in Figure 41.18b. The grasses then die, sink, and decompose, a process that uses the dissolved oxygen in the water. Organisms living in the water can no longer survive when oxygen is depleted, so they either

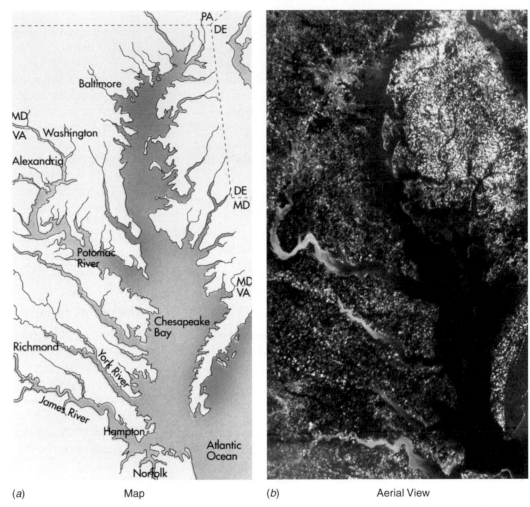

(a) Map  (b) Aerial View

**Figure 41.18 Chesapeake Bay.** Chesapeake Bay has more than 11,300 kilometers (7020 miles) of shoreline and drains more than 166,000 square kilometers (64,100 square miles) in one of the most densely populated and heavily industrialized areas in North America. (a) The large metropolitan areas noted on the map and shipping facilities make the bay one of the busiest natural harbors anywhere. (b) Uncontrolled erosion from certain agricultural practices, pesticides, and increases in nutrients block the light needed for photosynthesis and upset the delicate ecological balance on which the productivity of the bay depends. This material can be seen in some of the rivers that empty into the bay in this aerial view.

leave the area or they die. This situation caused a 90% decrease in the yield of fish from the area in the 1980s.

In the late 1990s, the primary problem of the Chesapeake Bay area continued to be the abundance of nutrients in the water. Due to the implementation of agricultural best management practices, however, these levels steadily declined during the 1990s. Maryland and Virginia, the states that border the bay, along with the Environmental Protection Agency, are working to bring the Chesapeake Bay estuary back to its former productivity. In late 2003, the governors of Maryland and Virginia called on the federal government to elevate Chesapeake Bay restoration to a rank equal to the Florida Everglades, a move that would establish federal funding for the project.

## 41.13 The oceans provide a variety of habitats.

Although only 2% of the Earth is covered by fresh water, nearly three-quarters of the Earth's surface is covered by ocean. These seas have an average depth of 4 kilometers (approximately 2 1/2 miles), and they are, for the most part, cold and dark. The concentration of oxygen, as well as the availability of light and food, is a factor that limits life in the ocean. Although cold water is able to "hold" more oxygen than warm water, the warmer seawater near the surface of the ocean mixes with the oxygen in the atmosphere. Therefore, oxygen is present in its highest concentrations in the upper 200 meters (650 feet) or so of the sea. Light is most abundant in the top 100 meters (325 feet).

The marine environment provides a variety of habitats that can be classified into three major life zones:

- The *intertidal zone*, the area between the highest tides and the lowest tides.
- The *neritic zone* (nih-RIT-ik), the area of shallow waters along the coasts of the continents, which extends from the low tide mark to waters down to 200 meters deep.
- The *open-sea zone*, comprising the remainder of the ocean.

## The intertidal zone

The wind-swept shoreline is a harsh place for organisms to live. As the tide rolls in and out, environmental conditions change from hour to hour: wet to dry, sun protected to sun parched, and wave battered to calm. Nevertheless, life abounds in the intertidal zone.

**Figure 41.19** shows the rocky shore at low tide on the Big Sur coastline in California. The exposed rocks are teeming with life, but these organisms vary along a continuum from the driest areas that are least often covered with water to the wettest areas that are most often covered with water. Highest up on the rocks, in areas that the high tide sometimes does not reach, grow certain lichens and algae. Somewhat lower on the rocks grow barnacles. Then oysters, blue mussels, and limpets (mollusks that have conical shells) take over, followed by brown algae and red algae. "Forests" of brown algae, or kelp, take over in areas that are exposed for only short periods of time. All of these organisms have adaptations such as hard shells or

gelatinous coverings that keep them from drying out, and they are either anchored within the sand or stick to the surfaces of rocks so that they will not be washed away.

In contrast to the rocky shore, the sandy shore (**Figure 41.20**) may look as though no life is present. Life is plentiful, however, within the sand and mud. Because organisms have no large surfaces to which they can attach, they are adapted to burrowing into the sand during low tide. Copepods (KO-peh-pods), tiny "micro" crustaceans, are predominant organisms. In addition, worms, crabs, and mollusks such as clams burrow to safety when the tides roll out. In areas near the low tide mark, sea anemones, sea urchins, and sea stars make their home.

The intertidal zone has plentiful light and is home to a variety of producers. Along with the algae, phytoplankton float in the water and are used for food by the zooplankton and many other consumers. In addition, the heterotrophs of the intertidal zone have the waves to thank for bringing fresh organic material to them and for washing away their wastes.

## The neritic zone

Surrounding the continents of the world is a shelf of land that extends out from the intertidal zone usually 50 to 100 kilometers (30 to 60 miles), sloping to a depth of about 200 meters (approximately 650 feet) beneath the sea. This margin of land is called the **continental shelf**. The waters lying above it make up the neritic zone, which is derived from a Greek root referring to the sea. Because light reaches the waters of most of this zone, it supports an abundant array of plant and therefore animal life.

One outstanding community in the neritic zone is the *coral reef* (see Figure 39.1). The term *reef* refers to a mass of rocks or coral in the ocean lying at or near the surface of the water. Coral reefs are built by corals (phylum *Cnidaria*) that secrete calcium carbonate, a hard, shell-like substance. With the help of algae that reside in their bodies, the coral build on already shallow portions of the continental shelf or on submerged volcanoes in the ocean. Complex and fascinating

**Figure 41.19** A rocky shore on the Big Sur coastline in California. Barnacles, such as the clusters of goose barnacles shown in the inset photo, are only one of myriad organisms growing on and around the rocks.

**Figure 41.20** Sandy beach and sand dune at Cape Cod National Seashore, Massachusetts. The inset photos show a blue crab and a sand dollar.

ecosystems, coral reefs provide habitats for a variety of invertebrates and fishes. Along with the tropical rain forests, coral reefs are the most highly productive ecosystems in terms of biomass.

### The open-sea zone

Beyond the continental shelf lies the great expanse of the open ocean. This open-sea zone, along with the neritic zone, is often referred to as the **pelagic zone**, a term derived from another Greek word meaning "ocean" (**Figure 41.21**). Within the huge ecosystem of the open sea exist many diverse forms of life. Some are familiar, such as squid and various species of fishes (**Figure 41.22a**). Other forms of life are unfamiliar—even bizarre—such as the flashlight fish shown in Figure 41.22b.

Organisms live in the vast expanse of the ocean in relationship to available light and food. Temperature, salinity, and water pressure also play roles in creating the various habitats of the ocean. Light is available to organisms from the water's surface only to an approximate depth of 200 meters (650 feet). This area of the open ocean is called the *photic zone*. (Photic means "light.") Phytoplankton, which are small plantlike organisms that float or drift in the water, thrive in this well-lighted layer of the ocean, especially within the upper 100 meters (325 feet), drifting freely with the ocean currents and serving as the base of oceanic food webs. Zooplankton, which are small

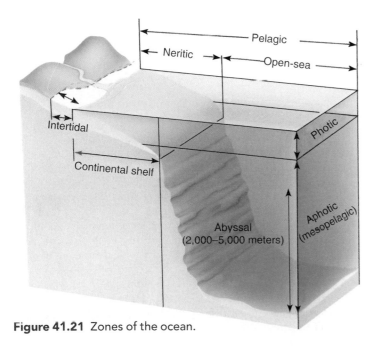

**Figure 41.21** Zones of the ocean.

(a)

(b)

**Figure 41.22** The open ocean. Many diverse forms of life exist in the open-sea zone including the (a) squid and the (b) flashlight fish. Notice the organ of illumination under the eye.

animallike organisms that float or drift in the water, float with the phytoplankton and are first-order consumers. Other typical heterotrophs of this zone are most air-breathing mammals, such as whales, porpoises, dolphins, seals, and sea lions, and fishes such as herring, tuna, and sharks. These fishes and mammals of the sea, which are called *nekton* (NEK-ton), feed on the plankton and on one another. Together, the organisms that make up the plankton and the nekton provide all of the food for those that live below.

Little light penetrates the ocean from 200 to 1000 meters (650 to 3250 feet). Therefore, no photosynthetic organisms live in this zone known as the *aphotic* ("without light") or *mesopelagic zone* (MEZ-oh-puh-LAJ-ik; this word means "middle ocean"). At this depth, temperatures remain somewhat steady throughout the year but are cooler than the water above; water pressure increases steadily with depth. Under these conditions, many bizarre organisms have evolved, such as those that exhibit bioluminescence (Figure 41.22*b*), which they use to communicate with one another or to attract prey. The organisms common to middle-ocean life are fish with descriptive names such as swordfish, lanternfish, and hatchetfish; certain sharks and whales; and cephalopods such as octopi and squid.

Peculiar creatures also live in the ocean depths of 1000 or more meters—a mile or so and more beneath the surface. This region of the ocean is called the *abyssal zone* (uh-BISS-uhl), meaning "bottomless," because it is so deep that it seems to be bottomless. The water at this tremendous depth contains high concentrations of salt, is under immense pressure from the water above, and is very cold. The organisms living at this level cannot make trips into the photic zone to capture food as some mesopelagic organisms do, but must feed on material that settles from above.

Organisms collectively referred to as *benthos* (BEN-thos), or bottom dwellers, live on the ocean floor. Sea cucumbers and sea urchins crawl around eating detritus. Various species of clams and worms burrow in the mud, feeding on a similar array of decaying organic material. Bacteria are also rather common in the deeper layers of the sea, playing important roles as decomposers as they do on land and in fresh water habitats. Chemosynthetic bacteria make their own food using energy derived from inorganic substances such as sulfur and hydrogen spewed from deep sea hot springs called hydrothermal vents. These bacteria live symbiotically with organisms such as tube worms and clams, and form the base of the hydrothermal vent community's food chain.

### CONCEPT CHECKPOINT

**5.** Why would organisms that inhabit the shoreline of a small lake or pond have very different adaptations from those that inhabit the shoreline of an ocean in the intertidal zone?

## CHAPTER REVIEW

## Summary of Key Concepts and Key Terms

### What are biomes, and how does climate define them?

**41.1 Biomes** (p. 721) are large recognizable terrestrial ecosystems; they occur over wide areas and within specific climatic regions.

**41.1** The characteristics of biomes are a direct result of temperature and rainfall patterns.

**41.2** The climate of a region is determined primarily by its latitude and wind patterns.

**41.2** These factors interact with the surface features of the Earth, resulting in particular rainfall patterns.

**41.2** The temperature, rainfall, and altitude of an area provide conditions that result in the growth of vegetation characteristic of that area.

**41.3** Seven categories of biomes, arranged by distance from the equator, are (1) tropical rain forests, (2) savannas, (3) deserts, (4) temperate grasslands, (5) temperate deciduous forests, (6) taiga, and (7) tundra.

### What are the features of the Earth's major biomes?

**41.4 Tropical rain forests** (p. 725) occur near the equator, receive an enormous amount of rain year round, and are characterized by the growth of tall trees and lush vegetation.

**41.5 Savannas** (p. 726) also lie near the equator but experience less rain than tropical rain forests and sometimes have prolonged dry spells.

**41.5** Savannas are characterized by open grasslands with scattered trees and shrubs.

**41.6 Deserts** (p. 726) are extremely dry biomes; hot deserts are hot year round, whereas cool deserts are hot only in the summer.

**41.6** Deserts are of great biological interest due to the extreme behavioral, morphological, and physiological adaptations of the plants and animals that live there.

**41.7 Temperate grasslands** (p. 729) receive less rainfall than savannas but more than deserts; the soil in temperate grasslands is rich, so they are well suited to agriculture.

**41.8 Temperate deciduous forests** (p. 729) receive moderate precipitation that is well distributed throughout the year.

**41.8** The climate of temperate deciduous forests differs from tropical forests in that they receive less rainfall, are found at higher and cooler latitudes, and experience cold winters.

**41.8** The trees in temperate deciduous forests lose their leaves and remain dormant throughout the winter.

**41.9** The **taiga** (TIE-gah, p. 729) is the coniferous forest of the north.

**41.9** The taiga consists primarily of cone-bearing evergreen trees, which are able to survive long, cold winters and low levels of precipitation.

**41.10** Even farther north than the taiga is the **tundra** (p. 731), which covers about 20% of the Earth's land surface and consists largely of open grassland, often boggy in summer, which lies over a layer of permafrost.

### How are the Earth's aquatic life zones distributed?

**41.11** Fresh water ecosystems make up only about 2% of the Earth's surface; most of them are ponds and lakes.

**41.11** Ponds and lakes have a **shore zone** (p. 732) and an **open-water zone** (p. 732), while lakes also have a **deep zone** (p. 732).

**41.11** Rivers and streams differ from ponds and lakes because they contain moving water that mixes with the air to provide high levels of oxygen for its fish and invertebrate inhabitants.

**41.12** **Estuaries** (ESS-choo-ER-eez, p. 732) are places where fresh water meets saltwater as rivers empty into the ocean.

**41.12** Estuaries receive nutrients from the surrounding land and usually support a large number of organisms.

**41.13** The marine environment consists of three major life zones: the intertidal zone, between the highest tides and the lowest tides; the neritic zone, the area of shallow water that lies over the **continental shelf** (p. 734); and the open-sea zone.

**41.13** Together, the open-sea zone and the neritic zone are often called the **pelagic zone** (p. 735).

**41.13** The ocean supports the most life in areas that have light and sufficient quantities of dissolved oxygen.

---

## Level 1  Learning Basic Facts and Terms

**Multiple Choice**

1. Biomes
   a. are large terrestrial ecosystems occurring within specific climatic regions.
   b. are often classified by their geographic location and dominant plant community.
   c. have characteristics that usually result from temperature and rainfall patterns.
   d. have patterns of distribution that are governed by climate and geography.
   e. All of the above.
2. Which of the following factors determine the climate of a region?
   a. Latitude
   b. Altitude
   c. Prevailing wind direction
   d. Local geography
   e. All of the above.
3. The oceanic life zone that extends over the continental shelf to depths of around 200 meters is the
   a. intertidal zone.
   b. abyssal zone.
   c. neritic zone.
   d. open-sea zone.
   e. estuary.

**Matching**

For items 4–9, match the terrestrial biome (a–f) associated with each description:

4. ____ Dominant trees such as oak and beech allow enough light so that herbs, ferns, and mosses grow on the ground.
5. ____ Because of permafrost, when surface ice melts, the ground cannot absorb the water.
6. ____ Although located near the equator, these regions are mostly open grasslands.
7. ____ These coniferous forests receive most of their precipitation during the summer.
8. ____ Herds of North American bison once roamed these areas, grazing on large expanses of perennial grasses.
9. ____ These regions radiate heat rapidly at night, causing wide temperature differences between day and night.

a. Tundra
b. Taiga
c. Temperate deciduous forest
d. Savanna
e. Desert
f. Temperate grassland

---

## Level 2  Learning Concepts

1. Why is rainfall or water availability such an important factor in determining the biotic characteristics of a biome?
2. January is a winter month in the United States but a summer month in Australia. Why?
3. For each of the fresh water and ocean ecosystems, discuss how photosynthetic activity varies with depth.
4. Explain why we find plants (and/or algae) that are adapted to low light levels in both terrestrial and aquatic ecosystems.
5. What factors limit rainfall in a desert?
6. Distinguish between taiga and tundra.
7. Describe the three life zones found in fresh water. What types of organisms live in each?
8. What is an estuary? Summarize how its inhabitants are adapted to its conditions.
9. Why are estuaries and fresh water regions so vulnerable to human pollution?
10. Identify and briefly describe the three major life zones of the ocean.

**Critical Thinking and Life Applications**

1. Much of the world's most productive agriculture is carried out on soil of temperate grasslands. Agriculture in the tropics is far less productive. Why do you think temperate grassland soil is so much richer than soil in a tropical rain forest?

2. There is a thin strip of desert along the central west coast of South America. Immediately to the east of this desert land lie the Andes Mountains. In the United States, the Mojave Desert lies east of the Sierra Madre Mountains. Explain why these deserts do not occur on the same side of the mountain ranges to which they are adjacent.

3. Analyze the graph to the right, which gives climate information for a particular city in North America. In which biome is this city located? Explain the reasons for your answer.

4. In which biome do you live? List the characteristics of your biome and describe how they match the characteristics of your biome as described in this chapter.

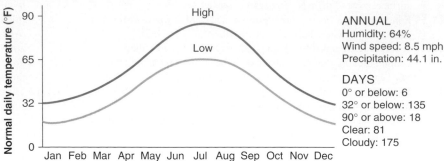

ANNUAL
Humidity: 64%
Wind speed: 8.5 mph
Precipitation: 44.1 in.

DAYS
0° or below: 6
32° or below: 135
90° or above: 18
Clear: 81
Cloudy: 175

5. If a river or stream were dammed to form a pond or lake, what changes would you expect to see in the plant and animal life in that aquatic environment?

## In The News | **Critical Thinking**

### FOOD FOR THOUGHT

Now that you understand more about biomes, reread this chapter's opening story about research into the effects of climate change on soybean and corn crop production. To better understand this research, it may help you to follow these steps:

1. Review your immediate reaction to Lobell and Asner's data on the effects of environmental temperature on soybean and corn crop yields that you wrote when you began reading this chapter.

2. Based on your current understanding, again summarize the main point of the research in a sentence or two.

3. What questions do you now have about this research that this chapter's opening story does not answer?

4. Collect new information about the research. Visit the *In The News* section of this text's companion Web site at www.wiley.com/college/alters and watch the "Food for Thought" video. Then use the "summary" link to read the accompanying story and access related links. Use this information, the links provided, and other online and library resources to answer your questions and find updates about this research topic. State the sources of your information. Explain why you think the information is accurate. Also determine whether the information expresses a particular point of view or is biased in any way.

5. What in your view are the most significant aspects of this research? Give reasons for your opinions and for any changes in your ideas based on the additional information you have collected and the analysis you have done.

# THE BIOSPHERE: TODAY AND TOMORROW

## In The News | Shark Test

In some restaurants that serve Asian cuisine, a bowl of shark fin soup can cost $90 or more. If people who enjoy this soup knew how the fins were obtained, they might not order the expensive item from the menu. After the shark is caught, the top (dorsal) fin is sliced off, and the shark, which often is still alive, is thrown back into the ocean to die. According to the nonprofit international organization WildAid, trade in shark fins increased by about 300% between 1980 and 2000. Each year, an estimated 100 million sharks are killed. In the northwest Atlantic Ocean, the populations of most shark species have declined by at least 50% since the late 1980s.

The demand for the fins of the great white shark, especially in Hong Kong and Taiwan, has decimated the population of this spectacular predator. In addition to killing great whites for their fins, fishermen catch the animal for its tasty meat, and its teeth and jaws are sold as curios throughout the world. There is also a demand for great white shark carcasses that are suitable for being preserved for public or private displays. As a result of overfishing, the population of great white sharks in the northwest Atlantic Ocean declined by about 80% over the past 20 years.

Great white sharks are protected in the United States, Australia, South Africa, and some other countries. Nevertheless, enforcing bans against killing these marine predators presents a challenge for conservationists. Illegal practices are difficult to control over the vast expanses of oceans. Furthermore, great white shark meat may be concealed by mixing it with the meat of other sharks, making it difficult to distinguish from the meat of unprotected shark species.

Ellen Pikitch, director of Ocean Strategy for New York-based Wildlife Conservation Society (WCS), and a team of scientists from Nova Southeastern University in Florida, used DNA technology to develop a relatively simple test that will enable conservationists to distinguish the tissues of one species of shark from that of others. The test uses a technique called *polymerase chain reaction* (*PCR*). As described in Chapter 11, PCR quickly and accurately amplifies DNA. The Pikitch PCR process is unique, however, in that it uses both nuclear and mitochondrial DNA, and is so sensitive that it can detect the presence of white shark DNA in a mixture of DNA from up to ten shark species. Now that a definitive method of identifying great white shark tissues has been developed, the animal is more likely to be given protection under the Convention on International Trade in Endangered Species (CITES), a treaty that regulates trade in threatened and endangered animals. You can learn more about this test and the great white shark, and view the video "Shark Test" by visiting the *In The News* section of this text's companion Web site at www.wiley.com/college/alters.

---

*Write your immediate reaction to the illegal trade in great white shark tissues and the DNA test that may help this protected species: first, summarize the main point in a sentence or two; then suggest what you think the significance of this research is. You will have an opportunity to reflect on your responses and gather more information on this topic in the* In The News *feature at the end of this chapter. In this chapter, you will learn more about other factors that threaten the survival of various species, as well as information about the health of the planet on which we live.*

## What is the biosphere?

**42.1** The biosphere is the part of the Earth that supports life.

## What resources does the biosphere provide?

**42.2** Nonrenewable resources are formed more slowly than they are consumed.

**42.3** Renewable resources replace themselves naturally over time.

## What threatens the health of the biosphere?

**42.4** Pollution threatens the purity of surface and ground water.

**42.5** Burning fossil fuels pollutes the air.

**42.6** Polluted air causes acid precipitation.

**42.7** Chlorofluorocarbons damage the ozone layer.

**42.8** Destruction of tropical rain forests threatens this resource and the habitats it provides.

**42.9** Carbon dioxide buildup in the atmosphere produces a greenhouse effect.

**42.10** Habitat destruction, habitat pollution, and illegal trade threaten the survival of many species.

**42.11** Many factors interact to threaten the overall health of the biosphere.

## What is the biosphere?

### 42.1 The biosphere is the part of the Earth that supports life.

Just imagine that you could hold the world in your hands and make a difference in its future. Well, imagining isn't even necessary. All people do hold the world in their hands in a figurative way. Everyone's actions have a direct impact on the Earth and on the quality of life that you and others will experience for generations to come.

First, you must ask yourself some crucial questions. How are you affecting the Earth? What environmental problems do people face today? And what can you do to deal with those problems to be sure that the quality of life on Earth will be enhanced for yourself and your children? The answers to these questions are the focus of this chapter and will help you understand more about how you shape this fragile planet on which you live.

Life on Earth is confined to a region called the **biosphere**, the global ecosystem in which all other ecosystems exist. It is the layer of Earth that contains all living organisms. The biosphere extends from approximately 9000 meters (30,000 feet) above sea level to about 11,000 meters (36,000 feet) below sea level. You can think of it as extending from the tops of the highest mountains, such as Mount Everest in the Himalayas, to the depths of the deepest oceans, such as the Mariana Trench of the Pacific Ocean. The biosphere is the part of our Earth in which the land, air, and water come together to help sustain life (**Figure 42.1**).

The biosphere is often spoken of as the **environment**. This general term refers to everything around you—not only the

land, air, and water, but also other living things. You can speak, for example, of the environment of an ant, a water lily, or all the peoples of the world. Your particular environment can change during the day from a home environment, to a classroom environment, and then to an office environment. The environment can include a great deal—or very little—of the total biosphere and its living things.

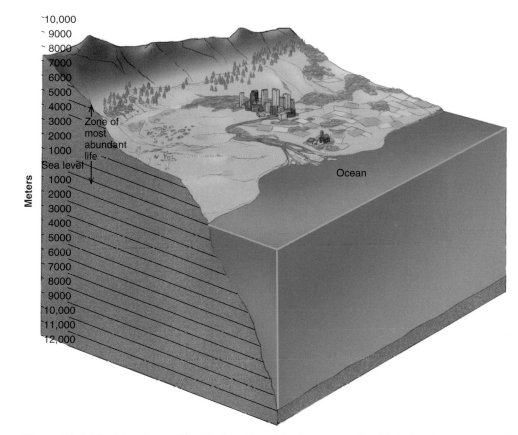

**Figure 42.1** **The biosphere.** The Mariana Trench in the western Pacific is the deepest part of the ocean. It is 11,022 meters below sea level. The deepest dive by an unmanned submersible was to 10,920 meters, and bottom fish and crustaceans were found at that level. Mount Everest in the eastern Himalayas is the highest mountain in the world, rising to 8848 meters. Spiders and insects have been found as high as 6100 meters.

## What resources does the biosphere provide?

### 42.2 Nonrenewable resources are formed more slowly than they are consumed.

The biosphere gives humans many things: fossil fuels, wood, food, and minerals. Water, too, is an important natural resource. Some of these resources—fossil fuels and minerals—are finite or **nonrenewable resources**; they are formed at a rate much slower than

their consumption. Coal, oil, copper, and iron are examples of nonrenewable resources.

#### Fuel Resources

Many of the fuels used to heat homes and run cars are **fossil fuels**. These substances—coal, oil, and natural gas—are formed over time, acted on by heat and pressure, from the undecomposed carbon compounds of organisms that died millions of years ago. Once these fuels are used up, they are gone forever. Currently, 77% of the world's energy is derived from coal (20%), oil (36%), and natural gas (21%).

Burning fossil fuels releases the stored carbon to the atmosphere. This carbon, in the form of carbon dioxide, is referred to as a greenhouse gas, which can result in global warming. Greenhouse gases and global warming are discussed in Section 42.9. Burning coal also releases sulfates and nitrates into the atmosphere, compounds that can result in air pollution and acid precipitation. These topics are discussed in Sections 42.5 and 42.6.

**Nuclear power**, once considered a viable alternative energy source, contributes no carbon, sulfates, or nitrates to the atmosphere. Although nuclear power is dependent on uranium, a finite but abundant natural resource, comparatively small amounts of uranium are needed to produce electricity. For example, 2.2 pounds of Uranium-235 can yield as much energy as 2200 tons of coal. The problems of nuclear power lie primarily with the cost of building nuclear power plants, the difficulty of disposing of highly radioactive wastes, and the public fear regarding safety. Because of these problems, nuclear power is minimally used and contributes only about 9% of the world's energy.

## Mineral resources

Minerals are inorganic substances that occur naturally within the Earth's crust. Zinc, lead, copper, aluminum, and iron are the minerals that humans use in large quantities. Other minerals, however, are also mined, such as gold, silver, and mercury. These minerals are present in the Earth in fixed amounts; once they are used, they are gone. **Table 42.1** lists some important minerals and their uses. Researchers agree that increased use of plastics and technology such as microelectronics will lower the demand for certain minerals.

## Disposal and conservation of nonrenewable resources

The American culture has been called the throwaway society because Americans use, and then dispose of, a large amount and variety of consumables. The average American uses six times more zinc, lead, and iron, 11 times more copper and 14 times more alu-

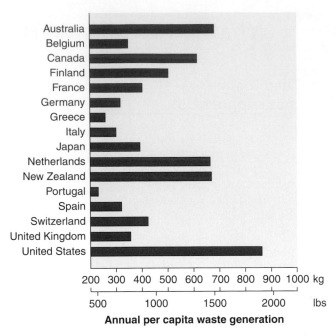

**Figure 42.2** Municipal solid waste. The graph shows the annual municipal waste generated per person in selected developed nations.

minum than does a person living outside the United States. The amount of municipal solid waste Americans generate per person is one of the highest rates of any industrialized nation in the world—almost one ton per person every year (**Figure 42.2**). **Municipal solid waste** includes paper, yard waste, food waste, plastics, metals, rubber, leather, textiles, glass, and wood—the typical items that you would find in your household garbage and trash plus materials from institutions and businesses. Municipal solid waste does not include agricultural and industrial wastes.

There are three strategies to conserve our nonrenewable natural resources: Reduce, reuse, and recycle. The first line of attack is to simply *reduce* consumption of energy and material goods. A second method of conservation is *reuse*. Items that cannot be directly reused can often be used again in a similar or different manner: A plastic bread bag, for example, can be used to carry your lunch to work; and paper

| Table 42.1 | Some Important Minerals and Their Uses |
| --- | --- |
| **Mineral** | **Some Uses** |
| Aluminum (Al) | Structural materials (airplanes, automobiles), packaging (beverage caps, toothpaste tubes), fireworks |
| Copper (Cu) | Alloy ingredient in gold jewelry, silverware, brass, bronze, electrical wiring, pipes, cooking utensils |
| Gold (Au) | Jewelry, money, dentistry, alloys |
| Iron (Fe) | Basic ingredient of steel (buildings, machinery) |
| Lead (Pb) | Lead pipes, solder, battery electrodes, pigments |
| Mercury (Hg) | Thermometers, barometers, dental inlays, electrical switches, streetlights, medicine |
| Platinum (Pt) | Jewelry, delicate instruments, electrical equipment, cancer chemotherapy, industrial catalyst |
| Silicon (Si) | Electronics, solar batteries, ceramics, silicones |
| Silver (Ag) | Jewelry, silverware, photography, alloys |
| Zinc (Zn) | Brass, metal coatings, electrodes in batteries, medicine (zinc salts) |

printed on one side only can be cut up and used as notepaper. Many of the items we use regularly can be put to creative reuse. This method avoids the consumption of energy necessary to produce recycled items.

*Recycling* is a third way to reduce the consumption of energy and material goods. Paper, glass, and aluminum soda cans are useful items to recycle. For example, aluminum cans are produced from the claylike mineral bauxite, an aluminum ore that must be processed to produce aluminum metal. Producing cans from recycled aluminum uses 95% less energy than producing them from bauxite, saves mineral resources, and saves waste disposal costs and problems.

Plastics, on the other hand, are not easily recyclable and are not "truly" recyclable. That is, plastics cannot be refashioned into the product from which they were claimed; plastic milk jugs cannot be made into plastic milk jugs but must be made into items such as park benches or parking lot curbs. Grocery store plastic bags, made from used plastic bags and other additives, are close to being truly recycled plastic products. Nevertheless, purchasing goods in recyclable containers such as paper, glass, and aluminum, and recycling those containers is a wise choice. Scrap metal and old automobiles are also recyclable, as are automobile tires and batteries.

An important part of recycling is buying recycled goods so that a market is maintained for them. Additional recycling behaviors are purchasing products in recyclable containers, avoiding the purchase of "overpackaged" products, and reusing products when feasible.

## 42.3 Renewable resources replace themselves naturally over time.

Maybe you live in a region of the country where part of your electricity is generated by the wind, or possibly you have solar panels on your house. Environmental scientists suggest that instead of our depending on nonrenewable fossil fuels for energy—fuels that are finite—people should work toward a *sustainable society* that uses renewable sources of fuel. In a sustainable society, the needs of the society are satisfied without compromising the ability of future generations to survive and without diminishing natural resources.

*Renewable resources* are those produced by natural systems that replace themselves naturally over time. These resources include trees, fishes, fertile agricultural soil, and fresh water. The primary renewable sources of power are the sun, the wind, moving water, geothermal energy, and bioenergy. Currently, renewable resources collectively supply about 14 percent of the world's energy.

### Solar power

*Solar power* is the use of the sun for heating or to produce electricity. Solar cells can collect energy from the sun, convert it to electricity, and store it in large batteries for later use (**Figure 42.3a**). Currently under development is a solar power satellite system, which would collect solar energy in space, convert it to electricity, transmit it to Earth, and then convert it to a form that could be delivered to the existing electric power distribution system. Scientists expect solar-powered satellite systems to be a major source of energy by 2030. Solar power can also power automobiles. In March 1999, Royal Dutch Shell opened its first stations in Germany and the Netherlands to charge electric cars with solar power.

Passive solar architecture is another way to use the energy from the sun. Homes and businesses can be built so that windows are located where the rays of the winter sun, low in the sky, enter the building, heating it. In the summer, an overhang protects the windows from the sun, which is high in the summer sky.

### Water power

Water has been used as an energy source for decades. Figure 42.3*b* shows the Hoover Dam, a typical hydroelectric plant. The dam is located 30 miles southeast of Las Vegas, Nevada, and provides energy to Arizona, Nevada, and California. Turbines located at the base of the dam rotate by the force of the falling water. The turbines drive generators, which produce electricity.

A new form of hydropower called *wave power* uses the vertical motion of sea waves to produce electricity. The world's first commercial wave power station was recently built on the island of Islay, Scotland. Although various technologies exist for wavepower

(a)

(b)

(c)

(d)

**Figure 42.3** Renewable energy sources. (*a*) An array of solar panels at a solar power station near Sacramento, California. (*b*) Hoover Dam in Arizona. (*c*) Tidal power plant in the Bay of Fundy. (*d*) Modern windmills in southern California.

stations, the Islay facility is a shoreline chamber in which seawater enters and leaves naturally with the waves. As the seawater enters, the air in the chamber is pushed through a hole that houses a turbine, making the turbine move. As the water recedes, the air is sucked back, keeping the turbine moving. The turbine drives a generator, which converts the energy into electricity.

In addition, *tidal power*, the use of the movement of the tides of the oceans, is being used in St. Malo, France, and in the Bay of Fundy, Canada. Figure 42.3*c* shows the Canadian facility. In 2003, the world's first offshore tidal energy turbine began producing electricity near Devon, England. The problems with producing electricity from tidal and wave sources are the high cost and the unsteady rate at which the electricity is produced. Therefore, neither method appears useful to fulfill the total energy demands of an area.

### Wind power

The wind has been used for centuries as a source of energy. Today, windmills are being used in developing countries to pump water to livestock and to irrigate the land. In developed countries, however, windmills are being used to generate electricity. This "new breed" of windmill has rigid blades fashioned from lightweight materials and is shown in Figure 42.3*d*. At present, the United States, Denmark, and China are the three countries that lead in the use of *wind power* to generate electricity. Some scientists predict that the United States will be generating 10 to 20% of its electricity from wind power by the year 2030.

### Geothermal energy

*Geothermal energy* refers to the use of heat deep within the Earth. In some places, reservoirs of hot water or steam exist that can be extracted from the Earth by drilling procedures much like those used to tap into the Earth's oil and natural gas reserves. Alternatively, dry, hot rock can be drilled and water flushed through it. After the water is heated within the Earth, it can be used directly for heating purposes or as part of a process to produce electricity. Currently, this technology is being used and further developed in the United States, Russia, England, Italy, New Zealand, Japan, China, Indone-

sia, Kenya, Mexico, and the Philippines. The largest plant in the world is near Geyserville, California, north of San Francisco.

### Bioenergy

*Bioenergy* refers to the use of living plants to produce energy. The most obvious type of bioenergy, the burning of wood, was first used by our ancestors more than 1 million years ago. Until the Industrial Revolution of the 1800s, wood, not coal, supplied most of the world's energy. Today, wood is an important energy source primarily in Latin America, Asia, India, and various African states. However, the world is experiencing a fuel wood crisis—demand is exceeding the supply. The reasons for this crisis are complex but include the high demand for wood by a growing world population, the degradation of woodlands without proper reforestation techniques, and the cutting and burning of huge areas of tropical rain forests (see Section 42.8).

In developing countries such as China, India, and Africa, the use of *biogas machines* is helping to ease the shortage of fuel wood. These stoves use microorganisms to decompose animal manure, harvest waste, wood waste, and even human sewage in a closed container. This process yields a methane-rich gas that can be used to fuel stoves, light lamps, and produce electricity. The substances left over after combustion can be used as fertilizer.

A newer form of bioenergy is the use of plants such as corn and sugar cane, producing fermented carbohydrate products such as ethanol that can be used as fuels or fuel additives. This technology, however, produces a product that appears to be an expensive alternative to fossil fuels. In addition, many cars do not function well on gasoline with ethanol additives.

---

CONCEPT CHECKPOINT

**1.** What are some advantages of using renewable energy sources such as wind, water, solar, and geothermal power over nonrenewable energy sources such as coal, oil, or nuclear power? Speculate on some potential environmental drawbacks to developing each of the renewable energy technologies on a large scale.

---

# What threatens the health of the biosphere?

## 42.4 Pollution threatens the purity of surface and ground water.

Have you ever traveled to a country in which you were warned not to drink the water? That warning also meant that you should not eat uncooked foods such as salads washed with the water. If you did not heed the warning, you likely developed a case of "traveler's diarrhea," caused by *infectious agents* (disease-causing microbes) usually not present in North American water supplies. In areas such as Africa, Asia, and Latin America, waterborne diseases are common.

Surface water—the fresh water found on Earth's surface in streams, rivers, lakes, and ponds—becomes polluted from untreated human wastes and from animal wastes, causing diseases such as hepatitis, polio, and cholera. The purity of surface water, groundwater (fresh water under the Earth's surface), and ocean water is threatened by various sources of **pollution**, substances that cause physical or chemical changes in the water that harm living and nonliving things. In addition to infectious agents, the types of contaminants that threaten the purity of surface water, groundwater, and ocean water are land runoff, mine drainage, hazardous waste dumps, organics, inorganics,

**Figure 42.4** A toxic waste dump in Chabarovice, Czechoslovakia. Toxic chemical dumps are serious threats to groundwater and surface water. Pollution occurs when the drums rust through and release their contents, which then enter the surface water and may eventually trickle down to the groundwater.

toxic substances, thermal pollution, and solid waste. These pollutants affect both aquatic organisms that live in the water and terrestrial organisms that drink the water.

## Land runoff, mine drainage, and hazardous waste dumps

Factories, power plants, and sewage treatment plants can pollute surface water by dumping waste chemicals, heated water, or human sewage into a lake, stream, or river. These sources of pollutants are called *point sources* because they enter the water at one or a few distinct places. Other types of pollutants may enter surface water at a variety of places and are called *nonpoint sources*. Examples of nonpoint sources of pollution are (1) sediments in land runoff caused by erosion from poor agricultural practices—a major type of water pollution; (2) metals and acids draining from mines; (3) poisons leaching from hazardous waste dumps (**Figure 42.4**); and (4) pesticides, herbicides, and fertilizers washing into surface waters after a rain. Land runoff of storm water from cities, agricultural areas, and roads is the single largest source of ocean pollution.

## Toxic substances

An array of *toxic substances* pollutes surface waters worldwide. Toxic substances include both organic compounds such as PCBs (polychlorinated biphenyls) and phenols, and inorganic substances such as metals, acids, and salts. These toxic, or poisonous, substances come from a di-

verse array of sources such as industrial discharge, mining, air pollution, soil erosion, old lead pipes, and many natural sources. The effects on humans from drinking these substances in water range from numbness, deafness, vision problems, and digestive problems to the development of cancers.

Most toxic pollutants do not degrade, or break down, and are therefore present in bottom sediments of surface waters for decades or longer. Some organisms do not excrete certain chemicals —often deadly ones—from their bodies, but accumulate them in their fatty tissues. This process is called **biological concentration**. Oysters, for example, accumulate heavy metals such as mercury; thus, these organisms might be highly toxic when living in waters with relatively low concentrations of this metal. Also, as organisms higher on the food chain eat organisms lower on the food chain, toxins accumulate in the predators in a high concentration, a concept called **biological magnification**. As discussed in Chapter 40, the relationships between trophic levels of food chains are pyramidal, with a smaller number of organisms at each level as you progress "up" the chain. **Figure 42.5** illustrates this concept. Scientists estimate that the concentration of a toxin in polluted water may be magnified from 75,000 to 150,000 times in humans who consume tainted fish.

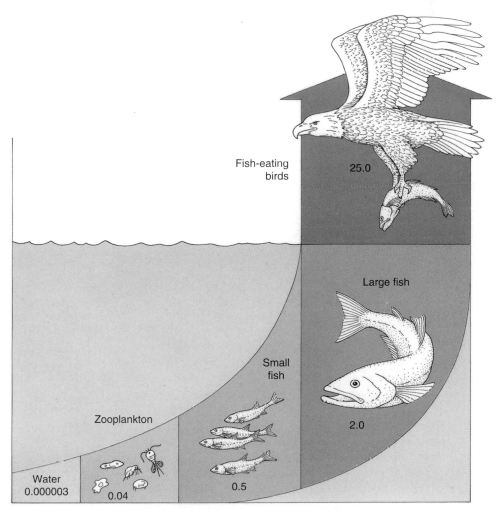

Amount of DDT (in ppm = parts per million)

**Figure 42.5** Biological magnification. The amount of DDT in an organism increases the higher the organism is on the food chain. This illustration depicts an 8 million-fold amplification of DDT in the tissues of the eagle.

## Organics from treatment plants, paper mills, and meat-packing plants

*Organic nutrients* are sometimes discharged into rivers or streams by sewage treatment plants, paper mills, and meat-packing plants. These "organics" are food for bacteria. If high amounts of organic nutrients are available to bacteria, their populations will grow exponentially. As they grow and reproduce, they use oxygen that fish need. Therefore, as the bacterial populations rise, the only organisms that survive are those that can live on little oxygen. Some fish such as carp can outsurvive other species such as trout and bass. If oxygen levels become extremely low, however, all the fish die, survived only by various worms and insects.

## Inorganics from croplands or laundry detergents

The accumulation of *inorganic nutrients* in a lake is called **eutrophication**, meaning "good feeding." Certain inorganics such as nitrogen and phosphorus, which come from croplands or laundry detergents, stimulate plant growth. Although heavy plant growth makes swimming, fishing, or boating difficult, it does not cause most of its problems until the autumn, when the plants die in regions of the world that have distinct seasons. At that time, bacteria decompose the dead plant material, and problems similar to those of organic nutrient pollution arise. In addition, the decomposed materials begin to fill the bottom of the lake. Eventually, the lake may become transformed into a marsh and then into a terrestrial community by an accelerated process of succession (see Chapter 39). *Sediment* that flows into lakes from erosion of the land caused by certain agricultural practices, mining, and road construction also fills in lakes and hastens natural succession.

## Thermal pollution

The electric power industry and various other industries such as steel mills, refineries, and paper mills use river water for cooling purposes, then discharge the heated water back into the river. Small levels of **thermal pollution** do not cause serious problems in aquatic ecosystems, but sudden, large temperature changes kill heat-intolerant plants and animals. Moreover, ecosystems that have slowly adjusted to artificially heated waters are damaged if the heat source is shut down, as when a power plant closes.

## Mismanagement of solid waste

Another threat to the purity of surface water and groundwater is the mismanagement of solid waste. What happens to your trash when you put it out for collection? The burial site for your throwaways is the **sanitary landfill**, an enormous depression in the ground where trash and garbage are dumped, compacted, and then covered with dirt (**Figure 42.6**). In 1983, the Federal Resource Conservation and Recovery Act forced the closing of all *open dumps* or required them to be converted to landfill sites. At an open dump, solid waste is heaped on the ground, periodically burned, and left uncovered. Landfills are considered superior to dumps because landfill wastes are covered, reducing the number of flying insects and rodents that are attracted to the site and reducing the odor produced by open, rotting organic material. In addition, wastes are not burned at landfills, decreasing the problem of air pollution. Furthermore, when the capacity of a landfill site is reached, it may be used as a building site or

recreational area. Examples of landfill reuse are Mount Trashmore recreational complex in Evanston, Illinois, and Mile High Stadium in Denver, Colorado.

Problems do exist with landfills, however. First of all, space for landfills is running out. In just one year, the population of New York City alone produces enough trash to cover more than 700 acres of land 10 feet deep. Another problem with landfills is that liquid waste can trickle down through a landfill, reaching and contaminating groundwater below. Liquids leaching from landfills can also pollute nearby streams, lakes, and wells; therefore, it is unwise to put batteries, paint solvents, drain cleaners, and pesticides in with the trash. In addition, as the organic material compacted in landfills is decomposed in the absence of oxygen, methane gas is produced. This highly explosive gas rises from landfills and can seep into buildings constructed on or near reclaimed sites.

### CONCEPT CHECKPOINT

2. After World War II, DDT, a persistent organic pesticide that does not occur naturally in the environment, was widely sprayed to control insects such as mosquitoes. It was not until the 1960s that scientists learned that this pesticide was linked to the death of many species of birds and fishes, and it was banned in United States in the early 1970s. How could a poison intended to be harmful only to insects also be harmful to birds and fishes?

## 42.5 Burning fossil fuels pollutes the air.

The Earth's atmosphere actually extends much higher than the portion within the biosphere, a part of the atmosphere more technically called the *troposphere*. The troposphere extends approximately 11 kilometers (36,000 feet) into the atmosphere. The word "troposphere" literally means "turning over," a name extremely descriptive of the atmosphere. As the sun heats the Earth, air rises

**Figure 42.6** A sanitary landfill in California.

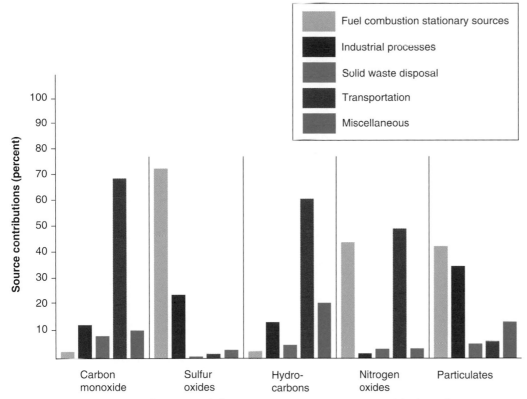

Figure 42.7 Major air pollutants and their sources. Transportation and fuel combustion at stationary sources are the main contributors to pollution.

**Visual Thinking:** Which of the pollutants shown here come primarily from transportation sources, such as cars, trucks, and SUVs? Which come primarily from stationary sources of fuel combustion, such as coal-fired power plants?

from its surface, cooling as it ascends. Cooler air then falls, resulting in a constant turnover of the air, aided by the prevailing winds (see Chapter 41).

About 99% of the clouds, dust, and other substances in the atmosphere are located in the troposphere. Some of these "other substances" are nitrogen and sulfur oxide pollutants. Other major air pollutants are carbon monoxide, hydrocarbons, and tiny particles called particulates. As **Figure 42.7** shows, sulfur oxides and particulates are produced primarily by the burning of coal in electricity-generating plants. Carbon monoxide and hydrocarbons are emitted primarily by cars, SUVs, buses, and trucks; all these sources spew out nitrogen oxides.

The type of air pollution in a city depends not only on which of the pollutants are in the air but also on the climate of the city. *Gray-air cities* (**Figure 42.8a**), such as New York and St. Louis, are located in relatively cold but moist climates and have an abundance of sulfur oxides and particulates in the air. The haze, or **smog**, that can be seen in the air is the result of the burning of fossil fuels. *Brown-air cities* (Figure 42.8b), such as Los Angeles, Denver, and Albuquerque, have an abundance of hydrocarbons and nitrogen oxides in the air. In these sun-drenched cities, the hydrocarbons and nitrogen oxides undergo photochemical reactions that produce "new" pollutants called *secondary pollutants*. The principal secondary pollutant formed is **ozone ($O_3$)**, a chemical that is extremely irritating to the eyes and the upper respiratory tract. Smog caused

Figure 42.8 Air pollution. (a) "Gray air" in New York City is caused by fossil fuel pollutants from power plants. (b) "Brown air" in Los Angeles is formed when sunlight reacts with the chemicals spewed from automobiles to create ozone. In the upper atmosphere, ozone protects you from the sun's harmful rays, but in the lower atmosphere it is poisonous.

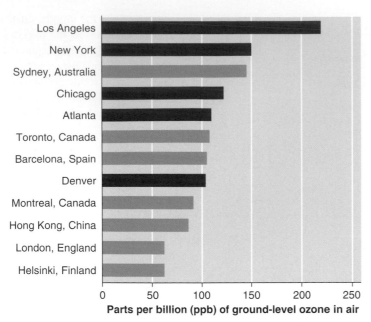

**Figure 42.9** Ground-level ozone in 12 cities around the world. Purple denotes U.S. cities.

by pollutants reacting in the presence of sunlight is called **photochemical smog. Figure 42.9** shows the average levels of ground-level ozone in 12 cities.

Solving the problem of air pollution and curtailing these emissions are not easily accomplished or inexpensive. One technique used to reduce sulfur dioxide is to put *scrubbers* on coal-burning power plants. This technology can remove up to 95% of the sulfur dioxide emissions produced. In addition, implementing international agreements to reduce emissions, employing energy-conservation measures, and increasing the use of public transportation all help reduce air pollution.

## 42.6 Polluted air causes acid precipitation.

Is acid precipitation a problem where you live? What is acid precipitation and how is it formed? Acid precipitation forms because of pollutants in the air. Two air pollutants, sulfur dioxide and nitrogen dioxide, form acids when combined with water. As this water vapor condenses and falls to the ground, it is commonly referred to as **acid rain**, although acid precipitation also falls as snow or as dry "micro" particles, mixing with water when it reaches surfaces on the ground.

How acid is acid rain? "Normal" rain has a slightly acidic pH of approximately 5.7, primarily because of dissolved carbon dioxide (see Chapter 3 for a discussion of pH). The pH

of acid rain is lower than 5.7 and usually ranges between 3.5 and 5.5. Rainfall samples taken in the eastern United States, however, have measured as low as 1.5, meaning the precipitation is more acidic than lemon juice and approaching the acidity of battery acid.

The map in **Figure 42.10** shows that the primary region of acid precipitation in the United States is the eastern half. Why? The eastern United States has more industry that emits sulfate and nitrate compounds, such as coal-fired power plants, industrial boilers, and large smelters that obtain metals from ores. In addition, nitrogen oxides are emitted by cars and trucks, and the eastern United States is heavily populated. Moreover, the eastern United States receives some of the emissions created in the Midwest as winds carry them eastward, while the wind takes emissions created in the northeast into Canada. As you can see, the situation is complicated because countries pollute both their own air and the air of other countries. Similarly, emissions produced in England move into the Scandinavian countries.

Acid rain results in many devastating effects on the environment. As it mixes with surface water, it acidifies lakes and streams, killing fish and other aquatic life. It seeps into groundwater, causing heavy metals such as cadmium and lead to leach out of the soil. The result is that these heavy metals enter the groundwater and surface water, posing health problems for humans as well as fish. Acid rain eats away stone buildings, monuments, and metal and painted surfaces (**Figure 42.11**). Acid rain also kills trees. It leaches many of the minerals essential to plant growth from the soil while killing or damaging microorganisms that live in symbiotic associations with forest trees. These microorganisms help trees extract water and needed minerals from the soil. Without the help of these organisms, the trees die (**Figure 42.12**).

The solutions to cutting down on acid precipitation are the same solutions to problems of smog formation and fossil fuel depletion: the use of coal-fired power plant scrubbers, energy conservation, and the recycling of resources leading to a reduction in energy use for manufacturing.

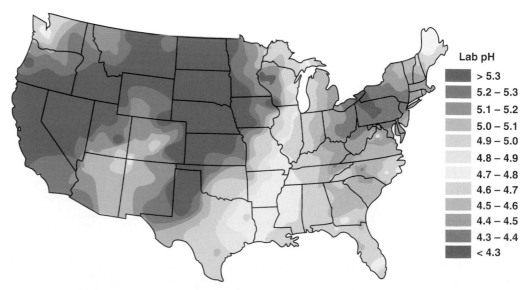

**Figure 42.10** The pH of precipitation in the United States in 2000.

Lab pH
> 5.3
5.2 – 5.3
5.1 – 5.2
5.0 – 5.1
4.9 – 5.0
4.8 – 4.9
4.7 – 4.8
4.6 – 4.7
4.5 – 4.6
4.4 – 4.5
4.3 – 4.4
< 4.3

**Figure 42.11** Acid rain damage. This statue has been eroded by acid precipitation.

 **Visual Thinking:** What pollutants lead to the formation of acid rain? What are the major sources of these pollutants? (See Figure 42.7.)

## 42.7 Chlorofluorocarbons damage the ozone layer.

Ironically, humans are producing ozone in the troposphere that is polluting the environment but are destroying it in the *stratosphere* where it is needed. The stratosphere is the layer of the atmosphere directly above the biosphere. It contains a layer of ozone that is formed when sunlight reacts with oxygen. Although ozone is harmful in air that is breathed, it is helpful in the stratosphere, acting as a shield against the sun's powerful ultraviolet (UV) rays. Excess exposure to UV rays can cause serious burns, cataracts (an opacity of the lens of the eye), and skin cancers, and can harm or kill bacteria and plants. This is also one explanation for the widespread declines in many amphibian populations. Scientists have measured "holes" in the ozone layer where it is thinnest, over the polar ice caps (**Figure 42.13**). These holes are huge—larger than the United States and Canada combined. In addition, the ozone layer is being reduced over much of the globe.

Ozone layer depletion was first noticed in 1974. The chemical chlorofluorocarbon (CFC) was discovered to be the main culprit. For example, freon, which is a refrigerant formerly used in air conditioners in the United States, is a CFC. In addition, CFCs are used in the manufacture of styrofoam and foam insulation. Today, several CFC-substitutes are already in use, and more are expected from manufacturers.

CFCs damage the ozone layer because they contain chlorine. At ground level, CFCs are stable, but when they reach the stratosphere, ultraviolet rays break them down, liberating chlorine gas. This gas reacts with ozone, producing chlorine oxide and oxygen. Chlorine oxide then breaks down, releasing molecules of chlorine gas that continue to destroy ozone. One molecule of chlorine gas can destroy up to 10,000 ozone molecules.

**Figure 42.12** Effects of acid rain. These balsam fir trees in North Carolina have been killed by acid rain. The chemicals in the acid rain, which are carried on prevailing winds, come from as far away as the Midwest.

In 1987, 57 nations signed the Montreal Protocol on Substances that Deplete the Ozone Layer. By the terms of the agreement, all production of CFCs in developed countries was halted on January 1, 1996, with the others to follow suit in 2006. In the United States, the Clean Air Act of 1990 and its subsequent amendments enforces the Montreal Protocol.

As a result of action taken under the terms of the Montreal Protocol, the ozone layer maybe showing signs of recovery (see Figure 42.13). Satellite-mounted instruments have collected data indicating that the amount of chlorine in the stratosphere peaked in 1997 and now may be on the decline. Nevertheless, experts estimate it may take until about 2050 before the effects of CFCs completely disappear.

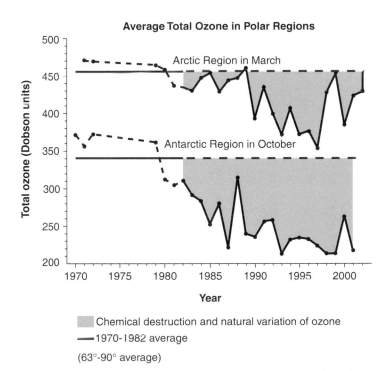

**Average Total Ozone in Polar Regions**

Arctic Region in March

Antarctic Region in October

Total ozone (Dobson units)

Year

Chemical destruction and natural variation of ozone

1970-1982 average

(63°-90° average)

**Figure 42.13** Depletion of the ozone layer over the South Pole.

 **Visual Thinking:** Describe the ozone layer over the South Pole from 1981 to 2001.

## *just wondering . . .*

### What effect does the depletion of the ozone layer have on our weather?

The thinning of the ozone layer in the stratosphere (discussed in Section 42.7) does not directly affect weather. It affects the amount of ultraviolet (UV) radiation that reaches the surface of the Earth. Although increased levels of UV radiation have no *direct* effect on weather, these rays penetrate bodies of water and appear to negatively affect photosynthesis in phytoplankton and algae. This has the potential to indirectly affect global climate. Climate differs from weather in that weather is a day-to-day occurrence, and climate is patterns of atmospheric conditions that remain relatively stable over long periods of time—decades, centuries, even millenia. Nonetheless, if climates change, then the weather in those changed climates will change as well.

To explain these connections, let's first look at how UV light affects photosynthesis in phytoplankton and algae. At the equator, these organisms already receive high levels of UV rays and appear to be adapted to this radiation. Antarctic phytoplankton and algae, however, normally receive low doses of UV light. When exposed to higher levels of UV radiation due to a hole in the ozone layer above them, they tend to sink deeper into the ocean where less light reaches. This protects them from damaging UV rays, but living deeper in the water means that these autotrophs receive less light for photosynthesis, and their productivity drops. Results of one large-scale field survey of Antarctic phytoplankton revealed a 6–12% drop in phytoplankton productivity in an ozone-hole area.

How can effects on photosynthesis affect climate? During photosynthesis, phytoplankton and algae incorporate at least 25 billion metric tons of carbon from carbon dioxide ($CO_2$) into carbohydrates each year as they make at least 35% of all food generated by photosynthesis on Earth. The carbon dioxide these autotrophic organisms incorporate into carbon compounds is removed from the atmosphere. Atmospheric carbon dioxide blocks outward heat radiation in a greenhouse effect. High levels of $CO_2$ in the atmosphere caused by burning tropical rain forests and fossil fuels enhance the greenhouse effect and may result in a worldwide temperature increase called global warming, with consequent changes in global climate patterns. Thus, depletion of the ozone layer could reduce the photosynthetic productivity of Antarctic phytoplankton and algae, which would reduce the amount of carbon dioxide they remove from the atmosphere, which could indirectly affect global warming, and thus global climates, and thus local weather. ●

*Are you wondering about a topic in biology and how it relates to your life? Submit your question by clicking the "Just Wondering" link in this text's companion Web site at www.wiley.com/college/alters.*

## 42.8 Destruction of tropical rain forests threatens this resource and the habitats it provides.

The destruction of tropical rainforests has been a topic in the popular press for many years, so it is likely that you have heard of this ecological problem. In general, the forests of the world are being depleted faster than they are being replanted or allowed to recover. The most severe crisis is that of **tropical rain forest deforestation**.

Tropical rain forests (see Chapter 41) are located in Central and South America, tropical Asia, and central Africa, forming a belt around the equatorial "waist" of the Earth. Although this belt of forest covers only 2% of the Earth's surface, it is home to more than *half* the world's species of plants, animals, and insects. These organisms contribute 25% of our medicines, along with fuel wood, rubber, charcoal, oils, and nuts. In addition, the tropical rain forests play an important role in the world climate.

Both population and poverty are high in rain forest countries. People with few resources move from towns and cities to the rain forest and cut the trees for sale as lumber or burn them to clear a patch of land to grow crops and raise cattle to sustain themselves and their families (**Figure 42.14**). Commercial ranchers also cut and burn the forests to make way for pastureland to feed beef-producing cattle. The soil of the rain forests is poor, however, with few nutrients, and does not support crops. Before it is cut, the forest sustains itself because of mutualistic relationships between the trees and microorganisms that quickly decompose dead and dying material on the forest floor. These "processed" nutrients are quickly reabsorbed by the tree roots. Few nutrients stay in the soil; most of the nutrients are in the vegetation. Cutting down these trees and burning them releases the nutrients from the trees into the atmosphere. Thus, the nutrients are not available in the soil to support food crops, which grow poorly on this land after it is

**Figure 42.14 Deforestation: a desperate effort to stave off poverty results in an environmental disaster.** These farmers live near the Andasibe reserve in Madagascar, an island where the per capita income is less than $250 a year. Clearing the rain forest allows these impoverished farmers to plant rice and graze cattle. But the environmental price may be too steep. In Madagascar alone, 84% of the rain forest has been destroyed.

**Figure 42.15 A rain forest reserve.** The World Wildlife Fund established a research program called the Biological Dynamics of Forest Fragments Project. The program isolated "islands" of rain forests out of the larger forest during conversion to pastureland. Scientists study these islands to determine the extent to which isolation and fragmentation alter the communities of organisms within them.

stripped. After a year or two, crops will not grow at all. The people move on, cutting and burning yet another portion of the forest.

Commercial logging also takes its toll on the tropical rain forests. Many of the trees are cut to supply fuel wood, paper, wood panels such as plywood, and charcoal and to supply furniture manufacturers with mahogany and other woods demanded by consumers around the world. At this time, the tropical forest is being slashed and burned at a devastating rate. The Food and Agriculture Organization of the United Nations estimates that an area of tropical forest large enough to cover North Carolina is currently deforested each year.

Research programs are currently being conducted to explore the concept of conserving forest "fragments" as reserves that will provide suitable habitats for species (**Figure 42.15**). Research results show, for example, that forest fragments with rounded corners, rather than with square corners, are more suitable for growing shade-loving trees. At corners, young trees are more exposed to sun, wind, and competition from invading species. To avoid the problem, reserve designers are considering planting sun-tolerant species of trees to round out the edges.

## 42.9 Carbon dioxide buildup in the atmosphere produces a greenhouse effect.

Many scientists agree that in addition to destroying a rich natural resource, the burning of the tropical forests is adding tremendous quantities of carbon dioxide ($CO_2$) to the air, in the same way that burning fossil fuels releases $CO_2$. Like the glass in a greenhouse or the windows in your car, $CO_2$ allows light to enter the Earth's atmosphere but preventing the resultant generated heat from leaving. This blocking of outward heat radiation from the atmosphere by high $CO_2$ levels is called the **greenhouse effect**. Rising levels of atmospheric $CO_2$ may therefore result in a worldwide temperature increase, a situation called **global warming**. Scientists are currently debating how greenhouse gases, which also include nitrous oxide and methane, will affect and are affecting global temperatures and climate. **Figure 42.16** shows atmospheric concentrations of three greenhouse gases. Warmer temperatures are predicted to cause changes in weather patterns, water supplies, ocean levels, growing conditions for crops (see Chapter 41's opening story), and the northward spread of tropical diseases.

A study conducted at the University of Massachusetts in 1999 concluded that the 1990s was the warmest decade of the millennium. This study adds to the growing body of evidence that global warming is occurring. Other studies show that portions of the Greenland ice sheet are melting, north temperate birds have earlier breeding dates, and both tropical and temperate species show latitudinal shifts in their ranges. These phenomena are further suggestions of global warming.

In a worldwide effort to reduce greenhouse gas emissions and mitigate their effects on global warming, representatives of more than 160 nations gathered in Kyoto, Japan, in December 1997 to discuss the issue. The outcome of the meeting was the Kyoto Protocol, an international treaty on global warming that sets limits on greenhouse gas emissions. Countries that ratify this protocol will commit to reduce their emissions of carbon dioxide and other greenhouse gases. As of this writing, the protocol had been approved by 120 countries, but had been rejected by the United States and Russia, two countries that are large emitters of greenhouse gases. The provisions of the Kyoto Protocol are controversial, and its ratification has become highly political. Although this treaty has not yet been ratified,

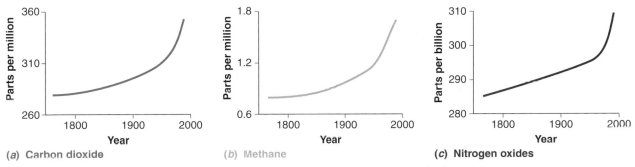

(a) Carbon dioxide  (b) Methane  (c) Nitrogen oxides

**Figure 42.16 Greenhouse gases.** Global greenhouse gas concentrations over the last 200 years.

 **Visual Thinking:** In approximately which decade did the concentration of all three greenhouse gases begin a dramatic rise?

many suggest that discussion about the protocol and greenhouse gas emissions has resulted in many countries taking significant steps on their own to reduce emissions that contribute to global warming.

## 42.10 Habitat destruction, habitat pollution, and illegal trade threaten the survival of many species.

Not only is the environment threatened, the richness and diversity of species on the Earth is threatened as well. Although scientists have classified approximately 1.7 million species of organisms, they estimate that approximately 40 million exist. Many of these species, however, are becoming extinct—dying out—and will never be seen again. An estimated 1000 extinctions are taking place each year, a number that translates into more than two species per day. Although such mass extinctions have occurred in the past, as with the dying out of the dinosaurs, these former extinctions were caused by climatic and geophysical factors. Much of today's species' extinction is caused by human activity.

One way in which humans destroy species is by destroying their habitats. Widespread habitat destruction is taking place as the tropical rain forests are cut down. This one factor alone will cause the extinction of one-third of the world's species.

Humans are also destroying coral reefs at an alarming rate. Because the base of the coral reef food chain is algae, the reef ecosystem depends on sunlight for its existence. In areas where soil erosion muddies the water, the reef dies from insufficient sunlight. In some cases, the water may become polluted with fertilizer runoff or with sewage, which causes the algae to overgrow, smothering the corals. In some cases, humans use dynamite to kill and harvest fish, a practice that obviously destroys the coral reef. In addition, coral is often harvested to sell to tourists.

Although habitat destruction and pollution are the most serious threats to the existence of certain species, exploitation of commercially valuable species threatens them as well. The Convention on International Trade in Endangered Species (CITES) regulates trade in live wildlife and products. Illegal trade still takes place, however, because of smuggling and the inability of law-enforcement inspectors to check all shipments of goods. Certain types of alligator, crocodile, sea turtle, snake, and lizard trade are illegal and endanger the existence of various species. Buying other products such as coral and ivory endangers the existence of coral reefs and African elephants. Even certain plants such as cacti and succulents should be purchased only if cultivated in greenhouses; they should not be taken from the wild. Organizations such as the World Wildlife Fund are invaluable sources of information regarding which products consumers should avoid so that they can help stop trade that threatens certain species.

The extinction of many species leads to a reduction in **biological diversity**, a loss in numbers of different kinds of species. Each species is not only the result of millions of years of evolution and can never be replaced once it is gone, but also is an intricate part of the interwoven relationships among organisms within the ecosystems of the world. The extinction of just one species can affect ecosystems in many unforeseen ways as well as affecting sources of food, medicine, and other substances.

Recently, the genetic diversity in crop plants has declined as scientists have used selective breeding techniques to produce plants

**Figure 42.17** A germ plasm bank at the U.S. Department of Agriculture. Researchers are in a cold storage room that contains seeds from around the world.

with specific characteristics, such as resistance to particular diseases or pests, hardiness, and other characteristics considered important. When only a few species or varieties of a species are cultivated or survive, however, the genetic diversity of the organism declines. Populations of species that have little diversity are more vulnerable to being wiped out by new diseases or climatic changes.

In the 1960s, the United Nations Food and Agriculture Organization (FAO) made recommendations that have led to the establishment of "banks" that store plant seeds and genetic material, or germ plasm (**Figure 42.17**). Zoos around the world are making similar efforts to preserve genetic diversity among animals by establishing and carrying out sophisticated breeding programs to increase genetic diversity among endangered species.

## 42.11 Many factors interact to threaten the overall health of the biosphere.

Some scientists hold that the enormous world population is the cause of the problems of pollution and diminishing natural resources. Others state that technology is the culprit. Most scientists, however, would agree that neither factor alone is the cause of these complex problems. Many factors interact to affect population size and the "health" of the environment.

The size of the human population, over 6.3 billion at the beginning of 2004, influences the environment because it puts a demand on resources. However, the social, economic, and technological development of a country affects the demand its population places on resources. A person living in a rural area of Kenya, for example, does not use the same amount and kind of environmental resources as a person living in a large city in the United States. The lifestyle and per capita consumption of a population make a big difference on the impact of that population on the environment.

Using resources and disposing of wastes in certain ways pollute the environment. The pollution of one aspect of the environment—such as the air—can lead to pollution of other aspects of the environment—such as the water through acid rain. Pollution, in turn, can limit population size by increasing the death rate. Living with pollution, however, can lead to attitudinal changes among members of a population, which may result in the development of laws to better manage the use of resources and to curb the pollution resulting from their use.

The impact of the human population on the environment is extremely complex and has multiple causes and effects. Many factors, such as population size, per capita consumption, technology, and politics, interact in complex ways, resulting not only in the prob-lems faced today but also in solutions for the future. Through research, environmental scientists will help everyone understand how each person can live on this planet without harming it or endangering everyone's future existence.

> ## CONCEPT CHECKPOINT
>
> **3.** Why should humans be concerned about extinctions taking place on the Earth today?

# CHAPTER REVIEW

## Summary of Key Concepts and Key Terms

### What is the biosphere?

**42.1** The **biosphere** (p. 741) is the part of the Earth that supports life; it is often spoken of as the **environment** (p. 741).

**42.1** The biosphere is the interface of the land, air, and water, extending from approximately 9000 meters (30,000 feet) above sea level to 11,000 meters (36,000 feet) below.

### What resources does the biosphere provide?

**42.2** Many of the natural resources of the Earth are finite and **nonrenewable resources** (p. 741); that is, they cannot be replaced.

**42.2** Coal, oil, and natural gas are **fossil fuels** (p. 742), nonrenewable resources formed over time from the remains of organisms that lived long ago.

**42.2** **Nuclear power** (p. 742) is a possible energy alternative, but people fear the danger of nuclear explosions, and it creates radioactive nuclear wastes.

**42.2** Solutions to the problem of **municipal solid waste** (p. 742) lie in reducing consumption, reusing, and recycling.

**42.3** **Renewable resources** (p. 743) replace themselves naturally over time.

**42.3** Alternative renewable energy sources are solar, water, wind, geothermal power, and bioenergy.

### What threatens the health of the biosphere?

**42.4** Preventing **pollution** (p. 744) is essential to ensuring a safe water supply.

**42.4** Underground reservoirs of water can be contaminated by the same sources as surface water when the contaminants trickle through the soil; however, the primary groundwater contaminants are toxic chemicals.

**42.4** Surface water is contaminated by factories, power plants, and sewage treatment plants that dump waste chemicals, heated water, and human sewage into lakes, streams, and rivers.

**42.4** Other sources of surface water pollution are sediment runoff during erosion and the leaching of chemicals from mines, hazardous waste dumps, and croplands.

**42.4** Accumulations of inorganic nutrients in a lake results in **eutrophication** (p. 744) which may eventually fill the lake with decomposed materials.

**42.4** Chemical pollutants affect aquatic life in different ways and can become concentrated in their bodies through **biological concentration** (p. 744) and **biological magnification** (p. 744).

**42.4** **Thermal pollution** (p. 744) may kill heat-intolerant plants and animals.

**42.4** In a **sanitary landfill** (p. 744), trash and garbage are dumped, compacted, and covered with dirt; liquids leaching from landfills can pollute nearby streams, lakes, and wells.

**42.5** The atmosphere has become polluted with sulfur and nitrogen oxides, primarily from the combustion of fossil fuels in automobiles and electricity-generating plants.

**42.5** Along with sulfur and nitrogen oxides, fuel combustion in cars, at electricity-generating plants, and in various industrial processes produces carbon monoxide, hydrocarbons, and particulates.

**42.5** As the nitrogen oxides and hydrocarbons react with sunlight, they form **smog** (p. 744), including **photochemical smog** (p. 744), an upper respiratory irritant that is a health hazard; a principal pollutant in photochemical smog is **ozone** (p. 744).

**42.6** As sulfur and nitrogen oxides mix with water vapor in the atmosphere, they form acid precipitation, commonly referred to as **acid rain** (p. 748), which harms both living and nonliving things.

**42.7** Ozone in the stratosphere acts as a shield against the sun's ultraviolet (UV) rays.

**42.7** Chlorofluorocarbons (CFCs) damage the ozone layer because they contain chlorine, which reacts with ozone, breaking it down.

**42.7** In 1987, 57 nations signed the Montreal Protocol, which required all production of CFCs in developed countries to be halted on January 1, 1996, with the others to follow suit in 2006.

**42.8** Approximately 100,000 square kilometers (62,000 square miles) of rain forest are being lost each year worldwide because of slashing and burning for agriculture and logging, and to create pastures for cattle; **tropical rain forest deforestation** (p. 750) is possibly the most severe environmental crisis that we face today.

**42.8** Scientists estimate that at the present rate, nearly all the tropical rain forest will be gone within the next 50 years.

**42.8** The atmospheric rise in carbon dioxide levels, which is caused by the burning of the tropical forests and the combustion of fossil fuels, is acting as a barrier against the escape of heat from the surface of the Earth.

**42.8** This **greenhouse effect** (p. 751) could affect global temperatures, rainfall patterns, and agricultural lands—in a situation called **global warming** (p. 751).

**42.9** Species are dying out at a rate of 1000 extinctions per year.

**42.9** A major cause of species extinction is habitat loss, which occurs when the tropical rain forests are cut or other habitats are destroyed.

**42.9** The extinction of many species leads to a reduction in **biological diversity** (p. 752), a loss of species richness.

**42.9** The loss of species also affects ecosystems and diminishes future sources of food, medicine, and other substances.

**42.10** Environmental problems are the result of many interacting factors.

**42.10** Some factors that modify the effects of a population on the environment include lifestyle, per capita consumption of natural resources, the manner of resource use, and the effects of resource use.

## Level 1    Learning Basic Facts and Terms

**Multiple Choice**

1. Which of the following resources are renewable and would contribute toward developing a more sustainable society?
   a. Fossil fuels
   b. Plastics
   c. Nuclear energy
   d. Geothermal energy
   e. None of the above

2. Which of the following correctly states the difference between a point and a nonpoint source of water pollution?
   a. Point sources are more toxic than nonpoint sources.
   b. Point sources pollute smaller lakes and ponds; nonpoint sources pollute larger bodies of water such as oceans or expansive estuaries.
   c. The entrance of point sources into the water can be localized to a distinct area, while nonpoint sources enter the water from a variety of places.
   d. Point sources require more immediate clean-up action than do nonpoint sources.
   e. The manner by which point sources affect living organisms is different from that of nonpoint sources.

3. Which of the following would be considered a point source of surface water pollution?
   a. Agricultural pesticides
   b. Thermal pollution from a power plant
   c. Mercury accumulation due to the burning of coal
   d. Sediment accumulation in the Mississippi River due to improper agricultural practices
   e. Acid rain

4. Which of the following types of pollution is correctly paired with its effect on the organisms or the environment?
   a. inorganic nutrients // eutrophication
   b. PCBs // cancer
   c. Sulfur and nitrogen dioxides // acid rain
   d. Chlorofluorocarbons // stratospheric ozone destruction
   e. All of the above are correctly paired.

**True–False**

5. _____ Because the soil in a tropical rainforest is so nutrient rich, farming of slashed and burned rainforests is considered a sustainable agricultural practice.

6. _____ Slashing and burning tropical rainforests may contribute to global warming because it adds to atmospheric $CO_2$ levels.

7. _____ Habitat destruction, pollution, and exploitation of commercially valuable wild species are the most significant threats to maintaining global biodiversity.

8. _____ The loss of a single species from an ecosystem is unlikely to affect the ecosystem as a whole.

9. _____ Ozone is an air pollutant when industrial practices release it into the lower atmosphere, but natural production of ozone in the stratosphere (upper atmosphere) shields the Earth from UV radiation.

## Level 2    Learning Concepts

1. Many scientists think that people should work toward creating a "sustainable society." What does this mean?

2. Name several energy sources that could be used to help create a sustainable society. Explain your answers.

3. What can be done to help prevent a mineral shortage in the future? Why are preventive measures important?

4. What is global warming? Why is it dangerous?

5. Define biological diversity. How are humans affecting the biological diversity of the world's species? Why is this serious?

6. How can the amount of solid waste be reduced and waste management improved?

7. Describe the different effects of pollution by organic nutrients, inorganic nutrients, infectious agents, and toxic substances in surface water.

8. How is acid rain produced? What are its effects?

9. When breathed, ozone is a dangerous pollutant in air. Why should people be concerned that ozone in the stratosphere is being depleted?

**Critical Thinking and Life Applications**

1. Suppose the president of the United States asked you to put together a plan of action for addressing environmental problems. What steps would you recommend?

2. Imagine that nothing was done to address the environmental problems you identified in your answer to question 1. Outline a scenario of the likely consequences—what would such a world be like in 20 years?

3. Why is the destruction of tropical rain forests a serious problem for everyone—not just the people living in tropical countries?

4. Measure how much solid waste you or your family produce each week. Sort your solid waste into categories of yard waste, trash, and garbage or food waste. How many pounds of each type of waste do you produce weekly? monthly? yearly? What proportion of the total is each? What steps could you take to reduce the amount of solid waste you contribute to the landfill?

5. Today the average American uses enough energy to release into the atmosphere about 50,000 pounds of carbon dioxide, the major contributor to global warming, per year. Describe 10 simple changes that you could make in your daily life that could help reduce $CO_2$ emissions to help combat global warming.

6. Do you think the development of new technologies alone will suffice to solve many of the human-caused environmental problems the Earth is experiencing today? Justify your answer.

7. Conserving habitat "fragments" as reserves that will provide suitable habitats for species is an important step in preserving biodiversity. However many ecologists today argue that this will be effective only if these fragments are connected by wide habitat corridors that allow for at least some migration and interbreeding with neighboring populations. Explain why they would advocate this. (*Hint:* See Chapter 17 and Figure 17.17 for more information.)

In The News | **Critical Thinking**

**SHARK TEST**

Now that you understand more about the biosphere, reread this chapter's opening story about the decline in the world's shark population, including great white sharks. To better understand this information, it may help you to follow these steps:

1. Review your immediate reaction to the decline of the great white shark that you wrote when you began reading this chapter.

2. Based on your current understanding, again summarize the main point of the research in a sentence or two.

3. What questions do you now have about this information that this chapter's opening story does not answer?

4. Collect new information about the research. Visit the *In The News* section of this text's companion Web site at www.wiley.com/college/alters and watch the "Shark Test" video. Then use the "summary" link to read the accompanying story and access related links. Use this information, the links provided, and other online and library resources to answer your questions and find updates about this research topic. State the sources of your information. Explain why you think the information is accurate. Also determine whether the information expresses a particular point of view or is biased in any way.

5. What in your view are the most significant aspects of this research? Give reasons for your opinion and for any changes in your ideas based on the additional information you have collected and the analysis you have done.

# APPENDIX A: A Scientific Journal Article Examined

ELSEVIER

**2** BIOLOGICAL CONSERVATION

**3** Biological Conservation 85 (1998) 63–68

**4** Experimental measurement of nesting substrate preference in Caspian terns, *Sterna caspia*, and the successful colonisation of human constructed islands

**5** James S. Quinn *, Jane Sirdevan

**6** *Biology Department, McMaster University, Hamilton, Ontario, Canada, L8S 4K1*

**7** Received 7 June 1997; accepted 30 September 1997

**8** **Abstract**

Caspian terns, *Sterna caspia*, recently bred in Hamilton Harbour, at the western end of Lake Ontario, on private property that is likely to be developed in the next decade. To reduce this land-use conflict and to promote the current level of biodiversity of colonial nesters in the area, artificial islands were built in the winter of 1995-1996 with different areas designated for a variety of nesting waterbirds including Caspian terns. In 1994, prior to island construction, we tested three substrate types for tern nesting preferences so that an appropriate substrate could be placed on the Caspian tern designated portion of the new islands. We found a preference for sand over pea-gravel and crushed stone, and indirect evidence for a preference favouring the experimental substrates over the pre-existing substrate of hard-packed ground. Based on these results, the small area of the island designed for Caspian tern nesting was surfaced with sand and was subsequently colonised successfully. The colony established and reproduced successfully on the designated site in 1996 and grew in numbers of nesting pairs in 1997. © 1998 Elsevier Science Ltd. All rights reserved

**9** *Keywords:* Habitat preference; Restoration; Colonial nesting waterbirds

**10** **1. Introduction**

Caspian terns, *Sterna caspia*, have probably never been abundant in North America, but they have nested in the Great Lakes since as early as 1896 (Blokpoel and Scharf, 1991). In 1986, a small colony of 48 nests was established in Hamilton Harbour (43° 16′N, 79° 46′W; Dobos et al.; 1988) at the western end of Lake Ontario. **11** Between 1990 and 1994, the main colony was located on the site of the experiment reported here and numbers in the harbour increased from 184 to 331 pairs, representing over 16% of Lake Ontario's Caspian terns (Moore et al., 1995). This immigration to Hamilton Harbour and subsequent growth of the colony, particularly the increase to 134 pairs in 1987 (Dobos et al., 1988), coincided with the decline and desertion of a Caspian tern colony site at the Eastern Headland, a man-made area extending into Lake Ontario from the Toronto waterfront (Morris et al., 1992). The success of this thriving new colony in Hamilton Harbour is encouraging given the questionable status of the species, ranging from rare to endangered (proposed) in Ontario and the States surrounding the Great Lakes (Blokpoel and Scharf, 1991).

Development plans for an area of the Hamilton Harbour shoreline included the Caspian tern main colony site. During the winter of 1995–1996 three islands were **12** constructed to provide nesting habitat for the six species of colonial nesters currently in the area, a major component of the remedial action plan for the rehabilitation of Hamilton Harbour (Quinn et al., 1996). The motivation for island construction was to maintain current levels of diversity of colonial nesters in the harbour and to reduce land-use conflict with the current property owners. We tested substrate preferences of Caspian terns on an experimental site on the mainland where the colony had been previously to facilitate the establishment of a colony on one of the new islands.

Little is known about the nesting substrate preferences of Caspian terns. **13** Caspian terns generally nest in dense colonies situated in open and largely unvegetated areas (Peck and James, 1983). Although descriptions

* Corresponding Author. Tel.: (905) 525-9140 ext. 23194; fax: (905) 522-6066; email: quinn@mcmaster.ca

0006-3207/98/$19.00 © 1998 Elsevier Science Ltd. All rights reserved
PII:S0006-3207(97)00142-0

---

**1** The name and logo of the publisher of the journal

**2** The journal title

**3** The journal title, volume number, year, and page numbers of the article

**4** TITLE OF THE PAPER. The title should be descriptive of the study and should have a clear focus. This title states the independent variable (IV—[nest] substrate) and the dependent variable (DV—nesting site preference [on the experimental grid]).

**5** AUTHORS OF THE STUDY. The author to whom queries should be addressed is noted by an asterisk, with contact information stated in a footnote at the bottom of the research page.

**6** Institutional affiliation of the authors

**7** This paper was received by the editor of the journal on June 7, 1997. It then was sent to peer reviewers (other scientists in this field). Peer reviewers give their opinion as to a paper's acceptability for publication. This paper was accepted for publication on September 30, 1997.

**8** ABSTRACT. An abstract is a concise summary of the research paper. Notice that it includes the statement of the problem, procedures, principal results, and conclusions. The hypothesis is also often stated in the abstract. The abstract gives readers an overview of the paper so they know whether it contains information they want or need. Additionally, it helps readers understand the paper by creating a context for the specific sections of the paper that follow.

**9** KEYWORDS. Keyword computer searches using scientific literature databases will yield titles, abstracts, and sometimes full texts of research papers related to the keywords. (Many such databases are available at university libraries.)

**10** INTRODUCTION. In this section, the investigators present the rationale (statements of their reasons) for their study and its foundation in the literature. The specific references cited here are listed at the end of the paper so that other researchers can find and read these papers. This section typically concludes with the purpose of the study and research hypotheses.

**11** The purpose of this paragraph is to provide the rationale for studying Caspian terns (i.e., why study conservation issues related to this species).

**12** This paragraph further explains the rationale for the study. The investigators discuss plans to develop the area in which a large colony of terns is nesting and the consequent construction of three islands to provide new places for the birds to nest. At the end of this paragraph, the investigators state that they tested nesting materials for use on these three islands' nesting sites.

**13** This paragraph gives background about tern preferences in nesting materials (substrates) and sites (habitats). The authors also mention the problems ring-billed gulls create with Caspian tern nesting.

**(14)** Statement of the "problem"; importance of the study

**(15)** PURPOSE OF THE STUDY. The investigators do not overtly state a hypothesis, but it is implied in the title and background given here: If the terns are provided with nesting substrates of gravel, sand, and stone, then the terns will favor one of these materials as a nesting substrate. The IV is the nesting substrate, and the DV is tern preference.

**(16)** METHODS. In the methods section, investigators describe the materials and procedures they used to conduct the study. Enough detail is included so that another researcher could repeat (replicate) the study.

**(17)** This section describes the selection of the study site and the design of the study.

**Photo A** Photograph of the distribution of substrates in the experimental grid. There are nine cells within the frame (see Figure 1).

**(18)** Here, the investigators cite references that further describe methods not described in detail in the paper.

**(19)** The investigators describe how they determined offspring (chick) survival. They wanted to determine whether chick survival would vary with substrate type. This was measured as it may provide insight as to why Caspian terns may prefer particular nesting substrates.

---

of nesting habitat are available (Peck and James, 1983; Quinn et al., 1996), experimental studies have not been reported and descriptive studies may not take habitat availability into account. For example, at sites where ring-billed gulls, *Larus delawarensis*, nest, only areas not used by gulls are available to terns as terns begin nesting after ring-billed gulls have become well established (Quinn et al., 1998). The later nesting terns are unable to take over areas that have become occupied by ring-billed gulls, a problem encountered also by common terns, *Sterna hirundo* (Morris et al., 1992). Ring-billed gulls began nesting as a colony of 17 pairs in Hamilton Harbour in 1978 (Dobos et al., 1998). Since then the colony has grown to about 40000 pairs when last counted in 1990 (Moore et al., 1995).

**(14)** The Hamilton Harbour Caspian tern colony is one of only five on the lower Great Lakes. In 1994, this colony was located on a site that is slated for development in the next 4 to 9 years. Caspian terns are sensitive to human disturbance and, unlike ring-billed gulls, will not nest in close proximity to human activity. The focus of **(15)** this study was to evaluate Caspian tern nesting preferences when given the opportunity to nest on one of three commercially available substrates, or on the hard-packed ground found on the colony site. The choice of materials for the experiment was based on examinations of Caspian tern nesting substrates in the Great Lakes and the Gulf of Mexico (J. Quinn, pers. obs.). Results of this experiment were used to determine substrate type placed on a designated area of recently built wildlife islands in Hamilton Harbour. The implementation resulted in the successful establishment of a colony of Caspian terns on the designated site. This colony has nested successfully for two seasons (1996 and 1997).

**(16)** ## 2. Methods

Our study site was Pier 26 in Hamilton Harbour, at the west end of Lake Ontario. Caspian terns nested at a main site in the midst of a large colony of ring-billed gulls, estimated at about 40 000 pairs in 1990 (Moore et al., 1995), and at a separate sub-colony site about 100m to the North-East. On 8 April 1994, prior to the arrival **(17)** of Caspian terns in the area, we established a $9 \times 9$ m frame subdivided into $3 \times 3$ m cells within the area of the 1993 Caspian tern colony site. Nest markers from the 1993 nesting season had been left in place and the number of clutches that had been initiated in each experimental cell was recorded (Fig. 1). We removed the old stakes and then placed the substrates within the grid, in a $3 \times 3$ Latin Square design, so that each substrate type was represented once in each row and once in each column, thus controlling for any effects due to position. The substrates tested were: (1) construction grade sand with a few small stones; (2) crushed stone

---

| | West | Middle | East | Total |
|---|---|---|---|---|
| **North** | Gravel<br><br>n=5(5) | Sand<br><br>n=17(7) | Stone<br><br>n=12(3) | 34 |
| **Middle** | Sand<br><br>n=18(10) | Stone<br><br>n=9(9) | Gravel<br><br>n=7(9) | 34 |
| **South** | Stone<br><br>n=6(3) | Gravel<br><br>n=3(9) | Sand<br><br>n=11(4) | 20 |
| **Total** | 29 | 29 | 30 | 88 |

Fig. 1. Distribution of substrates in the experimental grid. Each grid cell was 3x3m. Numbers of clutches initiated in each cell is given for 1994 and (1993)—prior to establishment of the grid. Row and column totals are for 1994 data.

**See Photo A.**

with sharp edges, approximately 1 cm in diameter; and (3) 'pea' gravel with rounded edges, approximately 1 cm in diameter. During the period of ring-billed gull nest building, prior to the arrival of terns, a field assistant visited the Caspian tern nesting area five times per day from 5 to 26 April 1994 and destroyed all ring-billed gull nests as they were started.

Upon tern clutch initiation we marked nests initially with numbered tongue depressors (wooden spatulas) placed about 15 cm from the edge of the nest scrape. Tongue depressors were replaced with 20 cm tall numbered wooden stakes on 17 May. Prior to 2 June we placed nest covers (Quinn, 1984) over most nests with eggs, particularly those near the periphery of the colony, **(18)** to protect them from the ring-billed gull depredation. We typically checked nests every second day until 2 June, and every sixth day thereafter, weather permitting (Fig. 2). We placed chick shelters (Burness and Morris, 1992) near where eggs had or were about to hatch to provide shelter from harsh weather or predators during parental absences.

To examine whether offspring survival varied with substrate types we followed the fates of eggs and chicks which were banded by age 7 days post-hatching. Caspian tern nestlings generally remain in the vicinity of the **(19)** nest until fledging. However, to restrict their movements when we checked the colony, we fenced the grid with 30 cm tall $6mm^2$ mesh hardware cloth on 8 June, before chicks became mobile, and used hatching and survival to at least age 24 days post-hatching as measures of breeding success.

On several occasions we observed a ring-billed gull that flew into the colony and removed an egg from an unattended clutch. Each time, the bird flew into a particular

NOTE: A clutch is a nest of eggs or newly hatched chicks. When chicks acquire the feathers necessary for flight, they are said to be fledged—they become fledglings.

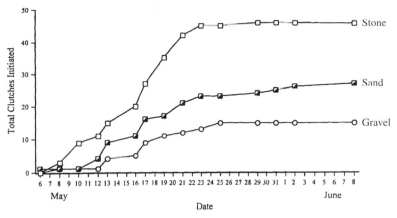

Fig. 2. Cumulative frequency of clutches initiated on three substrate types. Clutches from the three replicates of each substrate type were pooled. Data are presented for each nest census date.

**Photo B** Caspian terns nesting in the grid. The eggs or newly hatched nestlings are not visible in this photograph. The fence prevents them from leaving the area during nest checks by the researchers or during other disturbances. Caspian terns are nesting outside of the grid area also, but 20 of the first 21 eggs laid were laid on the grid.

part of the colony just outside the grid fence and ate the egg. Yolk stains on the gull's plumage and behavioural idiosyncrasies made it individually recognisable. Our observations from a blind located 13 m south of the edge of the colony suggest that the same individual took at least 30 eggs. On 15 June, this bird was shot under Canadian Wildlife Service permit and egg loss virtually stopped.

**(20)** During the winter of 1995–1996 three wildlife islands measuring about 100 by 20 m were constructed under the direction of the Hamilton Harbour Fish and Wildlife Habitat Restoration Project, part of the Hamilton Harbour Remedial Action Plan (Quinn et al., 1996). We covered the tern designated areas with heavy gauge plastic sheeting anchored with rocks prior to the beginning of gull nesting (to prevent nesting by gulls) and removed the sheeting shortly after the arrival of the terns. Nesting activities were monitored in 1996 and 1997 through twice-weekly visits to the islands which ceased when chicks became mobile.

We tested the spatial and temporal patterns of clutch initiations and reproductive parameters with the likelihood $L^2$ statistic (Hays, 1981). We employed a Spearman Rank Correlation (Siegel, 1988) on numbers of clutches initiated within the specific area of each of the cells between 1993 (prior to establishment of the experimental site; nest markers were left in place over winter) and 1994 (after establishment of the experimental site) to test for the possibility that pairs of terns established nests in the same location between years. To show that the frequencies of clutches initiated in the cells between 1993 and 1994 we used a multi-sample $\chi^2$ test with 2 degrees of freedom.

**(21) 3. Results**

See Photo B.

Between 6 May and 8 June 1994, 88 clutches were initiated on the experimental substrates. Although the available area of the Caspian tern colony was larger than that of the experimental grid (approximately 280 m² vs 81 m²), 20 of the first 21 clutches were initiated on the experimental substrates suggesting that **(22)** Caspian terns preferred the experimental substrates over the natural hard-packed ground at the colony site. Additionally, despite smaller numbers of nests initiated in the main colony site in 1994 ($n = 177$) compared with 1993 ($n = 242$), the number initiated on the experimental substrates in 1994 exceeded that on the same area in 1993 by about a third (Fig. 1). Patterns of nest placements were independent of nest positions in 1993 as the number of nests positioned within the location of each cell of the experimental grid in 1993 and 1994 were not correlated (Spearman Rank Correlation, $n_1 = 59$, $n_2 = 88$, $r_s = 0.15$, ns; Fig. 1). Furthermore, the nest frequencies in the cells (grouped by 1994 substrate-type) changed significantly between 1993 and 1994 (multi-sample $\chi^2 = 9.07$, df = 2, $p < 0.025$).

The number and timing of clutches initiated on each substrate were analysed using pooled values of the three replicates. Row position had an effect on the number of clutches initiated, with significantly fewer clutches initiated on the South row than on the middle or North rows of the experimental substrates ($\chi^2 = 4.76$, df = 1, $p < 0.050$). Because substrate preference patterns were constant for each row, replicates were pooled for analysis.

Clutch initiations were not at random with regard to substrate type ($L^2 = 16.8$, $p < 0.0005$). We found that

---

**(20)** After the preliminary nesting studies were completed, the Hamilton Harbour Remedial Action Plan's (HHRAP) Fish and Wildlife Habitat Restoration Project constructed the islands. The investigators were involved in the design of the topography, substrates, and vegetation of the islands.

**(21)** RESULTS. The results section of the paper present the data collected. The data are summarized in narrative form and often have accompanying charts, graphs, and/or tables. Statistical analyses of the data are presented; "raw" data are usually not presented unless it is manageable and useful, as in this study (i.e., Figure 1).

**(22)** First, the investigators describe the results of their preliminary testing of the three substrate types for tern nesting preferences.

NOTE: The article uses spelling and style conventions typical of Canadian and British publications. This includes a raised decimal point, as seen on this page.

Table 1

Reproductive parameters for caspian tern pairs nesting on experimental substrates in 1994

| Substrate type | Sand | Gravel | Stone |
|---|---|---|---|
| Clutches ($n$) | 46 | 15 | 27 |
| Clutch size (mean±1SD) | 2·17±0·74 | 2·00±0·63 | 2·07±0·87 |
| Clutches hatching ≥1 egg | 28 | 9 | 10 |
| Hatching success[a] | 0·61 | 0·60 | 0·37 |
| Clutches fledging ≥1 Chick | 21 | 6 | 7 |
| Fledging Success[b] | 0.75 | 0.67 | 0·70 |
| Chicks fledged/pair[c] (mean ± 1 SD) | 0·61±0·74 | 0·67±0·90 | 0·44±0·80 |

　a　Number of clutches hatching one or more eggs divided by total number of clutches initiated.

　b　Number of clutches fledging one or more chicks divided by total number of clutches hatching one or more eggs. Chicks were considered fledged if they survived to age 24 days.

　c　Number of clutches where one or more chicks survived to age 24 days divided by total number of clutches initiated.

more clutches were initiated on sand than on the other substrates (Fig. 2; $L^2=13\cdot3$, $p <0.0005$). There were no significant differences in the tendency to nest on stone compared with gravel ($L^2=3\cdot5$, ns). Nests initiated on sand were generally edged with a circle of pea gravel and/or crushed stone along the edge of the nest scrape. Although more pairs laid on sand, the timing clutch initiations did not differ among substrate types. The median clutch initiation date, 17 May, was the same for all three substrates.

Reproductive parameters were similar for pairs nesting on all three substrates (Table 1). Hatching success was defined as the proportion of successful clutches in which one or more eggs hatched, whereas fledging success was determined as the proportion of clutches where one or more eggs hatched and at least one chick fledged (survived to at least 24 days). Values for hatching and fledging success were highest on sand, but the differences were not statistically significant ($L^2=2\cdot1$ and $0\cdot38$, $p <0\cdot05$). Egg failures were categorised as: (1) failed to hatch; (2) cracked or broken; (3) disappeared. The proportion of eggs which disappeared or were cracked or

broken was the greatest in the South row of the experimental grid (Table 2). Chick loss prior to fledging was classified as death or disappearance. The most common source of chick loss was death (chicks found dead; sand 77%; stone 67%; gravel 71%). While we tried to minimise the effect of our nest chicks, we cannot dismiss the possibility that out infrequent nest visits affected these results. Notably, the activities of a family of red foxes (Vulpes vulpes) observed on Pier 26 in the area of the colony may have affected chick survival, however we have no direct data from 1994 to address this possibility. In 1995, fox activity was observed to cause colony desertion at night for periods of up to 42 min (J. Sirdevan, per. obs.).

During the winter of 1995 to 1996 three wildlife islands were constructed in Hamilton Harbour (Quinn et al., 1996). The design of the islands utilised the results reported here for the construction of one 200m$^2$ knoll surfaced with sand and a small amount of 1 cm diameter pea gravel for lining of nest rims. Most of the area was covered with plastic sheeting (held down with stones) to discourage nesting by ring-billed gulls until the terns arrived. Despite intentions to use decoys to attract Caspian terns to the site in 1996 (Quinn et al., 1996) they had already begun to lay eggs on the exposed sand that remained uncovered before we were able to remove the plastic and install decoys. After uncovering the sand covered habitat terns nested over the entire sandy knoll. In total there were 226 clutches (86% of clutches initiated in Hamilton Harbour) initiated on the designated Caspian tern site in 1996. About 37 clutches (14%) were initiated on a raft (H. Blokpoel, pers. comm.) which was positioned in a confined disposal facility near the original Caspian tern colony (Lampman et al., 1996). Again in 1997 we covered most of the sandy Caspian tern habitat with plastic. The first Caspian tern nests were found on 6 May, 7 days after the plastic was removed. There were 319 clutches initiated (80% of clutched initiated in Hamilton Harbour) on the Caspian tern site with a median initiation date of 21 May. A small proportion of pairs nested on the CDF tern raft (estimated

Table 2

Classification of egg failures on experimental substrates during the 1994 breeding season

| Substrate | Row[a] | Total eggs | Disappeared | Cracked/broken | Failed to hatch |
|---|---|---|---|---|---|
| | | | *Proportion of eggs* | | |
| Sand | N | 37 | 0.19 | 0.05 | 0.05 |
| | M | 37 | 0.51 | 0 | 0.03 |
| | S | 26 | 0.69 | 0.08 | 0 |
| Gravel | N | 10 | 0 | 0.10 | 0 |
| | M | 14 | 0.50 | 0 | 0 |
| | S | 6 | 0.83 | 0.17 | 0 |
| Stone | N | 28 | 0.39 | 0 | 0 |
| | M | 16 | 0.75 | 0 13 | 0 |
| | S | 12 | 0.92 | 0.08 | 0 |

　a　Row indicates position in experimental grid (N, north; M, middle; S, south)

N=15, 4%; H. Blokpoel, pers. Comm.) Sixty-three late-nesting or re-nesting pairs initiated clutches (16%) on another site on the wildlife islands with a median initiation date of 10 June. The secondary colony, on an island slightly to the south, was located on a site that had been targeting common terns. The area was open and covered with pea gravel (1 cm diameter).

**④ 4. Discussion** `See Photo C.`

Caspian terns demonstrated a statistical preference for nesting on sand over the other substrates. A preference for a particular substrate may be due to a variety of parameters. Substrates differed in colour, size and shape of grains. Construction grade sand is light brown in colour while crushed stone and pea gravel are shades of grey. Caspian tern eggs are light tan with dark brown spots. While they match each of the substrates quite well, they are relatively easy to see, even on sand. Chaniot (1970) noted changes in the frequency of tan-coloured relative to darker and lighter phase downy young Caspian terns in San Francisco Bay since 1943 and attributed this frequency change to natural selection favouring the colour morph which best matched the nesting substrate colour. Nesting substrate preference could reflect coloration of eggs or chicks. Nevertheless, the scrapes in sand were usually edged with crushed stone or pea gravel gathered by the birds from nearby cells with other substrates. This lining of nest edges made the nests more visible.

Sand may be marginally easier for terns to make scrapes. The nest depression is formed by resting the breast on the substrate while kicking substrate out behind. Additionally, sand, being more dense and darker in colour may carry thermal advantages.

The higher density of terns nesting on sand did not lead to diminished reproductive success as both hatching and fledging success values were comparable or slightly better than the alternative substrates. In contrast, Richards and Morris (1984) found that common **㉗** terns nesting on their preferred experimental substrate, which included structures providing shelter from predators, realised significantly improved fledging success. Larger sample sizes or different environmental conditions (e.g. absence of terrestrial predators) might be required to demonstrate differences in reproductive success if they exist. Fledging success reported here was low compared with previously reported fledging success rates for Hamilton Harbour Caspian terns (Ewins et al., 1994). Our sampling techniques differed from Ewins et al. (1994). Additionally, the presence of foxes in 1994 may have resulted in temporary desertions of the colony leaving chicks exposed to ambient conditions, particularly at night when foxes are active (Southern et al., 1995). Indeed, subsequent observations in 1995 revealed

that Caspian tern adults left the colony for periods of greater than 45 min when the foxes were active in the vicinity (J. Sirdevan, pers. obs.). Parental abandonment during periods of nocturnal disturbances by foxes caused ring-billed gull chick mortality due to prolonged exposure to low ambient temperatures (Southern et al., 1985). All Caspian terns nesting on the Hamilton Harbour mainland sites in 1995 eventually abandoned their eggs, probably due to the activity of six or more foxes observed on the site. Foxes remained active in the area in 1996 and 1997 and no nesting of Caspian terns was observed on the mainland at Hamilton Harbour.

The first eggs in 1996 were laid on the edge of the sandy knoll designed for Caspian terns on sand beside the plastic sheeting. Apparently the terns were attracted to the habitat, as no decoys or other efforts had been used to entice them. The colony grew from that nucleus of nests. The 200 m² Caspian tern area was the only part of the human-made island chain with a sand substrate and represented a very small area of the island chain (Quinn et al., 1996). It was one of the only two raised areas (about 1m higher than surrounding substrate) on the islands. The other, not included in the earlier design plan (Quinn et al., 1996), was a ridge placed on the middle island to encourage herring gull nesting. While it is possible that topography played a role, it should be noted that the first nests were at the edge of the plastic at the same elevation as the rest of the island. The substantial increase in the number of pairs nesting on the site in 1997 is a very good sign suggesting that the site will continue to be used. The Caspian tern nesting activity observed on the middle island in 1997 began after hatching had begun on the crowded main Caspian tern site. It is likely that the high density and the feeding activities at nests with chicks made the main colony site less attractive to the late nesters. The site used on middle island was one of two surfaced with a smaller pea gravel and intended for common terns. While not the most preferred of substrates, smaller-sized pea gravel was shown to be acceptable to Caspian terns in our experiment.

**㉘** This study shows a statistical preference for nesting on sand by Caspian terns. Additionally, we found that terns nesting at greater densities on sand were not disadvantaged reproductively compared with those nesting on other substrates. We concluded that construction grade sand is a suitable substrate for attracting nesting Caspian terns and recommended its use, with a small addition of pea gravel for nest lining, in a habitat creation project. The sandy raised knoll on the Northern end of the Northern-most wildlife island in Hamilton Harbour (Quinn et al., 1996) has attracted most of the Caspian terns nesting in Hamilton Harbour and this colony has grown in numbers from 1996 to 1997. The application of the results of our experiment to this habitat creation project has been a great success.

`See Photo D.`

Photo C The three types of substrates tested for nesting.

Photo D A new colony of Caspian terns on a human-made wildlife island. This site was designed for the terns based on the results of the experiment. The covers are chick shelters, which provide protection from weather or predators and provide a place into which the nestlings tend to settle when researchers perform nest checks.

**㉕ DISCUSSION.** The focus in this section is the interpretation of the data; that is, the investigators make sense of the data. They compare findings with other research, propose explanations, note discrepancies they may have found, and draw conclusions. Scientists compare their results with their hypotheses to determine if the data support or do not support the hypotheses. Sometimes researchers then make suggestions for improvements in their procedures if they are to continue study in this area and make recommendations for further study.

**㉖** The investigators discuss possible explanations for the terns' preference for nesting on sand.

**㉗** The investigators compare their finding with those of other researchers.

**㉘** The investigators draw conclusions from their study.

*J.S. Quinn, J. Sirdevan/Biological Conservation 85 (1998) 63-68*

**(29)** ACKNOWLEDGEMENTS. In this section, the authors of the paper (the investigators) give credit to all who helped them conduct their study or reviewed their paper before submission to the journal. Sources of funding for the study are usually noted here as well. They are noted in this paper.

**(30)** REFERENCES. The reference section lists all the books, papers, and journal articles that are listed in the paper. They are listed alphabetically by author last name. Authors of scientific papers give references whenever they cite prior work or make statements based on prior work. In this way, the reader can access the original source of that information.

**(29)** ### Acknowledgements

The authors thank John Hall and the Hamilton Harbour Remedial Action Plan team for getting us involved in this research and for incorporating our findings and ideas into their plans. They gratefully acknowledge the reliable assistance of Cynthia Pikerik, who prevented nesting by ring-billed gulls on the study site. Robert Dawson, Sarah Hopkin, Andrea Kirkwood, Vanessa Lougheed, Brent Murray, Joe Minor, Salmon Cuso, Cynthia Pekaric, Jennifer Startek, and Carole Yauk worked hard to help construct the substrate experimental grid for peanuts (and beer). Dedicated field assistance was provided by Carrie Rongits in 1994, Cynthia Pekaric and Angelo Nicassio in 1996 and Cheryl Fink and Cindy Anderson in 1997. The Hamilton Harbour Commissioners kindly provided access to their property and permission to construct the substrate grid. The authors also thank Chip Weseloh (CWS) for providing the services of Cynthia Peraric and the Canadian Wildlife Service for permission to carry on our studies of the colonial nesters. Finally, they are pleased to acknowledge the McMaster Eco-Research Project (MERP) for logistical and financial support. Funding for MERP was provided courtesy of the Tri-council Green Plan. Ralph D. Morris and two anonymous reviewers provided valuable comments on earlier versions of this paper.

68

**(30)** ### References

Blokpoel, H., Scharf, W.C., 1991. Status and conservation of seabirds nesting in the Great Lakes of North America. ICBP Technical Publication 11, pp. 17-41.

Burness, G.P., Morris, R.D., 1992. Shelters decrease gull predation on chicks at a common tern colony. Journal of Field Ornithology 63, 186-189.

Chaniot, G.E. Jr., 1970. Notes on color variation in downy Caspian terns. Condor 72, 460-465.

Dobos, R.Z., Struger, J., Blokpoel, H., Weseloh, D.V., 1988. The status of colonial waterbirds nesting at Hamilton Harbour, Lake Ontario, 1959-1987. Ontario Birds 6, 51-60.

Ewins, P.J., Weseloh, D.V., Norstrom, R.J., Legierse, K., Auman, H.J., Ludwig, J.P., 1994. Caspian terns on the Great Lakes: organochlorine contamination, reproduction, diet and population changes. Canadian Wildlife Service, Occasional paper No. 85.

Hays, W.L., 1981. Statistics. 3rd ed. Holt, Rinehart, and Winston, New York.

Lampman, K.P., Taylor, M.E., Blokpoel, H., 1996. Caspian terns, Sterna caspia, breed successfully on a nesting raft. Colonial Waterbirds 19, 135-138.

Moore, D.J., Blokpoel, H., Lampman, K.P., Weseloh, D.V., 1995. Status, ecology, and management of colonial waterbirds nesting in Hamilton Harbour, Lake Ontario, 1988-1994. Technical Report Series No. 213, Canadian Wildlife Service, Ontario Region.

Morris, R.D., Blokpoel, H., Tessier, G.D., 1992. Management efforts for the conservation of common tern Sterna hirundo colonies in the Great Lakes: two case histories. Biological Conservation 60, 7-14.

Peck, G.K., James, R.D., 1983. Breeding Birds of Ontario: Nidology and Distribution, Vol. 1: Nonpasserines. Royal Ontario Museum Life Science, Toronto.

Quinn, J.S., 1984. Egg predation reduced by nest covers during researcher activities in a Caspian tern colony. Colonial Waterbirds 7, 419-151.

Quinn, J.S., Morris, R.D., Blokpoel, H., Weseloh, D.V., Ewins, P.J., 1996. Design and management of bird nesting habitat: tactics for conserving colonial waterbird biodiversity on artificial islands in Hamilton Harbour Ontario. Canadian Journal of Fish and Aquatic Sciences 53 (Suppl.), 44-56.

Richards, M.R., Morris, R.D., 1984. An experimental study of nest site selection in common terns. Journal of Field Ornithology 55, 457-466.

Siegal, S., 1988. Nonparametric Statistics for Behavioural Sciences, 2nd ed. McGraw-Hill, New York.

Southern, W.E., Patton, S.R., Southern, L.K., Hanners, L., 1985. Effects of nine years of fox predation on two species of breeding gulls. Auk 102, 827-833.

Reprinted from *Biological Conservation*, Vol. 85, 1998, Quinn, J.S., Sirdevan, J. *Experimental measurement of nesting substrate preference in Caspian terns*, Sterna caspia, *and the successful colonisation of human-constructed islands*, pp. 63-68, 1998, with permission from Elsevier Science. Photographs courtesy of James S. Quinn, Ph.D.

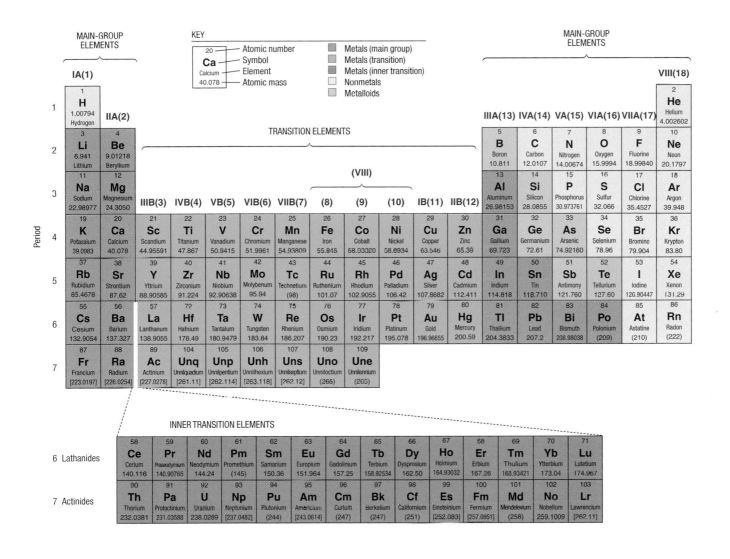

KEY

| 20 | Atomic number |
| **Ca** | Symbol |
| Calcium | Element |
| 40.078 | Atomic mass |

- Metals (main group)
- Metals (transition)
- Metals (inner transition)
- Nonmetals
- Metalloids

MAIN-GROUP ELEMENTS

MAIN-GROUP ELEMENTS

TRANSITION ELEMENTS

**IA(1)**

| 1 | | | |
| H | | | |
| 1.00794 | **IIA(2)** | | |
| Hydrogen | | | |

(VIII)

**VIII(18)**

| 2 |
| **He** |
| Helium |
| 4.002602 |

**IIIA(13) IVA(14) VA(15) VIA(16) VIIA(17)**

Period 1

| 3 | 4 |
| **Li** | **Be** |
| 6.941 | 9.01218 |
| Lithium | Beryllium |

| 5 | 6 | 7 | 8 | 9 | 10 |
| **B** | **C** | **N** | **O** | **F** | **Ne** |
| Boron | Carbon | Nitrogen | Oxygen | Fluorine | Neon |
| 10.811 | 12.0107 | 14.00674 | 15.9994 | 18.99840 | 20.1797 |

Period 2

| 11 | 12 |
| **Na** | **Mg** |
| Sodium | Magnesium |
| 22.98977 | 24.3050 |

**IIIB(3) IVB(4) VB(5) VIB(6) VIIB(7) (8) (9) (10) IB(11) IIB(12)**

| 13 | 14 | 15 | 16 | 17 | 18 |
| **Al** | **Si** | **P** | **S** | **Cl** | **Ar** |
| Aluminum | Silicon | Phosphorus | Sulfur | Chlorine | Argon |
| 26.98153 | 28.0855 | 30.973761 | 32.066 | 35.4527 | 39.948 |

Period 3

| 19 | 20 | 21 | 22 | 23 | 24 | 25 | 26 | 27 | 28 | 29 | 30 | 31 | 32 | 33 | 34 | 35 | 36 |
| **K** | **Ca** | **Sc** | **Ti** | **V** | **Cr** | **Mn** | **Fe** | **Co** | **Ni** | **Cu** | **Zn** | **Ga** | **Ge** | **As** | **Se** | **Br** | **Kr** |
| Potassium | Calcium | Scandium | Titanium | Vanadium | Chromium | Manganese | Iron | Cobalt | Nickel | Copper | Zinc | Gallium | Germanium | Arsenic | Selenium | Bromine | Krypton |
| 39.0983 | 40.078 | 44.95591 | 47.867 | 50.9415 | 51.9961 | 54.93809 | 55.845 | 58.9332 | 58.6934 | 63.546 | 65.39 | 69.723 | 72.61 | 74.92160 | 78.96 | 79.904 | 83.80 |

Period 4

| 37 | 38 | 39 | 40 | 41 | 42 | 43 | 44 | 45 | 46 | 47 | 48 | 49 | 50 | 51 | 52 | 53 | 54 |
| **Rb** | **Sr** | **Y** | **Zr** | **Nb** | **Mo** | **Tc** | **Ru** | **Rh** | **Pd** | **Ag** | **Cd** | **In** | **Sn** | **Sb** | **Te** | **I** | **Xe** |
| Rubidium | Strontium | Yttrium | Zirconium | Niobium | Molybenum | Technetium | Ruthenium | Rhodium | Palladium | Silver | Cadmium | Indium | Tin | Antimony | Tellurium | Iodine | Xenon |
| 85.4678 | 87.62 | 88.90585 | 91.224 | 92.90638 | 95.94 | (98) | 101.07 | 102.9055 | 106.42 | 107.8682 | 112.411 | 114.818 | 118.710 | 121.760 | 127.60 | 126.90447 | 131.29 |

Period 5

| 55 | 56 | 57 | 72 | 73 | 74 | 75 | 76 | 77 | 78 | 79 | 80 | 81 | 82 | 83 | 84 | 85 | 86 |
| **Cs** | **Ba** | **La** | **Hf** | **Ta** | **W** | **Re** | **Os** | **Ir** | **Pt** | **Au** | **Hg** | **Tl** | **Pb** | **Bi** | **Po** | **At** | **Rn** |
| Cesium | Barium | Lanthanum | Hafnium | Tantalum | Tungsten | Rhenium | Osmium | Iridium | Platinum | Gold | Mercury | Thallium | Lead | Bismuth | Polonium | Astatine | Radon |
| 132.9054 | 137.327 | 138.9055 | 178.49 | 180.9479 | 183.84 | 186.207 | 190.23 | 192.217 | 195.078 | 196.96655 | 200.59 | 204.3833 | 207.2 | 208.98038 | (209) | (210) | (222) |

Period 6

| 87 | 88 | 89 | 104 | 105 | 106 | 107 | 108 | 109 |
| **Fr** | **Ra** | **Ac** | **Unq** | **Unp** | **Unh** | **Uns** | **Uno** | **Une** |
| Francium | Radium | Actinium | Unniliquadium | Unnilpentium | Unnilhexium | Unnilseptium | Unniloctium | Unnilennium |
| [223.0197] | [226.0254] | [227.0278] | [261.11] | [262.114] | [263.118] | [262.12] | (265) | (265) |

Period 7

INNER TRANSITION ELEMENTS

6 Lanthanides

| 58 | 59 | 60 | 61 | 62 | 63 | 64 | 65 | 66 | 67 | 68 | 69 | 70 | 71 |
| **Ce** | **Pr** | **Nd** | **Pm** | **Sm** | **Eu** | **Gd** | **Tb** | **Dy** | **Ho** | **Er** | **Tm** | **Yb** | **Lu** |
| Cerium | Praseodymium | Neodymium | Promethium | Samarium | Europium | Gadolinium | Terbium | Dysprosium | Holmium | Erbium | Thulium | Ytterbium | Lutetium |
| 140.116 | 140.90765 | 144.24 | (145) | 150.36 | 151.964 | 157.25 | 158.92534 | 162.50 | 164.93032 | 167.26 | 168.93421 | 173.04 | 174.967 |

7 Actinides

| 90 | 91 | 92 | 93 | 94 | 95 | 96 | 97 | 98 | 99 | 100 | 101 | 102 | 103 |
| **Th** | **Pa** | **U** | **Np** | **Pu** | **Am** | **Cm** | **Bk** | **Cf** | **Es** | **Fm** | **Md** | **No** | **Lr** |
| Thorium | Protactinium | Uranium | Neptunium | Plutonium | Americium | Curium | Berkelium | Californium | Einsteinium | Fermium | Mendelevium | Nobelium | Lawrencium |
| 232.0381 | 231.03588 | 238.0289 | [237.0482] | (244) | [243.0614] | (247) | (247) | (251) | [252.083] | [257.0951] | (258) | 259.1009 | [262.11] |

| Metric Units | Metric Equivalents | Symbols | U.S. Equivalents |
|---|---|---|---|
| **Measures of length** | | | |
| 1 kilometer | = 1000 meters | km | 0.62137 mile |
| 1 meter | = 10 decimeters or 100 centimeters | m | 39.37 inches |
| 1 decimeter | = 10 centimeters | dm | 3.937 inches |
| 1 centimeter | = 10 millimeters | cm | 0.3937 inch |
| 1 millimeter | = 1000 micrometers | mm | |
| 1 micrometer | = 1/1000 millimeter or 1000 namometers | $\mu$ | |
| 1 nanometer | = 10 angstroms or 1000 picometers | nm | No U.S. equivalent |
| 1 angstrom | = 1/10,000,000 millimeter | Å | |
| 1 picometer | = 1/1,000,000,000 millimeter | pm | |
| **Measures of volume** | | | |
| 1 cubic meter | = 1000 cubic decimeters | $m^3$ | 1.308 cubic yards |
| 1 cubic decimeter | = 1000 cubic centimeters | $dm^3$ | 0.03531 cubic foot |
| 1 cubic centimeter | = 1000 cubic millimeters or 1 milliliter | $cm^3(cc)$ | 0.06102 cubic inch |
| **Measures of capacity** | | | |
| 1 kiloliter | = 1000 liters | kl | 264.18 gallons |
| 1 liter | = 10 deciliters | L | 1.0567 quarts |
| 1 deciliter | = 100 milliliters | dl | 0.4227 cup |
| 1 milliliter | = volume of 1 gram of water at standard temperature and pressure | ml | 0.3381 ounce |
| **Measures of mass** | | | |
| 1 kilogram | = 1000 grams | kg | 2.2046 pounds |
| 1 gram | = 100 centigrams or 1000 milligrams | g | 0.0353 ounce |
| 1 centigram | = 10 milligrams | cg | 0.1543 grain |
| 1 milligram | = 1/1000 gram | mg | |
| 1 microgram | = 1/1,000,000 gram | $\mu$g | |
| 1 nanogram | = 1/1,000,000,000 gram | ng | |
| 1 picogram | = 1/1,000,000,000,000 gram | pg | |

Note that a micrometer was formerly called a micron ($\mu$), and a nanometer was formerly called a millimicron (m$\mu$).

## Comparative Temperature Scales

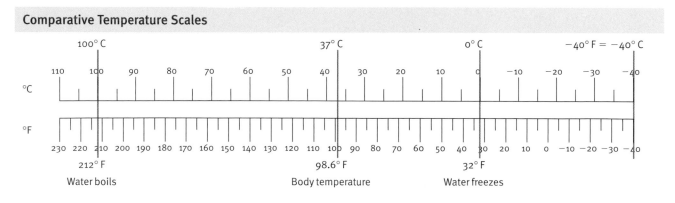

Temperature conversions:
Fahrenheit ($^\circ$ F) to Celsius ($^\circ$ C) $= \frac{5}{9} (^\circ F) - 32$
Celsius ($^\circ$ C) to Fahrenheit ($^\circ$ F) $= \frac{9}{5} (^\circ C) + 32$

# APPENDIX D: Answers

## CHAPTER 1

### Concept Checkpoint Answers

1. Cells form the tissues, organs, and organ systems of organisms.
2. populations, communities, ecosystems, and the biosphere.
3. Answers will vary, but should parallel the example of the CD cabinet, illustrating how parts, when separate entities, do not have the qualities of the whole they comprise.
4. DNA passes on information via a code, based on the sequence of its nucleotides. This code results in the formation of molecules that make up the organism.
5. Remove sunlight or producers (photosynthetic organisms) and the entire energy flow would eventually collapse. The reason is that sunlight ultimately provides the energy that supports all life, and only producers are able to harness this energy to create food. Remove consumers, and the populations of producers and decomposers would rise dramatically. Remove decomposers, and there would be no way of recycling nutrients back into other components of the ecosystem, and living organisms would eventually be enveloped in their own waste.
6. Answers will vary. An example answer is the teeth of humans. The front teeth function to bite and tear food such as meat; the form of these teeth is that they are sharp. The molars function to crush food such as plant material; these teeth are flattened. Humans eat both plant and animal foods, and our teeth are formed to chew, bite, and tear both plant and animal foods.
7. Answers will vary. An example answer is a dog and a flea. Both are made of cells, have DNA as the hereditary material, and eat food (that is, they do not make food via photosynthesis).
8. Answers will vary. For the organisms above: dogs have fur that insulates their bodies and canine teeth for tearing meat. Fleas have hard outer skeletons, which make them difficult to squash, and they can leap long distances, making it possible for them to leap onto your dog.
9. A phylum belongs to a single kingdom, so all the organisms in that phylum also are classified in that kingdom. Many families are included in a single phylum, so any two organisms in a phylum might be related closely enough to belong to the same family, but this cannot be assumed.
10. In the three-domain system, the kingdom Bacteria is split from the five-kingdom system into two domains—the Archaea and the Bacteria. All the organisms from the other four kingdoms of the five-kingdom system are classified in a single domain—the Eukarya. A good response to your classmate might be . . . "Biologists are not really debating about names when they propose new classification schemes. They are debating about relatedness among organisms. Taxonomic groupings reflect organisms' evolutionary history and evolutionary relationships."
11. All members of the domain Eukarya are eukaryotes. That is, each of their cells contains membrane-bounded organelles and a membrane-bounded nucleus.
12. **Animalia**—heterotrophs—eat other organisms. **Plantae**—autotrophs—make their own food by photosynthesis. **Protista**—autotrophs or heterotrophs. **Fungi**—heterotrophs—most are decomposers.

### Visual Thinking Answers

**Fig. 1.1** The cells look like columns, or rectangles, with central, dark circles. Some cells that have an indistinguishable shape are sandwiched in between the two layers of rectangular cells. **Fig. 1.7** Producers would no longer have energy to transform into chemical energy. They would die because they could not make their own food, and all the consumers dependent on them for food would die also. Consequently, the organisms dependent on the consumers who ate the producers would die. Populations of decomposers would likely increase because of the increase of dead material on which to feed. However, after this "feeding frenzy" was over and their food source was gone, even the decomposers would die. Life on Earth as we know it would end.

### End-of-Chapter Answers

*Level 1*

1. b. 2. c. 3. a. 4. d. 5. d. 6. T. 7. T. 8. F. 9. T. 10. F.

*Level 2*

1. The study of biology includes a wide variety of topics, such as studying hormonal changes in organisms, researching changes in populations over time, and learning how the human body protects itself against disease. You may have suggested other topics, but they should focus on the study of life.
2. Organisms have a cellular and hierarchical organization. Organisms interact with one another. Organisms are more than the sum of their parts. Organisms reproduce, passing on biological information. Organisms use and transform energy. The structures of organisms fit their functions. Organisms exhibit diversity and unity. Organisms alive today are descendants of those that lived long ago.
3. The levels are: (1) the cell (microscopic mass of protoplasm), (2) tissue (group of similar cells that work together to perform a function), (3) organ (group of tissues forming a structural and functional unit), and (4) organ system (group of organs that function together to carry out the principal activities of the organism).
4. The four levels are: (1) population (individuals of a given species that occur together at one place and time), (2) community (populations of different species that interact with each other), (3) ecosystem (community and its physical environment, including all the interactions among organisms and between organisms and the environment), and (4) biosphere (the part of the Earth where biological activity takes place).
5. Unlike all other organisms, people use many nonrenewable resources and fill the biosphere with wastes. However, scientists are trying to develop workable renewable energy sources for people to use.
6. Ultimately, most energy comes from the sun. Solar energy is captured by producers, which use photosynthesis to convert it into chemical energy. Other organisms (consumers) feed on the producers and on each other, passing the energy along. When you eat plants or animals, you are harvesting their stored energy.
7. DNA is the code of life—instructions that are capable of becoming translated into an organism. Using only four nucleotides, DNA codes for all the structural and functional components of an organism.
8. Evolution is descent with modification, which means that organisms alive today are descendants of organisms that lived long ago and that species have changed and diverged from one another over time. Fossils are the preserved remains of earlier cells and organisms; they provide a record of this change.
9. Bacteria, Archaea, and Eukarya. Humans belong to the domain Eukarya.
10. Domain, kingdom, phylum, class, order, family, genus, species.

*Level 3*

1. **a.** True. Classes are more inclusive than orders. That is, a single class contains many orders of organisms. Organisms of the same order are, by definition, of the same class, phylum, and kingdom. **b.** False. The two organisms might belong to the same genus, but this cannot be assumed. Many genera are included in a single order. **c.** False. Binomial nomenclature describes only the genus and species of an organism. These organisms may not belong to the same genus.
2. There are many characteristics of life; this textbook describes them as eight unifying themes. The scenario described in the question shows that computer-driven machines ostensibly have many of the characteristics of life; however, they do not carry out these activities in the same way as living things. For example, their "evolution" is driven by the marketplace and advances in technology, and is engineered by humans. Environmental selection pressures drive organismal evolution. Although humans may intervene to change environments, humans do not engineer organic change in the way they engineer new models of machines. Machines "reproduce" by manufacturing processes in factories, not by means of natural processes involving hereditary material. In addition, machines are not composed of cells, which are the smallest functional unit of living things. To be considered living, a machine would have to be composed of cells and carry on the basic processes of life in a naturalistic manner.
3. Answers will vary depending on the organisms chosen. However, all living things share the eight characteristics of life listed above in answer 2 of the Level 2 questions. Differences might concern such characteristics as the following: unicellular versus multicellular organisms; prokaryotes versus eukaryotes; and producers versus consumers.

4. Student answers will vary but should relate in some way to the following themes of life: ***Diversity and unity.*** Example observation: Although the different species of plants may look somewhat different from one another, there will be similarities, such as all of them having some green parts, which are important for photosynthesis (allowing the plants to make their own food). ***Hierarchical organization.*** Example observation: As the plants germinate from seeds, or the young plants mature, the development of different tissues within each plant will become apparent in the development of stems, leaves, and flowers. ***Interaction with the environment.*** Example observation: The plants may be seen to bend toward the light, depending on the availability of light in their environment. ***Transformation of energy.*** Example observation: The plants are growing, so must therefore be extracting energy and building blocks from nutrients in the soil and gases in the air (carbon dioxide). ***Forms that fit functions.*** Example observation: As the plants grow and mature, leaves develop that allow the light to hit these surfaces for photosynthesis to take place. ***Passing on of biological information.*** Example observation: Plants grown from seeds of the same species grow to look like one another and like the parent plants.

5. Viruses appear to have some of the characteristics of life, such as the ability to reproduce, based on everyday knowledge of the course of infection. They certainly interact with the environment—the human body, for example. In addition, they have biological information. However, they are not composed of cells, which is a key characteristic of life. Because they have many of the characteristics of life, some say they bridge the living and nonliving worlds. However, they lack at least one key characteristic and therefore could be considered nonliving. They are not classified by scientists as living things.

6. Rather than provide you with answers to this question, we urge you to think more about these examples as you learn more about biology in this course. Revisit this question at the end of the course and see if your answers have changed.

7. & 8. Biology-related issues include abortion, embryonic stem cell research, pollution, loss of natural habitat, large-scale extinction, global warming, genetically modified foods, overpopulation, and diseases like AIDS, mad cow, and SARS—and the list could go on and on. As you learn about the science of biology, you can use what you learn to better understand and make decisions about such issues.

# CHAPTER 2

## Concept Checkpoint Answers

1. Scientists study the natural world using a variety of scientific methods. These methods often include observation, hypothesis formation, and hypothesis testing.

2. You should gather information about your question. For example, you could observe whether any human activity is causing changes in the lake, such as building that results in soil entering the water, or farming that results in fertilizer runoff into the water. In addition, you might call your local conservation department for information. After obtaining information, you can develop your hypothesis.

3. One possible hypothesis might be: "If excess fertilizer results in conditions that lower the level of dissolved oxygen in the lake, then the fish in the lake will die."

4. One way to test this hypothesis would be to set up several large aquaria stocked with pond water, sediment, and native plants and algae from the pond in question. The aquaria would be supplied with various amounts of lawn and garden fertilizer. Another group of aquaria (the control group) would not receive any fertilizer. All other factors (temperature, light levels, light cycle, etc.) would be kept the same. Dissolved oxygen (D.O.) measurements would be taken every six hours for three weeks. D.O. levels in each aquarium would be compared to D.O. tolerance ranges of native species of fish.

5. Scientists use diverse processes for many reasons, such as: Not all sciences are based on the accumulation of observational data, and testing hypotheses by means of controlled experiments is difficult in some fields of scientific inquiry.

6. It depends on the question and your approach to answering it. First, the question would have to concern the natural world. (You can't use science to answer a question such as, "What is the meaning of life?") If you find out information about your natural-world question, develop a hypothesis with this information in mind, and test the hypothesis, then you are following a scientific approach.

7. A scientific theory is a well-substantiated explanation, not just a hunch or a guess. Theories explain existing data and consistently predict new data. Some theories are upheld over time so consistently that little doubt exists as to their validity. This is true of evolution.

8. The introduction.

9. An example is Peck and James (1983). This is background work on nesting substrate preferences of Caspian terns.

## Visual Thinking Answer

**Fig. 2.2** The cylinder at far the right has a wider diameter than the other cylinders. Therefore, the greater volume does not rise as high.

## End-of-Chapter Answers

*Level 1*

**1.** d. **2.** c. **3.** a. **4.** e. **5.** f. **6.** F. **7.** F. **8.** T. **9.** F. **10.** T

*Level 2*

1. Three key features of scientific inquiry are observation, hypothesis development, and hypothesis testing. Observation is not always present in scientific endeavor, but the hallmark of science is the testable hypothesis. Scientists test hypotheses in a variety of ways.

2. A hypothesis is a tentative explanation that guides inquiry. Scientists can only disprove hypotheses or support them with evidence. Future evidence may disprove a hypothesis that is supported with present evidence.

3. A scientific theory is a synthesis of hypotheses that are supported by so much evidence that the scientific community commonly accepts them, although conflicting theories occasionally exist. Scientific theories explain natural phenomena, whereas scientific laws describe them.

4. A variable is a factor that changes or varies. The independent variable is one that is manipulated, whereas the dependent variable changes in response to the independent variable. In a controlled experiment, researchers manipulate factors and observe other factors that change in response to the manipulated factors. Controlled variables are kept constant.

5. Quantitative data are based on numerical measurements, but qualitative data are descriptive.

6. Scientists in some disciplines find that conducting controlled experiments is difficult or impossible. Two examples are in paleontology, in which scientists hypothesize about past occurrences, and in biomedical research, in which all factors cannot be controlled when dealing with humans.

7. Using repeated trials increases the reliability of the results by reducing the effects of chance or error that may occur in a single trial.

8. An abstract is a concise summary of a paper. By reading the abstract, the reader knows whether the paper is likely to be useful for his or her purposes.

9. Scientists study the natural world and develop explanations of phenomena by means of testable hypotheses. Science does not employ psychic explanations, which depend on extrasensory perception (ESP)—the ability to acquire information without the benefit of the senses.

10. Is classical music more artistic than rock music?—**NO**. Do girls develop motor skills earlier in development than boys? **YES**. Does listening to Mozart affect learning in children?—**YES**. Is capital punishment just?—**NO**. Do nonhuman animals have self-awareness?—**YES**. How are whooping cranes able to migrate long distances?—**YES**.

11. Evidence that supports a hypothesis does not establish that further testing will also produce supporting evidence. Future testing may result in evidence that does not support or even refutes the original hypothesis. Therefore, a hypothesis can be supported with data but not proved. Nevertheless, some hypotheses are supported by such an overwhelming amount of evidence that they are accepted as well-substantiated explanations and scientifically valid statements.

*Level 3*

1. Your hypothesis could be the following: If the students walk one hour a day, four days a week, in addition to their usual activity, then they will each lose at least 5 pounds during the semester. The independent variable is the walking; the dependent variable is the change in their weights.

2. The control is the standard against which the subjects who go through the experiment's treatment would be compared. In this case, it would be a group of students who do not take the hour-long walks that the other students take.

3. Hypotheses may vary somewhat but should be in the format, "If (IV), then (DV)." For example, "If contact lens cleaner contains proteases, then it will degrade the protein in milk." You might test the hypothesis by placing increasing numbers of drops of lens cleaner on petri dishes containing milk agar. You would put no drops of lens cleaner on the control plate. Clear areas on the plates where the drops were placed would show protein degradation. Such data would support your hypothesis.

4. Deductive reasoning begins with a general statement and proceeds to a specific statement. For example, a scientist may think that fertilizer increases growth in houseplants, but would then conduct experiments to see if a particular fertilizer would increase growth in a particular houseplant. In a similar manner, a detective may know that if a married woman is found murdered in her home, the perpetrator of the crime is often her husband. The detective then conducts an investigation to see whether a particular female homicide victim was, in fact, killed by her husband.

5. Examples will vary. For example, knowledge of the biology underlying environmental issues helps people vote responsibly on environmentally related issues and also helps them incorporate environmentally sound practices into their daily routines.

# CHAPTER 3

## Concept Checkpoint Answers

1. Atoms are made up of three types of subatomic particles: protons, neutrons, and electrons. Protons and neutrons are found at the core of the atom, which is called the nucleus; electrons surround this core.
2. Protons; electrons
3. Isotopes of an atom have the same number of protons but different numbers of neutrons.
4. Molecules are atoms that interact by sharing electrons or by being attracted to one another because of opposing electrical charges.
5. Molecules made up of atoms of different elements are compounds.
6. (1) the tendency of electrons to occur in pairs, (2) the tendency of atoms to balance positive and negative charges, and (3) the tendency of the outer shell of electrons to be full
7. (b) chlorine. Radon has a complete outer shell and is unreactive. Carbon has a half-full outer shell and tends to share electrons.
8. The substance oxidized loses electrons, often to oxygen, as is the case in the rusting of iron.
9. An atom of carbon can bond with four atoms of hydrogen: Each hydrogen atom shares its one electron with the carbon atom, so the carbon atom completes its outer shell with eight electrons. The carbon atom shares one electron with each hydrogen atom, completing their outer shells with two electrons each. Similarly, a carbon atom can bond with two atoms of oxygen: Each oxygen atom shares two electrons with the carbon atom, completing its outer shell. The carbon atom shares two electrons with each oxygen atom, completing their outer shells.
10. **(a)** Water molecules "stick together" (are cohesive) because hydrogen bonds form between polar water molecules. Example: Water transport in plants. **(b)** Water molecules "stick to" (adhere to) other polar substances because hydrogen bonds form between polar water molecules and the molecules of other polar substances. Example: Water transport in plants. **(c)** The polarity of water molecules results in their surrounding and therefore dissolving other charged or polar molecules. Example: A variety of biological molecules can dissolve in water.
11. Because water ionizes spontaneously, it can act as either an acid or a base, accepting protons or donating protons. Many reactions that take place in living systems involve the transfer of a proton from an acid to a base. Since water can act as either, it plays a vital role in the chemistry of life as either a reactant or a product.
12. pH indicates the concentration of $H^+$ ions in a solution.
13. Low pH values indicate high hydrogen ion concentrations in the acid pH range. High pH values indicate low hydrogen ion concentrations in the alkaline pH range.

## Visual Thinking Answers

**Fig. 3.5** The arrow is showing that the sodium atom is losing an electron to the chlorine atom, changing those atoms to ions. **Fig. 3.8** You likely chose the shell model, for it shows electron configurations as well as the number of positive charges in the nucleus of each atom, visually showing the "pull" the oxygen atom has for the electrons of the hydrogen atoms. The space-filling model shows the space occupied by the electrons of each atom but does not provide as much information as the shell model. **Fig. 3.10** No. A nonpolar liquid, such as oil, does not have attractive forces to separate the salt ions. The attractive forces between the ions will keep the salt crystals together.

## End-of-Chapter Answers

*Level 1*

1. c.  2. a.  3. a.  4. d.  5. a.  6. b.  7. d.  8. c.  9. a.  10. e.

*Level 2*

1. **a.** See Figure 3.1. Your atom would be hydrogen and would have one proton, one electron, and from zero to two neutrons. **b.** The atom would still have one proton and one electron, but it would have a different number of neutrons from the atom in question 1.a., within the range mentioned.
2. Three factors that influence how an atom interacts with other atoms are: (1) the tendency of electrons to occur in pairs, (2) the tendency of atoms to balance positive and negative charges, and (3) the tendency of the outer shell (energy level) of electrons to be full. The third point is known as the octet rule because an atom with an unfilled outer shell tends to interact with other atoms to fill this outer shell. For many atoms, eight (an octet of) electrons fill the outer shell.
3. The atoms in nitrogen gas are bound together by covalent bonds, as are the hydrogen and oxygen atoms of water molecules. Both involve the sharing of electrons between molecules. However, nitrogen gas molecules have triple bonds, which are stronger bonds than the single covalent bonds in water. The bonds are different because the bond between the H and the O of water is a polar covalent bond, whereas that between the two nitrogen atoms is a nonpolar covalent bond.
4. Water is unusual because it is the only common molecule on Earth that exists as a liquid at the Earth's surface. This liquid enables other molecules dissolved or suspended in it to move and interact. A second important characteristic is its ability to form hydrogen bonds with itself and other molecules because it is a polar molecule. Water molecules are strongly attracted to ions and other polar molecules, giving water another important trait: It is an excellent solvent. Chemical interactions readily take place in water because so many molecules are water soluble.
5. This environment is slightly basic. Like all basic environments, the intestinal environment would have a relatively low level of free $H^+$ ions.
6. Water is a polar molecule. It quickly and easily forms hydrogen bonds, which, though not very strong, allow water to react with many other molecules.

*Level 3*

1. Oil is a nonpolar molecule, which means that it does not form hydrogen bonds with water and therefore does not dissolve. Answers will vary as to why an oil spill should be cleaned up, but one reason based on its nonpolarity is that the oil will form a film on the top of the water, affecting the amount of light entering the water. Lowered light levels would affect aquatic plant life. In addition, organisms that come to the surface of the water would get coated with this film of oil, which would probably affect their ability to survive.
2. Testing kits are available to test for radon levels in the home. Elevated radon levels are 4 picoCuries per liter (pCi/L) or higher. If such levels are found, the homeowner needs to find a qualified contractor to reduce the radon levels in the home and help the homeowner maintain lowered levels. The Environmental Protection Agency (www.epa.gov) has information on radon testing.
3. Acid precipitation is about 10 times more acidic than unpolluted rainwater. Reasons will vary why students think that acid precipitation might be harmful to living things. For discussion of this question, see Chapter 42.
4. NASA chose this site and designed the probe to accomplish such a goal because wherever we find life on Earth, we find water. Conversely, evidence of water on Mars in the past is evidence that life may have existed there. The chemical and physical properties of water make it a suitable medium in which the chemical reactions of life can occur.

# CHAPTER 4

## Concept Checkpoint Answers

1. No, because organic chemists study the compounds of life. Rocks and soil are not made up of these carbon-containing compounds.
2. the carbon core, or skeleton, and functional groups
3. carbohydrates, lipids, proteins, nucleic acids
4. Dehydration means to take out water. Synthesis means to build. Therefore, dehydration synthesis means to build by taking out water. Hydro means water. Lysis means to break. Therefore, hydrolysis means to break with water.
5. ready energy and structure
6. Monosaccharides are used to build carbohydrates. The general name for carbohydrate polymers is polysaccharides. An example of a monomer is glucose, and an example of a polymer is cellulose.
7. All fats are lipids, but not all lipids are fats.
8. Chicken, because it has the lowest percentage of saturated fat of the four foods listed.
9. Phospholipids are found in cell membranes, and sex hormones are steroids.
10. Letters = amino acids; words = polypeptides; sentences = proteins composed of multiple polypeptides (quaternary structure). The meaning of words and sentences could be analogous to the function of the protein.
11. Proteins recognize and interact with a wide array of other molecules. The specific shapes of proteins give them these capabilities. If the protein does not assume its correct shape, it cannot do these jobs.
12. Letters = nucleotides; words/sentences = genes. The meaning of words and sentences could be analogous to the protein the gene encodes.
13. DNA stores the information for making proteins, and RNA directs the production of proteins.

## Visual Thinking Answers

**Fig. 4.2** Hydroxyl, amino, sulfhydryl, and phosphate.   **Fig. 4.3** For the dehydration portion of the explanation, point to the water molecules being given off at the junction of two monomers being joined. For the synthesis portion of the explanation, point to the highlighted new bonds that join the monomers.   **Fig. 4.11**

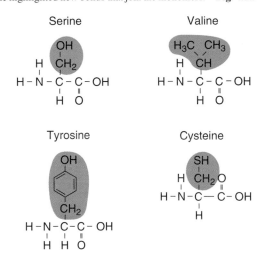

## End-of-Chapter Answers

*Level 1*

**1.** b.   **2.** a.   **3.** d.   **4.** b.   **5.** c.   **6.** T.   **7.** T.   **8.** F.   **9.** F.   **10.** T.

*Level 2*

1. Organic molecules tend to be large, covalently bonded, and carbon based. Inorganic molecules tend to be small, they do not usually contain carbon, and they interact by means of ionic bonding. If you wanted to learn about living things—such as humans—you would primarily study organic molecules.

2. Monosaccharides, disaccharides, and polysaccharides are all carbohydrates. Most organisms use carbohydrates as a primary fuel. Monosaccharides are among the least complex carbohydrates. Many organisms link monosaccharides to form disaccharides that are less readily broken down as they are transported within the organism. To store the energy from carbohydrates, organisms convert monosaccharides and disaccharides into polysaccharides, long insoluble polymers of sugars.
3. Plants and animals must store energy as insoluble polysaccharide compounds, and glucose is a very soluble sugar in either blood or water. Plants generally store sugar as starch, whereas animals store it as glycogen.
4. The three classes of macromolecules taken in as food energy by humans are carbohydrates, fats, and proteins. All three are made up of smaller units that, when disassembled, release energy. Because these molecules were all put together by dehydration synthesis, breaking them apart (digesting them) entails hydrolysis, an opposite process.
5. DNA is often called the code of life because it stores the information for making proteins in the sequence of its bases. Proteins play diverse roles in living things, including making up the bodies of organisms and speeding up chemical reactions.
6. The sequence of amino acids of a protein is its primary structure. Bonds form between amino acids in the chain. Thus, the sequence of amino acids determines the bonds that will form and the subsequent shape the protein will take.

*Level 3*

1. The macromolecules present in this food are lipids, carbohydrates, and proteins. This food could be a part of a heart-healthy diet because it supplies only 17% of its calories from fat. In addition, the amount of saturated fats as compared with the total fat is low—30%.
2. $C_{51}H_{68}O_6$. To obtain the formula, add the number of carbon, hydrogen, and oxygen atoms in the three fatty acids. To this, add the number of carbon, hydrogen, and oxygen atoms in the glycerol. Then subtract the number of hydrogen and oxygen atoms in three molecules of water, which would be released during the dehydration synthesis reaction that would bond the three fatty acids to the glycerol.
3. The triglyceride would be unsaturated. The fatty acid tails, which each contain 15 carbon atoms, have only 22 hydrogen atoms each. Saturated, each fatty acid tail would have 32 carbon atoms. So the triglyceride in the above question must have fatty acid tails with multiple double bonds between the carbon atoms.
4. In living systems, proteins are used as enzymes and as structural components. Because DNA codes for the sequence of amino acids in a protein, an error in DNA can result in an error in the sequence of amino acids in a protein. This sequence is called the primary structure of a protein, and it affects the formation of the secondary and tertiary levels of structure. If a protein does not have the correct sequence of amino acids and is not configured correctly, it cannot perform its job.
5. When each of these molecules is broken down in the body, the fats yield more energy due to the high number of energy-storing carbon-hydrogen bonds as compared to carbohydrates and proteins.
6. Answers will vary. Students may focus on the total number of calories as most important, grams of fat, or grams of sugar. Students involved in physical training and sports may consider the grams of protein as most important.

# CHAPTER 5

## Concept Checkpoint Answers

1. Here is an example of a corrected essay: A small cell has a *larger* surface area to volume ratio than a large cell, which *enhances* the uptake of molecules required for metabolism and the removal of accumulated wastes.
2. Mitochondria and chloroplasts are thought to have originated through endosymbiosis— an ancient larger pre-eukaryotic (prokaryotic cell) engulfing a smaller prokaryote. Today these bacterial endosymbionts are mitochondria and chloroplasts, and have retained DNA within their structures originally found in their prokaryotic ancestors.
3. Here are some examples of analogies. Your analogies may of course be different; check with your instructor to see if they are accurate. Membrane channel (or transport) proteins could be likened to a mouth. Membrane transport proteins are the "doors" through which molecules move into and out of a cell. A mouth is the "door" through which food moves into (and sometimes out of) the body.—The phospholipid bilayer of a cell could be likened to skin. The phos-

pholipid bilayer encloses/surrounds the cell; the skin encloses/surrounds the body.—Glycoproteins on the cell membrane surface can be likened to a face. Cells recognize one another by the structure of glycoproteins protruding from their surfaces. People recognize one another by the structures of their faces.

4. **a.** Engineering—analogous to the nucleus, where DNA acts as the blueprint for building proteins in the cell; **b.** Manufacturing—analogous to ribosomes, which are organelles that construct proteins; **c.** Waste—analogous to vacuoles, which store cellular waste products; **d.** Recycling center—analogous to lysosomes, which engulf and digest organic molecules and damaged organelles and then recycle their component parts; **e.** Sorting division—analogous to the endoplasmic reticulum, which receives proteins from the ribosomes, sorts and tags them, and packages them in membrane bound vesicles for transport to other organelles; **f.** Central receiving—analogous to the Golgi apparatus, which receives proteins from the ER and tags them with molecular labels that indicate their ultimate intracellular or extracellular destinations; **g.** Power plant—analogous to the mitochondria, which harness the energy from food molecules and supply the energy for all other processes that occur in the cell.

5. Widgets are analogous to proteins.

6. The blood cells became swollen with water and burst because the water is hypotonic relative to the cytoplasm of the red blood cells. The result is a net flow of water into the blood cells, which causes the osmotic pressure inside the cells to rise. Since the plasma membrane of these cells is relatively fluid and these animal cells lack a supportive, rigid cell wall, the rise in pressure ultimately causes the cells to rupture.

7. Because plant cells have a rigid cell wall surrounding their plasma membrane, the rise in osmotic pressure in the cells will cause the cells to swell and become turgid (filled with water), but the cells will not rupture.

8. **Active transport—(a)** Low to high concentration. **(b)** Small inorganic molecules like ions or larger organic molecules like sugars. **(c)** Example—$Na^+/K^+$ exchange. **Passive transport (diffusion or osmosis)—(a)** Low to high concentration. **(b)** Small molecules that diffuse across the lipid bilayer or through membrane channel proteins. **(c)** Example—$H_2O$, $O_2$, $CO_2$, some ions like $H^+$ ions. **Facilitated diffusion—(a)** High to low concentration. **(b)** Organic molecules required by the cell. **(c)** Example—Proteins, lipids (fatty acids). **Pinocytosis—(a)** Either direction. **(b)** Small organic or inorganic molecules dissolved in solution. **(c)** Example—nutrients taken up by the maturing egg cell. **Exocytosis—(a)** High to low concentration. **(b)** Organic molecules secreted by a cell. **(c)** Example—proteins such as digestive enzymes or hormones.

## Visual Thinking Answers

**Fig. 5.4** The illustration shows that bacteria (and other prokaryotes) have ribosomes in their cytoplasm. Ribosomes are the site of protein synthesis. **Fig. 5.8** The lipid bilayer has an inner core of nonpolar molecules. Highly charged particles such as ions do not interact with uncharged (nonpolar) molecules and so are repelled by the lipid bilayer. **Fig. 5.26** Energy for cell processes is derived from the breakdown of ATP to ADP and P. In (1), the illustration shows ATP near the sodium–potassium pump. The arrows emerging from the ATP molecule depict the splitting of ATP into ADP and P (black arrow) with the energy (orange arrow) being used by the pump.

## End-of-Chapter Answers

*Level 1*

**1.** c.  **2.** d.  **3.** c.  **4.** b.  **5.** F.  **6.** T.  **7.** T.  **8.** F.  **9.** F.

*Level 2*

1. They both involve passive transport of solute down a concentration gradient across a membrane; however, facilitated diffusion involves the use of membrane proteins, which facilitates the passive diffusion of specific solutes that bind to the protein.

2. The protein is produced by a ribosome bound to the rough ER. The protein then enters the rough ER, where it is modified and chemically tagged. It is then packaged into small membranous vesicles that bud off the ER and travel to the Golgi apparatus. Here the protein is further chemically tagged, sorted, and packaged into vesicles that will transport the protein to the plasma membrane, where its contents will be secreted from the cell through exocytosis.

3. Your diagram should show a cell surrounding a particle or large molecules with extensions or a vesicle fusing with the outer side of the plasma membrane, and

then spilling its contents inside the cell or pinching off a vesicle that moves into the cell. An example might be white blood cells engulfing foreign pathogens or damaged cells.

4. Exocytosis. Large molecules such as proteins cannot move through the lipid bilayer without being packaged in a vesicle that then fuses with the cell membrane for discharge from the cell.

5. The structure of the lipid bilayer, in part, makes the membrane differentially permeable. The hydrophobic interior of the bilayer, for example, prevents large polar or charged molecules from freely passing through the membrane. However, small nonpolar molecules like oxygen can freely diffuse across the membrane. Various proteins embedded in this membrane also regulate the transport of molecules across the membrane. For example, active transport proteins and proteins involved in facilitated diffusion only allow specific molecules to traverse the membrane. Differential permeability is essential for living cells because it allows cells to maintain an internal chemical environment that is favorable to the metabolic processes that sustain life.

6. **a.** Facilitated diffusion; passive. **b.** Diffusion; passive; solute = coffee crystals, solvent = water, solution = coffee. **c.** Active transport; active.

*Level 3*

1. A single nucleus can effectively control only a small cellular volume. Therefore, a large cell, such as a muscle cell, has numerous nuclei to coordinate the production of essential muscle proteins throughout this large volume.

2. See Table 5.2 for a summary of the differences. For example, the cell with chloroplasts would be the plant, the cell with mitochondria but no chloroplasts would be the animal, and the one with neither mitochondria nor chloroplasts would be the bacterium. The bacterium would be much smaller than the other two and would not have a membrane-bound nucleus or other membrane-bound intracellular organelles. Plant cells would have a large number of vacuoles and/or a large central vacuole that takes up most of the volume of the cell.

3. These protozoans live in a watery, hypotonic environment. That is, the solute concentration of the water in which they live is lower than that of their internal environment. Because of this difference in solute concentrations, water passively diffuses into the protozoan. Contractile vacuoles pump out the excess water acquired through osmosis.

4. Protein synthesis

## CHAPTER 6

## Concept Checkpoint Answers

1. **a.** Exergonic—energy releasing. Energy stored in the bonds of the tree's organic molecules is released as heat. **b.** Increases the entropy as the breakdown of complex organic molecules of the tree releases heat to the surroundings. **c.** The lightning provided the initial energy to start the reactions. The energy subsequently released from these initial reactions provided the activation energy for a cascade of future reactions. **d.** The reactants have higher energy. However, this energy was not lost or destroyed, but was simply converted to heat.

2. Some characteristics of enzymes: Most are proteins. They speed up (catalyze) biochemical reactions. They lower the energy of activation. They help bring substrates together, facilitating reactions. They have active sites where substrates bind. They may be altered during the course of a reaction, but in the end they revert to their original shape. They are reusable.

3. Temperature, ph, and chemicals

4. End-product inhibition in enzyme-catalyzed processes is a type of negative feedback: The end product feeds information back to the first enzyme in the pathway, shutting down the pathway when an additional end product is not needed. Similarly, a thermostat works by negative feedback: The end product feeds information back to the thermostat (analogous to the first enzyme in the pathway), which shuts down the furnace (the pathway) when the sufficient level of heat (end product) is reached and more heat (additional end product) is not needed.

5. Many of the trace elements in your diet act as cofactors, and many vitamins are used to synthesize coenzymes; cofactors and coenzymes both help enzymes catalyze reactions.

6. by eating food
7. the sun
8. Some of the things you may have listed are: contracting of heart muscle, which pumps blood; transporting molecules across a cell membrane via active transport; generating nerve impulses, which allows us to see, hear, smell, and read and understand this textbook; and synthesizing organic polymers such as polysaccharides, polypeptides, and DNA. All of these endergonic processes require energy from ATP.

## Visual Thinking Answers

**Fig. 6.4** The "excess" energy would be released as heat energy.    **Fig. 6.5** The difference is shown by the heights of the energy "hills" that must be climbed for each type of reaction. The difference in activation energy needed is the difference in the heights of the "hills."    **Fig. 6.7** Prior to binding the substrate, the enzyme's active site is shown as two curved depressions. After binding the substrate, these depressions in the enzyme conform to the shape of the substrate.    **Fig. 6.9** About 37°C—normal body temperature.    **Fig. 6.13** The substrates are ATP and Compound A. The products are ADP and phosphorylated Compound A.

## End-of-Chapter Answers

*Level 1*

**1.** d.   **2.** d.   **3.** d.   **4.** b.   **5.** d.   **6.** d.   **7.** c.   **8.** T.   **9.** F.   **10.** T.   **11.** T.

*Level 2*

1. **a.** substrates. **b.** product. **c.** enzymes. **d.** cofactors or coenzymes.
2. The "leftover" energy would be released as heat.
3. Thousands of biochemical reactions occur in the body each second, and each individual reaction requires a specific enzyme to catalyze it.
4. Food molecules, such as glucose (sugar), store considerable amounts of chemical energy, just as $500 bills represent a considerable amount of money. It would be very cumbersome to do your day-to-day transactions with $500 bills, just as it would be inefficient to "burn" a whole glucose molecule each time the cell needed to do some chemical work. So just as you might make change for $500 bills to conduct daily monetary transactions, your cells "make change" for glucose molecules by creating numerous ATP molecules that store small amounts of "useable" energy.
5. Both are molecules that will interact with enzymes to regulate a biochemical reaction (speed it up or slow it down). Inhibitors are the ones that play a key role in negative feedback.
6. Metabolism is the total of all chemical reactions that occur in a living organism.
7. "A" represents the activation energy. This is an overall exergonic reaction because the products have less energy than the reactants. In the presence of an enzyme, the activation energy would be represented by a lower hill, as shown in the following diagram.

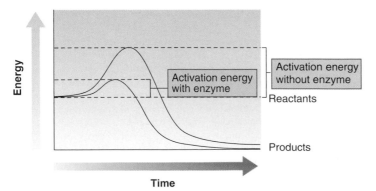

*Level 3*

1. "No, this is a misconception. Living organisms are not closed systems. Energy flows into the system, increasing its order. Living organisms create increased order within themselves at the expense of increased entropy in the surrounding universe."

2. This defect would likely affect endergonic chemical reactions in the organism. Depending on the severity of the defect, the genetic disorder could be fatal.
3. As temperature increases, enzyme activity correspondingly increases. Eventually, an optimal temperature is reached at which enzyme activity is at a maximum. This optimal temperature is probably about 37°C. Beyond this point, the enzyme begins to denature. One way of measuring enzyme activity would be to measure the change in the concentration of product molecules over time.
4. Photosynthetic organisms capture the sun's energy and convert it into the chemical energy stored in the bonds of organic molecules (sugars). Other living organisms eat plants (or other photosynthetic organisms) and/or are eaten by other organisms. When these organisms die, decomposers like bacteria and fungi derive energy from this dead organic material. In this way, the energy from the sun "flows" through life on Earth.
5. Most enzymes function best within a narrow range of pH. Water and ions enter the cells of the plants from their environment, possibly changing the internal environment of the plants. If the plant's internal environment becomes too acidic, enzymes could be affected, causing chemical reactions to stop, which ultimately would affect the metabolic pathways of organisms.

# CHAPTER 7

## Concept Checkpoint Answers

1. They are called pigments because they absorb some visible wavelengths of light and transmit or reflect others.
2. The antenna on your radio picks up a variety of radio signals. Photosynthetic pigments such as chlorophylls and carotenoids are much like antennas in that they allow plants to absorb and get energy from a variety of different wavelengths of light.
3. Plants are *autotrophs*, using energy from the *sun* to manufacture *glucose*, with *oxygen* as a byproduct. Animals are *heterotrophs*; they consume plants (or other organisms that consume plants). By the process of *cellular respiration*, animals unlock the energy stored in their food to make *ATP*, with *water* and *carbon dioxide* (the raw materials of photosynthesis) as byproducts.
4. **L**ose **E**lectron **O**xidation, **G**ain **E**lectron **R**eduction
5. ATP and NADPH
6. $CO_2$
7. The $H^+$ ions build up inside thylakoids, which is analogous to the buildup of water on the lake side of a dam. As $H^+$ ions flow down through ATP synthase, just as water flows over a dam, this energy is converted into another form. In a dam, the flow of water is converted into electrical energy. During chemiosmosis, $H^+$ ion flow is converted into chemical energy stored in ATP.
8. NADPH; high-energy electrons plus $H^+$ ions.

## Visual Thinking Answers

**Fig. 7.3** The leaves are the photosynthesizing parts of the plant. Chlorophylls, found in the leaves and used by green plants for photosynthesis, reflect mostly green light and absorb other wavelengths.    **Fig. 7.4** Photosynthesis evolved first. Autotrophs were able to make their own food and produced oxygen as a byproduct. Cellular respiration uses oxygen and is not possible without the availability of this gas.

## End-of-Chapter Answers

*Level 1*

**1.** b.   **2.** e.   **3.** a.   **4.** a.   **5.** e.   **6.** a and b.   **7.** c.   **8.** a, b, and c.   **9.** b.   **10.** b.   **11.** c.

*Level 2*

1. d—Remember that plants undergo both cellular respiration and photosynthesis.
2. Fungi do not belong because they are heterotrophs; all of the others are autotrophs. You also could have chosen cyanobacteria because they are the only prokaryotes in the group.

3. After a few days, none could occur except cellular respiration, which would continue as long as the plant produced ATP from its carbohydrate and fat (oil) reserves. Both cyclic and noncyclic electron flow would stop immediately in the absence of light. Although the Calvin cycle does not use light directly, it requires ATP and NADPH, which are produced during the light-dependent reactions and which would run out very quickly once the plant is placed in the dark.

4. $CO_2$ is being reduced to sugars, and water is being oxidized, resulting in the liberation of waste $O_2$. The former occurs during the Calvin cycle; the latter occurs during the light-dependent reactions. The oxidation of water provides the electrons to noncyclic electron flow.

5. light energy → electron flow → proton ($H^+$ ion) gradient → chemical energy stored in ATP and NADPH → chemical energy stored in sugars

*Level 3*

1. Mostly from $CO_2$, but some from plant nutrients that are taken up from the soil by the roots. As the plant grows, more water is taken up by the plant and held in its tissues, which would be reflected in its weight.

2. The roots do not manufacture ATP via photosynthesis but via cellular respiration. Sucrose would be transported to the roots from the photosynthetic parts of the plant, where it would be broken down during respiration, yielding ATP.

3. Deforestation removes vast numbers of huge trees, which are capable of taking up large amounts of $CO_2$ from the atmosphere and turning it into organic molecules via photosynthesis.

4. Calvin cycle

5. Since green cellophane transmits only green light to the plant, and plants do not use much green light in photosynthesis, you would expect the rate of photosynthesis to diminish drastically. With blue cellophane, you would not expect the rate of photosynthesis to diminish as drastically because plants use blue light extensively in photosynthesis. However, the rate of photosynthesis would drop a bit because other colors of light that plants use are being excluded.

6. There is evidence that this cataclysmic event was responsible for the extinction of many groups of organisms including the dinosaurs. The dust thrown into the atmosphere from the impact clouded out the sun and thereby may have reduced the levels of photosynthesis dramatically. The decreased food production by photosynthetic organisms may have resulted in the collapse of food chains.

# CHAPTER 8

## Concept Checkpoint Answers

1. **Plants**—sunlight; photosynthesis and cellular respiration (plants do both); aerobic (plants use oxygen in cellular respiration); plants produce $CO_2$ during cellular respiration but also take it up in photosynthesis. **Animals**—organic molecules such as sugars and fats; cellular respiration and fermentation (animal cells can use oxygen to make ATP, but they can also use fermentation—e.g., in muscle cells); both aerobic and anaerobic (some cells in animals undergo fermentation when deprived of $O_2$); animals produce $CO_2$ as a waste product of cellular respiration. **Baker's yeast**—organic molecules such as sugars; fermentation (these organisms don't use oxygen to make ATP from the food they eat); anaerobic (fermentation does not use $O_2$); as yeasts ferment sugars they produce $CO_2$ as a waste product. **Chlamydia**—ATP made by other organisms; these organisms live off the ATP they acquire from the cells they infect, so they don't use cellular respiration or fermentation.

2. **Molecules that enter into the reactions of glycolysis:** glucose (sugar); ATP; ADP; $NAD^+$ (electron carrier); $P_i$ (inorganic phosphate) **Molecules that are products of the reactions of glycolysis:** pyruvate, which will be used in reactions that produce acetyl-CoA for the Krebs cycle; ADP, which will be used again in glycolysis; ATP, which will provide energy to do cellular work, NADH, which will shuttle electrons to the electron transport chain.

3. **Molecules that enter into the reactions of the Krebs cycle:** acetyl-CoA; ADP (GDP); $P_i$; FAD; $NAD^+$; $H_2O$. **Molecules that are products of the reactions of the Krebs cycle:** $CO_2$, which is waste and will be released from the cell; ATP, which is used to do cellular work; NADH, which will shuttle electrons to the electron transport chain; $FADH_2$, which will shuttle electrons to the electron transport chain.

4.

| | ATP | NADH | FADH$_2$ |
|---|---|---|---|
| Glycolysis | 2 | 2 | 0 |
| Krebs cycle | 2 | 8 | 2 |
| Total net gain | 4 | 10 | 2 |

5. In addition to some energy stored within the bonds of $CO_2$ (a metabolic waste product), some energy is released as heat.

6. The pH would be more acidic in the outer mitochondrial compartment because $H^+$ ions (protons) are continually pumped there as energy is released from the electrons being passed along the electron transport chain. This proton gradient is maintained as long as oxygen and electrons are being supplied to the electron transport chains.

7. NAD and FAD return to the Krebs cycle, where they are "recycled" to NADH and $FADH_2$, and some NAD returns to the glycolytic pathway and is "recycled" to NADH.

8. The energy is stored in 36 ATP molecules and 6 waste $CO_2$ molecules. The rest is released as heat.

9. The energy is stored in 2 ATP molecules and 2 lactic acid molecules. The rest is released as heat. Once oxygen supply to the muscle is restored, the lactic acid will be metabolized aerobically.

## Visual Thinking Answers

**Fig. 8.2** Gluose is the fuel of cellular respiration; energy is extracted from this molecule and is ultimately stored in molecules of ATP. Oxygen ($O_2$) serves as the terminal electron acceptor, the last compound to accept electrons in the reactions of cellular respiration. The carbon molecules in carbon dioxide are from the breakdown of glucose. **Fig. 8.7** ATP is being generated. **Fig. 8.9** Photosynthetic eukaryotes (autotrophs) such as plants and algae have both kinds of cells. Heterotrophic eukaryotes do not have chloroplasts but do have mitochondria. The metabolic machine works between these two types of organisms. Heterotrophs breathe out carbon dioxide and release water to the environment. Photosynthetic organisms take up these substances from the environment. Similarly, heterotrophs eat autotrophs (and organisms that eat autotrophs), thereby taking in the carbohydrates autotrophs produce. In addition, autotrophs release oxygen to the environment, which heterotrophs breathe in.

## End-of-Chapter Answers

*Level 1*

1. d.  2. d.  3. a.  4. $CO_2$.  5. The sun.  6. a and b.  7. d.  8. c.  9. d.
10. a.

*Level 2*

1. Oxygen is the final acceptor of electrons in the electron transport chain.

2. Your body metabolizes foods by breaking down complex molecules into simple ones. The digestive system breaks down the proteins (hamburger meat) into amino acids, the carbohydrates (bun and fries) into simple sugars, and the lipids (within hamburger meat and fried potatoes) into fatty acids and glycerol. The carbohydrates (now glucose) are further broken down to release energy within the cells by means of glycolysis, the Krebs cycle, and the electron transport chain. The amino acids, fatty acids, and glycerol enter the Krebs cycle without undergoing glycolysis.

3. As oxygen concentration increases, more oxygen is being supplied to support the aerobic phases of metabolism. This results in more ATP production via the electron transport chain. This of course would be essential during times of strenuous activity when your cells require more ATP. The body could naturally increase oxygen concentration by increasing breathing rate and increasing heart rate.

4. During reactions of the electron transport chain, oxygen accepts hydrogen ions and electrons. Water is formed as a result. It is an important end product of cellular respiration; one water molecule is produced for each NAD or $FADH_2$ molecule that enters the electron transport chain.

5. Krebs cycle vs. Calvin cycle: In the Krebs cycle, $CO_2$ is released; in the Calvin cycle $CO_2$, is required. In the Krebs cycle, ATP is produced; in the Calvin cycle, ATP is required. In the Krebs cycle, electrons are stripped from organic

molecules to make NADH; in the Calvin cycle, NADPH provides electrons to make organic molecules like sugar.

6. Electron transport chain (ETC) versus noncyclic light-dependent reactions (LDR): **a.** In the ETC, oxygen accepts electrons and is reduced to water; in the noncyclic LDR, water supplies electrons and is oxidized, resulting in the production of the byproduct oxygen. **b.** In the ETC, electron carriers (i.e., NADH) supply electrons; in the noncyclic LDR, the electron carrier (NADPH) is the final acceptor of electrons. **c.** In the ETC, H$^+$ ions are pumped out of the matrix during chemiosmosis; in the noncyclic LDR, electrons are pumped into the thylakoid space during chemiosmosis.

*Level 3*

1. d
2. a
3. Most of the molecular weight of fats is in the carbon and oxygen bonds of the fatty acids. When these bonds were broken and the energy within them used, the carbon and oxygen remaining diffused from the cells to the blood as $CO_2$, was transported through the bloodstream to the lungs, and was exhaled.
4. True. The breakdown of glucose during cellular respiration is exergonic in that energy is released from the glucose being cleaved. The energy is then used endergonically to produce molecules of ATP.
5. This chapter shows how the mitochondria are the place in which a more complete breakdown of food molecules occurs than in the cytoplasm. It seems likely that a symbiotic relationship between certain cells and bacteria could result millions of years later in eukaryotic cells having bacterialike organelles. The structure of the mitochondrion is discussed more fully in this chapter in terms of its function—the inner membrane (possibly the original bacterial membrane) is the site of biochemical activity, as it is in bacteria. The outer membrane (which is relatively inactive in cellular respiration) could be derived from the plasma membrane of the cell and not be bacterial in origin.

# CHAPTER 9

## Concept Checkpoint Answers

1. DNA is coiled around histone proteins in the nucleus, which are then coiled further to form chromatin. The chromatin fibers are then supercoiled, and these supercoils are condensed with other proteins into chromosomes.
2. Morgan—discovered that genes are on chromosomes.—Chargaff's rule—in a living cell, the amount of adenine is equal to the amount of thymine, and the amount of cytosine is equal to the amount of guanine.—Mendel—suggested that traits were inherited as discrete factors.—Nucleotides—the subunits that make up DNA; each contains three parts: a sugar, a phosphate group, and a nitrogenous base.—Nucleus—the organelle in the cell where DNA is located.—Purines and pyrimidines—pyrimidines are single-ringed nitrogen bases (cytosine and thymine in DNA, cytosine and uracil in RNA); purines are double-ringed nitrogen bases (adenine and guanine).—Ribonucleic acid—another kind of nucleic acid (in addition to DNA) found in the cell.
3. The rigid rungs represent the nitrogen bases (adenine bonding to thymine and guanine bonding to cytosine). The flexible rope portion of the ladder represents the repeating sugars and phosphates on adjacent nucleotides.
4. Genes that code for polypeptides; genes that code for certain types of RNA; promoters that signify the start of a gene; introns, which are sequences that set off portions of a gene and are not expressed; exons, which are the expressed portions of genes; repetitive DNA sequences whose functions are mostly unknown; telomeres, which are repetitive sequences that protect the ends of DNA strands as they shorten during DNA replication.
5. Cellular events that can mutate DNA: Environmental factors such as high-energy radiation, low-energy radiation, and certain chemicals can cause changes in the nucleotide sequence of DNA.—Some chromosome breaks occur spontaneously.—Transposons are DNA sequences that can move from one position to another in the genome. Insertions of transposable elements are the cause of many visible mutations.

## Visual Thinking Answers

**Fig. 9.2** Just prior to cell division. At that time, the cell must give half of its duplicated hereditary material to each cell so that each retains a full complement. If DNA were in long threads, division of the chromatin would be impossible; it would get

tangled and break. **Fig. 9.7** The strands of DNA "run" in opposite directions. Each has a 3′ and a 5′ end, and replication can proceed only from the 5′ to the 3′ end on each growing strand. Replication, therefore, is able to take place continuously on the lower strand but must take place "piecemeal" on the upper.

## End-of-Chapter Answers

*Level 1*

**1.** c. **2.** g. **3.** e. **4.** d. **5.** a. **6.** f. **7.** b. **8.** h. **9.** d. **10.** b.

*Level 2*

1. The chemical instructions that determine our specific personal characteristics are located within chromosomes. The DNA within our chromosomes directs the millions of complex chemical reactions that govern our growth and development.
2. The two types of nucleic acid are DNA (deoxyribonucleic acid) and RNA (ribonucleic acid). Both consist of nucleotides that are made up of three molecular parts: a sugar, a phosphate group, and a base. The sugar in RNA is ribose, whereas the sugar in DNA is deoxyribose. DNA contains the bases adenine, thymine, cytosine, and guanine; RNA contains uracil instead of thymine.
3. Double helix refers to the DNA molecule, which is shaped like a double-stranded helical ladder. The bases of each strand together form rungs of uniform length, and alternating sugar-phosphate units form the ladder uprights.
4. Chargaff's experiments showed that the proportions of bases varied in the DNA of different types of organisms. This suggested that DNA has the ability to be used as a molecular code, because its base composition varies as its code varies from organism to organism. Chargaff also noted that the amount of adenine present in DNA molecules is always roughly equal to the amount of thymine, and the amount of guanine is always roughly equal to the amount of cytosine (A = T and G = C). This discovery was significant in determining the structure of DNA.
5. Watson and Crick understood that when a cell divides, each cell must get a complete set of the genetic instructions. This implies that the DNA in the nucleus must be copied prior to cell division. The pairing they are referring to is the regular pairing of adenine with thymine and guanine with cytosine to form the "rungs" of the DNA double helix. They were suggesting that perhaps somehow the two strands of DNA were separated and new nucleotides were added according to base-pairing rules. This would in their estimation produce an exact replica of the original strand of DNA.
6. We now know that genes code for polypeptides, which either alone, or in combination with other polypeptides forming proteins, have a variety of functions in cells besides acting as enzymes. For example, proteins are used for structure, for transport functions, and as hormones. In addition, some genes code for certain types of RNA.
7. You may choose from the work of Martha Chase, Barbara McClintok, and Rosalind Franklin, all of whom have made significant contributions to our understanding of molecular genetics.

*Level 3*

1. As cells continue to divide, they lose some of the nucleotides at the ends of each strand of DNA. Eventually this could pose a problem because this loss of nucleotides could damage or delete genes required for the normal functioning of a cell. Each strand of DNA has repetitive sequences called telomeres that act as "caps" on the ends of each strand, so that nucleotides that are lost are not part of important genes.
2. This *might* have had something to do with the shortened telomeres of the DNA from the original donor nucleus. Since this cell was acquired from an adult and was already the product of numerous prior cell divisions, the telomeres of this cell's DNA may have been significantly shortened, resulting in premature cell death of many of the subsequent daughter cells.
3. Introns. Much of our DNA is made up of intervening sequences that are not expressed. Some biologists have hypothesized that these introns come from past viral infections.
4. Science is an ongoing process of discovery. Scientists use observations and prior findings, synthesize this knowledge, and develop testable hypotheses. As more research is done, our understanding of nature is continuously refined to account for new findings. In fact, occasionally research conclusions that were once thought to be implausible are later found to be substantiated. The story of discovery in molecular genetics has all of these aspects.

# CHAPTER 10

## Concept Checkpoint Answers

1. **(1)** tRNA. **(2)** mRNA. **(3)** rRNA.
2. mRNA—CUAAUGUCUAAUGUGGAUCGAUAG
   Amino acid sequence—Leu-Met-Ser-Asn-Val-Asp-Arg-*Stop*
3. mRNA—CUAAAUGUGGAUCGAUAG
   Amino acid sequence—Leu-Asn-Val-Asp-Arg- *Stop*
4. The anticodon GGG is complementary to the mRNA codon CCC, which codes for proline.
5. Prokaryotes are single-celled organisms that have no membrane-bounded nucleus or organelles. The most familiar prokaryotes are the bacteria. These organisms turn genes on (induction) or off (repression) in response to the presence or absence of various substrates (food sources). These inducible or repressible genes encode enzymes that metabolize the substrates, with certain enzymes needed for the metabolism of certain subtrates. In positive control mechanisms, the products of regulator genes—genes that code for products that control other genes—turn on genes that code for enzymes. In negative control mechanisms, the products of regulator genes turn off genes that code for enzymes.
6.

| Point of Regulation | Mechanism | Brief Description |
|---|---|---|
| Transcription | *Transcription factors | *Transcription factors that turn on gene expression. |
| Transcription | Chromosome structure | Molecules that make the DNA physically accessible to binding of transcription factors. |
| mRNA processing | Alternate splicing of mRNA transcripts | Changing which introns are removed, a single gene sequence can code for a variety of polypeptides. |
| Translation | Degradation of mRNA | Breaking down mRNA transcripts in the cytoplasm can slow down or turn off polypeptide production. |

* Occurs in prokaryotes.

7. This discovery suggests that these homeotic genes date back to the common ancestor of fruit flies and mice.
8. Peptide hormones cannot enter a target cell but instead bind to a receptor protein, which triggers an intracellular cascade of molecular signaling that results in changes in gene expression. Steroid hormones are lipid soluble and can pass through membranes, thereby allowing them to enter the nucleus of a cell and more directly control gene expression in their target cells.

## Visual Thinking Answers

**Fig. 10.3** The splicing pattern would be 12678. The polypeptide would not be functional for sex determination because it would not follow the functional pattern shown in the illustration. Translation takes place in the cytoplasm in both eukaryotes and prokaryotes. The cytoplasm is also the location of the tRNA and amino acids. **Fig. 10.7** Transcription and mRNA processing take place in the nucleus of eukaryotes. Because prokaryotes do not have a nucleus, these processes take place at the DNA, in the cytoplasm of the cell, as the DNA is not separated by a membrane from the rest of the cell contents.

## End-of-Chapter Answers

*Level 1*

1. b.  **2.** c.  **3.** b.  **4.** a.  **5.** e.  **6.** a.  **7.** e.  **8.** f.  **9.** d.  **10.** c.  **11.** b. **12.** g.

*Level 2*

1. The type of RNA found in ribosomes is ribosomal RNA (rRNA). Transfer RNA (tRNA) is in the cytoplasm; during polypeptide synthesis, tRNA molecules transport amino acids to the ribosomes and position each amino acid at the correct place on the elongating polypeptide chain. Messenger RNA (mRNA) brings information from the DNA in the nucleus to the ribosomes in the cytoplasm to direct which polypeptide is assembled.
2. Regulation of gene expression. Different cell types express different sets of genes.
3. DNA is wound around proteins called histones, which form nucleosomes. Strings of nucleosomes then are coiled on a protein scaffold forming chromatin fibers. During cell division these chromatin fibers are themselves further coiled to form chromosomes. Because of the supercoiled nature of the large amount of DNA in the nucleus of a cell, most of the genes are physically inaccessible to transcription.
4. **a.** This substitution would produce a mRNA with the following sequence (boldface indicates changes): CUA**U**UUCUAAUGUGGAUCGAUAG, which would result in the following amino acid sequence: Leu-**Ile**-Ser-Asn-Val-Asp-Arg-*Stop*. **b.** This substitution would produce a mRNA with the following sequence (boldface indicates changes): CUAAUG**G**UCUAAUGUGGAUCGAUAG. This of course shifts the reading frame of the mRNA transcript, which would result in the following amino acid sequence: Leu-Met-**Val—Stop** (the rest would not be translated). The second mutation changes the polypeptide structure the most.

*Level 3*

1. There are many, but probably most important is that bacteria and humans share the same basic genetic code. The matching of codons and amino acids during translation is the same.
2. This means that for some reason the target cells for insulin lose their ability to respond when insulin binds to a membrane receptor protein. In other words, the binding of insulin does not trigger the normal intracellular cascade of events that trigger the cell to take up sugars.
3. All of these could be outcomes depending on how the mutation affects the function of the protein. If the repressor can't bind to the DNA, the result will be a. If the mutation causes the repressor to itself be prevented from binding to the DNA, then both b and c will result.

# CHAPTER 11

## Concept Checkpoint Answers

1. Classical biotechnology involves manipulating organisms to produce organisms with specific biochemical, morphological, and/or growth characteristics, to produce useful products, or to gain new knowledge about an organism or tissue. Examples include producing vaccines; using yeast to ferment sugars for beer, wine, and bread making; and selective breeding of livestock. Molecular biotechnology involves manipulating the genes of organisms to do essentially the same kinds of things as classical biotechnology. Examples include genetic engineering of bacteria so they produce human insulin and DNA fingerprinting.
2. Important discoveries include the basic laws of inheritance (Mendel), the recognition that DNA rather than protein is the genetic material (Hershey and Chase), the structure of the DNA molecule (Watson and Crick), and the nature of the genetic code, including the fact that this code is virtually identical for all of life.
3. **Transformation**—Naked DNA from ruptured bacteria. Bacteria take up DNA through their outer membrane and sometimes incorporate it into their chromosome. **Transduction**—Pieces of bacterial DNA combined with viral DNA. Viruses that infect bacteria occasionally take up fragments of host DNA as they replicate and transfer that DNA to other bacterial cells. **Conjugation**—Plasmids or sometimes whole bacterial chromosomes. Bacteria form sex pili and transfer DNA from one cell to another.

4. The correct order is: **c.** Cut DNA containing gene of interest with restriction enzyme. **f.** Cut plasmids (cloning vector) with the same restriction enzyme. **a.** Mix restriction fragments with plasmids. **d.** Seal bonds between sticky ends of restriction fragments and plasmids with DNA ligase. **g.** Insert hybrid plasmids into bacteria by transformation. **b.** Culture bacterial colonies. **e.** Screen the colonies to find the bacteria that took up the gene of interest.

5. Polymerase chain reaction (PCR).

6. DNA vaccines are sequences of DNA that code for proteins found on pathogens. These sequences are inserted into a human cell, which then produce the protein, which stimulates antibody production and immunity to the real pathogen. Genetically engineered vaccines are produced by splicing genes for the antigens of a pathogen into a nonpathogenic organism or virus.

7. A cDNA microarray lets researchers determine which genes are being expressed in a particular cell. It allows molecular biologists to monitor the interaction of all of the expressed genes in a cell when the cell is, for example, exposed to different conditions. The microarrays from a heart muscle cell and a pancreatic cell would be different because the two cells require the production of different sets of proteins that are specific to the functions of each cell type. To make this happen, they must of course express different sets of genes.

8. Knowing the human genome sequence does not tell us what this sequence means. The meaning of the sequence is conveyed by the protein products of the genes, their function in the body, and the control of gene expression in the cell. We still have a long way to go in comprehending these aspects of the "book of life."

9. The gene must be inserted into the stem cells (or parent cells) that divide to give rise to the fully differentiated cells that make up the affected tissue or organ.

10.

| Benefits | Risks |
|---|---|
| • Increase food production by producing disease-resistant or more environmentally tolerant organisms. | • Transgenic organisms could reproduce with or outcompete natural populations, resulting in severe disruptions of natural ecosystems. |
| • Produce medicines (hormones, enzymes, vaccines). | • Inadvertently creating transgenic viruses or bacteria that could cause disease if accidentally released. |
| • Produce other substances useful to humans (spider silk—Biosteel) | • It is virtually impossible to predict and plan for the long-term effects of transgenic organisms on complex and dynamic ecosystems. |

## Visual Thinking Answers

**Fig. 11.6** The F pilus is a hollow tube that serves as a mating bridge. The F plasmid, or the entire bacterial genome including the F plasmid, is transferred through the bridge from the F$^+$ bacterium to the F$^-$ bacterium. **Fig. 11.11** Media is a nutritive material on which bacteria grow. *Selective* media contains nutrients on which only *certain* bacteria will grow; the media, in effect, "selects" these bacteria from others. In this case, selective media is prepared on which only bacteria having the recombinant plasmid are able to grow, so they are "selected" in that they are the only cells that will reproduce. **Fig. 11.13** No. The banding patterns in the bloodstain specimen do not match the banding patterns in the victim, nor do they match patterns in suspects 2 and 3. There is a clear match, even beyond the bands denoted by the arrows, with suspect 1. **Fig. 11.19** *Agrobacterium* has the natural ability to transfer DNA to plant cells when it infects them, forming tumorlike growths. Scientists have removed *Agrobacterium*'s disease-causing genes while leaving intact its ability to transfer DNA.

## End-of-Chapter Answers

*Level 1*

1. f.  2. a.  3. e.  4. c.  5. b.  6. b.  7. a.  8. d.  9. c.  10. e.

*Level 2*

1. The use of scientific and engineering principles to manipulate organisms, producing one or more of the following: organisms with specific biochemical, morphological, and/or growth characteristics; organisms that produce useful products; information about an organism or tissue that would otherwise not be known. Genetic engineering is a form of molecular biotechnology in which the genes of organisms are manipulated or changed in some way for the purposes expressed above.

2. Classical biotechnology involves manipulating organisms to develop specific characteristics or synthesize chemical products that are useful to humans. Unlike classical biotechnology, molecular biotechnology manipulates the genes of living organisms for similar purposes. Both are indeed being practiced today. Examples of classical biotech include pasteurization, traditional means of developing vaccines, and selective breeding of organisms to bring out valuable characteristics. Examples of molecular biotech include transgenic bacteria that synthesize Humulin and human gene therapy to treat SCID.

3. Scientists induce bacteria to take up DNA by transformation.

4. Restriction enzymes allow biotechnologists to cut genes from the chromosomes of an organism. They also allow for the production of RFLPs, which can be used to construct detailed genetic maps for use in positional cloning. Positional cloning is a procedure of identifying and then isolating genes using information about their location in the genome.

5. A molecular probe is a short sequence of single-stranded DNA constructed with radioactive isotopes. These probes will hybridize to a known complementary sequence in a gene. Exposure of the probe to transgenic bacteria will allow the biotechnologists to identify the bacterial colonies that contain the gene of interest.

6. PCR allows very small samples of DNA to be amplified into hundreds of thousands of identical sequences. This allows biotechnologists to work with initially very small amounts of DNA and create copies of it to be used in a wide variety of applications, one of which is DNA fingerprinting.

7. In human gene therapy, the correct version of a gene is inserted into the cells that express an abnormal or mutant form of that gene. In theory, this technique can be used to treat inherited genetic disorders such as diabetes or hemophilia, or it can be used to treat cancer. The technique is still in clinical trials, and numerous problems need to be resolved. However, HGT has shown much promise—for example, the work done in Drs. Blaese and Anderson's labs has shown that HGT can successfully restore immune system function in a small number of SCID patients.

8. Shotgun cloning involves cutting an organism's DNA with restriction enzymes and inducing bacteria to take up *all* of these fragments. The bacterial colonies must then be screened to find the bacteria that took up the gene of interest. In cDNA cloning, DNA fragments are created from mRNA transcripts of a cell that strongly expresses the gene of interest. In this technique, only genes copied from mRNA transcripts synthesized by that cell are produced. This results in less screening. Gene synthesis cloning requires no screening at all because only the gene of interest is made synthetically.

9. Each child inherits different sets of genes (or alleles) from their parents as a result of mutations, crossing over, and independent assortment of chromosomes that occur during sex-cell formation in their parents. You are already aware of the concept of mutations; in Chapter 12 you will learn about their inheritance and contribution to genetic variation in offspring, and the shuffling of genes (crossing over and independent assortment) that occurs through sexual reproduction.

*Level 3*

1. Scientists must insert human genes into the bacterial genome so that bacteria can express those genes, producing human protein. Gold and other precious metals are not produced by living things and thus are not coded for by DNA. Although it sounds like a great idea to have bacteria produce gold, it won't work.

2. Regulator genes could be inserted that turn off the gene at some level, or targeted gene repair could be used.

3. Some bacteria may not have been transformed at all, some bacteria may have taken up a nonhybridized plasmid, some bacteria may have taken up a hybrid plasmid that contains fragments of human DNA but not the insulin gene, and some may have taken up a plasmid containing the insulin gene.

4. You would not want to use BamHI or HbaI. Both of these restriction enzymes would cut right in the middle of the gene and therefore would not yield a fragment that contains the whole gene.

ATCCATAGTCAAATTTT**GGATCC**GGATCCGG**GCGC**AAAATGGGGC

Instead, you would want to use a restriction enzyme that cut on either side of the gene.

5. Genetically engineered insulin is chemically identical to human insulin and therefore could not elicit an allergic reaction like animal-derived insulin could.

6. The DNA fingerprint analysis shows that Sally belongs to Mr. and Mrs. T and Ruth belongs to Mr. and Mrs C. Lane, however, is another issue—she may be the child of Mrs. R and Mr. C!

# CHAPTER 12

## Concept Checkpoint Answers

1. The ploidy of a cell does not change when it divides by mitosis because the DNA replicates just prior to cell division, and then the DNA is divided equally between the two resulting cells. The ploidy of a cell does change when it divides by meiosis because the resulting cells each contain just one member of each chromosome pair in the parent cell.

2.

### Description of cellular events

| | | |
|---|---|---|
| Interphase | G1 | Cell growth and normal daily operation |
| | S | DNA replication |
| | G2 | DNA begins condensation into chromosomes (sister chromatids); cell division "machinery" is assembled |
| Prophase | | DNA condensation into sister chromatids is completed; nucleolus disappears as rRNA genes become inaccessible; spindle fibers form and migrate until they extend to opposite poles of the cell; nuclear membrane breaks down; spindle fibers attach to kinetochores; aster forms |
| Metaphase | | Chromosomes are arranged in a circle at the cell's equator, forming the metaphase plate |
| Anaphase | | Sister chromatids separate at their centromere and begin to migrate toward opposite poles |
| Telophase | | Spindle fibers are disassembled; nuclear membrane reforms; nucleolus reforms; cytokinesis begins |

3. **Initiation**—Caused by mutations in proto-oncogenes and/or in tumor-suppressor genes, resulting in damage to the DNA; mutations can be inherited or can be caused by one or more mutagenic agents, including viruses, chemical, and radiation. **Promotion**—Caused by increased damage to the DNA through the action of agents called promoters, resulting in partially transformed cells; promoters may be the same agents that cause initiation. **Progression**—Caused by partially transformed cells becoming malignant, due to further damage to the DNA.

4. Prophase of mitosis and prophase I of meiosis: **Similarities**—DNA condensation into chromosomes is completed; nuclear membrane breaks down; spindle fibers form **Differences**—During Prophase I, homologous chromosomes undergo synapsis, and crossing over occurs. These processes do not occur during mitosis. Anaphase of mitosis and anaphase I of meiosis: **Similarities**—DNA begins to be separated into future daughter cells. **Differences**—Mitosis: sister chromatids separate and migrate toward opposite poles. Meiosis I: homologous chromosomes separate and migrate toward opposite poles. Telophase of mitosis and telophase I of meiosis: **Similarities**—Separated nuclear material gathers near each pole; nuclear membrane forms; cytokinesis begins. **Differences**—In mitosis, the separated nuclear material consists of separated sister chromatids; in meiosis, the separated nuclear material consists of separated homologues. Prophase of mitosis and prophase II of meiosis:

**Similarities**—DNA condensation into chromosomes is completed; nuclear membrane breaks down; spindle fibers form. **Differences**—These two stages are virtually identical, mitosis involves a single cell while meiosis involves two daughter cells. Anaphase of mitosis and anaphase II of meiosis: **Similarities**—Sister chromatids separate and migrate toward opposite poles. **Differences**—These two stages are virtually identical, except that mitosis involves a single cell while meiosis involves two daughter cells. Telophase of mitosis and telophase II of meiosis: **Similarities**—Separated sister chromatids gather near each pole; nuclear membrane forms; cytokinesis begins. **Differences**—These two stages are virtually identical, except that mitosis involves a single cell, with cytokinesis resulting in two daughter cells, while meiosis involves two cells, with cytokinesis resulting in four daughter cells.

## Visual Thinking Answers

**Fig. 12.2** In both animals and plants, mitosis is used for the growth and development of the diploid organism. However, in plants, mitosis is also used for two other purposes: (1) the growth and development of the haploid organism and (2) the production of eggs and sperm by mitosis in cells that are already haploid. **Fig. 12.4** The S phase (DNA replication) takes up about one-third of the cell cycle, so in this example, it would take about two hours. Mitosis and cytokinesis appear to take up about one-tenth or a bit more of the cell cycle, or a bit less than three-quarters of an hour. **Fig. 12.12** In the normal cells, the nuclei (dark dots within cells) are quite regular in size and shape; in the dysplastic cells, the nuclei are much larger and irregular in shape. In addition, the normal cells are all about the same size, while the dysplastic cells vary in size, and they appear to stain less evenly than the normal cells.

## End-of-Chapter Answers

*Level 1*

1. c. 2. e. 3. a. 4. e. 5. a. 6. d. 7. T. 8. F. 9. T. 10. F. 11. T. 12. T. 13. T.

*Level 2*

1. Independent assortment of chromosomes and crossing over during meiosis I. These events effectively shuffle genetic information on and among each homologous pair of chromosomes. This recombination produces offspring with unique combinations of genes, resulting in variation.

2. c → a → d → b. This describes either mitosis or meiosis II.

3. c → d → a → b → e. This describes meiosis I.

4. The cells that result from mitosis are diploid and are genetically identical. The cells that result from meiosis are haploid; they have one chromosome from each of the original homologous pairs. Because of crossing over, they are not genetically identical.

5. Sister chromatids would be most similar because they are the result of DNA replication, so their DNA sequence would be identical.

6. 4 (AB, AD, BC, and CD)

*Level 3*

1. Since these cells are from the root tip, they are somatic cells (non-sex cells). They are in various stages of the cell cycle: A cell with 20 pg of DNA could be at any stage from G2 of interphase to telophase of mitosis prior to cytokinesis. A cell with 10 pg of DNA could be at any stage from telophase of mitosis after cytokinesis to the end of G1 of interphase. A cell with 15 pg of DNA is in the S phase of interphase at a point where half its DNA has been replicated.

2. These cells are germ cells (reproductive cells). They are in various stages of the cell cycle: A cell with 40 pg of DNA could be at any stage from telophase I after cytokinesis to telophase II before cytokinesis, or it could be in the G1 stage of interphase of a germ cell prior to DNA replication. A cell with 49.3 pg of DNA is in the S phase of interphase at a point where slightly less than a quarter of its DNA has been replicated. A cell with 20 pg of DNA could be at any stage from telophase II after cytokinesis to interphase of a fully formed sex cell (sperm or egg). A cell with 80 pg of DNA could be at any stage from G2 of interphase to telophase I of meiosis prior to cytokinesis.

3. AB, Ab, aB, ab

4. Metaphase, the stage in which microtubules move the chromosomes to the metaphase plate.

5. If homologous chromosomes are arranged next to one another on each side of the metaphase plate, the cell is in metaphase I of meiosis. But if the chromosomes are lined up, not paired with homologues, on the metaphase plate, the cell could be in metaphase of mitosis or metaphase of meiosis II.

6. During anaphase II of meiosis, one pair of sister chromatids in one cell did not separate. As a result, one daughter cell got an extra chromosome, and the other daughter cell didn't get the same chromosome. If each of the four sperm cells were to fertilize an egg, two normal zygotes, one zygote with 47 chromosomes, and one zygote with 45 chromosomes would result. (This is what occurs in the disorder known as Down Syndrome [trisomy-21]. An individual inherits an extra chromosome 21 as a result of an error in separation of sister chromatids during meiosis.)

7. In order for tumors to grow and invade surrounding tissues, they require a blood supply, just as other tissues of the body do. The use of drugs that prevent formation of new blood vessels in tumors can starve tumors of needed blood, preventing growth. These drugs would be most useful during progression, when tumors are metastasizing and infiltrating tissues of the body.

# CHAPTER 13

## Concept Checkpoint Answers

1. An individual is true-breeding for a given trait if self-fertilization always produces offspring that are the same as the parent for that trait. Hybrids are offspring that are the product of a cross between two different varieties of true-breeding individuals for a given trait (i.e., flower color).

2. Whenever two varieties of true-breeding individuals for a given trait are crossed, only one of the parental traits appears in the offspring.

3. **Traits are inherited as genes:** Mendel observed that the plants exhibiting the traits he studied did not produce progeny of intermediate appearance when crossed. These observations did not uphold the theory of blending inheritance but suggested instead that traits are inherited as discrete factors that are either present or absent in a particular generation. **Alleles are alternative forms of genes:** For each pair of traits that Mendel examined, one alternative form was not expressed in the $F_1$ hybrids, although it reappeared in some $F_2$ individuals. He inferred from these observations that each individual, with respect to each trait, contains two factors. Each pair of factors may contain information for (be a code for) the same form of a trait, or each member of the pair may code for an alternative form of a trait. **Gametes receive only one allele of each pair:** Mendel hypothesized that the two factors (alleles) that coded for a trait remained "uncontaminated," not blending with one another as the theory of blending inheritance predicted, and the results obtained from experiments on various traits showed similar results; therefore, pea hybrids must form egg cells and pollen cells (gametes) in which the factors (alleles) for each trait separate from one another "in equal shares" during gamete formation. **Alleles may differ or may be identical:** Mendel realized that plants exhibiting the dominant trait in his monohybrid crosses of pea plants, when self-fertilized, would breed true or would produce plants exhibiting either the dominant or the recessive form of the characteristic in a 3:1 ratio, respectively. He observed that plants exhibiting the recessive trait, when self-fertilized, would always breed true. These data suggested to Mendel that true-breeding plants receive *only* the dominant factors or the recessive factors from each of their parents and that non-true-breeding plants were hybrids, which received the dominant and the recessive factors in equal shares. Therefore, individuals may have two identical alleles for a trait or two different alleles for a trait.

4.

| | Phenotypic ratio (dominant: recessive) | Genotypic ratio (CC:Cc:cc) |
|---|---|---|
| Cc x Cc | 3:1 | 1:2:1 |
| CC x Cc | 1:0 | 1:1:0 |
| CC x cc | 1:0 | 0:1:0 |
| Cc x cc | 1:1 | 0:1:1 |

5.

| | Phenotypic ratio (Yellow-round: green-round: yellow-wrinkled: green-wrinkled) | Probability |
|---|---|---|
| GGWW x ggww | 1:0:0:0 | $^{16}/_{16}$ yellow-round |
| GgWW x ggww | 1:1:0:0 | $^8/_{16}$ yellow-round, $^8/_{16}$ green-round |
| GgWw x ggww | 1:1:1:1 | $^4/_{16}$ yellow-round, $^4/_{16}$ green-round, $^4/_{16}$ yellow-wrinkled, $^4/_{16}$ green-wrinkled |
| GGWw x ggww | 1:0:1:0 | $^8/_{16}$ yellow-round, $^8/_{16}$ yellow-wrinkled |
| ggWw x ggww | 0:1:0:1 | $^8/_{16}$ green-round, $^8/_{16}$ green-wrinkled |
| ggWW x ggww | 0:1:0:0 | $^{16}/_{16}$ green-round |
| GGww x ggww | 0:0:1:0 | $^{16}/_{16}$ yellow-wrinkled |
| Ggww x ggww | 0:0:1:1 | $^8/_{16}$ yellow-wrinkled, $^8/_{16}$ green-wrinkled |

6. The phenotypic ratios for each cross and the corresponding probabilities are unique. We can use the ratios from a testcross to deduce the genotype.

7. The father would have to have white eyes (have the allele for white eyes on his X chromosome), and the mother would have to be heterozygous or homozygous for the allele.

8. In every case the female parent passed the allele to her male progeny. The male parent cannot pass the allele to the male progeny because he gives them the Y chromosome.

## Visual Thinking Answers

**Fig. 13.2** Sperm are located within the pollen grains on the anther.  **Fig. 13.4** Cross-pollination was used with the P generation. Pollination refers to the pollen grain landing on the stigma. Fertilization refers to sperm fertilizing an egg. Pollination leads to fertilization.

## End-of-Chapter Answers

*Level 1*

**1.** b. **2.** d. **3.** a. **4.** d. **5.** a. **6.** d. **7.** F. **8.** T. **9.** T. **10.** T.

*Level 2*

1. All of the offspring have long leaves and are heterozygous (Ss).

| | s | s |
|---|---|---|
| **S** | Ss | Ss |
| **S** | Ss | Ss |

**2.**

| | S | s |
|---|---|---|
| **S** | SS Long | Ss Long |
| **s** | Ss Long | ss Short |

**3.** $\frac{1}{4}$

**4.** Alleles of two different genes assort independently into sex cells. Today we know that for this to be true the genes must be located on separate, nonhomologous chromosomes.

**5.** $\frac{1}{16}$, $\frac{9}{16}$

**6.** Sutton and Morgan provided a framework that explained the equal genetic role of both egg and sperm in the hereditary material of offspring. Sutton suggested the presence of hereditary material within the nuclei of the gametes, whereas Morgan gave the first clear evidence that genes reside on chromosomes.

**7.** All seven genes are on separate, nonhomologous chromosomes. (The genes only behaved as if they were on separate chromosomes. We will discover why this is the case later when we explore linked genes.)

*Level 3*

**1.** $\frac{1}{2}$, $\frac{1}{2}$, $\frac{1}{4}$

**2.** Both are easy to raise, matings can easily be manipulated, the organisms are readily available, and it is relatively easy to observe different phenotypes.

**3.** This is difficult to do, since virtually all traits have some basis in heredity. For example, the development of freckles may not be directly hereditary; they are caused by exposure to the sun. However, the tendency to develop freckles most certainly is hereditary. For example, people with dark skin are less likely to develop freckles.

**4.**

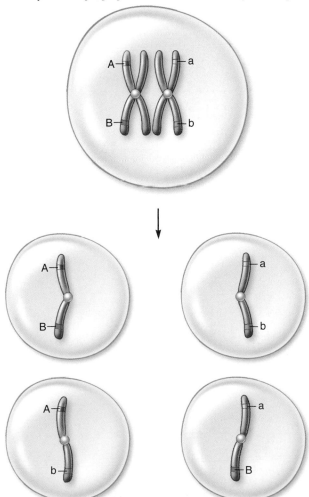

A crossover event would have to occur to produce the bottom two gametes.

**5.** All the F1 offspring would be light purple. The F2 offspring would be $\frac{1}{4}$ purple, $\frac{1}{2}$ light purple, and $\frac{1}{2}$ white.

*Genetics Problems*

**6.** Some of your bulls are heterozygous for the polled trait. To fix this problem, you would have to mate each polled bull with several of the horned cows. Those bulls who produce offspring with horns are the heterozygotes.

**7.** $\frac{1}{4}$

**8.** 50 Long-Gray, 50 Long-Black, 50 Dumpy-Gray, 50 Dumpy-Black

**9.** Just a cursory look at these results would indicate that they are close enough to the expected results. Differences are most likely due to chance.

**10.** Both of the parents are heterozygous.

**11.** $(\frac{1}{4})^5$

**12.** $(\frac{3}{4})^5$

## CHAPTER 14

### Concept Checkpoint Answers

**1.** *AA* and *Aa*

**2.** No. If both parents had attached earlobes, their genotypes would be *aa*. At least one parent would have to donate an *A* allele for any offspring to have unattached earlobes.

**3.** Yes. In incomplete dominance, alternative alleles are not dominant over or recessive to other alleles governing the same trait. Heterozygotes are phenotypic intermediates.

**4.** With codominance, neither allele is dominant; in a heterozygote, each allele contributes equally. The AB blood type is an example of a heterozygote, with the *A* and *B* allele each contributing equally. With incomplete dominance, alternative alleles are not dominant over or recessive to other alleles governing a particular trait. Instead, heterozygotes are phenotypic intermediates. A person with wavy hair is an example of a heterozygote of parents who are curly haired and straight haired.

**5.** A single gene codes for traits caused by multiple alleles. The gene is represented by more than two alleles, and inheritance patterns such as dominance, recessiveness, and codominance are evident. These patterns have a discrete nature, such as the presence of four distinct blood types in a population. In polygenic inheritance, more than one gene codes for a single trait. This type of inheritance has a continuous nature, such as the variations of height in a population.

**6.** The differences are likely due to a combination of polygenic inheritance and environmental factors. Tamika probably received more alleles from her parents that contribute to skin color than did Keisha. In addition, Tamika may be outdoors more than Keisha due to a job or to recreational activities, which would cause her skin to darken even more if she did not use sunblock.

**7.** Environmental factors, such as exposure to ultraviolet light, affect skin color. The environmental effects on phenotypes would result in a more continuous line of variation, as shown by the bell curve, rather than a less continuous representation, as shown by the bar graph.

**8.** During meiosis, the alleles for the disorder will be segregated into different gametes. The dominant allele will be present in 50% of the gametes, and the recessive allele, which does not produce the disorder, will be present in the other 50%. Therefore, each child has a 50% chance of acquiring the dominant gene.

**9.** The probability of parents who are both heterozygous for the sickle cell allele of having a child with sickle cell disease (genotype *SS*) is 1 out of 4.

| | | Father | |
|---|---|---|---|
| | | S | s |
| **Mother** | **S** | SS | Ss |
| | **s** | Ss | ss |

**10.** The alleles for sex-linked disorders are generally found on the X chromosome, not the Y chromosome. The human Y chromosome is much smaller than the X chromosome and carries few genes. Y-linked disorders are rare.

**11.** The trait is autosomal because it is expressed equally in males and females. The trait is recessive because the people expressing the trait have parents who do not express the trait.

12. Due to the age of their eggs, women over the age of 35 have a higher probability of bearing children with errors in chromosome number. Therefore, a woman over age 35 may choose to have information on the genetic health of her fetus. In this situation, the karyotype is used to determine whether the chromosome number is "normal" in the fetus and whether any gross abnormalities in its chromosome structure exist.

13. Inheriting an improper number of sex chromosomes is due to errors in meiosis in one or both parents. Nondisjunction of the sex chromosomes in females could lead to gametes with two X chromosomes or no X chromosomes at all. Nondisjunction of the sex chromosomes in males could lead to gametes with both an X and a Y chromosome or no sex chromosomes at all. When gametes with improper numbers of sex chromosomes join during fertilization with other gametes that are normal or abnormal, the fetus inherits an improper number of sex chromosomes.

14. A duplication is the doubling of a section of a chromosome; it lengthens the chromosome. A deletion is the loss of a section of a chromosome; it shortens the chromosome.

15. In general, genetic tests are performed to look for genes implicated in diseases or disorders or to determine the identity of a deceased person or the perpetrator of a crime.

16. A physician performs amniocentesis by inserting a sterile hypodermic needle into the amniotic cavity and withdrawing a small sample of amniotic fluid. The position of the needle in relationship to the fetus is observed by means of ultrasound. Amniocentesis is not a genetic test, but it is a way to harvest cells from the fetus for the genetic test.

## Visual Thinking Answers

**Table 14.2** O blood has neither A nor B antigens. Therefore, it can be given to persons with A, B, AB, or O blood without fear that it will stimulate and A B antigen/antibody reaction because the recipient would not be receiving these "foreign" antigens. However, persons with O blood have both A and B antibodies present in their plasma. If given A, B, or AB blood, their antibodies will react with these foreign antigens, causing a serious reaction. **Fig. 14.13 1.** During meiosis II, the sister chromatids in one of the cells produced during meiosis I do not separate. Therefore, the product of meiosis II is one cell with both sister chromatids and one cell with no chromatids. **2.** If each of these cells were to combine with a sperm, one zygote would have an extra chromosome and one would have one less chromosome than the normal complement. **Fig. 14.17** Prior to meiosis, DNA replicates, forming an XX chromosome (XX sister chromatids) and a YY chromosome (YY sister chromatids). These XX and YY chromosomes are homologues. During meiosis I, homologues separate, resulting in an XX cell and a YY cell. During normal meiosis II, sister chromatids separate. If nondisjunction occurs in the YY cell, however, one resulting cell would contain both sister chromatids (YY) and one would have no Y chromosome. **Fig. 14.18** A duplication and a deletion would be visible because each would result in a change in the length of the chromosome arm as measured from the centromere. A translocation would result in changes in the lengths of chromosome arms of nonhomologous chromosomes, unless the translocated pieces were of the same length. If so, then changes in banding patterns might reveal the translocation. An inversion could be determined if banding patterns were changed enough to show differences between homologues.

## End-of-Chapter Answers

*Level 1*

**1.** b. **2.** c. **3.** a. **4.** F. **5.** T. **6.** F. **7.** c. **8.** d. **9.** a. **10.** e.

*Level 2*

1. People who inherit abnormal numbers of sex chromosomes often have abnormal features and may be mentally retarded. Examples are: (**a**) triple X females (XXX zygote), underdeveloped females who may have lower-than-average intelligence, (**b**) Klinefelter syndrome (XXY zygote), sterile males with some female characteristics, (**c**) Turner syndrome (XO zygote), sterile females with immature sex organs, and (**d**) XYY males, fertile males of normal appearance.

2. **a.** X-linked. **b.** dominant. **c.** recessive. **d.** autosomal.

3. Most human genetic disorders are recessive because those genes are able to persist in the population among carriers; people carrying lethal dominant disorders are more likely to die before reproducing. Recessive disorders include cystic fibrosis, sickle cell anemia, and Tay-Sachs disease.

4. In incomplete dominance, alternative forms of an allele are neither dominant nor recessive; heterozygotes are phenotypic intermediates. In codominance, alternative forms of an allele are both dominant; thus, heterozygotes exhibit both phenotypes.

5. Couples who suspect that they may be at risk for genetic disorders can undergo genetic counseling to determine the probability of this risk. When a pregnancy is diagnosed as high risk, a woman can undergo amniocentesis (analysis of a sample of amniotic fluid) or chorionic villus sampling to test for many common genetic disorders.

*Level 3*

1. Familial hypercholesterolemia is a dominant autosomal disorder. The chances of inheriting this disorder from an affected father and unaffected mother is 1 out of 2, or 50%.

2. Answers will vary but will include a family history of genetic diseases and disorders or suspected genetic diseases and disorders.

3. **a.** Each parent is heterozygous for the condition: GL/gl. **b.** Your possible genotypes are GL/GL or GL/gl. **c.** No. She will pass on the gl allele, but unless her children's father passes on a gl allele as well, her children will remain unaffected.

4. The fact that a genetic abnormality such as trisomy 21 can exist shows that there are often groups of genes located near each other on a chromosome that may be transferred or altered as a group and cause multiple alterations that occur as a discrete group.

5. Their children's genotypes would be 1:2:1 (AA [type A]: AB [type AB]: BB [type B]).

|   | A | B |
|---|---|---|
| **A** | AA | AB |
| **B** | AB | BB |

6.

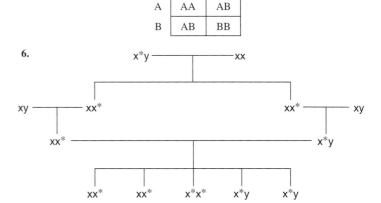

7. No. A type O child (ii) is possible, since dad can be AO ($I^A$i) and mom can be BO ($I^B$i).

8. 45 (44 autosomes, one X)

9. $I^A$i (blood type A)

10. The chance that offspring will be totally color blind is 1/4, as shown by the following pedigree:

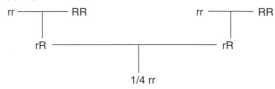

11. e. Y-linked

# CHAPTER 15

## Concept Checkpoint Answers

1. **a.** Organisms of the past and present are related to one another. **b.** Something more than just climate must play a role in the diversity of life. **c.** These geographically distinct populations must be related, and the ancestors of the off-

shore populations must have somehow traveled there from the mainland. **d.** The age of the Earth allows enough time for evolution to have taken place. **e.** There is natural variation in populations, and much of this variation is heritable. Over successive generations, species can change dramatically if breeders choose organisms with specific traits and selectively breed them. **f.** There is a struggle for existence. Organisms that have favorable characteristics (adaptations) are more likely to survive to reproduce than organisms that do not.

2. A scientific theory is a powerful concept that helps scientists make specific predictions about the world and explain particular phenomena. It can incorporate facts, laws, inferences, and tested hypotheses. In science, theories can be considered factual if they have been repeatedly confirmed and never refuted. This is true of evolution.

## Visual Thinking Answers

**Fig. 15.12** Merychippus. **Fig. 15.13** No, the wing of a bat is not homologous to the wing of an insect. From the illustration, it is clear that the bat wing contains bones, as do the homologous structures shown. The wing of an insect does not contain bones. The wings of bats and insects are analogous structures. **Fig. 15.19** Humans and yeast are most distantly related: they have a larger number of differences in their base sequences of cytochrome C oxidase than any other two organisms. The duck and the rattlesnake are most closely related: they have the smallest number of differences in their base sequences of cytochrome C oxidase than any other two organisms. Of the organisms shown on the tree, the pig is most closely related to humans because it has the smallest number of differences from humans in its base sequences of cytochrome C oxidase.

## End-of-Chapter Answers

*Level 1*

**1.** b. **2.** e. **3.** d. **4.** e. **5.** e. **6.** convergent evolution.
**7.** analogous. **8.** adaptive radiation. **9.** acquired.

*Level 2*

1. Species are not perfectly adapted to their environment, mainly because the environment is not static—it is always changing. This statement would also imply that extinction could never occur (except for the occurrence of catastrophic events), which is not at all supported by evidence from the fossil record.

2. The two dogs look so different today because of artificial selection; people who bred them selected for desired characteristics, so that over time the breeds changed. Artificial selection is based on the principle that all organisms within a sexually reproducing species exhibit variation.

3. **Molecular evidence:** DNA sequences of closely related species are more similar than are DNA sequences of distantly related species. **Embryological evidence:** Groups of organisms that differ as adults (e.g., vertebrates) all are similar in their early embryological development. **Homologies:** Closely related species that live in different environments have different-looking but structurally similar anatomical features that perform different functions. Example: the forearms of different species of vertebrates. **Analogies:** Distantly related species that live in similar environments have similar-looking but structurally different anatomical features that perform similar functions. **Fossil evidence:** The fossil record shows that current forms of life are different from but also similar in many respects to extinct organisms preserved in the fossil record. **Age of Earth:** Radiometric dating has shown that the Earth is old enough for evolution to be the cause of past and represent diversity of life.

4. (a) The development of analogous structural similarities through convergent evolution. (b) The development of homologous structural similarities through evolution.

5. Darwin had no understanding of genes or of how heredity works, so it was impossible for him to explain completely how evolution occurs. If he had had the benefit of Mendel's insights, Darwin might have understood that natural selection acts on the phenotype, which reflects the underlying genotype. In most populations, more offspring are produced than can be supported by available resources. The individuals that survive long enough to reproduce, or that reproduce the most, are those with alleles and combinations of alleles that confer an adaptive advantage.

6. The carbon-14 method can help establish a date by estimating the relative amounts of the different isotopes of carbon present in a fossil. The half-life of

carbon-14 is 5,730 years (the amount of time for half of the $^{14}$C to decay into nitrogen). Scientists can estimate the length of time that the carbon-14 has been decaying, which means the time that has elapsed since the organism died.

7. According to radioactive dating, the Earth is about 4.6 billion years old. This is significant because it allows sufficient time for evolution to occur.

8. Vestigial structures are structures that are present in an organism of today but are no longer useful. The tiny leg bones of a python are an example.

9. Studies of comparative anatomy show that many organisms have groups of bones, nerves, muscles, and organs with the same anatomical plan but with different functions. These homologous structures imply evolutionary relatedness. Studies in comparative embryology show that many organisms have early developmental stages that are similar. This again implies evolutionary relatedness.

*Level 3*

1. A population composed of only a few organisms has limited genetic variation and mating possibilities. Should conditions for survival change and should some of these organisms die due to their lack of characteristics that could accommodate that change, the population would become smaller over time and could eventually die out. Larger populations generally have wider genetic variation and are better able to withstand changes that affect the survival of the population.

2. All organisms are subject to the pressures of natural selection—those pressures differ from species to species and depend on the environment in which they live. Although humans shape their environments, they are still subject to environmental pressures that affect the reproductive fitness of the species. For example, in some countries humans experience environmental pressures such as famine.

3. No, evolution cannot be considered a ladder that organisms can climb because there is no goal to reach. Each surviving species is successful simply because it continues to survive in the face of environmental pressures. In addition, survival under differing environmental pressures does not imply that one species is better than another.

4. When a population of a particular species of bacterium is exposed to an antibiotic to which that species is susceptible, most bacteria are killed. However, some bacteria may have a heritable trait that confers resistance to that antibiotic. Even if only one bacterium were to survive, if given the proper growth conditions, that bacterium could reproduce asexually, resulting in a new population of resistant bacteria.

5. Tall, handsome, and rich DBM with two children, because he is the only one currently able to have children. He has the most offspring of the group and the potential to have more.

6. Individuals who have inherited a minimized nonuseful structure may actually be at a selective advantage over those who have the full-size nonuseful structure because just maintaining a full-size nonuseful structure requires energy that could be better used in other ways; and full-size nonuseful structures may be potential sites of infection or attract parasites. Therefore, individual moles that have inherited the genes for reduced eye development may actually stand a better chance of surviving and reproducing than their big-eyed neighbors.

7. Organisms are indeed very complex, much like a precision watch; however, evolution fashions organisms blindly, without any forethought or plan.

## CHAPTER 16

## Concept Checkpoint Answers

1. Counting all those with the recessive phenotype and those without will provide only the frequency of that phenotype in the population. Heterozygous individuals carry one recessive allele, which must be included in determining the frequency of the recessive allele in the population, even though those individuals do not express the recessive phenotype.

2. An allele that is adaptive may become detrimental if a significant change takes place in the salmon's environment. Alleles are only adaptive (or detrimental or neutral) in the context of environmental pressures put on the organism.

3. A mutation is always random and rare. A mutation might be beneficial, neutral, nonadaptive, heritable (if it occurs in the gametes), adaptive, and lethal.

4. Malaria is widespread in Africa but not in North America. Individuals who are heterozygous for the sickle cell gene have resistance to a particularly deadly species of the malaria parasite and have a selective advantage over

those individuals who are homozygous. Those who are homozygous have sickle cell disease, which reduces their fitness. Consequently, the frequency of heterozygous individuals and the recessive allele they carry increases in the population. Because of the relative absence of malaria in North America, there is no selective advantage to being heterozygous. This situation, combined with strong selection against those who have sickle cell anemia (homozygous recessive individuals), has resulted in a relatively low frequency of the sickle cell anemia allele in North American populations.

5. The few marbles that came out of the bottle represent a small population (founder population) that has separated from the large population of marbles in the bottle. The founder effect is an extreme type of genetic drift that occurs in founder populations: the frequencies of the alleles (colors, in this case) in the small population are quite different from the frequencies of the alleles (colors) in the larger population. The frequency of each color in the original population was 1/3, or 0.33. In the founder population, the frequency of red is 1/6 (0.17), the frequency of white is 2/6 (0.33), and the frequency of blue is 3/6 (0.50).

6. 1. Mating is random, meaning that individuals of different sexes can pair without limitations. (Nonrandom mating)
   2. The population is infinitely large. (Genetic drift)
   3. Genes are not added from outside the population—that is, there is no gene flow. (Gene flow between populations)
   4. Genes do not change from one allelic state to another—that is, no mutation occurs. (Mutations)
   5. All individuals have equal probabilities of survival and of reproduction—that is, there is no selection of certain alleles. (Natural selection)

## Visual Thinking Answers

**Fig. 16.4** Initially, as shown in graph (4), the birds with small beaks and the birds with large beaks are at the extremes of a single population. As selection for these two beak types progresses over generations (5), the population begins to split apart (is *disrupted*), and eventually (6) the one species has *diversified* into two.  **Fig, 16.5** Most students reading this book live in North America, where malaria is not prevalent. Therefore, the sickle cell allele is likely to have a low prevalence as well, because it has no adaptive advantage. Those with the disease die early, so they are less likely to reproduce and pass on the gene than those without the disease. Therefore, the frequency of the gene slowly becomes smaller in the gene pool. If you live where malaria is prevalent, then the sickle cell allele is likely to have a high prevalence because it confers an adaptive advantage. Those heterozygous for the trait have a higher level of fitness than others living in your country because they are less likely to die from malaria and will not get sickle cell disease.

## End-of-Chapter Answers

*Level 1*

1. d.   2. b.   3. e.   4. a.   5. T.   6. F.   7. F.   8. F.   9. T.

*Level 2*

1. A natural population can rarely meet all five conditions proposed by Hardy and Weinberg; thus, it is rare for a natural population to be in genetic equilibrium. Natural selection, nonrandom mating, mutations, and gene flow are always occurring in natural populations. Even very large populations undergo some genetic drift, although its effect on overall gene frequencies is probably minimal.
2. There would have to be *no* selection involving that allele. In other words, mates would not be chosen based on the phenotypic expression of that allele, and survival would not be affected by it. In addition, the population would have to be isolated, and any mutations in the allele would have to result in nonviable offspring.
3. The founder effect can occur when a small population separates from a much larger population and begins a new population elsewhere. In this case, genetic drift can occur over the course of one or a few generations rather than in many generations, because the gene pool of the founder population comprises the alleles of only a small portion of the original population, which means that a large sampling error is likely. The gene pool of the founder population is likely to contain only a small portion of the total genetic variation of the larger popula-

tion, and over generations, allelic frequencies in the gene pool based on the founder population are likely to deviate greatly from the allelic frequencies in the original population. The founder effect probably occurred on the Galapagos Islands, which were colonized by small populations of various species of plants and animals. These small populations had separated from much larger populations of the mainland.

4. A population bottleneck occurs when an event such as the founder effect, or some other natural or human-caused event, significantly reduces the size of a population. Genetic drift can then result in the fixation of alleles and the loss of genetic variability, which is critical to the long-term survival of the population.

*Level 3*

1. Yes, because with incompletely dominant alleles, heterozygotes express a phenotype that is distinguishable from (or intermediate to) that of homozygous individuals. Determining frequencies is then a matter of counting up individuals, remembering that heterozygotes have one of each allele, and then dividing by the total number of alleles in the population.
2. Heterozygous individuals would have little selective advantage, and homozygous recessive individuals would be strongly selected against (because they get sickle cell anemia). Therefore, the frequency of the allele would begin to diminish in the population.
3. The dramatic reduction in the wild condor population probably caused severe genetic drift and the loss of alleles. If the population rebounds, future survival will be more difficult for this species even if its habitat is restored, because of the reduced genetic variability in the population.
4. Small, isolated populations in fragmented patches of habitat are like the small, separated populations that result from the founder effect. Sampling error and genetic drift result in reduced genetic variation, making a population less able to withstand environmental changes, such as the introduction of a new pathogen or parasite.
5. **a.** Disruptive selection. **b.** Directional selection. **c.** Directional selection. **d.** Stabilizing selection.

# CHAPTER 17

## Concept Checkpoint Answers

1. Gene flow between apple flies and hawthorn flies is likely to be extremely low or nonexistent within populations of these flies. Therefore, the two types of flies are likely to diverge genetically from one another over time, and they probably have diverged genetically over the past 200 years. They may become separate species in the future. This type of speciation is termed sympatric speciation, referring to members of a population that occupy different ecological zones with the same geographical area.
2. **a.** Mechanical isolation. **b.** Temporal isolation. **c.** Habitat isolation. **d.** Temporal isolation. **e.** Maintenance of geographical isolation.
3. Example "e" would be the clearest case of evolution via punctuated equilibrium.
4. This depends on what examples you provide, but it is possible that all of them, either directly or indirectly, could be the result of human activity.

## Visual Thinking Answers

**Fig. 17.12** The slanted lines show a gradual change in morphology, the vertical lines show periods of stasis, or equilibrium, with little change in morphology.  **Fig. 17.17** Wildlife corridors increase animal movement between wildlife patches, and thus, their interactions with plants in various habitat fragments. This increases plants' opportunities for cross-pollination and for seed dispersal.

## End-of-Chapter Answers

*Level 1*

1. a.   2. e.   3. a.   4. d.   5. b.   6. F.   7. F.   8. T.   9. T.

*Level 2*

1. Biological species concept.
2. A supporter of punctuated equilibrium would argue that this is not surprising, since most species stay relatively unchanged for long periods, and that most change occurs very quickly. Therefore, it is unlikely that these transitional fossils would be found very often, since transitional forms only exist for short periods of time if at all. However, a supporter of gradualism may counter by arguing that the fossil record is an incomplete record of the history of life, since fossils are usually only the impressions or remains of hard parts of the organism. Many changes in organisms, such as behaviors, coloration, or soft body parts, aren't recorded in the fossil record. Therefore, it's not surprising that transitions are rare; however, this is no indication that gradualism is not at work.
3. Allopatric speciation. The Galapagos Islands provide an excellent opportunity for geographic separation of populations, both from the mainland populations and between islands in the archipelago. Genetic drift and the effects of different selection pressures on islands that have significantly different environments all led to speciation.
4. Extinctions open up habitats that were previously occupied. These habitats can then be exploited by those species that remain, presenting populations with new selection pressures.
5. Genetic drift can play a role, particularly if the subpopulation is very small. Initial differences in *allelic* frequencies and subsequent random fluctuations in *allelic* frequencies can result in microevolution. In addition, the new isolated subpopulation may experience different environmental selection pressures that will cause change. Finally, separated populations experience different mutations.

*Level 3*

1. Allopatric speciation, environmental change, temporal isolation
2. **a.** Gradualism. **b.** Punctuated equilibrium. **c.** Punctuated equilibrium.
3. There are greater opportunities for separation of populations in fresh water (i.e., lakes drying up into smaller separated lakes), which has led to higher levels of allopatric speciation.
4. Allopatric speciation
5. Some reasons might be as follows. Hawaii is an attractive place for humans to live, so there has been much loss of natural habitat due to agriculture and development.—Species on the Hawaiian Islands became highly specialized as they adapted to the conditions on each island. Minor changes in the environment, such as the natural or accidental introduction of new species could pose a significant threat to a genetically uniform population. Moreover, humans have introduced a wide variety of nonnative species to the islands.—Overfishing, hunting, and trade of local species have also taken their toll.
6. There could be a number or reasons for this: The environment of the Florida Keys is very similar to that of mainland Florida. Therefore, organisms that migrated to the Keys experienced selection pressures similar to those experienced by mainland populations.—The Keys are too close to mainland Florida for island populations to be reproductively isolated from those on the mainland.—The islands that make up the Keys are too close to each other to effectively prevent interbreeding between groups of organisms inhabiting different islands.
7. Their concerns are probably legitimate. Many examples here on Earth have shown that accidental (or sometimes purposeful) introduction of nonnative species or pathogens into a habitat can harm the native populations. There is a small chance that if microorganisms from Mars were accidentally released on Earth, they could cause similar environmental problems. Visit NASA's web site at www.nasa.gov to find out how they intend to deal with this concern.

# CHAPTER 18

## Concept Checkpoint Answers

1. Miller and Urey were asking whether simple organic molecules could have arisen in a primitive atmosphere composed of methane, ammonia, water vapor, and hydrogen gas if a source of energy existed, such as lightning. Fox, Wächtershäuser, Huber wondered about the conditions under which even more complex organic molecules, such as proteins, might have formed from simpler organic molecules, such as amino acids.

2. The presence of DNA suggests that perhaps at one time the "ancestors" of these organelles were free-living prokaryotes, which took up residence in a larger host cell.
3. One hypothesis is that they originated through the invagination, or infolding, of the plasma membrane. Ultimately these infoldings separated from the outer membrane and formed intracellular membrane-bound organelles. Some of these infoldings may have surrounded the nucleoid of a pre-eukaryote to form a nucleus.
4. Bony structures that support the pharyngeal arches evolved into jaws with teeth in vertebrates.—The fins of lobe finned fishes were used as appendages in terrestrial vertebrates.—Lung-like structures in fish were used as lungs in terrestrial vertebrates.—Feathers in reptiles possibly used as insulation and for trapping insects were used for flight in birds.
5. Preadaptations are inherited traits used for new functions for which they are later selected. The structure of the human hand developed as part of our ancestors' arboreal life and the brachiation they exhibited in the trees. This same structure is adaptive in humans for a variety of other uses.

## Visual Thinking Answers

**Fig. 18.11** As adults, humans exhibit only a single, hollow nerve cord (spinal cord). The notochord becomes the bony vertebral column, the pharyngeal gill arches (also called branchial arches) give rise to various blood vessels and structures of the throat, and the "tail bones" of the vertebral column only hint at a tail, which does not extend beyond the anus. **Fig. 18.27** chimpanzees **Fig. 18.31** No, none are considered to be apes or monkeys. *S. tchadensis, A. ramidus, A. anamensis, A. afarensis,* and *H. erectus* are ancestors of humans. The one species on the tree living today is modern *Homo sapiens.*

## End-of-Chapter Answers

*Level 1*

1. d.  2. c, currently inconclusive, but the oldest fossil evidence is from ~3.5 bya.  3. e.  4. b.  5. e.  6. d.  7. b.  8. d.  9. c.  10. a.

*Level 2*

1. The primordial soup theory attempts to explain how life began on Earth. It states that life arose in the sea as elements and simple compounds in the atmosphere interacted to form simple organic molecules. One alternative to this theory is the hypothesis that life might have arisen in hydrothermal vents in the oceans.
2. Initial invaginations may simply have increased the absorptive surface area of the outer membrane. Recall how important this outer membrane is for transporting materials into and out of the cell.
3. Cyanobacteria are single-celled organisms that are similar to the first oxygen-producing bacteria. These evolved during the Archean era and played an important evolutionary role by gradually oxygenating the atmosphere and oceans.
4. The Cambrian period was the oldest period in the Paleozoic era (roughly 590 to 500 million years ago). All of the main phyla and divisions of organisms that exist today (except the land plants) evolved by the end of the Cambrian period.
5. During the Carboniferous period, reptiles developed from amphibians and arthropods moved from the sea onto land. Fungi evolved during the late Carboniferous period. Much of the land was low and swampy with extensive forests; the worldwide climate was warm and moist.
6. The Cenozoic era, which began about 65 million years ago, saw the rapid evolution and growth of mammals, including primates.
7. Primates are the order of mammals with characteristics reflecting an arboreal lifestyle. They have developed two especially helpful characteristics: depth perception resulting from an overlapping visual field, and grasping hands with opposable thumbs.
8. These are both types of primates. Prosimians (lower primates such as lemurs) are small animals, usually nocturnal, with a well-developed sense of smell. Anthropoids (higher primates such as apes and humans) have larger brains, flatter faces, eyes that are closer together, and relatively long front and hind limbs. Anthropoids are also diurnal and possess color vision.
9. Statement (a) is correct; statement (b) is false. All hominins are hominids.

10. *Homo sapiens sapiens* had smaller heads, brow ridges, teeth, jaws, and faces than earlier species did. They also fashioned sophisticated tools from stone and from other substances such as bone and ivory. Over time, groups of *Homo sapiens sapiens* developed agriculture, complex social structures, and civilization.

*Level 3*

1. **a.** Myth—Humans share a common ancestor with apes but did not evolve from apes. **b.** Reality—The other branches of this "bush" contain extinct hominins and other primates that are either extinct or still present today. **c.** Myth—Neanderthals coexisted with *Homo sapiens* from about 230,000 to 30,000 years ago in Europe and the Middle East. **d.** Myth—Hominins did not evolve until some 6–7 million years ago, about 60 million years after the dinosaurs were extinct. **e.** Myth—Paleontologists and anthropologists are always receptive to new scientific data that could change how we understand human ancestry and evolution.
2. Current hypotheses regarding the origin of life on Earth all begin with the premise that life arose on this planet and did not come from another planet. Therefore, all these hypotheses take into account the environment of Earth at the time and how molecules might evolve into organisms. If life arose on Mars and were transported to Earth, that information would change that initial premise of the major hypotheses currently held.
3. This observation supports the endosymbiotic theory of the origin of these organelles because the initial ingestion of the mitochondrial precursor likely involved phagocytosis, in which the host cell surrounded, and engulfed, a smaller prokaryote. If the membrane of the mitochondrial precursors remained intact, it is not surprising that they retained their original prokaryote proteins.
4. If you recall, sexual reproduction contributes to variation in offspring. Natural selection then acts on this variation to choose organisms with characteristics that are most adaptive at the time. This probably led to enormous and rapid diversification.
5. Once oxygen became prevalent, organisms evolved the ability to use its oxidative powers in metabolism. The use of oxygen allows the production of more ATP than is possible without oxygen. These metabolic gains allowed organisms to exploit entirely new and more active modes of existence, leading to diversification.

# CHAPTER 19

## Concept Checkpoint Answers

1. Although viruses do have their own genetic material, they are incapable of replicating without infecting and using the transcription and translation machinery of a host cell. In addition, viruses have no cellular structure.
2. During the lytic cycle, the phage genetic material is actively used to assemble new viruses. During the lysogenic cycle, the phage genetic material is integrated into the host cell's chromosome and is replicated along with the rest of the host cell's DNA prior to cell division. The ability of the integrated phage genes to initiate assembly of new viruses remains latent unless the phage enters the lytic cycle.
3. Cancer-causing viruses may be transmitted by sexual contact, by contact with blood or blood products, by transfusion with contaminated blood, or through the placenta or breast milk from mother to child.
4. AIDS patients suffer from a depressed immune system due to their HIV infections and are therefore vulnerable to infection from other viruses.
5. Bacteria, unlike the others, have a cellular structure and are capable of reproducing on their own.

## Visual Thinking Answers

**Fig. 19.6** The abnormal white blood cells in (*a*) all have large round nuclei that take up most of the cell, while the normal white blood cells in (*b*) have differences in the shapes of their nuclei. In one, the nucleus takes up most of the cell, but in the other it does not.   **Fig. 19.7** The level is approximately the same as it was in 1983, 1988, and 1994. Although there has been a steady rise in the disease since 1966, there have been many upsurges and dips along the way. The year 2001 showed a continued decline from a high in 1999.

## End-of-Chapter Answers

*Level 1*

1. e.   2. c.   3. a.   4. d.   5. T.   6. T.   7. T.   8. F.   9. F.

*Level 2*

1. **Pervasiveness**—bacteria are everywhere, there are few places on the Earth where one can go and not find bacteria. **Diverse environments**—bacteria are capable of inhabiting virtually any environment on the planet, including inside living organisms, the soil, water, around hot springs, undersea volcanoes, and deep in caves. **Metabolic diversity**—bacteria are the only group of organisms that includes photoautotrophs, chemoautotrophs, and heterotrophs (this is one reason why they can live in so many different types of environments). **Adaptability**—bacteria, because of their reproductive potential, are capable of evolving very rapidly in response to environmental change. An example of this is the ever-increasing resistance of bacteria to various antibiotics. **Length of existence**—bacteria were among the first forms of life to emerge, some 3.5 billion years ago. Their ancestors have existed on this planet ever since. **Reproductive potential**—bacteria are capable of reproducing very rapidly; populations can double in size in a matter of hours!
2. In the lytic cycle, viruses enter a cell, replicate, and cause the cell to burst and release new viruses. In the lysogenic cycle, viruses enter into a long-term relationship with the host cells, with their nucleic acid replicating as the cells multiply.
3. Bacteria perform many vital functions, such as decomposing organic materials and recycling inorganic compounds. They were also largely responsible for creating the properties of the atmosphere and soil that are present today.
4. No. Sexually transmitted infections are not caused by poor personal hygiene. They are communicable illnesses transmitted between persons during sexual contact.
5. The simplicity of bacterial structure—small size, lack of discrete organelles, lack of a membrane-bound nucleus—seems to support the idea that they evolved before eukaryotes.
6. See Figure 19.4.
7. These diseases may be acquired without the knowledge of the person or partner. Infected people may not have symptoms of infection and thus may unwittingly pass on these diseases.
8. The only completely effective protection against sexually transmitted infections is abstinence. There is decreased risk if latex condoms are used correctly.

*Level 3*

1. It takes 10 days for the antibiotic your physician prescribed to kill all the bacteria causing your infection. After only 4 days of the antibiotic, some cells may still be living and able to reproduce. These cells could trigger another strep infection.
2. A cold sore results from cell damage when the herpes simplex virus emerges from the lysogenic cycle. The virus remains latent in nervous tissues until something triggers it, such as a cold. This virus could be transmitted to a partner's genitals during oral sex if the person was shedding virus at the time.
3. Answers will vary.
4. Viruses are capable of inserting genetic material into a host cell, and that genetic information can be incorporated into a host cell's chromosome. Viruses are host specific, which means the virus can be chosen to deliver the human genes only to human cells.

# CHAPTER 20

## Concept Checkpoint Answers

1. These names refer to protists' metabolism and mode of acquiring energy and nutrition. Animal-like protists are heterotrophs and get their energy and nutrition by eating (or parasitizing) other organisms. Plant-like protists are photoautotrophs. Fungus-like protists are also heterotrophs but get their energy and nutrition primarily by feeding on dead and decaying organic material.
2. Unicellular means these protists are single cells. Colonial means they are an assemblage of individual cells with little or no cell specialization. Multicellular means they are made up of many cells in which cell specialization is common (e.g., cells specialized for feeding, gas exchange, or reproduction).

3. **Similarities**—Both have stages in which they behave like amebas; both produce spores that are involved in sexual reproduction; both produce structures that can become dormant for a period of time; both aggregate in some way; both can reproduce asexually. **Differences**—Aggregation is acellular in Myxomycota and cellular in Acrasiomycota; cell division in Acrasiomycota produces separate cells; Myxomycota replicate nuclei, which are not separated by cell walls.

4. **Zygomycota**—The haploid stage of hyphae growing on a substrate is more conspicuous. Opposite mating strains of hyphae fuse to form diploid zygospores, which then produce haploid spores by meiosis. **Ascomycota**—Fungal mycelium is haploid, and the more conspicuous stage. Opposite mating strains fuse to form di-nucleated hyphae, which form a cup or sac. Nuclei fuse within the cup to form true diploid cells, which undergo meiosis and germinate haploid spores. **Basidiomycota**—Underground mycelium is haploid and is the more conspicuous stage. Hyphae of separate mating strains merge to from di-nucleated hyphae. These hyphae grow into a mushroom (or other club fungus), in which specialized cells in the gills fuse to form true diploid cells, which undergo meiosis to produce and germinate haploid spores.

## Visual Thinking Answers

**Fig. 20.17** Volvox lives in colonies. It is the only organism among the three in which multiple cells are visible, making up a "unit." The "units" are the large spheres, and the cells that make them up look like tiny dots. Daughter colonies can be seen within the spheres.   **Fig. 20.25** In this generalized life cycle, the fusion of hyphae of different mating types produces a diploid zygote. Meiosis then takes place to produce spores, which restores the haploid state. The haploid phase by far is the most prominent portion of the life cycle of fungi.   **Fig. 20.27** No, you would not have removed all of the mold. The illustration shows the mycelium embedded within the bread, which is the food source of the mold. Scraping mold from the food would not remove these structures.

## End-of-Chapter Answers

*Level 1*

**1.** e.  **2.** b.  **3.** d.  **4.** b.  **5.** c.  **6.** d.  **7.** f.  **8.** a.  **9.** e.

*Level 2*

1. Protists are single-celled eukaryotic organisms; this kingdom contains organisms that are animal-like (protozoa), plant-like (algae, including diatoms), and fungus-like (slime molds).
2. Mutualistic relationships are those in which two species of organisms live together, with both benefiting from the relationship. An example of this is *Tricho nympha*, a protist that lives in the guts of termites. Parasitic relationships are those in which two organisms live together, but one benefits from the relationship, while the other is harmed. Examples include most disease-causing protists or fungi, such as *Plasmodium*, the apicomplexan that causes malaria.
3. Both flagellates and ciliates are protozoans, and both have hairlike cellular processes that help them move and obtain food. However, flagellates have flagella (long hairlike processes), whereas ciliates have cilia (short hairlike processes). Also, ciliates have a more complex internal organization than flagellates do.
4. Sporozoans are nonmotile protozoans that parasitize animals. They have complex life cycles with asexual and sexual phases and may require several species of host.
5. Dinoflagellates and diatoms are both phyla of plant-like protists. Dinoflagellates have stiff outer coverings, and their flagella beat in two grooves. The diatoms look like microscopic boxes made up of top and bottom shells.
6. This classification stems from the characteristic pigmentation of most members of each group. These differences result from different photosynthetic pigments used by each of the different groups of algae. Since the photosynthetic pigments of red algae absorb mainly the green, blue, and violet light, which penetrates deeper into the water, red algae are capable of living at greater depths than the green or brown algae.
7. All fungi could be described as helpful, because as decomposers they are crucial to the cycling of materials in the environment. More specifically, helpful fungi include those that produce antibiotic drugs. Yeasts are fungi that are crucial in the production of bread, cheese, and beer. Other fungi, however, cause diseases in plants and animals, such as Dutch elm disease, wheat rust, and yeast infections.

*Level 3*

1. It is difficult to give a definitive answer to this question. As groups of cells within colonial organisms specialize, taking on particular jobs, many would suggest that they should be considered multicellular organisms. However, in this case, only one group of cells is specialized; if the cells could exist independently, the group might still be considered a colonial organism.
2. *Euglena* is difficult to place using this frame of reference. Clearly, classifying protists on the basis of such criteria is difficult and, in some cases, may be impossible. Protists are a varied group; classifying these organisms is the subject of debate among scientists. Using a simplistic approach such as "plant-like" and "animal-like" may not appropriately reflect evolutionary closeness or distance among organisms, and it may be an arbitrary way to group organisms.
3. Eventually, the exponential growth of the algae will slow as nutrients are used and the surface of the water becomes covered. Some algae will then begin to die. As the algae die, bacteria and other decomposers feed on it. As they do, they use oxygen dissolved in the water. As the oxygen level decreases, other organisms dependent on the oxygen begin to die also, providing yet more food for decomposers. Poisoning the algae would have a devastating effect on the lake because a great deal of food would become available to decomposers. Unless oxygen is returned to the system, as happens with a rapidly flowing stream splashing over rocks, the lake will die.
4. Both fungi and bacteria are decomposers and often compete for space and nutrients. A mold that could inhibit bacterial competitors from growing near it would have a survival and reproductive advantage over molds that could not suppress the growth of bacterial competitors.
5. Wet or moist environments are necessary for food getting and food absorption in all protists. Some protists are motile and can move toward nutrients by chemotaxis. Some sessile protists have cilia that create currents in the water to channel food to the organism. In others, nutrients flow past the organism carried by water currents. Food absorption takes place across cell surfaces in protists, sometimes helped by means of specialized organelles. In addition, most protists have little protection against drying out and could therefore not live in a dry environment.
6. DDT is an insecticide that is effective against mosquitoes. Mosquitoes are the intermediate hosts of the apicomplexan *Plasmodium*, a protozoan that causes malaria. Spraying of DDT dramatically reduced the mosquito population, which prevented the malaria parasite from completing its life cycle. The number of deaths from malaria dropped precipitously during the years when spraying of DDT was widespread in Africa. Today DDT is banned because of its harmful effects on other organisms such as birds.

## CHAPTER 21

### Concept Checkpoint Answers

1. **Patterns of embryonic development**—Biologists compare the patterns of cell division and the organization of those cells into tissues. Organisms that share similar patterns of embryonic development are thought to be more closely related. **Levels of organization**—In some animals cells are not organized into distinct tissues; in others they are. Similarly, some animals with tissues have organs formed from there tissues; and some animals have organ systems. The similarities and differences in levels of organization can be used to establish evolutionary relatedness. **Symmetry**—The simplest animals are asymmetrical (sponges) or radially symmetrical (cnidarians). More complex animals are bilaterally symmetrical. **Body plan**—The simplest animals (sponges, cnidarians, flatworms) lack a body cavity. They are acoelomates (without a coelom). More complex animals have a body cavity. They are either pseudocoelomates (having a false coelom, like roundworms) or coelomates (having a true coelom, like insects and humans).
2. Porifera: **Levels of organization**—no true tissues. **Symmetry**—asymmetrical. **Feeding**—flagellated collar cells create a flow of water from the outside of the sponge, through pores in the matrix, to the interior of the sponge, bringing in nutrients. **Reproduction**—asexual through fragmentation or branches; sexual through hermaphroditism (sponges' sperm fertilize other sponges' eggs), resulting in free-swimming larvae that then settle on rocks and develop into adults.
   Cnidaria: **Levels of organization**—tissues; some primitive organs (ovaries, testes); no organ systems. **Symmetry**—radial. **Feeding**—tentacles with cnidocytes that capture prey and draw it into the mouth. **Reproduction**—asexual through budding in many species; sexual through hermaphroditism in some,

separate sexes in others; larvae either develop directly into medusae or first set-tle as sessile polyps, which then produce medusae.

Platyhelminthes: **Levels of organization**—tissues; organs; organ systems; no coelom. **Symmetry**—bilateral. **Feeding**—flukes and tapeworms are parasitic; turbellarians are free living and feed on other organisms or particulate organic material. **Reproduction**—in some genera, asexual through fragmentation; sex-ual through hermaphroditism, in which two flatworms fertilize each other's eggs; eggs may hatch into miniature adults or may hatch into larvae that go through a succession of forms before becoming adults.

3. All phyla have tissues, organs, and organ systems. All phyla are bilaterally sym-metrical, except the echinoderms, which are bilaterally symmetrical as larvae and radially symmetrical as adults. All phyla have a body cavity, although the roundworm cavity is a pseudocoelom.

4. These trends are seen in Figure 21.2 and the accompanying text:
no tissues → distinct tissues → organs and organ systems
no symmetry → radial symmetry → bilateral symmetry
no body cavity → pseudo body cavity (false coelom) → true body cavity (true coelom)
protostome pattern of embryological development → deuterostome pattern of embryological development

## Visual Thinking Answers

**Fig. 21.9** The sponge is a sessile organism; that is, it cannot move from place to place. It relies on the flow of water for two main reasons: (1) It feeds by trapping and eating food that flows by its collar cells. (2) Water carries sperm from one sponge to the eggs of other sponges, promoting genetic diversity within sponge species. **Fig. 21.14** Turbellarians are acoelomates. In the cross section of the organism, no coelom (body cavity) is shown. Organs such as the testis are embedded directly in the organ-ism's tissues. **Fig. 21.21** A coelom is a fluid-filled enclosure lined with connective tissue within which body organs are suspended. The intestine is an organ of the di-gestive system and is one of the body organs suspended within the coelom. The in-testinal cavity (i.e., the cavity inside the intestine) is therefore not a coelom.

## End-of-Chapter Answers

*Level 1*

**1.** a.  **2.** d.  **3.** b.  **4.** e.  **5.** c.  **6.** e.  **7.** b.  **8.** c.  **9.** b.  **10.** d.

*Level 2*

1. Both you and jellyfish are animals. An important taxonomic difference is that jellyfish are invertebrates (they lack a backbone), whereas you are a vertebrate (with a backbone and dorsal nerve cord).
2. The cells of eukaryotic organisms have a distinct nucleus and a structure differ-ent from that of bacteria. "Multicellular" means that animals have more than one cell. As heterotrophs, animals are unable to make their own food; they must eat other organic matter for food. Beyond these basic similarities, however, ani-mals differ in many other characteristics (symmetry, body plan, etc.).
3. Sponges are called the simplest animals because they lack organs and true tissues.
4. Cnidarians exist as either polyps (cylindrically shaped animals that anchor to rocks) or medusae (free-floating, umbrella-shaped animals).
5. The acoelomate flatworms are the most primitive bilaterally symmetrical ani-mals. They have organs and some organ systems, including a primitive brain, but their organs are embedded within body tissues, not suspended within a body cavity.
6. Roundworms are cylindrical pseudocoelomates encased in a flexible outer cuti-cle; they have primitive excretory and nervous systems. They move by using muscles attached to the cuticle that push against the pseudocoel. Roundworms can be dangerous if they parasitize humans because they absorb digested food from their hosts.
7. They have a digestive cavity that branches throughout the body of the animal, ensuring that nutrients don't have to diffuse far to get to any tissue of the body.
8. Insects do not utilize their circulatory system to transport respiratory gases to their tissues. Instead they use a tracheal system that transports respiratory gases directly to their tissues.

9. Insects belong to a group of arthropods that have jaws (mandibles). Spiders be-long to a different group of arthropods that lack mandibles.
10. Echinoderms are deuterostomes, whereas the other invertebrates are proto-stomes. This makes echinoderms more similar to the chordates, which implies that echinoderms and chordates evolved from a common ancestor.

*Level 3*

1. Most hermaphroditic invertebrates still reproduce sexually, although each or-ganism has both male and female reproductive glands. Sponges produce both sperm and eggs, which float through the water; fertilization generally takes place between gametes of different organisms.
2. Answers may vary. Example: Malpighian tubules in terrestrial arthropods such as grasshoppers are slender projections from the digestive tract (see Figure 21.26). These projections, which function in waste removal, provide a large surface area over which fluid is passed to and from the blood in which the tubules are bathed.
3. Echinoderms have bilaterally symmetrical larval forms and are deuterostomes: They cleave in a radial pattern during early embryological development (see Fig-ure 21.3). These characteristics connect them evolutionarily with the chordates and suggest that they evolved from a common, bilaterally symmetrical ancestor.
4. Insects are the only arthropods (actually, the only invertebrates) that can fly. Hy-potheses vary as to why insects are so abundant on Earth but should center on the idea that insect species have adapted to a wide variety of environments, both aquatic and terrestrial.
5. Flatworms have a much simpler body plan than earthworms do and have no coelom. Earthworms are coelomates with a tube-within-a-tube body plan. The structural units of the earthworm are repeated in its segments, giving organization to its long body. Also, segmentation allows for greater flexibility and motility.
6. **Bilateral symmetry and cephalization**—The development of an anterior re-gion (head) where the sense organs are located led to bilateral symmetry. This cephalization allowed for the development of more elaborate sensory organs, which may have facilitated a more active lifestyle. This allowed animals to search for food rather than wait for it to come to them, as is the case in the more primitive sessile and radially symmetrical invertebrates. **Body cavity**—The de-velopment of more complex and elaborate internal organs probably led to the evolution of a fluid-filled body cavity that protects, supports, and cushions these important internal structures. **Complete digestive tract**—A complete digestive tract allows for differentiation and specialization of function along the digestive tube, making digestion more efficient.

## CHAPTER 22

### Concept Checkpoint Answers

1. Notochord, dorsal hollow nerve chord, pharyngeal (gill) arches and slits, post-anal tail.
2. Adult humans do not have a notochord, pharyngeal arches and slits, or an exter-nally visible post-anal tail. However, human embryos have all of these charac-teristics.
3. Lampreys are parasitic as adults; fertilization is external, and larvae develop from fertilized eggs shed into the water; their skeleton is cartilaginous. Sharks are primarily carnivorous predators, but some are plankton feeders; fertilization is internal, with some species giving birth to live young (viviparous), while oth-ers retain fertilized eggs in the reproductive tract (ovoviviparous), and still oth-ers are egg layers (oviparous); their skeleton is cartilaginous. Trout are bony fish, so their skeleton is primarily composed of bone. Like most bony fish, fertil-ization is external, and larvae develop from fertilized eggs; you may be aware from your own fishing experiences that trout feed on a variety of food sources, including larval fish and adult and larval insects.
4. Bony fish have opercula or gill flaps, covering their gills on each side of their head. Movements of the opercula enhance the flow of water over their gill sur-faces while the fish remain motionless, so the gills receive adequate oxygen. In addition, the presence of a gas-filled swim bladder allows fish to alter their den-sity and maintain buoyancy at any depth without swimming. Most sharks must swim continuously to move water over their gills and to maintain buoyancy.
5. Most amphibians are skin-breathers in addition to using gills or lungs. Gas ex-change occurs over the skin, which must be moist for gases to diffuse across it. In

addition, fertilization in amphibians is external, and their eggs lack a shell or membrane to prevent them from drying out. Therefore, their eggs must be laid in water or a moist environment. Also, the larval stage of many amphibians is aquatic.

6. Reptiles produce a shelled egg that allows the egg to be laid on land and prevents it from drying out. Also, reptiles have dry skins with scales that retard water loss.

7. Flight adaptations in birds include (1) lighter bones; (2) efficient lungs that increase oxygenation of the blood, supporting the high metabolic demands of flight; (3) feathers that form aerodynamic wing surfaces and provide insulation necessary to maintain elevated body temperatures; (4) a four-chambered heart in which deoxygenated blood from the body is kept separate from oxygenated blood from the heart, allowing for high levels of oxygen to be supplied to flight muscles and other tissues; (5) hollow bones and lightweight beaks.

## Visual Thinking Answers

**Fig. 22.4** The larval tunicate is structured quite similarly to the lancelet. Both have three principal features of a chordate: hollow nerve cord, rod-shaped notochord between the nerve cord and the gut, and pharyngeal (gill) slits. The larval tunicate looks much like a tadpole, while the lancelet looks like a tiny, scaleless fish. **Fig. 22.6** The shark is darker on its dorsal side and lighter on its ventral side. Therefore, it will blend in with the dark ocean bottom as a predator (or prey) looks down on it and will blend in with the light above as a predator (or prey) looks up at it. The ray and skate blend in with the bottom material over which they swim. **Fig. 22.7** Water will tend to move out of the cells of marine fish because their solute concentration is lower than that of seawater. Water with a higher concentration of solutes (in the fish) will move to water with a lower concentration of solutes (in the ocean). The opposite will happen with fresh water fish. Water will tend to move into the cells of fresh water fish because their solute concentration is higher than that of fresh water. Water with a higher concentration of solutes (fresh water) will move to water with a lower concentration of solutes (fish).

## End-of-Chapter Answers

*Level 1*

**1.** d. **2.** d. **3.** d. **4.** c. **5.** c. **6.** c. **7.** e.

*Level 2*

1. Hagfish retain the primitive characteristics of lacking jaws and paired fins. Both of these traits are present in all sharks.

2. Although some sharks give birth to their young live, the young are nourished from a yolk sac. In most mammals, the embryo receives nourishment directly from the mother through a placenta or by nursing (marsupials).

3. "Cold-blooded" implies that reptile body temperature is always cold. However, the body temperature of reptiles can be quite warm, as when they bask in the sun. "Warm-blooded" is also not an appropriate term because sometimes the body temperature of mammals can be lower than the surrounding environment, as on a very hot day.

4. Sharks maintain an internal solute concentration that is nearly the same as that of the surrounding water. They do this by retaining nitrogenous waste in the form of urea, not by losing water to the surrounding saltwater. Bony fishes use their gill epithelia to actively pump solutes from their blood or actively take up solutes. Marine bony fish are hyposomotic relative to the surrounding saltwater, so they tend to lose water. They drink lots of water and actively pump excess salt from their gills. Fresh water fishes are hyperosmotic relative to the surrounding fresh water, so they drink very little, excrete lots of very dilute urine, and actively take up salt across their gills.

5. Ovoviviparous organisms (some fish, reptiles, many insects) retain fertilized eggs within their oviducts until the young hatch; the young receive nourishment from the egg. Oviparous animals (birds) lay eggs, and their young hatch outside the mother. Viviparous organisms (humans) bear their young alive, and the young are nourished by the mother, not the egg.

6. Some salamanders spend their whole lives in the water, such as mud puppies.

7. Reptiles have dry skins covered with scales, retarding water loss. They also lay amniotic eggs, which contain nutrients and water for the embryos and protect the embryos from drying out.

8. Mammals have hair; female mammals secrete milk from mammary glands to feed their young.

9. The fish heart has two chambers (an atrium and a ventricle). The atrium receives blood from the body, and the ventricle pumps it out to the gills. From the gills, the blood continues on to the rest of the body without another push. The amphibian heart has three chambers (two atria and a ventricle)—blood returns to the heart through the second atrium after picking up oxygen at the lungs (or gills) and is then pumped out to the body. However, in amphibians, oxygenated and deoxygenated bloods mix in the ventricle. The reptile heart also has three chambers, with an incomplete septum (complete in most crocodiles) subdividing the ventricle and lessening the mixing of the bloods. The bird heart has four separate chambers (two atria, two ventricles). Birds have the most efficient heart of these groups of vertebrates because blood is pumped to the lungs and then pumped to the body without any mixing of the bloods.

*Level 3*

1. Rockfish, skate, sea squirt, sea star, octopus, flatworm, jellyfish, sponge.

2. Reptiles are ectothermic and regulate their body temperature by taking in heat from the environment. In the arctic, snakes would not be able to maintain a body temperature high enough to survive.

3. Bony fishes have considerable osmoregulatory capabilities, making life in waters of different salinities possible.

4. Birds and reptiles.

# CHAPTER 23

## Concept Checkpoint Answers

1. Mosses—gametophyte is dominant; Ferns—sporophyte is dominant. Mosses—the gametophyte exists as a free-living plant; a transient sporophyte grows from it. Ferns—the gametophyte is a small free-living plant that degenerates soon after giving rise to a much larger free-living sporophyte. Spores disperse both mosses and ferns. In both mosses and ferns, the sperm must swim through water to get to the archegonia containing the egg cells.

2. Conifers produce wind-blown pollen. When pollen lands on a female cone, a pollen tube grows into the female gametophyte and a sperm cell is discharged which fertilizes the egg.

3. Flowers house the reproductive structures of angiosperms; they may also attract pollinators.

4. **Flowers**, which may attract animal pollinators. **Fruits**, which may aid in dispersal of seeds. **Double fertilization**, which produces endosperm, a tissue that nourishes the embryo prior to seed germination.

5. Animals may eat fruit and carry away the seeds. Some seeds attach to the fur or feathers (or socks) of animals that brush against them. Some are dispersed by the wind. Still other seeds or fruits float on water and can be dispersed by ocean currents or rivers and streams.

## Visual Thinking Answers

**Fig. 23.3** The sporophyte stalk grows out of the tissues of the gametophyte, which is shown in (1) in the illustration. The sporophyte contains no chlorophyll and nutrients can reach the sporophyte only via the parent gametophyte. Conversely, gametophytes germinate as independent plants from spores. They undergo photosynthesis and may derive nutrition from the rootlike structures shown at (3) in the illustration. **Fig. 23.6** The archegonium protects the embryo for awhile, which is one function of a seed. It does not, however, contain stored food like a seed does. Instead, the sporophyte grows within gametophyte tissues until it becomes free-living. **Fig. 23.8** In angiosperms, the ovary completely encloses and protects the ovules. In the pine seed, the ovules are not surrounded and protected by an ovary. The ovules become the seeds, and in angiosperms, the ovary ripens into a fruit. **Fig. 23.11** Green peppers are considered fruits because they are the ripened ovary of an angiosperm, containing protected seeds. From a botanical perspective, tomatoes, squash, and cucumbers are also fruits.

## End-of-Chapter Answers

*Level 1*

**1.** d. **2.** c. **3.** a. **4.** b. **5.** mitotic, meiotic. **6.** eggs, sperm.
**7.** endosperm. **8.** ovary. **9.** anther, ovary. **10.** e. **11.** b, d, e, a, c.

*Level 2*

1. Dispersal of young is important because it prevents the young from growing next to their parents and competing for essential resources. The more primitive nonvascular and seedless vascular plants use haploid spores for dispersal. Gymnosperms produce a naked seed, while angiosperms produce a seed surrounded by fruit. Seeds are adaptations that provide nourishment to the growing embryo prior to its ability to carry out photosynthesis. In addition, a seed coat provides protection for the embryo. "Wings" on some seeds aid in disperal by wind. Edible fruit and burrs surrounding the seeds of angiosperms allow them to utilize animals to very effectively disperse the young from the parents.
2. See Figure 23.8.
3. In conifers and flowering plants, the microspore develops into the male gametophyte (pollen), while the megaspore develops into the female gametophyte. Plants such as ferns produce a bisexual gametophyte that contains both archegonia and antheridia and therefore produces both egg and sperm.
4. Both conifers and flowering plants have reduced gametophytes that are not separate and independent of the sporophyte; both produce pollen; pollen tubes; seeds with seed coats; and seeds that can remain dormant; and both can reproduce asexually through vegetative propagation.

*Level 3*

1. Both of these groups of plants rely on water for sexual reproduction. Sperm must swim to the egg cell, so these plants are tied to habitats that are moist or wet during some part of the year. You'll never see a moss or a fern growing in a desert!
2. Each Aspen grove is made up of genetic clones of many trees descended by vegetative reproduction. This is why they all change to the same color at the same time.
3. It is energy expensive to produce these energy-rich fruits, but fruit attracts animals that eat the fruit (and the seeds). They transport the seeds away from the parent plant. The alternative is to have the young growing near the parent plant, competing with it for essential resources (nutrients, water, light . . .).
4. The yellowish "dust" is pine pollen produced by the male cones that grow at the tips of the branches.
5. The capsule is the sporophyte stage of the moss, which germinates from the archegonia of the gametophyte after fertilization. The capsule of the sporophyte produces numerous haploid spores (the "puffs of smoke"), which are released and blown by the wind, thereby dispersing the new gametophytes away from their parents.
6. By attracting only one or a few species of pollinators, a plant insures that its pollen is efficiently transported to another plant of the same species, since its pollinators are likely to visit other plants of the same species. The pollinator also has a monopoly on a food source, which increases their survival chances.
7. Seeds that are only released and germinate after a forest fire (1) supply a new pool of seed to replace trees that may have been killed by the fire; (2) are likely to have a good chance of surviving because few underbrush plants will remain to compete with the seedling for available nutrients and water.

# CHAPTER 24

## Concept Checkpoint Answers

1. Meristematic tissue
2. **a.** Vascular tissue, xylem, tracheids and vessel elements. **b.** Ground tissue, sclerenchyma, fibers. **c.** Meristematic tissue, apical and lateral meristem. **d.** Ground tissue, parenchyma. **e.** Dermal tissue, epidermis. **f.** Meristematic tissue, lateral meristem.
3. Epidermis → root cortex (parenchyma) → endodermis→ pericycle → xylem

**4. a.** Pith, primary xylem and phloem, vascular cambium, cortex **b.** Secondary xylem and phloem. **c.** Secondary xylem and phloem **d.** Primary and secondary xylem
**5.** Soil → root epidermis → root cortex → endodermis→ pericycle → xylem → leaf mesophyll → leaf epidermis (through stomata) → atmosphere
**6.** Leaf mesophyll → phloem companion cell → phloem sieve-tube member → root pericycle → endodermis→ root cortex (parenchyma).

## Visual Thinking Answers

**Fig. 24.9** The root hair is a fine, threadlike, and delicate extension of a single cell. If root hairs were to grow from the root cap, they would be quickly sheared off as the root cap penetrated the soil. **Fig. 24.16** Between 10 and 100 years—a rough count of the rings indicates that the tree is about 50 years old. **Fig. 24.20** When the solute concentration in the guard cells is high, then the concentration of water is lower in the guard cells than in the surrounding cells. Water moves from an area of higher concentration (surrounding cells) to an area of lower concentration (guard cells) in a process termed osmosis. (See Chapter 5.) Similarly, when the solute concentration in the guard cells is low, then the concentration of water is higher in the guard cells than in the surrounding cells. Water moves from an area of higher concentration (guard cells) to an area of lower concentration (surrounding cells).

## End-of-Chapter Answers

*Level 1*

**1.** c. **2.** b. **3.** a. **4.** d. **5.** e. **6.** c. **7.** a. **8.** b. **9.** c. **10.** a.

*Level 2*

1. **a.** Xylem transport is always upward, from the roots through the stems to the leaves. Phloem sap transport is bidirectional but always flows from a sugar source to a sugar sink. **b.** The fluid transported in xylem is composed mainly of water and inorganic nutrients, phloem sap is composed mainly of water, organic molecules such as carbohydrates produced by photosynthesis, and some inorganic nutrients. **c.** Transport through xylem occurs because transpiration of water from leaf stomata results in a "pull" on a continuous column of water that is supported by the cohesion of water molecules (due to hydrogen bonding) and the adhesion of water molecules to the walls of xylem vessels. Phloem sap transport occurs because of a hydrostatic pressure gradient in phloem. High pressure at the sugar source end due to active loading of sugar into sieve-tube members, and low pressure at the sugar sink due to unloading of sugars result in pressure-driven bulk flow of phloem sap.
2. Primary growth occurs mainly at the tips of roots and shoots (apical meristems), making plants taller; secondary growth occurs in the lateral meristem and makes plants larger in diameter.
3. See Figure 24.9.
4. A dandelion has a taproot, a single major root that can grow deep as a firm anchor. Some taproots also store food for the plant. An ivy plant has adventitious roots that develop from the lower part of the stem and help anchor the plant. Grass has fibrous roots, an extensive root system without one major root; this type of system anchors the plant and absorbs nutrients over a large surface area.
5. Annual rings provide a clue to a tree's age by illustrating the annual pattern of rapid and slow growth in the cambium (lateral meristem tissue). This results in secondary growth.
6. See Figure 24.22.
7. In general, nonvascular plants grow closer to the ground than vascular plants because they lack the sophisticated transport systems of vascular plants.

*Level 3*

1. Air pressure still pushes on water in the ground, pushing water up the thin xylem tubes of the plant. In addition, the water molecules adhere to the walls of the xylem tubes and stick to one another (cohesion), which helps to maintain the column.
2. The fungi, because they are at the roots, probably function in making inorganic nutrients in the soil available to the plants.

**3.** Answers may vary. Here are some examples.
**Differences:**

| "Circulatory System" of Plants | Circulatory System of Animals |
|---|---|
| • Has no circulating pump | • Has a heart—a circulating pump |
| • Circulation is not a continuous loop; water moves from the roots to the leaves | • Circulation is a continuous loop throughout the body |
| • Fluid is water with dissolved | • Fluid is blood inorganic nutrients |

**Similarities:** In both, the "circulating" fluid is enclosed in vessels and is essential to the survival of the organism. Death results if too much of the fluid is lost.

**4.** You would look at the same spot—4.5 feet up the tree. The tree grows taller from the apical meristem, which would have been above the initials.

**5.** Answers may vary. Examples are: Plants contain fibrous material, which is beneficial in the diet. Dietary fiber helps substances move through the large intestine and gives the muscles of the large intestine bulk against which to push. Plants also contain inorganic nutrients, which supply the human body with minerals it needs for proper nutrition.

**6.** Xylem vessels are coated with small pores called pits that allow water to divert laterally around these air pockets into adjacent xylem vessels. Even so, over several winters, these air pockets can accumulate in xylem vessels. Each spring, however, the vascular cambium lays down new secondary xylem that is free of these air pockets.

**7.** Since water is pulled up the plant by transpiration, even a rootless plant can absorb a significant amount of water through the stem. Florists often recommend that you cut the stems of your roses under water to prevent air bubbles from forming in the xylem vessels. This would disrupt the continuous column of water needed to help pull more water up to the leaves and flowers.

# CHAPTER 25

## Concept Checkpoint Answers

**1.** Growth patterns are not nearly as genetically and physiologically predetermined in plants as they are in animals. For example, plants may respond to environmental signals by altering their pattern of growth to maximize light, water, or nutrient acquisition.

**2.** **a.** Auxins. **b.** Ethylene. **c.** Auxins. **d.** Brassinosteroids.

**3.** **a.** Extrinsic signal—light; intercellular control mechanism—phytochrome triggers the seed to imbibe water. **b.** Extrinsic signal—light; intercellular control mechanism—auxin on the side of the stem that is receiving less light causes cells to elongate, resulting in stems that grow toward light coming in through the window. **c.** Extrinsic signal—gravity; intercellular control mechanism—auxin on the lower surface of the root inhibits cell elongation, resulting in stems that grow downward in the same direction as the pull of gravity.

**4.** **a.** Avoidance. **b.** Acclimation. **c.** Senescence.

**5.** Photoperiodism and vernalization—Winter rye requires a cold treatment over the winter to render the seedling sensitive to photoperiodic control of flowering.

## Visual Thinking Answer

**Fig. 25.5** The cells on the sunny side of the stem should be shorter than the cells on the dark side of the stem.

## End-of-Chapter Answers

*Level 1*

**1.** e. **2.** e. **3.** b. **4.** e. **5.** F. **6.** T. **7.** F. **8.** T. **9.** F. **10.** F.

*Level 2*

**1. Gravity**—auxin diffuses to the lower side of the organ, increasing its concentration relative to the upper side of the organ. In negatively gravitropic organs, such as shoots, auxin causes greater elongation of the cells on the side of the shoot closer to the ground, resulting in the shoot curving upward. The effect is the opposite in positively gravitropic organs such as roots. **Light**—Phytochromes change form in response to different wavelengths of light and respond to the duration of dark periods between light. These changes are likely to affect the production of plant hormones through the action of gene promoters and inhibitors. **Availability of water**—Abscisic acid production is stepped up in response to water loss. This stimulates closure of stomata, thus conserving water.

**2.** Adaptation refers to the heritable modifications (encoded in a plant's genes) in structure or function that increase an organism's fitness in a stressful environment. Acclimation refers to nonheritable physiologic or growth modifications that occur during the life of a plant that help it deal with an environmental stress. Of course, the capacity to acclimate is heritable, but the specific physiologic changes that occur during a plant's life are not.

**3.** Plants continue to grow throughout their lifetimes. It is essential that meristem cells be able to differentiate into any of the cell types required to grow various tissues comprising a plant organ (such as xylem, phloem, and parenchyma) as it responds to external environmental signals.

**4.** Light exposure, abrasion of seed coat, availability of water—When phytochromes in the seed detect certain wavelengths of red light, it is likely that the seed has been dispersed far enough away from a dense population of trees or shrubs that would compete for soil nutrients, sunlight, or water. The subsequent change in the phytochromes due to absorption of red light triggers the events of seed germination.

**5.** Water uptake by the seed is necessary for metabolic breakdown of nutrients. Water helps activate enzymes already in the seed and provides a fluid medium in which chemical reactions can take place, such as the synthesis of new enzymes to help digest and use the food stored in the seed. With the intake of water, the seed coat ruptures, allowing oxygen to enter. Anaerobic metabolism becomes aerobic.

**6.** This refers to nastic movements in plants such as the opening and closing of flowers or the closing of the leaves of the Venus flytrap around an insect. Nastic movements generally result from changes in the sizes of motor cells, caused by movement of water into or out of them. The movement of water follows the movement of ions, which pass into or out of cells as ion channels open or close as a result of response to the stimulus. In many plants, shrinkage of the motor cells due to water loss causes the overall movement of the plant.

*Level 3*

**1. a. Avoidance**—the seeds are avoiding low sunlight, nutrient, or water levels that may be present if thick shrubs are present on the forest floor. Periodic low-intensity fires that sweep through these communities clear away the underbrush, providing a more favorable environment for germination of seeds. **b. Avoidance**—these plants are restricting their active life cycle to the season in which water is most available and survival chances are greatest. **c. Acclimation**—these plants acquire the ability to withstand freezing temperatures over their lifetimes.

**2.** Apples that have spoiled release large amounts of ethylene, which will rapidly cause other apples nearby to ripen more quickly and rot.

**3.** Tulips require a vernalization period, a continuous long stretch of cold prior to germination, for flowering to occur. Since the winter in the South is too warm to provide this cold period, gardeners dig up their bulbs and put them in the refrigerator to simulate the cold treatment that is a natural part of northern climates.

**4.** Poinsettias are short-day flowering plants that require a long period of uninterrupted darkness at night. In the spring, place the plants in a dark room every evening where it will experience approximately 12 uninterrupted hours. In the morning, remove the plant from the dark room so that it receives the essential light that it needs for photosynthesis. Continue to do this over several weeks until flowering is induced.

# CHAPTER 26

## Concept Checkpoint Answers

**1. Simple epithelium:** diffusion, filtration, secretion, excretion, absorption; squamous, cuboidal, or columnar cells; lining blood vessels. **Stratified epithelium:** protection squamous, cuboidal, or columnar cells; skin. **Pseudo-stratified**

**epithelium:** protection, secretion of mucus; columnar-type cells; airways. **Transitional epithelium:** stretching; cuboidal cells; bladder.

2. Blood has a fluid matrix (plasma), which allows it to be pumped to the various tissues of the body. Dissolved in this fluid are various nutrients, hormones, waste products, and salts, which are carried in the blood to various parts of the body. The cells of the blood carry respiratory gases, provide protection from pathogens, and clean up cellular debris. Bone is comprised of osteoblasts that secrete an extracellular matrix of collagen fibers embedded in calcium salts. The mineralized part of the matrix makes bone rigid, while the collagen keeps it from being brittle. Tendons are comprised of rows of fibroblasts that secrete collagen fibers, which allow for strong connections between bones and muscles.

3. All three muscle types are made up of contractile cells containing specialized filaments organized in fibers (myofibrils). Interaction of these fibers causes muscle cells to contract. However, the way in which they are stimulated to contract, and the general way in which the tissue contracts, vary among the three types of muscle. Skeletal muscle is stimulated to contract by voluntary nerve impulses. All muscle fibers within a cell contract simultaneously upon stimulation. Smooth muscle cells can be stimulated to contract by involuntary nerve impulses, hormones, or stretching. In addition, impulses may be able to pass directly from one smooth muscle cell to another, resulting in a self-propagating wave of contraction throughout a layer of the muscle. Cardiac muscle cells are organized into fibers that branch and interconnect, forming a latticework, which allows an entire portion of the heart to contract at one time.

4. Neurons have projections from the cell body called dendrites, which allow them to receive nerve impulses from many other nerve cells, including sensory cells. They also have an axon, a long extension from the cell body that allows them to transmit impulses to other nerve cells and effector cells such as muscles or glands.

5. **a.** Negative. **b.** Positive. **c.** Negative.

## Visual Thinking Answers

**Fig. 26.8** A tissue is a group of cells working together to perform a function. The illustration shows muscle tissue—groups of muscle fibers (cells) that work together to perform the job of contracting. Connective tissue is also shown. **Fig. 26.12** The feedback is considered negative because the response is opposite to the initial trigger. In the loop at the upper right, the response is the lowering of body temperature to a normal level, which is opposite to the initial rise in body temperature. In the loop at the lower left, the response is the raising of body temperature to a normal level, which is opposite to the initial decrease in body temperature.

## End-of-Chapter Answers

*Level 1*

**1.** a. **2.** e. **3.** d. **4.** b. **5.** c. **6.** a. **7.** e. **8.** d. **9.** a. **10.** b. **11.** c.

*Level 2*

1. (Examples will vary.) A cell is the smallest structure in an organism that is capable of performing all the functions necessary for life. An example is a cardiac muscle cell. The cells of the body are organized into tissues (groups of similar cells that work together to perform a function, such as cardiac muscle tissue). Several different tissues group together to form a structural and functional unit—an organ (for example, the heart). An organ system is a group of organs that works together to carry out the body's principal activities (such as the circulatory system).

2. (Examples will vary.) The four types of tissue in the body are: (1) epithelial (lining of the body cavities), (2) connective (tendons), (3) muscle (skeletal muscles such as leg muscles), and (4) nervous (the brain).

3. (Examples will vary.) See Figure 26.3. Squamous cells are found in the air sacs of the lungs, cuboidal cells line ducts in glands, and columnar cells line much of the digestive tract.

4. Lymphocytes, erythrocytes, and fibroblasts are all cells found in connective tissue, but they have different functions. In addition, lymphocytes and erythrocytes are found in the blood, but fibroblasts are found in other types of connective tissue. Lymphocytes defend the body against infection; erythrocytes pick up and deliver gases in the blood; and fibroblasts produce fibers that occur in various types of connective tissue.

5. Oxygen delivery by means of red blood cells is necessary for cells to use the aerobic process of cellular respiration.

6. Yes, fat cells have their uses. In addition to storing fuel, fat tissue helps shape and pad the body and insulate it against heat loss.

7. Skeletal muscle; cardiac muscle.

8. See Table 26.1.

9. Homeostasis is the maintenance of a stable environment inside your body despite varying conditions in the external environment. The molecules, cells, tissues, organs, and organ systems in your body must all work together to maintain this internal equilibrium.

*Level 3*

1. Like most living tissues, bone is constantly renewing itself. Blood vessels within the bone bring calcium and other substances necessary to heal a break to the area. In addition, the bone cells are able to lay down new matrix to help repair a break.

2. This is part of the body's inflammation response to injury. Mast cells produce histamine, which dilates blood vessels and increases blood flow to the area. This makes the area turn red and feel warm.

3. The connective tissue matrix may provide a nutrient pool for connective tissue cells, allow for waste removal, and provide an environment enhancing cell mobility. Proteins secreted into the matrix may provide a fibrous, stronger structure to matrix, such as that found in bone and cartilage.

4. (Answers may vary, for the chapter gave only the basics on which to base an answer.) Nervous tissue is specialized for conducting electrochemical impulses. The brain is specialized, in addition, for interpreting those impulses. This suggests that neurons are connected to one another in particular pathways, unlike other tissues such as skin. Nervous tissue in the brain and spinal cord might be incapable of regeneration because of this complex specialization. Other tissues are not as a specialized.

# CHAPTER 27

## Concept Checkpoint Answers

1. **a.** Vitamins are organic molecules; minerals are inorganic substances. **b.** Micronutrients (vitamins and minerals) are required only in small quantities. Macronutrients (carbohydrates, proteins, lipids) are required in substantially higher quantities. **c.** Essential amino acids must be supplied in the diet because they cannot be synthesized in the body; Conditionally, essential amino acids can be synthesized in the body if the diet supplies other materials needed to build them; Nonessential amino acids can be manufactured from other amino acids in the body. **d.** Unsaturated fats are certain plant fats whose fatty acids are not saturated with hydrogen. These are the healthiest fats. Saturated fats are mainly animal fats whose fatty acids are saturated with hydrogen. These fats are relatively unhealthy. Trans fats have had their fatty acids hydrogenated to make them more solid. These appear to be the unhealthiest of all the fats.

2. No specialized digestive organs (intracellular digestion) → digestive cavity combined with intracellular digestion → complete digestive tract with specialized organs that break down, store, digest, and absorb food; and eliminate wastes.

3. Tapeworms are parasitic organisms that live within the digestive tract of vertebrates. They simply absorb food that their host has already digested.

4. mouth (mechanical breakdown of food; salivary amylase hydrolyzes starch) → pharynx → peristalsis in the esophagus → through the lower esophageal sphincter → stomach (mechanical churning; secretes HCl and pepsin which digests proteins) → through the pyloric sphincter → small intestine

5. Carbohydrates—Pancreatic amylase secreted into the duodenum by the pancreas breaks down starch and glycogen into maltose. Disaccharidases produced in the small intestine digest maltose, sucrose and lactose into monosaccharides. Proteins are digested initially in the stomach through the action of pepsin. Trypsin, chymotrypsin, and carboxypeptidase produced by the pancreas break down polypeptides into shorter chains and then to amino acids. Peptidases produced by the small intestine also break down polypeptides.

6. Nearly all of lipid digestion occurs in the small intestine. Because fats are hydrophobic and insoluble in water, globules of fats are first emulsified, or separated into small droplets, by bile salts. This increases the surface area of contact with pancreatic lipase, which breaks the fats into fatty acids and glycerol.

7. The small intestine is relatively long and the epithelial cells lining the lumen have villi and microvilli, which together provide a huge surface area for the absorption of the products of digestion. The walls of the small intestine are also lined with smooth muscle, which move food through the intestine via waves of muscular contraction (peristalsis).

## Visual Thinking Answers

**Fig. 27.3** The prefix sub- means "under" or "beneath." Therefore, the submandibular gland is located under the mandible (lower jaw). The sublingual gland is located under the tongue. **Fig. 27.5** The pattern of teeth in humans is characteristic of an omnivore; thus, it has characteristics of both the herbivore and the carnivore. The canine teeth are pointed for biting and tearing, although modified from the carnivore. The molars (back teeth) are flattened much like those of the herbivore for grinding. The front incisors, also similar to those of an herbivore, are used for biting. **Fig. 27.13** As food passes along the small intestine and nears the large intestine, it pushes open the valve, which is pointed in the direction of flow. Food on the large intestine side of the valve, however, provides backpressure on the flaps of the valve, which are pointed away from that backward direction of flow. This pressure pushes the flaps of the valve together, closing the "door" to the small intestine.

## End-of-Chapter Answers

*Level 1*

1. b.  2. b.  3. d.  4. a.  5. c, ww.  6. b, ww.  7. c, ww.  8. a, ww.
9. c, vv.  10. a, uu.

*Level 2*

1. The six classes of nutrients are carbohydrates, fats, proteins, vitamins, minerals, and water. The first three are organic compounds that your body uses as a source of energy; they are also used as building blocks for growth, repair, maintenance, or reproduction. Vitamins, minerals, and water help body processes take place and allow vital chemical reactions. Some minerals are also part of body structures.

2. Three types of enzymes help to digest the energy nutrients. Proteases break down proteins into peptides and amino acids. Amylases break down carbohydrates (starches and glycogen) to sugars. Lipases break down the triglycerides in lipids to fatty acids and glycerol.

3. The process of digestion begins in the mouth, as the teeth crush the food, and the salivary amylase in salvia breaks down starches into maltose.

4. The liver and the pancreas perform other functions in addition to secreting substances used in digestion. In addition, they are not part of the digestive pathway that begins at the mouth and ends at the anus. They are accessory to it.

5. Stomach acid (HCl) reacts with sodium bicarbonate ($NaHCO_3$) to form carbonic acid ($H_2CO_3$) and salt (sodium chloride, $NaCl$). The carbonic acid quickly decomposes to carbon dioxide ($CO_2$) and water ($H_2O$).

6. Hydrochloric acid in the stomach converts pepsinogen to the active form of the enzyme pepsin, softens connective tissue in food, denatures large proteins, and kills bacteria. The stomach wall is protected by thick mucus, gastric juice is not secreted unless there is food in the stomach.

7. The duodenum is the first part of the small intestine; digestion is completed there. Starch and glycogen are broken down to disaccharides and then to monosaccharides. Proteins and polypeptides are broken down to shorter peptides and then to amino acids. Triglycerides are digested to fatty acids and glycerol.

8. These are hormones that help regulate digestion. Gastrin controls the release of gastric juice in the stomach. Secretin stimulates the release of sodium bicarbonate which neutralizes acid in the chyme, and increases bile secretion in the liver. Cholecystokinin stimulates the gallbladder to release bile into the small intestine and stimulates the pancreas to release digestive enzymes.

9. Increasing the internal surface area of the small intestine allows more absorption of nutrients from a given amount of food. Three features accomplish this:

inner folds, projections (villi) on the folds, and additional projections on the projections (microvilli).

10. Glucose is broken down by glycolysis, the Krebs cycle, and the electron transport chain. Amino acids and triglycerides are converted to substances that can be metabolized by two of these pathways. Fatty acids are metabolized by another pathway.

*Level 3*

1. Meat and animal products provide protein in the diet and some fat. To replace animal protein, a vegetarian could eat legumes, such as dried peas and beans, soy-based food, grains, nuts, and seeds. To ensure ingestion of the essential amino acids, legumes should be combined with any grain, nut, or seed supplemented with small amounts of milk, cheese, yogurt, or eggs. Persons who are not strict vegetarians might also consider eating small amounts of fish or poultry.

2. The shape of the teeth on the skull would indicate the type of food eaten by the animal. Herbivores (plant-eaters) have flat, grinding teeth that look like human molars. The student might also look for sharp front incisors that could have been used to bite off plants.

3.

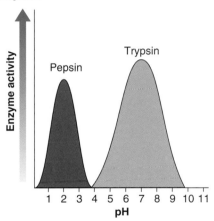

Pepsin is secreted into the stomach along with HCl. Because of the very acidic environment in the stomach, one would expect the maximum activity of pepsin to be at a pH close to that of the interior of the stomach. Trypsin, however, is secreted into the duodenum of the small intestine after the chyme has been neutralized, so one would expect its maximum activity to be at a pH closer to 7.

4. This prevents pepsin from hydrolyzing the proteins in the cells that secrete it into the stomach.

5. Answers will vary. See Section 27.2 and the *How Science Works* box. Comparisons should be made to percentage intake of fats, proteins, and carbohydrates, or to numbers of servings of categories of foods recommended.

6. Choking occurs when a piece of food gets lodged in the trachea, or windpipe. During swallowing a small muscular flap called the epiglottis covers the glottis (the opening to the windpipe); food is prevented from entering the windpipe and continues into the esophagus.

# CHAPTER 28

## Concept Checkpoint Answers

1. Most of these organisms "breathe" through their skin. They are small, and/or have fairly large surface area to volume ratios by being elongated, branched, or fringed. Larger animals such as amphibians breathe through skin that is infiltrated with a rich supply of capillaries.

2. Gills have a very large surface area for diffusion of gases. Countercurrent flow of water through the gills and blood through the gill capillaries maximizes oxygenation of the blood.

3. nasal cavities → pharynx → larynx → glottis → trachea → bronchi → bronchioles → alveoli → alveolar capillaries

4. The nervous system senses pH related to levels of $CO_2$ in the blood, and responds by increasing breathing rate as pH decreases, signaling an increase in blood $CO_2$ levels.

5. This could be any capillary bed in the body, except for the capillaries surrounding the alveoli.
6. This must be the alveolar capillaries because the blood is picking up $O_2$ and giving up $CO_2$.

## Visual Thinking Answers

**Fig. 28.6** Arteries, arterioles, and eventually capillaries at the aveoli; vein; the heart (left side), to be pumped to the body. **Fig. 28.9** When the diaphragm is pulled down, the air pressure within the jar drops. Air moves from an area of higher pressure outside the jar to the area of reduced pressure inside the jar.

## End-of-Chapter Answers

*Level 1*

**1.** a.  **2.** b.  **3.** c.  **4.** c.  **5.** a.  **6.** e.  **7.** F.  **8.** F.  **9.** T.  **10.** T.

*Level 2*

1. Organisms without special organs of respiration exchange respiratory gases over cell and skin surfaces. These organisms are either single-celled, multicellular with a large surface-to-volume ratio, or large with thin, moist skins supplied with abundant blood vessels.
2. *Respiration* is the uptake of oxygen and the release of carbon dioxide by the body. *Cellular respiration* is the process at the cellular level that uses the oxygen you breathe in and produces the carbon dioxide you breathe out. *Internal respiration* is the exchange of oxygen and carbon dioxide between the blood and tissue fluid; the exchange of the two gases between the blood and alveoli is *external respiration*.
3. Your vocal cords produce the sound for speech as air rushes by and makes them vibrate. Your lungs serve as a power supply and volume control for your voice, and your lips and tongue form the sounds into words.
4. As the volume of the thoracic cavity increases, the air pressure within it decreases. Outside air flows in, equalizing the pressures inside and outside the thoracic cavity. As the volume of the thoracic cavity decreases, the air pressure inside it increases and forces air out, equalizing the pressures.
5. At the alveoli, carbon dioxide diffuses out of the blood into the alveoli because it moves down its pressure gradient (its pressure in the blood is greater than its pressure within the alveoli). Meanwhile, oxygen diffuses from the alveoli into the blood because its pressure within the lungs is greater than in the blood.
6. Hemoglobin is known as the oxygen-carrying molecule, but it also carries almost 25% of circulating carbon dioxide in the bloodstream.
7. E, A, B
8. At the capillaries, oxygen moves down a pressure gradient to diffuse from the blood into the tissue fluid. Meanwhile, carbon dioxide moves down its own pressure gradient and diffuses from the tissue fluid into the blood.
9. At high altitudes, the pressure of the oxygen molecules in the air is lower than at sea level. This means that the pressure gradient is lower at the alveoli and that less oxygen diffuses into the blood. This can cause you to feel dizzy and short of breath.
10. Increased hemoglobin levels allow the blood to carry more oxygen to the tissues of the body to support aerobic metabolism. Remember from Chapter 8 that oxygen is necessary for ATP production and that when muscles are not being supplied with enough $O_2$, they revert to more inefficient anaerobic metabolism (fermentation). Increased hemoglobin levels in the blood ensure that during strenuous activity, muscle cells can stay in the aerobic phase longer.

*Level 3*

1. During hibernation, the metabolism of the frog slows, reducing the amount of oxygen it uses. Buried in the mud, the frog breathes through its thin skin, which has an abundant supply of blood vessels just beneath the skin surface. Air is available in the mud, dissolved in the water surrounding the microscopic spaces between particles of dirt and sand.
2. The sound of the hiccup occurs when the suddenly inhaled air causes the opening and quick closing of the vocal cords, producing a loud gasp.

3. The circulatory system provides a transport system that distributes gases throughout the body. Without it, you would not survive because it would take too long for oxygen to diffuse from the lungs to the rest of the body.
4. One of the primary symptoms of respiratory diseases is that patients tire easily. This is because their decreased lung capacity does not allow sufficient oxygenation of blood; less oxygen reaches the brain and other organ systems.
5. Water and blood flow over the gill surface in opposite directions, a mechanism called countercurrent flow. As oxygen diffuses out of the water, its partial pressure drops steadily. As oxygen diffuses into the blood, its partial pressure rises steadily. Since the blood is moving in a direction opposite to that of the water, the partial pressure of oxygen in the water never drops below that in the blood. If these two fluids moved in the same direction, diffusion would soon stop as soon as the partial pressure of oxygen in the blood equaled that in the water.
5. Emphysema results in loss of elasticity in the lungs. As such, one might expect the magnitudes of A, C, D, and E to decrease, while the magnitude of B would increase as more residual air would be left in the lungs due to their reduced elastic resilience.

# CHAPTER 29

## Concept Checkpoint Answers

1. **ameba**—no circulatory system; relies on diffusion to and from the surrounding environment; **flatworm**—a gastrovascular cavity branches throughout the body bringing nutrients to the body tissues, gas exchange is through diffusion to and from the surrounding environment; **oyster**—open circulatory system; **housefly**—open circulatory system carries nutrients; gas exchange occurs through a tracheal system; **octopus**—closed circulatory system; **salmon**—closed circulatory system with a two-chambered heart and a single circulatory loop; **salamander**—closed circulatory system with a three-chambered heart and a double circulatory loop with mixing of blood from each loop in a single ventricle; **hawk**—closed circulatory system with a four-chambered heart and a separate double circulatory loop; **wolf**—closed circulatory system with a four-chambered heart and a separate double circulatory loop.
2. vena cava, right atrium, right atrioventriular valve, right ventricle, pulmonary semilunar valve, pulmonary artery, lungs, pulmonary vein, left atrium, left atrioventricular valve, left ventricle, aortic semilunar valve, aorta
3. vena cava, right atrium, right ventricle, pulmonary artery; Pressure would be highest in the aorta.
4. hemoglobin, an oxygen-binding protein found in red blood cells
5. Formed elements of the blood
   **A. Erythrocytes.** (1) Structure—lack nuclei; small, concave, and round; contain hemoglobin. (2) Function—carry respiratory gases to and from tissues. **B. Leukocytes.** (1) Granulocytes. **a.** neutrophils—phagocytize microorganisms and foreign particles. **b.** basophils—rupture and release chemicals that enhance the body's response to injury or infection; play a role in causing allergic responses. **c.** eosinophils—involved in allergic reactions and act against certain parasitic worms. (2) Agranulocytes. **a.** monocytes—attracted to the sites of injury or infection, where they are converted into amebalike macrophages that phagocytize microorganisms and particles of foreign matter, and clean up any bacteria and dead cells after neutrophils. **b.** lymphocytes—recognize and react to substances that are foreign to the body, sometimes producing a protective immunity to disease. Also involved in inflammation and allergic responses. **C. Platelets**—cellular fragments pinched from megakaryocytes; produce proteins and enzymes involved in blood clotting.

## Visual Thinking Answers

**Fig. 29.4** In the vein, blood flows in the direction the valve is pointing. As blood moves in this direction, it easily opens the valve. However, if the blood flows back, then it presses on the curved portions of the valve, shutting the valve tightly. **Fig. 29.7** The left ventricle pumps blood to the body, while the right ventricle pumps blood to the lungs. Pumping blood throughout the body requires a stronger push than pumping blood to the lungs which surround the heart. Thus, the musculature of the left ventricle is thicker than the musculature of the right ventricle. **Fig. 29.10** Diastole is a period of relaxation of the heart chambers. Systole is a period of contraction. The atria and the ventricles cannot pump blood at the same time because ventricular contraction would push blood back into the atria if the atrioventricular valves opened when the atria contracted. Conversely, if the ventricular pumping forced these valves shut, then atrial pumping would do no good. The atria and ventricles must pump in succession for proper blood flow through the heart.

## End-of-Chapter Answers

*Level 1*

**1.** e.  **2.** b.  **3.** b.  **4.** b.  **5.** d.  **6.** F.  **7.** T.  **8.** F.  **9.** T.  **10.** F.

*Level 2*

1. The blood carries sugars, amino acids, and fatty acids to the liver, where some of the molecules are converted to glucose and released into the bloodstream. Excess energy molecules are stored for later use. Essential amino acids and vitamins pass through the liver into the bloodstream. The cells release their metabolic waste products into the blood, which carries them to the kidneys.
2. As blood flows through tissue, from capillaries to venules, it loses gases (primarily oxygen) and nutrients to the tissues. The blood in the venules is deoxygenated and carries waste products from the tissues.
3. When people are scared (or cold), the walls of some arterioles can contract and the blood flow through them decreases, routing more blood from the skin to other areas of the body such as the muscles. This can cause light-skinned people to "turn pale."
4. The walls of arteries have more elastic tissue and smooth muscle than those of veins; this helps arteries accommodate the pulses and high pressures of blood pumped from the heart. Veins contain one-way valves to keep blood moving toward the heart as it is pushed along by contractions of skeletal muscle. The walls of capillaries are only one cell thick, which permits the exchange of substances between the cells and the blood.
5. The heart has left and right sides, which both serve as pumps, yet are not directly connected with one another. Oxygenated blood from the lungs enters and is pumped from the left side, while deoxygenated systemic blood passes through the right side of the heart into the lungs. These two types of blood must be separated—hence the two-pump system.
6. Valves are necessary in heart function to prevent the backflow of blood.
7. The SA node is the pacemaker of the heart; this cluster of cells initiates the excitatory impulse that causes the atria to contract and the impulse to be passed along to other cardiac cells. The AV node conducts the impulse from the SA node to other cardiac cells, initiating the contraction of the ventricles.
8. See Figure 29.11. Plasma is the fluid in which various substances, such as nutrients and ions, are dissolved. Plasma carries these dissolved substances and the formed elements throughout the body and provides the fluid in which chemical reactions and the transport of substances takes place.
9. **a.** True—During diastole, both ventricles are relaxing. **b.** False—During diastole both ventricles are relaxing. **c.** True—The semilunar valves are closed, thereby preventing blood from the aorta and pulmonary arteries from flowing back into the ventricles when they are relaxing. **d.** False—The SA node stimulates the simultaneous contraction of the atria.

*Level 3*

1. The student should list family members who have (or have had) heart disease. The student should also list lifestyle factors that may contribute to heart disease such as whether they consume a high-fat diet, are overweight, smoke cigarettes, or exercise regularly. See Table 29.1.
2. Obviously, one cannot change one's genetics. The other factors are behavioral and can be changed. Answers will vary depending on the answer to question 1.
3. The first heart sound occurs at the start of ventricular contraction (systole), so it should be shown on your diagram just after the "R" peak. It represents the closing of the atrioventricular valves. The second heart sound occurs at the end of ventricular contraction and should coincide with the end of ventricular systole (during the T wave). It represents the closing of the semilunar valves.
4. Insects transport respiratory gases directly to and from their tissues via a tracheal system rather than a circulatory system.

5. Lymph nodes filter the lymph as it passes through. When cancer spreads, some cells break away from the original tumor and travel to other parts of the body in the blood and lymph. If cancer cells are traveling from the original site of the cancer (the breast), some of these cells may become trapped by the lymph nodes nearest to that site (nodes under the arms).
6. In the first case, the patient has a leaky atrioventricular valve because the first lub sound is the result of the turbulence surrounding the closing of these valves. As the ventricles then contract, blood is leaking through these valves back into the atria, which makes the hiss sound. In the second case, it is likely that one of the semilunar valves is leaking.
7. Bone marrow contains the precursor cells that give rise to all blood cells, including leukocytes, which play an important role in defending the body against invading microorganisms. The patient must be protected from contagious disease until the transplanted bone marrow begins to produce white blood cells.
8. Fluid would accumulate in the blocked lymphatic vessels leading to severe swelling (also known as edema).

## CHAPTER 30

### Concept Checkpoint Answers

1. Nonspecific immunity works to keep out any foreign invader while specific immunity responds specifically to each particular type of microbe. The process of inflammation defends against foreign microbes and substances in general by bringing an increased blood supply to the injured area, which brings defensive substances and cells there, and dilutes any toxins that may be present. Blood clots wall off the area, preventing the spread of the microbes or substances.
2. **Cell-mediated immune response:** T cells carry out the cell-mediated immune response, during which cytotoxic T cells recognize and destroy body cells infected with certain bacteria, viruses, and fungi. In addition, they destroy transplanted cells and cancer cells. Helper T cells initiate the response, attracting macrophages and activating cytotoxic T cells, inducer T cells, and, as the infection subsides, suppressor T cells. **Antibody-mediated immune response:** B cells carry out the antibody-mediated immune response in which B cells recognize foreign antigens and, if activated by helper T cells, produce large quantities of antibody molecules directed against the antigen. The antibodies bind to the antigens they encounter and mark them for destruction.
3. This analogy is not quite accurate. A better analogy would be a comparison of the immune system to purchasing a suit from a department store in the sense that you choose a suit off the store rack that fits your measurements and specifications. The suit is already in the store waiting for you to purchase it! This is like the immune response because when a pathogen invades the body, the immune system is generally pre-prepared for that attack—the body already has a population of antibodies specific to antigens on the pathogen. The immune system is prepared for a certain antigen to appear just as your suit was in the store when you arrived.
4. "Live" or active vaccines like the polio vaccine are attenuated versions of the virus that are unable to infect cells, but still possess the antigens that stimulate an immune response and production of memory cells.
5. Because of the central role of helper T cells in arming both components of specific immunity, the attack on these cells by HIV effectively cripples the immune response.

### Visual Thinking Answers

**Fig. 30.3** Mechanical, chemical and biological barriers are a "first line" of defense because they work to exclude any foreign invader from entry into the body. If these mechanisms are unsuccessful, other nonspecific mechanisms such as inflammation hold growth of the pathogen in check until the body's specific resistance mechanisms have time to mount a defense. **Fig. 30.7** Specific immunity is triggered when macrophages engulf bacteria and display parts of the bacteria on their surface that act as an antigen. Interaction with the helper T cells stimulates both "arms" of the immune system: the cell-mediated immune response and the antibody-mediated immune response. **Fig. 30.13** Helper T cells are key to stimulating both the cell-mediated immune response and the antibody-mediated immune response.

# End-of-Chapter Answers

*Level 1*

**1.** c.  **2.** d.  **3.** d.  **4.** b.  **5.** i.  **6.** d.  **7.** h.  **8.** f.  **9.** c.

*Level 2*

1. See Table 30.1 and descriptions in Section 30.3.
2. **a.** Corrected—When a virus *re-invades* the body, there are usually memory T cells in the body that are prepared to destroy virally infected body cells. **b.** Corrected—When a virus invades the body and infects cells, generally only one component of specific immunity is armed: *the cell-mediated immune response.* **c.** This choice is correct!
3. The transplanted cell would be identified as nonself; therefore, it would probably act as an antigen and produce an immune response in the new environment.
4. The five principal types of T cells are: (1) helper T cells that initiate the specific immune response, (2) cytotoxic T cells that break apart infected and foreign cells, (3) inducer T cells that oversee the development of T cells in the thymus, (4) suppressor T cells that limit the immune response, and (5) memory T cells that respond quickly and vigorously if the same foreign antigen reappears.
5. During the cell-mediated immune response, cytotoxic T cells recognize and destroy infected body cells; they may also destroy transplanted and cancer cells. Helper T cells initiate the response, activating cytotoxic T cells, macrophages, inducer T cells, and, finally, suppressor T cells.
6. During the antibody-mediated response, B cells recognize foreign antigens and, if activated by helper T cells, produce antibodies. The antibodies bind to the antigens and mark them for destruction.
7. Antibodies have highly specific binding sites that fit only one specific epitope with the exact amino acid composition necessary for complete binding.
8. Injection with antigens causes B cells to produce antibody and generate memory cells. A booster shot induces these memory cells to differentiate into antibody-producing plasma cells as well as to form still more memory cells. If the child is exposed to tetanus, the memory cells will again launch an antibody response to kill the pathogen.
9. The human immunodeficiency virus (HIV), which is responsible for AIDS, is dangerous because it destroys helper T cells, thus destroying the immune system's ability to mount a specific defense against infection.
10. An allergic reaction occurs when the immune system mounts a defense against a harmless antigen. When class E (IgE) antibodies bind to the antigens, they cause a strong inflammatory response that can dilate blood vessels and lead to symptoms ranging from uncomfortable to life-threatening.

*Level 3*

1. The cowpox virus had antigens that were nearly identical to smallpox antigens. They stimulated the production of memory cells that were also effective against smallpox.
2. Memory T cells were developed by your immune system as part of your immune response to the measles. These cells give you an accelerated response during subsequent exposures to measles, promptly defending you against infection.
3. This is a case of passive immunity, in which an organism is protected not by its own immune response, but is protected by antibodies from another source (in this case, the maternal immune system). This protection is temporary because these cells will eventually die, and the baby has produced no memory cells.
4. Antihistamines suppress the action of histamines, which are released during an allergic reaction. Histamines cause a variety of responses, such as the dilation of blood vessels. An antihistamine would block the production of histamine, diminishing its effects.
5. This strain of the flu virus probably had antigens that were different from anything that humans had ever experienced. As a result, the immune systems of most people lacked sufficient populations of T lymphocytes that were able to recognize these antigens. Therefore, the body's immune response to the flu was very weak. The virus was probably quite virulent as well.
6. Europeans had been exposed to European pathogens for many generations and had evolved immune systems capable of providing protection from these diseases. The native Indians, however, had never encountered European pathogens, and therefore had not evolved immune systems that could develop a significant immune response to them. At the cellular/molecular level, the immune systems of native Indians lacked sufficient populations of B and T cells that could respond to antigens on the newly introduced pathogens.

# CHAPTER 31

## Concept Checkpoint Answers

1. **Cell membrane**—simple diffusion, osmosis and active transport; bacteria, fungi, plants, plantlike and funguslike protists, cnidarians, and echinoderms. **Contractile vacuoles**—water taken up by osmosis in some animals that live in fresh water is pumped out by vacuoles that contract; fresh water protozoans and sponges. **Protonephridia**— cilia move the water and dissolved substances (filtrate) from surrounding fluids into the flame cell tubules and out excretory pores. Flame cells may modify the composition of the filtrate before it is excreted; flatworms. **Metanephridia (kidneys)**— similar to protonephridia except that selective tubular secretion and reabsorption of some substances to and from filtrate occurs, which helps maintain osmotic balance; mollusks. **Gills and antennal gland**—Gills excrete nitrogenous waste; the antennal glands are responsible for water and salt balance through filtration, tubular secretion, and reabsorption; Crustaceans. **Malpighian tubules**—Tubes within the blood-filled body cavity collect water, salts and wastes by diffusion or active transport, and empty into the intestine. Selective reabsorption and tubular secretion adjust the composition of the filtrate; spiders and insects. **Vertebrate kidneys**—composed of nephrons that carry out filtration, selective reabsorption, and tubular secretion to produce liquid waste called urine; humans.
2. **Glucose:** Afferent arteriole → glomerulus → glomerular capsule → proximal convoluted tubule → tubular capillaries
   **Urea:** Afferent arteriole → glomerulus → glomerular capsule → proximal convoluted tubule → loop of Henle → distal convoluted tubule → collecting duct → renal pelvis → ureter.
3. **Glucose**—filtration and reabsorption. **Urea**—filtration, some reabsorption of urea into the renal medulla also occurs. $H^+$ **ions**— filtration and tubular secretion. $Na^+$ **ions**—filtration and reabsorption. **Water**—filtration and reabsorption.
4. **a.** ADH secretion would increase, resulting in increased permeability to water of the collecting duct and distal tubules. **b.** The concentration of water in the urine would decrease. **c.** The concentration of urea in the urine would increase, mostly because of the decreased concentration of water due to its increased reabsorption. **d.** Collecting duct permeability to water would increase, and there would be an increase in the reabsorption of water. **e.** Aldosterone production would increase, promoting the retention of blood sodium and a corresponding reabsorption of water into the blood.

## Visual Thinking Answers

**Fig. 31.3** Skin—small amounts of salts and water; lungs—carbon dioxide and water; liver—bile pigments; large intestine—salts of some minerals; kidneys—ions of salts, water, and nitrogenous wastes. **Fig. 31.8** The walls of the descending limb of the loop of Henle are freely permeable to water. The longer the descending limb, the greater the surface area over which water can move out of the filtrate and back into the blood. **Fig. 31.10** A diuretic suppresses production of ADH. Decreased ADH levels cause distal convoluted tubules and collecting ducts to become impermeable to water. Thus, water is not reabsorbed from the filtrate and is excreted. Taking diuretics results in a large volume of dilute urine and can cause dehydration.

## End-of-Chapter Answers

*Level 1*

**1.** e.  **2.** c.  **3.** b.  **4.** a.  **5.** d, e.  **6.** e.  **7.** e.  **8.** a.  **9.** b.  **10.** e.

*Level 2*

1. **Structured:** no specialized excretory and water balance structures (osmosis and diffusion through cell membrane) → intracellular structures (contractile vacuoles) → specialized structures and organs (protonephridia/metanephridia/Mal-

phigian tubules/kidneys). **Function:** increased ability to regulate the composition of filtrate through reabsorption and tubular secretion.

2. These organisms are typically very small, single-celled organisms whose surface area-to-volume ratio is large enough to allow diffusion and osmosis to eliminate toxic waste.

3. Ammonia results from deamination of amino acids and is excreted by the kidneys. Ammonia combines with $CO_2$, a byproduct of cellular respiration, to form urea, and it is excreted by the kidneys. $CO_2$ is primarily excreted by the lungs. Creatinine is derived from creatine in muscles and is excreted by the kidneys. Uric acid is excreted by the kidneys.

4. All are nitrogen-containing waste products. Urea is diluted in the urine; uric acid forms a paste. Urea and uric acid are less toxic than ammonia.

5. See Figure 31.4.

6. Kidneys help conserve water by reabsorbing it from the filtrate. By establishing a high concentration of salts and urea in the renal medulla, the osmotic gradient results in water moving from the filtrate into the surrounding tissue fluid.

7. Homeostasis is the maintenance of constant physiological conditions in the body. ADH and aldosterone are two hormones that help regulate the water and ion balance necessary for homeostasis. ADH regulates that amount of water reabsorbed at the collecting ducts; aldosterone promotes the retention of sodium (and therefore water), while promoting the excretion of potassium.

*Level 3*

1. Diuretics inhibit the secretion or action of ADH and therefore decrease reabsorption of water into the blood by the collecting ducts. More water is excreted with the urine. This lowers blood volume, which in turn reduces blood pressure.

2. For reptiles, excreting uric acid conserves water; little water is used to dilute or excrete this waste product. For birds, excreting uric acid allows them to be lighter in flight; urea would need to be dissolved in water and stored in a bladder.

3. Normally, insulin allows glucose to enter the cells and leave the blood. Persons with diabetes mellitus have reduced levels of insulin or their tissues do not respond to insulin. Either situation leads to increased levels of glucose in the blood. In the kidneys, glucose is filtered out of the blood and returned to only a threshold level; the excess glucose present in the blood of diabetics is excreted in the urine. Because the urine contains excess glucose, the osmotic pressure in the urine results in less water leaving the urine through the collecting ducts. Urine is therefore produced in larger volumes, fills the bladder quicker, and more frequent urination results.

4. The mitochondria in the region of contractile vacuoles generate the ATP necessary for the contraction of the vacuole.

5. Urine can reveal the presence of drugs in the body because the kidney secretes the breakdown products of substances such as drugs into the filtrate.

# CHAPTER 32

## Concept Checkpoint Answers

1. Sensory neurons receive information and transmit it to the central nervous system; one long axon brings messages from receptor cells. Motor neurons transmit commands away from the central nervous system and to the muscles and glands; one long axon, usually branched at the end, brings messages to a muscle or a gland. Interneurons are located completely within the brain or spinal cord and integrate incoming information with outgoing messages; a highly branched system of dendrites, receives input from many different neurons converging on a single interneuron.

2. **A.** A stimulus causes $Na^+$ to enter the cell, resulting in depolarization. **B.** If enough $Na^+$ ions enter, $Na^+$-gated channels open, and $Na^+$ ions rush through, depolarizing the membrane to +30mV. **C.** $Na^+$-gated channels close, and $K^+$ channels open allowing $K^+$ to enter the cell. **D.** Continued influx of $K^+$ into the cell repolarizes the cell to –70 mV; $Na^+$-gated channels are still closed.

3. Interneurons integrate information they receive from sensory neurons by "keeping score" of the sum of all excitatory and inhibitory impulses they receive at one time. If excitatory impulses outweigh inhibitory impulses, the result may be enough of a depolarization of the postsynaptic membrane to generate an action potential in the interneuron. If the inhibitory impulses outweigh excitatory impulses, then the interneuron may be prevented from generating an impulse. This mechanism permits a great deal of flexibility of response to sensory input.

4. **a.** Cerebellum. **b.** Cerebrum. **c.** Hypothalamus. **d.** Cerebellum and basal ganglia **e.** Thalamus. **f.** Pons. **g.** Medulla.

5.

6. **a.** Inhibitory. **b.** Excitatory. **c.** Inhibitory.

7. **a.** No contraction results. **b.** Acetylcholine continuously stimulates muscles to contract resulting in muscle paralysis due to constant contraction. **c.** No contraction results.

## Visual Thinking Answers

**Fig. 32.10** At the synapse, neurotransmitter is stored only at the terminal ends of presynaptic neurons; receptors for neurotransmitters are found only in the postsynaptic neurons. The signal must travel from presynaptic neurons where neurotransmitter is stored to postsynaptic neurons where receptors bind with it, opening sodium ion channels and causing an action potential in the postsynaptic neuron. **Fig. 32.16** The incoming sensory message would enter the spinal cord via sensory neurons at about the level of the hand. These neurons would synapse directly in the spinal cord with motor neurons, which would carry messages to the muscles of the hand and arm to pull away from the stove. Since this is a simple reflex action, the spinal cord is the location of the integration. The brain plays no role in the integration.

## End-of-Chapter Answers

*Level 1*

1. e.  2. a.  3. e.  4. a.  5. d.  6. F.  7. F.  8. T.  9. F.

*Level 2*

1. Resting potential refers to the difference in electrical charge along the membrane of the resting neuron, that is, a neuron that is not conducting an impulse. The action of sodium—potassium pumps and ion-specific membrane channels results in more positive ions outside the membrane, while negative proteins remain in the cell. This separation of charges creates the resting potential, which is the basis for the transmission of nerve impulses.

2. A nerve impulse travels because voltage-gated channels respond to the electrical changes that accompany depolarization of nearby cell membrane. The adjacent section of membrane depolarizes, followed by another section, leading to a wave of depolarization. In a myelinated axon, channels respond to depolarization in adjacent nodes leading to rapid transmission along the axon.

3. The decision to move your finger is made in the brain. A nerve impulse travels down the spinal cord where it excites a motor neuron extending to the muscles in your finger. At the neuromuscular junction, acetylcholine is released from the presynaptic vesicles of the motor neuron. This neurotransmitter crosses the synapse and binds to receptors in the postsynaptic muscle cell membrane, opening sodium channels. A depolarization of the muscle cell membrane follows, which ultimately triggers muscle contraction.

4. Neurons in the CNS receive both excitatory impulses and inhibitory impulses. They "sum up" (integrate) these opposing impulses and respond accordingly.

5. A ganglion is an aggregation of nervous tissue. It is not really a brain because it is not capable of conscious thought or emotion, but simply processes incoming and outgoing impulses.

6. The vertebrate central nervous system consists of the brain and spinal cord and is the site of information processing. The peripheral nervous system includes sensory and motor nerves that shuttle messages to and from the central nervous system.

7. The brain stem includes the midbrain, pons, and medulla. It contains tracts of nerve fibers that carry messages to and from the spinal cord. Nuclei located in the brain stem control important body functions such as heart rate and breathing.

8. Sensory receptors react to stimuli by producing nerve impulses. The nerve fibers of sensory neurons carry this information to the central nervous system, where interneurons interpret them and direct a response—in this case, turning your head. The axons of motor neurons conduct these impulses to the appropriate muscles.

9. These systems act in opposition to each other to help maintain homeostasis as you respond to internal and external stimuli. The sympathetic nervous system generally mobilizes the body for greater activity (faster heart rate, increased respiration), whereas the parasympathetic nervous system stimulates normal body functions such as digestion. Together, they can fine tune physiological parameters.

*Level 3*

1. An action potential is not a chemical reaction because no molecules are changed from one substance to another. It consists simply of the movement of ions along the length of the neuron membrane.

2. Intensity of a sensation such as pain varies from occurrence to occurrence because (1) the number of neurons carrying the message differs, (2) the frequency of the impulses varies with the strength of the stimulus and (3) the brain interprets these impulses modified by a person's emotional and psychological response to the particular situation.

3. Infants and toddlers are quickly developing nervous system connections and expanding motor skills. The myelin sheath of many nerve fibers, particularly in the cerebral white matter and motor neurons, is critical for nerve signal transmission and normal development.

4. As the spinal cord extends down the back from the brain, sensory and motor nerves emerge along its length relative to their destination in the body. The higher the point of an injury that severs the spinal cord from the brain, the greater the number of sensory and motor nerves that are no longer connected to the brain and the more regions of the body that are affected.

5. Strawberry sundaes will not affect the action of neurotransmitters at your synapses; you will not develop a dependence on strawberry sundaes; and if you stop eating strawberry sundaes, you will not experience withdrawal symptoms.

# CHAPTER 33

## Concept Checkpoint Answers

1. An action potential is generated by a depolarization caused by the release of neurotransmitter by the presynaptic cell. The neurotransmitter stimulates the opening of voltage-gated channels in the neuron. A generator potential however, is a depolarization that results directly from a change in the environment of the sensory cell—a stimulus.

2. Body temperature, chemical concentrations (i.e., $CO_2$, $O_2$, and $H^+$ ion concentration in the blood [pH]), and mechanical change (i.e., stretching of cells).

3. **Mechanoreceptors:** (1) physical deformation due to pressure; (2) Meissner's corpuscles in the lips, genitals, tongue, and nipples; Pacinian corpuscles in the skin, joints, tendons and muscle; lateral line of fish; vertebrate ears; echolocation organs in whales; (3) sense touch and to interpret sound for hearing and echolocation. **Thermoreceptors:** (1) temperature changes (heat and cold); (2)

skin; pit organs of vipers; (3) provide information about external temperature. **Electroreceptors:** (1) electricity; (2) head of sharks; lateral line of some fish scattered over body of weakly electric fish; (3) sense electrical fields emitted by other organisms or detect objects in the environment. **Magnetic field receptors:** (1) magnetism; (2) salmon, salamanders, turtles, hornets, honeybees, homing pigeons, some bacteria; (3) allow organisms to orient themselves with respect to the earth's magnetic field. **Chemoreceptors:** (1) chemicals; (2) taste buds on the tongue or on the bodies of aquatic organism and olfactory receptors in nasal cavity; (3) allows detection of chemicals in solution or in the air.

4. **Photoreceptors:** (1) visible light, ultraviolet light, infrared light; (2) eyespots of protozoans; photoreceptors on earthworm epidermis; compound eyes of arthropods; vertebrate retina and cephalopod eyes; (3) allow organisms to detect motion, light intensity; wavelength of light (colors). **Gravity receptors:** (1) sense the pull of gravity; (2) invertebrate statocysts; vestibule of human inner ear; (3) pull of gravity allows organisms to detect their body position and movement

5. The lens focuses peripheral light on the periphery of the retina, which is predominantly populated by rods. Since rod cells are incapable of sensing different wavelengths of light that produce color, it is difficult to distinguish colors using peripheral vision.

6. **a.** This would result in hearing loss because sound waves cause vibrations in the tympanic membrane, which are then amplified and transmitted to the inner ear through the action of the stapes, incus, and malleus. **b.** Loud noises damage the hair cells in the organ of Corti. Since it is the bending of these hairs that allow one to sense sound, damage to these cells will result in hearing loss. **c.** Unless accompanied by a ruptured eardrum, this would have little effect on hearing. A ruptured vestibule, however, would result in a loss of balance because of leakage of jellylike material in the utricle, saccule, and semicircular canals. Movement of this material, in response to changes in the position of the head, is responsible for helping detect body position.

## Visual Thinking Answers

**Fig. 33.4** The muscle spindles are specialized muscle cells around which the ends of sensory neurons are wrapped. When the muscle is stretched and the muscle spindle gets longer, the stretched nerve ending is repeatedly stimulated. The sensory nerve sends a message to the brain that the muscle is being stretched—that a load is being put on the muscle. The CNS sends a response to the muscle via the motor nerve, which ends at the muscle fibers. **Fig. 33.14** The green color in these *Euglena* is due to the photosynthetic pigment chlorophyll. A critical part of the photosynthetic process needs light for the reactions to occur. Eyespots allow *Euglena* to detect light and orient toward it. **Fig. 33.23** The waves in the fluid of the vestibular canal push on the vestibular membrane. The push transfers through the cochlear duct and causes the basilar membrane to vibrate, which in turn, causes the hair cells in the organ of Corti to move up and down. These hairs are bent back and forth with the vibration, generating action potentials along neurons of the auditory nerve. The brain interprets these impulses as sound.

## End-of-Chapter Answers

*Level 1*

1. a, vv. 2. c, uu. 3. b, ww or zz. 4. f, xx. 5. e, yy, or ww. 6. d, ww. 7. b. 8. a. 9. a. 10. e.

*Level 2*

1. For you to sense a stimulus, it must be of sufficient magnitude to open ion channels within the membrane of the receptor cell. This depolarizes the membrane, creating a generator potential that leads to an action potential or nerve impulse in the sensory neurons with which the receptor synapses. Nerve fibers conduct the impulse to the central nervous system.

2. The vestibular organs and semicircular canals would be highly stimulated as a person's spatial orientation and acceleration change quickly during a roller coaster ride.

3. Satiation is a sensation involving the body's information about the internal environment, including levels of nutrients in the blood and signals from the digestive system.

4. Olfactory receptors detect different smells because of specific binding of airborne molecules with the receptors located in the cilia of the nasal epithelium.

5. On its way to the olfactory area of the cerebral cortex, the nerve impulse travels through the limbic system, the area of the brain that is responsible for many of your drives and emotions. Thus, certain odors become linked in your memory with emotions and events.

6. Taste buds are taste receptors concentrated on the tongue. They detect chemicals in food and register an overall taste that consists of different combinations of sweet, salty, sour, and bitter. The sense of taste interacts with the sense of smell to produce a taste sensation.

7. See Figure 33.16.

8. Rod cells, which can detect low light levels, are the primary cell type stimulated in a dark room, but in a well-lit setting, the cone cells would be equally activated, allowing sensation of color.

9. Your outer ear funnels sound waves toward the eardrum, which vibrates in response. These vibrations are amplified and transmitted by the bones of the middle ear through the oval window to fluids in the inner ear. Receptor cells in the inner ear possess hairs that are bent as the membranes of the inner ear move up and down. This results in nerve impulses carried to the brain by the cochlear nerve.

10. You can detect movements of your head because of the otoliths embedded in the jellylike layer that covers the cilia in the utricle, saccule, and semicircular canals in your inner ear. When you move your head, the otoliths are pulled by gravity and bend the underlying cilia. This generates signals to the brain, which interprets the type and degree of movement.

*Level 3*

1. The birds oriented in some way to the direction of migration, but that orientation was disrupted by an orientation to a magnet. Because the birds were placed in a covered cage, they would be unable to see movements of the sun or moon for navigational purposes. Possibly these birds have a sensory system in which they can detect and orient to the Earth's magnetic field, which is their means of navigation. Placing a magnet near them would provide a stronger stimulus than that of the Earth's magnetic field, so the birds would orient to it.

2. The inability to feel pain is a disadvantage to a person's well-being. We can avoid many injuries because we detect potentially dangerous stresses placed on our bodies because of the pain they cause, and we can avoid them. For example, a child who jumps from a staircase and slightly twists an ankle will not continue to walk on the ankle, further damaging it. In addition, the child will probably not jump from that height again. If the child feels no pain, however, he will continue to walk and jump, injuring the ankle further, and might not be deterred from this dangerous behavior in the future.

3. The drops affected the set of smooth muscles that cause the iris to open, enlarging the pupil.

4. Answers will vary.

5. During a middle ear infection, fluid builds up in the middle ear surrounding the malleus, incus, and stapes, pressing on the eardrum and filling the eustachian tube. Hearing can be affected because the tympanic membrane cannot vibrate as usual, and the bones of the middle ear cannot move as usual.

# CHAPTER 34

## Concept Checkpoint Answers

1. A hydroskeleton uses fluid under pressure as scaffolding for an organism or body part. Movement results when the muscles attached to the body wall pull in on one part of the body, which compresses the fluid, causing it to push out another part of the body.

2. Most of the length of the femur is made up of compact bone. The central core of long bones is hollow and lined with spongy bone, which is filled with yellow bone marrow. Spongy bone makes up most of the bone tissue of ankle bones. Spongy bone does not show a regular concentric ring structure like compact bone and is filled with red bone marrow. The spongy bone tissue can be surrounded by compact bone.

3. Thumb articulation—phalange-metacarpal, saddle joint, movement in right angles; Wrist articulation—carpals and radius; ellipsoidal joint, nearly hinge movement with some rotation; Elbow articulation—radius and ulna-humerus, hinge joint hinge movement; Shoulder articulation—humerus-scapula, ball-and-socket joint, wide range of movement.

4. Having the origin so close to the joint allows for a wider range of motion of the joint. If the origin were farther away from the joint, the range of motion would be dramatically decreased.

5. Infection barrier—skin. Body support—skin and skeleton. Movement—skeleton and muscle. Gas exchange—skin. Protection from predators—skin and skeleton; also, muscles help organisms move and avoid predators. Thermoregulation—skin; also smooth muscle in arteries constrict or dilate carrying blood to skin for conservation or release of heat. Water conservation—skin and skeleton.

## Visual Thinking Answers

**Fig. 34.4** When the vertebrae are stacked one upon the other, the foramen of each vertebra lines up with those invertebrae above and below creating a tunnel. The spinal cord passes through this tunnel. **Fig. 34.13** Mitochondria produce ATP, which is used to power muscle contraction.

## End-of-Chapter Answers

*Level 1*

**1.** e. **2.** a. **3.** b. **4.** e—all the other bones are found in the skull. **5.** a—all the other bones are part of the appendicular skeleton. **6.** c. **7.** b. **8.** a. **9.** c. **10.** e. **11.** d.

*Level 2*

1. Your movement resulted from the contraction of your skeletal muscles, which are anchored to bones. The muscles use bones like levers to direct force against an object. When you raised your hand, your skin stretched to accommodate the change in position.

2. The role of skin as a sensory surface is of great use to small children. As we age, we tend to rely less heavily on such input and more on visual perception.

3. Bone is a type of connective tissue consisting of living cells that secrete collagen fibers into a surrounding matrix. The bones of the skeletal system support the body and permit movement by serving as points of attachment and acting as levers against which muscles can pull. They also protect delicate internal structures, store important minerals, and produce red and white blood cells.

4. Bone marrow transplants involve the removal and transfer of red bone marrow from spongy bone found in flat bones like the sternum, or in the ends of long bones. In fact, red bone marrow is generally harvested from the crest of the hip bone where a large quantity of bone marrow is located. Bone marrow provides stem cells that divide to produce new blood cells, including the various classes of white blood cells that are involved in fighting off infections. The patient would have to be shielded from infection because his or her own white blood cells have been killed, and the transferred marrow will need time to re-populate the supply in the patient.

5. Spongy bone makes up most of the tissue of the bones of the axial skeleton.

6. Excess calcium taken in by the body and not needed at the moment is stored in bone tissue until it is needed for cellular activities.

7. A "slipped disk" refers to one of the intervertebral disks of fibrocartilage that separate vertebrae from each other. These disks act as shock absorbers, provide the means of attachment between vertebra, and allow the vertebral column to move.

8. These are all types of synovial (freely movable) joints. They differ in the type of movement they allow. A hinge joint allows movement in one plane only, a ball-and-socket joint allows rotation, and a pivot joint permits side-to-side movement.

*Level 3*

1. Answers may vary. With age, people generally exhibit wrinkling, sagging skin having less elasticity as well as a lower touch sensation. Bones often become brittle, causing more fractures and breaking more easily. Joints often stiffen, and osteoarthritis is common.

2. Bone is a type of connective tissue that protects organs, gives the body some of its shape and strength, and provides a place of attachment for muscles. Certain minerals, such as calcium, give hardness to bone but the bone also provides a storage place for them. In addition, the bone marrow is located within spaces of bones. Red bone marrow, located within spongy bone, produces red and white blood cells. Artificial bones can offer protection, help to impart shape to the body and act as levers in movement. Artificial joints can allow movement where bones meet bones. However, artificial bones cannot be storehouses for minerals, nor can they provide bone marrow.

3. The trunk of the elephant has a hydrostatic skeleton. Muscles are attached to the flexible walls of the elephant's trunk. When muscles pull on one part of the trunk, they compress body fluid, causing it to push out another part.

4. Human skin provides a relatively dry environment with little food for bacteria because the top layer of keratinized skin is dead, hardened cells. When the skin is broken, however, an environment opens that has fluid and tissue on which bacteria can reproduce in great numbers. In addition, the normal microbiota colonize the skin and therefore, disease-producing bacteria have no place to live.

5. Answers will vary, but moderate amounts of weight-bearing exercise help maintain bone density. Excessive strenuous exercising, irregular or sudden impacts to the limbs, and application of heavy loads to the limbs can damage the cartilage in joints.

# CHAPTER 35

## Concept Checkpoint Answers

1. Control by the endocrine system is slower than control by the nervous system. Its effects may also be longer lasting and more widespread.

2. **a.** Reproduction. **b.** Growth and development. **c.** Response. **d.** Regulation.

3. Releasing hormones are produced in the hypothalamus; enter the blood, and travel directly to the anterior pituitary, where they control the release of hormones by the anterior pituitary. Tropic hormones such as ACTH are hormones released by anterior pituitary that control the release of hormones by other endocrine glands in the body, like the adrenal cortex.

4. The adrenal medulla is stimulated to release hormones such as epinephrine and norepinephrine in response to nervous signals from the sympathetic portion of the autonomic nervous system.

5. c

## Visual Thinking Answers

**Fig. 35.3** In the negative feedback loop, the effect produced at 5 would slow down or stop the production of hormone from the endocrine gland at 2. **Fig. 35.11** They are both negative feedback loops. In the PTH feedback loop, the release of PTH causes an increase in calcium ions in the bloodstream. This information is fed back to the parathyroids, resulting in the secretion of less PTH. Similarly, the release of CT causes a decrease in calcium ions in the bloodstream. This information is fed back to the parathyroids, resulting in the secretion of less CT. The result—less secretion of the hormone—is opposite to the initial increase in secretion of the hormone.

## End-of-Chapter Answers

*Level 1*

1. e.   2. e.   3. d.   4. c.   5. a.   6. T.   7. T.   8. F.   9. T.   10. F.

*Level 2*

1. Only the target cells for insulin have the proper membrane receptor protein that will bind insulin and stimulate the activity of an intracellular second messenger such as cAMP.

2. The nervous system releases neurotransmitters at the synapse, affecting only nearby neurons, glands, or muscle cells, regulating glandular secretion and muscular contraction. Endocrine hormones are released into the bloodstream and may affect distant parts of the body. Endocrine effects are typically slower, but potentially longer-lasting and more widespread. They may affect any cell with specific, hormone receptors.

3. After a hormone is secreted and has exerted its effect on the target cell, the body feeds back information to the endocrine gland. In a positive feedback loop, the feedback causes the gland to produce more hormone; in a negative feedback loop, it causes the gland to slow down or stop hormone production.

4. Blood solute concentration would be monitored and controlled by the hypothalamus. This structure produces "releasing hormones" that trigger ADH production and release from the posterior pituitary. ADH would cause more water to be reabsorbed in the kidney, thus restoring blood volume.

5. **Similarities**—Tropic hormones are released by the anterior pituitary and control the release of hormones by other endocrine glands in the body. In a similar sense, releasing hormones secreted by the hypothalamus control the release of hormones by the anterior pituitary. **Differences**—Besides being secreted by different glands, the release of tropic hormones is under the control of endocrine signals from the hypothalamus; the release of releasing hormones is under the control of the nervous system.

6. Parathyroid hormone (PTH) and calcitonin (CT) work antagonistically to maintain appropriate calcium levels. If the level becomes too low, PTH stimulates osteoclasts to liberate calcium from the bones and stimulates the kidneys and intestines to reabsorb more calcium. When levels grow too high, more CT is secreted, which inhibits the release of calcium from bones and speeds up its absorption.

7. Over a prolonged period of stress, the body reacts in three stages: (1) The alarm reaction, characterized by increased heart rate and breathing rate, and increased blood flow to the skeletal muscles; triggered by adrenaline and noradrenaline. (2) Resistance, accompanied by glucose production and rise in blood pressure; hormones involved are ACTH, GH, TSH, mineralocorticoids and glucocorticoids. (3) Exhaustion characterized by loss of potassium and glucose; organs become weak and may stop functioning.

8. The pancreatic islets of Langerhans secrete two hormones that act antagonistically to one another to regulate glucose levels. Glucagon raises the glucose level by stimulating the liver to convert glycogen and other nutrients into glucose, whereas insulin decreases glucose levels in the blood by helping cells transport it across their membranes.

*Level 3*

1. People living far from the ocean faced little chance of receiving their iodide from seafood, a primary source of iodide. Until this century, iodized salt was not freely available. This lack of dietary iodide caused goiter.

2. **a.** Long-term stress can induce hypertension because of the continued secretion of mineralocorticoids such as aldosterone by the adrenal cortex. Aldosterone stimulates increased retention of sodium and water by the blood. This elevates blood volume and therefore blood pressure. Diuretics inhibit the secretion of aldosterone, thereby allowing the kidneys to reabsorb less sodium and water. More water is thus excreted from the body as urine so urine volume increases. Blood volume is diminished, and correspondingly blood pressure is reduced. **b.** ADH secreted by the posterior pituitary. ADH binds to the collecting ducts of nephrons in the kidneys and stimulates the increased reabsorption of water. Diuretics would inhibit the secretion of ADH, thereby diminishing reabsorption of water, and increasing urine volume.

3. Most youngsters of middle school age undergo puberty. During this time, the sex hormones cause the development of the primary and secondary sexual characteristics. In boys, these include a deepening of the voice and the growth of facial hair. In girls, these include the onset of menstruation and development of the breasts. Because of the hormonal changes that occur during puberty, adolescents undergo a wide variety of both physiological and psychological changes, which are often accompanied by behavioral and mood "swings."

# CHAPTER 36

## Concept Checkpoint Answers

1. Types of asexual and sexual reproduction in animals

| | Bacteria | Proto-zoans | Fungus-like Protists | Plants | Animals |
|---|---|---|---|---|---|
| Sexual | | X | | | X |
| Alternating sexual and asexual | | | X | X | |
| Binary fission | X | X | | | |
| Multiple fission | | X | | | |
| Budding | | | X | | |
| Fragmentation | | | | | X |
| Parthenogenesis | | | | | X |

2. In external fertilization, the female sheds eggs and the male fertilizes them after they leave her body. External fertilization requires a moist external environment to keep the eggs from drying out and to provide a medium in which sperm can swim. In internal fertilization, the female retains eggs within her body. The male deposits sperm in or near the female reproductive tract. Fertilization takes place within the moist internal environment of the female. Thus, the external environmental conditions need not be moist.

3.

| | Spermatogenesis | Oogenesis |
|---|---|---|
| Time frame of gamete maturation | continuously from the onset of puberty | a few each cycle |
| Site of production | testes | ovaries |
| Ploidy of gamete | haploid | haploid |
| Splitting of the cytoplasm during meiosis | equal | unequal, forming polar bodies |

4. The development of primary oocytes in the ovary is triggered by FSH. After oocyte stimulation, the follicles (developing oocytes and their supporting cells) produce estrogens. The hypothalamus senses this rise in estrogens and triggers the pituitary to secrete a surge of LH, which triggers ovulation. A small surge of FSH also occurs. The ruptured follicle (corpus luteum), left behind in the ovary after ovulation, secretes estrogens and progesterone, which maintain the uterine lining. If fertilization does not take place, the corpus luteum begins to degenerate, and its secretion of hormones stops. The endometrial lining sloughs off.

5. IUDs and emergency contraceptive pills.

6. None of them. Condoms are probably the most likely to prevent transmission of disease, but they are no guarantee.

7. The blastocyst secretes HCG, which is a hormone that maintains the corpus luteum. The corpus luteum continues to secrete estrogens and progesterone during pregnancy, which maintains the endometrium.

8. Fetal blood receives oxygen from the mother across the placenta. To prevent blood from circulating to the lungs, a hole (the foramen ovale) between the left and right atria allows blood to avoid the right ventricle and thus pulmonary circulation. In addition, any blood flowing into the right ventricle is shunted to the aorta, rather than to the pulmonary artery, through the *ductus arteriosus*.

## Visual Thinking Answers

**Fig. 36.12** An erection of the penis occurs when blood flow to the penis increases, causing blood to collect within the spaces of the corpus cavernosa and the corpus spongiosum. Pressure from the increased volume of blood within the spaces causes the spongy erectile tissue to become distended and compresses the veins that normally drain blood from the penis. This makes the penis become erect and rigid. Men who have inadequate erections or no erections due to blood flow problems are helped by enhancement of blood flow as a result of the action of the drug. **Fig. 36.13** The tube is a relatively thin-walled, narrow structure that cannot stretch to accommodate a developing embryo/fetus. Furthermore, the tube does not have the thick lining of the uterus necessary for implantation and development of the structures necessary to support and nourish the embryo/fetus. A tubal pregnancy is a serious condition and must be removed surgically or arrested by means of drugs if development does not arrest naturally. **Fig. 36.31** If the foramen ovale did not close, some of the deoxygenated blood from the right side of the heart would flow directly to the left side of the heart without going to the lungs to be oxygenated. This condition results in blood with a deep red color rather than the bright red of oxygenated blood. Such babies are termed "blue babies," because the blood has a bluish cast as seen through the skin. This abnormality must be repaired by surgery.

## End-of-Chapter Answers

*Level 1*

1. a. 2. f. 3. c. 4. b. 5. d. 6. e. 7. f. 8. e. 9. b. 10. a. 11. c. 12. d. 13. b. 14. b.

*Level 2*

1. Sexual reproduction, as in human reproduction, is a process in which a male and a female sex cell combine to form the first cell of a new individual. Asexual reproduction, as in growing a new plant from a cutting, is the generation of a new individual without the union of gametes.

2. A spermatozoon has a head that contains the hereditary material. Located at its leading tip is an acrosome containing enzymes that help the sperm penetrate an egg's membrane. The sperm also has a flagellum that propels it and mitochondria that produce ATP, from which the sperm derives energy to power the flagellum.

3. The reproductive cycle of females occurs roughly every 28 days. The primary oocyte matures, is released from the ovary during ovulation, and journeys through the uterine tube to the uterus. The endometrial lining of the uterus has thickened to prepare for implantation; if fertilization does not occur, the lining sloughs off during menstruation. The hormones FSH, LH, estrogens, and progesterone orchestrate these events.

4. The embryo secretes human chorionic gonadotropin (HCG), a hormone that acts like LH in that it maintains the corpus luteum. The corpus luteum continues to produce estrogens and progesterone, which in concert act to maintain the endometrium.

5. Asexual reproduction essentially produces clones of the original parent. During optimal times, this allows for rapid population increase. However, during drought or other extreme environmental change, sexual reproduction enhances genetic variability among offspring. Genetic variability increases the chance that some members of the population will inherit adaptations that will help them survive environmental changes.

6. If Mary knows when she ovulates, then fertilization can take place for about 16 hours, the approximate life of the egg after ovulation. However, Mary can have sexual intercourse 6 days or so before ovulation and have fertilization take place, because sperm can live in the female reproductive tract for that amount of time.

*Level 3*

1. Answers will vary. Students will answer the question based on self-knowledge of their behavior. If the student thinks that he or she will use a birth control method unreliably, then the actual data should carry more weight. If he or she is meticulous with use, the theoretical data might carry more weight in decision making.

2. Sea stars can reproduce asexually by fragmentation. Any piece that contains part of the central ring of a sea star has the ability to grow into a new organism.

3. This early movement of cells begins to put cells in the correct places for them to begin to develop into the three primary germ layers (ectoderm, endo-

derm, and mesoderm), from which all other tissues and organs are derived. If morphogenesis occurs improperly, the very first tissues layers develop improperly, which results in a profoundly malformed embryo.

4. During pregnancy high levels of estrogens and progesterone secreted by the corpus luteum (and later by the embryo) inhibit the hypothalamus from releasing the hormones that stimulate the production of LH and FSH by the pituitary gland. Since these hormones are maintained at low levels, follicular development is inhibited, as is ovulation (which is triggered by a surge in LH). Essentially, a woman is incapable of producing and ovulating new ova during pregnancy.

5. Birth control pills inhibit the hypothalamus from releasing the hormones that stimulate the production of LH and FSH by the pituitary gland. Since these hormones are maintained at low levels, follicular development is inhibited, as is ovulation (which is triggered by a surge in LH). Essentially, a woman is incapable of producing and ovulating new ova while she is on the pill.

6. In a woman's normal menstrual cycle, the level of estrogens and progesterone diminish as the corpus luteum degenerates. This allows for normal menstruation (shedding of the endometrium) that occurs each cycle. By coming off the pill late in the cycle, normal menstruation occurs. If a woman stayed on the hormones throughout her cycle, she would not menstruate. In fact, some women who get severe migraines or PMS (premenstrual syndrome) during the period of significant hormonal changes around menstruation are sometimes encouraged by their doctors to stay on the pill throughout this time. In addition, a new birth control pill has been formulated that a woman takes continually for three months, thereby menstruating only four times a year.

7. If an animal is sessile, it can only mate with those individuals that settle around it. By being both male and female at the same time, an animal can increase its reproductive success because it will be able to mate with every individual around it, not just those that are of the opposite sex.

8. HCG is human chorionic gonadotropin, a hormone that maintains the corpus luteum so that it will continue to secrete progesterone and estrogens. Progesterone is needed for the maintenance of the uterine lining and the continued attachment of the fetus to this lining. This substance would be filtered from the blood in the kidney during filtration and become part of the urine. Although some may be reabsorbed at the nephron, reabsorption would take place only to a threshold level.

9. The only completely effective protection against sexually transmissible diseases is abstinence. There is decreased risk if latex condoms are used correctly along with the spermicide nonoxynol-9.

# CHAPTER 37

## Concept Checkpoint Answers

1. Both of the cuckoo chick's behaviors are fixed action patterns.
2. **a.** Imprinting, because there is a critical period in which a sparrow must hear its species' song to sing it correctly as an adult. **b.** Possibly insight, because this problem engenders a variety of solutions, and the one demonstrated by the raven on the first try suggests that it was able to perceive relationships and manipulate concepts in its mind to solve the problem. (*Note*: Results of experiments show that prior learning in birds is important in their ability to show a response such as the one described in the question. Still unanswered is whether birds are capable of conscious, rational thought. Nevertheless, birds clearly are capable of "higher" learning to a greater extent than previously thought.)
3. Dominance hierarchies reduce the chance that individuals will be killed or injured through aggression in social groups. A healthy subordinate has a better chance of reproducing than an injured or dead one.
   Altruism—Many individuals in insect societies sacrifice their own reproductive success by being nonreproductive. These individuals share many genes in common with other related individuals in the colony. Enhancing the reproductive success of the queen enhances the chance that some of its own genes (those that it shares in common with the beneficiaries of the altruistic act) will be passed to the next generation.

## Visual Thinking Answers

**Fig. 37.3** Fish with gravity-detecting organ would orient to both stimuli, swimming at a 45° angle to each—like this:

Fish without a gravity-detecting organ would orient only to the light, so their dorsal sides would face the light—like this:

**Fig. 37.6** Lorenz was the first moving object the birds saw on hatching and imprinted on him as they would on their mother. Imprinting occurs during a short time early in the bird's life, when the young animal forms a learned attachment to a moving object, usually its mother. **Fig. 37.8** Long-distance migrants like the American golden plover are triggered to migrate by their biological clocks, which produce annual physiological and behavioral rhythms, just like biological clocks that produce daily rhythms. Their internal clocks are programmed in such a way that the animals perform the behavior for a set amount of time—whatever time is necessary to complete their journeys.

## End-of-Chapter Answers

*Level 1*

**1.** c. **2.** b. **3.** c. **4.** e. **5.** e. **6.** F. **7.** T. **8.** F. **9.** F. **10.** T.

## Level 2

1. This statement expresses a fundamental idea in the study of animal behavior—that animal behaviors are evolutionary adaptations that make an organism more fit to survive in and adjust to its environment.
2. Innate behaviors protect an animal from environmental hazards without its having learned to do so; many social behaviors also depend on innate behaviors. Organisms with simple nervous systems rely primarily on innate behaviors for survival. Learned behaviors, however, allow an animal to adjust its behavior on the basis of its experiences, helping it to adapt better to its environment and to environmental changes.
3. **a.** Taxis (phototaxis). **b.** Habituation. **c.** Trial-and-error conditioning (operant conditioning).
4. Operant conditioning is a form of learning in which an animal associates something that it does with a reward or punishment. In classical conditioning, an animal learns to associate a new stimulus with a natural stimulus that normally evokes a particular response; eventually, the animal will respond to the new stimulus alone.
5. Insight (reasoning) is the most complex form of learning, in which an animal recognizes a problem and solves it mentally before trying out the solution. Insight allows an animal to perform a correct behavior the first time it tries; it helps the animal adjust and respond to new situations and perhaps determine better ways to deal with its environment.
6. **a.** Locality imprinting (migration). **b.** Reflex. **c.** Fixed action pattern.
7. Competitive behavior results when two or more individuals are striving to obtain the same resource (such as food, territories, or mates).
8. **a.** Threat display. **b.** Submissive behavior. **c.** Parenting behavior. **d.** Courtship ritual (reproductive behavior). **e.** Territorial behavior.
9. A honeybee colony consists of three castes: (*a*) the queen, which lays the eggs, (*b*) the drones, male bees that fertilize the queen's eggs, and (*c*) the workers, female bees that develop from fertilized eggs and do almost everything around the hive except lay eggs and fertilize them. (This includes feeding larvae, guarding the hive, and foraging for food.)
10. A rank order is the social hierarchy that appears in groups of fishes, reptiles, birds, and mammals. This hierarchy helps reduce aggression and fighting within social groups, focusing these behaviors on a short time when the rank order develops.

## Level 3

1. Answers will vary. However, the evolutionary basis of group social behaviors is that the advantages of a particular social trait must outweigh the disadvantages for the individual and its kin. Individuals within colonies of bees, for example, are closely related. Mammals do not produce large numbers of offspring, as occurs in bee and ant colonies; large groups of mammals are generally not closely related. Research has shown, however, that naked mole rats live within colonies and are closely related—an unusual situation for mammals.
2. Answers will vary, but hypotheses should focus on imprinting and other means of animal migration mentioned in the chapter. Research shows, in fact, that Pacific salmon become imprinted with the smells of their native stream. Because they can discriminate between the waters of two streams coming together at a fork, they know which way to swim.
3. Answers will vary. Human behavior appears to result from a combination of hereditary and environmental factors.
4. A pheromone is a chemical produced by one individual that alters the physiology or behavior of other individuals of the same species. Since the chemicals on the pads of the first group of women altered the physiology of the second group of women, these chemicals fit the definition of a pheromone.
5. The genes that encode social behaviors are retained in the gene pool because they increase the reproductive fitness of the species.
6. They are taking advantage of a predator's ability to learn through operant conditioning, or "trial-and-error learning," and classical conditioning. For the former, as a predator attempts to eat a noxious prey organism several times it eventually learns to avoid the prey. For the latter, the bright coloration helps the predator to associate this color pattern with the noxious taste.

## CHAPTER 38

### Concept Checkpoint Answers

1. Uniform
2. The growth curve for newly established populations would show very slow initial growth, and then growth would occur more rapidly as the population size increased. This is because when population numbers are low, dune grasses would probably suffer higher mortality and lower successful seed germination than when there is a lot of dune grass nearby to stabilize the dune. This initial period of slow growth, or lag time would be much longer than for similar grasses that inhabit a more stable habitat.
3. It appears that both play a role. Seedlings are more successful when there is a higher density of dune grasses on the dune; however at very high densities, dune grasses can begin to compete for resources such as nutrients and water. An unpredictable storm, however, can wipe out a whole population of dune plants regardless of the number of dune grasses established on the dune.
4. The survivorship curve for dune grasses are likely to be somewhere between type II and type III. Seedling mortality is likely to be very high initially owing to the difficulty of becoming established in such a harsh and constantly changing environment. Once a dune is established, the chances of survival are better. The unpredictable nature of the ocean sand dune environment, however, results in continued high mortality from year to year.
5. One measure would simply to be to delay having children until later in life. For a demonstration of how this would decrease population growth, see Level 2 question 2 at the end of the chapter. This will help you to see how delaying childbirth results in slower overall population growth.

### Visual Thinking Answers

**Fig. 38.2** The lag phase is the first seven to eight hours of growth. Although bacteria are doubling, there are so few bacteria that the population growth as plotted on the graph rises very little; that is, the population is not yet growing dramatically. The term is appropriate because this time appears as a delay in growth before a virtual explosion in population numbers. **Fig. 38.5** In uniform distributions, individuals are evenly spaced; random distributions lack any definite order; in clumped distributions individuals form clusters or groups. Clumped distributions are the most prevalent in nature. Clumping occurs as a result of the interactions among animals, plants, and microorganisms, and may reflect unevenly distributed resources in an environment. **Fig. 38.9** The world population has not yet stabilized. It is still in a phase of exponential growth.

### End-of-Chapter Answers

*Level 1*

**1.** a. **2.** f. **3.** f. **4.** b. **5.** e. **6.** F. **7.** F. **8.** T. **9.** T. **10.** F.

*Level 2*

1. The population size is affected by immigration. The growth rate does not include factors of emigration and immigration.
2. Demographers study human populations, describing peoples and the characteristics of populations. They predict the ways in which the sizes of populations will change in the future, taking into account the age distribution of the population and its changing size through time. Population ecologists study how populations of any species grow and interact, investigating the factors that determine the distribution patterns of organisms and the factors that control their numbers in the locations in which they are distributed. Although both study populations, demographers are concerned only with human populations.
3. *Density* refers to the number of organisms per unit of area; *dispersion* refers to the way in which the individuals of a population are arranged. If individuals are too far apart, they may not be able to reproduce. If population becomes too dense, however, factors such as disease, predation, or starvation can limit the population's size.
4. The three patterns are (*a*) uniform (evenly spaced), (*b*) random, and (*c*) clumped (organisms are grouped in areas of the environment that have the necessary resources). Human populations are distributed in the clumped pattern.

5. The effect of density-dependent limiting factors such as competition varies with population size; density-independent factors such as weather have the same effect regardless of population size.

6. As the predator population increases, more members of the prey population are captured and eaten, reducing the size of the prey population. As the prey population continues to diminish, its predator population begins to decline, as fewer prey individuals are available to support the feeding requirements of a large predator population. The corresponding rebound in the prey population results from a diminishing number of predators. The interactions that limit population growth of each species are density dependent. Growth of the predator population is density-dependent because individuals compete with one another for food—the prey individuals. Growth of the prey population is density-dependent because a large population results in growth of a predator population that removes individuals for food.

7. Both predation and parasitism are density-dependent limiting factors. Predators are organisms that kill and eat organisms of another species; parasites live on or in other species and derive nourishment from them but do not necessarily kill them.

8. See Fig. 38.7. Type I is typical of organisms such as humans that provide a great deal of parental care to their offspring and tend to survive past their reproductive years. Type II characterizes organisms such as hydra that reproduce asexually and are equally likely to die at any age. Type III is characteristic of organisms such as oysters that produce abundant offspring that must survive on their own and therefore die in large numbers when young.

9. Developing countries tend to have a population pyramid with a broad base, reflecting a rapidly growing population with large numbers of individuals entering their reproductive years. Industrialized countries tend to have populations that have similar proportions of their populations in each age group and that are growing slowly or not at all.

*Level 3*

1. People are already putting pressure on our natural resources including land, water, forests, and atmosphere. Increasing industrialization leads to rising levels of pollution and solid waste. Even though some experts suggest that enough food can be produced to feed the world's population, many people live—and die—in poverty and hunger.

2. Answers will vary.

3. In a population pyramid of a stable population, births should equal deaths, and the number of females of each age group within the population should be similar. Therefore, the pyramid would not be a pyramid at all but would be "squared off" similar to Figure 38.8*b*, United States, only with a wider "top."

4. Such a population probably does not include a high percentage of women in their childbearing years, and deaths probably outnumber births. The fertility rate is probably lower than the replacement rate. The population pyramid would have a narrow base; this population would be one of negative growth.

5. Parasites may not kill their hosts, but they do weaken their hosts and make them less fit to compete for space and food. As the density of a population increases, competition for these resources increases, and the effects of parasites increase. The reproductive fitness of parasitized organisms therefore diminishes as density of the population increases.

6. A likely survivorship curve for lobsters:

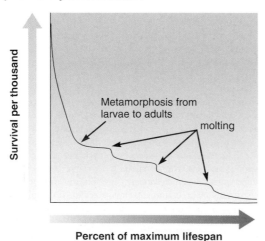

7. This population of monkeys seems to exhibit type I survivorship since mortality at early age is low, and most of the mortality occurs later in life, after reproduction. The low mortality of young is indicative of a great deal of parental care, as is exhibited in most primates.

# CHAPTER 39

## Concept Checkpoint Answers

1. In a scientific sense, competition refers to organisms that live near one another and strive to obtain the same limited resources. In an everyday sense, the term *competition* refers to striving against one another, so in a very general sense the terms are similar. However, in an everyday sense, competition among humans usually refers to a rivalry, such as between businesspersons who compete to gain customers, or between contestants who compete to win a prize. In nature, organisms that use the same limited resource for survival are said to "compete" for that resource, but they do not use any higher order thinking processes to do that and are generally unaware that a competition is taking place. Plants, for example, may produce chemicals that prevent other plants from growing nearby. Complex animals such as vertebrates compete by using genetically "hardwired" behaviors to compete for mates and territory.

2. When a predator limits the size of prey populations, it is less likely that competition between prey populations for limited resources will lead to extinction of one of the populations. Thus, predators can be important for maintaining species diversity within a community.

3. Toxic species that mimic each other receive the benefit of having more toxic individuals in the environment. This probably results in predators learning much more quickly to avoid prey with that coloration. If a predator only happens on a toxic individual very occasionally, it may take more encounters with the toxic species for the predator to learn to avoid it. In other words, the less time it takes for a predator to learn to avoid eating you, the better the chance that you will avoid being eaten!

4. This is not an example of mutualism because the two species of mimics are not living together in a symbiotic relationship. The two species of mimics may occasionally encounter one another in their habitat, but they do not live in association with one another in ways that are characteristic of symbiosis (living on or inside another organism, sharing or acquiring resources, etc. . . ).

## Visual Thinking Answers

**Fig. 39.5** The realized niche of *Chthamalus* is more restricted than its fundamental niche. This situation is usually the case because organisms compete for resources in overlapping fundamental niches. The organisms coexist only because their realized niches differ somewhat. **Fig. 39.19** Primary succession occurs where there is no soil. Volcanic activity kills the organisms in an area, depositing lava rock over the soil; creating conditions for primary succession.

## End-of-Chapter Answers

*Level 1*

1. b. 2. e. 3. c. 4. f. 5. b. 6. F. 7. T. 8. T.

*Level 2*

1. A population consists of individuals of the same species. Populations that interact with one another form a community. An ecosystem is a community of organisms along with the abiotic factors with which the community interacts.

2. The niche occupied by a species is its realized niche—the role it actually plays in the ecosystem. It is distinguished from its fundamental niche, the niche that it might occupy if competitors were not present. The realized niche of a species may be much narrower than its fundamental niche because of competition. In some species because of strong competitors, the realized and fundamental niches may be the same.

3. Gause's experiments showed that predator–prey relationships follow a cyclical pattern: As the number of prey increases, the number of predators increases. When the prey decrease because of predation, the predator population also declines until an upsurge in the prey population starts the cycle again.

4. The balance between predator and prey populations is complex because interactions depend on many factors: interactions with other organisms, movement of new organisms into or out of the community, and abiotic factors in the ecosystem.

5. Coevolution is the long-term, mutual evolutionary adjustment of characteristics of members of biological communities in relation to one another. Cattle are one example: Their digestive systems are adapted (grinding teeth, strong jaws, a rumen) to digesting silica in certain grasses.

6. **a.** Warning coloration. **b.** Camouflage.

7. Primary succession takes place on surfaces not previously supporting organisms and on which there is no soil. It could begin, for instance, with lichens growing on bare rock, producing acids that break down the rock to form soil. This gives rise to a pioneer community, which in turn leads to more vegetation, eventually making the area able to support other forms of life. At some point, the mix of plants and animals becomes somewhat stable, forming a climax community.

*Level 3*

1. Often what seems to be a relationship in which one organism benefits but the other is neither harmed nor benefited is the result of our lack of knowledge of the biology of the association between the two species. More research often reveals that what was thought to be a neutral relationship for one of the partners is actually negative or positive in some way.

2. Our towns and cities undergo successionary changes much like natural communities. Small towns initially attract a certain subset of the population, but the very presence of these individuals and their daily actions modify the towns. Roads are created, stores open, housing developments sprout to accommodate the needs of the population. These changes then attract people seeking to live in larger, more developed townships. This is similar to what happens in natural succession, as organisms that initially inhabit an area alter the biotic and abiotic aspects of the community through their normal biological processes. Natural communities, however, often reach a fairly stable, long-lasting climax community. In order to preserve long-lasting, more stable human communities that retain those aspects desirable to the population, it may be wise to set limits on growth and development. Just as there are natural limits to growth in nature, local comprehensive community plans that set limits on the growth of townships can help to preserve stable communities with desirable living arrangements that are maintained for long periods.

3. Much of the evidence that competition structures communities comes from "removal" experiments similar to the one involving the barnacles (*Balanus* and *Cthamalus*) in the rocky intertidal area. Removing competitors and then assessing each species' fundamental niche, in comparison to its realized niche, allow one to infer that competition played a role in structuring the community.

4. The most successful predator would feed on those organisms past reproductive age. In this way, the predator's food source continues to reproduce, ensuring food for the future.

5. The factors are: climate (including annual precipitation), types of available habitats, types of predators, and level and type of human activity.

6. When an herbivore eats plants but does not destroy them, allowing them to continue to grow, the relationship is considered parasitism. If, however, the herbivore destroys the plants and they are unable to continue to grow, the relationship is considered predation.

# CHAPTER 40

## Concept Checkpoint Answers

1. Since the ultimate source of energy for most ecosystems is the sun, ecosystems will continue to be provided with energy, as long as there are photosynthetic producers to capture that energy. Eventually the sun will burn out, and then, of course, those ecosystems that ultimately derive their energy from the sun will run out of energy. Fortunately, this event is many millions of years in the future!

2. Calvin cycle—sink, Krebs cycle—source. Burning of fossil fuels—source. Plants—sink. Animals—source. Phytoplankton—sink. Decomposers—source. Forest fires—source.

3. Nitrogen is generally the nutrient that most limits plant growth. Although nitrogen is very common in our atmosphere, it is in a form ($N_2$) that is unusable by most organisms for incorporation into organic molecules. $N_2$ must first be converted into ammonia or nitrate before it can by used by plants. Plants rely on the

activity of soil dwelling and/or symbiotic nitrogen fixing bacteria to do this, so usable nitrogen is quickly depleted from the soil. Phosphorus is also a limiting nutrient in many areas because it is not abundant in many types of rocks which, when weathered, produce soil. This is why most agricultural fertilizers have nitrates and phosphates as main components.

## Visual Thinking Answers

**Fig. 40.2** Both you and the lizard are consumers.   **Fig. 40.5** Energy enters the chain at the producers. Most of the energy comes from the sun, captured by photosynthetic organisms.   **Fig. 40.7** Small heterotrophs are any consumers smaller than smelt, a small fish. Examples are zooplankton, some arthropods, annelids, larvae, and fish smaller than smelt.

## End-of-Chapter Answers

*Level 1*

1. d.   2. b.   3. e.   4. c.   5. a.   6. d.   7. d.   8. b.   9. c.   10. a.

*Level 2*

1. An ecosystem is a community of organisms and the nonliving factors of their environments through which energy flows and within which nutrients cycle. Producers are autotrophic organisms that can make their own food. Consumers and heterotrophic organisms that eat other organisms for food. Decomposers are consumers sometimes called detritovores; they eat dead matter.

2. Within each trophic level most of the energy acquired is used for metabolism or lost with feces, metabolic waste, or heat, which makes it unavailable to the next trophic level.

3. A trophic level is a feeding level within a food chain. A food chain is a linear relationship among organisms that feed one on another. Food webs are food chains that interweave with one another.

4. **Carbon**—increased burning of fossil fuels and deforestation are disrupting the natural balance of this cycle and contributing to global warming. **Water**—pollution from industrial wastes, pesticides, and fertilizers are polluting our fresh water ecosystems and groundwater. **Nitrogen**—Agricultural fertilizers and crop rotation are used to increase the availability of this nutrient to crops. **Phosphorus**—Agricultural fertilizers are used to increase the availability of this nutrient to crops.

5. Crop rotation can improve soil by alternating a nonleguminous crop, which consumes nitrogen from the soil, with a leguminous crop. Legumes can enrich the nitrate level of depleted soil by forming mutualistic associations with nitrogen-fixing bacteria. This relates to the nitrogen cycle.

*Level 3*

1. Student answers will vary. However, the ecosphere must include producers to serve as food for the consumers. Decomposers are necessary to break down dead material, making the breakdown products available for reuse. The photosynthesizing producers will generate oxygen for the nonphotosynthesizing consumers. Both producers and consumers will generate carbon dioxide through cell respiration, which is needed by the producers. A moderate amount of water must be present in a terrestrial environment to be continually cycled throughout the system. Inorganic nutrients and nitrogen-fixing bacteria must be present in the soil. The specific organisms would, of course, need to be carefully chosen for their ability to survive in such a system. An aquatic self-sustaining system might include snails and elodea plants.

2. Plants transpire, or release water vapor into the atmosphere. This water vapor is a critical part of the water cycle. As the rain forest is cut down, the plants are lost and less water is available in the atmosphere to return to the ground as precipitation.

3. Energy "flows through" ecosystems, and once it makes its way through the various trophic levels, it is ultimately lost to that ecosystem. Therefore, the sun will have to continually supply energy to the producers of the ecosystem to meet its energy demands. When the sun burns out many millions of years in the future, so too will the ultimate source of energy for Earth's ecosystems. In the same sense, our energy demands are also limited by the availability of fossil fuels, whose energy flows through our industrial economy. Although there may be many yet untapped supplies of fossil fuels, it is incorrect to think that this supply will be limitless.

4. The dramatic increase in the level of $CO_2$ in the atmosphere, coinciding with the increased burning of fossil fuels and large-scale deforestation in the last two centuries, suggests that the cycle is out of balance.

5. This one species of owl is connected to many other species through feeding relationships. The removal of this species can ripple throughout the ecosystem, affecting abundance and diversity of other species. The spotted owl, for example, helps keep the small mammal population in check. An explosion in the population of small mammals could significantly affect the abundance and diversity of local vegetation; effects that could significantly impact other primary consumers.

# CHAPTER 41

## Concept Checkpoint Answers

1. Seasonal variation occurs because of the change in the angle at which sunlight strikes the northern and southern hemispheres of the Earth during its orbit around the sun. If the Earth's axis became perpendicular to the line between the Earth and the Sun, it would result in the elimination of seasonal variations.

2. Because the combination of the prevailing atmospheric circulation patterns along with the shape of ocean basins affects oceanic circulation patterns, this event probably had a significant impact on global climate patterns. It is also likely that the interior of this large continent had limited rainfall because of its distance from the oceans. Hemispheric seasonal variation in climate would probably have changed little since it is the Earth's orientation to the sun that causes this phenomenon.

3. Biome—tropical rain forest; Location—Central America, Amazon Basin of South America, central and west Africa, Southeast Asia; Climate—high rainfall (> 200 cm), high temperature (average 25°C); Adaptations—large trees form dense canopy, shallow-roots for quickly capturing nutrients from fallen leaves, shade tolerant forest floor plants. Biome—savanna; Location—between the tropical rain forests and deserts, cover much of central and southern Africa, western India, northern Australia, large areas of northern and east-central South America, and some of Malaysia; Climate—hot and more arid than tropical rain forests; Adaptations—open grassland supports herbivorous vertebrates and invertebrates. Large carnivores feed on grazing mammals. Biome—desert; Location—occur in regions 20 to 30 degrees from the equator, also in the interiors of continents far from the moist sea air, especially in Africa (the Sahara Desert), Eurasia (Europe and Asia), and Australia, the leeward side of mountain ranges, such as in the Great Basin of Nevada and Utah; Climate—<25 cm of annual rainfall, substantial temperature differential between day and night; Adaptations—Annual plants with seeds abundant to avoid dry season; succulent plants, trees and shrubs with deep roots that reach deep groundwater; woody plants either lose their leaves, or are evergreen with hard, reduced leaves; animals burrow, migrate to transient food/water sources or store water like camels.

4. Biome—temperate grasslands; Location—occur at higher latitudes than savannas but are often found bordering deserts; Climate—25 to 75 centimeters of rainfall annually; Adaptations—burrowing rodents and herds of grazing mammals feed on grasses. Biome—temperate deciduous forests; Location—in the northern hemisphere it occurs in the eastern United States and Canada and an extensive region in Eurasia. Climate—regions having warm summers, cold winters, and moderate amounts of precipitation; Adaptations—Animals abundant on ground and in trees; forest canopy not as dense as rain forests; smaller trees and shrubs grow on forest floor. Biome—taiga; Location—latitudes north of the temperate across vast areas of Eurasia and North America; Climate—long, cold winters with little precipitation; Adaptations—plants grow rapidly during short warm seasons, trees (alders) with nitrogen-fixing bacteria in nodules on their roots; large herbivores and carnivores. Biome—tundra; Location—farthest north in Eurasia, North America, between the taiga and the permanent ice, not in Antarctica; Climate—low annual of less than 25 centimeters that freezes for most of the year; during brief Arctic summers, surface ice melts and puddles; permafrost; Adaptations—plants grow rapidly during brief warm season, large grazing mammals migrate from the taiga, birds and waterfowl migrate to tundra to nest.

5. Although both environments are at the interface between a terrestrial and aquatic environment, the shoreline of an ocean has a variety of abiotic environmental factors that are not present along the shoreline of small lake or pond. These include a shifting shoreline due to ocean tides, high salinity, wave action, and stronger currents. Therefore, ocean intertidal organisms have adaptations that allow them to withstand these environmental pressures (burrowing, shells, gelatinous covering, etc.).

## Visual Thinking Answers

**Fig. 41.1** (1) The equator. The sun's rays arrive almost perpendicular to regions near and at the equator. These units of solar energy fall on a smaller surface of the Earth than they do farther north or south, so this energy is concentrated in a smaller area. Thus, the equator receives more solar energy than do other areas of the Earth. (2) In the winter, the northern hemisphere is tilted away from the sun; in the summer, the northern hemisphere is tilted toward the sun. (3) When the northern hemisphere is tilted toward the sun in the summer, its rays cover a smaller area. Therefore, in the summer a person will have greater exposure to UV rays because they are more concentrated than in the winter, when the northern hemisphere is tilted away from the sun and its rays cover a larger area. **Fig. 41.3** Regions at 60° latitude would have abundant precipitation because air rises and is cooled, thus holding less moisture. **Fig. 41.15** In winter, lichens are the chief food for the caribou. They also eat horsetails, sedges, and willow and birch twigs. Their food intake is much reduced in winter, and they lose weight.

## End-of-Chapter Answers

*Level 1*

**1.** e. **2.** e. **3.** c. **4.** c. **5.** a. **6.** d. **7.** b. **8.** f. **9.** e.

*Level 2*

1. Water is essential for plants to undergo photosynthesis. Plants have evolved a variety of adaptations to deal with different moisture levels. The availability of water affects or limits the growing season; affects the timing of flowering and reproduction; and affects overall plant structure (size, root structure, plant shape and texture etc.). For these reasons, water availability is one of the primary forces that govern the makeup, abundance, distribution, and physical characteristics of the plant community.

2. Australia is in the southern hemisphere, whereas the United States is in the northern hemisphere. Thus, when the United States is tilted away from the sun, it is winter in the United States but summer in Australia. When the United States is tilted toward the sun, it is summer in the United States, but winter in Australia.

3. Since light is capable of penetrating to the bottom of a pond and shallow estuarine creeks, photosynthesis is capable of occurring throughout the body of water by phytoplankton and along the bottom by bottom-dwelling plants and algae. Lakes and oceans, however, can be far deeper than ponds; the upper photic zone receives enough sunlight to support photosynthesis. Below this zone, light levels are too low to support photosynthetic production of food.

4. As implied in question 3, light is capable of penetrating only so far through water, and therefore light intensity diminishes with depth. Plants and algae that live deeper in the photic zone are adapted to low light levels, just as terrestrial plants that live under the forest canopy are adapted to lower light levels.

5. Desert rainfall is limited by latitude because the air, which has released its moisture over the rain forests, is dry over desert latitudes. Deserts can also occur in the interior of continents, far from the ocean, and on the leeward side of mountains that block rainfall.

6. The taiga is the northern coniferous forest with long, cold winters and little precipitation. The tundra lies farther north and is open, often boggy, with desertlike levels of precipitation.

7. The three fresh water life zones are: (a) shore zone (shallow water in which plants with roots may grow; some consumers live here), (b) open-water zone (the main body of water through which light penetrates; floating plants, microscopic floating animals, and fish live here), and (c) deep zone (light does not penetrate, so there are no producers; inhabited mainly by decomposers or organisms that feed on organic material that filters down).

8. An estuary is an environment where fresh water and saltwater meet. All organisms living here are adapted to living in an area of moving water and changing salt concentrations. Some are bottom dwellers that attach to bottom material or burrow into the mud; others spawn in less salty water.

9. Estuaries are ecosystems where fresh water rivers and streams meet the ocean. They are vulnerable to pollutants that flow from upstream, as well as to pesticides that leach from the land and soil.

10. The three major life zones of the ocean are: (a) intertidal (area between high and low tides), (b) neritic (shallow waters along the coasts of the continents), and (c) open sea (the rest of the ocean).

*Level 3*

1. Tropical rain forests make poor farmland because most nutrients are found in the vegetation, not the soil. Root systems are poor and shallow; the soil itself is thin, often only a few centimeters deep. Cutting down and burning the trees removes the nutrients and breaks down the organic matter to carbon dioxide. In two to three years, the land becomes barren. Conversely, in grasslands, root systems are extensive and deep. Much of the nutrients are held in the root systems and in the soil rather than in the vegetation above ground.

2. The windward sides of mountains are much wetter than the leeward sides. In fact, the leeward sides can have desertlike conditions as cited in this question. In the northern hemisphere, winds blow from the west over the Sierra Madre Mountains, so that desert conditions exist on the eastern side of this mountain range. In the southern hemisphere, however, winds blow from the east, so desertlike conditions exist on the western side of the Andes.

3. This city is located in temperate deciduous forest, in which the annual rainfall is from 30 to 60 inches, and summers are warm and winters are cold. Based on the temperatures shown and the above information, this city is probably in the northeastern United States. (Just so you won't wonder, this city is Hartford, Connecticut.)

4. Answers will vary.

5. Answers will vary, but the changes in plant and animal life should follow from the differences between that in a flowing river and a pond. For example, the water in a pond may not be aerated as much as in a stream, so with less oxygen, the numbers and types of fish might diminish.

# CHAPTER 42

## Concept Checkpoint Answers

1. *Advantages*: All renewable energy sources (wind, water, solar, geothermal) can be used to produce electricity with no resulting air pollution emissions. They are "clean" sources of energy. Each of these technologies, if developed wisely, could provide large amounts of energy. *Potential drawbacks*: Although these sources of energy would have less impact on habitats than, for example, strip mining for coal or drilling for oil, their very presence on a large scale could impact ecosystems. To adequately meet the energy needs of this country, large and numerous solar collection panels, wind turbines, or tidal or wave generators would have to be developed. These could affect populations of migrating birds or insects, and marine organisms. In addition, vast areas of land covered with wind turbines, or solar panels, or tidal/wave electric generators may be aesthetically displeasing.

2. Although DDT in small concentrations is probably not harmful to most species of birds and fishes, it can be harmful at high concentrations. Biological magnification resulted in the accumulation of high levels of DDT in the tissues of animals as they fed on numerous organisms with small quantities of DDT in their tissues. At these high levels DDT can be toxic to the larger organisms.

3. The extinctions taking place on Earth today concern humans for a variety of reasons. Here are two: (1) Humans, like any other species on this planet, rely on other species of organisms for our survival. We are all part of a global ecosystem that relies on the intricate and interwoven relationships among organisms to maintain the health and delicate balance of our biosphere. (2) Many consequences of the loss of biological diversity are completely unpredictable.

## Visual Thinking Answers

**Fig. 42.7** Transportation contributes primarily carbon monoxide, hydrocarbons, and nitrogen oxides to the air. Stationary sources contribute primarily sulfur oxides. **Fig. 42.11** Sulfur and nitrogen oxides are key in the formation of acid precipitation. Sulfur oxides are produced primarily by stationary sources; nitrogen oxides are produced by both stationary sources and transportation. **Fig. 42.13** As measured in October, the ozone hole became substantially thinner between 1981 and 2001. March valves appear to show some recovery of the ozone layer form 1999 to 2001. **Fig. 42.16** Greenhouse gases began to rise dramatically after World War II; from approximately the 1950s onward.

## End-of-Chapter Answers

*Level 1*

1. d. 2. c. 3. b. 4. e. 5. F. 6. T. 7. T. 8. F. 9. T.

*Level 2*

1. A sustainable society uses renewable sources of energy. Society's needs are satisfied without compromising future generations and natural resources.

2. People could use solar, wind, and water power, as well as the Earth's geothermal energy. All of these energy sources are renewable and contribute little to waste and pollution problems.

3. People can reduce, reuse, and recycle paper, glass, and metal products to help preserve mineral resources. This is important because mineral supplies are finite; once used up, they are gone forever.

4. Global warming refers to a worldwide increase in temperature due mainly to the production of greenhouse gases. Even a small increase in temperature could melt ice at the poles, raising the sea level and flooding one fifth of the world's land area. Higher temperatures also affect rain patterns and agriculture.

5. Biological diversity refers to the number, abundance, and variety of species. Humans are reducing the world's biological diversity through habitat destruction, pollution, exploitation of commercially valuable species, and selective breeding. Reducing diversity makes species more vulnerable to being wiped out by disease or environmental changes.

6. Humans can improve solid waste management by increasing and improving recycling, using safe incineration, engineering better landfill sites, and practicing waste-to-energy reclamation.

7. Organic nutrients are food for bacteria; as the bacteria multiply, they use up oxygen in the water, killing the fish. Inorganic nutrients can stimulate plant growth in the water. Bacteria decompose the plants after they die; again, when the bacteria multiply, they consume the water's oxygen. The decomposed materials also begin to fill in the body of water, eventually turning it into a terrestrial community. Pollution with infectious agents can cause serious diseases. Toxic substances in water can poison the organisms that take in this water; sometimes organisms accumulate these substances in their bodies.

8. Acid rain results from gases such as sulfur dioxide and nitrogen dioxide that combine with water in the atmosphere. When this acidic rain falls, it kills aquatic life, pollutes groundwater, damages plants, and eats away at stone, metal, and painted surfaces.

9. Ozone in the stratosphere helps shield you from the Sun's ultraviolet rays. Depletion of this layer can damage bacteria and plants and cause burns and skin cancers.

*Level 3*

1. Answers will vary.

2. Answers will vary.

3. Tropical rain forests are home to more than half the world's species of plants, animals, and insects from which people get many medicines as well as wood, charcoal, oils, and nuts. Destroying the forests destroys the habitat for many species and leads to extinctions. Burning the forests adds carbon dioxide to the air, which may raise temperatures worldwide.

4. Answers will vary.

5. Here are some simple steps: ride your bike or walk, use cold water when washing your clothes, run your dishwasher only with a full load or wash dishes by hand, line-dry your clothes when possible, clean or replace air conditioning/heating air filters regularly, use energy-efficient compact fluorescent bulbs, install low-flow shower heads, drive a fuel-efficient or hybrid car, and find out what waste you produce is recyclable in your town and recycle.

6. New technologies can help, but they will not suffice. Often new technologies bring with them unforeseen environmental consequences. In addition, new technologies can do little to mitigate the environmental effects of excessive consumption of natural resources and rapid population growth. It will also take substantial changes to the way each individual in our society lives to curb these problems.

7. Habitat fragments isolate small populations. The loss of genetic diversity due to isolation can make populations more susceptible to extinction if a significant environmental perturbation occurs. Keeping corridors between fragments can help preserve genetic diversity by allowing gene flow between populations occupying nearby habitat fragments.

# GLOSSARY

**abiotic factors** (AYE-bye-OT-ik) nonliving factors within the environment, such as air, water, and rocks.

**abscisic acid** a plant hormone that is found predominantly in mature, green leaves and regulates the opening and closing of stomata, inhibits shoot growth, induces seeds to synthesize storage proteins, and has some effect on inducing and maintaining dormancy.

**acclimation** (ACK-lih-MAY-shun) the ability of a plant to become accustomed to new conditions.

**acid** any substance that dissociates to form $H^+$ ions when it is dissolved in water.

**acid rain** precipitation having a low pH (usually between 3.5 and 5.5) produced when sulfur dioxide and nitrogen dioxide pollutants combine with water in the atmosphere.

**acoelomates** animals that have no body cavity; flatworms, cnidarians and sponges are acoelomates.

**acquired immunodeficiency syndrome** (AIDS) the final stage of HIV infection in which the helper T cell count falls below 200 cells/mm$^3$ and a person exhibits certain serious diseases typical of this end stage.

**actin** (AK-tin) a protein that makes up the thin myofilaments in a muscle fiber.

**action potential** the rapid change in a membrane's electrical potential caused by the depolarization of a neuron to a certain threshold.

**active immunity** a kind of immunity in which a vaccination or the contraction of a disease causes the body to build up antibodies against that particular disease.

**active sites** the grooved or furrowed locations on the surface of an enzyme where reactions are catalyzed.

**active transport** the movement of a solute across a membrane against the concentration gradient with the expenditure of energy; this process requires the use of a transport protein specific to the molecule(s) being transported.

**adaptations** naturally occurring, inheritable traits present in a population of organisms that confer reproductive advantages on organisms that possess them.

**adaptive behaviors** behaviors that enhance the ability of members of a population to live to reproductive age and that tend to occur at an increased frequency in successive generations.

**adaptive radiation** the phenomenon by which a population of a species changes as it is dispersed within a series of different habitats within a region.

**adaptive traits** traits that allow individual organisms to survive to reproductive age and bear offspring.

**adenosine diphosphate** (ADP) (uh-DEN-o-seen dye-FOS-fate) the molecule that remains after ATP has been used to drive an endergonic reaction.

**adenosine triphosphate** (ATP) (uh-DEN-o-seen try-FOS-fate) the primary molecule used by cells to capture energy and later release it during chemical reactions.

**adrenal cortex** (uh-DREE-nul KOR-tecks) the outer, yellowish portion of each adrenal gland that secretes a group of hormones known as corticosteroids in response to the hormone ACTH, or adrenocorticotropic hormone.

**adrenal glands** (uh-DREE-nul) two triangular glands located above the kidneys; each adrenal gland consists of two parts: the adrenal cortex and the adrenal medulla.

**adrenal medulla** (uh-DREE-nul muh-DULL-uh) the inner, reddish portion of each adrenal gland; the adrenal medulla is surrounded by the cortex and secretes the hormones adrenaline and noradrenaline.

**aerobic cellular respiration** the chemical process by which cells break down fuel molecules using oxygen, producing carbon dioxide, and releasing energy.

**age distribution** the proportion of individuals in the different age categories of a population.

**agranulocytes** (ay-GRAN-yuh-low-sites) one of the two major groups of leukocytes; they have neither cytoplasmic granules nor lobed nuclei.

**aldosterone** (al-DOS-ter-own) a hormone that regulates the level of sodium ions and potassium ions in the blood; it promotes the retention of sodium and the excretion of potassium.

**algae** (AL-jee) plantlike protists that are photosynthetic autotrophs; algae contain chlorophyll and manufacture their own food using energy from sunlight.

**allele** (uh-LEEL) each member of a factor pair containing information for an alternative form of a trait that occupies corresponding positions on paired chromosomes.

**allergic reaction** a specific immune response that results from the immune system mounting a major defense against a harmless antigen.

**allopatric speciation** speciation resulting from the separation of members of a population by a physical barrier that blocks gene flow.

**alternation of generations** a type of life cycle that has both a multicellular haploid phase and a multicellular diploid phase.

**alternative splicing** a process in which an mRNA transcript can have its introns cut out and its exons spliced together in different ways.

**altruistic behavior** (AL-troo-ISS-tik) a kind of behavior that benefits one at the cost of another; parenting behaviors are a type of altruistic behavior.

**alveoli** (al-VEE-uh-lye) microscopic air sacs in the lungs where oxygen enters the blood and carbon dioxide leaves.

**amebas** (uh-MEE-buhs) a phylum (Rhizopoda) of protists that have changing shapes brought about by cytoplasmic streaming, which forms cell extensions called pseudopodia.

**amniotic egg** (am-nee-OT-ik) an egg, characteristically produced by birds, reptiles, and monotremes, that protects the embryo from drying out, nourishes it, and enables it to develop outside of water.

**amphibians** (am-FIB-ee-uns) a class of vertebrates capable of living on land and in the water; amphibians depend on water during their early stages of development.

**amylases** (AM-uh-lace-es) digestive enzymes that break down starches and glycogen to sugars.

**anabolic reactions** (AN-uh-BOL-ick) chemical reactions that use energy to build complex molecules from simpler molecules.

**analogous** (uh-NAL-eh-gus) of differing evolutionary origins and basic anatomy, now similar in function.

**anaphase** (ANN-uh-faze) the third phase of mitosis, meiosis I, and meiosis II; during anaphase, sister chromatids separate and move to opposite poles of the cell.

**angiosperms** (AN-jee-oh-spurms) vascular plants with protected seeds and flowers that act as their organs of sexual reproduction.

**animal behavior** a scientific discipline that was formed when the fields of ethology and behavioral psychology merged; the field of animal behavior applies both the physiological perspective of the ethologist and the psychological perspective of the behaviorist to animal study; today, the field of animal behavior is composed of researchers from a variety of disciplines.

**annelids** (AN-ul-idz) a phylum (Annelida) of invertebrate animals—worms characterized by a soft, elongated body composed of a series of ringlike segments

**antennal gland** (*green gland*) paired excretory glands of crustaceans (lobsters) that consist of a sac, a long coiled tube, and a bladder.

**antheridia** (singular, antheridium) (an-thuh-RID-ee-uh / AN-thuh-RID-ee-um) specialized structures in which sperm are produced in nonvascular plants and several divisions of vascular plants.

**anthropoids** (AN-thruh-poyds) a suborder of primates that includes the monkeys, apes, gorillas, chimpanzees, and humans; they differ from prosimians in the structure of the teeth, brain, skull, and limbs.

**antibodies** proteins produced by plasma cells (B lymphocytes) that recognize foreign antigens and prevent them from causing disease.

**antibody-mediated immune response** one of the two branches of the specific immune response; it is initiated by helper T cells that have been activated by interleukin-1 and the presence of antigens, and results in B cells producing antibodies; the antibodies bind to the antigens they encounter and mark them for destruction.

**anticodon** a portion of a tRNA molecule with a sequence of three base pairs complementary to a specific mRNA codon.

**antidiuretic hormone (ADH)** (AN-ti-dye-yoo-RET-ik) the hormone that regulates the rate at which water is lost or retained by the body; it is secreted by the pituitary gland at the base of the brain.

**antigens** (ANT-ih-jens) foreign molecules that induce the formation of antibodies, which specifically bind to the foreign substance and mark it for destruction.

**anus** the opening of the rectum for the elimination of feces.

**aorta** (ay ORT-uh) the largest artery in the body; it receives blood from the left ventricle and has many vessels branching from it that carry blood throughout the body (with the exception of the lungs).

**appendicular skeleton** (AP-en-DIK-you-lur) the portion of the human skeleton, consisting of 126 bones, that forms the bones of the appendages (arms and legs) and the bones that help attach the appendages to the axial skeleton.

**aqueous humor** (AYK-wee-us HYOO-mur) the watery fluid that fills the chamber behind the cornea.

**Archaea** a domain of single-celled prokaryotic organisms; archaeans differ from bacteria in their cell walls, cell membranes, and ribosomal RNA; they are often found in extreme environments, such as the Antarctic, the ocean depths, and volcanic areas.

**archegonia** (singular, archegonium) (AR-kih-GO-nee-uh; AR-kih-GO-nee-um) specialized structures in which eggs are produced in nonvascular plants and several phyla of vascular plants.

**arteries** (ART-uh-rees) blood vessels that carry blood away from the heart.

**arterioles** (are-TEER-ee-oles) small arteries that lead from arteries to capillaries.

**arthropods** a phylum (Arthropoda) of invertebrate animals with jointed appendages and a rigid exoskeleton; crabs, lobsters, insects, centipedes, millipedes, spiders, horseshoe crabs, mites, and ticks are arthropods.

**articulation** (ar-TIK-you-LAY-shun) another term for a joint.

**asci** (AS-kye) saclike structures that enclose the sexual spores of sac fungi.

**asexual reproduction** the generation of a new individual without the union of gametes.

**atherosclerosis** (ATH-uh-ROW-skluh-ROW-sis) a disease in which the inner walls of the arteries accumulate fat deposits, narrowing the passageways.

**atoms** submicroscopic particles that make up all matter.

**atrioventricular (AV) node** (AY-tree-oh-ven-TRIK-yuh-lur) a group of specialized cardiac muscle cells located in the base of the atria that receives the impulses initiated by the sinoatrial node and conducts them to the atrioventricular bundle in the heart septum.

**auditory canal** (AWD-uh-tore-ee) a 1-inch long canal that receives sound waves funneled from the pinna and carries them directly to the eardrum.

**autonomic nervous system** (awe-tuh-NOM-ik) the branch of the peripheral nervous system consisting of motor neurons that control the involuntary and automatic responses of the glands and the nonskeletal muscles of the body.

**autosomes** (AW-tuh-somes) chromosomes that carry the majority of an individual's genetic information but do not determine gender.

**autotroph** (AW-tuh-TROFE) a self-feeder; an organism that produces its own food by photosynthesis or chemosynthesis.

**auxins** (AWK-sinz) plant hormones primarily found in actively growing regions of the plant that induce cell elongation in stems and affect other plant processes as well, such as the development of roots.

**axial skeleton** (AK-see-uhl) the central column of the skeleton, from which the appendages (arms and legs) of the appendicular skeleton hang; the axial skeleton consists of 80 bones, including the skull, vertebral column, and rib cage.

**axon** (AK-son) a single, long cell extension of a neuron that often makes distant connections; it may give out branches, which usually split off at right angles, and usually has a myelin covering, a type of insulation.

**B cells or B lymphocytes** (LIM-foe-sites) white blood cells that provide the antibody-mediated immune response; they develop and mature in the bone marrow; as they mature, B cells develop the ability to identify bacterial antigens and respond by developing into plasma cells, which are specialized to produce antibodies.

**Bacteria** a domain of single-celled prokaryotic organisms; they perform essential functions as decomposers of organic material and are natural recyclers of nitrogen and other organic compounds in ecosystems.

**basal ganglia** masses of gray matter located deep within the cerebrum that play important roles in the coordination of slow sustained movements, such as maintaining posture and the suppression of useless patterns of movement.

**base** any substance that combines with $H^+$ ions when it is dissolved in water.

**basidia** (buh-SID-ee-uh) in club fungi, club-shaped structures from which unenclosed sexual spores are produced.

**basophils** (BAY-soh-filz) one of three kinds of granulocytes; basophils contain granules that rupture and release chemicals that enhance the body's response to injury or infection; they also play a role in causing allergic responses.

**behavioral ecology** (or **sociobiology**) a science that applies the concepts of evolutionary biology to the study of social behavior to develop general laws of the biology and evolution of social behavior.

**behaviorism** a school of thought that suggests that learning takes place in a stimulus/response fashion, possibly reinforced by some type of reward that may or may not be readily apparent.

**behaviors** the patterns of movement, sounds (vocalizations), and body positions (postures) exhibited by an animal; behaviors also include any type of change in an animal, such as a change in coloration or the releasing of a scent, that can trigger certain behaviors in another animal.

**benign tumors** (buh NINE) growths or masses of cells that are made up of partially transformed cells, are confined to one location, and are encapsulated, shielding them from surrounding tissues.

**Big Bang** an explosion from a single-point of super-condensed matter about 15 billion years ago that was the start of our universe.

**bilateral symmetry** having two sides in which the right side is a mirror image of the left side.

**bile pigments** substances produced by the liver from the breakdown of old, worn-out red blood cells; bile pigments enter the small intestine with the bile and are the cause of the characteristic color of the feces.

**binary fission** (BYE-nuh-ree FIZH-un) a type of asexual reproduction in which one cell divides into two with no exchange of genetic material among cells; bacteria reproduce in this way.

**binomial nomenclature** (bye-NO-mee-uhl NO-men-clay-chur) literally, "two-name naming"; the system of determining the scientific name of an organism using its genus and species classifications; thus, an organism's genus becomes its first name, and its species becomes its last name.

**biological clocks** the governing mechanisms of circadian rhythms

**biological concentration** a process by which some organisms accumulate within their bodies certain harmful or deadly chemicals that are present in their environments or in the food they eat.

**biological diversity** the richness and variety of species on Earth.

**biological magnification** a process by which toxins accumulate in organisms in high concentrations when they consume tainted organisms lower on the food chain; this effect increases with progression up the food chain.

**biological species concept** recognizes species as a group of interbreeding natural populations that are reproductively isolated from other such groups.

**biomes** (BYE-omes) ecosystems of plants and animals that occur over wide areas of land within specific climatic regions easily recognized by their overall appearance.

**biosphere** (BYE-oh-sfear) the global ecosystem of life on Earth that extends from the tops of the tallest mountains to the depths of the deepest seas; the part of the Earth where biological activity exists.

**biotechnology** (BYE-oh-tek-NOL-uh-jee) the use of scientific and engineering principles to manipulate organisms.

**biotic factors** (bye-OT-ik) living factors within the environment, such as plants, animals, and microorganisms.

**birds** a class (Aves) of vertebrate animals with wings and feathers; birds have reptilian-like scales on their legs and lay amniotic eggs as reptiles do, but bird eggs have hard shells.

**blade** the flattened portion of a leaf.

**blastocyst** (BLAS-tuh-sist) a stage of development in which the preembryo is a hollow ball of cells; its center is filled with fluid.

**blastula** (BLAS-chuh-luh) the general term used to describe the saclike blastocyst of mammals and, in other animals, the embryonic stage that develops a similar fluid-filled cavity.

**bone** a type of connective tissue consisting of widely separated bone cells embedded in a matrix of collagen fibers and mineral salts; it forms the vertebrate skeleton.

**bony fishes** a class of vertebrates whose members have skeletons made of bone.

**boom and bust cycle** a pattern of exponential growth of a population followed by a sudden die-off.

**brain** one of the two main parts of the central nervous system; the brain is a complicated maze of interconnected neurons linked to the hormone-producing glands, the muscles, and other tissues; it consists of four main parts: the cerebrum, the cerebellum, the diencephalon, and the brainstem.

**brainstem** the part of the brain consisting of the midbrain, pons, and medulla that brings messages to and from the spinal cord and controls important body reflexes such as the rhythm of the heartbeat and rate of breathing.

**brassinosteroids** plant hormones that have a chemical structure similar to the steroid hormones in animals; they cause a wide array of responses in plants, such as stimulating stem and pollen tube elongation, stimulating ethylene production, and inhibiting root growth and development.

**breathing** the movement of air into and out of the lungs.

**brown algae** a phylum (Phaenophyta) of multicellular algae formerly classified with the plants; brown algae are found predominantly on northern, rocky shores and can grow to enormous sizes.

**bulbourethral glands** (BULL-bo-yoo-REE-thrul) a set of tiny glands lying beneath the prostate that secrete an alkaline fluid into the semen.

**bulk flow** a pressure-flow system of fluid transport in plants; sucrose is produced at a source, is actively transported into sieve-tube members by companion cells, and is then transported to a "sink" (e.g., roots) where it is used.

**Burgess Shale** one of the most diverse and well-preserved fossil localities in the world, found in an area of the Canadian Rocky Mountains known as the Burgess Pass; sedimentary rocks there contain many fossils of Cambrian organisms.

**Calvin cycle** another name for the light-independent reactions that take place during photosynthesis.

**capillaries** (KAP-uh-LARE-ees) fine latticeworks of microscopic blood vessels that permeate tissues.

**capsid** (KAP-sid) a protein "overcoat" that covers the nucleic acid core of a virus.

**carbohydrates** molecules that contain carbon, hydrogen, and oxygen, with the concentration of hydrogen and oxygen atoms in a 2:1 ratio.

**carbon cycle** the cycling of carbon between the atmosphere, the Earth, and the Earth's organisms.

**carbon fixation** a process by which organisms use the ATP and NADPH produced by light-dependent photosynthetic reactions to build organic molecules from atmospheric carbon dioxide.

**carboxypeptidase** (kar-BOK-see-PEP-ti-dace) an enzyme produced by the pancreas, which, together with trypsin and chymotrypsin, completes the digestion of proteins in the small intestine.

**carcinogens** (kar-SIN-uh-jens) cancer-causing substances.

**cardiovascular system** (KAR-dee-oh-VAS-kyuh-lur) the heart and blood vessels, the "plumbing" of the circulatory system.

**cardiovascular system** the heart and blood vessels of the body.

**carotenoids** (kuh-ROT-uh-noids) pigments that absorb photons of green, blue, and violet wavelengths and reflect red, yellow, and orange; they are second only to the chlorophylls in importance in photosynthesis.

**carrying capacity** the number of individuals within a population that can be supported within a particular environment for an indefinite period.

**cartilaginous fishes** (kar-tuh-LAJ-uh-nus) a class of vertebrates whose members have skeletons made of cartilage rather than bone that includes the sharks, skates, and rays.

**catabolic reactions** (CAT-uh-BOL-ick) chemical reactions that release energy by breaking down complex molecules into simpler molecules.

**cell** a microscopic mass of protoplasm; the unit of structure of all living things.

**cell body** the region of the neuron around the nucleus; the main organelles of the cytoplasm are in this portion of the cell.

**cell culture techniques** methods that enable scientists to remove cells from a parent plant and grow them into new plants.

**cell cycle** the time from the generation of a new cell until it reproduces; the cell cycle includes interphase, nuclear division, and cytoplasmic division.

**cell theory** a statement regarding the nature of living things, which holds that all living things are made up of cells; that the smallest living unit of structure and function of all organisms is the cell; and that all cells arise from preexisting cells.

**cell wall** a rigid structure that surrounds some cells' plasma membranes.

**cell-mediated immune response** one of the two branches of the specific immune response; it is initiated by helper T cells that have been activated by interleukin-1 and the presence of antigens, and results in cytotoxic T cells recognizing and destroying body cells.

**cellular respiration** a series of chemical reactions in humans and plants in which the sugars that make up carbohydrates are broken down.

**cellular slime molds** a phylum (Acrasiomycota) of protists that look and behave like amebas during much of their life cycle but can group together and form dormant, cystlike spores when food is scarce; these spores then revert to the amebalike form when conditions are favorable again.

**central nervous system** one of the two main parts of the vertebrate nervous system; the site of information processing within the nervous system; the central nervous system is made up of the brain and spinal cord.

**centrioles** pairs of cylindrical microtubular structures found in the cytoplasm of animal cells that play a role in cell division.

**cephalization** a characteristic of bilaterally symmetrical animals in which sense organs and nervous tissue are concentrated in the head end.

**cerebellum** (SER-uh-BELL-um) the part of the brain located below the occipital lobes of the cerebrum that coordinates subconscious movements of the skeletal muscles.

**cerebral cortex** (suh-REE-brul KOR-tecks) a thin layer of tissue, called gray matter, that forms the outer surface of the cerebrum, and within which most of the activity of the cerebrum occurs, including higher cognitive processes such as learning and memory.

**cerebrum** (suh-REE-brum) the largest and most dominant part of the human brain, which is divided into two halves, or hemispheres, connected by the corpus callosum; the cerebrum governs motor, sensory, and association activity.

**cervix** (SUR-viks) the narrower, bottom part of the uterus that opens into the vagina.

**chemical bonds** forces that hold atoms together.

**chemiosmosis** (KEM-ee-oz-MOH-sis) the synthesis of ATP using the potential energy of a hydrogen ion gradient across a membrane to phosphorylate ADP; chemiosmosis fuels most ATP synthesis in cells.

**chemoautotrophs** (KEE-mo-AWE-toe-trofes) organisms that make their own food by deriving energy from inorganic molecules.

**chiasmata** (keye-AZ-muh-tuh) in meiosis, the point of crossing over where parts of chromosomes have been exchanged during synapsis; under a light microscope, a chiasma appears as an X-shaped structure.

**chlorophyll *a*** the name given to special molecules of chlorophyll that absorb wavelengths of light in the 680- to 700-nanometer range, in the far-red portion of the light spectrum.

**chlorophylls** (KLOR-uh-fils) pigments that absorb photons of violet-blue and red wavelengths and reflect green and yellow; chlorophylls are the primary light gatherer in all plants and algae and in almost all photosynthetic bacteria.

**chloroplasts** (KLOR-oh-plasts) energy-capturing organelles that are found in the cells of plants and algae; these eukaryotic cell organelles manufacture carbohydrates during photosynthesis using energy from the sun and carbon dioxide.

**cholecystokinin** (CCK) (KOL-uh-SIS-tuh-KINE-un) one of the hormones that control digestion in the small intestine.

**Cholodny-Went hypothesis** the hypothesis that the growth curvatures seen as a plant bends toward the light are due to an unequal distribution of auxin between the two sides of the curving organ.

**chordates** (KOR-dates) animals having a hollow dorsal nerve cord, a dorsal rod-shaped notochord that forms during development, and gill arches at some stage of life; includes fishes, amphibians, reptiles, birds, mammals, and humans.

**choroid** (KOR-oyd) a thin, dark-brown membrane that lines the sclera of the eye and contains blood vessels that nourish the retina, and a dark pigment that absorbs light rays so they will not be reflected within the eyeball.

**chromatin** (KRO-muh-tin) the complex of deoxyribonucleic acid (DNA) and protein that makes up the chromosomes of eukaryotes.

**chromosomes** (KROW-muh-somes) DNA coiled tightly around proteins and condensed to form shortened, thickened structures.

**chronic bronchitis** (bron-KYE-tis) an inflammation of the bronchi and bronchioles that lasts for at least 3 months each year for 2 consecutive years with no accompanying disease as a cause; one of the disorders commonly included in chronic obstructive pulmonary disease.

**chronic obstructive pulmonary disease** (COPD) a term used to refer to disorders that block the airways and impair breathing.

**chymotrypsin** (kye-moe-TRIP-sin) an enzyme produced by the pancreas, which, together with trypsin and carboxypeptidase, completes the digestion of proteins in the small intestine.

**cilia** (SILL-ee-uh) hairlike extensions that often cover the surface of some eukaryotic cells and allow them to move or to move substances across their surfaces.

**ciliary muscle** (SILL-ee-err-ee) a tiny circular muscle that slightly changes the shape of the eye's lens by contracting or relaxing.

**ciliates** (SIL-ee-uts *or* SIL-ee-ates) a phylum (Ciliophora) of protists characterized by fine, short, hairlike extensions called cilia.

**circadian rhythms** (sir-KAY-dee-un) 24-hour cycles of physiological activity and behavior.

**circulation** the movement of a fluid through an organism in a regular manner.

**circulatory system** a transport system that uses a fluid to move substances such as nutrients, wastes, and gases throughout an organism.

**class** a taxonomic subcategory of phyla; as organisms appear to be increasingly related to each other, they are more and more narrowly classified within taxonomic subcategories.

**classical biotechnology** the use of scientific and engineering principles to manipulate organisms, relying on the traditional techniques of selection, mutation, and hybridization.

**classical conditioning** a form of learning in which an animal is taught to associate a new stimulus with a natural stimulus that normally evokes a response in the animal.

**classification** the categorization of organisms into a coherent scheme.

**climax community** a community in which the mix of plants and animals becomes stable; the last stage of succession.

**clitoris** (KLIT-uh-ris) a small mass of erectile and nervous tissue in the female genitalia that responds to sexual stimulation.

**clones** copies.

**cloning vector** a plasmid, virus, or other piece of self-replicating DNA that can carry an inserted DNA fragment into a bacterial, yeast, or human cell.

**closed circulatory system** a transport system in which blood is enclosed within vessels as it travels throughout an organism.

**club fungi** a phylum (Basidiomycota) of fungi that have club-shaped structures from which unenclosed spores are produced; mushrooms, puffballs, and shelf fungi are all types of club fungi.

**cnidarians** (neye-DARE-ee-uhnz) a phylum (Cnidaria) of radially symmetrical invertebrate animals with tentacles having stinging cells and with two basic body plans—medusae and polyps; jellyfish, hydras, sea anemones, and corals are cnidarians.

**cochlea** (KOCK-lee-uh) a winding, snail-shaped tube that contains the organ of hearing and forms a portion of the inner ear.

**codominant** (KO-DOM-uh-nunt) a term referring to traits in which the alternative forms of an allele are both dominant, and both characteristics are exhibited in the phenotype.

**codons** sequences of three nucleotide bases in transcribed mRNA that code for specific amino acids, which are the building blocks of polypeptides.

**coelom** (SEE-lum) a body cavity found within most bilaterally symmetrical organisms and the echinoderms that is a fluid-filled enclosure lined with connective tissue; in humans, the coelom is divided into two parts: the thoracic cavity and the abdominal cavity.

**coelomates** animals with a coelom—a fluid-filled body cavity lined with connective tissue; mollusks, annelids, arthropods, echinoderms, and chordates (including humans) are coelomates.

**coenzyme** (KO-ENN-zyme) cofactor that is an organic molecule.

**cofactors** nonprotein molecules that bind to enzymes and help them catalyze chemical reactions.

**cognitivism** (KOG-nih-tiv-izm) a school of thought that suggests that individuals acquire and then store information in memory; learning takes place as new information builds and merges with the old, leading to changes in behavior.

**coleoptile** (KO-lee-OP-tle) in grasses, a straight sheath that offers protection to newly emerging shoots.

**colon** the large intestine; its function is to absorb sodium and water, and to eliminate wastes.

**commensalism** a relationship in which an organism of one species benefits from its interactions with another, whereas the other species neither benefits nor is harmed.

**communication** the transfer of information from one individual to another, which is a stimulus for the other's reaction.

**community** a grouping of populations of different species living together in a particular area at a particular time.

**competition** a situation in which organisms that live near one another strive to obtain the same limited resources.

**competitive behavior** a type of behavior that occurs when two or more animals strive to obtain the same needed resource, such as food, water, nesting sites, or mates.

**competitive exclusion** a principle that states that if two species are competing with one another for the same limited resource in a specific location, the species able to use that resource most efficiently will eventually eliminate the other species in that location.

**complement** proteins that circulate in the blood and are part of both nonspecific and specific immunity; when triggered by an antigen, these proteins poke holes in the cell membranes of pathogens.

**complementary DNA (cDNA) cloning** the process of synthesizing DNA from mRNA and then making copies of this DNA using yeast or bacterial cells.

**compounds** molecules or ionically bonded substances made up of atoms of different elements.

**cones** light receptors located within the retina at the back of the eye that function in bright light and detect color.

**conjugation** (CON-juh-GAY-shun) one of the ways in which bacteria transfer genetic material; conjugation occurs as a donor and a recipient bacterium make contact, and the DNA from the donor is transferred to the recipient cell.

**connective tissue** groups of similar cells that provide a framework for the body, join its tissues, help defend it against foreign invaders, and act as a storage site for specific substances.

**constitutive genes** in prokaryotes, genes that encode gene products that are needed at all times to maintain life.

**consumers** organisms that cannot make their own food and must feed on producers or other consumers.

**continental shelf** the margin of land that extends out from the intertidal zone usually 50 to 100 kilometers (30 to 60 miles) and slopes to a depth of about 200 meters (approximately 650 feet) beneath the sea.

**contractile vacuoles** cellular organelles found in protists and sponges that collect wastes and excess water, squeezing them out of the cell when the vacuole fills.

**control** the standard against which experimental effects may be checked to establish their validity.

**controlled experiment** a method of testing a hypothesis in which a factor is changed and other factors that respond due to this change are observed; all other factors are held constant.

**convergent evolution** (kun-VUR-junt EV-uh-LOO-shun) the development of similar structures having similar functions in different species as the result of the same kinds of selection pressures.

**cornea** (KOR-nee-uh) the rounded, transparent portion of the eye's outer layer that permits light to enter the eye.

**corticosteroids** (KORT-ik-oh-STARE-oydz) a group of hormones, secreted by the adrenal cortex in response to ACTH, that act on the nuclei of target cells, causing the cell's hereditary material to produce certain proteins.

**cotyledon** (COT-ul-LEED-un) in plant seeds, a structure that absorbs, stores, and distributes nutrients.

**courtship** or **courtship ritual** patterns of behavior, usually consisting of a series of fixed action patterns of movement, designed to lead to mating; courtship rituals are unique to each species and occur after the formation of a male-female pair.

**covalent** (ko-VAY-lent) a type of chemical bond that is caused by the electrical attraction created by atoms sharing electrons.

**cranial nerves** any of the 12 pairs of nerves that enter the brain through holes in the skull.

**creatinine** (kree-AT-uh-neen) a nitrogenous waste found in small amounts in the urine that is formed from a nitrogen-containing molecule in muscle cells called creatine.

**Cro-Magnons** (KRO-MAG-nuns) early members of *H. sapiens sapiens* whose anatomical features were similar to modern humans; they made sophisticated tools, hunted, and created elaborate cave paintings.

**crossing over** a complex series of events during meiosis in which homologous nonsister chromatids cross over one another; these crossed-over pieces break away from the chromatids to which they are attached and reattach to the nonsister chromatid.

**cytokinesis** (sye-toe-kuh-NEE-sis) the physical division of the cytoplasm of a cell into two daughter cells.

**cytokinins** plant hormones that stimulate cell division.

**cytoplasm** (SYE-toe-PLAZ-um) a viscous fluid within a cell that contains all cell organelles except the nucleus.

**Darwinian evolution** Darwin's theory of evolution by natural selection.

**decomposers** (DEE-kum-POE-zurs) organisms such as bacteria and fungi that obtain their energy by breaking down organic material in dead organisms and contribute to the recycling of nutrients to the environment.

**deep zone** the area of pond or lake water into which light does not penetrate.

**dehydration synthesis** the process by which monomers are put together to form polymers.

**demography** (dih-MOG-ruh-fee) the statistical study of human populations.

**dendrite** a cell extension of a neuron that is usually much shorter than an axon, has no myelin covering (insulation), and is specialized to receive impulses from sensory cells or from axons of other neurons; most neurons contain multiple dendrites.

**density** the number of organisms or individuals in a population per unit of area.

**density-dependent limiting factors** environmental factors that result from the growth of a population but act to limit its subsequent growth; density-dependent limiting factors increase in effectiveness as population density increases.

**density-independent limiting factors** environmental factors that operate to limit a population's growth, regardless of its density.

**deoxyribonucleic acid** (DNA) (de-OK-see-RYE-boh-new-KLAY-ick) the hereditary material; DNA controls all cell activities and determines all of the characteristics of organisms.

**dependent variable** the factor that varies in response to changes in the independent variable during a controlled experiment.

**dermal tissue** the outer protective covering of virtually all plants and animals.

**desert** a biome that occurs around 20 to 30 degrees north and south latitude and in other areas that have 25 centimeters (10 inches) or less of precipitation annually; desert life is somewhat sparse, but exhibits adaptations to life in a dry environment.

**detritus** (dih-TRITE-us) the refuse or waste material of an ecosystem.

**deuterostomes** (DOOT-uh-row-stowmz) one of two distinct evolutionary lines of coelomates that includes the echinoderms and chordates; their embryological development is characterized by the anus developing from the first indentation of the gastrula and a radial pattern of cleavage.

**development** the changes in an organism throughout its life cycle.

**dialysis** (dye-AL-uh-sis) a method of treating renal failure in which blood is filtered through a machine called an artificial kidney or by using the peritoneum.

**diaphragm** (DYE-uh-fram) (1) a sheet of muscle that forms the horizontal partition between the thoracic cavity and the abdominal cavity; (2) a rubber dome that is inserted immediately before intercourse to cover the cervix and prevent the entry of sperm into the uterine tubes.

**diastole** (dye-AS-tl-ee) the time of relaxation of a heart chamber, when it is filling.

**diatoms** (DIE-uh-toms) a phylum (Chrysophyta) of protists—unicellular algae that look like microscopic pillboxes because they are made up of top and bottom shells that fit together.

**dicots** one of two groups of angiosperms that differ in the placement of stored food in their seeds; dicots store food in their cotyledons, or seed leaves.

**diencephalon** (DYE-un-SEF-uh-lon) the part of the brain consisting of the thalamus and hypothalamus.

**diffusion** the net movement of molecules from a region of higher concentration to a region of lower concentration, eventually resulting in a uniform distribution of the molecules.

**digestion** a process in which food particles are broken down into small molecules that can be absorbed by the body.

**digestive system** a series of organs specialized for breaking down food and ridding the body of waste.

**dinoflagellates** (dye-no-FLAJ-uh-luts *or* DYE-no-FLAJ-uh-lates) a phylum (Dinoflagellata) of protists that are unicellular algae characterized by stiff outer coverings; their flagella beat in two grooves, one encircling the cell like a belt and the other perpendicular to it.

**diploid** (DIP-loyd) having a full complement of genetic material for that species.

**directional selection** a type of natural selection in which an extreme phenotype is the most fit, with the average phenotype moving toward this extreme.

**disaccharidases** (dye-SACK-uh-rye-DAYS-is) enzymes, produced by specialized epithelial cells of the small intestine, that break down the disaccharides maltose, sucrose, and lactose to the monosaccharides glucose, fructose, and galactose.

**disaccharide** (dye-SACK-uh-ride) two monosaccharides linked together.

**dispersion** the way in which the individuals of a population are arranged within their environment.

**disruptive selection** (also called *diversifying selection*) a type of natural selection that occurs when two or more extreme phenotypes are fitter than the intermediates; each extreme becomes an average, and the population separates into increasingly different groups.

**DNA microarrays** (also called *DNA chips*) a technology for screening thousands of genes to determine which genes or gene products are present in a sample.

**domain** the most inclusive level of organism classification.

**dominance hierarchy** a rank order among social groups of animals, including many species of fishes, reptiles, birds, and mammals; dominance hierarchies reduce the costs of aggression within a social groups.

**dominant** in an organism carrying a pair of contrasting alleles for a particular trait, the form of the trait (the allele) that will be expressed.

**drug addiction** a compulsive urge to continue using a psychoactive drug, physical and/or psychological dependence on the drug, and a tendency to increase the dosage of the drug.

**duodenum** (DOO-oh-DEE-num *or* doo-ODD-un-um) the initial short segment of the small intestine that is actively involved in digestion and absorption of nutrients.

**echinoderms** a phylum (Echinodermata) of invertebrate animals that are bilaterally symmetrical as larvae but radially symmetrical as adults; they differ from other invertebrates in being deuterostomes, not protostomes, and thus are more closely related to the chordates; sea lilies, sea stars, brittle stars, sea urchins, sand dollars, and sea cucumbers are echinoderms.

**ecology** (eh-KOL-uh-gee *or* ee-KOL-uh-gee) the study of the interactions among organisms and between organisms and their environments.

**ecosystem** (EH-koe-SIS-tem *or* EE-koe-SIS-tem) all the biotic and abiotic factors within a certain area; a community consisting of plants, animals, and microorganisms that interact with one another and with their environments and are interdependent on one another for survival.

**ectoderm** (EK-toe-durm) the outer layer of cells formed during the early development of the embryos of all bilaterally symmetrical animals; the primary germ layer that gives rise to the outer layer of skin, the nervous system, and portions of the sense organs.

**ectothermic** (EK-toe-THUR-mik) a term referring to animals such as reptiles, amphibians, and fishes that regulate body temperature by taking in heat from the environment and by their behavior; their internal body temperature fluctuates.

**electrocardiogram** (ECG) (ih-LEK-trow-KARD-ee-uh-GRAM) a recording of the electrical impulses that pass throughout the heart as it contracts and relaxes.

**electron transport chain** a group of electron carriers located on the inner mitochondrial membrane and on the thylakoid membranes that, during redox reactions, release energy used to make ATP during photosynthesis and cellular respiration.

**electrons** particles that surround the core of an atom; electrons carry a negative electrical charge.

**electrophoresis** process in which DNA fragments are drawn through a gel using an electrical field, which separates the fragments into bands of different lengths.

**elements** pure substances that are made up of a single kind of atom and cannot be separated into different substances by ordinary chemical methods.

**elimination** a process whereby unabsorbed digestive wastes leave the body during defecation.

**embryo** (EM-bree-oh) the early stage of development in humans, from the third to eighth weeks.

**emigration** (EM-uh-GRAY-shun) the movement of organisms out of a population.

**emphysema** (em-fi-SEE-muh) a chronic obstructive pulmonary disease in which mucus plugs various bronchioles, trapping air within alveoli and often causing them to rupture; in this condition, the lungs lose their elasticity and the ability to recoil during exhalation, and instead stay filled with air.

**endangered species** those that face a very high risk of extinction in the wild in the near future.

**endergonic reaction** (ENN-der-GON-ick) a chemical reaction in which the products contain more energy than the substrates; energy must be supplied for this kind of reaction to proceed.

**endocrine glands** (ENN-doe-krin) ductless glands that secrete hormones and spill them directly into the bloodstream.

**endocrine system** the collective term for the 10 different endocrine glands of the human body, which secrete over 30 different hormones.

**endocytosis** (ENN-doe-sye-TOE-sis) a process in which cells engulf large molecules or particles and bring these substances into the cell packaged within vesicles.

**endoderm** (EN-doe-durm) the inner layer of cells formed during the early development of the embryos of all bilaterally symmetrical animals; the primary germ layer that gives rise to the digestive tract lining, the digestive organs, the respiratory tract, the lungs, the urinary bladder, and the urethra.

**endometrium** (en-do-MEE-tree-um) the inner lining of the uterus; this lining has two layers: one functional, transient layer that is in contact with the uterine cavity, and an underlying permanent layer; the functional layer is shed each month an embryo is not present in the uterus.

**endoplasmic reticulum** (EN-doe-PLAZ-mik ri-TIK-yuh-lum) an extensive system of interconnected membranes that forms flattened channels and tubelike canals within the cytoplasm of a cell.

**endoskeleton** a rigid support system within an animal body.

**endosperm** nutritive material that surrounds the embryo in some seeds ; these nutrients feed the embryo until the seed germinates.

**endosymbiotic theory** (EN-do-SIM-bye-OT-ik) the most widely accepted theory regarding how eukaryotes arose; according to this theory, bacteria became attached to or engulfed by host prokaryotic cells; mitochondria are thought to have originated from aerobic bacteria and chloroplasts from photosynthetic bacteria.

**endothermic** (EN-doe-THUR-mik) a term referring to organisms such as birds and mammals that maintain a stable internal body temperature by means of internal regulatory mechanisms.

**energy level** (or **shell**) the volume of space around an atom's nucleus where an electron is most likely to be found.

**entropy** (ENN-truh-pee) the energy lost to disorder; entropy is a measure of the disorder of a system.

**envelope** a chemical layer over the capsid of many viruses that is rich in proteins, lipids, and carbohydrate molecules.

**environment** a general term for the biosphere, encompassing the land, air, water, and every living thing on Earth.

**environmental contamination** the presence of toxic substances in a habitat.

**enzymes** (ENN-zymes) biological catalysts that reduce the amount of free energy of activation needed for a chemical reaction to take place, thus speeding up the reaction.

**eosinophils** (EE-oh-SIN-oh-filz) one of three kinds of granulocytes; eosinophils are believed to be involved in allergic reactions and also act against certain parasitic worms.

**epicotyl** the portion of a shoot above the cotyledons in most monocots and some dicots.

**epididymis** (ep-ih-DID-uh-mis) a long, coiled tube that sits on the back side of the testes and in which sperm mature.

**epiglottis** (ep-ih-GLOT-iss) a flap of tissue that folds back over the opening to the larynx, thus preventing food or liquids from entering the airway.

**epithelial tissue** (ep-uh-THEE-lee-uhl) groups of similar cells that cover body surfaces and line body cavities.

**epochs** (EP-uks or EE-poks) subdivisions of geological periods.

**eras** (EAR-uhs or AIR-uhs) geological time periods; scientists divide the time from the formation of the Earth until the present day into five eras.

**erythrocytes** (ih-RITH-row-sites) or red blood cells, one type of formed element of the blood that resembles flat disks with central depressions and that are packed with the oxygen-carrying molecule hemoglobin.

**esophagus** (ih-SOF-uh-gus) the food tube that connects the pharynx to the stomach.

**essential amino acids** amino acids that humans cannot manufacture and therefore must obtain from proteins in the food they eat.

**estrogens** (ES-truh-jens) various hormones that develop and maintain the female reproductive structures, such as the ovarian follicles, the lining of the uterus, and the mammary glands.

**estuary** (ESS-choo-ER-ee) a place where the fresh water of rivers and streams meets the salt water of oceans.

**ethology** (ee-THOL-uh-jee) the study of animal behavior in the natural environment; ethology examines the biological basis of the patterns of movement, sounds, and body positions of animals.

**ethylene** a gaseous plant hormone that causes fruit to ripen and that affects flower blossoming.

**euglenoids** (yoo-GLEE-noyds) a phylum (Euglenida) of flagellate protists that have chloroplasts and make their own food by photosynthesis.

**Eukarya** (you-CARE-yuh or you-CARE-ee-uh) the domain that consists of all eukaryotic organisms; these organisms can be unicellular (the protists and some fungi) or multicellular (most fungi and all plants and animals).

**eukaryotes** (you-KARE-ee-oats) cells that each have organelles and a membrane-bounded nucleus.

**eukaryotic cells** (YOO-kare-ee-OT-ick) cells with a membrane-bounded nucleus and membrane-bounded organelles that compartmentalize the cell; eukaryotic cells are much more complex in structure than prokaryotic cells.

**eustachian tube** (yoo-STAY-kee-un or yoo-STAY-shun) a structure that connects the middle ear with the nasopharynx (the upper throat); it equalizes air pressure on both sides of the eardrum when the outside air pressure is not the same as the pressure in the middle ear.

**eutrophication** (yoo-TROWF-uh-KAY-shun) the accumulation of inorganic nutrients in a lake, which stimulate plant growth; when the plants die, the decomposition process takes oxygen from the water, leading to the death of other organisms.

**evolution by means of natural selection** (ev-uh-LOO-shun) the descent, with modification, of different lineages from common ancestors.

**evolutionary psychology** the science that attempts to explain the behavior of our prehistoric hunter–gatherer ancestors as a direct result of adaptive evolution.

**evolutionary synthesis** (or **modern synthesis**) brings together Charles Darwin's theory of the evolution of species by natural selection with Gregor Mendel's theory of genetics as the basis for biological inheritance; recognizes that mechanisms in addition to natural selection are integral to evolution.

**evolutionary tree** a depiction of the evolutionary history of an organism or organisms.

**excretion** (ex-SKREE-shun) a process whereby metabolic wastes and excess water and salt are removed from the blood and passed out of the body.

**exergonic reaction** (EK-sur-GON-ick) a chemical reaction in which energy is released and, therefore, the products contain less energy than the substrate.

**exocrine glands** (EK-so-krin) glands whose secretions reach their destinations by means of ducts.

**exocytosis** (EK-so-sye-TOE-sis) the reverse of endocytosis; the discharge of material from a cell by packaging it in a vesicle and moving the vesicle to the cell surface.

**exoskeleton** (EK-so-SKEL-uh-tun) a rigid support system that covers an animal body.

**expiration** expelling air from the lungs.

**exponential growth** (ek-spo-NEN-shul) a rapid growth in the number of individual organisms in a population.

**external auditory canal** the passage from the pinna of the ear to the eardrum.

**external fertilization** a process in which a female sheds eggs and the male deposits sperm on them after they leave her body; the union of a male gamete (sperm) and a female gamete (egg) outside the body of the female.

**external genitals** (JEN-uh-tuls) sexual organs, such as the penis, that are located on the outside of the body.

**external respiration** the exchange of carbon dioxide and oxygen gases at the alveoli in the lungs.

**extinction** death of all members of a species.

**facilitated diffusion** a type of transport process in which molecules move across the cell membrane by means of a carrier protein, but down the concentration gradient without an input of energy by the cell.

**fact** proposition that is so uniformly upheld that little doubt exists as to its accuracy.

**family** a taxonomic subcategory of order.

**fats** large molecules made up of carbon, hydrogen, and oxygen, with a hydrogen-to-oxygen ratio higher than 2 : 1; one type of lipid.

**feedback loops** mechanisms by which information regarding the status of a physiological situation or system is fed back to the system so that appropriate adjustments can be made.

**fermentation** the chemical process by which cells extract energy from glucose without using oxygen.

**fertilization** the union of a male gamete (sperm) and a female gamete (egg).

**first filial ($F_1$) generation** (FIL-ee-uhl) the hybrid offspring of the parental (P) generation.

**first law of thermodynamics** a law stating that energy cannot be created or destroyed; it can only be changed from one form or state to another.

**fixed action patterns** sequences of innate behaviors in which the actions follow an unchanging order of muscular movements.

**flagella** (fluh-JELL-uh) whiplike extensions from some cells that allow them to move.

**flagellates** (FLAJ-uh-lates *or* FLAJ-uh-luts) protists that are characterized by fine, long, hairlike cellular extensions called flagella.

**flatworms** a phylum (Platyhelminthes) of invertebrate animals; flatworms have the simplest body plan among bilaterally symmetrical animals; they have organs and some organ systems, but no coelom; turbellarians, flukes, and tapeworms are flatworms.

**fluid mosaic model** a well-accepted explanation of the structure of the cell membrane as consisting of a fluid lipid bilayer studded with a mosaic of proteins, lipoproteins, and glycoproteins.

**food chain** a series of organisms from each trophic level that feed on one another.

**food web** the interwoven and interconnected feeding relationships of an ecosystem; interwoven food chains.

**formed elements** the solid portion of blood that is suspended in blood plasma, composed principally of erythrocytes, leukocytes, and platelets.

**fossil fuels** substances such as coal, oil, and natural gas that are formed over time (acted on by heat and pressure) from the undecomposed carbon compounds of organisms that died millions of years ago.

**fossil** preserved remains or impression of a dead organism.

**fossil record** the history of life as preserved in fossils.

**founder effect** extreme genetic drift that occurs if a small population becomes isolated from a larger population.

**fovea** (FOE-vee-uh) a spot on the retina that has the highest concentration of cones; the lens focuses images on this spot, resulting in sharp vision.

**fragmentation** a form of asexual reproduction in which an organism detaches parts of its body, which then grow into new individuals.

**fruit** mature ovary of an angiosperm, consisting of seeds and their surrounding tissues.

**functional group** a group of atoms with definite chemical properties that is attached to the carbon-based core of an organic molecule.

**fundamental niche** the role that an organism might play in an ecosystem if competitors were not present.

**fungi** (FUN-jye or FUN-ghee) a kingdom that consists of eukaryotic heterotrophs that digest food outside their bodies.

**gallbladder** a sac attached to the underside of the liver, where bile is stored and concentrated.

**gametes** (GAM-eets) sex cells; the female gamete is the egg, and the male gamete is the sperm.

**gametophyte (gamete-plant) generation** the haploid phase of a plant life cycle; this phase tends to dominate the life cycles of nonvascular plants.

**gametophyte** (guh-MEE-toe-fite) plant that produces haploid gametes by mitosis; also known as the gamete-plant generation.

**gastric glands** glands dotting the inner surface of the stomach that secrete a gastric juice of hydrochloric acid and pepsinogen.

**gastrin** a digestive hormone of the stomach that controls the production of gastric juice.

**gastrula** (GAS-truh-lah) three-layer embryo.

**gastrulation** (GAS-truh-LAY-shun) during prenatal development, the process by which various cell groups of the inner cell mass migrate, divide, and differentiate resulting in a three-layer embryo.

**gene** a unit of heredity formed of a sequence of nucleotides that codes for the amino acid sequences of polypeptides or for RNA.

**gene expression** the process by which genes produce polypeptides and RNA and by which these products exert their effects on an organism.

**gene flow** the exchange of genes among neighboring populations due to immigration and emigration.

**gene library** a collection of clones of DNA fragments, which together represents the entire genome of an organism.

**gene pool** the genetic composition of a population; all the alleles of all the genes of all the individuals in a population.

**gene synthesis cloning** the process of synthesizing DNA in the laboratory based on knowledge of its nucleotide sequence and then making copies of this DNA using yeast or bacterial cells.

**gene therapy** the treatment of a genetic disorder by the insertion of "normal" genes into the cells of a patient.

**general senses (somatic senses)** senses that have more than one location in the body, such as sense receptors distributed throughout the body for heat, cold, pain, and pressure.

**genetic counseling** the process in which a geneticist discusses a couple's genetic history early in a pregnancy or before conception to determine if their offspring may be at risk for a variety of genetic disorders.

**genetic engineering** (also called *recombinant DNA technology*) techniques of molecular biology that involve the manipulation of genes.

**genetic recombination** the exchange of genetic material between homologous chromosomes in meiosis; the exchange of DNA sequences between molecules.

**genetic testing** the examination of DNA for genes implicated in diseases or disorders.

**genetic variation** heritable differences among individuals in a population.

**genetics** (juh-NET-iks) the branch of biology dealing with the principles of heredity and the sources, nature, and implications of heritable variation in organisms.

**genome** (GEE-nome) an organism's total complement of genetic material.

**genomic library** a collection of clones of DNA fragments, which together represent the entire genome of an organism.

**genomics** subdiscipline of genetics that involves mapping, sequencing, and analyzing the functions of entire genomes.

**genotype** (JEEN-uh-type) an organism's allelic (genetic) makeup.

**genus** (GEE-nus) a taxonomic subcategory of family.

**germination** the sprouting of a seed, which begins when it receives water and has appropriate environmental conditions.

**gibberellins** plant hormones that promote growth through cell elongation.

**gills** evaginations of the body surface that form a respiratory surface in large, multicellular aquatic animals and that are confined to one part of the body as in bony fishes or are distributed throughout an organism as in aquatic worms.

**global warming** a worldwide temperature increase that could result from the greenhouse effect.

**glottis** (GLOT-iss) the space between the vocal cords; the opening to the larynx and trachea.

**glycogen** (GLYE-ko-jen) highly branched polysaccharides that are the storage form of sugar in animals.

**glycolysis** (glye-KOL-uh-sis) the first of the three series of chemical reactions of cellular aerobic respiration, in which glucose is broken down to pyruvate.

**Golgi apparatus** (GOL-gee) an organelle that collects, modifies, and packages molecules that are made at different locations within the cell and prepares them for transport.

**gonadotropins** (GON-ah-duh-TROP-inz) 2 of the 4 tropic hormones; gonadotropins affect the male and female sex organs.

**gonads** (GO-nads) the male and female reproductive organs that produce the sex cells, or gametes.

**gradient** differences in concentrations, pressures, and electrical charges that result in the net movement of molecules in a particular direction.

**grana** (GRA-nuh) stacks of thylakoid membranes within chloroplasts.

**granulocytes** (GRAN-yuh-low-sites) one of the two major groups of leukocytes distinguished by their cytoplasmic granules and lobed nuclei.

**gravitotropism** directional growth of a plant organ in response to gravity.

**green algae** a phylum (Chlorophyta) of multicellular algae including both unicellular and multicellular forms; most are aquatic, but some species live in moist places on land.

**greenhouse effect** the blocking of outward heat radiation from the Earth by carbon dioxide in the atmosphere.

**ground tissue** stores the carbohydrates the plant produces; it forms the substance of the plant and is the tissue in which vascular tissue is embedded.

**growth rate** the number of individuals added to a population during a given time; the growth rate of a population is determined by subtracting its death rate from its birth rate.

**guard cells** a pair of cells that brackets a stoma and regulates its opening and closing.

**gymnosperms** (JIM-no-spurms) vascular plants with naked seeds; four phyla of plants fall into this category: the conifers, cycads, ginkgos, and gnetophytes.

**habitat** a place where an organism lives or grows.

**habitat loss** the shrinking and fragmentation of the ranges in which species live.

**habituation** (huh-BICH-yoo-AY-shun) the ability of animals to "get used to" certain types of stimuli that they perceive as nonthreatening.

**haploid** (HAP-loyd) having half the usual number of chromosomes for that species.

**Hardy-Weinberg equilibrium** genotype frequencies and all allele frequencies will remain constant across generations when certain conditions are met.

**heart** the muscular pump of the circulatory system.

**hermaphrodites** (hur-MAF-rah-dytes) an animal or plant having both male and female reproductive organs.

**heterotroph** (HET-uhr-uh-trofe) an organism that cannot produce its own food; a consumer in an ecosystem or a food chain.

**heterozygous** (het-uhr-uh-ZYE-gus) a term referring to an individual who has two different alleles for a trait.

**high-glycemic index foods** carbohydrates that cause higher rises in blood sugar levels when eaten than do low-glycemic index foods; they have been linked to an increased risk for diabetes and heart disease.

**hinge joint** a kind of joint that allows movement in one plane only.

**homeostasis** (HOE-mee-oh-STAY-sis) the maintenance of a stable internal environment despite what may be a very different external environment.

**homeotic genes** (master control genes) genes that coordinate the action of a number of other genes, which, in turn, determine the development of a large region or body segment.

**hominids** (HOM-uh-nidz) one of three broad types of anthropoids, which includes all living and extinct ape and human species.

**hominins** (HOM-uh-ninz) formerly called hominids; all living and extinct human species, including immediate nonhuman ancestors; the only living hominin is *Homo sapiens sapiens.*

*Homo erectus* (HOE-moe ih-REK-tus) an extinct species of hominins whose fossil record dates back 1.8 million years; members of this species were fully adapted to upright walking, made sophisticated tools, and probably communicated with language.

*Homo habilis* (HOE-moe HAB-uh-lus) an extinct species of hominins whose fossil record dates back about 2 million years; members of this species are thought to have been the first humans and have been given a name that emphasizes their intelligence (*Homo habilis* means "skillful human").

*Homo sapiens* (HOE-moe SAY-pee-unz) an early or archaic form of modern humans (*Homo sapiens sapiens*) that is extinct; their fossil record dates back about 500,000 years.

*Homo sapiens sapiens* modern humans; the subspecies of hominins who made their appearance 10,000 years before the Neanderthal subspecies died out and whose early members are called Cro-Magnons.

**homologous** (hoe-MOL-eh-gus) of the same evolutionary origin and basic anatomy, now differing in function.

**homologous chromosomes** (also called *homologues*) (hoe-MOL-uh-gus KRO-muh-somes/HOME-uh-logs) pairs of chromosomes that all organisms produced by sexual reproduction receive; half of these chromosomes are from one parent organism and half from the other; homologous chromosomes each contain genes that code for the same inherited traits.

**homozygous** (hoe-muh-ZYE-gus) a term referring to an individual who has two identical alleles for a trait.

**hormone** (HORE-mone) a chemical messenger secreted and sent by a gland to other cells of the body.

**hornworts** a phylum (Anthocerophyta) of nonvascular plants having an elongated sporophytes that protrude like horns from the surface of the creeping gametophytes.

**Human Genome Project** (HGP) a worldwide scientific project that deciphered the DNA code of all 46 human chromosomes.

**human immunodeficiency virus** (HIV) (IM-yoo-no-de-FISH-un-see VYE-rus) a deadly virus that weakens the ability of the immune system to mount a defense against infection because it attacks and destroys helper T cells.

**hybrids** the offspring of members of two different species, genera, breeds, or varieties.

**hydrogen bond** weak electrical bond formed through the attraction between an oxygen or nitrogen atom and the H of an -NH group or -OH group.

**hydrolysis** (hi-DROL-uh-sis) the process by which polymers are disassembled into monomers.

**hydrophilic** (HI-droe-FIL-ick) characteristic of polar molecules that form hydrogen bonds with water.

**hydrophobic** (hi-dro-FO-bick) characteristic of nonpolar molecules that cannot form hydrogen bonds with water.

**hydroskeleton** a support system that uses a fluid under pressure as a scaffolding for an organism or body part.

**hypertension** (also called high blood pressure) characteristic of people with a blood pressure of 140/90 mm Hg or above.

**hypertonic** characteristic of a solution having a higher solute concentration than another solution to which it is being compared.

**hypocotyl** the shoot below the cotyledon.

**hypothalamus** (HYE-poe-THAL-uh-muss) a mass of gray matter located beneath the thalamus that regulates vital body functions such as respiration and the heartbeat and directs the hormone secretions of the pituitary gland.

**hypothesis** (hi-POTH-uh-sis) a tentative explanation of a phenomenon that guides scientific inquiry.

**hypotonic** characteristic of a solution having a lower solute concentration than another solution to which it is being compared.

**immigration** (IM-uh-GRAY-shun) the movement of organisms into a population.

**immune system** the body structures that perform specific defense responses; the organs of the immune system are the red bone marrow and the thymus gland (in which disease-fighting cells [lymphocytes] arise and/or develop) and the spleen and lymph nodes (in which mature lymphocytes accumulate and function).

**immunity** (ih-MYOON-ih-tee) protection from disease, particularly infectious disease; the two types of immunity are nonspecific and specific.

**imperfect fungi** a catchall group of fungi (Deuteromycota) whose sexual stages of reproduction have not been observed, making classification into one of the phyla of fungi impossible.

**implantation** (IM-plan-TAY-shun) the embedding of the developing blastocyst into the posterior wall of the uterus approximately 1 week after fertilization.

**imprinting** a rapid and irreversible type of stimulus/response learning that takes place during an early developmental stage of some animals.

**incomplete dominance** a term referring to traits in which alternative alleles are neither dominant over nor recessive to other alleles governing a particular trait; heterozygotes for incomplete dominance traits are phenotypic "intermediates."

**incus** (ING-kus) one of the three bones of the middle ear that amplify sound vibrations and carry them from the outer ear to the inner ear; the incus is also known as the anvil.

**independent variable** the factor that is manipulated during a controlled experiment.

**infectious diseases** those caused by pathogens that enter organisms and disrupt their normal functioning.

**inferior vena cava** a large vein that collects blood from the lower body and returns it to the right atrium of the heart.

**innate behaviors** those behaviors resulting from genetically determined neural programs that are part of the nervous system at the time of birth or develop at an appropriate point in maturation.

**insect societies** social groups that are formed by many species of insects, particularly bees and ants, and are characterized by a division of labor.

**insight** (also called *reasoning*) the capability of recognizing a problem and solving it mentally before ever trying out a solution.

**inspiration** taking air into the lungs.

**integument** (in-TEG-you-ment) in vertebrates, the skin, hair, and nails.

**integumentary system** (in-TEG-you-MEN-tuh-ree) another name for the vertebrate body's integument: the skin, hair, and nails.

**intercostal muscles** (inn-ter-KOS-tul) literally, "between-the-rib muscles"; muscles that extend from rib to rib and assist the diaphragm in the breathing process.

**internal fertilization** a process in which the male deposits sperm in or near the female reproductive tract and fertilization (the union of a male and a female gamete) takes place within the body of the female.

**internal respiration** the exchange of oxygen and carbon dioxide between the blood and the tissue fluid.

**interneurons** nerve cells located within the brain or spinal cord that integrate incoming information with outgoing information.

**internodes** the portions of a plant's stem that lie between the nodes.

**interphase** (IN-tur-faze) the portion of the cell cycle preceding cell division in which the cell grows and carries out life functions; during this time the cell also doubles in size and produces an exact copy of its hereditary material, DNA, as it prepares for cell division.

**intimidation displays** (or **threat displays**) a form of aggressive behavior that scares other animals away or causes them to back down before fighting takes place.

**ionic** (eye-ON-ick) a type of chemical bond between atoms that is caused by the attraction of oppositely charged particles formed by the gain or loss of electrons.

**ionizing radiation** (EYE-uh-nye-zing RAY-dee-AY-shun) a form of electromagnetic energy that can cause chromosomes to break or can cause changes in the nucleotide structure of DNA; X-rays and nuclear radiation are kinds of ionizing radiation.

**iris** (EYE-rus) a diaphragm lying between the cornea and lens that controls the amount of light entering the eye.

**islets of Langerhans** (EYE-lets of LANG-ur-HANZ) the separate types of cells within the exocrine cells of the pancreas that produce the hormones insulin and glucagon.

**isotonic** (EYE-so-TAWN-ick) refers to solutions having equal solute concentrations to one another.

**isotopes** (EYE-suh-topes) atoms that have the same number of protons but different numbers of neutrons.

**jawless fishes** tubular, scaleless, jawless vertebrates that live in the sea or in brackish water; they have no paired fins.

**joint** a place within the skeletal system where bones, or bones and cartilage, come together; also known as an articulation.

**karyotype** (KAIR-ee-oh-type) the condensed chromosomes of a eukaryote as they are seen at metaphase, arranged according to size, shape, and other characteristics.

**kidney stones** crystals of certain salts that can develop in the kidney and block urine flow.

**kidneys** along with the lungs, the primary organs of excretion; kidneys excrete the ions of salts and the nitrogenous wastes urea, creatinine, and uric acid, along with small amounts of other waste products.

**kinesis** (kih-NEE-sis) the change in the speed of the random, nondirected movements of an animal with respect to changes in certain environmental stimuli.

**kinetic energy** (kuh-NET-ick) energy actively doing work; the energy of motion.

**kingdom** a broad category in which taxonomists group living things; the classification scheme used increasingly today is domain, kingdom, phylum, class, order, family, genus, species.

**Krebs cycle** the second series of chemical reactions of aerobic cellular respiration, in which pyruvate, the end product of glycolysis, is oxidized to carbon dioxide; also called the *citric acid cycle*.

**labia majora** (LAY-bee-uh muh-JORE-uh) two longitudinal folds of skin that run posteriorly from the mons in the exterior genitals of the female.

**labia minora** (LAY-bee-uh mu-NORE-uh) folds of skin covered by the labia majora in the exterior genitals of the female.

**lancelets** (LANS-lets) a subphylum of the chordates; tiny, scaleless, fishlike marine organisms that are just a few centimeters long and pointed at both ends; their adult forms exhibit chordate characteristics.

**larynx** (LAIR-inks) the voice box, which is located at the beginning of the trachea.

**lateral buds** tiny undeveloped side shoots that develop at the angles between a plant's leaves and its stem.

**lateral line system** a complex system of mechanoreceptors possessed by all fishes and amphibians that detects mechanical stimuli such as sound, pressure, and movement.

**laws of thermodynamics** two laws that govern all of the changes in energy that take place in the universe.

**learning** an alteration in behavior based on experience.

**leaves** the parts of a plant's structure where most photosynthesis takes place.

**left atrium** (AY-tree-um) the upper left chamber of the heart, into which oxygenated blood enters from the lungs.

**left ventricle** (VEN-truh-kul) a chamber in the heart into which blood flows from the left atrium; the left ventricle then pumps blood through the aorta into the arteries.

**lens** a structure in the eye lying just behind the aqueous humor that plays a major role in focusing the light that enters the eye onto the retina at the back of the eye.

**leukocytes** (LOO-ko-sites) or white blood cells, one type of formed element of the blood that is larger than red blood cells and is essentially colorless; there are several kinds of leukocytes, including macrophages and lymphocytes, but all of them function in defending the body against invading microorganisms and foreign substances.

**lichens** (LIE-kins) associations between sac fungi and either cyanobacteria or green algae.

**life cycle** the progression of stages an organism passes through from its conception until it conceives another similar organism.

**ligaments** (LIG-uh-munts) bundles or strips of dense connective tissue that hold bones to bones.

**light-dependent reactions** the reactions of photosynthesis that produce ATP and NADPH, which can occur only in the presence of light.

**light-independent reactions** the reactions of photosynthesis that use ATP and NADPH to provide energy for the formation of sucrose from carbon dioxide; to occur, these reactions do not require light.

**lipases** (LYE-pays-es) digestive enzymes that break down the triglycerides in lipids to fatty acids and glycerol.

**lipids** composite molecules made up of glycerol and fatty acids (in the case of oils and fats) or carbon rings (in the case of steroids).

**liver** a large, complex organ weighing over 3 pounds, lying just under the diaphragm, that performs more than 500 functions in the body, including the secretion of bile, which aids in the digestion of lipids.

**liverworts** a phylum (Hepatophyta) of nonvascular plants; a well-known example is *Marchantia*.

**lungs** invaginations of the body surface, which are confined to one part of the body and form a respiratory surface in large, multicellular land animals.

**lymph** (LIMF) the name given to tissue fluid in the vessels of the lymphatic system.

**lymph nodes** small, ovoid, spongy structures located in various places in the body along the routes of the lymphatic vessels, which filter lymph as it passes through them.

**lymphatic system** (lim-FAT-ik) the body's one-way, passive circulatory system, which collects and returns to the blood tissue fluid that does not return directly from the body's tissues.

**lymphocytes** (LIM-foe-sites) white blood cells that are formed in the bone marrow, circulate in the blood and lymph, and reside in lymph tissue such as the lymph nodes, spleen, and thymus; lymphocytes recognize and react to substances that are foreign to the body, sometimes producing a protective immunity to disease.

**lysogenic cycle** (lye-suh-JEN-ik) a pattern of viral replication in which a virus integrates its genetic material with that of a host and is replicated each time the host cell replicates.

**lysosomes** (LYE-so-somes) membrane-bounded organelles that are essentially bags of many different digestive enzymes; lysosomes break down old cell parts or materials brought into the cell from the environment and are extremely important to the health of a cell.

**lytic cycle** (LIT-ik) a pattern of viral replication in which a virus enters a cell, replicates, and then causes the cell to burst, releasing new viruses.

**macroevolution** large-scale evolution; evolution above the species level.

**macrophages** (MAK-row-FAY-djus) enlarged, amebalike cells that entrap microorganisms and particles of foreign matter by phagocytosis.

**malignant** (muh-LIG-nunt) a term describing cancerous tumors that have the ability to invade and kill other tissues and move to other areas of the body.

**malleus** (MAL-ee-us) a very small bone, connected to the internal side of the eardrum, that works with two other small bones, the incus and stapes, to amplify sound vibrations and carry them from the outer ear to the inner ear; the malleus is also known as the hammer.

**Malpighian tubules** excretory organs of insects and spiders that consist of blind-ended tubes lying in the blood-filled body cavity and emptying into the intestine.

**mammals** (MAMM-uhls) warm-blooded vertebrates that have hair and whose females secrete milk from mammary glands to feed their young.

**mammary glands** milk-producing glands that lie over the chest muscles in female mammals.

**marsupials** (mar-SOO-pee-uhls) a subclass of mammals that gives birth to immature young that are carried in a pouch.

**mating** male-female behaviors that result in fertilization, regardless of whether copulation occurs.

**medulla** (mih-DULL-uh) the lowest portion of the brainstem, continuous with the spinal cord below; the medulla is the site of neuron tracts, which cross over one another, delivering sensory information from the right side of the body to the left side of the brain and vice versa.

**medusae** (meh-DOO-see *or* meh-DOO-zee) free-floating and often umbrella-shaped aquatic animals with the mouth usually located on the underside of the umbrella shape and tentacles hanging down around the umbrella's edge.

**meiosis** (my-OH-sis) a type of cell division by means of which the sex organs of a mature animal produce gametes; during meiosis, one parent cell produces four sex cells; each gamete produced during this process contains half the number of chromosomes of the original parent cell; consists of meiosis I and meiosis II.

**meiosis I and meiosis II** (my-OH-sis) the two-staged process of nuclear division in which the number of chromosomes in cells is halved during gamete formation; both meiosis I and meiosis II can be further divided into four stages: prophase, metaphase, anaphase, and telophase.

**Mendel's law of independent assortment** the concept that the distribution of alleles for one trait into the gametes does not affect the distribution of alleles for other traits.

**Mendel's law of segregation** the concept that each gamete receives only one of an organism's pair of alleles; chance determines which member of a pair of alleles becomes included in a gamete.

**meninges** (muh-NIN-jeez) the three layers of protective membranes covering both the brain and the spinal cord.

**menopause** (MEN-uh-pawz) the permanent cessation of menstrual activity in a woman, usually between the ages of 50 and 55; the end of the menses.

**menstruation** (MEN-stroo-AY-shun) the monthly sloughing of the blood-enriched lining of the uterus when pregnancy does not occur; the lining degenerates and causes a flow of blood, tissue, and mucus from the uterus out through the vagina.

**meristematic tissue** (MER-uh-stuh-MAT-ik) an undifferentiated type of tissue in a vascular plant in which cell division occurs during growth.

**mesoderm** (MEZ-oh-durm) the middle layer of cells formed during the early development of the embryos of all bilaterally symmetrical animals; the primary germ layer that gives rise to the skeleton, muscles, blood, reproductive organs, connective tissue, and the innermost layer of the skin.

**messenger RNA** (mRNA) a type of RNA that brings information from the DNA within the nucleus to the ribosomes in the cytoplasm and directs polypeptide synthesis.

**metabolic pathways** chains of reactions that move, store and release energy.

**metamorphosis** a process of development in insects that involves change in form; in complete metamorphosis, the stages of change are egg, larva, pupa, and adult; in incomplete metamorphosis, the stages are egg, nymph and adult.

**metaphase** (MET-uh-faze) the second phase of mitosis, meiosis I, and meiosis II; metaphase begins when the chromosomes align themselves equidistantly from the two poles of the cell.

**metastasis** (muh-TAS-tuh-sis) one of the characteristics of cancer cells: the ability to spread to multiple sites throughout the body.

**microevolution** small-scale change in the frequency of alleles in a gene pool over time.

**midbrain** the top part of the brainstem; it contains nerve tracts that connect the upper and lower parts of the brain, and nuclei that act as reflex centers.

**middle ear** the portion of the ear that contains three bones that amplify the force of sound vibrations as they conduct them from the tympanic membrane to the oval window of the inner ear.

**migrations** long-range, two way movements by animals, often occurring yearly with the change of seasons.

**mitochondria** (MITE-oh-KON-dree-uh) oval, sausage-shaped, or threadlike cellular organelles approximately the size of bacteria that have their own DNA; these eukaryotic cell organelles break down fuel molecules, releasing energy for cell work.

**mitosis** (my-TOE-sis or mih-TOE-sis) a process of cell division that produces two identical cells from an original parent cell.

**modern synthesis** (or **evolutionary synthesis**) brings together Charles Darwin's theory of the evolution of species by natural selection with Gregor Mendel's theory of genetics as the basis for biological inheritance; recognizes that mechanisms in addition to natural selection are integral to evolution.

**molecule** two or more atoms held together by shared electrons.

**mollusks** a phylum (Mollusca) of invertebrate animals with a soft body and a muscular foot; in most cases, the entire organism is covered with a hard shell; clams, oysters, squid, and snails are mollusks.

**molting** in arthropods, a process in which the organism outgrows its exoskeleton, which splits open and is shed; a new, soft exoskeleton lies underneath, which then hardens.

**monocots** one of two groups of angiosperms that differ in the placement of stored food in their seeds; monocots store most of their extra food in extraembryonic tissue called endosperm.

**monocytes** (MON-oh-sites) a group of agranulocytes that circulates as the granulocytes do; monocytes are attracted to the sites of injury or infection, where they mature into macrophages and engulf any bacteria or dead cells that neutrophils may have left behind.

**monosaccharides** (MON-o-SACK-uh-rides) simple sugars.

**monotremes** (MON-oh-treems) a subclass of mammals that lays eggs with leathery shells similar to those of reptiles.

**mons pubis** (monz PYOO-bis) the mound of fatty tissue that lies over the place of attachment of the two pubic bones in females.

**morphogenesis** (MORE-foe-JEN-uh-sis) the early stage of development in a vertebrate when cells begin to move, or migrate, thus shaping the new individual.

**morphological species concept** recognizes species as a group of organisms that share certain morphological (anatomical) and biochemical traits.

**mortality** the death rate of a population.

**morula** (MORE-yuh-luh) a stage of development in which the embryo consists of about 16 densely clustered cells and is still the same size as a newly fertilized ovum.

**mosses** the largest phylum (Bryophyta) of nonvascular plants; the gametophytes of most mosses have small, simple leaflike structures often arranged in a spiral around stemlike structures.

**motor neurons** nerve cells that are specialized to transmit information from the central nervous system to the muscles and glands.

**multiple alleles** (uh-LEELS) a system of more than two alleles that govern certain traits.

**municipal solid waste** refuse from homes, institutions, and businesses (including paper, yard waste, food waste, plastics, metals, textiles, glass, wood, etc.), but not including agricultural or industrial wastes.

**muscle tissue** groups of similar cells that are capable of contraction.

**mutations** (myoo-TAY-shuns) permanent changes in the genetic material that alter the original expression of a gene or genes; mutations can affect single genes, pieces of chromosomes, whole chromosomes, or entire sets of chromosomes.

**mutualism** a relationship in which two species live together in close association, both benefiting from the relationship.

**mycorrhizae** (MY-kuh-RYE-zuh) fungi that form mutually beneficial relationships with the roots of higher plants.

**myelin sheath** (MY-eh-len) the fatty wrapping surrounding some axons and long dendrites of sensory neurons created by multiple layers of many Schwann cell membranes.

**myosin** (MY-uh-sin) a protein that makes up the thick myofilaments in muscle fiber.

**nasal cavities** two hollow areas, located above the oral cavity and behind the nose, that are bordered by projections of bone covered with moist epithelial tissue.

**nastic movements** (NAS-tick) plant movements in which the direction of the response is not related to the direction of the stimulus as it is in tropisms.

**natural selection** the process in which organisms with adaptive traits survive in greater numbers than organisms without such traits.

**Neanderthal** (nee-AN-dur-thol or nee-AN-dur-tal) a subspecies of *Homo sapiens* named after the Neander Valley in Germany, where their fossils were first found; Neanderthals were short and powerfully built, with large brains; they made diverse tools, took care of the sick and injured, and buried their dead.

**negative feedback loop** a feedback loop in which the change that takes place is opposite to the condition that prompted the change; most of the body's regulatory mechanisms work by means of negative feedback loops.

**neo-Darwinian theory of evolution** a genetics-based view of natural selection developed in the early twentieth century.

**nephridia** (neh-FRID-ee-uh) excretory tubes found in the platyhelminthes (flatworms) and annelids (earthworms).

**nephrons** (NEF-rons) the microscopic filtering systems of the vertebrate kidneys in which urine is formed.

**nerve cord** a single, hollow cord along the back that carries sensory and motor impulses and that is a principal feature of chordates; in vertebrates, the nerve cord differentiates into a brain and spinal cord.

**nerve impulses** the electrochemical signals sent by nerves.

**nerve net** a system of interconnecting nerve cells with no central controlling area, or brain.

**nervous system** the quick message system of the body, the nervous system in all complex animals is made up of nerve cells that transmit electrochemical signals throughout the body.

**nervous tissue** groups of similar cells, along with supporting cells, that are specialized to conduct electrochemical impulses in animals.

**neuromuscular junction** (NER-oh-MUS-kyuh-lur) a synapse between a neuron and a skeletal muscle cell.

**neurons** (NER ons) nerve cells.

**neurotransmitters** (NER-oh-TRANS-mit-urs) chemicals released when nerve impulses reach the axon tip of a nerve cell; neurotransmitters then cross the synaptic cleft to combine with receptor molecules on the target cell.

**neurulation** (NOOR-oo-LAY-shun) the development of a hollow nerve cord.

**neutrons** (NOO-trons) particles found at the core of an atom; neutrons carry no electrical charge.

**neutrophils** (NOO- truh-filz) one of three kinds of granulocytes; neutrophils migrate to the site of an injury and engulf microorganisms and other foreign particles.

**niche** (nich) the role each organism plays within an ecosystem.

**nitrogen cycle** the cycling of nitrogen from the Earth's atmosphere to nitrogen-fixing and nitrifying bacteria, to other organisms, to deniftrifying bacteria, and then back to the atmosphere.

**nitrogen-fixing bacteria** microorganisms with the ability to convert nitrogen gas into ammonia, which other living things use to synthesize proteins, nucleic acids and other necessary nitrogen-containing compounds.

**nitrogenous wastes** (nye-TROJ-uh-nus) nitrogen-containing molecules that are produced as waste products from the body's breakdown of proteins and nucleic acids.

**nodes** locations on the stem of a plant where leaves form.

**nodes of Ranvier** (ron-VYAY) uninsulated spots on the myelin sheath between two Schwann cells.

**nondisjunction** (non-dis-JUNK-shun) the failure of homologous chromosomes to separate after synapsis during meiosis, resulting in gametes with abnormal numbers of chromosomes.

**nonnative species** organisms that are alien to a particular ecosystem.

**nonrandom mating** inbreeding, or the selection of mates for specific characteristics, both of which reduce genetic variability in a gene pool.

**nonrenewable resources** those formed at a rate much slower than their consumption; they are thus finite in supply; fossil fuels and minerals are nonrenewable resources.

**nonspecific immunity** a set of defenses the body has to keep out any foreign invader; these defenses include the skin and mucous membranes.

**nonvascular plants** (non-VAS-kyuh-ler) plants that lack specialized transport tissues.

**normal microbiota** bacteria that usually inhabit the human body in places such as the mouth, throat, colon, vagina, and skin; they normally do not cause disease and prevent foreign bacteria from invading the body.

**notochord** (NO-toe-kord) a rod-shaped structure that forms between the nerve cord and the gut (stomach and intestines) during the development of all chordates; the notochord forms the midline axis along which the vertebral column (backbone) develops in all vertebrate animals.

**nuclear power** (NOO-klee-ur) an energy source that derives its power from the splitting apart of the nuclei of large atoms (nuclear fission) or from the combining of the nuclei of certain small atoms (nuclear fusion); all nuclear reactors in use today use nuclear fission reactions to produce energy.

**nucleic acid** (noo-KLAY-ick) a long polymer of repeating subunits called nucleotides; the two types of nucleic acid within cells are deoxyribonucleic acid (DNA) and ribonucleic acid (RNA); nucleic acids store information about the structure of proteins.

**nucleolus** (noo-KLEE-oh-lus) a darkly staining region within the nucleus of a cell that contains a special area of DNA that directs the synthesis of ribosomal ribonucleic acid, or rRNA.

**nucleotide** (NOO-klee-o-tide) a single unit of nucleic acid, consisting of a five-carbon sugar bonded to a phosphate group and a nitrogen-containing base.

**nucleus** (NOO-klee-us) the control center of the cell, which is made up of an outer, double membrane that encloses the cell's chromosomes and one or more nucleoli.

**nutrients** carbohydrates, lipids, proteins, water, vitamins, and minerals in the diet that provide raw materials and other substances needed for growth, repair, maintenance, reproduction, and energy.

**octet rule** (OCK-tet ROOL) one of three factors that influence whether an atom will interact with other atoms; the octet rule states that an atom with an unfilled outer shell has a tendency to interact with another atom or atoms in ways that will complete this outer shell.

**olfactory receptors** (ole-FAK-tuh-ree) neurons whose cell bodies are embedded in the nasal epithelium; they detect smells when different airborne chemicals bind with receptor chemicals in their ciliated dendrite endings.

**oncogenes** (ON-ko-jeens) cancer-causing genes.

**one-gene-one-enzyme theory** a scientific explanation that states that the production of a given enzyme is under the control of a specific gene; if the gene mutates, the enzyme will not be synthesized properly or will not be made at all; therefore, the reaction it catalyzes will not take place, and the product of the reaction will not be produced.

**open circulatory system** a transport system in which blood flows in vessels leading to and from the heart but through irregular channels called blood sinuses in many parts of an organism; in an open circulatory system, blood mixed with tissue fluid bathes the organs and tissues directly.

**open-water zone** the main body of pond or lake water through which light penetrates.

**operant conditioning** (OP-uh-runt) a process in which the frequency of occurrence of a behavior is modified by the consequences of the behavior.

**operculum** (oh-PUR-kyuh-lum) a flap in a bony fish that extends posteriorly from the head over the gills, protects the gills, and enhances water flow over the gills.

**operons** in the bacterial genome, genes with related functions grouped in units.

**optic nerve** the nerve that carries impulses from the retina to the brain.

**order** a taxonomic subcategory of class.

**organ of Corti** (KORT-ee) the organ of hearing; the collective term for the hair cells, the supporting cells of the basilar membrane, and the overhanging tectorial membrane.

**organ system** a group of organs that function together to carry out the principal activities of an organism.

**organ** two or more tissues grouped together to form a structural and functional unit.

**organic compounds** the carbon-containing molecules that make up living things.

**organisms** living things; organisms can be either multicellular or unicellular.

**osmoregulation** (OZ-mo-reg-you-lay-shun or OS-mo-reg-you-LAY-shun) the control of water movement within an organism.

**osmosis** (os-MOE-sis) a special form of diffusion in which water molecules move from an area of higher concentration to an area of lower concentration across a differentially permeable membrane.

**osmotic pressure** the pressure that water exerts on a cell as it diffuses into the cell.

**oval window** the entrance to the cochlea of the inner ear; the stirrup fits into this membrane-covered opening.

**ovaries** (OH-vah-reez) the female gonads, in which secondary oocytes develop.

**overspecialization** characteristic of a species that is adapted to a narrow range of environmental conditions.

**oviducts** commonly referred to as the fallopian tubes; they lead from the ovaries to the uterus.

**oviparous** (oh-VIP-uh-rus) a term that describes an organism that lays eggs in which the embryo develops after egg laying.

**ovoviviparous** (OH-vo-vye-VIP-uh-rus) a term that describes an organism that retains fertilized eggs within the oviducts until the young hatch.

**ovulation** (OV-yuh-LAY-shun) the monthly process by which a secondary oocyte (potential egg) is expelled from the ovary.

**oxidation** (OK-si-day-shun) a reaction that involves an atom losing an electron.

**ozone** ($O_3$) (OH-zone) a chemical air pollutant formed as a secondary pollutant when hydrocarbons and nitrogen oxides undergo photochemical reactions; it is extremely irritating to the eyes and the respiratory tract; in the stratosphere, ozone protects the Earth from UV rays from the sun.

**pancreas** (PANG-kree-us or PAN-kree-us) a long gland that lies beneath the stomach and is surrounded on one side by the curve of the duodenum; it secretes a number of digestive enzymes and the hormones insulin and glucagon.

**pancreatic amylase** (pang-kree-AT-ick or pan-kree-AT-ick AM-uh-lace) a digestive enzyme secreted by the pancreas that works within the small intestine to break down starch and glycogen to maltose.

**pancreatic lipase** (pang-kree-AT-ick LYE-pace or LIP-ace) a digestive enzyme secreted by the pancreas that works in the small intestine to break down triglycerides to fatty acids and glycerol.

**parasites** organisms that feed on other, living organisms.

**parasitism** (PARE-uh-suh-tiz-um) an interaction in which an organism of one species (the parasite) lives in or on another (the host).

**parasympathetic nervous system** (PARE-uh-SIM-puh-THET-ik) a subdivision of the autonomic nervous system that generally stimulates the activities of normal internal body functions and inhibits alarm responses; acts in opposition to the sympathetic nervous system.

**parathyroid glands** (PARE-uh-THIGH-royd) four small glands embedded in the posterior side of the thyroid that produce parathyroid hormone.

**parental (P) generation** the members of a cross between pure-breeding organisms, giving rise to offspring called the $F_1$.

**parthenogenesis** (PAR-theh-no-JEN-eh-sis) a form of asexual reproduction in a variety of invertebrates and some vertebrates in which eggs of the female develop into adults without being fertilized.

**passive immunity** a kind of specific immunity conferred on an individual when he or she receives antibodies from another source; an example of passive immunity is a baby receiving antibodies during breastfeeding.

**passive transport** molecular movement down a gradient but across a cell membrane; three types of passive transport are diffusion, osmosis, and facilitated diffusion.

**pectoral girdle** (PEK-tuh-rul) the part of the appendicular skeleton that is made up of two pairs of bones: the clavicles, or collarbones, and the scapulae, or shoulder blades; the pectoral girdle is also known as the shoulder girdle.

**pedigrees** (PED-uh-greez) diagrams of genetic relationships among family members over several generations.

**pelagic zone** (puh-LAJ-ik) the open ocean.

**pelvic girdle** the part of the appendicular skeleton that is made up of the two bones called coxal bones, pelvic bones, or hip bones; the pelvic girdle is also known as the hip girdle.

**penis** (PEE-nis) a cylindrical organ of the male urogenital system that transfers sperm from the male reproductive tract to the female reproductive tract; it also functions in urination.

**pepsin** an enzyme of the stomach that digests only proteins, breaking them down into short peptides.

**peptidases** (PEP-teh-DACE-is) enzymes produced by cells in the intestinal epithelium that work as a team with trypsin, chymotrypsin, and carboxypeptidase to break down polypeptides into shorter chains and then to amino acids.

**peptide hormones** one of the two main classes of endocrine hormones; peptide hormones are made of amino acids and are unable to pass through cell membranes; instead, they bind to receptor molecules embedded in the membranes of target cells.

**peptides** short chains of amino acids.

**periods** shorter time units into which eras are subdivided.

**peripheral nervous system** (puh-RIF-uh-rul) one of the two main parts of the vertebrate nervous system; it consists of an information highway of nerves that bring messages to and from the brain and spinal cord.

**peristalsis** (PEAR-ih-STALL-sis) successive waves of contractions of the esophagus and small intestine that move food along these parts of the digestive system.

**petiole** (PET-ee-ole) the slender stalk of a leaf.

**pH scale** a succession of values (scale) based on the slight degree of spontaneous ionization of water that indicates the concentration of $H^+$ ions in a solution.

**phagocytosis** (FAG-oh-sye-TOE-sis) a type of endocytosis in which a cell ingests an organism or some other fragment of organic matter.

**pharyngeal (gill) arches** (fuh-RIN-jee-uhl or FAR-in-JEE-uhl) a principal feature of the embryos of all chordates; pharyngeal arches develop into the gill structures of fish, and the ear, jaw, nose, and throat structures of terrestrial vertebrates.

**pharynx** (FAIR-inks) the upper part of the throat that extends from behind the nasal cavities to the openings of the esophagus and larynx.

**phenotype** (FEE-nuh-type) the outward appearance or expression of an organism's genes.

**pheromone** (FARE-uh-moan) a chemical produced by one individual that alters the physiology or behavior of other individuals of the same species.

**phloem** (FLO-em) a type of vascular tissue that conducts carbohydrates the plant uses as food, along with other needed substances.

**phosphorous cycle** the movment of the element phosphorous from the soil, to plants, to animals that eat those plants, and then back to the soil when those animals die and decompose.

**phosphorylation** (FOS-for-ih-LAY shun) the bonding of a phosphate group to a molecule; the conversion of a molecule into a phosphorus compound.

**photoautotrophs** (FOTE-oh-AWE-toe-trofes) organisms that make their own food by photosynthesis, using the energy of the sun.

**photochemical smog** a type of air pollution caused by reactions between hydrocarbons and nitrogen oxides taking place in the presence of sunlight.

**photoperiodism** (FOE-toe-PEER-ee-oh-dizm) responses of plants to the relative lengths of light and dark cycles.

**photosynthesis** (foe-toe-SIN-thuh-sis) the process whereby energy from the sun is captured by organisms and used to produce carbohydrates and other organic molecules.

**photosystem I** the first photosynthetic pathway to evolve; in it, energy is transferred to a molecule of chlorophyll *a* called P700, which then acts as the photosystem's reaction center; photosystem I, along with hydrogen ions, reduces $NADP^+$ to NADPH.

**photosystem II** a photosynthetic pathway in which energy is transferred to a molecule of chlorophyll *a* called P680, which then acts as the photosystem's reaction center; energized electrons ejected from photosystem II are passed down an electron transport chain, triggering events of ATP production.

**phototropism** (foe-TOE-trah-PIZ-um) a plant growth response to light.

**phyla** (FYE-luh) subcategories of kingdoms into which taxonomists place organisms having similar evolutionary histories.

**phyletic gradualism** (fye-LET-ik GRADJ-oo-uh-LIZ um) a theory that states that new species develop slowly and gradually as an entire species changes over time.

**phylogenetic tree** (also called *evolutionary tree*) a depiction of the evolutionary history of an organism or organisms.

**phylogeny** the evolutionary history of an organism.

**phytochromes** plant photoreceptor pigments that absorb red and far-red light.

**pigments** molecules that absorb some visible wavelengths of light and transmit or reflect others.

**pineal gland** (PIN-ee-uhl) a gland resembling a tiny pine cone embedded deep within the brain; though its exact function remains a mystery, it is believed to be the site of the control center that regulates the body's daily rhythms; it may also stimulate the onset of puberty.

**pinna** (PIN-uh) the ear flap located on the outside of the head.

**pinocytosis** (PIE-no-sye-TOE-sis) a type of endocytosis in which a cell ingests liquid material containing dissolved molecules.

**pistil** in angiosperms, the female sex organ.

**pituitary** (puh-TOO-ih-tare-ee) a small but powerful gland hanging from the underside of the brain that is controlled by the hypothalamus and secretes 9 major hormones.

**pivot joint** a kind of joint that allows side-to-side movement.

**placenta** (pluh-SEN-tuh) a flat disk of tissue that grows into the uterine wall, made up of both chorionic and maternal tissues, through which the embryo and fetus are supplied with food, water, and oxygen and through which wastes are removed.

**placental mammals** (pluh-SENT-uhl) a subclass of mammals that nourishes their developing young within the body of the mother.

**plant embryo** an undeveloped plant housed within the seed; one end of the embryo forms the shoot when the seed germinates, while the other end gives rise to the roots.

**plants** multicellular, eukaryotic, photosynthetic autotrophs (producers).

**plaques** (PLAKS) masses of cholesterol and other lipids that can build up within the walls of large and medium-sized arteries in a disease called atherosclerosis; the accumulation of such masses impairs the arteries' proper functioning.

**plasma membrane** a thin, flexible lipid bilayer that encloses the contents of a cell.

**plasmids** (PLAZ-mids) extrachromosomal pieces of DNA that replicate independently of the main chromosome in bacteria.

**plasmodial (acellular) slime molds** (plaz-MOE-dee-uhl) a phylum (Myxomycota) of protists that has an ameboid stage in their life cycle; they spend much of their time as a nonwalled, multinucleated mass of cytoplasm called a plasmodium, but can form spores when food or moisture is in short supply.

**plasmodium** a nonwalled, multinucleate mass of cytoplasm that resembles a moving mass of slime; plasmodial slime molds spend much of their life cycle in this form.

**platelets** cell fragments present in blood that play an important role in blood clotting.

**point mutation** (also called *gene mutation*) a change in the genetic message of a chromosome caused by alterations of molecules within the structure of the chromosomal DNA.

**polar molecules** molecules (two or more atoms sharing electrons) that have opposite partial charges at either end because electrons are shared unequally.

**pollution** substances that cause physical or chemical changes in the environment that harm living and nonliving things; pollution can affect aqauatic and terrestrial environments as wells as the atmosphere.

**polygenic inheritance** the determination of a trait by alleles of two or more genes.

**polymerase chain reaction** (PCR) a technique for amplifying DNA fragments (i.e., producing large numbers of copies of the fragments).

**polypeptides** chains of amino acids linked by peptide bonds.

**polyploidy** having more than the usual amount of chromosomes.

**polyps** (POL-ups) aquatic, cylindrical animals with a mouth at one end that is ringed with tentacles.

**polysaccharides** (POL-ee-SACK-uh-rides) long, insoluble polymers composed of sugar.

**pons** (ponz) a part of the brainstem that consists of bands of nerve fibers that act as bridges, connecting various parts of the brain to one another; the pons also brings messages to and from the spinal cord.

**population** a group that consists of the individuals of the same species that occur together at one place and at one time.

**population bottleneck** an evolutionary event in which a significant percentage of a population dies before reaching reproductive age or is otherwise prevented from reproducing; this results in a dramatic loss of genetic variation within the population.

**population ecologists** (ih-KOL-uh-jists *or* ee-KOL-uh-jists) scientists who study how populations grow and interact.

**population genetics** the science of the frequency and distribution of genes in populations of organisms.

**population pyramid** a bar graph that shows the composition of a population by age and sex.

**positive feedback loop** a feedback loop in which the response of the regulating mechanism is intensified with respect to the output.

**positive phototaxis** (foe-toe-TAK-sis) an innate movement of an organism toward a source of light.

**post-anal tail** a tail posterior to the anus of an organism; it is one of the four principal features that characterize chordates.

**postmating reproductive isolating mechanisms** structures, functions, or developmental abnormalities that, once mating has occurred, prevent organisms of two species or diverging populations from producing vigorous, fertile offspring.

**potential energy** energy not actively doing work but having the capacity to do so.

**preadaptations** inherited traits used for new functions for which they are later selected.

**predation** a relationship in which an organism of one species kills and eats an organism of another.

**predator** an organism of one species that kills and eats organisms of another.

**premating reproductive isolating mechanisms** physical barriers, physiological functions, behaviors, or anatomical structures that prevent organisms from mating.

**prenatal development** life before birth; the gradual growth and progressive changes in a developing human from conception until the time the fetus leaves the mother's womb.

**prey** an organism killed and eaten by a predator.

**primary bronchi** (BRON-keye) the two airways branching immediately from the trachea into each lung.

**primary consumers** organisms that feed directly on green plants.

**primary germ layers** three layers of cells that develop from the inner cell mass of the blastocyst and from which all the organs and tissues of the body develop.

**primary growth** occurs mainly at the tips of the roots and shoots; during primary growth, the plant grows taller and its roots grow deeper into the ground.

**primary succession** succession that takes place in areas not previously supporting organisms.

**primates** placental mammals that have characteristics reflecting an arboreal, or tree-dwelling, lifestyle.

**prions** infectious proteins that do not contain DNA or RNA; believed to be the smallest infectious particles.

**probe** a molecule that binds to a specific gene or nucleotide sequence.

**producers** organisms that can make their own food.

**products** substances obtained as the result of chemical reaction; substrates that have undergone a chemical change.

**progesterone** hormone that increases the thickening of the lining of the uterus and dramatically increases its blood supply.

**prokaryotes** (pro-KARE-ee-oats) unicellular organisms having no membrane-bounded organelles or nucleus.

**prokaryotic cells** (PRO-kare-ee-OT-ick) the first type of cells to exist as life arose on Earth; unlike eukaryotic cells, they have no membrane-bounded nucleus (rather, they have a region of DNA concentration called a *nucleoid*) and no membrane-bounded organelles that compartmentalize the cell.

**prophase** (PRO-faze) the first stage of mitosis, meiosis I, and meiosis II; prophase begins when the chromosomes have condensed; they become visible under a light microscope.

**proprioceptors** (PRO-pree-oh-SEP-turz) receptors located within skeletal muscles, tendons, and the inner ear that give the body information about the position of its parts relative to each other and to the pull of gravity.

**prosimians** (pro-SIM-ee-uns) a suborder of primates that includes lemurs, indris, aye-ayes, and lorises; they are small animals, mostly nocturnal, with large ears and eyes, and elongated snouts and rear limbs.

**prostate** a chestnut-sized gland in males that adds an alkaline fluid to the semen.

**proteases** (PRO-tee-ACE-es) digestive enzymes that break down proteins to smaller polypeptides, and polypeptides to amino acids.

**proteins** long, complex chains of amino acids linked end to end.

**protists** members of the Kingdom Protista, which includes unicellular eukaroytes as well as some multicellular forms.

**protons** particles found at the core of an atom; protons carry a positive electrical charge.

**proto-oncogenes** genes that can mutate into oncogenes, causing cancer; unmutated, these genes play important roles in the regulation of biochemical activities within cells, including activities related to cell division.

**protostomes** (PRO-toe-stowmz) one of two distinct evolutionary lines of coelomates that includes the mollusks, annelids, and arthropods; their embryological development is characterized by the mouth developing from the first indentation of the gastrula and a spiral pattern of cleavage.

**protozoans** (PRO-tuh-ZOE-unz) animallike protists that are heterotrophs; they take in and use organic matter for energy.

**pseudocoelomates** organisms with a so-called false coelom (pseudocoel)—a fluid-filled cavity that houses the organs but is not lined completely by connective tissue; roundworms and rotifers are pseudocoelomates.

**psychoactive drugs** those that affect the action of neurotransmitters in the brain or spinal cord.

**pulmonary circulation** (PUHL-muh-NARE-ee sur-kyuh-LAY-shun) the part of the human circulatory system that carries blood to and from the lungs.

**pulmonary trunk** a pulmonary artery that branches into two arteries that carry deoxygenated blood to the lungs.

**pulmonary veins** (PUHL-muh-NARE-ee VAYNZ) large blood vessels that carry oxygenated blood from the lungs to the left atrium of the heart.

**punctuated equilibrium** (PUNGK-choo-AY-ted EE-kwuh-LIB-ree-um) a theory that states that new species arise suddenly and rapidly as small subpopulations of a species split from the populations of which they were a part.

**Punnett square** (PUN-et) a simple diagram that provides a way to visualize the possible combinations of genes in a cross and illustrates their expected ratios; male gametes are shown along one side of the square and female gametes are shown along the other; each possible combination of gametes is shown in cells within the square.

**pupil** the opening in the center of the iris through which light passes.

**purines** (PYOOR-eens) double-ring compounds that are components of nucleotides; both DNA and RNA contain the purines adenine and guanine.

**pyrimidines** (pih-RIM-uh-deens *or* pye-RIM-uh deens) single-ring compounds that are components of nucleotides; both DNA and RNA contain the pyrimidine cytosine; in addition, DNA contains the pyrimidine thymine, and RNA contains the pyrimidine uracil.

**qualitative data** descriptive results not based on numerical measurements.

**quantitative data** results based on numerical measurements.

**radial symmetry** having parts that emerge, or radiate, from a central point, much like spokes on a wheel.

**radicle** a young root, usually the first portion of the embryo to emerge from a germinating seed.

**random genetic drift** a change in the frequencies of alleles in the gene pool of a finite population due to chance events.

**reactants** substances entering into a chemical reaction.

**realized niche** the role an organism actually plays in the ecosystem.

**reasoning** (or **insight**) the most complex form of learning; it is the ability to recognize a problem and solve it mentally before trying out the solution.

**recessive** in an organism carrying a pair of contrasting alleles for a particular trait, the form of the trait (the allele) that recedes or disappears entirely.

**rectum** the lower portion of the colon, which terminates at the anus.

**red algae** a phylum (Rhodophyta) of multicellular algae that plays an important role in the formation of coral reefs and produces gluelike substances that make them commercially useful.

**redox reaction** a term used to describe the occurrence of oxidation and reduction reactions, which always happen together.

**reduced genetic variability** characteristic of a species that has a narrow range of genotypes.

**reduction** a reaction that involves an atom gaining an electron.

**reflex** an automatic response to nerve stimulation.

**reflex arcs** pathways that integrate responses to certain types of stimuli within the spinal cord.

**regulator genes** in prokaryotes, genes that code for products that control the expression of other genes.

**releasers** sign stimuli communicated by members of the same species.

**renal failure** a disease that occurs when the filtration of the blood at the glomerulus either slows or stops; in acute renal failure, filtration stops suddenly, while in chronic renal failure, the filtration of blood at the glomerulus slows gradually.

**repeated trials** performing experiments more than one time to increase the reliability of the results.

**reproductive fitness** the comparative ability of an organism to survive to reproductive age in a particular environment and to produce viable offspring, thereby passing on its genes.

**reptiles** a class (reptilia) of vertebrates characterized by dry skin covered with scales that help retard water loss.

**respiration** the uptake of oxygen and the release of carbon dioxide by the body.

**respiratory bronchioles** (BRON-kee-olz) the last division of bronchi; bronchioles have thousands of tiny air passageways whose walls have clusters of tiny pouches, or alveoli.

**respiratory system** collectively, all the organs of respiration in an organism.

**resting potential** an electrical potential difference, or electrical charge, along the membrane of a resting neuron.

**restriction enzymes** proteins that recognize certain nucleotide (base) sequences in a DNA strand and break the bonds between the nucleotides at those points.

**restriction fragments** the pieces of DNA that have been cut from larger pieces of DNA by restriction enzymes.

**restriction site** a particular nucleotide (base) sequence in DNA that is acted upon by a specific restriction enzyme.

**retina** (RET-uh-nuh) tissue at the back of the eye that is composed of rod and cone cells and that is sensitive to light.

**rhizoids** (RYE-zoyds) in nonvascular plants, slender, rootlike projections that anchor the plant to its substrate; unlike roots, however, they consist of only a few cells and do not play a major role in the absorption of water or minerals.

**rhizomes** (RYE-zomz) underground stems along which some plants produce new plants.

**ribonucleic acid** (RNA) (RYE-boh-noo-KLAY-ick) one of two types of nucleic acid found in cells; the nucleotides of RNA contain the 5-carbon sugar ribose.

**ribosomal ribonucleic acid** (rRNA) (RYE-buh-SO-mull RYE-boh-new-KLAY-ick) a type of ribonucleic acid that is manufactured by the DNA within a eukaryotic cell's nucleolus; rRNA molecules are structural components of ribosomes.

**ribosomes** (RYE-buh-somes) minute, round structures found on the endoplasmic reticulum of eukaryotic cells or free in the cytoplasm of both prokaryotes and eukaryotes; ribosomes are the places where proteins are manufactured.

**right atrium** the upper right chamber of the vertebrate heart.

**right ventricle** the lower right chamber of the vertebrate heart.

**rods** sensory cells in the retina of the eye that are stimulated by low levels of light and detect only white light.

**root** the part of a vascular plant that exists below the ground; the root penetrates the soil and absorbs water and various ions that are crucial for plant nutrition; it also anchors the plant.

**rough ER** a kind of endoplasmic reticulum that is covered with ribosomes and hence resembles long sheets of sandpaper; in tandem with its ribosomes, rough ER manufactures and transports proteins designed to leave the cell.

**roundworms** a phylum (Nematoda) of bilaterally symmetrical invertebrate animals with a pseudocoelomate body plan; tapeworms and many other parasitic worms are roundworms.

**sac fungi** a phylum (Ascomycota) of fungi that live in both aquatic and terrestrial environments and are characterized by sexual spores borne in saclike structures.

**saccule** (SAK-yool) the smaller of two membranous sacs within the vestibule of the inner ear, containing both ciliated and nonciliated cells, that functions in the maintenance of bodily equilibrium and coordination.

**salivary amylase** (SAL-i-VER-ee AM-uh-lace) a digestive enzyme found in saliva that breaks down starch into molecules of the disaccharide maltose.

**salivary glands** (SAL-ih-VER-ee GLANDZ) the paired glands of the mouth that secrete saliva.

**sampling error** a factor that can cause random genetic drift; it refers to the fact that a random sample of alleles from a population is unlikely to show the same relative frequency of alleles as the entire population.

**sanitary landfill** an enormous depression in the ground where trash and garbage are dumped, compacted, and then covered with dirt.

**saprophytic** (SAP-roe-FIT-ik) a term describing an organism that feeds on dead or decaying organic matter; most fungi are saprophytic.

**sarcomere** (SAR-koe-mere) the repeating bands of actin and myosin myofilaments that appear between two Z lines in a muscle fiber; the contractile unit of muscles.

**saturated fats** a group of "bad" fats; along with the trans fats, they tend to raise blood cholesterol levels, greatly increasing the risk of heart disease; foods with high levels of saturated fats include whole milk, cheese, and red meat.

**savanna** a biome found near the equator but in areas having less annual rainfall than tropical rain forests; this climate supports grasslands with only scattered trees and shrubs.

**Schwann cells** special neurological cells that are wrapped around many of the long cell processes of sensory and motor neurons of the peripheral nervous system.

**scientific laws** descriptions of patterns of regularity with respect to natural phenomena.

**scientific theory** a rigorously tested explanation of a phenomenon.

**sclera** (SKLEAR-uh) the tough outer layer of connective tissue that covers and protects the eye.

**scrotum** (SKRO-tum) a sac of skin, located outside the lower pelvic area of the male, which houses the testicles, or testes.

**second filial ($F_2$) generation** (FIL-ee-uhl) the offspring of the $F_1$ generation.

**second law of thermodynamics** a scientific law stating that disorder in a closed physical system tends to increase—that is, disorder is more likely than order.

**secondary bronchi** (BRON-keye) outbranchings of each of the two primary bronchi; three secondary bronchi serve the three right lobes of the lungs, whereas two secondary bronchi serve the two left lobes of the lungs.

**secondary consumers** meat eaters that feed on the primary consumers (herbivores).

**secondary growth** occurs at the stem; during secondary growth, a plant grows wider (thicker).

**secondary succession** succession that takes place in areas originally occupied by organisms that have been disturbed and where there is soil.

**secretin** (sih-KRE-tin) one of the hormones that controls digestion in the small intestine.

**seed dormancy** an evolutionary adaptation that allows seeds to germinate only when growing conditions are good.

**seed germination** the process by which a plant embryo begins to grow and the radicle emerges through the seed coat.

**seedless vascular plants** plants that reproduce by means of spores rather than seeds; a familiar member of this group is the fern.

**seeds** structures from which new sporophyte plants grow; seeds protect the embryonic plant from drying out or being eaten when it is at its most vulnerable stage.

**semicircular canals** three looped fluid-filled canals in the inner ear that detect the direction of the body's movement.

**seminal vesicles** (SEM-un-uhl VES-uh-kuls) two accessory glands that secrete a thick, clear fluid, primarily composed of fructose, that forms a part of the semen.

**seminiferous tubules** (SEM-uh-NIF-uhr-us TOO-byools) tightly coiled tubes within each testis, where sperm cells develop.

**senescence** (sih-NEH-sense) the final stage in development of a plant, in which parts of the plant begin to die.

**sensory neurons** nerve cells that are specialized to change specific stimuli into nerve impulses and transmit this information to the central nervous system.

**septum** tissue that separates the two sides of the heart.

**serial endosymbiotic theory** the scientific theory that eukaryotes arose through a succession of symbiotic relationships among various prokaryotic cells in which some prokaryotic cells became attached to or were engulfed by other, pre-eukaryotic cells and thus became precursors of present-day organelles of eukaryotic cells.

**sex chromosomes** chromosomes that determine the gender of an individual as well as certain other characteristics.

**sex-linked alleles** alleles found on an X or Y chromosome; almost all sex-linked alleles that cause sex-linked disorders are found on the X chromosome.

**sexual reproduction** a process whereby a male sex cell and female sex cell combine to form the first cell of a new individual.

**sexual selection** a form of natural selection that depends on a struggle between the individuals of one sex, generally the males, for mates of the other sex.

**shell** (or **energy level**) the volume of space around an atom's nucleus where an electron is most likely to be found.

**shoot** the part of a vascular plant that exists above the ground; the stem, leaves, and other plant structures are formed at the shoot.

**shore zone** the shallow water near the edges of a lake or pond in which plants with roots, such as cattails and water lilies, may grow.

**shotgun cloning** the process of synthesizing DNA by cutting the DNA of an entire genome into pieces, isolating and purifying the genomic DNA, and then making copies of this DNA using yeast or bacterial cells.

**sigmoid growth curve** the exponential growth of a population and its subsequent stabilization at the level of its environment's carrying capacity.

**sign stimulus** an environmental cue that triggers a fixed action pattern (a recognizable "unit" of behavior).

**signaling molecules** in eukaryotes, substances that induce gene expression; these substances include hormones, growth factors, and related compounds.

**sinoatrial (SA) node** (sye-no-AY-tree-uhl NODE) a small cluster of specialized cardiac muscle cells that are embedded in the upper wall of the right atrium and automatically and rhythmically send out impulses that initiate each heartbeat.

**sister chromatids** (CROW-mah-tidz) the original chromosome and its copy, connected by a centromere; the copy is created during the DNA replication stage of interphase, preceding mitosis.

**skeletal system** another term for the skeleton.

**skeleton** the collective term for a vertebrate's internal bony scaffolding; the skeleton provides support for the body, helps the body move (along with the muscles), and protects delicate internal structures.

**skull** the framework of the head, which in humans consists of 8 cranial and 14 facial bones.

**small intestine** the tubelike portion of the digestive tract that begins at the pyloric sphincter and ends at its T-shaped junction with the large intestine.

**smog** a type of air pollution that results from the burning of fossil fuels.

**smooth ER** a kind of endoplasmic reticulum that does not have ribosomes attached to its surface and hence manufactures no proteins; instead, smooth ER helps produce carbohydrates and lipids.

**social behaviors** activities of animals that help members of the same species communicate and interact with one another, each responding to stimuli from others.

**sociobiology** (SO-see-oh-bye-OL-uh-gee or SO-shee-oh-bye-OL-uh-gee) the biology of social behavior; this science applies the knowledge of evolutionary biology to the study of animal behavior.

**sodium-potassium pump** membrane proteins that transport sodium ions out of the cell and potassium ions into the cell.

**solid waste** garbage and trash.

**soluble** (SOL-you-ble) able to be dissolved in water.

**solute** the substance that dissolves in a solvent to form a solution.

**solvent** the substance in a solution that is present in the greater amount.

**somatic nervous system** (soe-MAT-ik) the branch of the peripheral nervous system consisting of motor neurons that sends messages to the skeletal muscles and controls voluntary responses.

**special senses** those restricted to particular locations of the body.

**speciation** (SPEE-shee-AY-shun) the process by which new species are formed during the process of evolution.

**species** (SPEE-shees or SPEE-sees) a group of interbreeding natural populations that are reproductively isolated from other such groups; the most narrow and specific taxonomic classification; a taxonomic subcategory of family.

**specific immunity** a set of cellular and molecular defenses that protect the body from each particular microbe that may invade it.

**spermatozoon** (also called *sperm*) (spur-MAT-uh-ZOO-un) a male sex cell.

**spinal nerves** nerves by which the spinal cord receives and transmits information; spinal nerves include sensory and motor neurons.

**spiracles** small openings for breathing; in tracheal systems, spiracles are muscular openings that can open and close, directed by nerves and triggered by carbon dioxide levels.

**sponges** a phylum (Porifera) of asymmetrical, invertebrate, aquatic animals; sponges are considered simpler than other animals because they have no tissues, no organs, and no coelom.

**sporangia** (spoh-RAN-jee-uh) spore cases.

**spores** reproductive bodies formed by cell division (mitosis or meiosis) in the parent organism; spores are formed by meiosis in a diploid parent and mitosis in a haploid parent; the spores are always haploid.

**sporophyte (spore-plant) generation** the diploid phase of a plant life cycle; this generation tends to dominate the life cycles of vascular plants.

**sporophyte** (SPOR-ih-fite) a plant that produces haploid spores by meiosis; also known as the spore-plant generation.

**sporozoans** (SPOR-uh-ZOH-uns) a phylum (Apicomplexa) of protists that is nonmotile, spore-forming parasites of vertebrates, including humans.

**stabilizing selection** a type of natural selection that occurs when the environment does not change; the averge phenotype does not change over time, but the range of variation in the population diminishes.

**stable population** a population whose size remains the same through time.

**stamens** in angiosperms, the male sex organs.

**stapes** (STAY-peez) one of the three bones of the middle ear that amplify sound vibrations and carry them from the outer ear to the inner ear; the stapes is also known as the *stirrup*.

**starches** polysaccharides that are the storage form of sugar in plants.

**starch-statolith hypothesis** the explanation that gravity sensing in plants involves the movement of starch granules within cells.

**stem** a part of a plant's structure that serves as a framework for the positioning of its leaves.

**steroid hormones** one of the two main classes of endocrine hormones; steroid hormones are made of cholesterol and are able to pass through cell membranes.

**stolons** (STOW-lunz) in some plants, horizontal stems that grow above the ground, also called runners.

**stomach** a muscular sac in which food is collected and partially digested by hydrochloric acids and proteases.

**stomata** (singular, stoma) (STOW-muh-tuh/STOW-muh) openings in the epidermis of a leaf bordered by guard cells.

**stretch receptors** sensory neuron endings that are stimulated to fire when tension is placed on them.

**stroma** (STROH-muh) within a chloroplast, a fluid that surrounds the thylakoids and contains the enzymes of the light-independent reactions.

**structural genes** in prokaryotes, genes that code for gene products that are produced only when needed.

**submissive behavior** behavior that animals often exhibit in response to a threat display in order to avoid fighting; submissive behavior may include making the body appear smaller or withdrawing fangs or claws.

**substrates** substances on which enzymes act.

**succession** a process of change during which a sequence of communities replaces one another in an orderly and predictable way.

**superior vena cava** (VEE-nuh KAY-vuh or VAY-nuh KAH-vuh) a large vein that drains blood from the upper body and returns it to the right atrium of the heart.

**survivorship** the proportion of an original population that lives to a certain age.

**suture** (SOO-chur) a type of immovable joint.

**symbiosis** types of interactions in which one organism lives with another in an intimate association; the three types are commensalism, mutualism, and parasitism.

**sympathetic nervous system** (SIM-puh-THET-ik) a subdivision of the autonomic nervous system that generally mobilizes the body for greater activity; acts in opposition to the parasympathetic nervous system.

**sympatric speciation** may begin when members of a population occupy different ecological zones.

**synapse** (SIN-aps) a place where a neuron communicates with another neuron or effector cell.

**synapsis** (sih-NAP-sis) a process during prophase I of meiosis in which homologous chromosomes line up side-by-side, initiating the process of crossing over.

**synaptic cleft** (sin-AP-tik KLEFT) the space or gap between two adjacent neurons, which the nerve impulse must cross.

**synovial joints** (suh-NO-vee-uhl) a freely movable joint in which a fluid-filled space exists between the articulating bones.

**systematics** (sis-teh-MAH-ticks) the study of the diversity of organisms, focusing on their evolutionary relationships as well as their comparative relationships.

**systemic circulation** the pathway of blood vessels to the body regions and organs other than the lungs.

**systole** (SIS-tl-ee) the time of contraction of a heart chamber, in which blood is forced out of the chamber.

**T cells or T lymphocytes** white blood cells that provide the cell-mediated immune response; they develop in the bone marrow but migrate to the thymus where they mature and develop the ability to identify body cells that have been infected with fungi, some viruses, and certain bacteria; they also recognize foreign body cells.

**taiga** a biome that consists of evergreen, cone-bearing trees; the climate of this biome is characterized by long, cold winters with little precipitation; it is also called the northern coniferous forest.

**taste buds** microscopic receptors embedded within the papillae of the tongue that work with olfactory receptors to produce taste sensations; humans have taste buds specialized to respond to four different kinds of sensations: salty, sweet, sour, and bitter.

**taxis** (TAK-sis) a directed movement by an animal toward or away from a stimulus, such as light, chemicals, or heat.

**taxon** (plural *taxa*) a named group of organisms, such as a phylum, family, or order.

**taxonomy** (tack-SAHN-uh-me) a method of classification of the diverse array of species based on their common ancestry.

**telophase** (TEL-uh-faze) the last phase of mitosis, meiosis I, and meiosis II; during telophase, the mitotic apparatus assembled during prophase is disassembled, the nuclear envelope is reestablished, and the normal use of the genes present in the chromosomes is reinitiated.

**temperate deciduous forest** a biome that occurs in areas having warm summers, cold winters, and moderate amounts of precipitation; the trees of this forest lose their leaves and remain dormant throughout the winter.

**temperate grasslands** a biome that experiences a greater amount of rainfall than deserts but a lesser amount than savannas; they occur at higher latitudes than savannas but, like savannas, are characterized by perennial grasses and herds of grazing mammals.

**tendons** cords of dense connective tissue that attach muscles to bones.

**territorial behaviors** behaviors an animal may exhibit that involve marking off an area as its own and defending it against the same-gender members of its species.

**testes** (TES-teez) the male gonads, where sperm production occurs.

**thalamus** (THAL-uh-muss) paired oval masses of gray matter at the base of the cerebellum, lying close to the basal ganglia, that receive sensory stimuli, interpret some of these stimuli, and send the remaining sensory messages to appropriate locations in the cerebrum.

**thermal pollution** a change in water temperature that occurs when industries release heated water back into rivers after using the water for cooling purposes.

**thoracic cavity** (thu-RASS-ick) the chest cavity in humans, located within the trunk of the body and extending above the diaphragm and below the neck.

**threat displays/intimidation displays** a form of aggressive behavior that animals may exhibit during a competitive situation; the purpose of these displays is to scare other animals away or cause them to "back down" before fighting takes place.

**threatened species** likely to become endangered in the near future.

**threshold potential** The level at which a stimulus results in an action potential along the nerve cell membrane.

**thylakoids** (THIGH-luh-koidz) flattened, saclike membranes within chloroplasts; these membranes contain the enzymes necessary for the light-dependent reactions.

**thyroid** the gland responsible for the secretion of hormones that determine the rate of the chemical reactions of the body's cells (the rate of metabolism).

**tidal volume** the amount of air inspired or expired with each breath.

**tissue differentiation** a prenatal developmental process in which groups of cells become distinguished from other groups of cells by the jobs they will perform in the body.

**tissues** groups of similar cells that work together to perform a function.

**trachea** (TRAY-kee-uh) the windpipe; the air passageway that runs down the neck in front of the esophagus and that brings air to the lungs.

**tracheal system** a branching network of microscopic air tubes found throughout the bodies of most spiders and insects.

**traits** distinguishing features or characteristics.

**trans fats** "bad" fats that are processed; created by adding hydrogen to oil to harden it, as for margarine; they tend to raise blood cholesterol levels, greatly increasing the risk of heart disease.

**transcription** (trans-KRIP-shun) the first step in the process of polypeptide synthesis and gene expression, in which a gene is copied into a strand of messenger RNA.

**transcription factors** in eukaryotes, proteins that interact with DNA, regulating the production of polypeptides coded for by that DNA.

**transduction** (tranz-DUK-shun) one of the ways in which bacteria transfer genetic material; occurs as DNA from a donor bacterium is transferred to a recipient bacterium by a virus.

**transfer RNA** (tRNA) a type of ribonucleic acid found in the cytoplasm that transports amino acids, used to build polypeptides, to the ribosomes; tRNA molecules also position each amino acid at the correct place on the elongating polypeptide chain.

**transformation** one of the ways in which bacteria transfer genetic material; transformation occurs as DNA from a lysed donor cell is taken up from the surrounding medium by a competent cell.

**transgenic plants** (tranz-GEE-nik) plants that are genetically altered using the techniques of genetic engineering.

**translation** (trans-LAY-shun) the second step of gene expression, in which the mRNA, using its copied DNA code, directs the synthesis of a polypeptide.

**transpiration** the process by which water vapor leaves through the stomata of a leaf.

**transpirational pull** the "pulling" or "sucking" effect created when water evaporates from the leaf (transpiration) and water moves from cell to cell in response until the xylem is reached, resulting in a pull on the water in the xylem.

**transposable genetic elements** DNA sequences that can move from one position to another in a genome.

**triglyceride** (try-GLISS-er-ide) a fat molecule in which three fatty acids are attached at each of the three carbon atoms of a glycerol molecule.

**trisomy** (TRI-so-me) a condition in which an individual has inherited an extra chromosome, resulting in three copies of a particular chromosome instead of the usual two.

**trophic levels** (TROW-fik) the different feeding levels within an ecosystem.

**tropic hormones** (TROW-pik) 4 of the 7 hormones produced by the anterior pituitary gland; the tropic hormones turn on, or stimulate, other endocrine glands.

**tropical rain forest** a biome that occurs in regions of high temperature and rainfall, characterized by tall trees that support a variety of plant and animal life on their branches.

**tropical rain forest deforestation** the loss of the plants and animals of the tropical rain forest (a biome that occurs in regions of high temperature and rainfall) through cut-and-burn agriculture and logging.

**tropisms** plant responses such as bending or curving toward or away from an external stimulus.

**true-breeding** a term referring to plants that produce offspring consistently identical to the parent with respect to certain defined characteristics after generations of self-fertilization.

**trypsin** (TRIP-sin) an enzyme produced by the pancreas, which, together with chymotrypsin and carboxypeptidase, completes the digestion of proteins in the small intestine.

**tumor-suppressor genes** "off" switches in the development of cancer, signaling cells to slow their growth and increase their levels of differentiation.

**tundra** a biome that encircles the top of the world; the tundra is characterized by desertlike levels of precipitation, extremely long and cold winters, and short, warmer summers.

**tunicates** (TOO-nih-kits *or* TOO-nih-KATES) a subphylum of the chordates; a group of about 2500 species of marine animals, most of which look like living sacs attached to the floor of the ocean; the only chordate characteristic adult tunicates exhibit is their gill slits; however, larvae have a notochord and nerve cord.

**tympanic membrane** (tim-PAN-ik) a thin piece of fibrous connective tissue that is stretched over the opening to the middle ear; the tympanic membrane is also known as the eardrum.

**urea** (yoo-REE-uh) the primary excretion product from the deamination of amino acids.

**ureter** (YER-ih-ter) the tube that takes urine from the kidney to the urinary bladder.

**urethra** (yoo-REE-thruh) a muscular tube that brings urine from the urinary bladder to the outside; in men, the urethra also carries semen to the outside of the body during ejaculation.

**uric acid** (YOO-rik) a nitrogenous waste found in small amounts in the urine that is formed from the breakdown of nucleic acids found in the cells of food and from the metabolic turnover of nucleic acids and ATP.

**urinary bladder** a hollow muscular organ in the pelvic cavity that acts as a storage pouch for urine.

**urinary system** a set of interconnected organs that not only remove wastes, excess water, and excess ions from the blood but store this fluid until it can be expelled from the body.

**urine** an excretion product composed of water and the substances excreted by the kidneys.

**utricle** (YOO-trih-kul) the larger of two membranous sacs within the vestibule, containing both ciliated and nonciliated cells, that functions in the maintenance of bodily equilibrium and coordination.

**vaccination** (VAK-suh-NAY-shun) a procedure that involves the injection (or oral administration, in some cases) of a weakened or killed microbe, or laboratory-made antigen proteins into a person or animal in order to confer resistance to a disease-causing pathogen.

**vacuoles** (VACK-yoo-ole) membrane-bounded storage sacs within eukaryotic cells that hold such substances as water, food, and wastes; vacuoles are most often found within plant cells.

**vagina** (vuh-JINE-uh) an organ of the female genital system; its muscular, tubelike passageway to the exterior has three functions: it accepts the penis during intercourse, it is the lower portion of the birth canal, and it provides an exit for the menstrual flow.

**vas deferens** two muscular tubes through which sperm move; they ascend from the epididymis into the pelvic cavity, loop over the side of the urinary bladder, and then enter the penis.

**vascular plants** (VAS-kyuh-ler) plants that have systems of specialized tissues that transport water and nutrients; they are made up of underground roots and aboveground shoots.

**vascular tissue** forms the "circulatory system" of a vascular plant; vascular tissue conducts water and dissolved inorganic nutrients up the plant and carries the products of photosynthesis throughout the plant.

**vegetative propagation** an asexual reproductive process in which a new plant develops from a portion of a parent plant.

**veins** (VAYNS) blood vessels that carry blood to the heart.

**venules** (VAYN-yooles) small veins that connect capillaries to larger veins.

**vernalization** the process whereby plants must undergo prolonged exposure to cold prior to flowering.

**vertebral column** (VER-tuh-brul) the collection of 26 vertebrae, stacked one on top of the other along the midline of the back, that acts as a strong, flexible rod and supports the head in a human skeleton; the vertebral column is also known as the backbone.

**vertebrates** (VER-tuh-bruts *or* ver-tuh-braytes) a subphylum of chordates; vertebrates have a bony (vertebral) column surrounding a dorsal nerve cord.

**vestibule** (VES-tuh-byool) a structure within the inner ear that detects the effects of gravity on the body.

**villi** (VIL-eye) fine, fingerlike projections of the epithelium of the small intestine.

**viroids** small strands of RNA that are different from viruses but are replicated in host cells as viruses are; viroids can cause plant diseases.

**viruses** (VYE-russ-es) nonliving infectious agents that enter living organisms and cause disease.

**viviparous** (vye-VIP-uh-rus) a term that describes an organism that gives birth to live offspring.

**vocal cords** two pieces of elastic tissue, covered with a mucous membrane, that are stretched across the upper end of the larynx and are involved in the production of sound.

**vulva** (VUL-vuh) the collective term for the external genitals of a female.

**water cycle** (or **hydrologic cycle**) the cycling of water into the atmosphere from the ocean and other bodies of water (though evaporation) and from plants (through transpiration) and then back to the Earth through rain and other forms of precipitation, where it is taken up and used by living organisms.

**water molds** (also called *egg fungi*) a phylum (Oomycota) of funguslike protists that predigests and absorbs food as fungi do; water molds thrive in moist places and aquatic environments and parasitize plants and animals.

**xylem** (ZY-lem) a type of vascular tissue in plants that conducts water and dissolved inorganic nutrients.

**zygospores** (ZYE-go-SPORZ) sexual spores formed by the zygote-forming fungi.

**zygote** (ZYE-goat) a new cell that contains intermingling genetic material from both the sperm and egg cells; the fertilized ovum.

**zygote-forming fungi** a phylum (Zygomycota) of fungi having a distinct sexual phase of reproduction that is characterized by the formation of sexual spores called zygospores.

# PHOTO CREDITS

## Chapter 1

Ch Op 1: Copyright/owner ScienCentral, Inc. and/or its licensors; Fig. 1.1: Carolina Biological/Visuals Unlimited; Fig. 1.2: Kim Taylor/Bruce Coleman, Inc.; Fig. 1.3: Stephen Dalton/NHPA/Photo Researchers; Fig. 1.4: J.D. Cunningham/Visuals Unlimited; Fig. 1.5a: NASA/Galileo Imaging Team; Fig. 1.5b: PhotoDisc; Fig. 1.5c: Joe McDonald/ Visuals Unlimited; Fig. 1.5c: PhotoDisc, Inc.; Fig. 1.5f: Richard Gross/Biological Photography; Fig. 1.5g: Biophoto Associates/Photo Researchers, Inc.; Fig. 1.5h: PhotoDisc, Inc; Fig. 1.6: PhotoDisc, Inc.; Fig. 1.7: Richard Gross/ Biological Photography; Fig. 1.8a: Stewart Halperin; Fig. 1.8b: Gail Shumway/Bruce Coleman, Inc.; Fig. 1.8c: Tom McHugh/Photo Researchers; Fig. 1.9a: R. Arndt/Visuals Unlimited; Fig. 1.9b: Richard Gross/Biological Photography; Fig. 1.9c: Terry Hazen/Visuals Unlimited; Fig. 1.10: Nature-M.P.F.T./Getty Images; Fig. 1.12a: David M. Phillips/Visuals Unlimited; Fig. 1.12b: Dr. T.J. Beveridge/Visuals Unlimited; Fig. 1.12c: Abbey/Visuals Unlimited; Fig. 1.12d: Stewart Halperin; Fig. 1.12e: Stephen Sharnoff/Visuals Unlimited; Fig. 1.12f: J. Westrich/Zefa/Masterfile

## Chapter 2

Ch Op 2: Copyright/owner ScienCentral, Inc. and/or its licensors; Fig. 2.1: Andrew C. Revkin; Fig. 2.2–2.4: Stewart Halperin; Fig. 2A: Tozzer Library/Harvard University; Fig. 2B: Edward Owen/Art Resource; Fig. 2.5: Courtesy of Thomas Cirrito, Ph.D; Fig. 2.6: Charles M. Russell/Peter Arnold, Inc.

## Chapter 3

Ch Op 3: Copyright/owner ScienCentral, Inc. and/or its licensors; Fig. 3.A: SIU/Visuals Unlimited; Fig. 3.7: Michael and Patricia Fogden; Fig. 3.9: COLOR-PIC/Animals Animals/Earth Scenes; Fig. 3.11: AFP/Getty Images News and Sport Services; Fig. 3.13: Kenneth Murray/Photo Researchers

## Chapter 4

Ch Op 4: Copyright/owner ScienCentral, Inc. and/or its licensors; Fig. 4.1: Jan Hinsch/Photo Researchers; Fig. 4.6: *Electron Microscopy 2e* by John J. Bozzola & Lonnie D. Russell, reprinted with permission of Jones & Bartlett Publishers, Inc. Fig. 4.7: J.D. Litvay/Visuals Unlimited

## Chapter 5

Ch Op 5: Copyright/owner ScienCentral, Inc. and/or its licensors; Fig. 5.1a: M. Abbey/ Visuals Unlimited; Fig. 5.1b: Richard Gross/Biological Photography; Fig. 5.1c: Science Pictures Ltd/Science Photo Library/Photo Researchers; Fig. 5A: Leonard Lessin/Peter Arnold, Inc.; Fig. 5B: Bruce Iverson Photomicrography; Fig. 5C: Robert Brons/Biological Photo Service; Fig. 5D: Lester Lefkowitz/Corbis Stock Market; Fig. 5E: Biophoto Associates/Photo Researchers; Fig. 5F: S. Schmitt from *Electron Microscopy 2e* by John J. Bozzola and Lonnie D. Russell, Jones and Bartlett Publishers, Inc.; Fig. 5.4: David M. Phillips/Visuals Unlimited; Fig. 5.7a: Donald Fawcett/Visuals Unlimited; Fig. 5.9: B. Crandall-Statler from *Electron Microscopy 2e* by John J. Bozzola and Lonnie D. Russell, Jones and Bartlett Publishers, Inc.; Fig. 5.11: from *Electron Microscopy 2e* by John J. Bozzola and Lonnie D. Russell, Jones and Bartlett Publishers, Inc.; Fig. 5.12b: W. Dougherty from *Electron Microscopy 2e* by John J. Bozzola and Lonnie D. Russell, Jones and Bartlett Publishers, Inc.; Fig. 5.13: from *Electron Microscopy 2e* by John J. Bozzola and Lonnie D. Russell, Jones and Bartlett Publishers, Inc.; Fig. 5.14: Ed Reschke; Fig. 5.15: from *Electron Microscopy 2e* by John J. Bozzola and Lonnie D. Russell, Jones and Bartlett Publishers, Inc.; Fig. 5G: Kevin and Betty Collins/Visuals Unlimited; Fig. 5.19: from *Electron Microscopy 2e* by John J. Bozzola and Lonnie D. Russell, Jones and Bartlett Publishers, Inc.; Fig. 5.20: Dr. Henry C. Aldrich/Visuals Unlimited; Fig. 5.21: *Electron Microscopy 2e* by John J. Bozzola and Lonnie D. Russell, Jones and Bartlett Publishers, Inc.; Fig. 5.22: M. Gillot from *Electron Microscopy 2e* by John J. Bozzola and Lonnie D. Russell, Jones and Bartlett Publishers, Inc.; Fig. 5.28: Dr. Birgit H. Satir

## Chapter 6

Ch Op 6: Copyright/owner ScienCentral, Inc. and/or its licensors; Fig. 6.1: Nick Bergkessel/Photo Researchers; Fig. 6.6: Dr. Thomas A. Steitz; Fig. 6.14: Darwin Dale/Photo Researchers

## Chapter 7

Ch Op 7: Copyright/owner ScienCentral, Inc. and/or its licensors; Fig. 7.1: Charles Winters/Photo Researchers; Fig. 7.5: L. L. Rue, III/Animals Animals/Earth Scenes; Fig. 7.6a: Richard Gross/Biological Photography; Fig. 7.6b: M. Gillott from *Electron Microscopy 2e* by John J. Bozzola and Lonnie D. Russell, Jones and Bartlett Publishers, Inc.; Fig. 7.7: Elizabeth Gentt/Visuals Unlimited

## Chapter 8

Ch Op 8: Copyright/owner ScienCentral, Inc. and/or its licensors; Fig. 8.3: PhotoDisc, Inc./Getty Images; Fig. 8.10a: Richard Gross/Biological Photography; Fig. 8.10b: Stewart Halperin Photography

## Chapter 9

Ch Op 9: Copyright/owner ScienCentral, Inc. and/or its licensors; Fig. 9.1: CNRI/Science Photo Library/Photo Researchers, Inc.; Fig. 9.2a: Ada L. Olins/Biological Photo Service; Fig. 9.2b: Biofoto Associates/Science Source/Photo Researchers, Inc.; Fig. 9.5: Cold Spring Harbor Laboratory Archives/Photo Researchers, Inc.; Fig. 9A: Lee Simon/Science Photo Library/Photo Researchers, Inc.; Fig. 9.6: Cold Springs Harbor Laboratory Archives/ Photo Researchers, Inc.; Fig. 9.9: Peter Lansdorp/Visuals Unlimited; Fig. 9B: Cellmark Diagnostics, Inc. /Germantown, Maryland; Fig. 9.10: Dan Gair Photographic/Index Stock

## Chapter 10

Ch Op 10: Copyright/owner ScienCentral, Inc. and/or its licensors; Fig. 10.5a: E. Kifelva-D. Fawcett/Visuals Unlimited; Fig. 10.6a: Dennis Kunkel/Phototake; Fig. 10.6b: Alfred Pasieka/Photo Researchers; Fig. 10.6c: Eye of Science/Photo Researchers; Fig. 10.8: Dr. Walter J. Gehring/ Biozentrum University of Basel, Switzerland

## Chapter 11

Ch Op 11: Copyright/owner ScienCentral, Inc. and/or its licensors; Fig. 11.1: ©Photodisc; Fig. 11.3: National Library of Medicine/Photo Researchers; Fig. 11.4: John Durham/Photo Researchers, Inc.; Fig. 11.6: David P. Allison/ Oak Ridge National Laboratory/Biological Photo Service; Fig. 11.7: Geoff Tomkinson/Photo Researchers; Fig. 11.8: Maximilian Stock Ltd./Photo Researchers; Fig. 11.13: Courtesy Cellmark Diagnostics/Zeneca; Fig. 11.15: Robert Longuehaye, NIBSC/Photo Researchers; Fig. 11.17: Courtesy Cellmark Diagnostics/Zeneca; Fig. 11.18: B-D: Courtesy of Patrick O. Brown. See Joseph DeRisi et al., *Science* 278:680, 1997 and Tracy L. Ferea and Patrick O. Brown, Curr. Opin. Gen. Develop. 9:715, 1999; Fig. 11.20: Courtesy, Shiloh Creek Farm, Fresno, Ohio; Fig. 11.21: M. Baret/Science Photo Library/Photo Researchers; Fig. 11.22: Dr. James M. Robl/Reprinted with permission from Science, Vol. 280, May 22, 1998, Figure 2, page 1257 © 1999 American Association for the Advancement of Science, Photograph courtesy Dr. James M. Robl, University of Massachusetts; Fig. 11.23: Courtesy Aqua Bounty Technologies, Inc.

## Chapter 12

Ch Op 12: Copyright/owner ScienCentral, Inc. and/or its licensors; Fig. 12.3a: Richard Gross/Biological Photography; Fig. 12.3b: Ken Wagner/Visuals Unlimited; Fig. 12.3c: Jerome Wexler/Photo Researchers; Fig. 12.3d: Michael P. Gadomski/Photo Researchers; Fig. 12.5: Science Source/Photo Researchers; Fig. 12.6: M. Abbey/Photo Researchers, Inc.; Fig. 12.7: Dr. A. S. Bajer; Fig. 12.8: B.A. Palevitz and E.H. Newcomb/BPS/Tom Stack and Associates; Fig. 12.12a: Cabisco/Visuals Unlimited; Fig. 12.12b: Cabisco/ Visuals Unlimited; Fig. 12.15: James Kezer/University of Oregon

## Chapter 13

Ch Op 13: Copyright/owner ScienCentral, Inc. and/or its licensors; Fig. 13.1: E. Rohne Rudder; Fig. 13.2: Jean-Michel Labat/Ardea London; Fig. 13.11: Carolina Biological Supply Company/Phototake; Fig. 13A: Nigel Cattlin/Photo Researchers.

## Chapter 14

Ch Op 14: Copyright/owner ScienCentral, Inc. and/or its licensors; Fig. 14.1: Stewart Halperin Photography; Fig. 14.3-14.5: ©AP/Wide World Photos; Fig. 14.6: Courtesy of Billie Hetrick; Fig. 14.7: Stan Flegler/Visuals Unlimited; Fig. 14.8: Courtesy Adam Hart-Davis; Fig. 14.12: CNRI/Photo Researchers; Fig. 14.14a: Courtesy Colorado Genetics Laboratory, University of Colorado Health Science Center, Denver; Fig. 14.14b: J. Cancalosi/DRK Photo; Fig. 14.16: Lester V. Bergman/Corbis Images; Fig. 14.19b: Med. Illus. SBHS/Stone/Getty Images

**Chapter 15**

Ch Op 15: Copyright/owner ScienCentral, Inc. and/or its licensors; Fig. 15.3a: Frank B. Gill/VIREO; Fig. 15.3b: John S. Dunning/VIREO; Fig. 15.4: Eric Sander; Fig. 15.5a: Ralph A. Reinhold/Animals Animals; Fig. 15.5b: Gerald Lacz/Animals Animals; Fig. 15A: Rosemary and Peter Grant; Fig. 15B: Yann Arthus-Bertrand/Corbis Images; Fig. 15C: Allen Chartier; Fig. 15.8: The National Portrait Gallery, London; Fig. 15.9: Phil Degginger/Bruce Coleman, Inc.; Fig. 15.10: Richard Gross/Biological Photography; Fig. 15.11: James Lemass/Index Stock; Fig. 15.14: Don and Pat Valenti/Tom Stack & Associates; Fig. 15.15a: Arthur Gloor/Animals Animals/Earth Scenes; Fig. 15.15b: George H.H. Huey/Animals Animals/Earth Scenes

**Chapter 16**

Ch Op 16: Copyright/owner ScienCentral, Inc. and/or its licensors; Fig. 16.1: Bettmann/Corbis Images; Fig. 16.2: Allen Traber/Alamy Images; Fig. 16.3a: Martin B. Withers; Frank Lane Picture Agency/Corbis Images; Fig. 16.3b: Christopher Talbot Frank/Alamy Images; Fig. 16A: Renee Purse/Photo Researchers; Fig. 16.8: Lester V. Bergman/Corbis Images; Fig. 16.9: George D. Lepp/Corbis Images

**Chapter 17**

Ch Op 17: Copyright/owner ScienCentral, Inc. and/or its licensors; Fig. 17.1a: COLOR-PIC/Animals Animals/Earth Scenes; Fig. 17.1b: Paul A. Souders/Corbis Images; Fig. 17.2a: Volvox/Index Stock; Fig. 17.2b: Fred McConnaughey/Photo Researchers; Fig. 17.3a: Bob Krist/Corbis Images; Fig. 17.3b: Wolfgang Kaehler/Corbis Images; Fig. 17.3c: Fritz Prenzel/Animals Animals/Earth Scenes; Fig. 17.3d: B.G. Thomson/Photo Researchers; Fig. 17.3e: Chris McLennan/Alamy Images; Fig. 17.5: Robert Lee Snyder; Fig. 17.6a: Tui De Roy/Minden Pictures, Inc.; Fig. 17.6b: Kevin Schafer/Corbis Images; Fig. 17.7: Montes De Oca & Associates; Fig. 17.8a: Rod Planck/Photo Researchers; Fig. 17.8b: Dan Suzio/Photo Researchers; Fig. 17.9: Joe McDonald/Corbis Images; Fig. 17.10a: Don Enger/Animals Animals/Earth Scenes; Fig. 17.10b: Melissa Farlow/National Geographic Society; Fig. 17.11a: Kit Houghton/Corbis Images; Fig. 17.11b: George McCarthy/Corbis Images; Fig. 17.11c: Kent Foster Photographs/Bruce Coleman, Inc.; Fig. 17.13a: A.H. Rider/Photo Researchers; Fig. 17.13b: Mary Liz Austin/The Image Bank/Getty Images; Fig. 17.14c: Heather Angel/Natural Visions; Fig. 17.15: Renee Lynn/Photo Researchers; Fig. 17.A (1): A.J. van Loon, "The Dubious Role of Man in a Questionable Mass Extinction," *Earth Science Reviews* 62 (2003)177-186 (c) Elsevier Science B.V. All rights reserved: Fig. 17A (2): From: Brook, et.a l.,"Catastrophic extrinctions follow deforestation in Singapore," NATURE, vol. 424, July 23, 2003. Used with permission.; Fig. 17.16a: Dr. Dan Sudia; Fig. 17.16b: James Zipp/Photo Researchers; Fig. 17.16c: Robert Royce; Fig. 17.17: Fred Atwood; Fig. 17.18: Steve Hillebrand/U.S. Fish &Wildlife Service

**Chapter 18**

Ch Op 18: Copyright/owner ScienCentral, Inc. and/or its licensors; Fig. 18.1: NASA; Fig. 18.2b: Dr. Stanley Miller; Fig. 18.3: Dudley Foster/Woods Hole Oceanographic Institute; Fig. 18.5a: M. R. Walter/ MacQuarie University, Australia; Fig. 18.6: Andrew W. Knoll/Botanical Museum of Harvard University; Fig. 18.8a: David Bruton; Fig. 18.8b: Simon Conway Morris/University of Cambridge; Fig. 18.8c: Simon Conway Morris/University of Cambridge; Fig. 18.9a: Richard Gross/Biological Photography; Fig. 18.9b: Frans Lanting/Minden Pictures; Fig. 18.10a: Heather Angel/Biophotos; Fig. 18.10b: Courtesy of Smithsonian Institution; Fig. 18.13 & 18.14: Steve Martin/Tom Stack and Associates; Fig. 18.16: The Field Museum; Fig. 18.17a: John D. Cunningham/Visuals Unlimited; Fig. 18.17b: O. Louis Mazzatenta/National Geographic Society; Fig. 18.18: Douglas Peebles/Corbis Images; Fig. 18.19: Stouffer Productions Ltd/Animals, Animals; Fig 18.21: Sidney Bahri/Photo Researchers, Inc.; Fig. 18.22: Tom McHugh/Photo Researchers, Inc.; Fig. 18.23a: David A. Northcott/Corbis Images; Fig. 18.23b: S. R. Maglione/Photo Researchers, Inc.; Fig. 18.24: J. C. Carton/Bruce Coleman, Inc.; Fig. 18.25a: Russell A Mittermeier; Fig. 18.25b: Anup Shah/DRK Photo; Fig. 18.25c: K & K Ammann/Bruce Coleman, Inc.; Fig. 18.26: Alan E. Mann; Fig. 18.28: Michel Brunet/M.P.F.T.; Fig. 18.29a: The Living World, St. Louis Zoo, St. Louis, MO; Fig. 18.29b & 18.30: John Reader; Fig. 18.32: Douglas Waugh/Peter Arnold, Inc.; Fig. 18-33: ©AP/Wide World Photos

**Chapter 19**

Ch Op 19: Copyright/owner ScienCentral, Inc. and/or its licensors; Fig. 19.1a: Science Source /Photo Researchers; Fig. 19.1b: Biophoto Associates/Science Source/ Photo Researchers, Inc.; Fig. 19.3a: K.G. Murti /Visuals Unlimited; Fig. 19.3b: K. Namba and D.L.D. Caspar; Fig. 19.4a: Visuals Unlimited; Fig. 19.8: Science Photo Library/Photo Researchers; Fig. 19.9: Stephen J. DeArmond, University of California, San Francisco; Fig. 19.10a: Barry Fitzgerald/ARS, USDA; Fig. 19.10b: Kent Loeffler/Cornell University; Fig. 19.10c: Jack Kelly Clark, Courtesy University of California Statewide IPM Program; Fig. 19.11: Elizabeth Gentt/Visuals Unlimited; Fig. 19.12: Runk/Schoenberger/Grant Heilman/Photography, Inc.; Fig. 19.13a: David M. Phillips/Visuals Unlimited; Fig. 19.13b: Dr. T.J. Beveridge/Visuals Unlimited; Fig. 19.13c: Abbey/Visuals Unlimited; Fig. 19.13d: Stewart Halperin; Fig. 19.13e: Stephen Sharnoff/Visuals Unlimited; Fig. 19.13f: J. Westrich/Zefa/Masterfile; Fig. 19.15: Jo Handelsman and Steven A. Vicen;

Fig. 19.16: Photo Researchers/CDC; Fig. 19.17: Custom Medical Stock Photo, Inc.; Fig. 19.18a: Biophoto Associates/Photo Researchers, Inc.; Fig. 19.18b: Visuals Unlimited

**Chapter 20**

Ch Op 20: Copyright/owner ScienCentral, Inc. and/or its licensors; Fig. 20.1: Dwight R. Kuhn; Fig. 20.2a: David M. Phillips/Visuals Unlimited; Fig. 20.2b: Dr. T.J. Beveridge/Visuals Unlimited; Fig. 20.2c: Abbey/Visuals Unlimited; Fig. 20.2d: Stewart Halperin Photography; Fig. 20.2e: Stephen Sharnoff/Visuals Unlimited Fig. 20.2f: J. Westrich/Zefa/Masterfile; Fig. 20.3: Abbey/Visuals Unlimited; Fig. 20.4a: Richard Gross/Biological Photography; Fig. 20.4b: Manfred Kage/Peter Arnold, Inc.; Fig. 20.5: Diana Laulaien-Schein; Fig. 20.6a: Dr. Yuuji Tsukii, Hosei University; Fig. 20.6b: Michael Abbey/Visuals Unlimited; Fig. 20.6c: Dennis Kunkel/Phototake; Fig. 20.7: Biophoto Associates/Photo Researchers, Inc.; Fig. 20.8: Oliver Meckes/Photo Researchers, Inc.; Fig. 20.9: Stanley Erlandson; Fig. 20.10: K.G. Murti/Visuals Unlimited; Fig. 20.11a: Ken Wagner/Visuals Unlimited; Fig. 20.11b: Biodisc/Visuals Unlimited; Fig. 20.12: Hisashi Fujioka & Masamichi Aikawa/The Malaria Parasite and its Life Cycle (1999); Fig. 20.13a&b: Carolina Biological/Visuals Unlimited; Fig. 20.13c: Photo by Gerry Nash, Electron Microscopy Unit, Australian Antartic Division. (c) Commonwealth of Australia 2002; Fig. 20.14: Bill Bachman/Photo Researchers, Inc.; Fig. 20.15a: David M. Phillips/Visuals Unlimited; Fig. 20.15b: Richard Gross/Biological Photography; Fig. 20.16a: Philip Sze/Visuals Unlimited; Fig. 20.16b: Gregory Ochocki/Photo Researchers, Inc.; Fig. 20.17a: Biophoto Associates/Photo Researchers, Inc.; Fig. 20.17b&c: John D. Cunningham/Visuals Unlimited; Fig. 20.18a: William C. Jorgensen/Visuals Unlimited; Fig. 20.18b: Gary K. Robinson/Visuals Unlimited; Fig. 20.19: Higuchi Bioscience Laboratory; Fig. 20.20: E. S. Ross; Fig. 20.21: Richard Gross/Biological Phototgraphy; Fig. 20.22: Carolina Biological /Visuals Unlimited; Fig. 20.23a: Reno A. DiTullio; Fig. 20.23b: Andrew J. Storer; Fig. 20.24: Dwight R. Kuhn; Fig. 20.26a: Tom Volk/University of Wisconsin-LaCrosse; Fig. 20.26b: Courtesy of Joyce E. Longcore, University of Maine; Fig. 20.27a: Stewart Halperin Photography; Fig. 20.27b: Ed Reschke; Fig. 20.28: Stephen Sharnoff/Visuals Unlimited; Fig. 20.29a: Manfred Kage/Peter Arnold, Inc.; Fig. 20.30: Steven Rothfeld/Stone/Getty Images; Fig. 20.31a: Kerry T. Givens/Bruce Coleman, Inc.; Fig. 20.31b: Mark Custance/Peter Raven/Bruce Coleman, Inc.; 20.31c: Robert Pickett/Corbis Images; Fig. 20.31d: W. Wayne Lockwood, M.D./Corbis Images; Fig. 20.31e: Frank Young/Papilio/Corbis Images; Fig. 20.31f: Richard Gross/Biological Photography; Fig. 20.31g: Dayton Wild/Visuals Unlimited; Fig. 20.33: Stanley L. Flegler/Visuals Unlimited; Fig. 20.34a: David M. Phillips/Visuals Unlimited; Fig. 20.34b: Dennis Kunkel/Phototake; Fig. 20.35: Ken Greer/Visuals Unlimited; Fig. 20.36: Richard Gross/Biological Photography; Fig. 20.37a: John Gerlach/Visuals Unlimited; Fig. 20.37b: V. Ahmadjian/Visuals Unlimited; Fig. 20A (1): Lee Berger, James Cook University; Fig. 20A(2): Ribbit Photography; Fig. 20.38a: J. Westrich/Zefa/Masterfile; Fig. 20.38b: Peter Arnold, Inc.; Fig. 20.38c: ChromoSohm Media Inc./Photo Researchers; Fig. 20.38d: Super Stock; Fig. 20.38e: David Sieren/Visuals Unlimited

**Chapter 21**

Ch Op 21: Copyright/owner ScienCentral, Inc. and/or its licensors; Fig 21.2a: Lawson Wood/Corbis Images; Fig. 21.2b: Cabisco/Visuals Unlimited; Fig. 21.2c: Science Photo Library/Photo Researchers; Fig. 21.2d: Jeffrey L. Rotman/Corbis Images; Fig. 21.2e: Robert Yin/Corbis Images; Fig. 21.2f: Custom Medical Stock Photo, Inc.; Fig. 21.2g: Margarette Mead/The Image Bank/Getty Images; Fig. 21.2h: Steven Frink; Fig. 21.2i: Kelvin Aitken/Peter Arnold, Inc.; Fig. 21.5a: Ronald L. Sefton/Bruce Coleman, Inc./ PNI; Fig. 21.5b: Amos Nachoum/Corbis Images; Fig. 21.5c: Alex Kerstitch/Visuals Unlimited; Fig. 21.7a: Secret Sea Visions/Peter Arnold, Inc.; Fig. 21.7b: Peter Parks/imagequestmarine.com; Fig. 21.9: Lawson Wood/Corbis Images; Fig. 21.10a: Cabisco/Visuals Unlimited; Fig. 21.10b: Stephen Frink; Fig. 21.11: E. Jeff Rotman; Fig. 21.13a: Jim and Cathy Church; Fig. 21.13b: E.J.Cable/Tom Stack and Associates; Fig. 21.13c: Science Photo Library/Photo Researchers; Fig. 21.14: Tom Adams/Visuals Unlimited; Fig. 21.15: Custom Medical Stock Photo, Inc.; Fig. 21.A: Lester V. Bergman/Corbis; Fig. 21.16: R. Umesh Chandran/TDR,WHO/SPL/Photo Researchers, Inc.; Fig. 21.18a: Milton Rand/Tom Stack & Associates; Fig. 21.18b: Jeffrey L. Rotman/Corbis Images; Fig. 21.18c: Heather Angel; Fig. 21.18d: Gustav W. Verderber/Visuals Unlimited; Fig. 21.19a: Robert Yin/Corbis Images; Fig. 21.19b: Stephen Frink/Corbis Images; Fig. 21.20: Photo by Jeff Miller/University of Wisconsin-Miller; Fig. 21.22a: Margarette Mead/The Image Bank/Getty Images; Fig. 21.22b: Doug Sokell/Visuals Unlimited; Fig. 21.22c: John Shaw/Bruce Coleman, Inc.; Fig. 21.22d: Oliver Meckes/Photo Researchers; Fig. 21.23a: Kjell Sandved; Fig. 21.23b: Cleveland P. Hickman, Jr.; Fig. 21.24a: Jan Taylor/Bruce Coleman, Inc.; Fig. 21.24b: Andrew Syred/Photo Researchers; Fig. 21.25: Kjell Sandved; Fig. 21.28a: Breck P. Kent/Animals Animals/Earth Scenes; Fig. 21.28b: Rod Planck/Photo Researchers; Fig. 21.28c: Scott Tysick/Masterfile; Fig. 21.28d: J. Mitchell/Photo Researchers; Fig. 21.29: Dan Mott, Entomology Department, NCSU; Fig. 21.30a&b: Stephen Frink; Fig. 21.30c: Jeff Rotman

**Chapter 22**

Ch Op 22: Copyright/owner ScienCentral, Inc. and/or its licensors; Fig. 22.1: Anatomical Travelogue/Photo Researchers; Fig. 22.2b: Cabisco/Visuals Unlimited; Fig. 22.2c: Jim and Cathy Church; Fig. 22.2d: Custom Medical Stock Photo, Inc.; Fig. 22.2e: Jeffrey L.

Rotman/Corbis Images; Fig. 22.2f: Robert Yin/Corbis Images; Fig. 22.2g: Margarette Mead/The Image Bank/Getty Images; Fig. 22.2h: Stephen Frink; Fig. 22.2i&j: Rick Harbo; Fig. 22.2k: Breck Kent; Fig. 22.2l: Kelvin Aitken/Peter Arnold, Inc.; Fig. 22.2m: Mark Giovannetti/Stamm Photo; Fig. 22.2n: Gary Meszaros/Visuals Unlimited; Fig. 22.2o: John Gerlach/Visuals Unlimited; Fig. 22.2p: ©PhotoDisc; Fig. 22.2q: Stewart Halperin; Fig. 22.3 & 22.4: Rick Harbo; Fig. 22.5a: Breck Kent; Fig. 22.5b: Hans Reinhard/Bruce Coleman, Inc.; Fig. 22.5c: Tom McHugh/Photo Researchers; Fig. 22.6a: Kelvin Aitken/Peter Arnold, Inc.; Fig. 22.6b: David Hall/Photo Researchers, Inc.; Fig. 22.6c: David Wrobel/Visuals Unlimited; Fig. 22.8a: Mark Giovannetti/Stamm Photo; Fig. 22.8b: Dr. John D. Cunningham/Visuals Unlimited; Fig. 22.10a: Gary Meszaros/Visuals Unlimited; Fig. 22.10b: Tim Flach/Stone/Getty Images; Fig. 22.10c: Richard Gross/Biological Photography; Fig. 22.11: Michael and Patricia Fogden; Fig. 22A ©PhotoDisc; Fig. 22.12a: Betsy R. Strasser/Visuals Unlimited; Fig. 22.12b: Rob and Ann Simpson/Visuals Unlimited; Fig. 22.12c: John Gerlach/Visuals Unlimited; Fig. 22.12d: G and C Merker/Visuals Unlimited; Fig. 22.14a: Tom Vezo; Fig. 22.14b: ©PhotoDisc; Fig. 22.16a: Tom McHugh/Photo Researchers, Inc.; Fig. 22.16b: Theo Allofs; Fig. 22.17a: C. Andrew Henley/Biofotos; Fig. 22.17b: John Cancalosi/Peter Arnold, Inc.; Fig. 22.18a&b: Stewart Halperin Photography; Fig. 22.18c: John Cancalosi/Peter Arnold, Inc.

**Chapter 23**
Ch Op 23: Copyright/owner ScienCentral, Inc. and/or its licensors; Fig. 23.2a: Stewart Halperin Photography; Fig. 23.2b: Heather Angel/Biofotos; Fig. 23.3: Courtesy Ari Kornfeld (www.perspective.com/nature/); Fig. 23.4: Kirtley-Perkins/Visuals Unlimited; Fig. 23.5: William E. Ferguson; Fig. 23.6: Ed Reschke/Peter Arnold, Inc.; Fig. 23.7a: PhotoDisc, Inc.; Fig. 23.7b: Gerald & Buff Corsi/Visuals Unlimited; Fig. 23.8: Deborah Allen; Fig. 23.9: Heather Angel/Biofotos; Fig. 23.10a: Whit Bronaugh Photography; Fig. 23.10b: Michael and Patricia Fogden; Fig. 23.11a (1): Eric Anderson/Visuals Unlimited; Fig. 23.11a (2): Brand X Pictures/Alamy Images; Fig. 23.11b: Charles O"Rear/Corbis Images; Fig. 23.11c: Robert Levin/Corbis Images; Fig. 23.12a: Ron Spomer/Visuals Unlimited; Fig. 23.12b: D. Cavagnaro/Visuals Unlimited; Fig. 23.12c&d: Stewart Halperin Photography; Fig. 23.14a-d: Richard Gross/Biological Photography; Fig. 23.14f: Whit Bronaugh Photography; Fig. 23.16: Terry W. Eggers/Corbis Images; Fig. 23.17: Jack M. Bostrack/Visuals Unlimited; Fig. 23.18: E. Webber/Visuals Unlimited; Fig. 23.A (1): Courtesy of Yuzuru Shimizu, PhD., University of Rhode Island; Fig. 23.A (2): Scott Camazine/Photo Researchers

**Chapter 24**
Ch Op 24: Copyright/owner ScienCentral, Inc. and/or its licensors; Fig. 24.1: Peter French/Bruce Coleman, Inc.; Fig. 24.4: Michael Clayton/University of Wisconsin-Madison; Fig. 24.5: James D. Mauseth, University of Texas at Austin; Fig. 24.6a&b: George J. Wilder/Visuals Unlimited; Fig. 24.6c: Richard Gross/Biological Photography; Fig. 24.7a: Cabisco/Visuals Unlimited; Fig. 24.7b: from P. Linstead, L. Dolan, and K. Roberts in J. Bowman, Arabidopsis; An Atlas of Morphology and Development, Springer-Verlag, NY 1994 Fig. 24.9a : Phil Degginger/Bruce Coleman, Inc.; Fig. 24.9b: Science VU/Visuals Unlimited; Fig. 24.9c: Richard Gross/Biological Photography; Fig. 24.9d: Biodisc/Visuals Unlimited; Fig. 24.10: Kent Wood; Fig. 24.11a: John D. Cunningham/Visuals Unlimited; Fig. 24.11b&c: Richard Gross/Biological Photography; Fig. 24.12a: Ernst van Jaarsueld/Kirsten-bosch National Botanical Garden, South Africa; Fig. 24.12b: Darrell Gulin/DRK Photo; Fig. 24.13a&b: Richard Gross/Biological Photography; Fig. 24.15: Ray F. Evert; Fig. 24.16: Andrew Brown/Corbis Images; Fig. 24.17 &24.18a: Deborah Allen; Fig. 24.18b: Stewart Halperin; Fig. 24.20: Ed Reschke/Peter Arnold, Inc.; Fig. 24A: David M. Dennis/Animals Animals/Earth Scenes; Fig. 24.23: Heather Angel/Biofotos

**Chapter 25**
Ch Op 25: Copyright/owner ScienCentral, Inc. and/or its licensors; Fig. 25.1: Kingsley R. Stern; Fig. 25.2: R. Lyons/Visuals Unlimited; Fig. 25.3: Courtesy of Bruce Bugbee, Crop Physiology Laboratory, Utah State University; Fig. 25.4: Bruce Iverson Photomicrography; Fig. 25A: Courtesy Rob Frohne; Fig. 25.5: Cathlyn Melloan/Stone/Getty Images; Fig. 25.6a: L.E. Gilbert/Biological Photo Service; Fig. 25.7: ; Fig. 25.9: Randy Moore; Fig. 25.10: John Kaprielian/Photo Researchers; Fig. 25.12: Tom & Susan Bean, Inc.; Fig. 25.13: Dodie Ulery/Gladstar Images; Fig. 25.15: Arthur Zangrel; Fig. 25.16: Bruce Iverson Photomicrography

**Chapter 26**
Ch Op 26: Copyright/owner ScienCentral, Inc. and/or its licensors; Fig. 26A: Sherman Thomson/Visuals Unlimited; Fig. 26B: Rod Barbee/Visuals Unlimited; Fig. 26.3a-c: Ed Reschke; Fig. 26.3d: Biophoto Associates/Science Source/Photo Researchers; Fig. 26.3e & f: Ed Reschke; Fig. 26.4a-j: Ed Reschke; Fig. 26.4k: Carolina Biological Supply Company/Phototake; Fig. 26.5: Lennart Nilsson/*Behold Man*, Little, Brown and Co.; Fig. 26.6: Brian Eyden/Science Photo Library/Photo Researchers; Fig. 26.8a: M.I. Walker/Science Source/Photo Researchers; Fig. 26.8b&c: Ed Reschke; Fig. 26.9: Digital Stock; Fig. 26.11: Bol Daemmrich/Stock, Boston

**Chapter 27**
Ch Op 27: Copyright/owner ScienCentral, Inc. and/or its licensors; Fig. 27.1a-c: Stewart Halperin; Fig. 27.4a&b: Glenn Oliver/Visuals Unlimited; Fig. 27.8: Ed Reschke; Fig. 27.11: Michael Webb/Visuals Unlimited

**Chapter 28**
Ch Op 28: Copyright/owner ScienCentral, Inc. and/or its licensors; Fig. 28.1: Frank Balthis; Fig. 28.6a: Lennart Nilsson/*Behold Man*, Little, Brown and Co.; Fig. 28.6b: Art Siegel/University of Pennsylvania; Fig. 28.7: Ellen Dirkson/Visuals Unlimited; Fig. 28.10: David M. Grossman/Photo Researchers, Inc.; Fig. 28A: Oliver Meckes/Photo Researchers, Inc.; Fig. 28.13a&b: Stewart Halperin Photography

**Chapter 29**
Ch Op 29: Copyright/owner ScienCentral, Inc. and/or its licensors; Fig. 29.4a&b: Gladden Willis, MD/Visuals Unlimited; Fig. 29.4c: Science VU/Visuals Unlimited; Fig. 29.4d: Carolina Biological /Visuals Unlimited; Fig. 29.6: Dr. P. Mazzi/Photo Researchers; Fig. 29.8: Biophoto Associates/Science Source/Photo Researchers; Fig. 29.12: CNRI/Science Photo Library/Photo Researchers, Inc.; Fig 29.14a&b: Harry Ransom Humanities Research Center

**Chapter 30**
Ch Op 30: Copyright/owner ScienCentral, Inc. and/or its licensors; Fig. 30A: Virginia Weinland/Photo Researchers, Inc.; Fig. 30.1: Mike Abbey/Visuals Unlimited; Fig. 30.5: Lester V. Bergman/Corbis Images; Fig. 30.7: Prof. S.H.E. Kaufman & Dr. J.R. Golecki/SPL/Photo Researchers, Inc.; Fig. 30.8: Dr. A. Liepins/SPL/Photo Researchers, Inc.; Fig. 30B: Courtesy Mervyn BiBB, BBSRC, UK; Fig. 30.11a: Dr. Gopal Murti/Photo Researchers; Fig. 30.11b: Dr. Donald Fawcett/Visuals Unlimited; Fig. 30.14: Ken Eward/Biografx/Photo Researchers, Inc.; Fig. 30.16: Joe Raedle/Getty Images News and Sport Services; Fig. 30.17: Roger Ressmeyer/Corbis Images; Fig. 30.19a: Ed Reschke; Fig. 30.19b: David Scharf/Peter Arnold, Inc.

**Chapter 31**
Ch Op 31: Copyright/owner ScienCentral, Inc. and/or its licensors; Fig. 31.1a: Stewart Halperin; Fig. 31.1b: Michael Sewell/Peter Arnold, Inc.; Fig. 31.2a: David M. Phillips/Visuals Unlimited; Fig. 31.2b: A. Kerstitch/Visuals Unlimited; Fig. 31.4 (part1): SIU/Photo Researchers, Inc.; Fig. 31.4(part 2): Cabisco/Visuals Unlimited; Fig. 31.5: Courtesy Kenjiro Kimura; Fig. 31.8a: Stewart Halperin; Fig. 31.8b: Tom McHugh/Photo Researchers, Inc.; Fig. 31.11: Stephen J. Krasemann/Photo Researchers, Inc.; Fig. 31.12: Michael Heron/Corbis Stock Market; Fig. 31A: Andrew Syred/Science Photo Library/Photo Researchers, Inc.; Fig. 31B: Reuters/Corbis Images

**Chapter 32**
Ch Op 32: Copyright/owner ScienCentral, Inc. and/or its licensors; Fig. 32.4a: David M. Phillips/Visuals Unlimited; Fig. 32.4b: Don W. Fawcett/Visuals Unlimited; Fig. 32.10: Dr. John Heuser; Fig. 32.11: E. R. Lewis/Visuals Unlimited; Fig. 32.19: Courtesy of National Institute of Drug Abuse/NIH; Fig. 32.A: Dr. Robert Friedland/Photo Researchers

**Chapter 33**
Ch Op 33: Copyright/owner ScienCentral, Inc. and/or its licensors; Fig. 33.5: Heather Angel/Biofotos; Fig. 33.7: Joe McDonald/Bruce Coleman, Inc.; Fig. 33.8a&b: Heather Angel/Biofotos; Fig. 33.10: Doug Perrine/Seapics.com; Fig. 33.11: Dr. Diana Wheeler; Fig. 33.14: Tom E. Adams/Peter Arnold, Inc.; Fig. 33.15: Richard Walters/Visuals Unlimited; Fig. 33.18: Scott Mittman; Fig. 33.22: PhotoDisc, Inc.; Fig. 33A: Cathlyn Malloan/Stone/Getty Images; Fig. 33B: SEMs by Robert S. Preston, courtesy of Dr. J. E. Hawkins, Kresge Hearing Research Institute

**Chapter 34**
Ch Op 34: Copyright/owner ScienCentral, Inc. and/or its licensors; Fig. 34.9: Video Surgery/Photo Researchers, Inc.; 34.10: Dr. Michael Klein/Peter Arnold, Inc.; Fig. 34.13: Don W. Fawcett/Visuals Unlimited; Fig. 34A: Richard Gross; Fig. 34.16: Ed Reschke

**Chapter 35**
Ch Op 35: Copyright/owner ScienCentral, Inc. and/or its licensors; Fig. 35.6: Cirone/Photo Researchers, Inc.; Fig. 35.7: *American Journal of Medicine* 20 (1956). Fig. 35.9: Science Photo Library/Photo Researchers; Fig. 35.10: Martin Rotker/Phototake; Fig. 35.14: Ed Reschke; Fig. 35A: PhotoDisc, Inc.

**Chapter 36**
Ch Op 36: Copyright/owner ScienCentral, Inc. and/or its licensors; Fig. 36.1: CNRI/SPL/Photo Researchers, Inc.; Fig. 36.2: Carolina Biological Supply/Phototake; Fig. 36.3: Stuart Westmorland/Photo Researcers, Inc.; Fig. 36.4: Educational Images/Custom Medical Stock Photo, Inc. Fig. 36.5: David Crews; Fig. 36.6: David Dennis/Tom Stack and Associates; Fig. 36.7: Gregory G. Dimijian/Photo Researchers, Inc.; Fig. 36.8: Ed Reschke; Fig. 36.11: Lennart Nilsson, *Behold Man*, Little, Brown and Company; Fig. 36.17: Lester V.

Bergman/Corbis Images; Fig. 36.19a: Michael Newman/PhotoEdit; Fig. 36.19b: Michael Newman/PhotoEdit; Fig. 36.19c: Yoav Levy/Phototake; Fig. 36.19d: Custom Medical Stock Photo, Inc.; Fig. 36.19e: Chuck Swartzell/Visuals Unlimited; Fig. 36.19f: Jonathan A. Meyers/Photo Researchers; Fig. 36.19g: Courtesy of Organon USA, Inc.; Fig. 36.19h: Nourypharma/Sipa Press; Fig. 36.19i: Saturn Stills/Science Photo Library/Photo Researchers; Figs. 36.20, 36.21, 36.22, 36.24, 36.25, 36.27: Lennart Nilsson/Allbert Bonniers Publishing Company, *A Child is Born*, Dell Publishing Company; Fig. 36.28: James Stevenson/Photo Researchers; Fig. 36.29: Petit Format/Photo Researchers

### Chapter 37

Ch Op 37: Copyright/owner ScienCentral, Inc. and/or its licensors; Fig. 37.1: Norbert Wu; Fig. 37.2: Pat and Tom Leeson/Photo Researchers, Inc.; Fig. 37.4: Heather Angel/Biofotos; Fig. 37.6: Thomas McAvoy/Life Magazine, 1955. Time Warner, Inc.; Fig. 37.7: Omikron/Photo Researchers, Inc.; Fig. 37.8: David Weintraub/Photo Researchers, Inc.; Fig. 37.10: Kennan Ward/The Stock Market; Fig. 37.11: Merlin Tuttle/National Geographic Society; Fig. 37.12: Jason Edwards/National Geographic Society; Fig. 37.13: Jane Burton/Bruce Coleman, Inc.; Fig. 37.14: Rick and Nora Bowers/Visuals Unlimited; Fig. 37.15: Jeffrey Rich; Fig. 37.17: J. C. Stevenson/Animals, Animals; Fig. 37.18 & 37.19: E. S. Ross; Fig. 37.21a: David Dennis/Tom Stack & Associates; Fig. 37.21b: Brian Rogers/Biofotos

### Chapter 38

Ch Op 38: Copyright/owner ScienCentral, Inc. and/or its licensors; Fig. 38.1: Cliff Beittel; Fig. 38.3a&b: Photography reproduced with permission of Natural Resources, Queensland, Australia; Fig. 38.5a: Jack Dykinga Photography; Fig. 38.5b: Greg Vaughn/Stone/Getty Images; Fig. 38.5c: Fred Breummer/Peter Arnold, Inc.; Fig. 38A: Charles Cecil/Visuals Unlimited; Fig. 38.6: Tom McHugh/Photo Researchers

### Chapter 39

Ch Op 39: Copyright/owner ScienCentral, Inc. and/or its licensors; Fig. 39.1: Ralph Oberlander/Stock, Boston; Fig. 39.2: Ed Reschke; Fig. 39.3: Roger del Moral/University of Washington; Fig. 39.4: D. P. Wilson/FLPA - Images of Nature; Fig. 39.6a: Arthur Siegelman/Visuals Unlimited; Fig. 39.6b: M. Abbey/Visuals Unlimited; Fig. 39.7: Calgene, Inc.; Fig. 39.8: Stewart Halperin Photography; Fig. 39.9a: Alex Kerstich; Fig. 39.9b: E. S. Ross; Fig. 39.10a: Gary Meszaros/Visuals Unlimited; Fig. 39.10b: Arthur Morris/Visuals Unlimited; Fig. 39.11: E. S. Ross; Fig. 39.12a: Art Wolfe/Stone/Getty Images; Fig. 39.12b: Hal Beral/Visuals Unlimited; Fig. 39.13: Richard Gross/Biological Photography; Fig. 39.14a: Francois Gohier/Photo Researchers, Inc.; Fig. 39.14b: Frank S. Balthis; Fig. 39.15: Brain Rogers/Biofotos Fig. 39.16: Bill Wood/Bruce Coleman, Inc.; Fig. 39.17: Cabisco/Visuals Unlimited; Fig. 39.18: Richard Gross; Fig. 39.19: Gary Braasch

### Chapter 40

Ch Op 40: Copyright/owner ScienCentral, Inc. and/or its licensors; Fig. 40.1: J. Cancalosi/Peter Arnold, Inc.; Fig. 40.3: E. S. Ross; Fig. 40.4: Richard Gross/Biological Photography; Fig. 40.12: Wally Eberhart/Visuals Unlimited; Fig. 40.13a: Gerald A. Peters/Virginia Commonwealth University; Fig. 40.13b: Tom Stack and Associates; Fig. 40.13c: Stewart Halperin; Fig. 40.16: Norbert Wu; Fig. 40A: Nick Gordon/Ardea London; Fig. 40B: John D. Cunningham/Visuals Unlimited

### Chapter 41

Ch Op 41: Copyright/owner ScienCentral, Inc. and/or its licensors; Fig. 41.8a: M. P. L. Fogden/Bruce Coleman, Inc.; Fig. 41.8b: Stuart Westmorland/Stone/Getty Images; Fig. 41.8c: Gregory G. Dimijian/Photo Researchers, Inc.; Fig. 41.8d: PhotoDisc, Inc.; Fig. 41.9 main: Heather Angel/Biofotos; Fig. 41.9 inset (a): D. Robert & Lorri Franz/Corbis Images; Fig. 41.9 inset (b): Peter Johnson/Corbis Images; Fig. 41.10: M. Harvey/DRK Photos; Fig. 41.11a: Digital Vision/Getty Images; Fig. 41.11b: Robert Harding World Imagery /Getty Images; Fig. 41.11c: Brian Rogers/Biofotos; Fig. 41.12: S.

Maslowski/Visuals Unlimited; Fig. 41.13a: Jeremy Woodhouse/DRK Photos; Fig. 41.13 inset (a): David Boag/Alamy Images; Fig. 41.13 inset (b): Scott Camazine/Photo Researchers; Fig. 41.13 inset (c): Image State; Fig. 41.14: Charlie Ott/Photo Researchers, Inc.; Fig. 41.15: Daniel J. Cox/Stone/Getty Images; Fig. 41.16: Tom Stack and Associates; Fig. 41.16 inset (a): Joe McDonald/Bruce Coleman, Inc.; Fig. 41.16 inset (b): Lynn Stone/Index Stock; Fig. 41.18: the Chesapeake Bay Foundation; Fig. 41.19: Roy Ooms/Masterfile; Fig. 41.19 inset: Ken Lucas/Visuals Unlimited; Fig. 41.20: Vanessa Vick/Photo Researchers, Inc.; Fig. 41.20 inset (a): Gerlach Nature Photography/Animals Animals/Earth Scenes; Fig. 41.20 inset (b): Mark Bacon/Stock Connection/PictureQuest; Fig. 41.22a: Jeffrey L. Rotmann/Peter Arnold, Inc.; Fig. 41.22b: Norbert Wu/Stone/Getty Images

### Chapter 42

Ch Op 42: Copyright/owner ScienCentral, Inc. and/or its licensors; Fig. 42.3a: Martin Bond/Science Photo Library/Photo Researchers, Inc.; Fig. 42.3b: Grant Heilman/Grant Heilman Photography, Inc.; Fig. 42.3c: Keven Schafer/Peter Arnold, Inc.; Fig. 42.3d: Stand W. Elms/Visuals Unlimited; Fig. 42.6: Inga Spence/Visuals Unlimited; Fig. 42.8b: Bruno J. Zehnder/Peter Arnold, Inc.; Fig. 42.11: Gramma Liaison Network; Fig. 42.12: Thomas A. Scheider; Fig. 42.14: Frans Lanting/Minden Pictures; Fig. 42.15: J. M. Rankin/Worldwide Fund for Nature; Fig. 42.17: Agricultural Research Service/Courtesy USDA; Fig. 42.4: Shepard Sherbell/SABA; Fig. 42.8a: Michael J. Doolittle/The Image Works; Fig. 42.16: Sally A. Morgan/Corbis Images

### Line Art Credits

Fig. 10.9, Snustad & Simmons, *Genetics 3e*, Fig. 24.8, p. 607, John Wiley & Sons (2003). Fig. 11.22, SCIENCE Magazine, vol. 280, May 22, 1998, p. 1257 Fig. 2, American Association for the Advancement of Science. Fig. 12.10, US Mortality Public Use Data Tapes 1960-2000/American Cancer Society, Surveillance Research 2004. Fig. 16.5, BBC Maps. Fig. 18.15, Bill Ober. Fig. 18.31, S.J. Gould, *The Book of Life*, W.W. Norton (1993). Fig. 19.4b, Carlyn Iverson. Fig. 20.3, Carlyn Iverson. Fig. 20.27b, Carlyn Iverson. Fig. 20.28b, Carlyn Iverson. Fig. 20.32, Carlyn Iverson. Fig. 21.18, Bill Ober. Fig. 21.21, Nadine Sokol. Fig. 21.24, Bill Ober. Fig. 21.31, Bill Ober. Fig. 22.9, Bill Ober. Fig. 22.13, Bill Ober. Fig. 22.15, Raychel Ciemma. Fig. 23.13, Carlyn Iverson. Fig. 24.4, Bill Ober. Fig. 24.15a, Carlyn Iverson. Fig. 25.8, Hopkins *Plant Physiology 3e*, Fig. 18.10, p. 399, John Wiley & Sons (2004). Fig. 25.11, Hopkins *Plant Physiology 3e*, Fig. 21.1, p. 460, John Wiley & Sons (2004). Fig. 26.6, Lennart Nilsson, Behold Man, Little, Brown, and Co. (1978). Fig. 27.4, Nadine Sokol. Fig. 27.6, Nadine Sokol. Fig. 27.13, Tortora & Grabowski, *Principles of Anatomy and Physiology 10e*, Fig. 24.25, p. 888, John Wiley & Sons (2003). Fig. 27.B, www.hsph.harvard.edu/nutritionsource/pyramids.html Fig. 28.7, Lennart Nilsson, Behold Man, Little, Brown, and Co. (1978). Fig. 29.4, Ed Reschke. Fig. 29.5, Bill Ober. Fig. 31.9, Barbara Cousins. Fig. 32.1, Aidley, D.S., *The Physiology of Excitable Cells 3e*, Cambridge University Press (1989). Fig. 32.15, Nadine Sokol. Fig. 33.6, Barbara Cousins. Fig. 33.13a, Marsha J. Dohrmann. Fig. 33.13b, Christine Oleksyk. Fig. 33.16, Marsha J. Dohrmann. Fig. 33.18, Scott Mittman. Fig. 33.19, Marsha J. Dohrmann. Fig. 34.3, Laurie O'Keefe/John Daugherty. Fig. 34.13, Barbara Cousins. Fig. 34.16, Ronald J. Ervin. Fig. 37.5, E. Rohne Rudder. Fig. 37.9, Goodenough, McGuire, & Wallace, *Perspectives on Animal Behavior 2e*, Fig. 10.15, p. 229, John Wiley & Sons (2001). Fig. 37.16, Bill Ober. Fig. 38.6, Raven, Berg, & Aliff, *Environment 4e*, Fig, 8.10, p. 170, John Wiley & Sons (2004). Fig. 39.5, Barbara Cousins. Fig. 40.5, Raychel Ciemma. Fig. 40.6, Raychel Ciemma. Fig. 40.8, Raven, Berg, & Aliff, Environment 4e, Fig. 4.12, 4.13, 4.14, pp. 75-76, John Wiley & Son (2004). Fig. 40.9, Raven, Berg, & Aliff, *Environment 4e*, Fig, 6.6, p. 112, John Wiley & Son (2004). Fig. 40.10, Raven, Berg, & Aliff, *Environment 4e*, Fig, 6.2, p. 106, John Wiley & Son (2004). Fig. 40.14, Raven, Berg, & Aliff, *Environment 4e*, Fig, 6.3, p. 107, John Wiley & Son (2004). Fig. 40.15, Raven, Berg, & Aliff, *Environment 4e*, Fig, 6.4, p. 109, John Wiley & Son (2004). Fig. 41.4, Bill Ober. Fig. 41.21, Raychel Ciemma. Fig. 42.5, Raychel Ciemma.

# INDEX

Note: Page numbers with an *f* indicate that the reference appears in a figure or a figure caption; page numbers with a *t* indicate that the reference appears in a table.

# eGrade Plus

## www.wiley.com/college/alters
## Based on the Activities You Do Every Day

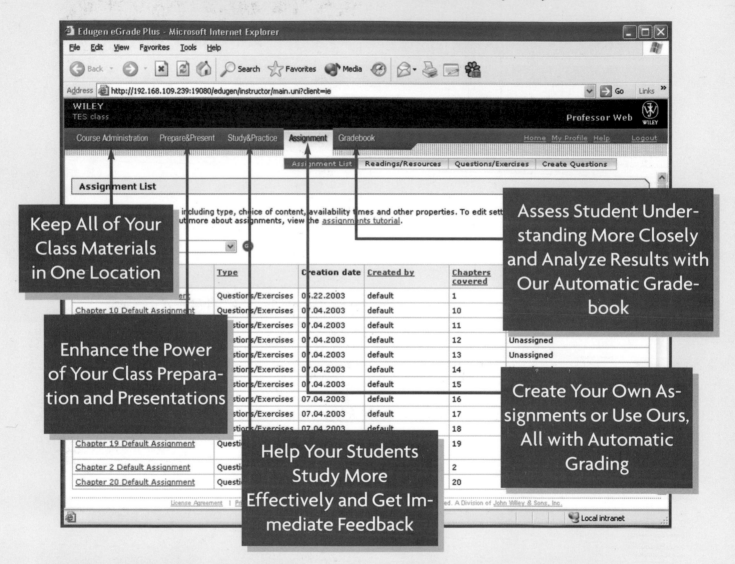

**Keep All of Your Class Materials in One Location**

**Enhance the Power of Your Class Preparation and Presentations**

**Help Your Students Study More Effectively and Get Immediate Feedback**

**Assess Student Understanding More Closely and Analyze Results with Our Automatic Gradebook**

**Create Your Own Assignments or Use Ours, All with Automatic Grading**

## All the content and tools you need, all in one location, in an easy-to-use browser format.

## Choose the resources you need, or rely on the arrangement supplied by us.

Now, many of Wiley's textbooks are available with eGrade Plus, a powerful online tool that provides a completely integrated suite of teaching and learning resources in one easy-to-use website. eGrade Plus integrates Wiley's world-renowned content with media, including a multimedia version of the text, PowerPoint slides, and more. Upon adoption of eGrade Plus, you can begin to customize your course with the resources shown here.

### See for yourself!
**Go to www.wiley.com/college/egradeplus for an online demonstration of this powerful new software.**